FROM CLASSICAL TO QUANTUM FIELDS

T0177641

From Classical to Quantum Fields

Laurent Baulieu

CNRS and Sorbonne Universités

John Iliopoulos

CNRS and École Normale Supérieure

Roland Sénéor

CNRS and École Polytechnique

OXFORD
UNIVERSITY PRESS

Great Clarendon Street, Oxford, OX2 6DP,
United Kingdom

Oxford University Press is a department of the University of Oxford.
It furthers the University's objective of excellence in research, scholarship,
and education by publishing worldwide. Oxford is a registered trade mark of
Oxford University Press in the UK and in certain other countries

First Edition published in 2017

Reprinted 2018 (with corrections)

Impression: 2

Published in the United States of America by Oxford University Press
198 Madison Avenue, New York, NY 10016, United States of America

British Library Cataloguing in Publication Data

Data available

Library of Congress Control Number: 2016935303

ISBN 978–0–19–878839–3 (hbk.)
ISBN 978–0–19–878840–9 (pbk.)

Printed and bound by
CPI Group (UK) Ltd, Croydon, CR0 4YY

The idea of this collaborative work is due to our late colleague Roland Sénéor. The book still carries part of his style with the quest for precision and rigour. He would have brought considerable improvements, had he had the chance to see the final version. We miss him.

During the preparation of this work, we have profited from the advice of many among our colleagues but we want to underline, in particular, the influence of Raymond Stora, a master and a friend, who first taught physicists the importance of fundamental mathematical concepts in the formulation of gauge theories. Together with Roland Sénéor, we had decided to dedicate this book to Stora's memory.

Contents

Prologue

This book originated in a course at the École Polytechnique which covered approximately the first third of it. The course was aimed at undergraduate students, some of whom were majoring in physics and some in mathematics. The problem we had to face was that, contrary to what happens in some other places, the education these two communities had received was rather unbalanced towards each other. The physicists had a good training in an advanced mathematical formalism, but mathematics was perceived as a mere tool to obtain a precise description of physical phenomena. Similarly, the mathematicians had used physics only as a demonstration of mathematics' ability to have a contact with the real world. Our aim was to show the deep entanglement of physics and mathematics and how these two disciplines through their mutual interactions over the past hundred years have enriched themselves and both have shaped our understanding of the fundamental laws of nature. Today, modern theoretical physics interacts with the mainstream research in pure mathematics and this interaction has resulted in the development of new concepts common to both. We chose an approach which attempts to reconcile the physicists' and the mathematicians', points of view. It is based, on two ingredients: the concept of field, as it appears primarily in Maxwells' equations, and that of the path integral, as formulated by Feynman. With these two ingredients we can make the connection between the classical and the relativistic quantum worlds. We can introduce the underlying symmetries, including those of general relativity, and show how some fundamental physical principles, such as relativistic invariance, locality of the interactions, causality, and positivity of the energy, can form the basic elements of a modern physical theory.

In this approach we were confronted with the fact that practically one can never directly give a mathematical meaning to the path integrals. This, however, can be done indirectly by the axiomatic link between our relativistic space–time and the Euclidean universe in which, under certain physical conditions, the existence of path integrals can be shown. This provides the foundation of what follows. We develop the standard theory of the fundamental forces which is a perfect example of the connection between physics and mathematics. Based on some abstract concepts, such as group theory, gauge symmetries, and differential geometry, it gives a detailed model whose agreement with experiment has been spectacular. This line of approach, from first principles, all the way to specific experimental predictions, has been the guiding line all along. As a consequence, we decided to leave out subjects, such as the attempts to obtain a quantum theory of gravity based on string theories, because, although they involve very beautiful modern mathematics, they have not yet been directly connected to concrete experimental results.

1

Introduction

1.1 The Descriptive Layers of Physical Reality

Figure 1.1 describes how the different degrees of approximation of the physical 'reality' fit into each other. This course aims to describe a possible progression from bottom to top.

We assume that the reader is familiar with classical mechanics, classical electromagnetic theory, and non-relativistic quantum mechanics. Although the students to whom the material of this book was first addressed had a good background in mathematics covering the standard fields such as analysis, geometry, and group theory, the techniques we shall use will not exceed what is usually taught in an advanced undergraduate course. We have added three appendices with some more specialised topics.

We start with a very brief reminder of the basic principles of classical mechanics. This book will develop the theory of relativistic quantum fields, so we have devoted the second chapter to the properties of the Lorentz group and, in particular, its spinorial representations. In the third chapter we present the first successful classical field theory, namely Maxwell's theory of electromagnetism formulated as a Lagrangian field theory. It will serve as a model throughout this book. The purpose is not to review classical electromagnetism, but rather to extract some features which will be useful in the discussion of more general field theories. In the fourth chapter we give a very brief review of general relativity, the other 'Classical' field theory. It shares with electrodynamics the property of gauge invariance, but it goes further because of the non-linearity of the transformations. These two classical theories have guided our intuition for the understanding of the fundamental forces of nature.

The main applications we have in mind will be in elementary particle physics, so in the fifth chapter we present the space of physical states as the Fock space built out of free particles. We use the invariance under the Poincaré group and no knowledge of second quantisation is required.

In the sixth chapter we present the simplest relativistic wave equations for fields of spin 0, 1/2, and 1. They are studied as classical differential equations and in the next chapter we attempt to construct out of them a relativistic version of quantum mechanics.

From Classical to Quantum Fields. Laurent Baulieu, John Iliopoulos and Roland Sénéor.

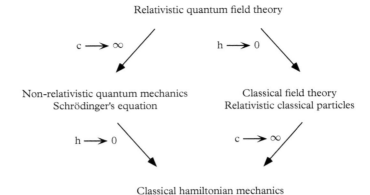

Figure 1.1 *The different approximations of the physical reality. h is the Planck constant, c is the speed of light and $\hbar = h/2\pi$.*

We show that all such attempts point unmistakably to a system with an infinite number of degrees of freedom, a quantum field theory.

This quantum theory is presented using the path integral approach which is gradually developed in the next chapters. Following Feynman's original presentation, we show how we can derive Schrödinger's equation in the non-relativistic case. The quantum mechanical harmonic oscillator is analysed and we establish the connection with the formalism of canonical quantisation. In order to establish a more concrete mathematical framework, we pass to the Euclidean version, in which some rigorous definitions can be given. A special chapter contains elements of the axiomatic formulation including the two basic theorems which form the foundations of relativistic quantum field theory, namely *PCT* and the connection between spin and statistics. Finally, we develop the asymptotic theory and the Feynman rules for scalar, spinor, and electromagnetic fields. The extension to Yang–Mills gauge theories is based on the invariance of the effective Lagrangian under BRST transformations which allow for a unified treatment of the Ward identities and the gauge independence of physical quantities.

The theory of perturbative renormalisation is developed separately, including its relation to the renormalisation group and the problem of anomalies. Several properties of the correlation functions in quantum field theory, such as the analyticity properties in the complex energy plane, the problem of the infrared divergences, and the classical limit through the use of the coherent state formalism, are presented in separate chapters. A special chapter summarises the rigorous results which show the existence of some simplified field theory models beyond the perturbation expansion. The last chapters show various applications to particle physics. We present the standard model of strong, electromagnetic, and weak interactions and the various attempts to go beyond.

A word about references and bibliography. This book touches so many subjects, without exhausting any of them, that a list of references is not practical. We decided to follow a minimal line and we shall give very few references. The choice is, to a large extent, arbitrary and the criteria vary in each case. They include: (i) books or articles whose arguments we follow very closely, (ii) works which are not widely known, and (iii) works whose reading is essential for the understanding of the results presented in the book.

1.2   Units and Notations

The basis for dimensional analysis is given by the mass $[M]$, the length $[L]$, and the time $[T]$. The dimensions of other physical quantities can be expressed as a combination of these three basic ones. For example,

The speed of light c: $[c] = \frac{[L]}{[T]}$

The Planck constant $\hbar$: $[\hbar] = \frac{[M][L]^2}{[T]}$

The force F: $[F] = \frac{[M][L]}{[T]^2}$

The charge e: $[e]^2 = \frac{[M][L]^3}{[T]^2}$

The fine-structure constant $\alpha = \frac{e^2}{4\pi\hbar c} = \frac{1}{137}$, which is dimensionless.

We will use from now on 'God-given' units, where

$$\hbar = c = 1. \tag{1.1}$$

In this system $[L] = [T], [M] = [L]^{-1}, [F] = [L]^{-2}$, and $[e^2] = [1]$.

The only unit left is $[L]$.

Spatial vectors will be denoted $\boldsymbol{x}$ or $\mathbf{x}$. Their components will generally be noted with Roman indices $\boldsymbol{x} = (x^1, x^2, x^3) = (\{x^j\})$.

Four-vectors will be denoted x and their component indices will be Greek letters: $x = (x^0, x^1, x^2, x^3) = (\{x^\mu\})$. The scalar product between x and y is denoted by xy or $x.y$.

We shall adopt the *Einstein convention* for the summation over repeated indices.

A word about complex conjugation: In physics books the notation is not uniform. We shall adopt the standard convention according to which $\bar{z}$ denotes the complex conjugate of the complex number z. For an operator A we shall write A^* or $A^\dagger$ for its Hermitian conjugate. Since for a c-number function the operations of complex and Hermitian conjugation coincide, we shall occasionally write f^* for the complex conjugate of f, especially in cases in which the same symbol may denote an operator in the following sections. Unfortunately, this simple rule has an exception: In the physics literature it has been established to use the 'bar' notation also in a different context. If Ψ represents a spinor, $\bar{\Psi}$ is not just the complex conjugate of it. We shall give the precise definition in the appropriate chapter and we shall make sure there is no confusion whenever the symbol is used.

1.3 Hamiltonian and Lagrangian Mechanics

We will give in this section a brief review of rational mechanics. The high-level conceptualisation of classical mechanics has played, since the end of the nineteenth century, an essential part in the development of physical theories.

1.3.1 Review of Variational Calculus

Given $q \in \mathbb{R}^n$ and let $\gamma = \{t, q \mid q = q(t), t_0 \le t \le t_1\}$ be a curve in $\mathbb{R}^n \times \mathbb{R}$ such that $q(t_0) = q_0$ and $q(t_1) = q_1$, and let the Lagrangian $L : \mathbb{R}^n \times \mathbb{R}^n \times \mathbb{R} \to \mathbb{R}$ be a sufficiently regular function of $2n + 1$ variables. We have the following.

Theorem 1. *The curve γ is extremal for the functional $\Phi(\gamma) = \int_{t_0}^{t_1} L(q, \dot{q}, t)dt$ in the space of the curves joining (t_0, q_0) to (t_1, q_1) if and only if the Euler–Lagrange equations are verified along γ.*

We recall that the *Euler–Lagrange equations* are

$$\frac{\mathrm{d}}{\mathrm{d}t}\left(\frac{\partial L}{\partial \dot{q}}\right) - \frac{\partial L}{\partial q} = 0. \tag{1.2}$$

The principle of least action links the Newton equations to the Euler–Lagrange equations.

Theorem 2 (Least Action Principle). *The extrema of the functional $\Phi(\gamma)$ where $L = T - U$, the difference between kinetic energy and potential energy, are given by the solutions of the equations of motion.*

The proof is obvious for $T = \frac{1}{2}m(\dot{q}, \dot{q})$ and $U = U(q)$. Introducing the generalised momentum $p = \frac{\partial L}{\partial \dot{q}}$, we have the following.

Theorem 3. *The Lagrange system of equations $\dot{p} = \frac{\partial L}{\partial q}$ and $p = \frac{\partial L}{\partial \dot{q}}$ is equivalent to the Hamilton system of $2n$ first-order equations*

$$\dot{p} = -\frac{\partial H}{\partial q} \tag{1.3}$$

$$\dot{q} = \frac{\partial H}{\partial p}, \tag{1.4}$$

where $H(p, q, t) = p\dot{q} - L(q, \dot{q}, t)$ is the Legendre transform of the Lagrangian.

Proof. The Legendre transform is used to pass from a function L of $\dot{q}$ and q to a function H of p and q. One can invert the equation describing the generalised momentum p from the Lagrangian to express $\dot{q}$ as a function of p and q: $\dot{q} = \dot{q}(p, q, t)$. Therefore, we have (we suppose for simplicity that L does not depend explicitly on time)

$$\mathrm{d}H = \frac{\partial H}{\partial p}\mathrm{d}p + \frac{\partial H}{\partial q}\mathrm{d}q \tag{1.5}$$

$$= \left(\dot{q} + p\frac{\partial \dot{q}}{\partial p} - \frac{\partial L}{\partial \dot{q}}\frac{\partial \dot{q}}{\partial p}\right)\mathrm{d}p + \left(p\frac{\partial \dot{q}}{\partial q} - \frac{\partial L}{\partial \dot{q}}\frac{\partial \dot{q}}{\partial q} - \frac{\partial L}{\partial q}\right)\mathrm{d}q \tag{1.6}$$

$$= \dot{q}\mathrm{d}p - \frac{\partial L}{\partial q}\mathrm{d}q \tag{1.7}$$

using Hamilton's equations.

We deduce that a function $f(p, q, t)$ varies with t like

$$\frac{\mathrm{d}f}{\mathrm{d}t} = \frac{\partial f}{\partial t} + \frac{\partial f}{\partial q}\frac{\partial H}{\partial p} + \frac{\partial f}{\partial p}\left(-\frac{\partial H}{\partial q}\right) = \frac{\partial f}{\partial t} + \{H, f\}, \tag{1.8}$$

where by definition, the *Poisson bracket* of two functions f and g is given, component-wise, by

$$\{f, g\} = \frac{\partial f}{\partial p^i}\frac{\partial g}{\partial q^i} - \frac{\partial f}{\partial q^i}\frac{\partial g}{\partial p^i}. \tag{1.9}$$

If f does not depend explicitly on time, saying that f is a constant of motion is equivalent to saying that its Poisson bracket with H vanishes.

Hamilton's equations also result from a stationary principle. Indeed, consider the variation of the action

$$I = \int_{t_0}^{t_1} L(q, \dot{q})\mathrm{d}t = \int_{t_0}^{t_1} p\mathrm{d}q - H(p, q)\mathrm{d}t \tag{1.10}$$

with $\tilde{q}(t) = q(t) + \delta q(t)$ and $\tilde{p}(t) = p(t) + \delta p(t)$ under the condition that $\delta q(t_1) = \delta q(t_0) = 0$. We find that δI is null at first order if Hamilton's equations are verified.

We will now give a global formulation of a Lagrangian system in mechanics.

Given M is a differentiable manifold, $T(M)$ its tangent fibre space, and $L : T(M) \to \mathbb{R}$ a differentiable function, a trajectory of the Lagrangian system (M, L), of configuration space M and Lagrangian L, is an extremal curve of the functional

$$\Phi(\gamma) = \int_{t_0}^{t_1} L(\dot{\gamma})\mathrm{d}t, \tag{1.11}$$

where $\dot\gamma \in T_{\gamma(t)}(M)$ is the velocity vector. The system of local coordinates $\boldsymbol{q} = (q_1,\ldots,q_m)$ for a point of $\gamma(t)$ subject to a Lagrangian system verifies the Euler–Lagrange equations, $L(\boldsymbol{q},\dot{\boldsymbol{q}})$ being the expression, in local coordinates, of the function L.

If the manifold is Riemannian, the quadratic form on the tangent space $T = \frac{1}{2}(v,v), v \in T_q(M)$, is the kinetic energy. The potential energy is a differentiable function $U: M \to \mathbb{R}$. A Lagrangian system is natural if $L = T - U$.

These few formulae make it possible to solve problems of mechanics with constraints, by using the Euler–Lagrange equations.

1.3.2   Noether's Theorem

We will now formulate an invariance theorem. For each one-parameter group of diffeomorphisms of the configuration manifold which preserves the Lagrangian there is a corresponding prime integral of the equations of motion, i.e. a conserved quantity.

Given $h : M \to M$, a differentiable map, and $T(h) : T(M) \to T(M)$, the induced map on speeds, a Lagrangian system (M,L) is invariant if $\forall v \in T(M)$:

$$L(T(h)v) = L(v).\tag{1.12}$$

We then have the following.

Theorem 4 (E. Noether). *If the Lagrangian system (M,L) is invariant under the one-parameter group of diffeomorphisms $h_s : M \to M, s \in \mathbb{R}$, then the system of the Euler–Lagrange equations admits a prime integral $I : T(M) \to \mathbb{R}$.*

In a local chart

$$I(\boldsymbol{q},\dot{\boldsymbol{q}}) = \frac{\partial L}{\partial \dot{\boldsymbol{q}}} \frac{\mathrm{d} h_s(\boldsymbol{q})}{\mathrm{d} s}\bigg|_{s=0}.\tag{1.13}$$

Proof. We only treat the case $M = \mathbb{R}^m$. Let $f : \mathbb{R} \to M$ be such that $\boldsymbol{q} = f(t)$ is a solution of the Euler–Lagrange equations. As L is invariant by $T(h_s)$, $T(h_s)f : \mathbb{R} \to M$ satisfies also the same equations for any s. Consider the map $\Phi : \mathbb{R} \times \mathbb{R} \to \mathbb{R}^m$, given by $\boldsymbol{q} = \Phi(s,t) = h_s(f(t))$. By hypothesis

$$0 = \frac{\partial L(\Phi,\dot\Phi)}{\partial s} = \frac{\partial L}{\partial \boldsymbol{q}}\Phi' + \frac{\partial L}{\partial \dot{\boldsymbol{q}}}\dot\Phi',\tag{1.14}$$

where the $'$ indicate the derivation with respect to s, the dots with respect to t, and where the derivatives are calculated at the points $\boldsymbol{q} = \Phi(s,t)$ and $\dot{\boldsymbol{q}} = \dot\Phi(s,t)$. By hypothesis, for all constant s, the mapping

$$\Phi\big|_{s=constant} : \mathbb{R} \to \mathbb{R}^m\tag{1.15}$$

verifies the Euler–Lagrange equations

$$\frac{\partial}{\partial t}\left[\frac{\partial L(\varPhi,\dot{\varPhi})}{\partial \dot{q}}\right] = \frac{\partial L(\varPhi,\dot{\varPhi})}{\partial q}.$$ (1.16)

By using this equality in (1.14), we have

$$0 = \left(\frac{\mathrm{d}}{\mathrm{d}t}\frac{\partial L}{\partial \dot{q}}\right)q' + \frac{\partial L}{\partial \dot{q}}\left(\frac{\mathrm{d}}{\mathrm{d}t}q'\right) = \frac{\mathrm{d}}{\mathrm{d}t}\left(\frac{\partial L}{\partial \dot{q}}q'\right) = \frac{\mathrm{d}I}{\mathrm{d}t}.$$ (1.17)

1.3.3   Applications of Noether's Theorem

Let us apply the theorem to the Lagrangian of N particles $L = \sum_{i=1}^{N}\frac{1}{2}m_i(\dot{q}^i)^2 - U(q^1,\cdots,q^N)$, $q^i \in \mathbb{R}^3$, $i = 1,\cdots,N$ in the following two cases:

1. The Lagrangian is invariant under translations

$$h_s : q^i \rightarrow q^i + s\boldsymbol{a}, \quad \boldsymbol{a} \in \mathbb{R}^3, \quad i = 1,\cdots,N.$$ (1.18)

2. The Lagrangian is invariant under rotations

$$h_s : q^i \rightarrow (R(\boldsymbol{n},\theta)q^i), \quad i = 1,\cdots,N,$$ (1.19)

where $R(\boldsymbol{n},\theta)$ is a rotation of direction $\boldsymbol{n}$ and angle θ.

Setting $\boldsymbol{q} = (q^1,\cdots,q^N)$, $\boldsymbol{q} \in M = (\mathbb{R}^3)^{\otimes N}$, we can define obviously in both cases an extension of h_s to M.

Note that in both cases it is a property of the potential that is expressed, since the kinetic energy term is invariant under both translations and rotations.

1.3.3.1   *Invariance under Translations*

Given $h_s(q^i) = q^i + s\boldsymbol{a}$, therefore

$$\frac{\mathrm{d}h_s(q^i)}{\mathrm{d}s} = \boldsymbol{a},$$ (1.20)

so

$$\frac{\partial L}{\partial \dot{q}}q' = \sum_i \frac{\partial L}{\partial \dot{q}^i}(q^i)' = \sum_i m_i\dot{q}^i\boldsymbol{a} = \left(\sum_i p^i\right)\boldsymbol{a} = \boldsymbol{pa}$$ (1.21)

and

$$0 = \frac{\mathrm{d}I}{\mathrm{d}t} = \frac{\mathrm{d}\boldsymbol{p}\boldsymbol{a}}{\mathrm{d}t} = \boldsymbol{a}\frac{\mathrm{d}\boldsymbol{p}}{\mathrm{d}t} \tag{1.22}$$

since $\boldsymbol{a}$ is a constant.

We thus see that the invariance under translation of the Lagrangian implies the *conservation of the total momentum*.

1.3.3.2   Invariance under Rotations

Let us fix the direction of the rotation and take as parameter s the angle of rotation. It is a one-parameter group of diffeomorphisms. Now with

$$h_\theta(\boldsymbol{q}^i) = R(\boldsymbol{n}, \theta)\boldsymbol{q}^i = \boldsymbol{q}^i(\theta) \tag{1.23}$$

we have

$$\frac{\mathrm{d}}{\mathrm{d}\theta} h_\theta(\boldsymbol{q}^i) = \frac{\mathrm{d}}{\mathrm{d}\theta} \boldsymbol{q}^i(\theta). \tag{1.24}$$

The conserved quantity is

$$I = \sum_{i=1}^{N} m_i \dot{\boldsymbol{q}}^i(\theta) \cdot \frac{\mathrm{d}}{\mathrm{d}\theta} \boldsymbol{q}^i(\theta)|_{\theta=0},$$

and since

$$\boldsymbol{q}\cos\theta + \boldsymbol{n}(1 - \cos\theta)\boldsymbol{n}\cdot\boldsymbol{q} + \sin\theta\,\boldsymbol{n}\wedge\boldsymbol{q} \tag{1.25}$$

this shows that

$$\frac{\mathrm{d}}{\mathrm{d}\theta} \boldsymbol{q}^i(\theta) = \sin\theta\,(\boldsymbol{n}\,\boldsymbol{n}\cdot\boldsymbol{q}^i - \boldsymbol{q}^i) + \cos\theta\,\boldsymbol{n}\wedge\boldsymbol{q}^i \tag{1.26}$$

and therefore

$$I = \sum_{i=1}^{N} m_i \dot{\boldsymbol{q}}^i \cdot (\boldsymbol{n}\wedge\boldsymbol{q}^i) = \sum_{i=1}^{N} m_i \boldsymbol{n}\cdot(\boldsymbol{q}^i\wedge\dot{\boldsymbol{q}}^i) = \boldsymbol{n}\cdot\left(\sum_{i=1}^{N} \boldsymbol{q}^i\wedge\boldsymbol{p}^i\right). \tag{1.27}$$

As I is conserved for any $\boldsymbol{n}$, what follows is the   *conservation of the total angular momentum*:

$$L = \sum_{i=1}^{N} \boldsymbol{q}^i\wedge\boldsymbol{p}^i\,. \tag{1.28}$$

2
Relativistic Invariance

2.1 Introduction

Until the beginning of the twentieth century, physicists postulated the existence of an absolute time, a Euclidean space, and *'Galileo's Principle of Relativity'*, that is to say the invariance of the laws of physics by the *Galilean group of transformations*. The transformations generating this group are space rotations, constant translations in space and time, and the passage from one reference frame to any other in uniform motion with regard to the first. Some reference frames, the *reference frames of inertia*, are privileged. They are such that the laws of physics are the same there as in absolute space.

The principle of Galilean relativity and the idea of an absolute time are at the origin of the *Principia*, written by Newton in 1686. Newton justifies them by the predictive power of the physical theories that come from them. However, he perceives their limits by the following consequences: the unobservability of space by any experiment, the instantaneous propagation of signals, and the interaction between corpuscles. He writes to Richard Bentley in 1687:

> 'That one body may act upon another at a distance through a vacuum, without the mediation of something else, by and through which their action and force may be conveyed from one to another, is to me so great an absurdity, that I believe no man, who has in philosophical matters a competent faculty of thinking, can ever fall into it.'

More than any other, with the range of mathematical knowledge of his time, Newton knew that he could not go any further in his analysis of space–time. Many of his writings state his faith in the future generations to deepen the understanding of natural phenomena, theoretically and experimentally.

The theory of electrodynamics presented by Maxwell in 1864 and the Michelson experiment mark the abandon of the Newtonian theory of space and time. The Maxwell equations predict that the speed of light in the vacuum is a universal constant c (equal, by *definition*, since 1983 to $299,792,458 \, \mathrm{m \, s^{-1}}$). Because of the Newtonian law on composition of speeds, this prediction cannot be true in two frames in relative motion defined by a Galilean transformation. The existence of an ether serving as support to the propagation

From Classical to Quantum Fields. Laurent Baulieu, John Iliopoulos and Roland Sénéor.
© Laurent Baulieu, John Iliopoulos and Roland Sénéor, 2017. Published 2017 by Oxford University Press.

of light has allowed for a certain time to admit that only the laws of mechanics obeyed the Galilean principle of relativity, the laws of electromagnetism being true only for some particular inertial frames, those at rest with respect to the ether. Faced by the failure of the experiments to measure the speed of motion of the earth in the ether, and particularly the Michelson experiment, the proposition from Poincaré, Lorentz, and Einstein to replace the Galilean transformations with the Lorentz transformations led Einstein to introduce in 1905 a new mechanics in which electrodynamics and mechanics satisfy a unique principle, the *principle of special relativity*. This new principle postulates the invariance of the equations of physics by Lorentz transformations rather than by those of Galileo. One of the spectacular consequences of this new mechanics is the violation of the law of composition of velocities predicted by Newton's theory. The principle of equivalence of the forces of inertia and of gravitation, discovered by Einstein in 1916 makes it possible to incorporate gravitation within the frame of a relativistic theory. This principle is the basis for the *theory of general relativity*. It is equivalent to the statement that no experiment confined to an infinitely small region of space–time could allow us to distinguish an inertial frame from another one.

Today the hypothesis on space–time is as follows. It should be possible to write the laws of physics geometrically in a Riemannian manifold of four dimensions, locally equivalent to a Minkowskian space with a signature $(+, -, -, -)$. The microscopic particles are described by local fields. If we introduce a system of local coordinates, the classical equations of motion are differential equations, deriving by a principle of least action from an action invariant under local Lorentz transformations. The quantum theory is founded using the formalism of functional integration, leading back within certain limits to the classical theory.

Einstein's theory allows coherent classical descriptions of gravity and electrodynamics. Although we will not study the quantum theory of gravitation in this book, it is still necessary to understand how the force of gravity is linked to the local properties of space–time. We shall give a brief summary of the classical theory of general relativity in Chapter 4; here we want only to point out that a simple consequence of Einstein's equations is that the gravitational attraction between corpuscles is linked to the curvature of space, itself induced by the corpuscles. Let us take the simplified case of a two-dimensional curved space, for example a sphere. Let two test particles exist in a neighbourhood considered as small compared to the curvature of the sphere. According to Einstein's principle, they must move at a constant speed on geodesics, that is to say on big circles. If we observe the particles for a sufficiently short time, their trajectories appear to be parallel and the Newtonian principles are respected. After a certain time, the observer will see the particles to converge. He will then be able to conclude that there exists an attractive force between the two particles.

Einstein's analysis on the nature of space–time is confirmed by many experiments in classical physics. However, the quantum theory of gravitation still escapes our comprehension despite the common efforts of physicists and mathematicians. Moreover, because of the orders of magnitude of physical constants and perhaps because of our lack of imagination, there do not exist experiments putting into evidence a gravitational quantum effect.

In the following, unless stated otherwise, we will systematically neglect the action of gravity. We will try to give an insight of the techniques that have made it possible to give a sufficiently satisfactory description of microscopic physical interactions, other than gravity, between particles.

In the approximation where we neglect gravitation, everything is as if space–time is not curved. We therefore postulate that the physical scene is the Minkowski space M^4, that is $\mathbb{R}^4$, with the metric tensor $g_{\mu\nu}\mathrm{d}x^\mu \otimes \mathrm{d}x^\nu = \mathrm{d}s^2$, with components

$$G = \{g_{\mu\nu}\} = \begin{pmatrix} 1 & 0 & 0 & 0 \\ 0 & -1 & 0 & 0 \\ 0 & 0 & -1 & 0 \\ 0 & 0 & 0 & -1 \end{pmatrix}, \tag{2.1}$$

where the indices μ and ν take the values $0, 1, 2, 3$.

The rest of this chapter is devoted to the study of the Lorentz transformations, i.e. to the group leaving the quadratic form $g_{\mu\nu}x^\mu x^\nu$ invariant; it is the invariance group of the Maxwell equations. Since we intend to write later the Dirac equation, which generalises the Maxwell equations, to include particles of spin 1/2, we will introduce the notion of spinors, which is otherwise fundamental in rotation theory. We will consider the spinors firstly in the simplest case, that of the three-dimensional Euclidean space, then those linked to the four-dimensional relativistic space.

2.2 The Three-Dimensional Rotation Group

Let us consider the rotation group in the three-dimensional Euclidean space $\mathbb{R}^3$. The vectors x of this space form a representation space which is easy to visualise. The action of a rotation on the vectors is characterised by a rotation axis, which is a unitary vector n, and an angle θ defined modulo 2π. It is represented by a real 3×3 matrix R, such that the length of the transformed vector $x' = Rx$ is equal to that of x. If G is the metric of $\mathbb{R}^3$ (in Cartesian coordinates, $G_{ij} = \delta_{ij}$, the Kronecker symbol), we have the relation

$$R^{tr} G R = G, \tag{2.2}$$

where R^{tr} is the transposed matrix. By taking the determinant of both sides of this equality we find $\det R = \pm 1$. We deduce that the group of rotations $O(3)$ is made of two connected components: one containing the identity 1, the other -1. We go from one to the other by the action of the inversion operator, -1, which by definition changes the sign of every coordinate.[1] Let us study, without loss of generality, $SO(3)$, the connected component of $O(3)$'s identity, defined by $\det R = 1$.

The elements of $SO(3)$ are made of the elements of the group of 3×3 orthogonal matrices with determinant equal to 1. This group is connected but not *simply connected*.

---

[1] In physics we often call the operation of space inversion *parity (P)*.

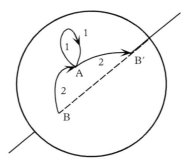

The points of intersection with the sphere, B and B′, are identified.

Figure 2.1 *SO(3) is not simply connected.*

Simply connected means that if we have two elements of the group g_1 and g_2 and two curves continuously joining these two elements, $g(t)$ and $\tilde{g}(t)$, $0 \le t \le 1$ with $g_1 = g(0) = \tilde{g}(0)$ and $g_2 = g(1) = \tilde{g}(1)$, there exists a continuous function $g(t,s)$ on the group such that $g(t,0) = g(t)$; $g(t,1) = \tilde{g}(t)$.

To prove that the rotation group is not simply connected, we can proceed as follows: see Fig. 2.1. Any rotation is characterized by the direction n and the angle ϕ of the rotation, $0 \le \phi \le \pi$. We represent a rotation by a vector, at the origin of coordinates, of direction n and length ϕ. The manifold of the group is then a ball of radius π such that 2 antipodal points of the edge are identified (this means that a rotation around an axis n with an angle $\pi + \varepsilon$ is represented by a rotation of axis $-n$ and angle $\pi - \varepsilon$). We see in this representation of the group manifold that this group is compact (more precisely, it is the image of a compact set in the space of parameters). Moreover, given a rotation of angle θ, $\theta \ne 2k\pi$, there exist two types of laces (continuous families of one-parameter rotations) linked to this element that are not continuously deformable one into the other: the laces that stay inside the ball and those that cross two antipodal points of the edge.

Let us study the group $SO(3)$ without losing the general nature of the exposition. The theory of Lie groups enables us to write any element $R \in SO(3)$ as an exponential matrix:

$$R(n,\theta) = R = e^{-i\theta n \cdot J} \tag{2.3}$$

with $J = (\mathcal{J}^1, \mathcal{J}^2, \mathcal{J}^3)$ and

$$[\mathcal{J}^k, \mathcal{J}^l] = i\varepsilon^{klm} \mathcal{J}^m. \tag{2.4}$$

The three matrices J are the three generators of the Lie algebra of the rotation group $O(3)$ and we deduce from the fact that the elements of the group have a unit determinant that the Lie algebra is made of Hermitian matrices with trace equal to 0. They define, with their commutation relations, a three-dimensional representation, the vectorial representation of the Lie algebra, also called the associated representation. The commutation relations are by definition independent of the chosen representation.

Let us find the expression of the generators in Cartesian coordinates. Consider first the case of a rotation $R_3(\theta)$, of angle θ, around the third axis, the z axis. We have

$$R_3(\theta) = \begin{pmatrix} \cos\theta & -\sin\theta & 0 \\ \sin\theta & \cos\theta & 0 \\ 0 & 0 & 1 \end{pmatrix}. \tag{2.5}$$

As $R_3(\theta) = e^{-i\theta \mathcal{J}^3}$ for all θ, we find $\mathcal{J}^3$ by comparing (2.5) to the derivative of $e^{-i\theta \mathcal{J}_3}$ with respect to θ, calculated at 0. Similarly, we can consider the rotations around x and y axes. We come up with the following expression of the three generators $\mathcal{J}^i$:

$$\mathcal{J}^1 = \begin{pmatrix} 0 & 0 & 0 \\ 0 & 0 & -i \\ 0 & i & 0 \end{pmatrix}; \quad \mathcal{J}^2 = \begin{pmatrix} 0 & 0 & i \\ 0 & 0 & 0 \\ -i & 0 & 0 \end{pmatrix}; \quad \mathcal{J}^3 = \begin{pmatrix} 0 & -i & 0 \\ i & 0 & 0 \\ 0 & 0 & 0 \end{pmatrix}. \tag{2.6}$$

Note that the matrix elements of the three generators $\mathcal{J}_i$ are identified with the structural coefficients $i\varepsilon_{ijk}$ of the group. The composition law of the rotations on the vectors is automatically taken into account by the formula expressing an element R as an exponential of matrices of the Lie algebra.

Let us now consider a differentiable function $f : \mathbb{R}^3 \to \mathbb{C}$ and study its variation for an infinitesimal rotation of the argument. This variation is described by the derivative

$$\frac{\mathrm{d}}{\mathrm{d}\theta} f(e^{-i\boldsymbol{n} \cdot \boldsymbol{J}\theta} \boldsymbol{x})\big|_{\theta=0} = \frac{\partial f(\boldsymbol{x})}{\partial x^i}(-i\boldsymbol{n} \cdot \boldsymbol{J}_{ij})x^j = i\boldsymbol{n} \cdot (\boldsymbol{x}, \boldsymbol{J}\nabla)f(\boldsymbol{x}). \tag{2.7}$$

The three differential operators

$$L_i = -(\boldsymbol{x}, \mathcal{J}^i\nabla) = -x^k \mathcal{J}^i_{kl}\frac{\partial}{\partial x^l} \tag{2.8}$$

are the components of the angular momentum $\boldsymbol{L}$. We can verify by using formula (2.6) that

$$L_i = i\varepsilon_{ijk}x^j\nabla^k = [\vec{r} \wedge i\vec{\nabla}]_i \tag{2.9}$$

and therefore at the leading order in θ, for θ small,

$$f(e^{-i\boldsymbol{n} \cdot \boldsymbol{J}\theta}\boldsymbol{x}) \simeq f(\boldsymbol{x}) - i\theta \boldsymbol{n} \cdot \boldsymbol{L}f(\boldsymbol{x}). \tag{2.10}$$

Of course, the differential operators L_i verify the commutation relations (2.4) of the Lie algebra of the rotation group.

2.3 Three-Dimensional Spinors

Experimentally, we know there exist in $\mathbb{R}^3$ geometric objects other than vectors. A deeper understanding of the action of the rotation group in three dimensions, and consequently of the Lorentz group, is obtained by the introduction of spinor representations.[2] These representations are different from those that transform vectors and tensors. The spinors and their applications in relativistic physics have been revealed by detailed studies of rotations in the three-dimensional Euclidean space and in space–time. Already in the nineteenth century, mathematicians and theoretical physicists had foreseen some subtleties of the rotation group that we cannot understand if we only consider the action of this group on vectors of the Euclidean $\mathbb{R}^3$ space. Their detailed motivations present only a historical interest today but, among them, a simple one is the following: it is well known that complex numbers are associated with rotations around an axis. This is due to the local isomorphy between the $O(2)$ and $U(1)$ groups. The natural question is now 'Are there "numbers" which are similarly associated with $O(3)$ rotations?' It seems that the answer was found by Sir William Rowan Hamilton in 1843.[3] On October 16th of that year, after some years of thought, and during a walk with his wife along the Royal Canal at Dublin, Hamilton engraved with a knife on a stone of the Brougham Bridge the formula

$$i^2 = j^2 = k^2 = ijk = -1. \tag{2.11}$$

He had just discovered quaternions which today we represent using the Pauli matrices

$$\sigma_1 = \begin{pmatrix} 0 & 1 \\ 1 & 0 \end{pmatrix} = \mathrm{i}i \qquad \sigma_2 = \begin{pmatrix} 0 & -i \\ i & 0 \end{pmatrix} = \mathrm{i}j \qquad \sigma_3 = \begin{pmatrix} 1 & 0 \\ 0 & -1 \end{pmatrix} = \mathrm{i}k. \tag{2.12}$$

We see that in order to find the 'numbers' sought by Hamilton, we must give up commutativity.[4] Using quaternions, the rotation formula (2.3) becomes

$$R(\boldsymbol{n}, \theta) = \cos\frac{\theta}{2} - \mathrm{i}\sin\frac{\theta}{2}\,\boldsymbol{n}\cdot\boldsymbol{\sigma}. \tag{2.13}$$

The law of composition of rotations is obtained by matrix multiplication:

$$R(\boldsymbol{n}_3, \theta_3) = R(\boldsymbol{n}_1, \theta_1)R(\boldsymbol{n}_2, \theta_2). \tag{2.14}$$

---

[2] The terminology 'three-dimensional spinors' may cause confusion. More precisely it refers to the two-dimensional representation of the Lie algebra of $O(3)$. In fact, a three-dimensional spinor has two components.

[3] As usual the actual history of the discovery is more complicated. Already Euler in the previous century had obtained the so-called 'four squares formula'. In 1840 the French banker and mathematician Benjamin Olinde Rodrigues applied this formula to describe rotations in three-space, but he did not write the algebraic properties of quaternions. It seems also that quaternions were known to Gauss as early as 1819, but his work remained unpublished until 1900.

[4] We can extend the argument and find octonions, but then we must also give up associativity.

The remarkable point is the dependence on $\theta/2$, which shows the existence of geometric objects needing a rotation around an axis of 4π rather than 2π to return to their original state.

The law of transformation discovered by Rodrigues and Hamilton is that of a rotation acting on a new geometric object, a *spinor*, the properties of which we will study next.[5]

We will introduce the spinors simply, beginning with the transformation laws under rotation of vectors belonging to $\mathbb{R}^3$. For each vector $x \in \mathbb{R}^3$, represented by the triplet of numbers x^i, we can associate the 2×2 matrix

$$\tilde{x} = \sigma_i x^i = x \cdot \boldsymbol{\sigma}, \tag{2.15}$$

where the σ_i are the Pauli matrices. The $x^{i\prime}$ values being real, the matrix $\tilde{x}$ is Hermitian. Furthermore, it has a null trace because the Pauli matrices have a null trace. From the identity

$$\sigma_i \sigma_j = \delta_{ij} + i\epsilon_{ijk}\sigma_k, \tag{2.16}$$

we easily show that

$$x^i = \frac{1}{2}\text{Tr}(\sigma_i \tilde{x}) \quad \text{and} \quad \det \tilde{x} = - <x, x> = -x^{tr}Gx = -x^i x_i = -|x|^2, \tag{2.17}$$

and there is a one-to-one correspondence between $\tilde{x}$ and x.

As we know the laws of transformation by rotations of the components of the vector x, we can calculate the manner in which the matrix $\tilde{x}$ transforms. For any 3×3 matrix of rotation R, such that $x' = Rx$, we will associate a 2×2 matrix with complex coefficients $A(R)$, noted A, such that

$$\tilde{x}' = A\tilde{x}A^*. \tag{2.18}$$

This matrix is not unique, as we can see from the previous formula, because if $A(R)$ satisfies it, so does $aA(R)$, with a being a complex number with modulus 1. The transformation is not linear but quadratic, and the form of the equation has been chosen in such a way as to preserve hermiticity. The conservation by rotation of the norm of a vector, $<x', x'> = <x, x>$, implies that $|\det A| = 1$; we can, therefore, choose $a = (\det A)^{-1/2}$ and suppose, without loss of generality, that $\det A = 1$. We will now determine the $A(R)$ with the condition that these matrices have a group structure. We will, in fact, show that A is an element of $SU(2)$, the *unimodular group*, that is to say the group of 2×2 unitary matrices with unit determinant, and that there exists a homomorphism from the $SU(2)$ group to the rotation group.

The unitarity property results from the preservation of the property of nullity of the trace

---

[5] The modern theory of spinors was developed by Elie Cartan in 1910. In physics the spin of the electron was introduced by Samuel Goudsmit and George Uhlenbeck in 1925, following suggestions by Pauli, who also introduced the Pauli matrices in 1927.

$$\text{Tr}\tilde{x}' = \text{Tr}A\tilde{x}A^* = \text{Tr}\tilde{x}A^*A = 0 \qquad \forall x, \tag{2.19}$$

which means that $A^*A = \alpha\mathbf{1}$ and the condition on the determinant forces the constant α to equal 1. This is proved by using the fact that the matrix A^*A is Hermitian and that any Hermitian matrix is a linear combination with real coefficients of the three Pauli matrices and of the identity.

The 2×2 identity matrix $\mathbf{1}$ is a natural neutral element for the group of matrices that we are looking for. It corresponds to the identity of the rotation group and its determinant is 1. To calculate the explicit form of any matrix A, we will use the group structure. We will therefore consider transformations close to the identity, a general transformation being obtained by the composition of infinitesimal transformations.

We can characterise an infinitesimal rotation by its three angles $\delta\theta^i$ and

$$x_i' = x_i + \varepsilon_{ijk}\delta\theta^j x^k. \tag{2.20}$$

By continuity, the matrix A is infinitesimally different from the unit matrix and can be written $A = 1 + \delta A$. As A has a unitary determinant, δA's trace is null, and we can express it using the Pauli matrices. Therefore, $A = 1 + \boldsymbol{\lambda} \cdot \boldsymbol{\sigma}$, where the components $\lambda_1, \lambda_2, \lambda_3$ of $\boldsymbol{\lambda}$ are infinitesimal. Thus we have

$$\tilde{x}' = \tilde{x} + (\boldsymbol{\lambda} \cdot \boldsymbol{\sigma})\tilde{x} + \tilde{x}(\bar{\boldsymbol{\lambda}} \cdot \boldsymbol{\sigma}). \tag{2.21}$$

By comparison of the two equations giving x' and by using the multiplication law of the Pauli matrices, we find easily that the components λ^j of the vector $\boldsymbol{\lambda}$ are pure imaginary numbers, with $\lambda^j = -i\delta\theta^j/2$. Finally, for an infinitesimal rotation, we have

$$A = 1 - \frac{i}{2}\delta\boldsymbol{\theta} \cdot \boldsymbol{\sigma}. \tag{2.22}$$

As the matrices A form a group, this formula is integrable as an exponential. We find for a finite rotation, with angle θ and direction $\boldsymbol{n}$:

$$A = \exp\{-i\theta\boldsymbol{n} \cdot \frac{\boldsymbol{\sigma}}{2}\}. \tag{2.23}$$

Remark: The correspondence which for $R \in SO(3)$ associates a matrix A with $\det A = 1$ is not bijective because the matrices A and $-A$ give the same rotation as $\det(-\mathbf{1}) = \det\mathbf{1} = 1$.

Comparing this formula to (2.3), we see that the matrices $\boldsymbol{\sigma}/2$ have played for A the same role as J for three-dimensional rotations. In fact, from the properties of the Pauli matrices, the commutation relation

$$[\frac{1}{2}\sigma_i, \frac{1}{2}\sigma_j] = i\varepsilon_{ijk}\frac{1}{2}\sigma_k \tag{2.24}$$

follows.

These matrices form a two-dimensional representation of the Lie algebra of the rotation group. The fact that the groups $SO(3)$ and $SU(2)$ have the same Lie algebra shows that they are very close to each other. In fact, the group $SU(2)$ is a simply connected Lie group; it is the covering group of the rotation group. Simple connectedness is obtained by associating with each element of $SO(3)$ two elements of $SU(2)$: A and $-A$. The preceding arguments yield the existence of a group homomorphism

$$\phi : SU(2) \to SO(3), \tag{2.25}$$

given by $\phi(A(R)) = R$ and with kernel $\{-1, 1\}$. Using the exponential development of matrices and the properties of Pauli matrices, we can rewrite (2.23) as

$$A = \cos\frac{\theta}{2} - \mathrm{i}\, \boldsymbol{n} \cdot \boldsymbol{\sigma} \sin\frac{\theta}{2}. \tag{2.26}$$

Using the property $\|\boldsymbol{n}\| = 1$, it is easy to verify that $\det A = 1$ and that the matrix A is unitary.

It is possible to make the matrices $A \in SU(2)$ act as linear operators on the vectors ξ of the $\mathbb{C}^2$-vectorial space and on the components ξ^α as

$$\xi'^\alpha = A^\alpha_\beta \xi^\beta. \tag{2.27}$$

The preceding transformation law defines ξ as a three-dimensional spinor. What distinguishes a spinor from a vector is the dependence of the transformation matrix A to half the angle of rotation.

In modern language, the set of matrices $A \in SU(2)$ and the vectorial space $\mathbb{C}^2$ constitute the representation of spin $\frac{1}{2}$—that we note $D^{1/2}$—of the rotation group. This representation is not reducible (as we shall see in the next section).

Later on, we will see that the physically interesting objects, the wave functions, solutions of a spin-$\frac{1}{2}$ non-relativistic equation, and the Pauli equation,[6] transform as the spinors of this representation.

As in the case of vectors, we can easily define a scalar product invariant under the action of the rotations. For this, we introduce the spinor's metric tensor ζ:

$$\zeta_{\alpha\beta} = \begin{pmatrix} 0 & 1 \\ -1 & 0 \end{pmatrix} = \mathrm{i}\sigma_2. \tag{2.28}$$

The scalar product of two arbitrary spinors ξ and ϕ is then defined by the antisymmetric quadratic form

$$< \phi, \xi > = \phi^{tr}\zeta\xi = \phi^\alpha \zeta_{\alpha\beta}\xi^\beta = \phi^1\xi^2 - \phi^2\xi^1. \tag{2.29}$$

---

[6] The Pauli equation is a generalization of the Schrödinger equation for the spin-$\frac{1}{2}$ particles that we will study in Chapter 6.

The invariance of the scalar product under rotation results from the property $\det A = 1$. We can observe that the scalar product of any spinor with itself is zero:

$$< \xi, \xi >= \xi^1 \xi^2 - \xi^2 \xi^1 = 0. \tag{2.30}$$

We therefore cannot use the invariant scalar product $< \ , \ >$ to define a norm for spinors.

The introduction of the metric ζ makes it possible nevertheless to define a tensor calculus according to the usual rules of rising and lowering indices. Thus, the components ξ^α of a spinor are called contravariant and the components $\xi_\alpha = \zeta_{\alpha\beta} \xi^\beta$ are called covariant, and we have $\zeta_{\alpha\beta} = -\zeta^{\alpha\beta}$. We can rewrite the scalar product of two spinors under the form

$$< \phi, \xi >= -\phi_\alpha \xi^\alpha = \phi^\alpha \xi_\alpha. \tag{2.31}$$

The action of a rotation of matrix A on the covariant components is therefore expressed by the matrix $\zeta A \zeta^{-1}$, which is also unimodular. A simple calculation shows that

$$\zeta A \zeta^{-1} = -\zeta A \zeta = \bar{A}, \tag{2.32}$$

and since $\zeta \in SU(2)$, the representations given by A or by $\bar{A}$ are equivalent. We recall that in these formulae 'bar' means complex conjugation. Furthermore, we can see that by rotation, the covariant components ξ_α of a spinor ξ transform in an identical manner as the complex conjugate components $\bar{\xi}^\alpha$ of the contravariant components ξ^α.

The preceding remark leads to the observation that the initial representation of the rotation group defined by vectors of $\mathbb{R}^3$ can be obtained from the tensor product $D^{1/2} \otimes D^{1/2}$ of two spin $\frac{1}{2}$ representations, a phenomenon described by the formula $\tilde{x}' = A\tilde{x}A^*$, which is written in components

$$\tilde{x}'^{\alpha\beta} = A^\alpha_\gamma \bar{A}^\beta_\delta \tilde{x}^{\gamma\delta}. \tag{2.33}$$

As the matrices A are unitary, we can define, without using the scalar product defined earlier, the invariant norm of a spinor in the following manner:

$$\| \xi \| = \xi^{tr} \bar{\xi} = \xi^1 \bar{\xi}^1 + \xi^2 \bar{\xi}^2. \tag{2.34}$$

In quantum mechanics, this bilinear and positively defined scalar form is used to define the probability density of the presence of non-relativistic particles of spin $\frac{1}{2}$.

The identity of the transformation laws by rotation of $(\bar{\xi}^1, \bar{\xi}^2)$ and $(\xi^2, -\xi^1)$ is closely related to the symmetry with respect to the inversion of time.

2.4   Three-Dimensional Spinorial Tensors

The notion of spinors of higher rank is introduced according to the ordinary procedure of tensor formalism. Thus, the spinors which have been introduced earlier are called rank 1 spinors. A contravariant spinor of rank 2 is a quantity made of four complex components $\Psi^{\alpha\beta}$ that transform by rotation as the products $\xi^{\alpha}\phi^{\beta}$ of the contravariant components of two rank 1 spinors. By using the metric tensor $\zeta_{\alpha\beta}$, we can consider the covariant $\Psi_{\alpha\beta}$ or mixed Ψ_{β}^{α} components, transforming as the products $\xi_{\alpha}\phi_{\beta}$ or $\xi^{\alpha}\phi_{\beta}$. We have

$$\Psi_{\alpha\beta} = \zeta_{\alpha\gamma}\zeta_{\beta\delta}\Psi^{\gamma\delta} \qquad \Psi_{\beta}^{\alpha} = \zeta_{\beta\gamma}\Psi^{\alpha\gamma}. \tag{2.35}$$

Therefore, $\Psi_{12} = -\Psi_{1}^{1} = -\Psi^{21}, \Psi_{11} = \Psi_{1}^{2} = \Psi^{22}$, etc. We define similarly tensors of arbitrary rank, the rank of a tensor being the number of its indices. Spinor algebra, and particularly the operations of symmetrisation and anti-symmetrisation, is relatively simple to master, given that each index can take only two values.

The quantities $\zeta_{\alpha\beta}$ form themselves an invariant antisymmetric tensor of rank 2. Because an index can only take 2 values, there exist only rank 2 antisymmetric tensors, and any antisymmetric tensor is the product of a scalar by the metric tensor $\zeta_{\alpha\beta}$. We can easily verify that the product $\zeta_{\alpha\beta}\zeta^{\beta\gamma}$ is, as it should be, the 'unit spinor', that is to say, the rank 2 symmetric spinor $\delta_{1}^{1} = \delta_{2}^{2} = 1, \delta_{2}^{1} = \delta_{1}^{2} = 0$, with

$$\zeta_{\alpha\beta}\zeta^{\beta\gamma} = \delta_{\alpha}^{\gamma}. \tag{2.36}$$

In spinor algebra, as in tensor algebra, we have two fundamental operations: the product of two spinorial tensors and the contraction of two indices of the same type (either upper or lower) by the metric tensor. The multiplication of two tensors of ranks n and m gives a tensor of rank $n + m$, and the contraction on two indices lowers the rank by two units. Therefore, the contraction of the rank 6 spinor $\Psi_{\lambda\mu}^{\nu\varrho\sigma\kappa}$ on the indices μ and ν gives the rank 4 spinor $\Psi_{\lambda\mu}^{\mu\varrho\sigma\kappa}$; the contraction of the rank 2 spinor Ψ_{μ}^{ν} gives the scalar (or rank 0 spinor) Ψ_{μ}^{μ}, etc. Because of the antisymmetry of the metric tensor, it is clear that the contraction of a tensor on two symmetric indices gives the null tensor.

Contraction being the only covariant operation that makes it possible to lower the rank of a tensor, we can deduce that spinors which are totally symmetric on all their indices form the spaces of the irreducible representations of the rotation group. Consider a spinor u of rank n, element of $(\mathbb{C}^{2})^{\otimes n}$, on which acts the tensor representation $(D^{1/2})^{\otimes n}$ by

$$(D^{1/2})^{\otimes n} : u^{\alpha_{1}\cdots\alpha_{n}} \to A^{\alpha_{1}}_{\ \beta_{1}} \cdots A^{\alpha_{n}}_{\ \beta_{n}} u^{\beta_{1}\cdots\beta_{n}}. \tag{2.37}$$

If $n \geq 1$, the representation defined by (2.37) is not irreducible. To show this let us consider the contraction of two indices of u:

$$v^{\alpha_{3}\cdots\alpha_{n}} = \zeta_{\alpha_{1}\alpha_{2}}u^{\alpha_{1}\cdots\alpha_{n}} = u^{12\cdots\alpha_{n}} - u^{21\cdots\alpha_{n}}. \tag{2.38}$$

The space of the v's is an invariant sub-space of the representation defined by (2.37). This shows that only the completely symmetric tensor spaces do not have invariant sub-spaces.

With a simple count, we find that the space of representations of totally symmetric spinors of rank n has dimension $n + 1$, as only the components having 0 times, 1 times, ... or n times the index 1 are distinct.

The decomposition of any rank n spinor $\Psi^{\alpha\beta\gamma\cdots}$ with respect to the space of irreducible representations (made of totally symmetric spinors) gives a set of symmetric tensors of rank n, $n-2$, $n-4$, etc. The algorithm of decomposition is the following: symmetrisation of $\Psi^{\alpha\beta\gamma\cdots}$ with respect to all its indices which gives a tensor of the same rank; then, contraction of the various pairs of indices of the initial spinor gives rank $n-2$ tensors of the form $\Psi^{\alpha\gamma\cdots}_{\alpha}$ that, once symmetrised, give symmetric spinors of rank $n-2$. Symmetrising the spinors obtained by contraction of two pairs of indices, we obtain symmetric spinors of rank $n-4$, etc. Symbolically, we can write

$$\Psi^n = \Psi^{\{n\}} + \zeta \Psi^{\{n-2\}} + \zeta\zeta \Psi^{\{n-4\}} + \cdots + \zeta\zeta\zeta \Psi^{\{n-6\}} + \cdots. \tag{2.39}$$

As an example, for $n = 2$, $\Psi^{\{2\}}$ has for components $\frac{1}{2}(\Psi^{\alpha\beta} + \Psi^{\beta\alpha})$, and $\Psi^{\{0\}}$ is a scalar whose value is $\frac{1}{4}(\Psi^{\alpha\beta} - \Psi^{\beta\alpha})\zeta_{\alpha\beta}$. The space of representations of $\Psi^{\{2\}}$ is of dimension three. We can see that this spinor corresponds to the vector $\boldsymbol{a}$ whose components are

$$a_1 = \frac{i}{\sqrt{2}}(\Psi_{11} - \Psi_{22}), \qquad a_2 = -\frac{1}{\sqrt{2}}(\Psi_{11} + \Psi_{22}), \qquad a_3 = -i\sqrt{2}\Psi_{12}. \tag{2.40}$$

From spinors, we can generate, in fact, all the tensors with integer spin that are usually obtained by tensor products of vectors.

Let us end this section by pointing out that the components of the spinor $\bar{\Psi}_{\alpha\beta\cdots}$, the complex conjugates of $\Psi_{\alpha\beta\cdots}$, transform as the components of a contravariant spinor $\Phi^{\alpha\beta\cdots}$ and vice versa. The sum of the squares of the moduli of the components of any spinor is therefore an invariant.

2.5   The Lorentz Group

When we neglect the effects of gravitation, the Lorentz group is the fundamental group of invariance of any physical theory. Physically, it is clear that this group contains two subgroups, with six generators in total: the group of spatial rotations, already studied, with three generators corresponding to rotations in the planes xy, yz, and zx, and the group of pure Lorentz transformations, with three generators corresponding to changes, at constant speed, of the reference frame along the axes x, y, and z, transformations that we can represent formally as rotations of imaginary angles in the planes xt, yt, and zt.

We, therefore, define the *Lorentz group* as the set of all real linear transformations that leave invariant the quadratic form $(x, y) = g_{\mu\nu}x^{\mu}y^{\nu} = x^{\mu}y_{\mu} = x_{\mu}y^{\mu}$.

The fact that these transformations form a group is obvious. The neutral element of this group is the identity $\mathbf{1}$. Let Λ be a Lorentz transformation: $x^{\mu} \mapsto \Lambda^{\mu}_{\nu}x^{\nu}$; the invariance of the quadratic form $(x, y) = (\Lambda x, \Lambda y)$ leads to the equality

$$g_{\mu\nu} = \Lambda^\varrho{}_\mu \Lambda^\sigma{}_\nu g_{\varrho\sigma} = (\Lambda^{tr})_\mu{}^\varrho \Lambda^\sigma{}_\nu g_{\varrho\sigma}, \tag{2.41}$$

which in matrix notation can be written as

$$\Lambda^{tr} G \Lambda = G, \tag{2.42}$$

where G is given by (2.1). By taking the determinant of the two members of the previous equality, we find that

$$\det \Lambda^{tr} \det G \det \Lambda = \det G (\det \Lambda)^2 = \det G. \tag{2.43}$$

Therefore, $\det \Lambda = \pm 1$. Taking the component $(0, 0)$ of that equality we find that

$$(\Lambda^0_0)^2 - \sum_{i=1}^{3} (\Lambda^i_0)^2 = 1, \tag{2.44}$$

which implies that $|\Lambda^0_0| \geq 1$. We can deduce that the Lorentz group is formed out of four connected components corresponding to the different signs of $\det \Lambda$ and of Λ^0_0.

These four connected components respectively contain the identity matrix **1**, the inversion of space I_s given by the matrix G, the inversion of time I_t given by $-G$, and the product $I_s I_t$ of the inversions of space and time, given by $-\mathbf{1}$ (see Fig. 2.2).

We will now focus on one subgroup of the Lorentz group, *the restricted Lorentz group*, $L^\uparrow_+$, that is such that $\det \Lambda = 1$, represented by the sign +, and $\Lambda^0_0 \geq 1$, represented by a time arrow axed up. This subgroup is the connected component of the identity. The benefit of its study is that the knowledge of an infinitesimal neighbourhood of the identity suffices to totally determine the subgroup.

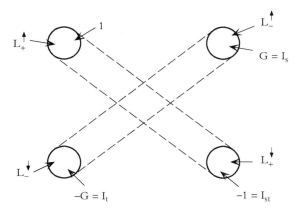

Figure 2.2 *The four connected components of the Lorentz group.*

The other subgroups are the proper group $L_+ = L_+^\uparrow \cup L_+^\downarrow$, the orthochronous group $L^\uparrow = L_+^\uparrow \cup L_-^\uparrow$, and $L_+^\uparrow \cup L_-^\downarrow$.

Two types of transformations are particularly important for physics: the pure Lorentz transformations of direction n, $n \cdot n = 1$, and speed $v = c \ \tanh \xi$,

$$(\Lambda x)^0 = x^0 \cosh \xi + x \cdot n \sinh \xi$$
$$\Lambda x = x - (x \cdot n)n + (x \cdot n \cosh \xi + x^0 \sinh \xi)n, \tag{2.45}$$

and the spatial rotations of angle θ around an axis n,

$$(\Lambda x)^0 = x^0$$
$$\Lambda x = x \cos \theta + n(n \cdot x)(1 - \cos \theta) + n \wedge x \sin \theta. \tag{2.46}$$

We can prove that any element of the restricted Lorentz group can be written as the product of a rotation and of a pure Lorentz transformation.

The spatial rotations are given by 4×4 matrices of the form

$$\begin{pmatrix} 1 & 0 \\ 0 & r \end{pmatrix}, \tag{2.47}$$

where r is a 3×3 rotation matrix of the underlying Euclidean space $\mathbb{R}^3$. We can easily check that these matrices verify the relation (2.42), that they contain the identity, and that this form is preserved by multiplication: therefore, that these matrices form a subgroup of the Lorentz group.

We shall now recall some essential facts of relativistic geometry.

The transformations (2.46) are usual rotations of the underlying three-dimensional space; we shall, therefore, focus strictly on pure Lorentz transformations.

Let us consider two frames K and $\bar{K}$, $\bar{K}$ in motion with constant speed v along an axis Ox_1 with respect to K. Let us write, using the general formula (2.45), the coordinates of a point x of universe in a frame $\bar{K}$ in terms of its coordinates in K. We find that

$$\bar{x}^0 = \frac{x^0 - \frac{v}{c}x^1}{\sqrt{1 - \frac{v^2}{c^2}}}; \quad \bar{x}^1 = \frac{x^1 - \frac{v}{c}x^0}{\sqrt{1 - \frac{v^2}{c^2}}}; \quad \bar{x}^2 = x^2; \quad \bar{x}^3 = x^3, \tag{2.48}$$

where we used the fact that $\cosh \xi = 1/\sqrt{1 - \frac{v^2}{c^2}}$ and $\sinh \xi = -(v/c)\sqrt{1 - \frac{v^2}{c^2}}$.

Given a four-vector x, the sign of $x^2 = (x, x)$ splits the Minkowski space into three disjoint subsets:

- The time-like vectors defined by $x^2 > 0$
- The light-like vectors defined by $x^2 = 0$
- The space-like vectors defined by $x^2 < 0$.

Two events, characterised by two space–time points x and y, are in causal relation if $x-y$ is time-or light-like. If $x^0 > y^0$, x is said to be in the future of y, or that y is in the past of x. We can check that for time-like vectors the notions of past or future are Lorentz invariant.

Finally, it is useful to enlarge the Lorentz group by including the translations. Thus, we define the *Poincaré group*, or the non-homogeneous Lorentz group as the group of transformations[7] (a, Λ), $a \in M^4$, such that $((a, \Lambda)x)^\mu \mapsto \Lambda^\mu{}_\nu x^\nu + a^\mu$, satisfying the composition law

$$(a_1, \Lambda_1)(a_2, \Lambda_2) = (a_1 + \Lambda_1 a_2, \Lambda_1 \Lambda_2). \tag{2.49}$$

Just as the Lorentz group, the Poincaré group has four connected components that we can determine. The Poincaré group is the group of invariance of the equations of relativistic physics. It is the most general real group leaving invariant the metric ds^2. It will play an important role later.

2.6 Generators and Lie Algebra of the Lorentz Group

Each Lorentz transformation, pure or rotation, depends on three parameters: two for the direction n and one for the angle or the speed. We will focus on infinitesimal transformations.

Let ε be an angle or a speed and $\Lambda(\varepsilon)$ an associated Lorentz transformation (also depending on the direction n) that we normalise so that when $\varepsilon \to 0$, the transformation $\Lambda(\varepsilon) \to 1$. So, at first order in ε, $\Lambda(\varepsilon) = 1 + \varepsilon\lambda$, and

$$\Lambda^{tr} G \Lambda = (1 + \varepsilon\lambda^{tr})G(1 + \varepsilon\lambda) = G; \tag{2.50}$$

that is,

$$\lambda^{tr} G + G\lambda = 0. \tag{2.51}$$

For the sake of simplicity, we will write all the matrices in covariant notation; we will be able, with the help of the metric tensor, to lower or raise the indices. Thus, in this notation, the preceding relation is written as $\lambda^{\mu\nu} = -\lambda^{\nu\mu}$; that is, λ is antisymmetric. We have

$$\lambda = \frac{d}{d\varepsilon} \Lambda(\varepsilon) \Big|_{\varepsilon=0}. \tag{2.52}$$

Let us find a basis for the λ' values. For this purpose, let us suppose that the pure Lorentz transformations are of the form[8]

---

[7] It is the semi-direct product of the Lorentz group and of the group of translations in Minkowski space.

[8] We can verify that for any λ such that $\lambda^{tr} G + G\lambda = 0$, and any real ε, $e^{\varepsilon\lambda}$ is a Lorentz transformation belonging to the restricted group.

$$e^{i\xi n \cdot N} \tag{2.53}$$

and the pure rotations

$$e^{-i\theta n \cdot J}. \tag{2.54}$$

Note: The notation $a \cdot A$, with a being a three-vector and A three matrices A_1, A_2, and A_3, means that $\sum_{j=1}^{3} a^j A_j$ and, if the three matrices A_j do not commute among each other, $e^{ia \cdot A}$ is generally different from $e^{ia^1 A_1} e^{ia^2 A_2} e^{ia^3 A_3}$, the exponential being defined by its series expansion.

N and J are six antisymmetric (4×4) matrices (with mixed indices, i.e. as operators of $M^4 \to M^4$, they verify the relation (2.51) that all the generators of the Lorentz group must satisfy).

By computing the derivatives around the identity of the Lorentz transformations and by comparing them to $in \cdot N$ for the pure Lorentz transformations and to $-in \cdot J$ for the rotations we find the relations

$$[\mathcal{J}^k, \mathcal{J}^l] = i\varepsilon^{klm} \mathcal{J}^m \tag{2.55}$$

$$[N^k, N^l] = -i\varepsilon^{klm} \mathcal{J}^m \tag{2.56}$$

$$[\mathcal{J}^k, N^l] = i\varepsilon^{klm} N^m. \tag{2.57}$$

The six matrices J and N are the six *generators* of the *Lie algebra* of the Lorentz group, algebra defined by the preceding relations. The first of these relations is nothing else than the relation that must be satisfied by the generators of the rotation group. The generators N do not form a closed sub-algebra since their commutators are expressed in terms of the J' values. The structure of the Lie algebra of the Lorentz group can be understood in a better way by complexifying it:

$$J_1 = \frac{1}{2}(J + iN) \quad \text{and} \quad J_2 = \frac{1}{2}(J - iN). \tag{2.58}$$

One then finds that

$$[\mathcal{J}_\alpha^k, \mathcal{J}_\alpha^l] = i\varepsilon^{klm} \mathcal{J}_\alpha^m \quad \text{with} \quad \alpha = 1 \text{ or } 2$$

$$[J_1, J_2] = 0. \tag{2.59}$$

One has, thus, brought to the fore that the Lie algebra of the Lorentz group is generated by two copies of the Lie algebra of the spatial rotation group.

One can form with J_1 and J_2 two quadratic invariant operators (the so-called 'Casimir operators'): J_1^2 and J_2^2, where $J^2 = \mathcal{J}^1 \mathcal{J}^1 + \mathcal{J}^2 \mathcal{J}^2 + \mathcal{J}^3 \mathcal{J}^3$. These operators commute with J_1 and J_2, thus with all the elements of the algebra and with each element of the restricted Lorentz group. In the case of an irreducible representation, they are, from Schur's lemma, proportional to the unit matrix

$$J_1^2 = j_1(j_1 + 1)\mathbf{1} \text{ and } J_2^2 = j_2(j_2 + 1)\mathbf{1}, \tag{2.60}$$

j_1 and j_2 being integers or half-integers.

2.7   The Group *SL*(2, ℂ)

The Lorentz group is a Lie group which has properties close to those of the rotation group that was previously studied. The theory of four-dimensional spinors had been worked out in the 1920s by Van der Waerden, Uhlenbeck, and Laporte, by a generalisation of the theory of three-dimensional spinors.

To introduce the four-dimensional spinors, we will follow the same approach as for those of the three-dimensional rotations. We will start from the laws of transformation of the four-vectors, obtained by asking for the invariance of $x^2 = (x^0)^2 - \mathbf{x}^2$, and show that they can be represented by the matrices of $SL(2,\mathbb{C})$, that is the group of 2×2 matrices with complex coefficients and with unit determinant. By generalising the three-dimensional spinors, the four-dimensional spinors will be introduced as the space of representations of minimal dimension of $SL(2,\mathbb{C})$. An important difference between the three-dimensional and the four-dimensional cases is that in this last case, two representations, one being a complex conjugate of the other, are not equivalent.

Let us denote by $\sigma^\mu = (\sigma^0, \boldsymbol{\sigma})$ the quadruplet of the matrix 1 and of the σ^i.

Each 2×2 matrix with complex elements can be written as a linear combination with coefficients in ℂ of the unit matrix $\sigma_0 = \mathbf{1}$ and of the three Pauli matrices.

Each 2×2 Hermitian matrix can be written as a linear combination with real coefficients of these four matrices.

To each vector x of M^4, there corresponds a 2×2 matrix

$$\tilde{x} = \sigma_\mu x^\mu = \begin{pmatrix} x^0 - x^3 & -x^1 + ix^2 \\ -x^1 - ix^2 & x^0 + x^3 \end{pmatrix}. \tag{2.61}$$

$\tilde{x}$ is Hermitian if and only if x is real. It can be checked that

$$x^\mu = \frac{1}{2}\text{Tr}(\sigma_\mu \tilde{x}) \qquad \text{and} \qquad \det \tilde{x} = x^\mu x_\mu. \tag{2.62}$$

There exists a group homomorphism (i.e. respecting the group structure) of $SL(2,\mathbb{C})$ on the restricted Lorentz group, the kernel of which is $\mathbb{Z}_2$. The mapping is given by

$$A \in SL(2,\mathbb{C}) \mapsto \Lambda(A) \tag{2.63}$$

with

$$\Lambda(A)^\mu{}_\nu = \frac{1}{2}\text{Tr}(\sigma_\mu A \sigma_\nu A^*), \tag{2.64}$$

where A^* is the conjugate transposed of A. If $A = A^*$, that is A is Hermitian, then $\Lambda(A)$ is symmetric.[9] The action of an element of $SL(2,\mathbb{C})$ is given by

$$\tilde{x} \mapsto A\tilde{x}A^*. \tag{2.65}$$

As is right and proper, the dimension of $SL(2,\mathbb{C})$ is identical to that of the restricted Lorentz group. An element of $SL(2,\mathbb{C})$ is indeed parameterised by $8-2 = 6$ independent parameters.

Note that as far as the action on four-vectors is concerned, the matrices $\bar{A}$, complex conjugate of A, and A do not correspond to the same Lorentz transformation.

One sees also on the formula (2.64) that the kernel of the mapping is $\mathbb{Z}_2$ since $\pm A$ give the same $\Lambda(A)$.

The group $SL(2,\mathbb{C})$ is the covering group of the Lorentz group. It is a simply connected group.

One shows easily that $\Lambda(A)$ is a rotation if and only if A is unitary. It is a pure Lorentz transformation if and only if A is Hermitian.

We will now introduce two important automorphisms of $SL(2,\mathbb{C})$ (a group automorphism is a bijective and bicontinuous mapping of the group on itself):

$$A \mapsto (A^{tr})^{-1} \quad \text{and} \quad A \mapsto \bar{A}. \tag{2.66}$$

The first of these automorphisms is an internal automorphism[10]

$$(A^{tr})^{-1} = \zeta A \zeta^{-1}, \quad \text{where} \quad \zeta = -(\zeta)^{-1} = \begin{pmatrix} 0 & 1 \\ -1 & 0 \end{pmatrix}. \tag{2.67}$$

The operation of complex conjugation on a matrix $\tilde{x}$ is the same as changing the component y to $-y$ and is therefore identical to a symmetry I_2 with respect to the hyperplane orthogonal to the two axes.

The second automorphism is such that

$$\Lambda(\bar{A}) = I_2 \Lambda(A) I_2. \tag{2.68}$$

Since I_2 does not belong to the restricted Lorentz group, the second automorphism is not internal, and, consequently, we must distinguish between representations which are complex conjugate of one another (in contradistinction with the representations of three-dimensional rotations).

---

[9] We remind the reader that for a matrix $\Lambda : \mathbb{R}^4 \to \mathbb{R}^4$, of components $\Lambda^\mu{}_\nu$, we define its transposed and its Hermitian conjugate by $\Lambda^{tr\mu}{}_\nu = \Lambda_\mu{}^\nu$ and $\Lambda^{*\mu}{}_\nu = (\bar{\Lambda}^{tr})^\mu{}_\nu$.

[10] An internal automorphism is a conjugation by an element of the group. If G is a group, $g \in G$, one defines an internal automorphism τ_g by $\tau_g : h \mapsto ghg^{-1}, \quad \forall h \in G$.

2.8 The Four-Dimensional Spinors

It follows from the preceding discussion that each element of the restricted Lorentz group can be represented by a 2×2 matrix with determinant 1 which we have denoted by A. The action of this element on a four-vector is quadratic in A. As a consequence, A and $-A$ define the same transformation. We will now introduce objects, the spinors, which remove this degeneracy.

By definition, a four-dimensional spinor is an object with two complex components $\xi^\alpha, \alpha = 1, 2$, which transform as

$$\xi \to A\xi. \tag{2.69}$$

This representation of $SL(2, \mathbb{C})$ in $\mathbb{C}^2$ is denoted $D^{[1/2,0]}$. This law of transformation defines the four-dimensional spinors as objects on which acts the lowest dimensional representation of $SL(2, \mathbb{C})$.

The matrix $\bar{A}$, complex conjugate to the matrix A, and A are therefore not equivalent: there are no matrices U such that for all A, $\bar{A} = U^{-1}AU$. One can thus introduce a second space of representation, made also of two component complex vectors $\eta^{\dot\alpha}, \dot\alpha = 1, 2$, called pointed, or dotted, spinors. By definition, the law of transformation corresponding to the element A of $SL(2, \mathbb{C})$ is

$$\eta \to \bar{A}.\eta \tag{2.70}$$

One has, therefore,

$$\eta^{\dot\alpha} \sim \bar{\xi}^\alpha, \tag{2.71}$$

the symbol $\sim$ meaning 'transforming' as.

The manipulation of dotted and undotted indices is, as in three-dimensions, done by the tensor $\zeta_{\alpha\beta}$. Thus, by definition,

$$\xi_1 = \xi^2 \qquad \xi_2 = -\xi^1 \tag{2.72}$$

$$\eta_{\dot 1} = \eta^{\dot 2} \qquad \eta_{\dot 2} = -\eta^{\dot 1}. \tag{2.73}$$

The tensor $\zeta_{\alpha\beta}$ makes it possible also to define an invariant scalar product for undotted and dotted spinors. The reasoning is the same as for three-dimensional spinors and uses the unimodularity property of A and $\bar{A}$. One gets

$$< \xi, \xi' > = \xi^{tr} \zeta \xi' = \xi^\alpha \xi'_\alpha = \xi^1 \xi'^2 - \xi^2 \xi'^1 \tag{2.74}$$

$$< \eta, \eta' > = \eta^{tr} \zeta \eta' = \eta^{\dot\alpha} \eta'_{\dot\alpha} = \eta^{\dot 1} \eta'^{\dot 2} - \eta^{\dot 2} \eta'^{\dot 1}. \tag{2.75}$$

For a three-dimensional spinor, the positive combination

$$\varrho = \xi^1 \bar{\xi}^1 + \xi^2 \bar{\xi}^2 \tag{2.76}$$

is an invariant which can be interpreted, if ξ is a wave function, as a density of probability. In four dimensions, this is not an invariant, because the matrix of transformation A is unitary only in the case of spatial rotation. However, this is not a problem for the interpretation of ϱ. In relativistic theory, the density of probability need not be a scalar, but rather the time component of a four-vector. We will see when studying the Dirac equation how to construct a conserved probability current, precisely based on ϱ.

Before studying the higher order representations, it is important to render explicit the form of a matrix A corresponding to an arbitrary Lorentz transformation. The method is similar to that followed in the three-dimensional case. It consists in identifying the Lorentz transformations in their usual form with the expression $\tilde{x}' = A\tilde{x}A^*$. Let us consider first the spatial rotations. We have for an infinitesimal rotation of angle $\delta\theta\, \boldsymbol{n}$, with $\|\boldsymbol{n}\| = 1$,

$$t' = t$$
$$\boldsymbol{x}' = \boldsymbol{x} + \delta\theta\, \boldsymbol{n} \wedge \boldsymbol{x}, \qquad (2.77)$$

where the symbol $\wedge$ means the external product of three-dimensional vectors. Writing $A = 1 + \delta A$, and expanding to first order $\tilde{x}' = A\tilde{x}A^*$, we find the equation

$$\delta A(t + \boldsymbol{\sigma} \cdot \boldsymbol{x}) + (t + \boldsymbol{\sigma} \cdot \boldsymbol{x})\delta A^* = \delta\theta\,(\boldsymbol{n} \wedge \boldsymbol{x}) \cdot \boldsymbol{\sigma}. \qquad (2.78)$$

We deduce from this equality that $\delta A + \delta A^* = 0$ and therefore that δA is an anti-Hermitian matrix of null trace, that is of the form $i\boldsymbol{\sigma} \cdot \boldsymbol{u}$, with finally $\boldsymbol{u} = -\frac{1}{2}\delta\theta\boldsymbol{n}$. Consequently, we find for a pure infinitesimal rotation that

$$A = 1 - \frac{i}{2}\delta\theta\, \boldsymbol{n} \cdot \boldsymbol{\sigma}. \qquad (2.79)$$

As for the rotations around an axis, the rotation angle is an additive parameter of the group. This formula is integrated as an exponential for finite angle transformation. One gets the unitary matrix

$$A = \exp -\frac{i}{2}\delta\theta\, \boldsymbol{n} \cdot \boldsymbol{\sigma} = \cos\frac{\theta}{2} - i\boldsymbol{n} \cdot \boldsymbol{\sigma} \sin\frac{\theta}{2}. \qquad (2.80)$$

Of course, this formula is identical to the one we obtained for the three-dimensional spinors and we find again that the generators $\boldsymbol{J}$ are here represented by the matrices $\boldsymbol{\sigma}/2$.

Let us now consider a pure infinitesimal Lorentz transformation, that is the law of transformation between two reference frames in uniform relative motion of speed v, with $\|\boldsymbol{v}\| \ll 1$. If we proceed as for the space rotations, we are tempted to write that a four-vector $x^\mu = (t, \boldsymbol{x})$ transforms in $x'^\mu = (t', \boldsymbol{x}')$, with

$$t' = t + \boldsymbol{v} \cdot \boldsymbol{x}$$
$$\boldsymbol{x}' = \boldsymbol{x} + \boldsymbol{v}t, \qquad (2.81)$$

where we neglect the terms of order $||v||^2$. This is, however, not useful because, according to formulae (2.45) and (2.46) giving the law of transformation of a four-vector by a pure Lorentz transformation, the components of the speed v cannot be considered as group parameters. This is simply the physical assertion according to which one must abandon the Galilean law of addition of speeds when the speeds are not negligible with respect to the speed of light. In reality, a pure Lorentz transformation acting on a four-vector must be compared (because of the conservation of the pseudo-Euclidean norm) to a complex rotation, mixing space and time components, around the axis of the unit vector $n = v/||v||$ and of angle $i\xi$ such that $\tanh\xi = v/c$, where $v = ||v||$. For successive Lorentz transformations along the same axis, one checks easily that the additive parameter of the group is ξ.

Since the idea is to generate the matrices A from their infinitesimal forms, it is natural to write the infinitesimal Lorentz transformations as

$$t' = t + \delta\xi n \cdot x$$
$$x' = x + \delta\xi n t. \tag{2.82}$$

Writing $A = 1 + \delta A$, and working at the lower order in $||v||$, one gets by comparison with $\tilde{x}' = A\tilde{x}A^*$:

$$\delta A(t + \sigma \cdot x) + (t + \sigma \cdot x)\delta A^* = \delta\xi\, (n \cdot x + t\sigma \cdot n)\,. \tag{2.83}$$

The matrices 1 and σ being a basis, one deduces that δA is a Hermitian matrix, $\delta A^* = \delta A$ and $\delta A = \delta\xi n \cdot \sigma/2$. Consequently, for an infinitesimal pure Lorentz transformation,

$$A = 1 + \frac{\delta\xi}{2}n \cdot \sigma. \tag{2.84}$$

This formula is integrated as an exponential for a finite transformation. Indeed, performing a sequence of infinitesimal Lorentz transformations along n one finds the matrix A corresponding to the finite Lorentz transformation of axis n and parameter ξ to be

$$A = \exp\frac{\xi}{2}\, n \cdot \sigma = \cos\frac{\xi}{2} + n \cdot \sigma \sin\frac{\xi}{2}. \tag{2.85}$$

One sees that A is a non-unitary Hermitian matrix. The appearance of the $\frac{1}{2}$ factor, as for the formulae found in the three-dimensional spinor case, is related to the fact that the matrix A is quadratically linked to the matrix Λ.

The fact that the restricted Lorentz group is locally isomorphic to $SU(2) \times SU(2)$ is obvious in the way we describe it. One can also check the commutation relations for the generators that we gave earlier.

Higher order spinors are defined using tensor products of fundamental representations of undotted and dotted spinors, that is order-1 spinors. For a higher order spinor, it is important to distinguish dotted indices from undotted indices. For example, there exist three order-2 spinors:

$$\xi^{\alpha\beta} \sim \xi^\alpha \, \Xi^\beta$$
$$\zeta^{\alpha\dot\beta} \sim \xi^\alpha \eta^{\dot\beta}$$
$$\eta^{\dot\alpha\dot\beta} \sim \eta^{\dot\alpha} \zeta^{\dot\beta}. \tag{2.86}$$

We have already seen the correspondence between four-vectors x^μ and order-2 spinors of mixed indices $\zeta^{\alpha\dot\beta}$. More generally, the order of a spinor is given by a couple of numbers (k, l) representing the number of dotted and undotted indices. The dotted and undotted indices cannot be mixed through a restricted Lorentz transformation; *it is not, therefore, necessary to specify the order of the indices*. To be covariant, any spinorial equality must have in its two sides the same number of dotted and undotted indices. This remark is the key point in the construction of the Dirac equation. It is evident that a complex conjugation means the reciprocal exchange of dotted and undotted indices. Thus, a relation like $\eta^{\alpha\dot\beta} = \bar\xi^{\dot\alpha\beta}$ is covariant.

The contraction of spinors by the tensor $\zeta_{\alpha\beta}$ is only meaningful on indices of the same species. It is indeed easy to check that the contraction of indices of different nature cannot be covariant. More generally, one gets a sequence of lower rank tensors starting from a given tensor and contracting pairs of its indices. We saw, when studying three-dimensional spinors, that the contraction of a couple of symmetrical indices gives 0. This result is still valid for pairs of dotted or undotted indices. Consequently, the higher order irreducible representations of $SL(2, \mathbb{C})$ are given by the spinors

$$\xi^{\alpha_1\alpha_2...\alpha_k\dot\beta_1\dot\beta_2...\dot\beta_l}, \tag{2.87}$$

separately symmetrical on the k indices of α and the l indices of $\dot\beta$. Indeed, it is impossible to pass by using linear combination of such tensors to non-vanishing tensors of strictly lower ranks. An irreducible representation is, thus, characterised by a pair of numbers (k, l).

Since each dotted or undotted index can take only two values, one easily finds that a completely symmetrical spinor of rank (k, l) has only $(k + 1)(l + 1)$ independent components; it is the dimension of the irreducible representation realised by this spinorial tensor.

2.9   Space Inversion and Bispinors

For three-dimensional spinors, space inversion is obvious because it commutes with the rotations. For four-dimensional spinors, the space inversion, which we shall denote by I_s, is not so obvious because it does not commute with the pure Lorentz transformations. We will now see that this operation makes it possible to link complex conjugate representations and in particular to transform undotted spinors to dotted spinors. It also makes it possible to introduce the bispinors which are necessary to the theory of massive half-integer spin particles, such as the electron or the nucleons.

In the three-dimensional theory, one defines, without contradiction, the action of space inversion on a spinor by

$$\xi \rightarrow I_s\xi = P\xi, \qquad (2.88)$$

where P is a constant coefficient. For a three-dimensional spinor, the operator I_s is thus diagonal. This is a consequence of Schur's lemma, since space inversion commutes with the rotations.

The eigenvalue P is defined up to some degree of freedom. By two successive space inversions we go back to the initial state. However, such a way back can be interpreted in two different manners: as a rotation of angle 0 or as a rotation of angle 2π. For a spinor, these two possibilities are not equivalent, since for a rotation of angle 2π the spinor changes sign. We, thus, have two possibilities for P. In one case

$$P^2 = 1 \qquad \Rightarrow \qquad P = \pm 1, \qquad (2.89)$$

and in the other one

$$P^2 = -1 \qquad \Rightarrow \qquad P = \pm i. \qquad (2.90)$$

We will choose, once and for all, the second possibility. Let us remark that the relative parity of two spinors, defined as the parity of the scalar formed by these two spinors $< \xi, \psi >$, is independent of the convention taken for the parity of a spinor.

Let us now consider the case of the Minkowski space. For a vector, the inversion law is $(t, x) \rightarrow (t, -x)$. It is easy to see that the inversion I_s does not commute with the Lorentz transformations. Indeed, if we denote by L_v the pure Lorentz transformation of speed v, it is easy to check that

$$I_sL_v = L_{-v}I_s \Rightarrow [L_v, I_s] \neq 0. \qquad (2.91)$$

It thus follows that the components of a four-dimensional spinor cannot be transformed into themselves. If this property had been true, since space inversion must commute with all spatial rotations, that is with all matrices of the type $\cos \frac{\theta}{2} + i n \cdot \sigma \sin \frac{\theta}{2}$, Schur's lemma would have implied that I_s has to be represented by a matrix proportional to the identity matrix, as for three-dimensional spinors. But then, I_s will have to commute with the pure Lorentz transformations, which is not true.

The inversion thus must transform the components of ξ^α into other quantities. We can guess the right answer by considering the action of I_s on the generators of the Lie algebra of the group. We see immediately that the generators for rotations, the components of the angular momentum, are pseudovectors and do not change sign under space inversions: $J \rightarrow J$. On the contrary, the generators of pure Lorentz transformations are vectors and change sign: $N \rightarrow -N$. As a result I_s interchanges the generators J_1 and J_2 of

Eqs. (2.58) and (2.59). It follows that space inversion should interchange dotted and undotted indices. Therefore, a possible choice is to identify the components of ξ^α with the components of another dotted spinor $\eta^{\dot\alpha}$, which by definition do not coincide regarding their transformation laws with ξ^α. With our convention that the inversion applied twice gives minus the identity, we define the action of I_s in the following way:

$$\xi^\alpha \to i\eta_{\dot\alpha} \qquad \eta_{\dot\alpha} \to i\xi^\alpha. \tag{2.92}$$

One has also

$$\xi_\alpha \to -i\eta^{\dot\alpha} \qquad \eta^{\dot\alpha} \to -i\xi_\alpha; \tag{2.93}$$

thus, $I_s^2 = -1$.

Therefore, to combine in a coherent way, space inversion and the Lorentz group, one must consider spaces of representations in which there are simultaneously undotted and dotted spinors. At least we need to have in a minimal representation the set of all pairs $(\xi^\alpha, \eta_{\dot\alpha})$, called bispinors of order 1. Starting from this minimal representation, it is then possible to construct by tensor products the irreducible representations of the extended Lorentz group.

The scalar product of two bispinors can be defined in two different ways. Either

$$\xi^\alpha \xi'_\alpha + \eta_{\dot\alpha} \eta'^{\dot\alpha}, \tag{2.94}$$

which gives a Lorentz scalar, or

$$\xi^\alpha \xi'_\alpha - \eta_{\dot\alpha} \eta'^{\dot\alpha}, \tag{2.95}$$

which gives a Lorentz pseudoscalar, that is a quantity changing its sign under space inversions.

Let us give some examples of order-2 spinors. We have two possibilities for the laws of transformation. The first consists in writing

$$\zeta^{\alpha\dot\beta} \sim \xi^\alpha H^{\dot\beta} + \Xi^\alpha \eta^{\dot\beta}, \tag{2.96}$$

and through space inversion $\zeta^{\alpha\dot\beta} \to \zeta_{\dot\alpha\beta}$. In this case, using the properties of the metric tensor, it is easy to check that the four-vector a^μ equivalent to such a rank 2 spinor is a true vector, that is transforming by inversion as $(a^0, \mathbf{a}) \to (a^0, -\mathbf{a})$. The other possibility is to impose the law of transformation

$$\zeta^{\alpha\dot\beta} \sim \xi^\alpha H^{\dot\beta} - \Xi^\alpha \eta^{\dot\beta}. \tag{2.97}$$

In this case, the space inversion gives $\zeta^{\alpha\dot\beta} \to -\zeta_{\dot\alpha\beta}$, and the corresponding four-vector is such that by inversion $(a^0, \mathbf{a}) \to (-a^0, \mathbf{a})$. Such a four-vector, $\mathbf{a}$ being axial, is called a four-pseudovector.

Let us now consider the order-2 symmetrical spinors

$$\xi^{\alpha\beta} \sim \xi^{\alpha} \, \Xi^{\beta} + \xi^{\beta} \, \Xi^{\alpha}$$

$$\eta_{\dot{\alpha}\dot{\beta}} \sim \eta_{\dot{\alpha}} H_{\dot{\beta}} + \eta_{\dot{\beta}} H_{\dot{\alpha}}. \tag{2.98}$$

The couple $(\xi^{\alpha\beta}, \eta_{\dot{\alpha}\dot{\beta}})$ forms a bispinor of order 2. The number of its independent components is $3 + 3 = 6$ and one can prove that this bispinor is equivalent to a rank 2 antisymmetric four-tensor.

2.10 Finite-Dimensional Representations of *SU*(2) and *SL*(2, $\mathbb{C}$)

Let us recall from non-relativistic quantum mechanics the form of the irreducible representations of $SU(2)$. They are indexed by a non-negative number, j, integer or half-odd integer. The representation, denoted D^j, is defined by its matrix elements $D^j_{mm'}$ in the basis formed by the homogeneous polynomials in 2 variables ξ^1, ξ^2, of degree $2j$; ξ^1 and ξ^2 are the 2 components of a spinor $\xi \in C^2$. The representation is of dimension $2j + 1$.

If

$$\begin{pmatrix} \xi^1 \\ \xi^2 \end{pmatrix} \to A \begin{pmatrix} \xi^1 \\ \xi^2 \end{pmatrix}$$

then

$$|j; m > \to \sum_{m'} D^j_{mm'}(A) \, |j; m' >,$$

where $m = -j, -j + 1, \ldots, j$ and

$$|j; m >= \frac{(\xi^1)^{j+m}(\xi^2)^{j-m}}{\sqrt{(j + m)!(j - m)!}}. \tag{2.99}$$

These representations are all the irreducible unitary representations of $SU(2)$.

The tensor product of two representations is in general reducible. Using the preceding formulae one gets

$$D^{j_1} \otimes D^{j_2} = \sum_{j = |j_1 - j_2|}^{|j_1 + j_2|} D^j. \tag{2.100}$$

This is the well-known formula describing the quantum mechanical rule for the addition of two angular momenta.

In Eq. (2.59) we saw that the Lie algebra of the Lorentz group is generated by two copies of the algebra of $SU(2)$. It follows that the finite-dimensional irreducible representations $D^{[j_1,j_2]}$ of $SL(2,\mathbb{C})$ (in general non-unitary) are characterised by two non-negative numbers, j_1 and j_2, integers or half-odd integers. They are of dimension $(2j_1 + 1)(2j_2 + 1)$ and given on a basis of the homogeneous polynomials of degree $2j_1 + 2j_2$.

If

$$\begin{pmatrix} \xi^1 \\ \xi^2 \end{pmatrix} \to A \begin{pmatrix} \xi^1 \\ \xi^2 \end{pmatrix}, \qquad \begin{pmatrix} \xi^3 \\ \xi^4 \end{pmatrix} \to \bar{A} \begin{pmatrix} \xi^3 \\ \xi^4 \end{pmatrix},$$

then

$$|j_1 j_2; m_1 m_2 > \; \to \; \sum_{m_1' m_2'} D^{[j_1,j_2]}_{m_1 m_1', m_2 m_2'} (A) \, |j_1 j_2; m_1' m_2' >$$

with $m_k = -j_k, -j_k + 1, \ldots, j_k$, $k = 1, 2$

$$|j_1 j_2; m_1 m_2 > = \frac{(\xi^1)^{j_1+m_1} (\xi^2)^{j_1-m_1} (\xi^3)^{j_2+m_2} (\xi^4)^{j_2-m_2}}{\sqrt{(j_1 + m_1)!(j_1 - m_1)!(j_2 + m_2)!(j_2 - m_2)!}}. \qquad (2.101)$$

One deduces from the preceding formulae that

$$D^{[j_1,j_2]}(A) = D^{[j_1,0]}(A) \otimes D^{[0,j_2]}(A).$$

Moreover,

$$D^{[0,j]}(A) = D^{[j,0]}(\bar{A})$$

and the representations

$$D^{[0,j]}(A)$$

and

$$D^{[j,0]}(A)$$

are inequivalent. If A is unitary, $\bar{A} = (A^{tr})^{-1}$, and the two representations are equivalent since one goes from A to $\bar{A}$ through an internal automorphism of the group. They are equivalent to the representation of $SU(2)$ that we already built and we identify them with D^j.

One has

$$D^{[0,0]}(A) = 1$$
$$D^{[\frac{1}{2},0]}(A) = A$$
$$D^{[0,\frac{1}{2}]}(A) = \bar{A}$$
$$D^{[\frac{1}{2},\frac{1}{2}]}(A) = A \otimes \bar{A}.$$

Note: The multispinors $a^{\alpha_1 \alpha_2 \cdots \alpha_i \dot\beta_1 \dot\beta_2 \cdots \dot\beta_j}$ are the elements of $\mathbb{C}^{2 \otimes i} \otimes \mathbb{C}^{2 \otimes j}$ on which the tensor product of the representation $(D^{[1/2,0]})^{\otimes i} \otimes (D^{[0,1/2]})^{\otimes j}$ acts through

$$(D^{[1/2,0]})^{\otimes i} \otimes (D^{[0,1/2]})^{\otimes j} : a^{\alpha_1 \cdots \alpha_i \dot\beta_1 \cdots \dot\beta_j} \rightarrow A^{\alpha_1}_{\gamma_1} \cdots A^{\alpha_i}_{\gamma_i} \bar{A}^{\dot\beta_1}_{\delta_1} \cdots \bar{A}^{\dot\beta_j}_{\delta_j} a^{\gamma_1 \cdots \gamma_i \delta_1 \cdots \delta_j} \qquad (2.102)$$

For $i + j \geq 1$, these representations are not irreducible; the irreducible representations are always given by the same formula restricted to spinors symmetrised separately with respect to undotted and dotted indices.

2.11 Problems

Problem 2.1 Show that the matrix elements of the $O(3)$ generators are defined by

$$(\mathcal{J}^k)_{ij} = i\epsilon_{kij}.$$

Problem 2.2 Prove Equ. (2.32).

Problem 2.3 Prove that any Hermitian 2×2 matrix A can be written in a unique way as

$$A = a + \boldsymbol{b} \cdot \boldsymbol{\sigma},$$

where a and $\boldsymbol{b}$ are real.

Problem 2.4 Prove (formula (2.40)) that $\boldsymbol{a}$ with components

$$a_1 = \frac{i}{\sqrt{2}}(\Psi_{11} - \Psi_{22}), \qquad a_2 = -\frac{1}{\sqrt{2}}(\Psi_{11} + \Psi_{22}), \qquad a_3 = -i\sqrt{2}\Psi_{12}$$

transforms as a vector.

Problem 2.5 Show that if A is unitary, then $\Lambda(A)$ is a rotation, and if A is Hermitian, then $\Lambda(A)$ is a pure Lorentz transformation.

Hint: A, Hermitian and of determinant 1, can be written as

$$A = U \begin{pmatrix} e^{\frac{\xi}{2}} & 0 \\ 0 & e^{-\frac{\xi}{2}} \end{pmatrix} U^*,$$

where U is some unitary matrix.

Problem 2.6 Prove the following identities between Pauli matrices:

$$\sigma_i \sigma_j = \delta_{ij} + i\epsilon_{ijk}\sigma_k$$
$$\sigma_i \sigma_j \sigma_k = \sigma_i \delta_{jk} - \sigma_j \delta_{ik} + \sigma_k \delta_{ij} + i\epsilon_{ijk}$$
$$\sigma_i \sigma_j \sigma_k \sigma_l = \delta_{ij}\delta_{kl} - \delta_{ik}\delta_{jl} + \delta_{il}\delta_{jk} + i[\epsilon_{ijm}\delta_{kl} - \epsilon_{ikm}\delta_{jl} + \epsilon_{ilm}\delta_{jk}]\sigma_m.$$

Problem 2.7 Use formula (2.64)

$$\Lambda(A)^{\mu}{}_{\nu} = \frac{1}{2}\mathrm{Tr}(\sigma_{\mu}A\sigma_{\nu}A^*)$$

to compute the matrix elements of $\Lambda(A)$ as a function of

$$A = a^0 + \boldsymbol{a} \cdot \boldsymbol{\sigma} \in SL(2, \mathbb{C}).$$

Hint: Show that

$$\Lambda^0{}_0 = 4|a^0|^2$$
$$\Lambda^k{}_0 + \Lambda^0{}_k + i\epsilon_{ijk}\Lambda^i{}_j = 4\bar{a}^0 a^k.$$

Problem 2.8 The purpose of this exercise is to study the automorphisms of $SL(2, \mathbb{C})$:

Show that ζ is in $SL(2, \mathbb{C})$ and is a rotation.

Show that $\Lambda(\zeta) = I_1 I_3$ where I_1 and I_3 are the symmetries with respect to the three-dimensional planes orthogonal to the 1st and the 3rd axis, respectively.

Show that there are no elements U of $SL(2, \mathbb{C})$ such that for $A \in SL(2, \mathbb{C})$:

$$UAU^{-1} = \bar{A}$$

Problem 2.9 Show that any antisymmetric four-tensor, $F_{\mu\nu} = -F_{\nu\mu}$, $\mu, \nu = 0, 1, 2, 3$ can be expressed in terms of the elements of a bispinor of order 2 ($\xi^{\alpha\beta}, \eta_{\dot{\alpha}\dot{\beta}}$) as given by formula (2.98).

3

The Electromagnetic Field

3.1 Introduction

Maxwell's equations are among the most important discoveries of the nineteenth century. They are at the origin of special relativity. Efforts to quantise Maxwell's theory have resulted in the relativistic quantum field theory with its numerous applications.

Maxwell's theory and its quantum version can be conceived as the prototype of modern theories for the description of microscopic physics. They are based on three fundamental principles: locality of interactions, invariance under Poincaré's transformations and its generalisation to curved space, and invariance under a gauge symmetry, i.e. a symmetry for which its parameters depend on the space–time point considered.

The purpose of this chapter is not to present a complete current survey of the electromagnetic theory. It is rather to emphasise some of its properties within the modern formalism in order to help understand the generalisations to more complex theories, such as those unifying the weak, the electromagnetic, and the strong interactions.

In classical Newtonian mechanics, particle interactions are depicted using the force field concept. The interaction between two particles is expressed by saying that one of the particles is in the other's force field.

Such an analysis becomes meaningless in special relativity if we view particles as small rigid spheres. The notion of a particle itself conceived as an elementary object is questionable. Indeed, a particle cannot be an extended absolutely solid body, i.e. an object that retains its shape and size independently of the frame of reference. This kind of solid cannot exist due to the finite speed of propagation of the interactions. So, within the classical relativistic mechanical framework, an elementary particle is necessarily a point-like object. In terms of mathematics, this implies a description of the particles by distributions. Thus, a charge density e located at the centre of coordinates is written as $e\delta(\mathbf{x})$.

When dealing with interactions, due to the finite propagation speed, the forces acting at a given moment on a particle rely on the other particles' states at the preceding moments. One can then describe the forces exerted by a set of particles on one particle located in a space–time point x by a field at this same point.

The effect of the forces on the particle, *the coupling*, is *local*; i.e. it is expressed in terms of quantities evaluated at the same point. The notion of a point-like interaction of a particle with the resultant fields created by other particles replaces that of a force

From Classical to Quantum Fields. Laurent Baulieu, John Iliopoulos and Roland Sénéor.
© Laurent Baulieu, John Iliopoulos and Roland Sénéor, 2017. Published 2017 by Oxford University Press.

exerted by particles on a test particle. The interaction is then instantaneous and local. This principle of *locality of the interactions* is one of the fundamental principles of the modern relativistic quantum field theory.

A difficulty inherent in the notion of fields must be pointed out. The existence of a particle is translated into the existence of a field at each point of the entire space. This field can interact with the particle and one can end up with the paradox that an isolated particle can self-accelerate itself. This self-interaction has been the source of technical and conceptual difficulties that have been understood only during the past sixty years by the theory of *renormalisation*. The issue was that of a link between fields and particles, which only finds a relevant answer within the quantum framework.

From a historical point of view, the field notion comes out from the study of electromagnetism. Ampère, Gauss, Weber, Faraday, and Maxwell were inspired by the non-instantaneousness of interactions, which was already underlying Newton's work. The contribution of Maxwell, who understood the unifying role of mathematics, is exceptional. After writing the equations bearing his name, he gave the interpretation of light as an electromagnetic phenomenon by comparing equations he obtained for the vector potential to those of a wave propagation in an elastic medium. Among the two theories of light proposed at this time (corpuscular and wave), Maxwell supported the wave theory. In contrast, a quarter of a century later, Planck and Einstein showed that in certain cases the corpuscular interpretation was necessary. It is within the synthesis of those two currents of thought that the modern relativistic quantum field theory arose.

Four pages from Maxwell's *Treatise on Electricity and Magnetism* (written around 1873) are given in Appendix E. They illustrate well the state of mind of the epoch concerning certain issues referred to in this book.

3.2 Tensor Formulation of Maxwell's Equations

Maxwell's equations were established through a long historical journey. At the end of the nineteenth century, it was commonly assumed that electric and magnetic fields were indissociable. The equations of electromagnetism proposed by Maxwell are the following, using the units $\mu_0 = \epsilon_0 = 1$,

$$
\begin{aligned}
&\mathrm{div}\boldsymbol{B} = 0 && \mathrm{div}\boldsymbol{E} = \varrho \\
&\mathrm{rot}\boldsymbol{E} = -\dot{\boldsymbol{B}} && \mathrm{rot}\boldsymbol{B} = \boldsymbol{j} + \dot{\boldsymbol{E}}
\end{aligned}
\tag{3.1}
$$

$\boldsymbol{E}$ is the electric field vector, $\boldsymbol{B}$ is the magnetic field pseudovector, ϱ is the scalar electric charge density, and $\boldsymbol{j}$ is the electric current density vector. These differential equations express the various possibilities that electric and magnetic fields transform into each other and their interaction with charged matter.

A more symmetrical form of these equations is obtained if we assume the existence of a pseudo-scalar magnetic charge density $\varrho_{\mathrm{magn.}}$, with $\mathrm{div}\boldsymbol{B} = \varrho_{\mathrm{magn.}}$. Such charges, the magnetic monopoles, are very interesting from the theoretical point of view because of their implication on the topology of the space and the quantisation of electric charge.

We shall study some of their properties in Chapter 26. On the other hand, there is no experimental evidence of their existence and nature seems to rule in favour of the dissymmetry by allowing only one kind of charge.

After the discovery of the invariance of Maxwell's equations under Lorentz transformations and the development of the tensor formalism, the E and B fields were quickly identified as components of an irreducible Lorentz group representation: the antisymmetric tensor of rank 2, with

$$\{F^{\mu\nu}\} = \begin{pmatrix} 0 & -E^1 & -E^2 & -E^3 \\ E^1 & 0 & -B^3 & B^2 \\ E^2 & B^3 & 0 & -B^1 \\ E^3 & -B^2 & B^1 & 0 \end{pmatrix}. \tag{3.2}$$

The first index of $F^{\mu\nu}$ is the row index. So we have $F^{i0} = E^i$.

By convention, Latin indices $i, j, k, \ldots$ are used for space indices. We also use the convention that boldface vectors like x, E, B, j, etc., are the contravariant triplets (x^1, x^2, x^3), (E^1, E^2, E^3), (B^1, B^2, B^3), etc.

The assembly of the electric and magnetic fields into the $F^{\mu\nu}$ entity explains their transformations through changes of reference frame. Maxwell's equations can be written in a compact form

$$\partial_\mu {}^*F^{\mu\nu} = 0 \qquad\qquad \partial_\mu F^{\mu\nu} = j^\nu, \tag{3.3}$$

where ${}^*F^{\mu\nu} = \frac{1}{2}\varepsilon^{\mu\nu\varrho\sigma} F_{\varrho\sigma}$, $j^\mu = (\varrho, j)$, $\partial_\alpha = \frac{\partial}{\partial x^\alpha}$, and $\varepsilon^{\mu\nu\varrho\sigma}$ is the completely antisymmetric tensor in the four indices so that $\varepsilon_{0123} = 1$.

Using the tensor notation, it's easy to see that Maxwell's equations imply current conservation. Indeed, as $F^{\mu\nu} = -F^{\nu\mu}$, we have $\partial_\nu \partial_\mu F^{\mu\nu} = 0$ and hence

$$\partial_\nu j^\nu = 0. \tag{3.4}$$

This equation can also be written as

$$\frac{\partial \varrho}{\partial t} + \text{div} j = 0, \tag{3.5}$$

which is the familiar current conservation equation.

The first couple of Maxwell's equations $\partial_\mu^* F^{\mu\nu} = 0$ is equivalent to the Bianchi identities satisfied by $F^{\mu\nu}$, $\varepsilon^{\mu\nu\varrho\sigma}\partial_\nu F_{\varrho\sigma} = 0$. These equations are equivalent to the property that the electromagnetic field is obtained locally from a potential:

$$F_{\mu\nu} = \partial_\mu A_\nu - \partial_\nu A_\mu. \tag{3.6}$$

So this expression of F in terms of A resolves the homogeneous sector of Maxwell's equations.

The four-vector A_μ has the electrostatic potential A_0 as time component and the vector potential A with A_i as space components. In terms of these quantities, the electromagnetic fields are written as

$$E^i = F^{i0} = \partial^i A^0 - \partial^0 A^i = -\partial_i A^0 - \partial_0 A^i \qquad\qquad B^i = -\varepsilon_{ijk} F^{jk}; \qquad (3.7)$$

hence,

$$E = -\frac{\partial A}{\partial t} - \nabla A^0 \qquad\qquad B = \text{rot} A. \qquad (3.8)$$

Equations (3.8) show how magnetic and electric fields are transformed by space inversion. B is a pseudo-vector, whereas E is a vector.

The transformation laws of A^0 and A under the action of the Lorentz group can be easily deduced from the definition of A_μ as a four-vector. Similarly, the transformation laws by Lorentz transformation of E and B can be deduced from the fact that $F_{\mu\nu}$ is a rank 2 antisymmetric tensor.

Let us now make explicit the modern formalism using the differential geometry. The main concern in doing this is not to introduce excessively heavy formalism, but, in fact, this somewhat abstract approach has applications in the quantisation of the theory.

3.3   Maxwell's Equations and Differential Forms

The rules of differential calculus are set forth briefly. There exists an intrinsic way of formulating what follows, i.e. without choosing local coordinates. One can find more details in Appendix B.

Given a manifold M_n of dimension n, we can locally define coordinates $\{x^\mu\} = \{x^1 \ldots x^n\}$. We introduce the exterior derivatives in the tangent plane defined in every point of M_n. These objects are built on real or complex functions of points and 1-form basis $\{dx^\mu\} = (dx^1, \ldots, dx^n)$ on the manifold. The dx^μ form a real or complex vector space. They can be multiplied by exterior product, an operation symbolised by $\wedge$, which permits the construction of a form of a higher degree. The rules of the exterior calculus are the following:

(i) $dx^\mu \wedge dx^\nu = -dx^\nu \wedge dx^\mu$.

(ii) A p-form f is defined as

$$f = \frac{1}{p!} f_{\mu_1 \ldots \mu_p}(x) dx^{\mu_1} \wedge \ldots \wedge dx^{\mu_p}. \qquad (3.9)$$

The tensor $f_{\mu_1 \ldots \mu_p}$ is a non-singular object that takes real or complex values and is fully antisymmetric in its indices. A consequence of this definition is that any form of rank higher than or equal to n, n being the manifold's dimension, vanishes identically.

(iii) The p-forms constitute a vector space (on $\mathbb{R}$ or $\mathbb{C}$ if they are real or complex). Given two p-forms f and g, $af + bg$, a and b in $\mathbb{R}$ or $\mathbb{C}$, is a p-form:

$$af + bg = \frac{1}{p!}(af_{\mu_1\ldots\mu_p} + bg_{\mu_1\ldots\mu_p})dx^{\mu_1} \wedge \ldots \wedge dx^{\mu_p}. \qquad (3.10)$$

(iv) Given a p-form f and a q-form g, their exterior product is a $(p + q)$-form:

$$f \wedge g = \frac{1}{p!q!}f_{\mu_1\ldots\mu_p}g_{\mu_{p+1}\ldots\mu_{p+q}}dx^{\mu_1} \wedge \ldots \wedge dx^{\mu_{p+q}}.$$

(v) Let $\varepsilon_{\mu_1\ldots\mu_n}$ be the fully antisymmetric tensor with value 1 if $\mu_1\ldots\mu_n$ is an even substitution of $1\ldots n$ and -1 otherwise. The dual $*f$ of a p-form f is an $(n-p)$-form defined as

$$*f = \frac{1}{p!(n-p)!}f_{\nu_1\ldots\nu_p}g^{\nu_1\mu_1}\ldots g^{\nu_p\mu_p}\varepsilon_{\mu_1\ldots\mu_n}dx^{\mu_{p+1}} \wedge \ldots \wedge dx^{\mu_n}.$$

The introduction of the dual necessitates the existence of a metric $g_{\mu\nu}$ to lower and raise the indices.

(vi) The exterior derivative operator d is a differential operator with the following properties: d satisfies the graded Leibniz rule, with f a p-form and g a q-form $d(fg) = (df)g \pm fdg$, where the minus sign arises if p is odd. On the 0-forms and 1-forms, we have $df = dx^\mu \partial_\mu f$ and $d(dx^\mu) = 0$. From these rules, one can show easily that the d operator raises the rank of the forms by one unit, and can be written as $d = dx^\mu \partial_\mu$. We have the property

$$d^2 = 0, \qquad (3.11)$$

true for any form. Reciprocally, if a rank p-form satisfies $df = 0$, one can construct locally a form g, of rank $p-1$, so that $f = dg$.

(vii) The volume element $d^n x = \frac{1}{n!}\varepsilon_{\mu_1\ldots\mu_n}dx^{\mu_1} \wedge \ldots \wedge dx^{\mu_n}$ is invariant under change of coordinates; it is usually written as $dx^1 \ldots \wedge dx^n$. Given a 0-form f, the integration over a sub-manifold M_p of M_n is defined by $\int_{M_p} d^p x f$. An important property of the integration is the Stokes formula

$$\int_M d\omega = \int_{\partial M} \omega, \qquad (3.12)$$

where ∂M is the border of the manifold M, and ω is a form whose rank is equal to the dimension of the border of M. A well-known example of application is the magnetic flux theorem, $\oint_\Gamma \boldsymbol{A} \cdot d\boldsymbol{l} = \int_\Sigma \boldsymbol{B} \cdot d\boldsymbol{s}$, where Γ is the contour of the Σ surface.

(viii) Given a vector field ξ^μ, there exists a differential operator i_ξ, called contraction operator, which reduces the rank of a form by one unit. It satisfies the Leibniz rule. It is sufficient to define it on the 0-forms and on the 1-forms. By definition, we have for f a function $i_\xi f = 0$ and $i_\xi \, dx^\mu = \xi^\mu$.

(ix) Once d and i_ξ are given, we may construct an algebra of operators acting on forms by successive commutations. Thus, the Lie derivative along a vector ξ gives

$$L_\xi = [i_\xi, d] = i_\xi d + d i_\xi.$$

Note the plus sign because both i_ξ and d are odd elements. We have the properties $[L_\xi, d] = 0$, $[L_\xi, i_{\xi'}] = i_{\{\xi,\xi'\}}$, $[i_\xi, i_{\xi'}] = 0$. All this notation is explained in Appendix B.

Let us show how the differential form formalism applies to electromagnetism. Given the electromagnetic tensor, we construct the 2-form

$$F = \frac{1}{2} F_{\mu\nu} dx^\mu \wedge dx^\nu. \tag{3.13}$$

So we have explicitly

$$F = dt \wedge dx^i E^i + dx^1 \wedge dx^2 B_3 + dx^3 \wedge dx^1 B_2 + dx^2 \wedge dx^3 B_1 \tag{3.14}$$

A form being an invariant object, one can see from the preceding formula that the components of the electric field, i.e. F_{0i}, form a space vector which is odd with respect to time reversal, whereas the components of the magnetic field, i.e. $-F_{ij}$, form a space pseudovector, even with respect to time reversal.

Maxwell's equations are written under the form

$$dF = 0 \qquad d^*F = -^*\mathcal{J}. \tag{3.15}$$

By introducing the exterior coderivative $\delta = *d*$ one can rewrite Eq. (3.15) as

$$\delta F = \mathcal{J} \tag{3.16}$$

using the fact that on a p-form $** = -1(-1)^{p(n-p)}$ and $\mathcal{J} = j_\mu dx^\mu$.

The equation $dF = 0$ indicates that the 2-form F is closed.[1]

In Minkowski space, which is a star-shaped space (see Appendix B), one can deduce that the form is exact and that there exists a 1-form A so that

$$F = dA. \tag{3.17}$$

From $A = A_\mu dx^\mu$, one can deduce that

$$dA = \partial_\nu A_\mu dx^\nu \wedge dx^\mu = -\partial_\mu A_\nu dx^\nu \wedge dx^\mu = \frac{1}{2}(\partial_\mu A_\nu - \partial_\nu A_\mu) dx^\mu \wedge dx^\nu, \tag{3.18}$$

---

[1] The precise definitions for an exact and a closed form are given in Appendix B.

where

$$F_{\mu\nu} = \partial_\mu A_\nu - \partial_\nu A_\mu. \tag{3.19}$$

Global properties correspond to the local versions (3.15) of the homogeneous and inhomogeneous Maxwell equations.

From the Stokes theorem, $dF = 0$ is equivalent to $\int_{\partial N_3} F = 0$, where N_3 is any sub-manifold of dimension 3. So, if we take as N_3 the ball $N_3 = \{(t, \boldsymbol{x}) \in \mathbb{R}^4 \mid t = t_0 \in \mathbb{R}, |\boldsymbol{x}| \leq R\}$, its surface is ∂N_3, a sphere of radius R in the plane $t = t_0$ and $0 = \int_{\partial N_3} F = \int_{|\boldsymbol{x}|=R} \boldsymbol{B}.d\boldsymbol{S}$, which expresses the absence of magnetic charges. We also have $\int_{N_2} F = \int_{\partial N_2} A$, where N_2 is a sub-manifold of dimension 2.

The importance of the fact that the space in which we are solving the equation $dF = 0$ by $F = dA$ is star-shaped can be seen in the following example. Consider the field created by a magnetic charge μ (a magnetic monopole), $\boldsymbol{E} = 0$ and $\boldsymbol{B} = \dfrac{\mu \boldsymbol{x}}{4\pi |\boldsymbol{x}|^3}$. It follows immediately that $\text{div}\boldsymbol{B} = \mu \delta^{(3)}(\boldsymbol{x})$. As $\boldsymbol{B}$ is singular at the origin, F is defined only for $M = \{(t, \boldsymbol{x}) \in \mathbb{R}^4 \mid x \neq 0\}$ where it satisfies $dF = 0$, but there exists no globally defined vector potential from which F is obtained. This situation will be analysed in Chapter 26.

As $\text{d}^2 = 0$ we deduce from the inhomogeneous equation the conservation of the 3-form $^*\mathcal{J}$:

$$\delta \mathcal{J} = 0 \quad \text{so} \quad \text{d}^*\mathcal{J} = 0. \tag{3.20}$$

Note:

- If N_4 is a manifold of dimension 4 and if $^*\mathcal{J}$ is a 3-form with a compact support over ∂N_4, the Stokes theorem gives $\int_{\partial N_4} {}^*\mathcal{J} = 0$.
- Reciprocally, from the current conservation, we deduce the existence of a 2-form, $^*F$.
- If N_3 exists, a section $t = 0$ of the physical space that is compact and without frontier, then $Q = \int_{N_3} {}^*\mathcal{J} = \int_{\partial N_3} {}^*F = 0$, because $\partial N_3 = \emptyset$. A closed universe must have a total charge null.

3.4   Choice of a Gauge

An important consequence of $F = dA$ is that the vector potential is not uniquely defined by F. Indeed, we can add any exact 1-form $d\phi$ to A since $dd\phi = 0$. We express this non-uniqueness by saying that the theory, expressed in terms of the vector potential, is gauge invariant.[2]

The transformation $A \to A + d\phi$ which is written in local coordinates $A_\mu(x) \to A_\mu(x) + \partial_\mu \phi(x)$ is called a gauge transformation, due to the fact that the scalar function ϕ

---

[2] The origin of this strange terminology goes back to the late twenties in the work of Hermann Weyl. Since the concept of gauge invariance plays a central role in the theory of all fundamental forces, we shall give in section 14.8 a brief historical account.

can vary depending on the point of space–time we are considering. An essential property of the theory, at the classical and quantum levels, is that the experimentally measured quantities, such as electric and magnetic fields or scattering cross sections in which the electromagnetic force intervenes, are left unchanged under a redefinition of the gauge of the vector potential.

Let us write the inhomogeneous equations $\mathrm{d}^*F = -^*\mathcal{J}$ in terms of the vector potential

$$\partial_\mu F^{\mu\nu} = \partial_\mu(\partial^\mu A^\nu - \partial^\nu A^\mu) = j^\nu. \tag{3.21}$$

Introducing

$$\partial_\mu \partial^\mu = \partial_t^2 - \Delta = \Box, \tag{3.22}$$

we obtain

$$\Box A^\nu - \partial^\nu(\partial_\mu A^\mu) = j^\nu, \tag{3.23}$$

which is the usual form of Maxwell's inhomogeneous equations in terms of the vector potential. One can see easily that these expressions are *gauge invariant*; i.e. they remain unchanged under $A_\mu \to A_\mu + \partial_\mu \phi$ transformations where ϕ is an arbitrary 0-form, and they imply current conservation $\partial_\mu j^\mu = 0$.

If we express the components, we obtain

$$\rho = -\Delta A^0 - \frac{\partial}{\partial t}\mathrm{div}\boldsymbol{A} \tag{3.24}$$

$$\boldsymbol{j} = \Box \boldsymbol{A} + \boldsymbol{\nabla}\left(\frac{\partial}{\partial t}A^0 + \mathrm{div}\boldsymbol{A}\right). \tag{3.25}$$

The observable quantities, the magnetic and electric fields, are gauge invariant. It follows that the four-component A_μ field contains redundant degrees of freedom, since we can change them without changing the physical quantities. We shall see later that this has a deep physical origin, but, for the moment, we proceed by trying to lift this degeneracy, by imposing a condition on the vector potential. This operation is called a gauge choice. Some gauges are more or less adapted to the type of problem one wants to solve. Some examples are as follows.

We call *Lorenz gauge*[3] the choice of A_μ which satisfies $\partial_\mu A^\mu = 0$. This gauge is Lorentz covariant. The *Coulomb gauge* is the gauge for which $\mathrm{div}\boldsymbol{A} = 0$. This gauge breaks Lorentz invariance, but this will not be seen in the results of calculations of physical quantities. The *axial gauge* is a gauge in which the component $A_0 = 0$.

Hence, in the Coulomb gauge, Eq. (3.24) becomes

$$-\Delta A^0 = \varrho. \tag{3.26}$$

---

[3] This condition was introduced as early as 1867 by the Danish mathematical physicist Ludvig Lorenz.

It is straightforward to establish the relation

$$-\triangle \frac{1}{|\mathbf{r} - \mathbf{r}'|} = 4\pi \delta (\mathbf{r} - \mathbf{r}'). \tag{3.27}$$

Using this equation we can obtain the solution of (3.26) as

$$A^0 (t, \boldsymbol{x}) = \frac{1}{4\pi} \int d^3 y \frac{\varrho (t, \boldsymbol{y})}{|\boldsymbol{y} - \boldsymbol{x}|}, \tag{3.28}$$

which shows that the time component of A is entirely determined by the external sources. In the Lorenz gauge, the inhomogeneous equations (3.25) are written as

$$\Box A^\mu = j^\mu. \tag{3.29}$$

3.5   Invariance under Change of Coordinates

To say that $F_{\mu\nu}$ is a tensor is equivalent to saying how it transforms under a change of coordinates (those coordinate changes are those that preserve the structure of the Minkowski space, i.e. Lorentz transformations). Using the fact that the electromagnetic tensor is a 2-form, we are going to make this transformation explicit.

Let $\Lambda : x \mapsto \bar{x} = \Lambda x$ be a Lorentz transformation. Then let us write that the intrinsic F object does not depend on a change of coordinates. We will denote by $F_{\mu\nu}$ the coefficients in the initial system and by $\bar{F}_{\mu\nu}$ those in the final one. So we obtain

$$\begin{aligned} F_{\mu\nu}(x) dx^\mu \wedge dx^\nu &= \bar{F}_{\varrho\sigma} (\bar{x}) d\bar{x}^\varrho \wedge d\bar{x}^\sigma \\ &= \bar{F}_{\varrho\sigma} (\bar{x}) \Lambda^\varrho{}_\mu \Lambda^\sigma{}_\nu dx^\mu \wedge dx^\nu. \end{aligned}$$

Then we deduce for the components in matrix terms

$$F = \Lambda^{tr} \bar{F} \Lambda. \tag{3.30}$$

Using the identities $\Lambda^{tr} G \Lambda = G$ and $g_{\mu\varrho} g^{\varrho\nu} = \delta^\nu_\mu$, we find that

$$F = G\Lambda^{-1} G\bar{F} G (\Lambda^{tr})^{-1} G; \tag{3.31}$$

hence,

$$G\bar{F} G = \Lambda G F G \Lambda^{tr}. \tag{3.32}$$

So, in terms of components,

$$\bar{F}_{\mu\nu}(\bar{x}) = (\Lambda^{-1})^{tr}_{\mu}{}^{\rho}(\Lambda^{-1})^{tr}_{\nu}{}^{\sigma}F_{\rho\sigma}(x) \tag{3.33}$$

$$\bar{F}^{\mu\nu}(\bar{x}) = \Lambda^{\mu}{}_{\rho}\Lambda^{\nu}{}_{\sigma}F^{\rho\sigma}(x). \tag{3.34}$$

Thus we can see that the components of $F^{\mu\nu}$ transform like those of a twice contravariant tensor,

$$F^{\mu\nu}(x) \rightarrow \Lambda^{\mu}{}_{\rho}\Lambda^{\nu}{}_{\sigma}F^{\rho\sigma}(\Lambda^{-1}x). \tag{3.35}$$

Similarly, A is a 1-tensor and its components transform as

$$A^{\mu}(x) \rightarrow \Lambda^{\mu}{}_{\nu}A^{\nu}(\Lambda^{-1}x). \tag{3.36}$$

We interpret the relation (3.36) by saying that the four-vector potential transforms as a four-dimensional representation of the Lorentz group.

The four-vector potential is, for us, the prototype of what we will call later *a classical or quantum field*. In general, a field ϕ will be indexed by a finite number of indices $\{\alpha\}_{\alpha \in I}$ and its components ϕ_{α} will transform according to

$$\phi^{\alpha}(x) \rightarrow S(\Lambda)^{\alpha}{}_{\beta}\phi^{\beta}(\Lambda^{-1}x), \tag{3.37}$$

where the square matrix with $|I|$ components, $S(\Lambda)$, is an $|I|$-dimensional representation of the Lorentz group.

Let us look for the invariants under the Lorentz group that we may construct from the electromagnetic tensor. They will be the coefficients of 4-forms, indeed after a Lorentz transformation,

$$d\bar{x}^0 \wedge d\bar{x}^1 \wedge d\bar{x}^2 \wedge d\bar{x}^3 = (\det \Lambda)dx^0 \wedge dx^1 \wedge dx^2 \wedge dx^3,$$

and for an element Λ of the special group we have $\det \Lambda = 1$. We can construct two gauge invariant 4-forms from F: on the one hand, $F \wedge {}^*F$, and on the other, $F \wedge F$. The gauge invariance is evident since the 2-form F is gauge invariant itself.

First, let us consider $F \wedge {}^*F$. We have

$$F \wedge {}^*F = \frac{1}{2}F_{\mu\nu}dx^{\mu} \wedge dx^{\nu} \wedge \frac{1}{4}\varepsilon_{\gamma\delta\rho\sigma}g^{\gamma\gamma}g^{\delta\delta}F_{\gamma\delta}dx^{\rho} \wedge dx^{\sigma}$$

$$= \frac{1}{4}F_{\mu\nu}F^{\mu\nu}\varepsilon_{\mu\nu\rho\sigma}dx^{\mu} \wedge dx^{\nu} \wedge dx^{\rho} \wedge dx^{\sigma}$$

$$= F_{\mu\nu}F^{\mu\nu}dx^0 \wedge dx^1 \wedge dx^2 \wedge dx^3 \tag{3.38}$$

and

$$F_{\mu\nu}F^{\mu\nu} = 2(\boldsymbol{B}^2 - \boldsymbol{E}^2). \tag{3.39}$$

In a similar way

$$F \wedge F = \frac{1}{2} F_{\mu\nu} \mathrm{d}x^\mu \wedge \mathrm{d}x^\nu \wedge \frac{1}{2} F_{\varrho\sigma} \mathrm{d}x^\varrho \wedge \mathrm{d}x^\sigma$$

$$= -\frac{1}{4} F_{\mu\nu} F_{\rho\sigma} \varepsilon^{\mu\nu\varrho\sigma} \varepsilon_{\mu\nu\varrho\sigma} \mathrm{d}x^\mu \wedge \mathrm{d}x^\nu \wedge \mathrm{d}x^\varrho \wedge \mathrm{d}x^\sigma$$

$$= -\frac{1}{2} F_{\mu\nu}{}^* F^{\mu\nu} \mathrm{d}x^0 \wedge \mathrm{d}x^1 \wedge \mathrm{d}x^2 \wedge \mathrm{d}x^3 \tag{3.40}$$

and

$$F_{\mu\nu}{}^* F^{\mu\nu} = -4 \boldsymbol{E} \cdot \boldsymbol{B}. \tag{3.41}$$

Under a particular Lorentz transformation, the space inversion I_s, the tensor $F^{\mu\nu}(t, \boldsymbol{x})$ transforms into

$$\bar{F}^{\mu\nu}(\boldsymbol{E}, \boldsymbol{B})(t, \boldsymbol{x}) = F^{\mu\nu}(-\boldsymbol{E}, \boldsymbol{B})(t, -\boldsymbol{x}), \tag{3.42}$$

i.e.

$$\boldsymbol{E}(t, \boldsymbol{x}) \rightarrow -\boldsymbol{E}(t, -\boldsymbol{x}) \tag{3.43}$$

$$\boldsymbol{B}(t, \boldsymbol{x}) \rightarrow \boldsymbol{B}(t, -\boldsymbol{x}), \tag{3.44}$$

which expresses the fact that $\boldsymbol{E}$ is a vector and $\boldsymbol{B}$ is a pseudovector (or axial vector).

Therefore, we can see that $F_{\mu\nu} F^{\mu\nu}$ defines a scalar (i.e. a space parity invariant), whereas $F_{\mu\nu}{}^* F^{\mu\nu}$ defines a pseudoscalar.

3.6   Lagrangian Formulation

The theory of electromagnetism is a theory based on the locality principle and satisfying at the classical level the least action principle. Thus, it is possible to construct a Lagrangian density whose equations of motion are Maxwell's equations.

Only very few changes are necessary in order to adapt the variational calculus of mechanics to the case of a field theory.

In classical mechanics, it is easy to associate with the equations of motion a Lagrangian depending on positions $q_i(t)$ and on the time derivatives of the positions $\dot{q}_i(t)$. In relativistic field theories, the parameters t and i are replaced by a point $x \in M^4$. Thus, q becomes a field ϕ and can have indices or be a tensor, as the electromagnetic field A_μ.

The models of statistical mechanics on lattice give interesting examples of field theories. Let us consider, as an example, a system of spins σ placed on the vertices of a cubic lattice in a three-dimensional space and suppose that we are interested in the time evolution of the system under the action of the interaction

$$H = \sum_{\imath,\jmath,|\imath-\jmath|=\delta} \sigma_\imath(t)\sigma_\jmath(t) + \sum_\imath \sigma_\imath(t)B_\imath(t), \tag{3.45}$$

δ being the lattice spacing. The spins are coupled with their nearest neighbours as well as with the third component of a time-dependent magnetic field $B_\imath(t)$. Each spin depends on time and on its location on the lattice: $\sigma_\imath(t)$, where $\imath \in \delta\mathbb{Z}^3$. We can express this relation by saying that σ depends on the parameters $(t,\imath)$. If we imagine this lattice as a discretisation of $\mathbb{R}^3$, in the limit of vanishing lattice spacing, σ becomes a function $\sigma(t,x)$, that is, a classical field, and the Hamiltonian becomes the integral over $\mathbb{R}^3$ of a Hamiltonian density, which is a function of $\sigma(t,x)$ and of the derivatives $\nabla\sigma(t,x)$. We could have replaced the spin σ, a scalar quantity, by a vector $\boldsymbol{\sigma}$ and think about a coupling such as $\boldsymbol{\sigma}.\boldsymbol{B}$.

The instantaneity of the interaction requires that the fields can only interact locally in order not to violate the causality principle. This is what happened in the particular case of the preceding spin model. More generally, a model of field theory is given by a Lagrangian density $\mathcal{L}(\phi_i(x), \partial_\mu\phi_i(x), x) = \mathcal{L}(x)$, $x \in \mathbb{R}^4$, where $\{\phi_i(x)\}_{i\in I}$ is a family of fields indexed by I.

3.6.1   The Euler–Lagrange Equations and Noether's Theorem

Let us consider the action

$$S[\phi] = \int_\Omega \mathcal{L}(x)\mathrm{d}^4x, \tag{3.46}$$

where Ω is a regular region of space–time over which the Lagrangian density is being integrated (usually, it will be the whole space $\mathbb{R}^4$). Also for simplicity, we choose that the Lagrangian density $\mathcal{L}$ does not depend explicitly on x.

The change induced in the Lagrangian by an infinitesimal shift, $\delta\phi(x)$, of the fields $\phi(x) \to \phi(x) + \delta\phi(x)$ gives[4]

$$\delta\mathcal{L} = \frac{\delta\mathcal{L}}{\delta\phi_i}\delta\phi_i + \frac{\delta\mathcal{L}}{\delta(\partial_\mu\phi_i)}\delta\partial_\mu\phi_i$$

---

[4] One defines the functional derivative $\frac{\delta F(\phi)}{\delta\phi(x)}$ of a function $F(\phi)$ of ϕ by the limit, when it exists, as

$$\frac{\delta F(\phi)}{\delta\phi(x)} = \lim_{\varepsilon\to 0}\frac{F(\phi + \varepsilon\delta_x) - F(\phi)}{\varepsilon}, \tag{3.47}$$

where δ_x is the Dirac measure at the point x. If $F = \int(\phi(y))^n\mathrm{d}^4y$, then

$$\frac{\delta F(\phi)}{\delta\phi(x)} = n(\phi(x))^{n-1}. \tag{3.48}$$

$$= \frac{\delta \mathcal{L}}{\delta \phi_i} \delta \phi_i + \partial_\mu \left(\frac{\delta \mathcal{L}}{\delta (\partial_\mu \phi_i)} \delta \phi_i \right) - \partial_\mu \left(\frac{\delta \mathcal{L}}{\delta (\partial_\mu \phi_i)} \right) \delta \phi_i$$

$$= \left[\frac{\delta \mathcal{L}}{\delta \phi_i} - \partial_\mu \left(\frac{\delta \mathcal{L}}{\delta (\partial_\mu \phi_i)} \right) \right] \delta \phi_i + \partial_\mu \left(\frac{\delta \mathcal{L}}{\delta (\partial_\mu \phi_i)} \delta \phi_i \right). \tag{3.49}$$

The last term in (3.49) is a 4-divergence. When $\delta \mathcal{L}$ is integrated in Ω, there is no contribution of this last term because of the vanishing on the boundary $\partial \Omega$ of the fields and of their infinitesimal variations.

We will now examine this result according to two different points of view: on the one hand, it can be used to describe the fields when the action is extremal under an arbitrary variation of the fields, and on the other, if this variation describes a symmetry of the theory, it can generate some consequences of this symmetry.

3.6.1.1   *The Principle of Stationary Action*

This principle is a generalisation of what we know in classical mechanics: the equations of motion for a classical field can be derived from a Lagrangian through the principle of stationary action.

Theorem 5 (Euler–Lagrange's equations). *Let ϕ_i, $i \in I$, be a solution of the equations of motion. If we add to the solution an arbitrary infinitesimal function $\delta \phi_i$, $i \in I$, which vanishes on the boundary $\partial \Omega$, then the variation of the action S is of second order:*

$$S[\phi + \delta \phi] - S[\phi] \simeq \delta S[\phi] = 0. \tag{3.50}$$

From (3.49), we have that

$$\delta S[\phi] = \int_\Omega \left[\left[\frac{\delta \mathcal{L}}{\delta \phi_i} - \partial_\mu \left(\frac{\delta \mathcal{L}}{\delta (\partial_\mu \phi_i)} \right) \right] \delta \phi_i + \partial_\mu \left(\frac{\delta \mathcal{L}}{\delta (\partial_\mu \phi_i)} \delta \phi_i \right) \right] \mathrm{d}^4 x$$

$$= \int_\Omega \left[\frac{\delta \mathcal{L}}{\delta \phi_i} - \partial_\mu \left(\frac{\delta \mathcal{L}}{\delta (\partial_\mu \phi_i)} \right) \right] \delta \phi_i \mathrm{d}^4 x = 0. \tag{3.51}$$

Since the last equality is true for any variation $\delta \phi$ satisfying the boundary condition, we deduce that[5]

$$\frac{\delta \mathcal{L}}{\delta \phi_i(x)} - \partial_\mu \frac{\delta \mathcal{L}}{\delta [\partial_\mu \phi_i(x)]} = 0.$$

These are the Euler–Lagrange equations of motion of the classical field.

---

[5] There are numerous improper notations when using functional derivatives. For example, the Euler–Lagrange equations are sometimes written as

$$\frac{\delta \mathcal{L}(x)}{\delta \phi_i(x)} - \partial_\mu \frac{\delta \mathcal{L}(x)}{\delta [\partial_\mu \phi_i(x)]} = 0 \quad \text{or} \quad \frac{\delta \mathcal{L}}{\delta \phi_i}(x) - \partial_\mu \frac{\delta \mathcal{L}}{\delta [\partial_\mu \phi_i]}(x) = 0. \tag{3.52}$$

It must be noted that under the conditions we imposed for the derivation of the Euler–Lagrange equations, the action is not related in a unique way to the Lagrangian density. We can, in fact, add to the Lagrangian any divergence (since, by Gauss theorem, its integral on the boundary of Ω will give zero).

3.6.1.2   Noether's Theorem for Classical Fields

We saw in section 1.3.2 that any continuous transformation of the coordinates leaving invariant a Lagrangian generates a conserved quantity. We will now prove that any continuous change of the fields, leaving invariant the action $S[\phi]$, generates a conserved current.

We start with some definitions. A symmetry is called *internal* if the transformations act on the fields without affecting the space–time point x. If the parameters of the transformation are x-independent the symmetry will be called *global*. In the general case, in which the parameters of the transformation are arbitrary functions of x, the symmetry will be called *local*, or more often *gauge symmetry*.

Suppose now that the infinitesimal change of the fields, $\phi_i(x) \rightarrow \phi_i(x) + \delta\phi_i(x)$, $i \in I$, leave the action $S[\phi]$ invariant. The variation of the Lagrangian due to this change $\delta\mathcal{L}$ is still given by (3.49), but now the change $\delta\phi$ is not arbitrary. Since the action remained unchanged, this means that the change of the Lagrangian density must be the 4-divergence of some quantity,

$$\delta\mathcal{L} = \partial_\mu F^\mu, \tag{3.53}$$

and this is true for arbitrary configurations of the fields ϕ_i. This is often called an off-shell equality because the fields do not need to satisfy the Euler–Lagrange equations.

If now we suppose that the fields ϕ_i satisfy the Euler–Lagrange equations, then comparing (3.53) and (3.49), we get that

$$\delta\mathcal{L} = \partial_\mu F^\mu = \partial_\mu \left(\frac{\delta\mathcal{L}}{\delta(\partial_\mu\phi_i)} \delta\phi_i \right), \tag{3.54}$$

from which follows that

$$\mathcal{J}^\mu = \left(\frac{\delta\mathcal{L}}{\delta(\partial_\mu\phi_i)} \delta\phi_i \right) - F^\mu \tag{3.55}$$

is conserved (it is an on-shell property):

$$\partial_\mu \mathcal{J}^\mu = 0,$$

with an associated conserved charge

$$Q = \int_\Omega \mathcal{J}^0 \mathrm{d}^3 x.$$

This current, the Noether current, is conserved and is a consequence of the symmetries of the theory.

Remark that there is some freedom in redefinition of the Noether current. We can add and multiply by any constant without changing the fact that its 4-divergence vanishes.

3.6.2 Examples of Noether Currents

3.6.2.1 *Energy–Momentum Tensor*

In classical mechanics, invariance under spatial translations gives rise to the conservation of the momentum while invariance under time translation gives rise to conservation of energy. We will see something similar in field theories. We will show directly that the fact that the Lagrangian does not explicitly depend on x makes it possible to define, using the Euler–Lagrange equations, the existence of a conserved tensor. We choose Ω to be the four-dimensional space–time (a domain of integration which is invariant under translations) and assume that the fields decrease fast enough at infinity to ignore all surface terms.

Let us write in two different ways the derivative of $\mathcal{L}$ with respect to x:

$$
\begin{aligned}
\frac{\mathrm{d}\mathcal{L}(x)}{\mathrm{d}x^{\mu}} &= \frac{\delta\mathcal{L}(x)}{\delta\phi_i(x)}\partial_\mu\phi_i(x) + \frac{\delta\mathcal{L}(x)}{\delta[\partial_\nu\phi_i(x)]}\partial_\mu\partial_\nu\phi_i(x) \\
&= \partial_\nu\frac{\delta\mathcal{L}(x)}{\delta[\partial_\nu\phi_i(x)]}\partial_\mu\phi_i(x) + \frac{\delta\mathcal{L}(x)}{\delta[\partial_\nu\phi_i(x)]}\partial_\mu\partial_\nu\phi_i(x) \\
&= \partial_\nu\left(\frac{\delta\mathcal{L}(x)}{\delta[\partial_\nu\phi_i(x)]}\partial_\mu\phi_i(x)\right).
\end{aligned}
\tag{3.56}
$$

By equating the last equality to

$$
\frac{\mathrm{d}\mathcal{L}}{\mathrm{d}x^{\mu}} = \partial_\mu\mathcal{L} = \delta_\mu^\nu\partial_\nu\mathcal{L},
\tag{3.57}
$$

we find that

$$
\partial_\nu\tilde{T}_\mu^\nu = 0,
\tag{3.58}
$$

where

$$
\tilde{T}_\mu^\nu(x) = \frac{\delta\mathcal{L}(x)}{\delta[\partial_\nu\phi_i(x)]}\partial_\mu\phi_i(x) - \mathcal{L}(x)\delta_\nu^\mu
\tag{3.59}
$$

is the *energy–momentum* tensor.

We deduce from it that the four-vector P^μ given by

$$
P^\mu = \int \mathrm{d}^3x\,\tilde{T}^{0\mu}(\boldsymbol{x},t)
\tag{3.60}
$$

is time independent since

$$\dot{P}^\mu = \int d^3x \partial_0 \tilde{T}^{0\mu}(\boldsymbol{x}, t) = -\int d^3x \partial_i \tilde{T}^{i\mu}(\boldsymbol{x}, t) = 0 \qquad (3.61)$$

if the fields vanish fast enough at infinity as it was supposed.

The fact that P^μ is a four-vector can be seen through the integrals defining it. In fact, the integration which is performed on the surface $t = 0$ can be made on any space–time surface, the integration element d^3x being replaced by the associated covariant surface element directed according to the normal of the surface

$$P^\mu = \int d\sigma_\nu \, \tilde{T}^{\nu\mu}(\boldsymbol{x}, t). \qquad (3.62)$$

For reasons which will become clear later on, it is useful to replace this energy–momentum tensor by another tensor, symmetric with respect to the exchange of indices, and obtained from the first one by adding to it a divergence. In the following chapter we will show that the new energy–momentum tensor can be obtained as the result of a problem of variational calculus in which it is not the fields but the metric $g_{\mu\nu}$ which varies.

Remark that the conservation of the energy–momentum tensor can be obtained by applying directly Noether's theorem of section 3.6.1.2. The fact that the Lagrangian does not depend explicitly on x expresses the invariance under translations.

Let us consider an infinitesimal translation of x: $x \to x - \epsilon$.

This translation implies that $\phi_i(x) \to \phi_i(x) + \epsilon^\nu \partial_\nu \phi_i(x)$ and equivalently $\mathcal{L}(x) \to \mathcal{L}(x) + \epsilon^\nu \partial_\nu \mathcal{L}(x)$. Then from the definition (3.55) of the Noether current, we get, rewriting $\partial_\nu \mathcal{L}$ as $\delta^\mu_\nu \partial_\mu \mathcal{L}$,

$$(\mathcal{J}_\mu)_\nu = \frac{\delta\mathcal{L}}{\delta(\partial_\mu\phi_i)} \partial_\nu\phi_i - \delta^\mu_\nu \mathcal{L}, \qquad (3.63)$$

which is nothing else than the energy-momentum tensor $\tilde{T}^\mu_\nu$.

3.6.2.2   *Global Symmetries*

A global symmetry is a symmetry for which the parameters of the transformation are x-independent.

Let us consider a global internal symmetry of the Lagrangian and suppose, as it is practically always the case, that it is given by a unitary transformation,

$$\phi_j(x) \to (e^{i\alpha_a T^a})_{jk}\phi_k(x), \qquad (3.64)$$

where the α_a's denote a set of parameters defining the transformation, for example, the rotation angles for a group of rotations. Infinitesimally, the transformation can be written as

$$\phi_j(x) \to \phi_j(x) + i\alpha_a T^a_{jk}\phi_k(x) \qquad (3.65)$$

From the definition of the Noether current 3.55, one gets

$$\mathcal{J}_\mu = \frac{\delta \mathcal{L}(x)}{\delta(\partial^\mu \phi_j(x))} \delta \phi_j(x)$$

$$= i\alpha_a \frac{\delta \mathcal{L}(x)}{\delta(\partial^\mu \phi_j(x))} T^a_{jk} \phi_k(x) \qquad (3.66)$$

thus showing the existence of a conserved current

$$\mathcal{J}^a_\mu(x) = i \frac{\delta \mathcal{L}(x)}{\delta[\partial^\mu \phi_j(x)]} T^a_{jk} \phi_k(x) \qquad (3.67)$$

where we made use of the previous remark that we can add or multiply a current by any constant.

3.6.3 Application to Electromagnetism

Let us apply these results to the electromagnetic field. We will write the Lagrangian density as a function of the field A, the vector potential. Using the hypothesis of locality, Lorentz invariance and the requirements that the equations of motion are linear, second-order differential equations in A, the most general form of the electromagnetic part is the invariant quantity:

$$\mathcal{L}_{em} = aA^\mu A_\mu + b\partial_\mu A^\nu \partial_\nu A^\mu + c\partial_\mu A^\nu \partial^\mu A_\nu + d(\partial_\mu A^\mu)^2 \qquad (3.68)$$

where a, b, c, d are constants to be fixed. If one now asks the Euler–Lagrange equations to be gauge invariant, one gets $b + c + d = 0$ and $a = 0$. The Lagrangian can then be written as

$$\frac{b}{2} F^{\mu\nu} F_{\mu\nu} + 2d\, \partial_\mu (A^\nu \partial_\nu A^\mu)$$

Therefore, neglecting the divergence term which will not contribute to the integrated form,[6] and choosing the value of b which corresponds to an appropriate rescaling of the field A, the Lagrangian density becomes

$$\mathcal{L}_{em} = -\frac{1}{4} F^2, \qquad (3.69)$$

where $F^2 = F_{\mu\nu} F^{\mu\nu}$. The Euler–Lagrange equations of this density (in the absence of charged particles) are the Maxwell equations $\partial_\nu F^{\nu\mu} = 0$.

---

[6] In most cases studied in this Book we shall assume that the fields vanish at infinity, so surface terms can be set to zero. Whenever this is not the case, see for example the discussion in section 25.4.5, we shall state it explicitly.

To get the Maxwell equations coupled to a current, it is enough to change the free action as

$$\mathcal{L} = -\frac{1}{4}F^2 - j \cdot A = \mathcal{L}_{em} + \mathcal{L}_{coupl.} \tag{3.70}$$

By construction F^2 is gauge invariant. In contrast, a gauge transformation $A_\mu \to A_\mu + \partial_\mu \phi$ leads to $j \cdot A \to j \cdot A + j \cdot \partial \phi$. It is only if the current four-vector is conserved that the last term can be written as a divergence $\partial_\mu(j^\mu \phi)$, that is to say, a term which does not contribute to the action because of the conditions at infinity. Conversely, if the action is gauge invariant, the current must be conserved.

In the language of forms, the Lagrangian density $\mathcal{L}$ is a 4-form. The Lagrangian for the electromagnetic field is

$$\mathcal{L} = -\frac{1}{2}dA \wedge {}^*dA - A \wedge {}^*\mathcal{J}. \tag{3.71}$$

The equations of motion can be obtained easily. Let us consider indeed the variation of the action I on a manifold N_4 of dimension 4,

$$I = \int_{N_4} \mathcal{L}, \tag{3.72}$$

obtained by varying A to $A + \delta A$, δA vanishing on ∂N_4. We get

$$\delta I = \int_{N_4} \delta \mathcal{L} = -\frac{1}{2}\int_{N_4} d\delta A \wedge {}^*dA - \frac{1}{2}\int_{N_4} dA \wedge {}^*d\delta A - \int_{N_4} \delta A \wedge {}^*\mathcal{J}$$

$$= -\int_{\partial N_4} \delta A \wedge {}^*dA - \int_{N_4} \delta A \wedge (d\,{}^*dA + {}^*\mathcal{J}). \tag{3.73}$$

Thus, from the boundary conditions, $d\,{}^*dA + {}^*\mathcal{J} = 0$ if the action is stationary. We used $\delta dA = d\delta A$ et $dA \wedge {}^*\delta dA = \delta dA \wedge {}^*dA$.

Let us look for the form of the energy–momentum tensor. Using $\frac{\delta \mathcal{L}_{em}}{\partial_\mu A_\nu} = -F^{\mu\nu}$, we find that

$$\widetilde{T}^{\mu\nu} = -F^{\mu\varrho}\partial^\nu A_\varrho + \frac{1}{4}g^{\mu\nu}F^2 + g^{\mu\nu}j \cdot A = \widetilde{T}^{\mu\nu}_{em} + \widetilde{T}^{\mu\nu}_{coupl}. \tag{3.74}$$

Adding to $\widetilde{T}$ the divergence $\partial_\varrho(F^{\mu\varrho}A^\nu)$, we make the purely electromagnetic part symmetric and gauge invariant,

$$\widetilde{T}^{\mu\nu}_{em} \to T^{\mu\nu}_{em}, \tag{3.75}$$

where

$$T^{\mu\nu}_{em} = -F^{\mu\varrho}F^{\nu}_{\ \varrho} + \frac{1}{4}g^{\mu\nu}F^2. \tag{3.76}$$

We can verify that

$$T^{00}_{em} = -F^{0\varrho}F^{0}_{\ \varrho} + \frac{1}{4}F^2 = E^2 + \frac{1}{2}(B^2 - E^2) = \frac{1}{2}(E^2 + B^2) \tag{3.77}$$

$$T^{i0}_{em} = [E \wedge B]^i = S^i, \tag{3.78}$$

where T^{00}_{em} is the energy density of the electromagnetic field and $S = E \wedge B$ is the Poynting vector. In the presence of charges, the energy–momentum tensor has a divergence which opposes the Lorentz force, $-j_\sigma F^{\sigma\nu}$,

$$\partial_\mu T^{\mu\nu}_{em} = j_\sigma F^{\sigma\nu}. \tag{3.79}$$

We can then deduce that

$$-\frac{\mathrm{d}}{\mathrm{d}t}\int_\Omega \mathrm{d}^3x\, T^{00}_{em} = \int_{\partial\Omega} E \wedge B \cdot \mathrm{d}\boldsymbol{\sigma} + \int_V \mathrm{d}^3x\, \boldsymbol{j} \cdot \boldsymbol{E} = 0, \tag{3.80}$$

which expresses the conservation of the total four-momentum as a sum of two terms, one corresponding to the momentum carried by the electromagnetic field and a second one associated with the particles.

Obviously, with charges, the energy–momentum tensor is not gauge invariant. We must add to the system what corresponds to the dynamics of these charges in such a way that the system is closed.

A consequence of the symmetry of the energy–momentum tensor is that the third-order tensor,[7]

$$M^{\mu\nu\varrho} = T^{\mu\nu}x^\varrho - T^{\mu\varrho}x^\nu, \tag{3.81}$$

a natural generalisation of the angular momentum density, is conserved,

$$\begin{aligned}\partial_\mu M^{\mu\nu\varrho} &= (\partial_\mu T^{\mu\nu})x^\varrho + T^{\mu\nu}\partial_\mu x^\varrho - (\partial_\mu T^{\mu\varrho})x^\nu - T^{\mu\varrho}\partial_\mu x^\nu \\ &= T^{\varrho\nu} - T^{\nu\varrho} = 0,\end{aligned} \tag{3.82}$$

where we used the conservation of $T^{\mu\nu}$.

---

[7] This conserved tensor is obtained, up to a divergence, by writing that the Lagrangian is rotation invariant.

3.7 Interaction with a Charged Particle

Let us consider now the interaction with a charged particle that we assume to be point-like. If $\mathbf{v}$ is the speed of the particle, its space–time trajectory is described by $x^\mu(s)$, where s is the proper time. We recall that $\mathrm{d}s = \sqrt{\mathrm{d}x^\mu \mathrm{d}x_\mu} = \sqrt{1 - v^2}\mathrm{d}t$ with $\mathrm{d}t = \mathrm{d}x^0(s)$.

The relativistic current of the particle is given by

$$
j^\mu(t, \mathbf{y}) = e\frac{\mathrm{d}x^\mu}{\mathrm{d}t}\delta^3(\mathbf{y} - \mathbf{x}(s))\,|_{t=x^0(s)}
$$

$$
= e\int_{-\infty}^{+\infty}\mathrm{d}s\frac{\mathrm{d}x^\mu}{\mathrm{d}s}\delta^4(y - x(s)). \tag{3.83}
$$

This current is conserved, $\partial_\mu j^\mu = 0$. The trajectory of the particle $x^\mu(s)$ is a time-like curve, that is causal, a fact which is defined by saying that at each point of the curve, the tangent vector $u^\mu = \dfrac{\mathrm{d}x^\mu}{\mathrm{d}s}$ is a time-like vector.

The Lagrangian $L_{\text{part.}}$ of a particle of mass m is

$$
L_{\text{part.}} = -m\int \mathrm{d}s. \tag{3.84}
$$

The Lagrangian of the system is the sum of the electromagnetic Lagrangian, of the Lagrangian of the coupling of the particle with the field, and of the Lagrangian of the particle (the additivity of the Lagrangians is an empirical rule which is not justified by any principle other than the adequacy with the laws of nature):

$$
L = L_{\text{em}} + L_{\text{part.}} + L_{\text{coupl.}}. \tag{3.85}
$$

The coupling term can be rewritten as

$$
\int \mathrm{d}^4 y\, j^\mu(y) A_\mu(y) = e\int \mathrm{d}^4 y\, A_\mu(y)\left(\int \mathrm{d}s\frac{\mathrm{d}x^\mu}{\mathrm{d}s}\delta^4(y - x(s))\right)
$$

$$
= e\int A_\mu(x(s))\mathrm{d}x^\mu. \tag{3.86}
$$

We have

$$
L = \int \mathrm{d}^4 x\, \mathcal{L}_{em}(x) - m\int \mathrm{d}s - e\int \mathrm{d}x^\mu A_\mu(x(s))
$$

$$
= \int \mathrm{d}^4 x\, \mathcal{L}_{em}(x) + \int \mathrm{d}t(-m\sqrt{1 - v^2} - eA^0 + e\mathbf{v}\mathbf{A}). \tag{3.87}
$$

We will first consider the interaction of a point-like charge with an electromagnetic field fixed by external conditions, which means that in the Lagrangian we will ignore the pure electromagnetic contribution. Thus,

$$
L = -m\int \mathrm{d}s - e\int A_\mu \mathrm{d}x^\mu. \tag{3.88}
$$

Let us apply the least action principle in order to get the equations of motion. By a variation $x^\mu \to x^\mu + dx^\mu$,

$$\delta ds = \frac{(\delta dx^\mu) dx_\mu}{ds} = u_\mu \delta dx^\mu \tag{3.89}$$

and

$$\delta(A_\mu dx^\mu) = \delta A_\mu dx^\mu + A_\mu \delta dx^\mu = \partial_\nu A_\mu \delta x^\nu dx^\mu + A_\mu d\delta x^\mu$$

$$= \partial_\nu A_\mu \delta x^\nu dx^\mu - dA_\nu \delta x^\nu + d(A_\nu \delta x^\nu)$$

$$= (\partial_\nu A_\mu - \partial_\mu A_\nu) \delta x^\nu dx^\mu + d(A_\nu \delta x^\nu). \tag{3.90}$$

Then

$$\delta L = m \int ds \delta x^\nu \frac{du_\nu}{ds} - e \int ds F_{\mu\nu} u^\nu \delta x^\mu, \tag{3.91}$$

and therefore the equations of motion are

$$m \frac{du_\mu}{ds} = e F_{\mu\nu} u^\nu. \tag{3.92}$$

Going back to the usual coordinates, these equations are equivalent to (we used the fact that the covariant vector u can be written as $(\frac{1}{\sqrt{1-v^2}}, \frac{\mathbf{v}}{\sqrt{1-v^2}})$)

$$\frac{d}{dt} \frac{m\mathbf{v}}{\sqrt{1-v^2}} = e\mathbf{E} + e\mathbf{v} \wedge \mathbf{B} \tag{3.93}$$

$$\frac{d}{dt} \frac{m}{\sqrt{1-v^2}} = e\mathbf{v} \cdot \mathbf{E}. \tag{3.94}$$

The first equation is the Lorentz force equation. The second equation expresses the time variation of the particle's kinetic energy; we see that it depends only on the electric field.

It is useful to check this result from the Lagrangian expressed in terms of t and $\mathbf{v}$: $L = \int dt(-m\sqrt{1-v^2} - eA^0 + e\mathbf{v} \cdot \mathbf{A})$.

If we compute now the conjugate momentum $\mathbf{p} = \frac{\partial \mathcal{L}}{\partial \mathbf{v}}$, we find that

$$\mathbf{p} = \frac{m\mathbf{v}}{\sqrt{1-v^2}} + e\mathbf{A}, \tag{3.95}$$

and the Hamiltonian is

$$\mathcal{H} = \mathbf{v} \cdot \mathbf{p} - \mathcal{L} = [m^2 + (\mathbf{p} - e\mathbf{A})^2]^{\frac{1}{2}} + eA^0. \tag{3.96}$$

We call *minimal coupling* the fact that under the action of an electromagnetic field the conjugate momentum $\boldsymbol{p}$ becomes $\boldsymbol{p} - e\boldsymbol{A}$. This terminology is related to the extremely simple form $j \cdot A$ of the coupling between the field and the current.

3.8   Green Functions

With the introduction of the vector potential, Maxwell's equations reduce to the study of a propagation equation with a source term. The typical equation is the equation of the vector potential in the Lorenz gauge.

To avoid inessential complications due to the vectorial character of the initial equation, we will consider first a simpler version of this equation: the massive *Klein–Gordon equation* with a source term:

$$(\Box + m^2)\phi(x) = j(x). \tag{3.97}$$

In our usual system of units, the parameter m^2 has dimensions of $[\text{mass}]^2$. We will see later that in the quantum theory it describes indeed the mass of a particle, but, for the classical differential equation we are studying here, it is just a parameter. We will show shortly that choosing $m \neq 0$ avoids a singular behaviour of the solution at infinity.

The Klein–Gordon equation is the most general linear, second-order differential equation for a scalar function which is invariant under Lorentz transformations. It is an equation of hyperbolic type. As a second-order differential equation, its solutions are determined by the Cauchy data on a surface, called the Cauchy surface. We will now characterise the Cauchy surface of the Klein–Gordon equation.

A sufficiently regular hypersurface $N_3 \subset \mathbb{R}^4$ is space-like if its normal derivatives are time-like. It splits space into two half-spaces, the future and the past. We define the past area of influence of a space-like hypersurface N as the set of all points in the future which are at least on one time-like curve[8] which crosses N: it is the closure of the interior $D^+(N)$ of the light cone open towards the future and sitting on N. We define similarly the future area of influence $D^-(N)$. The influence area of the Cauchy data restricted to N is $D = D^+ \cup D^-$. The hypersurface N of a manifold M is a Cauchy surface if $D(N) = M$: see Fig. 3.1. In practice every time we shall use local coordinates, we will take as a Cauchy surface, the hyperplane $\mathbb{R}^3 = \{(t, \boldsymbol{x}) \in M^4 \,|\, t = 0\}$ or its translations. By relativistic invariance, the quantities that we look at are independent of this particular choice.

Thus, the current 3-form $*\mathcal{J}$ can be integrated on a space-like surface. On the particular surface we choose,

$$\int_{t=0} *\mathcal{J} = \int_{t=0} \mathrm{d}^3 x \mathcal{J}_0 \tag{3.98}$$

---

[8] A curve is time-like or causal if at each of its points, the tangent is time-like.

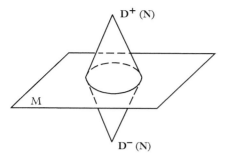

Figure 3.1 *The Cauchy surface for the Klein–Gordon equation.*

Since it is an invariant, it can also be written on an arbitrary space-like 3-surface N_3

$$\int_{N_3} d\sigma^\mu \mathcal{J}_\mu, \tag{3.99}$$

where $d\sigma^\mu$ is the μth component of the normal element of the 3-surface.

In the case of electromagnetism, if N is a Cauchy surface, F is uniquely determined in $D(N)$ by the data $\mathcal{J}$, $F|_N$, and $*F|_N$. We can be convinced that if these data determined F in points outside of $D(N)$, we would have a speed of propagation larger than the speed of light.

Let us go back to the Klein–Gordon equation. We obtain the solutions of this equation if we know the solutions of the equation

$$(\Box + m^2)\, G(x, x') = \delta^4(x - x'). \tag{3.100}$$

In physics, these solutions are the so-called elementary solutions, or *Green functions*.

Since we are interested only in the solutions which are translation invariant, we look for the Green functions which are translation invariant. A general solution will then be

$$\phi(x) = \phi_0(x) + \int d^4x\; G(x - x')j(x'), \tag{3.101}$$

where ϕ_0 is a solution of the homogeneous equation $(\Box + m^2)\,\phi_0 = 0$, which is fixed by the Cauchy data.

3.8.1   The Green Functions of the Klein–Gordon Equation

By Fourier transform, the equation for the elementary solution becomes

$$(-p^2 + m^2)\, \tilde{G}(p) = 1. \tag{3.102}$$

In other words, the general Green function is given by the inverse of the differential operator on the left-hand side. This is very general and we shall use it often in what follows. Let us consider a linear, homogeneous differential equation of the form

$$\mathcal{D}\Psi(x) = 0, \tag{3.103}$$

where the function $\Psi(x)$ may be an n-component vector and the differential operator $\mathcal{D}$ an $n \times n$ matrix. The associated general Green function is given by $[\mathcal{D}]^{-1}$, where the inverse is understood both in the sense of the differential operator and that of the matrix.

Coming back to the Klein–Gordon equation (3.102), we see that if $p^2 \neq m^2$, $\tilde{G}(p) = (p^2 + m^2)^{-1}$. It follows that $\tilde{G}$ is completely fixed up to a distribution whose support is on $p^2 = m^2$. We conclude that we can obtain several Green functions, depending on the choice of this distribution. We shall describe them in this section.

In order to satisfy the causality principle, that is, such that the value of the solution at a point depends only on the values at points belonging to the past, we shall look at the so-called *retarded Green function* $G_{\text{ret}}(x - x')$, whose support is in the past cone of x.

Claim   The retarded elementary solution G_{ret} of the Klein–Gordon equation is given by

$$G_{\text{ret}}(x - x') = -\frac{1}{(2\pi)^4} \lim_{\varepsilon \to 0} \int d^4 p \frac{e^{-ip\cdot(x-x')}}{(p_0 + i\varepsilon)^2 - \mathbf{p}^2 - m^2}. \tag{3.104}$$

Proof.   In what follows, the limit, $\varepsilon \geq 0$, ε going to 0, will be implicit.

To prove this result, let us introduce $\omega = \omega(\mathbf{p}) = \omega_p = \sqrt{\mathbf{p}^2 + m^2}$. Reordering the integrations, we get

$$G_{\text{ret}}(x - x') = -\frac{1}{(2\pi)^4} \int \int \int d^3 p \, e^{i\mathbf{p}\cdot\mathbf{x}} \left\{ \int_{-\infty}^{+\infty} dp^0 \frac{e^{-ip^0(x^0 - x^{0\prime})}}{(p^0 + i\varepsilon)^2 - \omega^2} \right\}. \tag{3.105}$$

Using the expression of the Heaviside θ-function, which can be easily proved using the Cauchy theorem,

$$\int_{-\infty}^{+\infty} dp^0 \frac{e^{-ip^0 x^0}}{p^0 + i\varepsilon + a} = -2i\pi \, e^{iax^0} \theta(x^0), \tag{3.106}$$

the expression in the brackets can be written as

$$-\frac{i\pi}{\omega(\mathbf{p})} \theta(x^0 - x^{0\prime})(e^{-i\omega(x^0 - x^{0\prime})} - e^{i\omega(x^0 - x^{0\prime})})$$

and finally

$$G_{\text{ret}}(x - x') =$$
$$\frac{i\theta(x^0 - x^{0\prime})}{(2\pi)^3} \int \frac{d^3p}{2\omega(p)} e^{ip \cdot (x-x')} \left\{ e^{-i\omega(x^0 - x^{0\prime})} - e^{i\omega(x^0 - x^{0\prime})} \right\}. \quad (3.107)$$

Let $V^+ = \{x \in \mathbb{R}^4 \,|\, x^0 \geq 0, x^2 \geq 0\}$ and $V^- = -V^+$ be the past and future half-cones. The function G_{ret} has its support in the variable x' in $\{x' \in \mathbb{R}^4 \,|\, x' \in x + V^-\}$. In fact, the Heaviside function limits its support to the x' values in the past cone of x and it is Lorentz invariant.

Let us consider the integral giving G_{ret} as a limit of Riemann discrete sums on a lattice defined on $\mathbb{R}^3$ with a lattice spacing $1/L$,

$$\frac{1}{L^3} \sum_{p = \frac{2\pi n}{L}, n \in \mathbb{Z}^3} \rightarrow \frac{1}{(2\pi)^3} \int d^3p.$$

Introducing plane waves

$$\phi_{\pm,p}(x) = \frac{e^{\pm i\omega(p)x^0 + ip \cdot x}}{L^{\frac{3}{2}} \sqrt{2\omega(p)}} \quad (3.108)$$

we have that

$$G_{\text{ret}}(x - x') = \lim_{L \to \infty} i\theta(x^0 - x^{0\prime}) \sum (\phi_{+,p}(x)\phi_{+,p}^*(x') - \phi_{-,p}(x)\phi_{-,p}^*(x')).$$

This can be interpreted as contributions to the Green function of plane waves of positive and negative energy propagating towards the future (in the exponent $-ip \cdot x$ is written $-ip^0 x^0 + ip \cdot x$, where $p^0 = p_0$ is the energy, and one has either $p^0 = \omega$ or $p^0 = -\omega$).

It is straightforward to repeat this analysis and obtain the *advanced Green function* given by the expression

$$G_{\text{adv}}(x - x') = -\frac{1}{(2\pi)^4} \lim_{\varepsilon \to 0} \int d^4p \frac{e^{-ip(x-x')}}{(p_0 - i\varepsilon)^2 - p^2 - m^2}, \quad (3.109)$$

which has support in the past light cone. $G_{\text{adv}}(x - x')$ propagates both positive and negative energy solutions towards the past.

These two Green functions are given by real distributions and they are the ones often used in the classical electromagnetic theory.

Another elementary solution will play a fundamental role in relativistic quantum field theory, the Feynman Green function G_F. It is a complex distribution given by

$$G_F = -\frac{1}{(2\pi)^4} \int d^4p\, e^{-ip.x} \frac{1}{p^2 - m^2 + i\varepsilon}. \tag{3.110}$$

It splits into two parts

$$G_F(x - x')$$
$$= i \lim_{L \to \infty} \{\theta(x^0 - x^{0\prime}) \sum \phi_{+,p}(x)\phi^*_{+,p}(x') + \theta(x^{0\prime} - x^0) \sum \phi_{-,p}(x)\phi^*_{-,p}(x')\}$$

and one sees from the above expression that the positive energy solutions propagate towards the future while the negative energy solutions propagate towards the past. These last ones will later be interpreted as the propagation of antiparticles.

We can explicitly compute the case where $m = 0$. Since $\omega(p) = |p| = p$ we have[9]

$$\begin{aligned}
G_{\text{ret}}(x) &= \frac{1}{(2\pi)^3}\theta(x^0) \int d^3p \frac{\sin px^0}{p} e^{ip.x} \\
&= \frac{2\theta(x^0)}{(2\pi)^2} \int_0^{+\infty} dp \frac{\sin px^0 \sin p|x|}{|x|} \\
&= \frac{\theta(x^0)}{(2\pi)^2} \int_0^{+\infty} dp \frac{1}{2|x|} \{e^{ip(x^0 - |x|)} + e^{-ip(x^0 - |x|)} - e^{ip(x^0 + |x|)} - e^{ip(x^0 + |x|)}\} \\
&= \frac{\theta(x^0)}{(2\pi)} \delta(x^2).
\end{aligned}$$

We thus see that in the massless case the function G_{ret} has its support on the boundary of the future light cone.

Before closing this section, we want to notice that the three Green functions we introduced, G_{ret}, G_{adv}, and G_F, all have a simple geometrical interpretation. We saw that they are boundary values of the same analytic function $f = 1/(p^2 - m^2)$. Considered as a function of p^0, f has two poles in the complex plane corresponding to the values $p^0 = \pm\omega$. In taking the Fourier transform we must integrate over p^0 from $-\infty$ to $+\infty$.

---

[9] We recall that for a function $f(x)$ with a finite number of distinct zeros x_i

$$\delta(f(x)) = \sum_i \frac{1}{|f'(x_i)|} \delta(x - x_i). \tag{3.111}$$

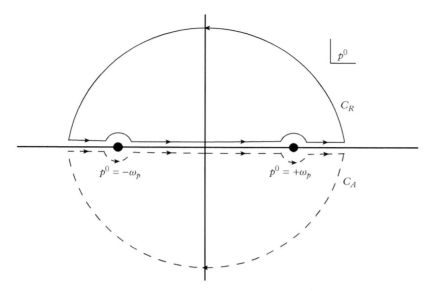

Figure 3.2 *Two of the integration contours in the complex p^0 plane. For the retarded Green function, we can close the contour in the upper half plane and find that G_{ret} vanishes if $x^0 < 0$. The opposite is true for G_{adv}.*

Since the singularities are on the real axis we must deform the contour, going above or below the singularities, the way it is shown in Figs. 3.2 and 3.3. It is clear that there exist four possible deformations, denoted in the Figures as C_R, C_A, and C_F together with its complex conjugate $\bar{C}_F$. It is easy to verify that they give rise to G_{ret}, G_{adv}, G_F, and $\bar{G}_F$, respectively.

3.8.2 The Green Functions of the Electromagnetic Field

Let us apply this formalism to the case of the electromagnetic potential $A^\mu(x)$. It satisfies the Maxwell equation which we can write as:

$$[\Box g_{\mu\nu} - \partial_\mu \partial_\nu]A^\mu(x) = j_\nu(x) \tag{3.112}$$

According to our previous discussion, the Green functions should be given by the inverse of the differential operator $\mathcal{D}_{\mu\nu} = \Box g_{\mu\nu} - \partial_\mu \partial_\nu$. However, as it is straightforward to verify, this operator has no inverse. The reason is that it has a zero eigenvalue which we can easily exhibit. Given any scalar function $\phi(x)$ we get $\mathcal{D}_{\mu\nu}\partial^\mu\phi(x) = 0$. This is the old problem of gauge invariance which we have faced already in a previous section. It is due to the fact that the vector potential contains redundant degrees of freedom and one should impose special constraints among the four components of A^μ. We have called this procedure *choice of a gauge*, or, in the usual jargon, *gauge fixing*. What constitutes a

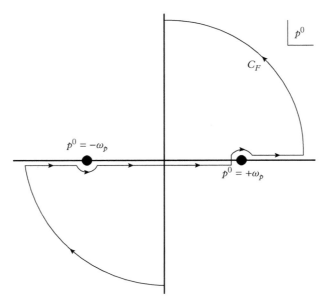

Figure 3.3 *The integration contour in the complex p^0 plane which gives rise to the Feynman propagator.*

legitimate gauge fixing, i.e. one which does not affect the values of physically measurable quantities, will be the subject of a special chapter in this book, so here we present only a particular family of such gauges and we leave the proof of their legitimacy for later. We choose the family which is parametrised by a real parameter α and results in replacing the Maxwell equation (3.112) with

$$[\Box g_{\mu\nu} - (1 - \alpha^{-1})\partial_\mu\partial_\nu]A^\mu(x) = j_\nu(x). \qquad (3.113)$$

This equation can be viewed as coming from a gauge fixed Lagrangian density of the form

$$\mathcal{L} = -\frac{1}{4}F_{\mu\nu}F^{\mu\nu} - \frac{1}{2\alpha}[\partial_\mu A^\mu]^2 - A^\mu j_\mu. \qquad (3.114)$$

The term $\frac{1}{2\alpha}[\partial_\mu A^\mu]^2$ breaks gauge invariance and it represents the gauge fixing term in the Lagrangian. The simplest choice is $\alpha = 1$, in which case (3.113) reduces to (3.29). If A^μ vanishes at infinity, a solution of this equation is given by

$$A^\mu(\bar{x}) = (G_{ret} * j^\mu)(\bar{x}) = \frac{1}{2\pi}\int d^4x\,\theta(\bar{x}^0 - x^0)\delta((\bar{x} - x)^2)j^\mu(x)$$

$$= \frac{1}{4\pi}\int d^3x\frac{1}{|\bar{x} - x|}j^\mu(\bar{x}^0 - |\bar{x} - x|, x), \qquad (3.115)$$

where we used the fact that

$$\theta(x^0)\delta(x^2) = \frac{1}{2|x^0|}\delta(x^0 - |\boldsymbol{x}|).$$

Because of the conservation of the current, this solution satisfies $\partial_\mu A^\mu = 0$. We see the retarded effects on the potential through the dependence on $\bar{x}^0 - |\bar{\mathbf{x}} - \mathbf{x}|$ of the time variable of the current. Of course, if j is the current defined by the static charges, then $j = (\varrho, 0)$ and we find again the Coulomb potential formula since ϱ does not depend on time.

For an arbitrary value of α the Green functions of Eq. (3.113) are given, in momentum space, by

$$\tilde{G}_{\mu\nu}(k) = [\tilde{\mathcal{D}}^{-1}]_{\mu\nu} = \frac{1}{k^2}\left[g^{\mu\nu} - (1-\alpha)\frac{k^\mu k^\nu}{k^2}\right]. \tag{3.116}$$

Compared with the corresponding expressions we found in the study of the Klein–Gordon equation, we see that the singularity in the complex k^2 plane has moved to $k^2 = 0$. This was expected since Maxwell's equations do not contain any parameter with the dimensions of a mass. The $i\epsilon$ prescription remains the same. Depending on our choice, we can obtain the advanced, the retarded, or the Feynman Green functions, which have the same support properties as those we found in the case of the Klein–Gordon equation.

3.9   Applications

3.9.1   The Liénard–Wiechert Potential

We will now compute the electromagnetic field created by a pointlike charge, the so-called *radiation field*.

Carrying the definition of the current of a charged pointlike particle into formula (3.115), we get

$$\begin{aligned}
A^\mu(\bar{x}) &= \frac{e}{2\pi}\int_{-\infty}^{+\infty}\mathrm{d}s\int\mathrm{d}^4x\frac{\mathrm{d}x^\mu}{\mathrm{d}s}\delta^4(x-x(s))\theta(\bar{x}^0-x^0)\delta((\bar{x}-x)^2)\\
&= \frac{e}{2\pi}\int_{-\infty}^{+\infty}\mathrm{d}s\frac{\mathrm{d}x^\mu}{\mathrm{d}s}\theta(\bar{x}^0-x^0(s))\delta((\bar{x}-x(s))^2).
\end{aligned} \tag{3.117}$$

Let $g(s)$ be a function of s and let us compute

$$I_1 \stackrel{\text{def}}{=} \int_{-\infty}^{+\infty}\mathrm{d}s\,g(s)\theta(\bar{x}^0-x^0(s))\delta((\bar{x}-x(s))^2).$$

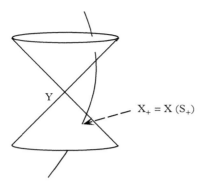

$X_+ = X(S_+)$

Figure 3.4 *The intersection of $x(s)$ with the light cone.*

Since $(\bar{x} - x(s))^2$ is a quadratic function of x we can write, using formula (3.111),

$$\delta((\bar{x} - x(s))^2) = \frac{1}{\bar{x}^0 - x^0(s) + \sqrt{(\bar{x} - x(s))^2}} \delta(\bar{x}^0 - x^0(s) - \sqrt{(\bar{x} - x(s))^2}$$
$$+ \frac{1}{\bar{x}^0 - x^0(s) - \sqrt{(\bar{x} - x(s))^2}} \delta(\bar{x}^0 - x^0(s) + \sqrt{(\bar{x} - x(s))^2}).$$

Moreover since the θ function implies that $\bar{x}^0 > x^0(s)$, for generic values of $x(s)$,

$$I_1 = \int_{-\infty}^{+\infty} \mathrm{d}s g(s) \frac{1}{\bar{x}^0 - x^0(s) + \sqrt{(\bar{x} - x(s))^2}} \delta(\bar{x}^0 - x^0(s) - \sqrt{(\bar{x} - x(s))^2}). \qquad (3.118)$$

Let us call $f(s)$ the argument of the delta function Eq. (3.118). Since the curve $x(s)$ is time-like, there exists a unique value s_+ such that $f(s_+) = 0$ (see Fig. 3.4). Thus, according to (3.111),

$$I_2 \overset{\text{def}}{=} \int_{-\infty}^{+\infty} \mathrm{d}s g(s) \frac{1}{\bar{x}^0 - x^0(s) + \sqrt{(\bar{x} - x(s))^2}} \delta(f(s))$$
$$= \int_{-\infty}^{+\infty} \mathrm{d}s g(s) \frac{1}{\bar{x}^0 - x^0(s) + \sqrt{(\bar{x} - x(s))^2}} \frac{1}{|f'(s_+)|} \delta(s - s_+)$$

Using the value of f

$$f'(s) = \frac{\mathrm{d}f}{\mathrm{d}s} = -\frac{\mathrm{d}x^0(s)}{\mathrm{d}s} + \frac{\mathrm{d}\boldsymbol{x}}{\mathrm{d}s} \cdot \frac{(\bar{\boldsymbol{x}} - \boldsymbol{x})}{\sqrt{(\bar{\boldsymbol{x}} - x(s))^2}}. \qquad (3.119)$$

Since $\mathrm{d}x^\mu/\mathrm{d}s$ is a time-like vector, $f'(s)$ is negative, and we have for $s = s_+$ and $x_+ = x(s_+)$

$$\left(\bar{x}^0 - x^0(s_+) + \sqrt{(\bar{x} - x(s_+))^2}\right) |f'(s_+)| = 2\frac{\mathrm{d}x_+}{\mathrm{d}s} \cdot (\bar{x} - x_+), \qquad (3.120)$$

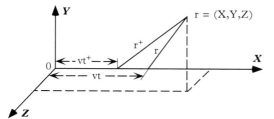

Figure 3.5 *The charge in motion.*

and thus

$$I_1 = \frac{g(s_+)}{2\frac{dx_+}{ds} \cdot (\bar{x} - x_+)}.$$ (3.121)

Applying this result to the case where $g(s) = \frac{dx^\mu}{ds}$, we get

$$A^\mu(\bar{x}) = \frac{e}{4\pi} \frac{dx^\mu_+}{ds} \frac{1}{\frac{dx_+}{ds} \cdot (\bar{x} - x_+)} = \frac{e}{4\pi} \left(\frac{dx^\mu}{ds} \frac{1}{\frac{dx}{ds} \cdot (\bar{x} - x)} \right) \Big|_{\text{ret}},$$ (3.122)

where the index 'ret' means that the expression instead of being evaluated at the proper time s, corresponding to the time t, is evaluated at a later time s_+. Formula (3.122) is, in fact, at the origin of the discovery by Lorentz of the transformations which bear his name. Let us consider the four-vector potential created at a point $\bar{x} = (t, r)$, with $r = (x, y, z)$, by a charge at uniform velocity v, along the x-axis (see Fig. 3.5). The point of universe $x(s)$ has for components $(t, vt, 0, 0) = (1 - v^2)^{-1/2}(s, vs, 0, 0)$.

At time 0, the charge is at O of space coordinates $(0, 0, 0)$; at time t, it is at the point with space coordinates $(vt, 0, 0)$. Let r be the distance between the position of the charge at time t and the point of space where we will evaluate the potential. The calculus of the vector potential shows us that because of the finite speed of propagation of the interaction, it is the effect of the charge at time $t_+ = (1 - v^2)^{-1/2}s_+$, defined by

$$t - t_+ = \sqrt{(x - vt_+)^2 + y^2 + z^2},$$ (3.123)

which counts. We thus introduce, the vector r_+ between the retarded position of the charge and $\bar{x}$, of components $(x - vt_+, y, z)$. Solving Eq. (3.123), we find that

$$t - t_+ = r_+ = \frac{v}{1 - v^2}(x - vt) + \frac{1}{1 - v^2}\sqrt{(x - vt)^2 + (1 - v^2)(y^2 + z^2)}.$$ (3.124)

Since $dx^0/dt = 1$ (because $c = 1$) and $dx/dt = v$,

$$\frac{dx}{dt} \cdot (\bar{x} - x) \Big|_{ret} = r_+ - v \cdot r_+,$$ (3.125)

we finally get

$$A^0(t, x, y, z) = \frac{1}{4\pi} \frac{e}{\sqrt{1 - v^2}} \frac{1}{\sqrt{\left(\frac{x - vt}{\sqrt{1 - v^2}}\right)^2 + y^2 + z^2}}$$

$$A(t, x, y, z) = vA^0.$$ (3.126)

Everything happens as if the solution of the problem is obtained by writing the usual formulae, giving the vector potential, in terms of the Coulomb potential. The common point is replaced by

$$x \rightarrow \frac{(x - vt)}{\sqrt{1 - v^2}}$$

$$y \rightarrow y$$

$$z \rightarrow z.$$

The factor $1/\sqrt{1 - v^2}$ expresses the fact that A is a four-vector. More precisely, we get formulae (3.126) by computing the vector potential created by the charge in the frame where it is at rest and then applying a Lorentz transformation to return to the initial frame.

The Lorentz transformations appear therefore naturally when we want to find the field created by a particle in motion. This observation, insignificant today since the modern idea is to obtain the Maxwell equations assuming Lorentz invariance, has marked the physics at the beginning of the previous century and the abandonment of the principle of Galilean invariance.

We will now compute, at the lowest order in v, the electric and magnetic fields created by the charge in motion. For this purpose, we must take into account the condition which forces $x(s)$ to be on the light cone of vertex $\bar{x}$. It makes it possible to define s_+ as a function of $\bar{x}$. Indeed by taking the derivative of the relation

$$\bar{x}^0 - x^0(s) = |\bar{x} - \bar{x}(s)|$$

we deduce

$$\frac{ds}{d\bar{x}^0} = \frac{r}{r - v \cdot r} \simeq 1 + \frac{v \cdot r}{r}$$ (3.127)

$$\frac{ds}{d\bar{x}} = -\frac{r}{r - v \cdot r} \simeq -\frac{r}{r}\left(1 + \frac{v \cdot r}{r}\right),$$

where all equations should be understood with $s = s_+$, $r = r_+$, etc.

Starting from

$$\mathbf{A} \simeq \frac{\mathbf{e}}{4\pi \mathbf{r}} \boldsymbol{v} \qquad\qquad \mathbf{A}^0 \simeq \frac{\mathbf{e}}{4\pi \mathbf{r}} \left(1 + \frac{\boldsymbol{v} \cdot \boldsymbol{r}}{\boldsymbol{r}} \right)$$

we find at the leading order in v

$$E = \frac{e}{4\pi r} \left(\frac{\dot{\boldsymbol{v}} \cdot \boldsymbol{r}\, \boldsymbol{r}}{r^2} - \dot{\boldsymbol{v}} \right) + \frac{e}{4\pi r^2} \left(\frac{\boldsymbol{r}}{r} - \dot{\boldsymbol{v}} + 3 \frac{\boldsymbol{r}\boldsymbol{v} \cdot \boldsymbol{r}}{r^2} \right) \tag{3.128}$$

$$B = \frac{e}{4\pi} \frac{\dot{\boldsymbol{v}} \wedge \boldsymbol{r}}{r^2} + \frac{e}{4\pi} \frac{1}{r^2} \frac{\boldsymbol{v} \wedge \boldsymbol{r}}{r}; \tag{3.129}$$

therefore,

$$B = \frac{\boldsymbol{r} \wedge E}{r}. \tag{3.130}$$

(As it turns out, this last relation is true to all orders in v: the magnetic field is at any point perpendicular to the electric field.)

We see that the electric field is made of two parts: one depending on v which behaves at large distances as $1/r^2$ and a second one depending on the acceleration, which shows a $1/r$ behaviour. Every accelerated charge radiates[10] and we are interested in the field of radiation

$$E_{rad} = \frac{e}{4\pi r^3} \left[\boldsymbol{r} \wedge \left(\boldsymbol{r} \wedge \frac{\mathrm{d}\boldsymbol{v}}{\mathrm{d}t} \right) \right] \qquad B_{rad} = \frac{e}{4\pi r^2} \left(\boldsymbol{r} \wedge \frac{\mathrm{d}\boldsymbol{v}}{\mathrm{d}t} \right) \tag{3.131}$$

If we have several charges $\{e_i\}$, v is replaced in the preceding formula by $\sum_i e_i \boldsymbol{v}_i$, which is nothing else than the derivative of the dipole moment. At this order of approximation, we say that we are in presence of a dipole radiation.

3.9.2 The Larmor Formula

Let us observe the radiation of an accelerated charge in a frame such that the velocity of the charge is negligible with respect to the velocity of light. The instantaneous flux of electromagnetic energy is given by the Poynting vector

$$\mathbf{S} = E_{rad} \wedge B_{\mathrm{rad}} = \frac{e}{4\pi r^2} \boldsymbol{r} \wedge \frac{\mathrm{d}\boldsymbol{v}}{\mathrm{d}t}. \tag{3.132}$$

The power radiated per unit of solid angle is thus

$$\frac{\mathrm{d}P}{\mathrm{d}\Omega} = \frac{e^2}{(4\pi)^2 r^2} \left| \boldsymbol{r} \wedge \frac{\mathrm{d}\boldsymbol{v}}{\mathrm{d}t} \right|^2 = \frac{e^2}{(4\pi)^2} \left(\frac{\mathrm{d}\boldsymbol{v}}{\mathrm{d}t} \right)^2 \sin^2 \theta, \tag{3.133}$$

---

[10] The charges in uniform motion do not radiate; in fact, we can choose a frame in which the charge is at rest and therefore it does not radiate.

where θ is the angle between r and $\frac{d\boldsymbol{v}}{dt}$. The total radiated power is given by the *Larmor formula*, obtained by integrating the preceding expression over the solid angle:

$$P = \frac{2}{3}\frac{e^2}{4\pi}\left(\frac{d\boldsymbol{v}}{dt}\right)^2. \tag{3.134}$$

In order to derive Larmor formula we assumed that the velocity of the charge is small with respect to the velocity of light. Actually, the power, which is the derivative of the radiated energy with respect to time ($dE_{\text{rad}} = Pdt$), is Lorentz invariant since the energy transforms as the fourth component of a four-vector under a Lorentz transformation. To write the relativistic generalisation we have one four-vector at our disposal: the momentum four-vector p and the formula must be quadratic in p/m since the Larmor formula is quadratic in $\boldsymbol{v}$. It is, therefore, easy to write the most general Lorentz invariant expression using this four-vector which reduces to the Larmor formula for non-relativistic velocities. We get the following formula, valid whatever the speed may be,

$$P = \frac{2}{3}\frac{e^2}{4\pi m^2}(dp^{\mu}/ds).(dp_{\mu}/ds) = \frac{2}{3}\frac{e^2}{4\pi m^2}\frac{dp}{ds}\cdot\frac{dp}{ds}, \tag{3.135}$$

where p is the momentum four-vector, m is the mass of the particle, and s is the proper time.

3.9.3   The Thomson Formula

A charged particle hit by an electromagnetic wave is set in motion. This motion creates in turn a radiation. It is said that there is a diffusion of light by the charge. We will calculate this effect in the case of a monochromatic incident plane wave of low frequency arriving on a fixed charge and describe it by a *diffusion cross section*. The plane wave is characterised by an electric field

$$\boldsymbol{E} = \boldsymbol{E}_0\cos(\boldsymbol{k}\cdot\boldsymbol{x} - \omega t + \phi), \tag{3.136}$$

with $\boldsymbol{E}.\boldsymbol{k} = 0$.

We admit that the speed of oscillation of the particle is small compared to the speed of light, and thus to a good approximation the electromagnetic force acting on the particle is reduced to the electric field term. Moreover, this field is correctly represented by the field at the origin of the coordinates where the particle starts its motion. The equation of motion of the particle is therefore

$$\frac{d\boldsymbol{v}}{dt} = \frac{e}{m}\boldsymbol{E}. \tag{3.137}$$

The diffusion cross section in the solid angle dΩ is given by

$$\frac{d\sigma}{d\Omega} = \frac{\text{energy radiated per second and per unit of solid angle}}{\text{energy density of the incident flux}}. \tag{3.138}$$

From the Larmor formula, the numerator is

$$\frac{e^2}{(4\pi)^2} \left(\frac{dv}{dt} \right)^2 \sin^2 \theta \tag{3.139}$$

and the density of the incident flux, given by the modulus of the Poynting vector, is $|S| = E^2$. Thus

$$\frac{d\sigma}{d\Omega} = \left(\frac{e^2}{4\pi m} \right)^2 \sin^2 \theta, \tag{3.140}$$

where θ is the angle between the direction of diffusion and the direction of the incident electric field. We get the total cross section by integrating over the solid angle, and we obtain the Thomson formula

$$\sigma_{tot} = \frac{8\pi}{3} \left(\frac{e^2}{4\pi m} \right)^2 = \frac{8\pi}{3} r_e^2, \tag{3.141}$$

where we introduce the electromagnetic radius of the particle (reintroducing c)

$$r_e = \frac{e^2}{4\pi mc^2}.$$

This result supposes that the frequency ω is small, i.e. $\omega \ll mc^2/\hbar$. Quantum and relativistic effects appear for larger frequencies or wavelengths smaller than the Compton wave length $\frac{\hbar}{mc}$.

3.9.4 The Limits of Classical Electromagnetism

The system made of charged particles in an electromagnetic field shows the appearance of some difficulties. They are of two types: ultraviolet and infrared.

The ultraviolet problems are related to short distance effects. They lead to infinities in some of the fundamental parameters characterising a particle, such as its mass or its charge. One can see this problem immediately by computing the electrostatic potential created by an ensemble of fixed charged particles. Because of the $1/r$ nature of the Coulomb potential, we see that the hypothesis of pointlike structure of the particles leads to infinite contributions. We recover the usual formula if we forbid the self-interactions of particles. This is an example of what we shall call later renormalisation of the energy. Refining this argument, we could deduce that the mass of a charged particle receives an infinite contribution of electromagnetic origin. These difficulties haunted the theory of charged particles during the first half of the twentieth century and were finally solved by the *theory of renormalisation*, which we shall present in Chapter 16.

Infrared problems are another type of challenge. They are related to the fact that light propagates at the speed of 'light' or in a more modern way that the associated quantum, the photon, is massless. The effects are related to large distances or equivalently to small frequencies. We have already encountered these problems in non-relativistic quantum mechanics when we studied the scattering of a charged particle from a Coulomb potential. They are due to the long range of the electromagnetic interactions which is manifest by the too slow decrease of the potential at large distances. As we shall see later, these difficulties can be controlled by a more refined analysis of the physical processes.

4

General Relativity: A Field Theory of Gravitation

4.1 The Equivalence Principle

4.1.1 Introduction

In the last chapter we presented Maxwell's theory as a classical field theory of electrodynamics. In this chapter we shall present Einstein's general relativity as a classical field theory of gravitation. We warn the reader again that this chapter is not meant to replace a regular course on general relativity any more than Chapter 3 was meant to do that for electrodynamics.

We have seen that a relativistic theory fundamentally relies on the invariance under the Lorentz transformations, namely changes of the reference frame from one inertial frame K to another one K'. The restriction to *inertial* frames is crucial. We also saw that the Lorentz symmetry is, in fact, a subgroup of the Poincaré symmetry, which includes the space and time translations. Other internal symmetries may exist and, in the case of electrodynamics, we found a very special such symmetry, namely the transformations which change the vector potential A_μ by adding to it the derivative of an arbitrary scalar function: $A_\mu(x) \rightarrow A_\mu(x) + \partial_\mu\theta(x)$. What is special with this transformation is that it depends on a parameter, θ, which is an arbitrary function of the space–time point x. We called this transformation *gauge transformation* and we saw that it plays an important role in electrodynamics. In this chapter we want to show that a similar principle of gauge invariance will allow us to construct a classical field theory of gravitation.

The invariance under Lorentz transformations implies that one cannot define an absolute reference frame as in the Newton paradigm. For instance, the observed time interval of a physical process, such as the periodicity of the pulses of an atomic clock, the lifetime of a radioactive particle or a living creature, varies depending on the reference frame of the observer. The most popular illustration of this property is the Langevin 'twin paradox'. We showed that these phenomena are mathematically described by the

From Classical to Quantum Fields. Laurent Baulieu, John Iliopoulos and Roland Sénéor.
© Laurent Baulieu, John Iliopoulos and Roland Sénéor, 2017. Published 2017 by Oxford University Press.

invariance under the Lorentz transformations of the Minkowski distance between two events, P_a and P_b, with coordinates x_i, y_i, z_i, t_i $i = a, b$:

$$(P_a - P_b)^2 = c^2(t_b - t_a)^2 - (x_b - x_a)^2 - (y_b - y_a)^2 - (z_b - z_a)^2$$
$$= \eta_{\mu\nu}(P_a - P_b)^\mu(P_a - P_b)^\nu. \tag{4.1}$$

In fact, the Lorentz transformations leave invariant the flat Minkowski metric

$$\eta_{\mu\nu} = \begin{pmatrix} 1 & 0 & 0 & 0 \\ 0 & -1 & 0 & 0 \\ 0 & 0 & -1 & 0 \\ 0 & 0 & 0 & -1 \end{pmatrix}. \tag{4.2}$$

In Chapter 2 we saw that the Lorentz invariance of a physical theory is enforced by postulating Lorentz covariant equations of motion. This, in turn, is achieved by expressing all physical quantities in terms of Lorentz tensors, i.e. quantities belonging to representation spaces of the Lorentz group. It is probably Einstein's greatest contribution to realise that in order to describe the effects of a gravitational field in a relativistically invariant framework, we must go beyond the requirement of Lorentz invariance. The theory of general relativity, which he invented in 1915, is a revolutionary step that changed profoundly our views of space and time.

Right after having established the principles of special relativity, Einstein addressed the following problem: the Newton equations for (slowly moving) particles with trajectories $X(t)$ and mass density $\mu(X)$ are

$$\Delta\varphi(x) = \mu(x) \quad \frac{d^2X}{dt^2} = -\text{grad}\varphi(X), \tag{4.3}$$

where φ is the gravitational potential they produce. These equations are not relativistically invariant, exactly as those of pure electrostatics in the absence of magnetism. No 'naive' attempt to make them invariant, such as interpreting the gravity potential as an element of a tensor, can be achieved without leading to theories in contradiction with experiment.

4.1.2 The Principle

In order to formulate a relativistically invariant gravity theory, Einstein raised to the level of a fundamental principle of nature a property which is true in Newtonian mechanics: he postulated that all fundamental interactions must be such that one can locally cancel the effects of gravitation by changing the reference system and choosing a *non-Galilean* coordinate frame with an opposite compensating acceleration. This principle is known as *the equivalence principle*. For instance, in a non-relativistic framework, if an observer is in a box in free fall within a gravitational field, he cannot possibly experimentally detect the presence of gravity; analogously, inside a spaceship which accelerates far away from any gravitational fields, the effect of the propulsion simulates a gravity field, by pushing the passengers on their seats as if there was an equivalent gravity field, by making fall an object as if it were subject to the equivalent gravity, etc. So Einstein postulated that the coupling of gravity with all other interactions, whatever they are (today we would say electromagnetism, electroweak, strong interactions), must be such that their effects remain locally the same as if gravity were decoupled and locally replaced by an inertia force obtained by making all experiments and observations within an accelerated frame. He called this the principle of equivalence.

The postulate of this principle has profound consequences. We see immediately that it establishes a relation between gravity and the geometric properties of space and time. As we shall show in the following sections, it allowed Einstein to conclude that the classical effects due to gravity can be viewed as consequences of a non-trivial geometry, such as a non-zero curvature, of the manifold describing space–time. In this manifold the value of the non-uniform metric will replace the notion of a gravity potential and the idea of complicated trajectories implied by the gravitational forces in non-relativistic Newtonian mechanics will be replaced by that of geodesics. In other words, Einstein showed that the gravity effects for a relativistic particle can only be described consistently in a curved manifold. For achieving a correct construction, some non-trivial mathematics must be gradually introduced, namely the principles of Riemannian geometry, which cannot be avoided if we want to describe curved spaces.

The theory of general relativity relies on clear physical hypothesis. In particular, we admit the existence of identical universal clocks (atomic clocks are an example) that measure their proper time in the frame where they are at rest and where the system of coordinates has been chosen so that the metric is locally flat, as in Eq. (4.2). In this frame, everything works in such a way that the effects of gravity cannot be seen, and the clocks indicate to their local observers their proper time by markers, such as series of emitted regular beeps. By counting their number, an observer can measure what he calls the duration between two local events. Moreover, in this frame, all other possible interactions should respect the invariance under the residual Lorentz invariance of the frame, as if we were in a flat space with no gravity. In turn, universal clocks allow us to define rulers that measure space-like intervals, for instanced by sending signals that come back using mirrors suitably located, and measuring the time for the reflected signal to come back. Given two events $P = (t, x, y, z)$ and $P + dP = (t + dt, x + dx, y + dy, z + dz)$ in a small region where the gravity fields have been cancelled, the fundamental property is the Lorentz invariance of the distance:

$$ds^2 = |P, P + dP|^2 = dx^\mu \eta_{\mu\nu} dx^\nu. \tag{4.4}$$

When we change the frame to recover gravity, the metric becomes coordinate dependent, but ds^2 remains the same, and physics just looks more complicated because we see the effects of gravity.

Before going to the formal description of this theory, we want to present two important consequences which have been verified experimentally.

4.1.3   Deflection of Light by a Gravitational Field

Consider a ray of light such as that of a laser beam that propagates through a constant gravitational field g. According to the principle of equivalence, its trajectory must be identical to that observed in a system with no gravity, but in which the observer is sitting in a spaceship accelerating with acceleration $-g$. For another observer sitting outside in an inertial frame, the light ray obviously goes along a straight line. For the observer sitting in the accelerating spaceship, its trajectory will be seen as a parabola, due to the uniformly accelerated trajectory of the spaceship. Experiments show indeed that a light ray is deviated accordingly in a gravitational field, and the postulates of special relativity alone cannot explain this phenomenon. Within the context of the quantum theory, we must admit that the massless photons that make the laser beam are attracted by gravity, as massive particles are. The attraction of light by gravity was predicted by Einstein in 1911 and demonstrated by the famous experiment of Eddington in 1919. It was observed during a solar eclipse that the sun deviates the light coming from distant stars, slightly away from the straight line trajectory it has when measured far away from the sun. We shall see that the laws of general relativity do provide equations which accurately describe this phenomenon.

4.1.4   Influence of Gravity on Clock Synchronisation

In special relativity, the Lorentz contraction phenomenon implies that when one observes in a given frame two identical clocks moving one with respect to the other at a constant speed, they get desynchronised. We will show, using the equivalence principle, that two identical clocks which do not move with respect to each other, but are located at two distinct points where the gravity potential is different, get also desynchronised.

Consider the following experiment: one puts two identical clocks in a uniform gravity field g at two locations A and B, with altitudes z_A and z_B such that $z = z_A - z_B > 0$. Suppose they are atomic clocks. The way they function is by emitting monochromatic waves and we count the number of emitted periods between two events located in the vicinity of the clock. For an observer located near the clock A, a unit τ_0 of proper time has passed when he counted a given number N_0 of pulses emitted by the clock. The question is to find out if, when the clock located in A has emitted N_0 pulses towards B, the observer located at B sees these N_0 pulses passing by for a time τ_0 measured on his own clock. If it is not the case, he concludes that the two steady clocks, structurally identical but located at positions with different values of the gravity potential, get desynchronised.

To find out, we simulate once more the gravity field by putting both clocks in an accelerated spaceship in an empty space with no gravity. The experiment goes as follows. The observer in A sends a first light signal at $t = 0$ measured in universal time and another one at $t = \tau_0$. The first signal arrives at B at a moment T with

$$z - \frac{1}{2}gT^2 = cT. \tag{4.5}$$

The second one arrives at $T + \delta T$ with

$$\left(z + \frac{1}{2}g\tau_0^2\right) - \frac{1}{2}g(T + \delta T)^2 = c(T + \delta T - \tau_0). \tag{4.6}$$

These formula rely on the fact that the speed of propagation of light is c in the universal frame, and that the speed of the spaceship has increased for the time it takes the light to go from A to B, so that the second signal has less distance than the first signal to go and connect A to B. By combining the two equations, we find that

$$\delta T \sim \tau_0 \left(1 - \frac{gz}{c^2}\right). \tag{4.7}$$

In principle, to compare the universal time with the proper time in the spaceship with speed v, we should have taken in consideration the Lorentz dilatation factor, but it gives a negligible correction $\sim 1 - \frac{1}{2}\frac{v^2}{c^2} \sim 1 - g^2 z^2/2c^4$ of order $1/c^2$.

δT is what measures the observer in B for both signals emitted by A with a time separation in A equal to τ_0. Therefore, it looks as if the clock in B runs late as compared to that in A, although they are identical. We conclude that the principle of equivalence implies that clocks at rest with respect to each other but located at points with different gravity potential will be desynchronised.

The first experiment proving this prediction was done in 1960 and is based on the Mössbauer effect, with two identical clocks located down and up in a Harvard University tower. Nowadays, this effect must be taken into account to ensure a correct behaviour of GPS and cell phones, due to the variations of the gravity field on earth and on satellites! In fact, the formula generalises to the following one, for a general gravity potential $\varphi(z)$:

$$\tau(z) = \tau_o \left(1 + \frac{\varphi(z)}{c^2}\right). \tag{4.8}$$

All these effects must be predicted by a theory that describes well the behaviour of light and matter in the presence of a gravitational field, whether or not the theory is classical or quantum.

4.2 Curved Geometry

4.2.1 Introduction

The conclusion of the discussion in the previous section is that we must write the laws of physics in a manifold allowing for the metric to depend on the space–time point in order to accommodate the possibility of going to an accelerated frame where the gravity is cancelled. This way the laws of physics in the rest frames of objects will be the same as those we established from the principle of Lorentz invariance. The guideline is that we know the laws of physics in flat space where we can measure the invariant element

$ds^2 = c^2 dt^2 - dx^2 - dy^2 - dz^2$. If we find the way of writing the laws of physics in a manifold with arbitrary metric instead of a constant metric, we will be able to see the effects of gravity from the principle of equivalence. By doing so, we will find, by consistency of the formalism, that the energy–momentum of matter actually determines the metric of the universe, which in turns produces gravity. This will be a complete change of paradigm, since the notion of gravity force and gravity potential will be replaced by the notion of a curved space where objects basically follow geodesics.

Mathematicians understood already during the nineteenth century that a manifold with a point-dependent metric is generally curved, the most simple examples being a sphere and a torus in the case of manifolds with dimension 2. In fact, for a manifold with coordinate-dependent metric, we find that, depending on the values of the second derivatives of its metric, it cannot always be represented globally as a flat space. It is only locally that we can reduce, by a suitable change of coordinates, a metric whose components depend on the coordinates of the point, to a flat metric. This is easily understood in the theory of a 2-surface, where we can visualise the way the tangent space tilts when it runs over the points of the surface, if it has curvature.

However, we cannot visualise the curvature of a manifold with dimension larger than 2, since it would mean looking at it by an embedding in a Euclidean flat space with dimension larger than 3 and we have no intuition to do so. For dimensions larger than 2, we must content ourselves to define the curvature mathematically, a work that was achieved by mathematicians following the work of Riemann.

This section explains the needed material, which is actually necessary to describe Einstein's theory of gravity.

4.2.2   Tensorial Calculus for the Reparametrisation Symmetry

We suppose that phenomena are described by a local field theory, with fields defined over a 4-manifold with metric $g_{\mu\nu}(x)$, where x is a given system of coordinates. We must define a tensorial formalism for the symmetry under change of coordinates of all fields, and find the way to write Lagrangians that are invariant under this symmetry. For this purpose, we generalise the method we followed to obtain field theories in flat space that are Lorentz invariant.

(1) We must be able to classify the fields as elements of representation spaces, which express all possible ways they can transform under change of coordinates.

(2) We must find the way to write equations that take the same form, for any given choice of a coordinate system.

This is called covariance under general changes of coordinates. Above each non-singular point of a manifold with coordinates x^μ, there is a tangent plane: it is a vector space with the same dimension as the manifold, and is spanned by the differential 1-forms dx^μ. The dx^μ can be interpreted as small variations of the coordinates x^μ when the point moves to another point with coordinates $x^\mu + dx^\mu$ within a small enough domain of the manifold. In other words, we can locally identify a small domain of the manifold around a point with its tangent space at this point.

We shall call a general coordinate transformation $x^\mu \to x'^\mu = x'^\mu(x)$ a *reparametrisation*. The transformation law for the differentials is accordingly

$$\mathrm{d}x'^\mu = \frac{\partial x'^\mu}{\partial x^\nu}\mathrm{d}x^\nu. \tag{4.9}$$

By definition of the metric $g_{\mu\nu}$, the squared line element

$$\mathrm{d}s^2 = g_{\mu\nu}\mathrm{d}x^\mu\mathrm{d}x^\nu \tag{4.10}$$

is invariant under changes of coordinates. Thus, to counter-back the transformations of the $\mathrm{d}x^\mu$, $g_{\mu\nu}$ must transform as

$$g_{\mu\nu} \to g'_{\mu\nu} = \frac{\partial x^\rho}{\partial x'^\mu}\frac{\partial x^\sigma}{\partial x'^\nu}g_{\rho\sigma}. \tag{4.11}$$

If an observer measures the components of the metric as $g_{\mu\nu}$ in the system with coordinates x^μ, he will measure it as $g'_{\mu\nu}$ in the system with coordinates x'^μ.

Given the definition (4.11), the invariance of the line-element length $\mathrm{d}s^2$ is due to

$$\frac{\partial x^\mu}{\partial x'^\rho}\frac{\partial x'^\rho}{\partial x^\nu} = \delta^\mu_\nu. \tag{4.12}$$

We wish to classify all possible local fields by the way they change under reparametrisation. Later on, we will interpret the group of these transformation laws as a gauge symmetry.

In other words, we want to organise the local fields as tensors with so-called world indices μ, according to the way they transform with respect to the group of the coordinate transformations. Thus, if the fields are known in a given system of coordinates, they will be determined by well-defined transformations in any other system, as it is the case for the differentials dx^μ or $g_{\mu\nu}$.

By definition, a contravariant vector A^μ and a covariant vector B_μ are objects that transform as follows under change of coordinates, at the same point on the manifold:

$$A'^\mu = \frac{\partial x'^\mu}{\partial x^\nu}A^\nu \tag{4.13}$$

$$B'_\mu = \frac{\partial x^\nu}{\partial x'^\mu}B_\nu. \tag{4.14}$$

More generally, a p,q-tensor $T^{\mu_1\dots\mu_p}_{\nu_1\dots\nu_q}$ with p contravariant indices and q covariant indices is an object whose $4p \times 4q$ components transform as

$$T^{\mu_1\dots\mu_p}_{\nu_1\dots\nu_q} \to T'^{\mu'_1\dots\mu'_p}_{\nu'_1\dots\nu'_q} = \frac{\partial x'^{\mu'_1}}{\partial x^{\mu_1}}\cdots\frac{\partial x'^{\mu'_p}}{\partial x^{\mu_p}}\frac{\partial x^{\nu_1}}{\partial x'^{\nu'_1}}\cdots\frac{\partial x^{\nu_q}}{\partial x'^{\nu'_q}}T^{\mu_1\dots\mu_p}_{\nu_1\dots\nu_q}. \tag{4.15}$$

A scalar is invariant. Examples of scalars are the length $\mathrm{d}s^2 = g_{\mu\nu}\mathrm{d}x^\mu\mathrm{d}x^\nu$ as well as the exterior derivative $\mathrm{d} = \mathrm{d}x^\mu\frac{\partial}{\partial x^\mu}$; given a vector A_μ, the 1-form $A = A_\mu\mathrm{d}x^\mu$ is a scalar, etc. The metric $g_{\mu\nu}$ is a rank 2 symmetric covariant tensor.

We have the following properties that are useful.

- If a tensor vanishes in a given system of coordinates, it vanishes in all other systems.
- The product of two tensors is a tensor. If the product of a tensor by an object with indices is a tensor, the object is a tensor.
- The properties of symmetry and antisymmetry of the indices of tensor remain true by a change of coordinates.
- If $T^{\mu_1...\mu_p}_{\nu_1...\nu_q}$ is a (p,q) tensor, $g_{\mu_1\mu_2} T^{\mu_1\mu_2...\mu_p}_{\nu_1...\nu_q}$ is a $(p-2,q)$ tensor. If A^μ is a $(1,0)$ vector, then $g_{\mu\nu}A^\mu$ is a $(0,1)$ vector. In fact, the knowledge of the metric $g_{\mu\nu}$ allows us to identify contravariant and covariant indices.
- The inverse $g^{\mu\nu}$ of the metric is $g_{\mu\nu}$, and

$$g^{\mu\rho}g_{\rho\nu} = \delta^\mu_\nu, \tag{4.16}$$

where δ^μ_ν is the invariant unit tensor. So $g^{\mu\nu}$ is a $(0,2)$ tensor.

- One defines the moving frame e^a_μ over each point of the manifold, where a is a Lorentz index. As a 4×4 matrix, it has an inverse e^ν_b such that $e^{a\mu} e^b_\nu = \delta^\mu_\nu \eta^{ab}$, where η^{ab} is the flat Minkowski metric and $e^a_\mu e^b_\nu \eta_{ab} = g_{\mu\nu}$ or $e^a_\mu e_{a\nu} = g_{\mu\nu}$. e^ν_b is a contravariant four-vector which takes its values in the same Lorentz representation as e^a_μ. The introduction of the moving frame e^a_μ is necessary to express the transformation of half-integer spin fields. As we shall see later, the description of such fields in flat space requires the introduction of a set of numerical matrices called γ matrices. The moving frame formalism will enable us to consistently describe γ matrices in curved space.
- In contrast to what happens in flat space, the antisymmetrisation symbol $\epsilon^{\mu\nu\rho\sigma}$ that equals 1 or -1 depending on the parity of the permutation of indices $\mu\nu\rho\sigma$, or zero if two indices are equal, is not a tensor, since we have generally $-\det(g_{\mu\nu}) \neq 1$. Rather, $\sqrt{-g}\epsilon^{\mu\nu\rho\sigma}$ is a rank 4 antisymmetric tensor. In fact, $\mathrm{d}^4x\sqrt{-g} = \frac{1}{16}\sqrt{-g}\epsilon^{\mu\nu\rho\sigma}\mathrm{d}x^\mu \wedge \mathrm{d}x^\nu \wedge \mathrm{d}x^\rho \wedge \mathrm{d}x^\sigma$ is a scalar. It is the invariant measure for defining integrals over the manifold: if f is a scalar function, the integral $\int \mathrm{d}^4x\sqrt{-g}f(x)$ is a number whose value is independent of the choice of the coordinate system.
- Equations that are covariant with respect to changes of coordinates imply that they equate tensors of the same rank.

We now show that the definition of the derivation must be improved in order to obtain an operation that is covariant.

4.2.3   Affine Connection and Covariant Derivation

Let $f(x)$ be a scalar function and $x + \mathrm{d}x$ the coordinates of a point $P + \mathrm{d}P$ that differs slightly from that of the point P with coordinates x. By a change of coordinates $x \to x'$, we have

$$f(x' + \mathrm{d}x') - f(x') = f(x + \mathrm{d}x) - f(x) = \mathrm{d}f(x). \tag{4.17}$$

By Taylor expansion, we get

$$\frac{\partial f}{\partial x^\mu} = \frac{\partial x'^\nu}{\partial x^\mu} \frac{\partial f}{\partial x'^\nu}. \tag{4.18}$$

Thus, if $f(x)$ is a scalar function, $\partial_\mu f(x)$ is a vector.
However, if we consider a vector B_μ, we have at point P

$$B'_\mu(x') = \frac{\partial x^\nu}{\partial x'^\mu} B_\nu(x). \tag{4.19}$$

If we compare this equation to the analogous one that holds at the neighbouring point $P + \mathrm{d}P$ with coordinates $x + \mathrm{d}x$, where $\mathrm{d}x^\mu$ is infinitesimal, we have

$$\frac{\partial B'_\mu}{\partial x'^\nu} = \frac{\partial x^\rho}{\partial x'^\mu} \frac{\partial x^\sigma}{\partial x'^\nu} \frac{\partial B_\rho}{\partial x^\sigma} + B_\rho \frac{\partial^2 x^\rho}{\partial x'^\mu \partial x'^\nu}. \tag{4.20}$$

The last term indicates that $\frac{\partial B_\mu}{\partial x^\nu}$ does not transform as a tensor, except if we restrict ourselves to affine transformations, i.e. when the x's are linear functions in x, so that $\frac{\partial^2 x^\rho}{\partial x'^\mu \partial x'^\nu}$ vanishes.

Thus, we see that $B_\mu(x + \mathrm{d}x) - B_\mu(x)$ is not a tensor. This in fact is geometrically obvious. We cannot consistently interpret the difference $B_\mu(x + \mathrm{d}x) - B_\mu(x)$, since it amounts to comparing two vectors that live in different tangent planes, which are generally tilted due to the transport between the points with coordinates x and $x + \mathrm{d}x$. To compare two objects over different points, we must transport one of them at the point over which the other is defined in such a way that the transported object transforms tensorially like the other one. The definition of this operation must be geometrical, that is, independent of the system of coordinates.

This can be illustrated very simply in the case of vectors drawn in a plane. If we wish to compare two vectors in Cartesian coordinates, we can subtract their components. But if we use polar coordinates, it is geometrically absurd to subtract their angular and radial coordinates. More generally, saying that two vectors at different points are the same if they have equal coordinates cannot be generally true: when the metric depends on the point, this definition would imply that they have different lengths.

We thus need to add a new object to the manifold called the affine connection, in such a way that the derivation can be generalised into a covariant operation. For the type of manifolds that one uses in physics, the affine connection can be computed as a function of the metric.

We thus consider again two points with coordinates x and $x + \mathrm{d}x$ and a vector field A^μ. We have seen that $A^\mu(x) + \partial_\nu A^\mu(x)\mathrm{d}x^\nu$ doesn't transform as a vector. The idea of parallel transport is to add a quantity $\delta_\parallel A^\mu$ to $A^\mu(x + \mathrm{d}x)$ that is linear in $\mathrm{d}x^\mu$ and A^ν, such that

$$A^\mu(x + \mathrm{d}x) + \delta_\parallel A^\mu = A^\mu(x) + \partial_\nu A^\mu(x)\mathrm{d}x^\nu + \delta_\parallel A^\mu \tag{4.21}$$

transforms as a vector at $x + dx$. By hypothesis, we can express $\delta_{\parallel}A\mu$ as

$$\delta_{\parallel}A^\mu = \Gamma^\mu_{\nu\rho}dx^\nu A^\rho. \tag{4.22}$$

$\Gamma^\mu_{\nu\rho}$ is not a tensor and is called the affine connection. Its transformation law will be computed shortly. We also define

$$\Gamma_{\mu\nu\rho} = g_{\mu\tau}\Gamma^\tau_{\nu\rho} \tag{4.23}$$

$$\Gamma^\mu_{\nu\rho} = g^{\mu\tau}\Gamma_{\tau\nu\rho}. \tag{4.24}$$

The requirement that $A^\mu(x) + \Gamma(x)^\mu_{\nu\rho}dx^\nu A^\rho(x)$ be a tensor at point $x + dx$ implies that

$$A'^\mu(x' + dx') + \Gamma'(x')^\mu_{\nu\rho}dx'^\nu A'^\rho(x) = (A^\sigma(x + dx) + \Gamma^\sigma_{\nu\rho}(x + dx)dx^\nu A^\rho)\left(\frac{\partial x'^\mu}{\partial x^\sigma}\right)_{x+dx}.$$

Using

$$\left(\frac{\partial x'^\mu}{\partial x^\sigma}\right)_{x+dx} = \left(\frac{\partial x'^\mu}{\partial x^\sigma}\right)_x + \frac{\partial^2 x'^\mu}{\partial x^\sigma \partial x^\tau}dx^\tau \tag{4.25}$$

and

$$A^\mu dx^\nu = \frac{\partial x^\mu}{\partial x'^\sigma}\frac{\partial x^\nu}{\partial x^\tau}A'^\sigma dx'^\tau \tag{4.26}$$

we get

$$\Gamma'^\mu_{\nu\rho} = \frac{\partial x'^\mu}{\partial x^\sigma}\frac{\partial x^\alpha}{\partial x'^\nu}\frac{\partial x^\beta}{\partial x'^\rho}\Gamma^\sigma_{\alpha\beta} + \frac{\partial^2 x^\alpha}{\partial x'^\nu \partial x'^\rho}\frac{\partial x^\mu}{\partial x^\alpha}. \tag{4.27}$$

This transformation law defines the connection Γ. The last term is non-homogeneous and shows that Γ is not a tensor. On the other hand, the antisymmetric part in $\nu\rho$ of $\Gamma^\mu_{\nu\rho}$ transforms as a tensor. It is called the torsion of the manifold.

So, we invented the transformation laws of Γ such that

$$D_\nu A^\mu = \frac{\partial}{\partial x^\nu}A^\mu + \Gamma^\mu_{\nu\rho}A^\rho \tag{4.28}$$

is a tensor.

Γ is called the affine connection. If the manifold has no metric, Γ must be given as a defining property of the manifold (this is an abstract possibility we do not consider). If the manifold has a metric, we will see that its components can be computed in terms of those of the metric.

$D = dx^\mu D_\mu$ is called the exterior covariant differential and $D_\nu A^\mu$ the covariant derivative of A^μ.

Once Γ has been introduced, we can define the parallel transport along a curve for consistently comparing two tensors at different points.

Given two neighbouring points on a curve with coordinates x and $x + dx$, we say that A has been parallel transported from x to $x + dx$ if its components satisfy

$$A^\mu(x + dx) = A^\mu(x) + \Gamma^\mu_{\nu\rho} A^\nu dx^\rho, \tag{4.29}$$

that is,

$$dx^\nu D_\nu A^\mu = 0. \tag{4.30}$$

Consider now an arc of a curve that is parametrised with a parameter λ. Its tangent vector is $\xi^\rho = \frac{dx^\rho}{d\lambda}$ at every point. Saying that the vector A is parallel transported along this arc means that its covariant derivative satisfies everywhere on the arc

$$\xi \cdot DA^\mu \equiv \xi^\nu D_\nu A^\mu = 0. \tag{4.31}$$

When the coordinate system is such that the connection vanishes all along the curve, the parallel transport is done with no modification of components. We can always find a system of coordinates such that the connection vanishes along a given curve, a property that disappears as soon as we take a curve that differs from the initial one, even slightly, as an effect of the curvature of the manifold.

4.2.4   Parallel Transport and Christoffel Coefficients

When the manifold has a metric (which is always the case in physics), it requires that the parallel transport conserves the angles and the length; that is, it preserves the scalar product $A \cdot B \equiv g_{\mu\nu} A^\mu B^\nu = A^\mu B_\nu$, for all A and B. This request determines $\Gamma^\rho_{\mu\nu}$ as a function of $g_{\mu\nu}$. Indeed, given a curve with parametrisation λ, it implies that

$$\frac{d}{d\lambda}(g_{\mu\nu} A^\mu B^\nu) = 0. \tag{4.32}$$

This reads as

$$\frac{\partial g_{\mu\nu}}{\partial x^\rho} \frac{dx^\rho}{d\lambda} A^\mu B^\nu + g_{\mu\nu} \frac{dA^\mu}{d\lambda} B^\nu + g_{\mu\nu} A^\mu \frac{dB^\nu}{d\lambda} = 0. \tag{4.33}$$

Since, for all vectors A and B, we have $dA^\mu/d\lambda = -\Gamma^\mu_{\rho\nu} A^\nu dx^\rho/d\lambda$, and an analogous equation for B, it follows that

$$\frac{\partial g_{\mu\nu}}{\partial x^\rho} - g_{\sigma\nu} \Gamma^\sigma_{\rho\mu} - g_{\mu\sigma} \Gamma^\sigma_{\nu\rho} = 0. \tag{4.34}$$

We can invert this equation, using some cyclic permutations of indices, and obtain

$$\Gamma^\sigma_{\mu\nu} = \frac{1}{2} g^{\sigma\rho} \left(\frac{\partial g_{\mu\rho}}{\partial x^\nu} + \frac{\partial g_{\rho\nu}}{\partial x^\mu} - \frac{\partial g_{\mu\nu}}{\partial x^\rho} \right). \tag{4.35}$$

The upper index $^\sigma$ is tensorial. Thus, we can define

$$\Gamma_{\rho\mu\nu} = \frac{1}{2}\left(\frac{\partial g_{\mu\rho}}{\partial x^\nu} + \frac{\partial g_{\rho\nu}}{\partial x^\mu} - \frac{\partial g_{\mu\nu}}{\partial x^\rho}\right). \tag{4.36}$$

$\Gamma_{\rho\mu\nu}$ is called the Christoffel symbol:

$$\Gamma^\sigma_{\mu\nu} = g^{\sigma\rho}\Gamma_{\rho\mu\nu}. \tag{4.37}$$

We have the following useful properties. Since the parallel transport preserves the length, we have

$$0 = \delta_{||}(A^\mu B_\mu) = \delta_{||}(A^\mu)B_\mu + A^\mu\delta_{||}B_\mu. \tag{4.38}$$

Thus, for any given covariant vector B, we have

$$\delta_{||}B_\mu = -\Gamma^\nu_{\mu\rho}B_\nu dx^\rho \tag{4.39}$$

and its covariant derivative is

$$D_\nu B_\mu = \frac{\partial B_\mu}{\partial x^\nu} - \Gamma^\rho_{\mu\nu}B_\rho. \tag{4.40}$$

For a scalar, we have

$$D_\mu f = \frac{\partial f}{\partial x^\mu}. \tag{4.41}$$

The concept of parallel transport applies to arbitrary tensors $T^{\mu_1\mu_2\cdots}_{\nu_1\nu_2\cdots}$ as

$$\begin{aligned}
D_\nu T^{\mu_1\mu_2\cdots}_{\nu_1\nu_2\cdots} = \frac{\partial T^{\mu_1\mu_2\cdots}_{\nu_1\nu_2\cdots}}{\partial x^\nu} &+ \Gamma^{\mu_1}_{\rho_1\nu}T^{\rho_1\mu_2\cdots}_{\nu_1\nu_2\cdots} \\
&+ \Gamma^{\mu_2}_{\rho_2\nu}T^{\mu_1\rho_2\cdots}_{\nu_1\nu_2\cdots} + \ldots \\
&- \Gamma^{\rho_1}_{\nu_1\nu}T^{\rho_1\mu_2\cdots}_{\rho_1\nu_2\cdots} \\
&- \Gamma^{\rho_2}_{\nu_2\nu}T^{\mu_1\mu_2\cdots}_{\nu_1\rho_2\cdots} - \ldots
\end{aligned} \tag{4.42}$$

Equation (4.34) means that the covariant derivative of the metric tensor vanishes identically

$$D_\rho g_{\mu\nu} = 0. \tag{4.43}$$

We thus have

$$g_{\mu\mu_1}D_\nu T^{\mu_1\mu_2\cdots}_{\nu_1\nu_2\cdots} = D_\nu\left[g_{\mu\mu_1}T^{\mu_1\mu_2\cdots}_{\nu_1\nu_2\cdots}\right] = D_\nu T^{\mu_2\cdots}_{\mu\nu_1\nu_2\cdots}. \tag{4.44}$$

Under this form, the tensorial character of the covariant derivative is explicit.

By using covariant derivatives D_μ, instead of non-covariant partial derivatives, we get differential equations that are automatically covariant under change of coordinates. It is thus quite easy to generalise the classical laws of physics in curved space, and, similarly, to write them in flat space using a system of curved coordinates.

4.2.5 Geodesics

Let P and P' be two points in the manifold. There is a very special trajectory among all curves that connect P and P', namely the trajectory that minimises its invariant length.

For a flat Euclidean space, such special curves are straight lines, for a 2-sphere they are pieces of grand-circles, etc. In special relativity, we have seen that a particle that satisfies its equations of motion between two points follows the trajectory that minimises the amount of proper time it needs to go from one point to the other.

In fact, for a general curved manifold, the geometrical definition of a geodesic does not primarily refer to its length, and it is expressed only in terms of the connection Γ. The definition is that for any two points P and P', the geodesic that connects P and P' is such that its tangent vector ξ is parallel transported from P to P'. This property is easily checked to be true for a straight line in a Euclidean flat space.

Consider therefore a geodesic and let us choose a parametrisation λ, so that the co-ordinates of the points of the geodesic are $x^\mu(\lambda)$. By definition, $\xi^\mu(\lambda)$ satisfies for all values of λ

$$\frac{\mathrm{d}\xi^\mu}{\mathrm{d}\lambda} + \Gamma^\mu_{\nu\rho}\frac{\mathrm{d}x^\nu}{\mathrm{d}\lambda}\xi^\rho = 0. \tag{4.45}$$

We can chose the tangent vector as $\xi^\mu = \mathrm{d}x^\mu/\mathrm{d}\lambda$. But we have the freedom of local dilatations of this vector by a factor $f(\lambda)$. So we can write the geodesic equation as

$$\frac{\mathrm{d}}{\mathrm{d}\lambda}\left(f(\lambda)\frac{\mathrm{d}x^\mu}{\mathrm{d}\lambda}\right) + \Gamma^\mu_{\nu\rho}\frac{\mathrm{d}x^\nu}{\mathrm{d}\lambda}f(\lambda)\frac{\mathrm{d}x^\rho}{\mathrm{d}\lambda} = 0. \tag{4.46}$$

We can change the parametrisation $\lambda \to s = s(\lambda)$ such that $\mathrm{d}\lambda/\mathrm{d}s = f(\lambda)$. We then obtain the geodesic equation in the so-called 'normal form':

$$\frac{\mathrm{d}^2x^\mu}{\mathrm{d}s^2} + \Gamma^\mu_{\nu\rho}\frac{\mathrm{d}x^\nu}{\mathrm{d}s}\frac{\mathrm{d}x^\rho}{\mathrm{d}s} = 0. \tag{4.47}$$

Affine transformations on s stand for redefinitions of length units. We will see that (i) $\mathrm{d}s$ can be identified with the square of the norm $\mathrm{d}s^2 = g_{\mu\nu}\mathrm{d}x^\mu\mathrm{d}x^\nu$ and (ii) that a geodesic between two points is the curve with extremal length connecting these two points.

Consider therefore a manifold with a metric. Since by definition of a geodesic its tangent vector is parallel transported, it keeps the same length all along the curve. Choosing $\xi^\mu = \mathrm{d}x^\mu/\mathrm{d}s$ as the tangent vector, we have on each point of the geodesic

$$|\xi|^2 = g_{\nu\rho}\frac{\mathrm{d}x^\nu}{\mathrm{d}s}\frac{\mathrm{d}x^\rho}{\mathrm{d}s} = \mathrm{cte}. \tag{4.48}$$

It follows that $ds^2 = cte^{-1}g_{\mu\nu}dx^\mu dx^\nu$ and we verify that s is truly the parametrisation of the length measured on the curve, modulo the multiplication by a global constant.

We can express Γ as a function of the metric for the geodesic equation

$$\frac{d^2x^\sigma}{ds^2} + \frac{1}{2}g^{\sigma\rho}\left(\frac{\partial g_{\mu\rho}}{\partial x^\nu} + \frac{\partial g_{\rho\nu}}{\partial x^\mu} - \frac{\partial g_{\mu\nu}}{\partial x^\rho}\right)\frac{dx^\mu}{ds}\frac{dx^\nu}{ds} = 0. \tag{4.49}$$

This is a second-order differential equation which looks exactly as an equation of motion. Given a point, there is only one geodesic that passes through it with a given direction for its tangent vector at this point. It follows that a manifold can be described by a series of geodesics. The latter cannot cross each other in a small enough domain, which generalises the definition of parallelism. The rate at which geodesics tend to separate themselves depends on the curvature.

Going back to physics, we can better understand the Euler–Lagrange equations in non-relativistic mechanics as stemming from the minimisation of the proper time and a deformation of space by a point-dependent metric. Consider the manifold with coordinates $x^\mu = (ct, x^i)$ where t is the Newtonian absolute time and x^i the Cartesian coordinates of space. We can use the universal Newton time t as the parameter of all trajectories, so $ct = s$. The Newton equations are

$$\frac{d^2x^i}{ds^2} + \frac{dV}{dx^i} = 0$$
$$\frac{d^2t}{ds^2} = 0. \tag{4.50}$$

They can be rewritten, modulo terms in $1/c^4$, as

$$\frac{d^2x^\mu}{ds^2} + \Gamma^\mu_{\nu\rho}\frac{dx^\nu}{ds}\frac{dx^\rho}{ds} = 0 \tag{4.51}$$

with

$$\Gamma^\sigma_{\mu\nu} = \frac{1}{2}g^{\sigma\rho}\left(\frac{\partial g_{\mu\rho}}{\partial x^\nu} + \frac{\partial g_{\rho\nu}}{\partial x^\mu} - \frac{\partial g_{\mu\nu}}{\partial x^\rho}\right) \tag{4.52}$$

and

$$g_{ij} = \delta_{ij}$$
$$g_{00} = 1 + \frac{V}{c}$$
$$g_{0i} = 0. \tag{4.53}$$

If curved coordinates are used for the space, we replace δ_{ij} by the expression of the metric in the new system of coordinates. So, in general, Newton's equation can be rewritten as a geodesic equation (4.51).

It shows that in non-relativistic mechanics, we cannot distinguish locally between an internal force obtained by an accelerated frame and an external force that derives from a potential. We also found this relation when we detailed the deviation of light by gravity by using the equivalence principle, replacing the gravity field by that of the acceleration of the frame. Here, we understand that the light follows a geodesic in a curved space, where the connection is determined by the gravity potential.

We now show that it is equivalent to define a geodesic by the demand that its tangent vector moves by parallel transport and that its length is extremal. The length is defined as follows, when $\lambda_i \leq \lambda \leq \lambda_f$:

$$s[\lambda_i, \lambda_f] = \int_{\lambda_i}^{\lambda_f} d\lambda \sqrt{g_{\mu\nu} \frac{dx^\mu}{d\lambda} \frac{dx^\nu}{d\lambda}}. \tag{4.54}$$

The trajectory which extremises s is such that

$$\delta \int_{\lambda_i}^{\lambda_f} \sqrt{L(x^\mu, \dot{x}^\mu)} d\lambda = 0 \tag{4.55}$$

with

$$L(x^\mu, \dot{x}^\mu) = \frac{1}{2} g_{\mu\nu} \frac{dx^\mu}{d\lambda} \frac{dx^\nu}{d\lambda}. \tag{4.56}$$

We can solve this equation by making it more general, with

$$\delta \int_{\lambda_i}^{\lambda_f} F[L(x^\mu, \dot{x}^\mu)] d\lambda = 0, \tag{4.57}$$

where $F[L]$ is an arbitrarily given function of L that is monotonous and differentiable. By using the methods of the calculus of variations, we have

$$\frac{d}{d\lambda} \left(\frac{\partial F}{\partial \dot{x}^\mu} \right) = \frac{\partial F}{\partial x^\mu} \tag{4.58}$$

and thus

$$\frac{d}{d\lambda} \left(F'[L] g_{\mu\nu} \frac{dx^\nu}{d\lambda} \right) = \frac{1}{2} F'[L] \frac{\partial g_{\nu\rho}}{\partial x^\mu} \frac{dx^\nu}{d\lambda} \frac{dx^\rho}{d\lambda}. \tag{4.59}$$

We can choose λ as the arc length s on the curve, with $ds^2 = g_{\mu\nu} dx^\mu dx^\nu$, so that $L(s) = \frac{1}{2} g_{\mu\nu} \frac{dx^\mu}{ds} \frac{dx^\nu}{ds} = \frac{1}{2}(ds/ds)^2 = \frac{1}{2}$, and thus L is constant along the curve. For any given function F, $F(L)$ and $F'(L)$ are then fixed along the curve, independently of the value of s.

We end up therefore with the following expression of the Euler–Lagrange equation that expresses the condition of extremal length:

$$\frac{\mathrm{d}}{\mathrm{d}s}\left(g_{\mu\nu}\frac{\mathrm{d}x^{\nu}}{\mathrm{d}s}\right) = \frac{1}{2}\frac{\partial g_{\nu\rho}}{\partial x^{\mu}}\frac{\mathrm{d}x^{\nu}}{\mathrm{d}\lambda}\frac{\mathrm{d}x^{\rho}}{\mathrm{d}\lambda}. \tag{4.60}$$

This equation is nothing but the geometrical equation (4.49) for a geodesic, defined from the property that its tangent vectors are parallel transported along it.

The variation principle gives the same equation for the geodesics, independently of the choice of the function F. If we choose $F(L) = \sqrt{L}$ we see that Eq. (4.49) for a geodesic is truly the constraint that its length, defined, in Eq. (4.54), is extremal.

We have thus demonstrated the identification of geodesics as curves of extremal length between different points. Both definitions are useful. They offer different ways of visualising and drawing geodesics.

The following equation for a geodesic

$$\delta\int_{s_{i}}^{s_{f}}\frac{1}{2}g_{\mu\nu}\frac{\mathrm{d}x^{\mu}}{\mathrm{d}s}\frac{\mathrm{d}x^{\nu}}{\mathrm{d}s}\mathrm{d}s = 0 \tag{4.61}$$

gives again a link between mechanics and geodesics. Indeed in a non-relativistic limit, when $s \sim ct$, we find with $g_{00} = 1$, $L = c^2 - \frac{1}{2}\dot{q}^2$, that is the well-know Lagrange function for a free non-relativistic particle.

4.2.6 The Curvature Tensor

On a curved manifold, many 'obvious' properties of Euclidean geometry get modified. The sum of the angles of a triangle drawn with three geodesics (three grand circle arcs on a sphere) is not 2π. A square made by drawing four geodesics starting from a point and, making three successive 90° angles, does not close; the ratio of the circumference of a circle divided by its radius measured on the surface is not 2π; etc.

These facts boil down to the property that if we consider points in a manifold and the closed path C obtained by connecting them in a given order by geodesics, and if we parallel transport a vector along this curve, the vector will not come back identical to itself after a cycle, because of the successive jumps of direction at each corner of the curve of the angles between the tangent vector and the geodesics. We get an angular deficit between the vector we started from and the vector it became at the end of the cycle, when it returns to the starting point. This angle is non-zero in general and makes it possible for us to define precisely the curvature as a tensor.

We have seen that the equations of geodesics depend on the connection. We also suggested that the rate at which geodesics separate one from the other depends on the curvature. We now show the precise link between the connection and the curvature.

Consider a vector A_{μ}, which is parallel transported on a closed curve C. When it goes from the point x to the point $x + \mathrm{d}x$ on the curve, its coordinates change from A_{μ} to $A_{\mu} + \delta_{\parallel}A_{\mu}$, with $\delta_{\parallel}A_{\mu} = -\Gamma_{\mu\nu}^{\sigma}A_{\sigma}\mathrm{d}x^{\nu}$. The $\delta_{\parallel}A_{\mu}$'s are not the components of a vector at x, but the integrated quantity

$$\Delta_{C}A_{\mu} = -\oint_{C}\Gamma_{\mu\nu}^{\sigma}A_{\sigma}\mathrm{d}x^{\nu} \tag{4.62}$$

does express the components of a vector, since it is nothing but the difference at a given point (arbitrarily chosen on $\mathcal{C}$) between A_μ and the transformation of A_μ after having been parallel transported once around the closed path Γ. If one changes the contour $\mathcal{C}$, the values of ΔA_μ generally change.

To define the curvature, the correct idea is to consider a small enough contour $\mathcal{C}$ and to use Eq. (4.62) to define a tensor that does not depend on the details of the contour and characterises the curvature of the manifold.

So, by doing an expansion of ΔA_μ at the lowest order in the surface defined by $\Sigma^{\rho\sigma} = dx^\rho \wedge dx^\sigma$ that is delimited by the infinitesimal contour $\mathcal{C}$, we will find a formula for the curvature tensor $R^\nu_{\mu\rho\sigma}$ such that

$$\Delta A_\mu = \frac{1}{2} R^\nu_{\mu\sigma\tau} A_\nu \Sigma^{\sigma\tau}. \tag{4.63}$$

ΔA_μ and $\Sigma^{\sigma\tau}$ being tensors, $R^\nu_{\mu\rho\sigma}$ is a tensor of rank 4. It is called the Riemann curvature tensor.

To compute R we apply the Stokes theorem to the right-hand side of Eq. (4.62) for ΔA_μ, and we use the property that A is parallel transported to itself along the contour,

$$\frac{\partial A_\mu}{\partial x^\nu} = \Gamma^\sigma_{\mu\nu} A_\sigma. \tag{4.64}$$

We get

$$\Delta A_\mu = \frac{1}{2} \left(\frac{\partial(\Gamma^\nu_{\mu\tau} A_\nu)}{\partial x^\sigma} - \frac{\partial(\Gamma^\nu_{\mu\sigma} A_\nu)}{\partial x^\sigma} \right) \Sigma^{\sigma\tau}. \tag{4.65}$$

By comparison with (4.63), we find that R is

$$R^\nu_{\mu\sigma\tau} = \frac{\partial \Gamma^\nu_{\mu\tau}}{\partial x^\sigma} - \frac{\partial \Gamma^\nu_{\mu\sigma}}{\partial x^\tau} + \Gamma^\nu_{\rho\sigma} \Gamma^\rho_{\mu\tau} - \Gamma^\nu_{\rho\tau} \Gamma^\rho_{\mu\sigma}. \tag{4.66}$$

The Riemann tensor is thus calculable in terms of the metric, using the expression of Γ as a function of $g_{\mu\nu}$. We see that the Riemann tensor depends on the second derivatives of the metric. We also have

$$R_{\mu\nu\rho\sigma} = g_{\mu\tau} R^\tau_{\nu\rho\sigma}. \tag{4.67}$$

We find analogous formulae for a contravariant vector B^μ. To do so, either we do a direct computation or we use the fact that $\Delta(A^\mu B_\mu) = 0$ since $A^\mu B_\mu$ is a scalar. We get by both methods

$$\Delta B^\mu = \oint_\Gamma \Gamma^\mu_{\sigma\nu} B^\sigma \, dx^\nu = -\frac{1}{2} R^\mu_{\nu\sigma\tau} B^\nu \Sigma^{\sigma\tau}. \tag{4.68}$$

The generalisation of these formulae for computing ΔT with T a tensor of arbitrary rank is obvious.

The following formulae express the non-commutativity of the covariant derivatives in curved space. The result is proportional to the Riemann tensor:

$$(D_\sigma D_\tau - D_\tau D_\sigma) B_\mu = R^\nu_{\mu\sigma\tau} B_\nu \tag{4.69}$$

$$(D_\sigma D_\tau - D_\tau D_\sigma) A^\mu = R^\mu_{\nu\sigma\tau} A^\nu \tag{4.70}$$

$$(D_\sigma D_\tau - D_\tau D_\sigma) B_{\mu\rho} = R^\nu_{\mu\sigma\tau} B_{\nu\rho} + R^\nu_{\rho\sigma\tau} B_{\mu\nu}. \tag{4.71}$$

We can show that the definition of flat space is that its Riemann tensor globally vanishes. We have the following symmetries for its indices:

$$R_{\mu\nu\rho\sigma} = -R_{\nu\mu\rho\sigma}$$
$$R_{\mu\nu\rho\sigma} = -R_{\mu\nu\sigma\rho}$$
$$R_{\rho\sigma\mu\nu} = R_{\mu\nu\sigma\rho}. \tag{4.72}$$

Thus, the only rank 2 tensor we can extract from the Riemann tensor is by contracting its first and third indices. This gives the Ricci tensor.

$$R_{\nu\sigma} = g^{\mu\rho} R_{\mu\nu\rho\sigma}. \tag{4.73}$$

Then, the scalar curvature is

$$R = g^{\nu\sigma} R_{\nu\sigma}. \tag{4.74}$$

The notion of a curvature explains well the twin Langevin paradox, taking correctly into account the effect of the accelerations on the trajectories of both twins. The twin who stays on Earth follows a given geodesic between the two moments when his brother takes off from Earth and comes back near him. The second one follows another geodesic, computed by using a metric expressing all the forces he is submitted to. Whatever the form of the geodesics, the one who remains on Earth will have aged more than the other one, who travelled along other geodesics, when the latter comes back.

4.3 Reparametrization Gauge Symmetry and Einstein's General Relativity

4.3.1 Reparametrisation Invariance as a Gauge Symmetry

In the previous chapter on classical electrodynamics we showed that Maxwell's theory can be obtained from an action which is invariant under the gauge transformation $A_\mu(x) \rightarrow A_\mu(x) + \partial_\mu \theta(x)$. In this section we want to show that reparametrisation invariance can also be interpreted as a gauge symmetry acting on the fields at the same point. This approach de-emphasises the geometrical idea of parallel transport, since the symmetry is expressed at a given point, but we get a unified description of gauge transformations which will be used again later in the formulation of the Yang–Mills theory in flat space.

The idea is to interpret the transformation laws of tensors under a change of coordinates as a symmetry acting on the dynamical variables. Imagine that we have a theory whose dynamical variables are a set of fields which we denote collectively by $\varphi(x)$. We assume that they transform as tensors under a change of coordinates. We build a Lagrangian density $L(\varphi, D_\mu\varphi, g_{\mu\nu})$ as a polynomial in the fields and their covariant derivatives. All indices in every monomial of L are supposed to be fully contracted using the metric g. It follows that the action given by

$$I[\varphi, D_\mu\varphi, g_{\mu\nu}] = \int \mathrm{d}^4x\sqrt{-g}L(\varphi, D_\mu\varphi, g_{\mu\nu}) \tag{4.75}$$

is invariant under reparametrisations because both L and $\mathrm{d}^4x\sqrt{-g}$ are scalars. By minimising this action, we get Euler–Lagrange equations that are automatically covariant under changes of coordinates. We will show that the action I is determined by a gauge symmetry associated with the reparametrisation invariance.

We consider the infinitesimal change of coordinates $x \to x'$ with

$$x'^\mu = x^\mu - \xi^\mu(x), \tag{4.76}$$

where all components of the vector ξ^μ are infinitesimal and depend on x.

Let us start by considering a scalar function f. By change of coordinates, we have

$$f'(x') = f(x). \tag{4.77}$$

The idea is to identify the way $f(x)$ transforms into $f'(x')$ as a change δf by a local transformation acting on f without changing its argument x. δf is defined as

$$\begin{aligned}
\delta_\xi f(x) &\equiv f'(x') - f(x') \\
&= f(x) - f(x - \xi) = \xi^\mu \partial_\mu f(x).
\end{aligned} \tag{4.78}$$

These transformations form an algebra, since given two transformations with parameters ξ_1 and ξ_2, we have

$$\left(\delta_{\xi_1}\delta_{\xi_2} - \delta_{\xi_2}\delta_{\xi_1}\right)f = \delta_{\xi_3}f \tag{4.79}$$

with

$$\xi_3^\mu = \xi_1^\nu \partial_\nu \xi_2^\mu - \xi_2^\nu \partial_\nu \xi_1^\mu. \tag{4.80}$$

This relation shows that the infinitesimal transformations δ build a Lie algebra, at least when they act on scalars.

As a second example, let us consider a vector A^μ and proceed as for the scalar f. By change of coordinates, we have

$$A'^\mu(x') = \frac{\partial x'^\mu}{\partial x^\nu} A^\nu(x). \tag{4.81}$$

We thus define

$$\delta_\xi A^\mu(x) \equiv A'^\mu(x') - A^\mu(x'), \tag{4.82}$$

so that

$$\delta_\xi A^\mu(x) = \xi^\nu \partial_\nu A^\mu - A^\nu \partial_\nu \xi^\mu. \tag{4.83}$$

More generally, given a tensor T we can compute δT such that

$$\delta_\xi T(x) \equiv T'(x') - T(x') \tag{4.84}$$

with $x' = x - \xi$.

Generally, at first order in ξ, the result $\delta_\xi T(x)$ can be written as

$$\delta_\xi T(x) = \mathcal{L}_\xi T(x). \tag{4.85}$$

The operation $\mathcal{L}_\xi$ acts on any kind of tensor. It is generically called the Lie derivative of T along the vector ξ. The explicit expression of the action of $\mathcal{L}_\xi$ is tedious to write, because its expression gets more and more lengthy when the rank of the tensor T increases; however, we find that for all tensors, independently of the rank, we have the Lie algebra structure

$$[\mathcal{L}_\xi, \mathcal{L}_{\xi'}] = \mathcal{L}_{\{\xi,\xi'\}} \tag{4.86}$$

with

$$\{\xi, \xi'\}^\mu = \xi^\nu \partial_\nu \xi'^\mu - \xi'^\nu \partial_\nu \xi^\mu. \tag{4.87}$$

It will be useful to consider, as another example, the case of the metric g. We find that

$$\delta_\xi g_{\mu\nu} = \mathcal{L}_\xi g_{\mu\nu} = g_{\mu\rho} \partial_\nu \xi^\rho + g_{\nu\rho} \partial_\mu \xi^\rho + \xi^\rho \partial_\rho g_{\mu\nu}. \tag{4.88}$$

By using the chain rule and the definition of a determinant, we find that the density $\sqrt{-g}$ transforms as

$$\delta_\xi \sqrt{-g} = \partial_\mu (\xi^\mu \sqrt{-g}). \tag{4.89}$$

Thus, assuming that the Lagrangian density is a scalar, $\delta_\xi L = \xi^\mu \partial_\mu L$ we have

$$\delta_\xi(\sqrt{-g}L) = \partial_\mu(\xi^\mu\sqrt{-g})L + \sqrt{-g}\xi^\mu\partial_\mu L = \partial_\mu(\xi^\mu\sqrt{-g}L). \tag{4.90}$$

The Lagrangian is invariant modulo a pure derivative. If all fields, as well as ξ^μ, vanish at the boundary, the action $I = \int d^4x\sqrt{-g}L$ is invariant under the action of the transformation δ_ξ,

$$\delta_\xi \int \sqrt{-g}\mathrm{d}^4xL = 0. \tag{4.91}$$

We say that the transformation defines a symmetry which is local, since its parameters ξ depend on the coordinates. So reparametrisation acts as a gauge symmetry, in a way analogous to the one we found for electrodynamics.

4.3.2 Reparametrisation Invariance and Energy–Momentum Tensor

In flat space, we have shown that the conservation of energy–momentum follows from the translation invariance of the action

$$I = \int \mathrm{d}xL(\varphi, \partial_\mu\varphi). \tag{4.92}$$

We used the proof that combines

$$\partial_\mu L = \frac{\delta L}{\delta\varphi}\partial_\mu\varphi + \frac{\delta L}{\delta\partial_\nu\varphi}\partial_\nu\partial_\mu\varphi \tag{4.93}$$

and the Euler–Lagrange equations

$$\frac{\delta L}{\delta\varphi} = \partial_\nu\frac{\delta L}{\delta\partial_\nu\varphi}. \tag{4.94}$$

It gives

$$\partial_\mu\left(\frac{\delta L}{\delta\partial_\mu\varphi}\partial_\nu\varphi - \delta_\nu^\mu L\right) = 0. \tag{4.95}$$

The tensor

$$T_{\mu\nu} = \frac{\delta L}{\delta\partial_\mu\varphi}\partial_\nu\varphi - \delta_\nu^\mu L \tag{4.96}$$

is thus conserved modulo the equations of motion. This is a possible expression of the energy–momentum tensor.

Although this derivation is correct, the resulting expression for the energy–momentum tensor is not the most convenient one. For example, depending on the field representations, it does not give a tensor which is automatically symmetric in the indices μ and ν. In fact, this expression must be often improved by the addition of terms with vanishing divergence in order to get a symmetric expression. Even worse, for electromagnetism with the Lagrangian $L = -\frac{1}{4}(\partial_\mu A_\nu - \partial_\nu A_\mu)$, we must modify $T_{\mu\nu}$, not only to make it symmetrical, but also to make it gauge invariant at the same time. Furthermore, it is challenging to generalise this definition of $T_{\mu\nu}$ in curved space, and get a covariant definition.

The right thing to do is to use the invariance under reparametrisation of the action, which imposes that it depends on the metric (which we can always take to be constant at the end of the computation). In doing so we shall obtain a symmetric second-rank tensor which is covariantly conserved, as the natural candidate for the energy–momentum tensor.

This is quite simple. Consider the action

$$I = \int dx \sqrt{-g} L(\varphi, \partial_\mu \varphi, g_{\mu\nu}), \tag{4.97}$$

which now depends on the metric. In this equation, the metric $g_{\mu\nu}$ is considered as an external field, so the variational principle only applies to the fields φ and their derivatives. We look for a covariantly conserved tensor, modulo the equations of motion. The action is reparametrisation invariant. When the equations of motion hold, the action is invariant under any given local variation of the fields, which include variations under reparametrisation as a particular case. Therefore, we have that, modulo the equations of motion, the reparametrisation invariance implies that

$$\int dx \left(\left(\frac{\delta \sqrt{-g} L}{\delta g_{\mu\nu}} - \partial_\sigma \left(\frac{\delta \sqrt{-g} L}{\delta \partial_\sigma g_{\mu\nu}} \right) \right) \mathcal{L}_\xi g_{\mu\nu} \right) = 0, \tag{4.98}$$

where $\mathcal{L}_\xi g_{\mu\nu}$ is the Lie derivative of $g_{\mu\nu}$ along the infinitesimal vector ξ of an infinitesimal diffeomorphism. Let us define

$$\frac{1}{2} \sqrt{-g} T_{\mu\nu} = \frac{\delta \sqrt{-g} L}{\delta g_{\mu\nu}}. \tag{4.99}$$

Using the definition of $\mathcal{L}_\xi g_{\mu\nu}$, and performing partial integrations, Eq. (4.98) can be expressed as

$$\int dx \sqrt{-g} \, \xi^\mu D_\rho T^\rho_\mu = 0. \tag{4.100}$$

Therefore, $T_{\mu\nu}$ is covariantly conserved and is the right definition for the energy–momentum tensor. Just two examples:

For the action of a scalar field, we have

$$I = \int \mathrm{d}x \sqrt{-g} \left(\frac{1}{2} g^{\mu\nu} \partial_\mu \varphi \partial_\nu \varphi \right). \tag{4.101}$$

By derivation with respect to the metric, we obtain

$$T_{\mu\nu} = \partial_\mu \varphi \partial_\nu \varphi - \frac{1}{2} g_{\mu\nu} \partial_\rho \varphi \partial^\rho \varphi. \tag{4.102}$$

For the action of electromagnetism,

$$\int \mathrm{d}x \sqrt{-g} \left(-\frac{1}{4} F^{\mu\nu} F_{\mu\nu} \right) \tag{4.103}$$

(this Lagrangian is reparametrisation invariant since $D_\mu A_\nu - D_\nu A_\mu = \partial_\mu A_\mu - \partial_\nu A_\mu$), we find that

$$T^{\alpha\beta} = \frac{2}{\sqrt{-g}} \frac{\delta L}{\delta g_{\alpha\beta}} = \frac{1}{4} g^{\alpha\beta} F^{\mu\nu} F_{\mu\nu} - F^{\alpha\nu} F^\beta_\nu. \tag{4.104}$$

$T^{\alpha\beta}$ is both gauge invariant and symmetric. For a flat metric, we can check that $T_{00} = E^2 + B^2$ is the energy density and $T_{0i} = (E \wedge B)_i$ is the Poynting momentum–flux vector.

4.3.3   The Einstein–Hilbert Equation

We found the remarkable result that the energy–momentum tensor of a system with local fields satisfying their Euler–Lagrange field equations is the response of the action to a variation of the metric.

Up to now the metric has been treated as an external field. We want to remind, however, that, in various static cases we have discussed, we have found relations between the metric and the gravitational field as consequences of the equivalence principle. We have given examples suggesting that the presence of a gravity field can be absorbed into a redefinition of the metric of space–time. In this section we want to generalise these results and obtain an equation of motion expressing the reciprocal interactions between matter and gravity in a geometrical way.

A mathematically simple way to do so is to add to the matter Lagrangian $L(\varphi, g_{\mu\nu})$ a Lagrangian for pure gravity L_{gravity}, depending only on the metric $g_{\mu\nu}$ and its derivatives, so that $g_{\mu\nu}$ becomes a dynamical field.

We thus consider the action

$$\int \mathrm{d}^4 x \sqrt{-g} \left(L(\varphi, g_{\mu\nu}) + L_{\mathrm{gravity}}(g_{\mu\nu}) \right). \tag{4.105}$$

The simplest scalar invariant Lagrangian density that depends on $g_{\mu\nu}$ and is of the lowest possible order on the derivatives of $g_{\mu\nu}$ is the scalar curvature R, which we obtained in contracting in all possible ways the four indices of the Riemann tensor.

$$L_{\text{gravity}} = \kappa^{-1} R = \kappa^{-1} g_{\mu\nu} R^{\mu\nu}. \tag{4.106}$$

κ is a dimensionful constant. It will be related to the Newton constant in the non-relativistic limit of the theory.

A rather simple computation shows that for a local variation of the metric $\delta g_{\mu\nu}$, the variation of the action $\int \mathrm{d}^4 x \sqrt{-g} L_{\text{gravity}}$ is

$$\int \mathrm{d}^4 x \sqrt{-g} \delta g_{\mu\nu} \left(R^{\mu\nu} - \frac{1}{2} g^{\mu\nu} R \right). \tag{4.107}$$

When we vary the part of the action that depends on the matter fields, we find, by definition of the energy–momentum tensor of the matter $T_{\mu\nu}$, that

$$\int \mathrm{d}^4 x \sqrt{-g} T^{\mu\nu} \delta g_{\mu\nu}. \tag{4.108}$$

Thus, when we look for the metric configuration that extremises the action, the Euler–Lagrange equations give the following relation between the curvature of the space and the energy–momentum tensor of matter:

$$R_{\mu\nu} - \frac{1}{2} g_{\mu\nu} R = \kappa \, T_{\mu\nu}. \tag{4.109}$$

This equation expresses how the matter determines the space–time curvature through its energy–momentum tensor. It is called the Einstein–Hilbert equation.

The matter dynamics is itself determined by the Euler–Lagrange equation of matter fields

$$\frac{\delta L}{\delta \varphi} = D_\mu \frac{\delta L}{\delta D_\mu \varphi}. \tag{4.110}$$

The system of the two coupled non-linear differential equations (4.109) and (4.110) expresses the way gravity and matter classically interact.

Contact with physics is simple when the observed matter is sufficiently dilute to leave unperturbed the background geometry given by a solution of the Einstein–Hilbert equation. This is the concept of test particles in a gravitational background. Then, the action of a pointlike particle with trajectory $X^\mu(\tau)$ is

$$I[X^\mu] = \frac{1}{2} \int \mathrm{d}\tau \, g_{\mu\nu} \frac{\mathrm{d}X^\mu}{\mathrm{d}\tau} \frac{\mathrm{d}X^\nu}{\mathrm{d}\tau}. \tag{4.111}$$

This value of the action is extremal when $X^\mu(\tau)$ satisfies its Euler–Lagrange equation

$$\frac{\mathrm{d}^2 X^\mu}{\mathrm{d}\tau^2} + \Gamma^\mu_{\rho\sigma} \frac{\mathrm{d}X^\rho}{\mathrm{d}\tau} \frac{\mathrm{d}X^\sigma}{\mathrm{d}\tau} = 0. \tag{4.112}$$

This is the equation for a geodesic. We thus recover in full generality the results we found on examples, using the equivalence principle.

In the Newtonian approximation, $g_{\mu\nu}$ is a flat metric, and τ is the universal time

$$I[X^\mu] = I[x, t] \sim \int d\tau \left(\frac{1}{2} - \frac{1}{2} \left(\frac{dx}{dt} \right)^2 \right) \qquad (4.113)$$

We can expand all coefficients of the metric $g_{\mu\nu}$ in a Taylor expansion in $1/c^2$. At first non-trivial order in $1/c^2$, the general solution of the Einstein–Hilbert equation (4.109) is nothing but $g_{ij} = \delta_{ij}$, $g_{i0} = 0$, and

$$g_{00} = 1 + \frac{2V}{c^2}. \qquad (4.114)$$

For this parametrisation, the Einstein–Hilbert equation implies that V satisfies the Poisson equation, so that V can be identified with a Newton potential. The connection between the constant κ and the Newton constant can be obtained by considering the Einstein–Hilbert equation for the static energy–momentum tensor of a massive particle localised at a point.

We found earlier this approximate relation between the metric component g_{00} and the non-relativistic gravity potential by using the equivalence principle. The Einstein–Hilbert equation explains this relation in full depth.

The Einstein–Hilbert equation is the fundamental equation for gravity. It implies that gravity deforms the particle trajectories into curved trajectories, which, in the previous formulation, were interpreted in the non-relativistic approximation as coming from the existence of forces which deviate them from straight lines. In general relativity they simply follow the geodesics for the curved metric which describe the curvature of space–time through the Einstein–Hilbert equation.

General relativity is thus a gauge theory for a spin-2 field, the metric $g_{\mu\nu}$, while electromagnetism is a gauge theory for a spin-1 field, the gauge potential A_μ. Both fields satisfy wave equations which, for the metric, is nothing but the Einstein equation. The two theories share some common features, but gravity also presents serious additional complications.

The common part is that they are both gauge invariant. We saw in the case of the electromagnetic field that this invariance has two consequences which are related to each other: (i) all four components of A_μ are not independent dynamical degrees of freedom, and (ii) in order to solve for the Green functions of the theory we must apply a procedure which we called 'gauge fixing'. Both features appear also in the equation for gravity: (i) the ten components of the rank 2 symmetric tensor $g_{\mu\nu}$ are not independent degrees of freedom. In fact, we shall show in the next chapter, that, as it is the case for the electromagnetic field, only two degrees of freedom are physical. The others can be eliminated by an appropriate choice of gauge. (ii) We need to proceed to a gauge fixing in order to solve for the Green functions. For example, the analogue of the Lorenz gauge condition $\partial^\mu A_\mu = 0$ for the electromagnetic field is the 'de Donder' gauge condition for the gravitational field:

$$\partial^\mu A_\mu = 0 \quad \Rightarrow \quad \partial^\mu g_{\mu\nu} = 0. \tag{4.115}$$

The additional complications we alluded to stem, essentially, from two problems. (i) The Einstein equation is a non-linear equation, so simple methods of solutions based on Fourier transform do not apply. (ii) The weak field approximation, in which all components of the field can be considered infinitesimal, does not apply either, since the determinant of the metric cannot vanish. As a result, in general relativity non-perturbative phenomena are the most interesting ones, and they have to do with the structure of space. On the other hand, some features are similar; for example, it is simple to reach the conclusion that planar gravitational waves can propagate in empty space. They are solutions of the linear approximation of the Einstein action and are expected to be produced by strongly accelerating massive bodies, such as strong gravitational bursts in the early Universe. We shall briefly mention them again in a later chapter.

Exploring all classical consequences of the equations of gravitation we just derived is not a subject that we want to pursue as a main topic in this book. It is a fact that general relativity has passed all experimental tests that were possible to do for the past decades to check its specific predictions, such as the motion of stars in the post-Newtonian approximation, the formation, and the dynamics of Galaxies, but also the existence of black holes, the gravitational red shift, and many cosmological observations such as the spectrum of the microwave background radiation.

However, this chapter makes us understand with no artifice how the relativistic quantum field theories we can construct in flat space can be generalised in the presence of classical gravity. We have seen that the mathematical recipe is just to change the Minkowski space into a curved manifold, as it is required for sustaining the principle of equivalence. All actions we may write in flat space can be easily written in a coordinate invariant way, and this defines the interaction between spin 0, $\frac{1}{2}$, and 1 fields and the gravitational field $g_{\mu\nu}$.

In the subsequent chapters we shall present the methods to go from a classical to the corresponding quantum field theory. In principle, the method could be applied to general relativity and we shall sketch it briefly, after we have established and proved it for flat space gauge theories. It is indeed gratifying to realise that the method is universal and we shall understand why it fails to give calculable results for quantum gravity.

4.4   The Limits of Our Perception of Space and Time

A deep assumption of both classical and quantum physics is that space and time are described by a four-dimensional Riemannian (or pseudo-Riemannian) manifold. Going from non-relativistic mechanics to special relativity and, finally, to general relativity, imply only different assumptions on the properties of the underlying metric. But the basic properties of the manifold, such as the number of dimensions or the property of being continuous and differentiable, are not challenged. In this section we want to present the experimental evidence which supports this assumption, together with some speculations concerning possible deviations.

4.4.1 Direct Measurements

A first guess would be that the limits on possible defects[1] are given by the resolution of our measuring instruments. They are clocks for time and microscopes for space measurements.

Let us start with the time variable. We saw in this chapter that an observer sees that identical clocks (i.e. cloned perfect atomic clocks) get desynchronised in a well-defined way, when they are at rest in different frames with relative speeds and/or accelerations. This is the lesson of general relativity. This fact generalises to an analogous desynchronisation for clocks relatively at rest but situated at different locations in a classical gravitational field. These counter-intuitive phenomena have been accurately verified experimentally. They have deep consequences in cosmology. At the experimental side we can observe nowadays phenomena with time-scales running over an enormous range, with impressive accuracy. State-of-the-art atomic clocks rely on the transition process between two states of a caesium atom, where the frequency of the emitted light is on the order of 10^{-13} seconds. The great precision of this transition makes it possible to measure reliably time intervals as short as 10^{-18} seconds, in periods on the order of a day. As a result we could claim that the hypothesis of a continuous real time is well supported experimentally for time periods as short as 10^{-18} seconds and probably as long as $10^{10} - 10^{11}$ years, the age of the Universe. Note that even without any other measurement, this result could be translated into a distance down to which space can be assumed to have a continuous structure: the distance covered by a light signal in 10^{-18} seconds. This is 3×10^{-8} cm, on the order of the typical size of an atom.

For the space part direct measurements can do better. Light diffraction experiments can have a resolution better than 10^{-13} cm (the electromagnetic form factor of the proton was measured this way), and high energy accelerators can probe the structure of space at distances as short as $10^{-16} - 10^{-17}$ cm. The discovery of the composite nature of nucleons as bound states of quarks was the result of such scattering experiments, in the same spirit that Rutherford discovered the composite nature of atoms. An immediate conclusion could be that the same measurements show indirectly the absence of any space defects down to these distances, although, as we shall show presently, this conclusion is not as solid as it sounds.

Before discussing the significance of these limits, let us remark that they are expressed in terms of space and time units, the centimetre and the second, which are related to our macroscopic environment. Let us first try to find a more abstract system. At the beginning of this book we introduced the notion of *natural units*, i.e. units related to fundamental physical constants. Let us complete this analysis here.

---

[1] Here by defect we mean any structure which is not described by the usual four-dimensional smooth metrics we have been considering. Some special examples will be given below.

The first 'fundamental constant' which was introduced in physics was Newton's constant G_N, which, in our earth-related units, equals $6.67384(80) \times 10^{-11} \text{kg}^{-1}\text{m}^3\text{s}^{-2}$. In the late nineteenth century the constant c was introduced through Maxwell's equations, $c = 299792458$ m s^{-1}, which represents the speed of light in a vacuum. Finally, with quantum theory came the constant $\hbar = 1.054571726(47) \times 10^{-34} \text{kg m}^2\text{s}^{-1}$. It seems that it was Planck himself who first realised that the set of these three independent physical constants offers a way to define a system of *natural* units, i.e. units which are dictated by the laws of nature and not by our earthly environment. It is the system in which $c = \hbar = G_N = 1$ and we shall call them *Planck units*. Given the values of c, $\hbar$, and G_N we have just provided, we find that the Planck units of time τ_P, distance l_P, and mass M_P (we often use the unit of energy E_P) are

$$1\tau_P = \sqrt{\frac{G_N \hbar}{c^5}} \sim 0.52 \times 10^{-43} \text{s} \quad ; \quad 1l_P = \sqrt{\frac{G_N \hbar}{c^3}} \sim 1.6 \times 10^{-35} \text{m} \qquad (4.116)$$

$$1M_P = \sqrt{\frac{\hbar c}{G_N}} \sim 2.12 \times 10^{-8} \text{kg} \quad ; \quad 1E_P \sim 1.2 \times 10^{19} \text{GeV}. \qquad (4.117)$$

We see that our measurement capabilities for space and time are many orders of magnitude away from Planck units. Where do we fail? In our experiments, in particle physics, the typical velocities are on the order of the speed of light and we are in the quantum regime in which the typical action is on the order of $\hbar$. So, setting $c = \hbar = 1$ is normal, it describes phenomena for which quantum mechanics and special relativity are important. This is the system of units we presented in the Introduction and it is the one we adopt in this book. The point of failure is the relation $G_N = 1$. Together with the other two, it describes conditions under which we have, in addition, strong gravitational forces. The second of the relations (4.117) shows that this is expected to happen at extremely high energies, beyond the reach of existing, or imaginable, accelerators. For comparison, the maximum energy of LHC is on the order of a few times 10^3 GeV. If whichever new structure of space and time exists appears only at Planck units, it may well remain undetectable for any foreseeable future. If our current ideas on the evolution of the Universe are correct, conditions such that $c = \hbar = G_N = 1$ prevailed only during the first 10^{-43}s after the Big Bang. In order to describe physics under these conditions a classical theory is not sufficient and we need a fully quantised theory of gravity. Although such a complete theory is still missing, most speculations studied so far predict a breakdown of the usual assumptions corresponding to a smooth four-dimensional manifold of space and time.

4.4.2   Possible Large Defects

Let us now come back to the existing experimental evidence for the structure of space and time. In the previous section we presented the current precision of direct measurements. We want here to argue that they do not necessarily guarantee the absence of any 'defect' down to the corresponding accuracy. The reason is obvious: our capability to detect any kind of defect in the structure of space depends on the interaction this defect may have with the probes we are using, usually light, or other elementary particles such as electrons or protons. In the absence of such interactions a defect will remain invisible to scattering experiments. We want here to give an example.

The example is inspired by the study of quantum string theories which offer the most promising candidate to obtain a quantum theory of gravity. They are formulated in a ten-dimensional space (nine space and one time) and the usual assumption is that six space dimensions are compact. The natural size of compactification is given by the Planck length l_P of (4.116). However, it is interesting to study an alternative possibility, namely the one of *large extra dimensions*. So, leaving any theoretical prejudice aside, let us assume that there exist $3 + d$ spatial dimensions with the d forming a compact space of size R. How large can R be without contradicting existing experiments? As we noted earlier, we expect the answer to depend on the specific assumptions concerning the relation of our probes with the extra dimensions. We shall consider two extreme cases.

Let $z_M = (x_\mu, y_i)$ $M = 0, 1, ..., 3 + d$, $\mu = 0, 1, 2, 3$ and $i + 1, ..., d$ denote a point in the $(4 + d)$-dimensional space. The metric tensor is written as $g_{MN}(z)$. In the first case let us assume that a field, such as the electromagnetic field, depends on both x and y. Consider, for simplicity, that the compact space is a sphere of radius R. In this case the d y's can be chosen to be angular variables and we can take the field A to satisfy periodic boundary conditions. Expanding in a Fourier series, we obtain

$$A(z) = \sum_{n=-\infty}^{\infty} A_n(x) e^{in\frac{y}{R}}, \tag{4.118}$$

where we have suppressed the vector index for A and we wrote the formula for $d = 1$. We see that the field A in five dimensions is equivalent to an infinity of fields in four dimensions. The equation of motion (3.29) in the vacuum will be generalised to

$$\partial_M \partial^M A(z) = 0 \quad \Rightarrow \quad \partial_\mu \partial^\mu A_n(x) = \frac{n^2}{R^2} A_n(x); \tag{4.119}$$

in other words, we obtain an infinite tower of modes, the so-called *Kaluza–Klein modes*, with equal spacing given by n/R. In this case light scattering experiments will detect these modes, if the resolution is better than R, and our previous assumption that the limiting size of the defect will be given by the resolution of our microscope will be correct.

In the opposite extreme case, let us assume that the field A, as well as any other physical field with the exception of the metric tensor, is independent of y. This means that all our fields and particles other that the gravitational field live in the four-dimensional space. In this case only A_0 survives and a diffraction experiment will see nothing. The compact space can only be detected by gravitational experiments.

In our four-dimensional space–time the gravitational potential of a mass M at a distance r is given by Newton's formula

$$V(r) = \frac{G_N M}{r}, \tag{4.120}$$

and, in the units we are using, $G_N \sim (E_P)^{-2}$. In the $(4 + d)$-dimensional space and for $r < R$, the corresponding formula is

$$V(r) = \frac{KM}{(E_*)^{2+d}} \frac{1}{r^{1+d}}, \tag{4.121}$$

where E_* is the new value of the 'Planck energy' in $4 + d$ dimensions and K is a numerical factor which takes into account the fact that force lines spread over more dimensions: $K \sim 8\pi/[(2 + d)\Omega_{(2+d)}]$. For $r > R$ we must obtain back the four-dimensional potential (4.120), so by matching the two we obtain

$$\left(\frac{E_P}{E_*}\right)^2 = \frac{1}{K}(E_* R)^d. \tag{4.122}$$

R cannot be larger than 0.1 mm, because Newton's law has been tested to distances of that order. In our system of units ($c = \hbar = 1$), $E_* R$ is dimensionless and the conversion factor with ordinary units is given by the approximate relation 10^{-13} cm $\sim (200$ MeV$)^{-1}$. We see that assuming, for example, $d = 2$, we can have a compact sphere at every point in space with radius as large as some fraction of a millimetre, provided E_* is on the order of a few TeV, in perfect agreement with all existing experiments. The reason why such a 'defect' could have escaped detection is that it interacts only gravitationally with our usual probes. Note also that in such a world gravitational interactions would become strong at an energy scale of a few TeV. An accelerator reaching this scale would discover entirely new and spectacular strong interaction phenomena associated with gravitation, such as multi-graviton radiation, mini black hole production, etc. All these phenomena become detectable at an energy of several TeV which corresponds to a space resolution smaller than 10^{-17} cm, although the compact sphere of extra dimensions has a size of 0.1 mm. In this simple example we see that the limits in our perception of the properties of space are not necessarily identical with the resolution of our microscopes.

5

The Physical States

5.1 Introduction

In establishing the symmetry properties of a physical theory it is important to know which are the associated physical states. Indeed, all observable consequences of the theory are expressed in terms of relations among such states.

Classical physics is formulated in terms of the dynamical variables $q_i(t)$ and $p_i(t)$, $i = 1, ..., N$, where N is the number of degrees of freedom of the system. For example, if the system consists of point particles, the q and the p values could be the positions and momenta of the particles. The physical states of such a system are just the possible values of these variables at a given time, so they do not require any special care in order to be specified. The situation does not change radically when we take the large N limit and consider the index i taking values in a continuum, for example, the points in the three-dimensional space. The theory becomes a classical field theory, such as the electromagnetic theory or the theory of general relativity, which we reviewed briefly in the previous chapters. The physical states are still determined by the values of the fields and their first-time derivatives at each point in space, which, for the electromagnetic field, correspond to a given configuration of the electric and magnetic fields.

Although the subject of this book is the theory of quantum fields, we will see that the associated physical states we will use will be states of particles. This connection between fields and particles, already present under the form of wave–particle duality in the old quantum theory, will be fully developed in the next chapters. However, we know already that in quantum mechanics positions and momenta cannot be specified simultaneously and the states of a non-relativistic quantum mechanical particle are determined by the possible wave functions, i.e. vectors in a certain Hilbert space. Special relativity brings a further, very important, complication. Already in classical physics, special relativity establishes a connection between energy and mass with the famous relation $E = mc^2$. But it is in quantum physics that this relation reveals all its significance. We expect therefore that establishing a relativistic quantum formalism will not be a simple exercise. In fact, such a formalism must include in an essential way the non-conservation of the number of particles and, as we will see in this book, the fact that to each particle is associated an antiparticle. In the old books this is often called *the formalism of second quantisation*, although it is, in fact, the theory of quantum fields, the subject of this book.

From Classical to Quantum Fields. Laurent Baulieu, John Iliopoulos and Roland Sénéor.

From these remarks, we see that a necessary step for understanding relativistic quantum phenomena is a good knowledge of the states consisting of an arbitrary number of particles. In the following chapters we shall show that this is, in fact, the space of physical states appropriate to a quantum field theory. However, studying the most general n-particle state, with arbitrary n, will turn out to be an impossible task for any physically interesting theory. Fortunately, for most applications we are going to consider in this book, it will be enough to restrict ourselves to a particular class of n-particle states, the states of non-interacting particles. They can be viewed as the states of n particles lying far away one from the other and having only short-range interactions. If the distances among them are much larger than the interaction range, the particles can be considered as free. The interactions among them will be treated separately. Since the Poincaré group is the invariance group of any physical system we are going to consider, a good description of these states will be provided by the irreducible representations of this group.[1]

5.2 The Principles

We note by $\mathcal{H}$ the Hilbert space of the physical states of a system. The set of physical states describes the results of all the experiments done during the whole history of the system. A physical state is represented by a ray in this space and a ray is a vector $\Psi \in \mathcal{H}$ defined up to a multiplicative constant.

We will only consider unit vectors, i.e. $\Psi \in \mathcal{H}$ such that $(\Psi, \Psi) = 1$. To simplify, we will assume that we deal only with a part of the space which is *coherent*, namely one in which the superposition principle can be applied: a linear combination of physically realisable states is a physically realisable state. In real life the Hilbert space of all physical states is not coherent. The linear superposition of two physical states, although it is a vector in the Hilbert space, does not always correspond to a physical state. For example, a state with one proton and a state with one electron are both physical states. The first has electric charge +1 and the second −1. A linear superposition of the two will not have a definite value of the electric charge, so it will not be an eigenstate of the charge operator. It is an empirical fact that there are operators such that all physical states are eigenstates of them with well-defined eigenvalues. We say that such an operator defines *a superselection rule*. These operators split the Hilbert space into sub-spaces corresponding to definite eigenvalues. We will restrict our discussion to one of these subspaces, which are said to be coherent and for which all the states are always physically realisable. In the example above, the electric charge operator Q defines a superselection rule: every physical state $|\Phi >$ must be an eigenstate of Q; in other words, it must have a definite value of the electric charge. There are other operators with the same property. For example, the operator of baryon number B seems to be one of them. Let us consider the one proton state $|p >$ which satisfies

$$Q|p >= |p > \quad ; \quad B|p >= |p > . \tag{5.1}$$

[1] In this chapter, a large part of the content was inspired by the presentation given by A. S. Wightman in his lecture notes 'L'invariance dans la mécanique quantique relativiste', in *Relations de dispersion et particules élémentaires*, Les Houches (1960), Hermann, Paris.

Similarly, the one positron state $|e^+>$ satisfies

$$Q|e^+>= |e^+ > \quad ; \quad B|e^+ >= 0, \tag{5.2}$$

because the positron has a baryon number equal to 0. A linear superposition of the form $|\Psi> = C_1|p> +C_2|e^+ >$ would still be an eigenstate of Q with eigenvalue $+1$, but not an eigenstate of B. It has never been observed experimentally and this is interpreted as an indication of the existence of a superselection rule associated with the baryon number. A superselection rule is always associated with a conservation law but the inverse is not true. The z-component of angular momentum is an absolutely conserved quantity but all physical states are not necessarily eigenstates of it. Note also that if tomorrow we discover that baryon number is not absolutely conserved, for example if we discover that protons may decay, we will have to abandon the corresponding superselection rule. We will present the current experimental situation in Chapter 26.

5.2.1 Relativistic Invariance and Physical States

The Poincaré group, which is the invariance group of a relativistic physical theory, keeps invariant the 2-form $g_{\mu\nu}dx^\mu \otimes dx^\nu$ (i.e. ds^2). An element of the Poincaré group $\mathcal{P}$ is a couple $\{a, \Lambda\}$, where a is a translation of vector $a \in \mathbb{R}^4$ and Λ is a Lorentz transformation, acting on $\mathbb{R}^4$ by

$$x \in \mathbb{R}^4 \rightarrow x' = \{a, \Lambda\}x = \Lambda x + a. \tag{5.3}$$

It obeys the multiplication law of the group

$$\{a_1, \Lambda_1\}\{a_2, \Lambda_2\} = \{a_1 + \Lambda_1 a_2, \Lambda_1 \Lambda_2\} \tag{5.4}$$

with the operations acting from right to left. The group identity is $\{0, \mathbf{1}\}$. The Poincaré group possesses two distinguished subgroups: the translation subgroup and the Lorentz subgroup whose elements are respectively $\{a, \mathbf{1}\}$ and $\{0, \Lambda\}$. As for the Lorentz group, the Poincaré group consists of four connected components and we will only consider the group built over the elements Λ of the restricted Lorentz group.

We will now show that physics imposes to study the *unitary* representations of the Poincaré group. In the study of these representations we will encounter a difficulty: the fact that apart from the trivial representation of dimension 1, there exist no unitary irreducible representations of finite dimension.

Let us first consider the relativistic invariance properties.

The relativistic invariance can be expressed in two different ways.

5.2.1.1 The Relativistic Symmetry

This property expresses the fact that to each state Ψ and to each Poincaré transformation $\{a, \Lambda\}$, there corresponds a state $\Psi_{\{a,\Lambda\}}$ such that $\Psi_{\{0,1\}} = \Psi$ and such that the transition probabilities are invariant:

$$| (\Phi, \Psi) |^2 = | (\Phi_{\{a,\Lambda\}}, \Psi_{\{a,\Lambda\}}) |^2. \tag{5.5}$$

It results from eq. (5.5) that there exists a linear mapping $U(a, \Lambda)$ on $\mathcal{H}$ which is unitary,

$$(U(a, \Lambda)\Phi, U(a, \Lambda)\Psi) \equiv (\Phi_{\{a,\Lambda\}}, \Psi_{\{a,\Lambda\}}) = (\Phi, \Psi), \tag{5.6}$$

or antiunitary

$$(U(a, \Lambda)\Phi, U(a, \Lambda)\Psi) = \overline{(\Phi, \Psi)}. \tag{5.7}$$

If we are only interested in the restricted Poincaré group, $U(a, \Lambda)$ is unitary, unique up to a phase, and such that

$$U(a_1, \Lambda_1)U(a_2, \Lambda_2) = \pm U(a_1 + \Lambda_1 a_2, \Lambda_1 \Lambda_2). \tag{5.8}$$

These are the so-called representations up to a phase.

The representation of inversions can be unitary or antiunitary. In practice, to the space inversion I_s will correspond a unitary representation and to the time reflection I_t will correspond an antiunitary representation. A theorem by E. Wigner asserts that each unitary, continuous representation, up to a phase, of the restricted Poincaré group arises from a unitary continuous representation of its covering group; this means that $U(a, \Lambda)$ is indeed $U(a, A)$ with $\Lambda = \Lambda(A)$.

5.2.1.2 The Relativistic Equivalence

Two unitary representations U_1 and U_2 are equivalent according to the relativistic point of view if there exists a unitary operator V acting on $\mathcal{H}$ such that

$$U_2(a, \Lambda) = VU_1(a, \Lambda)V^{-1}. \tag{5.9}$$

This expresses the fact that if a state Ψ_1 transforms under the representation U_1 and if there is a corresponding state Ψ_2 given by $\Psi_2 = V\Psi_1$, then Ψ_2 transforms under U_2. Moreover, if the operators or the observables which act on the states transform in the same way, i.e. to each operator O_1, there corresponds an operator O_2 such that $O_2 = VO_1V^{-1}$; we will speak about physical equivalence. The relativistic equivalence is therefore expressed through the equivalence of the representations.

5.2.1.3 *The Vacuum*

We define a particular state, the *vacuum* Ψ_0 or $\mid \Omega >$, which is the state with no particles. Since this state must be the same for all the observers

$$U(a, \Lambda)\Psi_0 = \Psi_0. \tag{5.10}$$

5.2.1.4 *The Particle States*

We shall call an *elementary system* a system whose states transform according to an irreducible representation $\{a, \Lambda\} \rightarrow U(a, \Lambda)$ of the restricted Poincaré group. A stable elementary particle is, at this level of discussion, assimilated to an elementary system as will be the case for an atom in its fundamental state. So we see that this notion of elementarity does not mean that the system is not a bound state of constituents, which may, or may not, be themselves elementary systems. A simple example is given by a stable nucleus, such as a deuteron. According to our definition it is an elementary system. It is also a bound state of a proton and a neutron. Although we know that both of them are in fact composite from quarks and gluons, the proton is an elementary system, but the neutron, which in a free state is unstable under β-decay, is not.

5.3 The Poincaré Group

5.3.1 The Irreducible Representations of the Poincaré Group

We present here some of the basic notions concerning the classification of useful representations of the restricted Poincaré group (up to a unitary equivalence). We study these representations in the Hilbert space $\mathcal{H}$ of the one-particle states, i.e. the space of square integrable functions with respect to some measure which will be defined later.

Let us first consider a subgroup, the group of space–time translations. Since it is an Abelian group, its finite-dimensional irreducible representations are of dimension 1. Since they are unitary, they are like phase factors and

$$U(a, \mathbf{1}) = \mathrm{e}^{ia.p}, \tag{5.11}$$

where p is a real four-vector of components p_μ and $a.p = a^\mu p_\mu$, . We then have the following theorem (which will be partially justified in the next section), which expresses the fact that unitary representations of the translation group are direct integrals of irreducible representations.

Theorem 6. *Any continuous unitary representation of the space–time translation group can be written as*

$$(U(a, \mathbf{1})\Phi)(p) = \mathrm{e}^{ipa}\Phi(p), \tag{5.12}$$

where $\Phi(p) \in \mathcal{H}_p$ is the space of the physical states of momentum p, the space $\mathcal{H}$ being a direct integral[2] over the momentum space

$$\mathcal{H} = \int \mathrm{d}\mu(p) \mathcal{H}_p. \tag{5.14}$$

Since $U(0, A)U(a, 1)U(0, A)^{-1} = U(\Lambda(A)a, 1)$, $U_1(a, 1) = U(\Lambda(A)a, 1)$ is a representation equivalent to $U(a, 1)$, it is natural to take an invariant measure $\mathrm{d}\mu$, $\mathrm{d}\mu(\Lambda(A)(p)) = \mathrm{d}\mu(p)$ and to identify $\mathcal{H}_{\Lambda p} = \mathcal{H}_p$.

The most general positive invariant measure can be written as

$$\mu = c\delta^4(p) + \int_0^\infty \mathrm{d}\rho_+(m)\theta(p_0)\mathrm{d}\Omega_m(p) + \int_0^\infty \mathrm{d}\rho_-(m)\theta(-p_0)\mathrm{d}\Omega_m(p) + \mu_-, \tag{5.15}$$

where c is a constant, $c \geq 0$, and $\mathrm{d}\Omega_m$ is the invariant measure on the hyperboloid $p^2 = m^2$, with $p_0 \geq 0$ for the second term and $p_0 \leq 0$ for the third term. For the positive energy branch it is given by

$$\mathrm{d}\Omega_m = \frac{\mathrm{d}^4 p}{(2\pi)^4}(2\pi)\delta(p^2 - m^2)\theta(p^0) = \frac{\mathrm{d}^3 p}{2(2\pi)^3\sqrt{p^2 + m^2}}. \tag{5.16}$$

$\mathrm{d}\rho_+(m)$ and $\mathrm{d}\rho_-(m)$ are positive measures on $\mathbb{R}_+$. Finally, μ_- is a positive measure, the support of which consists of space-like four-vectors p, which are of no interest since we will only consider light- or time-like four-vectors p (there is no physical meaning for states corresponding to a purely imaginary energy).

It is easy to check that the measure $\mathrm{d}\Omega_m$ is Lorentz invariant.

We then prove the following theorem.

Theorem 7. *Each continuous unitary representation of* $\mathcal{P}$ *is equivalent to a representation in* $\mathcal{H}$ *of the form*

$$(U(a, A)\Phi)(p) = \mathrm{e}^{ipa}Q(p, A)\Phi(\Lambda(A^{-1})p), \tag{5.17}$$

where $Q(p, A)$ *satisfies*

$$Q(p, A_1)Q(\Lambda(A_1^{-1})p, A_2) = Q(p, A_1 A_2). \tag{5.18}$$

---

[2] It suffices to understand the definition for functions: let $\mathcal{H}$ be the Hilbert space of the functions of p which are square integrable with respect to the measure $\mathrm{d}\mu$; if Φ and Ψ are two elements of $\mathcal{H}$, their scalar product is defined by

$$(\Phi, \Psi) = \int \mathrm{d}\mu(p)\Phi^*(p)\Psi(p) \tag{5.13}$$

and $\Phi(p)$ and $\Psi(p)$ are elements of $\mathcal{H}_p \simeq \mathbb{C}$. Here $*$ means complex conjugation.

If $\mathcal{H}_p$ has a dimension larger than 1 (this is the case for states corresponding to particles with non-zero spin) and $\Phi(p)$ or $\Psi(p)$ are vectors of $\mathcal{H}_p$, their product in the integral is replaced by the scalar product in this space.

Remark: $U(0,1)$ is the identity on $\mathcal{H}$.

Given the four-vectors p, which characterise the representation of the translation subgroup, we see, following the previous formula, that the matrices A, such that $\Lambda(A)p = p$, play a special role. They form a subgroup L_p of $SL(2, \mathbb{C})$: the *little group* of p. We see that the mapping $A \in L_p \to Q(p, A)$ is a representation of the little group of p. To classify these representations, it can be shown that two representations giving rise to equivalent representations of their translation subgroups are equivalent if and only if they have equivalent representations of their little groups.

The little groups belong to four different classes corresponding to the various possible types of p: time-like, light-like, identically null, and space-like.

We will only consider the first three classes since they are the only ones which have a physical meaning. If, in addition, we want to identify p with the energy-momentum four-vector, then we need only to consider the first two classes.

Let us consider

1. **time-like p.**
   One can choose a particular p since all the representations associated with this type of p are equivalent. Let us take $p = (1, 0, 0, 0)$. Then[3] $\tilde{p} = \mathbf{1}$ and $A\tilde{p}A^* = AA^* = \mathbf{1}$, and the little group L_p is the *special unimodular group* $SU(2, \mathbb{C})$. The unitary representations of $SU(2, \mathbb{C})$ are D^s, s integer or half-odd integer. The representations of the Poincaré group of this type are therefore characterised by two numbers (m, s), where $m = \sqrt{p^2}$ is the mass and s is the spin. Obviously, from the physical point of view, only representations corresponding to positive values of the energy ($p_0 \geq 0$) must be considered. We denote $[m, s]$ as an irreducible representation of this type. It is possible to show in this case that we can take

$$Q(p, A) = D^s((\sqrt{\tilde{p}})^{-1} A \sqrt{\widetilde{\Lambda(A^{-1})p}}). \tag{5.19}$$

2. **light-like p.**
   We can choose $p = (1, 0, 0, 1)$. The elements of the little group are the A's of the form

$$A(\phi, z) = \begin{pmatrix} e^{\frac{i\phi}{2}} & e^{-i\frac{\phi}{2}} z \\ 0 & e^{-i\frac{\phi}{2}} \end{pmatrix}, \tag{5.20}$$

   where $z = x_1 + ix_2$. With this parametrisation

$$A(\phi_1, z_1) A(\phi_2, z_2) = A(\phi_1 + \phi_2, z_1 + e^{i\phi_1} z_2),$$

---

[3] In the notation of Eq. (2.61), $\tilde{p} = \sigma_\mu p^\mu$.

where one recognises the law of composition of the *Euclidean group of motion in the plane*: translation of vector (x_1, x_2) and rotation of angle ϕ around the axis x_3 perpendicular to the plane. There exists an invariant Abelian subgroup, the translation subgroup given by $A(0, z)$. Again we can do the same analysis as for the first case. We first seek the representations of the translation subgroup. They are of dimension 1 and characterised by two real numbers $p = (p_1, p_2)$. Thus, there are two cases: either $p^2 > 0$ or $p^2 = 0$. The first case is excluded since the representation is characterised by a continuous number p^2 which must be an invariant associated with a particle; it could be a spin value but no continuous spin has been observed. The second case corresponds to $p_1 = p_2 = 0$. The little group consists of all the $A(\phi, 0)$'s. Since the $A(\phi, 0)$'s commute, their representations are of dimension 1 and, the $A(\phi, 0)$'s being unitary, are of the form $e^{i\phi s}$. Since $A(4\pi, 0) = 1$, this means $s = 0, \pm\frac{1}{2}, \pm 1, \pm\frac{3}{2}, \ldots$. The parameter s is called the *helicity* of the massless particle. We denote by $[0, s]$ an irreducible representation of this type.

3. $p \equiv 0$.

The corresponding representations are generally of infinite dimension. Their associated fields have an infinite number of components. For this reason, they are not considered, except for one of them which plays an important role in physics. This representation, the only one which has a finite dimension, is the *trivial representation*. It corresponds to $U(a, \Lambda) = 1$. It is used for the vacuum state.

5.3.2   The Generators of the Poincaré Group

One of the aims of this section is to show how the study of the infinitesimal generators of the Poincaré group makes it possible to recover some of the characteristics of its irreducible representations. Let $U(a, \Lambda)$ be a unitary irreducible representation of the Poincaré group.

Let us first consider the space–time translation subgroup whose elements are $U(a, 1)$.

Since the translations along different axes of coordinates commute, it suffices to study the translations on **R**. Let $T(a)$ be a translation of length a. Acting on a function $f(x)$, it is defined by

$$(T(a)f)(x) = f(x - a). \tag{5.21}$$

For a infinitesimal, $f(x - a) = f(x) - a\frac{\mathrm{d}}{\mathrm{d}x}f(x) + O(a^2)$ and $T(a) = 1 + a\delta T + O(a^2)$, δT being the infinitesimal generator of the transformation. In this case, we see that

$$\delta T = -\frac{\mathrm{d}}{\mathrm{d}x}, \tag{5.22}$$

and the global transformation is given formally by the exponential

$$T(a) = \mathrm{e}^{-a\frac{\mathrm{d}}{\mathrm{d}x}}, \qquad (5.23)$$

which yields the Taylor expansion when the transformation is applied to infinitely differentiable functions. Now if we consider the functions f as elements of the Hilbert space of square integrable functions with scalar product

$$(f, g) = \int \mathrm{d}x \overline{f(x)} g(x), \qquad (5.24)$$

the representation is automatically unitary, i.e. such that $(T(a)f, T(a)g) = (f, g)$ or equivalently $T(a)^* T(a) = \mathbf{1}$. Indeed, writing the infinitesimal form, the consequence of this last identity is $\delta T^* + \delta T = 0$, showing that δT is an anti-Hermitian operator. Since $\delta T = -\frac{\mathrm{d}}{\mathrm{d}x} = \mathrm{i}(\mathrm{i}\frac{\mathrm{d}}{\mathrm{d}x})$ this shows that $\mathrm{i}\frac{\mathrm{d}}{\mathrm{d}x}$ is Hermitian in this space as it can also be checked directly. We will now show that the unitary operator $T(a)$ is bounded and of norm 1 although its generator is an unbounded operator. To prove this assertion we use the Fourier transform

$$(T(a)f)(x) = \frac{1}{2\pi} \int \mathrm{d}p \mathrm{e}^{-\mathrm{i}px} \widetilde{(T(a)f)}(p) = \frac{1}{2\pi} \int \mathrm{d}p (T(a)\mathrm{e}^{-\mathrm{i}px}) \tilde{f}(p) \qquad (5.25)$$

$$= \frac{1}{2\pi} \int \mathrm{d}p \mathrm{e}^{-\mathrm{i}p(x-a)} \tilde{f}(p),$$

the third equality resulting from the linearity of T and the last one being the definition of the translation. We deduce from it the action of $T(a)$ on the functions ϕ of p:

$$(T(a)\phi)(p) = \mathrm{e}^{\mathrm{i}ap} \phi(p). \qquad (5.26)$$

This shows that the representation of the group of translations on the space of square integrable functions is the direct integral of irreducible representations, each of them being characterised by a real vector p.

In the case of the translations in $\mathbb{R}^4$, this gives

$$U(a, \mathbf{1}) = \mathrm{e}^{\mathrm{i}P.a}, \qquad (5.27)$$

where $P = (P_0, P_1, P_2, P_3)$, $P_\mu = \mathrm{i}\partial_\mu$, are the four generators of the group. Moreover, using the fact that U is a continuous representation of the Abelian group of translations in $\mathbb{R}^4$

$$U(a, \mathbf{1}) = \int \mathrm{e}^{\mathrm{i}a.p} \mathrm{d}E(p), \qquad (5.28)$$

where $E(p)$ is a spectral measure[4] and

$$P = \int p \, dE(p). \tag{5.32}$$

Let us show that the operator $P^\mu P_\mu = P^2$ is Lorentz invariant. To prove it, let us consider the relation

$$U(0, \Lambda)U(a, 1)U(0, \Lambda)^{-1} = U(\Lambda a, 1),$$

from which we deduce for a infinitesimal

$$U(0, \Lambda)P_\mu U(0, \Lambda)^{-1} = (\Lambda^{tr} P)_\mu, \tag{5.33}$$

which proves the assertion by an obvious calculation.

This shows also that the spectral measure $E(p)$ is Lorentz invariant.

Continuous bases are of common use in physics.[5] In the momentum space, projectors on states of given momentum p are given by $|p><p|$ from which we get a very simple expression of the spectral measure :

$$dE(p) = d\Omega_m(p)\,|p><p|\,.$$

Since, on the other hand, P^2 commutes with translations, it commutes with all the transformations $U(a, \Lambda)$ of the Poincaré group and by Schur's lemma, P^2 is proportional to the identity. Let us note m^2 the coefficient of proportionality. On the spaces $\mathcal{H}_p$, the operators P_μ are proportional to the identity with coefficients p_μ and $p^2 = m^2$.

To an infinitesimal Lorentz transformation $\Lambda = 1 + \lambda$, with components $\Lambda^\mu_\nu = \delta^\mu_\nu + \lambda^\mu_\nu$ with $\lambda_{\mu\nu} = -\lambda_{\nu\mu}$, corresponds, because of continuity, $U(\Lambda) = 1 - \frac{1}{2}M_{\mu\nu}\lambda^{\mu\nu}$ (one

---

[4] A spectral measure $\{E(\lambda)\}$, $\lambda \in \mathbb{R}$, is a family of orthogonal projections such that

1. $E(\lambda) \le E(\lambda')$ if $\lambda < \lambda'$

2. the following strong limits exist:

$$s - \lim_{\lambda \to -\infty} E(\lambda) = 0 \tag{5.29}$$

$$s - \lim_{\lambda \to +\infty} E(\lambda) = 1. \tag{5.30}$$

3. The family is right continuous: at each point $-\infty < \lambda < +\infty$

$$s - \lim_{\varepsilon \to +0} E(\lambda + \varepsilon) = E(\lambda) \tag{5.31}$$

The integral is a Stieltjes integral. For more details see, for example, M. Reed and B. Simon, *Methods of Mathematical Physics*, vols I–IV, Academic Press (1972).

[5] See section 8.5.1.2.

has set $U(\Lambda) \equiv U(0, \Lambda)$). The operators $M_{\mu\nu}$ are a representation of the infinitesimal generators of the Lorentz group.

Remark: For the representation of the Lorentz group in $\mathbb{R}^4$, we computed (see section 2.6) the matrix representation of the infinitesimal generators, $\boldsymbol{J}$ and $\boldsymbol{N}$. We can define, through these operators, a 2-tensor $\mathcal{M}_{\mu\nu}$, whose coefficients are matrices, by setting $\boldsymbol{J} = (\mathcal{M}^{23}, \mathcal{M}^{31}, \mathcal{M}^{12})$ and $\boldsymbol{N} = (\mathcal{M}^{01}, \mathcal{M}^{02}, \mathcal{M}^{03})$. For pure Lorentz transformations of speed $\tanh \xi$ in the direction $\boldsymbol{n}$ and rotations of angle θ around $\boldsymbol{n}$ the values of the parameters are given by $\lambda^{0i} = \xi n^i$ and $\lambda^{ij} = \theta \varepsilon^{ijk} n^k$.

This tensor satisfies the commutation relations

$$[\mathcal{M}^{\varrho\sigma}, \mathcal{M}^{\mu\nu}] = \mathrm{i}(g^{\sigma\mu}\mathcal{M}^{\varrho\nu} + g^{\varrho\nu}\mathcal{M}^{\sigma\mu} - g^{\varrho\mu}\mathcal{M}^{\sigma\nu} - g^{\sigma\nu}\mathcal{M}^{\varrho\mu}). \tag{5.34}$$

The $M_{\mu\nu}$'s, which are a representation of the $\mathcal{M}_{\mu\nu}$'s, satisfy obviously the same commutation relations.

From the relation

$$U(\Lambda)P_\mu U(\Lambda)^{-1} = \Lambda^\nu{}_\mu P_\nu \tag{5.35}$$

we deduce that

$$-\lambda^{\varrho\sigma}\frac{\mathrm{i}}{2}[M_{\varrho\sigma}, P_\mu] = \lambda^\nu{}_\mu P_\nu = g_{\mu\varrho}\lambda^{\sigma\varrho}P_\sigma. \tag{5.36}$$

Thus, identifying the antisymmetric part of the coefficients of $\lambda^{\varrho\sigma}$, we find the commutation relations between the $M_{\mu\nu}$'s and the P_μ's:

$$[M_{\varrho\sigma}, P_\mu] = \mathrm{i}(g_{\sigma\mu}P_\varrho - g_{\varrho\mu}P_\sigma). \tag{5.37}$$

These commutation relations show that the operator P_μ transforms like a four-vector under Lorentz transformations. It also explains why P^2 was found to be an invariant.

Let us now introduce a new operator which commutes with P

$$W^\mu = \frac{1}{2}\epsilon^{\mu\nu\varrho\sigma}P_\nu M_{\varrho\sigma}. \tag{5.38}$$

It is interesting to have such an operator since it is a multiple of the identity in any irreducible representation of the Poincaré group. This operator can be written in terms of the generators of the group

$$W^0 = P^1 \mathcal{J}^1 + P^2 \mathcal{J}^2 + P^3 \mathcal{J}^3 = \boldsymbol{P}\boldsymbol{J} \qquad \boldsymbol{W} = P_0\boldsymbol{J} - \boldsymbol{P} \wedge \boldsymbol{N}, \tag{5.39}$$

where $(\boldsymbol{P} \wedge \boldsymbol{N})^i = \varepsilon_{ijk}P^j N^k$.

One indeed finds that $W^2 = W^\mu W_\mu$ commutes with P_ϱ and with $M_{\varrho\sigma}$. For an irreducible representation characterised by $[m, s]$, mass and spin (a non-zero mass representation), we have $W^2 = -m^2 s(s + 1)\mathbf{1}$.

We have thus recovered as a function of the generators the characteristic parameters of the irreducible representations of interest for us. The operator-valued four-vector W^μ has the meaning of a relativistic spin.

In the case $m \neq 0$, we have seen that an irreducible representation is partially charac-terised by p. To complete the characterisation, we choose for each eigenvalue p_μ of P_μ, three four-vectors $n^{(i)}$, $i = 1, 2, 3$, orthogonal to p, of square -1 (since they are space-like) and pairwise orthogonal. Introducing $n^{(0)} = p/m$, the $n^{(\alpha)}$'s form a basis and we can define $S^{(\alpha)} = -W.n^{(\alpha)}/m$. We have $S^{(\alpha)} = (0, S)$ with $S = (S^{(1)}, S^{(2)}, S^{(3)})$. The $S^{(i)}$'s are the three generators of the little group of p isomorphic to SO_3. In the representation $[m, s]$, $S^2 = s(s + 1)$ and since $W^2 = -m^2 S^2$, we obtain the claimed value.

Note that in the reference frame at rest for p, $p = (m, 0)$, following (5.39), we have $S = W/m = J$.

In the case $m = 0$, since $W^2 = p^2 = W.p = 0$, we find that $W^\mu = \pm s p^\mu$, where s takes again integer or half-odd-integer positive values. We shall call s *helicity* and we see that for $s \neq 0$, all representations are two dimensional. We say often that massless particles with $s \neq 0$ have only two helicity states.

5.4   The Space of the Physical States

We described in the preceding section the action of the Poincaré group on the one-particle states. The aim of this section is to generalise this analysis and formulate a general description of many particle states, a framework which will be useful in the study of quantum field theories.

5.4.1   The One-Particle States

In the massive scalar case, a one-particle state Ψ is given by a complex valued function $\Psi(p)$ in a Hilbert space $\mathcal{H}^{(1)}$ equipped with the norm

$$\|\Psi\|^2 = (\Psi, \Psi) = \int |\Psi(p)|^2 \mathrm{d}\Omega_m(p) \tag{5.40}$$

and which transforms[6] under the Poincaré group as

$$(U(a, \Lambda)\Psi)(p) = \mathrm{e}^{\mathrm{i}p.a}\Psi(\Lambda(A)^{-1}p). \tag{5.42}$$

---

[6] We can use the bra and ket formalism by introducing a continuous basis $|p>$, $p^2 = m^2$, such that $<q|p> = \delta(q - p)p^0$ with $\Psi(p) = <p|\Psi>$, the transformation law of $|p>$ being

$$U(a, \Lambda)|p> = \mathrm{e}^{\mathrm{i}\Lambda p.a}|\Lambda^{-1}p>. \tag{5.41}$$

For a particle with mass m and spin s, the state Ψ is described by a set of complex-valued functions depending on $2s$ spinorial indices: $\Psi_{\alpha_1\cdots\alpha_{2s}}(p)$. Their transformation laws are given by

$$(U(a,A)\Psi)_{\alpha_1\cdots\alpha_{2s}}(p) = e^{ip.a} \sum_{\beta_1\cdots\beta_{2s}} (\prod_{i=1}^{2s} A_{\alpha_i}{}^{\beta_j})\Psi_{\beta_1\cdots\beta_{2s}}(\Lambda(A)^{-1}p) \tag{5.43}$$

and the scalar product by

$$(\Phi,\Psi) = \int d\Omega_m(p)\bar{\Phi}_{\alpha_1\cdots\alpha_{2s}}(p) \prod_{i=1}^{2s} (\tilde{p}/m)^{\alpha_i\beta_i} \Psi_{\beta_1\cdots\beta_{2s}}(p). \tag{5.44}$$

The massless case is more complicated. We saw that the representations corresponding to $p^2 = 0$ are characterised by the value of helicity which takes integer or half-integer values. For zero helicity the representation is one dimensional and for all other values it is two dimensional. However, since the corresponding particle moves always with the speed of light, we cannot assign square integrable wave functions to it. This is known as *the infrared problem* and has a simple physical origin. Let us take a one-photon state. Following what we said previously, it has a momentum p with $p^2 = 0$. However, since every measurement has a finite resolution Δp, there is no conceivable experiment which can distinguish this single-particle state from an infinity of others which will contain in addition an arbitrary number of photons, provided their total energy is smaller than Δp because, since the mass is zero, there is no lower limit to the energy of a photon and therefore no upper limit to the number of photons. We shall come back to this problem when we develop better tools to tackle it. For the moment we restrict ourselves to the description of a world containing only massive particles.

5.4.2   The Two- or More Particle States without Interaction

Let us consider a system made of two scalar particles without interaction, i.e. described by plane waves. Their state Ψ is given by a complex-valued function of two variables $\Psi(p_1,p_2)$, such that $p_1^2 = m_1^2$, $p_2^2 = m_2^2$, $p_1^0 > 0$, and $p_2^0 > 0$. This function transforms according to

$$(U(a,\Lambda)\Psi)(p_1,p_2) = e^{i(p_1+p_2).a}\Psi(\Lambda^{-1}p_1, \Lambda^{-1}p_2) \tag{5.45}$$

and is an element of the Hilbert space $\mathcal{H}^{(2)}$ with the scalar product

$$(\Phi,\Psi) = \int\int d\Omega_{m_1}(p_1)d\Omega_{m_2}(p_2)\overline{\Phi(p_1,p_2)}\Psi(p_1,p_2). \tag{5.46}$$

This shows that the transformation law of the state is the same as that for $[m_1,0] \otimes [m_2,0]$. If the two particles are identical, $m_1 = m_2 = m$, Ψ is symmetrical with respect to the exchange of the variables and transforms like $([m, 0] \otimes [m, 0])_S$. The scalar product defined in Eq. (5.46) is invariant by the transformations of $([m, 0])^{\otimes 2}$.

The states of two non-interacting particles are described by the mass M of the system they form, given by $M^2 = (p_1 + p_2)^2$, a number which varies from $m_1 + m_2$ to infinity, and by the angular momentum of the system, characterised by an integer L. In terms of our previous analysis of elementary systems, it is parameterised by $[M, L]$. Therefore, there exists a continuum of possible states.

One can obviously do a similar analysis for a system of more than two[7] particles with or without spin (in the case of identical particles, a rule of symmetrisation or antisymmetrisation, depending on whether the spin is integer or half-odd integer, is understood). One represents the space of all these states as a Hilbert space, $\mathcal{F}$, the Fock space.

5.4.3 The Fock Space

The Fock space describes the states of an arbitrary number of non-interacting particles. To simplify the presentation we will assume that we have only one kind of particle: a scalar particle of mass m.

The Fock space is built as a direct sum of spaces, each of which corresponds to a fixed number of particles

$$\mathcal{F} = \oplus_n \mathcal{H}^{(n)}, \tag{5.47}$$

where $\mathcal{H}^{(n)}$ is the Hilbert space of n-particle states. The elements of $\mathcal{H}^{(n)}$ are the symmetric functions $\Phi^{(n)}(p_1, \cdots, p_n)$ and the scalar product, invariant under $([m, 0])_S^{\otimes n}$, is given by

$$(\Phi^{(n)}, \Psi^{(n)})$$
$$= \int \cdots \int d\Omega_m(p_1) \cdots d\Omega_m(p_n) \overline{\Phi^{(n)}(p_1, \cdots, p_n)} \Psi^{(n)}(p_1, \cdots, p_n). \tag{5.48}$$

The space $\mathcal{H}^{(0)}$ is the one-dimensional space of complex numbers.

An element of $\mathcal{F}$ is a series

$$\Phi = (\Phi^{(0)}, \Phi^{(1)}, \Phi^{(2)}, \cdots)$$

and the scalar product is given by

$$(\Phi, \Psi) = \sum_0^\infty (\Phi^{(n)}, \Psi^{(n)}); \tag{5.49}$$

thus, $\Psi \in \mathcal{H}$ if $(\Psi, \Psi) = \|\Psi\|^2 < \infty$. Fock spaces of more than one species of particles can be built in a similar way.

---

[7] If there are more than two particles, there exists not only one but an infinite number of states characterised by the values of the mass and the angular momentum.

In the case of particles with spin s, the elements of $\mathcal{H}^{(n)}$ are the functions $\Psi^{(n)}_{\alpha_1 \cdots \alpha_n}(p_1, \cdots, p_n)$, which are, according to the value of the spin, integer or half-odd integer, symmetric or antisymmetric in the exchanges of pairs (p_i, α_i) and (p_j, α_j). The scalar product is given by

$$
(\Phi^{(n)}, \Psi^{(n)}) = \int \cdots \int d\Omega_m(p_1) \cdots d\Omega_m(p_n) \overline{\Phi^{(n)}_{\alpha_1 \cdots \alpha_n}(p_1, \cdots, p_n)}
$$
$$
\cdot \prod_{j=1}^{n} D^{[s,0]}(\tilde{p}_j/m)^{\alpha_j \beta_j} \Psi^{(n)}_{\beta_1 \cdots \beta_n}(p_1, \cdots, p_n), \tag{5.50}
$$

where α_i and β_j represent groups of $2s$ undotted spinorial indices.

5.4.4 Introducing Interactions

The construction of the Fock space can be extended to include interacting particles provided, as we discussed already, the interactions are short ranged and the particles are far apart from each other. For example, two initially well-separated particles will feel the effects of the interaction, i.e. diffusion, creation, or absorption of particles, only when they come close enough to each other. There are two possibilities: either they form a bound state (its mass is less than the sum of the masses, the difference being the binding energy) or they separate, forming a set of non-interacting particles moving away from each other. All these processes are realised in accordance with the fundamental laws of physics such as conservation of energy–momentum, of angular momentum, and of any quantum numbers preserved by the interaction. Bound states are new elementary systems generated by the interactions and must be included in the space of possible states. Moreover, by reversing the velocities of the particles, the initial and the final states can be interchanged, which means that they must be of the same nature. This analysis leads us to introducing the notion of the incoming (and outgoing) Fock space: the space of initial or incoming states $\mathcal{H}_{\text{in}}$. It consists of non-interacting states composed of wave packets well separated and evolving independently. Symmetrically, we introduce the Fock space of final or outgoing states $\mathcal{H}_{\text{out}}$. It is identical to the previous one. Finally, we set that these two spaces are identical to the space of the physical states $\mathcal{H}$, i.e.

$$
\mathcal{H}_{\text{in}} = \mathcal{H}_{\text{out}} = \mathcal{H}. \tag{5.51}
$$

This is the hypothesis of *asymptotic completeness*. From our construction of Fock spaces as spaces on which the action of the Poincaré group is well defined (by the action of $U(a, \Lambda)$), the last equality means that even during the collisions, states transform in a well-defined way under the actions of the Poincaré group. This point of view reinforces the ideas on the principle of preservation of covariance we sketched out in Chapter 2.

A central problem in this book will be the one we studied also in quantum mechanics, namely the computation of a transition probability between initial and final states. Let us consider an initial state Ψ_{in}; then the probability to obtain a final state Φ_{out} is obtained from $(\Phi_{\text{out}}, \Psi_{\text{in}})$. We introduce a unitary operator S such that

$$(\Phi_{\text{out}}, \Psi_{\text{in}}) = (\Phi_{\text{in}}, S\Psi_{\text{in}}).\tag{5.52}$$

This operator, which for historical reasons is called the *S-matrix* (*S* for scattering), will become an important object of study all throughout this book.

5.5   Problems

Problem 5.1 The Bargmann–Michel–Telegdi equation.
We like to give a relativistic generalisation of the spin. There are many mathematical quantities which can reduce at rest to the three-dimensional spin. We saw in this chapter and in the previous ones some of them. Bargmann, Michel, and Telegdi proposed a relativistic equation for the time evolution of the spin in an electromagnetic field. Their idea was to take the simplest relativistic generalisation of the spin at rest, i.e. a four-vector s_μ. The description of a spinning pointwise particle is given in the rest frame of the particle by its angular momentum. The relativistic generalisation is to describe now the particle by two four-vectors, p^μ and s^μ,

$$p^\mu = mu^\mu = m\frac{\mathrm{d}x^\mu}{\mathrm{d}s} \qquad \text{with} \qquad p^2 = m^2,$$

with the condition $s.p = 0$ which expresses the fact that the spin is a space-like vector (no time component in the frame at rest). In the rest frame, the dimensional spin s (from $s = (0, \boldsymbol{s})$) is supposed to satisfy the classical equation (generalisation to a spin of the angular momentum time evolution):

$$\frac{\mathrm{d}\boldsymbol{s}}{\mathrm{d}t} = g\frac{e}{2m}\boldsymbol{s} \wedge \boldsymbol{B}.$$

Then they wrote the most general covariant equation, linear in s, one can build with at our disposal two four-vectors and one antisymmetric tensor, s_μ, u_μ, and $F_{\mu\nu}$

$$\frac{\mathrm{d}s^\mu}{\mathrm{d}s} = aF^{\mu\nu}s_\nu + b(u^\lambda F_{\lambda\nu}s^\nu)u^\mu.$$

The inputs are the Lorentz equation

$$\frac{\mathrm{d}mu^\mu}{\mathrm{d}s} = eF^\mu{}_\nu u^\nu$$

and

$$s.u = 0.$$

Show that

$$a = g\frac{e}{2m} \qquad \text{and} \qquad b = (\frac{g-2}{2})\frac{e}{m}.$$

$(g-2)/2$ is known as the particle's magnetic moment anomaly.

Problem 5.2 Properties of W.

Prove the equalities

$$W^\mu P_\mu = 0 \qquad [W^\mu, P^\nu] = 0$$
$$[W^\mu, W^\nu] = -i\epsilon^{\mu\nu\rho\sigma} P_\rho W_\sigma$$
$$[W^\mu, M^{\nu\rho}] = i(g^{\mu\nu} W^\rho - g^{\mu\rho} W^\nu).$$

6

Relativistic Wave Equations

6.1 Introduction

Maxwell's equations were at the origin of the discovery of special relativity. They were the first relativistically invariant wave equations. In modern notation they describe the propagation of a vector field. In this chapter we want to introduce wave equations for fields of any spin.

Historically, the Dirac equation was found as a response to a physical problem, the need to write a relativistic equation for the wave function of the electron. In doing so, Paul Adrien Maurice Dirac was the first to find in 1928, by trial and error, the spinorial representations of the Lorentz group. Today we do not have to follow this historical route. We can use our knowledge of group theory and write down directly the relativistic, lowest order, linear differential equations for fields belonging to various representations of the Lorentz group.

To write a relativistic wave equation means, in particular, to write an equality between quantities of the same type: equality between scalars, between four-vectors, etc. On the other hand, the equation we would like to write, which is a differential equation, should be of the lowest possible order so as to restrict the number of initial data. In this chapter we apply this recipe to fields belonging to some among the low-dimensional representations of the Lorentz group.

6.2 The Klein–Gordon Equation

We start with the simplest case, the equation for a real, scalar field. In this case the elements which are at our disposal are the field itself Φ and the four-vector operator of derivation ∂_μ. It is clear that the lowest order, non-trivial, relativistically invariant equation which can be built with these quantities is[1]

$$(\partial_\mu g^{\mu\nu} \partial_\nu + m^2)\Phi(x) = 0. \tag{6.1}$$

---

[1] Strictly speaking, there exists a simpler one, of the form $\partial_\mu \Phi(x) = 0$, but it is trivial.

From Classical to Quantum Fields. Laurent Baulieu, John Iliopoulos and Roland Sénéor.
© Laurent Baulieu, John Iliopoulos and Roland Sénéor, 2017. Published 2017 by Oxford University Press.

It is the Klein–Gordon equation whose Green functions we studied in section 3.8.1. We remind the reader that m^2 is a parameter which, in our usual system of units $\hbar = c = 1$, has the dimensions of [mass]^2. We shall often call this equation *the massive Klein–Gordon equation*, although, at this stage, we have no real justification for this name. m^2 is just a parameter which can take any real value.

This equation can be derived from a variational principle applied to the quantity

$$S[\Phi] = \frac{1}{2} \int d^4x \left(\partial_\mu \Phi(x) g^{\mu\nu} \partial_\nu \Phi(x) - m^2 \Phi(x)^2\right) = \int d^4x \mathcal{L}(x), \tag{6.2}$$

where the local Lagrangian and Hamiltonian densities are given, respectively by

$$\mathcal{L} = \frac{1}{2} \left(\partial_\mu \Phi(x) g^{\mu\nu} \partial_\nu \Phi(x) - m^2 \Phi(x)^2\right) \tag{6.3}$$

and

$$\mathcal{H} = \frac{1}{2} \left[(\partial_0 \Phi(x))^2 + (\partial_i \Phi(x))^2 + m^2 \Phi(x)^2\right]. \tag{6.4}$$

The action S is quadratic in the field Φ and gives rise to the linear equation (6.1). Using the diagonal form of the metric, and restoring the dependence on c and $\hbar$, we get

$$\left(\frac{1}{c^2}\left(\frac{\partial}{\partial t}\right)^2 - \left(\frac{\partial}{\partial x}\right)^2 + \frac{m^2 c^2}{\hbar^2}\right) \Phi(x) = 0. \tag{6.5}$$

It is a linear equation which admits solutions of the plane wave type

$$e^{-i\frac{(p_0 t - \boldsymbol{p} \cdot \boldsymbol{x})}{\hbar}} \tag{6.6}$$

with

$$p_0 = \pm c\sqrt{\boldsymbol{p}^2 + m^2 c^2}. \tag{6.7}$$

Thus, in Fourier transform, the Klein–Gordon equation can be written as

$$(p_0^2 - c^2 \boldsymbol{p}^2 - m^2 c^4) \tilde{\Phi}(p_0, \boldsymbol{p}) = 0, \tag{6.8}$$

where $\tilde{\Phi}(p_0, \boldsymbol{p})$ is the Fourier transform of $\Phi(x)$.

In the system of units where $\hbar = c = 1$

$$(p^2 - m^2) \tilde{\Phi}(p) = 0. \tag{6.9}$$

The most general solution can be easily obtained by looking at (6.9) which gives

$$\tilde{\Phi}(p) = F(p)\delta(p^2 - m^2), \tag{6.10}$$

where $F(p)$ is an arbitrary function of p provided it is sufficiently regular at $p^2 = m^2$. We see that the solution has as support the two branches of the hyperboloid $p^2 = m^2$. We have already introduced the invariant measure on the positive energy branch of the hyperboloid in Eq. (5.16), which we wrote as

$$d\Omega_m = \frac{d^4 p}{(2\pi)^4} (2\pi)\delta(p^2 - m^2)\theta(p^0) = \frac{d^3 p}{(2\pi)^3 2E} \tag{6.11}$$

with $E = +\sqrt{p^2 + m^2}$. Using this expression we can write the most general x-space solution of the Klein–Gordon equation as an expansion in plane waves

$$\Phi(x) = \int d\Omega_m \left[a(p)e^{-i(Et - p \cdot x)} + a^*(p)e^{i(Et - p \cdot x)} \right], \tag{6.12}$$

where a^* is the complex conjugate of a. This complex function $a(p)$ is arbitrary and should be determined by the initial data. Indeed, since the Klein-Gordon equation is a second-order differential equation, the initial data at $t = 0$ involve the value of the function and its first-time derivative $\Phi(0, x)$ and $\dot{\Phi}(0, x)$. Inverting the expression (6.12) we obtain

$$a(p) = \int d^3 x e^{-ip \cdot x} \left[E\Phi(0, x) + i\dot{\Phi}(0, x) \right], \tag{6.13}$$

which shows that a depends only on p.

The total Hamiltonian H, i.e. the space integral of the Hamiltonian density (6.4), has a very simple expression in terms of the function $a(p)$:

$$H = \int d\Omega_m \frac{1}{2} E \left[a(p)a^*(p) + a^*(p)a(p) \right]. \tag{6.14}$$

For later convenience we have kept the peculiar ordering $aa^* + a^*a$, although the two terms are equal for the classical theory we are discussing here.

The Green functions we derived in Chapter 3 allow us to write the general form of the solution for the inhomogeneous Klein–Gordon equation (3.97).

An obvious generalisation is to consider the equation for a multiplet of N scalar fields $\Phi_i(x)$, $i = 1, \ldots, N$. We can write an $O(N)$ invariant Lagrangian density as

$$\mathcal{L} = \frac{1}{2} \sum_{i=1}^{N} \left(\partial_\mu \Phi_i(x) \partial^\mu \Phi_i(x) - m^2 \Phi_i(x)\Phi_i(x) \right). \tag{6.15}$$

Of particular interest will turn out to be the case $N = 2$. Taking into account the fact that $O(2)$ is locally isomorphic to $U(1)$, we can rewrite the two-component real scalar field as a complex field $\Phi = (\Phi_1 + i\Phi_2)/\sqrt{2}$ and the Lagrangian becomes

$$\mathcal{L} = \partial_\mu \Phi(x) \partial^\mu \Phi^*(x) - m^2 \Phi(x) \Phi^*(x), \tag{6.16}$$

which is invariant under phase transformations of the field: $\Phi \to e^{i\theta} \Phi$. We can compute the corresponding conserved current and we obtain

$$j_\mu(x) = \Phi(x) \partial_\mu \Phi^*(x) - \Phi^*(x) \partial_\mu \Phi(x). \tag{6.17}$$

Since Φ satisfies the Klein–Gordon equation, the current is conserved $\partial^\mu j_\mu(x) = 0$.

We can still expand the complex field Φ in terms of plane waves, but now the coefficients of the positive and negative frequency terms will not be complex conjugates of each other. To keep the notation as close as possible to the real field case, we shall write the solution as

$$\Phi(x) = \int d\Omega_m \left[a(p) e^{-i(Et - \boldsymbol{p} \cdot \boldsymbol{x})} + b^*(p) e^{i(Et - \boldsymbol{p} \cdot \boldsymbol{x})} \right] \tag{6.18}$$

and similarly, for Φ^* in terms of the coefficient functions a^* and b.

Naturally, this equation can be made more complicated by adding non-linear terms in the field Φ and its derivatives. In particular, since Φ is a scalar, any power of it will be a Lorentz scalar as well. Of course, for the non-linear cases, we lose the plane wave solutions and the power of the Fourier transform.

Although we obtained the Klein–Gordon equation from purely group theory considerations, we can think of possible physical applications. In classical physics this equation, or its non-linear generalisations, can be applied to problems of relativistic fluid dynamics. Another application, which we shall encounter in this book, involves the addition of a cubic term in the field and it is known as *the Landau–Ginzburg* equation. It describes the order parameter in various problems of phase transitions. In the next chapters we shall study possible applications in the quantum physics of elementary particles.

6.3   The Dirac Equation

The scalar field belongs to the simplest, in fact trivial, representation of the Lorentz group. The next interesting case is the spinor representation.

We will look for a wave equation, invariant under the action of the Lorentz group and the parity operation[2] which, at Dirac's time, seemed to be one of the discrete symmetries leaving unchanged the laws of physics.

---

[2] As we announced in the Introduction, when studying the Dirac equation, people use a notation for the complex conjugation acting on a spinor field which is different from that used everywhere else: the complex conjugate of Φ is denoted Φ^* and the transposed conjugate $\Phi^\dagger$. It is unfortunate, but it is firmly established.

We saw that there exist two irreducible inequivalent representations of minimal dimension of the restricted Lorentz group, one of them corresponding to spinors ξ^α with undotted indices and the other one corresponding to spinors $\eta^{\dot\alpha}$ with dotted indices. These representations do not mix by the transformations of the restricted Lorentz group but transform into each other by space inversion, which, in particle physics jargon, is often called *parity operation P*. We introduced in a previous chapter bispinors, noted $(\xi^\alpha, \eta_{\dot\alpha})$, to build the minimal dimensional representation of the Lorentz group extended by parity. From the action of the spatial rotations on each of the components of a bispinor, we note that its spin is $\frac{1}{2}$.

Let us thus consider a field which is a bispinor $\Psi \sim (\xi^\alpha, \eta_{\dot\alpha})$. In the following, we will note Ψ^α, $\alpha = 1, 2, 3, 4$, the four independent components of Ψ, and we will accept the possibility of combining these components through unitary transformations. According to the rules of spinor calculus, to the four-momentum operator $P_\mu = i\hbar\dfrac{\partial}{\partial x^\mu} = i\hbar\partial_\mu$ corresponds the operatorial spinor

$$\tilde{P} = P^\mu \sigma_\mu = P^0 + \boldsymbol{\sigma} \cdot \boldsymbol{p}. \tag{6.19}$$

The σ_i's are the Pauli matrices, $\boldsymbol{\sigma} \cdot \boldsymbol{p} = \sum \sigma_j P^j$ and $\tilde{P}$ is a 2×2 matrix, the elements of which $P^{\alpha\dot\beta}$ depend linearly on the components of P_μ. As in the scalar case, the requirement of relativistic covariance dictates the most general form of the minimal degree wave equation which can be written for the bispinor (we will omit in the sequel the $\hbar$ and c dependences);

$$P^{\alpha\dot\beta}\eta_{\dot\beta} = m\xi^\alpha \tag{6.20}$$
$$P_{\dot\beta\alpha}\xi^\alpha = m\eta_{\dot\beta}$$

with m a constant with the dimensions of a mass. Multiplying the second equation by $P^{\gamma\dot\beta}$ and using the fact that $P^{\alpha\dot\beta}P_{\dot\beta\gamma} = P^2\delta^\alpha_\gamma$, we get

$$(P^2 - m^2)\xi^\alpha = 0 \tag{6.21}$$
$$(P^2 - m^2)\eta_{\dot\beta} = 0.$$

The two spinors satisfy separately the Klein–Gordon equation.

The space parity, which corresponds to $P_{\dot\beta\alpha} \to P^{\beta\dot\alpha}$, $\xi^\alpha \to i\eta_{\dot\alpha}$, and $\eta_{\dot\beta} \to i\xi^\beta$, exchanges the two equations (6.20). For $m \neq 0$, we have thus introduced a parity invariant system of equations.

Using the relation $\zeta_{\gamma\alpha}\zeta_{\dot\delta\dot\beta}P^{\alpha\dot\beta} = P_{\gamma\dot\delta}$, we can transcribe (6.20) as

$$(P_0 + \boldsymbol{p} \cdot \boldsymbol{\sigma})\eta = m\xi \tag{6.22}$$
$$(P_0 - \boldsymbol{p} \cdot \boldsymbol{\sigma})\xi = m\eta,$$

that is, with matrix notation:

$$\begin{pmatrix} 0 & P_0 + \boldsymbol{p} \cdot \boldsymbol{\sigma} \\ P_0 - \boldsymbol{p} \cdot \boldsymbol{\sigma} & 0 \end{pmatrix} \begin{pmatrix} \xi \\ \eta \end{pmatrix} = m \begin{pmatrix} \xi \\ \eta \end{pmatrix}. \tag{6.23}$$

Introducing the 2×2 matrices whose elements are themselves 2×2 matrices (that is, altogether 4×4 matrices)

$$\gamma^0 = \begin{pmatrix} 0 & 1 \\ 1 & 0 \end{pmatrix} \qquad \gamma^i = \begin{pmatrix} 0 & -\sigma_i \\ \sigma_i & 0 \end{pmatrix}, \quad i = 1, 2, 3, \tag{6.24}$$

the matrix equation (6.23) is written in its usual form, the *Dirac equation*

$$(\not{P} - m)\Psi = 0 \tag{6.25}$$

with

$$\Psi = \begin{pmatrix} \xi \\ \eta \end{pmatrix} \tag{6.26}$$

and

$$\not{P} = \gamma^\mu P_\mu = \gamma P = \gamma^0 P_0 - \boldsymbol{\gamma} \cdot \boldsymbol{p} = i\gamma^0 \frac{\partial}{\partial t} + i\boldsymbol{\gamma} \cdot \boldsymbol{\nabla}. \tag{6.27}$$

The matrices $\gamma^\mu = (\gamma^0, \boldsymbol{\gamma})$ are called *gamma (or Dirac) matrices*. We can verify by explicit calculation that the set of matrices (6.24) satisfy the anticommutation relations

$$\{\gamma^\mu, \gamma^\nu\} = \gamma^\mu \gamma^\nu + \gamma^\nu \gamma^\mu = 2g^{\mu\nu} \mathbf{1}. \tag{6.28}$$

From now on, we will use indistinguishably p for the four-vector or for the four-operator, whenever there is no risk of confusion.

Before closing this section, let us make a few remarks which will be useful later. We see from Eq. (6.20) that for $m = 0$, the two equations decouple. So, the simplest relativistically invariant equation for a spinor field is to consider one of the two spinorial representations, either $(\frac{1}{2}, 0)$ or $(0, \frac{1}{2})$, and write an equation of the form $P^{\alpha\dot{\beta}} \eta_{\dot{\beta}} = 0$, or, similarly, for ξ. As we shall see later, such a theory with no mass parameter could be adequate to describe particles with spin $\frac{1}{2}$ and zero mass. These equations were first written by Hermann Weyl in 1929 and they were dismissed because, as we explained earlier, they were not invariant under parity (space inversion), which, at that time, was believed to be an exact symmetry of nature. So, in order to satisfy the requirements of invariance under parity and the appearance of a mass parameter, it seems that we are forced to use both a ξ and an η spinor which transform into each other by parity. It is the choice which gave us the Dirac equation. However, as Ettore Majorana first observed, there exists a more economical solution: we remember that for the representations of the

Lorentz group, the operation of complex conjugation is not always an equivalence. In particular, a member of $(\frac{1}{2}, 0)$ is transformed by complex conjugation into a member of $(0, \frac{1}{2})$. Therefore, we could choose in the system (6.20) $\eta \sim \xi^*$. The resulting system will contain half as many variables as that introduced by Dirac. We shall study the properties of both the Weyl and the Majorana spinors in a later section.

6.3.1   The γ Matrices

We obtained the Dirac equation from our knowledge on spinors and the form we found is particular in the sense that it naturally exhibits the relativistic invariance. However, in the applications, it may be more useful to change the representation, that is to apply unitary linear transformations on the four components of the bispinor Ψ, $\Psi \to \Psi' = U\Psi$, where U is a 4×4 unitary matrix. From the Dirac equation we found that $(\gamma.p - m)\Psi = 0$; it is easy to see that the transformed equation is $(\gamma'.p - m)\Psi' = 0$, with $\gamma' = U\gamma U^\dagger$. Here $U^\dagger$ denotes the Hermitian conjugate of U and, since U is unitary, we have that $U^\dagger = U^{-1}$.

The new γ matrices satisfy again the relation (6.28):

$$\{\gamma'^\mu, \gamma'^\nu\} = 2g^{\mu\nu} \, \mathbf{1}. \tag{6.29}$$

Alternatively, we can also prove the opposite property, namely, given a set of four 4×4 matrices satisfying (6.28), there exists a unitary, 4×4 matrix U, such that the transformed matrices $\gamma' = U\gamma U^\dagger$ are equal to our first choice (6.24). It follows that any such set can be used to write the Lorentz invariant Dirac equation in a particular basis. In other words, (6.28) is equivalent to the property of Lorentz covariance.

The algebra defined by the anticommutation relation (6.28) is called *a Clifford algebra*. It is the necessary and sufficient condition for a set of γ matrices to give, through (6.25), a Lorentz covariant Dirac equation. It is also the necessary relation for the Dirac equation $(\gamma \cdot p - m)\Psi = 0$ to give the Klein–Gordon equation by left (or right) multiplication by the operator $\gamma \cdot p + m$.

The γ matrices satisfy the hermiticity relations

$$(\gamma^0)^\dagger = \gamma^0 \qquad \text{and} \qquad (\gamma^i)^\dagger = -\gamma^i \quad i = 1, 2, 3. \tag{6.30}$$

It is easy to check that these relations are independent of the chosen representation. It will be useful to introduce a fifth 4×4 matrix, called γ^5,

$$\gamma^5 = \gamma_5 = \mathrm{i}\gamma^0\gamma^1\gamma^2\gamma^3 = \frac{\mathrm{i}}{4!}\varepsilon_{\mu\nu\varrho\sigma}\gamma^\mu\gamma^\nu\gamma^\varrho\gamma^\sigma \tag{6.31}$$

and

$$(\gamma_5)^2 = 1 \qquad \{\gamma_5, \gamma^\mu\} = 0 \tag{6.32}$$

and the matrices

$$\sigma^{\mu\nu} = \frac{i}{2}[\gamma^\mu, \gamma^\nu]. \tag{6.33}$$

Finally, note that

$$\frac{1}{4}\gamma^\mu\gamma_\mu = 1 \qquad \mathrm{Tr}\gamma^\mu = 0 \quad \mathrm{Tr}\mathbf{1} = 4 \quad \mathrm{Tr}\gamma_5 = 0. \tag{6.34}$$

The following identities are often used with $\not{a} = a_\mu\gamma^\mu$:

$$\not{a}^2 = a^2\mathbf{1} \qquad \not{a}\not{b} + \not{b}\not{a} = 2a \cdot b\,\mathbf{1}. \tag{6.35}$$

A useful exercise is the computation of the trace of a product of γ matrices. Using the Clifford algebra anticommutation relations and the cyclic property of the trace, we can easily obtain the following results: if $T^{\mu_1\cdots\mu_n}$ is the trace of a product of n γ matrices, we have

$$\begin{aligned}
T^{\mu_1\mu_2} &= 4g^{\mu_1\mu_2} \\
T^{\mu_1\cdots\mu_{2n+1}} &= 0 \\
T^{\mu_1\cdots\mu_{2n}} &= \sum_{k=1}^{2n-1}(-)^k g^{\mu_1\mu_k} T^{\mu_2\cdots\mu_{k-1}\mu_{k+1}\cdots\mu_{2n}}.
\end{aligned} \tag{6.36}$$

The last relation expresses the $2n$-trace in terms of the $2n-2$ ones. With the help of these relations we can compute, by induction, the trace of the product of any number of γ matrices.

6.3.2   The Conjugate Equation

Let us apply a complex conjugation on the two spinorial equations at the beginning of this chapter. Since $p_\mu = i\partial_\mu$, $p_\mu^* = -p_\mu$

$$(\not{p} - m)\Psi = 0 \rightarrow (\not{p}^* - m)\Psi^* = ((\gamma^0)^{tr}p_0^* - (\gamma^i)^{tr}p_i^* - m)\Psi^* = 0 \tag{6.37}$$

or $(\Psi^*)^{tr}(-\gamma^0p_0 + \gamma^ip_i - m) = 0$, the differential operators acting on the left. Here Ψ^* is the four-component spinor whose elements are the complex conjugates of the elements of Ψ. Multiplying this equation by γ^0 and using $\gamma^0\gamma^0 = 1$ and $\gamma^0\gamma^i\gamma^0 = -\gamma^i$ we have

$$(\Psi^*)^{tr}\gamma^0(-\gamma^0p_0 - \gamma^ip_i - m) = -\overline{\Psi}(\not{p} + m) = 0, \tag{6.38}$$

where we have introduced the Dirac conjugate spinor

$$\overline{\Psi} = (\Psi^*)^{tr}\gamma^0. \tag{6.39}$$

Note, that, when referring to Dirac spinors, $*$ means complex conjugation and 'bar' the Dirac conjugation.

6.3.3   The Relativistic Invariance

Although we know that, by construction, the Dirac equation transforms covariantly under the extended Lorentz transformations, it is instructive to verify this property by explicit computation. Let us first look at how the equation transforms under space inversion $I_s : x \rightarrow \bar{x} = (x^0, -\mathbf{x})$. We check that by parity $\Psi(x) \rightarrow i\gamma^0 \Psi(\bar{x})$ and $\overline{\Psi}(x) \rightarrow -i\overline{\Psi}(\bar{x})\gamma^0$, since γ^0 exchanges ξ and η. Therefore,

$$(i\gamma^0 \frac{\partial}{\partial x^0} + i\gamma^k \frac{\partial}{\partial x^k} - m)i\gamma^0\Psi(x) = i\gamma^0(\gamma^0 \frac{\partial}{\partial \bar{x}^0} + i\gamma^k \frac{\partial}{\partial \bar{x}^k} - m)\Psi(\bar{x}) = 0. \qquad (6.40)$$

Multiplying on the left by $-i\gamma^0$ we recover the Dirac equation (we knew that already by construction).

The field Ψ (a 1-bispinor) transforms under $D^{[\frac{1}{2},0]} \oplus D^{[0,\frac{1}{2}]}$, that is under a Lorentz transformation $\Lambda(A) : x \rightarrow x' = \Lambda x$, $\Psi(x) \rightarrow \Psi'(x') = S(\Lambda(A))\Psi(x)$. A spinor ξ^α transforms as A, a dotted spinor $\eta^{\dot\alpha}$ transforms as A^*, thus $\eta_{\dot\alpha}$ transforms as $\zeta A^* \zeta^{-1} = (A^\dagger)^{-1}$; therefore,[3]

$$S(\Lambda(A)) = \begin{pmatrix} A & 0 \\ 0 & (A^\dagger)^{-1} \end{pmatrix}. \qquad (6.41)$$

Once again, claiming that the equation is form invariant, one deduces that

$$S(\Lambda)\gamma^\nu S(\Lambda)^{-1} = (\Lambda^{-1})^\nu{}_\mu \gamma^\mu. \qquad (6.42)$$

We shall often need to consider expressions of the form $\overline{\Psi}_r \Gamma^{rs} \Psi_s$, where Γ is a general four-by-four matrix. A convenient basis of sixteen such matrices is given by: $\mathbb{1}, \gamma^5, \gamma_\mu, \gamma_\mu\gamma^5, 1/2[\gamma_\mu, \gamma_\nu]$, $\mu, \nu = 0, 1, 2, 3$. It is an easy exercise to show that under Lorentz transformations, these expressions transform as

$\overline{\Psi}\Psi$: a scalar
$\overline{\Psi}\gamma^5\Psi$: a pseudo-scalar
$\overline{\Psi}\gamma_\mu\Psi$: a vector
$\overline{\Psi}\gamma_\mu\gamma^5\Psi$: a pseudo-vector
$\overline{\Psi}[\gamma_\mu, \gamma_\nu]\Psi$: a tensor.

---

[3] This is not exactly the expression of $S(\Lambda(A))$ given at the end of Chapter 3. Indeed, the formula was written for upper dotted spinors $\eta^{\dot\beta}$ and not for lower dotted spinors.

6.3.4 The Current

Multiplying the Dirac equation on the left by $\overline{\Psi}$ and the equation for the conjugate spinor on the right by Ψ and adding the two quantities, we get

$$0 = \overline{\Psi}(\not{p}\Psi) + (\overline{\Psi}\ \not{p})\Psi = \overline{\Psi}\gamma^{\mu}(p_{\mu}\Psi) + (p_{\mu}\overline{\Psi})\gamma^{\mu}\Psi, \tag{6.43}$$

that is,

$$p_{\mu}(\overline{\Psi}\gamma^{\mu}\Psi) = 0, \tag{6.44}$$

which shows that the current

$$j^{\mu} = \overline{\Psi}\gamma^{\mu}\Psi \tag{6.45}$$

is conserved. It is the Noether current corresponding to the invariance of the Dirac equation under global phase transformations of Ψ: $\Psi \rightarrow e^{i\alpha}\Psi$.

The charge density $j^{0} = \overline{\Psi}\gamma^{0}\Psi = \Psi^{*}\Psi$ is positive.

6.3.5 The Hamiltonian

The Dirac equation, which is linear in the time derivative, can be put in a Hamiltonian form. Multiplying the equation on the left by γ^{0}, gives (after the reintroduction of c and $\hbar$)

$$i\hbar\frac{\partial}{\partial t}\Psi = H\Psi, \tag{6.46}$$

with

$$H = c\boldsymbol{\alpha} \cdot (-i\hbar\boldsymbol{\nabla}) + \beta mc^{2} = c\boldsymbol{\alpha} \cdot \boldsymbol{p} + \beta mc^{2}, \tag{6.47}$$

where

$$\alpha^{i} = \gamma^{0}\gamma^{i} = \begin{pmatrix} -\sigma_{i} & 0 \\ 0 & \sigma_{i} \end{pmatrix} \qquad \beta = \gamma^{0} = \begin{pmatrix} 0 & 1 \\ 1 & 0 \end{pmatrix}. \tag{6.48}$$

6.3.6 The Standard Representation

The representation of the wave function as a bispinor (ξ, η) is called the spinorial representation. It is the most natural one from the mathematical point of view and it is the one we found when we first constructed the equation using only group theory considerations. We saw also that we can transform this representation by applying a unitary transformation $\Psi \to U\Psi$. Depending on the physical application we have in mind, some particular forms of these transformed spinors will turn out to be useful. We shall present here those which are most commonly used.

In order to discuss the non-relativistic limit we use the so-called *standard, or Dirac,* representation. It can be deduced from the spinorial representation by the following unitary transformation on ξ and η:

$$\phi = \frac{1}{\sqrt{2}}(\xi + \eta) \quad \kappa = \frac{1}{\sqrt{2}}(\xi - \eta). \tag{6.49}$$

The corresponding γ matrices have the form

$$\gamma^i = \begin{pmatrix} 0 & \sigma_i \\ -\sigma_i & 0 \end{pmatrix} \quad \gamma^0 = \begin{pmatrix} 1 & 0 \\ 0 & -1 \end{pmatrix} \tag{6.50}$$

and the matrices operating in the Hamiltonian are

$$\alpha^i = \gamma^0 \gamma^i = \begin{pmatrix} 0 & \sigma_i \\ \sigma_i & 0 \end{pmatrix} \quad \beta = \gamma^0 = \begin{pmatrix} 1 & 0 \\ 0 & -1 \end{pmatrix}. \tag{6.51}$$

In this representation, the parity transforms ϕ and κ into themselves, according to the formula

$$\phi \to i\phi \quad \kappa \to -i\kappa. \tag{6.52}$$

The γ^5 matrix is given by

$$\gamma^5 = \begin{pmatrix} 0 & 1 \\ 1 & 0 \end{pmatrix}, \tag{6.53}$$

an expression to be compared with the corresponding one in the spinorial representation in which γ^5 is diagonal:

$$\gamma^5 = \begin{pmatrix} 1 & 0 \\ 0 & -1 \end{pmatrix}. \tag{6.54}$$

The Dirac equation implies the following equations for ϕ and κ,

$$p_0 \phi - \boldsymbol{p} \cdot \boldsymbol{\sigma} \kappa = m\phi$$
$$p_0 \kappa - \boldsymbol{p} \cdot \boldsymbol{\sigma} \phi = -m\kappa; \tag{6.55}$$

thus,

$$i\frac{\partial}{\partial t}\phi = -i\boldsymbol{\sigma} \cdot \frac{\partial}{\partial \boldsymbol{x}}\kappa + m\phi$$

$$i\frac{\partial}{\partial t}\kappa = i\boldsymbol{\sigma} \cdot \frac{\partial}{\partial \boldsymbol{x}}\phi - m\kappa, \tag{6.56}$$

The energy being positive, we see that in the reference frame $\boldsymbol{p} = 0$, which we shall call *the rest frame* $\kappa = 0$, and the four-dimensional spinor Ψ reduces to the three-dimensional spinor ϕ. In the non-relativistic approximation, we will see that the norm of κ remains small compared to ϕ. For this reason, we call κ the *small* components of the four-dimensional spinor and ϕ the *large* components. We will see that thanks to this representation the Dirac equation has a non-relativistic limit: the Pauli equation.

6.3.7 The Spin

We know already that, by construction, the Dirac equation is associated with the representations $[m, \frac{1}{2}]$ of the Poincaré group, since the little group corresponds to $D^{\frac{1}{2}}$. Let us check it directly from the generators.

According to the general analysis done in the first section of this chapter, under the action of the Poincaré group, Ψ transforms as a unitary representation of $\{a, \Lambda\} \rightarrow U(a, \Lambda)$ which maps Ψ into Ψ' and as we saw

$$\Psi'(x) = S(\Lambda)\Psi(\Lambda^{-1}(x-a)). \tag{6.57}$$

Let us take $a \equiv 0$. Then an infinitesimal transformation $\Lambda = 1 + \lambda$ induces infinitesimal transformations $U(0, \Lambda) = 1 - \frac{i}{2}M_{\mu\nu}\lambda^{\mu\nu}$ and $S(\Lambda) = 1 - \frac{i}{2}T_{\mu\nu}\lambda^{\mu\nu}$. Let us find the value of $T_{\mu\nu}$. We know explicitly $S(\Lambda)$ in terms of A and we know that the A's are of two types, corresponding to either rotations and $A = \exp\{-i\theta\frac{\boldsymbol{\sigma}}{2} \cdot \boldsymbol{n}\}$ or pure Lorentz transformations and $A = \exp\{\xi\frac{\boldsymbol{\sigma}}{2} \cdot \boldsymbol{n}\}$. After some manipulations that we leave to be checked by the reader, we find that $T_{\mu\nu} = \frac{1}{2}\sigma_{\mu\nu}$.

Therefore, at first order

$$S(\Lambda)\Psi(\Lambda^{-1}x) = (1 - \frac{i}{4}\sigma_{\mu\nu}\lambda^{\mu\nu})\Psi(x - \lambda x) \tag{6.58}$$

$$= (1 - \frac{i}{4}\sigma_{\mu\nu}\lambda^{\mu\nu})(1 - \lambda^{\nu\mu}x_\mu\partial_\nu)\Psi$$

$$= (1 - \frac{i}{4}\sigma_{\mu\nu}\lambda^{\mu\nu} + \frac{1}{2}\lambda^{\mu\nu}(x_\mu\partial_\nu - x_\nu\partial_\mu))\Psi$$

and by identification the angular momentum tensor (the generators of the Lorentz group) is given by

$$M_{\mu\nu} = \frac{1}{2}\sigma_{\mu\nu} + (x_\mu (\mathrm{i}\partial_\nu) - x_\nu (\mathrm{i}\partial_\mu)). \tag{6.59}$$

The second term is identified with what we would call the orbital angular momentum. We see that a supplementary term has been added to it which, in the quantum version of the theory, will be interpreted as the intrinsic angular momentum of the particle, its *spin*. From the relativistic point of view it is a 2-tensor.

In terms of the little group generators, we have

$$W^\mu = \frac{1}{4}\varepsilon^{\mu\nu\varrho\sigma}\sigma_{\nu\varrho}P_\sigma, \tag{6.60}$$

because the contribution to the kinetic term is 0 by symmetry.

6.3.8   The Plane Wave Solutions

We are looking for the plane wave solutions of the Dirac equation. The interest of plane waves lies in their simplicity; we can then write a general solution as a superposition of plane waves.

We are interested in the plane waves of positive or negative energy. Let us note that

$$\Psi^+(x) = \mathrm{e}^{-ik.x}u(k)$$

is a positive energy solution and

$$\Psi^-(x) = \mathrm{e}^{ik.x}v(k)$$

is a negative energy solution. u and v are 4-component spinors u_r and v_r ($r = 1, \ldots, 4$), and k^0 is positive.

These solutions satisfy the Klein–Gordon equation; thus, $k^2 = m^2$. Therefore, $k_0 = \omega(k) \equiv \sqrt{k^2 + m^2}$. We will note in the sequel E_k the energy $\omega(k)$.

From $(\slashed{p} - m)\Psi^\pm = 0$, we get

$$(\slashed{k} - m)u = 0 \tag{6.61}$$
$$(\slashed{k} + m)v = 0.$$

Let us choose the standard representation. In the rest reference frame $k = (m, \mathbf{0})$, the equations become

$$(\gamma^0 - 1)u(m, \mathbf{0}) = 0 \tag{6.62}$$
$$(\gamma^0 + 1)v(m, \mathbf{0}) = 0;$$

thus, $u_2 = u_3 = v_1 = v_2 = 0$. A possible basis of the solutions is given by

$$\hat{u}^{(1)}(m,0) = \begin{pmatrix} 1 \\ 0 \\ 0 \\ 0 \end{pmatrix} \quad \hat{u}^{(2)}(m,0) = \begin{pmatrix} 0 \\ 1 \\ 0 \\ 0 \end{pmatrix} \quad \hat{v}^{(1)}(m,0) = \begin{pmatrix} 0 \\ 0 \\ 1 \\ 0 \end{pmatrix} \quad \hat{v}^{(2)}(m,0) = \begin{pmatrix} 0 \\ 0 \\ 0 \\ 1 \end{pmatrix}. \tag{6.63}$$

We have now to find the expressions of $u^{(\alpha)}$ and $v^{(\beta)}$ in an arbitrary reference frame. Let us consider, for example, the spinor u. We remark that in order for u to satisfy $(\not{k} - m)u = 0$, since

$$k^2 - m^2 = (\not{k} - m)(\not{k} + m) = 0, \tag{6.64}$$

it is enough to look for $u(k)$ of the form $u(k) = (\not{k} + m)f(k)\hat{u}(m,0)$ with $f(k) \to 1/\sqrt{2m}$ when $k \to (m,0)$. A convenient choice resulting from the normalisation conditions we shall use[4] is $1/f(k) = \sqrt{(m + E_k)}$. A similar analysis holds for v, so we find that

$$u^{(\alpha)}(k) = \frac{\not{k} + m}{\sqrt{(m + E_k)}} \hat{u}^{(\alpha)}(m,0) \tag{6.65}$$

$$v^{(\alpha)}(k) = \frac{-\not{k} + m}{\sqrt{(m + E_k)}} \hat{v}^{(\alpha)}(m,0).$$

we thus get

$$u^{(1)}(k) = \sqrt{E_k + m} \begin{pmatrix} 1 \\ 0 \\ \dfrac{\boldsymbol{\sigma} \cdot \boldsymbol{k}}{E_k + m} \begin{pmatrix} 1 \\ 0 \end{pmatrix} \end{pmatrix} \tag{6.66}$$

$$u^{(2)}(k) = \sqrt{E_k + m} \begin{pmatrix} 0 \\ 1 \\ \dfrac{\boldsymbol{\sigma} \cdot \boldsymbol{k}}{E_k + m} \begin{pmatrix} 0 \\ 1 \end{pmatrix} \end{pmatrix}$$

and similar expressions for the $v^{(\alpha)}$'s. They can be rewritten as

$$u^{(\alpha)}(k) = \sqrt{E_k + m} \begin{pmatrix} \phi^{(\alpha)}(m,0) \\ \dfrac{\boldsymbol{\sigma} \cdot \boldsymbol{k}}{E_k + m} \phi^{(\alpha)}(m,0) \end{pmatrix} \tag{6.67}$$

$$v^{(\alpha)}(k) = \sqrt{E_k + m} \begin{pmatrix} \dfrac{\boldsymbol{\sigma} \cdot \boldsymbol{k}}{E_k + m} \kappa^{(\alpha)}(m,0) \\ \kappa^{(\alpha)}(m,0) \end{pmatrix} \tag{6.68}$$

---

[4] The conventions regarding the normalisation of the spinor solutions are not standard and they are dictated by the physical applications we have in mind. In some treatments the choice $f(k) \to 1/2m$ is used, which makes the non-relativistic limit simpler. In present high-energy physics experiments, however, the zero-mass limit is more often appropriate and our normalisation makes it simpler. Naturally, the final results of physically measurable quantities are independent of normalisation conventions.

with an explicit use of the non-vanishing 2-component spinors $\phi^{(\alpha)}$ and $\kappa^{(\alpha)}$ of, respectively, $\hat{u}^{(\alpha)}(m, 0)$ and $\hat{v}^{(\alpha)}(m, 0)$. We get analogous expressions for the conjugate spinors

$$\bar{u}^{(\alpha)}(k) = \bar{\hat{u}}^{(\alpha)}(m, 0) \frac{\slashed{k} + m}{\sqrt{(m + E_k)}} \tag{6.69}$$

$$\bar{v}^{(\alpha)}(k) = \bar{\hat{v}}^{(\alpha)}(m, 0) \frac{-\slashed{k} + m}{\sqrt{(m + E_k)}},$$

where 'bar' denotes the Dirac conjugate spinors introduced in Eq.(6.39). We have the orthogonality relations

$$\bar{u}^{(\alpha)}(k) u^{(\beta)}(k) = 2m\delta^{\alpha\beta} \qquad \bar{u}^{(\alpha)}(k) v^{(\beta)}(k) = 0 \tag{6.70}$$

$$\bar{v}^{(\alpha)}(k) v^{(\beta)}(k) = -2m\delta^{\alpha\beta} \qquad \bar{v}^{(\alpha)}(k) u^{(\beta)}(k) = 0.$$

Let us consider the 'three-vector' matrix $\mathbf{\Sigma}$ of components $\Sigma^i = \frac{1}{2}\varepsilon_{ijk}\sigma^{jk}$

$$\Sigma^i = \begin{pmatrix} \sigma_i & 0 \\ 0 & \sigma_i \end{pmatrix}. \tag{6.71}$$

We check easily that $u^{(1)}(m, 0)$ and $v^{(1)}(m, 0)$ are eigenvectors of Σ^3 with eigenvalue $+1$, while $u^{(2)}(m, 0)$ and $v^{(2)}(m, 0)$ are eigenvectors with eigenvalue -1. More generally, we find that $\mathbf{\Sigma} \cdot \mathbf{n} = \sum_i \Sigma^i n^i$ has eigenvectors on a basis generated by the $u^{(\alpha)}(k)$'s and the $v^{(\alpha)}(k)$'s only if the vector $\mathbf{n}$ is parallel to $\mathbf{k}$. In this case $\mathbf{\Sigma} \cdot \hat{\mathbf{k}}$, where $\hat{\mathbf{k}} = \frac{\mathbf{k}}{|\mathbf{k}|}$, has for eigenvectors $u^{(1)}(k)$ and $v^{(1)}(k)$ with eigenvalue $+1$ and $u^{(2)}(k)$ and $v^{(2)}(k)$ with eigenvalue -1. The operator $\mathbf{\Sigma} \cdot \hat{\mathbf{k}}$ is the *helicity* operator. The matrices

$$Q_\pm(k) = \frac{1}{2}(1 \pm \mathbf{\Sigma} \cdot \hat{\mathbf{k}}) \tag{6.72}$$

are the projectors on the two helicity states.

To completely determine the eigenvectors we will build the projectors on the states of positive, respectively negative, energy. The matrices

$$P_+(k) = \frac{\slashed{k} + m}{2m} \tag{6.73}$$

$$P_-(k) = \frac{-\slashed{k} + m}{2m}$$

are the projectors respectively on u and v. We have

$$P_\pm^2(k) = P_\pm(k) \qquad P_+ + P_- = 1 \qquad \mathrm{Tr} P_\pm(k) = 2. \tag{6.74}$$

Let us compute the charge density $\varrho = j^0 = \overline{\Psi}\gamma^0\Psi$ for a plane wave of positive energy

$$\varrho = \overline{\Psi}^+\gamma^0\Psi^+ = \bar{u}\gamma^0 u \tag{6.75}$$
$$= \bar{u}\{\not{k}, \gamma^0\}u = 2E_k.$$

As expected, it is positive.

As we did with the Klein–Gordon equation, we can expand an arbitrary solution of the Dirac equation on the basis of plane waves. In analogy with Eq. (6.18) for a complex scalar field, we write

$$\Psi = \int \frac{\mathrm{d}^3 k}{(2\pi)^3}\frac{1}{2E_k}\sum_{\alpha=1}^{2}\left[a_\alpha(k)u^{(\alpha)}(k)\mathrm{e}^{-ikx} + b_\alpha^*(k)v^{(\alpha)}(k)\mathrm{e}^{ikx}\right] \tag{6.76}$$

$$\overline{\Psi} = \int \frac{\mathrm{d}^3 k}{(2\pi)^3}\frac{1}{2E_k}\sum_{\alpha=1}^{2}\left[a_\alpha^*(k)\bar{u}^{(\alpha)}(k)\mathrm{e}^{ikx} + b_\alpha(k)\bar{v}^{(\alpha)}(k)\mathrm{e}^{-ikx}\right], \tag{6.77}$$

where a_α and b_α, $\alpha = 1, 2$, are two pairs of arbitrary complex functions and $\star$ means 'complex conjugation'. With our normalisation conventions the integration measure in the expansions (6.76) and (6.77) is the invariant measure on the mass hyperboloid $\mathrm{d}\Omega_m$.

6.3.9 The Coupling with the Electromagnetic Field

The coupling with the electromagnetic field is the minimal coupling introduced in section 3.7. This leads us to replace ∂_μ by the *covariant derivative*

$$D_\mu = \partial_\mu + ieA_\mu. \tag{6.78}$$

The Dirac equation in the presence of an external electromagnetic field can therefore be rewritten as

$$(i\not{\partial} - e\not{A} - m)\Psi(x) = 0. \tag{6.79}$$

It is invariant under the local *gauge transformation* (i.e. depending on the point x)

$$\Psi(x) \quad \rightarrow \quad \Psi(x)\mathrm{e}^{ie\alpha(x)} \tag{6.80}$$
$$A_\mu(x) \quad \rightarrow \quad A_\mu(x) - \partial_\mu\alpha(x).$$

The conjugate spinor transforms as

$$\overline{\Psi}(x) \rightarrow \overline{\Psi}(x)\mathrm{e}^{-ie\alpha(x)}. \tag{6.81}$$

The conserved current has, as it can be easily shown, the same form as in the absence of the electromagnetic field, since, as we have noticed earlier, it is the Noether current of the invariance under the global transformations with constant phase.[5]

To obtain the Hamiltonian form, let us multiply the Dirac equation by the γ^0 matrix. This gives

$$i\frac{\partial}{\partial t}\Psi = [\boldsymbol{\alpha} \cdot (-i\boldsymbol{\nabla} - e\boldsymbol{A}) + \beta m + eA_0]\Psi \tag{6.82}$$

$$= [\boldsymbol{\alpha} \cdot (\boldsymbol{p} - e\boldsymbol{A}) + \beta m - eA_0]\Psi \tag{6.83}$$

$$= (\boldsymbol{\alpha} \cdot \boldsymbol{p} + \beta m)\Psi - e(\boldsymbol{\alpha} \cdot \boldsymbol{A} - A_0)\Psi \tag{6.84}$$

$$= (H_0 + H_{\text{int}})\Psi,$$

where H_0 is the non-interacting Hamiltonian, that is with $e = 0$.

Let us conclude with the operator relation

$$[D_\mu, D_\nu] = ieF_{\mu\nu}, \tag{6.85}$$

which will be generalised to more complicated gauge theories.

6.3.10   The Constants of Motion

Let us look for operators which commute with the Hamiltonian.[6] To start with, let us consider the Dirac Hamiltonian in the absence of an electromagnetic field.

Since $H_0 = c\boldsymbol{\alpha} \cdot \boldsymbol{p} + \beta mc^2$, $\boldsymbol{p}$ commutes with H_0 and is therefore a constant of the motion.

Consider $\boldsymbol{L} = \boldsymbol{r} \wedge \boldsymbol{p}$, the angular momentum operator. It commutes with β and

$$[H_0, \boldsymbol{L}] = [c\boldsymbol{\alpha} \cdot \boldsymbol{p}, \boldsymbol{L}] = -ic\boldsymbol{\alpha} \wedge \boldsymbol{p}. \tag{6.86}$$

Consider then the operator $\boldsymbol{\Sigma}$. We check easily that

$$[H_0, \boldsymbol{\Sigma}] = 2ic\boldsymbol{\alpha} \wedge \boldsymbol{p}; \tag{6.87}$$

consequently, since $\boldsymbol{p}$ commutes with H_0, the helicity $\boldsymbol{\Sigma} \cdot \boldsymbol{p}$ and the angular momentum

$$J = L + \frac{1}{2}\boldsymbol{\Sigma} \tag{6.88}$$

are constants of motion.

---

[5] This justifies the terminology we used earlier when we called the parameter e which appears in the phase transformation *charge*. It is the parameter which determines the strength of the coupling with the electromagnetic field.

[6] The terminology we are using is slightly misleading. Up to this point, the Dirac equation is just a differential equation for a classical spinor field. So, when we speak about operators which commute with the Hamiltonian, we really mean quantities whose Poisson brackets with the Hamiltonian vanish. For this we must consider the Dirac field as a canonical dynamical system. However, the notation, which we borrow from quantum mechanics, is more convenient and it will make the transition to the quantum system easier.

The study of the Dirac equation with an electromagnetic field will be useful for the sequel. In the case of a static, spherically symmetric field, we will have to consider the Hamiltonian H_0 to which is added a potential $V(r)$, that is $H = H_0 + V(r)$. Since $\boldsymbol{L}$ and all the more so $\boldsymbol{\Sigma}$, which does not depend on r, commute with V, $\boldsymbol{J}$ is still a constant of the motion. The helicity is not conserved. Intuitively, we expect to be able to specify whether the spin of the electron is parallel or anti-parallel to the total angular momentum. By an explicit calculation, we can see that the quantity which commutes with H is

$$K = \beta(\boldsymbol{\Sigma} \cdot \boldsymbol{L} + \hbar) = \beta \boldsymbol{\Sigma} \cdot \boldsymbol{J} - \beta \frac{\hbar}{2}. \tag{6.89}$$

We also check that

$$[\mathcal{J}, K] = 0. \tag{6.90}$$

The eigenstates of the Dirac Hamiltonian with a spherically symmetric potential can therefore be simultaneously the eigenstates of $H, K, \mathcal{J}^2$, and $\mathcal{J}_z$.

6.3.11 Lagrangian and Green Functions

The Lagrangian density which corresponds to the free Dirac equation is

$$\mathcal{L}_D = \frac{\mathrm{i}}{2}(\overline{\Psi}\gamma^\mu \partial_\mu \Psi - \partial_\mu \overline{\Psi}\gamma^\mu \Psi) - m\overline{\Psi}\Psi. \tag{6.91}$$

Indeed, we obtain the Dirac equation for Ψ and $\overline{\Psi}$ as a consequence of the stationarity of the action $S = \int \mathcal{L}_D \mathrm{d}^4 x$ under independent variation of $\overline{\Psi}$ and Ψ. Because of the linear dependence in Ψ or $\overline{\Psi}$, the action has neither a minimum nor a maximum. Thus, the overall sign of the action can be chosen at will. Provided that the field vanishes at infinity, we can rewrite the action as

$$S = \int \mathrm{d}^4 x \left(\overline{\Psi}(\mathrm{i}\,\slashed{\partial}\Psi) - m\overline{\Psi}\Psi \right). \tag{6.92}$$

We can easily extend the action (6.92) to describe the interaction between the complex spinor and an electromagnetic field A_μ. As usual, we use the principle of minimal coupling, that is, the coupling deduced from gauge invariance. This yields the action

$$S = \int \mathrm{d}^4 x \left(\overline{\Psi}((\mathrm{i}\,\slashed{\partial} - e\,\slashed{A})\Psi) - m\overline{\Psi}\Psi \right) = \int \mathrm{d}^4 x \, \mathcal{L}_D + \int \mathrm{d}^4 x \, \mathcal{L}_{\mathrm{int}}. \tag{6.93}$$

Here $\mathcal{L}_{\mathrm{int}} = -j^\mu A_\mu$ and $j^\mu = e\overline{\Psi}\gamma^\mu \Psi$.

The previous action assumes that A_μ is an external fixed field. In order to describe the full interacting system of the spinor and the electromagnetic field, we must add to the spinor Lagrangian the Maxwell Lagrangian. We thus obtain the combined Lagrangian density:

$$S = \int d^4x \left(\overline{\Psi}((i\,\slashed{\partial} - e\,\slashed{A})\Psi) - m\overline{\Psi}\Psi \right) - \frac{1}{4}F_{\mu\nu}F^{\mu\nu}. \tag{6.94}$$

By construction the whole action is gauge invariant and yields a system where both the spinor field and the electromagnetic field can exchange energy.

The Green functions of the Dirac equation are obtained following the same steps we used for the Klein–Gordon theory. We must find the inverse of the differential operator in the quadratic part of the Lagrangian (6.91). Therefore, the general expression for the Green functions is $S = (i\,\slashed{\partial} - m)^{-1}$, where the inverse is understood with respect to the differential operator, as well as the 4×4 matrix.

As usual, the expression simplifies in momentum space. Taking into account the fact that $(\slashed{k}+m)(\slashed{k}-m) = (k^2-m^2)$ we can write the general expression for the Green function as

$$\tilde{S} = \frac{\slashed{k}+m}{k^2 - m^2}. \tag{6.95}$$

Depending on which $i\epsilon$ prescription we choose for the singularities in the complex k^0 plane, we will obtain the retarded, the advanced, or the Feynman Green function.

As for the Klein–Gordon theory, in order to obtain the corresponding expressions in the configuration space, we must determine a convenient basis. For example, the retarded Green function can be written as

$$S_{ret}(x_2 - x_1) = \tag{6.96}$$
$$\theta(t_2 - t_1)\frac{1}{(2\pi)^3} \int d^3k \frac{1}{2E_k}[(\slashed{k} + m)e^{-i\mathbf{k}\cdot(\mathbf{x_2}-\mathbf{x_1})} + (\slashed{k} - m)e^{i\mathbf{k}\cdot(\mathbf{x_2}-\mathbf{x_1})}].$$

Here $k^0 = E_k$, and we have

$$(i\,\slashed{\partial} - m)S_{ret}(x - y) = i\delta^4(x - y). \tag{6.97}$$

6.4   Relativistic Equations for Vector Fields

Up to now we have obtained relativistic wave equations for scalar and spinor fields. In addition we have written Maxwell's equations for the electromagnetic field. In this section we want to write the general relativistically invariant wave equation for fields belonging to the $(\frac{1}{2}, \frac{1}{2})$ representation of the Lorentz group.

Let $A_\mu(x)$ denote a general vector field. In writing a differential equation we can use the differential operator ∂_μ, which has also a vector index, and the field itself. So the most general Lorentz invariant, linear, second-order differential equation for A_μ is of the form

$$\Box A_\mu(x) - \frac{1}{\alpha} \partial_\mu \partial_\nu A^\nu(x) + m^2 A_\mu(x) = 0. \tag{6.98}$$

It involves two arbitrary parameters, m^2, which has the dimensions of mass square, and α, which is dimensionless. Maxwell's equations in the vacuum are obtained for the particular choice $m^2 = 0$ and $\alpha = 1$. It is only for these values that the equation is gauge invariant. Indeed, the transformation $A_\mu(x) \rightarrow A_\mu(x) + \partial_\mu \theta(x)$ applied to (6.98) gives an extra term of the form $[(1 - \alpha^{-1})\Box + m^2]\partial_\mu\theta$, which, for arbitrary θ, vanishes only when $m^2 = 0$ and $\alpha = 1$.

Equation (6.98) can be obtained from a Lagrangian density:

$$\mathcal{L} = -\frac{1}{2}\left[\partial_\mu A^\nu \partial^\mu A_\nu - \frac{1}{\alpha}\partial_\mu A^\mu \partial_\nu A^\nu - m^2 A^\mu A_\mu\right]. \tag{6.99}$$

It is straightforward to obtain the corresponding Green functions. They are given by the inverse of the differential operator and they satisfy the equation

$$\left[k^2 g_{\mu\nu} - \frac{1}{\alpha}k_\mu k_\nu - m^2 g_{\mu\nu}\right]\tilde{G}^{\mu\rho} = g_\nu^\rho, \tag{6.100}$$

which gives

$$\tilde{G}^{\mu\rho}(k) = \frac{1}{k^2 - m^2}\left[g^{\mu\rho} - \frac{k^\mu k^\rho}{k^2(1-\alpha) + \alpha m^2}\right] \tag{6.101}$$

with, always, the appropriate $i\epsilon$ prescription.

As expected, the limit of (6.101) for $m^2 = 0$ and $\alpha = 1$ does not exist. The reason is gauge invariance. As we have seen, for this choice of the values of the parameters the differential operator in the wave equation satisfies the condition $\mathcal{D}_{\mu\nu}\partial^\mu\theta = 0$, which means that $\mathcal{D}$ has a zero mode and, consequently, it has no inverse.

It is instructive to study the general equation (6.98) for particular values of the parameters:

(i) For $m^2 = 0$ we obtain the equation for the electromagnetic field in the family of gauges parametrised by α which we introduced in section 3.8.2. $\alpha \rightarrow \infty$ gives the Feynman gauge, in which the Green function reduces to four independent components: $\tilde{G}^{\mu\rho}(k) = g^{\mu\rho}/k^2$. $\alpha = 0$ gives the Landau gauge in which the Green function is transverse in four dimensions: $k_\mu \tilde{G}^{\mu\rho}(k) = 0$.

(ii) For $m^2 \neq 0$ the equation describes a theory which, as we shall show later, is appropriate for describing the fields of massive particles with spin equal to 1. Of particular interest is the choice $\alpha = 1$ for which the equation is known as *the Proca equation*. We shall use it extensively in this book:

$$\Box A_\mu(x) - \partial_\mu \partial_\nu A^\nu(x) + m^2 A_\mu(x) = 0. \tag{6.102}$$

By taking the partial derivative ∂^μ of this equation we obtain

$$m^2 \partial^\mu A_\mu(x) = 0, \qquad (6.103)$$

which, for $m^2 \neq 0$, implies a condition among the four components of the vector field. Two remarks are in order here: first, for the Proca equation with the term proportional to m^2, this condition is a consequence of the equation of motion, in contradistinction with what happens in the $m = 0$ case in which this, or any similar, condition had to be imposed by hand. Second, note that this condition remains valid even if we introduce a source term $j_\mu(x)$ on the right-hand side of Eq. (6.102), provided the source is a conserved current. It follows that out of the four components of the field, only three are independent dynamical variables. We shall come back to these points in more detail in Chapter 12.

Up to the $i\epsilon$ prescription, the Green function of the Proca equation takes the simple form

$$\tilde{G}^{\mu\rho}(k) = \frac{1}{k^2 - m^2}\left[g^{\mu\rho} - \frac{k^\mu k^\rho}{m^2}\right]. \qquad (6.104)$$

As we did for the Dirac equation, we can find plane wave solutions which we can use as basis to expand any other solution.

We start by defining four basic vectors $\epsilon_\mu^{(\lambda)}(k)$, with λ running from 0 to 3. We can use any system of linearly independent vectors and a particularly convenient choice is the following: we choose $\epsilon_\mu^{(0)}(k)$ to be a unit vector in the time direction with $\epsilon_\mu^{(0)}(k)\epsilon^{(0)\mu}(k) = 1$ and $\epsilon_0^{(0)}(k) > 0$. Then we choose a three-vector $\boldsymbol{k}$ and choose the z-axis along its direction. For the $m = 0$ case the natural choice is the direction of propagation. $\epsilon_\mu^{(3)}(k)$ will be a unit vector in this direction. The other two vectors $\epsilon_\mu^{(1)}(k)$ and $\epsilon_\mu^{(2)}(k)$ are chosen in the plane perpendicular to that formed by the other two and orthogonal to each other. In our particular reference frame they take the simple form

$$\epsilon^{(0)} = \begin{pmatrix} 1 \\ 0 \\ 0 \\ 0 \end{pmatrix} \qquad \epsilon^{(1)} = \begin{pmatrix} 0 \\ 1 \\ 0 \\ 0 \end{pmatrix} \qquad \epsilon^{(2)} = \begin{pmatrix} 0 \\ 0 \\ 1 \\ 0 \end{pmatrix} \qquad \epsilon^{(3)} = \begin{pmatrix} 0 \\ 0 \\ 0 \\ 1 \end{pmatrix}. \qquad (6.105)$$

The four ϵ vectors are called *polarisation vectors* and, when $\boldsymbol{k}$ is chosen along the direction of propagation, we can call $\epsilon^{(3)}$ longitudinal, $\epsilon^{(1)}$ and $\epsilon^{(2)}$ transverse, and $\epsilon^{(0)}$ scalar. In an arbitrary reference frame they satisfy the orthonormality relations

$$\sum_{\lambda=0}^{3} \frac{\epsilon_\mu^{(\lambda)}(k)\epsilon_\nu^{(\lambda)*}(k)}{\epsilon^{(\lambda)\rho}(k)\epsilon_\rho^{(\lambda)*}(k)} = g_{\mu\nu} \qquad \epsilon^{(\lambda)\rho}(k)\epsilon_\rho^{(\lambda')*}(k) = g^{\lambda\lambda'}, \qquad (6.106)$$

where, again, $*$ means 'complex conjugation'.

With the help of these unit vectors, an arbitrary solution of the wave equation for a real vector field can be expanded in plane waves. For the simple case of $m = 0$ in a linear gauge, the expansion reads

$$A_\mu(x) = \int \frac{\mathrm{d}^3 k}{(2\pi)^3 2\omega_k} \sum_{\lambda=0}^{3} \left[a^{(\lambda)}(k)\epsilon_\mu^{(\lambda)}(k)\mathrm{e}^{-ikx} + a^{(\lambda)*}(k)\epsilon_\mu^{(\lambda)*}(k)\mathrm{e}^{ikx} \right] \qquad (6.107)$$

with $a^{(\lambda)}(k)$ four complex functions. Here $\omega_k = k^0$. As we shall see in a later chapter, depending on the choice of gauge, the number of independent polarisation vectors can be reduced. Note the Minkowski metric $g^{\lambda\lambda'}$ in the second of the relations (6.106), which is necessary to reproduce the correct transformation properties of the field A_μ.

We can obtain a similar expansion for the solutions of the Proca equation. The four-dimensional transversality condition (6.103) can be used to eliminate the polarisation in the zero direction. For the general $m \neq 0$ equation, the expansion involves all four polarisation vectors.

We can obtain relativistically invariant equations for fields belonging to higher representations of the Lorentz group, but we shall not use any of them in this book.

7

Towards a Relativistic Quantum Mechanics

7.1 Introduction

In the previous chapter we derived relativistically invariant wave equations for fields of low spin. Although the derivation was purely mathematical, these equations have important physical applications. The obvious case is Maxwell's equation which accurately describes the dynamics of a classical electromagnetic field. In this chapter we will study the possibility of a particular application, namely that of a relativistic extension of Schrödinger's equation, the equation for the quantum mechanical wave function of a particle. Indeed, we may wonder why Schrödinger, who wrote his equation in 1926, twenty-one years after the discovery of special relativity, choose to write a non-relativistic equation for the electron. In fact, historically, it seems that he first tried to find a relativistic equation and wrote the equivalent of the Klein–Gordon equation (the degrees of freedom associated with the spin of the electron were not used). In this chapter we want to see how far we can go in this direction using everything we know today about the representations of the Lorentz group.

7.2 The Klein–Gordon Equation

We start with the simplest wave equation, the Klein–Gordon equation. From our study of the Lorentz group, we know that it should correspond to the wave function of a spinless particle. The wave function in quantum mechanics is complex valued, so we write the equation for a complex field (6.16).

Remember that the free Schrödinger equation can be obtained from the correspondence principle by substituting to the non-relativistic relation $p_0 = \frac{\boldsymbol{p}^2}{2m}$, the operator $i\hbar \frac{\partial}{\partial t}$ for the energy p_0, and the operator $-i\hbar \frac{\partial}{\partial \boldsymbol{x}}$ for the momentum $\boldsymbol{p}$. We see similarly that the Klein–Gordon equation, which was introduced from the relativistic invariance criterion,

From Classical to Quantum Fields. Laurent Baulieu, John Iliopoulos and Roland Sénéor.
© Laurent Baulieu, John Iliopoulos and Roland Sénéor, 2017. Published 2017 by Oxford University Press.

can also be deduced by the same relativistic invariance criterion from the Einstein relation $p_0^2 = m^2c^4 + c^2\boldsymbol{p}^2$. In this derivation we see that m does correspond to the mass of the particle.

In the non-relativistic limit, $c \to +\infty$, the Klein–Gordon equation gives back the Schrödinger equation. To see this property, it is enough to extract the mass from the energy dependence of the field. This is obtained by parameterising Φ:

$$\Phi(t, \boldsymbol{x}) = \exp\left(-\frac{imc^2}{\hbar}t\right)\Psi(t, \boldsymbol{x}). \tag{7.1}$$

The Klein–Gordon equation gives then

$$\left(\frac{1}{c^2}\left(\frac{\partial}{\partial t}\right)^2 - \frac{i}{\hbar}2m\frac{\partial}{\partial t} - \left(\frac{\partial}{\partial \boldsymbol{x}}\right)^2\right)\Psi(x) = 0. \tag{7.2}$$

In the non-relativistic approximation, $c \to +\infty$, the first term can be neglected, and we recover the free Schrödinger equation:

$$i\hbar\frac{\partial}{\partial t}\Psi(x) = -\frac{\hbar^2}{2m}\triangle\Psi(x). \tag{7.3}$$

So, at first sight, the Klein–Gordon equation seems to have the right properties to give the relativistic generalisation of the Schrödinger equation. We suspect that this could not be right because, if it were that simple, Schrödinger would have written directly this more general relativistic equation. Let us show that, indeed, this interpretation does lead to physical incoherences.

Without loss of generality we can assume that m is real. The field satisfies Eq. (6.1). The complex conjugate field Φ^* satisfies the complex conjugate equation. If Φ is to be considered as a wave function, there must exist a conserved probability current, as for the Schrödinger equation. In the previous chapter we have constructed this current. We saw that the two equations derive from a variational principle applied to the real action (6.16). The invariance of this action under phase transformations of Φ implies the conservation of the current (6.17). Let us separate the zero component and the three space components:

$$\varrho = \frac{ie\hbar}{2m}\left(\Phi^*\left(\frac{\partial}{\partial t}\Phi\right) - \left(\frac{\partial}{\partial t}\Phi^*\right)\Phi\right)$$

$$\boldsymbol{j} = -\frac{ie\hbar}{2m}\left(\Phi^*(\nabla\Phi) - (\nabla\Phi^*)\Phi\right). \tag{7.4}$$

We see that ϱ satisfies the conservation equation which would make it possible to interpret it as a probability density. This interpretation is, however, impossible since the expression (7.4) is not positive definite. Indeed, the Klein–Gordon equation has solutions of the form $e^{\frac{iEt}{\hbar}}u(\boldsymbol{x})$ for all $E^2 \geq m^2c^4$. It follows that whatever the sign of

E, ϱ can be negative and therefore cannot be interpreted as a density. Technically, the problem arises from the fact that the energy dependence of the Klein–Gordon equation is quadratic. This is the clue that gave to Dirac the idea of his equation.

A second problem, as fundamental as the first one and linked to the probabilistic interpretation of the wave function and independent from the fact that we consider a real field or a complex field, is that the energy spectrum p_0 is not bounded from below, since $p_0 = \pm c\sqrt{p^2 + m^2 c^2}$. Thus, it seems that the system is capable of furnishing spontaneously to the environment, under the effect of small perturbations, arbitrarily large quantities of energy. We conclude that a complex field $\Phi(x)$, which satisfies the Klein–Gordon equation, cannot be interpreted as a wave function of a relativistic particle.

7.3   The Dirac Equation

Historically, the Dirac equation was derived as early as 1928, as a relativistic generalisation of the Schrödinger equation for a spinning electron. It was a most remarkable achievement because, at that time, the spinorial representations of the Lorentz group were not known. Dirac looked for a first-order differential equation in order to avoid the problem of the non-positivity of the probability density which plagued the Klein–Gordon equation. He wrote his famous equation with the aim of describing a relativistic electron, that is, a particle with mass m, electric charge $e = -|e|$, and spin $\frac{1}{2}$, which satisfies Pauli's exclusion principle. Seen with today's knowledge, almost a century later, his formulation of the problem was rather primitive. For instance, the Pauli principle was imposed by hand. Today we know that it is not 'a principle' but a consequence of a theorem that asserts that particles with half-integer spin must satisfy odd statistics, the so-called Fermi–Dirac statistics. This theorem relies on some basic physical principles, namely that the electron is described in the context of a local quantum theory which is Lorentz invariant and unitary.

Dirac met his first objective: compared to the Klein–Gordon equation, his equation has an enormous advantage, namely it has a conserved current with positive-definite density. We have obtained this result in the previous chapter when we derived the conserved current (6.45). Therefore, the first objection to the interpretation of the solution as a probability amplitude is no more valid and we could be tempted to consider the Dirac equation as a relativistic equation for the quantum mechanical wave function of a spin-$\frac{1}{2}$ particle. We shall see, however, that if we want to follow consistently this road, we will be led unambiguously to the formulation of a relativistic quantum field theory.

7.3.1   The Non-relativistic Limit of the Dirac Equation

The first step is to check that we obtain the correct non-relativistic limit. We can do it for the case of an electron interacting with an external electromagnetic field.

The energy of a relativistic particle of mass m includes its rest energy mc^2; we can eliminate it by setting

$$\psi = \Psi \, e^{imc^2 t}, \tag{7.5}$$

which satisfies the equation (we have reintroduced the dependence on $\hbar$ and c)

$$\left(i\hbar \frac{\partial}{\partial t} + mc^2 \right) \psi = \left[c\boldsymbol{\alpha} \cdot \left(-i\boldsymbol{\nabla} - \frac{e}{c}\mathbf{A} \right) + \beta mc^2 + eA_0 \right] \psi. \tag{7.6}$$

In the standard representation where $\psi = \begin{pmatrix} \phi \\ \kappa \end{pmatrix}$, this equation can be written in terms of the components

$$\left(i\hbar \frac{\partial}{\partial t} - eA^0 \right) \phi = c\boldsymbol{\sigma} \cdot \left(\boldsymbol{p} - \frac{e}{c}A \right) \kappa \tag{7.7}$$

$$\left(i\hbar \frac{\partial}{\partial t} - eA^0 + 2mc^2 \right) \kappa = c\boldsymbol{\sigma} \cdot \left(\boldsymbol{p} - \frac{e}{c}A \right) \phi. \tag{7.8}$$

Let us consider the non-relativistic approximation where all the energies are small with respect to the rest energy. This means, for example, that in the second equation, the leading term on the left-hand side is $2mc^2$ and therefore that

$$\kappa = \frac{1}{2mc} \boldsymbol{\sigma} \cdot \left(\boldsymbol{p} - \frac{e}{c}A \right) \phi. \tag{7.9}$$

Substituting this quantity in the first equation gives

$$\left(i\hbar \frac{\partial}{\partial t} - eA^0 \right) \phi = \frac{1}{2m} \left(\boldsymbol{\sigma} \cdot \left(\boldsymbol{p} - \frac{e}{c}A \right) \right)^2 \phi. \tag{7.10}$$

Using the fact that

$$\boldsymbol{\sigma} \cdot \boldsymbol{a}\, \boldsymbol{\sigma} \cdot \boldsymbol{b} = \boldsymbol{a} \cdot \boldsymbol{b} + i\boldsymbol{\sigma} \cdot \boldsymbol{a} \wedge \boldsymbol{b} \tag{7.11}$$

we get

$$\left(\boldsymbol{\sigma} \cdot \left(\boldsymbol{p} - \frac{e}{c}A \right) \right)^2 = \left(\boldsymbol{p} - \frac{e}{c}A \right)^2 + i\boldsymbol{\sigma} \cdot \left(\boldsymbol{p} - \frac{e}{c}A \right) \wedge \left(\boldsymbol{p} - \frac{e}{c}A \right) \tag{7.12}$$

and since $\boldsymbol{p}$ and A do not commute, the last term is non-zero and is equal to

$$-\frac{e\hbar}{c} \boldsymbol{\sigma} \cdot \left(\boldsymbol{\nabla} \wedge A + A \wedge \boldsymbol{\nabla} \right) = -\frac{e\hbar}{c} \boldsymbol{\sigma} \cdot \mathrm{rot} A = -\frac{e\hbar}{c} \boldsymbol{\sigma} \cdot \boldsymbol{B}. \tag{7.13}$$

We thus obtain the Pauli equation

$$i\hbar \frac{\partial}{\partial t}\phi = \left[\frac{1}{2m}\left(p - \frac{e}{c}A\right)^2 + eA_0 - \frac{e\hbar}{2mc}\sigma \cdot B \right]\phi. \tag{7.14}$$

This equation is different from the Schrödinger equation by the last term which is a term representing the coupling between the electromagnetic field and a magnetic moment μ given by

$$\mu = \frac{e\hbar}{2mc}\sigma = \frac{g\mu_B S}{\hbar},$$

where $\mu_B = 9.27 \times 10^{-24}$J/T, $S = \frac{\hbar}{2}\sigma$ is the spin $\frac{1}{2}$, and g is the gyromagnetic ratio.

The Dirac equation is therefore the theory of a massive particle of charge e and spin $\frac{1}{2}$ and predicts its gyromagnetic ratio[1] to be equal to 2.

Let us consider for simplicity a stationary state of the initial equation of energy E. The next order of approximation, under the hypothesis in which the energy E is close to the rest energy $(E - mc^2)/mc^2 \ll 1$ and $|eA^0| \ll mc^2$, can be obtained by writing that

$$(E - eA_0 + 2mc^2)\kappa = c\,\sigma \cdot \left(p - \frac{e}{c}A\right)\phi. \tag{7.15}$$

At first order in $1/c$ this equation gave

$$\kappa = \frac{1}{2mc}\sigma \cdot \left(p - \frac{e}{c}A\right)\phi. \tag{7.16}$$

To second order it gives

$$\kappa = \frac{1}{2mc}\left(1 - \frac{E + eA_0}{2mc^2}\right)\sigma \cdot \left(p - \frac{e}{c}A\right)\phi. \tag{7.17}$$

We find at this order the appearance of a spin–orbit coupling term $L.S$, where L is the operator of orbital angular momentum.

7.3.2 Charge Conjugation

If, following Dirac, one interprets his equation as that giving the wave function of a free particle, the first task is to search for the spectrum of the Hamiltonian. It yields a continuous energy spectrum $\pm E_k$, with

---

[1] We find experimentally that $g = 2,00233184$. This value of the 'anomalous' magnetic moment $\frac{g-2}{2}$ is known up to 10^{-7}. The theoretical computations predict $g = 2[1 + (\frac{e^2}{4\pi\hbar c})\frac{1}{2\pi} + \ldots]$. They come from corrections given by quantum electrodynamics (the quantum field version of the Dirac equation) and coincide with the same precision to this numerical value.

$$(-\infty, -mc^2] \cup [mc^2, +\infty).$$

This spectrum remains of the same type if one introduces in the Dirac equation a potential that vanishes at spatial infinity. Moreover, if one considers that this electron is bound to an atom, it must interact with the electromagnetic field of the nucleus. Whichever theory is used, the electron will be able to lower its energy by photon emission. Since its spectrum is not bounded from below, there is no limit for such a process, and we reach the conclusion that an atom cannot be stable. To bypass this contradiction, Dirac proposed in 1930 that under normal conditions all states with negative energy are occupied, which of course forbids the unbounded fall of the energy of the electron. Indeed, because of the Pauli principle, no electron with positive energy can 'fall' in a state of negative energy, which is already occupied, and this explains the observed stability of all atoms. This hypothetical infinite set of negative energy electrons has been called the *Dirac sea* and determines the ground state of the Dirac theory. It can be interpreted as a Fermi gas with infinite density. All observables of the full system (energy, electric charge, etc.) must take into account all possible interactions between the Dirac sea and the electrons with positive energy. In Dirac's point of view, the energy of an electron must be computed as the energy of the system [one electron+Dirac sea] minus the energy of the system [zero electron+Dirac sea]. In the Dirac theory, when a photon with energy $\hbar\omega > 2mc^2$ is absorbed by an electron of the sea, this electron can become an electron with positive energy and a hole is created in the sea. The electron and the hole become an observable system. The energy of the hole is positive, its electric charge is opposite that of an electron, and its mass is identical to that of the electron. Dirac thought for a while that he could identify the holes with the protons, but it was quickly understood that, were it the case, the hydrogen atom would decay into two photons with an unrealistic lifetime of $\sim 10^{-10}$ s. The great triumph of the Dirac theory, apart from predicting the value $g = 2$ for the electron gyromagnetic ratio, came with the experimental discovery of the positron by Anderson in 1932. This discovery was the cornerstone in the establishment of the existence of *antiparticles*. After that, instead of being a mathematical artefact, a hole was to be understood as a particle with a positive energy, and all puzzles could be resolved by understanding that the Dirac equation describes in a unified way either a positron or an electron, each of them having positive energy larger than the rest mass m. For instance, if we put together the spectrum of the electron and that of the positron, which run between m and ∞, we formally get a bijection with the above-mentioned unphysical spectrum that ranges from $-\infty$ and ∞, with no states between $-m$ and $+m$. In its early interpretation, Dirac was invoking the following mechanism for the absorption of a photon by a hole:

$$e^-_{E \ll 0} + \gamma \rightarrow e^-_{E > 0}. \tag{7.18}$$

In the interpretation with particles and antiparticles, this process is naturally interpreted as a decay process of a photon in an electron–positron pair:

$$\gamma \rightarrow e^-_{E > 0} + e^+_{E > 0}. \tag{7.19}$$

Such a process can occur if we send a light ray in the Coulomb field of an atom or a molecule (it would be forbidden for an isolated photon because of energy–momentum conservation, which necessitates an external source that brings momentum).

We see that although the Dirac theory was originally constructed for describing a single charged particle, in a straightforward generalisation of the Schödinger picture, a thorough analysis of this equation implies the introduction of another particle, called an antiparticle. Eventually, it implies a multi-particle interpretation, due to the possibility of creating pairs of electrons and positrons, provided there is energy and momentum conservation. This non-trivial generalisation of the idea of a wave equation came as a great shock. This new paradigm in which the number of particles is not conserved was originally called second quantisation. Nowadays, it is just called quantum field theory, and it became very clear with the introduction of the so-called *path integral*, which we will describe later on. Nevertheless, although the traditional one-particle interpretation of quantum mechanics is not consistent for a relativistic theory, the Dirac equation can still be used as an approximation in such a picture, provided the energies are low enough so that the effects of electron–positron pair creation can be neglected.

The apparent asymmetry between electrons and holes completely disappears if we reinterpret all phenomena in terms of electrons and positrons. We will explain this by establishing the existence of a symmetry of the Dirac equation, called charge conjugation.

To understand the multiparticle interpretation of the theory, we must couple the electrons to the electromagnetic field A. In order that positrons play a role symmetric to that of electrons, the former should satisfy the same Dirac equation as the latter, but with an opposite electric charge, $-e$. Thus, we look for a transformation $\Psi \to \Psi^c$ such that

$$(\mathrm{i}\,\slashed{\partial} - e\,\slashed{A} - m)\Psi = 0 \tag{7.20}$$

$$(\mathrm{i}\,\slashed{\partial} + e\,\slashed{A} - m)\Psi^c = 0.$$

In order to satisfy the basic principles of quantum mechanics, this transformation must be local and involutive, up to a phase. In this way, the norms of Ψ and Ψ^c, which measure the probability of presence of the particle, will be the same.

Assuming that the spinor Ψ is a solution of the Dirac equation, its conjugate $\overline{\Psi}(-\mathrm{i}\,\slashed{\partial} - e\,\slashed{A} - m) = 0$ satisfies

$$\left(-(\gamma^\mu)^{tr}(\mathrm{i}\partial_\mu - eA_\mu) - m\right)\gamma^0\Psi^* = 0. \tag{7.21}$$

Suppose now the existence of a matrix C such that

$$C(\gamma^\mu)^{tr}C^{-1} = -\gamma^\mu. \tag{7.22}$$

Then

$$\Psi^c \equiv C\overline{\Psi}^{tr} \tag{7.23}$$

is a solution of the Dirac equation with an opposite electric charge, $-e$. Using the standard representation, we get the following solution for C:

$$C = i\gamma^2\gamma^0 = \begin{pmatrix} 0 & -i\sigma_2 \\ -i\sigma_2 & 0 \end{pmatrix}. \tag{7.24}$$

C satisfies

$$-C = C^{-1} = C^{tr} = C^+. \tag{7.25}$$

Note that for the spinor field that represents an electron at rest, we have

$$\Psi = e^{imt} \begin{pmatrix} 0 \\ 0 \\ 0 \\ 1 \end{pmatrix} \rightarrow \Psi^c = e^{-imt} \begin{pmatrix} 1 \\ 0 \\ 0 \\ 0 \end{pmatrix}. \tag{7.26}$$

We see that what Dirac called an electron with negative energy and spin projection $-\frac{1}{2}$ is transformed by charge conjugation into an electron with positive energy, positive electric charge, and spin projection $+\frac{1}{2}$. This is identified to the positron. Note that the matrix C is anti-unitary. However, if we want to interpret C as an operator in the Hilbert space of one-particle states, electrons or positrons, it will be a unitary operator, because it involves also the operation of complex conjugation of the wave function. We will show this property explicitly in Chapter 12.

7.3.3 PCT Symmetry

We have seen that two symmetries exist for the Dirac equation, the spatial parity and the charge conjugation. We want to investigate here the consequences of time reversal I_t. We can easily check that $\gamma^0\gamma^5\Psi^c(I_tx)$, which is an anti-unitary operation, also satisfies the Dirac equation.

Consider now the product of the three symmetries:

$$\begin{aligned} P &: \ \Psi(x) \rightarrow i\gamma^0\Psi(I_sx) \\ C &: \ \Psi(x) \rightarrow \Psi^c(x) = C\gamma^0\Psi^*(x) \\ T &: \ \Psi(x) \rightarrow \gamma^0\gamma^5\Psi^c(I_tx). \end{aligned} \tag{7.27}$$

We find that the product PCT of the three operators taken in any order is an invariant of the theory

$$\begin{aligned} PCT\Psi(x) &= i\gamma^0(C\gamma^0(\gamma^0\gamma^5 C\gamma^0\Psi^*(I_sI_tx))^*) \\ &= i\gamma^0(C\gamma^0(\gamma^0\gamma^5 C^{-1}\gamma^0\Psi(-x))) \\ &= i\gamma^5\Psi(-x). \end{aligned} \tag{7.28}$$

Let us ask the following question: given a wave function $\Psi(x)$ which satisfies the Dirac equation in an external electromagnetic field $A_\mu(x)$, can we find the equation satisfied by its PCT-transformed wave function $i\gamma^5\Psi(-x)$? The answer is simple;

$$
\begin{aligned}
0 &= [\mathrm{i}\,\not\partial_x - e\,\not\!A(x) - m]\,\Psi(x) \\
&= \mathrm{i}\gamma^5\,[\mathrm{i}\,\not\partial_x - e\,\not\!A(x) - m]\,\Psi(x) \\
&= [-\mathrm{i}\,\not\partial_x + e\,\not\!A(x) - m]\,\mathrm{i}\gamma^5\Psi(x) \\
&= [\mathrm{i}\,\not\partial_x + e\,\not\!A(-x) - m]\,PCT\Psi(x),
\end{aligned}
\tag{7.29}
$$

where, in order to obtain the last equation, we have performed a change of variable $x \to -x$.

This equation shows that the PCT-transformed wave function $i\gamma^5\Psi(-x)$ satisfies the same Dirac equation *provided* at the same time we replace the external electromagnetic field $A_\mu(x)$ by $PCTA_\mu(x) = -A_\mu(-x)$. This is not surprising. We saw already that the operation of charge conjugation relates an electron to a positron whose coupling to the electromagnetic field has the opposite sign.

At this level the invariance under PCT looks like a definition. It is the consequence of our assumption that the electromagnetic potential changes sign under this transformation. We shall see in Chapter 12 that, in fact, this is much more general. It is possible to prove, using only general principles based on locality and Lorentz invariance, that every relativistically invariant quantum theory with local interactions is invariant under the product, taken at any order, of three appropriately defined operators: P for space inversion, T for time reversal, and C for particle-antiparticle conjugation.

7.3.4   The Massless Case

When the mass m vanishes, the equations for the dotted and undotted spinors in (6.20) are decoupled. The spinors ξ and η independently satisfy the Dirac equation. This can be also directly shown at the level of the Dirac equation with 4-spinor notation, independently of the chosen gamma matrix representation.

Indeed, by left multiplication of the equation $\not p\Psi = 0$ by $\gamma^5\gamma^0$, we get

$$
\gamma^5 P^0 \Psi = \boldsymbol{\Sigma} \cdot \boldsymbol{P}\,\Psi.
\tag{7.30}
$$

Thus, γ^5 and the helicity operator have the same eigenvectors. γ^5 is suggestively called the *chirality operator*. Note that since γ^5 anticommutes with $\not p$, if Ψ satisfies the massless Dirac equation, so does $\gamma^5\Psi$.

It is customary to introduce $P_L = (1 - \gamma^5)/2$ and $P_R = (1 + \gamma^5)/2$, the projectors on states of chirality $+1$ and -1. We also define

$$
\Psi_L = P_L\Psi \qquad \Psi_R = P_R\Psi.
\tag{7.31}
$$

Here, the indices L and R stand for 'left' and 'right'. The notation is borrowed from optics, where it denotes the two states of polarisation of light.

The equation $\not{p}\Psi_L = 0$ is nothing else but the first equation in (6.20), and the analogous equation involving Ψ_R is the second one.

Particles with spin equal to $\frac{1}{2}$ and mass equal to 0 do not seem to exist in nature. However, there exist particles, the *neutrinos*, which have masses so tiny that, for them, the zero-mass approximation is very often adequate. The $m \rightarrow 0$ limit of the Dirac theory should apply to them. They should be described as Dirac spinors with four components, provided their two possible states of polarisation exist, and we can pass from one polarisation to the other by a space parity.

But in 1957, Maurice Goldhaber and co-workers showed in a very beautiful experiment that the neutrinos produced in nuclear beta-decay, the only ones known at the time, were only left-polarised (helicity -1). Anti-neutrinos were right-polarised.

The neutrino wave function must therefore be represented by a complex spinor Ψ_L. In the spinor representation, only two components of Ψ_L are non-zero. Then the Dirac equation of a neutrino can be reduced to a 2×2 matrix equation:

$$(i\partial_t - i\boldsymbol{\sigma} \cdot \boldsymbol{\nabla})\Psi = 0. \tag{7.32}$$

This equation was originally written by Hermann Weyl as early as 1929. It was discarded because it lacks parity invariance. Obviously, it was resurrected in 1957 after the great discovery of T. D. Lee and C. N. Yang (theory) and C. S. Wu (experiment) that the weak β decay, $n \rightarrow p + e^- + \bar{\nu}$, violates the conservation of parity (here $\bar{\nu}$ stands for an anti-neutrino).

The description of the neutrino by a two-component spinor only makes sense if its mass is strictly zero. Otherwise, a Lorentz transformation would transform a state with positive helicity into a state with negative helicity. Thus, when one attempts to put experimental limits on the smallness of the mass of a neutrino, we must use a four-component Dirac equation.

The charge conjugation associates with the neutrino its antiparticle, called the *anti-neutrino*. In the model for elementary particle physics which we shall present in Chapter 25, known as *the standard model*, the antineutrinos are different particles from the neutrinos, although this may change radically in the near future. In contrast, the photon is its own anti-particle. Note that if neutrinos and anti-neutrinos turn out to be identical, we would have to use a different equation, the *Majorana equation*, to describe them. We shall present it briefly in the next section. The experimental situation will be briefly discussed in Chapter 25. In the standard model neutrinos and anti-neutrinos have opposite polarisations. The charge conjugation and the space-parity are not invariances of the Weyl equation, but the product CP is. As a consequence of the PCT theorem, the time reversal T is thus also a symmetry; this can be shown also directly by inspection of the Lagrangian or of its equations of motion.

It is often convenient to use the projectors P_L and P_R, even for massive spinors. Indeed, this decomposition projects any given spinor Ψ on eigenvectors of the chirality operator which are interchanged by space parity. Using the relations

$$\overline{\Psi}_L = (P_L \Psi)^\dagger \gamma^0 = \Psi^\dagger P_L \gamma^0 = \overline{\Psi} P_R \tag{7.33}$$

and

$$\overline{\Psi}_R = \overline{\Psi} P_L, \tag{7.34}$$

we get

$$\overline{\Psi} \gamma^\mu \Psi = \overline{\Psi}_L \gamma^\mu \Psi_L + \overline{\Psi}_R \gamma^\mu \Psi_R \tag{7.35}$$

$$\overline{\Psi} \Psi = \overline{\Psi}_L \Psi_R + \overline{\Psi}_R \Psi_L; \quad \overline{\Psi}_L \Psi_L = \overline{\Psi}_R \Psi_R = 0. \tag{7.36}$$

We see that the vector current is the sum of two currents that separately correspond to states with distinct helicities. On the other hand, the term $\overline{\Psi} \Psi$ couples states with different helicities.

7.3.5   Weyl and Majorana Spinors

In the spinorial representation a four-dimensional spinor is formed out of the two two-dimensional spinors ξ and η. They transform according to the $(0, \frac{1}{2})$ and $(\frac{1}{2}, 0)$ representation of $SU(2) \times SU(2)$, respectively. Each one has two complex components, so a Dirac spinor depends on four complex variables. Note that if ξ transforms as a member of $(0, \frac{1}{2})$, its complex conjugate ξ^* transforms as a member of $(\frac{1}{2}, 0)$. It was Ettore Majorana who first realised that one can use this property in order to build a four-component spinor with half as many variables as a Dirac spinor. It is sufficient to choose in the Dirac equation (6.20) $\eta \sim \xi^*$. Equation (6.20) still contains a mass parameter but in the four-dimensional spinor (6.26) the two lower components are given by the complex conjugates of the two upper ones:

$$\Psi_M = \begin{pmatrix} \xi \\ \xi^* \end{pmatrix}. \tag{7.37}$$

Therefore, we can find a unitary matrix U_M to make all four components of Ψ_M real. Ψ_M is called *a Majorana spinor* and (7.37) shows that it is equivalent to a single Weyl spinor. Formally, a Majorana spinor still satisfies a Dirac-type equation, but it cannot be interpreted as the equation for the wave function of a particle since the components of Ψ_M are real valued. In particular, there is no $U(1)$ phase symmetry and, hence, no conserved current. It is easy to verify that the transformation of charge conjugation acts trivially on a Majorana spinor which implies that if a particle is described by such an equation, it must be identical with its own anti-particle.

As we shall see, Majorana spinors will turn out to be very useful later on, so we ask, as an exercise, to compute the unitary matrix U_M, which relates a Majorana and a Weyl spinor, as well as the associated γ matrices.

7.3.6 Hydrogenoid Systems

The extremely precise analysis of the hydrogen spectrum played an essential role in the development of quantum mechanics in the first half of the previous century. A major success of the Dirac equation was the accurate determination of the relativistic corrections to the Schrödinger theory. This came in two steps, first at the level of the one-particle approximation and, second, at that of the complete picture, where one takes into account the effects of particle–anti-particle creation and annihilation.

7.3.6.1 Hydrogenoid Atoms in Non-relativistic Quantum Mechanics

The states of an electron bound to a nucleus with Z protons are described by the Schrödinger equation

$$\left(-\frac{\hbar^2}{2m} \Delta - \frac{Ze^2}{4\pi r} - E \right) \psi(r) = 0, \tag{7.38}$$

where m is the reduced mass

$$m = \frac{m_e m_N}{m_e + m_N} \simeq m_e, \tag{7.39}$$

m_e stands for the electron mass, and m_N is the nucleus mass. Using the rotational symmetry of the Coulomb potential, we can parameterise the solutions in terms of the eigenfunctions of the angular momentum operator $\psi_{l,m}(r) = Y_l^m(\theta, \phi) \Phi(r)$. Then, the Schrödinger equation reads

$$\left(-\frac{\hbar^2}{2m} \Delta - \frac{Ze^2}{4\pi r} - E_{n,l} \right) \psi_{n,l,m}(r) = 0, \tag{7.40}$$

where $\psi_{n,l,m}$ satisfies

$$\begin{aligned} L^2 \psi_{n,l,m} &= l(l+1) \psi_{n,l,m} & l &= 0, 1, 2, \ldots \\ L_z \psi_{n,l,m} &= m \psi_{n,l,m} & |m| &\leq l, \end{aligned} \tag{7.41}$$

and the Laplacian is

$$-\Delta = -\frac{\partial^2}{\partial r^2} - \frac{2}{r} \frac{\partial}{\partial r} + \frac{L^2}{r^2}. \tag{7.42}$$

The radial Schrödinger equation is

$$\left(-\frac{\hbar^2}{2m}\left(-\frac{\partial^2}{\partial r^2} - \frac{2}{r}\frac{\partial}{\partial r} + \frac{l(l+1)}{r^2} \right) - \frac{Ze^2}{4\pi r} - E \right)\Phi = 0. \tag{7.43}$$

Its normalisable solutions are

$$\Phi(r) = r^l e^{-\frac{\sqrt{2mE}}{\hbar}r} P_{n'}(r), \tag{7.44}$$

where $P_{n'}$ is a polynomial with degree n',

$$E = -\frac{mc^2(Z\alpha)^2}{2(n'+l+1)^2}, \tag{7.45}$$

and α is the fine structure constant

$$\alpha = \frac{e^2}{4\pi\hbar c} \sim \frac{1}{137}. \tag{7.46}$$

By defining the quantum number $n = n' + l + 1$ we now see that the solutions of the Schrödinger equations are functions $\psi_{n,l,m}$ that are the eigenfunctions of L^2 and L_z with eigenvalues

$$E_n = -\frac{mc^2(Z\alpha)^2}{2n^2} \qquad n = 1, 2, \cdots. \tag{7.47}$$

n and l are such that $n' = n - (l+1)$ cannot be negative. Formula (7.47) indicates that the energy levels that are independent of l are degenerate. For a given n, we have

$$\sum_{l=0}^{n-1}(2l+1) = n^2 \tag{7.48}$$

degenerate levels.

7.3.6.2 *Hydrogenoid Atoms in Relativistic Quantum Mechanics*

In order to conveniently solve the Dirac equation in the radial case, we multiply the left-hand side of

$$(\not{p} - e\not{A} - m)\psi = 0 \tag{7.49}$$

by $\not{p} - e \not{A} + m$, which gives

$$(\not{p} - e \not{A} + m)(\not{p} - e \not{A} - m)\psi$$
$$= [(\not{p} - e \not{A})^2 - m^2]\psi$$
$$= \left[(i\partial - eA)^2 + \frac{1}{4}[\gamma^\mu, \gamma^\nu][i\partial_\mu - eA_\mu, i\partial_\nu - eA_\nu] - m^2 \right]\psi$$
$$= \left[(i\partial - eA)^2 - \frac{e}{2}\sigma^{\mu\nu}F_{\mu\nu} - m^2 \right]\psi, \tag{7.50}$$

since $A = 0$ and $A_0 = -\frac{Ze}{4\pi r}$, $F_{0i} = -F_{i0} = -\partial_i A_0 = E^i$. We use the spinorial representation

$$\sigma^{\mu\nu}F_{\mu\nu} = \begin{pmatrix} 2i\sigma \cdot \boldsymbol{E} & 0 \\ 0 & -2i\sigma \cdot \boldsymbol{E} \end{pmatrix} \tag{7.51}$$

$$2i\sigma \cdot \boldsymbol{E} = -iZ\alpha \frac{\sigma \cdot \hat{r}}{r^2}, \tag{7.52}$$

where $\hat{r}$ is the unit vector along r. Since the other terms are diagonal, we naturally write the solution as

$$\psi = \begin{pmatrix} \psi_+ \\ \psi_- \end{pmatrix}, \tag{7.53}$$

where ψ_+ and ψ_- stand for two-component spinors. The Dirac equation now reduces to a pair of spinorial equations for the stationary solutions with energy E

$$\left[-\left(\frac{\partial^2}{\partial r^2} + \frac{2}{r}\frac{\partial}{\partial r} \right) + (L^2 - Z^2\alpha^2 \mp iZ\alpha\sigma \cdot \hat{r})\frac{1}{r^2} - \frac{2Z\alpha E}{r} - (E^2 - m^2) \right]\psi_\pm = 0. \tag{7.54}$$

The idea is to compare this equation with that of the non-relativistic case by simply writing the second term in parentheses as an operator analogous to L^2, the eigenvalues of which are parameterised as $\lambda(\lambda + 1)$. Now we use the fact that J commutes with H, the Dirac Hamiltonian in the Coulomb field potential, as well as with L^2. Thus at a given eigenvalue j, $j = \frac{1}{2}, \frac{3}{2}, \cdots$, of J such that $\mathcal{J}^2 = j(j + 1)I$, the integer eigenvalues of L^2, $l(l + 1)$, must be written as $l_\pm(l_\pm + 1)$ and satisfy

$$l_+ = j + \frac{1}{2}$$
$$l_- = j - \frac{1}{2}.$$

For each one of the equations in (7.54), we can expand the solutions on the two-dimensional basis of solutions given by $|l_+ >$ and $|l_- >$:

$$L^2|l_\pm >= l_\pm(l_\pm + 1)|l_\pm > . \tag{7.55}$$

L^2 is a diagonal operator in this basis. The operator $\sigma \cdot \hat{r}$ is Hermitian and traceless, and its square is the unit operator **1**. Moreover, we have

$$< l_{\pm} | \sigma \cdot \hat{r} | l_{\pm} > = 0, \tag{7.56}$$

since $\sigma . \hat{r}$ is odd under space parity, $r \to -r$. Thus,

$$\sigma \cdot \hat{r} = \begin{pmatrix} 0 & \tau \\ \bar{\tau} & 0 \end{pmatrix},$$

where τ is a phase. We finally get that $L^2 - Z^2\alpha^2 \mp iZ\alpha\sigma \cdot \hat{r}$ can be represented in this basis by a 2×2 matrix

$$\begin{pmatrix} L^2 - Z^2\alpha^2 & \mp\tau iZ\alpha \\ \mp\bar{\tau}iZ\alpha & L^2 - Z^2\alpha^2 \end{pmatrix},$$

which gives on the $\binom{l_+}{l_-}$ basis

$$\begin{pmatrix} \left(j + \frac{1}{2}\right)\left(j + \frac{3}{2}\right) - Z^2\alpha^2 & \mp\tau iZ\alpha \\ \mp\bar{\tau}iZ\alpha & \left(j + \frac{1}{2}\right)\left(j - \frac{1}{2}\right) - Z^2\alpha^2 \end{pmatrix}.$$

This matrix has the eigenvalues

$$\left(\frac{j + \frac{1}{2}}{2}\right)^2 - Z^2\alpha^2 \pm \sqrt{\left(\frac{j + \frac{1}{2}}{2}\right)^2 - Z^2\alpha^2}. \tag{7.57}$$

They can be written under the form $\lambda(\lambda + 1)$ provided that

$$\lambda = \sqrt{\left(\frac{j + \frac{1}{2}}{2}\right)^2 - Z^2\alpha^2} \quad \text{and} \quad \lambda = \sqrt{\left(\frac{j + \frac{1}{2}}{2}\right)^2 - Z^2\alpha^2} - 1. \tag{7.58}$$

We have, thus, reduced the problem to that of finding the eigenstates of

$$\left(-\frac{\hbar^2}{2m}\left(-\frac{\partial^2}{\partial r^2} - \frac{2}{r}\frac{\partial}{\partial r} + \frac{\lambda(\lambda + 1)}{r^2}\right) - \frac{2Z\alpha E}{r} - (E^2 - m^2)\right)\Phi = 0. \tag{7.59}$$

This operator is formally the same as that of the non-relativistic case, (7.43) with the substitutions

$$l(l+1) \rightarrow \lambda(\lambda+1)$$
$$\alpha \rightarrow \frac{\alpha E}{m}$$
$$E \rightarrow \frac{E^2 - m^2}{2m}.$$
(7.60)

Let us define a new orbital quantum number λ by

$$\lambda = \left(j \pm \frac{1}{2}\right) - \delta_j,$$
(7.61)

where

$$\delta_j = j + \frac{1}{2} - \sqrt{\left(\frac{j+\frac{1}{2}}{2}\right)^2 - Z^2\alpha^2} \simeq \frac{Z^2\alpha^2}{2j+1} + O(Z^4\alpha^4).$$
(7.62)

In order to get the condition of normalisability of the solutions, we shift the energy quantum number n by δ_j, in such a way that $(n-\delta_j) - \lambda - 1$ is a non-negative integer. In turn, $j \leq n - \frac{3}{2}$ for $\lambda = j + \frac{1}{2} - \delta_j$ and $j \leq n - \frac{1}{2}$ for $\lambda = j + \frac{1}{2} - \delta_j$ implies that for a given n, we have a double degeneracy of each level for $j \leq n - \frac{3}{2}$ and no degeneracy for $j = n - \frac{1}{2}$. The substitutions indicate immediately that the new energy levels are such that[2]

$$\frac{E_{nj}^2 - m^2}{2m} = -\frac{mZ^2\alpha^2}{2} \frac{E_{nj}^2}{m^2} \frac{1}{(n-\delta_j)^2}.$$
(7.63)

This gives finally

$$E_{nj} = \frac{m}{\sqrt{1 + \left(\frac{Z^2\alpha^2}{n-\delta_j}\right)^2}} \simeq m - \frac{mZ^2\alpha^2}{2n^2} - \frac{mZ^2\alpha^2}{n^3(2j+1)} + \frac{3}{8}\frac{mZ^4\alpha^4}{n^4} + O(\alpha^4)$$
(7.64)

for $n = 1, 2, \cdots$ and $j = \frac{1}{2}, \frac{3}{2}, \cdots, n - \frac{1}{2}$.

The difference between the non-relativistic and the relativistic cases is the occurrence of a *fine structure* that lifts the degeneracy of the j dependent levels, for a given value of n. Using the atomic physics notations, and for $Z = 1$, we have

---

[2] Obviously, in discussing Eq. (7.62) we have assumed that $Z\alpha < 1$, which is true for all known stable atoms. It is, however, of some interest, even if only academic, to see what happens when Z becomes very large. Formally Eqs. (7.62) and (7.63) imply that, in this case, the energy of the state becomes complex. We know in the theory of non-relativistic scattering that considering complex values of the energy is a formal way to take into account the effects of unstable states such as resonances. Later, in section 20.4, we will study this phenomenon in quantum field theory. Here we only note that nuclei with very large Z tend to be unstable because of the Coulomb repulsion among protons, a phenomenon unrelated to the one we discuss here. However, there are speculations in nuclear physics regarding the possible existence of a new stability region for nuclei with very large Z and A. If such nuclei can be produced, it would be interesting to study the corresponding highly ionised atoms and see the possible effects of the instability given by Eq. (7.62).

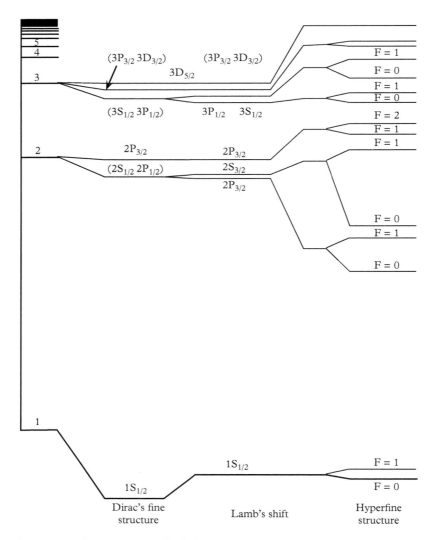

Figure 7.1 *The spectrum of the hydrogen atom.*

$$E(2P_{\frac{3}{2}}) - E(2P_{\frac{1}{2}}) =\simeq \frac{m\alpha^4}{32} = 4,53 \times 10^{-5} eV \qquad (7.65)$$

for $n = 2$, $P \to l = 1$, and $j = \frac{3}{2}$ or $\frac{1}{2}$.

We can refine the relativistic correction by including the effect of the interaction between the spin of the nucleus and the spin of the electron. This means adding to the Hamiltonian a term of the form

$$H = -\frac{e}{2m_e}\boldsymbol{\sigma}_e.\boldsymbol{B}. \tag{7.66}$$

Since $\boldsymbol{B}$ is proportional to the magnetic moment of the proton $\mu_p \simeq g_p\frac{e}{2m_p}$, we get an additional contribution which lifts further the degeneracy of the fundamental level, the $1S_{\frac{1}{2}}$:

$$\Delta E = \frac{4}{3}m_e\alpha^4\frac{m_e}{m_p}g_p = 5,89 \times 10^{-6}\ eV. \tag{7.67}$$

This new effect is called the *hyperfine structure*. The new correction is proportional to the mass ratio $\frac{m_e}{m_p}$, which makes it smaller by an order of magnitude than the previous effect. Figure 7.1 indicates the values of the correction. Figure 7.1 also indicates another correction, the *Lamb shift*, due to the quantum field effect, which cannot be computed within the framework of first quantisation (we left this computation as an exercise for the next chapters).

7.4 Problems

Problem 7.1 Show for bilinear forms the following equalities under a proper group transformation $L_+ = L_+^\uparrow \cup L_-^\downarrow$, $x'^\mu = \Lambda^\mu_{\ \nu}x^\nu$:

$$\begin{aligned}
\overline{\psi}'(x')\psi'(x') &= \overline{\psi}(x)\psi(x) & \text{scalar} \\
\overline{\psi}'(x')\gamma^\mu\psi'(x') &= \Lambda^\mu_{\ \nu}\overline{\psi}(x)\gamma^\nu\psi(x) & \text{vector} \\
\overline{\psi}'(x')\sigma^{\mu\nu}\psi'(x') &= \Lambda^\mu_{\ \alpha}\Lambda^\mu_{\ \beta}\overline{\psi}(x)\sigma^{\alpha\beta}\psi(x) & \text{antisymmetric tensor} \\
\overline{\psi}'(x')\gamma_5\gamma^\mu\psi'(x') &= \det\Lambda\ \Lambda^\mu_{\ \nu}\overline{\psi}(x)\gamma_5\gamma^\nu\psi(x) & \text{pseudovector} \\
\overline{\psi}'(x')\gamma_5\psi'(x') &= \det\Lambda\ \overline{\psi}(x)\gamma_5\psi(x) & \text{pseudoscalar.}
\end{aligned}$$

$$\tag{7.68}$$

Problem 7.2 Compute the transformation properties of the bilinear expressions of Problem 7.1 under the transformations of time reversal T and charge conjugation C.

Problem 7.3 Prove that the infinitesimal generator $T_{\mu\nu}$ of the representation of the Lorentz group acting on the Dirac equation is given by the formula

$$T_{\mu\nu} = \frac{1}{2}\sigma_{\mu\nu}.$$

Problem 7.4 Check that $\Sigma.\boldsymbol{p}$ and $\mathcal{J} = L + \frac{1}{2}\Sigma$ are constants of motion of the free Dirac equation.

Compute $[H, \mathcal{J}]$ when H is the Dirac Hamiltonian with an electromagnetic field.

Problem 7.5 The current.

Consider a relativistic particle of charge e and mass m under the action of a magnetic field. Its dynamics is described by the Dirac equation. This problem has for purpose to compare in some case the Dirac equation to the Schrödinger equation.

(1) Show that the current density $j^\mu(x) = e\bar\Psi(x)\gamma^\mu\Psi(x)$ is conserved.

(2) Show that $j^\mu = j^\mu_{(1)} + j^\mu_{(2)}$, where

$$j^\mu_{(1)} = \frac{e}{2m}\left(\partial^\mu\bar\Psi(x)\Psi(x) - \bar\Psi(x)\partial^\mu\Psi(x)\right) - \frac{e^2}{2m}A^\mu(x)\bar\Psi(x)\Psi(x) \quad (7.69)$$

$$j^\mu_{(2)} = -\frac{e}{2m}\partial^\nu\left(\bar\Psi(x)\sigma^{\nu\mu}\Psi(x)\right). \quad (7.70)$$

(3) Assume that all the energies are small with respect to the rest energy of the particle. Give the non-relativistic limit of the two currents $j_{(1)}$ and $j_{(2)}$. Interpret the results (one can use the fact that, classically, a magnetic dipole M generates an effective density of current $J_{\text{eff}} = \text{rot}M$).

(4) Express the part of the electromagnetic coupling $\int j_{(2)} \cdot A\,\mathrm{d}^4x$ in terms of the electromagnetic tensor $F_{\mu\nu}$. Give the non-relativistic limit.

Problem 7.6 The Zitterbewegung.

We will set in the Heisenberg representation and therefore suppose that the operators are time dependent. If Q is an operator in the Schrödinger representation, we remember that $Q_H = \exp(itH)Q\exp(-itH)$ and the Schrödinger representation is equivalent to the Heisenberg equations of motion;

$$\frac{\mathrm{d}Q_H}{\mathrm{d}t} = i[H, Q_H] + \frac{\partial Q_H}{\partial t},$$

the last term only exists if Q is explicitly time dependent. The operator H is the Dirac Hamiltonian.

We will not in the following give an index H to the operators which are in the Heisenberg representation.

(1) Let $\pi^\mu = p^\mu - eA^\mu$ the conjugate momentum. Show that

$$\frac{\mathrm{d}\pi}{\mathrm{d}t} = e(E + \alpha \wedge B). \quad (7.71)$$

(2) Give an interpretation to α in the preceding equation. One can compute the mean value of this operator on the plane waves of positive energy:

$$< \alpha >= \int \Psi^{(+)\dagger}(x)\alpha\Psi^{(+)}(x)\mathrm{d}^3x. \tag{7.72}$$

(3) Compute

$$\frac{\mathrm{d}x^k}{\mathrm{d}t} = \mathrm{i}[H, x^k]. \tag{7.73}$$

Is it possible to simultaneously measure all the components of the speed? Let us suppose now that there is no electromagnetic field.

(4) Write the evolution equation satisfied by α and show using the fact that p and H are constants of motion that

$$\alpha(t) = \frac{p}{H} + \left(\alpha(0) - \frac{p}{H}\right)\mathrm{e}^{-2\mathrm{i}Ht} \tag{7.74}$$

is the solution of this equation.

(5) Deduce from it that $x(t)$ is made of two parts, one classical and the other rapidly oscillating. Comment?

Problem 7.7 The Landau levels.
Solve the Dirac equation for an electron moving in a space with a constant magnetic field $B = (0, 0, B_z)$.

(1) Find the energy levels, the relativistic version of the Landau levels we studied in non-relativistic quantum mechanics.
(2) Find the degree of degeneracy for each one and interpret this degeneracy in terms of the symmetries of the problem.
(3) Show that the two components of the velocity of the electron v_x and v_y do not commute.

We can go one step further and show that a free electron in a space with non-commutative geometry reproduces the Landau results.

Hint: The method is similar to that we followed in solving the same problem using the Schrödinger equation. We choose a gauge in which the magnetic field is given by a vector potential of the form $A = (0, B_z y, 0)$ and reformulate the problem in terms of an equation for a harmonic oscillator.

8

Functional Integrals and Probabilistic Amplitudes

8.1 Introduction

The functional integral is one of the most remarkable inventions of contemporary theoretical physics. The formal and practical simplifications that it has generated have had important consequences in many discoveries, particularly in the theory of elementary particles and in statistical physics. More recently, this tool was proven to be interesting in the reformulation of various problems encountered in pure mathematics.

Historically, functional integrals, or *path integrals*, were introduced in the 1920s by N. Wiener to solve some diffusion and Brownian motion problems. They were rediscovered in the 1940s by R. Feynman in his approach to reformulate first quantum mechanics and then quantum electrodynamics. Today, the path integral is considered to be one of the rigorous methods possible from a classical theory expressed by a Lagrangian to define the corresponding quantum theory.

In comparison with the standard method of operator quantisation, the functional integral quantisation shows many advantages, especially for constrained systems with an infinite number of degrees of freedom like field theories. It is also one of the most powerful tools for studying the preservation, after quantisation, of the symmetries of the classical Lagrangian.

The main physical applications are related to the description of the interaction among elementary particles by means of quantum field theories as well as that of many problems in statistical physics such as superfluidity, superconductivity, and plasma physics.

It is important to distinguish between the basic ideas of functional integration and the technical difficulties we may encounter in manipulating objects having an infinite number of degrees of freedom. Many among these technical difficulties are out of the scope of this book. The main problems related to the infinite number of degrees of freedom will be treated only in the special case of perturbation theory. For most physically interesting theories a non-perturbative rigorous approach is still missing, since, in depth, there are unsolved technical problems. In this chapter and the next one we will establish the foundations of the formalism of the functional integral with an emphasis on

From Classical to Quantum Fields. Laurent Baulieu, John Iliopoulos and Roland Sénéor.
© Laurent Baulieu, John Iliopoulos and Roland Sénéor, 2017. Published 2017 by Oxford University Press.

the applications to non-relativistic quantum mechanics. Then in a subsequent chapter we will show how we can extend the definition of the functional integral to relativistic theories and to include fermions.

Before deriving the basic results, we will show how simple arguments based on the principles of quantum mechanics lead naturally to the idea of path integrals. We will mainly follow the original approach of R. P. Feynman.[1]

8.2   Brief Historical Comments

The idea of describing a physical phenomenon through the concept of summing over its possible past histories is quite old and, although it has been applied to various problems, its precise history has not yet been written. The correct mathematical tools were developed in probability theory during the first half of the twentieth century, but the less rigorous applications in physics are much older. In this short notice we do not pretend to cover the field and we shall only mention some of the essential steps which led to Feynman's path integral formulation of quantum mechanics.

In this book we often derived Lagrange's equations by using the principle of least action. The calculus of variations was developed to solve problems of this kind and the ideas go back to the eighteenth century. A very instructive example is the connection between *physical* and *geometrical* optics.

A simple model to illustrate Huygens' principle of physical optics is given by a generalisation of the massless Klein–Gordon equation with a variable index of refraction,

$$\nabla^2 \phi - \frac{n^2}{c^2} \frac{\partial^2 \phi}{\partial t^2} = 0, \tag{8.1}$$

where c is the speed of light in a vacuum and n the index of refraction, i.e. the ratio of c to the speed of light in the medium. n may be a function of x and t. Equation (8.1) describes the propagation of wave fronts, thus exemplifying Huygens' principle. It is easy to show (see Problem 8.1) that in the *eikonal approximation*, i.e. the approximation in which the wave length of the propagated wave is short compared to the distance over which n changes appreciably, Eq. (8.1) is equivalent to the equation of geometrical optics; in other words, Huygens' principle gives Fermat's principle in the eikonal approximation. The same result can be obtained by starting from the wave equations of classical electrodynamics. We can understand why the two pre-Maxwellian descriptions of light propagation, Newton's light particles and Huygens' waves, could both describe reflection and refraction phenomena.

This connection was known already to Hamilton as far back as 1834. He realised that the Hamilton–Jacobi formulation of classical mechanics could be considered

---

[1] The arguments presented in the first sections of this chapter follow very closely Feynman's original approach: R. P. Feynman, *Rev. Mod. Phys.* **20**, 367 (1948); *see also* R. P. Feynman and A. R. Hibbs, *Quantum Mechanics and Path Integrals*, (Mc Graw-Hill, New York, 1965).

as the eikonal approximation of some kind of wave mechanics. Today we know that Schrödinger's wave equation has precisely this property; see Problem 8.2. There is no evidence that Hamilton ever attempted to write this wave equation and, in fact, he had no reason to do so. In the early nineteenth century there was absolutely no indication that a wave propagation should be associated with the motion of a classical point particle.

It seems that this kind of connection was first formulated by Peter Debye in discussions he had with Arnold Sommerfeld in 1910. Although we do not know whether de Broglie was aware of the Debye–Sommerfeld discussion, his approach makes use of the analogy between Fermat's principle and Maupertuis' formulation of the least action principle. Schrödinger was certainly aware of these developments and the story goes that he attempted to write the wave equation which bears his name trying to answer a question by Debye. The idea was to give a precise meaning to de Broglie's wave–particle duality conjecture. After struggling for a while with the problem of the negative energy solutions of a relativistic equation, presumably the Klein–Gordon equation, Schrödinger proposed his equation as its non-relativistic limit. The correct classical, or $\hbar \to 0$, limit was an important consideration in his approach and, as we show in Problem 8.2, this is equivalent to the eikonal approximation. There is a beautiful paper, published by Schrödinger in 1926, in which he summarises his approach to quantum mechanics.[2]

In Heisenberg's matrix formulation of quantum mechanics the classical limit is obtained by replacing matrix commutators by Poisson brackets. The canonical transformation which relates Schrödinger's with Heisenberg's pictures was found by Dirac at the end of 1926. He was the first to understand fully the classical limit, so it is not surprising that he was also the first to propose the path integral in order to exemplify the eikonal approximation. Dirac published a short paper (eight pages) on this subject in a Soviet journal in 1933[3] (he later called it 'the little paper'), in which he writes the expression we know today of the path integral and argues that summing over all paths is *analogous* to the quantum theory. In Dirac's formulation this was only an analogy, not a complete theory. A historical puzzle, not easy to solve given Dirac's rather introvert character,[4] is why he stopped there and did not proceed to the next logical step, namely the complete formulation of the quantum theory as the sum over classical trajectories. This was done by Feynman in 1948 and it is the subject of this and the following chapters. Feynman relates a discussion he had with Dirac in 1946, before he had completed his own formulation, in which he tried to find out what Dirac meant with the term analogous. Feynman suggested the correct term should be *proportional*. Dirac: 'Are they?' Feynman: 'Yes.' Dirac: 'This is very interesting.' End of the conversation!

The difference is not trivial. As we shall see in this and the following chapters, it is only the ratio of two path integrals, with and without external sources, which can be defined, because the infinite constant of proportionality cancels.[5] Coming back to

---

[2] E. Schrödinger, *Physical Review* **28**, 1049 (1926).
[3] P.A.M. Dirac, *Physicalische Zeitschrift der Sowjetunion* **3**, 64 (1933).
[4] For a recent Dirac biography see: Graham Farmello, *The Strangest Man* (Faber and Faber, 2009)
[5] According to Freeman Dyson, Feynman said later, 'I don't know what all the fuss is about. Dirac did it all before me'; see Graham Farmello, cited earlier.

Dirac, it is quite possible that he was discouraged by the difficulties encountered when one attempts to give a rigorous mathematical definition of the quantum mechanical path integral. Two points should be mentioned here: (i) in the early years the connection with the formulation in Euclidean space, which, as we show in Chapter 10, is mathematically better defined, had not been fully understood, and (ii) Dirac had always been reluctant to accept prescriptions, such as the theory of renormalisation, as valid ways to define a physical theory. It is conceivable that such considerations played a role in his final negative attitude towards quantum field theory in general and the path integral formulation in particular, subjects on whose formulation he played a fundamental role and which, to a large extent, were initiated by him.

8.3   The Physical Approach

The principles of non-relativistic quantum mechanics can be elucidated by the diffraction experiments of non-luminous bodies. We will describe an experiment which was initially a *gedanken* experiment, but which is now a real experiment thanks to the progress made on the control of sources of slow electrons and the ability to detect the impacts of isolated electrons. This experiment makes it possible to understand the physical origin of the description of the quantum reality in terms of functional integrals.

Let us consider a source of slow, non-relativistic electrons emitting, inside a cone, wave packets spaced out in time with a nearly uniform angular distribution. The trajectory of each wave packet is assimilated to a straight line. The source is supposed to be 'monochromatic'; i.e. the electron speeds are approximatively the same.

If we put in front of the source a sensor plate, we will see after some time the appearance of a disk-like uniform density of impacts. Let us pierce now the plate with two small holes and let us put behind them a second sensor plate. What will be seen on the second plate?

Reasoning according to the principles of classical mechanics, the answer is obvious. Either an electron pulse is directed towards one of the holes, goes through it, and reaches the second plate or the electron pulse crashes into the first plate. After some time, two spots must appear on the second plate, concentrated around the intersection with the plate of the straight lines joining the source to the holes. If the rhythm of emission is sufficiently slow, it is in principle possible to imagine an experimental device able to identify for each impact the hole which the electron went through. If we close up one of the holes, after some time the figure drawn on the second plate is a unique spot. This reasoning is in principle independent of the size of the holes, provided we take into account the probability for an electron to be deflected by the border of the hole. For a hole of small but finite size, this probability is expected to be very small since it is a border effect.

Reality is completely different from what was described earlier if the size of the hole is on the order of $\lambda = \hbar/m_e c$, the Compton wave length of the electrons. If we observe the plate, we see the slow appearance of impacts, apparently randomly and often far away from the position predicted by classical reasoning. If we wait long enough, we will see

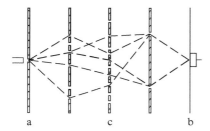

Figure 8.1 *The possible paths.*

a   c   b

the formation of an interference pattern, with fringes comparable to those of the Young interferences in optics. If we speed up the emission of electrons, the interference figure is obtained more rapidly. If we close up one of the holes, the result is a unique uniform spot. Repeating the experiment with the two holes and adding a detection apparatus capable of reporting which hole each electron went through, we see that the figure has no more fringes and we observe on the second plate the two spots predicted by the classical reasoning. Everything happens as if we were studying the emission of two incoherent sources.

We can improve the device by adding a third plate pierced with two holes and then close up the holes in all possible ways. The figures obtained after some time will always be identical to the optical diffraction patterns created by a light source whose wavelength equals the Compton wavelength of the electrons. Obviously what is against intuition is the fact that these interference patterns were created in a progressive way by the particles emitted at relatively slow rhythm and that it is impossible, although the interference pattern is built by successive impacts, to identify through which hole each one of the electrons crossed.

This experiment can be interpreted in the framework of ordinary quantum mechanics by allocating to each electron a wave function satisfying the Schrödinger equation and interpreted as the probability amplitude of presence of the electron. Of course, there is also some arbitrariness in the expression of the Schrödinger equation, obtained by the principle of correspondence and confirmed by its predictive capacity. We need also to set the traditional quantum mechanics postulates concerning the measurement process and the preparation of the wave packet. In that way, all the results of the experiment are perfectly interpretable.

We will show now how the notion of path integral can be deduced from this experiment. Let us forget about the Schrödinger equation, but keep in mind the idea that there exists a ket formalism to describe the physical states. Since in some configurations, we have interference fringes, this means that we must introduce the notion of the amplitude of probability $A(b, a)$, of having, at time t_b, at the point q_b of the plate an electron which has been emitted by the source, at time t_a, at the point q_a. This amplitude is a function of the trajectories of each electron. Piercing new holes, adding plates, and closing up holes are various manipulations whose aim is to modify the paths that the electrons may follow, making some of them possible and others impossible. From the experiments, we see that

$A(b, a)$ is a function, in fact, a functional, of the possible trajectories, and in fact the electrons can follow all possible paths $q_{pos.}$ joining the source at q_a to the point of detection of the plate q_b. Of course, a trajectory $q_{pos.}$ is not necessarily a real trajectory, i.e. such that it corresponds to a solution of the classical equations of motion. The word 'possible' refers only to the part of space in which the electron can be localised. Hence, we expect that the amplitude $A(b, a)$ is a sum over all possible paths of elementary amplitudes $A[q_{pos.}]$:

$$A(b, a) = \sum_{q_{pos.}} A[q_{pos.}]. \qquad (8.2)$$

$A[q_{pos.}]$ is a functional of the path $q_{pos.}$ defined by a trajectory $t \rightarrow q_{pos.}[t]$, with $q_{pos.}[t_a] = q_a$ and $q_{pos.}[t_b] = q_b$. The meaning of such a formula is well defined in the discrete approximation of space. The path integral is defined as its continuous limit; we will make this statement more rigorous later on in this chapter. In the following, we will omit to specify that only the possible trajectories, i.e. physically realisable trajectories, must be taken into account.

The problem we face is to find the expression of the functional $A[q]$. Let us remark that the arbitrariness governing the choice of this functional is similar to that concerning the wave function in traditional quantisation. We *postulate* that

$$A[q] \sim e^{\frac{i}{\hbar} S[q]}. \qquad (8.3)$$

The functional $S[q]$ is the value of the classical action of the electron on the trajectory $t \rightarrow q[t]$. We thus have

$$A(b, a) = \sum_{q} e^{\frac{i}{\hbar} S[q]}. \qquad (8.4)$$

Going back to the analysis of what happened in interposing new plates with holes between the source and the final plate, we can write for the amplitudes of probability a formula of summation over intermediate states. Let indeed t_c be an intermediate time between t_a and t_b, each possible trajectory from q_a to q_b will cross at some point q_c a plate P_c located between the source and the point where the impacts are observed. We can describe the set of all possible trajectories $t \rightarrow q[t]$, $t \in [t_a, t_b]$, linking q_a at time t_a to q_b at time t_b as the union, for all points $q_c \in P_c$, of the possible trajectories $t \rightarrow q[t]$, $t \in [t_a, t_b]$, such that $q[t_a] = q_a$, $q[t_b] = q_b$, and $q[t_c] = q_c$ for some time $t_c \in [t_a, t_b]$. Since the amplitudes are functionals of the trajectories and since the action is additive, $S(b, a) = S(b, c) + S(c, a)$, the phase factors factorise and the total amplitude is obtained by summing independently on the paths from (q_a, t_a) to (q_c, t_c) and on the paths from (q_c, t_c) to (q_b, t_b), then finally on the position of the intermediate point q_c. It is therefore reasonable to postulate the *composition formula*

$$A(b, a) = \int dq_c A(b, c) A(c, a). \qquad (8.5)$$

We will see that from these two postulates follow all the results of quantum mechanics, the most spectacular being the *proof* of the Schrödinger equation.

8.4   The Reconstruction of Quantum Mechanics

To avoid unnecessary notational complications due to space dimensions, we suppose from now on that *physics is one dimensional.*

To deduce a useful expression of the probability amplitude, we will make use of the composition formula (8.5). We can iterate this identity and rewrite $A(b, a)$, introducing $n - 1$ intermediate times $t_0 = t_a < t_1 < t_2 \cdots < t_n = t_b$ and $n - 1$ intermediate points q_i, $i = 1, \ldots, n - 1$. We find then that

$$A(b, a) = \int \int \ldots \int dq_1 \ldots dq_{n-1} A(b, n-1) A(n-1, n-2) \ldots A(1, a), \qquad (8.6)$$

where we used again the same notation, namely i stands for '$q = q_i$ at $t = t_i$'. This last identity is interesting because it is of direct use. It implies indeed that the trajectory is specified more precisely since it must go through q_1 at time t_1, $\ldots$, through q_i at time t_i, $\ldots$, through q_{n-1} at time t_{n-1}. Let us take a regular time spacing by dividing the time interval into n intervals of length ε, $n\varepsilon = t_b - t_a$, corresponding to the $n - 1$ intermediate times $t_1, t_2, \ldots, t_{n-1}$. Since the expression does not depend on the number of intermediate times, by taking n large enough, i.e. ε sufficiently small, we can expect to extract the dominant contribution to the formula of the amplitudes $A(j + 1, j)$.

We remark first that we can approximate the value of the action between two neighbouring points. Thus in the interval (t_{i+1}, t_i), we can write, assuming the Lagrangian sufficiently regular and setting $q_0 = q_a$ and $q_n = q_b$,

$$\int_{t_i}^{t_{i+1}} dt\, L(\dot{q}, q, t) \simeq \varepsilon L \left(\frac{q_{i+1} - q_i}{\varepsilon}, \frac{q_{i+1} + q_i}{2}, \frac{t_{i+1} + t_i}{2} \right), \qquad (8.7)$$

where $\dot{q}$ has been replaced by $\dfrac{(q_{i+1} - q_i)}{\varepsilon}$, q by $(q_{i+1} + q_i)/2$, and t by $(t_{i+1} + t_i)/2$.

We can then set

$$A(i + 1, i) \simeq \frac{1}{N} e^{\left(\frac{i\varepsilon}{\hbar} L \left(\frac{q_{i+1} - q_i}{\varepsilon}, \frac{q_{i+1} + q_i}{2}, \frac{t_{i+1} + t_i}{2} \right) \right)}, \qquad (8.8)$$

where $1/N$ is the number of trajectories satisfying the required boundary conditions. At first glance N depends on the Lagrangian and on the characteristics imposed to the trajectories, we will, however, see that it is enough to take $N = N(\varepsilon)$.

In fact, we will see that there exists a suitable $N(\varepsilon)$, such that

$$\lim_{\varepsilon \to +0} \frac{1}{N(\varepsilon)^n} \int \int \ldots \int dq_1 dq_2 \ldots dq_{n-1} e^{\frac{i}{\hbar} S(b, a)} \qquad (8.9)$$

exists with $S(b, a) = S(b, n-1) + \cdots + S(2, 1) + S(1, a)$, each of the actions satisfying the boundary conditions associated with the intervals. This limit will be noted

$$\int_{\substack{q(t_b)=q_b \\ q(t_a)=q_a}} e^{\frac{i}{\hbar} S(b,a)} \mathcal{D}(q(t)). \tag{8.10}$$

8.4.1   The Quantum Mechanics of a Free Particle

As an example we will quantise a free particle, a case for which the functional Integral can be explicitly computed. The Lagrangian is given by $L = (\frac{1}{2}) m \dot{q}^2$.

Let us first define what is a Gaussian oscillatory integral. If a is a complex number whose real part, Re a, is positive,

$$\int_{-\infty}^{+\infty} dx e^{-ax^2} = \sqrt{\frac{\pi}{a}}, \tag{8.11}$$

the square root being defined as the usual square root for positive numbers and the origin as the branching point.

By continuity, we define, for $b > 0$, that

$$\int_{-\infty}^{+\infty} dx e^{ibx^2} = e^{i\frac{\pi}{4}} \sqrt{\frac{\pi}{b}} = \sqrt{\frac{i\pi}{b}}, \tag{8.12}$$

an expression which must be understood as $\lim_{\delta \to +0} \int_{-\infty}^{+\infty} dx\, e^{(ib-\delta)x^2}$.

Let us compute the amplitude $A(b, a)$ using formula (8.6). With n intermediate points for the approximation, we define A as

$$A^{(n)}(b, a) = \frac{1}{N^{n+1}} \int \int \cdots \int dq_1 \ldots dq_n e^{\frac{im}{2\hbar\varepsilon} \sum_{i=1}^{n+1} (q_i - q_{i-1})^2}, \tag{8.13}$$

where $N = \sqrt{\frac{2\pi i \hbar \varepsilon}{m}}$.

Performing the integrations one after the other, starting from the integration over q_1, then over q_2, ..., up to the integration over q_n, then making a repeated use of the following formula, for $C > 0$,

$$\frac{1}{\sqrt{j}} \int_{-\infty}^{+\infty} dq_j e^{\frac{iC}{j}(q_0-q_j)^2 + iC(q_j-q_{j+1})^2} = e^{i\frac{\pi}{4}} \sqrt{\frac{\pi}{C(j+1)}} e^{\frac{iC}{j+1}(q_0-q_{j+1})^2}, \tag{8.14}$$

we obtain with $C = m/2\hbar\varepsilon$

$$A^{(n)}(a, b) = \sqrt{\frac{m}{2\pi i \hbar \varepsilon (n+1)}} e^{\frac{im}{2\hbar\varepsilon(n+1)}(q_0-q_n)^2}. \tag{8.15}$$

But $(n + 1)\varepsilon = t_b - t_a$, thus finally

$$A^{(n)}(b, a) = \sqrt{\frac{m}{2\pi i\hbar(t_b - t_a)}} \exp \frac{im(q_b - q_a)^2}{2\hbar(t_b - t_a)}. \tag{8.16}$$

The resulting expression *is independent of the number of intermediate points* and is therefore equal to the limit value. This is a remarkable property which is due to the fact that the Lagrangian is quadratic in q.

We thus prove that

$$A(b, a) = \sqrt{\frac{m}{2\pi i\hbar(t_b - t_a)}} \exp \frac{im(q_b - q_a)^2}{2\hbar(t_b - t_a)}. \tag{8.17}$$

8.4.2   A Particle in a Potential

The general case can be treated in the same way as the free particle case. For a general Lagrangian

$$L(q, \dot{q}, t) = \frac{1}{2}m\dot{q}^2 - V(q) \tag{8.18}$$

we introduce an approximate expression of the amplitude

$$A^{(n)}(b, a) = \frac{1}{N(\varepsilon)^n} \int \int \cdots \int dq_1 dq_2 \ldots dq_{n-1} e^{\frac{i\varepsilon}{\hbar} \sum_{k=1}^{n} (\frac{m}{2}(\frac{q_k - q_{k-1}}{t_k - t_{k-1}})^2 - V(\frac{q_k + q_{k-1}}{2}))} \tag{8.19}$$

with the same $N(\varepsilon)$ as in the free case.

If the limit of $A^{(n)}(b, a)$ exists for $n \to \infty$, we write it under the form (8.10).

The existence of the limit is a difficult mathematical problem, strongly depending on the potential term. We can completely compute the integral when V is a quadratic function of q.

We will consider in the following that (8.10) is the limit of (8.19) whether this limit exists or not.

8.4.3   The Schrödinger Equation

From the amplitude $A(b, a)$ previously defined and which we will note in the sequel $G(q_b, t_b; q_a, t_a)$ in order to make explicit the parameter dependence, we can define, for $t > s$, a wave function

$$\Psi(q, t) = \int G(q, t; y, s)\phi(y, s)dy, \tag{8.20}$$

the function ϕ being the initial function. Using the composition formula (8.5), this wave function satisfies

$$\Psi(q,t) = \int_{-\infty}^{+\infty} \mathrm{d}q' \, G(q,t;q',t') \Psi(q',t') \tag{8.21}$$

for all times t', $t > t' > s$. This equation is the basic equation that we will take to define the wave function. It shows indeed that the wave function at a point (q, t) can be expressed as the action of the amplitude on a wave function defined at a previous time, this function summarising all the knowledge we have on the past of the system.

In the case of a Lagrangian $L = \frac{1}{2} m \dot{q}^2 - V(q, t)$, we will now derive a partial differential equation that is satisfied by the wave function. To do it let us compute the time derivative of $\Psi(q, t)$:

$$\frac{\partial \Psi(q,t)}{\partial t} = \lim_{\varepsilon \to +0} \frac{\Psi(q, t+\varepsilon) - \Psi(q,t)}{\varepsilon}. \tag{8.22}$$

Using formula (8.21), we compute $\Psi(q, t+\varepsilon) - \Psi(q, t)$ up to $O(\varepsilon^2)$. Let us first rewrite the difference

$$\Psi(q, t+\varepsilon) - \Psi(q,t) = \int \mathrm{d}y \, G(q, t+\varepsilon; y, t) \Psi(y, t) - \Psi(q, t) \tag{8.23}$$

$$= \int \mathrm{d}y \left(\frac{1}{N} \mathrm{e}^{\frac{im}{\hbar\varepsilon}(q-y)^2} \mathrm{e}^{-\frac{i\varepsilon}{\hbar} V(\frac{q+y}{2}, t+\frac{\varepsilon}{2})} \right) \Psi(y, t) - \Psi(q, t).$$

We must compute $\frac{1}{N(\varepsilon)} \int \mathrm{d}y \, \mathrm{e}^{\frac{im}{2\hbar\varepsilon}(q-y)^2} f(y)$. We find that[6]

$$\frac{1}{N(\varepsilon)} \int \mathrm{d}y \, \mathrm{e}^{\frac{im}{2\hbar\varepsilon}(q-y)^2} f(y) = f(q) - \frac{\hbar\varepsilon}{2im} f''(q) + \mathcal{O}(\varepsilon^2). \tag{8.24}$$

Applying this formula to the function

$$f(y) = \mathrm{e}^{-\frac{i\varepsilon}{\hbar} V(\frac{q+y}{2}, t+\frac{\varepsilon}{2})} \Psi(y, t) \tag{8.25}$$

and keeping the terms of order at most ε, we find that

$$\frac{1}{N} \int \mathrm{d}y \, \mathrm{e}^{\frac{im}{2\hbar\varepsilon}(q-y)^2} \mathrm{e}^{-\frac{i\varepsilon}{\hbar} V(\frac{q+y}{2}, t+\frac{\varepsilon}{2})} \Psi(y, t) \tag{8.26}$$

$$= \Psi(q, t) - i\frac{\varepsilon}{\hbar} V(q, t+\frac{\varepsilon}{2}) \Psi(q, t) + \frac{i\hbar\varepsilon}{2m} \frac{\partial^2}{\partial q^2} \Psi(q, t).$$

---

[6] This shows that $\lim_{\varepsilon \to +0} \frac{1}{\sqrt{i\pi\varepsilon}} \mathrm{e}^{\frac{i}{\varepsilon} x^2}$ converges as a distribution to $\delta(x)$.

Collecting all the result we find that

$$i\hbar\frac{\partial}{\partial t}\Psi(q,t) = H\Psi(q,t) \tag{8.27}$$

with

$$H = -\frac{\hbar^2}{2m}\frac{\partial^2}{\partial q^2} + V(q,t). \tag{8.28}$$

We proved the following.

Theorem 8. *The wave function* $\Psi(q,t)$, *defined by Eq. (8.20), is a solution of the Schrödinger equation.*

This relatively intuitive approach could have been the starting point to define quantum mechanics, one postulate being replaced by another one.

This formalism gives back all the properties of the Schrödinger equation. We can, for example, recover the property of conservation of probability.

Let $\Psi(q_2,t_2) = \int dq_1 G(q_2,t_2;q_1,t_1)\Psi(q_1,t_1)$; we must prove that

$$\int dq_2\Psi^*(q_2,t_2)\Psi(q_2,t_2) = \int dq_1\Psi^*(q_1,t_1)\Psi(q_1,t_1), \tag{8.29}$$

which implies that

$$\int dq_2 G^*(q_2,t_2;q_1',t_1)G(q_2,t_2;q_1,t_1) = \delta(q_1'-q_1). \tag{8.30}$$

We can prove this relationship directly by infinitesimal steps using the law of composition of amplitudes. It is, thus, enough to show it for $t_2 = t_1 + \varepsilon$. In this case

$$\int dq_2\, G^*(q_2,t_2;q_1',t_1)\, G(q_2,t_2;q_1,t_1)$$

$$= \frac{m}{2\pi\hbar\varepsilon}\int dq_2\, e^{\frac{i\varepsilon}{\hbar}L(\frac{q_2-q_1}{\varepsilon},\frac{q_2+q_1}{2},t_1+\frac{\varepsilon}{2})}e^{-\frac{i\varepsilon}{\hbar}L(\frac{q_2-q_1'}{\varepsilon},\frac{q_2+q_1'}{2},t_1+\frac{\varepsilon}{2})}$$

$$= \frac{m}{2\pi\hbar\varepsilon}\int dq_2\, e^{\frac{im}{\hbar\varepsilon}q_2(q_1'-q_1)}e^{\frac{im}{2\hbar\varepsilon}(q_1^2-q_1'^2)}e^{\frac{i\varepsilon}{\hbar}(V(\frac{q_2+q_1}{2},t_1+\frac{\varepsilon}{2})-V(\frac{q_2+q_1'}{2},t_1+\frac{\varepsilon}{2}))}.$$

At order ε, the integration over q_2 gives

$$\int dq_2\, e^{\frac{im}{\hbar\varepsilon}q_2(q_1'-q_1)} = \frac{2\pi\hbar\varepsilon}{m}\delta(q_1-q_1'),$$

which implies the stated result.

8.5 The Feynman Formula

We place ourselves in a non-relativistic framework. Let us consider a translation invariant system consisting of a scalar particle moving on a straight line under the action of a time-independent potential $V(q)$. The classical dynamics is described by a Hamiltonian $H = p^2/2m + V(q)$ or equivalently, after a Legendre transformation, by a Lagrangian $L(q, \dot{q}) = m\dot{q}^2/2 - V(q)$.

The laws of classical mechanics are such that if this particle moves from the point q_i at the initial time t_i to the point q_f at a later time t_f, its trajectory is given by a function $\bar{q}(t)$, solution of the Euler–Lagrange equations. These equations are obtained by writing that the action of the system $S(f, i)[q] = \int_{t_i}^{t_f} dt L(\dot{q}, q)$ is stationary and satisfies the boundary conditions: $q_i = \bar{q}(t_i)$, $q_f = \bar{q}(t_f)$.

To obtain the quantum theory of the system, the usual starting point is to write a causal equation of the Schrödinger type defining the evolution of the wave function of the particle, this wave function being interpreted as the probability amplitude to find the particle at a particular point. The quantisation in terms of a path integral requires a reinterpretation of the corresponding physical quantities.

8.5.1 The Representations of Quantum Mechanics

The probability density for the presence of the particle at a point in space is given by the square of the modulus of the wave function. It follows that the overall phase of the latter is not an observable quantity. There exists therefore an equivalence between all the physical theories described by unitary equivalent wave functions. The action of a unitary transformation on a wave function is called a change of representation. We can construct infinitely many such representations, but, in this book, we shall consider two different classes: one related to the time dependence of states and the other related to the expression of quantised operators.

8.5.1.1 The Different Time Dependences

Three representations play a key role: the Schrödinger representation, the Heisenberg representation, and the interaction representation, depending on whether the emphasis is put on the time dependence of the states or of the operators.[7]

- The **Schrödinger representation**: the states are time dependent and the operators are time independent. We note $|\Psi(t)>_S$ such states and the Hamiltonian in the Schrödinger representation is the usual Hamiltonian $\hat{H}_S = \hat{H} = \hat{H}_0 + \hat{H}_{\mathrm{int}}$, which is written as the sum of a free and an interacting part. In this representation, the time evolution of the states is given by the Schrödinger equation

$$i\hbar\partial_t|\Psi(t)>_S = H_S|\Psi(t)>_S .\tag{8.31}$$

---

[7] In what follows, we will adopt the rule according to which a hat above a quantity means that it is an operator.

- The **Heisenberg representation:** the states are time independent and the operators depend on time. We denote by $\hat{Q}_H(t)$ the operators and $|\Psi>_H$ the states. The time evolution is given by

$$\partial_t |\Psi>_H = 0 \quad \text{and} \quad -i\hbar \dot{\hat{Q}}_H(t) = [\hat{H}_H, \hat{Q}_H(t)], \tag{8.32}$$

where $\hat{H}_H$ is the Hamiltonian in the Heisenberg representation.

- The **interaction representation:** one splits the Hamiltonian into $\hat{H}_I = (\hat{H}_0)_I + (\hat{H}_{\text{int}})_I$, where H_0 is the free part. The time evolution of the states $|\Psi>_I$ and of the operators $\hat{Q}_I$ is given by

$$i\hbar\partial_t |\Psi>_I = (\hat{H}_{int})_I |\Psi>_I \quad \text{and} \quad -i\hbar\dot{\hat{Q}}_I(t) = [(\hat{H}_0)_I, \hat{Q}_I(t)]. \tag{8.33}$$

These different representations are linked by the fact that the expectation values of operators between different states are physically measurable quantities and cannot depend on the choice of the representation:

$$<\Psi(t)|\hat{Q}|\Phi(t)>_S = <\Psi|\hat{Q}(t)|\Phi>_H = <\Psi(t)|\hat{Q}(t)|\Phi(t)>_I . \tag{8.34}$$

We link up the Schrödinger and the Heisenberg representations by setting

$$|\Psi(t)>_H = |\Psi(0)>_H = |\Psi(0)>_S .$$

Since

$$|\Psi(t)>_S = e^{-i\frac{tH_S}{\hbar}} |\Psi(0)>_S \tag{8.35}$$

we find that

$$|\Psi(t)>_H = e^{i\frac{tH_S}{\hbar}} |\Psi(t)>_S . \tag{8.36}$$

It follows that the correspondence between an operator $\hat{Q}_S$ in the Schrödinger representation and its expression in the Heisenberg representation is given by (omitting from now on the index S for 'Schrödinger representation' for $\hat{H}$ and for $\hat{H}_0$)

$$\hat{Q}_H(t) = e^{i\frac{t\hat{H}}{\hbar}} \hat{Q}_S e^{-i\frac{t\hat{H}}{\hbar}} \tag{8.37}$$

and thus $\hat{H} = \hat{H}_S = \hat{H}_H$,
   Similarly

$$|\Psi(t)>_I = e^{i\frac{t\hat{H}_0}{\hbar}} |\Psi(t)>_S, \tag{8.38}$$

and we define operators in the interaction representation by

$$\hat{Q}_I(t) = e^{i\frac{t\hat{H}_0}{\hbar}} \hat{Q}_S e^{-i\frac{t\hat{H}_0}{\hbar}}, \tag{8.39}$$

which leads to $(\hat{H}_0)_I = \hat{H}_0$. Let us write now the equation satisfied by $|\Psi_I(t)>$.

$$
\begin{aligned}
i\hbar\partial_t|\Psi_I(t) > &= -\hat{H}_0|\Psi_I(t) > +e^{i\frac{t\hat{H}_0}{\hbar}}H|\Psi_I(t) > \\
&= -\hat{H}_0|\Psi_I(t) > +\hat{H}_0|\Psi_I(t) > +e^{i\frac{t\hat{H}_0}{\hbar}}\hat{H}_{\text{int}}e^{-i\frac{t\hat{H}_0}{\hbar}}|\Psi_I(t) > \\
&= e^{i\frac{t\hat{H}_0}{\hbar}}\hat{H}_{\text{int}}e^{-i\frac{t\hat{H}_0}{\hbar}}|\Psi_I(t) >= (\hat{H}_{\text{int}})_I(t)|\Psi_I(t) >,
\end{aligned}
$$

which leads to the definition

$$
(\hat{H}_{\text{int}})_I(t) = e^{i\frac{t\hat{H}_0}{\hbar}}\hat{H}_{\text{int}}e^{-i\frac{t\hat{H}_0}{\hbar}}. \tag{8.40}
$$

It is easy to check that the equations of motion are equivalent in these representations.

Remark that in the interaction representation, $|\Psi_I(t)>$ coincides with $|\Psi_H>$ in the case where there is no interaction. This property enables the interaction representation to play a key role in problems in which the weak coupling approximation is justified.

8.5.1.2   *The Different Operator Representations*

In quantum mechanics, the phase space coordinates p and q are replaced by operators $\hat{q}$ and $\hat{p}$, satisfying the commutation relations

$$
[\hat{q}, \hat{p}] = \hat{q}\hat{p} - \hat{p}\hat{q} = i\hbar. \tag{8.41}
$$

In the space representation[8] where the operator $\hat{q}$ is diagonal and acts as the multiplication by q, $\hat{q} = q1$, the momentum operator is the derivative with respect to q: $\hat{p} = -i\frac{d}{dq}$. To each function $f(p, q)$, We associate an operator by replacing p and q with the corresponding operators. This method, known as canonical quantisation, is not without ambiguities. In fact since the operators $\hat{p}$ and $\hat{q}$ do not commute, to each ordering of p and q in a product corresponds a different operator. Such a problem is not manifest if there is an additive dependence in these variables. This is the case for the Hamiltonian $H(p, q) = p^2/2m + V(q)$ to which is associated in a unique way a unique Hamiltonian operator $\hat{H}$.

It is useful to introduce continuous basis associated with each of these operators. They are often called the p-basis, for momentum, and the q-basis, for position, representations of quantum mechanics. They are particularly interesting for the description of quantum systems having a classical equivalent.

The q-basis, noted $|q>$, is the basis formed by the eigenstates of the operator $\hat{q}$:

$$
\hat{q}|q >= q|q > . \tag{8.42}
$$

It is formally orthonormal (this is an improper basis)

$$
< q|q' >= \delta(q - q'). \tag{8.43}
$$

---

[8] The name representation has here a meaning different from the preceding one. By the choice of a representation, we mean the choice of a particular functional space in which the operators $\hat{p}$ and $\hat{q}$ act.

Moreover, it is supposed to be complete; i.e. the following closure relation is satisfied:

$$1 = \int dq \, |q><q| \, . \tag{8.44}$$

Similarly, the eigenfunctions of the momentum operator $\hat{p}$ form an orthonormal basis (improper). We thus have

$$1 = \int dp \, |p><p| \, . \tag{8.45}$$

The position ket basis is related to the momentum ket basis by the Fourier transform

$$<p|q> = \overline{<q|p>} = \frac{1}{\sqrt{2\pi\hbar}} e^{-i\frac{pq}{\hbar}} \, . \tag{8.46}$$

It will be useful to introduce, in addition, the time-dependent continuous basis[9]:

$$|q, t> = e^{it\hat{H}} |q> \, . \tag{8.47}$$

This basis is built using the eigenvectors of the position operator in the Heisenberg representation

$$\hat{q}(t)|q, t> = q|q, t> \, . \tag{8.48}$$

We have obviously a closure relation

$$1 = \int dq \, |q, t><q, t| \qquad \forall t. \tag{8.49}$$

To each state $|\Psi>$ is associated a function of the position $<q|\Psi> = \Psi(q)$ and a function of the momentum $<p|\Psi> = \Psi(p)$. Obviously $\Psi(q)$ and $\Psi(p)$ are related by Fourier transforms. Moreover, if $|\Psi>$ is a solution of the Schrödinger equation, its wave function $\Psi(q)$ is also a solution of this equation, the operators $\hat{p}$ and $\hat{q}$ being, respectively, replaced by the operator of derivation with respect to q and by the operator of multiplication by q.

8.5.2 The Feynman Formula for Systems with One Degree of Freedom

Let us consider the Schrödinger representation. The description of a physical phenomenon is made by computing the transition probability between an initial state $|\Phi_i>$ and a final state $|\Psi_f>$, a probability which can be computed from the scalar product $<\Psi_f|\Phi_i>$. To compute such a quantity, it is enough to know the Green function $<q'|e^{-i(t'-t)\hat{H}}|q>$ describing the evolution of the particle from its position q at time t to its position q' at time t'. Knowing in fact the Green function, we can express the solution of the Schrödinger equation at time t' in terms of the initial value at time t:

---

[9] Be careful: the kets $|q, t>$ are not the states $|q(t)> = e^{-it\hat{H}}|q>$ of the Schrödinger representation.

$$\Psi(q',t') = <q'|\Psi(t')> = <q'|e^{-i(t'-t)H}|\Psi(t)> \tag{8.50}$$

$$= \int_{-\infty}^{+\infty} dq <q'|e^{-i(t'-t)H}|q><q|\Psi(t)>$$

$$= \int_{-\infty}^{+\infty} dq <q'|e^{-i(t'-t)H}|q> \Psi(q,t). \tag{8.51}$$

Comparing this formula to (8.21), We see that we can set

$$<q'|e^{-i(t'-t)H}|q> = G(q',t';q,t). \tag{8.52}$$

We will give a deeper justification of this identification by showing, through an analysis similar to that made at the beginning of this chapter, that the Green functions have exactly the same formulation in terms of functional integrals as the amplitudes of probability.

As before, we will first treat the free case.

We must calculate $<q'|e^{-\frac{i}{\hbar}(t'-t)\hat{H}_0}|q>$. Since $\hat{H}_0$ is a function of $\hat{p}$, it is useful to choose the momentum representation and we have

$$<q'|e^{-\frac{i}{\hbar}(t'-t)\hat{H}_0}|q>$$

$$= \int <q'|p><p|e^{-\frac{i}{\hbar}(t'-t)\hat{H}_0}|q> dp = \frac{1}{\sqrt{2\pi\hbar}} \int e^{\frac{i}{\hbar}pq'} <p|e^{-\frac{i}{\hbar}(t'-t)\hat{H}_0}|q> dp$$

$$= \frac{1}{2\pi\hbar} \int e^{\frac{i}{\hbar}p(q'-q)} e^{-\frac{i}{\hbar}(t'-t)\frac{p^2}{2m}} dp = \sqrt{\frac{m}{2i\pi\hbar(t'-t)}} e^{\frac{i}{\hbar}\frac{1}{2m}\frac{(q'-q)^2}{(t'-t)}}. \tag{8.53}$$

We recognise in the last equality the expression of the functional integral in the free case. We have thus shown

$$<q'|e^{-\frac{i}{\hbar}(t'-t)\hat{H}_0}|q> = \int_{\substack{q(t')=q' \\ q(t)=q}} e^{\frac{i}{\hbar}\int_t^{t'} \frac{\dot{q}^2(\tau)}{2m} d\tau} \mathcal{D}(q). \tag{8.54}$$

The general case, with a potential V, can be treated in the same way. We divide the time interval into n intervals of size ε and we write

$$<q'|e^{-\frac{i}{\hbar}(t'-t)\hat{H}}|q> \tag{8.55}$$

$$= \int <q'|e^{-\frac{i}{\hbar}\varepsilon\hat{H}}|q_{n-1}> dq_{n-1} \cdots <q_2|e^{-\frac{i}{\hbar}\varepsilon\hat{H}}|q_1> dq_1 <q_1|e^{-\frac{i}{\hbar}\varepsilon\hat{H}}|q>.$$

We compute each of these terms at the leading approximation in ε using[10]

$$e^{-\frac{i}{\hbar}\varepsilon\hat{H}} = e^{-\frac{i}{\hbar}\varepsilon\hat{H}_0} e^{-\frac{i}{\hbar}\varepsilon\hat{V}} + O(\varepsilon^2), \tag{8.58}$$

---

[10] We must be careful: if two operators $\hat{A}$ and $\hat{B}$ do not commute, then

$$e^{\hat{A}}e^{\hat{B}} \neq e^{\hat{A}+\hat{B}}. \tag{8.56}$$

If the commutator $[\hat{A}, \hat{B}]$ is small, a good approximation is given by

$$e^{\hat{A}}e^{\hat{B}} = e^{\hat{A}+\hat{B}}(1 + \mathcal{O}([\hat{A}, \hat{B}])). \tag{8.57}$$

H_0 being the free Hamiltonian. Thus,

$$< q_{j+1} | e^{-\frac{i}{\hbar}\varepsilon \hat{H}} | q_j > = < q_{j+1}, t_{j+1} | q_j, t_j > \tag{8.59}$$

$$= \int \mathrm{d}r < q_{j+1} | e^{-\frac{i}{\hbar}\varepsilon \hat{H}_0} | r > < r | e^{-\frac{i}{\hbar}\varepsilon \hat{V}} | q_j >$$

and since

$$< r | e^{-\frac{i}{\hbar}\varepsilon \hat{V}} | q_j > = e^{-\frac{i}{\hbar}\varepsilon V(q_j)} \delta(r - q_j) \tag{8.60}$$

it follows that

$$< q_{j+1} | e^{-\frac{i}{\hbar}\varepsilon H} | q_j > = \sqrt{\frac{m}{2i\pi \hbar \varepsilon}} e^{\frac{i}{\hbar}(\frac{m}{2\varepsilon}(q_{j+1}-q_j)^2 - \varepsilon V(q_j))} \tag{8.61}$$

$$= \sqrt{\frac{m}{2i\pi \hbar \varepsilon}} e^{\frac{i}{\hbar}(\frac{m}{2\varepsilon}(q_{j+1}-q_j)^2 - \varepsilon V(\frac{q_{j+1}+q_j}{2}))} + O(\varepsilon^{3/2}).$$

The factor $\frac{3}{2}$ in the error estimate comes from the fact that for V we have replaced q_j by $(q_{j+1} + q_j)/2$ and that $q_{j+1} - q_j = O(\varepsilon^{1/2})$ because of the oscillation factor $e^{i(q_{j+1}-q_j)^2/2\varepsilon}$.

We can thus write

$$< q' | e^{-\frac{i}{\hbar}(t'-t)\hat{H}} | q > = \lim_{n \to \infty} < q' | e^{-\frac{i}{\hbar}(t'-t)\hat{H}} | q >^{(n)}, \tag{8.62}$$

where the approximation of order n is defined by

$$< q' | e^{-\frac{i}{\hbar}(t'-t)\hat{H}} | q >^{(n)}$$

$$= \left(\sqrt{\frac{m}{2i\pi \hbar \varepsilon}}\right)^n \int \mathrm{d}q_1 \cdots \mathrm{d}q_{n-1} e^{\frac{i}{\hbar}\varepsilon \sum_{i=1}^{n-1}(\frac{m}{2}(\frac{q_i - q_{i-1}}{\varepsilon})^2 - V(\frac{q_i + q_{i-1}}{2}))}$$

$$= \frac{1}{N(\varepsilon)^n} \int \mathrm{d}q_1 \cdots \mathrm{d}q_{n-1} e^{\frac{i}{\hbar}\varepsilon \sum_{i=1}^{n-1}(\frac{m}{2}(\frac{q_i - q_{i-1}}{\varepsilon})^2 - V(\frac{q_i + q_{i-1}}{2}))}. \tag{8.63}$$

We recognise in this last expression of Eq.(8.63) the formula (8.19). Thus, if the limit exists, we have shown that

$$< q' | e^{-\frac{i}{\hbar}(t'-t)\hat{H}} | q > = \int_{\substack{q(t')=q' \\ q(t)=q}} \mathcal{D}(q) e^{\frac{i}{\hbar} \int_t^{t'} \mathcal{L}(q(\tau), \dot{q}(\tau)) \mathrm{d}\tau}. \tag{8.64}$$

This formula is called the *Feynman formula*. It shows the equivalence between an operator expression in terms of the Hamiltonian with a functional expression in terms of the Lagrangian; the advantage of this last expression is the fact that it depends only on the classical Lagrangian and is expressed only in terms of commuting variables.

Note the dependence in $\hbar$. We will see that in perturbation theory $\hbar$ is a natural parameter of expansion.

Observe also the similarity between the probability amplitude, expressed by (8.64), and the partition functions in statistical mechanics. Thus, the probabilistic nature of quantum mechanics and that of statistical mechanics, very different in principle, give rise to very similar mathematical descriptions.

It is possible to rewrite the approximate expression of the Green function in a form hiding the appearance of the singular normalisation factor $N(\varepsilon)$. Indeed, with the identity

$$\frac{1}{2\pi\hbar}\int_{-\infty}^{\infty}\mathrm{d}p\,\mathrm{e}^{\frac{\mathrm{i}}{\hbar}\varepsilon(p\dot{q}-\frac{p^2}{2m})}=\sqrt{\frac{m}{2\mathrm{i}\pi\hbar\varepsilon}}\mathrm{e}^{\frac{\mathrm{i}}{\hbar}\varepsilon\frac{m}{2}\dot{q}^2}\tag{8.65}$$

We can rewrite (8.62) as

$$<q_{j+1}\,|\,\mathrm{e}^{-\frac{\mathrm{i}}{\hbar}\varepsilon H}\,|\,q_j>=\frac{1}{2\pi\hbar}\int\mathrm{d}p_j\,\mathrm{e}^{\frac{\mathrm{i}}{\hbar}[p_j(q_{j+1}-q_j)-\varepsilon(\frac{p_j^2}{2m}+V(\frac{q_{j+1}+q_j}{2}))]}\tag{8.66}$$

and thus

$$<q'\,|\,\mathrm{e}^{-\frac{\mathrm{i}}{\hbar}(t'-t)\hat{H}}\,|\,q>=G(q',t';q,t)=<q',t'|q,t>=$$

$$=\lim_{n\to+\infty}\int\frac{\mathrm{d}p_1\,\mathrm{d}q_1}{2\pi}\cdots\frac{\mathrm{d}p_{n-1}\mathrm{d}q_{n-1}}{2\pi}\,\mathrm{e}^{\frac{\mathrm{i}}{\hbar}\sum_{i=0}^{n-1}[p_i(q_{i+1}-q_i)-\varepsilon(\frac{p_i^2}{2m}-V(\frac{q_{i+1}+q_i}{2}))]}$$

$$=\lim_{n\to+\infty}\int\frac{\mathrm{d}p_1\,\mathrm{d}q_1}{2\pi}\cdots\frac{\mathrm{d}p_{n-1}\mathrm{d}q_{n-1}}{2\pi}\,\mathrm{e}^{\frac{\mathrm{i}}{\hbar}\sum_{i=0}^{n-1}[p_i(q_{i+1}-q_i)-\varepsilon H(p_i,\frac{q_{i+1}+q_i}{2})]}$$

$$=\int\mathcal{D}(p,q)\,\mathrm{e}^{\frac{\mathrm{i}}{\hbar}\int_t^{t'}(p\dot{q}-H(p,q))\mathrm{d}s},\tag{8.67}$$

the symbol $\mathcal{D}(p,q)$ being defined by the last equality.

Taking into account the definition of the symbol $\int\mathcal{D}(q)$ in (8.64), we have the convolution property

$$\int_{\substack{q(t_f)=q_f\\q(t_i)=q_i}}\mathcal{D}(q)\exp\frac{\mathrm{i}}{\hbar}S(t_i;t_f)\tag{8.68}$$

$$=\int_{-\infty}^{+\infty}\mathrm{d}\bar{q}\int_{\substack{q(t)=\bar{q}\\q(t_i)=q_i}}\mathcal{D}(q)\exp\frac{\mathrm{i}}{\hbar}S(t_i;t)\int_{\substack{q(t_f)=q_f\\q(t)=\bar{q}}}\mathcal{D}(q)\exp\frac{\mathrm{i}}{\hbar}S(t;t_f)$$

valid for any time t in the interval $[t_i,t_f]$.

If we are in a case where the action is large with respect to $\hbar$ for the majority of the trajectories linking q_i to q_f between times t_i and t_f, all the trajectories except those minimising the action, i.e. the classical trajectories, give to G rapidly oscillating factors, and finally the main contribution comes from the classical trajectories. If the classical trajectory from q_i to q_f is unique, we thus get, in the limit $\hbar \to 0$,

$$< q_f | e^{-\frac{i}{\hbar}(t_f - t_i)H} | q_i > \sim e^{\frac{i}{\hbar} S_{\text{clas.}}(q_i, t_i; q_f, t_f)}, \tag{8.69}$$

where $S_{\text{clas.}}$ is the value of the action on the classical trajectory linking q_i to q_f. Thus quantum mechanics appears as implying fluctuations around the classical trajectory. More precisely, we can assert that the classical action appears as the leading exponent in the weight of the trajectories implied in the functional integral. This domination appears more clearly in the Euclidean space formalism, a formalism obtained by time complexification, i.e. by changing t into $-it$. By this change of variables, the rapidly oscillating effect is replaced by an exponential decrease, much easier to analyse.

8.6   The Harmonic Oscillator

We will now compare, by a 2-point function calculation, the operator formalism of quantisation to the path integral formalism in the case of the harmonic oscillator. This case can be treated explicitly because the Hamiltonian is a quadratic function of p and q.

Let us first recall the operator formalism associated with the one-dimensional harmonic oscillator.

The Lagrangian L_0 and the Hamiltonian H_0 are given by

$$L_0 = \frac{m}{2}\dot{q}^2 - \frac{m\omega^2}{2}q^2 \qquad H_0 = \frac{p^2}{2m} + \frac{m\omega^2}{2}q^2. \tag{8.70}$$

Let $\Psi \in L^2(\mathbb{R})$ be a solution of the Schrödinger equation

$$i\hbar \frac{\partial \Psi}{\partial t} = H_0 \Psi = \left(-\frac{\hbar^2}{2m} \frac{d^2}{dq^2} + \frac{m\omega^2}{2}q^2 \right) \Psi. \tag{8.71}$$

Since H_0 does not depend on time, the time dependence of Ψ can be written as $\Psi = e^{-iEt/\hbar} \Phi$ with $H_0 \Phi = E\Phi, \quad \Phi \in L^2(\mathbb{R})$.

We find that the possible eigenvalues are $E_n = \hbar\omega(n + 1/2)$, $n = 0, 1, 2 \cdots$ and that the corresponding wave functions Φ_n, of unit norm, are given by

$$\Phi_n(q) = \frac{1}{\sqrt{2^n n!}} \left(\frac{m\omega}{\hbar\pi} \right)^{1/4} e^{-\frac{m\omega}{2\hbar}q^2} H_n(q\sqrt{\frac{m\omega}{\hbar}}), \tag{8.72}$$

where the H_n's are Hermite polynomials given by

$$H_n(x) = (-1)^n e^{x^2} \left(\frac{\mathrm{d}}{\mathrm{d}x} \right)^n e^{-x^2}. \tag{8.73}$$

The generating function of the Hermite polynomials is

$$e^{-t^2 + 2tx} = \sum_{n=0}^{+\infty} H_n(x) \frac{t^n}{n!}. \tag{8.74}$$

By derivation with respect to t, we get the main properties of these polynomials, in particular the orthonormality property

$$\int H_n(x) H_m(x) e^{-x^2} \, \mathrm{d}x = \sqrt{\pi} \, 2^n n! \delta_{nm}, \tag{8.75}$$

from which results the orthonormality of the harmonic oscillator wave functions

$$\int \overline{\Phi_n(q)} \Phi_m(q) \, \mathrm{d}q = \delta_{nm}. \tag{8.76}$$

The energy of the fundamental state is $E_0 = \frac{\hbar \omega}{2}$ and its wave function is $\Phi_0 = (\frac{m\omega}{\hbar \pi})^{1/4} e^{-\frac{m\omega}{2\hbar} q^2}$.

Let us set $q = \sqrt{\frac{\hbar}{m\omega}} x$. As a function of x, H can be written as

$$H_0 = \hbar \omega \left(-\frac{1}{2} \frac{\partial^2}{\partial x^2} + \frac{1}{2} x^2 \right). \tag{8.77}$$

Let us introduce the following operators acting on functions of $L^2(\mathbb{R})$:

$$a = \frac{1}{\sqrt{2}} \left(x + \frac{\mathrm{d}}{\mathrm{d}x} \right) = \sqrt{\frac{m\omega}{2\hbar}} q + \sqrt{\frac{\hbar}{2m\omega}} \frac{\mathrm{d}}{\mathrm{d}q} \tag{8.78}$$

$$a^\dagger = \frac{1}{\sqrt{2}} \left(x - \frac{\mathrm{d}}{\mathrm{d}x} \right) = \sqrt{\frac{m\omega}{2\hbar}} q - \sqrt{\frac{\hbar}{2m\omega}} \frac{\mathrm{d}}{\mathrm{d}q}. \tag{8.79}$$

From the preceding expressions, we easily check that

$$a\Phi_0 = 0$$
$$a\Phi_n = \sqrt{n} \Phi_{n-1}$$
$$a^\dagger \Phi_n = \sqrt{n+1} \Phi_{n+1}. \tag{8.80}$$

Moreover, we have the commutation relations

$$[a, a] = 0 \qquad [a^\dagger, a^\dagger] = 0 \qquad [a, a^\dagger] = 1. \tag{8.81}$$

In the ket formalism, the fundamental state Φ_0 is denoted $|0>$, which is a vacuum state; the n-quantum state, or n particle state according to the usual terminology, given by the wave function Φ_n is denoted $|n>$ and given by

$$|n> = \frac{(a^\dagger)^n}{\sqrt{n!}} |0> \tag{8.82}$$

with

$$\Phi_n(x) = \frac{1}{\pi^{1/4}} \frac{1}{\sqrt{2^n n!}} H_n(x)\, e^{-\frac{x^2}{2}} = <x \mid n>.$$

The operators a and $a^\dagger$ are respectively the *annihilation and creation operators*. The Hamiltonian H_0 and the particle number operator N can be written as

$$H_0 = \hbar\omega \left(N + \frac{1}{2} \right) \qquad N = a^\dagger a. \tag{8.83}$$

The kets $|n>$ are the eigenvectors of H_0 and N:

$$N|n> = n|n> \qquad H_0|n> = \hbar\omega \left(n + \frac{1}{2} \right) |n> .$$

They form a complete orthonormal basis of $L^2(\mathbb{R})$. This means that any element of $L^2(\mathbb{R})$ can be expanded over the kets $|n>$. In particular, we have the closure relation

$$\sum_n |n><n| \equiv 1. \tag{8.84}$$

Remark: $H_0\Phi_0 = \frac{1}{2}\hbar\omega\Phi_0$, which means that for an oscillator the energy of the fundamental state is different from 0. We call the value $\frac{1}{2}\hbar\omega$ the *zero-point energy* . If we imagine a system made of a very large number of oscillators, the energy of the fundamental state will be very large or even infinite if we have an infinite number of oscillators. This behaviour is not acceptable because we would like the energy of the ground state to be 0, or at least finite. Such a difficulty is solved by modifying the Hamiltonian in such a way that its lower eigenvalue is 0, $H_{\text{ren}}\Phi_0 = 0$. The new Hamiltonian is, for one oscillator, $H_{\text{ren}} = H_0 - 1/2\hbar\omega$. The change from H_0 to H_{ren} is called *a renormalisation*. For a finite system it is just an additive constant. In the absence of gravitational interactions, which couple to any kind of energy, only energy differences are important, so the modification is not physically observable. The important point is that the resulting

Hamiltonian is positive definite. As we shall see later, this kind of renormalisation pre-scription will present some technical difficulties for systems which behave as if they have an infinite number of oscillators.

To link the creator and annihilator formalism to the functional integral formalism, we must introduce the notion of chronological product or T-product. The natural frame-work for this analysis is the interaction representation. We define for two operators of this representation (thus, time dependent) $\mathcal{O}_1(t_1)$ and $\mathcal{O}_2(t_2)$

$$T(\mathcal{O}_1(t_1)\mathcal{O}_2(t_2)) = \theta(t_1 - t_2)\mathcal{O}_1(t_1)\mathcal{O}_2(t_2) + \theta(t_2 - t_1)\mathcal{O}_2(t_2)\mathcal{O}_1(t_1). \tag{8.85}$$

Let us consider the simple case with $\mathcal{O} = \hat{q}$, the position operator of the harmonic os-cillator, and consider the vacuum expectation value of the T-product of two q at different times; this expectation value is called a *propagator*,

$$< 0| T(\hat{q}(t_1)\hat{q}(t_2))|0 > . \tag{8.86}$$

We first need to compute the value of the operator $\hat{q}(t)$ in terms of annihilators and creators.

From $\hat{q} = \sqrt{\frac{\hbar}{2m\omega}}(a + a^\dagger)$, we get

$$\begin{aligned}
\hat{q}(t) &= e^{it\hat{H}_0/\hbar}\hat{q}e^{-it\hat{H}_0/\hbar} \\
&= \hat{q} + \frac{it}{\hbar}[\hat{H}_0, \hat{q}] + \frac{(it)^2}{2!\hbar^2}[\hat{H}_0, [\hat{H}_0, \hat{q}]] + \dots \\
&= \sqrt{\frac{\hbar}{2m\omega}}(ae^{-it\omega} + a^\dagger e^{it\omega}),
\end{aligned} \tag{8.87}$$

where we used

$$[\hat{H}_0, a] = -\hbar\omega a \qquad [\hat{H}_0, a^\dagger] = \hbar\omega a^\dagger \tag{8.88}$$

We thus get

$$< 0|\hat{q}(t_1)\hat{q}(t_2)|0 > \equiv < \hat{q}(t_1)\hat{q}(t_2) > = \frac{\hbar}{2m\omega}e^{-i\omega(t_1-t_2)}. \tag{8.89}$$

Therefore,

$$\begin{aligned}
< T(\hat{q}(t_1)\hat{q}(t_2)) > &= \frac{\hbar}{2m\omega}\left(\theta(t_1 - t_2)\,e^{-i\omega(t_1-t_2)} + \theta(t_2 - t_1)\,e^{i\omega(t_1-t_2)}\right) \\
&= \frac{i\hbar}{2\pi m}\int\frac{e^{-ik_0(t_1-t_2)}}{k_0^2 - \omega^2 + i\varepsilon}dk_0 \\
&= -iG_F(t_1, t_2).
\end{aligned} \tag{8.90}$$

This is a Green function similar to that shown as being interesting in the study of the Klein–Gordon equation. The present case corresponds to the case[11] of 0 space-dimension and 1 time-dimension. It is called the *Feynman propagator*.

We will link this expression to the path integrals.

Up to now, we have defined the path integral by fixing the end points of the path over which we perform the integration, a condition which can be read directly on the Feynman formula (8.64). Suppose now that we consider an integral corresponding to paths going from q at time t to q' at time t'.

Let us consider, formally, the eigenstates of the position operator $\hat{q}$: $\hat{q}|q> = q|q>$.[12]

Since the kets $|q>$ form a basis, we can define path integrals between any states; in particular, we can choose the vacuum states as the initial and final states. Using the fact that

$$\int_{-\infty}^{+\infty} |q><q|0> \, \mathrm{d}q = |0>, \tag{8.91}$$

We define

$$\int \mathcal{D}(q)\mathrm{e}^{\mathrm{i}\int_t^{t'} L_0 \mathrm{d}\tau} = \int \mathrm{d}q'\mathrm{d}q <0|q'> \int_{\substack{q(t')=q' \\ q(t)=q}} \mathcal{D}(q)\mathrm{e}^{\mathrm{i}\int_t^{t'} L_0 \mathrm{d}\tau} <q|0>, \tag{8.92}$$

where $<q|0>$ is the wave function $\Phi_0(q)$ of the fundamental state of the harmonic oscillator. With this definition,[13] from the Feynman formula it follows that

$$\int \mathcal{D}(q)\mathrm{e}^{\mathrm{i}\int_t^{t'} L_0 \mathrm{d}\tau} = <0|\mathrm{e}^{-\mathrm{i}(t'-t)\hat{H}_0/\hbar}|0> = \mathrm{e}^{-\frac{\mathrm{i}\omega}{2}(t'-t)}, \tag{8.93}$$

where the last equality can be checked using the operator formalism.

We now define

$$\int \mathcal{D}(q)q(t_1)q(t_2)\mathrm{e}^{\mathrm{i}\int_t^{t'} L_0 \mathrm{d}\tau} \tag{8.94}$$

---

[11] Warning: by dimension we mean the dimension of the space of parameters. Quantum mechanics corresponds to a one-parameter theory: the time. The basic objects can be scalar functions $q(t)$ for the quantum mechanics on the line or three functions $q(t) = (q_1(t), q_2(t), q_3(t))$ for the quantum mechanics in the usual three-dimensional space.

[12] In standard quantum mechanics the position operator has a continuous spectrum; therefore, the eigenstates are not square integrable functions. All formulae remain formally the same if what orthogonality conditions the Kronecker delta is replaced by the Dirac delta function and summations over indices are replaced by integrals. If we wish to give a precise meaning to the states $|q>$ we must go through the lattice approximation.

[13] We omit to indicate the boundary conditions on the trajectories in the case it is given by the vacuum state.

for $t_1, t_2 \in [t, t']$. Let us suppose that $t_1 > t_2$. Using the factorisation property of the probability amplitude (8.5), we interpret $q(t_i)$ as the multiplication by q at time t_i and the preceding expression can be written as

$$\int \mathrm{d}q_1 \mathrm{d}q_2 \left\{ \left(\int_{q(t)=q_2} \mathcal{D}(q) \mathrm{e}^{\mathrm{i} \int_t^{t_2} L_0 \mathrm{d}\tau} \right) q_2 \right. \tag{8.95}$$

$$\left. \cdot \left(\int_{\substack{q(t_1)=q_1 \\ q(t_2)=q_2}} \mathcal{D}(q) \mathrm{e}^{\mathrm{i} \int_{t_2}^{t_1} L_0 \mathrm{d}\tau} \right) q_1 \left(\int_{q(t_1)=q_1} \mathcal{D}(q) \mathrm{e}^{\mathrm{i} \int_{t_1}^{t'} L_0 \mathrm{d}\tau} \right) \right\},$$

namely, thanks to the Feynman formula

$$\int \mathrm{d}q' \mathrm{d}q < 0|\mathrm{e}^{-\mathrm{i}(t'-t_1)\hat{H}_0}|q> q < q|\mathrm{e}^{-\mathrm{i}(t_1-t_2)\hat{H}_0}|q'> q' < q'|\mathrm{e}^{-\mathrm{i}(t_2-t)\hat{H}_0}|0>$$

$$= \int \mathrm{d}q' \mathrm{d}q < 0|\mathrm{e}^{-\mathrm{i}(t'-t_1)\hat{H}_0} q \mathrm{e}^{-\mathrm{i}(t_1-t_2)\hat{H}_0} q' \mathrm{e}^{-\mathrm{i}(t_2-t)\hat{H}_0}|0>$$

$$= < 0|\mathrm{e}^{-\mathrm{i}t'\hat{H}_0} q(t_1) q'(t_2) \mathrm{e}^{\mathrm{i}t\hat{H}_0}|0>$$

$$= \mathrm{e}^{-\frac{\mathrm{i}\omega}{2}(t'-t)} < 0|q(t_1)q(t_2)|0> . \tag{8.96}$$

Doing this calculation again but now reversing the time ordering, we get

$$\int \mathcal{D}(q) q(t_1) q(t_2) \mathrm{e}^{\mathrm{i} \int_t^{t'} L_0 \mathrm{d}\tau}$$

$$= \mathrm{e}^{-\frac{\mathrm{i}\omega}{2}(t'-t)} (\theta(t_1-t_2) < 0|q(t_1)q(t_2)|0> + \theta(t_2-t_1) < 0|q(t_2)q(t_1)|0>)$$

$$= \mathrm{e}^{-\frac{\mathrm{i}\omega}{2}(t'-t)} < 0|T(q(t_1)q(t_2))|0> \tag{8.97}$$

under the condition that t_1 and t_2 are between t' and t .

We have therefore rigorously proved

$$< 0|T(q(t_1)q(t_2))|0> = \lim_{t' \to +\infty, t \to -\infty} \frac{\int q(t_1)q(t_2)\mathrm{e}^{\mathrm{i} \int_t^{t'} L_0 \mathrm{d}\tau} \mathcal{D}(q)}{\int \mathrm{e}^{\mathrm{i} \int_t^{t'} L_0 \mathrm{d}\tau} \mathcal{D}(q)}, \tag{8.98}$$

the limit being obvious since as soon as $t < t_1, t_2 < t'$ the functional integral is independent of t and t'.

In the next chapter, we will show more systematically how the path integral makes it possible to reformulate important properties of quantum mechanics.

8.7 The Bargmann Representation

8.7.1 The Coherent States

In this section, we will introduce another representation of quantum mechanics which will be useful in the sequel.

We define, for any $z \in \mathbb{C}$, *the coherent state* $|z>$ by[14]

$$|z >= e^{za^\dagger}|0 > . \tag{8.99}$$

The scalar product between two coherent states is given by

$$< z_1|z_2 >= e^{\bar{z}_1 z_2}. \tag{8.100}$$

To each vector $\psi \in L^2(\mathbb{R})$ corresponds an antiholomorphic function

$$\psi(\bar{z}) =< z|\psi >, \qquad z \in \mathbb{C}. \tag{8.101}$$

This is the so-called *Bargmann representation*.

The scalar product between two vectors in the Bargmann representation is given by

$$(\psi|\chi) = \int \overline{\psi(\bar{z})} \chi(\bar{z}) e^{-|z|^2} \frac{d\bar{z}dz}{2\pi i}$$

$$= \int < \psi|z >< z|\chi > e^{-|z|^2} \frac{d\bar{z}dz}{2\pi i}. \tag{8.102}$$

The last equality is nothing else than

$$\int |z >< z|e^{-|z|^2} \frac{d\bar{z}dz}{2\pi i} = 1, \tag{8.103}$$

an identity that we now prove.

Writing $z = x + iy$, we have $\dfrac{d\bar{z}dz}{2\pi i} = \dfrac{dxdy}{\pi}$.

From (8.82), we get

$$|z >= \sum_{n=0}^{+\infty} \frac{\bar{z}^n (a^\dagger)^n}{n!}|0 >= \sum_{n=0}^{+\infty} \frac{\bar{z}^n}{\sqrt{n!}}|n >,$$

---

[14] The coherent states are introduced here as an alternative description of the quantum mechanical space of states. Because of their simple evolution law, they offer a convenient basis for formulating the path integral. In Chapter 22 we will come back to this definition and show that the coherent states have remarkable physical properties which make it possible to study the classical limit of a quantum mechanical system.

from which it follows, using (8.84), that

$$
\int |z><z| e^{-|z|^2} \frac{d\bar{z}dz}{2\pi i} = \int \sum_{n=0}^{+\infty} \sum_{m=0}^{+\infty} \frac{\bar{z}^m z^n}{\sqrt{m!}\sqrt{n!}} |n><m| e^{-|z|^2} \frac{dxdy}{\pi}
$$

$$
= \sum_{n=0}^{+\infty} \sum_{m=0}^{+\infty} \frac{1}{\sqrt{m!}\sqrt{n!}} |n><m| \int \bar{z}^m z^n e^{-|z|^2} \frac{dxdy}{\pi}
$$

$$
= \sum_{n=0}^{+\infty} |n><n| = 1 \tag{8.104}
$$

since

$$
\int \bar{z}^m z^n e^{-|z|^2} \frac{dxdy}{\pi} = \frac{1}{\pi} \int_0^{+\infty} e^{-\rho^2} \rho^{(n+m+1)} d\rho \int_0^{2\pi} e^{i\theta(n-m)} d\theta = n!\delta_{nm}.
$$

Any function $f \in L^2(\mathbb{R})$ can be expanded over the orthonormal basis of the harmonic oscillator wave functions $\Phi_n(x) = <x|n>$,

$$
f(x) = \sum_{n=0}^{\infty} f_n \Phi_n(x), \tag{8.105}
$$

where

$$
f_n = \int f(x) \Phi_n(x).
$$

We can check that

$$
f \in L^2(\mathbb{R}) \iff \sum_{n=0}^{+\infty} |f_n|^2 < +\infty. \tag{8.106}
$$

To each function $f \in L^2(\mathbb{R})$, there corresponds a function f in the Bargmann space, its Bargmann transform, given by

$$
f(\bar{z}) = <z|f> = \int_{-\infty}^{+\infty} B(z,x) f(x) dx = \int_{-\infty}^{+\infty} <z|x><x|f> dx
$$

$$
= \int_{-\infty}^{+\infty} <z|x> f(x) dx; \tag{8.107}
$$

thus, the kernel of the Bargmann transform is

$$
B(z,x) = <z|x> .
$$

The function $f(\bar{z})$ is an entire antiholomorphic function (of type $\frac{1}{2}$) of z. It has an expansion

$$f(\bar{z}) = \sum_{n=0}^{+\infty} <z|n><n|f> = \sum_{n=0}^{+\infty} \Phi_n(\bar{z}) <n|f> = \sum_{n=0}^{\infty} \frac{1}{\sqrt{n!}} f_n \bar{z}^n, \qquad (8.108)$$

where the functions

$$\Phi_n(\bar{z}) = <z|n> = \frac{\bar{z}^n}{\sqrt{n!}}$$

are the eigenfunctions of the Hamiltonian H_0 and of the particle number operator N.

In the Bargmann representation, the creation and annihilation operators, a and $a^\dagger$, are, respectively, represented by z and $\frac{d}{dz}$. They obviously satisfy the commutation relations and are formally adjoint with respect to the scalar product we introduced in Eq. (8.102).

In fact,

$$a|z> = z|z>, \qquad a^\dagger|z> = \frac{d}{dz}|z>.$$

Thus,

$$<z|a|\psi> = <0|e^{\bar{z}a}a|\psi> = \frac{d}{d\bar{z}}<0|e^{\bar{z}a}|\psi>$$

or

$$(a\psi)(\bar{z}) = \frac{d}{d\bar{z}}\psi(\bar{z}).$$

Similarly, We find that

$$(a^\dagger\psi)(\bar{z}) = \bar{z}\psi(\bar{z}).$$

The complete space of vectors on which $a^\dagger$ and a act is obtained as the closure of linear combinations of monomials with respect to the norm defined by the scalar product (8.102).

We now look for the expression of linear operators on $L^2(\mathbb{R})$ in the Bargmann representation.

Let A be a linear operator on $L^2(\mathbb{R})$. There will be two ways of representing operators in the Bargmann space.

Let us first consider an expression related to the decomposition of the operator according to the eigenfunction basis. A can be written as

$$A = \sum_{n,m=0,1,2\cdots} |n> A_{nm} <m|,$$

which suggests to define a kernel for A in the Bargmann space

$$A(\bar{z}, z) = \sum_{n,m} A_{nm} \frac{\bar{z}^n}{\sqrt{n!}} \frac{z^m}{\sqrt{m!}}. \tag{8.109}$$

The obvious action on an arbitrary function $f(\bar{z})$ is given by

$$(Af)(\bar{z}) = \int A(\bar{z}, \xi) f(\bar{\xi}) e^{-|\xi|^2} \frac{d\bar{\xi}\, d\xi}{2\pi i} \tag{8.110}$$

and the product of two operators is given by

$$(A_1 A_2)(\bar{z}, z) = \int A_1(\bar{z}, \xi) A_2(\bar{\xi}, z) e^{-|\xi|^2} \frac{d\bar{\xi}\, d\xi}{2\pi i} \tag{8.111}$$

It is clear from these formulae that $\bar{z}$ and z have to be considered as two independent variables.

Let us check that formula (8.110) is satisfied by our definition of the kernel of A given by formula (8.109). In fact

$$(Af)(\bar{z}) = <z|A|f> = \sum_{n,m} <z|n><n|A|m><m|f>$$

$$= \sum_{n,m} <z|n> A_{nm} \int <m|\xi><\xi|f> e^{-|\xi|^2} \frac{d\bar{\xi}\, d\xi}{2\pi i}.$$

$$= \int \sum_{n,m} A_{nm} \frac{\bar{z}^n}{\sqrt{n!}} \frac{\xi^m}{\sqrt{m!}} f(\bar{\xi}) e^{-|\xi|^2} \frac{d\bar{\xi}\, d\xi}{2\pi i}$$

$$= \int A(\bar{z}, \xi) f(\bar{\xi}) e^{-|\xi|^2} \frac{d\bar{\xi}\, d\xi}{2\pi i}.$$

Formula (8.111) follows immediately by iteration.

We will now choose to privilege the algebraic structure built by the annihilation and creation operators.

Let us expand A as

$$A = \sum_{n,m} \mathcal{A}_{nm} (a^\dagger)^n a^m. \tag{8.112}$$

Any linear operator on $L^2(\mathbb{R})$ can be expanded as a power series in $a^\dagger$ and a. Since the creation and the annihilation operators do not commute, it is usual to write products in a and $a^\dagger$ in a special order, the *normal order* or *normal product* (it will be also called *Wick order* later on). The normal ordering is defined by the fact that in any monomial made of creation and annihilation operators, the annihilation operators must be written to the right of the creation operators. The normal order of an expression is usually represented by writing the expression between two double-dots:

$$\text{Normal order of } A(a^\dagger, a) =: A(a^\dagger, a):$$

In particular

$$: a^{n_1}(a^\dagger)^{m_1} \cdots a^{n_N}(a^\dagger)^{m_M} := (a^\dagger)^{\sum_{j=1}^M m_j} a^{\sum_{i=1}^N n_i}. \tag{8.113}$$

The expression given by formula (8.112) is the normal ordered expansion of A. We define the normal kernel of A expressed as in (8.112) to be

$$\mathcal{A}(\bar{z}, z) \overset{\text{def}}{=} \sum_{n,m} A_{nm}(\bar{z})^n z^m \tag{8.114}$$

We claim that the link between the kernel of $A(\bar{z}, z)$ and its normal kernel $\mathcal{A}(\bar{z}, z)$ is given by

$$A(\bar{z}, z) = e^{\bar{z}z} \mathcal{A}(\bar{z}, z). \tag{8.115}$$

To prove this identity, it is enough to choose A to be a monomial in $a^\dagger$ and a:

$$A = (a^\dagger)^k a^l.$$

Thus, since for this operator $A_{nm} = \delta_{n,k} \delta_{m,l}$

$$\mathcal{A}(\bar{z}, z) = \bar{z}^k z^l. \tag{8.116}$$

Let us now compute $A(\bar{z}, z)$ using (8.80):

$$A_{nm} = <n|(a^\dagger)^k a^l|m>$$

$$= \sqrt{n(n-1)\cdots(n-k+1)}\sqrt{m(m-1)\cdots(m-l+1)}\theta(n-k)$$

$$\cdot \theta(m-l) <n-k|m-l>$$

$$= \sqrt{n(n-1)\cdots(n-k+1)}\sqrt{m(m-1)\cdots(m-l+1)}\theta(n-k)$$

$$\cdot \theta(m-l)\delta_{n-k,m-l}, \tag{8.117}$$

where the θ-functions ensure that it is only when $n \geq k$ and $m \geq l$ that the kernel is different from 0.

Therefore,

$$
\begin{aligned}
A(\bar{z}, z) &= \sum_{n,m} A_{nm} \frac{\bar{z}^n}{\sqrt{n!}} \frac{z^m}{\sqrt{m!}} \\
&= \bar{z}^k z^l \sum_{n,m} \theta(n-k)\theta(m-l) \frac{\bar{z}^{n-k}}{\sqrt{(n-k)!}} \frac{z^{(m-l)}}{\sqrt{(m-l)!}} \delta_{n-k,m-l} \\
&= \bar{z}^k z^l \sum_p \frac{\bar{z}^p z^p}{p!} = \bar{z}^k z^l e^{\bar{z}z} = \mathcal{A}(\bar{z}, z) e^{\bar{z}z}.
\end{aligned}
\tag{8.118}
$$

Remark that this result shows that, as expected,

$$
< z \mid A \mid z > = A(\bar{z}, z)
\tag{8.119}
$$

since

$$
\begin{aligned}
< z \mid A \mid z > &= < z \mid (a^\dagger)^k a^l \mid z > = \bar{z}^k z^l < z \mid z > = \bar{z}^k z^l e^{\bar{z}z} \\
&= \mathcal{A}(\bar{z}, z) e^{\bar{z}z} = A(\bar{z}, z),
\end{aligned}
$$

where the last two equalities result from formulae (8.116) and (8.118).

By linearity, this result proves formulae (8.115) and (8.119) in the general case.

8.7.2 The Path Integral Formula in the Bargmann Space

We want to prove in this section a path integral formula similar to the Feynman formula (8.64).

Having in mind the harmonic oscillator model, we will take, for Hamiltonian function H, a normal ordered function of $a^\dagger$ and a. Therefore,

$$
H(a^\dagger, a) =: H(a^\dagger, a) : .
$$

Note that for the Hamiltonian of the harmonic oscillator normal ordering is just the renormalisation we introduced previously, namely $: H := H_{ren}$.

From the previous discussion, we find, with $\varepsilon = \frac{t_f - t_i}{N}$, $\bar{z}_N = \bar{z}$ and $z_0 = \zeta$,

$$
\begin{aligned}
< z \mid e^{-i(t_f - t_i)H} \mid \zeta > &= < z \mid (e^{-i\varepsilon H})^N \mid \zeta > \\
&= \int \prod_{j=1}^{N-1} \frac{d\bar{z}_j dz_j}{2\pi i} e^{-|z_j|^2} \prod_{i=0}^{N-1} < z_{i+1} \mid e^{-i\varepsilon H} \mid z_j > \\
&\simeq \int \prod_{j=1}^{N-1} \frac{d\bar{z}_j dz_j}{2\pi i} e^{-|z_j|^2} \prod_{i=0}^{N-1} < z_{i+1} \mid 1 - i\varepsilon H \mid z_i > \\
&\simeq \int \prod_{j=1}^{N-1} \frac{d\bar{z}_j dz_j}{2\pi i} e^{-|z_j|^2} \prod_{i=0}^{N-1} e^{\bar{z}_{i+1} z_i - i\varepsilon \mathcal{H}(\bar{z}_{i+1}, z_i)}.
\end{aligned}
\tag{8.120}
$$

We can rewrite this last expression in a way suitable to take the limit when $\varepsilon \to 0$:

$$\lim_{\varepsilon \to 0} \int \prod_{j=1}^{N-1} \frac{d\bar{z}_j dz_j}{2\pi i} e^{-|z_j|^2} \prod_{i=0}^{N-1} e^{\bar{z}_{i+1} z_i - i\varepsilon \mathcal{H}(\bar{z}_{i+1}, z_i)}$$

$$= \lim_{\varepsilon \to 0} \int \prod_{j=1}^{N-1} \frac{d\bar{z}_j dz_j}{2\pi i} e^{\frac{1}{2}(\bar{z}_1 \zeta + \bar{z} z_{N-1})} e^{\frac{1}{2} \sum_{i=1}^{N-1} [(\bar{z}_{i+1} - \bar{z}_i) z_i + \bar{z}_i (z_{i-1} - z_i)]} e^{-i\varepsilon \sum_{i=0}^{N-1} \mathcal{H}(\bar{z}_{i+1}, z_i)}$$

$$\overset{\text{def}}{=} \int_{\substack{\bar{z}(t_f) = \bar{z} \\ z(t_i) = \zeta}} \mathcal{D}\bar{z}\mathcal{D}z \, e^{\frac{1}{2}(\bar{z}(t_i)\zeta + \bar{z} z(t_f)) + i \int_{t_i}^{t_f} ds[\frac{1}{2i}(\dot{\bar{z}}(s) z(s) - \bar{z}(s)\dot{z}(s)) - \mathcal{H}(\bar{z}(s), z(s))]}, \tag{8.121}$$

where the last integral is defined by the limit. In this definition of the path integral, $\bar{z}(s)$ and $z(s)$ are independent variables, $\bar{z}(s)$ is not the complex conjugate of $z(s)$, and the integral is defined as the sum over all the complex trajectories,

$$s \in [t_i, t_f] \to \bar{z}(s)$$
$$s \in [t_i, t_f] \to z(s)$$

with the boundary conditions $\bar{z}(t_f) = \bar{z}$ and $z(t_i) = \zeta$.

Remark that in the final formula, the term

$$\int_{t_i}^{t_f} \frac{1}{2i}(\dot{\bar{z}}(s) z(s) - \bar{z}(s)\dot{z}(s)) - \mathcal{H}(\bar{z}(s), z(s)) ds \tag{8.122}$$

is nothing else than the action in terms of the complex coordinates.

To conclude this section we will extend this analysis to some time-dependent Hamiltonians. More specifically, we will compute the functional integral for the harmonic oscillator in interaction with a time-dependent external source since, being Gaussian, it can be exactly solved.

A good approximation of the functional integral will be given, as explained in section 8.5, by the value of the integrand at the trajectories for which the action is stationary (there is an analogous oscillatory argument if we reintroduce the $\hbar$ dependence).

Let us consider, following (8.122), a general action

$$\int_{t_i}^{t_f} [\frac{1}{2i}(\dot{\bar{z}}(s) z(s) - \bar{z}(s)\dot{z}(s)) - \mathcal{H}(\bar{z}(s), z(s); s)] ds. \tag{8.123}$$

This action is extremal if the change of the action

$$\delta\left(\int_{t_i}^{t_f} [\frac{1}{2i}(\dot{\bar{z}}(s) z(s) - \bar{z}(s)\dot{z}(s)) - \mathcal{H}(\bar{z}(s), z(s); s)] ds\right) = 0$$

under an infinitesimal change $\bar{z}(s) \to \bar{z}(s) + \delta\bar{z}(s)$ and $z(s) \to z(s) + \delta z(s)$ with the given boundary conditions, i.e. $\delta\bar{z}(t_f) = \delta z(t_i) = 0$. This is the case if the paths $\bar{z}(s)$ and $z(s)$ satisfy the equations for stationarity

$$\dot{\bar{z}}(s) = i\frac{\partial \mathcal{H}}{\partial z(s)}$$

$$\dot{z}(s) = -i\frac{\partial \mathcal{H}}{\partial \bar{z}(s)}.$$
(8.124)

If we now restrict our attention to the harmonic oscillator Hamiltonian with external time-dependent sources $H = \omega a^* a + \bar{\xi}(t)a + a^*\xi(t)$, these equations are now

$$\dot{\bar{z}}(s) = i\omega\bar{z}(s) + i\bar{\xi}(s)$$

$$\dot{z}(s) = -i\omega z(s) - i\xi(s)$$
(8.125)

with the boundary conditions $\bar{z}(t_f) = \bar{z}$ and $z(t_i) = \zeta$. They have unique solutions given by

$$\bar{z}(t) = \bar{z}e^{-i\omega(t_f - t)} - ie^{i\omega t}\int_t^{t_f} e^{-i\omega s}\bar{\xi}(s)ds$$

$$z(t) = \zeta e^{-i\omega(t - t_i)} - ie^{-i\omega t}\int_{t_i}^t e^{i\omega s}\xi(s)ds.$$
(8.126)

Expressing the exponent of the functional integral in (8.121) in terms of these solutions, we find

$$\bar{z}\zeta e^{-i\omega(t_f - t_i)} \quad -i\bar{z}\int_{t_i}^{t_f} e^{-i\omega(t_f - s)}\xi(s)ds$$

$$-i\zeta\int_{t_i}^{t_f} e^{-i\omega(s - t_i)}\bar{\xi}(s)ds - \int_{t_i}^{t_f}ds\int_{t_i}^s e^{-i\omega(s-u)}\bar{\xi}(s)\xi(u)du$$

$$= \bar{z}\zeta e^{-i\omega(t_f - t_i)} \quad -i\bar{z}\int_{t_i}^{t_f} e^{-i\omega(t_f - s)}\xi(s)ds - i\zeta\int_{t_i}^{t_f} e^{-i\omega(s - t_i)}\bar{\xi}(s)ds$$

$$-\frac{1}{2}\int_{t_i}^{t_f}\int_{t_i}^{t_f} e^{-i\omega|u-s|}\bar{\xi}(s)\xi(u)duds.$$
(8.127)

The value of the functional integral is in fact equal to the exponential of the above expression. This is a general feature that the path integral of the exponential of a quadratic form is given, up to some normalisation factor, by the exponential of this quadratic form computed on the stationary trajectories.

Let us give a proof of this assertion in the discrete case.

Let Q be the quadratic form

$$Q(\bar{z}, z) = \sum_{i,j}\bar{z}_i A_{ij}z_j + \sum_k(\bar{\xi}_k z_k + \xi_k\bar{z}_k).$$

Its critical or stationary points $\bar{u}$ and u are the solutions of the equations

$$\frac{\partial}{\partial \bar{z}_i} Q(\bar{z}, z) = \frac{\partial}{\partial z_i} Q(\bar{z}, z) = 0 \quad \forall i.$$

We find that

$$u_i = -\sum_j (A^{-1})_{ij} \xi_j \tag{8.128}$$

$$\bar{u}_i = \sum_j \bar{\xi}_j (A^{-1})_{ji} \tag{8.129}$$

and thus

$$Q(\bar{u}, u) = \bar{\xi} A^{-1} A A^{-1} \xi - \bar{\xi} A^{-1} \xi - \bar{\xi} A^{-1} \xi = -\bar{\xi} A^{-1} \xi$$
$$= -\sum_{i,j} \bar{\xi}_i (A^{-1})_{ij} \xi_j. \tag{8.130}$$

We can prove that

$$\int_{\mathcal{C}^N} e^{-Q(\bar{z}, z)} \prod_{i=1}^N \frac{d\bar{z}_i dz_i}{2\pi i} = (\det A)^{-1} e^{-Q(\bar{u}, u)} \tag{8.131}$$

provided $A + A^\dagger > 0$.

8.8   Problems

Problem 8.1 The eikonal approximation of the generalised Klein–Gordon equation.

Prove that Eq. (8.1) in the eikonal approximation describes the propagation according to geometrical optics, namely along rays perpendicular to wave fronts.

Problem 8.2 Classical mechanics as the eikonal limit of Schrödinger's equation. Prove that for the Schrödinger equation the eikonal approximation is equivalent to the $\hbar \to 0$ classical limit.

Problem 8.3 Completeness of the set of eigenfunctions of the harmonic oscillator. Let $f \in L^2(\mathbb{R})$. Prove that if $\forall n \in \mathbb{Z}$, $(f|\Phi_n) = 0$, then $f \equiv 0$. Deduce from this result that $\sum_n |n><n| = 1$.

Hint: Use the generating function of Hermite polynomials to show that the Fourier transform of f is identically 0.

Problem 8.4 Orthonormality of the Hermite polynomials.

Prove by using the generating function of Hermite polynomials that

$$\int H_n(x)H_m(x)\,dx = \sqrt{\pi}\,(\frac{d}{dt})^n(\frac{d}{ds})^m e^{2ts}\Big|_{t=s=0}$$

Problem 8.5 Bargmann transforms
Prove that the Bargmann transform kernel $B(z,x)$ is explicitly given by

$$B(z,x) = \frac{1}{\pi^{\frac{1}{4}}}e^{-\frac{\bar{z}^2}{2}-\sqrt{2}\bar{z}x-\frac{x^2}{2}}$$

Hint: Compute $< z|x > = \sum_n < z|n > < n|x >$.
Give the explicit form of the inverse transform.

Problem 8.6 Show that

$$|0 > < 0| =: e^{-a^\dagger a}.$$

Hint: Show that $a|0 > < 0| = |0 > < 0|a^\dagger = 0$.

9

Functional Integrals and Quantum Mechanics: Formal Developments

---

9.1 T-Products

9.1.1 General Definition

In the preceding chapter we introduced the notion of *chronological product* or T-*product* of two operators $q(t)$ in order to establish the link between the creator-annihilator formalism and the functional integral. This notion can be generalised to any time-dependent operators and to an arbitrary number of them. For two time-dependent operators $\mathcal{O}_1(t_1)$ and $\mathcal{O}_2(t_2)$, we defined

$$T(\mathcal{O}_1(t_1)\mathcal{O}_2(t_2)) = \theta(t_1 - t_2)\mathcal{O}_1(t_1)\mathcal{O}_2(t_2) + \theta(t_2 - t_1)\mathcal{O}_2(t_2)\mathcal{O}_1(t_1). \tag{9.1}$$

More generally, we define the T-product of n operators $\mathcal{O}_1(t_1), \mathcal{O}_2(t_2), \cdots, \mathcal{O}_n(t_n)$, at distinct times, by

$$T(\mathcal{O}_1(t_1), \mathcal{O}_2(t_2), \cdots, \mathcal{O}_n(t_n)) = \mathcal{O}_{\sigma(1)}(t_{\sigma(1)})\mathcal{O}_{\sigma(2)}(t_{\sigma(2)}) \cdots \mathcal{O}_{\sigma(n)}(t_{\sigma(n)}), \tag{9.2}$$

where σ is a permutation of $(1, 2, \cdots, n)$ such that the times $t_{\sigma(j)}$, $j = 1, \cdots, n$, are ordered:

$$t_{\sigma(1)} > \cdots > t_{\sigma(n)}. \tag{9.3}$$

It results obviously from this formula that the T-product is a symmetric function of the operators, i.e. such that for any permutation σ

$$T(\mathcal{O}_1(t_1), \mathcal{O}_2(t_2), \cdots, \mathcal{O}_n(t_n)) = T(\mathcal{O}_{\sigma(1)}(t_{\sigma(1)})\mathcal{O}_{\sigma(2)}(t_{\sigma(2)}) \cdots \mathcal{O}_{\sigma(n)}(t_{\sigma(n)})).$$

These products have been defined only for distinct times. It is possible, as in the case of the T-product of two operators, to rewrite the general definition by introducing Heaviside (or step) functions:

From Classical to Quantum Fields. Laurent Baulieu, John Iliopoulos and Roland Sénéor.
© Laurent Baulieu, John Iliopoulos and Roland Sénéor, 2017. Published 2017 by Oxford University Press.

$$T(\mathcal{O}_1(t_1), \mathcal{O}_2(t_2), \cdots, \mathcal{O}_n(t_n)) \tag{9.4}$$
$$= \sum_{\sigma} \theta(t_{\sigma(1)} - t_{\sigma(2)}) \cdots \theta(t_{\sigma(n-1)} - t_{\sigma(n)}) \times \mathcal{O}_{\sigma(1)}(t_{\sigma(1)}) \mathcal{O}_{\sigma(2)}(t_{\sigma(2)}) \cdots \mathcal{O}_{\sigma(n)}(t_{\sigma(n)}).$$

The definition of Eq. (9.4) is the source of some of the infinite quantities which appear in field theories. T-products will have to be handled with care when applied to quantum fields because then we will have the difficulties which may appear when taking products of distributions: the Heaviside functions and quantum fields which are distribution-valued operators. In all the physically interesting cases, either these products of distributions exist and will be defined by continuity[1] from the values at non-coinciding times or there exists a difficulty at coinciding times, the expressions being ill-defined or infinite when some of the arguments coincide. Thus, T-products are *a priori* defined up to local terms, i.e. delta functions or derivatives of delta functions whose arguments are the difference of coordinates of pairs of space–time points: finding these local terms is typically what is called a renormalisation problem. This problem will be solved by introducing more global physical constraints which complete the definition of T-products and determine without ambiguity the nature of the local terms.

However, in this chapter we will limit ourselves to a formal definition where the time ordering is given explicitly by products of Heaviside θ functions. Thus, for example, the T-product of three operators will be written as

$$T(\mathcal{O}_1(t_1)\mathcal{O}_2(t_2)\mathcal{O}_3(t_3)) = \theta(t_1 - t_2)\theta(t_2 - t_3)\mathcal{O}_1(t_1)\mathcal{O}_2(t_2)\mathcal{O}_3(t_3)$$
$$+ \theta(t_1 - t_3)\theta(t_3 - t_2)\mathcal{O}_1(t_1)\mathcal{O}_3(t_3)\mathcal{O}_2(t_2) + \theta(t_2 - t_1)\theta(t_1 - t_3)\mathcal{O}_2(t_2)\mathcal{O}_1(t_1)\mathcal{O}_3(t_3)$$
$$+ \theta(t_2 - t_3)\theta(t_3 - t_1)\mathcal{O}_2(t_2)\mathcal{O}_3(t_3)\mathcal{O}_1(t_1) + \theta(t_3 - t_1)\theta(t_1 - t_2)\mathcal{O}_3(t_3)\mathcal{O}_1(t_1)\mathcal{O}_2(t_2)$$
$$\theta(t_3 - t_2)\theta(t_2 - t_1)\mathcal{O}_3(t_3)\mathcal{O}_2(t_2)\mathcal{O}_1(t_1). \tag{9.5}$$

The vacuum expectation value of the T-product of n operators is usually called an *n-point function*.

9.1.2   Application to the Harmonic Oscillator

We choose to apply these ideas to the one-dimensional harmonic oscillator defined in Chapter 8. Let us compute the vacuum expectation value of the T-product of an arbitrary number of operators $q(t)$ using the formalism of creation and annihilation operators. Let us start with three $q(t)$'s. Suppose $t_1 > t_2 > t_3$; then

$$< 0| T(q(t_1)q(t_2)q(t_3))|0 >$$
$$= < 0|q(t_1)q(t_2)q(t_3)|0 >$$

---

[1] In the case of quantum mechanics, where the space dimension is 0, we should have delta functions $\delta(t_i - t_j)$, but since the theory is very regular, there is no problem in defining T-products by the formal multiplication by θ functions.

$$= (\frac{\hbar}{2m\omega})^{3/2} < 0|(ae^{-it_1\omega} + a^\dagger e^{it_1\omega})(ae^{-it_2\omega} + a^\dagger e^{it_2\omega})(ae^{-it_3\omega} + a^\dagger e^{it_3\omega})|0 >$$

$$= (\frac{\hbar}{2m\omega})^{3/2} e^{-it_1\omega} e^{it_3\omega} < 1|(ae^{-it_2\omega} + a^\dagger e^{it_2\omega})|1 >= 0. \qquad (9.6)$$

In fact, from the orthogonality properties of n-quanta states

$$< 1|a|1 >=< 1|a^\dagger|1 >= 0, \qquad (9.7)$$

from which follows that the vacuum expectation value (v.e.v.) of the T-product of three $q(t)$'s vanishes identically. This result can be generalised to an arbitrary *odd* number n of $q(t)$'s. In fact, expanding the product of $q(t)$ in terms of creator and annihilator operators, we get a sum of homogeneous monomials of a and $a^\dagger$. Using the commutation relations, let these monomials act on the ket $|0 >$; we then get a ket proportional to a state $|m >$, $m \geq 0$, or 0. The only possibility for the scalar product with $< 0|$ to be different from 0 is that $m = 0$. This is only possible if the initial monomial had as many creation as annihilation operators, i.e. an even number of operators.

Let us now treat the problem of the computation of the v.e.v. of the T-product of four operators:

$$< 0|T(q(t_1)q(t_2)q(t_3)q(t_4))|0 > . \qquad (9.8)$$

There exist 4! possible time orderings and each corresponding term written in terms of creation and annihilation operators generates 16 monomials in a and $a^\dagger$. Thus, if $t_1 > t_2 > t_3 > t_4$

$$< 0|T(q(t_1)q(t_2)q(t_3)q(t_4))|0 >$$
$$= < 0|q(t_1)q(t_2)q(t_3)q(t_4)|0 >$$
$$= e^{-i\omega(t_4-t_1)/2} < 0|qe^{-it_1 H/\hbar}q(t_2)q(t_3)e^{it_4 H/\hbar}q|0 >$$
$$= \frac{\hbar}{2m\omega}e^{-i\omega(t_4-t_1)/2}e^{i3\omega(t_4-t_1)/2} < 1|q(t_2)q(t_3)|1 >$$
$$= (\frac{\hbar}{2m\omega})^2 e^{-i\omega(t_1-t_4)}e^{-i\omega(t_2-t_3)}(< 1|aa^\dagger|1 > +e^{-i\omega(t_3-t_2)} < 1|a^\dagger a|1 >)$$
$$= 2(\frac{\hbar}{2m\omega})^2 e^{-i\omega(t_1-t_4)}e^{-i\omega(t_2-t_3)} = (\frac{\hbar}{2m\omega})^2 e^{-i\omega(t_1-t_4)}e^{-i\omega(t_3-t_2)}. \qquad (9.9)$$

We note that the result is the value, for the given time ordering, of

$$< 0|T(q(t_1)q(t_2))|0 >< 0|T(q(t_3)q(t_4))|0 >$$
$$+ < 0|T(q(t_1)q(t_3))|0 >< 0|T(q(t_2)q(t_4))|0 >$$
$$+ < 0|T(q(t_1)q(t_4))|0 >< 0|T(q(t_2)q(t_3))|0 > .$$

Repeating the analysis for all possible time orderings, we have thus proved that

$$< 0|T(q(t_1)q(t_2)q(t_3)q(t_4))|0 >$$
$$= < 0|T(q(t_1)q(t_2))|0 >< 0|T(q(t_3)q(t_4))|0 >$$
$$+ < 0|T(q(t_1)q(t_3))|0 >< 0|T(q(t_2)q(t_4))|0 >$$
$$+ < 0|T(q(t_1)q(t_4))|0 >< 0|T(q(t_2)q(t_3))|0 > . \qquad (9.10)$$

More generally, seeing how tedious the proof would be, we have

$$< 0|T(q(t_1)q(t_2) \cdots q(t_{2n}))|0 >$$
$$= \sum_{pairs} < 0|T(q(t_{i_1})q(t_{i_2}))|0 > \cdots < 0|T(q(t_{i_{2n-1}})q(t_{i_{2n}}))|0 > . \qquad (9.11)$$

In this formula, the sum is carried out over the $(2n-1)(2n-3)\cdots = \frac{2n!}{n!2^n}$ distinct pairs of variables (t_i, t_j). The preceding results, i.e. the formula (9.11) for an even number of $q(t)$ and the vanishing for an odd number, express the fact that the operator $q(t)$ is Gaussian. This is the consequence of the fact that its dynamics is described by a Lagrangian (or Hamiltonian) quadratic in q and $\dot{q}$ (or p). A characteristic of this Gaussian (or quadratic) nature is the result that the v.e.v. of the T-product of an arbitrary number of these operators can be solely expressed as a function of the v.e.v.'s of the T-products of two operators.

Formula (8.98) of Chapter 8 is easily generalised

$$< 0|T(q(t_1)q(t_2) \cdots q(t_{2n}))|0 > = \lim_{t' \to +\infty, t \to -\infty} \frac{\int q(t_1) \cdots q(t_n) e^{i \int_t^{t'} L_0 d\tau} \mathcal{D}(q)}{\int e^{i \int_t^{t'} L_0 d\tau} \mathcal{D}(q)}. \qquad (9.12)$$

The right-hand side is indeed independent of t and t' as soon as $(t_1, \cdots, t_n) \in [t, t']$.

Equation (9.12) expresses the fact that $< 0|T(q(t_1)q(t_2) \cdots q(t_{2n}))|0 >$ are the moments of the Gaussian measure[2]

$$d\nu_0(q) = \frac{e^{i \int_t^{t'} L_0 d\tau} \mathcal{D}(q)}{\int e^{i \int_t^{t'} L_0 d\tau} \mathcal{D}(q)}. \qquad (9.13)$$

As a measure, this means that $d\nu_0(q)$ is a Gaussian measure of mean zero and of covariance (or propagator)

$$< 0|T(q(t_1)q(t_2))|0 > = -iG_F(t_1 - t_2). \qquad (9.14)$$

The fact that the mean is 0, i.e. $\int q(t) d\nu_0(q) = 0$, indicates that it is even; therefore, all its odd moments are 0. This last property can be directly seen in the definition of

---

[2] It is a complex measure.

path integrals. Indeed, since the Lagrangian is even in q, we can directly see on the finite approximations that they are invariant by the change, for all s, of $q(s)$ into $-q(s)$.

It is useful, whenever possible, to associate with a measure a generating functional from which the moments with respect to this measure can be obtained through functional derivatives. We thus define

$$Z_0[j] = \int e^{i \int_{-\infty}^{+\infty} j(s)q(s)\,ds}\,dv_0(q) \tag{9.15}$$

and

$$< 0|T(q(t_1)q(t_2)\cdots q(t_n))|0 > = i^{-n} \frac{\delta}{\delta j(t_1)} \cdots \frac{\delta}{\delta j(t_n)} Z_0[j]\big|_{j=0}, \tag{9.16}$$

or equivalently

$$Z_0[j] = < 0|T(e^{i \int_{-\infty}^{+\infty} j(s)q(s)\,ds})|0 > . \tag{9.17}$$

The fact that the odd moments vanish and that the other ones can be expressed in terms of G_F allows us to guess the possible form of $Z_0[j]$:

$$Z_0[j] = e^{i/2(j,G_F j)} \tag{9.18}$$

with

$$(j, G_F j) = \int_{-\infty}^{+\infty} \int_{-\infty}^{+\infty} j(u)\,G_F(u-v)j(v)\,du\,dv. \tag{9.19}$$

We can understand intuitively this formula by going back to the definition of the measure $dv_0(q)$. Indeed,

$$\frac{1}{\hbar} \int L_0\,d\tau = \frac{m}{2\hbar} \int q(\tau)\{-\frac{d^2}{d\tau^2} - \omega^2\}q(\tau)\,d\tau = \frac{m}{2\hbar} \int \left((\frac{dq(\tau)}{d\tau})^2 - \omega^2 q(\tau)^2\right)d\tau$$

$$= \frac{m}{4\pi\hbar} \int |\tilde{q}(k_0)|^2(k_0^2 - \omega^2)\,dk_0 = \frac{1}{2}(q, G^{-1}q), \tag{9.20}$$

where it is expected that the functions $q(\tau)$ have zero boundary values (or vanish at infinity) and where we have introduced the Fourier transform $\tilde{q}$ of q. The quadratic form G^{-1} is formally the Fourier transform $\tilde{G}^{-1}(k_0) = \frac{m}{2\hbar}(k_0^2 - \omega^2)$. Thus, by the change of variables, $q(\tau) - G * j(\tau) \to q(\tau)$,

$$\int e^{\frac{i}{2}(q,G^{-1}q)+i(q,j)}\mathcal{D}(q) = e^{\frac{i}{2}(j,Gj)}\int e^{\frac{i}{2}(q,G^{-1}q)}\mathcal{D}(q). \tag{9.21}$$

We see that this expression is practically the one obtained by combining (9.15) and (9.18):

$$\int e^{i \int_{-\infty}^{\infty} L_0 d\tau + i \int_{-\infty}^{\infty} q(\tau) j(\tau) d\tau} \mathcal{D}(q) = Z_0[j] \int e^{i \int_{-\infty}^{\infty} L_0 d\tau} \mathcal{D}(q), \tag{9.22}$$

the only difference coming from the fact that, in the formula we are looking for, G has been replaced by G_F. We can explain this replacement by noting that in our intuitive vision, we did not care about convergence problems. But the replacement of G by G_F is equivalent to replace the Fourier transform $\tilde{G}^{-1}$ by $\tilde{G}_F^{-1} = \frac{m}{2\hbar}(k_0^2 - \omega^2 + i\varepsilon)$. The small imaginary part has the effect of replacing $i \int L_0$ by $i \int L_0 - \varepsilon \int q^2$ and hence the convergence becoming exponential because of this last term. We intuitively note that the definition we choose for Green functions can be understood as to add what is necessary to make the functional integrals converge.

A better justification of this choice will be given later on. A similar analysis can be also made for expectations of T-products on kets $|q, t>$, i.e. for path integrals with boundary conditions $q(t) = q$ and $q(t') = q'$.

In the next section, we will show that the S-matrix, an essential object for calculating transition probabilities, can be naturally expressed in terms of T-products.

9.2   *S*-Matrix and *T*-Products

A scattering process can be described by an operator, S, acting in the space of the physical states and transforming an initial state vector into a final state vector. Ideally, the initial state is given at time $t = -\infty$ and the final state at time $t = +\infty$. The diffusion amplitude, or scattering amplitude, is the S-matrix element between these two states.

Let us now define the S-matrix.

Let $|\Psi(t)>$ be a state in the *interaction representation* and let $U(t, t_0)$ be the Green function describing the evolution of an initial state $|\Psi(t_0)>$ towards $|\Psi(t)>$:

$$|\Psi(t)> = U(t, t_0)|\Psi(t_0)> .$$

It is the solution of the equation

$$i \frac{d}{dt} U(t, t_0) = H_{\text{int}}(t) U(t, t_0) \tag{9.23}$$

with initial condition $U(t_0, t_0) = 1$. We have introduced $(H_{\text{int}})_I = H_{\text{int}}(t)$ with the definition

$$H_{\text{int}}(t) = e^{itH_0} H_{\text{int}} e^{-itH_0}, \tag{9.24}$$

a formula in which H_{int} represents the interaction operator in the Schrödinger representation, i.e. in terms of time-independent q. If the Hamiltonian does not explicitly depend on time, H_{int} satisfies the equation

$$-i\frac{d}{dt}H_{\text{int}}(t) = [H_0, H_{\text{int}}(t)].$$ (9.25)

The conjugate transposed of U obeys

$$-i\frac{d}{dt}U^\dagger(t, t_0) = U^\dagger(t, t_0)H_{\text{int}}(t).$$ (9.26)

Combining Eqs. (9.23) and (9.26), we get

$$-i\frac{d}{dt}[U^\dagger(t, t_0)U(t, t_0)] = 0$$ (9.27)

and therefore

$$U^\dagger(t, t_0)U(t, t_0) = 1,$$ (9.28)

an equation which expresses the fact that $U(t, t_0)$ is unitary.

We call S-matrix the limit

$$S = \lim_{\substack{t_0 \to -\infty \\ t \to +\infty}} U(t, t_0).$$ (9.29)

We can understand this expression in a more operatorial form. Indeed, let us introduce $V(t, t_0) = e^{-iH_0 t}U(t, t_0)$. From (9.23),

$$e^{iH_0 t}(-H_0 V + i\partial_t V) = H_{\text{int}}U(t, t_0) = e^{iH_0 t}H_{\text{int}}V = e^{iH_0 t}(H - H_0)V;$$ (9.30)

thus,

$$i\frac{d}{dt}V(t, t_0) = HV(t, t_0)$$ (9.31)

with initial condition $V(t_0, t_0) = e^{-iH_0 t_0}$. This equation has for solution

$$V(t, t_0) = e^{-iHt}e^{iHt_0}e^{-iH_0 t_0};$$ (9.32)

thus,

$$U(t, t_0) = e^{iH_0 t}e^{-iH(t-t_0)}e^{-iH_0 t_0}.$$ (9.33)

We will now give the expression of $U(t, t_0)$ in terms of T-products. To do this, let us integrate Eq. (9.23)

$$U(t, t_0) = 1 - \mathrm{i} \int_{t_0}^{t} \mathrm{d}t_1 H_{\text{int}}(t_1) U(t_1, t_0). \tag{9.34}$$

By an indefinite iteration, we obtain

$$U(t, t_0) = 1 - \mathrm{i} \int_{t_0}^{t} \mathrm{d}t_1 H_{\text{int}}(t_1) + (-\mathrm{i})^2 \int_{t_0}^{t} \mathrm{d}t_1 \int_{t_0}^{t_1} \mathrm{d}t_2 H_{\text{int}}(t_1) H_{\text{int}}(t_2) + \cdots$$

$$+ (-\mathrm{i})^n \int_{t_0}^{t} \mathrm{d}t_1 \cdots \int_{t_0}^{t_{n-1}} \mathrm{d}t_n H_{\text{int}}(t_1) \cdots H_{\text{int}}(t_n) + \cdots . \tag{9.35}$$

Each term can be rewritten in another way. Let us consider, for example, the third one. It can be rewritten as

$$\int_{t_0}^{t} \mathrm{d}t_1 \int_{t_0}^{t_1} \mathrm{d}t_2 H_{\text{int}}(t_1) H_{\text{int}}(t_2)$$

$$= \int_{t_0}^{t} \mathrm{d}t_2 \int_{t_0}^{t_2} \mathrm{d}t_1 H_{\text{int}}(t_2) H_{\text{int}}(t_1)$$

$$= \frac{1}{2} \left(\int_{t_0}^{t} \mathrm{d}t_1 \int_{t_0}^{t_1} \mathrm{d}t_2 H_{\text{int}}(t_1) H_{\text{int}}(t_2) + \int_{t_0}^{t} \mathrm{d}t_2 \int_{t_0}^{t_2} \mathrm{d}t_1 H_{\text{int}}(t_2) H_{\text{int}}(t_1) \right)$$

$$= \frac{1}{2} \left(\int_{t_0}^{t} \mathrm{d}t_1 \int_{t_0}^{t} \mathrm{d}t_2 (\theta(t_1 - t_2) H_{\text{int}}(t_1) H_{\text{int}}(t_2) + \theta(t_2 - t_1) H_{\text{int}}(t_2) H_{\text{int}}(t_1)) \right)$$

$$= \frac{1}{2!} \int_{t_0}^{t} \mathrm{d}t_1 \int_{t_0}^{t} \mathrm{d}t_2 T(H_{\text{int}}(t_1) H_{\text{int}}(t_2)). \tag{9.36}$$

This expression in terms of T-products can be generalised to all orders and we get

$$U(t, t_0) = \sum_{0}^{+\infty} \frac{(-\mathrm{i})^n}{n!} \int_{t_0}^{t} \mathrm{d}t_1 \int_{t_0}^{t} \mathrm{d}t_2 \cdots \int_{t_0}^{t} \mathrm{d}t_n T(H_{\text{int}}(t_1) H_{\text{int}}(t_2) \cdots H_{\text{int}}(t_n)). \tag{9.37}$$

This formula is often rewritten in a more symbolic form

$$U(t, t_0) = T[\exp(-\mathrm{i} \int_{t_0}^{t} \mathrm{d}s H_{\text{int}}(s))]. \tag{9.38}$$

This result shows that the S-matrix can be formally written as

$$S = \lim_{t' \to +\infty, t \to -\infty} T[\exp(-\mathrm{i} \int_{t}^{t'} \mathrm{d}s H_{\text{int}}(s))] = T[(\exp(-\mathrm{i} \int_{-\infty}^{+\infty} \mathrm{d}s H_{\text{int}}(s))]. \tag{9.39}$$

Let us now give three examples.

9.2.1 Three Examples

We first give a direct application of formula (9.39).

Let us consider the S-matrix element between two states, one of momentum p and the other of momentum p' (we deal with a one-dimensional case; the extension to three dimensions is straightforward):

$$< p' | S | p > = \lim_{t' \to +\infty, t \to -\infty} < p' | e^{iH_0 t'} e^{-iH(t'-t)} e^{-iH_0 t} | p > . \tag{9.40}$$

Suppose H to be the Hamiltonian of a particle of mass m in the potential V: $H = \frac{p^2}{2m} + V(q)$. Introducing the kets of the position representation

$$< p' | S | p > = \lim_{t' \to +\infty, t \to -\infty} e^{i(E_{p'} t' - E_p t)} \int_{\substack{q(t')=q' \\ q(t)=q}} dq' dq \, e^{-i(p'q'-pq)} < q' | e^{-iH(t'-t)} | q > \tag{9.41}$$

and using the Feynman formula, we obtain

$$< p' | S | p > \tag{9.42}$$

$$= \lim_{t' \to +\infty, t \to -\infty} e^{i(E_{p'} t' - E_p t)} \int dq' dq \, e^{-i(p'q'-pq)} \int_{\substack{q(t')=q' \\ q(t)=q}} e^{i \int_t^{t'} L d\tau} \mathcal{D}(q),$$

where L is the associated Lagrangian.

The second example is an application of Section 8.7.

In this section, we have given a closed expression for the matrix element of the operator e^{-itH} in the Bargmann space. In this case, the Hamiltonian was quadratic and more precisely it represented a harmonic oscillator with external sources. We will show now how, from this expression, we get a simple form for the S-matrix element in the Bargmann space.

The inputs are

$$H_0 = \omega a^\dagger a$$
$$H = H_0 + \bar{\xi}(t) a + a^\dagger \xi(t).$$

With the notations of Section 8.7, we first compute $< z | e^{-itH_0} | \zeta >$. It is given by the exponential of formula (8.127) with $\bar{\xi} = \xi = 0$, i.e.

$$< z | e^{-itH_0} | \zeta > = e^{\bar{z} \zeta e^{i\omega t}} \tag{9.43}$$

Using the fact that

$$f(\bar{z}) = \int_C e^{\bar{z}u} f(\bar{u}) e^{-|u|^2} \frac{d\bar{u} du}{2\pi i}, \tag{9.44}$$

which is nothing else than

$$< z \mid f > = \int < z \mid u >< u \mid f > e^{-|u|^2} \frac{d\bar{u}du}{2\pi i},$$

we have that for any operator A

$$< z \mid e^{it_2 H_0} A e^{-it_1 H_0} \mid \zeta >$$

$$= \iint < z \mid e^{it_2 H_0} \mid u >< u \mid A \mid v >< v \mid e^{-it_1 H_0} \mid \zeta > e^{-|u|^2-|v|^2} \frac{d\bar{u}du}{2\pi i} \frac{d\bar{v}dv}{2\pi i}$$

$$= \iint e^{\bar{z}ue^{i\omega t_2}} < u \mid A \mid v > e^{\bar{v}\zeta e^{-i\omega t_1}} e^{-|u|^2-|v|^2} \frac{d\bar{u}du}{2\pi i} \frac{d\bar{v}dv}{2\pi i}$$

$$= A(\bar{z}e^{i\omega t_2}, \zeta e^{-i\omega t_1}). \tag{9.45}$$

We are now ready to compute the matrix element of the evolution operator

$$< z \mid U(t, t_0) \mid \zeta > = < z \mid e^{itH_0} e^{-i(t-t_0)H} e^{-it_0 H_0} \mid \zeta >$$

$$= < ze^{-i\omega t} \mid e^{-i(t-t_0)H} \mid \zeta e^{-i\omega t_1} > \tag{9.46}$$

$$= e^{\bar{z}\zeta - i\bar{z}\int_{t_0}^{t} e^{i\omega s}\xi(s)ds - i\zeta\int_{t_0}^{t} e^{-i\omega s}\bar{\xi}(s)ds - \frac{1}{2}\int_{t_0}^{t}\int_{t_0}^{t} e^{-i\omega|u-s|}\bar{\xi}(u)\xi(s)duds},$$

from which results a close expression for the S-matrix element in the Bargmann space in the case of a harmonic oscillator with external sources:

$$< z \mid S \mid \zeta > = S(\bar{z}, \zeta) = \lim_{\substack{t \to +\infty \\ t_0 \to -\infty}} < z \mid U(t, t_0) \mid \zeta > \tag{9.47}$$

$$= e^{\bar{z}\zeta - i\bar{z}\int_{-\infty}^{+\infty} e^{i\omega s}\xi(s)ds - i\zeta\int_{-\infty}^{+\infty} e^{-i\omega s}\bar{\xi}(s)ds}$$

$$- \frac{1}{2}\iint_{-\infty}^{+\infty} e^{-i\omega|u-s|}\bar{\xi}(u)\xi(s)duds.$$

The third example is a direct computation of the S-matrix in the case of a free quantum mechanical particle in an external field j. If

$$H = H_0 + H_{int} \tag{9.48}$$

where $H_{int} = -\int_{-\infty}^{+\infty} j(s)q(s)ds$, or equivalently

$$L = L_0 + L_{int} \tag{9.49}$$

with

$$L_{int} = -H_{int} \tag{9.50}$$

then, according to formula (9.39), we have

$$S = T(e^{i\int_{-\infty}^{+\infty} j(s)q(s)ds}). \tag{9.51}$$

9.3 Elements of Perturbation Theory

The goal of this section is to show, using the evolution operator U, how the expectation values of quantities whose evolution is given by the complete Hamiltonian H, can be expressed with the help of quantities evolving in the free dynamics defined by the Hamiltonian H_0. The presentation is very general. Even if the notation is that of quantum mechanics, the final results can be transposed to field theories, by replacing $q(t)$ by $\phi(x)$.

Let us compute the vacuum-to-vacuum transition. We need first of all to calculate $< 0|U(t', t)|0 >$. From the value of U given by (9.33), we get

$$
\begin{aligned}
< 0|U(t', t)|0 > &= < 0|e^{it'H_0}e^{-i(t'-t)H}e^{-itH_0}|0 > = e^{i(t'-t)E_0} < 0|e^{-i(t'-t)H}|0 > \\
&= e^{i(t'-t)E_0} \int \mathcal{D}(q)e^{i\int_t^{t'} L} = e^{i(t'-t)E_0} \int \mathcal{D}(q)e^{i\int_t^{t'} L_{\text{int}}}e^{i\int_t^{t'} L_0} \\
&= \frac{\int \mathcal{D}(q)e^{i\int_t^{t'} L_{\text{int}}}e^{i\int_t^{t'} L_0}}{\int \mathcal{D}(q)e^{i\int_t^{t'} L_0}} = < 0|T[e^{i\int_t^{t'} L_{\text{int}}}]|0 >,
\end{aligned} \tag{9.52}
$$

where E_0 is the energy of the ground state of H_0.

Let us now, for $t_1 > t_2$, compute the quantity

$$
A(t', t) = < 0|U(t', t_1)q(t_1)U(t_1, t_2)q(t_2)U(t_2, t)|0 >, \tag{9.53}
$$

where the $q(t)$'s are given in the interaction representation and $|0 >$ is the ground state of the Hamiltonian H_0. Replacing the U's with their values (9.33), we can rewrite it as a functional integral

$$
\begin{aligned}
A(t', t) &= e^{i(t'-t)E_0} < 0|e^{-i(t'-t_1)H}qe^{-i(t_1-t_2)H}qe^{-i(t_2-t)H}|0 > \\
&= e^{i(t'-t)E_0} \int \mathrm{d}q' \int \mathrm{d}q < 0|e^{-i(t'-t_1)H}|q' > q' \\
&\quad \times < q'|e^{-i(t_1-t_2)H}|q > q < q|e^{-i(t_2-t)H}|0 > \\
&= e^{i(t'-t)E_0} \int \mathcal{D}(q)q(t_1)q(t_2)e^{i\int_t^{t'} L} \\
&= \frac{\int \mathcal{D}(q)q(t_1)q(t_2)e^{i\int_t^{t'} L_{\text{int}}}e^{i\int_t^{t'} L_0}}{\int \mathcal{D}(q)e^{i\int_t^{t'} L_0}} \\
&= < 0|T[q(t_1)q(t_2)e^{i\int_t^{t'} L_{\text{int}}}]|0 >
\end{aligned} \tag{9.54}
$$

since

$$
e^{i(t'-t)E_0} = \int \mathcal{D}(q)e^{i\int_t^{t'} L_0}. \tag{9.55}
$$

Using the group property of the evolution operator

$$
U(t_1, t_2)U(t_2, t_3) = U(t_1, t_3), \tag{9.56}
$$

which results from formula (9.33), we can re-express $A(t', t)$ in another way[3]

$$
\begin{aligned}
A(t', t) &= <0|U(t',0)U(0,t_1)q(t_1)U(t_1,0)U(0,t_2)q(t_2)U(t_2,0)U(0,t)|0> \\
&= <0|U(t',0)q_H(t_1)q_H(t_2)U(0,t)|0>
\end{aligned} \tag{9.57}
$$

using that

$$
U(0,t)q(t)U(t,0) = e^{itH}qe^{-itH} = q_H(t) \tag{9.58}
$$

is the operator $q(t)$ in the Heisenberg representation. We now consider the limits when $t' \to +\infty$ and $t \to -\infty$ of the various expressions of A. For this purpose, we introduce the ground state $|\Omega>$, of the complete Hamiltonian H. We suppose it is unique and normalised to 1. Let $P_\Omega = |\Omega><\Omega|$ be the projector on that state. Let us suppose that $<\Omega|0> \neq 0$. We *assert* that

$$
\lim_{t \to -\infty} \frac{U(0,t)|0>}{<0|U(0,t)|0>} = \frac{P_\Omega|0>}{<0|P_\Omega|0>}. \tag{9.59}
$$

It is beyond the scope of this book to give a rigorous proof of this statement, so we will present only a heuristic argument.

We want to estimate the expression $U(0,t)|0>$ at the limit when t goes to $-\infty$. We write

$$
e^{iHt}|0> = \sum_q e^{iHt}|q><q|0>, \tag{9.60}
$$

where the states $|q>$ are the eigenstates of the Hamiltonian H with eigenvalues E_q: $H|q> = E_q|q>$. We assume that H has a unique ground state $|\Omega>$ with eigenvalue E_Ω, which is separated by a finite amount from the rest of the spectrum of H. So Eq. (9.60) can be written as

$$
e^{iHt}|0> = e^{iE_\Omega t}|\Omega><\Omega|0> + \sideset{}{'}\sum_q e^{iHt}|q><q|0>, \tag{9.61}
$$

where by $\sum'$ we denote the sum over all other states of H except the state $|\Omega>$. Since E_Ω is the lowest eigenvalue of H, we can isolate it by taking the limit $t \to -\infty$ with a small positive imaginary part. A justification of this procedure will be given in the next chapter. The contribution of all other states will go to 0 faster and we shall obtain the dominant contribution

$$
U(0,t)|0> = e^{iE_\Omega t}P_\Omega|0> + \cdots, \tag{9.62}
$$

---

[3] In the case of a non-Hermitian operator like U, the expression of the expectation $<\Phi|U|\Psi>$ must be understood as the scalar product between $|\Phi>$ and $|U\Psi>$, that is, to say U is always understood as acting on its right.

where the dots stand for terms which are sub-dominant when $t \to -\infty$. Taking the matrix element $< 0|U(0,t)|0 >$, we can eliminate the exponential factor $e^{iE_\Omega t}$ and we obtain the result we announced in (9.59).

We can take the limits in formulae (9.54) and (9.57). Starting from (9.57), we obtain

$$A(+\infty,-\infty) \tag{9.63}$$

$$= < \Omega|U(0,t_1)q(t_1)U(t_1,0)U(0,t_2)q(t_2)U(t_2,0)|\Omega > \frac{|< 0|U(0,-\infty)|0 >|^2}{|< \Omega|0 >|^2}.$$

We can also look at

$$< 0|U(t',t)|0 > = < 0|U(t',0)U(0,t)|0 > \tag{9.64}$$

in the same limits and find that

$$< 0|U(+\infty,-\infty)|0 > = \frac{|< 0|U(0,-\infty)|0 >|^2}{|< \Omega|0 >|^2}. \tag{9.65}$$

Finally, thanks to (9.64) we can write that

$$A(+\infty,-\infty) = < \Omega|q_H(t_1)q_H(t_2)|\Omega > < 0|U(+\infty,-\infty)|0 > . \tag{9.66}$$

Comparing this expression to (9.54) and using (9.52) in the infinite limit, we finally show for $t_1 > t_2$ that

$$< \Omega|q_H(t_1)q_H(t_2)|\Omega > = \frac{\int \mathcal{D}(q)q(t_1)q(t_2)e^{i\int_t^{t'} L_{\text{int}}}e^{i\int_{-\infty}^{+\infty} L_0}}{\int \mathcal{D}(q)e^{i\int_t^{t'} L_{\text{int}}}e^{i\int_{-\infty}^{+\infty} L_0}} \tag{9.67}$$

and thus more generally that

$$< \Omega|T[q_H(t_1)q_H(t_2)]|\Omega > = \frac{\int \mathcal{D}(q)q(t_1)q(t_2)e^{i\int_{-\infty}^{+\infty} L_{\text{int}}}e^{i\int_{-\infty}^{+\infty} L_0}}{\int \mathcal{D}(q)e^{i\int_{-\infty}^{+\infty} L_{\text{int}}}e^{i\int_{-\infty}^{+\infty} L_0}}. \tag{9.68}$$

Introducing the measure $d\nu_0(q)$ that was defined in (9.13), with the index 0 to indicate that it comes from the free Lagrangian, we can rewrite Eq. (9.68) as

$$< \Omega|T[q_H(t_1)q_H(t_2)]|\Omega > = \frac{\int d\nu_0(q)q(t_1)q(t_2)e^{i\int_{-\infty}^{+\infty} L_{\text{int}}}}{\int d\nu_0(q)e^{i\int_{-\infty}^{+\infty} L_{\text{int}}}}$$

$$= \int q(t_1)q(t_2)d\nu(q) \tag{9.69}$$

with the measure

$$d\nu(q) = \frac{d\nu_0(q)e^{i\int_{-\infty}^{+\infty} L_{\text{int}}}}{\int d\nu_0(q)e^{i\int_{-\infty}^{+\infty} L_{\text{int}}}}. \tag{9.70}$$

This identity can be extended to an arbitrary number of q's.

Formula (9.69) is very important because it gives a recipe for calculating the physical vacuum expectation values of operators as a function of quantities measured with respect to the free measure (that is to say, the measure generated by the quadratic part of the Lagrangian).

Let us now consider an important application of the functional integral, namely the establishment of perturbation theory. In practice, most of the potentials taking place in physically relevant Lagrangians are such that the functional integral expressing the Green functions has no analytical expression. On the other hand, for the free theory, the calculation reduces to Gaussian integrals which can be explicitly computed. To be more explicit, we will suppose that the Hamiltonian splits into two parts: H_0, at most quadratic, and H_{int}, small with respect to H_0. For example, H_{int} can be a function of the 'fields' times a dimensionless quantity, negligible with respect to 1 (we may think of the dimensionless constant $\alpha = 1/137$ of electrodynamics). To the Hamiltonian corresponds a Lagrangian L that splits into two parts: L_0, generating in the functional formalism a Gaussian measure, and $L_{int} = -H_{int}$ corresponding to the interaction. Since we have assumed that the interaction is small, we can expand the exponential in formula (9.69) in a power series. Let us add, for the sake of simplicity, in front of the interaction a parameter λ that will be set equal to 1 at the end of the calculation. It will help us to keep track of the order of the various terms in the expansion. Writing

$$e^{i\lambda \int_{-\infty}^{+\infty} L_{int}} = \sum_{0}^{+\infty} \frac{(i\lambda)^n}{n!} \left(\int_{-\infty}^{+\infty} L_{int} \right)^n \tag{9.71}$$

we find formally (the question of convergence is not taken into consideration)

$$< \Omega | T[q_H(t_1) q_H(t_2) \cdots q_H(t_n)] | \Omega > \tag{9.72}$$

$$= \frac{\sum_{0}^{+\infty} \int d\nu_0(q) q(t_1) q(t_2) \cdots q(t_n) \frac{(i\lambda)^p}{p!} \left(\int_{-\infty}^{+\infty} L_{int} \right)^p}{\sum_{0}^{+\infty} \int d\nu_0(q) \frac{(i\lambda)^p}{p!} \left(\int_{-\infty}^{+\infty} L_{int} \right)^p}.$$

The right-hand side is the ratio of two series in λ. We can perform the quotient of the two series if the constant term in the denominator is different from 0 (it is the case here since the first term of the denominator is exactly equal to 1) and we find a formal series in λ:

$$< \Omega | T[q_H(t_1) q_H(t_2) \cdots q_H(t_n)] | \Omega > \tag{9.73}$$

$$= \sum_{0}^{+\infty} \frac{\lambda^p}{p!} < \Omega | T[q_H(t_1) q_H(t_2) \cdots q_H(t_n)] | \Omega >^{(p)}.$$

Remark that as in the free case, we can introduce a generating functional

$$Z[j] = \int e^{i \int_{-\infty}^{+\infty} j(s) q(s) ds} d\nu(q) \tag{9.74}$$

and

$$< \Omega | T[q_H(t_1)q_H(t_2) \cdots q_H(t_n)]|\Omega > = i^{-n} \frac{\delta}{\delta j(t_1)} \cdots \frac{\delta}{\delta j(t_n)} Z[j]\big|_{j=0}. \qquad (9.75)$$

The expression $< \Omega | T[q_H(t_1)q_H(t_2) \cdots q_H(t_n)]|\Omega >^{(p)}$ is called the pth order of the perturbation series. Since the coefficients of the numerator and denominator series in (9.72) are the expectation values of T-products in a vacuum of the free Hamiltonian, i.e. explicitly computable Gaussian expectation values, the pth order of perturbation is expressed in terms of such quantities.

The expansion in Feynman diagrams is nothing other than the development of all these quantities in terms of 2-point functions. Of course, as it was said at the beginning of this chapter, some of the expressions which appear in the expansion can be, because of the T-products, ill-defined. Some kind of regularisation process is therefore necessary in order to have a better understanding of the nature of these singularities. This will be made more transparent after we perform a complex rotation of the time, as we will see in the next chapter.

Before concluding this section, we want to introduce the notion of a truncated vacuum expectation value (t.v.e.v.). It is an operation which has the effect, when taking the vacuum expectation value of a product of $q_H(t)$, of eliminating all the factorisations due to the insertion of a vacuum state between any two $q_H(t)$'s. We will also see later on that when expressing the vacuum expectation values as diagrams, then the t.v.e.v.s correspond to connected diagrams.

More precisely, consider n observables $q_H(t_1) \cdot q_H(t_n)$. For any subset

$$I = \{i_1, i_2, \cdot, i_k\}, \qquad i_1 < i_2 < \cdots < i_k$$

of $(1, 2, \cdot, n)$, we define

$$q_H(I) = q_H(t_{i_1})q_H(t_{i_2}) \cdots q_H(t_{i_k}) \qquad (9.76)$$

and the t.v.e.v. as

$$< \Omega | q_H(I) | \Omega >^T = < \Omega | q_H(t_1)q_H(t_2) \cdots q_H(t_n) | \Omega >^T$$
$$= \sum_{m=1}^{|I|} \sum_{\{I_\nu\}} (-1)^{m-1}(m-1)! \prod_{\nu=1}^{m} < \Omega | q_H(I_\nu) | \Omega >, \qquad (9.77)$$

where the subsets I_ν, $\nu = 1, \cdots, m$, have no intersection with each other, their union is I, and the second summation is over all such partitions. $|I|$ is the number of elements in I.

For example:

$$
\begin{aligned}
&< \Omega|q_H(t)|\Omega >^T =< \Omega|q_H(t)|\Omega >\\
&< \Omega|q_H(t_1)q_H(t_2)|\Omega >^T\\
&=< \Omega|q_H(t_1)q_H(t_2)|\Omega > - < \Omega|q_H(t_1)|\Omega >< \Omega|q_H(t_2)|\Omega >\\
&< \Omega|q_H(t_1)q_H(t_2)q_H(t_3)|\Omega >^T\\
&=< \Omega|q_H(t_1)q_H(t_2)q_H(t_3)|\Omega > - < \Omega|q_H(t_1)q_H(t_2)|\Omega >< \Omega|q_H(t_3)|\Omega >\\
&- < \Omega|q_H(t_2)q_H(t_3)|\Omega >< \Omega|q_H(t_1)|\Omega > - < \Omega|q_H(t_1)q_H(t_3)|\Omega >< \Omega|q_H(t_2)|\Omega >
\end{aligned}
\tag{9.78}
$$

and so on.

Conversely, we can express any vacuum expectation value as a sum of products of truncated expectation values:

$$
\begin{aligned}
< \Omega|q_H(I)|\Omega > &= < \Omega|q_H(t_1)q_H(t_2)\cdots q_H(t_n)|\Omega >\\
&= \sum_{m=1}^{|I|}\sum_{\{I_\nu\}}\prod_{\nu=1}^{m} < \Omega|q_H(I_\nu)|\Omega >^T .
\end{aligned}
\tag{9.79}
$$

This definition is also valid for vacuum expectation values of T-products.

Formally, if we have a generating functional $Z[j]$ whose functional derivatives are vacuum expectation values, as for example the one given by formula (9.74), then the generating functional of their truncated values is given by

$$
G[j] = \log Z[j].
\tag{9.80}
$$

9.4   Generalizations

9.4.1   Three-Dimensional Quantum Mechanics

The usual quantum mechanics is over a three-dimensional space. Consequently, $q(t)$ must be replaced by $\boldsymbol{q}(t) = (q_1(t), q_2(t), q_3(t))$. Let us consider a system in the three-dimensional Euclidean space given by the Lagrangian

$$
L(\dot{\boldsymbol{q}}, \boldsymbol{q}) = \frac{1}{2}m\dot{\boldsymbol{q}}^2 - V(\boldsymbol{q}).
\tag{9.81}
$$

If $V \equiv 0$, we can repeat the analysis carried out in Chapter 8. Since the kinetic part is the sum of three independent parts and since the exponential of the action factorises into the product of three one-dimensional exponentials, we define the path integral as the product of three path integrals. Given $\boldsymbol{q}_a$ and $\boldsymbol{q}_b$ at times t_a and t_b, we find that

$$S(b, a) = \sum_{i=1}^{3} S_i(b, a) = \sum_{i=1}^{3} \int_{t_a}^{t_b} \frac{1}{2} m (\dot{q}_i)^2 \tag{9.82}$$

and

$$G(\boldsymbol{q}_b, t_b; \boldsymbol{q}_a, t_a) = \prod_{i=1}^{3} \int_{\substack{q_i(t_b)=(q_b)_i \\ q_i(t_a)=(q_a)_i}} \mathrm{e}^{\frac{i}{\hbar} S_i(b,a)} \mathcal{D}(q_i) \overset{\text{def}}{=} \int_{\substack{q(t_b)=q_b \\ q(t_a)=q_a}} \mathrm{e}^{\frac{i}{\hbar} S(b,a)} \mathcal{D}(\boldsymbol{q}). \tag{9.83}$$

This expression generalises in an obvious way when $V \neq 0$, but then the integral does not factorise into the product of three integrals:

$$
\begin{aligned}
&G(\boldsymbol{q}_b, t_b; \boldsymbol{q}_a, t_a) \\
&= \int_{\substack{q_1(t_b)=(q_b)_1 \\ q_1(t_a)=(q_a)_1}} \mathrm{e}^{\frac{i}{\hbar} \int_{t_a}^{t_b} \frac{1}{2} m (\dot{q}_3)^2} \left(\int_{\substack{q_2(t_b)=(q_b)_2 \\ q_2(t_a)=(q_a)_2}} \mathrm{e}^{\frac{i}{\hbar} \int_{t_a}^{t_b} \frac{1}{2} m (\dot{q}_2)^2} \right. \\
&\quad \times \left. \left(\int_{\substack{q_3(t_b)=(q_b)_3 \\ q_3(t_a)=(q_a)_3}} \mathrm{e}^{\frac{i}{\hbar} \int_{t_a}^{t_b} \frac{1}{2} m (\dot{q}_1)^2} \mathrm{e}^{\frac{i}{\hbar} V(q_1, q_2, q_3)} \mathcal{D}(q_1) \right) \mathcal{D}(q_2) \right) \mathcal{D}(q_3) \\
&\overset{\text{def}}{=} \int_{\substack{q(t_b)=q_b \\ q(t_a)=q_a}} \mathrm{e}^{\frac{i}{\hbar} S(b,a)} \mathcal{D}(\boldsymbol{q}).
\end{aligned}
\tag{9.84}
$$

Taking into account the analysis carried out in the previous sections, these last equalities show that the path integral formalism is equivalent to ordinary quantum mechanics. (This is not exactly true since from the path integral formalism we can recover the Schrödinger equation, the reverse being not obvious!) Even more generally, it is possible to extend the path integral formalism to arbitrary dimensions. Two extensions will be particularly useful in the next sections: the first is that for which the potential is a local function of the positions, the second that describing an infinite number of degrees of freedom located at each point in space. These two extensions will correspond to the quantum theory of fields.

9.4.2   The Free Scalar Field

Let us apply this analysis to the case of a field. For us, a field will be a natural generalisation of $q(t)$. The original meaning of the quantity $q(t)$ is that of a position at time t but, already in classical mechanics, it often has an abstract interpretation in which t is a parameter and q is a quantity which has a physical meaning. The field, in analogy with the potential four-vector, is parameterised by a space–time point x and represented by a function q (we rather note it ϕ) of these parameters. This function will have a meaning at the microscopic physical level. It will be possible to do with the field the same type of manipulations which were done with the $q(t)$'s of quantum mechanics.

This statement becomes more precise if we formulate the theory on a space lattice. The field $\phi(x, t)$ becomes a set of functions $\phi_i(t)$. For a lattice with a finite number of points N, we obtain quantum mechanics of N degrees of freedom. The limit $N \to \infty$, with the lattice spacing going to 0, is expected to reproduce field theory. The time, as a parameter, will play a particular role[4] as compared to the other parameters and it is at the level of the physical interpretation that the difference will be relevant.

Let us thus consider a free massive scalar field $\phi(x, t)$, x being the space variable, in a one-dimensional space. The Lagrangian density for such a field is

$$\mathcal{L}_0(\phi) = \frac{1}{2}\left((\frac{\partial\phi(x, t)}{\partial t})^2 - (\frac{\partial\phi(x, t)}{\partial x})^2 - m^2(\phi(x, t))^2\right) \tag{9.85}$$

of Lagrangian $L_0 = \int \mathcal{L}_0(\phi(x, t))\mathrm{d}x$. From this Lagrangian, we can define an amplitude $G(\phi, t'; \psi, t)$ as the sum of $\mathrm{e}^{\frac{i}{\hbar}S}$ over all paths, the paths having to coincide with the function $\phi(x)$ at time t' and with $\psi(x)$ at time t.

To control such an expression, we approximate the Lagrangian by a Riemann sum

$$L_0 = \lim_{\delta \to 0} \delta \sum_i \mathcal{L}_i(\phi), \tag{9.86}$$

where

$$\mathcal{L}_i(\phi) = \frac{1}{2}\left((\frac{\dot{\phi}(x_i, t) + \dot{\phi}(x_{i-1}, t)}{2})^2\right.$$
$$\left. -(\frac{\phi(x_i, t) - \phi(x_{i-1}, t)}{\delta})^2 - m^2(\frac{\phi(x_i, t) + \phi(x_{i-1}, t)}{2})^2\right) \tag{9.87}$$

and, for example, $x_i = i\delta$, $i \in \mathbb{Z}$. It is possible to argue as before and to speak for each x_i of paths $\phi(x_i, t)$ interpolating between $\phi(x_i)$ and $\psi(x_i)$. To each x_i is associated a measure $\mathcal{D}(\phi(x_i, t))$ and

$$A(\phi, t'; \psi, t) = \lim_{\delta \to 0} \int \mathrm{e}^{\frac{i}{\hbar}\delta \sum_i \int \mathrm{d}t \mathcal{L}_i} \prod_i \mathcal{D}(\phi(x_i, t)) \tag{9.88}$$

that we note, if the limit exists,

$$\int_{\substack{\phi(x, t') = \phi(x) \\ \phi(x, t) = \psi(x)}} \mathrm{e}^{\frac{i}{\hbar}\int \mathrm{d}t \mathrm{d}x \mathcal{L}_0} \mathcal{D}(\phi). \tag{9.89}$$

Remark: The approximation, with a fixed spacing δ, by Riemann sums, can be interpreted as a linear chain of atoms, x_j measuring the departure from equilibrium of

---

[4] The time direction plays a particular role in relativistic physics because of the causality principle. In Euclidean physics, all the parameters are on equal footing.

the jth atom, these atoms interacting through a potential depending only on their distance. At first approximation, we can suppose that it is different from 0 only between nearest neighbours and that at equilibrium, this potential is minimum; we can develop the potential around its equilibrium value and keep the lowest order, i.e. the quadratic term.

The previous studies can be easily extended to the cases where the field is defined in a higher dimensional space.

By analogy with formula (9.18), we can define the generating functional of T-products of free fields,[5] for regular enough functions f, by

$$Z_0[f] = e^{\frac{i}{2}(f, Gf)}, \tag{9.90}$$

where

$$(f, Gf) = \int f(x)\, G_F(x-y) f(y)\, d^4x\, d^4y \tag{9.91}$$

with the Feynman propagator G_F

$$G_F(x-y) = -\frac{1}{(2\pi)^4} \int \frac{e^{-ip.(x-y)}}{p^2 - m^2 + i\varepsilon}\, d^4p. \tag{9.92}$$

As before, we identify the propagator with a 2-point function of the free field

$$-iG_F(x-y) = <0|T(\phi(x)\phi(y))|0>. \tag{9.93}$$

In analogy with what we have just shown in the harmonic oscillator case, we set

$$<0|T(\phi(x_1)\phi(x_2)\cdots\phi(x_n))|0> = i^{-n}\frac{\delta}{\delta f(x_1)}\cdots\frac{\delta}{\delta f(x_n)} Z_0[f]\big|_{f=0} \tag{9.94}$$

and write

$$Z_0[f] = <0|T(e^{i\int \phi(x)f(x) d^4x})|0> \tag{9.95}$$

Remark that applying formula (9.80) to the functional $Z_0[f] = e^{\frac{i}{2}(f, Gf)}$, we get

$$G[j] = \frac{i}{2}(f, Gf); \tag{9.96}$$

thus, the only truncated expectation value of a T-product of a free field is

$$<0|T(\phi(x)\phi(y))|0>^T = <0|T(\phi(x)\phi(y))|0> = -iG_F(x-y).$$

---

[5] Here, we give the definition for free scalar fields in four dimensions.

The fact that the only t.v.e.v. of free fields is the 2-point function is characteristic of the free fields. For free fields, the *n*-point functions, for $n > 2$, split into a sum of products of 2-point functions, and therefore their truncated values are identically equal to 0.

We must now justify these identifications.

In the problem of the harmonic oscillator we proved that computing the functional integral reproduces the quantum mechanical amplitudes we would have obtained by writing $q(t)$ in terms of creation and annihilation operators, satisfying the canonical commutation relations (8.81). We just noted that a free scalar field is equivalent to an infinite number of harmonic oscillators, one at each point in space. Therefore, the computation of Feynman's path integral will be equivalent to imposing canonical commutation relations to the corresponding creation and annihilation operators. We remind that a free field $\phi(x, t)$ can be expanded as

$$\phi(t, x) = \frac{1}{(2\pi)^3} \int \frac{d^3 k}{2\omega_k} [a(k)e^{-ikx} + a^*(k)e^{ikx}] \tag{9.97}$$

with $\omega_k = \sqrt{k^2 + m^2}$.

We conclude that $a(k)$ and $a^*(k)$ should be promoted to operators satisfying the canonical commutation relations

$$[a(k), a(k')] = 0 [a^*(k), a^*(k')] = 0$$
$$[a(k), a^*(k')] = (2\pi)^3 \, 2\omega_k \delta^3(k - k'), \tag{9.98}$$

which generalise those written in Chapter 8 in the case of the harmonic oscillator. These relations express the connection we have often alluded to between a quantum field and a particle. It is straightforward to verify that the quanta created by the operators $a^*(k)$ have a minimum energy $k_0 = m$ and represent a spinless particle of mass m. Similarly, we can write the expression for the Hamiltonian operator as a formal sum over the energies of an infinite number of harmonic oscillators. Using the renormalised expression $H_{ren}(k)$ we introduced previously for each oscillator, we find that

$$H_{ren} = \frac{1}{(2\pi)^3} \int \frac{d^3 k}{2\omega_k} H_{ren}(k) = \frac{1}{2(2\pi)^3} \int d^3 k a^*(k) a(k), \tag{9.99}$$

which is positive definite.[6]

Naturally, these relations generalise to the case of N independent scalar fields ϕ^i, $i = 1, 2, ..., N$. The commutation relations read

$$[a^i(k), a^{*j}(k')] = \delta^{ij}(2\pi)^3 \, 2\omega_k \delta^3(k - k') \tag{9.100}$$

with all other commutators vanishing.

---

[6] Had we used the unrenormalised form of the Hamiltonian for each oscillator $H(k)$ we would have obtained an expression which would differ from H_{ren} of (9.99) by an infinite constant, proportional to the volume of the mass hyperboloid. As we noted previously, this constant is unobservable in the absence of a gravitational field, but we will come back to this point when we discuss the quantisation of the electromagnetic field.

In particular, if we consider a complex scalar field with its real and imaginary part, as we did in section 6.2, we find the commutation relations

$$[a(\boldsymbol{k}), a^*(\boldsymbol{k}')] = [b(\boldsymbol{k}), b^*(\boldsymbol{k}')] = (2\pi)^3 2\omega_p \delta^3(\boldsymbol{k} - \boldsymbol{k}') \tag{9.101}$$

with all other commutators vanishing.

The free quantum fields ϕ, which as operators are linear combinations of creation and annihilation operators, are not commutative. From the canonical commutation relations, it follows that the commutator of two fields is not an operator but a number (in fact, a number times the identity operator)

$$[\phi(x), \phi(y)] = i\Delta(x - y), \tag{9.102}$$

where

$$
\begin{aligned}
\Delta(x) &= \frac{i}{(2\pi)^3} \int \frac{d^3k}{2\omega_k} e^{ik.x} - \frac{i}{(2\pi)^3} \int \frac{d^3k}{2\omega_k} e^{-ik.x} \\
&= -\frac{i}{(2\pi)^3} \int d^4k \delta(k^2 - m^2) \varepsilon(k_0) e^{-ik.x}.
\end{aligned}
\tag{9.103}
$$

The last identity shows that Δ is a Lorentz scalar. We check easily that if $x^0 = 0$, the two terms contributing to Δ are equal[7] and thus that $\Delta(0, x) = 0$. Since Δ is a Lorentz invariant, this means that $\Delta(x) = 0$ for space-like four-vectors x, that is to say, such that $x^2 < 0$. It follows that fields at different space-like separated points commute. This is the manifestation of causality for a quantum field theory.

The way causality is expressed is in fact dependent on the nature of the fields. More precisely, we show that fields corresponding to integer spin, i.e. associated with to a representation $[m, s]$ of the Poincaré group with s an integer, commute for space-like separated points of space–time. If s is a half-odd integer, causality is expressed by the fact that it is the anticommutator of the fields which vanishes. We will accept here without proof that there exists a link between the expression of causality and the spinor nature of the fields. This property is known under the name of *spin-statistics theorem*: integer spin particles satisfy the Bose statistic and particles with half-odd integer spin satisfy the Fermi statistics. We shall come back to this point shortly.

We have studied here the case of a Lagrangian of free fields, that is to say given by a quadratic form in the fields. It is possible to give a rigorous meaning to this case. It is also possible to write a formal similar expression in the case of a general Lagrangian. It is then more difficult to give a precise meaning to the expressions which appear and we will consider this problem only in the Euclidean framework.

---

[7] It is enough to note that the integrals depend only on $|x|$.

9.5   Problems

Problem 9.1 Derive for free fields:

$T(\phi(x_1)\phi(x_2)) =: \phi(x_1)\phi(x_2) : -iG(x_1, x_2)$

$T(\phi(x_1)\phi(x_2)\phi(x_3))$

$=: \phi(x_1)\phi(x_2)\phi(x_3) : -i[\phi(x_1)G(x_2, x_3) + \phi(x_2)G(x_1, x_3) + \phi(x_3)G(x_1, x_2)]$

$\quad T(\phi(x_1)\phi(x_2)\phi(x_3)\phi(x_4))$

$=: \phi(x_1)\phi(x_2)\phi(x_3)\phi(x_4) : -i[: \phi(x_1)\phi(x_2) : G(x_3, x_4) + : \phi(x_1)\phi(x_3) : G(x_2, x_4)$

$+ : \phi(x_1)\phi(x_4) : G(x_2, x_3) + : \phi(x_2)\phi(x_3) : G(x_1, x_4) + : \phi(x_2)\phi(x_4) : G(x_1, x_3)$

$+ : \phi(x_3)\phi(x_4) : G(x_1, x_2)] - [G(x_1, x_2)G(x_3, x_4) + G(x_1, x_3)G(x_2, x_4)$

$+ G(x_1, x_4)G(x_2, x_3)]$

with $x_i = (t_i, x_i)$.

What is the general formula?

Give a proof of formula (9.80) by using formula (9.11) which gives the expectation values of T-product of harmonic oscillators.

10

The Euclidean Functional Integrals

10.1 Introduction

One of the main difficulties in the Feynman approach to path integral lies in the fact that the expressions which appear are only conditionally convergent. Indeed, in the simplest cases, we must deal with imaginary quadratic exponential integrals which were only defined by introducing a 'regulator'. This regulator was defined by giving to the exponent a small imaginary part ε and by taking the limit, the effect of it being to replace oscillating integrands by exponentially decreasing ones. The general case is even more complicated since in general we have no analytical expression of the result of the integration and a priori estimates that could help to control the convergence are very difficult to obtain.

This difficulty is solved formally by a complex rotation by $90°$ of the time axis $t \rightarrow -it$. This operation is called a *Wick rotation*. The effect of this rotation is to replace the Minkowski space $\mathbb{M}^4$ with the metric $g_{\mu\nu}$, by the Euclidean space $\mathbb{R}^4$ with the usual metric. An equivalence theorem makes it possible to link by analytic continuation the quantum theories of relativistic fields to the quantum theory of Euclidean fields (more precisely, we can give axiomatic definitions to field theories in Minkowski and Euclidean space and this equivalence is a correspondence between axioms). To the invariance under Lorentz transformations corresponds in Euclidean space the invariance under four-dimensional rotations, that is to say the invariance under O_4. The benefit is a complete symmetry of the expressions, no direction playing a particular role.[1] If the universe of four-dimensional Euclidean theories has no obvious physical reality, it can be identified without difficulty with a generalised statistical mechanics. In contrast, three-dimensional Euclidean theories are a natural generalisation of usual statistical mechanics.

Let us explain in more detail the technical advantage of the change to Euclidean space. In its operator version, the analysis made in the previous chapters has produced quantities such as e^{-itH}, which can be continued into e^{-tH}. For $t > 0$, this operator is bounded if the Hamiltonian H corresponds to a realistic physical system. Indeed, in

---

[1] In fact, one of the axioms in the Euclidean space which makes it possible to return to Minkowski space expresses the fact that in the Euclidean theory, we can choose a particular direction, satisfying a condition that makes it possible to perform the analytic continuation. We will explain this point further in Chapter 24.

this case, the spectrum of the Hamiltonian, i.e. the energy spectrum of the system, is bounded from below (there is always a finite minimal energy) but unbounded from above.

Remark: If the Hamiltonian of a system is time independent, the change to imaginary time will not modify it. The Euclidean Hamiltonian is thus the same as the usual Hamiltonian. Theories in Minkowski space and Euclidean theories coincide at time $t = 0$.

Let us consider how some of the Lagrangians we met in the preceding chapters are transformed.

For the three-dimensional quantum mechanics

$$L(\dot{q}, q) = \frac{1}{2} m\dot{q}^2 - V(q) \tag{10.1}$$

becomes

$$-\frac{1}{2}\dot{q}^2 - V(q) \tag{10.2}$$

and

$$iS = i \int dt L \rightarrow - \int dt \left(\frac{1}{2}\dot{q}^2 + V(q) \right). \tag{10.3}$$

Similarly, the Lagrangian density of the free scalar field (Klein–Gordon field)

$$\mathcal{L}(\phi) = \frac{1}{2}\left(\left(\frac{\partial \phi(x,t)}{\partial t}\right)^2 - \left(\frac{\partial \phi(x,t)}{\partial x}\right)^2 - m^2(\phi(x,t))^2 \right) \tag{10.4}$$

becomes

$$\mathcal{L}(\phi) = -\frac{1}{2}\left(\left(\frac{\partial \phi(x,t)}{\partial t}\right)^2 + \left(\frac{\partial \phi(x,t)}{\partial x}\right)^2 + m^2(\phi(x,t))^2 \right)$$
$$= -\frac{1}{2}\left((\nabla \phi(x))^2 + m^2(\phi(x))^2 \right), \tag{10.5}$$

where x, in the last term, is a point of $\mathbb{R}^4$ and the gradient is the usual gradient with respect to the four components. Thus e^{iS} becomes

$$e^{-\frac{1}{2} \int d^4x \left((\nabla \phi(x))^2 + m^2(\phi(x))^2 \right)}, \tag{10.6}$$

which is a quantity bounded by 1 and exponentially decreasing for large ϕ.

The Euler–Lagrange equation related to this Lagrangian is

$$(-\Delta + m^2)\phi = 0. \tag{10.7}$$

This is thus an elliptic system instead of being a hyperbolic one as in Minkowski space. The price to be paid by this passage to the Euclidean framework is the disappearance of the notion of causality (related to the hyperbolicity of the equation of motion). By an analysis similar to that done in Chapter 3, we can seek elementary solutions of this equation. A natural boundary condition is that they vanish at infinity. We then find (the scalar products being with respect to the Euclidean metric) that

$$G(x - x') = \frac{1}{(2\pi)^4} \int_{\mathbb{R}^4} d^4 p \frac{e^{-ip.(x-x')}}{p^2 + m^2}, \tag{10.8}$$

where G is a solution of the equation

$$(-\Delta + m^2)G = \delta. \tag{10.9}$$

This solution vanishes at infinity and is a function of x^2, the square of the Euclidean length of x.

10.1.1 The Wiener Measure

The Feynman analysis can be repeated for systems in which time has been complexified. Let $\mathcal{W}(q, t; q', t')$ be the set of continuous paths linking q at time t to q' at time t', we will define the kernel (or the associated amplitude) of the operator e^{-tH}, $t > 0$, by a formula similar to that we used in the case of real time. Formally, we have (we set $m = 1$ and we use the language of quantum mechanics)

$$< q' | e^{-itH} | q > = \int_{\mathcal{W}(q,-t/2;q',t/2)} \prod_{-t/2 < \tau < t/2} dq(\tau) e^{-\int_{-t/2}^{t/2} d\tau [\frac{1}{2}\dot{q}(\tau)^2 + V(q(\tau))]}, \tag{10.10}$$

to which we will give a precise meaning.

Let us first consider the case where $V \equiv 0$. In this case, the kernel of e^{-tH_0}, where $H_0 = \frac{1}{2}p^2$, is the elementary solution of the heat equation

$$\frac{\partial}{\partial t}\psi(q, t) = \frac{1}{2}\Delta\psi(q, t) = -H_0\psi(q, t) \tag{10.11}$$

and is given, for example in three dimensions, by

$$< q' | e^{-tH_0} | q > = (2\pi t)^{-3/2} e^{-\frac{(q-q')^2}{2t}} = \mathcal{K}_t^0(q, q'), \tag{10.12}$$

which is a well-defined expression.

$\mathcal{K}_t^0(\boldsymbol{q}, \boldsymbol{q}')$ is positive, normalised to 1 (its integration with respect to $\boldsymbol{q}$ or $\boldsymbol{q}'$ gives 1) and satisfies the composition law

$$\mathcal{K}_{t+s}^0(\boldsymbol{q}, \boldsymbol{q}') = \int \mathrm{d}^3 r \mathcal{K}_t^0(\boldsymbol{q}, \boldsymbol{r}) \mathcal{K}_s^0(\boldsymbol{r}, \boldsymbol{q}'). \tag{10.13}$$

These properties will allow us to define a probability measure, the conditioned *Wiener measure* (by the conditions imposed to the path ends). It is enough for this purpose to define it on some subsets of $\mathbb{R}^3$. We will take as subsets the products I of finite intervals of $\mathbb{R}$ (the correct notion is that of Borelian subsets). We thus define, for $-t/2 < t_1 \cdots < t_n < t/2$, the measure of the set of continuous paths

$$\{\boldsymbol{q}(\tau) \mid \boldsymbol{q}(-t/2) = \boldsymbol{q}; \boldsymbol{q}(t_1) \in I_1, \ldots, \boldsymbol{q}(t_n) \in I_n; \boldsymbol{q}(t/2) = \boldsymbol{q}\} \tag{10.14}$$

starting from $\boldsymbol{q}$ at time $-t/2$, going through $I_j \in \mathbb{R}^3$ at time t_j and ending at $\boldsymbol{q}'$ at time $t/2$, by

$$\int_{I_1} \mathrm{d}^3 q_1 \ldots \int_{I_n} \mathrm{d}^3 q_n \mathcal{K}_{t_1+t/2}^0(\boldsymbol{q}, \boldsymbol{q}_1) \mathcal{K}_{t_2-t_1}^0(\boldsymbol{q}_1, \boldsymbol{q}_2) \ldots \mathcal{K}_{t/2-t_n}^0(\boldsymbol{q}_n, \boldsymbol{q}'). \tag{10.15}$$

A probability theorem shows then that this defines a conditional measure on the space of continuous paths. Its element of integration is $\mathrm{d}W_{q,q'}^t$. If we consider regular and bounded functions, $A_i(\boldsymbol{q})$, $i = 1, 2, \ldots, p$, of $\boldsymbol{q}$, we can, by taking small enough intervals I_k, consider these functions as constant on these intervals and, by going through the limit, we can show the existence of

$$\int \prod_{i=1}^p A_i(\boldsymbol{q}(t_i)) \mathrm{d}W_{q,q'}^t, \tag{10.16}$$

with the $A_i(q)$'s as multiplication operators in $L^2(\boldsymbol{q}, \mathrm{d}^3 q)$. The interpretation in terms of kernels gives, for $-t/2 < t_1 \cdots < t_n < t/2$,

$$\int \prod_{i=1}^p A_i(\boldsymbol{q}(t_i)) \mathrm{d}W_{q,q'}^t$$
$$= < \boldsymbol{q}' | \mathrm{e}^{-(t_1+t/2)H_0} A_1 \mathrm{e}^{-(t_2-t_1)H_0} A_2 \ldots A_n \mathrm{e}^{-(t/2-t_n)H_0} | \boldsymbol{q} > . \tag{10.17}$$

In particular, we recover that if $A_i \equiv 1$ then

$$\int \mathrm{d}W_{q,q'}^t = \mathcal{K}_t^0(\boldsymbol{q}, \boldsymbol{q}'). \tag{10.18}$$

Remark: The support of this measure is something quite complicated. For example, the differentiable paths are of measure 0.

In the case where $V \neq 0$, we get the *Feynman-Kac formula*

$$
\begin{aligned}
\mathcal{K}_t(q, q') &\equiv\ <q'|\mathrm{e}^{-tH}|q> \\
&= \int \mathrm{e}^{-\int_{-t/2}^{t/2} V(q(\tau))\mathrm{d}\tau}\,\mathrm{d}W_{q,q'}^t.
\end{aligned}
\tag{10.19}
$$

This formula is proved like the Feynman formula; the basic input, which can be made rigorous for sufficiently regular operators (in fact, self-adjoint and bounded from below operators), is the Trotter formula

$$
\mathrm{e}^{-t(H_0+V)} = \lim_{n\to\infty}\left(\mathrm{e}^{-tH_0/n}\mathrm{e}^{-tV/n}\right)^n.
\tag{10.20}
$$

The limit is a strong limit, that is to say like the convergence of vectors obtained by applying the operator on any vector belonging to its domain of definition.

In reality, this form of path integral is not, as in the Minkowskian case, the form that can be used for the calculation of quantities which appear in field theories. The conditioning to points q and q' has no physical meaning when the objects in question are fields since these points are replaced by the fields themselves. The procedure we follow will be the same as what was done at the end of Chapter 8.

We take as unperturbed Hamiltonian H_0, the harmonic oscillator Hamiltonian, renormalised in such a way that the ground state has energy zero. We thus take

$$
H_0 = -\frac{1}{2}\Delta + \frac{1}{2}q^2 - \frac{1}{2},
\tag{10.21}
$$

which, by an analysis similar to what we already did, make it possible to define a measure $\mathrm{d}U_{q,q'}^t$, the so-called Ornstein–Uhlenbeck measure, less singular than the Wiener measure. We obtain a Feynman–Kac formula

$$
\begin{aligned}
\mathcal{K}_t(q, q') &\equiv\ <q'|\mathrm{e}^{-tH}|q> \\
&= \int \mathrm{e}^{-\int_{-t/2}^{t/2} V(q(\tau))\mathrm{d}\tau}\,\mathrm{d}U_{q,q'}^t,
\end{aligned}
\tag{10.22}
$$

which differs from the preceding ones by the definition of what we call $V \equiv H - H_0$.

Finally, we show for this measure a formula similar to that which was obtained for the T-products. If $-t/2 < t_1 \cdots < t_n < t/2$,

$$
\int \prod_{i=1}^{p} A_i(q(t_i))\mathrm{d}\mu_0 =\ <0|A_1\mathrm{e}^{-(t_2-t_1)H_0}A_2\,\mathrm{e}^{-(t_3-t_2)H_0}A_3 \ldots A_n|0>,
\tag{10.23}
$$

where $|0>$ is the ground state of the harmonic oscillator, and $\mathrm{d}\mu_0$ is the measure defined by

$$\mathrm{d}\mu_0 = \int_{\mathbb{R}^3 \times \mathbb{R}^3} \phi_0(\boldsymbol{q})\phi_0(\boldsymbol{q}')\mathrm{d}U_{q,q'}^t, \tag{10.24}$$

ϕ_0 being the wave function of the ground state of the harmonic oscillator. This measure is the Euclidean analogue of

$$\mathrm{d}\nu = \frac{\mathrm{e}^{\mathrm{i}\int L\mathrm{d}\tau}\mathcal{D}(\boldsymbol{q})}{\int \mathrm{e}^{\mathrm{i}\int L\mathrm{d}\tau}\mathcal{D}(\boldsymbol{q})}, \tag{10.25}$$

with L being the Lagrangian of the harmonic oscillator.

The measure $\mathrm{d}\mu_0$ is a Gaussian measure, i.e. it is completely fixed by (we write the formulae for the one-dimensional harmonic oscillator of frequency $\omega = 0$)

$$\int q(t)\mathrm{d}\mu_0 = \; <0|q|0> \; = 0 \tag{10.26}$$

and by the *covariance*

$$
\begin{aligned}
\int q(t_1)q(t_2)\mathrm{d}\mu_0 &= \; <0|q(t_1)q(t_2)|0> \\
&= \; <0|q\mathrm{e}^{-|t_1-t_2||H_0}q|0> \; = \mathrm{e}^{-|t_1-t_2|} \; <0|q^2|0> \\
&= \frac{1}{2}\mathrm{e}^{-|t_1-t_2|} \\
&= \frac{1}{2\pi}\int_{-\infty}^{+\infty}\mathrm{d}k_0\frac{\mathrm{e}^{-\mathrm{i}k_0(t_1-t_2)}}{k_0^2+1}.
\end{aligned}
\tag{10.27}
$$

The first equality results from the preceding formula applied for $t_1 > t_2$ and then in the reverse case. The second and third equalities are consequences of our choice of H_0 (the factor $-\frac{1}{2}$) and of the expression of the different quantities in terms of creators and annihilators. The last one is important because it generalises and expresses the fact that the covariance is in fact the kernel[2] of the operator $(-\frac{d^2}{dt^2}+1)^{-1}$ between t_1 et t_2: this is the *Euclidean propagator*.

---

[2] The operator $A = -\frac{\mathrm{d}^2}{\mathrm{d}t^2}+1$, invariant under translations, acting on sufficiently regular functions $f(t)$ can be written using the Fourier transform $\tilde{f}$:

$$(Af)(t) = \int A(t-s)f(s)\mathrm{d}s = \frac{1}{2\pi}\int (p^2+1)\mathrm{e}^{\mathrm{i}pt}\tilde{f}(p)\mathrm{d}p. \tag{10.28}$$

We use the following terminology: $A(t-s)$ is the kernel of A; this kernel has for Fourier transform p^2+1 and formally (for convergence reasons)

$$A(t-s) = \frac{1}{2\pi}\int (p^2+1)\mathrm{e}^{\mathrm{i}p(t-s)}\mathrm{d}p. \tag{10.29}$$

The fundamental properties of a Gaussian measure of mean zero are

$$\int q(t_1) \ldots q(t_{2n+1}) \mathrm{d}\mu_0 = 0 \tag{10.31}$$

$$\int q(t_1) \ldots q(t_{2n}) \mathrm{d}\mu_0 = \sum_{pairs} \int q(t_{i_1}) q(t_{i_2}) \mathrm{d}\mu_0 \ldots \int q(t_{i_{2n-1}}) q(t_{i_{2n}}) \mathrm{d}\mu_0 \tag{10.32}$$

with $i_1, i_2, \ldots, i_{2n}$ a permutation of $1, 2, \ldots, 2n$, the sum running over the $(2n-1)!!$ pairings of $1, 2, \ldots, 2n$. We recall that $n!! = n(n-2)(n-4)\ldots 1$.

Formally, everything happens as if

$$\mathrm{d}\mu_0 \simeq \mathrm{e}^{-\int q(\tau)(-\frac{\mathrm{d}^2}{\mathrm{d}t^2}+1)q(\tau)\mathrm{d}\tau} \prod_s \mathrm{d}q(s), \tag{10.33}$$

which is what we were expecting.

The change to three dimensions, which corresponds to ordinary quantum mechanics, is simply done by considering $q = (q_1, q_2, q_3)$, and a measure product of the measures $\mathrm{d}\mu_0(q_i)$ with a measure for each component q_i.

10.2   The Gaussian Measures in Euclidean Field Theories

We saw in the introduction of this chapter that a free massive Euclidean scalar field in four dimensions (i.e. 1 dimension of 'time' and $4 - 1$ dimensions of space) has a Lagrangian density

$$\mathcal{L}(\phi) = -\frac{1}{2}\left((\nabla\phi(x,t))^2 + m^2(\phi(x,t))^2\right), \tag{10.34}$$

where the gradient is a gradient in 4 dimensions, all directions playing the same role. We could have clearly defined such an object in an arbitrary d-dimensional space with $d-1$ space dimensions, the gradient becoming the gradient in d dimensions.

We might then repeat the analysis of the preceding section concerning quantum mechanics, the time parameter being replaced by d parameters x and q being the field ϕ. Since the free Lagrangian is the sum of d squared gradients and of the mass term, the infinitesimal contributions to the measure over the paths factorise and add up in the exponential, $\mathrm{d}^2/\mathrm{d}t^2$ being replaced by $\triangle$.

We check easily that the inverse operator A^{-1} has for kernel

$$A^{-1}(t-s) = \frac{1}{2\pi} \int \frac{\mathrm{e}^{ip(t-s)}}{p^2+1} \mathrm{d}p. \tag{10.30}$$

10.2.1 Definition

We now define a Gaussian measure of mean zero and covariance the kernel of $(-\triangle + m^2)^{-1}$. Its existence relies on the following theorem.

Theorem 9 (Minlos' theorem). *Let C be a bilinear, bicontinuous positive form on $\mathcal{S}(\mathbb{R}^d) \times \mathcal{S}(\mathbb{R}^d)$, where $\mathcal{S}(\mathbb{R}^d)$ is the space of C^∞ functions rapidly decreasing at infinity. Then there exists a unique normalised measure $d\mu_C$, of mean zero, such that $\forall f \in \mathcal{S}(\mathbb{R}^d)$*

$$S(f) = e^{-<f,Cf>/2} = \int e^{i\phi(f)} d\mu_C, \tag{10.35}$$

where

$$< f, Cg > = \int d^d x d^d y f(x) C(x,y) g(y) \tag{10.36}$$

and

$$\phi(f) = \int d^d x \phi(x) f(x). \tag{10.37}$$

We will not give the proof of this theorem but just make some comments about it. $S(f)$ is a generating functional. Indeed,

$$\int \phi(f_1) \ldots \phi(f_n) d\mu_C \tag{10.38}$$

is equal to

$$i^{-n} \frac{d}{d\lambda_1} \cdots \frac{d}{d\lambda_n} S(\lambda_1 f_1 + \cdots + \lambda_n f_n)\Big|_{\lambda_1 = \cdots = \lambda_n = 0} \quad .. \tag{10.39}$$

Remark: We will often use the following form of the functional derivation:

$$\int \phi(x_1)\phi(x_2)\cdots\phi(x_n) d\mu_C = i^{-n} \frac{\delta}{\delta f(x_1)} \cdots \frac{\delta}{\delta f(x_n)} S(f)|_{f=0} \tag{10.40}$$

We easily check the Gaussian properties of $d\mu_C$:

1.

$$\int \phi(f) d\mu_C = 0. \tag{10.41}$$

2.

$$\int \phi(f_1)\phi(f_2) d\mu_C = < f_1, Cf_2 > . \tag{10.42}$$

3.

$$\int \phi(f_1)\dots\phi(f_{2n})\mathrm{d}\mu_C = \sum_{\text{pairs}} <f_{i_1}, Cf_{i_2}> \cdots <f_{i_{2n-1}}, Cf_{i_{2n}}> . \tag{10.43}$$

4.

$$\int \phi(f_1)\dots\phi(f_{2n+1})\mathrm{d}\mu_C = 0, \tag{10.44}$$

with, in particular,

$$\int \phi(f)^n \mathrm{d}\mu_C = \left(-\mathrm{i}\frac{\mathrm{d}}{\mathrm{d}\lambda}\right)^n S(\lambda f)\Big|_{\lambda=0} \tag{10.45}$$

$$= \begin{cases} 0 & n \text{ odd} \\ (n-1)!! <f, Cf>^{n/2} & n \text{ even} \end{cases}.$$

The support of the measure, that is to say the ϕ's over which we are integrating, is the space of tempered distributions, i.e. the dual space of S. This is a very singular infinite-dimensional space. A measure in a space of infinite dimensions is defined by its compatibility with respect to restrictions. Suppose we are interested in a finite-dimensional subspace K and let us consider the measure $\mathrm{d}\mu|_K$ restricted to this subspace. If we take a subspace L of K, the restriction of $\mathrm{d}\mu|_K$ to L is equal to $\mathrm{d}\mu|_L$.

More precisely, suppose we want to integrate only functions $\phi(f_j)$, $j = 1, 2, \dots, n$, we then consider

$$S\left(\sum_i \lambda_i f_i\right) = \mathrm{e}^{-<\sum_i \lambda_i f_i, C\sum_j \lambda_j f_j>} \tag{10.46}$$

$$= \mathrm{e}^{-\sum_{i,j} \lambda_i \lambda_j <f_i, Cf_j>}$$

Applying Minlos' theorem to the last expression, since

$$\Gamma(n)_{ij} = <f_i, Cf_j> \tag{10.47}$$

is a positive matrix $\Gamma(n)$ $n \times n$, we see that there exists a measure $\mathrm{d}\mu_{\Gamma(n)}$ which is the restriction of the measure $\mathrm{d}\mu_C$ to the finite-dimensional subspace formed by $\phi(f_i)$. If we set $q_i = \phi(f_i)$, the measure $\mathrm{d}\mu_{\Gamma(n)}$ is given by

$$\mathrm{d}\mu_{\Gamma(n)} = (\det \Gamma(n))^{1/2} (2\pi)^{-n/2} \mathrm{e}^{-\frac{1}{2}\sum_{i,j} q_i \Gamma(n)_{ij}^{-1} q_j} \prod_{i=1}^n \mathrm{d}q_i \tag{10.48}$$

We could check the restriction property: suppose that $f_n = 0$. Minlos tells us that there exists a measure of covariance $\Gamma(n-1)$ such that $\mathrm{d}\mu_{\Gamma(n-1)}$ is the integral over q_n of the measure $\mathrm{d}\mu_{\Gamma(n)}$.

We can understand also the Gaussian measure in another way, a way not too far from the original Feynman idea that we developed in the preceding chapters. We can indeed approximate the action

$$\frac{1}{2} \int [(\nabla \phi(x))^2 + m^2 (\phi(x))^2] \mathrm{d}^d x = \frac{1}{2} \int \left(\phi(x), (-\triangle + m^2) \phi(x) \right) \mathrm{d}^d x \qquad (10.49)$$

by approximating the integral by Riemann sums. Let us consider a d-dimensional cubic lattice with lattice spacing a. The action is approximated by

$$a^{-d} \sum_{i,j \in Z^d} (\phi(ai), (-\triangle + m^2)_{ij} \phi(aj)), \qquad (10.50)$$

where $(-\triangle + m^2)_{ij} = (-\triangle)_{ij} + m^2 \delta_{ij}$ and $(-\triangle)_{ij}$ is a finite difference approximation of the Laplacian.[3] If now we sum over finite space volumes Λ, we can consider the measure

$$\mathrm{d}\mu(a, \Lambda) \qquad (10.52)$$
$$= \frac{1}{N(a, \Lambda)} \exp\left(-a^{-d} \sum_{i,j \in Z^d \cap \Lambda} (\phi(ai), (-\triangle + m^2)_{ij} \phi(aj)) \right) \prod_{k \in Z^d \cap \Lambda} \mathrm{d}\phi(ak),$$

where $N(a, \Lambda)$ is a normalisation factor such that the integral is of measure 1. This measure is of the same type as those we considered previously. We can easily understand that

$$\lim_{a \to 0} \lim_{\Lambda \to R^d} \mathrm{d}\mu(a, \Lambda) = \mathrm{d}\mu_C, \qquad (10.53)$$

since the finite difference operator converges to C^{-1}.

To conclude this section we give some properties of the covariance.

The covariance that we considered is the inverse of $-\triangle + m^2$. The kernel of this covariance can be represented by

$$C(x - y) = \frac{1}{(2\pi)^d} \int \mathrm{d}^d p \frac{1}{p^2 + m^2} e^{ip(x-y)}, \qquad (10.54)$$

where p^2 is the Euclidean square of the d-vector p and $p.(x - y)$ the scalar product between p and $x - y$. This is the Euclidean propagator of a free massive scalar field. It propagates from x to y or from y to x since it is a symmetric function of the argument. Graphically, it is represented by a segment joining the points x and y. We use often the momentum space description. In this case, it is said that the propagator, of momentum p, is given by $1/(p^2 + m^2)$.

As a function, $C(x - y)$ is positive for all values of x and y. We check without difficulty that when $x = y$, $C(0)$ is infinite except if the dimension d is smaller than 2. On the other hand for large Euclidean distances, $|x - y| \to +\infty$, the kernel decreases as $\exp(-m|x - y|)$;

---

[3] We can take, for example,

$$(-\triangle f)(ai) = a^{-2}[2df(ai) - \sum_{|i-j|=1} f(aj)], \qquad (10.51)$$

where the sum on $j \in Z^d$ is over the $2d$ nearest neighbours of i.

this decrease expresses the fact that a particle of mass m influences its neighbourhood only over distances smaller than $1/m$. It is the translation in field theory language of the well-known formula, giving the Yukawa potential created by the exchange of a particle of mass m.

10.2.2   The Integration by Parts Formula

In this section, we will prove an integration by parts formula with respect to the measure $\mathrm{d}\mu_C$. There is no hope to have an integration by parts formula with respect to the underlying Lebesgue measures since the functional measure is the product of an infinity of such measures.

Let us give the formula.

Let $F(\phi)$ be a function of the field ϕ. Then

$$\int \phi(x)F(\phi)\mathrm{d}\mu_C = \int_{R^d} \mathrm{d}^d y\, C(x-y) \int \frac{\delta}{\delta\phi(y)} F(\phi)\mathrm{d}\mu_C. \tag{10.55}$$

We first prove this formula for the case where the measure is the restricted measure (10.48). Let us consider

$$I = \int q_i F(q)\mathrm{d}\mu_{\Gamma(n)}. \tag{10.56}$$

Since $\Gamma(n)$ is positive, there exists an orthogonal matrix U such that $U^\dagger \Gamma(n)U = \Delta$ with Δ diagonal, with eigenvalues δ_j. Let us set $p = U^\dagger q$; thus, $U^\dagger C^{-1}U = \Delta^{-1}$ and

$$\sum_{i,j} q_i C_{ij}^{-1} q_j = \sum_k \delta_k^{-1} p_k^2. \tag{10.57}$$

Let us make the change of variables $q \to p$,

$$\begin{aligned}
I &= \sqrt{\prod_k \frac{\delta_k}{2\pi}} \int U_{ij} p_j F(U^\dagger p)\mathrm{e}^{-\frac{1}{2}\sum_k \delta_k^{-1} p_k^2} \prod \mathrm{d}p_l \\
&= -\sqrt{\prod_k \frac{\delta_k}{2\pi}} \sum_j U_{ij} \int F(U^\dagger p)\mathrm{e}^{-\frac{1}{2}\sum_{k\neq j} \delta_k^{-1} p_k^2} \prod_{l\neq j} \mathrm{d}p_l \delta_j \mathrm{d}\mathrm{e}^{-\delta_j^{-1} p_j^2} \\
&= \sqrt{\prod_k \frac{\delta_k}{2\pi}} \sum_j U_{ij} \int \delta_j \frac{\delta}{\delta p_j} F(U^\dagger p)\mathrm{e}^{-\frac{1}{2}\sum_k \delta_k^{-1} p_k^2} \prod \mathrm{d}p_l \\
&= \int U_{ij} \delta_j U_{kj} \frac{\delta}{\delta_k} F(q)\mathrm{d}\mu_{\Gamma(n)} \\
&= \int \Gamma(n)_{ik} \frac{\delta}{\delta_k} F(q)\mathrm{d}\mu_{\Gamma(n)},
\end{aligned} \tag{10.58}$$

where we used the fact that $\frac{\delta}{\delta p_j} = U_{kj}\frac{\delta}{\delta q_k}$ and $U_{kj} = U_{jk}^\dagger$. The replacement of two fields, in the integration by parts formula, by the propagator linking their ends is called a *contraction*.

This formula makes it possible to recover immediately the basic formulae of Gaussian integration:

$$\int \phi(x_1)\phi(x_2)\mathrm{d}\mu_C = \int \mathrm{d}^d y\, C(x_1 - y)\int \frac{\delta}{\delta\phi(y)}\phi(x_2)\mathrm{d}\mu_C$$

$$= \int\int C(x_1 - y)\delta(y - x_2)\mathrm{d}^d y\, \mathrm{d}\mu_C$$

$$= \int C(x_1 - x_2)\mathrm{d}\mu_C = C(x_1 - x_2). \qquad (10.59)$$

Similarly,

$$\int \phi(x_1)\phi(x_2)\phi(x_3)\phi(x_4)\mathrm{d}\mu_C$$

$$= \int C(x_1 - y)\frac{\delta}{\delta\phi(y)}(\phi(x_2)\phi(x_3)\phi(x_4))\mathrm{d}\mu_C\mathrm{d}^d y$$

$$= C(x_1 - x_2)\int \phi(x_3)\phi(x_4)\mathrm{d}\mu_C + C(x_1 - x_3)\int \phi(x_2)\phi(x_4)\mathrm{d}\mu_C$$

$$+ C(x_1 - x_4)\int \phi(x_2)\phi(x_3)\mathrm{d}\mu_C \qquad (10.60)$$

$$= C(x_1 - x_2)C(x_3 - x_4) + C(x_1 - x_3)C(x_2 - x_4) + C(x_1 - x_4)C(x_2 - x_3).$$

Remark: It is obvious that the result is independent of the choice of the field which has been taken to initiate the contractions. In the previous proof, it would have been possible to start with $\phi(x_2)$ or $\phi(x_3)$ or $\phi(x_4)$.

The integration by parts formula makes it possible to perform completely the functional integration of polynomial expressions.

10.2.3 The Wick Ordering

According to formula (10.59), $\int \phi(x)^2 \mathrm{d}\mu_C = C(0)$ is infinite when the dimension of the underlying space is greater than or equal to 2. This implies that $\phi(x)^2$ is not a measurable function with respect to the measure $\mathrm{d}\mu_C$. More generally, we can check that all powers of $\phi(x)$ are not measurable. We remedy this difficulty by introducing an algebraic combination, the *Wick ordering*, formally defined by[4]

---

[4] We can make these formulae rigorous by regularising the covariance $C \to C_\kappa$ with

$$C_\kappa(x - y) = \frac{1}{(2\pi)^d}\int \frac{\eta_\kappa(p)}{p^2 + m^2}\mathrm{d}^d p,$$

$$: \phi(x) : = \phi(x)$$
$$: \phi(x)^2 : = \phi(x)^2 - C(0)$$
$$: \phi(x)^3 : = \phi(x)^3 - C(0)\phi(x)$$
$$: \phi(x)^4 : = \phi(x)^4 - 6C(0)\phi(x)^2 + 3C(0)^2$$
$$: \phi(x)^n : = \sum_0^{[n/2]} \frac{(-1)^j n!}{(n-2j)! j! 2^j} C(0)^j \phi(x)^{n-2j}. \tag{10.61}$$

Algebraically, the introduction of the Wick ordering[5] for field monomials at the same points means that these fields cannot contract between themselves. Moreover, Wick powers behave with respect to derivation as usual powers

$$\frac{\mathrm{d}}{\mathrm{d}\phi} : \phi^n : = n : \phi^{n-1} : \qquad \forall n \in Z_+. \tag{10.62}$$

It follows from the definition that

$$\int : \phi(x)^n : \mathrm{d}\mu_C = 0 \tag{10.63}$$

and

$$\int : \phi(x)^p :: \phi(y)^q : \mathrm{d}\mu_C = \delta_{pq} \, p! \, [C(x-y)]^p \tag{10.64}$$

if p and q are positive integers such that $p \neq q$.

Remark: We should remember that the Wick ordering is relative to a given covariance.

10.3   Application to Interacting Fields

We will apply the analysis of the preceding sections to the concrete case of a given theory and show how the integration by parts method, or the contraction, generates a perturbation expansion particularly easy to be described in terms of diagrams.

---

where $\eta_\kappa(p)$ is a cutoff function which is 0 if $|p| > \kappa$ and such that $\lim_{\kappa \to +\infty} \eta_\kappa(p) = 1$. In this case $C_\kappa(0)$ is finite and the formulae must be understood as the limits of the formulae written in terms of the fields ϕ_κ of covariance C_κ.

[5] We had used previously the term *Wick ordering* to denote the normal ordering in the product of operators in which creation operators are placed on the left and annihilation operators on the right. The use of the same term here is not accidental. It is easy to prove that the two operations are related by the same analytic continuation which relates Euclidean and Minkowski theories. For example, the infinite constant $C(0)$ corresponds to the infinite subtraction we found in the previous chapter.

We consider the theory of a massive scalar field interacting with itself through a fourth-order interaction. Physically, this theory can represent the dynamics of meson fields, such as those corresponding to the neutral pions π^0.

We fix the dimension $d = 4$ and consider the Lagrangian density

$$\mathcal{L}(x) = \frac{1}{2}[(\nabla\phi(x))^2 + m^2\phi(x)^2] + \lambda\phi(x)^4, \tag{10.65}$$

where λ is the coupling constant (supposed to be small).

The action can be written as

$$S = \int_{R^4} \mathcal{L}(x)\mathrm{d}^4x \tag{10.66}$$

and, in order to compute an n-point function, we apply the Euclidean version of formula (9.12).

Let us give as examples the 2-point and the 4-point functions.

10.3.1   The 2-Point Function

By definition

$$< \phi(x)\phi(y) > = \frac{\int \phi(x)\phi(y)\mathrm{e}^{-S}\mathcal{D}(\phi)}{\int \mathrm{e}^{-S}\mathcal{D}(\phi)}. \tag{10.67}$$

To give a rigorous meaning to this expression, we write

$$\mathcal{L}(x) = \mathcal{L}_0(x) + \mathcal{L}_{\mathrm{int}}(x) \tag{10.68}$$

and thus

$$S = S_0 + S_{\mathrm{int}}, \tag{10.69}$$

from which

$$
\begin{aligned}
< \phi(x)\phi(y) > &= \frac{\int \phi(x)\phi(y)\mathrm{e}^{-S_{\mathrm{int}}}\mathrm{e}^{-S_0}\mathcal{D}(\phi)}{\int \mathrm{e}^{-S_{\mathrm{int}}}\mathrm{e}^{-S_0}\mathcal{D}(\phi)} \\
&= \frac{\int \phi(x)\phi(y)\mathrm{e}^{-S_{\mathrm{int}}}\mathrm{d}\mu_C}{\int \mathrm{e}^{-S_{\mathrm{int}}}\mathrm{d}\mu_C}.
\end{aligned} \tag{10.70}
$$

We take this last equality, in which the measure $\mathrm{d}\mu_C$ is perfectly defined (which is equivalent to formula (8.98) of Chapter 8), as the definition of the 2-point function in Euclidean space. In particular, we can, if necessary, regularise the expression by replacing C by C_κ and by restricting the integration in S_{int} to a finite volume, i.e. by applying an ultraviolet and an infrared regularisation.

We saw in Eq. (10.70) that the action which is expressed by using the fourth power of the field is not measurable with respect to the measure. We can try to remedy this problem by replacing ϕ^4 by $: \phi^4 :$. This shows that our initial Lagrangian density (10.65) is not correct and must be replaced by

$$\mathcal{L}(x) = \frac{1}{2}[(\nabla \phi(x))^2 + m^2 \phi(x)^2] + \lambda : \phi(x)^4 : . \tag{10.71}$$

Indeed, using the properties of the Wick ordering given in Eqs. (10.61), we find that the 2-point function to first order in λ is proportional to

$$< \phi(x)\phi(y) > \sim \int \phi(x)\phi(y) : \phi(z)^4 : \mathrm{d}\mu_C \mathrm{d}^4 z, \tag{10.72}$$

which vanishes. However, this will not solve all problems in higher orders.

Since λ is small, let us compute the 2-point function to second order in λ.

We start by an evaluation of the numerator. To do this we contract one of the fields, for example $\phi(x)$,

$$\int \phi(x)\phi(y)\mathrm{e}^{-S_{\text{int}}} \mathrm{d}\mu_C = \int \phi(x)\phi(y)\mathrm{e}^{-\int \mathcal{L}_{\text{int}}(x)\mathrm{d}^4 x} \mathrm{d}\mu_C$$

$$= \int C(x-x_1)\frac{\delta}{\delta\phi(x_1)}\left(\phi(y)\mathrm{e}^{-\int \mathcal{L}_{\text{int}}(x)\mathrm{d}^4 x}\right)\mathrm{d}^4 x_1 \mathrm{d}\mu_C$$

$$= \int C(x-x_1)\delta^4(x_1-y)\mathrm{e}^{-\int \mathcal{L}_{\text{int}}(x)\mathrm{d}^4 x}\mathrm{d}^4 x_1 \mathrm{d}\mu_C$$

$$- 4\lambda \int C(x-x_1)\delta^4(x_1-z) : \phi(z)^3 : \phi(y)\mathrm{e}^{-\int \mathcal{L}_{\text{int}}(x)\mathrm{d}^4 x}\mathrm{d}^4 z\mathrm{d}^4 x_1 \mathrm{d}\mu_C$$

$$= C(x-y)\int \mathrm{e}^{-\int \mathcal{L}_{\text{int}}(x)\mathrm{d}^4 x}\mathrm{d}\mu_C \tag{10.73}$$

$$-4\lambda \int C(x-x_1) : \phi(x_1)^3 : \phi(y)\mathrm{e}^{-\int \mathcal{L}_{\text{int}}(x)\mathrm{d}^4 x}\mathrm{d}^4 x_1 \mathrm{d}\mu_C$$

$$= C(x-y)Z - 4\lambda \int C(x-x_1) : \phi(x_1)^3 : \phi(y)\mathrm{e}^{-\int \mathcal{L}_{\text{int}}(x)\mathrm{d}^4 x}\mathrm{d}^4 x_1 \mathrm{d}\mu_C$$

with Z given by

$$Z = \int \mathrm{e}^{-\int \mathcal{L}_{\text{int}}(x)\mathrm{d}^4 x}\mathrm{d}\mu_C. \tag{10.74}$$

Let us consider the second term of the preceding equality. Since we are interested in terms up to the second order only, it can be written to order $O(\lambda^3)$ as

$$-4\lambda \int C(x-x_1) : \phi(x_1)^3 : \phi(y)(1 - \int \mathcal{L}_{\text{int}}(z)d^4 z)d^4 x_1 \mathrm{d}\mu_C$$

$$= -4\lambda \int C(x-x_1) : \phi(x_1)^3 : \phi(y)d^4 x_1 \mathrm{d}\mu_C$$

$$+ 4\lambda^2 \int C(x-x_1) : \phi(x_1)^3 : \phi(y) \int \mathcal{L}_{\text{int}}(z)d^4 z d^4 x_1 \mathrm{d}\mu_C + O(\lambda^3)$$

$$= A + B + O(\lambda^3). \tag{10.75}$$

The term A vanishes because $\int \,\, : \phi^3 : \phi \,\, \mathrm{d}\mu_C = 0$. Contracting $\phi(y)$ and using the symmetry $C(x-y) = C(y-x)$, we obtain

$$
\begin{aligned}
B &= +4\lambda^2 \int C(x-x_1) C(x_2-y) \frac{\delta}{\delta\phi(x_2)} \left(:\phi(x_1)^3: \int \mathcal{L}_{\text{int}}(z)\mathrm{d}^4 z \right) \mathrm{d}^4 x_2 \mathrm{d}^4 x_1 \mathrm{d}\mu_C \\
&= 12\lambda^2 \int C(x-x_1) C(x_2-y) :\phi(x_1)^2: \delta(x_1-x_2) \int \mathcal{L}_{\text{int}}(z)\mathrm{d}^4 z \mathrm{d}^4 x_2 \mathrm{d}^4 x_1 \mathrm{d}\mu_C \\
&\quad +16\lambda^2 \int C(x-x_1) C(x_2-y) :\phi(x_1)^3 :: \phi(z)^3: \delta(x_2-z)\mathrm{d}^4 z \mathrm{d}^4 x_2 \mathrm{d}^4 x_1 \mathrm{d}\mu_C \\
&= B_1 + B_2.
\end{aligned}
\tag{10.76}
$$

Again B_1 is 0 and we have

$$
\begin{aligned}
B_2 &= 16\lambda^2 \int C(x-x_1) C(x_2-y) :\phi(x_1)^3 :: \phi(x_2)^3: \mathrm{d}^4 x_2 \mathrm{d}^4 x_1 \mathrm{d}\mu_C \\
&= 96\lambda^2 \int C(x-x_1) C(x_1-x_2)^3 C(x_2-y)\mathrm{d}^4 x_2 \mathrm{d}^4 x_1.
\end{aligned}
\tag{10.77}
$$

This last equality comes from

$$
\begin{aligned}
&\int : \phi(x_1)^3 :: \phi(x_2)^3 : \mathrm{d}\mu_C \\
&= 3 \int C(x_1-u) : \phi(x_1)^2 :: \phi(x_2)^2 : \delta^4(x_1-u)\mathrm{d}^4 u \mathrm{d}\mu_C \\
&= 3 \int C(x_1-x_2) : \phi(x_1)^2 :: \phi(x_2)^2 : \mathrm{d}\mu_C \\
&= 3 C(x_1-x_2) \int : \phi(x_1)^2 :: \phi(x_2)^2 : \mathrm{d}\mu_C \\
&= 6 C(x_1-x_2)^2 \int \phi(x_1)\phi(x_2)\mathrm{d}\mu_C \\
&= 6 C(x_1-x_2)^3.
\end{aligned}
\tag{10.78}
$$

We have thus shown that

$$
\begin{aligned}
&< \phi(x)\phi(y) > \\
&= \frac{1}{Z} \left(C(x-y)Z + 96\lambda^2 \int C(x-x_1) C(x_1-x_2)^3 C(x_2-y)\mathrm{d}^4 x_2 \mathrm{d}^4 x_1 + O(\lambda^3) \right) \\
&= C(x-y) + 96\lambda^2 \int C(x-x_1) C(x_1-x_2)^3 C(x_2-y)\mathrm{d}^4 x_2 \mathrm{d}^4 x_1 + O(\lambda^3),
\end{aligned}
\tag{10.79}
$$

because $Z = 1 + O(\lambda^2)$. Indeed, expanding the exponential of the interaction, the first term is 1 and the second one is

$< \varphi(x)\varphi(y) >=$ 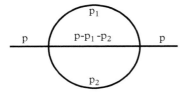   **Figure 10.1** *Expansion of the 2-point function.*

Figure 10.2 *A 2-loop contribution to the 2-point function.*

$$\int \left(\int : \phi(z)^4 : d^4 z \right) d\mu_C = 0. \tag{10.80}$$

Diagrammatically, we represent the expansion of the 2-point function as shown in Fig. 10.1.

Let us examine the 2-loop diagram (Fig. 10.2).

As we said previously, a propagator $C(x_1 - x_2)$ going from x_1 to x_2 (or the other way around since C is symmetric) is represented by a segment joining the points x_1 and x_2. The end points are either starting points or are integrated. If they are integrated, and we call them internal, then they are part of a vertex which links four propagators. This last property is the trace of the fact that the interaction is ϕ^4.

It is often useful to express the diagrams which have been drawn in Fig. 10.1 in terms of Fourier transforms. We have already seen that

$$C(x-y) = \frac{1}{(2\pi)^4} \int \frac{e^{-ip(x-y)}}{p^2 + m^2} d^4 p. \tag{10.81}$$

A propagator is therefore represented by $1/(p^2 + m^2)$. The second diagram corresponds to

$$I = \int C(x-x_1)C(x_1-x_2)^3 C(x_2-y)d^4x_2 d^4 x_1$$

$$= \frac{1}{(2\pi)^4} \int \frac{e^{-ip(x-x_1)}}{p^2 + m^2} d^4 p \tag{10.82}$$

$$\times \prod_{i=1}^{3} \left(\frac{1}{(2\pi)^4} \int \frac{e^{-ip_i(x_1-x_2)}}{p_i^2 + m^2} d^4 p_i \right) \frac{1}{(2\pi)^4} \int \frac{e^{-iq(x_2-y)}}{q^2 + m^2} d^4 q d^4 x_1 d^4 x_2.$$

Integrating over x_1 and x_2, the exponential factors give $(2\pi)^4\delta^{(4)}(p-\sum_1^3 p_j)$ and $(2\pi)^4\delta^{(4)}(q-\sum_1^3 p_j)$ and since

$$\delta^{(4)}\left(p-\sum_1^3 p_j\right)\delta^{(4)}\left(q-\sum_1^3 p_j\right) = \delta^{(4)}\left(p-\sum_1^3 p_j\right)\delta^{(4)}(p-q)$$

we get

$$\frac{1}{(2\pi)^{12}}\int \delta^{(4)}(p-q)\frac{1}{p^2+m^2}\prod\frac{1}{p_i^2+m^2}$$

$$\times\delta^{(4)}\left(p-\sum_1^3 p_j\right)\frac{1}{p^2+m^2}e^{-ip(x-y)}\,\mathrm{d}^4p\,\mathrm{d}^4q\,\mathrm{d}^4p_1\,\mathrm{d}^4p_2\,\mathrm{d}^4p_3$$

$$=\frac{1}{(2\pi)^{12}}\int \frac{1}{p^2+m^2}\prod\frac{1}{p_i^2+m^2}$$

$$\times\delta^{(4)}\left(p-\sum_1^3 p_j\right)\frac{1}{p^2+m^2}e^{-ip(x-y)}\,\mathrm{d}^4p\,\mathrm{d}^4p_1\,\mathrm{d}^4p_2\,\mathrm{d}^4p_3$$

$$=\frac{1}{(2\pi)^4}\int e^{-ip(x-y)}\frac{1}{(p^2+m^2)^2}\Pi_2(p^2)\,\mathrm{d}^4p \tag{10.83}$$

with

$$\Pi_2(p^2) = \frac{1}{(2\pi)^8}\int \delta^{(4)}\left(p-\sum_i p_i\right)\prod_1^3\frac{\mathrm{d}^4p_i}{p_i^2+m^2}$$

$$=\frac{1}{(2\pi)^8}\int \frac{1}{p_1^2+m^2}\frac{1}{(p-p_1-p_2)^2+m^2}\frac{1}{p_2^2+m^2}\mathrm{d}^4p_1\,\mathrm{d}^4p_2. \tag{10.84}$$

It is easy, by an $O(4)$ invariance argument, to justify the fact that the right-hand side of (10.84) is a function of p^2.

The expression (10.84) is still quadratically divergent. We see that using the Wick ordering has not cured the divergences beyond the lowest order. We shall present the complete solution to all such problems order by order in perturbation theory in Chapter 16. For the moment expressions such as that of Eq. (10.84) should be considered as formal. For example, we can assume that the propagator $C(x-y)$ is replaced by the regularised one C_k.[6]

---

[6] If we had used the original ϕ^4 interaction instead of the Wick ordered one, we would have obtained an extra term in first order of perturbation theory corresponding to the diagram of Fig. 10.3. It is also quadratically divergent. In the jargon of particle physics such diagrams are called 'tadpoles'.

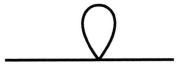

Figure 10.3 *A 1-loop divergent diagram for the 2-poin function in the ϕ^4 theory which we obtain in the absence of Wick ordering.*

10.3.2 The 4-Point Function

Following the same type of analysis as it was done for the 2-point function, the Euclidean 4-point function is given by

$$< \phi(x_1)\phi(x_2)\phi(x_3)\phi(x_4) >= \frac{\int \phi(x_1)\phi(x_2)\phi(x_3)\phi(x_4)\mathrm{e}^{-S_{\text{int}}}\mathrm{d}\mu_C}{\int \mathrm{e}^{-S_{\text{int}}}\mathrm{d}\mu_C}. \tag{10.85}$$

We can compute this expression up to the third order in the coupling constant and get

$$
\begin{aligned}
& < \phi(x_1)\phi(x_2)\phi(x_3)\phi(x_4) > \\
&= -24\lambda \int \prod C(x_i - y)\mathrm{d}^4 y \\
&\quad + C(x_1 - x_2)C(x_3 - x_4) + C(x_1 - x_3)C(x_2 - x_4) + C(x_1 - x_4)C(x_2 - x_3) \\
&\quad + 96\lambda^2 C(x_1 - x_2) \int C(x_3 - y)(C(y - z))^3 C(z - x_4)\mathrm{d}^4 y \mathrm{d}^4 z \\
&\quad + 96\lambda^2 C(x_1 - x_3) \int C(x_2 - y)(C(y - z))^3 C(z - x_4)\mathrm{d}^4 y \mathrm{d}^4 z \\
&\quad + 96\lambda^2 C(x_1 - x_4) \int C(x_2 - y)(C(y - z))^3 C(z - x_3)\mathrm{d}^4 y \mathrm{d}^4 z \\
&\quad + 96\lambda^2 C(x_2 - x_3) \int C(x_1 - y)(C(y - z))^3 C(z - x_4)\mathrm{d}^4 y \mathrm{d}^4 z \\
&\quad + 96\lambda^2 C(x_2 - x_4) \int C(x_1 - y)(C(y - z))^3 C(z - x_3)\mathrm{d}^4 y \mathrm{d}^4 z \qquad (10.86) \\
&\quad + 96\lambda^2 C(x_3 - x_4) \int C(x_1 - y)(C(y - z))^3 C(z - x_2)\mathrm{d}^4 y \mathrm{d}^4 z \\
&\quad + 288\lambda^2 \int C(x_1 - y)C(x_2 - y)(C(y - z))^2 C(z - x_3)C(z - x_4)\mathrm{d}^4 y \mathrm{d}^4 z \\
&\quad + 288\lambda^2 \int C(x_1 - y)C(x_3 - y)(C(y - z))^2 C(z - x_2)C(z - x_4)\mathrm{d}^4 y \mathrm{d}^4 z \\
&\quad + 288\lambda^2 \int C(x_1 - y)C(x_4 - y)(C(y - z))^2 C(z - x_2)C(z - x_3)\mathrm{d}^4 y \mathrm{d}^4 z + O(\lambda^3)
\end{aligned}
$$

or diagrammatically as shown in Fig. 10.4.

The first three diagrams represent the free contribution and we recover, as expected, the formula (10.61). The fourth term is the first manifestation of the fact that there is a

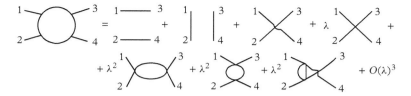

Figure 10.4 *The 4-point function.*

local point-wise interaction of strength λ which couples four fields. The fifth, sixth, and seventh terms express the first-order (in λ) correction to the coupling strength.

The connected part, up to order 2, of the 4-point function is nothing else than the truncated expectation value of the 4-point function as given algebraically by formula (9.77). Diagrammatically, it corresponds to the last four diagrams of the expansion (Fig. 10.4).

Introducing the Fourier transform of the propagator we get

$$\lambda (\frac{1}{(2\pi)^4})^3 \int \prod [e^{-ip_j x_j} \frac{1}{p_j^2 + m^2} d^4 p_j] \delta^{(4)}(p_1 + p_2 + p_3 + p_4)$$
$$\times [1 - 12\lambda(G(p_1 + p_2) + G(p_1 + p_3) + G(p_1 + p_4)], \qquad (10.87)$$

where $G(p)$ is given by

$$G(p) = \frac{1}{(2\pi)^4} \int \frac{1}{q^2 + m^2} \frac{1}{(p - q)^2 + m^2} d^4 q, \qquad (10.88)$$

which corresponds to the 1-loop 4-point function diagram shown in Fig. 10.5.

The diagram in Fig. 10.5 has two vertices. Each vertex links four lines, because it is a ϕ^4-theory. Two of these lines are external; from them flows a momentum p which is the vectorial sum of the momenta carried by each line. This momentum is split inside the loop in two contributions q and $p - q$ whose vectorial sum is again p because of the conservation of momenta at each vertex. Here q is the momentum of the loop and there is an integration over the whole space on this variable.

The function $G(p)$ represents the diagram of Fig. 10.5, i.e. a diagram with no contributions from the external lines. If we multiply $G(p)$ by

$$\prod [\frac{1}{p_j^2 + m^2}] \delta^{(4)}(p_1 + p_2 + p_3 + p_4)$$

we get the Fourier transform of one of the terms in (10.87).

Figure 10.5 *A 1-loop contribution to the 4-point function.*

$G(p)$ is not integrable for large $|q|$. If we introduce a momentum cutoff Λ, it can be seen immediately that

$$G^{(\Lambda)}(0) = \frac{1}{(2\pi)^4} \int_{|q|<\Lambda} \left(\frac{1}{q^2 + m^2}\right)^2 d^4q$$

behaves as $\log \Lambda$ since for large q it is like the integration

$$\int_{1<|q|<\Lambda} \frac{d^4q}{|q|^4} \simeq \int_{1<|q|<\Lambda} \frac{dq}{|q|} \simeq \log \Lambda. \tag{10.89}$$

This type of analysis is a power-counting argument: we integrate in a d-dimensional space 2 propagators which are quadratic functions of the integration variable. More generally, an integral

$$\int_{1<|q|<\Lambda} f(q) d^d q,$$

where $f(q) \simeq |q|^m$, is finite for $\Lambda \to \infty$ if $-m > d$. It is thus natural to introduce an index, the superficial degree of divergence D, $D = d + m$, whose values measure the integrability at infinity of the integral. The value of the integral is finite if $D < 0$. The superficial degree of divergence of $G(p)$ is $D = d - 4 = 0$ since $d = 4$; thus, this diagram is logarithmically divergent as can be seen from (10.89).

A similar 1-loop contribution for a ϕ^4 theory in d-dimension

$$\frac{1}{(2\pi)^4} \int \frac{1}{q^2 + m^2} \frac{1}{(p-q)^2 + m^2} d^d q \tag{10.90}$$

will have $D = d - 4$, i.e. will be convergent for any dimension $d < 4$. We will expand on this power-counting argument in Chapter 16.

10.3.3   The General Feynman Rules

From these two examples, it is possible to state the rules of computation of the (Euclidean) Feynman diagrams:

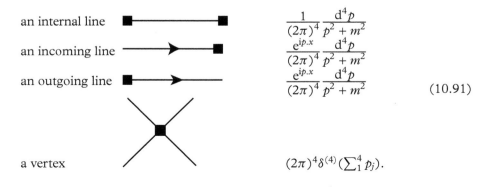

an internal line	$\dfrac{1}{(2\pi)^4} \dfrac{d^4p}{p^2 + m^2}$
an incoming line	$\dfrac{e^{ip.x}}{(2\pi)^4} \dfrac{d^4p}{p^2 + m^2}$
an outgoing line	$\dfrac{e^{ip.x}}{(2\pi)^4} \dfrac{d^4p}{p^2 + m^2}$ $\qquad$ (10.91)
a vertex	$(2\pi)^4 \delta^{(4)}\left(\sum_1^4 p_j\right).$

Because of the delta function at each vertex, each internal loop carries one momentum which is integrated over the whole space.

Thus, we obtain the following expression for the diagram of Fig. 10.2:

$$\frac{1}{(2\pi)^{12}}\int e^{-ip(x-y)}\frac{1}{(p^2+m^2)^2}\frac{1}{p_1^2+m^2}\frac{1}{p_2^2+m^2}\frac{1}{(p-p_1-p_2)^2+m^2}\,\mathrm{d}^4p\,\mathrm{d}^4p_1\mathrm{d}^4p_2$$

$$= \frac{1}{(2\pi)^4}\int e^{-ip(x-y)}\frac{1}{(p^2+m^2)^2}\Pi_2(p). \tag{10.92}$$

p_1 and p_2 are the two internal momenta over which the integration is performed. Again if we multiply $\Pi_2(p)$ by

$$\frac{1}{(p^2+m^2)^2},$$

we get the Fourier transform of (10.83).

10.4   Problems

Problem 10.1 What can be said about the convergence or the divergence of the diagrams of a theory with interaction : ϕ^m :, $m = 2, 4, 6$, and 8, in dimension d, d taking the values 2, 3, or 4?

Problem 10.2 Suppose the interaction is $(\nabla\phi.\nabla\phi)^2$. What are the divergent diagrams of this theory in two, three, or four dimensions?

11

Fermions and Functional Formalism

11.1 Introduction

Let us summarise the results we obtained in the previous chapters.

Consider a classical system of N degrees of freedom represented by the variables $q_i(t)$, $i = 1, 2, ..., N$. We assume that the dynamics is described by a Lagrangian $L(q_i, \dot{q}_i)$, or, equivalently, by a Hamiltonian $H(q_i, p_i)$, where $p_i(t)$ are the corresponding conjugate momenta. Starting from this classical system, we saw that we have at least two ways to obtain the corresponding quantum system.

The first is to follow Heisenberg's recipe and postulate that the q_i's and the corresponding p_i's are operators satisfying canonical commutation relations

$$[q_i, p_j] = i\hbar\delta_{ij} \quad ; \quad [q_i, q_j] = [p_i, p_j] = 0. \tag{11.1}$$

The classical theory is obtained at the limit $\hbar \to 0$, i.e. when the variables become commuting classical functions.

The second quantisation recipe is given by Feynman's path integral formalism: we consider the classical action S, which is a functional of the variables q_i. Summing $\exp(iS/\hbar)$ over all trajectories $q_i(t)$ joining an initial and a final configuration reproduces the quantum mechanical amplitude for the transition between the corresponding initial and final states.

Formally, we can take the large N limit and obtain a system with an infinite number of degrees of freedom. At the classical level such a system may be a classical field theory in a d-dimensional space with variables $\phi(x, t)$ and conjugate momenta $\pi(x, t)$. Either of the two recipes will give the corresponding quantum field theory. In a previous chapter we showed that a free real scalar field can be expanded in terms of an infinite set of creation and annihilation operators satisfying canonical commutation relations; see Eqs. (9.96) and (9.97). Two, seemingly unrelated properties follow from them. The first is that the corresponding Hamiltonian given by Eq. (9.99) is positive definite. The second is that a state which contains two excitations, one with momentum k_1 and a second with momentum k_2, is symmetric under the exchange of the two momenta. Indeed, the state is given by $a^\dagger(k_1)a^\dagger(k_2)|0>$ and the symmetry is a direct consequence of the fact

From Classical to Quantum Fields. Laurent Baulieu, John Iliopoulos and Roland Sénéor.
© Laurent Baulieu, John Iliopoulos and Roland Sénéor, 2017. Published 2017 by Oxford University Press.

that the two creation operators commute. This symmetry property can be obviously generalised to an arbitrary number of excitations and reflects the property we have mentioned already, namely that scalar fields describe bosons. At the level of the path integral this is reflected in the fact that we integrate in a space of classical commuting functions.

In this chapter we want to see how to extend these considerations when quantising a Dirac field $\psi(x)$. We wrote the expansion in plane waves in section 6.3.8, Eqs. (6.76) and (6.77). Heisenberg's canonical quantisation recipe tells us that we should interpret the complex functions a and b as two pairs of creation and annihilation operators satisfying canonical commutation relations. At first sight the differences between the Dirac and the scalar field expansions seem to be unimportant. The fact that we have two pairs of creation and annihilation operators is just a consequence of the fact that the Dirac field is complex. But then we see immediately that we shall run into trouble. The first comes from the Hamiltonian. A straightforward calculation gives, for the normal ordered form,

$$: H := \int \mathrm{d}\Omega_m E_k \sum_{\alpha=1}^{2} \left[a_\alpha^\dagger(k) a_\alpha(k) - b_\alpha^\dagger(k) b_\alpha(k) \right]. \tag{11.2}$$

The trouble comes from the minus sign. No matter how we arrange an overall sign, this Hamiltonian will be unbounded from below. The two kinds of excitations, those created by the $a^\dagger$ or the $b^\dagger$ operator, have opposite energies. This is not a new problem. It is the one we have found in analysing the energy spectrum of the Hamiltonian when we were studying the Dirac equation as the wave equation for an electron. The solution we adopted there was to assume that electrons are fermions and obey the Pauli exclusion principle. We thus assumed, following Dirac, that all negative energy states are occupied. This was meant to be an *ad hoc* principle inspired by the phenomenology of atomic spectra. But here we do not have that choice. If the a and b operators satisfy canonical commutation relations, the same argument we developed for the scalar field shows that a two-excitation state is always symmetrical; in other words, the excitations are bosons.

Can we modify the commutation relations in order to describe fermions? Let us consider a particular two-excitation state. We must have

$$|k_1, \alpha; k_2, \alpha > = a_\alpha^\dagger(k_1) a_\alpha^\dagger(k_2)|0> = -a_\alpha^\dagger(k_2) a_\alpha^\dagger(k_1)|0> \tag{11.3}$$

This suggests that the $a^\dagger$'s must satisfy an *anticommutation* relation of the form $a_\alpha^\dagger(k_1) a_\beta^\dagger(k_2) + a_\beta^\dagger(k_2) a_\alpha^\dagger(k_1) = 0$. The entire Heisenberg recipe should be rewritten with anticommutators replacing commutators. Note that this fixes also the energy crisis because in defining the normal ordering we get an extra minus sign: $: bb^\dagger := -b^\dagger b$.

We have just shown an example of what we called earlier *the spin-statistics theorem*. Positivity of the energy requires that half-odd integer spin particles obey the Fermi statistics.

Their quantisation requires the use of anticommutation relations instead of commutation relations which are used for integer spin particles. This rule is not obvious either from the point of view of classical physics or from the elementary formalism of quantum mechanics called the first quantisation. However, it appeared to become a necessity when Dirac discovered the equation describing spin-$\frac{1}{2}$ particles and tried to quantise it.

Note that the use of anticommutation relations solves also another apparent contradiction, namely the fact that the vacuum expectation value of the commutator of two half-odd integer spin fields does not vanish for space-like separations. The difference between the two types of particles, namely the fermions with anticommutation rules and the bosons with commutation rules, is crucial for multiparticle systems. In quantum mechanics, it results in the Pauli principle which says that the wave functions of fermions are antisymmetric under the exchange of any two fermions while it is symmetric for bosons. Both the microscopic and the macroscopic consequences of this fact are spectacular: Mendeleyev classification in chemistry, superconductivity, Bose condensation, etc.

In every relativistic theory, the energy can be transformed into mass and, as a result, there is no conservation of the number of particles. We have, therefore, from the beginning a multiparticle system and it is very important to have at one's disposal a formalism able to distinguish bosons from fermions.

We will see that it is hopeless to use the same definition of functional integral for fermions and for bosons, since, as we saw, the functional formalism reproduces in extenso the results of the canonical quantisation of bosons. This is not surprising. The functional integral recipe states that one should integrate over all classical configurations. Here 'classical' means what we obtain at the limit $\hbar = 0$. When we use commutators this is the space of c-number functions. However, when we use anticommutators the classical limit is a set of anticommuting objects. It is therefore necessary to find a generalisation which can take into account the fermionic nature of the fields.

We conclude that the extension of the path integral quantisation method to fermions requires, at the price of some abstractions, to define a summation on paths by developing an integration formalism for anticommuting variables. This necessity to appeal to some abstract notions in the case of fermions is not totally unexpected: we have already seen that half-odd integer spin particles had some non-obvious properties such as their difference of behaviour with respect to rotations by 2π or 4π.

The change from quantum mechanics to quantum field theory is conceptually simple. It consists of passing from a system of several degrees of freedom to a system with infinitely many degrees of freedom. The functional integral formalism of quantum mechanics generalises directly. Thus, the action $\int dt \mathcal{L}(t)$ is replaced by $\int dx dt \mathcal{L}(x, t)$, and the measure $\mathcal{D}(q)$ by the measure $\mathcal{D}(\phi)$, which, formally, is nothing else than the product for each value of x of measures $\mathcal{D}\phi(x, \cdot)$. We get an explicitly relativistic system and, apart from technical complications due to the introduction of an infinite number of degrees of freedom, there are no major differences with elementary quantum mechanics.

In the case of fermions, there is the same phenomenon. Once the path integral is defined for the anticommuting variables of quantum mechanics, the way to field theory, for example spinorial electrodynamics, is conceptually straightforward.

11.2   The Grassmann Algebras

We will develop in this section a differential and integral calculus on finite-dimensional algebras: the *Grassmann algebras*.

A Grassmann algebra $\mathcal{G}_n$ with n generators is an algebra generated by n anticommuting variables $\theta_1, \cdots, \theta_n$ satisfying the relations

$$\{\theta_i, \theta_j\} = \theta_i\theta_j + \theta_j\theta_i = 0 \qquad i, j = 1, \cdots, n. \tag{11.4}$$

A basis of the vector space, of dimension 2^n, underlying $\mathcal{G}_n$, is given by the monomials

$$1, \theta_1, \cdots, \theta_n, \theta_1\theta_2, \cdots, \theta_{n-1}\theta_n, \cdots, \theta_1\theta_2\cdots\theta_n \tag{11.5}$$

since from (11.4)

$$\theta_i\theta_i = \theta_i^2 = 0 \qquad \forall i. \tag{11.6}$$

Each element f of the algebra can thus be written as

$$f(\theta) = f_0^{(0)} + f_{i_1}^{(1)}\theta_{i_1} + \frac{1}{2!}f_{i_1 i_2}^{(2)}\theta_{i_1}\theta_{i_2} + \cdots + \frac{1}{n!}f_{i_1\cdots i_n}^{(n)}\theta_{i_1}\cdots\theta_{i_n} \tag{11.7}$$

with the convention of summation over repeated indices. From the anticommutativity of the θ's follows the fact that we can restrict the coefficients $f^{(p)}$, $p > 1$, to be completely antisymmetric under permutations of p indices; the function is entirely determined by the coefficients $f^{(p)}$, $p \geq 0$. It follows that every function is, in fact, a polynomial.

We now define the main operations that can be done in the differential and integral calculus on Grassmann algebra. Unless stated otherwise, we will suppose that these operations are defined for the algebra $\mathcal{G}_n$.

11.2.1   The Derivative

Grassmann algebras being non-commutative algebras, we expect that derivation is also a non-commutative operation. In particular, we can define two types of derivatives, the left and the right derivatives.

Let us consider first the left derivative. It is defined by its action on the monomials of the basis

$$\frac{\partial}{\partial\theta_k}\theta_{i_1}\cdots\theta_{i_p} = \delta_{ki_1}\theta_{i_2}\cdots\theta_{i_p} - \delta_{ki_2}\theta_{i_1}\theta_{i_3}\cdots\theta_{i_p} + \cdots + (-1)^{p-1}\delta_{ki_p}\theta_{i_1}\cdots\theta_{i_{p-1}}, \tag{11.8}$$

which shows that $\frac{\partial}{\partial\theta_p}f(\theta)$ is an element of the algebra. We similarly define the right derivative by

$$\theta_{i_1} \cdots \theta_{i_p} \frac{\overset{\leftarrow}{\partial}}{\partial \theta_k} = \delta_{ki_p} \theta_{i_1} \cdots \theta_{i_{p-1}} - \delta_{ki_{p-1}} \theta_{i_1} \cdots \theta_{i_{p-2}} \theta_{i_p} + \cdots + (-1)^{p-1} \delta_{ki_1} \theta_{i_2} \cdots \theta_{i_p}. \qquad (11.9)$$

We can apply the usual rules of composition of derivatives. Let us introduce linear combinations of θ, i.e. new anticommuting variables η:

$$\eta_k = A_{kp}(s)\theta_p. \qquad (11.10)$$

Let us consider now a function $f(\eta) = g(\theta)$, then

$$\frac{\partial}{\partial \theta_p} f(\eta(\theta)) = \frac{\partial}{\partial \theta_p} g(\theta) = [\frac{\partial}{\partial \eta_k} f(\eta)]\,|_{\eta=\eta(\theta)} A_{kp}(s) \qquad (11.11)$$

and

$$\frac{d}{ds} f(\theta) = \frac{d\theta_k}{ds} \frac{\partial}{\partial \theta_k} f(\theta). \qquad (11.12)$$

It is, however, necessary to exercise caution because Leibniz rule is not valid and some among the usual rules are modified. Thus,

$$\frac{\partial}{\partial \theta_p} (\frac{\partial}{\partial \theta_q} g(\theta)) = -\frac{\partial}{\partial \theta_q} (\frac{\partial}{\partial \theta_p} g(\theta)). \qquad (11.13)$$

Equation (11.13) shows that the derivatives are anticommuting operators:

$$\frac{\partial}{\partial \theta_p} \frac{\partial}{\partial \theta_q} + \frac{\partial}{\partial \theta_q} \frac{\partial}{\partial \theta_p} = 0. \qquad (11.14)$$

11.2.2   The Integration

To define integration, we introduce the differential symbols $d\theta_1, d\theta_2 \cdots d\theta_n$, which satisfy

$$\{d\theta_i, d\theta_j\} = \{d\theta_i, \theta_j\} = 0. \qquad (11.15)$$

Since, as we remarked previously, a Grassmann algebra $\mathcal{G}_n$ spans a finite-dimensional space, every function is a polynomial. Therefore, it is sufficient to define the integration for monomials. For a Grassmann algebra $\mathcal{G}_n$ the 'Table of Integrals' has only 2^n entries!

Following F. A. Berezin,[1] we remark that when we compute in bosonic theories, we usually encounter integrals from $-\infty$ to $+\infty$ for which the most useful property is translational invariance. So, taking the simple case of just one variable, we *define* the integration by imposing the property

---

[1] F. A. Berezin, *The Method of Second Quantization*, trans. N. Mugibayashi and A. Jeffrey (Academic Press, New York, 1966).

$$\int f(\theta)\mathrm{d}\theta = \int f(\theta + \eta)\mathrm{d}\theta \tag{11.16}$$

with η being a constant element of the algebra. The most general function $f(\theta)$ of one anticommuting variable is of the form $f = c_0 + c_1\theta$, with the c's being two arbitrary complex numbers. Imposing (11.16) we find that

$$(c_0 + c_1\eta)\int \mathrm{d}\theta = 0, \tag{11.17}$$

while $\int \theta\mathrm{d}\theta$ remains arbitrary. The property (11.17) should be an identity, so we are led to the *definitions*

$$\int \mathrm{d}\theta_i = 0 \tag{11.18}$$

$$\int \theta_i\mathrm{d}\theta_j = \delta_{ij}. \tag{11.19}$$

We check easily that the integral of the function $f \in \mathcal{G}_n$ given by (11.7) is equal to

$$\int f(\theta)\mathrm{d}\theta_n \cdots \mathrm{d}\theta_1 = f_1^{(n)} \cdots n. \tag{11.20}$$

The following integration by parts formula is also easy to prove. For all $f, g \in \mathcal{G}_n$

$$\int f(\theta)\left(\frac{\partial}{\partial\theta_p}g(\theta)\right)\mathrm{d}\theta_n \cdots \mathrm{d}\theta_1 = \int \left(f(\theta)\frac{\overleftarrow{\partial}}{\partial\theta_p}\right)g(\theta)\,\mathrm{d}\theta_n \cdots \mathrm{d}\theta_1. \tag{11.21}$$

This formula extends to the cases when we integrate only over some of the variables of the Grassmann algebra.

In particular, it shows that

$$\int \frac{\partial}{\partial\theta_i}f(\theta)\,\mathrm{d}\theta_i = 0. \tag{11.22}$$

Thus, splitting each function f of the Grassmann algebra

$$f(\theta) = f_+(\theta) + f_-(\theta), \tag{11.23}$$

where $f_+(\theta)$ is a sum of even polynomials in the θ_i's and $f_-(\theta)$ a sum of odd polynomials, we get

$$\int f_\pm(\theta)\,\frac{\partial g(\theta)}{\partial\theta_i}\,\mathrm{d}\theta_i = \mp\int \frac{\partial f_\pm(\theta)}{\partial\theta_i}g(\theta)\,\mathrm{d}\theta_i. \tag{11.24}$$

We define the 'Fourier transform' by

$$\tilde{f}(\eta) = \int e^{(\eta,\theta)} f(\theta) \prod_i d\theta_i, \tag{11.25}$$

where $(\eta, \theta) = \eta_i \theta_i$, η and θ being the $2n$ generators of $\mathcal{G}_{2n}$, and $\prod_i d\theta_i = d\theta_1 \cdots d\theta_n$, and the δ function by

$$\delta(\eta) = \int e^{(\eta,\theta)} \prod_i d\theta_i = \prod_i (-\eta_i). \tag{11.26}$$

We prove that

$$f(\theta) = \int \delta(\theta - \eta) \prod_i' d\eta_i f(\eta), \tag{11.27}$$

where $\prod_i' d\eta_i = d\eta_1 \cdots d\eta_n$.

It is also possible to define the scalar product of two functions f and g. To fit with the usual notation we consider these functions as functions of $\bar{\theta}$

$$(g,f) = \int e^{-\sum_i \bar{\theta}_i \theta_i} g(\bar{\theta})^* f(\bar{\theta}) d\bar{\theta}_n d\theta_n \cdots d\bar{\theta}_1 d\theta_1, \tag{11.28}$$

where $g(\bar{\theta})^* = \bar{g}_0^{(0)} + \bar{g}_{i_1}^{(1)} \theta_{i_1} + \bar{g}_{i_1 i_2}^{(2)} \theta_{i_2} \theta_{i_1} + \cdots + \bar{g}_{i_1 i_2 \cdots i_n}^{(n)} \theta_{i_n} \cdots \theta_{i_2} \theta_{i_1}$. We then check that the following formula holds:

$$(f,g) = f_0^{(0)} \bar{g}_0^{(0)} + f_{i_1}^{(1)} \bar{g}_{i_1}^{(1)} + f_{i_1 i_2}^{(2)} \bar{g}_{i_1 i_2}^{(2)} + \cdots + f_{i_1 i_2 \cdots i_n}^{(n)} \bar{g}_{i_1 i_2 \cdots i_n}^{(n)}. \tag{11.29}$$

Thus,

$$(f,f) = |f_0^{(0)}|^2 + |f_{i_1}^{(1)}|^2 + |f_{i_1 i_2}^{(2)}|^2 + \cdots + |f_{i_1 i_2 \cdots i_n}^{(n)}|^2. \tag{11.30}$$

It remains to see how this integral calculus behaves under the change of the integration variables.

Let U be a linear transformation of the variables θ

$$\tilde{\theta}_k = \tilde{\theta}_k[\theta] = U_{kp} \theta_p. \tag{11.31}$$

From

$$\tilde{\theta}_1 \cdots \tilde{\theta}_n = \det U \, \theta_1 \cdots \theta_n \tag{11.32}$$

and

$$d\tilde{\theta}_1 \cdots d\tilde{\theta}_n = \det U^{-1} \, d\theta_1 \cdots d\theta_n \tag{11.33}$$

we deduce that

$$\int f(\tilde{\theta}) \, d\tilde{\theta}_n \cdots d\tilde{\theta}_1 = \det U^{-1} \int f(\tilde{\theta}[\theta]) \, d\theta_n \cdots d\theta_1. \tag{11.34}$$

We remark that this formula is different from the usual formula of change of variables in the commutative case since it is the inverse of the determinant which appears.

To close this section, we now apply the integration formula to the computation of Gaussian integrals.

Let A be an $n \times n$ antisymmetric matrix. Let us consider the quadratic form[2] $(\theta, A\theta) = \theta_i A_{ij} \theta_j$. We want to prove the following.

Theorem 10. *Let A be an antisymmetric $n \times n$ matrix and $(\theta, A\theta)$ the quadratic form defined by A on the Grassmann algebra $\mathcal{G}_n$, then*

$$Z(A) \equiv \int e^{\frac{1}{2}(\theta, A\theta)} \, d\theta_n \cdots d\theta_1 = \sqrt{\det A}. \tag{11.35}$$

This theorem is characteristic of the difference of behaviour between commuting and anticommuting variables. We see indeed that the Gaussian integration in the anticommutating case gives an inverse result of what we get in the commuting case (ordinary theory of integration). The result, $Pf(A)$ of the gaussian integration over anticommuting variables of the antisymmetric matrix A is often called the Pfaffian of A and we have $Pf(A)^2 = \det(A)$.

Proof. Let us suppose first that the matrix A is real. There exists an orthogonal transformation U such that UAU^{-1} is of the form

$$\begin{pmatrix} 0 & \lambda_1 & 0 & 0 & 0 \cdots \\ -\lambda_1 & 0 & 0 & 0 & 0 \cdots \\ 0 & 0 & 0 & \lambda_2 & 0 \cdots \\ 0 & 0 & -\lambda_2 & 0 & 0 \cdots \\ \cdots\cdots\cdots \end{pmatrix}. \tag{11.36}$$

If we perform the change of variables $\eta = U\theta$, $Z(A)$ becomes

$$Z(A) = \int e^{\lambda_1 \eta_1 \eta_2 + \lambda_2 \eta_3 \eta_4 + \cdots + \lambda_m \eta_{2m-1} \eta_{2m}} \prod_i d\eta_i \tag{11.37}$$

if $n = 2m$ and

$$Z(A) = \int e^{\lambda_1 \eta_1 \eta_2 + \lambda_2 \eta_3 \eta_4 + \cdots + \lambda_m \eta_{2m-1} \eta_{2m}} \prod_i d\eta_i \tag{11.38}$$

if $n = 2m + 1$.

---

[2] Since we are interested in the quadratic form on $\mathcal{G}_n$ defined by A, it is in full generality that we can take an antisymmetric A. Indeed, for an arbitrary A, the symmetric part of A will not contribute to the quadratic form.

From the integration rules, $Z(A) \equiv 0$ for odd n since the integrand is independent of the variable η_n.

Let us consider the case where n is even. Since the variables η are anticommuting, the variables $\eta_{2i+1}\eta_{2(i+1)}$ and $\eta_{2j+1}\eta_{2(j+1)}$ commute for $i \neq j$. Thus the exponential of the sum in (11.37) can be written as the product of the exponential of each term and since $\eta_i^2 = 0$, $\forall i$,

$$e^{\lambda_p \eta_{2p-1} \eta_{2p}} = 1 + \lambda_p \eta_{2p-1} \eta_{2p} \tag{11.39}$$

and according to the integration rules

$$Z(A) = \int \lambda_1 \eta_1 \eta_2 \lambda_2 \eta_3 \eta_4 \ldots \lambda_m \eta_{2m-1} \eta_{2m} \prod_i d\eta_i = \lambda_1 \lambda_2 \ldots \lambda_m = \sqrt{\det A}. \tag{11.40}$$

This formula extends to the case where the matrix A is complex, the square root being defined as the determination given by taking the positive root for a positive determinant. We remark that

- Gaussian integrals are different from 0 for algebras $\mathcal{G}_n$ with n even.
- No convergence argument is necessary since each function on $\mathcal{G}_n$ including the exponential ones is in fact a polynomial.

From the change of variable property and from the preceding theorem we get the following.

Theorem 11. *Let A be an $n \times n$ matrix with a non-zero determinant, $\theta_1, \cdots, \theta_n, \bar{\theta}_1, \cdots, \bar{\theta}_n$ and $\eta_1, \cdots, \eta_n, \bar{\eta}_1, \cdots, \bar{\eta}_n$, two sets of Grassmannian variables, then*

$$\int e^{(\bar{\theta}, A\theta) + \bar{\theta}.\eta + \bar{\eta}.\theta} \prod_{i=1}^n d\bar{\theta}_i d\theta_i = (-1)^n \det A \, e^{-(\bar{\eta}, A^{-1}\eta)}, \tag{11.41}$$

where $(\bar{\theta}, A\theta) = \sum_{ij} \bar{\theta}_i A_{ij} \theta_j$ and $\bar{\theta}.\eta = \sum_i \bar{\theta}_i \eta_i$, $\bar{\eta}.\theta = \sum_i \bar{\eta}_i \theta_i$.

Proof. The proof is easy. We reduce the exponent to a quadratic form by performing the change of variables $\theta \to \theta - A^{-1}\eta$ and $\bar{\theta} \to \bar{\theta} - A^{-1}\bar{\eta}$, which gives the quadratic form of the right-hand side. It remains to prove that

$$\int e^{(\bar{\theta}, A\theta)} \prod_{i=1}^n d\bar{\theta}_i d\theta_i = (-1)^n \det A. \tag{11.42}$$

It is enough to compute this integral for a diagonalisable matrix A since the two sides are polynomials of the coefficients. The identity is then the consequence of

$$\int e^{\sum_j \lambda_j \bar{\theta}_j \theta_j} \prod_{i=1}^{n} d\bar{\theta}_i d\theta_i = \prod_i \int e^{\lambda_i \bar{\theta}_i \theta_i} d\bar{\theta}_i d\theta_i = \prod_i (-\lambda_i) = (-1)^n \prod_i \lambda_i. \qquad (11.43)$$

This theorem will be used in the following.

11.3   The Clifford Algebras

We have already introduced this concept when studying the Dirac equation. Here we give some general properties.

We call Clifford algebra[3] an algebra Γ_n with n generators $\gamma_1, \cdots, \gamma_n$ satisfying the relations

$$\{\gamma_i, \gamma_j\} = \gamma_i \gamma_j + \gamma_j \gamma_i = 2\delta_{ij} \quad i, j = 1, \cdots, n. \qquad (11.44)$$

To each Grassmann algebra $\mathcal{G}$ with n generators $\theta_1, \cdots, \theta_n$, we can associate a Clifford algebra Γ_{2n}. Let us, indeed, consider the operators $\hat{\theta}_i$, the operator of multiplication by θ_i, and $\frac{\partial}{\partial \theta_j}$ the left derivative with respect to θ_j, acting on functions $f(\theta)$. These operators satisfy

$$\{\theta_i, \frac{\partial}{\partial \theta_j}\} = \delta_{ij}. \qquad (11.45)$$

We can then define new operators

$$Q_i = \frac{1}{\sqrt{2}}(\hat{\theta}_i + \frac{\partial}{\partial \theta_i}) \qquad (11.46)$$

$$P_i = \frac{1}{i\sqrt{2}}(\hat{\theta}_i - \frac{\partial}{\partial \theta_i}), \qquad (11.47)$$

$i = 1, \cdots, n.$

We check easily that these operators satisfy the relations

$$\{P_i, Q_j\} = \delta_{ij} \qquad \{P_i, P_j\} = \{Q_i, Q_j\} = 0. \qquad (11.48)$$

The operators P_i and Q_i are representations of the generators of the Clifford algebra Γ_{2n}, in the space of functions on $\mathcal{G}_n$.

Looking at the expressions (11.46), we note a remarkable analogy between these operators and the position and momentum operators for the harmonic oscillator.

---

[3] The fact that the metric is Euclidian instead of Minkowskian is not important. This can be shown by setting for the Dirac matrices: $\hat{\gamma}_0 = \gamma^0$ and $\hat{\gamma}_j = i\gamma^j$.

11.4 Fermions in Quantum Mechanics

11.4.1 Quantum Mechanics and Fermionic Oscillators

We saw that it was possible to express the usual quantum mechanics in terms of creators and annihilators by means of which we can construct the space of physical states. The commutation rules satisfied by these operators are the ones expected from quantities obeying Bose statistics.

The characteristic properties of Fermi statistics, particularly the Pauli principle, are naturally expressed by introducing creation and annihilation operators satisfying anticommutation rules.

Let us restrict our attention to the one-dimensional case (one mode) and let us introduce creation and annihilation operators a and $a^\dagger$ such that

$$\{a, a^\dagger\} = 1 \tag{11.49}$$

and

$$\{a, a\} = \{a^\dagger, a^\dagger\} = 0 \tag{11.50}$$

or equivalently

$$a^2 = (a^\dagger)^2 = 0. \tag{11.51}$$

If, by analogy with the bosonic case, we define the particle number operator

$$N = a^\dagger a \tag{11.52}$$

we see, thanks to the anticommutation relations,

$$N^2 = a^\dagger a a^\dagger a = a^\dagger (1 - a^\dagger a) a = a^\dagger a = N, \tag{11.53}$$

showing that the operator N has only for eigenvalues 0 and 1, which is in agreement with the Pauli principle.

As in the bosonic case, we assume the existence of a state, $|0>$, the vacuum state, such that

$$a|0> = 0. \tag{11.54}$$

We then check that

$$Na^\dagger|0> = a^\dagger|0> \tag{11.55}$$

from which we may define the one-particle state as $|1 > \overset{\text{def}}{=} a^\dagger|0 >$. We then check that

$$a^\dagger|1 >= 0. \tag{11.56}$$

The space of physical states is therefore a two-dimensional space generated by $|0 >$ and $|1 >$. The operators and the space of states we introduced are the elementary bricks allowing the quantum description of a spin-$\frac{1}{2}$ particle.

We can generalise these expressions to systems with n modes by introducing for each mode creation and annihilation operators satisfying

$$\{a_i, a_j^\dagger\} = \delta_{ij} \tag{11.57}$$

$$\{a_i^\dagger, a_j^\dagger\} = \{a_i, a_j\} = 0. \tag{11.58}$$

With each of these modes are associated two states, the vacuum and one-particle states, the vacuum $|0 >$ being characterised by $a_i|0 >= 0, \forall i$. We build in this way a space with 2^n states:

$$|0 >, a_1^\dagger|0 >, \cdots, a_n^\dagger|0 >, a_2^\dagger a_1^\dagger|0 >, \cdots .$$

By analogy with the quantum mechanics of the harmonic oscillator, we can introduce a Hamiltonian

$$H = \omega a^\dagger a. \tag{11.59}$$

It follows from the anticommutation rules that the states $|n >$, $n = 0, 1$, are the eigenstates

$$H|n >= n\omega|n > . \tag{11.60}$$

We can 'represent' the operators a and $a^\dagger$ of the Clifford algebra Γ_2 as operators on $\mathcal{G}_1$ with generator θ. Defining the representation of an operator O as functions of θ by[4]

$$< \theta|O|n >= O(\theta) < \theta|n > \tag{11.61}$$

we find that

$$H(\theta) = \omega \frac{\partial}{\partial\theta}\theta = \omega - \omega\theta\frac{\partial}{\partial\theta} \tag{11.62}$$

and from now on we will omit to distinguish an operator from its representation in the space of functions on the algebra.

---

[4] We write the ket $O|n >$ as a function of θ.

The ground state $f_0(\theta) = <\theta|0>$ satisfies $Hf_0 = 0$ and, if we impose the condition $(f_0, f_0) = 1$, we find that

$$f_0(\theta) = \theta. \tag{11.63}$$

Equivalently, the eigenfunction $f_1(\theta) = <\theta|1>$, with eigenvalue ω, orthogonal to $<\theta|0>$ and of norm 1, is given by

$$f_1(\theta) = 1. \tag{11.64}$$

If we set $<0|\theta> = -1$ and $<1|\theta> = \theta$, then

$$<m|n> = \int <m|\theta> \, d\theta \, <\theta|n> = \delta_{mn} \tag{11.65}$$

so formally,

$$\int |\theta> \, d\theta \, <\theta| = 1 \tag{11.66}$$

and we have the closure relation

$$\sum_{n=0,1} <\eta|n><n|\theta> = \theta - \eta = \delta(\eta - \theta). \tag{11.67}$$

These results extend to the case of N generators.

If the Hamiltonian is given by $H = \sum \omega_i a_i^\dagger a_i$, we have a representation of the states as functions on the Grassmann algebra $\mathcal{G}_N$ with generators $\theta_1, \cdots, \theta_N$. The ground state $f_0(\theta)$ is given by

$$f_0(\theta) = <\theta|0> = \theta_1 \theta_2 \cdots \theta_N \tag{11.68}$$

and the n-particle states, $n = (n_1, n_2, \cdots, n_N)$, are given by

$$<\theta|n> = (\frac{\partial}{\partial \theta_N})^{n_N} \cdots (\frac{\partial}{\partial \theta_1})^{n_1} <\theta|0> . \tag{11.69}$$

To satisfy the relations

$$\int <m|\theta> \prod_i d\theta_i <\theta|n> = \delta_{mn} \tag{11.70}$$

we introduce the functions $<n|\theta> = \pm \theta_N^{n_N} \cdots \theta_1^{n_1}$, where the sign is fixed by Eq. (11.70). As a consequence, we have the closure relation

$$\sum_n <\bar{\theta}|n><n|\theta> = \delta(\bar{\theta}_1 - \theta_1)\cdots\delta(\bar{\theta}_N - \theta_N) = \prod_i^N \delta(\bar{\theta}_i - \theta_i). \qquad (11.71)$$

11.4.2 The Free Fermion Fields

If we are interested in the description of the states of a fermionic particle, these states are characterised by their three momenta and by the state of their spin. It is natural to introduce as many creation and annihilation operators as there are distinct possible states. We then introduce a continuous set of operators $a_s(\boldsymbol{p})$ and $a_s^\dagger(\boldsymbol{p})$. At $\boldsymbol{p}$ and s given,[5] the space of physical states reduces to the vacuum state $|0>$ and to the one-particle states $a_s^\dagger(\boldsymbol{p})|0>$, of four-momentum $(\sqrt{\boldsymbol{p}^2 + m^2}, \boldsymbol{p})$, and helicity s. Relations (11.57) and (11.58) are written in the continuum case

$$\{a_s(\boldsymbol{p}), a_{s'}^\dagger(\boldsymbol{p}')\} = (2\pi)^3 2\omega_p \delta(\boldsymbol{p} - \boldsymbol{p}')\,\delta_{ss'} \qquad (11.72)$$

$$\{a_s(\boldsymbol{p}), a_{s'}(\boldsymbol{p}')\} = \{a_s^\dagger(\boldsymbol{p}), a_{s'}^\dagger(\boldsymbol{p}')\} = 0 \qquad (11.73)$$

They automatically imply that fermions satisfy the Pauli principle since

$$a_s^\dagger(\boldsymbol{p})a_{s'}^\dagger(\boldsymbol{p}')|0> = -a_{s'}^\dagger(\boldsymbol{p}')a_s^\dagger(\boldsymbol{p})|0>. \qquad (11.74)$$

Moreover when $\boldsymbol{p} = \boldsymbol{p}'$ and $s = s'$, the vector (11.74) is zero, showing that the particles satisfy the Pauli exclusion principle.

To conclude we give an expression of the field of a free particle with spin $\frac{1}{2}$. It is necessary to introduce a second series of creation and annihilation operators, $b_s^\dagger(\boldsymbol{p})$ and $b_s(\boldsymbol{p})$, satisfying

$$\{b_s(\boldsymbol{p}), b_{s'}^\dagger(\boldsymbol{p}')\} = (2\pi)^3 2\omega_p \delta(\boldsymbol{p} - \boldsymbol{p}')\,\delta_{ss'} \qquad (11.75)$$

$$\{b_s(\boldsymbol{p}), b_{s'}(\boldsymbol{p}')\} = \{b_s^\dagger(\boldsymbol{p}), b_{s'}^\dagger(\boldsymbol{p}')\} = 0 \qquad (11.76)$$

and anticommuting with a and $a^\dagger$. The states $a_s^\dagger(\boldsymbol{p})|0>$ and $b_s^\dagger(\boldsymbol{p})|0>$ represent, respectively, the one-electron state and the one-positron state with momentum $(\sqrt{\boldsymbol{p}^2 + m^2}, \boldsymbol{p})$ and helicity s.

We can thus define a Dirac quantum field by

$$\psi(x) = \int d\Omega_m \sum_{s=\pm 1/2} [a_s(k)u^{(s)}(k)e^{-ik.x} + b_s^\dagger(k)v^{(s)}(k)e^{ik.x}] \qquad (11.77)$$

$$\bar{\psi}(x) = \int d\Omega_m \sum_{s=\pm 1/2} [a_s^\dagger(k)\bar{u}^{(s)}(k)e^{ik.x} + b_s(k)\bar{v}^{(s)}(k)e^{-ik.x}], \qquad (11.78)$$

---

[5] Here, the helicity s represents the projection of the spin on the direction of propagation $\boldsymbol{p}$.

where we have introduced the notation $u^{(s)}$ with $s = \pm\frac{1}{2}$ in order to make the eigenstates of the helicity to appear instead of the $u^{(\alpha)}$, $\alpha = 1$ or 2, as they were defined in Chapter 6. Obviously, we have the same convention for the v's.

We see from these expressions that the field ψ annihilates an electron and creates a positron while the field $\bar{\psi}$ creates an electron and annihilates a positron. We can check the causality condition by taking an anticommutator of Dirac fields. We find that

$$
\begin{aligned}
&\{\psi_\alpha(x), \bar{\psi}_\beta(y)\} \\
&= \frac{1}{(2\pi)^3} \int \frac{d^3k}{2\omega_k} [(\not{k} - m)e^{ik.(x-y)} + (\not{k} + m)e^{-ik.(x-y)}]_{\alpha\beta} \\
&= (i\not{\partial} + m)_{\alpha\beta} i\Delta(x-y),
\end{aligned} \tag{11.79}
$$

where Δ is the causal Green function introduced in formula (9.103).

11.5   The Path Integrals

The formalism we developed in the previous section makes it possible to define for fermions a functional integral very similar to that developed for bosons.

11.5.1   The Case of Quantum Mechanics

Let us consider a system with N degrees of freedom given by the Hamiltonian

$$
H = \sum \omega_{\alpha\beta} a_\alpha^\dagger a_\beta. \tag{11.80}
$$

We first want to solve the Schrödinger equation

$$
i\frac{\partial}{\partial t}|\psi(t)> = H|\psi(t)>. \tag{11.81}
$$

Equation (11.81) can be written in the representation on functions of a Grassmann algebra

$$
i\frac{\partial}{\partial t} < \theta|\psi(t)> = < \theta|H\psi(t)> = H < \theta|\psi(t)> = \sum_{ij} \omega_{ij} \frac{\partial}{\partial\theta_i} \theta_j < \theta|\psi(t)>. \tag{11.82}
$$

The second equality defines the action of H on $\psi(t, \theta) = <\theta|\psi(t)>$.

Using the closure relations, we can rewrite the solution of the Schrödinger equation

$$
< \theta'|\psi(t')> = \int < \theta'|e^{-i(t'-t)H}|\theta> \prod_i d\theta_i < \theta|\psi(t)>. \tag{11.83}
$$

As in Chapter 6, we see that to solve this equation it is enough to know the Green function:

$$G(t', \theta'; t, \theta) = <\theta'|e^{-i(t'-t)H}|\theta> . \tag{11.84}$$

It is this quantity that we will write as a path integral. The method we follow is to cut the time interval into $n + 1$ equal intervals $\epsilon = (t' - t)/(n + 1)$ with $t_0 = t$, $t_n = t'$. At each intermediate time t_j, $t_j = t + \epsilon j$, we insert, with variables $\theta(j) = \theta(t_j)$, the identity operator given by

$$1 = \int |\theta(j)> \prod_i d\theta_i(j) <\theta(j)| \tag{11.85}$$

and we use

$$e^{-i(t'-t)H} = (e^{-i\varepsilon H})^{n+1}. \tag{11.86}$$

We therefore must compute $G(t_j + \varepsilon, \theta(j+1); t, \theta(j))$. We will do this calculation at first order in ε:

$$<\theta(j+1)|e^{-i\varepsilon H}|\theta(j)> = <\theta(j+1)|\theta(j)> -i\varepsilon <\theta(j+1)|H|\theta(j)> +\mathcal{O}(\varepsilon^2). \tag{11.87}$$

To make the mechanism clear, we will start with a simple, one-mode, case, i.e. such that $H = \omega a^* a$.

Let us compute

$$<\theta'|e^{-i\varepsilon H}|\theta>$$
$$= \sum_n <\theta'|n> e^{-i\varepsilon n\omega} <n|\theta>$$
$$= <\theta'|0><0|\theta> +e^{-i\varepsilon\omega} <\theta'|1><1|\theta>; \tag{11.88}$$

therefore, at first order

$$<\theta'|e^{-i\varepsilon H}|\theta> = -\theta' + e^{-i\varepsilon\omega}\theta = -\theta' + \theta - i\varepsilon\omega\theta + \mathcal{O}(\varepsilon^2), \tag{11.89}$$

which can be written as

$$<\theta'|e^{-i\varepsilon H}|\theta> = \int d\kappa e^{(\kappa,(\theta'-\theta))}(1 + i\varepsilon\omega\kappa\theta) + \mathcal{O}(\varepsilon^2). \tag{11.90}$$

Thus,

$$<\theta'|e^{-i\varepsilon H}|\theta> d\theta = \int d\kappa d\theta e^{(\kappa,(\theta'-\theta))}(1 + i\varepsilon\omega\kappa\theta) + \mathcal{O}(\varepsilon^2). \tag{11.91}$$

The interesting aspect of this last form is that it commutes with the Grassmann variables.

This result extends to the case of many variables and we can prove that

$$< \theta' |e^{-i\varepsilon H}| \theta > \prod_i d\theta_i = \int \prod_i d\kappa_i d\theta_i e^{(\kappa,(\theta'-\theta))} (1 + i\varepsilon\omega_{\alpha\beta}\kappa_\alpha\theta_\beta) + \mathcal{O}(\varepsilon^2). \qquad (11.92)$$

Let us write the result of this calculation in the form

$$< \theta(j+1)|e^{-i\varepsilon H}|\theta(j) > \prod_i d\theta_i(j)$$

$$= \int \prod_i d\bar{\theta}(j) d\theta(j) e^{(\bar{\theta}_\alpha(j),[\theta_\alpha(j+1)-\theta_\alpha(j)])+i\varepsilon\omega_{\alpha\beta}\bar{\theta}_\alpha(j)\theta_\beta(j)} \qquad (11.93)$$

and insert it in

$$G(\theta', t'; \theta, t)$$

$$= \int < \theta' |e^{-i\varepsilon H}|\theta(n) > \prod d\theta(n) < \theta(n)|e^{-i\varepsilon H}|\theta(n-1) >$$

$$\times \prod d\theta(n-1) \cdots < \theta(1)|e^{-i\varepsilon H}|\theta > .$$

Performing a right multiplication on the Green function by $\prod d\theta$ and using the commutativity of the expressions giving the value of the Green function at first order in ε, we get

$$G(\theta', t'; \theta, t) \prod d\theta = \int \prod_i d\bar{\theta}(i) d\theta(i) e^{i\varepsilon \sum_j [-\frac{i}{\varepsilon}(\bar{\theta}_\alpha(j),\theta_\alpha(j+1)-\theta_\alpha(j))+\omega_{\alpha\beta}\bar{\theta}_\alpha(j)\theta_\beta(j)]}. \qquad (11.94)$$

Taking the limit $\varepsilon \to 0$ and denoting the limit measure by $\mathcal{D}(\bar{\theta}, \theta)$, we show that

$$\psi(t', \theta') = \int \mathcal{D}(\bar{\theta}, \theta) e^{i \int_t^{t'} L(\bar{\theta}(s), \theta(s)) ds} \psi(t, \theta) \qquad (11.95)$$

with the Lagrangian and the Hamiltonian given by

$$L(\bar{\theta}(s), \theta(s)) = -i\bar{\theta}(s)\dot{\theta}(s) + H(\bar{\theta}(s), \theta(s)) \qquad (11.96)$$

$$H(\bar{\theta}(s), \theta(s)) = \sum_{\alpha,\beta} \omega_{\alpha\beta}\bar{\theta}_\alpha(s)\theta_\beta(s). \qquad (11.97)$$

Remark the striking analogy between this path integral and that obtained in the bosonic case. As in this last case, we integrate the exponential of the Lagrangian by replacing the creation and annihilation operators a and a^* with Grassmann integration variables θ and $\bar{\theta}$.

11.5.2   The Case of Field Theory

It is possible to extend the results of the preceding section to the case of free fermion fields, that is to say, to define a Gaussian Grassmannian measure for an infinite number of degrees of freedom. Proceeding as earlier, we replace the annihilators and creators with Grassmannian integration variables. The construction can be done by approximating the integrals over the momenta by Riemann sums. For example, introducing Grassmann variables of even indices θ_{2i} for a^* and of odd indices θ_{2i+1} for b, the integration is performed on pairs $d\bar{\theta}_j d\theta_j$, the final integral being defined by taking the limit. However, since the fields ψ and $\bar{\psi}$ are the real variables and are linear combinations of creation and annihilation operators, thus Grassmann variables, we change the integration variables to integrate over the fields $\bar{\psi}$ and ψ themselves taken as independent variables. We are therefore led to introduce a generating functional

$$Z(\bar{\eta}, \eta) = \frac{\int e^{i\int d^4 x \mathcal{L}_D(\bar{\psi},\psi)(x) + i\int \bar{\eta}(x)\psi(x)d^4 x + \int \bar{\psi}(x)\eta(x)d^4 x} \mathcal{D}(\bar{\psi}, \psi)}{\int e^{i\int d^4 x \mathcal{L}_D(\bar{\psi},\psi)(x)} \mathcal{D}(\bar{\psi}, \psi)}, \tag{11.98}$$

where $\mathcal{L}_D$ is the Lagrangian density of Dirac and $\bar{\eta}$ and η are the anticommuting sources. Formally, the integration measure can be understood as

$$\mathcal{D}(\bar{\psi}, \psi) = \prod_{x \in \mathbb{R}^4} d\bar{\psi}(x)d\psi(x).$$

The Lagrangian is a bilinear form in $\bar{\psi}$ and ψ which can be written as

$$\int d^4 x \mathcal{L}_D(\bar{\psi}, \psi)(x) = \int d^4 x d^4 y \bar{\psi}(x) S_F^{-1}(x-y)\psi(y) \tag{11.99}$$

and by analogy with the bosonic case, the Gaussian measure is defined by the 2-point function. We set

$$< 0|T(\psi_\alpha(x)\bar{\psi}_\beta(y))|0> = \frac{\int \psi_\alpha(x)\bar{\psi}_\beta(y) e^{i\int d^4 x \mathcal{L}(\bar{\psi},\psi)(x)} \mathcal{D}(\bar{\psi}, \psi)}{\int e^{i\int d^4 x \mathcal{L}(\bar{\psi},\psi)(x)} \mathcal{D}(\bar{\psi}, \psi)}$$

$$= \frac{1}{i^2} \frac{\delta}{\delta\bar{\eta}_\alpha(x)} \frac{\delta}{\delta\eta_\beta(y)} Z(\bar{\eta}, \eta)|_{\bar{\eta}=\eta=0}. \tag{11.100}$$

Since the variables η are anticommuting, by convention the functional derivatives with respect to $\bar{\eta}$ act always in the given order to the left[6] of Z, that is to say, from right to left, and the derivative with respect to η acts always in the given order to the right of Z, that is to say, from the left to the right.

---

[6] Note that the integrand or the integration measure involving pairs of anticommuting variables commutes with the Grassmann variables.

An explicit computation from the definitions (11.79) shows that

$$< 0| T(\psi_\alpha(x)\bar{\psi}_\beta(y))|0 >= iS_F(x-y) \tag{11.101}$$

and thus (since it is a Gaussian expression)

$$Z(\bar{\eta}, \eta) = e^{-i(\bar{\eta}, S_F \eta)} = e^{-i\int\int \bar{\eta}(x)S_F(x)y)\eta(y)d^4x d^4y}. \tag{11.102}$$

The last equality, which can be taken as a definition of the integration measure on the Dirac fields, gives back all the properties of the Gaussian measure. Thus,

$$< 0| T(\psi_{\alpha_1}(x_1)\psi_{\alpha_2}(x_2)\bar{\psi}_{\beta_1}(y_1)\bar{\psi}_{\beta_2}(y_2))|0 >$$

$$\frac{1}{i^4}\frac{\delta}{\delta\bar{\eta}_{\alpha_1}(x_1)}\frac{\delta}{\delta\bar{\eta}_{\alpha_2}(x_2)}\frac{\delta}{\delta\eta_{\beta_1}(y_1)}\frac{\delta}{\delta\eta_{\beta_2}(y_2)}Z(\bar{\eta}, \eta)|_{\bar{\eta}=\eta=0}$$

$$= \frac{\delta}{\delta\bar{\eta}_{\alpha_1}(x_1)}\frac{\delta}{\delta\eta_{\beta_1}(y_1)}\frac{\delta}{\delta\eta_{\beta_2}(y_2)}i(S_F\eta)(x_2)_{\alpha_2}e^{i(\bar{\eta},S_F\eta)}|_{\bar{\eta}=\eta=0}$$

$$= -\frac{\delta}{\delta\eta_{\beta_1}(y_1)}\frac{\delta}{\delta\eta_{\beta_2}(y_2)}i(S_F\eta)(x_2)_{\alpha_2}i(S_F\eta)(x_1)_{\alpha_1}e^{i(\bar{\eta},S_F\eta)}|_{\bar{\eta}=\eta=0}$$

$$= -\frac{\delta}{\delta\eta_{\beta_2}(y_2)}\{-iS_F(x_2-y_1)_{\alpha_2\beta_1}i(S_F\eta)(x_1)_{\alpha_1} \tag{11.103}$$

$$+ i(S_F\eta)(x_2)_{\alpha_2}iS_F(x_1-y_1)_{\alpha_1\beta_1}e^{i(\bar{\eta},S_F\eta)}|_{\bar{\eta}=\eta=0}$$

$$= S_F(x_1-y_1)_{\alpha_1\beta_1}S_F(x_2-y_2)_{\alpha_2\beta_2} - S_F(x_2-y_1)_{\alpha_2\beta_1}S_F(x_1-y_2)_{\alpha_1\beta_2}$$

The final expression can be reinterpreted by saying that it is the sum over all possible contractions of a ψ with a $\bar{\psi}$ with a factor -1 allocated to each circular permutation making them closest in the following order: first ψ then $\bar{\psi}$, each contraction generating a propagator iS_F.

Remark:

- The equality between T-products and moments of a Gaussian Grassmannian measure is justified by the fact that the two expressions have the same rules of transformation with respect to permutations of the fields.

- The convergence argument, valuable in bosonic field theories, justifying the choice of $i\varepsilon$ for the propagator has no meaning here since the variables of integration are bounded.

- The integral expressions in this chapter are limits of integrals over a finite number of variables, integrals which are reduced to the integration over polynomials of Grassmann variables. They only express algebraic relations. They have, however, the interest, up to anticommutation problems, to behave as usual integrals by changes of coordinates, Jacobians and determinants being replaced by inverses of Jacobians and determinants.

- We can define, as in the bosonic case, the integration of arbitrary Lagrangians, including fermionic variables that are behaving like bosons (sums of monomials having an even number of fermion fields), by considering that the non-quadratic part is a perturbation of the quadratic part.

- The choice of Dirac fields as integration variables can be understood by the fact that they satisfy equal time anticommutation relations. Using indeed the formula (11.79), we get

$$\{\psi_\alpha(t, \boldsymbol{x}), \psi_\beta^\dagger(t, \boldsymbol{y})\} = \{\psi_\alpha(t, \boldsymbol{x}), \bar{\psi}_\sigma(t, \boldsymbol{y})\}\gamma_{\sigma\beta}^0$$

$$= \frac{1}{(2\pi)^3}\int \frac{\mathrm{d}^3 k}{2\omega_k}[(\slashed{k} - m)\gamma^0 \mathrm{e}^{-i k \cdot (x-y)}$$

$$+ (\slashed{k} + m)\gamma^0 \mathrm{e}^{i k \cdot (x-y)}]|_{\alpha\beta}$$

$$= \delta_{\alpha\beta}\delta^{(3)}(\boldsymbol{x} - \boldsymbol{y}). \tag{11.104}$$

- Using the integration formula (11.41), we can show that

$$\int \mathrm{e}^{i\int \mathrm{d}^4 x \mathcal{L}_D(\bar{\psi}, \psi)(x)}\mathcal{D}(\bar{\psi}, \psi) \propto \det S_F^{-1} = \mathrm{e}^{-\mathrm{Tr}\log S_F}. \tag{11.105}$$

12

Relativistic Quantum Fields

12.1 Introduction

We studied interacting quantum fields in Euclidean space. This formalism is quite useful because it allows for the manipulation of well-defined quantities. The connection between quantum Euclidean quantities and physically meaningful relativistic quantities can be established via a complicated operation of analytic continuation which can be synthesised into an equivalence between a set of axioms that Euclidean theories must satisfy and the basic axioms that any 'reasonable' relativistic theory seems to obey.

We will show in this chapter how, starting from these general reasonable basic axioms, more or less independent from any specific theories, but which are admitted to be true by any theoretical physicist, we arrive to deep results shaping our understanding of physical reality.[1]

In some, admittedly oversimplified, cases it has been shown during the recent years that some of the Euclidean theories exist, satisfy the Euclidean axioms, and obey certain resummation properties of their perturbative expansions (Euler resummation property linked to analyticity in the coupling constant). This last point gives a solid basis to the use of perturbation expansions (and Feynman diagram analysis) to extract predictions and numerical values from the general models of quantum field theories. A brief exposition of these results, as well as of the mathematical methods used to obtain them, is presented in Chapter 24.

This makes it possible to assert that quantum relativistic theories have a real global mathematical existence, and therefore we can manipulate, with the usual precautions, interesting quantities such as vacuum expectation values of products of fields.

12.2 Relativistic Field Theories

We admit in what follows that the calculation of vacuum expectation values of the T-product of a product of fields is achieved by the following set of correspondences:

---

[1] The first three sections will follow closely the route traced in: R. F. Streater and A. S. Wightman *PCT, Spin & Statistics, and All That* (W. A. Benjamin, New York, 1964).

From Classical to Quantum Fields. Laurent Baulieu, John Iliopoulos and Roland Sénéor.
© Laurent Baulieu, John Iliopoulos and Roland Sénéor, 2017. Published 2017 by Oxford University Press.

Relativistic quantities	Euclidean quantities
t	$-it$
$\phi(t, \boldsymbol{x})$	$\phi_E(x) = \phi(-it, \boldsymbol{x})$
$\mathcal{L}(\phi, \partial_\mu \phi)$	$\mathcal{L}_E(\phi_E, \nabla \phi_E)$
$i \int \mathcal{L} \mathrm{d}^4 x$	$-\int \mathcal{L}_E \mathrm{d}^4 x$
$< \Omega\lvert T(\cdots)\rvert\Omega > = \dfrac{<0\lvert T((\cdots)(\mathrm{e}^{i\int \mathcal{L}\mathrm{d}^4 x}))\rvert 0>}{<0\lvert T(\mathrm{e}^{i\int \mathcal{L}\mathrm{d}^4 x})\rvert 0>}$	$< \Omega\lvert(\cdots)\rvert\Omega > = \dfrac{<0\lvert(\cdots)\,\mathrm{e}^{-\int \mathcal{L}_E\mathrm{d}^4 x}\rvert 0>}{<0\lvert \mathrm{e}^{-\int \mathcal{L}_E\mathrm{d}^4 x}\rvert 0>}$

However, to relate the vacuum expectations to physical quantities, it is necessary to define with more precision a certain number of basic properties of relativistic quantum fields. These properties have been grouped together in the form of axioms, the *Wightman axioms*, that any reasonable quantum field theory is supposed to satisfy.

12.2.1 The Axiomatic Field Theory

12.2.1.1 The Wightman Axioms

We give a short survey of what these axioms are for a massive scalar field.

- Axiom 0: *The space of states*
  The states of the theory are the unit rays of a Hilbert space $\mathcal{H}$ on $\mathbb{C}$.
- Axiom 1: *The regularised (or smeared) fields*
  If $f \in \mathcal{S}$, the space of C^∞ rapidly decreasing functions, then

$$\phi(f) = \int \phi(x) f(x) \, \mathrm{d}^4 x \tag{12.1}$$

is an unbounded operator, defined on a dense subset D of $\mathcal{H}$ and, for $\Phi, \Psi \in D$

$$(\Phi, \phi(f)\Psi) = (\phi(\bar{f})\Phi, \Psi), \tag{12.2}$$

which implies that $(\Phi, \phi(x)\Psi)$ is a tempered distribution.
Moreover $\phi(f)D \subset D$ and the products of smeared fields is meaningful.

- Axiom 2: *The Lorentz invariance*
  There exists a continuous unitary representation $U(a, \Lambda)$ of the restricted Poincaré group such that

$$U(a, \Lambda)\phi(f)U^{-1}(a, \Lambda) = \phi(f_{(a, \Lambda)}), \tag{12.3}$$

where $f_{(a, \Lambda)}(x) = f(\Lambda^{-1}(x - a))$; moreover $U(a, \Lambda)D \subset D$.

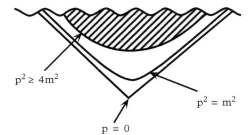

$p^2 \geq 4m^2$

$p^2 = m^2$

$p = 0$

Figure 12.1 *The spectrum of P.*

- Axiom 3: Stability of the vacuum state and the energy–momentum spectrum
  Let P be the energy–momentum operator. Its spectrum is identical to that of a free scalar field of mass m (see Fig. 12.1), that is to say given by[2] $\{p = 0\} \cup \{p|p_0 > 0, p^2 = m^2\} \cup V_+^{2m}$, where $V_+^{2m} = \{p \in V_+|p^2 \geq 4m^2\}$; this domain is Lorentz invariant. $p = 0$ is an isolated eigenvalue which corresponds to an eigenstate $|\Omega >$: the vacuum. Since this eigenvalue is invariant under transformations of the restricted Lorentz group, it follows that $U(a, \Lambda)|\Omega >= |\Omega >$.
  Since P is the generator of the translation subgroup, this means (in the notations of Chapter 5) that the support of $E(p)$ is $\{p = 0\} \cup \{p|p_0 > 0, p^2 = m^2\} \cup V_+^{2m}$.

- Axiom 4: *The locality*
  Let f_1 and $f_2 \in \mathcal{S}$, with compact supports, such that the supports of f_1 and of f_2 are space-like separated, then for all $\Phi \in D$

$$[\phi(f_1), \phi(f_2)]|\Phi >\equiv 0. \tag{12.4}$$

- Axiom 5: *The completeness hypothesis or the cyclicity of the vacuum state*
  The set of vectors $P(\phi)|\Omega >$, obtained by applying polynomials $\phi(f), f \in \mathcal{S}$, on the vacuum, is dense in $\mathcal{H}$. We say that the *vacuum is cyclic*.
  This last axiom implies that the fields form a complete set of operators in the Hilbert space.

Remark: The hypothesis according to which the vacuum state and the mass hyperboloid $p^2 = m^2$ are isolated in the spectrum P (that is to say, the mass $m \neq 0$) is fundamental for the proof we will give on the existence of asymptotic states. This hypothesis is not satisfied for the electromagnetic field (its mass m is 0) for which the spectrum of P is in V^+.

These axioms, which were written for a massive scalar field, can be extended to the case of several fields and, in particular, to the case of massive spinor fields. The introduction of spin requires some simple modifications. Axioms 0, 1, and 5 are unchanged. Axiom 2 is modified because of the presence of spinor indices. Indeed, U is a continuous unitary representation of the covering group of the Poincaré group, i.e.

---

[2] We define $V^+ = \{p \in \mathbb{R}^4|p_0 > 0, p^2 > 0\}$ and $\bar{V}^+ = \{p \in \mathbb{R}^4|p_0 \geq 0, p^2 \geq 0\}$.

of the inhomogeneous $SL(2,\mathbb{C})$ group. Thus if ϕ_j are the components of an irreducible representation of $SL(2,\mathbb{C})$ or of $SL(2,\mathbb{C})$ and of some reflections, then

$$U(a,A)\phi_j(f)U^{-1}(a,A) = \sum S_{jk}(A^{-1})\phi_k(f_{\{a,A\}}), \tag{12.5}$$

where $A \to S(A^{-1})$ is a finite-dimensional irreducible representation of $SL(2,\mathbb{C})$ of the form $D^{[j_1,j_2]}$. This implies for the fields themselves the following.

- Scalar field

$$U(a,A)\phi(x)U^{-1}(a,A) = \phi(\Lambda x + a) \tag{12.6}$$

- Dirac field

$$U(a,A)\psi_\alpha(x)U^{-1}(a,A) = \sum_{\beta=1}^{4} S_{\alpha\beta}(A^{-1})\psi_\beta(\Lambda x + a) \tag{12.7}$$

- Vector field

$$U(a,A)A_\mu(x)U^{-1}(a,A) = \Lambda_\mu^{\ \nu}(A^{-1})A_\nu(\Lambda x + a) \tag{12.8}$$

Axiom 4, axiom of locality, is replaced by the following:
One or the other of the following commutation relations is satisfied:

$$[\phi_j(f_1), \phi_k(f_2)]_\pm |\Phi> = 0 \tag{12.9}$$

if the supports of f_1 and f_2 are space-like separated, and

$$[A,B]_\pm = AB \pm BA \tag{12.10}$$

The choice of the commutator for integer spin fields and of the anticommutator for half-odd integer spin fields is a consequence of the spin-statistics theorem which will be proven from these axioms.

Most of the results in axiomatic field theories are proved starting from vacuum expectation values of products of fields. A theorem, due to A. Wightman, shows that there exists a complete equivalence between vacuum expectation values satisfying certain properties and the existence of fields satisfying the previous axioms.

12.2.1.2 *The Discrete Symmetries*

In a previous chapter we had introduced some important discrete transformations in the framework of the Dirac theory. Here we want to extend and generalise these notions to the case of a quantum field theory. The symmetries we have in mind are the following:

- *Space inversion I_s,* often called *parity P.* It is the transformation in which $x \to -x$. In classical physics all three-vectors, such as three-momenta, the electric field, etc., change sign under parity, while pseudo-vectors, such as the angular momentum, or the magnetic field, don't.

- *Time reversal I_t* or, simply, *T.* It is the transformation which changes $t \to -t$ and, thus, reverses the sign of velocities.

- *Charge conjugation C.* This transformation has no analogue in classical physics. It changes particles to anti-particles but leaves the space–time point unaffected.

Here we want to define these operators and study some of their properties in quantum field theory. In traditional quantum mechanics an operator is defined through its action on the physical states. Therefore, the action of a symmetry transformation on an observable Q, $Q \to \hat{Q}$, is supposed to be such that for all states $|\Phi>$

$$< \Phi | \hat{Q} | \Phi > = < \hat{\Phi} | Q | \hat{\Phi} >, \tag{12.11}$$

where $|\hat{\Phi}> = U|\Phi>$, which can be interpreted as a relation between the active and passive interpretations of the symmetries. This leads us to write that

$$\hat{Q} = U^{-1} Q U \tag{12.12}$$

if $\Phi \to \hat{\Phi}$ is unitary,

$$\hat{Q} = (U^{-1} Q U)^* \tag{12.13}$$

if $\Phi \to \hat{\Phi}$ is antiunitary.

In quantum field theory we can prove that it is sufficient to define these transformations through their action on the fundamental fields. Let us consider the example of space inversion I_s for a theory with a single scalar field $\phi(x)$. ϕ has no spinor indices, so we can write that

$$U(I_s)\phi(x)U^{-1}(I_s) = \eta_P \phi(I_s x) = \eta_P \phi(x^0, -x), \tag{12.14}$$

where we have allowed for a phase η_P. For boson fields we will adopt the convention that $P^2 = 1$, so $\eta_P = \pm 1$. Although the choice is, in principle, arbitrary, we will see that it is often dictated by physical considerations. We will call fields with $\eta_P = 1$ *scalars*, while those with $\eta_P = -1$ *pseudo-scalars*.

From this definition we can infer the action of $U(I_s)$ on the transformations of the Poincaré group as

$$U(I_s)U(a, \Lambda)U^{-1}(I_s) = U(I_s a, I_s^{-1} \Lambda I_s). \tag{12.15}$$

For simplicity, let us assume the unicity of the vacuum state.[3] So, up to a phase, we can write $U(I_s)|\Omega > \sim |\Omega >$.

It is not difficult to prove that these definitions fully determine the parity operator $U(I_s)$. For that we use the fact that the vacuum state is the only state invariant under the action of the Poincaré group and the fact that since the field is irreducible, the polynomials of the field applied on the vacuum form a dense set in the Hilbert space. Thus, any operator which commutes with the field is a multiple of the identity. The compatibility with the Poincaré group as it is expressed in (12.15) and the action on the vacuum determines the action on all states. For example, $|\Psi > = \phi(f_1) \cdots \phi(f_n)|\Omega >$ becomes

$$U(I_s)|\Psi > = U(I_s)\phi(f_1)U^{-1}(I_s) \cdots U(I_s)\phi(f_n)U^{-1}(I_s)U(I_s)|\Omega >$$
$$= \sim \phi(\hat{f}_1) \cdots \phi(\hat{f}_n)|\Omega >, \tag{12.16}$$

where $\hat{f}_j(x) = f_j(I_s x)$.

This discussion can be repeated for the other two discrete transformations, time reversal and charge conjugation. Staying always with the example of a scalar field, we write that

$$U(I_t)\phi(x)U^{-1}(I_t) = \eta_T \phi(I_t x) = \eta_T \phi(-x^0, \mathbf{x}) \tag{12.17}$$

$$U(C)\phi(x)U^{-1}(C) = \eta_C \phi^\dagger(x), \tag{12.18}$$

where we have introduced two new arbitrary phases, which we take as being equal to ± 1. Note also that $U(C)$ in (12.18) is assumed to be a linear unitary operator. For example, if the field ϕ is a free complex field with creation and annihilation operators a, b, $a^\dagger$, and $b^\dagger$, the action of charge conjugation is defined, up to a phase, as $a \leftrightarrow b$ and $a^\dagger \leftrightarrow b^\dagger$, leaving complex numbers unchanged. In contrast, $U(I_t)$ is assumed to be anti-unitary.

We can similarly define the relations that $U(I_t)$ and $U(C)$ must satisfy with respect to the Poincaré group:

$$U(I_t)U(a, \Lambda)U^{-1}(I_t) = U(I_t a, I_t^{-1}\Lambda I_t)$$
$$U(C)U(a, \Lambda)U^{-1}(C) = U(a, \Lambda).$$

The action of these operators on the vacuum gives back the vacuum up to a phase. We usually fixed this phase to be equal to 1 and we write that

$$U(I_s)|\Omega > = U(I_t)|\Omega > = U(C)|\Omega > = |\Omega >. \tag{12.19}$$

The action on any state follows as was done for the parity operator.

We thus see that the implementation of symmetries is linked to the *definition of their action* on the fields and this is independent of the fact that these symmetries are, or are not,

---

[3] We could relax this assumption and cover more general cases.

symmetries of the theory. It is therefore important to extend these definitions for fields with any spin. This can be done because our fields correspond to irreducible represent-ations of the Lorentz group, so their components have well-defined spinor indices. Con-sequently, we can use the results of section 2.9. In an obvious generalisation we write that

$$U(I_s)\xi^{(\alpha)(\dot\beta)}U(I_s)^{-1} = \eta_P \tau_P(j+k)(-1)^j \zeta \otimes \cdots \otimes \zeta \eta^{(\dot\alpha)(\beta)}$$

$$U(I_s)\eta^{(\dot\alpha)(\beta)}U(I_s)^{-1} = \eta_P \tau_P(j+k)(-1)^k \zeta \otimes \cdots \otimes \zeta \xi^{(\alpha)(\dot\beta)} \tag{12.20}$$

$$U(C)\xi^{(\alpha)(\dot\beta)}U(C)^{-1} = \eta_C \zeta \otimes \cdots \otimes \zeta \eta^{(\dot\alpha)(\beta)*}$$

$$U(C)\eta^{(\dot\alpha)(\beta)}U(C)^{-1} = \eta_C \zeta \otimes \cdots \otimes \zeta \xi^{(\alpha)(\dot\beta)*} \tag{12.21}$$

$$U(I_t)\xi^{(\alpha)(\dot\beta)}U(I_t)^{-1} = \eta_T \zeta \otimes \cdots \otimes \zeta \xi^{(\alpha)(\dot\beta)}$$

$$U(I_t)\eta^{(\dot\alpha)(\beta)}U(I_t)^{-1} = \eta_T \zeta \otimes \cdots \otimes \zeta \eta^{(\dot\alpha)(\beta)}, \tag{12.22}$$

$$\tag{12.23}$$

where (α) represents a set of j undotted indices and $(\dot\beta)$ a set of k dotted indices: $\xi^{(\alpha)(\dot\beta)} = \xi^{\alpha_1\cdots\alpha_j\dot\beta_1\cdots\dot\beta_k}$ and $\eta^{(\dot\alpha)(\beta)} = \eta^{\dot\alpha_1\cdots\dot\alpha_j\beta_1\cdots\beta_k}$, and $\tau_P(j+k)$ equal i if $j+k$ is odd and 1 if $j+k$ is even. The notation $\zeta \otimes \cdots \otimes \zeta$ must be understood as the action of one ζ on each of the indices. The η's are again arbitrary phases which should be chosen for every field. We check on the previous expressions that the antiunitary operator Θ corresponding to the product *PCT* has for action

$$\Theta\xi^{(\alpha)(\dot\beta)}\Theta^{-1} = \eta\, i^{F(j+k)}(-1)^j \xi^{(\alpha)(\dot\beta)*}$$

$$\Theta\eta^{(\dot\alpha)(\beta)}\Theta^{-1} = \eta\, i^{F(j+k)}(-1)^k \eta^{(\dot\alpha)(\beta)*}, \tag{12.24}$$

where η equals the product of the three η's and $F(j+k)$ equals 0 if $j+k$ is even and 1 if $j+k$ is odd.

We deduce for the expectation values that if the action of the operator PCT is well defined, for example by the formulae (12.24), on fields of components $\phi_\mu, \chi_\varrho, \cdots, \psi_\nu$, where $\mu, \varrho, \cdots, \nu$ are sets of undotted and dotted indices, then

$$
\begin{aligned}
(\Omega, \phi_\mu(x_1)\cdots\psi_\nu(x_n)\Omega) &= (\Omega, \Theta\Theta^{-1}\phi_\mu(x_1)\Theta\cdots\Theta^{-1}\psi_\nu(x_n)\Theta\Omega)\\
&= (\Omega, \Theta\phi_\mu^*(-x_1)\cdots\psi_\nu^*(-x_n)\Omega)\\
&= (\Theta\Omega, \phi_\mu^*(-x_1)\cdots\psi_\nu^*(-x_n)\Omega)\\
&= \overline{(\Omega, \phi_\mu^*(-x_1)\cdots\psi_\nu^*(-x_n)\Omega)}\\
&= (\Omega, \psi_\nu(-x_n)\cdots\phi_\mu(-x_1)\Omega).
\end{aligned}
\tag{12.25}
$$

Conversely, if this identity is verified, we show that there exists an operator Θ sat-isfying the relations (12.24). We have already mentioned the PCT theorem, which can be stated here by saying that for a Lorentz invariant quantum field theory with local in-teractions, there exists always a choice for the phases η of the various fields, usually in more than one way, such that the product PCT, taken in any order, is a symmetry of the theory. We shall present the precise conditions for the validity of this theorem in the next section.

12.2.1.3 The Wightman Functions

Most of the results in axiomatic field theories are proved starting from vacuum expectation values of products of fields. A theorem, due to A. Wightman, shows that there exists a complete equivalence between vacuum expectation values satisfying certain properties and the existence of fields satisfying the previous axioms. Numerous theorems follow directly from these axioms and, as a result, their validity is independent of the precise nature of the interaction.

It is not in the scope of this book to give a systematic presentation of axiomatic field theory. However, in the sequel, we will give the proof of two important consequences of these axioms, the PCT theorem and the spin-statistics theorem, because they still play a fundamental role in our description of reality.

Let us consider the vacuum expectation value

$$< \Omega, \phi_{\alpha_1}(x_1) \cdots \phi_{\alpha_n}(x_n)\Omega >= \mathcal{W}_{\alpha_1,\cdots,\alpha_n}(x_1,\cdots,x_n), \qquad (12.26)$$

where α_i represents a set of spinor indices (dotted or undotted). $\mathcal{W}(x_1,\cdots,x_n)$ is called a *Wightman function*. Let us study some consequences of the Wightman axioms on the structure of the Wightman functions.

We will list some of the main properties of the vacuum expectation values.

From axiom I, we get that

$$< \Omega, \phi_1(f_1)\phi_2(f_2),\ldots,\phi_n(f_n)\Omega >$$

exists and is a separately continuous multilinear functional of the arguments $f_1, f_2, \ldots, f_n$ which are elements of $\mathcal{S}(\mathbb{R}^4)$. It follows from this fact (through the so called nuclear theorem) that this functional can be uniquely extended to be a tempered distribution of the four-vectors $x_1, x_2, \ldots, x_n$. This tempered distribution, element of $\mathcal{S}'(\mathbb{R}^{4n})$, is the Wightman function $\mathcal{W}_{\alpha_1,\cdots,\alpha_n}(x_1,\cdots,x_n)$ introduced in Eq. (12.26). Some difficulties may result from this abuse of notation since it is only $< \Omega, \phi_{\alpha_1}(f_1) \cdots \phi_{\alpha_n}(f_n)\Omega >$ which is meaningful. Particularly, this is what happens if some variables are taken to be identical, in which case the pointwise multiplication by functions which are not sufficiently regular (e.g. θ functions) can result in quantities which are meaningless.

This last difficulty is the source of renormalisation problems.

We now state a theorem, without proof, which summarises all the properties of the Wightman functions resulting from the axioms.

Theorem 12.

(a) *Relativistic transformation law*

Since the vacuum state is invariant under the action of the Poincaré group and in particular of the translation group, we obtain

$$
\begin{aligned}
\mathcal{W}_{\alpha_1,\cdots,\alpha_n}(x_1,\cdots,x_n) &= <\Omega, U(a,1)\phi_{\alpha_1}(x_1)\cdots\phi_{\alpha_n}(x_n)U^{-1}(a,1)\Omega> \\
&= (<\Omega, U(a,1)\phi_{\alpha_1}(x_1)U^{-1}(a,1)U(a,1)\cdots U(a,1)\phi_{\alpha_n}(x_n) \\
&\quad U^{-1}(a,1)\Omega> \\
&= <\Omega, \phi_{\alpha_1}(x_1+a)\cdots\phi_{\alpha_n}(x_n+a)\Omega> = \mathcal{W}_{\alpha_1,\cdots,\alpha_n} \\
&\quad (x_1+a,\cdots,x_n+a),
\end{aligned}
\tag{12.27}
$$

which shows that the Wightman functions depend only on the difference of the arguments. Setting $\xi_j = x_j - x_{j+1}, j = 1,\cdots,n-1$, we define

$$
W(\xi_1,\cdots,\xi_{n-1}) = \mathcal{W}(x_1,\cdots,x_n).
\tag{12.28}
$$

Formula (12.27) can be rewritten by replacing the translation group by the Lorentz group. With $A \in SL(2,\mathbb{C})$, the Wightman functions transform according to

$$
\begin{aligned}
\mathcal{W}_{\alpha_1,\cdots,\alpha_n}(x_1,\cdots,x_n) &= <\Omega, U(0,A)\phi_{\alpha_1}(x_1)\cdots\phi_{\alpha_n}(x_n)U^{-1}(0,A)\Omega> \\
&= \sum_{\beta_1,\cdots\beta_n} S_{\alpha_1,\beta_1}(A^{-1})\cdots S_{\alpha_n,\beta_n}(A^{-1})<\Omega, \phi_{\beta_1}(\Lambda(A)x_1)\cdots \\
&\quad \phi_{\beta_n}(\Lambda(A)x_n)\Omega> \\
&= \sum_{\beta_1,\cdots\beta_n} S_{\alpha_1,\beta_1}(A^{-1})\cdots S_{\alpha_n,\beta_n}(A^{-1})\mathcal{W}_{\beta_1,\cdots,\beta_n}(\Lambda(A)x_1,\cdots, \\
&\quad \Lambda(A)x_n),
\end{aligned}
\tag{12.29}
$$

where $S(A)$ is a finite-dimensional irreducible representation of $SL(2,\mathbb{C})$.

(b) Spectral condition
Using the fact that $\phi(x) = U(x,1)\phi(0)U^{-1}(x,1) = U(x,1)\phi(0)U(-x,1)$ and the spectral representation of U, we obtain

$$
\begin{aligned}
&W_{\alpha_1,\cdots,\alpha_n}(x_1,\cdots,x_{n-1}) \tag{12.30} \\
&= <\Omega, U(x_1,1)\phi_{\alpha_1}(0)U(x_2-x_1,1)\cdots U(x_n-x_{n-1},1)\phi_{\alpha_n}(0)U(-x_n,1)\Omega> \\
&= <\Omega, \phi_{\alpha_1}(0)U(-x_1,1)\cdots U(-x_{n-1},1)\phi_{\alpha_n}(0)\Omega> \\
&= \int\int e^{-ip_1x_1\cdots-ip_{n-1}x_{n-1}}<\Omega, \phi_{\alpha_1}(0)\mathrm{d}E(p_1)\cdots\mathrm{d}E(p_{n-1})\phi_{\alpha_n}(0)\Omega> \\
&= \frac{1}{(2\pi)^{4(n-1)}}\int\int e^{-ip_1x_1\cdots-ip_{n-1}x_{n-1}}\widetilde{W}_{\alpha_1,\cdots,\alpha_n}(p_1,\cdots,p_{n-1})\,\mathrm{d}^4p_1\cdots\mathrm{d}^4p_{n-1}.
\end{aligned}
$$

The Fourier transform $\widetilde{W}$ of W is also a tempered distribution.

(c) Hermiticity condition

$$
<\Omega, \phi_1(x_1)\phi_2(x_2)\ldots,\phi_n(x_n)\Omega> = \overline{<\Omega, \phi_n^*(x_n)\ldots,\phi_2^*(x_2),\phi_1^*(x_1)\Omega>}.
\tag{12.31}
$$

(d) Local commutativity
Let π be a permutation of $\{1, 2, \ldots, n\} \rightarrow \{i_1, i_2, \ldots, i_n\}$, *then*

$$\mathcal{W}(x_1, x_2, \cdots, x_n) = \mathcal{W}_\pi(x_{i_1}, x_{i_2}, \cdots, x_{i_n}),$$

where

$$\mathcal{W}_\pi(x_{i_1}, x_{i_2}, \cdots, x_{i_n}) = (-1)^m < \Omega, \phi_{i_1}(x_{i_1})\phi_{i_2}(x_{i_2}) \ldots \phi_{i_n}(x_{i_n})\Omega >$$

if the differences $x_j - x_k$ are space-like for all indices j and k which are exchanged by the permutation and m is the number of exchanges of anti-commuting fields necessary to arrive to the new order.

(e) Positive definiteness condition
It is a set of formulas expressing in terms of the Wightman functions the fact that the norm of the vector

$$|\Psi > = f_0\Omega > + \int \phi_{11}(x_1)f_1(x_1)\mathrm{d}^4x_1\Omega >$$

$$+ \int \phi_{21}(x_1)\phi_{22}(x_2)f_2(x_1, x_2)\mathrm{d}^4x_1\mathrm{d}^4x_2\Omega > + \cdots$$

$$+ \int \phi_{k1}(x_1)\phi_{k2}(x_2) \ldots \phi_{kk}(x_k)f_k(x_1, x_2, \ldots, x_k)\mathrm{d}^4x_1 \cdots \mathrm{d}^4x_k\Omega >$$

is non-negative.

These five statements make it possible to reconstruct, starting from the vacuum expectation values, a field theory satisfying the Wightman axioms.

We need another result before going to the analytical properties of the Wightman functions. It is a consequence of the axioms, but it is not necessary for the reconstruction of the theory.

This result expresses the fact that when two systems at point x and y become separated by a large space-like distance, then the interaction between them falls off to 0. This is what has been observed for interactions driven by massive particles, the clustering property being on the order of $e^{-\text{Mass}\times\text{Distance}}$. A weaker result can be obtained, with more difficulties, when the particle describing the force is massless. This is, for example, the case for electromagnetism since the photons are massless. In that case, the clustering is no more exponential but power-like.

Theorem 13 (Cluster Decomposition Property).
Let a be a space-like vector. Then, as tempered distribution

$$\mathcal{W}(x_1, \cdots, x_j, x_{j+1} + \lambda a, \cdots, x_n + \lambda a)$$
$$\rightarrow \mathcal{W}(x_1, \cdots, x_j)\mathcal{W}(x_{j+1}, \cdots, x_n) \tag{12.32}$$

as $\lambda \rightarrow \infty$.

We will not give a complete proof of this theorem, only some plausibility arguments. We will also suppose that we are in the case described by the first axiom, i.e. a theory in which there are massive particles, therefore ensuring that in the energy–momentum spectrum, there is a mass gap.

We note that since $\phi(x + a) = U(a, 1)\phi(x)U^{-1}(a)$ and the vacuum is invariant under the action of the Poincaré group

$$
\begin{aligned}
&\mathcal{W}(x_1, \cdots, x_j, x_{j+1} + \lambda a, \cdots, x_n + \lambda a) \\
&= <\Omega\phi(x_1) \cdots \phi(x_j)U(\lambda a, 1)\phi(x_{j+1}) \cdots \phi(x_n)\Omega>.
\end{aligned}
\tag{12.33}
$$

Now the rest of the proof consists of showing that in some sense (in fact in a weak operator topology), the operator $U(\lambda a, 1)$ tends, when $\lambda \to \infty$, a being space-like, to the orthogonal projector on the vacuum state.

This also can be seen using physical arguments. Let us introduce a partition of the identity as in Chapter 8,

$$
\sum |n><n| = 1,
\tag{12.34}
$$

where the sum (or the integral) runs over a set of orthonormal states spanning the Hilbert space of physical states (which includes the vacuum state).

Then the left-hand side of (12.34) can be written as

$$
\begin{aligned}
&<\Omega\phi(x_1) \cdots \phi(x_j)U(\lambda a, 1)\phi_{j+1} \cdots \phi(x_n)\Omega> \\
&= \sum_n <\Omega\phi(x_1) \cdots \phi(x_j)|n> e^{i\lambda aP} <n|\phi_{j+1} \cdots \phi(x_n)\Omega> \\
&= <\Omega\phi(x_1) \cdots \phi(x_j)\Omega><\Omega\phi_{j+1} \cdots \phi(x_n)\Omega> \\
&+ \sum_{n \neq \Omega} <\Omega\phi(x_1) \cdots \phi(x_j)|n> e^{i\lambda aP} <n|\phi_{j+1} \cdots \phi(x_n)\Omega>,
\end{aligned}
\tag{12.35}
$$

where P is the momentum operator.

Now the end of the argument is that the second part of the right-hand side of (12.35) tends to 0 when $\lambda \to \infty$ by the Riemann–Lebesgue lemma. This uses the fact that in momentum space, all states $|n>$ have a spectrum whose closure does not include $p = 0$.

12.2.1.4   *Analytical Properties of the Wightman Functions*

Our goal is the proof of the two main theorems PCT and spin-statistics. We will see that because of the conditions imposed on the vacuum expectation values by the axioms, the Wightman function have some analyticity properties that we will now explore.

We remark that if $\eta \in V^+$, the scalar product in the Minkowski space $p.\eta$ is positive and the functions

$$
e^{-ip_1.(\xi_1 - i\eta_1) \cdots - ip_{n-1}.(\xi_{n-1} - i\eta_{n-1})}
\tag{12.36}
$$

are exponentially decreasing and in $\mathcal{S}(\mathbb{R}^{4(n-1)})$. We show easily that in the variables $\zeta_j = \xi_j - i\eta_j$,

$$W(\zeta_1, \cdots, \zeta_{n-1}) = W(\xi_1 - i\eta_1, \cdots, \xi_{n-1} - i\eta_{n-1}) \tag{12.37}$$

is an analytic function of ζ_j for $-\Im\zeta_j \in V^+$, which is usually described by saying that the variables $\zeta_1, \cdots, \zeta_{n-1}$ are in the *tube* $\mathcal{T}_{n-1}$ where $\mathcal{T}_{n-1} = \{\zeta_1, \cdots, \zeta_{n-1} | \Im\zeta_i \in V^+, i = 1, \ldots, n-1\}$. This result implies the following:

- The Wightman functions are the boundary values (from the analyticity domain) of the analytic functions:

$$W(\xi_1, \cdots, \xi_{n-1}) = \lim_{\substack{\eta_1, \ldots, \eta_{n-1} \to 0 \\ \eta_j \in V^+}} W(\xi_1 - i\eta_1, \cdots, \xi_{n-1} - i\eta_{n-1}).$$

- The points $\zeta_j = (-i\eta_j^0, \boldsymbol{\xi}_j)$ $j = 1, \cdots, n-1$, belong to the domain of analyticity of the Wightman functions. These points are the so called *Euclidean points*.

This last result clears up partly the preceding remarks showing the existence of a relation between Euclidean theories and relativistic theories. It justifies in particular the replacement of t by $-it$ as being the result of an analytic continuation.

We restrict our discussion to the case of one scalar field, the extension to more than one field or to spinor fields being relatively easy. Formula (12.29) becomes then

$$\mathcal{W}(x_1, \cdots, x_n) = \mathcal{W}(\Lambda(A)x_1, \cdots, \Lambda(A)x_n), \tag{12.38}$$

that is to say the Wightman functions are invariant under the action of the restricted Lorentz group or equivalently under the action of $SL(2, \mathbb{C})$. This property is obviously also true for the W's. This result makes it possible to enlarge the domain of analyticity of the Wightman functions.

We can indeed introduce the complex Lorentz transformations by replacing in the formula which links the restricted Lorentz group to $SL(2, \mathbb{C})$ $\bar{A}$ by B an element of $SL(2, \mathbb{C})$. We thus define $x \to \Lambda(A, B)x$ for $A, B \in SL(2, \mathbb{C})$ by

$$\tilde{x} \to A\tilde{x}B^{tr} \tag{12.39}$$

and therefore

$$\Lambda(A, B)^\mu{}_\nu = \frac{1}{2}\text{Tr}(\sigma_\mu A\sigma_\nu B^{tr}). \tag{12.40}$$

We check that the group law defined by Eq. (12.40) corresponds to the following law of multiplication for the elements of the complex Lorentz group $L_+(\mathbb{C})$

$$\Lambda(A_1, B_1)\Lambda(A_2, B_2) = \Lambda(A_1 A_2, B_1 B_2). \tag{12.41}$$

A multiplication law can be defined over pairs of elements $\{A, B\}$ of $SL(2, \mathbb{C}) \times SL(2, \mathbb{C})$ by

$$\{A_1, B_1\}\{A_2, B_2\} = \{A_1 A_2, B_1 B_2\} \tag{12.42}$$

and the representations $D^{[j,k]}$ of Chapter 2 extend to $L_+(C)$ by

$$\xi_{\alpha_1, \cdots, \alpha_{2j}, \dot\beta_1 \cdots \dot\beta_{2k}} \to A_{\alpha_1}^{\gamma_1} \cdots A_{\alpha_{2j}}^{\gamma_{2j}} B_{\dot\beta_1}^{\dot\delta_1} \cdots B_{\dot\beta_{2k}}^{\dot\delta_{2k}} \xi_{\gamma_1, \cdots, \gamma_{2j}, \dot\delta_1 \cdots \dot\delta_{2k}}. \tag{12.43}$$

The pairs $\{A, B\}$ and $\{-A, -B\}$ define the same element $\Lambda(A, B)$ and they are the only ones. The interest of the introduction of $L_+(\mathbb{C})$ comes from the fact that two elements of the group $\{1, 1\}$ and $\{1, -1\}$ can be continuously related. Thus, the complex Lorentz group connects the two components $L_+^\uparrow$ and $L_+^\downarrow$.

Coming back to formula (12.38), this means that we can rewrite it as

$$\mathcal{W}(x_1, \cdots, x_n) = \mathcal{W}(\Lambda(A, B)x_1, \cdots, \Lambda(A, B)x_n). \tag{12.44}$$

The left-hand term having an analytic continuation into the tube $\mathcal{T}_{n-1}$, the right-hand term is analytic for all the points $\Lambda(A, B)x_i$ with $\{x_1 - x_2, \cdots, x_{n-1} - x_n\} \in \mathcal{T}_{n-1}$. We call *extended tube* the set

$$\mathcal{T}'_{n-1} \tag{12.45}$$
$$= \{\Lambda(A, B)\zeta_1, \cdots, \Lambda(A, B)\zeta_{n-1} | (\zeta_1, \cdots, \zeta_{n-1}) \in \mathcal{T}_{n-1} \forall A, B \in SL(2, \mathbb{C})\}.$$

The Wightman functions are therefore analytic in the extended tube. Applying formula (12.44) for $A = -1$ and $B = 1$, which corresponds to $\Lambda = -1$, we find that

$$W(\xi_1, \cdots, \xi_n) = W(-\xi_1, \cdots, -\xi_n). \tag{12.46}$$

The result extends to the spinor case. Using the fact that $S(-1, 1) = (-1)^j 1$, j being the number of undotted spinors, we get the general formula

$$W_{\alpha_1, \cdots, \alpha_n}(\xi_1, \cdots, \xi_n) = (-1)^{\mathcal{J}} W_{\alpha_1, \cdots, \alpha_n}(-\xi_1, \cdots, -\xi_n), \tag{12.47}$$

where $\mathcal{J}$ is the total number of undotted indices (here with our hypothesis: only one spinor field, $\mathcal{J} = nj$). As we will see, this formula is one of the essential steps in the proof of the PCT theorem.

The last property which must be exploited is the local commutativity. We have seen that the vacuum expectation value of a product of fields, in a given order, makes it possible to define a function analytic in an extended tube of which it is the boundary value. What is the relation between this analytic function and that which can be obtained from the vacuum expectation value of a product of these fields but in another order?

Let us show first that the extended tube contains real points of analyticity. To be convinced, it is enough to consider the case of a four-vector. Therefore, let us assume that $\zeta \in T_1'$ and seek if there exist real ζ's. By definition ζ has the form $\zeta = \Lambda(\xi - i\eta)$ with $\eta \in V^+$; therefore $\zeta^2 = (\xi - i\eta)^2 = \xi^2 - \eta^2 - 2i\xi.\eta$. The reality of ζ implies that $\xi.\eta = 0$; thus, ξ is space-like since η is time-like; therefore, $\zeta^2 < 0$. It is easy to explicitly construct such points. For example, let us consider $\zeta = (\zeta^0, \zeta^1, 0, 0)$, ζ^0 with ζ^1 real and such that $|\zeta^0| < \zeta^1$; thus, $\zeta = \Lambda(\alpha)z$ with $z = (\zeta^0 \cos\alpha - i\zeta^1 \sin\alpha, -i\zeta^0 \sin\alpha + \zeta^1 \cos\alpha, 0, 0)$, where

$$\Lambda(\alpha) = \begin{pmatrix} \cos\alpha & i\sin\alpha & 0 & 0 \\ i\sin\alpha & \cos\alpha & 0 & 0 \\ 0 & 0 & 1 & 0 \\ 0 & 0 & 0 & 1 \end{pmatrix} \tag{12.48}$$

is an element of the extended tube since, if $\pi > \alpha > 0$, then $-\Im z \in V^+$. The real points of T_{n-1}', which are called *Jost points*, form a space-like convex set.[4] We show that an analytic function which vanishes at the Jost points is identically 0. This implies that if two analytic functions coincide on the Jost points (or an open subset of Jost points), then there exists a *unique analytic function* which coincides with each of them.

This last result shows, using local commutativity, that all the analytic functions obtained by the analytic continuation of vacuum expectation values of the product in any order of n fields are in fact the different pieces of a unique analytic function. We will prove this result.

Let us consider spinor fields with components ϕ_α where, as before, α represents a set of dotted or undotted indices. The axiom of local commutativity implies the following equality between Wightman functions,

$$\mathcal{W}_{\alpha_1,\cdots,\alpha_n}(x_1,\cdots,x_n) = (-1)^F \mathcal{W}_{\alpha_{i_1},\cdots,\alpha_{i_n}}(x_{i_1},\cdots,x_{i_n}), \tag{12.49}$$

if all the differences $x_i - x_j$ are space-like, $(i_1,\cdots,i_n)$ being a permutation of $(1,\cdots,n)$ and F being the number of half-integer spin fields which have been permuted. We would like to identify these two functions as boundary values of the same analytic function. This will be possible if they have in common an open set of Jost points. We will show that this is indeed the case in an elementary context, that is to say the permutation of two contiguous indices j and $j + 1$, since an arbitrary permutation is a sequence of local permutations, and thus the result extends to the general case.

For simplicity, let us consider two Wightman functions of a scalar field, $F_1 = \mathcal{W}(x_1,\cdots,x_j,x_{j+1},\cdots,x_n)$ and $F_2 = \mathcal{W}(x_1,\cdots,x_{j+1},x_j,\cdots,x_n)$.

---

[4] More precisely a real point $\zeta_1,\cdots,\zeta_n$ is in T_n' if and only if the vectors of the form

$$\sum_{i=1}^n \lambda_i \zeta_i \qquad \text{with} \qquad \lambda_i \geq 0, \quad \sum_i \lambda_i > 0$$

are space-like.

Corresponding to them there are the functions $W(\xi_1, \cdots, \xi_{j-1}, \xi_j, \xi_{j+1}, \cdots, \xi_{n-1})$ and $W(\hat{\xi}_1, \cdots, \hat{\xi}_{j-1}, \hat{\xi}_j, \hat{\xi}_{j+1}, \cdots, \hat{\xi}_{n-1})$ with

$$
\begin{aligned}
\hat{\xi}_k &= \xi_k \qquad 1 \le k < j-1 \quad \text{and} \quad j+1 < k \le n \\
\hat{\xi}_{j-1} &= \xi_{j-1} + \xi_j \\
\hat{\xi}_j &= -\xi_j \\
\hat{\xi}_{j+1} &= \xi_{j+1} + \xi_j.
\end{aligned}
\tag{12.50}
$$

These two functions are the boundary values of functions analytic in the extended tubes and are therefore analytic in the respective Jost points. Let us show that they have common Jost points. For this purpose, let us take $\xi_k = (0, b, 0, 0)$ for $k \neq j_1, j, j+1$, $\xi_{j-1} = (a, b, 0, 0)$, $\xi_j = (0, 0, c, 0)$ and $\xi_{j+1} = (-a, b, 0, 0)$ with $0 < |a| < b$. Then the vectors $A = \lambda_{j-1}\xi_{j-1} + \lambda_j\xi_j + \lambda_{j+1}\xi_{j+1}$ are space-like vectors for all $\lambda \ge 0$. It follows that

$$
\lambda_{j-1}\hat{\xi}_{j-1} + \lambda_j\hat{\xi}_j + \lambda_{j+1}\hat{\xi}_{j+1} = A + (\lambda_{j-1} + \lambda_{j+1} - 2\lambda_j)\xi_j
\tag{12.51}
$$

is also space-like. Adding the other components ξ_k, we see that these particular Jost points are common to the two tubes and form an open set. Since these points are real points of analyticity and because of local commutativity $F_1 = F_2$ coincide on them, they are the continuation of a unique analytic function.

12.2.1.5 *The PCT and Spin-Statistics Theorems*

The proof of these two theorems, which we will only sketch, requires the use of nearly all the Wightman axioms. This means that they are deep results of general principles which form the foundations of contemporary physical theories. For example, in the case of the PCT theorem, the link between processes involving particles and those involving anti-particles can only be understood using the large domain of analyticity in which physical quantities can be analytically continued.

Let us consider a theory with spinor fields $\phi_\mu, \cdots, \psi_\nu$, each index $\mu, \cdots, \nu$ representing a set of dotted or undotted indices $(\alpha)(\dot\beta)$, transforming according to irreducible representation of the restricted Lorentz group. We saw in the section about the discrete symmetries that this implies the existence of an antiunitary operator Θ such that

$$
\Theta^{-1}\phi_\mu\Theta = (-1)^j \mathrm{i}^{F(\phi)}\phi_\mu^*,
\tag{12.52}
$$

where j is the number of undotted indices in μ and $F(\phi)$ is 0 if the spin of the field is an integer and 1 if it is a half-odd integer. We can therefore state the following theorem, originally due to W. Pauli.

Theorem 14. *The PCT theorem: consider a relativistic field theory with spinor fields $\phi_\mu, \cdots, \psi_\nu$ transforming according to irreducible representations of the Lorentz group and satisfying the usual commutation rules. Then the following identities are satisfied,*

$$< \Omega, \phi_\mu(x_1) \cdots \psi_\nu(x_n)\Omega > = \mathrm{i}^F(-1)^{\mathcal{J}} < \Omega, \psi_\nu(-x_n) \cdots \phi_\mu(-x_1)\Omega >, \qquad (12.53)$$

where F is the number of integer-spin fields and $\mathcal{J}$ the total number of undotted indices.

A corollary of this theorem is the following.

Corollary 1. *If ϕ is a scalar field satisfying all the Wightman axioms, then the Wightman functions are invariant under the action of the operator PCT, appropriately defined.*[5]

The corollary results from the fact that for a scalar field $F(\phi) = 0$ and $\mathcal{J} = 0$.
Let us prove the theorem.
Let

$$\mathcal{W}_{\mu,\cdots,\nu}(x_1,\cdots,x_n) = < \Omega, \phi_\mu(x_1) \cdots \psi_\nu(x_n)\Omega > \qquad (12.54)$$

and $W_{\mu,\cdots,\nu}(\xi_1,\cdots,\xi_n)$ be the corresponding Wightman function. It transforms under $SL(2,\mathbb{C}) \times SL(2,\mathbb{C})$ as

$$\begin{aligned}
S^{(\phi)}_{\mu\mu'}(A,B) &\cdots S^{(\psi)}_{\nu\nu'}(A,B) W_{\mu'\cdots\nu'}(\zeta_1,\cdots,\zeta_{n-1}) \\
&= W_{\mu,\cdots,\nu}(\Lambda(A,B)\zeta_1,\cdots,\Lambda(A,B)\zeta_{n-1}) \qquad (12.55)
\end{aligned}$$

for ζ in the extended tube. It follows from (12.55) that if the total number of indices, dotted and undotted, is odd, the Wightman function vanishes. It is enough indeed to write this equation for $(A,B) = (\mathbf{1},-\mathbf{1})$ and $(A,B) = (-\mathbf{1},\mathbf{1})$ and to remark that

$$\begin{aligned}
S^{(\phi)}_{\mu\mu'}(\mathbf{1},-\mathbf{1}) &\cdots S^{(\psi)}_{\nu\nu'}(\mathbf{1},-\mathbf{1}) = \delta_{\mu'\mu} \cdots \delta_{\nu'\nu}(-1)^{\mathcal{J}} \\
S^{(\phi)}_{\mu\mu'}(-\mathbf{1},\mathbf{1}) &\cdots S^{(\psi)}_{\nu\nu'}(-\mathbf{1},\mathbf{1}) = \delta_{\mu'\mu} \cdots \delta_{\nu'\nu}(-1)^{K} \qquad (12.56)
\end{aligned}$$

and thus the first term in (12.55) changes sign while, since $\Lambda(\mathbf{1},-\mathbf{1}) = \Lambda(-\mathbf{1},\mathbf{1}) = -\mathbf{1}$, the second term remains invariant. Thus, according to formulae (12.55), we deduce that for $\zeta_1,\cdots,\zeta_{n-1}$ in the extended tube,

$$W_{\mu,\cdots,\nu}(\zeta_1,\cdots,\zeta_{n-1}) = (-1)^{\mathcal{J}} W_{\mu,\cdots,\nu}(-\zeta_1,\cdots,-\zeta_{n-1}). \qquad (12.57)$$

From local commutativity, for space-like points $x_1,\cdots,x_n$,

$$< \Omega, \phi_\mu(x_1) \cdots \psi_\nu(x_n)\Omega > = \mathrm{i}^F < \Omega, \psi_\nu(x_n) \cdots \phi_\mu(x_1)\Omega >, \qquad (12.58)$$

the factor i^F coming from the equality

$$(-1)^{(F-1)+(F-2)+\cdots+1} = (-1)^P = (-1)^{\frac{F(F-1)}{2}} = \mathrm{i}^F, \qquad (12.59)$$

---

[5] As this was explained in the previous section, this assertion results from the fact that we can always choose the phases for the various fields such that the invariance holds.

where P is the number of commutations of half-integer spin fields to reverse the order and F is the number of half-integer spin fields, the last equality being due to the preceding result which says that $j + k$ is even and therefore that F is even.

As we saw in the preceding section, the equality (12.58) extends to the extended tube[6]

$$W_{\mu,\cdots,\nu}(\zeta_1,\cdots,\zeta_{n-1}) = i^F W_{\nu,\cdots,\mu}(-\zeta_{n-1},\cdots,-\zeta_1). \tag{12.60}$$

Combining this relation with (12.57) we get

$$W_{\mu,\cdots,\nu}(\zeta_1,\cdots,\zeta_{n-1}) = i^F(-1)^{\mathcal{J}} W_{\nu,\cdots,\mu}(\zeta_{n-1},\cdots,\zeta_1). \tag{12.61}$$

Going to the boundary values, $\zeta_i \to \xi_i$ with $-\Im\zeta_i \in V^+$, we deduce the following equality for all the real points:

$$< \Omega, \phi_\mu(x_1)\cdots\psi_\nu(x_n)\Omega >= i^F < \Omega, \psi_\nu(-x_n)\cdots\phi_\mu(-x_1)\Omega >. \tag{12.62}$$

This identity is equivalent, according to (12.25), to the invariance of the theory under PCT.

To prove the PCT theorem we have assumed that the fields satisfy the normal commutation relations. We will show that it is indeed the case.

Let us consider formula (12.60). $\zeta_1,\cdots,\zeta_{n-1}$ being in the extended tubes, so it is the the case for $-\zeta_{n-1},\cdots,-\zeta_1$. We can therefore let the imaginary parts go to 0 and take the boundary values and obtain a relation between Wightman distributions valid for all real points. We restrict our interest to the case $n = 2$ with two fields ϕ and ϕ^* and we get, using (12.59),

$$W(\zeta) = (-1)^{P+\mathcal{J}}\hat{W}(-\zeta), \tag{12.63}$$

and from

$$\lim_{\eta\in V^+,\eta\to 0} W(x-y-i\eta) = < \Omega, \phi(x)\phi^*(y)\Omega >$$

$$\lim_{\eta\in V^+,\eta\to 0} \hat{W}(y-x-i\eta) = < \Omega, \phi^*(y)\phi(x)\Omega > \tag{12.64}$$

we get

$$< \Omega, \phi(x)\phi^*(y)\Omega >= (-1)^{P+\mathcal{J}} < \Omega, \phi^*(-y)\phi(-x)\Omega >. \tag{12.65}$$

---

[6] Be careful, (12.60) is not an equality on the real numbers but only at the Jost points! The equality which follows between analytic functions does not give an equality between vacuum expectation values by taking the boundary values since they are obtained as limit when $\Im\zeta \to 0$, $-\Im\zeta \in V^+$, and thus $\Im\zeta$ is not in V^+ but in V^-.

We integrate both sides by $f(x)\bar{f}(y)$ and get with $\hat{f}(x) = f(-x)$

$$\| \phi^*(f)\Omega >\|^2 = (-1)^{P+\mathcal{J}} \| \phi(\hat{f})\Omega >\|^2 \tag{12.66}$$

from which follows that $(P+\mathcal{J})$ must be even or the norms must be 0. If the norms must be 0, i.e. the vectors, whatever is the test function f this implies that $\phi^* = \phi = 0$.[7]

Now $\mathcal{J}$ is the sum of dotted and undotted indices of ϕ. $\mathcal{J}$ is even for integer spin representations and odd for half-integer spin representations, and P counts the number of permutation of half-integer spin field to reverse the order.

Then if we call a field ϕ a Bose field if its commutator vanishes for space-like distance and a Fermi field it is its anticommutator which vanishes:

Bose field	$[\phi(x), \phi^*(y)] = 0$	for	$(x-y)^2 < 0$	(12.67)
Fermi field	$[\phi(x), \phi^*(y)]_+ = 0$	for	$(x-y)^2 < 0.$	(12.68)

So for fields which anticommute at space-like distances $P = -1$ and for fields which commute at space-like distances $P = 1$.

To conclude, we need to clear off a situation. Can a field ϕ have different commutation relations with another field ψ and its adjoint ψ^*? The answer is given by a lemma due to Dell Antonio.

Theorem 15. *If in a field theory we have*

$$[\phi(x), \psi(y)]_{\pm} = 0 \qquad \textit{for} \qquad (x-y)^2 < 0 \tag{12.69}$$

and the opposite

$$[\phi(x), \psi^*(y)]_{\mp} = 0 \qquad \textit{for} \qquad (x-y)^2 < 0, \tag{12.70}$$

then either ϕ or ψ vanishes.

Let us give the proof.

Take two test functions f and g, C^∞ with compact supports and such that these supports are space-like separated. Then

$$< \Omega, \phi(f)^*\psi(g)^*\psi(g)\phi(f)\Omega > = \| \psi(g)\phi(f)\Omega >\|^2 \geq 0. \tag{12.71}$$

With the condition on the support of f and g and the commutation relations (12.69), the left-hand side of (12.71) is

$$- < \Omega, \psi(g)^*\psi(g)\phi(f)^*\phi(f)\Omega > . \tag{12.72}$$

---

[7] This results from axiom 5, 12.2.1, on the cyclicity of the vacuum. In fact, roughly speaking according to this axiom, $\prod \phi(f_i)\Omega >$ for arbitrary test functions f_i must span the whole physical Hilbert space.

If we now let the support of g tend to infinity in a space-like direction, then (12.72), using the cluster decomposition property (13), approaches

$$- < \Omega, \psi(g)^* \phi(g) \Omega > < \Omega, \phi(f)^* \phi(f) \Omega > = - \| \phi(g) \Omega > \|^2 \| \phi(f) \Omega > \|^2, \quad (12.73)$$

which is not positive.

Therefore, there is a contradiction with (12.71) and the only possibility for the limit is that the right-hand side of (12.73) is 0. This means that either $\phi(g)\Omega >$ or $\psi(f)\Omega >$ is 0. By the cyclicity of the vacuum, this means that either $\phi(g)$ or $\psi(f)$ vanish. If $\phi \neq 0$, then there exists a test function g such that $\phi(g) \neq 0$. This means that for all test functions f with compact support, $\psi(f) = 0$, which implies (because the space of all test functions with compact support $\mathcal{D}$ is dense in $\mathcal{S}$) that $\psi = 0$. Similarly, if $\psi \neq 0$, then $\phi = 0$.

We can now state the theorem.

Theorem 16. *The spin-statistics theorem. Fields belonging to integer spin representations are Bose fields and fields belonging to half-odd integer representation are Fermi fields. Let ϕ be a scalar field. Let us assume that for x, y space-like, we have*

$$[\phi(x), \phi^*(y)]_+ = 0, \quad (12.74)$$

then $\phi(x)|\Omega >= \phi^*(x)|\Omega >= 0$. In a field theory where ϕ and ϕ^* either commute or anticommute with all the other fields, this implies that

$$\phi = \phi^* = 0. \quad (12.75)$$

We only prove the first part of the theorem. The vanishing for space-like points of the vacuum expectation value of the commutator

$$< \Omega, \phi(x)\phi^*(y)\Omega) + (\Omega, \phi^*(y)\phi(x)\Omega >= 0 \quad (12.76)$$

expresses a relation between the boundary values of two analytic functions $W(\zeta)$ and $\hat{W}(\zeta)$ given by

$$< \Omega, \phi(x)\phi^*(y)\Omega > = \lim_{\eta \in V^+, \eta \to 0} W(x - y - i\eta)$$

$$< \Omega, \phi^*(y)\phi(x)\Omega > = \lim_{\eta \in V^+, \eta \to 0} \hat{W}(y - x - i\eta). \quad (12.77)$$

These two functions are analytic in the extended tube T_1'. The relation (12.76), true for the space-like points which are the Jost points, extends to all points of analyticity

$$W(\zeta) + \hat{W}(-\zeta) = 0. \quad (12.78)$$

But the invariance under the action of the complex group implies that

$$\hat{W}(\zeta) = \hat{W}(-\zeta) \tag{12.79}$$

and combining this relation with (12.78), we get

$$W(\zeta) + \hat{W}(\zeta) = 0; \tag{12.80}$$

thus taking the boundary value, we have for *all real points x and y*

$$< \Omega, \phi(x)\phi^*(y)\Omega > + < \Omega, \phi^*(-y)\phi(-x)\Omega >= 0. \tag{12.81}$$

Multiplying this equality by $f(x)$ and $\bar{f}(y)$ and integrating over x and y, we get

$$
\begin{aligned}
0 &= (< \Omega, \phi(f)\phi^*(\bar{f})\Omega > + < \Omega, \phi^*(\hat{\bar{f}})\phi(\hat{f})\Omega > \\
&= \| \phi(f)^*|\Omega > \|^2 + \| \phi(\hat{f})|\Omega > \|^2,
\end{aligned} \tag{12.82}
$$

where $\hat{f}(x) = f(-x)$ and using $\phi^*(\bar{f}) = \phi(f)^*$.

We get

$$\phi(f)^*|\Omega >= \phi(f)|\Omega >= 0; \tag{12.83}$$

the last part of the theorem, which will not be proved, results from the fact that the vacuum state is cyclic.

12.3   The Asymptotic States

12.3.1   Introduction

In 1909, Ernest Rutherford, along with Hans Geiger and Ernest Marsden, established experimentally the existence of the atomic nucleus. The prevailing conception of atomic structure at that time presented atoms as small balls full of 'matter', with no precise idea of the nature of the latter. Rutherford was inspired to investigate this question using a method which was, in principle, very simple: he sent a beam of α-particles to a thin foil of gold. Rutherford asked Geiger and Marsden to look at scattered α's at large deflection angles, something not expected from the theory of diffusion of particles from a soft medium. To their surprise they found that such scattering events did occur. More precisely, the picture they obtained was the following: most α-particles were going through the foil unaffected, but occasionally, some were deflected at very large angles. Rutherford interpreted these results as meaning that the atom was mostly empty space containing some small hard grains. Starting from this idea he formulated the first realistic atomic model with a hard nucleus and electrons orbiting around it. He also became the father of a new method of studying the structure of matter through the use of scattering experiments.

Rutherford's α-particles had a space resolution on the order of 10^{-11} cm, enough to discover the structure of atoms of 10^{-8} cm, but much too large to study the details of a nucleus whose typical size is 10^{-13} cm. Today's accelerators reach distances as low as 10^{-17} cm and it is through their use that we have unravelled most of what we know about the microscopic structure of matter. In this section we-will introduce the physical concepts necessary for the formulation and interpretation of scattering experiments.

A scattering experiment is ideally presented as follows: At time $t = t_1$, two particles move one against the other with a relative speed v. At $t = t_0$ they collide and at a later time $t = t_2$ we observe the results of the collision, typically particles moving away from each other. We want to compute the probability amplitude for observing such an event. In quantum mechanics this problem is easy to formulate: Let us represent the initial state at $t = t_1$ as $|\Psi(t = t_1) >_{\text{in}}$. Similarly, we represent the final state at $t = t_2$ by $|\Phi(t = t_2) >_f$. If H is the Hamiltonian of the system, the initial state $|\Psi >$ evolves with time according to.

$$|\Psi(t = t_2) >= U(t_1, t_2)|\Psi(t = t_1) > \quad ; \quad U(t_1, t_2) = e^{iH(t_1 - t_2)} \tag{12.84}$$

and the probability amplitude $A(t_1, t_2)$ we want to compute is given by the scalar product

$$A(t_1, t_2) =< \Psi(t = t_1)|U(t_2, t_1)|\Phi(t = t_2) > . \tag{12.85}$$

This is a typical problem in quantum mechanics and does not require any special formulation. In particular, the path integral formalism we presented in the previous chapters seems to be perfectly suitable for describing such a process. We should compute the Feynman integral over all paths joining the initial and the final states. The trouble is that, as we have noted already, we can have explicit results only with a Gaussian measure, which means that, in field theory, we can handle only theories which are quadratic in the field variables. We have called such theories 'free field theories' because they describe only non-interacting particles. For the interesting cases of interactions we must resort to perturbation theory. In this and the following sections we will present a formulation of scattering phenomena which will allow us to compute amplitudes of the form given in Eq. (12.85) order by order in the perturbation expansion.

The first step is to construct the space of states which are eigenstates of the unperturbed Hamiltonian, i.e. H_0, the Hamiltonian of non-interacting particles. We did that in Chapter 5 and in the next section we recall their properties. The physical justification for the use of this space in describing scattering experiments is based on the assumption that the interactions are short ranged. Let us give some orders of magnitude: experimentally, the range of the nuclear forces has been determined to be of order of one fermi, or 10^{-13} cm. Weak interactions have an even shorter range. By comparison, the size of an atom is of order of 10^{-8} cm. In scattering experiments the smallest separation among particles we will ever have to consider, in either the initial or the final state, is much larger than interatomic distances, so it is safe to assume that these particles are free.

This requirement of space separation can be translated into a similar requirement on the time interval $t_1 - t_2$ which separates the initial and the final states. Since in particle physics experiments the typical relative speed among particles is a sizable fraction of the speed of light, the characteristic time for a strong interaction collision is on the order of, or smaller than, 10^{-22} sec. Compared to that, even the lifetime of an unstable particle like a charged π-meson ($\sim 10^{-8}$ sec) can be considered as infinite. This led John Archibald Wheeler in 1937, and later W. Heisenberg in the 1940s, to introduce and develop the concept of the *S-matrix*, formally defined as the double limit of the evolution operator $U(-\infty, +\infty)$, see Eq. (5.52). We will study its properties in the following sections.

Before closing this introduction we want to repeat an important limitation of the formalism we are going to develop. As we have already noted, the assumption of short-range forces fails for the electromagnetic interactions. We called the resulting difficulties 'the infrared problem' and we will study it in Chapter 21.

12.3.2 The Fock Space

We introduced in Chapter 5 the Hilbert space $\mathcal{H}$ of the asymptotic states of scalar massive particles. This space of states, the Fock space, describes the possibilities of having no particles, of having one particle without interaction, two particles without interaction, etc.

The space $\mathcal{H}$ is the direct sum of spaces $\mathcal{H}^{(n)}$, Hilbert spaces of states with n particles, whose elements are functions $\Phi^{(n)}(p_1, \ldots, p_n)$, symmetrical because we must deal with only one species of particles of boson type, and square integrable with respect to the measure $\prod_i \mathrm{d}\Omega_m(p_i)$. We can define these elements as vectors in the Hilbert space built by the action of the creation operators on the vacuum state through

$$\Phi^{(n)} \to |\Phi^{(n)}> \tag{12.86}$$
$$= \frac{1}{\sqrt{n!}} \int \prod_i \mathrm{d}\Omega_m(p_i) \Phi^{(n)}(p_1, \cdots, p_n)\, a^*(p_1) \cdots a^*(p_n)|0>.$$

This construction is similar to that obtained when studying the eigenstates of the harmonic oscillator. We start from the vacuum state $|0>$ and we build with the action of the creation operators the states of n particles of momenta $p_1 \ldots p_n$:

$$\frac{1}{\sqrt{n!}} a^*(p_1) \ldots a^*(p_n)|0> = |p_1 \ldots p_n>. \tag{12.87}$$

We then check easily using the canonical commutation relations

$$[a(p), a^*(q)] = (2\pi)^3 2\omega(p)\, \delta^{(3)}(p-q) \tag{12.88}$$

that

$$<\Phi^{(n)}|\Phi^{(m)}> = \delta_{nm}||\Phi^{(n)}||^2. \tag{12.89}$$

With the notation

$$f\overleftrightarrow{\partial_0}g = f\partial_0 g - \partial_0 f g \tag{12.90}$$

and the definition

$$\phi_f(t) = i \int_{x_0=t} f \overleftrightarrow{\partial_0} \phi \, \mathrm{d}^3 x \qquad (12.91)$$

we can express the creation and annihilation operators in terms of the free fields (solutions of the Klein–Gordon equation $(\Box + m^2)\phi(x) = 0$)

$$a^*(\boldsymbol{k}) = -i \int_{x_0=t} \mathrm{d}^3 x \, [\mathrm{e}^{-ik.x} \overleftrightarrow{\partial_0} \phi(x)]$$

$$a(\boldsymbol{k}) = i \int_{x_0=t} \mathrm{d}^3 x \, [\mathrm{e}^{ik.x} \overleftrightarrow{\partial_0} \phi(x)]. \qquad (12.92)$$

Since both $\phi(x)$ and $\mathrm{e}^{\pm ik.x}$ are solutions of the Klein–Gordon equation, we check that the derivative with respect to t of each of the integrals is null and therefore the left-hand sides which are the creation and annihilation operators are independent of time.

We will be obliged in the sequel to use wave packets which correspond to states of the Fock space of the form

$$|f_1 \ldots f_n> = \int \prod_i (\mathrm{d}\Omega_m(\boldsymbol{p}_i)\tilde{f}_i(\boldsymbol{p}_i)) |\boldsymbol{p}_1 \ldots \boldsymbol{p}_n> \qquad (12.93)$$

with

$$f(x) = \int \mathrm{d}\Omega_m(\boldsymbol{p}) \, \mathrm{e}^{-ip.x} \tilde{f}(\boldsymbol{p}); \qquad (12.94)$$

thus

$$|f_1 \ldots f_n> = a^*(f_1) \cdots a^*(f_n)|0> = (-i)^n \phi_{f_1} \cdots \phi_{f_n}|0> \qquad (12.95)$$

with

$$a^*(f) = \int \mathrm{d}\Omega_m(\boldsymbol{p})\tilde{f}(\boldsymbol{p})a^*(\boldsymbol{p}). \qquad (12.96)$$

12.3.3   Existence of Asymptotic States

In Chapter 5 the space of asymptotic states was introduced on physical grounds. It is possible to propose a general axiomatic framework for the relativistic quantum field theories incorporating this description and leading to a certain number of results such as the spin-statistics theorem, the PCT theorem, or the existence of asymptotic states that these theories must satisfy. The advantage of this approach is that it is independent of the specific nature of the theories under consideration. We will use this framework to prove the existence of asymptotic states.

Let us introduce the notion of *regular solution* of the Klein–Gordon equation. f is such a solution: $(\Box + m^2)f(x) = 0$, if it is of the form

$$f(x) = \frac{1}{(2\pi)^3} \int_{R^4} \delta(p^2 - m^2)\theta(p_0) \{e^{-ip.x}h_+(p) + e^{ip.x}h_-(p)\} \, d^4p$$

$$= \int_{R^3} \{e^{-ip.x}f^+(\boldsymbol{p}) + e^{ip.x}f^-(\boldsymbol{p})\} \, d\Omega_m(\boldsymbol{p}), \qquad (12.97)$$

where $h_\pm$ are $\mathbb{C}^\infty$ functions with compact support.

Using formula (12.91), this notion of regular solution makes it possible to define, from any field $\phi(x)$, a regularised field $\phi_f(t)$.

We check easily that if ϕ is a free field, ϕ_f does not depend on time.

To construct the asymptotic fields, we need an auxiliary function $h(p)$ whose support contains a neighbourhood of the mass hyperboloid $p^2 = m^2$ and which does not intersect $\{p = 0\} \cup V_+^{2m}$. Moreover, we suppose that $h(p) = 1$ for $p \in \{p | p^2 = m^2\}$.

Finally, writing $\phi(x) = U(x)\phi(0)U^{-1}(x)$, we **suppose** that $\phi(0)|\Omega >= c_1|\psi_1 > +c|\psi >$, where $|\psi_1 >\in \mathcal{H}_1$ and $|\psi >$ belongs to the continuum. c and c_1 are two complex numbers to ensure the normalisation of the state and we assume that $c_1 \neq 0$. This hypothesis guarantees us that the field ϕ has a non-vanishing probability to create a one-particle state when applied to the vacuum state.

Starting from the field ϕ, we define the auxiliary field

$$B(x) = \phi * \tilde{h}(x) = \frac{1}{(2\pi)^4} \int e^{ip.x}\tilde{\phi}(p)h(p) \, d^4p \qquad (12.98)$$

and its regularisation $B_f(t)$. We check easily that $B(x)|\Omega >$ is a solution of the Klein–Gordon equation. Indeed

$$B(x)|\Omega >= \int \tilde{h}(x-y)\phi(y)|\Omega > \, d^4y = \int \tilde{h}(x-y)e^{ik.y}dE(k)\phi(0)|\Omega > \, d^4y$$

$$= \int e^{ip.y}h(p)dE(p)\phi(0)|\Omega >= \int e^{ip.y}h(p)dE(p)\psi_1|\Omega >, \qquad (12.99)$$

the last equality coming from the fact that h restricts the spectral measure $E(p)$ to the hyperboloid $p^2 = m^2$, therefore to the projector over the one-particle state.

Thus under the preceding hypothesis, we have the following.

Theorem 17 (Existence of asymptotic fields). *Let $f, f_1, \cdots, f_n$ be regular solutions of the Klein–Gordon equation. We first define the 'out' fields.*

The vectors

$$|\Psi(t) >= B_{f_1}(t)B_{f_2}(t) \cdots B_{f_n}(t)|\Omega > \qquad (12.100)$$

form a Cauchy sequence; i.e. given ε, *there exists* $T = T(\varepsilon)$ *such that for* $t_1, t_2 \geq T$

$$||\,|\Psi(t_1) > -|\Psi(t_2) > ||\, \leq \varepsilon. \tag{12.101}$$

If we note that

$$|\Psi^{\text{out}}(f_1, f_2, \ldots, f_n) >= \lim_{t \to \infty} B_{f_1}(t) B_{f_2}(t) \cdots B_{f_n}(t)|\Omega > \tag{12.102}$$

then by linear extension

$$\phi_f^{\text{out}}|\Psi^{\text{out}}(f_1, f_2, \ldots, f_n) >= |\Psi^{\text{out}}(f, f_1, f_2, \ldots, f_n) > \tag{12.103}$$

defines a real free scalar field ϕ^{out} *such that*

1.

$$\phi_f^{\text{out}} = \mathrm{i} \int_{x_0 = t} \mathrm{d}^3 x \, [f \overleftrightarrow{\partial_0} \phi^{\text{out}}] \tag{12.104}$$

2.

$$U(a, \Lambda) \phi^{\text{out}}(x) U^{-1}(a, \Lambda) = \phi^{\text{out}}(\Lambda x + a) \tag{12.105}$$

By

$$|\Psi^{\text{in}}(f_1, f_2, \ldots, f_n) >= \lim_{t \to -\infty} B_{f_1}(t) B_{f_2}(t) \cdots B_{f_n}(t)|\Omega > \tag{12.106}$$

We define the 'in' fields with properties identical to those of the 'out' fields. Using the properties of Θ, *the PCT operator, we thus show that*

$$\Theta^{-1} \phi^{\text{in}}(x) \, \Theta = \phi^{\text{out}}(-x) \tag{12.107}$$

We give the main steps leading to the proof of the theorem, writing

$$|\Psi(t_1) > -|\Psi(t_2) >= \int_{t_2}^{t_1} \frac{\mathrm{d}}{\mathrm{d}t} |\Psi(t) > \mathrm{d}t. \tag{12.108}$$

It is enough to prove that for $|t|$ large enough, there exists a constant C such that

$$||\frac{\mathrm{d}}{\mathrm{d}t} |\Psi(t) > ||^2 \leq C|t|^{-\frac{3}{2}}. \tag{12.109}$$

This bound is proved by splitting the vacuum expectation value of the left-hand side of (12.109) into a sum of product of truncated vacuum expectation values and using the following intermediate results.

1. Properties of the regular solutions of the Klein–Gordon equation
   Let $f(x) = f(t, \boldsymbol{x})$ be such a solution. Thus, for fixed t, it is an element of $\mathcal{S}(\mathbf{R}^3)$ and there exists a constant K such that for $n = 0, \frac{1}{2}, 1, \frac{3}{2}, \ldots$

$$|f(t, \boldsymbol{x})| < \frac{K}{(1 + |\boldsymbol{x}|^2 + |t|^2)^{\frac{3}{4}}} \left(\frac{1 + |t|}{1 + |\boldsymbol{x}|} \right)^n. \tag{12.110}$$

2. Properties of B_f
   The vector $B_f(t)|\Omega >$ is independent of t. Therefore,

$$(\Omega, B_f \frac{\mathrm{d}}{\mathrm{d}t} B_f \Omega) = (\frac{\mathrm{d}}{\mathrm{d}t} B_f^* \Omega, B_f \Omega) = 0. \tag{12.111}$$

3. Decrease of the truncated vacuum expectation values
   If $f_1, \ldots, f_n$ are the regular solutions of the Klein–Gordon equation, then

$$|(\Omega, B_{f_1} \cdots B_{f_n} \Omega)^T| |t|^{\frac{3(n-1)}{2}} \tag{12.112}$$

is bounded in t.

We will need later on a natural extension of this theorem. Let D^{in} be a dense subset of $\mathcal{H}^{in}$ on which will be defined the asymptotic 'in' fields.
Then for each $|\phi^{in} >\in D^{in}$

$$\lim_{t \to -\infty} B_{f_1}(t) B_{f_2}(t) \cdots B_{f_n}(t)|\phi^{in} >= B_{f_1}^{in} B_{f_2}^{in} \cdots B_{f_n}^{in} |\phi^{in} > \tag{12.113}$$

Obviously, we have a similar result for the 'out' case.

Remark: In the statement of the theorem and its extension, nothing was said on the nature of the convergence. The precise result is that if the test functions $f, f_1, \ldots, f_n$ are any of the regular solutions of the Klein–Gordon equation, then the convergence is a weak convergence. If, on the other hand, the supports of these functions do not overlap, that is to say if their supports are such that $\frac{\boldsymbol{p}}{\omega(p)} \neq \frac{\boldsymbol{p}_i}{\omega(p_i)} \neq \frac{\boldsymbol{p}_j}{\omega(p_j)}$, for $p \in \text{supp } f$, $p_i \in \text{supp } f_i$ and $p_j \in \text{supp } f_j$, $i \neq j$, then the convergence is a strong convergence.

These results show, in particular, that for positive energy solutions of the Klein–Gordon equation $\{f_i\}$, $i = 1, \ldots, n$, we have

$$|\Psi^{ex}(f_1, f_2, \ldots, f_n) >= |f_1, f_2, \ldots, f_n; ex >, \tag{12.114}$$

where '*ex*' stands for 'in' or 'out', the '*ex*' on the right-hand side stating that the state under consideration is in the Fock space $\mathcal{H}^{ex}$.

We saw in Chapter 5 on the physical states that it was natural to suppose that

$$\mathcal{H}^{\text{in}} = \mathcal{H}^{\text{out}} = \mathcal{H}. \tag{12.115}$$

In fact, from the Wightman axioms and from the previous theorem, it is enough to add

- Axiom 5': The completeness of the 'in' states

$$\mathcal{H}^{\text{in}} = \mathcal{H} \tag{12.116}$$

in order to prove the existence of an S-matrix. Indeed, from (12.107) we deduce that

$$\Theta \mathcal{H}^{\text{in}} = \mathcal{H}^{\text{out}} \tag{12.117}$$

and since $\Theta \mathcal{H} = \mathcal{H}$, (12.115) follows. Therefore, ϕ^{out} and ϕ^{in} act in the same Hilbert space and are unitarily equivalent. This leads to the existence of a unique unitary operator S such that

$$S\phi^{\text{out}}(x) = \phi^{\text{in}}(x)S. \tag{12.118}$$

Moreover, S satisfies

$$\begin{aligned} S|\Omega > &= |\Omega > \\ U(a, \Lambda)SU^{-1}(a, \Lambda) &= S \\ \Theta^{-1}S\Theta &= S. \end{aligned} \tag{12.119}$$

This result shows that a set of relatively natural hypotheses makes it possible to prove the existence of the diffusion operator. It is remarkable to get this result without being obliged to provide the details of the precise nature of the interaction. We will show now how to express the matrix elements of this operator in terms of basic quantities such as the vacuum expectation values of time-ordered products of fields.

12.4 The Reduction Formulae

We will show in this section how the elements of the S-matrix can be expressed as functions of the vacuum expectation values of T-products.

Let $f_1, \ldots, f_n$ and $g_1, \ldots, g_m$ be regular solutions of the Klein–Gordon equation corresponding to *positive frequencies* and let us consider the expression

$$\begin{aligned} I &= <f_1, \ldots, f_n; \text{in}|S|g_1, \ldots, g_m; \text{in} > = <f_1, \ldots, f_n; \text{out}|g_1, \ldots, g_m; \text{in} > \\ &= (\Phi^{\text{out}}(f_1, \ldots, f_n), \Phi^{\text{in}}(g_1, \ldots, g_m)), \end{aligned} \tag{12.120}$$

where we have introduced the usual notation of the scalar product in a Hilbert space and replaced the vectors $|g_1, \ldots, g_m; \text{in} >$ and $|f_1, \ldots, f_n; \text{out} >$ by the corresponding functions $\Phi^{\text{in}}(g_1, \ldots, g_m)$ and $\Phi^{\text{out}}(f_1, \ldots, f_n)$.

The derivation proceeds in steps. As a first step, let us choose one of the incoming particles, say that corresponding to g_1. According to (12.103) applied to the 'in' states

$$I = (\Phi^{\text{out}}(f_1,\ldots,f_n),\ \phi_{g_1}^{\text{in}}\Phi^{\text{in}}(g_2,\ldots,g_m))$$

$$= -i \lim_{t\to-\infty} \int_{x_0=t} d^3x (\Phi^{\text{out}}(f_1,\ldots,f_n),\ g_1\overleftrightarrow{\partial_0}B\ \Phi^{\text{in}}(g_2,\ldots,g_m))$$

$$= -i \lim_{t\to\infty} \int_{x_0=t} d^3x (\Phi^{\text{out}}(f_1,\ldots,f_n),\ g_1\overleftrightarrow{\partial_0}B\ \Phi^{\text{in}}(g_2,\ldots,g_m)) \qquad (12.121)$$

$$+ i \lim_{t_f\to\infty,\ t_i\to-\infty} \int_{t_i}^{t_f} dx_0 \int d^3x\,\partial_0 (\Phi^{\text{out}}(f_1,\ldots,f_n),\ g_1\overleftrightarrow{\partial_0}B\ \Phi^{\text{in}}(g_2,\ldots,g_m)).$$

Let us study separately the two terms of the last equality.
The first one shows an annihilator acting on the 'out' vector:

$$\lim_{t\to\infty} \int_{x_0=t} d^3x (\Phi^{\text{out}}(f_1,\ldots,f_n),\ g_1\overleftrightarrow{\partial_0}B\ \Phi^{\text{in}}(g_2,\ldots,g_m))$$

$$= (\Phi^{\text{out}}(f_1,\ldots,f_n),\ \phi_{g_1}^{\text{out}}\Phi^{\text{in}}(g_2,\ldots,g_m)). \qquad (12.122)$$

Indeed since g_1 is a solution with positive energy, $\phi_{g_1}^{\text{out}}$ is a creation operator $ia^*(g_1)$ and the expression that we study becomes, modulo coefficients,

$$(a(\bar{g}_1)\Phi^{\text{out}}(f_1,\ldots,f_n),\ \Phi^{\text{in}}(g_2,\ldots,g_m))$$

$$= \sum_{i=1}^{n} (g_1,f_i)(\Phi^{\text{out}}(f_1,\ldots,\widehat{f_i},f_n),\ \Phi^{\text{in}}(g_2,\ldots,g_m)), \qquad (12.123)$$

where $\widehat{f_i}$ means that f_i is left out and (g_1,f_i) is the scalar product of g_1 and of f_i in $\mathcal{H}_1$. Let us consider one of the terms of the sum in the formula (12.123). It can be interpreted by saying that the interaction process factorises into a coupling between the ingoing particle of index i and the outgoing particle of index 1 times the process having one initial particle less and one final particle less.

Due to this factorisation, we say that this is a *non-connected* process and we will see later that, diagrammatically, it corresponds to a disconnected figure. Physically it corresponds to a process in which an outgoing particle is in exactly the same momentum state as one among the incoming ones, so it is as if this particle has not participated in the scattering process.

The second term can be written as

$$i \int d^4x_1\,\partial_0 (\Phi^{\text{out}}(f_1,\ldots,f_n),\ (g_1\overleftrightarrow{\partial_0}B)(x_1)\ \Phi^{\text{in}}(g_2,\ldots,g_m)) \qquad (12.124)$$

$$= i \int d^4x_1 (\Phi^{\text{out}}(f_1,\ldots,f_n),\ (g_1\partial_0^2 B - B\partial_0^2 g_1)(x_1)\Phi^{\text{in}}(g_2,\ldots,g_m))$$

$$= i \int d^4x_1 (\Phi^{\text{out}}(f_1,\ldots,f_n),\ (g_1\partial_0^2 B - B(-\triangle + m^2)g_1)(x_1)\Phi^{\text{in}}(g_2,\ldots,g_m))$$

$$= \mathrm{i} \int \mathrm{d}^4 x_1 (\Phi^{\mathrm{out}}(f_1, \ldots, f_n), (g_1(\partial_0^2 - \triangle + m^2) B(x_1)) \Phi^{\mathrm{in}}(g_2, \ldots, g_m))$$

$$= \mathrm{i} \int \mathrm{d}^4 x_1 g_1(x_1)(\square_{x_1} + m^2)(\Phi^{\mathrm{out}}(f_1, \ldots, f_n), B(x_1) \Phi^{\mathrm{in}}(g_2, \ldots, g_m)),$$

where we used the fact that g_1 is a test function at the point x_1 to carry the action of the Laplacian from g_1 to B. This result deserves a comment: let us introduce the Fourier transform of g_1 and of the field B

$$\int \mathrm{d}^4 x_1 g_1(x_1)(\square_{x_1} + m^2)(\Phi^{\mathrm{out}}(f_1, \ldots, f_n), B(x_1) \Phi^{\mathrm{in}}(g_2, \ldots, g_m))$$

$$= \frac{1}{(2\pi)^4} \int \mathrm{e}^{-\mathrm{i} p_1 \cdot x_1} \tilde{g}_1(p_1) \mathrm{e}^{\mathrm{i} k \cdot x_1} (-k^2 + m^2)$$

$$\times (\Phi^{\mathrm{out}}(f_1, \ldots, f_n), \tilde{B}(k) \Phi^{\mathrm{in}}(g_2, \ldots, g_m)) \, \mathrm{d}^4 k \, \mathrm{d}\Omega(p_1) \, \mathrm{d}^4 x_1$$

$$= \frac{1}{(2\pi)} \int \mathrm{e}^{\mathrm{i}(k_0 - \omega(p_1))t_1} \tilde{g}_1(p_1)(-k_0^2 + p_1^2 + m^2)$$

$$\times (\Phi^{\mathrm{out}}(f_1, \ldots, f_n), \tilde{B}((k_0, p_1)) \Phi^{\mathrm{in}}(g_2, \ldots, g_m)) \, \mathrm{d}k_0 \, \mathrm{d}\Omega(p_1) \, \mathrm{d}t_1$$

$$= \frac{1}{(2\pi)} \int \delta(k_0 - \omega(p_1)) \tilde{g}_1(p_1)(-p_1^2 + m^2)$$

$$\times (\Phi^{\mathrm{out}}(f_1, \ldots, f_n), \tilde{B}((k_0, p_1)) \Phi^{\mathrm{in}}(g_2, \ldots, g_m)) \, \mathrm{d}k_0 \, \mathrm{d}\Omega(p_1)$$

$$= \frac{1}{(2\pi)^4} \int \theta((p_1)_0) \delta(p_1^2 - m^2) \tilde{g}_1(p_1)(-p_1^2 + m^2)$$

$$\times (\Phi^{\mathrm{out}}(f_1, \ldots, f_n), \tilde{B}(p_1) \Phi^{\mathrm{in}}(g_2, \ldots, g_m)) \, \mathrm{d}^4 p_1.$$

This last equality, to be meaningful, implies that

1. $(\Phi^{\mathrm{out}}(f_1, \ldots, f_n), \tilde{B}(p_1) \Phi^{\mathrm{in}}(g_2, \ldots, g_m))$; has a simple pole at $p_1^2 = m^2$;
2. We multiply it by $p_1^2 - m^2$ (operation which is called *amputation*);[8] and
3. The delta function which restricts p_1 to be on the mass shell tells us that what contributes to the diffusion is the residue of this pole and that this residue, as a function of p_1 must be evaluated at $(p_1)_0 = \omega(p_1)$. Since $\tilde{B}(p_1) = \tilde{\phi}(p_1) h(p_1)$, we see that the restriction to the mass shell implies that the contribution of this field is independent of the function h and we can replace, in the final expression, the regularised field B by ϕ.

---

[8] The existence of the pole at $p_1^2 = m^2$ is not surprising. Going back to the Feynman rules we obtained previously, we see that at least in perturbation theory, there exists a propagator $(p_i^2 - m^2)^{-1}$ associated with every one of the fields. On the other hand, we see that if, for some reason, this singularity is absent, the corresponding S-matrix element vanishes.

The result of this first step is summarised in

$$I = \text{non–connected terms} \tag{12.125}$$

$$+ i \int d^4x_1 g_1(x_1)(\Box_{x_1} + m^2)(\Phi^{\text{out}}(f_1, \ldots, f_n), \phi(x_1)\Phi^{\text{in}}(g_2, \ldots, g_m)).$$

Apart from the non-connected term, I is given by the matrix element of the field $\phi(x_1)$ between in and out states, but the in state has one particle less.

It is now easy to guess the next steps. For example, as a second step, let us choose to transform in the integral of the right-hand side of (12.125), one of the out fields into an interacting field:

$$\int d^4x_1 g_1(x_1)(\Box_{x_1} + m^2)(\Phi^{\text{out}}(f_1, \ldots, f_n), B(x_1)\Phi^{\text{in}}(g_2, \ldots, g_m)) \tag{12.126}$$

$$= \lim_{t \to \infty} \int d^4x_1 g_1(x_1)(\Box_{x_1} + m^2)(\Phi^{\text{out}}(f_2, \ldots, f_n), (B_{f_1}(t))^* B(x_1)\Phi^{\text{in}}(g_2, \ldots, g_m))$$

$$= i \lim_{t \to \infty} \int d^4x_1 g_1(x_1)$$

$$\times (\Box_{x_1} + m^2) \int_{y_1^0 = t} d^3y_1 \bar{f}_1(y_1) \overleftrightarrow{\partial}_{y_1^0} (\Phi^{\text{out}}(f_2, \ldots, f_n), B^*(y_1)B(x_1)\Phi^{\text{in}}(g_2, \ldots, g_m))$$

$$= i \lim_{t \to \infty} \int d^4x_1 g_1(x_1)(\Box_{x_1} + m^2) \int_{y_1^0 = t} d^3y_1$$

$$\times \bar{f}_1(y_1) \overleftrightarrow{\partial}_{y_1^0} (\Phi^{\text{out}}(f_2, \ldots, f_n), T(B^*(y_1)B(x_1))\Phi^{\text{in}}(g_2, \ldots, g_m)).$$

The introduction of the T-product in the last equality is justified by the fact that the quantity is computed at $y_1^0 = +\infty$. We can thus replace the last term of (12.127) to make the appearance of an 'in' field and we get

$$\lim_{t \to \infty} \int_{y_1^0 = t} d^3y_1 \bar{f}_1(y_1)(\Phi^{\text{out}}(f_2, \ldots, f_n), \overleftrightarrow{\partial}_{y_1^0} T(B^*(y_1)B(x_1))\Phi^{\text{in}}(g_2, \ldots, g_m))$$

$$= \lim_{t \to -\infty} \int_{y_1^0 = t} d^3y_1 \bar{f}_1(y_1)(\Phi^{\text{out}}(f_2, \ldots, f_n), \overleftrightarrow{\partial}_{y_1^0} T(B^*(y_1)B(x_1))\Phi^{\text{in}}(g_2, \ldots, g_m))$$

$$+ \int d^4y_1 \bar{f}_1(y_1) \partial_{y_1^0}(\Phi^{\text{out}}(f_2, \ldots, f_n), \overleftrightarrow{\partial}_{y_1^0} T(B^*(y_1)B(x_1))\Phi^{\text{in}}(g_2, \ldots, g_m)).$$

The first term can be written as

$$\lim_{t \to -\infty} (\Phi^{\text{out}}(f_2, \ldots, f_n), T((B_{f_1}(t))^* B(x_1))\Phi^{\text{in}}(g_2, \ldots, g_m))$$

$$= (\Phi^{\text{out}}(f_2, \ldots, f_n), B(x_1)(B_{f_1}^{\text{in}})^* \Phi^{\text{in}}(g_2, \ldots, g_m))$$

$$= \sum_{j=2}^{m} (f_1, g_j)(\Phi^{\text{out}}(f_2, \ldots, f_n), B(x_1)\Phi^{\text{in}}(g_2, \ldots, \widehat{g_j}, \ldots, g_m)) \tag{12.127}$$

and corresponds to a sum of non-connected contributions.

The second term can be handled as in the first step and

$$\int d^4 y_1 \bar{f}_1(y_1) \partial_{y_1^0} (\Phi^{\text{out}}(f_2, \dots, f_n), \overleftrightarrow{\partial}_{y_1^0} T(B^*(y_1)B(x_1))\Phi^{\text{in}}(g_2, \dots, g_m)) \qquad (12.128)$$

$$= \int d^4 y_1 \bar{f}_1(y_1)(\Box_{y_1} + m^2)(\Phi^{\text{out}}(f_2, \dots, f_n), T(B^*(y_1)B(x_1))\Phi^{\text{in}}(g_2, \dots, g_m))$$

$$= \int d^4 y_1 \bar{f}_1(y_1)(\Box_{y_1} + m^2)(\Phi^{\text{out}}(f_2, \dots, f_n), T(\phi(y_1)B(x_1))\Phi^{\text{in}}(g_2, \dots, g_m)),$$

where we used the property that on the mass shell the fields B and the fields ϕ coincide.
Finally, after these two steps, we have shown that

$$I \quad = \quad \text{non- connected terms} \qquad (12.129)$$
$$+ i^2 \int d^4 x_1 \, d^4 y_1 g_1(x_1) \bar{f}_1(y_1)(\Box_{x_1} + m^2)(\Box_{y_1} + m^2)$$
$$\times (\Phi^{\text{out}}(f_2, \dots, f_n), \, T(\phi(x_1)\phi(y_1))\Phi^{\text{in}}(g_2, \dots, g_m))$$

and we show in this way the general result for the connected part of the S-matrix element under consideration,

$$< f_1, \dots, f_n; \text{in}|S|g_1, \dots, g_m; \text{in} >^c$$
$$= i^{n+m} \int d^4 y_1 \cdots d^4 y_n \cdots d^4 x_m \bar{f}_1(y_1) \cdots g_1(x_1) \cdots g_m(x_m)$$
$$\times (\Box_{y_1} + m^2) \cdots (\Box_{x_m} + m^2) < T(\phi(y_1) \cdots \phi(y_n) \cdots \phi(x_m)) >, \qquad (12.130)$$

where $< T(\cdots) >$ represents the vacuum expectation value of the T-product.
For the momentum states, the formula can be written as

$$< p_1, \dots, p_n; \text{in}|S|q_1, \dots, q_m; \text{in} >^c$$
$$= i^{n+m} \int d^4 y_1 \cdots d^4 y_n \cdots d^4 x_m e^{i \breve{p}_1 \cdot y_1 \cdots + i \breve{p}_n \cdot y_n - i \breve{q}_1 \cdot x_1 - \cdots - i \breve{q}_m \cdot x_m}$$
$$\times (\Box_{y_1} + m^2) \cdots (\Box_{x_m} + m^2) < T(\phi(y_1) \cdots \phi(y_n) \cdots \phi(x_m)) >, \qquad (12.131)$$

where $\breve{p}$ and $\breve{q}$ mean that the four-vectors p and q are on the positive energy branch of the mass hyperboloid. This formula expresses the fact that the S-matrix element can be obtained by amputating the fields corresponding to the asymptotic fields introduced in the T-product and then replacing them by the plane-wave solutions of the corresponding Klein–Gordon equation.

In analogy with the Euclidean approach, we formally write that

$$< T(\phi(x_1) \cdots \phi(x_n)) > = \frac{\int \phi(x_1) \cdots \phi(x_n) \, e^{-i \int \mathcal{L}_{\text{int}}(x) \, d^4 x} \mathcal{D}\mu(\phi)}{\int e^{-i \int \mathcal{L}_{\text{int}}(x) \, d^4 x} \mathcal{D}\mu(\phi)} \qquad (12.132)$$

where $\mathcal{D}\mu(\phi)$ is a formal Gaussian measure, that is to say that

$$\int \phi(x)\phi(y)\mathcal{D}\mu(\phi) = < 0|T(\phi(x)\phi(y))|0 > = -iG_F(x-y) \qquad (12.133)$$

and

$$< 0|T(\phi(x_1)\cdots\phi(x_{2n}))|0 > \qquad\qquad\qquad\qquad (12.134)$$

$$= \int \phi(x_1)\cdots\phi(x_{2n})\mathcal{D}\mu(\phi)$$

$$= \sum_{\text{pairs}} < 0|T(\phi(x_{i_1})\phi(x_{i_2}))|0 > \cdots < 0|T(\phi(x_{i_{2n-1}})\phi(x_{i_{2n}}))|0 >$$

and

$$< 0|T(\phi(x_1)\cdots\phi(x_{2n+1}))|0 > = \int \phi(x_1)\cdots\phi(x_{2n+1})\mathcal{D}\mu(\phi) = 0, \qquad (12.135)$$

the formula (12.132) being interpreted as

$$< T(\phi(x_1)\cdots\phi(x_n)) >$$
$$= \frac{\sum_{n=0}^{+\infty} \frac{1}{n!} \int \phi(x_1)\cdots\phi(x_n)\,(-i\int \mathcal{L}_{\text{int}}(x)\,\mathrm{d}^4x)^n \mathcal{D}\mu(\phi)}{\sum_{k=0}^{+\infty} \frac{1}{k!} \int (-i\int \mathcal{L}_{\text{int}}(x)\,\mathrm{d}^4x)^k \mathcal{D}\mu(\phi)}. \qquad (12.136)$$

This last expression tells us that the vacuum expectation values of T-products of interacting fields can be written as the ratio of two infinite series whose elements can be written as Gaussian integrals of free-field polynomials. These Gaussian integrals are sums of products of Feynman propagators G_F. Each of these products forms what we call a Feynman diagram. We will study them in the next section.

We can also introduce a generating functional

$$Z[j] = < T(\mathrm{e}^{i\int \phi(x)j(x)\mathrm{d}^4x}) > = \frac{\int \mathrm{e}^{i\int \phi(x)j(x)\mathrm{d}^4x}\,\mathrm{e}^{-i\int \mathcal{L}_{\text{int}}(x)\,\mathrm{d}^4x}\mathcal{D}\mu(\phi)}{\int \mathrm{e}^{-i\int \mathcal{L}_{\text{int}}(x)\,\mathrm{d}^4x}\mathcal{D}\mu(\phi)} \qquad (12.137)$$

with

$$< T(\phi(x_1)\cdots\phi(x_n)) > = \frac{1}{i^n} \frac{\delta}{\delta j(x_1)} \cdots \frac{\delta}{\delta j(x_n)} Z[j]|_{j\equiv 0}. \qquad (12.138)$$

The connected processes, that is to say the processes which do not reduce to independent subprocesses, play a very important role in diffusion theory. They are given by formulae analogous to the truncated expectation values defined previously. Truncation is the algebraic way to extract connectivity from a given vacuum expression. The generating functional of the connected or truncated vacuum expectation values of T-products is given by

$$G[j] = \log Z[j]. \tag{12.139}$$

Sketch of proof: let us compute the functional derivative of $Z[j]$ with respect to $j(x)$:

$$\frac{\delta}{\delta j(x)} Z[j] = < T(\phi(x)e^{\int \phi(x)j(x)\mathrm{d}^4 x} > . \tag{12.140}$$

If we expand, formally, the right-hand side of this formula in powers of the function j, we get, after integration over the free measure, a sum of products of diagrams, some of them being connected to the point x. These diagrams are made of propagators $-iG_F$ having, except x, all their end points integrated over a function j or integrated as part of a vertex of the Lagrangian $\mathcal{L}_{int}$. Let us choose one of the terms in the sum. It is the product of two parts: a part (a diagram) connected to the point x and another part made of diagrams having no connected links to x. It is obvious that the coefficient of the connected term is made out of all diagrams, connected or not, generated by the infinite expansion in powers of j of the functional $Z(j)$.

Therefore, $\frac{\delta Z[j]}{\delta j(x)} = (\sum[\text{All diagrams connected to the point } x])Z(j)$. Now if we call $G[j]$ the generating functional of the connected diagrams, $\frac{\delta}{\delta j(x)} G[j]$ is precisely equal to $\sum[\text{All diagrams connected to the point x}]$. Therefore,

$$\frac{\delta}{\delta j(x)} Z[j] = \frac{\delta G[j]}{\delta j(x)} Z[j], \tag{12.141}$$

leading to

$$\frac{1}{Z[j]} \frac{\delta Z[j]}{\delta j(x)} = \frac{\delta \log Z[j]}{\delta j(x)} = \frac{\delta G[j]}{\delta j(x)}, \tag{12.142}$$

which gives by integration formula (12.138).

12.4.1 The Feynman Diagrams

If we leave aside very few exceptions, the results, which can be numerically exploitable, are obtained by perturbative methods. In particular, this is the case for the computation of cross sections. These methods allow us to give a concrete content to the diagrammatic analysis of particle interactions.

Schematically, a collision (more generally an interaction) process can be represented by a black box, the interaction area, which is drawn as a ball, where the particles which will interact arrive and from where the particles resulting from the interaction leave (see Fig.12.2a).

The perturbative expansions can be described in graphical terms as diagrams, the *Feynman diagrams* from which the associated expressions represent a possible contribution to the process (see Fig.12.2b). More precisely, we will see that to each diagram is associated a mathematical expression contributing to the probability amplitude of the process.

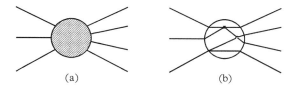

(a)   (b)   **Figure 12.2** *A diffusion diagram.*

A diagram is made of *internal lines*, the end points of which are vertices and of *external lines* for which only one of the end points is a vertex (with the exception of single propagators). A *vertex* is the intersection point of lines, it is generally the result of a contribution of the interaction Lagrangian and there corresponds to it an integration over the entire position space. It is usual to split the set of external lines into two subsets: the subset of *ingoing lines*, which are usually drawn at the left, and the subset of *outgoing lines*, which are usually drawn at the right of the figures. The whole process is then interpreted as a physical process in which the elements (particles) associated with the ingoing lines interact and produce the elements associated with the outgoing lines.

Since generally the diagrams are computed in Fourier transforms, we indicate the ingoing and outgoing lines and the internal lines with arrows on the lines. In general, an arrow on a line denotes the flow of some conserved quantity along this line. In the neutral scalar theory we are considering here, energy–momentum is the only conserved quantity, but we shall soon study more general theories with other conserved quantities, such as the electric charge.

The orientation from left to right corresponds to the flow of time direction. Each line corresponds to a propagator and is characterised by a four-momentum and indices associated with the nature of the particles which propagate (spinor indices, internal symmetry indices, ...). For an ingoing particle with momentum p with an orientation from left to right, p_0 is the energy, thus a positive number, and the exponential factor in the Fourier transform is chosen to correspond to a positive energy plane wave. An outgoing particle oriented from left to right will have a four-momentum p with $p_0 \geq 0$ or if it is oriented from right to left with $p_0 \leq 0$. Figure 12.3a shows the result of such a procedure. Figure 12.3b makes precise the content of the previous statement and corresponds, if the time components of the p_i are positive, to the reaction

$$1 + 2 + 3 \rightarrow 4 + 5 + 6 + 7, \tag{12.143}$$

where i indicates the particle with momentum p_i. It makes explicit a possible, but not unique, choice for the orientation of the internal lines.

Remark that with the above choice of orientation of the external lines, if for example p_2^0, p_3^0, p_4^0, p_5^0, and p_7^0 were positive and p_1^0 and p_6^0 negative, then the diagram would have been associated with the reaction

$$2 + 3 + 6 \rightarrow 1 + 4 + 5 + 7. \tag{12.144}$$

The choice of orientation for the internal lines is arbitrary; the only constraints must do with the four-momentum which, as we will see, are submitted to the condition

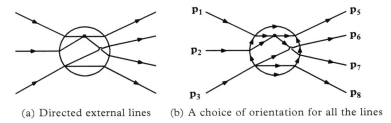

(a) Directed external lines   (b) A choice of orientation for all the lines

Figure 12.3 *Choices of orientations. (a) Direct external lines. (b) A choice of orientation for all the lines.*

of conservation of energy–momentum at each vertex[9] (the algebraic sum of the four-momentum vanishes).

In Minkowski space, the Feynman diagrams correspond to physical processes.

In Chapter 10 concerning the Euclidean space formulation, we have, in the framework of an interacting Lagrangian field theory, introduced the perturbative expansion of the vacuum expectation values. This expansion was written diagrammatically, each diagram appearing with a certain combinatoric weight and corresponding to a precise analytic expression obtained by rules assigning particular functions to each line or to each vertex.

The correspondences between Euclidean and relativistic theories extend obviously to perturbative elements and makes it possible to associate with the graphical diagrammatic expressions, analytical expressions in the Minkowski space with the same weight. From the important role played by the time or energy component, it is necessary to emphasise the orientation of the lines.

The correspondence between Euclidean diagrams and Minkowski space diagrams is given in the bosonic case by the following table, which completes the one given in Chapter 10.

Graphical element	Name	Euclidean factor	Relativistic factor
■———————■	Internal line	$\frac{1}{(2\pi)^4}\frac{1}{p^2+m^2}\mathrm{d}^4p$	$\frac{1}{(2\pi)^4}\frac{i}{p^2-m^2+i\varepsilon}\mathrm{d}^4p$
———▶———■	Ingoing line	$\frac{1}{(2\pi)^4}\frac{e^{ip.x}}{p^2+m^2}\mathrm{d}^4p$	$\frac{1}{(2\pi)^4}\frac{ie^{-ip.x}}{p^2-m^2+i\varepsilon}\mathrm{d}^4p$
■———▶———	Outgoing line	$\frac{1}{(2\pi)^4}\frac{e^{-ip.x}}{p^2+m^2}\mathrm{d}^4p$	$\frac{1}{(2\pi)^4}\frac{ie^{ipx}}{p^2-m^2+i\varepsilon}\mathrm{d}^4p$
or	Vertex	$\lambda(2\pi)^4\delta^{(4)}\left(\sum p_j\right)$	$-i\lambda(2\pi)^4\delta^{(4)}\left(\sum p_j\right)$

[9] This condition of conservation is due to the fact that the interaction is a local function of the fields and that the volume of integration is the whole space.

Let us illustrate these rules in the case of the theory with interaction $-\lambda\phi^4$. We saw, in Chapter 10, the form of the Euclidean perturbative expression at second order in λ of the 2-point function $< \phi(x)\phi(y) >$

$$B_2 = 96\lambda^2 \int \int C(x - x_1) C(x_1 - x_2)^3 C(x_2 - y) \, d^4x_1 \, d^4x_2. \tag{12.145}$$

The corresponding relativistic expression for $< T(\phi(x)\phi(y)) >$ is given by

$$\bar{B}_2 = i96\lambda^2 \int \int G_F(x - x_1) G_F(x_1 - x_2)^3 G_F(x_2 - y) \, d^4x_1 \, d^4x_2,$$

where

$$-iG_F(x - y) = < 0|T(\phi(x)\phi(y))|0 > = \frac{i}{(2\pi)^4} \int \frac{e^{-ip.(x-y)}}{p^2 - m^2 + i\varepsilon} d^4p. \tag{12.146}$$

Carrying the expression of G_F into $\bar{B}_2$, we get

$$\bar{B}_2 = -i96\lambda^2 \frac{1}{(2\pi)^{12}} \int \cdots \int e^{-ip.x} e^{iq.y} \delta^{(4)}\left(p - \sum_{i=1}^{3} p_i\right) \delta^{(4)}\left(q - \sum_{i=1}^{3} p_i\right)$$

$$\times \frac{1}{p^2 - m^2 + i\varepsilon} \frac{1}{q^2 - m^2 + i\varepsilon} \prod_{i=1}^{3} \frac{1}{p_i^2 - m^2 + i\varepsilon} \, d^4p \, d^4q \, d^4p_1 .. d^4p_3$$

$$= -i96\lambda^2 \frac{1}{(2\pi)^{12}} \int e^{-ip.(x-y)} \delta^{(4)}(p - q) \left(\frac{1}{p^2 - m^2 + i\varepsilon}\right)^2$$

$$\times \int \int \int \delta^{(4)}\left(p - \sum_{i=1}^{3} p_i\right) \prod_{i=1}^{3} \frac{1}{p_i^2 - m^2 + i\varepsilon} \, d^4p \, d^4q \, d^4p_1 \, d^4p_2 \, d^4p_3$$

$$= -i96\lambda^2 \frac{1}{(2\pi)^{12}} \int e^{-ip.(x-y)} \left(\frac{1}{p^2 - m^2 + i\varepsilon}\right)^2 \bar{B}_2(p) \, d^4p. \tag{12.147}$$

The last equality yields an expression made of three blocks of terms:

- A set of numerical coefficients including the combinatoric factor associated with the diagram, the coupling constants, and the various 2π factors we found previously;

- Expressions characterising external lines: propagators, exponential factors, and a delta function expressing the energy–momentum conservation of the set of external lines;

- An expression (here $\bar{B}_2$) which gives the value of the diagram, as a function of the four-momentum of the external lines, with the external lines amputated; and

- the integration is over the set of the four-momentum of the internal lines. In fact, we have one four-integration less than the number of internal lines because of the delta function of the overall energy–momentum conservation.

Let us consider another example originated from the ϕ^4 theory: the connected contributions to the first order in the coupling constant to $I = \langle T(\phi(x_1)\phi(x_2)\phi(x_3)\phi(x_4)) \rangle$. The Euclidean analysis gives

$$I = -4!\lambda \int \prod_{i=1}^{4} C(x - x_i)\, \mathrm{d}^4 x. \tag{12.148}$$

The corresponding expression in the physical space is

$$\bar{I} = -\mathrm{i}4!\lambda \int \prod_{i=1}^{4} G_F(x - x_i)\mathrm{d}^4 x. \tag{12.149}$$

To this order, the connected contribution to the scattering amplitude

$$\langle p_1, p_2; in|S|q_1, q_2; in \rangle$$

is given by

$$-\mathrm{i}(2\pi)^4 4!\lambda \delta^{(4)}(\check{p}_1 + \check{p}_2 - \check{q}_1 - \check{q}_2), \tag{12.150}$$

where we used the fact that $(\Box_{x_i} + m^2) G_F(x_i - x) = -\delta^{(4)}(x_i - x)$. Here again $\check{p}$ means that $p^2 = m^2$ and $p^0 > 0$.

12.5   The Case of the Maxwell Field

Introducing spin brings only minor modifications. We start with the case of the electromagnetic field.

12.5.1   The Classical Maxwell Field

In a previous chapter we introduced the vector potential $A_\mu(x)$, which transforms as a four-vector under Lorentz transformations, in order to describe the electromagnetic field. The corresponding field equation is the Maxwell equation

$$\left(\Box g_{\mu\nu} - \partial_\mu \partial_\nu\right) A^\mu(x) = j_\nu(x). \tag{12.151}$$

We noted that not all four components of A_μ can be considered as independent dynamical variables. In particular, Eq. (12.151) contains no time derivative for A_0, the zero component of the field. This is due to an invariance of this equation, which we called 'gauge invariance', which states that (12.151) remains invariant under the transformation $A_\mu(x) \rightarrow A_\mu(x) + \partial_\mu\phi(x)$, with $\phi(x)$ an arbitrary function. We noted that this invariance implies the non-existence of a Green function for the Maxwell equation. In order to overcome this problem we introduced the concept of 'gauge fixing' and we would like here to explain this concept a bit further.

Maxwell's equations are obtained, through a variational principle, from the action

$$S[A] = \int d^4x \left(-\frac{1}{4} F_{\mu\nu}(x) F^{\mu\nu}(x) - A^\mu(x) j_\mu(x) \right). \tag{12.152}$$

Since all components of the vector potential cannot be considered as independent dynamical variables, the variational method will not give a unique answer. For every solution satisfying (12.151), we can find an infinity of others by adding the divergence of an arbitrary scalar function $\phi(x)$. We want to impose a condition among the four components of A_μ to remove this arbitrariness. Therefore, *a gauge fixing condition* is a condition of the form $G(A) = 0$, where G is some suitably chosen functional of A. The obvious condition on G is the following: the equation

$$G[A + \partial\phi] = 0, \tag{12.153}$$

considered as an equation for ϕ, should yield a unique solution $\phi = \phi_0(x)$ for every vector potential $A_\mu(x)$. It is clear that this condition implies that G is not gauge invariant. The Lorenz condition $G = \partial_\mu A^\mu(x)$ satisfies this criterion if we impose on ϕ to vanish at infinity. Indeed, for $G = \partial_\mu A^\mu(x)$, (12.153) implies that $\Box\phi = 0$, which has a unique solution $\phi = 0$ for functions which vanish at infinity. In fact, an apparently much simpler choice is to choose one component of A_μ to vanish, for example to impose $A_0(x) = 0$. It is easy to check that this condition also satisfies the criterion expressed by (12.153). In the next section we examine the so-called 'Coulomb gauge' given by the choice $G = \text{div} A = 0$. The advantage of these non-covariant gauge conditions is that they can be solved explicitly and eliminate the redundant degrees of freedom. The disadvantage is that they break explicit Lorentz covariance. In this book, unless stated otherwise, we want to keep Lorentz covariance at every step of the calculation, so we shall choose gauge conditions, such as the Lorenz one, which we cannot explicitly solve. Therefore, we are led to use the method introduced by Lagrange in order to solve constrained problems in classical mechanics.

Maxwell's equations (12.151), together with the subsidiary condition $G = 0$, can be obtained from the Lagrangian density

$$\mathcal{L} = -\frac{1}{4} F_{\mu\nu}(x) F^{\mu\nu}(x) + b(x) G[A] - A^\mu(x) j_\mu(x), \tag{12.154}$$

where $b(x)$ is an auxiliary field called 'Lagrange multiplier'. Indeed, varying independently with respect to A_μ and b we obtain the system of equations

$$\partial^{\mu} F_{\mu\nu} + b \frac{\delta G}{\delta A^{\nu}} = j_{\nu}; \quad G[A] = 0. \tag{12.155}$$

For the particular example in which $G = \partial A$ this system is equivalent to

$$\Box A_{\nu} - \partial_{\nu} b = j_{\nu}; \quad \partial_{\mu} A^{\mu} = 0. \tag{12.156}$$

Since j is assumed to be a conserved current (otherwise, as we noted already, Maxwell's equations are inconsistent), these equations imply that $\Box b = 0$; in other words, the field $b(x)$ is a free field and does not participate in the dynamics. If it vanishes at infinity, the system (12.156) is equivalent to $\Box A_{\nu} = j_{\nu}$ and $b = 0$. For more complicated gauge conditions, for example if G depends non-linearly on A, the Lagrange multiplier field does not decouple and we should study the coupled system of A and b.

12.5.2   The Quantum Field: I. The Functional Integral

How much of the classical discussion in the previous section applies to the quantum field problem? We learned in the previous chapters that the transition from the classical to the quantum theory is conveniently encoded into the calculation of a Euclidean functional integral. For the electromagnetic field this amounts to the calculation of a generating functional of the form

$$Z[j] = \frac{\int \mathcal{D}[A] e^{-S_{\mathrm{inv}}[A] - \int d^4 x A j}}{\int \mathcal{D}[A] e^{-S_{\mathrm{inv}}[A]}}, \tag{12.157}$$

where, $S_{\mathrm{inv}}[A]$ is the Maxwell gauge invariant action proportional to the integral of F^2. However, we see immediately that this expression is ill-defined. The reason is again gauge invariance. If we denote, symbolically, the gauge transformation as $A \rightarrow^{\Omega} A$, with Ω an element of the gauge group, both the numerator and the denominator of (12.157) remain invariant. A simple way to visualise the problem is to consider an integral from $-\infty$ to $+\infty$ of a function of two variables x and y: $\int dx dy Q(x, y)$. If Q is translational invariant it is, in fact, a function of only the difference $x - y$. Therefore, the integral contains an infinite factor given by $\int d(x + y)$, the volume of the translation group. Similarly, the expressions in (12.157) contain the volume of the gauge group. We remind that although the underlying group of electromagnetic gauge invariance is $U(1)$, the gauge potentials take values in the Lie algebra of the group which is an infinite-dimensional space. We say often, in a formal sense, that the electromagnetic gauge group is $U(1)^{\infty}$, since it is as if we had a $U(1)$ factor at every point in space.[10]

We could argue that this is not a serious problem, since we expect the factors to cancel between the numerator and the denominator, and here we want to investigate under which conditions this is indeed the case. For this, let us try to impose a gauge fixing condition of the form we used before, namely, $G[A] = 0$. We want to separate

---

[10] Invariance under a compact group of global transformations does not create any difficulties, because the corresponding volume is finite. This is the case, for example, with the $O(4)$ invariance in Euclidean space.

the measure of the functional integral into a product of two factors, one $\mathcal{D}[\Omega]$ on the group and a second $\mathcal{D}[\tilde{A}]$ on the space of functions satisfying the gauge condition. It will be convenient to use a more general gauge condition of the form $G[A] = C(x)$, with C any given function of x, and then average over all functions C with a Gaussian measure. Consider one of the two integrals, for example the denominator D. We write

$$D = \int \mathcal{D}[A] \mathrm{e}^{-S_{\mathrm{inv}}[A]} \Rightarrow \int \mathcal{D}[A]\mathcal{D}[C]\mathcal{D}[G]\mathrm{e}^{-S_{\mathrm{inv}}[A]-\frac{1}{2}\int \mathrm{d}^4x C^2(x)}\delta[G-C], \qquad (12.158)$$

where we have introduced a functional delta function which has no effect, since we integrate over all gauge conditions given by functions G. The discussion is only formal, but we could make it more precise by formulating the theory on the points of a finite Euclidean space lattice.

Looking at the expression (12.158) we see that we can use the delta function to get rid of the integration over C. Then we change variables from G to Ω. This is, in principle, possible, since we assumed that the gauge condition has a unique solution as an equation for the element of the gauge group. Therefore, taking into account the invariance of the action under gauge transformations, we find that

$$D = \int \mathcal{D}[\Omega]\mathcal{D}[A]\mathrm{e}^{-S_{\mathrm{inv}}[A]-\frac{1}{2}\int \mathrm{d}^4x G^2[A]}\det\frac{\delta G}{\delta\Omega}. \qquad (12.159)$$

For the class of gauge conditions we are considering, such as $G = \partial A$, the Jacobi determinant is given by $\det\square$. It is a divergent quantity but we will never have to compute it. The only thing that matters here is that it is a constant, independent of the vector potential A. So it can be taken out of the integral. The final result is that the denominator D is given by

$$D = K \int \mathcal{D}[A]\mathrm{e}^{-S_{\mathrm{inv}}[A]-\frac{1}{2}\int \mathrm{d}^4x G^2[A]} \qquad (12.160)$$

with K some infinite constant. The same constant will appear in the expression for the numerator of (12.157). So, the net result of all this gauge fixing procedure is that, for the family of linear gauges, it is enough to change the Lagrangian density and use an *effective* one in which we have added the term G^2, in accordance with the formula we used in (3.114). What happens when we use more complicated gauge conditions will be discussed in Chapter 14.

We see that the effective field theory is gauge dependent, since the Lagrangian depends on the arbitrary gauge fixing function G. This is not a trivial dependence since, as we have seen already, the Green functions depend on it. Even if we restrict ourselves to the very special family of (3.114), the Green functions are given by (3.116) and they depend on the arbitrary parameter α. It seems that we are obtaining an infinite family of different field theories and this is indeed the case. In which sense do all these theories describe the quantum version of Maxwell's electrodynamics? We shall answer this question in two steps. In the next section we shall introduce the Fock space of states which is associated with each one of these field theories. In a later Chapter (Chapter 18) we will argue that observable quantities are indeed gauge independent.

12.5.3 The Quantum Field: II. The Particle Concept

The free electromagnetic field is supposed to describe free photons, i.e. neutral, massless particles with helicity equal to 1. As we have seen in the analysis of the one-particle states, photons have two degrees of polarisation corresponding to helicity equal ± 1. In other words, there are no longitudinally polarised free photons. This is the particle analogue of the familiar property of the classical electromagnetic radiation to be transversely polarised. Since we describe the electromagnetic field with the four-component vector potential, we expect that all components will not create physical photons. This is the redundancy we explained previously which is translated in the gauge invariance of the equations of motion. Therefore, we expect that the particle content of the electromagnetic field will depend on the choice of the gauge condition. In this section we want to show and explain this dependence.

12.5.3.1 The Coulomb gauge

We start with a non-relativistic choice in which the particle concept is particularly transparent. We choose the three-dimensional transverse gauge given by the gauge condition div $A=0$. It is often called the *Coulomb gauge*, although Coulomb never wrote this relation. As we showed in section 3.4, in this gauge the zero component of the field is uniquely determined by the external sources, Eq. (3.28). We see, in particular, that for a fixed point charge, A_0 reproduces the $1/r$ Coulomb potential, which explains the name attached to this gauge.

We are left with the three-component vector potential $A(x)$. We still must impose the gauge condition div $A = 0$ and this is easily done in momentum space. We start by writing the expansion in plane waves

$$A(x) = \int \mathrm{d}\Omega_0 \sum_{i=1}^{3} \left[a_i(k)\epsilon^{(i)}(k)\mathrm{e}^{-ikx} + a_i^*(k)\epsilon^{(i)*}(k)\mathrm{e}^{ikx} \right], \tag{12.161}$$

where $\epsilon^{(i)}(k)$ form a system of three orthonormal unit vectors. As usual, we choose $\epsilon^{(3)}(k)$ to be in the direction of k:

$$\epsilon^{(3)}(k) = \frac{k}{|k|}; \quad \epsilon^{(i)}(k) \cdot \epsilon^{(j)}(k) = \delta^{ij}; \quad k \cdot \epsilon^{(i)}(k) = 0 \quad i = 1, 2. \tag{12.162}$$

We shall call the third direction *longitudinal* and the first two *transverse* and we can rewrite the expansion (12.161) as

$$\begin{aligned} A(x) = \ & \int \mathrm{d}\Omega_0 \left\{ a_3^L(k)\epsilon^{(3)}(k)\mathrm{e}^{-ikx} + a_3^{*L}(k)\epsilon^{(3)*}(k)\mathrm{e}^{ikx} \right. \\ & \left. + \sum_{i=1}^{2} \left[a_i^T(k)\epsilon^{(i)}(k)\mathrm{e}^{-ikx} + a_i^{*T}(k)\epsilon^{(i)*}(k)\mathrm{e}^{ikx} \right] \right\}. \end{aligned} \tag{12.163}$$

We see that the gauge condition div $A = 0$ implies that $a_3^L(\mathbf{k}) = 0$; in other words, only the transverse degrees of freedom survive. In terms of them we can write the canonical commutation relations as

$$[a_i^T(\mathbf{k}), a_j^{*T}(\mathbf{k}')] = \delta_{ij}(2\pi)^3 2\omega_k \delta^3(\mathbf{k} - \mathbf{k}'), \qquad (12.164)$$

which allows us to interpret $a_i^T(\mathbf{k})$ and $a_j^{*T}(\mathbf{k}')$ as annihilation and creation operators, respectively. The resulting Fock space is given by $\mathcal{H}^T = \sum \mathcal{H}_n^T$, i.e. it is the direct sum of the Hilbert spaces made out of the states of n free photons with transverse polarisation. The renormalised Hamiltonian is given by the sum of two expressions of the form given in Eq. (9.99):

$$H_{\text{ren}} = \frac{1}{2(2\pi)^3} \int \mathrm{d}^3k \sum_{i=1}^{2} a_i^{*T}(\mathbf{k}) a_i^T(\mathbf{k}), \qquad (12.165)$$

and it is positive definite. The overall picture is consistent with what we expect from the classical theory.

Naturally, this analysis depends on the particular reference frame we have chosen and it is not Lorentz invariant. If we perform a Lorentz transformation, we mix transverse, longitudinal, and scalar components. For the free-field theory we are studying we can easily show that a Lorentz transformation can be compensated by a gauge transformation and the physical picture remains the same.

A final remark: It is obvious that this analysis becomes meaningless when $\mathbf{k} \to 0$, or, equivalently, $k_\mu \to 0$ since, in this limit we cannot separate the longitudinal and transverse directions. So, we expect to find a singularity when the photon momentum vanishes. These singularities are what we have often called *infrared singularities* and will be studied later, in Chapter 21.

12.5.3.2 A covariant gauge

Let us now analyse the picture in a covariant gauge belonging to the family we studied in the previous section. Let us choose the Lorenz condition which, for the classical field, reads $\partial A = 0$. The Lagrangian reduces to

$$\mathcal{L} = -\frac{1}{2}\partial_\mu A_\nu \partial^\mu A^\nu \qquad (12.166)$$

and the propagator is proportional to $g_{\mu\nu}/(k^2 + i\epsilon)$[11].

It, looks as if each component of the field satisfies an independent, massless, Klein-Gordon equation, but with one important difference. Because of the Minkowski metric, the term for A_0 in the Lagrangian has the opposite sign. If we ignore this problem for the moment, we proceed as we did with the Klein–Gordon field and we expand in plane waves using the equation we wrote in (6.107) with the normalisation of the four unit

---

[11] We recover here the infrared singularity when $k_\mu \to 0$.

vectors $\epsilon_\mu^{(\lambda)}(k)$ given by (6.106). The standard quantisation rule gives the canonical commutation relations of the form

$$[a^{(\lambda)}(k), a^{(\lambda')*}(k')] = -2k_0(2\pi)^3 g^{\lambda\lambda'} \delta(\boldsymbol{k} - \boldsymbol{k'}). \qquad (12.167)$$

We can easily show that this form is also necessary in order to satisfy the causality equation (12.173), which we shall derive shortly. The usual convention implies that $a^{(1)*}$ and $a^{(2)*}$ are the creation operators of transverse photons, and $a^{(0)*}$ and $a^{(3)*}$ are, respectively, the creation operators of scalar and longitudinal photons (and equivalent notions for the annihilation operators). The Fock space appears to be that of four types of excitations, twice as many as that we obtained in the Coulomb gauge.

So far, so good. However, we see immediately that a new problem arises: the commutation relations for the scalar photons, $\lambda' = \lambda = 0$, as written in Eq. (12.167), have the opposite sign from those of the other three. This in turn implies that some states have a negative norm! Indeed, the one-photon state

$$|\Psi_1 > = \frac{1}{(2\pi)^3} \int \frac{\mathrm{d}^3 k}{2k_0} f(k) a^{(0)*}(k)|0 > \qquad (12.168)$$

has the norm

$$< \Psi_1 | \Psi_1 > = -\frac{1}{(2\pi)^3} \int \frac{\mathrm{d}^3 k}{2k_0} |f(k)|^2 \le 0. \qquad (12.169)$$

We want to emphasise that this difficulty is again the reflection of the Minkowski metric of space–time and cannot be eliminated by any redefinition of the fields. As a consequence, the energy is not positive definite. We obtain

$$H_{\mathrm{ren}} = \int \mathrm{d}\tilde{k} E_k \left[\sum_{\lambda=1}^{3} a^{(\lambda)*}(k) a^{(\lambda)}(k) - a^{(0)*}(k) a^{(0)}(k) \right]. \qquad (12.170)$$

A formal way to describe this situation is to say that in the Fock space of states the scalar product is defined with an indefinite metric of the form

$$(\Psi, \Phi) \quad \Rightarrow \quad (\Psi, \eta\Phi) \; ; \quad \eta|\Phi > = (-)^s|\Phi >, \qquad (12.171)$$

where s is the number of scalar photons in the state $|\Phi >$. This metric is Hermitian, since it satisfies $\eta^2 = 1$ and $\eta = \eta^*$. Furthermore, the creation and annihilation operators for all four polarisations satisfy

$$(\Psi, a^{(\lambda)}(k)\Phi) = (a^{(\lambda)*}(k)\Psi, \Phi), \qquad (12.172)$$

which means that the field $A_\mu(x)$ is a self-adjoint operator.

This metric was introduced by S. N. Gupta and K. Bleuler in 1950 and it is called *the Gupta–Bleuler formalism*. We could give a more detailed mathematical description of the resulting Fock space with this indefinite metric, but, in this book, we shall never need anything beyond the straightforward application of formula (12.171).

We still have the problem of having too many states in the Fock space. From our experience with the Coulomb gauge we expect some of them to be unphysical. Can we separate them? In other words, can we find a subspace of the large Fock space which contains only physical, positive norm, states? And, *last but not least*, can we obtain this separation in a Lorentz covariant way? Of course, we can apply the gauge condition $\partial A = 0$ which is Lorentz covariant, but here we are facing a new problem: the canonical quantisation yields the usual commutation relation

$$[A_\mu(x), A_\nu(y)] = ig_{\mu\nu}\triangle(x-y), \qquad (12.173)$$

where $\triangle$ is given by the formula (9.103). This shows that, as we expect, the quantum field is not a classical function but an operator valued distribution. Can we apply the Lorenz condition $\partial A = 0$ as an operator equation? In other words, can we assume that

$$\partial A|\Psi> = 0 \qquad (12.174)$$

for *all* states $|\Psi>$ in the Fock space? The answer is no, as we can easily convince ourselves by applying ∂_x^μ to both sides of the commutation relation (12.173). We obtain

$$[\partial^\mu A_\mu(x), A_\nu(y)] = i\partial_\nu\triangle(x-y) \qquad (12.175)$$

and the right-hand side is, obviously, non-zero. There is therefore an inconsistency in willing to quantify the photon field in a Lorentz covariant way and at the same time imposing the Lorenz condition. Again, we overcome this problem by choosing to impose the Lorenz condition in a weaker form.[12] We split the ∂A operator into positive and negative frequency parts and we define the following.

A state $|\Psi>$ is physically acceptable if

$$\partial^\mu A_\mu^{(+)}(x)|\Psi> = 0, \qquad (12.176)$$

where (+) indicates that we only consider the positive frequency part of A_μ.

In a given reference frame this condition is equivalent to

$$(a^{(3)}(k) - a^{(0)}(k))|\Psi> = 0 \qquad (12.177)$$

on each physical state.[13] Note that this separation into 'physical' and 'unphysical' states is Lorentz covariant because it is based on the condition (12.176) which is valid in any frame.

---

[12] All these steps appear to be arbitrary and, in some sense, they are. Their final justification is based on the fact that, as we shall show, they lead to a mathematically consistent and physically correct theory.

[13] One may wonder whether for an interacting field, the splitting of frequencies into positive and negative ones is meaningful. The answer is yes because from the equations of motion, if the current is conserved, the divergence of the field $\partial.A$ satisfies the equation

$$\Box\partial.A = 0$$

and is therefore a free field.

The condition is obviously satisfied by the vacuum state. It is also satisfied by the one-particle states with transverse photons. In fact, it is satisfied by any state which has only transverse photons. So the states which we expect, intuitively, to be physical satisfy the condition. However, and this comes as a surprise, there are other states which satisfy the condition and are, according to this definition, 'physical'. A simple example is given by the one-photon state which is a superposition of the form $|\Psi> = (|1(s)> + |1(l)>)/\sqrt{2}$, i.e. one scalar and one longitudinal photon of the same momentum. In a similar way we can construct superpositions of states with two, three, etc. scalar and longitudinal photons which satisfy the condition. It is easy to show that any such state which has longitudinal and scalar photons has zero norm.

Let us summarise. We quantised the free electromagnetic field using various gauge conditions. In both the field theory picture and the particle picture we found totally different quantum field theories. The space of states is different; the Green functions are different. We recover the question we asked in the previous section: in which sense can we say that all these field theories describe the same underlying physical theory? We expect the answer to be in the following form: they all must give the same result when we compute 'physically measurable quantities'. So the question is to define which quantities are physically measurable. We cannot answer this question if we restrict ourselves to free fields, since any physical measurement implies an interaction of the system with the measuring apparatus. Therefore, we should postpone this discussion until the interacting theory is studied.

12.5.4   The Casimir Effect

Before closing this section we want to point out an interesting consequence of the quantisation procedure we have established for the electromagnetic field. Let us consider, for simplicity, the Coulomb gauge which involves only the physical degrees of freedom given by the transversely polarised photons. We showed that the free field is described by a double infinity of three-dimensional harmonic oscillators. The vacuum state $|0>$ is annihilated by all annihilation operators and the vacuum expectation value of the renormalised Hamiltonian given by (12.165) vanishes:

$$< 0|H_{\text{ren}}|0 >= 0. \tag{12.178}$$

This result is not a direct consequence of the formalism. It was obtained in section 8.6 by adjusting an additive constant with the effect of putting the zero-point energy of every harmonic oscillator equal to 0. We justified this choice by arguing that, in the absence of a gravitational field, only energy differences are measurable, so the absolute value of the vacuum expectation value in the expression (12.178) is arbitrary. The condition (12.178) *defines* in fact the Hamiltonian operator. It was H. Casimir who first realised that this procedure does lead to measurable effects.

We start with the electromagnetic field quantised in empty space. The Hamiltonian is given by (12.165). Let us now consider that we bring two conducting plates at a distance L apart, placed at the points $z = \pm L/2$. For simplicity, let us assume that in the corresponding (x, y) planes they extend to infinity and they are infinitely conducting.

We must now repeat the quantisation procedure restricted in the space between the two plates, with the boundary condition that the electromagnetic field must vanish for $z = \pm L/2$. The calculation is straightforward: the only difference is that the k-modes in the z direction are discrete. We can compute again the new renormalised Hamiltonian operator $\tilde{H}_{\text{ren}}$ using the condition that its value in the ground state vanishes. The new additive constant will depend on L; therefore, we can compute the difference in the ground state values $\Delta E(L)$ using H_{ren} and $\tilde{H}_{\text{ren}}$. The derivative $d\Delta E(L)/dL$ will give a force between the two plates. In a problem at the end of this chapter we ask the reader to perform this computation and calculate the resulting force per unit surface in the transverse (x, y) plane. This effect, which has been measured experimentally, shows the physical meaning of the vacuum fluctuations in quantum field theory. We will encounter further manifestations of such phenomena in this book.

12.6 Quantization of a Massive Field of Spin-1

The quantisation of a massive spin-1 field is much simpler than the one we encountered for the massless case. First, there is no infrared singularity. Indeed, the Green function we computed in section 6.4 shows that there is no singularity when $k_\mu \to 0$, Eq. (6.104). Second, there is no gauge invariance and, therefore, we have no extra condition to impose.

Going back to the notation of section 6.4, we write the Lagrangian as

$$\mathcal{L} = -\frac{1}{4}F^2 + \frac{1}{2}m^2A^2 - A_\mu j^\mu. \tag{12.179}$$

It gives the Proca equation of motion with the current on the right-hand side[14]:

$$\Box A_\mu(x) - \partial_\mu \partial_\nu A^\nu(x) + m^2 A_\mu(x) = j_\mu(x). \tag{12.180}$$

This equation implies that

$$m^2 \partial^\mu A_\mu(x) = \partial^\mu j_\mu(x). \tag{12.181}$$

If the current is independently conserved, this equation, for $m \neq 0$, implies that $\partial A = 0$. Note, that this is an equation of motion and not an external condition we should impose.

From the group theory point of view, a vector field has a dotted and an undotted index and, in three dimensions, it contains a spin 1 and a spin 0. If $\partial A = 0$, the spin 0 part is eliminated and we are left with a pure spin 1, therefore three degrees of freedom.

---

[14] In an interacting theory the current is a function of various fields. In deriving this equation we made an implicit assumption, namely that j_μ does not depend on A_μ; otherwise, we should include a term proportional to $\delta j/\delta A$. This assumption is verified for the electromagnetic field interacting with a charged Dirac field; in other words, quantum electrodynamics, but it is not for many other theories. In particular, there is an important class of vector field theories, the so-called Yang–Mills theories, for which this assumption does not hold. As a result, the quantisation of these theories is more complicated and it will be studied in a special chapter later (Chapter 14).

This leads naturally to the introduction of a free field

$$A_\mu(x) = \frac{1}{(2\pi)^3} \int \frac{\mathrm{d}^3 p}{2\omega(p)}$$
$$\times \sum_{i=1}^{3} [a^{(i)}(p)\epsilon_\mu^{(i)}(p)\mathrm{e}^{-ip.x} + a^{(i)*}(p)\epsilon_\mu^{(i)*}(p)\mathrm{e}^{ip.x}], \tag{12.182}$$

where we have taken into account the conditions related to the divergence by introducing three polarisation vectors $\epsilon^{(i)}(p)$ such that

$$p.\epsilon^{(i)}(p) = 0$$
$$\epsilon^{(i)}(p).\epsilon^{(j)}(p) = \delta_{ij}$$
$$\sum_i \epsilon_\varrho^{(i)}(p)\epsilon_\sigma^{(i)}(p) = -\left(g_{\varrho\sigma} - \frac{p_\varrho p_\sigma}{m^2}\right) \tag{12.183}$$

with the a's satisfying the usual commutation rules. The resulting Fock space has positive-definite metric.

The canonical commutation relations are given by

$$[a^{(i)}(p), a^{(j)*}(p')] = \delta_{ij}(2\pi)^3 2\omega_p \delta^{(3)}(p - p'). \tag{12.184}$$

We define the T-product by

$$< 0|T(A_\mu(x)A_\nu(y))|0 > = -\frac{i}{(2\pi)^4} \int \mathrm{e}^{-ip.(x-y)} \frac{g_{\mu\nu} - p_\mu p_\nu/m^2}{p^2 - m^2 + i\varepsilon} \mathrm{d}^4 p. \tag{12.185}$$

This description, which is convenient for massive spin-1 particles, is singular in the limit when the mass tends to 0. In fact, this should be expected. When the mass is 0, the quadratic form is no more invertible, because of the degeneracy due to gauge invariance. Physically, this corresponds to the fact that, at this limit, one degree of freedom must disappear, and we should be left with only two. If one wants to construct an expression having a limit for $m \to 0$, it is necessary to modify the initial Lagrangian and take the full Lagrangian we used in Eq. (6.99) with the propagator given in (6.101). It corresponds to the massive spin-1 theory of the Proca equation, but with the addition of an extra gauge fixing term.

Such a Lagrangian density gives the equations of motion

$$(\Box + m^2)A_\nu - \frac{1}{\alpha}\partial_\nu(\partial.A) = 0, \tag{12.186}$$

which implies, for the divergence of the field, that:

$$(\Box + \mu^2)\partial.A = 0 \tag{12.187}$$

with $\mu^2 = \frac{\alpha m^2}{\alpha - 1}$.

Note that this equation remains valid even if the field is coupled to a current with a term $A_\mu j^\mu$, provided the current is conserved. In other words, a conserved current does not affect the dynamics of the divergence of the field which is still given by the free-field equation (12.187).

The functional integral over the A's gives for propagator the one of Eq. (6.101).

If the current is conserved, we can introduce a field with a divergence equal to 0

$$A_\nu^T = A_\nu + \frac{1}{\mu^2}\partial_\nu\partial.A \tag{12.188}$$

since

$$\partial.A^T = \partial.A + \frac{1}{\mu^2}\Box\,\partial.A = \frac{1}{\mu^2}(\Box + \mu^2)\,\partial.A = 0. \tag{12.189}$$

The field can then be written as

$$A_\nu = A_\nu^T - \frac{1}{\mu^2}\partial_\nu\partial.A. \tag{12.190}$$

Remark that if this field is coupled to a conserved current j^ν, then, modulo a divergence,

$$j.A = j.A^T. \tag{12.191}$$

By the procedures of canonical quantisation, to Eq. (12.186) corresponds a free field

$$A_\mu(x)$$
$$= \int d\Omega_m \sum_{i=1}^{3}[a^{(i)}(p)\epsilon_\mu^{(i)}(p)e^{-ip.x} + a^{(i)*}(p)\epsilon_\mu^{(i)*}(p)e^{ip.x}]$$
$$+ \int d\Omega_\mu \frac{p_\mu}{\mu}[a^{(0)}(p)e^{-ip.x} + a^{(0)*}(p)e^{ip.x}] \tag{12.192}$$

with

$$[a^{(i)}(p), a^{(j)*}(q)] = \delta_{ij}(2\pi)^3 2\sqrt{p^2 + m^2}\delta^{(3)}(p - q)$$
$$[a^{(0)}(p), a^{(0)*}(q)] = -(2\pi)^3 2\sqrt{p^2 + \mu^2}\delta^{(3)}(p - q) \tag{12.193}$$

In the limit $m \to 0$, we get

$$< 0|T(A_\mu(x)A_\nu(y))|0 >$$
$$= -\frac{i}{(2\pi)^4}\int e^{-ip.(x-y)}\left(\frac{g_{\mu\nu}}{p^2 + i\varepsilon} + \frac{p_\mu p_\nu}{(p^2 + i\varepsilon)^2}(\alpha - 1)^{-1}\right)d^4p. \tag{12.194}$$

Note that now the resulting Fock space has an indefinite metric. These formulae are interesting because they have the correct massless limit but no infrared singularities for $m \neq 0$. So, they offer a 'regularised' form, namely expressions which depend on the parameter m, are regular when k goes to 0 as long as m is kept different from 0, and go to the corresponding expressions for the photon at the limit $m \to 0$. Therefore, they can be used in intermediate steps of a calculation. A convenient choice of gauge is the Feynman gauge corresponding to $\alpha \to \infty$. In section 16.5.4 we will see that quantum electrodynamics with a massive photon is in fact a perfectly acceptable physical theory.

12.7   The Reduction Formulae for Photons

We will now derive reduction formulae for processes involving photons. Let us consider an S-matrix element for which the 'out' state contains a transverse photon of momentum $k = (|\boldsymbol{k}|, \boldsymbol{k})$ and polarisation ϵ

$$< ...; k, \epsilon; \text{out}|\psi; \text{in} > = < ...; \text{out}|a^\epsilon(k)|\psi; \text{in} >$$
$$= -\mathrm{i} \int_t e^{\mathrm{i}k.x} \overleftrightarrow{\partial}_0 < ...; \text{out}|\epsilon.A^T(x)|\psi; \text{in} > \mathrm{d}^3 x, \qquad (12.195)$$

where we use the orthogonality relations of the polarisations. The matrix element can be written, as for the other reduction formulae, as

$$-\mathrm{i} \lim_{t_f \to \infty} \int_{t_f} e^{\mathrm{i}k.x} \overleftrightarrow{\partial}_0 < ...; \text{out}|\epsilon.A(x) + \frac{1}{\mu^2}\epsilon.\partial\partial.A(x)|\psi; \text{in} > \mathrm{d}^3 x \qquad (12.196)$$

$$= \text{non-connected terms}$$

$$-\mathrm{i} \int e^{\mathrm{i}k.x}(\Box_x + m^2) < ...; \text{out}|\epsilon.A(x) + \frac{1}{\mu^2}\epsilon.\partial\partial.A(x)|\psi; \text{in} > \mathrm{d}^4 x.$$

But from the equation of motion (with a conserved current), we have

$$(\Box + m^2)\frac{1}{\mu^2}\epsilon.\partial\partial.A = \frac{m^2 - \mu^2}{\mu^2}\epsilon.\partial\partial.A = -\alpha^{-1}\epsilon.\partial\partial.A, \qquad (12.197)$$

and the connected term, with sources, can be written as

$$-\mathrm{i} \int e^{\mathrm{i}k.x}(\Box_x + m^2)$$

$$\times < ...; \text{out}|\epsilon.\left((\Box_x + m^2)A(x) - \alpha^{-1}\partial\partial.A(x)\right)|\psi; \text{in} > \mathrm{d}^4 x$$

$$= -\mathrm{i} \int e^{\mathrm{i}k.x} < ...; \text{out}|\epsilon.j(x)|\psi; \text{in} > \mathrm{d}^4 x. \qquad (12.198)$$

This formula can be generalised and gives

$$< ...; k_f, \epsilon_f; \text{out}|k_i, \epsilon_i; ...; \text{in} >^c$$

$$= -\int e^{\mathrm{i}k_f.x - \mathrm{i}k_i.y} < ...; \text{out}|T(\epsilon_f.j(x)\epsilon_i.j(y))|...; in >^c \mathrm{d}^4 x \mathrm{d}^4 y. \qquad (12.199)$$

12.8   The Reduction Formulae for Fermions

We can also prove reduction formulae for theories involving Dirac fields. We only give the result, the proof being the same as in the bosonic case, the Klein–Gordon operator being replaced by the Dirac operator. The spinors $u^{(s)}$ and $v^{(s)}$ have been chosen as eigenstates of helicity and the creations and the annihilation operators create or annihilate states with well-defined helicity. Thus, $a_1^*(p)$ and $b_{-1}^*(p)$ create states of helicity $\frac{1}{2}$ when a_{-1}^* and b_1^* create states of helicity $\frac{-1}{2}$. An asymptotic state of well-defined momentum and helicity is of the form

$$a_{s_1}^*(\boldsymbol{p}_1) \cdots b_{s_1'}^*(\boldsymbol{p}_1') \cdots |0>= |\boldsymbol{p}_1, s_1; \cdots; \boldsymbol{p}_1', s_1'; \cdots; \text{in} >, \tag{12.200}$$

the a^*'s creating electrons and the b^* creating positrons.

Since, if ψ and $\bar{\psi}$ are free Dirac fields,

$$a_s^*(p) = \int_t \bar{\psi}(x) \gamma^0 e^{-ip.x} u^{(s)}(p) d^3x$$

$$b_s^*(p) = \int_t \bar{v}^{(s)}(p) e^{-ip.x} \gamma^0 \psi(x) d^3x$$

$$b_s(p) = \int_t \bar{\psi}(x) \gamma^0 e^{ip.x} v^{(s)}(p) d^3x$$

$$a_s(p) = \int_t \bar{u}^{(s)}(p) e^{ip.x} \gamma^0 \psi(x) d^3x, \tag{12.201}$$

the spinors with well-defined helicity can be defined satisfying

$$v^{(s)}(p) = C \bar{u}^{(s)tr}(p). \tag{12.202}$$

We get the reduction formula

$$< p_1, \alpha_1; \cdots; p_1', \alpha_1'; \cdots; \text{out}|q_1, \beta_1; \cdots; q_1', \beta_1'; \cdots; \text{in} >$$
$$= \text{non- connected   terms}$$
$$+(-i)^n i^{n'} \int d^4x_1 \cdots d^4y_1 \cdots e^{-iq_1.x_1-\cdots+ip_1.y_1+\cdots}$$
$$\times \bar{u}^{(\alpha_1)}(p_1)(i\overrightarrow{\not{\partial}}_{y_1} - m) \cdots \bar{v}^{(\beta_1')}(q_1')(i\overrightarrow{\not{\partial}}_{x_1'} - m)$$
$$\times < T(\cdots \bar{\psi}(y_1') \cdots \psi(y_1)\bar{\psi}(x_1) \cdots \psi(x_1') \cdots) > \cdot$$
$$\times (-i\overleftarrow{\not{\partial}}_{x_1} - m) u^{(\beta_1)}(q_1) \cdots (-i\overleftarrow{\not{\partial}}_{y_1'} - m) v^{(\beta_1')}(p_1') \cdots, \tag{12.203}$$

where n and n' are, respectively, the numbers of particles and antiparticles.

As for bosons, the amputated fields are replaced by plane wave solutions of the Dirac equation

$$u(q)\mathrm{e}^{-\mathrm{i}q.x} \quad \text{for an ingoing electron}$$
$$-\bar{v}(q')\mathrm{e}^{-\mathrm{i}q'.x'} \quad \text{for an ingoing positron}$$
$$\bar{u}(p)\mathrm{e}^{\mathrm{i}p.y} \quad \text{for an outgoing electron}$$
$$-v(p')\mathrm{e}^{\mathrm{i}p'.y'} \quad \text{for an outgoing positron}$$

$$(12.204)$$

12.9   Quantum Electrodynamics

We already saw that the Lagrangian density for electrodynamics is

$$\mathcal{L} = \mathcal{L}_{\mathrm{em}} + \mathcal{L}_D + \mathcal{L}_{\mathrm{int}}, \tag{12.205}$$

where

$$\mathcal{L}_{\mathrm{em}} = -\frac{1}{4}F^2 = -\frac{1}{4}(\partial_\mu A_\nu - \partial_\nu A_\mu)(\partial^\mu A^\nu - \partial^\nu A^\mu)$$
$$\mathcal{L}_D = \bar{\psi}\gamma^\mu(\mathrm{i}\partial_\mu - m)\psi$$
$$\mathcal{L}_{\mathrm{int}} = -e\bar{\psi}\gamma^\mu\psi A_\mu.$$

The interaction Lagrangian can be obtained from the Dirac Lagrangian by replacing the usual derivative by the covariant derivative

$$\mathrm{i}\partial_\mu \to \mathrm{i}D_\mu = \mathrm{i}\partial_\mu - eA_\mu \tag{12.206}$$

To this expression we should add a gauge fixing term, as explained previously.

Now we are ready to give the Feynman rules.

12.9.1   The Feynman Rules

The theory is quantised through the functional integral and the T-products are generated by the generating functional

$$Z[j,\bar{\eta},\eta] = \frac{\int \mathrm{e}^{\mathrm{i}\int j_\mu A^\mu + \mathrm{i}\int \bar{\eta}\psi + \mathrm{i}\int \bar{\psi}\eta}\mathrm{e}^{\mathrm{i}\int \mathcal{L}_{\mathrm{int}}}\mathrm{e}^{\mathrm{i}\int \mathcal{L}_{em} + \mathrm{i}\int \mathcal{L}_D}\mathcal{D}(\bar{\psi},\psi,A)}{\int \mathrm{e}^{\mathrm{i}\int \mathcal{L}_{\mathrm{int}}}\mathrm{e}^{\mathrm{i}\int \mathcal{L}_{em} + \mathrm{i}\int \mathcal{L}_D}\mathcal{D}(\bar{\psi},\psi,A)}, \tag{12.207}$$

where

$$\frac{\mathrm{e}^{\mathrm{i}\int \mathcal{L}_{em} + \mathrm{i}\int \mathcal{L}_D}\mathcal{D}(\bar{\psi},\psi,A)}{\int \mathrm{e}^{\mathrm{i}\int \mathcal{L}_{em} + \mathrm{i}\int \mathcal{L}_D}\mathcal{D}(\bar{\psi},\psi,A)} \tag{12.208}$$

defines in the variables A, $\bar{\psi}$, and ψ a 'Gaussian measure' characterised by the propagators (12.193) for the photons and by

$$< 0|T(\psi(x)\bar{\psi}(y))|0 >= \mathrm{i}S_{\mathrm{F}}(x-y) \tag{12.209}$$

for the fermions.

The Feynman rules, for non-amputated diagrams in momentum space, are the following:

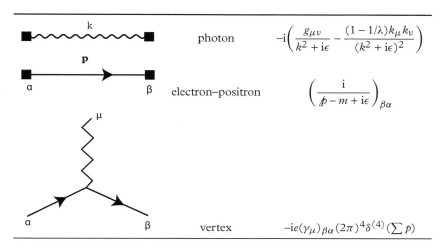

photon		$-\mathrm{i}\left(\dfrac{g_{\mu\nu}}{k^2 + \mathrm{i}\epsilon} - \dfrac{(1 - 1/\lambda)k_\mu k_\nu}{(k^2 + \mathrm{i}\epsilon)^2}\right)$
electron–positron		$\left(\dfrac{\mathrm{i}}{\not{p} - m + \mathrm{i}\epsilon}\right)_{\beta\alpha}$
vertex		$-\mathrm{i}e(\gamma_\mu)_{\beta\alpha}(2\pi)^4\delta^{(4)}\left(\sum p\right)$

Where λ is the gauge fixing parameter.

12.10   A Formal Expression for the *S*-Matrix

As was said Previously, the Bargmann representation is particularly suited to express *S*-matrix elements. We gave in Chapters 8 and 9 a rather complete analysis of the Bargmann representation for systems with one degree of freedom and explicit formulae in the case of quadratic Hamiltonians. The extension to harmonic oscillators living in the underlying three-dimensional space is obvious, as well as to an arbitrary number of such oscillators. Therefore, there are no real problems in extending these ideas to quantum fields. However, the Bargmann representation involves naturally Hamiltonians and not the Lagrangians.

Our first task is, therefore, to express quantum field theories in terms of Hamiltonians. We show in section 9.4 how the free scalar field is a natural generalisation of the position operator and how it can be expressed in terms of annihilation and creation operators which are obvious extensions of those defined for the dimensional harmonic oscillator.

We will extend to quantum fields the result given at the end of Chapter 8 on the *S*-matrix in the case of a free massive scalar field interacting with an external source.

The action is given by

$$S = \int \mathrm{d}^4x\, \mathcal{L}(x) = \int \mathrm{d}^4x\left[\frac{1}{2}\partial_\mu\phi(x)\partial^\mu\phi(x) - \frac{m^2}{2}\phi(x)^2 + j(x)\phi(x)\right], \qquad (12.210)$$

where $j(x)$ is the source term.

The corresponding Hamiltonian is

$$H_0(j) = \int d^3x \left[\frac{1}{2}\pi(x)^2 + \frac{1}{2}(\nabla\phi(x))^2 + \frac{m^2}{2}\phi(x)^2 - j(t,x)\phi(x) \right], \qquad (12.211)$$

where $\pi(x)$ is given, following the definition of the Legendre transform, by

$$\pi(x) = \frac{\partial}{\partial x_0}\phi(x)\big|_{x_0=0}.$$

We note that $H_0 = H_0(j)|_{j=0}$. The quantised fields are, according to formula (9.96), expressed in terms of the creation and annihilation operators $a^*(k)$ and $a(k)$ by

$$\phi(x) = \frac{1}{(2\pi)^3} \int \frac{d^3k}{2\omega_k}[a(k)e^{ik\cdot x} + a^*(k)e^{-ik\cdot x}]$$

$$\pi(x) = -\frac{i}{(2\pi)^3} \int \frac{d^3k}{2}[a(k)e^{ik\cdot x} - a^*(k)e^{-ik\cdot x}] \qquad (12.212)$$

and the Hamiltonian is therefore

$$H_0(j) = \frac{1}{(2\pi)^3} \int \frac{d^3k}{2\omega_k}\left[\omega(k)a^*(k)a(k) - \gamma(t,k)a^*(k) - \bar{\gamma}(t,k)a(k) \right] \qquad (12.213)$$

with

$$\gamma(t,k) = \int d^3x e^{-ik\cdot x}j(t,x).$$

We are therefore in a situation similar to that treated in sections 8.7.2 and 9.2.1 since the Hamiltonian is diagonal and is a sum of harmonic oscillators with one degree of freedom coupled to an external source. Thus, we can follow the steps leading to formula (9.48) and compute the matrix element of the evolution operator in the Bargmann space. States are of the form $| z >= e^{\int d^3k z(k)a^*(k)} | 0 >$ and we can compute

$$< z | e^{-i(t-t_0)H_0(j)} | \zeta >$$

$$= \int_{\substack{\bar{z}(t,k)=\bar{z}(k)\\z(t_0,k)=\zeta(k)}} \mathcal{D}(\bar{z})\mathcal{D}(z) \exp\left[\frac{1}{2}\frac{1}{(2\pi)^3} \int \frac{d^3k}{2\omega_k}\left\{ \bar{z}(k)z(t,k) + \bar{z}(t_0,k)\zeta(k) \right\} \right]$$

$$\times \exp\left[\int_{t_0}^t \left\{ \frac{1}{(2\pi)^3} \int \frac{d^3k}{2\omega_k}\left[\frac{1}{2}(\dot{\bar{z}}(s,k)z(s,k) - \bar{z}(s,k)\dot{z}(s,k)) \right.\right.\right.$$

$$\left.\left.\left. -i\omega_k\bar{z}(s,k)z(s,k) - i\bar{\gamma}(s,k)z(s,k) - i\bar{z}(s,k)\gamma(s,k) \right] \right\}ds \right]$$

$$= \exp\left[\frac{1}{(2\pi)^3} \int \frac{d^3k}{2\omega_k}\left\{ \bar{z}(k)\zeta(k)e^{-i\omega(t-t_0)} + i\bar{z}(k) \int_{t_0}^t e^{-i\omega(t-s)}\gamma(s,k)ds \right.\right.$$

$$\left.\left. +i\zeta(k)\int_{t_0}^t e^{-i\omega(s-t_0)}\bar{\gamma}(s,k)ds - \frac{1}{2}\int_{t_0}^t\int_{t_0}^t e^{-i\omega_k|u-s|}\bar{\gamma}(u,k)\gamma(s,k)duds \right\} \right]. \qquad (12.214)$$

The time evolution of $\mid z >$ follows from this formula by replacing $H_0(j)$ by H_0, i.e. by setting γ and $\bar{\gamma}$ equal to 0. The result is, as before, the replacement of $\bar{z}(\boldsymbol{k})$ and $z(\boldsymbol{k})$ by $\bar{z}(\boldsymbol{k})\mathrm{e}^{i\omega_k t}$ and $z(\boldsymbol{k})\mathrm{e}^{-i\omega_k t}$.

We are now ready to compute

$$\mathcal{S}_0(j)(\bar{z},\zeta) =< z \mid \mathcal{S}_0(j) \mid \zeta >= \lim_{\substack{t_f \to +\infty \\ t_i \to -\infty}} < z \mid U(t_f,t_i) \mid \zeta > .$$

From the previous discussion

$$\mathcal{S}_0(j)(\bar{z},\zeta) = \lim_{\substack{t_f \to +\infty \\ t_i \to -\infty}} =< z \mid \mathrm{e}^{it_f H_0}\mathrm{e}^{-i(t_f-t_i)H_0(j)}\mathrm{e}^{-it_i H_0} \mid \zeta >$$

$$= \lim_{\substack{t_f \to +\infty \\ t_i \to -\infty}} \int_{\substack{\bar{z}(t,\boldsymbol{k})=\bar{z}_t(\boldsymbol{k}) \\ z(t_0,\boldsymbol{k})=\zeta_{t_0}(\boldsymbol{k})}} \mathcal{D}(\bar{z})\mathcal{D}(z) \exp\left[\frac{1}{2}\frac{1}{(2\pi)^3}\int \frac{\mathrm{d}^3 k}{2\omega_k}\left\{\bar{z}_t(\boldsymbol{k})z(t,\boldsymbol{k}) + \bar{z}(t_0,\boldsymbol{k})\zeta_{t_0}(\boldsymbol{k})\right\}\right]$$

$$\times \cdot \exp\left[\int_{t_0}^{t}\left\{\frac{1}{(2\pi)^3}\int\frac{\mathrm{d}^3 k}{2\omega_k}\frac{1}{2}[(\dot{\bar{z}}(s,\boldsymbol{k})z(s,\boldsymbol{k}) - \bar{z}(s,\boldsymbol{k})\dot{z}(s,\boldsymbol{k})) - i\omega_k\bar{z}(s,\boldsymbol{k})z(s,\boldsymbol{k})\right.\right.$$

$$\left.\left.-i\bar{\gamma}(s,\boldsymbol{k})z(s,\boldsymbol{k}) - i\bar{z}(s,\boldsymbol{k})\gamma(s,\boldsymbol{k})]\right\}\mathrm{d}s\right]$$

$$= \exp\left[\frac{1}{(2\pi)^3}\int\frac{\mathrm{d}^3 k}{2\omega_k}\left\{\bar{z}(\boldsymbol{k})\zeta(\boldsymbol{k}) + i\bar{z}(\boldsymbol{k})\int_{-\infty}^{+\infty}\mathrm{e}^{i\omega_k s}\gamma(s,\boldsymbol{k})\mathrm{d}s\right.\right. \tag{12.215}$$

$$\left.\left.+i\zeta(\boldsymbol{k})\int_{-\infty}^{+\infty}\mathrm{e}^{-i\omega_k s}\bar{\gamma}(s,\boldsymbol{k})\mathrm{d}s - \frac{1}{2}\int_{-\infty}^{+\infty}\int_{-\infty}^{+\infty}\mathrm{e}^{-i\omega_k|u-s|}\bar{\gamma}(u,\boldsymbol{k})\gamma(s,\boldsymbol{k})\mathrm{d}u\mathrm{d}s\right\}\right]$$

with

$$\bar{z}_t(\boldsymbol{k}) = \bar{z}\mathrm{e}^{i\omega_k t} \qquad \zeta_{t_0}(\boldsymbol{k}) = \zeta(\boldsymbol{k})\mathrm{e}^{-i\omega_k t_0}$$

and the normal kernel is given by suppressing the first term of the exponent of the last equation in formula (12.216):

$$\mathcal{S}_0(j)(\bar{z},\zeta) = \exp\left[\frac{1}{(2\pi)^3}\int\frac{\mathrm{d}^3 k}{2\omega_k}\left\{i\bar{z}(\boldsymbol{k})\int_{-\infty}^{+\infty}\mathrm{e}^{i\omega_k s}\gamma(s,\boldsymbol{k})\mathrm{d}s\right.\right. \tag{12.216}$$

$$\left.\left.+i\zeta(\boldsymbol{k})\int_{-\infty}^{+\infty}\mathrm{e}^{-i\omega_k s}\bar{\gamma}(s,\boldsymbol{k})\mathrm{d}s - \frac{1}{2}\int_{-\infty}^{+\infty}\int_{-\infty}^{+\infty}\mathrm{e}^{-i\omega_k|u-s|}\bar{\gamma}(u,\boldsymbol{k})\gamma(s,\boldsymbol{k})\mathrm{d}u\mathrm{d}s\right\}\right].$$

We can now rewrite this formula in a more intrinsic way. The first two terms in the exponent can be recombined

$$\frac{i}{(2\pi)^3}\int\frac{\mathrm{d}^3 k}{2\omega_k}\int_{-\infty}^{+\infty}\left\{\bar{z}(\boldsymbol{k})\mathrm{e}^{i\omega_k s - i\boldsymbol{k}\cdot\boldsymbol{x}}\gamma(s,\boldsymbol{k}) + \zeta(\boldsymbol{k})\mathrm{e}^{-i\omega_k s - i\boldsymbol{k}\cdot\boldsymbol{x}}\bar{\gamma}(s,\boldsymbol{k})\right\}\mathrm{d}s$$

$$= i\int_{-\infty}^{+\infty}j(s,\boldsymbol{x})\left\{\frac{i}{(2\pi)^3}\int\frac{\mathrm{d}^3 k}{2\omega_k}[\bar{z}(\boldsymbol{k})\mathrm{e}^{i\check{k}\cdot x}\gamma(s,\boldsymbol{k}) + \zeta(\boldsymbol{k})\mathrm{e}^{-i\check{k}\cdot x}\bar{\gamma}(s,\boldsymbol{k})]\right\}\mathrm{d}s\mathrm{d}^3 x$$

$$= i\int j(x)\varphi(x)\mathrm{d}^4 x \tag{12.217}$$

by introducing

$$\varphi(x) = \frac{1}{(2\pi)^3} \int \frac{\mathrm{d}^3 k}{2\omega_k} \{\bar{z}(\boldsymbol{k})\mathrm{e}^{i\check{k}.x} + \zeta(\boldsymbol{k})\mathrm{e}^{-i\check{k}.x}\}. \tag{12.218}$$

The function $\varphi(x)$ has the form of a free field written in the complex coordinate system. It is a solution of the Klein–Gordon equation.

The last term in the exponent of formula (12.218) can also be transformed by reintroducing the source as a function of x:

$$-\frac{1}{2}\frac{1}{(2\pi)^3} \int \frac{\mathrm{d}^3 k}{2\omega_k} \iint_{-\infty}^{+\infty} \mathrm{e}^{-i\omega_k|u-s|}\bar{\gamma}(u,\boldsymbol{k})\gamma(s,\boldsymbol{k})\mathrm{d}u\mathrm{d}s$$

$$= -\frac{1}{2}\frac{1}{(2\pi)^3} \int \frac{\mathrm{d}^3 k}{2\omega_k} \iint_{-\infty}^{+\infty} \iint_{\mathbb{R}^3} j(u,\boldsymbol{x})\mathrm{e}^{i\boldsymbol{k}\cdot\boldsymbol{x}}\mathrm{e}^{-i\omega_k|u-s|}j(s,\boldsymbol{y})\mathrm{e}^{-i\boldsymbol{k}\cdot\boldsymbol{y}}\mathrm{d}^3 x\mathrm{d}^3 y\mathrm{d}u\mathrm{d}s$$

$$= -\frac{1}{2}\iint_{\mathbb{R}^4} j(x)\frac{1}{(2\pi)^3}\int\frac{\mathrm{d}^3 k}{2\omega_k}\left\{\mathrm{e}^{i\check{k}.(x-y)}\theta(y_0-x_0) + \mathrm{e}^{-i\check{k}.(x-y)}\theta(x_0-y_0)\right\}j(y)\mathrm{d}^4 x\mathrm{d}^4 y$$

$$= \frac{i}{2}\iint_{\mathbb{R}^4} j(x)\ G_F(x-y)j(y)\mathrm{d}^4 x\mathrm{d}^4 y \tag{12.219}$$

since

$$\frac{1}{(2\pi)^3}\int\frac{\mathrm{d}^3 k}{2\omega_k}\left\{\mathrm{e}^{i\check{k}.(x-y)}\theta(y_0-x_0) + \mathrm{e}^{-i\check{k}.(x-y)}\theta(x_0-y_0)\right\}$$

$$= \frac{1}{(2\pi)^4}\int\mathrm{d}^4 k\frac{1}{k^2-m^2+i\epsilon}\mathrm{e}^{-ik.(x-y)} = -G_F(x-y),$$

which is the expression of the Feynman propagator given in (9.91). Combining these results we have obtained a concise form for the normal kernel of the S-matrix:

$$S_0(j)(\bar{z},\zeta) = \mathrm{e}^{\left[i\int j(x)\varphi(x)\mathrm{d}^4 x + \frac{i}{2}\iint j(x)\,G_F(x-y)j(y)\mathrm{d}^4 x\mathrm{d}^4 y\right]}. \tag{12.220}$$

We are now in position to reverse the process, that is to say, to go from a normal kernel to the operator structure in the Fock space. This is done by replacing in the matrix element every $\bar{z}(\boldsymbol{k})$ and $\zeta(\boldsymbol{k})$ by, respectively, $a^*(\boldsymbol{k})$ and $a(\boldsymbol{k})$. We get the S-matrix of a free scalar field with an external source j

$$S_0(j) = :\mathrm{e}^{\{i\int j(x)\phi(x)\mathrm{d}^4 x + \frac{i}{2}\iint j(x)\,G_F(x-y)j(y)\mathrm{d}^4 x\mathrm{d}^4 y\}}:$$

$$= :\mathrm{e}^{i\int j(x)\phi(x)\mathrm{d}^4 x}:\ \mathrm{e}^{\frac{i}{2}\iint j(x)\,G_F(x-y)j(y)\mathrm{d}^4 x\mathrm{d}^4 y}, \tag{12.221}$$

where we have introduced the operator version $\phi(x)$ of $\varphi(x)$

$$\phi(x) = \frac{i}{(2\pi)^3} \int \frac{d^3k}{2\omega_k} \{a^*(k)e^{i\check{k}.x} + a(k)e^{-i\check{k}.x}\},$$

i.e. the free scalar field in terms of creation and annihilation operators.

We will now treat the general case of the theory of a scalar field in interaction. To be explicit, we will choose as an example the model we studied in Chapter 9, i.e. the $\lambda\phi^4$-theory of a massive scalar field with self-interaction.

Then the free Lagrangian density will be changed by introducing an interaction term $\mathcal{L}(\phi(x)) =: \frac{\lambda}{4!}\phi(x)^4 :$. We choose to Wick order this term in order to eliminate divergent diagrams, like those we called 'tadpoles' in a footnote 6 in section 10.3. This ordering will be as much necessary in the Minkowski space as it was in the Euclidean space. Then the Hamiltonian will become, with $V(\phi(x)) = \mathcal{L}(\phi(x))$,

$$
\begin{aligned}
H(j) &= \int d^3x \{\frac{1}{2}\pi(x)^2 + \frac{1}{2}(\nabla\phi(x))^2 + \frac{m^2}{2}\phi(x)^2 - V(\phi(x)) - j(t,x)\phi(x)\} \\
&= H_0(j) - V(\phi),
\end{aligned}
$$

where $V(\phi) = \int d^3x V(\phi(x))$. We also introduce that

$$V(\phi(t,\cdot)) = \int d^3x \mathcal{L}(\phi(x))$$
$$V(\phi) = V(a^*, a),$$

the last equality expressing the fact that ϕ is a free field, function of the creation and annihilation operators. Remark that $V(a^*, a)$ is normal ordered. We can thus perform all the analysis done in the case of the free Hamiltonian with an external source and leading to the functional integral in (12.217). We obtain, with $a^*(k)$ and $a(k)$ being replaced respectively by $\bar{z}(k)$ and $z(k)$

$$
\begin{aligned}
\mathcal{S}_I(j)(\bar{z},\zeta) &= \lim_{\substack{t_f \to +\infty \\ t_i \to -\infty}} = <z \mid e^{it_f H_0} e^{-i(t_f - t_i)H(j)} e^{-it_i H_0} \mid \zeta> \\
&= \lim_{\substack{t_f \to +\infty \\ t_i \to -\infty}} \int_{\substack{\bar{z}(t,k)=\bar{z}_t(k) \\ z(t_0,k)=\zeta_{t_0}(k)}} \mathcal{D}(\bar{z})\mathcal{D}(z) \exp\left[\frac{1}{2}\frac{1}{(2\pi)^3}\int\frac{d^3k}{2\omega_k}\left\{\bar{z}_t(k)z(t,k) + \bar{z}(t_0,k)\zeta_{t_0}(k)\right\}\right]. \\
&\quad \times \exp\left[\int_{t_0}^t \left\{\frac{1}{(2\pi)^3}\int\frac{d^3k}{2\omega_k}\left[\frac{1}{2}(\dot{\bar{z}}(s,k)z(s,k) - \bar{z}(s,k)\dot{z}(s,k)) - i\omega_k\bar{z}(s,k)z(s,k)\right.\right.\right. \\
&\quad \left.\left.\left. -i\bar{\gamma}(s,k)z(s,k) - i\bar{z}(s,k)\gamma(s,k)\right] - iV(\bar{z}(s,\cdot), z(s,\cdot))\right\}ds\right].
\end{aligned}
\tag{12.222}
$$

Since V is of order larger than 2 in $\bar{z}$ and z, (12.222) is no more a Gaussian functional integral and there is no closed form for its value.

We will now use formal identities.
We start with an obvious identity

$$\phi(x_1)\cdots\phi(x_n) = \prod_{i=1}^{n}\frac{1}{i}\frac{\delta}{\delta j(x_i)}e^{\left\{i\int d^4x j(x)\phi(x)\right\}}\Big|_{j=0}, \qquad (12.223)$$

which leads to

$$e^{\left\{-i\int ds V(\varphi(s,\cdot))\right\}} = e^{\left\{-i\int d^4x \mathcal{L}_I(\varphi(x))\right\}}$$

$$= e^{-i\int d^4x \mathcal{L}_I(\frac{1}{i}\frac{\delta}{\delta j(x)})}e^{\left\{i\int d^4x j(x)\varphi(x))\right\}}\Big|_{j=0} \qquad (12.224)$$

with the equalities to be taken as power series equalities.

We will apply this equality with φ, the free field expressed in terms of the complex coordinates

$$\varphi(x) = \frac{1}{(2\pi)^3}\int\frac{d^3k}{2\omega_k}\left\{\bar{z}(k)e^{ik.x} + z(k)e^{-ik.x}\right\}. \qquad (12.225)$$

Thus, we can rewrite (12.222) with $j = 0$

$$\mathcal{S}_I(\bar{z},\zeta) = \lim_{\substack{t_f\to+\infty \\ t_i\to-\infty}} = <z\,|\,e^{it_f H_0}e^{-i(t_f - t_i)H}e^{-it_i H_0}\,|\,\zeta>$$

$$= \lim_{\substack{t_f\to+\infty \\ t_i\to-\infty}}\int_{\substack{\bar{z}(t,k)=\bar{z}_t(k)\\ z(t_0,k)=\zeta_{t_0}(k)}}\mathcal{D}(\bar{z})\mathcal{D}(z)\exp\left[\frac{1}{2}\frac{1}{(2\pi)^3}\int\frac{d^3k}{2\omega_k}\left\{\bar{z}_t(k)z(t,k) + \bar{z}(t_0,k)\zeta_{t_0}(k)\right\}\right]$$

$$\times\exp\left[\int_{t_0}^{t}\left\{\frac{1}{(2\pi)^3}\int\frac{d^3k}{2\omega_k}\frac{1}{2}(\dot{\bar{z}}(s,k)z(s,k) - \bar{z}(s,k)\dot{z}(s,k)) - i\omega_k\bar{z}(s,k)z(s,k)\right]\right.$$

$$\left. -iV(\bar{z}(s,\cdot),z(s,\cdot))\}ds\right] \qquad (12.226)$$

$$= e^{-i\int d^4x \mathcal{L}_I(\frac{1}{i}\frac{\delta}{\delta j(x)})}\lim_{\substack{t_f\to+\infty \\ t_i\to-\infty}}\int_{\substack{\bar{z}(t,k)=\bar{z}_t(k)\\ z(t_0,k)=\zeta_{t_0}(k)}}\mathcal{D}(\bar{z})\mathcal{D}(z)\exp\left[\frac{1}{2}\frac{1}{(2\pi)^3}\int\frac{d^3k}{2\omega_k}\cdot\right.$$

$$\cdot\left\{\bar{z}_t(k)z(t,k) + \bar{z}(t_0,k)\zeta_{t_0}(k)\right\}\right]\exp\left[\int_{t_0}^{t}\left\{\frac{1}{(2\pi)^3}\int\frac{d^3k}{2\omega_k}\frac{1}{2}(\dot{\bar{z}}(s,k)z(s,k)\right.\right.$$

$$\left.-\bar{z}(s,k)\dot{z}(s,k)) - \bar{z}(s,k)\dot{z}(s,k)) - i\omega_k\bar{z}(s,k)z(s,k) - i\bar{\gamma}(s,k)z(s,k)\right.$$

$$\left.\left.-i\bar{z}(s,k)\gamma(s,k)\}ds\right]\right|_{j=0}$$

$$= e^{i\int d^4x \mathcal{L}_I\left(\frac{1}{i}\frac{\delta}{\delta j(x)}\right)} S_I(j)(\bar{z},\zeta)\Big|_{j=0} \tag{12.227}$$

$$= e^{-i\int d^4x \mathcal{L}_I\left(\frac{1}{i}\frac{\delta}{\delta j(x)}\right)} e^{\left[i\int j(x)\varphi(x)d^4x + \frac{i}{2}\iint j(x)\,G_F(x-y)j(y)d^4x d^4y\right]}\Big|_{j=0}.$$

We are now able to express the general S-matrix as an operator in the Fock space

$$S_I = e^{-i\int d^4x \mathcal{L}_I\left(\frac{1}{i}\frac{\delta}{\delta j(x)}\right)} : e^{\left[i\int j(x)\varphi(x)d^4x\right]} :$$

$$\times e^{\left[\frac{i}{2}\iint j(x)\,G_F(x-y)j(y)d^4x d^4y\right]}\Big|_{j=0}. \tag{12.228}$$

When restricted to the free-field case, i.e. no interaction, this formula can be written as

$$T\left(e^{-i\int d^4x \phi(x)j(x)}\right) =: e^{\left[i\int j(x)\varphi(x)d^4x\right]} : e^{\left[\frac{i}{2}\iint j(x)\,G_F(x-y)j(y)d^4x d^4y\right]}. \tag{12.229}$$

It is the generating function with respect to j of a set of identities expressing T-products of free fields as sums of normal products. This is the so-called Wick theorem.

On the other hand, the S-matrix can be developed on a normal product basis on the Fock space,

$$S_I = \left(\sum_{n=0}^{+\infty}\frac{1}{n!}\int S_n(x_1,\cdots,x_n):\phi(x_1)\cdots\phi(x_n): \prod_{i=1}^{n}d^4x_i\right) < 0|S_I|0 >, \tag{12.230}$$

where

$$< 0|S_I|0 >= e^{-i\int \mathcal{L}_I\left(\frac{1}{i}\frac{\delta}{\delta j(x)}\right)d^4x} e^{\left[\frac{i}{2}\iint j(x)\,G_F(x-y)j(y)d^4x d^4y\right]}\Big|_{j=0} \tag{12.231}$$

The $S_n(x_1,\cdots,x_n)$'s are the (distribution-valued) coefficients of the S-matrix. From formulae (12.228) and (12.231), we have

$$S_n(x_1,\cdots,x_n)$$

$$= e^{-i\int d^4x \mathcal{L}_I\left(\frac{1}{i}\frac{\delta}{\delta j(x)}\right)} i^n j(x_1)\cdots j(x_n) e^{\left[\frac{i}{2}\int j(x)\,G_F(x-y)j(y)d^4x d^4y\right]}\Big|_{j=0}$$

$$= e^{-i \int d^4x \mathcal{L}_I \left(\frac{1}{i} \frac{\delta}{\delta j(x)} \right)} \prod_{i=1}^{n} \frac{\delta}{\delta h(x_i)} e^{i \int j(x) h(x) d^4x} \int j(x) G_F(x-y) j(y) d^4x d^4y$$

$$= \prod_{i=1}^{n} \frac{\delta}{\delta h(x_i)} S_I^{ext}(h,j) \Big|_{h=0} \Big|_{j=0}, \tag{12.232}$$

where by definition

$$S_I^{ext}(h,j) = < 0|S_I|0 >^{-1} e^{-i \int \mathcal{L}_I(\frac{1}{i} \frac{\delta}{\delta j(x)}) d^4x} \left\{ e^{i \int h(x) j(x) d^4x} \right.$$

$$\left. \times e^{\left[\frac{i}{2} \int \int j(x) G_F(x-y) j(y) d^4x d^4y \right]} \right\} \Big|_{j=0} \tag{12.233}$$

We will now establish the link with the results obtained in Section 10.3.
First we have

$$S_I^{ext}(0,j) = Z(j) \tag{12.234}$$

with $Z(j)$ defined by formula (12.137).

In fact, by using the same reasoning as that used to prove formula (12.224), we shows
with $\mathcal{L}_{int}(x) = \mathcal{L}_I(\phi(x))$, that

$$\int e^{i \int \phi(x) j(x) d^4x} e^{-i \int \mathcal{L}_{int}(x) d^4x} \mathcal{D}\mu(\phi)$$

$$= e^{-i \int \mathcal{L}_I(\frac{1}{i} \frac{\delta}{\delta j(x)}) d^4x} \int e^{i \int \phi(x) j(x) d^4x} \mathcal{D}\mu(\phi)$$

$$= N^{-1} e^{-i \int \mathcal{L}_I(\frac{1}{i} \frac{\delta}{\delta j(x)}) d^4x} e^{\left[\frac{i}{2} \int \int j(x) G_F(x-y) j(y) d^4x d^4y \right]}.$$

The normalisation factor N^{-1} will cancel when the same recipe is applied to the
numerator and the denominator of (12.224) and we get

$$< T(\phi(x_1) \cdots \phi(x_n)) > = \prod_{j=1}^{n} \frac{1}{i} \frac{\delta}{\delta j(x_j)} S_I^{ext}(0,j) \Big|_0. \tag{12.235}$$

Now since

$$(\Box_x + m^2) G_F(x-y) = \delta^4(x-y)$$

and

$$(\Box_x + m^2) \frac{\delta}{\delta j(x)} e^{\left\{\frac{i}{2}\int\int j(x) G_F(x-y) j(y) d^4x d^4y\right\}} = i j(x) e^{\left\{\frac{i}{2}\int\int j(x) G_F(x-y) j(y) d^4x d^4y\right\}}$$

we get that if for all $i \neq j$, $x_i \neq x_j$, then

$$\prod_{j=1}^{n} (\Box_{x_j} + m^2) \frac{\delta}{\delta j(x_j)} S_I^{\text{ext}}(0,j)\Big|_{j=0} = \Big[\prod_{k=1}^{n} \frac{\delta}{\delta h(x_k)} S_I^{\text{ext}}(h,j)\Big|_{h=0}\Big]\Big|_{j=0} \tag{12.236}$$

and from formula (12.235) we finally obtains

$$S_n(\phi(x_1), \cdots, \phi(x_n)) = \prod_{i=1}^{n} (\Box_{x_i} + m^2) < T(\phi(x_1) \cdots \phi(x_n)) > . \tag{12.237}$$

12.11 Problems

Problem 12.1 *The Casimir effect:* Following the discussion of the Casimir effect in section 12.5, compute the force per unit surface between two conducting plates at a distance L apart. Show that it corresponds to attraction.

Hint: Start by quantising the electromagnetic field in a box $L \times R \times R$ with $R >> L$. In the sum over the momentum modes introduce a cut-off function to regulate the high modes.

13

Applications

13.1 On Cross Sections

In the previous chapters, we saw how to compute the vacuum-to-vacuum matrix elements of elementary processes and we understood how we can calculate from these quantities the transition amplitudes between the asymptotic ingoing and outgoing states. We now give the rules which make it possible to extract from these amplitudes the numbers which are directly confronted to experiments. We need for that to introduce the notion of *cross section*.

Although we are concerned with fields, the analysis which will follow favours the corpuscular interpretation (it is an excellent justification of the so-called wave–particle duality).

To simplify, we will consider the case of two interacting particles. Usually one of these particles is the target and the other is the projectile which will hit the target. In the laboratory reference frame, this corresponds to one particle, the first one, at rest and the other one in motion. Obviously, in the *centre-of-mass* reference frame, nothing distinguishes the target from the beam.[1]

Classically, a projectile, idealised as a ball of radius r in a linear motion, will hit the target, a ball of radius R, if the straight line followed by the centre of the projectile meets the disk of radius $r + R$ perpendicular to it and centred at the centre of the target.

The surface of this disk, $\sigma_{\text{tot}} = \pi (r + R)^2$, is the total cross section of this collision process.

Let us now consider the interaction of a unidirectional beam of monoenergetic particles I with a target built up of n_c centres of collision, for example C particles per unit volume and suppose that this target is a cylinder of volume $V = Sl$, where l is the length and S is the surface perpendicular to the direction of the ingoing particle. If the target is thin enough, the density n_c not too large, and the number of ingoing particles per unit time N_i large enough but not too large, then the study of this interaction yields

---

[1] The terminology has a purely historical origin. In the early experiments a beam of accelerated particles was hitting a target which was fixed in the laboratory, hence the name of the reference frame. Today, however, most accelerators are colliders in which two beams of particles are accelerated in opposite directions and are brought into a head-on collision. In these cases the 'laboratory' frame is in fact the centre-of-mass one.

From Classical to Quantum Fields. Laurent Baulieu, John Iliopoulos and Roland Sénéor.
© Laurent Baulieu, John Iliopoulos and Roland Sénéor, 2017. Published 2017 by Oxford University Press.

the knowledge of that of a process involving an ingoing particle and a centre of diffusion. The probability of a collision is then given by

$$P = \frac{\sigma_{\text{tot}}}{S} V n_c = \sigma_{\text{tot}} n_c l, \tag{13.1}$$

where the ratio σ_{tot}/S measures the fraction of the surface of the target that may generate a collision. We can reinterpret this equation by noting that the total number of scattered particles, N_{tot}, is proportional to the relative flux Φ_i of the ingoing particles, to the number of outgoing particles and to the number, $N_c = n_c S l$, of scattering centres, the proportionality coefficient being the cross section. The probability is, indeed, given by

$$P = \frac{N_{\text{tot}}}{N_i}, \tag{13.2}$$

where N_i is the number of ingoing particles reaching the target per unit time. Since $\Phi_i = N_i/S$ we have

$$N_{\text{tot}} = N_i n_c l \sigma_{\text{tot}} = \Phi_i n_c S l \sigma_{\text{tot}} = \Phi_i N_c \sigma_{\text{tot}}. \tag{13.3}$$

Clearly, this last relation does not make explicit the nature of the collisions. This nature depends on the level of description of the physical world we choose. If classically we represent the ingoing particles and the scattering centres as solid balls that keep their shapes, then the cross section is given by the geometric formula written at the beginning of this section. If, always classically, the scattering centres are potentials, the cross section is expressed in terms of the potentials. If now, our level of description is quantum physics, the particles are characterised by quantum numbers and we have two different types of collisions: the elastic collisions which preserve the quantum numbers and the inelastic collisions which may change them. Finally, if we accept (in the frame of quantum fields) that some particles can be created or annihilated, the nature and number of components of the result of the collision may sensitively differ from the initial state. This corresponds to one or more reactions of the type

$$I + C \longrightarrow A + B + \dots. \tag{13.4}$$

If we restrict the collisions to be elastic collisions,

$$I + C \longrightarrow I + C, \tag{13.5}$$

we can define, in the same way, the total elastic cross section $\sigma_{\text{tot}}^{\text{el}}$ by

$$N_{\text{tot}}^{\text{el}} = \Phi_i N_c \sigma_{\text{tot}}^{\text{el}}. \tag{13.6}$$

The difference $\sigma_{\text{tot}} - \sigma_{\text{tot}}^{\text{el}}$ is the total cross section of all the inelastic collisions.

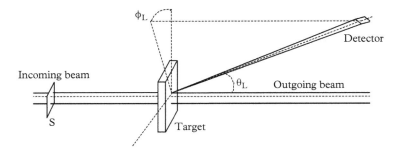

Figure 13.1 *The geometry of a collision.*

To study in more detail the interaction process, we must look at the scattering differ-ential cross section. The definition and the study of this quantity suppose one chooses a reference frame and generally it is either the laboratory reference frame or the centre-of-mass reference frame, the first one being often the natural system related to observation, the second one being the most convenient for computation.

Let us choose for simplification the laboratory reference frame and let us study the elastic scattering in a solid angle characterised by the two angles θ_L and ϕ_L (see Fig. 13.1).

If dN^{el} is the number of particles scattered in the solid angle $d\Omega_L = d\Omega(\phi_L, \theta_L)$ per unit time, we define the differential (elastic) cross section by

$$dN^{el} = \Phi_i N_c \frac{d\sigma_{el}(\phi_L, \theta_L)}{d\Omega_L} \, d\Omega_L. \tag{13.7}$$

Identically we define the differential cross sections corresponding to inelastic colli-sions. It is frequent to use $\sigma_{el}(\phi_L, \theta_L) = \sigma(\Omega)$ instead of $d\sigma_{el}(\phi_L, \theta_L)/d\Omega_L$.

Let us define the differential cross section in the centre-of-mass reference frame with $\sigma_{el}(\phi, \theta)$, ϕ and θ being the Euler angles in the centre-of-mass frame. It is easy to see that since the flux of ingoing particles is the flux relative to the target, then

$$\frac{d\sigma_{el}(\phi_L, \theta_L)}{d\Omega_L} \, d\Omega_L = \frac{d\sigma_{el}(\phi, \theta)}{d\Omega} \, d\Omega. \tag{13.8}$$

For the collisions between elementary particles, the basic units for the length are the **fermi**, $1 fm = 10^{-15} \ m$, and for the cross section the *barn*: $1b = 10^2 \ fm^2 = 10^{-28} \ m^2$.

We now apply these ideas to the study of the non-relativistic quantum interaction of two particles A and B.

Let m_A, m_B, $\mathbf{r}_A$, and $\mathbf{r}_B$ be the respective masses and positions of particles A and B. They are submitted to the action of a potential $V(\mathbf{r}_A - \mathbf{r}_B)$ and their dynamics is governed by the Schrödinger Hamiltonian:

$$H = \frac{\boldsymbol{p}_A^2}{2m_A} + \frac{\boldsymbol{p}_B^2}{2m_B} + V(\boldsymbol{r}_A - \boldsymbol{r}_B). \tag{13.9}$$

After introducing the centre-of-mass variables $r = r_A - r_B$ and $r = \dfrac{m_A r_A + m_B r_B}{m_A + m_B}$, the problem factorises into two parts: the free motion of the centre of mass and the relative motion of a 'particle' of mass $m = \dfrac{m_A m_B}{m_A + m_B}$ submitted to the potential $V(r)$. The initial problem reduces thus to the study of the scattering described by the Schrödinger equation

$$H\psi(r, t) = (-\frac{\hbar^2}{2m}\Delta + V(r))\psi(r, t) = i\hbar\frac{\partial}{\partial t}\psi(r, t). \tag{13.10}$$

We now suppose that the potential decreases sufficiently rapidly at infinity:

$$rV(r) \to 0 \tag{13.11}$$

for $|r| = r \to \infty$ (this condition excludes the Coulomb-type potentials).

Let us suppose that initially the particle has a well-defined momentum $p_i = \hbar k_i$, therefore an energy $E_i = p_i^2/2m$, and let us consider the corresponding stationary solution (we assume evidently that E_i belongs to the continuous spectrum)

$$\psi_i(r, t) = \psi_i(r)\, e^{-i\frac{E_i t}{\hbar}} \tag{13.12}$$

with

$$(-\frac{\hbar^2}{2m}\Delta + V(r))\psi(r) = E\psi(r). \tag{13.13}$$

We are interested in the solutions $\psi^{(+)}(r)$ satisfying the asymptotic condition

$$\psi_{k_i}^{(+)}(r) \underset{r\to\infty}{\to} A\,(e^{ik_i \cdot r} + f(\theta, \phi)\,\frac{e^{ikr}}{r}), \tag{13.14}$$

the angles θ and ϕ representing the polar angles of the vector r in the reference frame in which the origin is the origin of r and the polar axis is the direction of the initial momentum. A is a normalisation constant, independent of r, θ, and ϕ.

This asymptotic condition can be physically understood as a superposition of a plane wave of momentum k_i corresponding to the non-scattered initial wave and an outgoing spherical wave (we can check that the group speed is positive). The differential cross section $\sigma(\Omega)\,d\Omega$, which is the number of particles emitted per unit time and per unit of incident flux in the solid angle $d\Omega$, is therefore equal to the ratio of the flux of diffused particles crossing the element of surface $r^2\,d\Omega$, for $r \to \infty$, by the incident flux.

To the Schrödinger equation is associated a probability current

$$j(r) = \frac{\hbar}{2mi}\{\psi^*(r)(\nabla\psi(r)) - (\nabla\psi^*(r))\psi(r)\}, \tag{13.15}$$

which satisfies the continuity equation

$$\nabla \cdot j + \frac{\partial \rho}{\partial t} = 0, \tag{13.16}$$

where $\rho = |\psi|^2$.

If the ingoing wave is given by $\phi_{k_i}(r) = A\,e^{ik_i \cdot r}$, its flux through a unit of surface normal to the direction of propagation $\hat{k}_i$ is

$$\Phi_i = j_i \cdot \hat{k}_i = \frac{\hbar}{m} AA^* k_i = AA^* v_i. \tag{13.17}$$

The flux of the outgoing spherical wave, with the same normalisation, is given by

$$j \cdot \hat{r} = \Re[AA^* \frac{\hbar}{mi} f^*(\theta, \phi) \frac{e^{-ikr}}{r} \frac{\partial}{\partial r}(f(\theta, \phi)\frac{e^{ikr}}{r})]. \tag{13.18}$$

Consequently, apart from the direction of motion, i.e. for $\theta \neq 0$, and the limit $r \to \infty$ showing that the observation is done at sufficiently large distance from the interaction zone,

$$\frac{d\sigma(\theta, \phi)}{d\Omega} d\Omega = \lim_{r \to \infty} \frac{j \cdot \hat{r}}{j_i \cdot \hat{k}_i} r^2\, d\Omega = |f(\theta, \phi)|^2\, d\Omega, \tag{13.19}$$

thus,

$$\frac{d\sigma(\theta, \phi)}{d\Omega} = |f(\theta, \phi)|^2. \tag{13.20}$$

This result has its real justification in a more realistic formulation of the scattering process. Indeed, the initial state cannot be a plane wave with a well-defined momentum, but must be a wave packet, that is to say a superposition of plane waves centred around a value k_i. In this case, the asymptotic state is a superposition of asymptotic states (13.14). If the parameters which characterise the process are physically reasonable, it is straightforward to show that the differential cross section is effectively given by the formula (13.20).

This simple notion of cross section can be generalised to the framework of relativistic quantum field theories. The quantum aspect will be rendered by the fact that the cross section which measures a probability can be expressed as the square of a transition amplitude, the relativistic aspect being characterised by the way the cross section transforms under the action of the Lorentz group.

13.2   Formal Theory of Scattering in Quantum Mechanics

We saw in Chapter 8 that the solutions of the Schrödinger equation are easily expressed through the retarded Green function. We will show that the knowledge of this function makes it possible to write the transition probabilities. In order to do this, we will first write an integral equation which the Green function satisfies.

13.2.1   An Integral Equation for the Green Function

The Schrödinger equation for the Hamiltonian $H = H_0 + V$, with boundary conditions (q', q), is given by

$$G(q', t'; q, t) = \int_{\substack{q(t')=q' \\ q(t)=q}} \mathcal{D}(q) \exp \frac{\mathrm{i}}{\hbar} \int_t^{t'} \mathrm{d}s \mathcal{L}(q(s), \dot{q}(s)), \tag{13.21}$$

where $\mathcal{L}$ is the corresponding Lagrangian. We will compare this Green function to that of a free theory (that is to say without interaction) given by the Lagrangian $\mathcal{L}_0$ using

$$\mathcal{L} = \mathcal{L}_0 - V. \tag{13.22}$$

Let us write that

$$\mathrm{e}^{\frac{\mathrm{i}}{\hbar} \int_t^{t'} \mathrm{d}s \mathcal{L}(q(s), \dot{q}(s))}$$

$$= \mathrm{e}^{\frac{\mathrm{i}}{\hbar} \int_t^{t'} \mathrm{d}s \mathcal{L}_0(q(s), \dot{q}(s))} \, \mathrm{e}^{-\frac{\mathrm{i}}{\hbar} \int_t^{t'} \mathrm{d}s V(q(s), s)} \tag{13.23}$$

$$= \mathrm{e}^{\frac{\mathrm{i}}{\hbar} \int_t^{t'} \mathrm{d}s \mathcal{L}_0(q(s), \dot{q}(s))} \sum_{n=0}^{\infty} (\frac{\mathrm{i}}{\hbar})^n \frac{1}{n!} (\int_t^{t'} \mathrm{d}s V(q(s), s))^n$$

from which we can write

$$G(q', t'; q, t) = G_0(q', t'; q, t) + G^{(1)}(q', t'; q, t) + G^{(2)}(q', t'; q, t) + \cdots, \tag{13.24}$$

$G_0(q', t'; q, t)$ being the Green function of the free theory. Let us study the nature of the first few terms:

$$G^{(1)}(q', t'; q, t)$$

$$= -\frac{\mathrm{i}}{\hbar} \int_{\substack{q(t')=q' \\ q(t)=q}} \mathcal{D}(q) \mathrm{e}^{\frac{\mathrm{i}}{\hbar} \int_t^{t'} \mathrm{d}s \mathcal{L}_0(q(s), \dot{q}(s))} \int_t^{t'} \mathrm{d}s V(q(s), s)$$

$$= -\frac{\mathrm{i}}{\hbar} \int_t^{t'} \left(\int_{\substack{q(t')=q' \\ q(t)=q}} \mathcal{D}(q) \mathrm{e}^{\frac{\mathrm{i}}{\hbar} \int_t^{t'} \mathrm{d}s \mathcal{L}_0(q(s), \dot{q}(s))} V(q(s), s) \right) \mathrm{d}s$$

$$= -\frac{\mathrm{i}}{\hbar} \int_t^{t'} \mathrm{d}t_1 \int \mathrm{d}^3 q_1 \left(\int_{\substack{q(t')=q' \\ q(t_1)=q_1}} \mathcal{D}(q) \mathrm{e}^{\frac{\mathrm{i}}{\hbar} \int_t^{t'} \mathrm{d}s \mathcal{L}_0} V(q_1, t_1) \int_{\substack{q(t_1)=q_1 \\ q(t)=q}} \mathcal{D}(q) \mathrm{e}^{\frac{\mathrm{i}}{\hbar} \int_t^{t'} \mathrm{d}s \mathcal{L}} \right)$$

$$
= -\frac{i}{\hbar} \int_t^{t'} \left(\int G_0(\boldsymbol{q}', t'; \boldsymbol{q}_1, s_1) \, V(\boldsymbol{q}_1, s_1) \, G_0(\boldsymbol{q}_1, s_1; \boldsymbol{q}, t) \, \mathrm{d}^3 q_1 \right) \mathrm{d}s_1
$$

$$
= -\frac{i}{\hbar} \int G_0(q'; q_1) \, V(q_1) \, G_0(q_1; q) \, \mathrm{d}^4 q_1, \tag{13.25}
$$

where a point of position $\boldsymbol{q}$ at time s is denoted $q = (\boldsymbol{q}, s)$, $\mathrm{d}^4 q = \mathrm{d}^3 \boldsymbol{q} \, \mathrm{d}s$ and the fact that the retarded Green functions $G(\boldsymbol{q}', t'; \boldsymbol{q}, t)$ and $G_0(\boldsymbol{q}', t'; \boldsymbol{q}, t)$ vanish if $t > t'$ has been used.

Similarly,

$$
G^{(2)}(\boldsymbol{q}', t'; \boldsymbol{q}, t)
$$

$$
= \frac{1}{2} \left(\frac{i}{\hbar}\right)^2 \int_{\substack{q(t')=q' \\ q(t)=q}} \mathcal{D}(q) \mathrm{e}^{\frac{i}{\hbar} \int_t^{t'} \mathrm{d}s \mathcal{L}_0(q(s), \dot{q}(s))} \left(\int_t^{t'} \mathrm{d}s V(q(s), s) \right)^2
$$

$$
= \frac{1}{2} \left(\frac{i}{\hbar}\right)^2 \int_t^{t'} \int_t^{t'} \left(\int_{\substack{q(t')=q' \\ q(t)=q}} \mathcal{D}(q) \mathrm{e}^{\frac{i}{\hbar} \int_t^{t'} \mathrm{d}s \mathcal{L}_0(q(s), \dot{q}(s))} V(q(s_1), s_1) V(q(s_2), s_2) \right) \mathrm{d}s_1 \, \mathrm{d}s_2
$$

$$
= \frac{1}{2} \left(\frac{i}{\hbar}\right)^2 \int_t^{t'} \left(\int_{s_1}^{t'} \left(\int_{\substack{q(t')=q' \\ q(t)=q}} \mathcal{D}(q) \mathrm{e}^{\frac{i}{\hbar} \int_t^{t'} \mathrm{d}s \mathcal{L}_0(q(s), \dot{q}(s))} V(q(s_1), s_1) V(q(s_2), s_2) \right) \mathrm{d}s_2 \right) \mathrm{d}s_1
$$

$$
+ \frac{1}{2} \left(\frac{i}{\hbar}\right)^2 \int_t^{t'} \left(\int_t^{s_1} \left(\int_{\substack{q(t')=q' \\ q(t)=q}} \mathcal{D}(q) \mathrm{e}^{\frac{i}{\hbar} \int_t^{t'} \mathrm{d}s \mathcal{L}_0(q(s), \dot{q}(s))} V(q(s_1), s_1) V(q(s_2), s_2) \right) \mathrm{d}s_2 \right) \mathrm{d}s_1.
$$

By a change of variables, the last equality can be rewritten as

$$
G^{(2)}(\boldsymbol{q}', t'; \boldsymbol{q}, t)
$$

$$
= \frac{1}{2} \left(\frac{i}{\hbar}\right)^2 \int_t^{t'} \int_{s_1}^{t'} \left(\iint G_0(q'; q_2) V(q_2) G_0(q_2; q_1) V(q_1) G_0(q_1; q) \, \mathrm{d}^3 q_1 \, \mathrm{d}^3 q_2 \right) \mathrm{d}s_1 \mathrm{d}s_2
$$

$$
+ \frac{1}{2} \left(\frac{i}{\hbar}\right)^2 \int_t^{t'} \int_t^{s_1} \left(\iint G_0(q'; q_1) V(q_1) G_0(q_1; q_2) V(q_2) G_0(q_2; q) \, \mathrm{d}^3 q_1 \, \mathrm{d}^3 q_2 \right) \mathrm{d}s_1 \mathrm{d}s_2
$$

$$
= \left(\frac{i}{\hbar}\right)^2 \iint G_0(q'; q_2) V(q_2) G_0(q_2; q_1) V(q_1) G_0(q_1; q) \, \mathrm{d}^4 q_1 \, \mathrm{d}^4 q_2. \tag{13.26}
$$

This analysis can be repeated for the other terms of the expansion (13.24) and we find that

$$
G(q'; q) = G(\boldsymbol{q}', t'; \boldsymbol{q}, t)
$$

$$
= G_0(\boldsymbol{q}', t'; \boldsymbol{q}, t) - \frac{i}{\hbar} \int G_0(q'; q_1) V(q_1) G_0(q_1; q) \, \mathrm{d}q_1 + \cdots \tag{13.27}
$$

$$
+ \left(-\frac{i}{\hbar}\right)^n \iint G_0(q'; q_n) V(q_n) G_0(q_n; q_{n-1}) \cdots V(q_1) G_0(q_1; q) \, \mathrm{d}^4 q_n \cdots \mathrm{d}^4 q_1 + \cdots
$$

$$= G_0(q',t';q,t) + \int G_0(q';p)V(p)\left(G_0(p;q) + \int G_0(p;q_1)V(q_1)G_0(q_1;q)\,\mathrm{d}^4q_1\right.$$

$$+ \int\int G_0(p;q_1)\,V(q_1)\,G_0(q_1;q_2)\,V(q_2)\,G_0(q_2;q)\,\mathrm{d}^4q_2\,\mathrm{d}^4q_1 + \cdots\bigg).$$

$$= G_0(q';q) + \int G_0(q';p)\,V(p)\,G(p;q)\,\mathrm{d}^4q. \tag{13.28}$$

Given a wave function $\phi_0(q)$ at the initial time t, the wave function at time t' due to the complete dynamics, that is to say including the action of the potential V, is given by

$$\psi(q',t') = \int G(q',t';q,t)\phi_0(q)\,\mathrm{d}^3q. \tag{13.29}$$

Introducing the result at time t' of the evolution under the free dynamics, that is to say H_0, of the initial wave function,

$$\phi(q',t') = \int G_0(q',t';q,t)\phi_0(q)\,\mathrm{d}^3q \tag{13.30}$$

and using Eq. (13.28), we get

$$\psi(q') = \phi(q') + \int G_0(q';q)\,V(q)\,\psi(q)\,\mathrm{d}^4q. \tag{13.31}$$

Iterating this equation, we obtain the Born expansion

$$\psi(q') = \phi(q') - \frac{\mathrm{i}}{\hbar}\int G_0(q';q_1)\,V(q_1)\,\phi(q_1)\,\mathrm{d}^4q_1 + \cdots$$

$$+ (-\frac{\mathrm{i}}{\hbar})^n\int\int G_0(q';q_n)V(q_n)G_0(q_n;q_{n-1})\cdots V(q_1)\phi(q_1)\,\mathrm{d}^4q_n\cdots\mathrm{d}^4q_1 + \cdots.$$

Reduced to the first two terms, the last equality is known as the Born approximation.

It is easy to find the asymptotic formula (13.14) by inserting in the preceding equation an initial wave function ϕ of the plane wave type.

The physical problem that we consider is that of a beam of identical particles, for example electrons hitting a thin aluminium sheet. If the beam is collimated the experience shows that one part of the beam crosses the sheet without change while another part is diffused in all the directions. We are interested in the study of this scattering process.

By assuming that either the sheet is very thin or the interaction corresponding to each of the atoms very weak, we can reduce[2] the problem to that of the interaction of the electron with one of the atoms. This system can be described, approximately, by a

---

[2] Under this hypothesis, we can neglect the interaction of an electron with all the other atoms or with other electrons.

Schrödinger equation with Hamiltonian $H = H_0 + V$, V being a potential characterising the effect of the atom. This system remains nevertheless very complex: the atom can be excited or ionised, and the electron can lose energy. To simplify a little more, we will assume that we are dealing with an elastic scattering; that is to say, the atom remains in the same energy state before and after the scattering.

Without interaction, the particle evolves according to the free dynamics given by H_0 and, in order to simplify even further, we will assume that we have replaced the electron by a massive scalar particle in which case $H_0 = -\Delta/2m$. The scattering problem is then more simply analysed if one assumes that the range of the interaction V is finite. In this case, we can measure the effect of the diffusion by computing the probability that a free particle of momentum p_i becomes after a long enough time a free particle of momentum p_f. Since the energy is conserved, p_i and p_f have the same module and the scattering is characterised by the angle between these two vectors. In fact, with the hypothesis of finite range of the interaction, this probability could be measured by taking initial and final states localised out of the area of influence of the interaction.

As we have already discussed, the notion of initial and final times, i.e. the beginning of the experiment and the moment when the measurement is done, is translated mathematically into an infinitely remote time in the past $(t_i = -\infty)$ and an infinitely distant time in the future $(t_f = +\infty)$.

The physical picture is quite simple. The initial state $(t \to -\infty)$ is formed by two particles, far apart from each other, moving one towards the other. Since the particles are assumed to be localised, they must be described by two well-separated, wave packets. If the distance between them is very large and the interaction of short range, the two particles can be assumed to be free. Similarly, the final state after the collision $(t \to +\infty)$ is formed by two free particle wave packets moving away one from the other. This is the correct physical description. The trouble is that it is rather awkward to work with wave packets and we would like to replace them by the much simpler plane waves, but, by doing so, we lose the property of localisation. We go around this difficulty by a technical artefact, which will be quite useful in several other cases: we assume that at the beginning and at the end of the experiment, there is no interaction and that the interaction appears and then disappears gradually at the moment of the measurement. We still have to face the technical problem of the control of the double limit. We can solve this problem thanks to, for example, a mathematical device consisting in the replacement of V by $e^{-|\varepsilon|t}V$, ε going to 0 at the end of the calculation. This device simulates what is called *an adiabatic switching on and off of the interaction*. We shall come back to this point shortly.

So, let us consider a free state of momentum p with wave function

$$\phi_p(x, t) = \frac{1}{(2\pi)^{\frac{3}{2}}} e^{ip \cdot x - iEt}, \tag{13.32}$$

where $E = p^2/2m$ and $\phi(x, t)$ is a solution of the Schrödinger equation

$$i\frac{\partial}{\partial t}\phi(x, t) = H_0\phi(x, t) \tag{13.33}$$

and

$$\phi(\boldsymbol{x}',t') = \int G_0(\boldsymbol{x}',t';\boldsymbol{x},t)\,\phi(\boldsymbol{x},t)\,\mathrm{d}^3\boldsymbol{x}. \tag{13.34}$$

The solution of the Schrödinger equation with Hamiltonian H, equal to $\phi_i(\boldsymbol{x},t)$ at the time t, can be written as

$$\psi_i(\boldsymbol{x}',t') = \int G(\boldsymbol{x}',t';\boldsymbol{x},t)\,\phi_i(\boldsymbol{x},t)\,\mathrm{d}^3\boldsymbol{x}. \tag{13.35}$$

If the initial condition was a plane wave of momentum $\boldsymbol{k}_i$ at time $t = -\infty$, the solution at a later time t' can be written as

$$\psi_i^{(+)}(\boldsymbol{x}',t') = \lim_{t\to-\infty} \int G(\boldsymbol{x}',t';\boldsymbol{x},t)\,\phi_i(\boldsymbol{x},t)\,\mathrm{d}^3\boldsymbol{x}. \tag{13.36}$$

Therefore, the probability amplitude for the initial state ϕ_i at time $t = -\infty$ to become the final state ϕ_f at time t' is given by

$$
\begin{aligned}
(\phi_f, \psi_i^{(+)}) &= \int \phi_i(\boldsymbol{x}',t')^\dagger \psi_i^{(+)}(\boldsymbol{x}',t')\,d^3\boldsymbol{x}' \\
&= \lim_{t\to-\infty} \int \phi_f^\dagger(\boldsymbol{x}',t')\,G(\boldsymbol{x}',t';\boldsymbol{x},t)\,\phi_i(\boldsymbol{x},t)\,\mathrm{d}^3\boldsymbol{x}'\,\mathrm{d}^3\boldsymbol{x}.
\end{aligned}
\tag{13.37}
$$

Let us rewrite this last expression after introducing the bra and ket notations,

$$\phi_k(\boldsymbol{x},t) = <\boldsymbol{k}|\boldsymbol{x}>\,\mathrm{e}^{-\mathrm{i}Et}$$
$$G(\boldsymbol{x}',t';\boldsymbol{x},t) = <\boldsymbol{x}'|\mathrm{e}^{\mathrm{i}(t'-t)H}|\boldsymbol{x}>;$$

thus,

$$(\phi_f, \psi_i^{(+)}) \tag{13.38}$$
$$= \lim_{t\to-\infty} \int <\boldsymbol{k}_f'|\boldsymbol{x}'>\,\mathrm{e}^{\mathrm{i}t'E_f}\,<\boldsymbol{x}'|\mathrm{e}^{\mathrm{i}(t'-t)H}|\boldsymbol{x}><\boldsymbol{x}|\boldsymbol{k}_i>\,\mathrm{e}^{-\mathrm{i}tE_i}\,\mathrm{d}^3\boldsymbol{x}\,\mathrm{d}^3\boldsymbol{x}'$$
$$= \lim_{t\to-\infty} \int <\boldsymbol{k}_f'|\boldsymbol{x}'>\,\mathrm{e}^{\mathrm{i}t'E_f}\,<\boldsymbol{x}'|\mathrm{e}^{\mathrm{i}(t'-t)H}|\boldsymbol{k}_i>\,\mathrm{e}^{-\mathrm{i}tE_i}\,\mathrm{d}^3\boldsymbol{x}'$$
$$= \lim_{t\to-\infty} \int <\boldsymbol{k}_f'|\boldsymbol{x}'>\,\mathrm{e}^{\mathrm{i}t'E_f}\,<\boldsymbol{x}'|\mathrm{e}^{\mathrm{i}(t'-t)H}\,\mathrm{e}^{-\mathrm{i}tH_0}|\boldsymbol{k}_i>\,\mathrm{d}^3\boldsymbol{x}\,\mathrm{d}^3\boldsymbol{x}'$$
$$= \lim_{t\to-\infty} <\boldsymbol{k}_f|\mathrm{e}^{\mathrm{i}t'H_0}\mathrm{e}^{\mathrm{i}(t'-t)H}\,\mathrm{e}^{-\mathrm{i}tH_0}|\boldsymbol{k}_i> = \lim_{t\to-\infty} <\boldsymbol{k}_f|U(t',t)|\boldsymbol{k}_i>$$
$$= <\boldsymbol{k}_f|U(t',-\infty)|\boldsymbol{k}_i>. \tag{13.39}$$

It follows that

$$\lim_{t'\to\infty} (\phi_f, \psi_i^{(+)}) = <\boldsymbol{k}_f|U(\infty,-\infty)|\boldsymbol{k}_i> = <\boldsymbol{k}_f|S|\boldsymbol{k}_i> = S_{fi} \tag{13.40}$$

is the S-matrix element between two plane waves ϕ_i and ϕ_f. Applying the equality (13.31) for $\psi_i^{(+)}$ to the expression of the S-matrix element we get

$$S_{fi} = \delta^3(\boldsymbol{k}_f - \boldsymbol{k}_i) + \lim_{t' \to \infty} \int \phi_f^\dagger(\boldsymbol{x}', t') G_0(\boldsymbol{x}', t'; \boldsymbol{x}, t) V(\boldsymbol{x}, t) \psi_i^{(+)}(\boldsymbol{x}, t) \, \mathrm{d}^4 x \, \mathrm{d}^3 \boldsymbol{x}'. \quad (13.41)$$

The first term of this equality corresponds to the absence of interaction, the plane wave propagating without being perturbed. The second one corresponds to the scattering by the potential. Using the Born expansion, it can be developed in powers of V. The different terms which appear are what is called the perturbative expansion of the S-matrix element under consideration.

We will give a more rigorous justification of the introduction of the evolution operator $U(t', t)$ and its relation with the S-matrix.

The interest of the unitary operator U is that it is the evolution operator for the states in the interaction representation:

$$|\psi_I(t)> = U(t, t_0)|\psi_I(t_0)>. \quad (13.42)$$

The last equality of Eq. (13.39) has shown the appearance of the ket $U(t', -\infty)|\phi_i>$. We shall give a meaning to this notation through the notion of the adiabatic switching.

As we explained earlier, replacing the potential V by $\mathrm{e}^{-\varepsilon|t|} V$ does not mean that the physical potential depends effectively on time. It is just a formal way of taking into account the physical assumption that the range of the potential is sufficiently small so that the particle in the initial state has a very large probability to be far from the scattering area. $1/\varepsilon$ gives the time beyond which the effective interaction is nearly 0. We can show in fact the following result: if we solve the Schrödinger equation with potential $\mathrm{e}^{-\varepsilon|t|} V$ and the initial condition

$$\psi(\boldsymbol{x}, t) \underset{r \to -\infty}{\to} \mathrm{e}^{i\boldsymbol{k}\cdot\boldsymbol{x} - iE_k t}, \quad (13.43)$$

then at the limit $\varepsilon \to 0$, for each time t, $\psi(\boldsymbol{x}, t)$ is a solution of the stationary equation

$$H\psi = E\psi. \quad (13.44)$$

Thus, the adiabatic switching transforms each eigenfunction of H_0 into an eigenfunction of H with the same eigenvalue.

We now give a meaning to $U(t, s)$ when one or two of the arguments are infinite.

For this purpose, we rewrite the equation satisfied by U in the adiabatic approximation, replacing, since we are in the interaction representation, $V(t) = \mathrm{e}^{itH_0} V \mathrm{e}^{-itH_0}$ with $V(t)\mathrm{e}^{-\varepsilon|t|}$:

$$i\partial_t U_\varepsilon(t, t') = \mathrm{e}^{-\varepsilon|t|} V(t) U_\varepsilon(t, t'). \quad (13.45)$$

After integration we get

$$U_\varepsilon(t, t') = 1 - i \int_{t'}^t \mathrm{e}^{-\varepsilon|s|} V(s) U_\varepsilon(s, t') \, \mathrm{d}s. \quad (13.46)$$

Taking the complex conjugate of this equation and permuting t and t', we get

$$U_\varepsilon(t,t') = 1 - i \int_{t'}^{t} U_\varepsilon(t,s) V(s) e^{-\varepsilon|s|} \, ds = 1 + i \int_{t'}^{t} e^{-\varepsilon|s|} V(s) U_\varepsilon(s,t) \, ds. \qquad (13.47)$$

As for a usual evolution operator, $U_\varepsilon(t',t)$ satisfies the semi-group and unitarity conditions and we define the expressions at the large time limit by

$$U(t,\pm\infty) = 1 - i \lim_{\varepsilon \to 0^+} \int_{\pm\infty}^{t} U(t,s) V(s) e^{-\varepsilon|s|} \, ds \qquad (13.48)$$

and check that

$$U(t,-\infty) = U(t,s) U(s,-\infty). \qquad (13.49)$$

Introducing the orthonormal basis of the eigenstates of H_0 (we assume that H_0 has no bound states) given by the ϕ_k's and the fact that this basis is complete for L^2,

$$\int |\phi_k><\phi_k| \, d^3k = 1 \qquad (13.50)$$

we get

$$U(0,\pm\infty)$$
$$= 1 - i \lim_{\varepsilon \to 0^+} \int_{\pm\infty}^{0} e^{-\varepsilon|s|} e^{isH} e^{-isH_0} V(s) \left(\int |\phi_k><\phi_k| \, d^3k \right) ds$$
$$= 1 - i \lim_{\varepsilon \to 0} \int_{\pm\infty}^{0} \left(\int e^{-\varepsilon|s|} e^{isH} e^{-isE_k} V|\phi_k><\phi_k| \, ds \right) d^3k$$
$$= 1 + \lim_{\varepsilon \to 0^+} \int \frac{1}{E_k - H \mp i\varepsilon} V |\phi_k><\phi_k| \, d^3k$$
$$= \lim_{\varepsilon \to 0^+} \int \frac{\mp i\varepsilon}{E_k - H \mp i\varepsilon} |\phi_k><\phi_k| \, d^3k. \qquad (13.51)$$

Going back to the definition of $\psi_i^{(+)}(t)$, we find that

$$|\psi_i^{(+)}(t)> = e^{itH_0} e^{-itH} \lim_{\varepsilon \to 0^+} \int \frac{i\varepsilon}{E_k - H + i\varepsilon} |\phi_k><\phi_k|\phi_i> \, d^3k$$
$$= e^{itH_0} e^{-itH} \lim_{\varepsilon \to 0^+} \frac{i\varepsilon}{E_i - H + i\varepsilon} |\phi_i>$$
$$= e^{itH_0} e^{-itH} (|\phi_i> + \lim_{\varepsilon \to 0^+} \frac{1}{E_i - H + i\varepsilon} V |\phi_i>); \qquad (13.52)$$

thus, with the notation $\psi_i^{(\pm)}(0) = \psi_i^{(\pm)}$,

$$|\psi_i^{(\pm)}> = |\phi_i> + \lim_{\varepsilon \to 0^+} \frac{1}{E_i - H \pm i\varepsilon} V |\phi_i>; \qquad (13.53)$$

$$U(0, -\infty) = \int |\psi_k^{(+)}> <\phi_k| \, \mathrm{d}^3 k \qquad (13.54)$$

and

$$U(0, -\infty)|\phi_k> = |\psi_k^{(+)}> = \Omega^{(+)}|\phi_k>, \qquad (13.55)$$

where we introduced the Møller operator $\Omega^{(+)} = U(0, -\infty)$, which transforms the initial asymptotic states into scattering states. In fact, we can show, under very general conditions on the potential V, that if $H = H_0 + V$ and if $\psi(t)$ is a solution of the Schrödinger equation, with H for Hamiltonian, orthogonal to all the bound states (that is to say belonging to the continuous spectrum of H), then there exists a unique solution of the Schrödinger equation of Hamiltonian H_0 such that[3]

$$s - \lim_{t \to -\infty} (\psi(t) - \phi(t)) = 0 \qquad (13.56)$$

Conversely, to each solution of the free equation corresponds a scattering solution of the complete equation.

It thus follows that

$$s - \lim_{t \to -\infty} U(0, t) = s - \lim_{t \to -\infty} \mathrm{e}^{iHt} \, \mathrm{e}^{-itH_0} = \Omega^{(+)} \qquad (13.57)$$

and

$$H\Omega^{(+)} = \Omega^{(+)} H_0, \qquad (13.58)$$

where

$$H|\psi_i^{(+)}> = H\Omega^{(+)}|\phi_i> = \Omega^{(+)} H_0|\phi_i> = E_i \Omega^{(+)}|\phi_i> = E_i|\psi_i^{(+)}>. \qquad (13.59)$$

In the same way, we define

$$\Omega^{(-)} = U(0, \infty) \qquad (13.60)$$

which transforms the final asymptotic states into scattering states.

Using the semi-group properties of U and the limiting procedure given by (13.58)

$$U(\infty, -\infty) = U(\infty, 0)U(0, -\infty) = U(0, \infty)^\dagger U(0, -\infty) \qquad (13.61)$$

---

[3] We say that a sequence of vectors of a Hilbert space Φ_i converges strongly to 0: $s - \lim_i \Phi_i = 0$, if it converges in norm, that is to say if $\lim_i \|\Phi_i\| = 0$. It converges weakly, $w - \lim_i \Phi_i = 0$, if for any fixed vector Ψ, $\lim_i (\Phi_i, \Psi) = 0$. We have a similar definition for the operators. If A_i is a sequence of operators, we say that it converges strongly (or weakly) to A if for all vector Ψ in the domain of definition of the A_i's and of A, the sequence of vectors $A_i \Psi$ converge strongly (or weakly) to $A\Psi$.

we can express the S-matrix in terms of the Møller operator:

$$S = \Omega^{(-)\dagger}\Omega^{(+)}. \tag{13.62}$$

The Møller operators are isometric (and not unitary although they are strong limits of unitary operators), that is to say they preserve the norms, because

$$\Omega^{(+)\dagger}\Omega^{(+)} = \int\int |\phi_k><\psi_k^{(+)}|\psi_p^{(+)}><\phi_p|\,\mathrm{d}^3k\,\mathrm{d}^3p = \int |\phi_k><\phi_k|\,\mathrm{d}^3k = 1, \tag{13.63}$$

since, even at the limit, the U's remain unitary and thus

$$<\psi_k^{(+)}|\psi_p^{(+)}> = <\phi_k|\phi_p> \tag{13.64}$$

and

$$\Omega^{(+)}\Omega^{(+)\dagger} = \int\int |\psi_p^{(+)}><\phi_p|\phi_k><\psi_k^{(+)}|\,\mathrm{d}^3k\,\mathrm{d}^3p = \int |\psi_p^{(+)}><\psi_p^{(+)}|\,\mathrm{d}^3p. \tag{13.65}$$

But this last expression which projects on the eigenstates of H is the identity only if H has no bound states; otherwise,

$$\int |\psi_p^{(+)}><\psi_p^{(+)}|\,\mathrm{d}^3p = 1 - \sum_\alpha |\psi_\alpha^{(B)}><\psi_\alpha^{(B)}|, \tag{13.66}$$

the sum being on the bound states of H satisfying $H|\psi_\alpha^{(B)}> = E_\alpha|\psi_\alpha^{(B)}>$.

If ψ is an eigenstate of H of energy E

$$\Omega^{(+)\dagger}|\psi> = \lim_{\varepsilon\to 0^+}\varepsilon\int \mathrm{e}^{\mathrm{i}H_0 t}\,\mathrm{e}^{-\mathrm{i}Ht}|\psi>\,\mathrm{d}t = \lim_{\varepsilon\to 0^+}\frac{\mathrm{i}\varepsilon}{E-H_0+\mathrm{i}\varepsilon}|\psi>. \tag{13.67}$$

If ψ is a bound state of H, its energy is lower than the energy of the eigenstates of H_0 and therefore $E - H_0$ never vanishes. It follows from these remarks that we can take the limit $\varepsilon \to 0$ and we find 0. Thus,

$$\Omega^{(+)\dagger}|\psi_\alpha^{(B)}> = 0. \tag{13.68}$$

Evidently all the preceding properties extend to $\Omega^{(-)}$. This result makes it possible to prove the unitarity of the S-matrix

$$S^\dagger S = \Omega^{(+)\dagger}\Omega^{(-)}\Omega^{(-)\dagger}\Omega^{(+)} = \Omega^{(+)\dagger}[1 - \sum |\psi_\alpha^{(B)}><\psi_\alpha^{(B)}|]\Omega^{(+)} = 1$$

and

$$SS^\dagger = \Omega^{(-)\dagger}\Omega^{(+)}\Omega^{(+)\dagger}\Omega^{(-)} = 1 - \Omega^{(-)\dagger}\sum |\psi_\alpha^{(B)}><\psi_\alpha^{(B)}|\Omega^{(-)} = 1$$

as well as

$$[S, H_0] = 0, \tag{13.69}$$

which is obtained using (13.58) and an identical relation for $\Omega^{(-)}$.

We now give an expression for the elements of the S-matrix:

$$S_{fi} = < \phi_f|S|\phi_i > = < \phi_f|\Omega^{(-)\dagger}\Omega^{(+)}|\phi_i > = < \psi_f^{(-)}|\psi_i^{(+)} > . \tag{13.70}$$

Using the identity (13.52) for $t = 0$

$$
\begin{aligned}
S_{fi} &= < \psi_f^{(-)}|\psi_i^{(+)} > = < \psi_f^{(+)}|\psi_i^{(+)} > + < \psi_f^{(-)} - \psi_f^{(+)}|\psi_i^{(+)} > \\
&= \delta(\boldsymbol{k}_f - \boldsymbol{k}_i) + \lim_{\varepsilon \to 0^+} < \phi_f|V(\frac{1}{E_f - H + i\varepsilon} - \frac{1}{E_f - H - i\varepsilon})|\psi_i^{(+)} > \\
&= \delta(\boldsymbol{k}_f - \boldsymbol{k}_i) + \lim_{\varepsilon \to 0^+} < \phi_f|V(\frac{1}{E_f - E_i + i\varepsilon} - \frac{1}{E_f - E_i - i\varepsilon})|\psi_i^{(+)} > \\
&= \delta(\boldsymbol{k}_f - \boldsymbol{k}_i) + \lim_{\varepsilon \to 0^+} \frac{2i\varepsilon}{(E_f - E_i)^2 + \varepsilon^2} < \phi_f|V|\psi_i^{(+)} > \\
&= \delta(\boldsymbol{k}_f - \boldsymbol{k}_i) - 2i\pi \delta(E_f - E_i) < \phi_f|V|\psi_i^{(+)} > .
\end{aligned} \tag{13.71}
$$

We can show equivalently that

$$S_{fi} = \delta(\boldsymbol{k}_f - \boldsymbol{k}_i) - 2i\pi \delta(E_f - E_i) < \psi_f^{(-)}|V|\phi_i > . \tag{13.72}$$

Equations (13.71) and (13.72) show that an S-matrix element depends only on the matrix element $< \phi_f|V|\psi_i^{(+)} >$ computed on the *energy shell* $E_f = E_i$ and moreover on this energy shell we have

$$< \phi_f|V|\psi_i^{(+)} > = < \psi_f^{(-)}|V|\phi_i > . \tag{13.73}$$

In the expression giving the S-matrix element (13.71) the first term disappears in case the transition $i \to f$ involves different states. It corresponds to a purely elastic forward scattering, that is to say in the direction of the incident particle. The second term expresses the fact that transitions between different states, that is to say states having different quantum numbers, are possible if they have the same energy.

Remark that at lowest order in the Born expansion $\psi_i^{(+)} = \phi_i$ and therefore

$$< \phi_f|V|\psi_i^{(+)} > = \tilde{V}(\boldsymbol{p}_f - \boldsymbol{p}_i), \tag{13.74}$$

which is the Fourier transform of the potential.

Expressions like (13.72) of the S-matrix elements lead to the introduction of the T operator through the formula

$$S = 1 + \mathrm{i}T; \tag{13.75}$$

thus,

$$< \phi_f | T | \phi_i >= T_{fi} = -2\pi \delta(E_f - E_i) < \phi_f | V | \psi_i^{(+)} > . \tag{13.76}$$

The unitarity of the S-matrix leads to

$$TT^* = T^*T = -\mathrm{i}(T - T^*) = 2\Im T. \tag{13.77}$$

13.2.2 The Cross Section in Quantum Mechanics

13.2.2.1 The Lippmann–Schwinger Equation

It is possible to recover the results of the first section in this formalism and in particular to prove formula (13.14).

Starting from Eq. (13.53) and using the operator identity

$$\frac{1}{A} - \frac{1}{B} = \frac{1}{B}(B-A)\frac{1}{A} \tag{13.78}$$

with $A = E - H \pm \mathrm{i}\varepsilon$ and $B = E - H_0 \pm \mathrm{i}\varepsilon$ we find the *Lippmann–Schwinger equation*

$$
\begin{aligned}
|\psi_i^{(\pm)} > = & = |\phi_i > + \lim_{\varepsilon \to 0^+} \frac{1}{E_i - H \pm \mathrm{i}\varepsilon} V |\phi_i > \\
= & |\phi_i > + \lim_{\varepsilon \to 0^+} \{ \frac{1}{E_i - H_0 \pm \mathrm{i}\varepsilon} V |\phi_i > \\
& + \frac{1}{E_i - H_0 \pm \mathrm{i}\varepsilon} V \frac{1}{E_i - H \pm \mathrm{i}\varepsilon} V |\phi_i > \} \\
= & |\phi_i > + \lim_{\varepsilon \to 0^+} \frac{1}{E_i - H_0 \pm \mathrm{i}\varepsilon} V \{ |\phi_i > + \frac{1}{E_i - H \pm \mathrm{i}\varepsilon} V |\phi_i > \} \\
= & |\phi_i > + \lim_{\varepsilon \to 0^+} \frac{1}{E_i - H_0 \pm \mathrm{i}\varepsilon} V |\psi_i^{(\pm)} > .
\end{aligned}
\tag{13.79}
$$

Projecting the Lippmann–Schwinger equation on the position basis we get

$$\psi_i^{(\pm)}(\boldsymbol{r}) = \phi_i(\boldsymbol{r}) + \int < \boldsymbol{r} | \lim_{\varepsilon \to 0^+} \frac{1}{E_i - H_0 \pm \mathrm{i}\varepsilon} | \boldsymbol{r}' > V(\boldsymbol{r}') \psi_i^{(\pm)}(\boldsymbol{r}') \, \mathrm{d}\boldsymbol{r}' \tag{13.80}$$

and

$$
\begin{aligned}
&< r| \lim_{\varepsilon \to 0^+} \frac{1}{E_i - H_0 \pm i\varepsilon} |r' > \\
&= \int < r|p > \lim_{\varepsilon \to 0^+} < p|\frac{1}{E_i - H_0 \pm i\varepsilon}|p' >< p'|r' > \, dp' \, dp \\
&= \lim_{\varepsilon \to 0^+} \frac{1}{(2\pi)^3} \int \frac{1}{E_i - \frac{p^2}{2m} \pm i\varepsilon} \, e^{-ip.(r'-r)} \, dp \\
&= \lim_{\varepsilon \to 0^+} \frac{2m}{(2\pi)^3} \int \frac{1}{p_i - p^2 \pm i\varepsilon} \, e^{-ip.(r'-r)} \, dp \\
&= -2m\frac{1}{4\pi} \frac{e^{\pm ip_i|r'-r|}}{|r'-r|},
\end{aligned} \tag{13.81}
$$

where we used $E_i = p_i^2/2m$ and a redefinition of ε.

We now check that the solution of the integral equation

$$
\psi_i^{(+)}(r) = \phi_i(r) - 2m\frac{1}{4\pi} \int \frac{e^{ip_i|r'-r|}}{|r'-r|} \, V(r') \, \psi_i^{(+)}(r') \, dr' \tag{13.82}
$$

satisfies the asymptotic condition (13.14).

For this purpose we use

$$
|r'-r| \underset{r\to\infty}{\to} r - r'.\hat{r} + \frac{1}{2r}(\hat{r} \wedge r')^2 + \mathcal{O}(\frac{1}{r^2}) \tag{13.83}
$$

and therefore that

$$
\frac{e^{ip_i|r'-r|}}{|r'-r|} \underset{r\to\infty}{\to} \frac{e^{ip_i r}e^{-ip_i r'.\hat{r}}}{r}[1 + \frac{r'.\hat{r}}{r} + \frac{ip_i}{2r}(\hat{r} \wedge r')^2 + \mathcal{O}(\frac{1}{r^2})] \tag{13.84}
$$

If we now suppose that the potential has a finite range,

$$
V(r') \simeq 0 \quad \text{if} \quad r' \geq a \tag{13.85}
$$

and that

$$
r \gg a \quad \text{and} \quad r \gg p_i a^2, \tag{13.86}
$$

then

$$
\int \frac{e^{ip_i|r'-r|}}{|r'-r|} \, V(r') \, \psi_i^{(+)}(r') \, dr' \simeq \frac{e^{ip_i r}}{r} \int e^{-ip_i r'.\hat{r}} \, V(r') \, \psi_i^{(+)}(r') \, dr'
$$

and

$$\psi_i^{(+)}(\mathbf{r}) \;\to_{r\to\infty}\; \frac{1}{(2\pi)^{\frac{3}{2}}} e^{i\mathbf{p}_i \cdot \mathbf{r}} - \frac{2m}{4\pi} \frac{e^{ipr}}{r} \int e^{-i\mathbf{r}' \cdot \mathbf{p}_f} V(\mathbf{r}') \psi_i^{(+)}(\mathbf{r}') \, d\mathbf{r}'$$

$$= \frac{1}{(2\pi)^{\frac{3}{2}}} [e^{i\mathbf{p}_i \cdot \mathbf{r}} + f(p,\theta,\phi) \frac{e^{ipr}}{r}] \tag{13.87}$$

with the ingoing momentum $\mathbf{p}_f = p\hat{\mathbf{r}}$, $|\mathbf{p}_f| = |\mathbf{p}_i| = p$ and

$$f(p,\theta,\phi) = -\frac{(2\pi)^{\frac{3}{2}}}{4\pi} 2m \int e^{-i\mathbf{r}' \cdot \mathbf{p}_f} V(\mathbf{r}') \psi_i^{(+)}(\mathbf{r}') \, d\mathbf{r}', \tag{13.88}$$

thus reintroducing the dependence with respect to $\hbar$

$$f(p,\theta,\phi) = -\frac{(2\pi)^2 m}{\hbar^2} < \phi_f| V |\psi_i^{(+)} > \tag{13.89}$$

and therefore from (13.20)

$$\frac{d\sigma}{d\Omega} = \frac{(2\pi)^4 m^2}{\hbar^4} | < \phi_f| V |\psi_i^{(+)} > |^2. \tag{13.90}$$

13.2.2.2 *Cross Sections and T-Matrices*

We will now establish in a very general way the expression of cross sections in terms of T-matrix elements.

The transition probability from the initial state $|\phi_i >$ to the final state $|\phi_f >$ is given by (the two states are supposed to be different)

$$W_{fi} = | < \phi_f|S|\phi_i > |^2 = | < \psi_f^{(-1)}|\psi_i^{(+)} > |^2 = |T_{fi}|^2, \tag{13.91}$$

since from (13.20)

$$< \phi_f|S|\phi_i >\, = \delta_{fi} + iT_{fi} \tag{13.92}$$

$$= \delta_{fi} - 2i\pi \delta(E_f - E_i) < \phi_f| V |\psi_i^{(+)} >\, = \delta_{fi} - 2i\pi \delta(E_f - E_i)|T_{fi}|^2,$$

$\mathbf{T}_{fi}$ being the reduced T-matrix element (at given energy)

$$\mathbf{T}_{fi} =< \phi_f| V |\psi_i^{(+)} > . \tag{13.93}$$

We see when taking the square of this function that there will be an interpretation problem because of the delta function of conservation of energy. In fact (following J. Schwinger), we can interpret the product of two delta functions

$$(\delta(E_f - E_i))^2 = \frac{1}{2\pi\hbar} \int e^{i\frac{(E_f - E_i)t}{\hbar}} \, dt \delta(E_f - E_i) = \frac{1}{2\pi\hbar} \int dt \delta(E_f - E_i). \tag{13.94}$$

Therefore,

$$W_{fi} = \frac{2\pi}{\hbar} \delta(E_f - E_i) |<\phi_f|V|\psi_i^{(+)}>|^2 \int dt, \tag{13.95}$$

which has a logical interpretation: the probability of the transition is proportional to the total duration of the interaction. It is therefore natural that the probability of the transition per unit time is given by

$$w_{fi} = \frac{2\pi}{\hbar} \delta(E_f - E_i) |<\phi_f|V|\psi_i^{(+)}>|^2. \tag{13.96}$$

This intuitive derivation can be obtained more rigorously by saying that w_{fi} is the increase per unit time of the probability that a system initially in the state i is at the time t in the state f:

$$w_{fi} = \frac{d}{dt} \lim_{t' \to -\infty} |<\phi_f|U(t,0)U(0,t')|\psi_i(t')>|^2$$
$$= \frac{d}{dt} |<\phi_f|U(t,0)|\psi_i^{(+)}>|^2. \tag{13.97}$$

Now,

$$w_{fi} = \frac{d}{dt} [<\phi_f|U(t,0)|\psi_i^{(+)}> <\phi_f|U(t,0)|\psi_i^{(+)}>^*]$$
$$= <\phi_f|\frac{d}{dt}U(t,0)|\psi_i^{(+)}> <\phi_f|U(t,0)|\psi_i^{(+)}>^* + h.c., \tag{13.98}$$

thus using

$$\frac{d}{dt}U(t,0) = -iV(t)U(t,0) \tag{13.99}$$

$$w_{fi} = -i <\phi_f|V(t)U(t,0)|\psi_i^{(+)}> <\phi_f|U(t,0)|\psi_i^{(+)}>^* + h.c. \tag{13.100}$$

From the definition of $V(t)$ and of $U(t,0)$

$$<\phi_f|V(t)U(t,0)|\psi_i^{(+)}> = e^{itE_f} <\phi_f|Ve^{-itH}|\psi_i^{(+)}>$$
$$= e^{it(E_f - E_i)} <\phi_f|V|\psi_i^{(+)}> \tag{13.101}$$

and

$$<\phi_f|U(t,0)|\psi_i^{(+)}> = e^{it(E_f - E_i)} <\phi_f|\psi_i^{(+)}> \tag{13.102}$$

and therefore

$$w_{fi} = -i < \phi_f|V|\psi_i^{(+)}) > < \phi_f|\psi_i^{(+)}) >^* + h.c. \tag{13.103}$$

This last expression can be clarified using the Lippmann–Schwinger equation

$$w_{fi} \tag{13.104}$$

$$= -i < \phi_f|V|\psi_i^{(+)}) > \left(< \phi_f|\phi_i >^* + \lim_{\varepsilon \to 0^+} < \phi_f|\frac{1}{E_i - H_0 + i\varepsilon}V|\psi_i^{(+)}) >^* \right) + h.c.$$

$$= -i < \phi_f|V|\psi_i^{(+)}) > \delta_{fi}$$

$$\quad -i < \phi_f|V|\psi_i^{(+)}) > \lim_{\varepsilon \to 0^+} \frac{1}{E_i - E_f - i\varepsilon} < \phi_f|V|\psi_i^{(+)}) >^* + h.c.$$

$$= 2\Im < \phi_f|V|\psi_i^{(+)}) > \delta_{fi} \tag{13.105}$$

$$\quad -i| < \phi_f|V|\psi_i^{(+)}) >|^2 \lim_{\varepsilon \to 0^+} \left(\frac{1}{E_i - E_f - i\varepsilon} - \frac{1}{E_i - E_f + i\varepsilon}\right)$$

$$= 2\Im < \phi_f|V|\psi_i^{(+)}) > \delta_{fi} + 2\pi\delta(E_f - E_i)| < \phi_f|V|\psi_i^{(+)}) >|^2, \tag{13.106}$$

and thus, if the final state is different from the initial state,

$$w_{fi} = 2\pi\delta(E_f - E_i)| < \phi_f|V|\psi_i^{(+)}) >|^2 = 2\pi\delta(E_f - E_i)|\mathbf{T}_{fi}|^2. \tag{13.107}$$

In practice, we are always interested in the transitions from a given initial state to a bunch of final states whose energies are between E_f and $E_f + dE_f$ and of density $\rho(E_f)$.[4] The probability of transition per unit time is therefore

$$w = 2\pi \int df w_{fi} \tag{13.108}$$

$$= 2\pi \sum_{f'} \int_{E_f - \Delta E_f}^{E_f + \Delta E_f} \delta(E' - E_i)|\mathbf{T}_{fi}|^2 \rho_{f'}(E')dE' = 2\pi \sum_{f'} \rho_{f'}(E)|\mathbf{T}_{fi}|^2,$$

where the matrix element and the density are supposed to be slowly varying and in the final formula, the matrix element is calculated for the energy $E = E_f = E_i$ and the sum over the states f' excludes the states f.

Remark that if we sum the probability of transition w_{fi} over all the possible final states, we get from the closure relation

---

[4] $\rho_f(E)dE$ is the number of spin states of ϕ_f in the energy interval $(E, E + dE)$.

$$\int \mathrm{d}f w_{fi} = \lim_{t' \to -\infty} \frac{\mathrm{d}}{\mathrm{d}t} \int \mathrm{d}f < \phi_f |U(t,t')|\phi_i > < \phi_i |U(t',t)|\phi_f >$$

$$= \lim_{t' \to -\infty} \frac{\mathrm{d}}{\mathrm{d}t} \int \mathrm{d}f < \phi_i |U(t',t)|\phi_f > < \phi_f |U(t,t')|\phi_i >$$

$$= \lim_{t' \to -\infty} \frac{\mathrm{d}}{\mathrm{d}t} < \phi_i |U(t',t)U(t,t')|\phi_i >$$

$$= \lim_{t' \to -\infty} \frac{\mathrm{d}}{\mathrm{d}t} < \phi_i |\phi_i > = 0.$$

Applying this result to Eq. (13.106), we get

$$0 = 2\Im < \phi_i |V|\psi_i^{(+)}) > + 2\pi \int \mathrm{d}f\, \delta(E_f - E_i)\, | < \phi_f |V|\psi_i^{(+)}) > |^2. \tag{13.109}$$

The second term on the right-hand side is proportional (neglecting the contribution from the state f) to the total cross section

$$\sum_{f' \neq f} \sigma_{fi} \tag{13.110}$$

since the cross section σ_{fi} is calculated by dividing the probability w_{fi} by the ingoing state: $(2\pi)^{-3}v_i$, v_i being the speed of the ingoing state.

The equality (13.109) is known as the *optical theorem*.

13.3   Scattering in Field Theories

The preceding formalism extends to field theories. The calculation of cross sections is reduced to the computation of the T-matrix elements.

From the results of Chapter 10 it follows that the T-matrix can be written, f and i characterising the initial and final states of respective momenta $q_1, \cdots, q_l$ and $p_1, \cdots, p_n$, as

$$T_{fi} = (2\pi)^4 \delta^{(4)}(\sum_{i=1}^{n} p_i - \sum_{j=1}^{l} q_j)\mathbf{T}_{fi}. \tag{13.111}$$

We can always assume that the momenta of the initial states are known with precision.[5] The probability of a collision leading to a final state, the particles of which are characterised by spin states and by momenta denoted collectively by $p_1, \cdots, p_n$ up to $\mathrm{d}p_1, \cdots, \mathrm{d}p_n$ in a set Ω, is given by[6]

---

[5] This is, of course, an approximation. Any physical one-particle state should be described by a wave packet, but, in most experiments, the uncertainty of the initial momenta is very small.

[6] Unless it is necessary, we will omit to write explicitly the spin indices.

$$\mathrm{d}w'_{fi} = \int_{\Omega} |T_{fi}|^2 \mathrm{d}\tilde{\boldsymbol{p}}_1, \cdots, \mathrm{d}\tilde{\boldsymbol{p}}_n, \tag{13.112}$$

where

$$\mathrm{d}\tilde{\boldsymbol{p}}_i = \frac{1}{(2\pi)^3} \frac{d^3 p_i}{2\omega(p_i)} = d\Omega_{m_i}. \tag{13.113}$$

As in the case of quantum mechanics, the expression (13.112) is singular since it contains the square of a delta function. However, we can proceed as before and show that one of the delta functions can be replaced by

$$\frac{1}{(2\pi)^4} \int \mathrm{d}^4 x \tag{13.114}$$

and introduce the probability of transition per unit volume, the initial states being normalised to one particle per unit volume,

$$\mathrm{d}w_{fi} = \frac{\mathrm{d}w'_{fi}}{\prod_1^l \rho_i \int \mathrm{d}^4 x} = (2\pi)^4 \int_{\Omega} \delta^{(4)} (\sum_{i=1}^{n} p_i - \sum_{j=1}^{l} q_j) \frac{1}{\prod_1^l \rho_i} |\mathbf{T}_{fi}|^2 \mathrm{d}\tilde{\boldsymbol{p}}_1, \cdots, \mathrm{d}\tilde{\boldsymbol{p}}_n, \tag{13.115}$$

the ρ_i's representing the densities of the initial particles.

Formula (13.115) will be our starting point for the calculation of cross sections of l initial particles and n final particles. We will consider the two most interesting cases.

13.3.1   The Case of Two Initial Particles

In the case $l = 2$ and in order to simplify the discussion, let us assume that we have only massive scalar particles. The cross section is the number of transitions per unit time and unit volume divided by the incident flux and the number of particles (diffusion centres) of the target. Since we normalised the incoming states to contain one particle per unit volume, the division factor reduces to the incoming flux whose value is

$$v_i = |\mathbf{v}_1 - \mathbf{v}_2|, \tag{13.116}$$

where $\mathbf{v}_1$ and $\mathbf{v}_2$ are the velocities of the two incoming particles.

Therefore,

$$\mathrm{d}\sigma_{fi} = \frac{\mathrm{d}w_{fi}}{v_i}. \tag{13.117}$$

We will restrict once more our analysis by limiting the final states to two-particle final states and consider the reaction

$$A + B \rightarrow C + D, \tag{13.118}$$

where A and B are with four-moment q_1 and q_2 and C and D are with four-momenta p_1 and p_2.

The cross section can be written as

$$
\begin{aligned}
\mathrm{d}\sigma_{fi} &= \frac{(2\pi)^4}{(2\pi)^6 v_i \varrho_1 \varrho_2} \int \delta^{(4)}(p_1 + p_2 - q_1 - q_2)|\mathbf{T}_{fi}|^2 \frac{\mathrm{d}p_1}{2\omega(p_1)} \frac{\mathrm{d}p_2}{2\omega(p_2)} \\
&= \frac{1}{4(2\pi)^2} \frac{1}{\sqrt{(q_1 \cdot q_2)^2 - q_1^2 q_2^2}} \int \delta^{(4)}(p_1 + p_2 - q_1 - q_2)|\mathbf{T}_{fi}|^2 \frac{\mathrm{d}p_1}{2\omega(p_1)} \frac{\mathrm{d}p_2}{2\omega(p_2)},
\end{aligned}
$$

where we use the fact that the density of initial states is given by

$$
\varrho_i = 2\omega(q_i). \tag{13.119}
$$

The expression $\varrho_1 \varrho_2 v_i$ can be written as a relativistic invariant

$$
\begin{aligned}
(\varrho_1 \varrho_2 v_i)^2 &= 16\omega(q_1)^2 \omega(q_2)^2 |\frac{q_1}{\omega(q_1)} - \frac{q_2}{\omega(q_2)}|^2 = 16|q_1|^2(\omega(q_1) + \omega(q_2))^2 \\
&= 16\left((\omega(q_1)\omega(q_2) + q_1^2)^2 - \omega(q_1)^2\omega(q_2)^2 + q_1^2(\omega(q_1)^2 + \omega(q_2)^2) - q_1^4 \right) \\
&= 16\left((q_1 \cdot q_2)^2 - q_1^2 q_2^2 \right)
\end{aligned}
$$

and the last term is positive since q_1 and q_2 are in V^+.

We get the differential cross section by restricting the particle C (we could also have chosen D) to have its momentum p_1 in the solid angle $\mathrm{d}\Omega$. This constraint (with given q_1 and q_2), by the conservation of the energy–momentum imposed by the delta function, fixes the length of p_1, p_2, and thus the energies $p_i^0 = \omega(p_i)$. Let us choose the reference frame of the centre of mass $q_1 + q_2 = 0$. Then

$$
\begin{aligned}
\mathrm{d}\sigma_{fi} &= \frac{(2\pi)^4}{(2\pi)^6 v_i \varrho_1 \varrho_2} \int \delta(p_1^0 + p_2^0 - q_1^0 - q_2^0)\, \delta^{(3)}(p_1 + p_2)|\mathbf{T}_{fi}|^2 \frac{\mathrm{d}p_1}{2\omega(p_1)} \frac{\mathrm{d}p_2}{2\omega(p_2)} \\
&= \frac{1}{(2\pi)^2 v_i \varrho_1 \varrho_2} \int \delta(p_1^0 + p_2^0 - q_1^0 - q_2^0)|\mathbf{T}_{fi}|^2 \frac{\mathrm{d}p_1}{4\omega(p_1)\omega(p_2)} \\
&= \frac{1}{(2\pi)^2 v_i \varrho_1 \varrho_2} \int \delta(p_1^0 + p_2^0 - q_1^0 - q_2^0)\, |\mathbf{T}_{fi}|^2 \frac{1}{4\omega(p_1)\omega(p_2)}|p_1|^2 \,\mathrm{d}|p_1|\mathrm{d}\Omega \\
&= \frac{1}{(2\pi)^8 v_i \varrho_1 \varrho_2} \frac{|p_1|}{4(p_1^0 + p_2^0)})|\mathbf{T}_{fi}|^2 \mathrm{d}\Omega = \frac{1}{(2\pi)^8 v_i \varrho_1 \varrho_2} \frac{|p_1|^2}{4\omega(p_1)\omega(p_2)v_f}|\mathbf{T}_{fi}|^2 \mathrm{d}\Omega.
\end{aligned}
$$

For the integration of the delta function, we used that $p_1^0 + p_2^0 - q_1^0 - q_2^0$ is, through $p_i^0 = \omega(p_i)$, a function $f(|p_1|)$ and that $f(|p_1|) = 0$ for a value $|p_1| = p = p(q_1, q_2)$; thus

$$\delta(f(|\boldsymbol{p}_1|)) = \frac{1}{|f'(p)|}\,\delta(|\boldsymbol{p}_1| - p) \tag{13.120}$$

and since

$$f' = \frac{\mathrm{d}f(|\boldsymbol{p}_1|)}{\mathrm{d}|\boldsymbol{p}_1|} = \frac{p_1^0\boldsymbol{p}_1\cdot\boldsymbol{p}_2 - (\boldsymbol{p}_1)^2 p_2^0}{p_1^0 p_2^0 |\boldsymbol{p}_1|} = |\boldsymbol{p}_1|\frac{p_1^0 + p_2^0}{p_1^0 p_2^0} \tag{13.121}$$

we get in the centre-of-mass reference frame

$$\delta(p_1^0 + p_2^0 - q_1^0 - q_2^0)|\boldsymbol{p}_1|d|\boldsymbol{p}_1| = \frac{p_1^0 p_2^0}{p_1^0 + p_2^0}\delta(|\boldsymbol{p}_1| - p)\,\mathrm{d}|\boldsymbol{p}_1|. \tag{13.122}$$

Finally, since $\boldsymbol{p}_1\sqrt{1 - v_1^2} = m_C v_1$ and $\boldsymbol{p}_2\sqrt{1 - v_2^2} = m_D v_2$ where v_1 and v_2 are the velocities of the particles C and D

$$\boldsymbol{p}_i = v_i\omega(p_i)\qquad i = 1, 2. \tag{13.123}$$

We have used the fact that the relativistic velocity v_f is given by

$$\begin{aligned}
v_f^2 = |v_1 - v_2|^2 &= \frac{\boldsymbol{p}_1^2}{\omega(p_1)^2} + \frac{\boldsymbol{p}_2^2}{\omega(p_2)^2} - 2\frac{\boldsymbol{p}_1\cdot\boldsymbol{p}_2}{\omega(p_1)\omega(p_2)} \\
&= \frac{\boldsymbol{p}_1^2}{\omega(p_1)^2} + \frac{\boldsymbol{p}_2^2}{\omega(p_2)^2} + \frac{\boldsymbol{p}_1^2 + \boldsymbol{p}_2^2}{\omega(p_1)\omega(p_2)} = \boldsymbol{p}_1^2(\frac{1}{\omega(p_1} + \frac{1}{\omega(p_2})^2 \\
&= [|\boldsymbol{p}_1|\frac{p_1^0 + p_2^0}{p_1^0 p_2^0}]^2.
\end{aligned} \tag{13.124}$$

13.3.2   The Case of One Initial Particle

We consider the decay of one particle into many particles:

$$A \to B + C + \cdots \tag{13.125}$$

and to simplify, we assume that the decay generates two particles. The decay rate (the inverse of the life time τ) is

$$w = \frac{1}{\tau} = \frac{(2\pi)^4}{\varrho}\sum_{spin}\int \delta^{(4)}(q - p_1 - p_2)|\mathbf{T}_{fi}|^2\frac{\mathrm{d}\boldsymbol{p}_1}{2\omega(p_1)(2\pi)^3}\frac{\mathrm{d}\boldsymbol{p}_2}{2\omega(p_2)(2\pi)^3}, \tag{13.126}$$

the summation being over the spin states of the final particles.

In the centre of mass of the initial particle, whose mass is m_A,

$$\delta^{(4)}(q - p_1 - p_2) = \delta(m_A - p_1^0 - p_2^0)\delta^{(3)}(\boldsymbol{p}_1 + \boldsymbol{p}_2) \tag{13.127}$$

and

$$w = \frac{1}{\tau} = \frac{1}{\varrho(2\pi)^5} \sum_{\text{spin}} \int \delta \left(m_A - \sqrt{p^2 + m_B^2} - \sqrt{p^2 + m_C^2}\right) |\mathbf{T}_{fi}|^2 \mathrm{d}\boldsymbol{p}. \tag{13.128}$$

13.4 Applications

We will compute the scattering cross section of a photon by a free electron to lowest order of perturbation theory.

The amplitude is given by the matrix element

$$S_{fi} = \; < p_f, \alpha_f; k_f, \epsilon_f; \text{out} | q_i, \alpha_i; k_i, \epsilon_i; \text{in} > . \tag{13.129}$$

From formula (12.198), the connected part of this matrix element is given by (the S-matrix element is identical to the T-matrix element since the difference which corresponds to the propagation without interaction of the fermion and of the photon is given by the disconnected diagram)

$$S_{fi}^c = \mathrm{i} T_{fi}^c = - \int \mathrm{e}^{\mathrm{i} k_f . y - \mathrm{i} k_i . x} < p_f, \alpha_f; out | T(\epsilon_f . j(y) \epsilon_i . j(x)) | q_i, \alpha_i; i >^c \tag{13.130}$$

and the fermion current is given by

$$j_\mu(x) = e : \bar{\psi}(x) \gamma_\mu \psi(x) :, \tag{13.131}$$

where the Wick ordering indicates that there are no possible contractions between the two fields which appear in the current.

Let us apply formula (12.202)

$$S_{fi}^c = (-\mathrm{i})^2 \int \mathrm{d}^4 x_i \mathrm{d}^4 y_f \mathrm{d}^4 x \mathrm{d}^4 y \, \mathrm{e}^{-\mathrm{i} q_i . x_i + \mathrm{i} p_f . y_f} \, \mathrm{e}^{\mathrm{i} k_f . y - \mathrm{i} k_i . x} \tag{13.132}$$

$$\times \bar{u}^{(\alpha_f)}(p_1)(\mathrm{i} \overrightarrow{\slashed{\partial}}_{y_f} - m) < T(\psi(y_f)\epsilon_f . j(y)\epsilon_i . j(x)\bar{\psi}(x_i)) >^c (-\mathrm{i} \overleftarrow{\slashed{\partial}}_{x_i} - m) u^{(\alpha_i)}(q_i).$$

According to formulae (12.132) and (12.136) this expression can be calculated as a vacuum expectation value of the same T-product in which is inserted the exponential of the interaction. If we are interested in the lowest order of perturbation theory, given by the lowest power of the coupling constant e, we get

$$S_{fi}^c = (-\mathrm{i})^2 \int \mathrm{d}^4 x_i \mathrm{d}^4 y_f \mathrm{d}^4 x \mathrm{d}^4 y \, \mathrm{e}^{-\mathrm{i} q_i . x_i + \mathrm{i} p_f . y_f} \, \mathrm{e}^{\mathrm{i} k_f . y - \mathrm{i} k_i . x} \bar{u}^{(\alpha_f)}(p_1)(\mathrm{i} \overrightarrow{\slashed{\partial}}_{y_f} - m)$$

$$\times < 0 | T(\psi(y_f)\epsilon_f . j(y)\epsilon_i . j(x)\bar{\psi}(x_i)) | 0 >^c (-\mathrm{i} \overleftarrow{\slashed{\partial}}_{x_i} - m) u^{(\alpha_i)}(q_i).$$

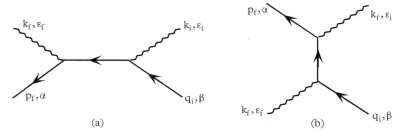

Figure 13.2 *The diagrams contributing to the S-matrix element.*

Let us compute the value of the connected part of the T-product

$$< 0|T(\psi(y_f)\epsilon_f.j(y)\epsilon_i.j(x)\bar\psi(x_i))|0 >^c$$
$$= ieS_F(y_f - y) < 0|T(\not\epsilon_f\psi(y)\epsilon_i.j(x)\bar\psi(x_i))|0 >^c$$
$$+ieS_F(y_f - x) < 0|T(\not\epsilon_i\psi(x)\epsilon_f.j(y)\bar\psi(x_i))|0 >^c$$
$$= e^2 S_F(y_f - y) < 0|T(\not\epsilon_f\psi(y)\bar\psi(x)\not\epsilon_i)|0 > S_F(x - x_i)$$
$$+e^2 S_F(y_f - x) < 0|T(\not\epsilon_i\psi(x)\bar\psi(y)\not\epsilon_f)|0 > S_F(y - x_i)$$
$$= ie^2 S_F(y_f - y)\not\epsilon_f S_F(y - x)\not\epsilon_i S_F(x - x_i)$$
$$+ie^2 S_F(y_f - x)\not\epsilon_i S_F(x - y)\not\epsilon_f S_F(y - x_i). \tag{13.133}$$

This result corresponds to the two diagrams given by Fig. 13.2.
Now since

$$(\mathrm{i}\,\overrightarrow{\not\partial}_x - m)S_F(x - y) = \delta^{(4)}(x - y) \tag{13.134}$$

$$S_F(x - y)(-\mathrm{i}\,\overleftarrow{\not\partial}_y - m) = \delta^{(4)}(x - y) \tag{13.135}$$

we find that the S-matrix element contribution is

$$S_{fi}^c = -\mathrm{i}(e)^2 \int \mathrm{d}^4x_i \mathrm{d}^4y_f \mathrm{d}^4x \mathrm{d}^4y\, e^{-\mathrm{i}q_i.x_i + \mathrm{i}p_f.y_f}\, e^{\mathrm{i}k_f.y - \mathrm{i}k_i.x} u^{(\alpha_f)}(p_f)$$

$$\times \Bigg\{ \delta^{(4)}(y_f - y)\not\epsilon_f S_F(y - x)\not\epsilon_i \delta^{(4)}(x - x_i)$$

$$+\delta^{(4)}(y_f - x)\not\epsilon_i S_F(x - y)\not\epsilon_f \delta^{(4)}(y - x_i) \Bigg\} u^{(\alpha_i)}(q_i)$$

$$= -\mathrm{i}(e)^2 \int \mathrm{d}^4x \mathrm{d}^4y\, u^{(\alpha_f)}(p_f) \Bigg\{ e^{-\mathrm{i}(q_i + k_i).x + \mathrm{i}(p_f + k_f).y}\not\epsilon_f S_F(y - x)\not\epsilon_i$$

$$+ e^{-\mathrm{i}(q_i - k_f).x_i + \mathrm{i}(p_f - k_i).x}\not\epsilon_i S_F(x - y)\not\epsilon_f \Bigg\} u^{(\alpha_i)}(q_i)$$

$$= -\mathrm{i}e^2 (2\pi)^4 \delta^{(4)}(p_f + k_f - q_i - k_i)$$

$$
\times \int d^4 p \, \delta^{(4)}(p - q_i - k_i) u^{(\alpha_f)}(p_f) \, \slashed{\epsilon}_f \frac{1}{\slashed{p} - m + i\epsilon} \, \slashed{\epsilon}_i u^{(\alpha_i)}(q_i)
$$

$$
-ie^2 (2\pi)^4 \delta^{(4)}(p_f + k_f - q_i - k_i) \cdot
$$

$$
\times \int d^4 p \, \delta^{(4)}(p - p_f + k_i) u^{(\alpha_f)}(p_f) \, \slashed{\epsilon}_i \frac{1}{\slashed{p} - m + i\epsilon} \, \slashed{\epsilon}_f u^{(\alpha_i)}(q_i)
$$

$$
= -ie^2 (2\pi)^4 \delta^{(4)}(p_f + k_f - q_i - k_i) u^{(\alpha_f)}(p_f) \{ \slashed{\epsilon}_f \frac{1}{\slashed{q}_i + \slashed{k}_i - m + i\epsilon} \, \slashed{\epsilon}_i
$$

$$
+ \slashed{\epsilon}_i \frac{1}{\slashed{q}_i - \slashed{k}_f - m + i\epsilon} \, \slashed{\epsilon}_f \} u^{(\alpha_i)}(q_i) \tag{13.136}
$$

and therefore

$$
\mathbf{T}^{fi} = -e^2 u^{(\alpha_f)}(p_f) \{ \slashed{\epsilon}_f \frac{1}{\slashed{q}_i + \slashed{k}_i - m + i\varepsilon} \, \slashed{\epsilon}_i + \slashed{\epsilon}_i \frac{1}{\slashed{q}_i - \slashed{k}_f - m + i\varepsilon} \, \slashed{\epsilon}_f \} u^{(\alpha_i)}(q_i)
$$

$$
= -e^2 u^{(\alpha_f)}(p_f) \{ \slashed{\epsilon}_f \frac{\slashed{q}_i + \slashed{k}_i + m}{(q_i + k_i)^2 - m^2 + i\varepsilon} \, \slashed{\epsilon}_i + \slashed{\epsilon}_i \frac{\slashed{q}_i - \slashed{k}_f + m}{(q_i - k_f)^2 - m^2 + i\varepsilon} \, \slashed{\epsilon}_f \} u^{(\alpha_i)}(q_i)
$$

$$
= -e^2 u^{(\alpha_f)}(p_f) \left(\frac{\slashed{\epsilon}_f \, \slashed{k}_i \, \slashed{\epsilon}_i}{2q_i.k_i} + \frac{\slashed{\epsilon}_i \, \slashed{k}_f \, \slashed{\epsilon}_f}{2q_i.k_f} + \frac{q_i.\epsilon_i}{q_i.k_i} \, \slashed{\epsilon}_f - \frac{q_i.\epsilon_f}{q_i.k_f} \, \slashed{\epsilon}_i \right) u^{(\alpha_i)}(q_i), \tag{13.137}
$$

where we used the zero mass of the photon to write $(q_i + k_i)^2 - m^2 = 2q_i.k_i$ and $(q_i - k_f)^2 - m^2 = -2q_i.k_f$ and that $(\slashed{p} - m)u(p) = 0$. If now we take the laboratory reference frame, that is to say the one in which the electron is at rest, then $q_i = (m, 0, 0, 0)$ and with polarisation vectors orthogonal to the hyperplane (k_i, q_i) for ϵ_i and to the hyperplane (k_f, q_i), for ϵ_f the expression reduces to

$$
\mathbf{T}^{fi} = -e^2 u^{(\alpha_f)}(p_f) \left(\frac{\slashed{\epsilon}_f \, \slashed{k}_i \, \slashed{\epsilon}_i}{2q_i.k_i} + \frac{\slashed{\epsilon}_i \, \slashed{k}_f \, \slashed{\epsilon}_f}{2q_i.k_f} \right) u^{(\alpha_i)}(q_i). \tag{13.138}
$$

If the initial and final spin states of the electron are unknown, one has to sum the square of the amplitude on all possible values of α_i and α_f.

The sum over α_i must be averaged over the number of possible spin states, namely 2 in this case. There are indeed $2s + 1$ possible initial spin states and we ignore which one it is, each state being weighted by its probability which is here $1/(2s + 1)$. Thus,

$$
dw_{fi} = \frac{1}{2} e^4 (2\pi)^4 \delta^{(4)}(p_f + k_f - q_i - k_i) \frac{1}{2s + 1} \sum_{\alpha_i \alpha_f} |\mathbf{T}^{fi}|^2. \tag{13.139}
$$

To compute the square of the T-matrix element, we use the identity

$$
\sum_{\alpha, \beta} |\bar{u}^{(\alpha)}(p) Q u^{(\beta)}(q)|^2 = Tr \left((\slashed{p} + m) Q (\slashed{q} + m) \gamma^0 Q^\dagger \gamma^0 \right). \tag{13.140}
$$

In fact

$$\sum_{\alpha,\beta} |\bar{u}^{(\alpha)}(p)Qu^{(\beta)}(q)|^2$$

$$= \sum_{\alpha,\beta} \bar{u}_r^{(\alpha)}(p)Q^{rs}u_s^{(\beta)}(q)u_t^{\dagger(\beta)}(q)Q^{\dagger tv}\bar{u}_v^{\dagger(\beta)}(q)$$

$$= \sum_{\alpha,\beta} \bar{u}_r^{(\alpha)}(p)Q^{rs}u_s^{(\beta)}(q)(u^{\dagger(\beta)}(q)\gamma^0)_t(\gamma^0 Q^\dagger \gamma^0)^{tv}(\gamma^0 \bar{u}^{(\alpha)})_v(q)$$

$$= \sum_{\alpha,\beta} \bar{u}_r^{(\alpha)}(p)Q^{rs}u_s^{(\beta)}(q)\bar{u}_t^{(\beta)}(q)(\gamma^0 Q^\dagger \gamma^0)^{tv}u_v^{(\alpha)}(q)$$

$$= \sum_{\alpha,\beta} u_v^{(\alpha)}(q)\bar{u}_r^{(\alpha)}(p)Q^{rs}u_s^{(\beta)}(q)\bar{u}_t^{(\beta)}(q)(\gamma^0 Q^\dagger \gamma^0)^{tv}$$

$$= (\not{p}+m)_{vr}Q^{rs}(\not{q}+m)_{st}(\gamma^0 Q^\dagger \gamma^0)^{tv}$$

$$= Tr\left((\not{p}+m)Q(\not{q}+m)\gamma^0 Q^\dagger \gamma^0\right).$$

where we used

$$\sum_\alpha u_r^{(\alpha)}(p)\bar{u}_s^{(\alpha)}(p) = (\not{p}+m)_{rs}. \tag{13.141}$$

Formula (13.140) makes it possible to compute the contribution of $\mathbf{T}^2$. Using the commutation rules of the gamma matrices and in particular the identities (6.35) the norms of the k's, $k_i^2 = k_f^2 = 0$ and of the polarisations $\epsilon_i^2 = \epsilon_f^2 = -1$, we get

$$\frac{1}{2}\sum_{\alpha_i \alpha_f} |\mathbf{T}_{fi}|^2$$

$$= \frac{1}{2}Tr\left[\left(\frac{\not{\epsilon}_f \not{k}_i \not{\epsilon}_i}{2q_i.k_i} + \frac{\not{\epsilon}_i \not{k}_f \not{\epsilon}_f}{2q_i.k_f}\right)(\not{q}_i + m)\left(\frac{\not{\epsilon}_i \not{k}_i \not{\epsilon}_f}{2q_i.k_i} + \frac{\not{\epsilon}_f \not{k}_f \not{\epsilon}_i}{2q_i.k_f}\right)(\not{p}_f + m)\right]$$

$$= \left[\frac{k_f^2 + k_i^2}{k_i k_f} + 4(\epsilon_f.\epsilon_i)^2 - 2\right] \tag{13.142}$$

with the following notation for a massless four-vector: $k = k_0 = |\mathbf{k}|$.

Taking into account the flux of incoming particles[7] $4p_i.k_i$, the cross section is given by

$$d\sigma = (2\pi)^4 \delta^{(4)}(p_f + k_f - q_i - k_i)\frac{1}{4p_i.k_i}\frac{1}{2}\sum_{\alpha_i,\alpha_f}|\mathbf{T}_{fi}|^2\frac{1}{(2\pi)^6}\frac{d^3 p_f}{2\omega(p_f)}\frac{d^3 k_f}{2k_f^0}. \tag{13.143}$$

---

[7] The density of photons is $\varrho_\gamma = 2k_i^0$, the density of electrons $\varrho_{el} = 2\omega(q_i)$, and the relative speed $v_i = 1$, so, in the laboratory reference frame,

$$\varrho_1 \varrho_2 v_i = 4p_i.k_i.$$

If we are interested in the photons scattered in a solid angle Ω modulo $d\Omega$, we must perform the integration over the electron momenta in their final state and over the photon energy. We use

$$\int \delta^{(4)}(p_f + k_f - q_i - k_i) \mathrm{d}^3 p_f \, \mathrm{d}^3 k_f = \int \delta(p_f^0 + k_f^0 - q_i^0 - k_i^0)|\boldsymbol{k}_f|^2 \mathrm{d}|\boldsymbol{k}_f| \mathrm{d}\Omega$$

$$= k_f^0 \frac{k_f^0 \omega(p_f)}{p_f . k_f} \mathrm{d}\Omega. \tag{13.144}$$

Thus, since in the laboratory reference frame $p_i . k_i = p_f . k_f = m k_i^0$, we have

$$\frac{\mathrm{d}\sigma}{\mathrm{d}\Omega} = \alpha^2 \frac{1}{(k_i^0)^2} \frac{1}{4m^2} \left[\frac{k_f^2 + k_i^2}{k_i k_f} + 4(\epsilon_f . \epsilon_i)^2 - 2 \right]$$

$$= \frac{\alpha^2}{4m^2} \left(\frac{k_f}{k_i}\right)^2 \left[\frac{k_f^2 + k_i^2}{k_i k_f} + 4(\epsilon_f . \epsilon_i)^2 - 2 \right] \tag{13.145}$$

after having reintroduced the dependence on c, $\hbar$ and the dimensionless fine structure constant α. This last formula is called the *Klein–Nishina formula*. If we introduce the angle between $\boldsymbol{k}_i$ and $\boldsymbol{k}_f$, the scattering angle, since

$$(k_f - k_i)^2 = -2k_i . k_f = -2k_i^0 k_f^0 (1 - \cos\theta)$$

$$= (p_i - p_f)^2 = 2m^2 - 2m\omega(p_f) = 2m(k_f^0 - k_i^0), \tag{13.146}$$

we get

$$k_f = \frac{k_i}{1 + (\frac{k_i}{m})(1 - \cos\theta)}, \tag{13.147}$$

which gives the frequency shift as a function of the scattering angle.

In the limit of low energies $k_i/m \to 0$, the preceding formula shows that $k_f/k_i \to 1$ and we obtain

$$\frac{\mathrm{d}\sigma}{\mathrm{d}\Omega} = \frac{\alpha^2}{m^2} (\epsilon_f . \epsilon_i)^2. \tag{13.148}$$

This expression is nothing else than the Thomson formula (3.141) of Chapter 3. Indeed, the Poynting vector of the radiation field is given by $\mathbf{S} = \mathbf{E}_{\mathrm{rad}} \wedge \mathbf{B}_{\mathrm{rad}} = |\mathbf{E}_{\mathrm{rad}}|^2 \boldsymbol{r}$ and leads to a Larmor formula proportional to $|\mathbf{E}_{\mathrm{rad}}|^2$. If we are now interested in the energy radiated in a state of polarisation ϵ_f, it is enough to replace $\mathbf{E}_{\mathrm{rad}}$ by $\epsilon_f . \mathbf{E}_{\mathrm{rad}}$. Similarly, if the incoming wave has a polarisation ϵ_i, then $\mathbf{E}_0 = \epsilon_i E_0$ and under the conditions of Chapter 3,

$$\left(\frac{\mathrm{d}\boldsymbol{v}}{\mathrm{d}t}\right)^2 \to \left(\epsilon_f . \frac{\mathrm{d}\boldsymbol{v}}{\mathrm{d}t}\right)^2 = E_0^2 \left(\frac{e}{m}\right)^2 |\epsilon_i . \epsilon_f|^2 = E_0^2 \left(\frac{e}{m}\right)^2 |\epsilon_i . \epsilon_f|^2 \tag{13.149}$$

with our choice of the polarisation four-vectors.

If the incoming beam of photons is not polarised, we average over the initial polarisations, that is to say over the $\boldsymbol{\epsilon}_i$'s orthogonal to $\boldsymbol{k}_i$, and we find that

$$\frac{\mathrm{d}\sigma}{\mathrm{d}\Omega} = \frac{\alpha^2}{m^2} \frac{1 + \cos\theta^2}{2}. \tag{13.150}$$

These equations tell us that scattering on free electrons can produce linearly polarised light. Let us choose the rest frame of the target electron and consider a beam of photons incident along the z axis. The initial polarisation vectors $\boldsymbol{\epsilon}_i$ lie on the x–y plane. If we look at a scattered photon at large angles, say along the x axis, its polarisation vector $\boldsymbol{\epsilon}_f$ will lie on the y–z plane. Formula (13.148) tells us that only the y component survives, i.e. even if the incident beam was unpolarised, the scattered photon will be linearly polarised.

There is an interesting and timely application of the Thomson scattering formula (13.148). According to the standard cosmological model, the Universe started very hot and dense and it cooled down as it expanded. During this process it went through several phase transitions. When the temperature was above a few keV, matter consisted of a hot plasma made out of electrons and protons with some light nuclei, mainly helium. Photons were trapped in the plasma and were not free. As the temperature dropped electrons and nuclei combined to form the first atoms. Matter became neutral and photons could travel freely through space and can be observed today. They carry precious information regarding the conditions that prevailed when they last interacted with free electrons, i.e. the moment of recombination. According to the model, this happened at a time around 380,000 years after the Big Bang. Because of the expansion, the photon wavelength has since been redshifted and today it is observed as cosmic microwave background (CMB) radiation. It is remarkably homogeneous and isotropic and this fact is best understood in the framework of the so-called *inflation* model which postulates that, at very early times, on the order of 10^{-33} sec after the Big Bang, the Universe went through a state of exponentially fast expansion. Thus, the present visible Universe results from a very small region of the early Universe and the study of the CMB radiation brings to us the earliest information of the world history. This information is of two kinds. First, we observe density and temperature fluctuations on the order of 10^{-5}. They are at the origin of the formation of the large structures we observe today, but here we want to concentrate on a second kind of information which concerns the polarisation of the CMB photons. Thomson scattering can produce such a polarisation; therefore, its properties will tell us something about the conditions during the last scattering. As we noted already, scattering at 90° of unpolarised photons produces linear polarisation. However, the same argument shows that if, in the electron rest frame, the incident radiations along the z and y axes have the same intensity, the polarisation of the photons scattered in the x direction cancels. It follows that a net polarisation will reveal the presence of anisotropies along perpendicular directions, i.e. *quadrupole anisotropies*, at the moments just before the last scattering. It is assumed that inflation has washed out any large anisotropies, so the polarisation is expected to be small. We can treat these anisotropies as perturbations and make a multipole expansion in scalar, vector, tensor, etc. We can imagine various sources of such perturbations. In the inflation model scalar perturbations come from density anisotropies, vector perturbations from the presence of magnetic vortices in the plasma, and tensor perturbations from the presence of gravitational waves created during the setup of inflation. These perturbations are expected to leave their imprint in the polarisation pattern of CMB radiation.

Like any vector field, a polarisation $\epsilon(x)$ of the CMB radiation can be decomposed into a pure gradient part and a pure curl part. They have different transformation properties under parity: the first, called *the E-mode*, is a pure vector and the second, *the B-mode*, a pseudo-vector. The scalar perturbations will contribute only to the E-mode because we cannot make a pseudo-vector out of the derivatives of a scalar. The E-mode polarisation has been measured already and it is well correlated with the observed temperature fluctuations. A B-mode can come only from vector or tensor perturbations. The first are expected to be very small because inflation has presumably diluted any primordial vortices in the plasma, so we are left with gravitational waves as the principal source of a B-mode. Furthermore, their presence is a generic prediction of all inflation models. Therefore, it is easy to understand the excitement caused by the recent announcement by the BICEP2 observatory in the Antarctica of the first detection of this B-mode with the magnitude and properties expected from inflation models. This observation has not been confirmed by the Planck collaboration but, if such a result is verified by independent measurements and if all other sources of contamination[8] are eliminated (note that this makes already many ifs), it will be the first, albeit indirect, observation of gravitational waves.[9]

As a second example of a process in quantum electrodynamics we choose to calculate the differential cross section for the creation of a pair $\mu^+\mu^-$ in electron–positron annihilation:

$$e^- + e^+ \rightarrow \mu^- + \mu^+. \tag{13.151}$$

It is a very important process for various reasons. First, electron–positron colliders have proven to be an extremely powerful tool for probing the structure of matter at very short distances. The reason is that the initial state, an electron and a positron, has the quantum numbers of the vacuum and, as such, it does not privilege any particular final state. Second, the amplitude for the reaction (13.151) can be computed in quantum electrodynamics, but also it can be accurately measured experimentally. Therefore, it provides a good test of the fundamental theory. Third, because of these properties, it can be used to calibrate the accelerator and to provide a measure with respect to which other processes can be compared.

The muons are spin-$\frac{1}{2}$ fermions, like the electrons, but their mass is almost 200 times larger. They are unstable with a lifetime of 2.2×10^{-6} sec. Their electromagnetic interactions are the same as those of the electrons, so the interaction Lagrangian density for the process (13.151) is given by

---

[8] An important source of error seems to be the scattering of the CMB photons from intergallactic dust whose effects are not easy to model.

[9] In 2016 the LIGO collaboration announced the direct detection of gravitational waves produced by a coalescence of two black holes.

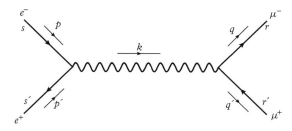

Figure 13.3 *The Feynman diagram for the reaction $e^- + e^+ \rightarrow \mu^- + \mu^+$. The arrows on the fermion lines indicate the charge flow, those next to the lines the momentum flow.*

$$\mathcal{L}_I = -e\left[\bar{\psi}(x)\gamma_\mu\psi(x) + \bar{\Psi}(x)\gamma_\mu\Psi(x)\right]A^\mu, \tag{13.152}$$

where ψ and Ψ are the Dirac quantum fields for the electron and the muon, respectively, A is the electromagnetic field, and e is the electric charge. There is only one diagram contributing to this process, that of Fig. 13.3. p and p' are the four-momenta of the electron and the positron, respectively, s and s' the spin indices, and (q, r) and (q', r') the same quantities for the final muons. Energy–momentum conservation gives $p + p' = k = q + q'$, where k is the momentum of the virtual photon. We shall use the Lorentz invariant quantities, the so-called *Mandelstam variables*:

$$(p + p')^2 = k^2 = s, \quad ; \quad (p - q)^2 = t, \quad ; \quad (p - q')^2 = u. \tag{13.153}$$

Note that s represents the square of the total energy in the centre-of-mass frame and t and u are the two invariant momentum transfers. These quantities satisfy $s + t + u = 2m^2 + 2m'^2$.

Using the Feynman rules we developed previously, we can compute the amplitude for this reaction,

$$\mathcal{M} = ie^2\left[\bar{v}^{s'}(p')\gamma_\mu u^s(p)\right]\frac{g^{\mu\nu}}{k^2}\left[\bar{U}^r(q)\gamma_\nu V^{r'}(q')\right], \tag{13.154}$$

where u and v are the spinor wave functions for the initial electron and positron and U and V those of the final μ^- and μ^+.

In most cases the electron and positron beams are unpolarised and we do not measure the polarisation of the final muons, (see, however, Problem 13.1 at the end of this chapter). So, in computing the square of the amplitude, we sum over final polarisations and average over the initial ones,

$$\frac{1}{4}\sum_{\text{spins}}|\mathcal{M}|^2 = \frac{e^4}{4s^2}Tr[(\not{p}' - m)\gamma_\mu(\not{p} + m)\gamma_\nu]Tr[(\not{q} + m')\gamma^\mu(\not{q}' - m')\gamma^\nu], \tag{13.155}$$

where m is the mass of the electron and m' that of the muon.

The traces of the γ matrices can be easily performed using the identities (6.36). For four matrices we obtain

$$\text{Tr } \gamma_\mu \gamma_\nu \gamma_\rho \gamma_\sigma = 4 \left(g_{\mu\nu} g_{\rho\sigma} - g_{\mu\rho} g_{\nu\sigma} + g_{\mu\sigma} g_{\nu\rho} \right). \tag{13.156}$$

The differential cross section is then given by

$$\frac{d\sigma}{d\Omega} = \frac{\alpha^2}{4s} \sqrt{\frac{s - 4m'^2}{s}} \left[\frac{s + 4m'^2}{s} + \frac{s - 4m'^2}{s} \cos^2\theta \right], \tag{13.157}$$

where we have neglected the electron mass m as compared to the muon mass and have introduced θ, the scattering angle, in the centre-of-mass frame (which, in this case, is also the laboratory frame). It is given, in terms of Lorentz invariant quantities, by the relation

$$\cos\theta = \frac{t - u}{\sqrt{s(s - 4m'^2)}}. \tag{13.158}$$

Following the same method, we can compute the cross section for any elementary reaction in quantum electrodynamics. In the list of exercises at the end of this chapter we propose the calculation of two other important processes, the *Bhabha scattering*, i.e. the electron–positron elastic scattering $e^- + e^+ \rightarrow e^- + e^+$, named after the Indian physicist Homi Jehangir Bhabha, and the *Møller scattering*, i.e. the electron–electron elastic scattering $e^- + e^- \rightarrow e^- + e^-$, named after the Danish physicist Christian Møller. The first receives contributions from two diagrams, the annihilation diagram, which is the analogue of that shown in Fig. 13.3, and the one-photon exchange diagram in which a photon line is exchanged between the electron and the positron lines. The second has no annihilation diagram but it has two exchange diagrams, in which the two electrons in the final state are interchanged. The resulting cross sections, which the reader is asked to verify, are given by

$$\frac{d\sigma}{d\Omega}\Big|_{\text{Bhabha}} = \frac{\alpha^2}{2s} \left[u^2 \left(\frac{1}{s} + \frac{1}{t} \right)^2 + \left(\frac{t}{s} \right)^2 + \left(\frac{s}{t} \right)^2 \right] \tag{13.159}$$

$$\frac{d\sigma}{d\Omega}\Big|_{\text{Møller}} = \frac{\alpha^2}{2s} \left[s^2 \left(\frac{1}{u} + \frac{1}{t} \right)^2 + \left(\frac{t}{u} \right)^2 + \left(\frac{u}{t} \right)^2 \right], \tag{13.160}$$

where we have neglected the electron mass compared to the centre-of-mass energy of the interaction: $s \gg m^2$.

Both expressions become singular when $t = 0$ and (13.160) also when $u = 0$. These singularities are due to the photon propagator in the exchange diagrams because of the zero photon mass. They are what we have called earlier *infrared singularities* and they will be studied in Chapter 21.

13.5   The Feynman Rules for the *S*-Matrix

From the preceding sections, we can extract the Feynman rules for the diagrams in momentum space which are involved in the computation of the *S*-matrix elements of quantum electrodynamics.

We find, up to combinatoric coefficients to be explained later, the following.

- For the external lines
    The external lines are amputated and put on the mass shell $p^2 = m^2$. They are allocated:
    - a spinor $u(p)$ for an incoming electron
    - a spinor $v(p)$ for an incoming positron
    - a spinor $\bar{u}(p)$ for an outgoing electron
    - a spinor $\bar{v}(p)$ for an outgoing positron
    - a polarisation vector $\epsilon^\lambda(p)$ for an incoming photon
    - a polarisation vector $\epsilon^{*\lambda}(p)$ for an outgoing photon
- For the internal lines
    The propagators are:
    - $(\mathrm{i}S_\mathrm{F}(p))_{\alpha\beta} = \left(\dfrac{\mathrm{i}}{\not{p} - m + \mathrm{i}\varepsilon}\right)_{\alpha\beta}$ for a fermion propagating from the index α to the index β
    - $(\mathrm{i}D_\mathrm{F}(p))_{\mu\nu} = -\mathrm{i}\left(\dfrac{g_{\mu\nu} + (\lambda^{-1} - 1)k_\mu k_\nu/k^2}{k^2 + \mathrm{i}\varepsilon}\right)$ for a photon
- The vertices are given by:
    - a delta function of conservation of energy and momentum
    - a factor $(\gamma_\mu)_{\alpha\beta}$ for a photon of index μ, an incoming fermion of index α and an outgoing fermion of index β.
- An overall sign. It comes from the fact that fermion operators under the normal ordering, or time ordering, symbol, anticommute. When we move the operators to bring them in the required order to obtain a given diagram, we may end up with minus signs. It is easy to check that the net effect can be summarised in the following two rules:
    - A minus sign for every closed fermion loop; and
    - A minus sign for every anti-fermion line going through the diagram from the initial to the final state.
- An overall numerical coefficient given by:
    - The factor coming from the expansion of the exponential. For QED, at the nth order of perturbation theory this factor equals $(-\mathrm{i}e)^n/n!$.
    - A combinatoric factor which tells us how many times this particular diagram appears in the Wick expansion. There is no close formula giving this factor

for an arbitrary diagram, so we must compute it every time. Here are some simple rules: (i) QED has only one type of vertex, so diagrams with two vertices interchanged are identical. This gives a factor $n!$, which cancels the one coming from the perturbation expansion. (ii) A symmetry factor coming from any interchange among external, or internal lines, which leaves the diagram unchanged.

In Appendix D. 3 at the end of this book we summarise all these rules and we apply them in some simple examples.

13.5.1   Feynman Rules for Other Theories

These rules can be easily generalised to any local quantum field theory described by a Lagrangian density which is polynomial in the fields and their derivatives. In perturbation theory a correlation function will be expressed, order by order, as a sum of Feynman diagrams. For the actual computation we must give the rules for the propagators and the vertices.

We have already derived the Feynman propagators for scalar, spinor, or vector fields. For the latter we distinguish the massless case, for which a gauge fixing is required, and the massive one. Although it is straightforward to derive propagators for fields of any spin, we shall never need them in this book.

So, the only new element in the rules for other field theories is the rule for the vertices. It is easily derived by going back to the perturbation expansion (9.72) and the Wick theorem.

Let us first consider a term in the interaction Lagrangian which is a monomial in the fields with no derivatives. It has the general form

$$\mathcal{L}_I = \Phi_{\{i_1\}} \Phi_{\{i_2\}} ... \Phi_{\{i_k\}} \Gamma^{\{\alpha_1 \beta_1\}}_{\{s_1\}} ... \Gamma^{\{\alpha_r \beta_r\}}_{\{s_r\}}, \tag{13.161}$$

where the Φ's are the various fields, the set of indices $\{i_j\}$ denote collectively the nature of the field, and they include Lorentz as well as internal symmetry indices. $\Gamma^{\{\alpha\beta\}}$ are numerical matrices with indices which may also refer to Lorentz or internal symmetry. The lower index $\{s\}$ indicates the type of the matrix, for example a γ matrix, and the upper indices the matrix element. We have included the coupling constant in these matrices.

This interaction term generates a vertex in the Feynman diagrams which has the following form:

- It has k lines, one for each field.
- Each line carries the set of $\{i_j\}$ indices.
- The term is proportional to the product of the matrix elements of the corresponding numerical matrices.

- There is an overall factor of $(2\pi)^4\delta^{(4)}(\Sigma p)$ with the sum of the momenta for every line in the vertex considered as incoming, ensuring energy–momentum conservation for the vertex.

In most cases the interaction Lagrangian is a Lorentz scalar, so the corresponding Lorentz indices are summed over. The same happens also for internal symmetry indices. These summations are also understood in the expression for the vertex.

Including derivatives in some of the fields is also straightforward. In the momentum space Feynman rules, a derivative ∂_μ on the field Φ will give a factor ip_μ, so the above set of rules is complemented by the following:

- A derivative $\partial_\mu \Phi_{\{i_j\}}$ gives a factor ip_μ, where p is the momentum of the line corresponding to the field $\Phi_{\{i_j\}}$.

- As we said for QED, there is no simple rule for computing the combinatoric factor for an arbitrary diagram. However, the cancellation of the $1/n!$ factor from the expansion of the exponential holds for any theory. Let us consider an interaction Lagrangian which is a sum of k terms:

$$\mathcal{L}_I = \mathcal{L}_{I,1} + \mathcal{L}_{I,2} + \cdots + \mathcal{L}_{I,k}. \tag{13.162}$$

A term contributing to the nth order of the perturbation expansion will be propositional to $1/(n_1!\, n_2!...n_k!)$ with $n_1 + n_2 + \cdots + n_k = n$. In other words, the interaction $\mathcal{L}_{I,i}$ will appear n_i times and a given diagram will have n_i vertices of this type. Permutation among these vertices will give a combinatoric factor $n_i!$ and will cancel the one coming from the expansion.

This completes the Feynman rules for arbitrary interaction Lagrangians which are given as a sum of monomials in the fields and their derivatives. We give below a list of those most commonly used.

- ϕ^n theory:

$$\mathcal{L}_I = -\lambda\, \phi^n,$$

where λ is a coupling constant with dimensions $[\text{mass}]^{4-n}$.

The vertex is represented by

and it is given by $-i\lambda(2\pi)^4\delta^{(4)}(\Sigma p)$.

- Scalar Yukawa theory:

$$\mathcal{L}_I = g\overline{\psi}\psi\phi.$$

The vertex is represented by

and it is given by $ig(2\pi)^4\delta^{(4)}(\Sigma p)$.

- Pseudo-scalar Yukawa theory:

$$\mathcal{L}_I = g\overline{\psi}\gamma_5\psi\phi.$$

The vertex is represented by

and it is given by $ig(\gamma_5)_{\alpha\beta}(2\pi)^4\delta^{(4)}(\Sigma p)$.

- $SU(2)$ pseudo-scalar Yukawa theory with a doublet of fermions and a triplet of bosons:

$$\mathcal{L}_I = g\overline{\psi}\gamma_5\tau\psi \cdot \phi$$

The vertex is represented by

and it is given by $ig(\tau^k)_{ij}(\gamma_5)_{\alpha\beta}(2\pi)^4\delta^{(4)}(\Sigma p)$. Here i, j, and k are the $SU(2)$ indices ($i, j=1, 2$ and $k=1, 2, 3$) and α and β the Dirac indices ($\alpha, \beta=1,..,4$). This is a model for the interaction between pions and nucleons.

- The interaction between the photon and a charged scalar particle, or, *scalar electrodynamics*. Gauge invariance and the minimal coupling prescription generate two couplings:

$$(i) \quad \mathcal{L}_I^{(1)} = \mathrm{i}eA_\mu(\phi\partial^\mu\phi^* - \phi^*\partial^\mu\phi).$$

The vertex is represented by

and it is given by $-\mathrm{i}e(p + p')_\mu(2\pi)^4\delta^{(4)}(\Sigma p)$.

$$(ii) \quad \mathcal{L}_I^{(2)} = e^2A_\mu A^\mu\phi\phi^*.$$

The vertex is represented by

and it is given by $2ie^2 g_{\mu\nu}(2\pi)^4 \delta^{(4)}(\Sigma p)$.

- Interaction between vector bosons. Of particular interest is the case of a self-interaction between the members of a multiplet of vector bosons belonging to the adjoint representation of a compact Lie group $\mathcal{G}$. We call A_μ^a the members of the multiplet. The index a runs between 1 and d_A, the dimension of the adjoint representation of $\mathcal{G}$. For example, for $SU(N)$ $d_A = N^2 - 1$. In the next chapter we shall derive the form of the interaction using a principle of gauge invariance, but for the purposes of obtaining the Feynman rules, it is sufficient to write the form of the interaction Lagrangian. Again, we obtain two couplings:

(i) $\mathcal{L}_I^{(1)} = \frac{1}{4} g f_{abc} (\partial_\mu A_\nu^a - \partial_\nu A_\mu^a) A^{\mu b} A^{\nu c}$, where f_{abc} are the structure constants of the Lie algebra of $\mathcal{G}$.

The vertex is represented by

- and it is given by

$$gf_{abc}[g_{\mu\nu}(p_1 - p_2)_\rho + g_{\nu\rho}(p_2 - p_3)_\mu \qquad (13.163)$$
$$+ g_{\rho\mu}(p_3 - p_1)_\nu](2\pi)^4 \delta^{(4)}(\Sigma p).$$

(ii) $\mathcal{L}_I^{(2)} = -\frac{1}{4} g^2 f_{abc} f_{ab'c'} A_\mu^b A_\nu^c A^{\mu b'} A^{\nu c'}$.

The vertex is represented by

and it is given by

$$-ig^2 [f_{eab} f_{ecd} (g_{\mu\rho} g_{\nu\sigma} - g_{\mu\sigma} g_{\nu\rho}) \qquad (13.164)$$
$$+ f_{eac} f_{edb} (g_{\mu\sigma} g_{\rho\nu} - g_{\mu\nu} g_{\rho\sigma}) + f_{ead} f_{ebc} (g_{\mu\nu} g_{\rho\sigma} - g_{\mu\rho} g_{\nu\sigma})](2\pi)^4 \delta^{(4)}(\Sigma p)$$

We shall see in the next chapter that the process of choosing a gauge for the interaction among massless vector bosons belonging to the adjoint representation of some non-Abelian group often generates new vertices which were absent in the classical Lagrangian. We shall complete the Feynman rules accordingly.

- The Fermi current–current model for weak interactions. It is an interaction between four Dirac fermions which gives a good low-energy approximation to processes such as the nuclear β-decay,

$$\mathcal{L}_I = \frac{G_F}{\sqrt{2}} \overline{\Psi}^{(1)} \gamma_\mu (1 + \gamma_5) \Psi^{(2)} \cdot \overline{\Psi}^{(3)} \gamma^\mu (1 + \gamma_5) \Psi^{(4)},$$

where G_F is a constant with dimensions $[\text{mass}]^{-2}$.

The vertex is represented by

and it is given by

$$i \frac{G_F}{\sqrt{2}} [\gamma_\mu (1 + \gamma_5)]_{\alpha\beta} [\gamma^\mu (1 + \gamma_5)]_{\alpha'\beta'} (2\pi)^4 \delta^{(4)} (\Sigma p).$$

As we can see, the vertex function is a matrix both in Minkowski space and in whichever internal symmetry space we have. Let us recall that propagators are also matrices in the same spaces. The correlation function is thus a product of matrices. When computing a scattering amplitude we saturate the indices with those of the external particles and we obtain a scalar.

It must be clear by now how one can obtain the Feynman rules for any given theory expressed in terms of a Lagrangian density which is polynomial in the fields and their derivatives. This is already an important step in our understanding of quantum field theory. Using these rules we can compute any correlation function, or any scattering amplitude, in lowest order of perturbation theory. By this we mean diagrams which contain no closed loops, those we called 'tree diagrams'. We still have two important questions to address: first, how to choose the right theory among the infinity of polynomial Lagrangians. Second, how to go beyond the lowest order and consider complicated diagrams involving loops. As it will turn out, the two questions are not unrelated. They will be the subject of the following chapters.

13.6 Problems

Problem 13.1 For some physical questions the use of polarised electron and positron beams in $e^- e^+$ colliders makes it possible to extract information which is washed out when we average over the initial spins. Repeat the calculation of the cross section for the process $e^- + e^+ \rightarrow \mu^- + \mu^+$ we presented earlier and find the dependence on the spins of the initial particles.

Problem 13.2 Compute the differential cross section for the elastic electron–positron scattering $e^- + e^+ \rightarrow e^- + e^+$. Show that at the limit when the mass of the electron is negligible compared to the centre-of-mass energy we obtain the formula (13.159).

Problem 13.3 Compute the differential cross section for the elastic electron–electron scattering $e^- + e^- \rightarrow e^- + e^-$. Show that at the limit of negligible electron mass we obtain the formula (13.160).

14

Geometry and Quantum Dynamics

14.1 Introduction. QED Revisited

In the first chapter we introduced the concept of symmetry. We saw that it restricts the possible form of interactions and leads to the appearance of conserved quantities. In this chapter we shall exploit this concept further. Let us choose the simplest example of a free Dirac field $\psi(x)$. In section 6.3.11 we saw that it is described by the Lagrangian density:

$$\mathcal{L} = \bar{\psi}(x)(i\partial\!\!\!/ - m)\psi(x). \tag{14.1}$$

It is invariant under a $U(1)$ group of phase transformations acting on the fields $\psi(x)$:

$$\psi(x) \rightarrow e^{i\theta}\psi(x). \tag{14.2}$$

It is this invariance which leads to the conservation of the Dirac current $j_\mu = \bar{\psi}(x)\gamma_\mu\psi(x)$. It is a *global* invariance, in the sense that the parameter θ in (14.2) is independent of the space–time point x. Is it possible to extend this invariance to a *local* one, namely one in which θ is replaced by an arbitrary function of x; $\theta(x)$? There may be various, essentially aesthetic, reasons for which we may wish to do that. In physical terms, we can argue that the formalism should allow for a local definition of the phase of the field, which is an unobservable quantity. If the phase is defined locally, we need a means to transfer the information from point to point. This requirement can be made more precise using a mathematical language. If we view the fields as sections in a fibre-bundle with base on the four-dimensional space–time, the global transformations of (14.2) imply only a trivial structure with vanishing curvature. A much richer geometry can be obtained by allowing the group action to depend on the fibre, in other words by allowing local, or *gauge*, transformations. Whatever our motivations may be, physical or mathematical, it is clear that (14.1) is not invariant under (14.2) with θ replaced by $\theta(x)$. The reason is the presence of the derivative term in (14.1), which gives rise to a term proportional to $\partial_\mu\theta(x)$. In order to restore invariance, we must modify (14.1), in which case it will no longer describe a free Dirac field; invariance under gauge transformations leads to the introduction of interactions. Both physicists and mathematicians know the

From Classical to Quantum Fields. Laurent Baulieu, John Iliopoulos and Roland Sénéor.
© Laurent Baulieu, John Iliopoulos and Roland Sénéor, 2017. Published 2017 by Oxford University Press.

answer to the particular case of (14.1) and we used it extensively in the first chapters of this book. We introduced a new field $A_\mu(x)$ and we replaced the derivative operator ∂_μ by a 'covariant derivative' D_μ which we found in (6.78)

$$D_\mu = \partial_\mu + ieA_\mu, \tag{14.3}$$

where e is an arbitrary real constant. D_μ is called 'covariant' because it satisfies

$$D_\mu[e^{i\theta(x)}\psi(x)] = e^{i\theta(x)}D_\mu\psi(x), \tag{14.4}$$

valid if, at the same time, the gauge field $A_\mu(x)$ undergoes the transformation

$$A_\mu(x) \to A_\mu(x) - \frac{1}{e}\partial_\mu\theta(x). \tag{14.5}$$

The Dirac Lagrangian density becomes now

$$\mathcal{L} = \bar\psi(x)(i\slashed{D} - m)\psi(x) = \bar\psi(x)(i\slashed{\partial} - e\slashed{A} - m)\psi(x). \tag{14.6}$$

It is invariant under the gauge transformations (14.2) and (14.5) and describes the interaction of a charged spinor field with an external electromagnetic field; For a physicist $A_\mu(x)$ is the electromagnetic field; for a mathematician the 1-form A is the connection on the fibre-bundle. We can complete the picture by including the degrees of freedom of the electromagnetic field itself and add to (14.6) the corresponding Lagrangian density. Again, gauge invariance determines its form uniquely and we are led to (3.6). A simple rule to derive the field tensor $F_{\mu\nu}$ is to note that it is given by the commutator of two covariant derivatives:

$$[D_\mu, D_\nu] = ieF_{\mu\nu}. \tag{14.7}$$

The constant e we introduced is the electric charge, the coupling strength of the field ψ with the electromagnetic field. Note that if we consider a second field ψ', it will be coupled with its own charge e' which does not have to be equal to e. We shall come back to this remark shortly.

Let us summarise: we started with a field theory invariant under a group $U(1)$ of global phase transformations. The extension to a local invariance can be interpreted as a $U(1)$ symmetry at each point x. In a qualitative way we can say that gauge invariance induces an invariance under $U(1)^\infty$. This extension, a purely geometrical requirement, implies the introduction of new interactions. The surprising result is that these 'geometrical' interactions describe the well-known electromagnetic forces.

A few remarks before closing this section:

1. We could as well have started from the free Schrödinger equation and derive the motion of a non-relativistic charged particle in an external electromagnetic field (see Problem 14.1).

2. Gauge invariance is already present in the classical Maxwell theory but its interpretation is physically more intuitive in the framework of quantum mechanics. Indeed, it is the invariance under phase transformations of the wave function of the electron which naturally extends to a local invariance.

3. The same reasoning can be applied to space–time transformations. Let us consider the example of translations. Classical mechanics is assumed to be invariant under $x_\mu \to x_\mu + a_\mu$, where a_μ is a constant four-vector. We can again attempt to extend this invariance to local translations, where a_μ is an arbitrary function of the space–time point x_μ. This can be done in a mathematically consistent way by extending the invariance to the group of diffeomorphisms, which can be defined on any manifold, and gives the correct mathematical definition of the 'equivalence principle'. We explained in Chapter 4 that this extension implies the introduction of new forces. The resulting gauge invariant theory turns out to be classical general relativity, which is a consistent relativistic theory and contains Newton's theory in the non-relativistic limit. Thus, gravitational interactions have also a geometric origin. In fact, and we must insist on this, the mathematical formulation of Einstein's original motivation to extend the principle of equivalence to accelerated frames is precisely the requirement of gauge invariance. Historically, many mathematical techniques used in today's gauge theories were developed in the framework of general relativity.

14.2 Non-Abelian Gauge Invariance and Yang–Mills Theories

The extension of the formalism of gauge theories to non-Abelian groups has not been a trivial task and it was first discovered by trial and error. At the end of this chapter we present a very brief history of the development of these ideas, both because of their fundamental importance in physics, but also because the history itself happens to be complex and not widely known. References to more complete studies are also given. Here we with restrict ourselves to internal symmetries which are simpler to analyse and they are the ones we shall apply to particle physics outside gravitation.

Let us consider a classical field theory given by a Lagrangian density $\mathcal{L}$. It depends on a set of N fields $\psi^i(x)$, $i = 1, ..., r$, and their first derivatives. The Lorentz transformation properties of these fields will play no role in this section. We assume that the ψ's transform linearly according to an r-dimensional representation, not necessarily irreducible, of a compact, simple, Lie group G which does not act on the space–time point x,

$$\Psi = \begin{pmatrix} \psi^1 \\ \vdots \\ \psi^r \end{pmatrix} \qquad \Psi(x) \to U(\omega)\Psi(x) \qquad \omega \in G, \qquad (14.8)$$

where $U(\omega)$ is the matrix of the representation of G. In fact, in this chapter we shall be dealing mainly with perturbation theory and it will turn out that we shall need only to look at transformations close to the identity in G. Therefore, it will be useful to exhibit the corresponding Lie algebra. Whenever the global group structure is relevant it will be explicitly stated,

$$\Psi(x) \rightarrow e^{i\Theta}\Psi \qquad\qquad \Theta = \sum_{a=1}^{m} \theta^a T^a, \qquad\qquad (14.9)$$

where the θ^a's are a set of m constant parameters, and the T^a's are m $r \times r$ matrices representing the m generators of the Lie algebra of G. They satisfy the commutation rules

$$\left[T^a, T^b\right] = \mathrm{i}f^{abc}T^c. \qquad\qquad (14.10)$$

The f's are the structure constants of G and a summation over repeated indices is understood. The normalisation of the structure constants is usually fixed by requiring that, in the fundamental representation, the corresponding matrices of the generators t^a are normalised such as

$$Tr\left(t^a t^b\right) = \frac{1}{2}\delta^{ab}. \qquad\qquad (14.11)$$

The Lagrangian density $\mathcal{L}(\Psi, \partial\Psi)$ is assumed to be invariant under the global transformations (14.9) or (14.8). As was done for the Abelian case, we wish to find a new $\mathcal{L}$ invariant under the corresponding gauge transformations in which the θ^a's of (14.9) are arbitrary functions of x. In the same qualitative sense, we look for a theory invariant under G^∞. This problem was first solved by trial and error for the case of $SU(2)$ by C. N. Yang and R. L. Mills in 1954.[1] They gave the underlying physical motivation and these theories have been called since 'Yang–Mills theories'. The steps are direct generalisations of those followed in the Abelian case. We need a gauge field, the analogue of the electromagnetic field, to transport the information contained in (14.9) from point to point. Since we can perform m independent transformations, the number of generators in the Lie algebra of G, we need m gauge fields $A_\mu^a(x)$, $a = 1, ..., m$. It is easy to show that they belong to the adjoint representation of G. Using the matrix representation of the generators we can cast $A_\mu^a(x)$ into an $r \times r$ matrix:

$$\mathcal{A}_\mu(x) = \sum_{a=1}^{m} A_\mu^a(x)T^a. \qquad\qquad (14.12)$$

The covariant derivatives can now be constructed as

$$\mathcal{D}_\mu = \partial_\mu + \mathrm{i}g\mathcal{A}_\mu \qquad\qquad (14.13)$$

with g an arbitrary real constant. They satisfy

$$\mathcal{D}_\mu e^{i\Theta(x)}\Psi(x) = e^{i\Theta(x)}\mathcal{D}_\mu\Psi(x) \qquad\qquad (14.14)$$

---

[1] See, however, the historical remarks at the end of this chapter.

provided the gauge fields transform as

$$\mathcal{A}_\mu(x) \to e^{i\Theta(x)} \mathcal{A}_\mu(x) e^{-i\Theta(x)} + \frac{i}{g}\left(\partial_\mu e^{i\Theta(x)}\right) e^{-i\Theta(x)}. \tag{14.15}$$

The Lagrangian density $\mathcal{L}(\Psi, \mathcal{D}\Psi)$ is invariant under the gauge transformations (14.9) and (14.15) with an x-dependent Θ, if $\mathcal{L}(\Psi, \partial\Psi)$ is invariant under the corresponding global ones (14.8) or (14.9). As was done with the electromagnetic field, we can include the degrees of freedom of the new gauge fields by adding to the Lagrangian density a gauge-invariant kinetic term. It turns out that it is slightly more complicated than $F_{\mu\nu}$ of the Abelian case. Yang and Mills computed it for $SU(2)$ but, in fact, it is uniquely determined by geometry plus some obvious requirements, such as absence of higher order derivatives. The result is the 2-form constructed out of $\mathcal{A}_\mu$ which corresponds to the curvature on the fibre-bundle:

$$\mathcal{F}_{\mu\nu} = \partial_\mu \mathcal{A}_\nu - \partial_\nu \mathcal{A}_\mu - ig\left[\mathcal{A}_\mu, \mathcal{A}_\nu\right]. \tag{14.16}$$

It is again given by the commutator of the two covariant derivatives:

$$[\mathcal{D}_\mu, \mathcal{D}_\nu] = ig\mathcal{F}_{\mu\nu}. \tag{14.17}$$

The full gauge invariant Lagrangian density can now be written as

$$\mathcal{L}_{\text{inv}} = -\frac{1}{2}\text{Tr}\mathcal{F}_{\mu\nu}\mathcal{F}^{\mu\nu} + \mathcal{L}(\Psi, \mathcal{D}\Psi). \tag{14.18}$$

By convention, in (14.16) the matrix $\mathcal{A}$ is taken to be

$$\mathcal{A}_\mu = A_\mu^a t^a, \tag{14.19}$$

where we recall that the t^a's are the matrices representing the generators in the fundamental representation. It is only with this convention that the kinetic term in (14.18) is correctly normalised. In terms of the component fields A_μ^a, $\mathcal{F}_{\mu\nu}$ reads as

$$\mathcal{F}_{\mu\nu} = F_{\mu\nu}^a t^a \qquad\qquad F_{\mu\nu}^a = \partial_\mu A_\nu^a - \partial_\nu A_\mu^a + gf^{abc}A_\mu^b A_\nu^c. \tag{14.20}$$

Under a gauge transformation $\mathcal{F}_{\mu\nu}$ transforms like a member of the adjoint representation.

$$\mathcal{F}_{\mu\nu}(x) \to e^{i\theta^a(x)t^a}\, \mathcal{F}_{\mu\nu}(x)\, e^{-i\theta^a(x)t^a}. \tag{14.21}$$

This completes the construction of the gauge-invariant Lagrangian. We add some remarks:

1. As it was the case with the electromagnetic field, the Lagrangian (14.18) does not contain terms proportional to $A_\mu A^\mu$. This means that under the usual quantisation rules developed earlier, the gauge fields describe massless particles.

2. Since $\mathcal{F}_{\mu\nu}$ is not linear in the fields $\mathcal{A}_\mu$, the $\mathcal{F}^2$ term in (14.18), besides the usual kinetic term which is bilinear in the fields, contains trilinear and quadrilinear terms. In perturbation theory they will be treated as coupling terms whose strength is given by the coupling constant g. In other words, the non-Abelian gauge fields are self-coupled while the Abelian (photon) field is not. A Yang–Mills theory, containing only gauge fields, is still a dynamically rich quantum field theory while a theory with the electromagnetic field alone is a trivial free theory.

3. The same coupling constant g appears in the covariant derivative of the fields Ψ in (14.13). This simple consequence of gauge invariance has an important physical application: if we add another field Ψ', its coupling strength with the gauge fields will still be given by the same constant g. Contrary to the Abelian case studied earlier, if electromagnetism is part of a non-Abelian simple group, gauge invariance implies charge quantisation.

4. The above analysis can be extended in a straightforward way to the case where the group G is the product of simple groups $G = G_1 \times ... \times G_n$. The only difference is that we should introduce n coupling constants $g_1, ..., g_n$, one for each simple factor. Charge quantisation is still true inside each subgroup, but charges belonging to different factors are no more related.

5. The situation changes if we consider non-semi-simple groups, where one or more of the factors G_i is Abelian. In this case the associated coupling constants can be chosen different for each field and the corresponding Abelian charges are not quantised.

14.3   Field Theories of Vector Fields

As we discussed in Chapter 2, spin-1 particles are described by vector, or sometimes tensor, fields. Poincaré invariance implies that a massive spin-1 particle has three physical degrees of freedom and a massless one has two. On the other hand, a Lorentz vector has four components, therefore, we expect our field theories to have redundant variables and we must find a way to impose some constraints in order to eliminate this redundancy.[2] For quantum electrodynamics this was achieved by gauge invariance. In this section we want to address this question more generally.

Let us start with the case of a massive spin-1 particle described by a vector field $A_\mu(x)$. A first guess would be to write a Lagrangian density for the free field proportional to $(\partial_\mu A_\nu)^2 - m^2 A_\mu A^\mu$, which implies that each component of the field satisfies the Klein–Gordon equation. We see immediately that this is unsatisfactory precisely because we want to end up with three independent degrees of freedom and not four. Furthermore,

---

[2] This counting applies to the quantum theory, where fields describe particles, but the study of constraint equations can be first done in classical field theory, although quantum corrections will bring additional complications. Notice also that this problem becomes more severe if we want to study higher spin fields since the mismatch between the number of components of the corresponding tensor field and that of the physical degrees of freedom of the particle becomes larger.

we see that the space and time components of the field have opposite signs in the kinetic energy, so the Hamiltonian will not be positive definite. A better choice is given by

$$\mathcal{L} = -\frac{1}{4}(\partial_\mu A_\nu - \partial_\nu A_\mu)^2 + \frac{1}{2}m^2 A_\mu A^\mu + j_\mu A^\mu, \tag{14.22}$$

where we have introduced a coupling with some external current $j_\mu(x)$. The equations of motion now become

$$\Box A_\mu - \partial_\mu \partial_\nu A^\nu + m^2 A_\mu + j_\mu = 0 \tag{14.23}$$

and they imply that

$$m^2 \partial_\mu A^\mu + \partial_\mu j^\mu = 0. \tag{14.24}$$

This consequence of the equations of motion is very interesting because it shows that the dynamics of the zero-spin content of the vector field is entirely determined by the properties of the current. In particular, if the latter is independently conserved, A_μ satisfies a four-dimensional transversality condition which eliminates the extra component, leaving three physical degrees of freedom. If, on the other hand, the current is not conserved, the theory contains automatically spin-1 and spin-0 excitations. We shall see in Chapter 16 that this conclusion survives quantisation and holds order by order in perturbation theory, but only the conserved current case gives a consistent quantum field theory. Note also that the vector field propagator, computed as the inverse of the operator of the quadratic part of (14.22), is proportional, in momentum space, to

$$D_{\mu\nu}(k) = \frac{1}{k^2 - m^2}\left(g_{\mu\nu} - \frac{k_\mu k_\nu}{m^2}\right). \tag{14.25}$$

As expected, it is singular when m^2 goes to 0. Indeed, we have already seen that in quantum electrodynamics the condition on $\partial_\mu A^\mu$ is not a consequence of Maxwell's equations and should be imposed independently. Physically, this follows from the fact that at $m \to 0$, one extra degree of freedom should decouple, leaving only two physical ones. In Chapter 16 we will see in which sense quantum electrodynamics can be obtained as the limit of a massive theory.

Let us now come to the massless case. In Chapter 3 we saw that Maxwell's equations can be obtained from a Lagrangian density

$$\mathcal{L} = -\frac{1}{4}(\partial_\mu A_\nu - \partial_\nu A_\mu)^2 + j_\mu A^\mu \tag{14.26}$$

by considering all components of the vector field as independent:

$$\Box A_\mu - \partial_\mu \partial_\nu A^\nu + j_\mu = 0. \tag{14.27}$$

Note that they imply the current conservation $\partial_\mu j^\mu = 0$; in other words, coupling a massless vector field to a non-conserved current gives inconsistent equations, already at the classical level. We also noted that the Lagrangian (14.26) does not allow us to compute the propagator of the photon field because the operator of the quadratic part is not invertible. To solve this problem we had imposed a gauge condition of the form $\partial_\mu A^\mu = \sqrt{\alpha}\, b(x)$, with $b(x)$ some given scalar function of x and α a constant, introduced only for convenience. Using a Lagrange multiplier, we can promote this condition to an equation of motion by considering the Lagrangian

$$\mathcal{L} = -\frac{1}{4}(\partial_\mu A_\nu - \partial_\nu A_\mu)^2 + \frac{1}{2}b^2 - \frac{1}{\sqrt{\alpha}}b\partial_\mu A^\mu + j_\mu A^\mu. \tag{14.28}$$

Varying with respect to b gives precisely the gauge condition. The Lagrangian (14.26) is equivalent to the usual one

$$\mathcal{L} = -\frac{1}{4}(\partial_\mu A_\nu - \partial_\nu A_\mu)^2 - \frac{1}{2\alpha}(\partial_\mu A^\mu)^2 + j_\mu A^\mu, \tag{14.29}$$

which gives a whole family of theories labelled by the parameter α. For a long time gauge invariance was seen as a mere mathematical curiosity and it is only in the past forty years that it was recognised as a most important guiding physical principle. We could say that it follows from the non-observability of the gauge potential in local phenomena where only the electric and magnetic fields are measurable. However, in the quantum theory, this is only partly true. A famous counter-example is the well-known Bohm–Aharonov effect in which a charged particle is forced to circulate around an obstruction, such as an external magnetic field. In this case a winding number can arise in the form of a phase difference, and be detected. We shall see more examples of such topological effects later.

We may worry whether imposing a gauge condition is legitimate and whether one has changed the physics. The answer requires a proof that physical quantities remain unchanged, in particular that they do not depend on α. This proof will be given in Chapter 18. Here we only note that the equations of motion for the gauge fixed Lagrangian (14.29), together with the conservation of the current, imply that

$$\Box\partial_\mu A^\mu = 0, \tag{14.30}$$

which means that the divergence of A_μ satisfies a free-field equation of motion, even in the presence of the interaction with the current j_μ. This is why we can consistently assign to it a given value $b(x)$. In the quantum theory it implies that although the complete Fock space of states associated with the Lagrangian (14.29) contains spin-0 photons, they will not be created by the interaction because the corresponding field is free. The argument presented here is only heuristic, but a more general result will be proven in Chapter 18. Note also that this is no more true if we use a non-linear gauge condition, for example one of the form $\partial_\mu A^\mu + A_\mu A^\mu = \sqrt{\alpha}\, b(x)$. Although there is nothing a priori wrong with such a condition, we do not expect it to be implementable at the quantum level by a procedure as simple as that of (14.29) because the combination $\partial_\mu A^\mu + A_\mu A^\mu$ does not satisfy a free-field equation of motion. We shall come back to this point shortly.

As a third example we look at the Yang–Mills Lagrangian (14.18). We see again that the quadratic part is not invertible because of the zero modes associated with gauge invariance. We see also that, as it was the case for quantum electrodynamics, a gauge condition does not follow from the equations of motion and must be imposed as a constraint equation. Again the simplest condition is a linear one of the form $\partial^\mu A_\mu^a = \sqrt{\alpha}\, b^a(x)$. Following the same steps as before we arrive at the analogue of the gauge fixed Lagrangian (14.29), which now reads

$$\mathcal{L} = \mathcal{L}_{\mathrm{inv}} - \frac{1}{\alpha} \mathrm{Tr}[(\partial^\mu \mathcal{A}_\mu)(\partial^\nu \mathcal{A}_\nu)] \tag{14.31}$$

with $\mathcal{L}_{\mathrm{inv}}$ given by (14.18). But here the analogy stops and trouble starts. Even in the absence of any other fields, keeping only the Yang–Mills part of the Lagrangian, the divergence of the gauge field $\partial^\mu A_\mu^a$ does not satisfy a free-field equation. This, of course, is due to the self-coupling terms among the gauge fields in the Yang–Mills Lagrangian. We conclude that the simple rule to solve the gauge ambiguity which we used in quantum electrodynamics, namely to add a $(\partial A)^2$ term in the Lagrangian, is not applicable in the non-Abelian theory, even for a linear gauge condition. Let us emphasise here that this problem will appear only at the quantum level. As a classical theory (14.31) is perfectly adequate. But the fact that ∂A is not a free field makes us believe that scalar degrees of freedom will be created at higher orders of perturbation theory.

In order to understand the origin of the problem, let us go back to the reasoning that led us to the gauge-fixing procedure (14.29) or (14.31). We want to describe the quantum properties of a system of massless, spin-1, bosons. We know that each one has two physical degrees of freedom. So our physical space of states is the Fock space constructed out of the creation and annihilation operators for transversely polarised gauge bosons. However, we want also to keep explicit Lorentz covariance at every step of the calculation. This forces us to use four-component vector fields. So it is natural to start by considering a quantum theory inside a larger Fock space containing, in addition, longitudinal and scalar excitations. This is the natural choice consistent with Lorentz invariance and it is the one dictated by classical arguments. For the quantum theory, however, we saw in quantum electrodynamics that some of these new states have negative norm. This means that all our computations will yield transition probabilities between states inside this large space, which, because of the negative metric, will not satisfy a unitarity relation. It follows that in the quantum theory we must impose a second requirement, on top of Lorentz invariance, namely a requirement related to unitarity. We must be able to prove that because of the gauge invariance of the original theory, our set of transition probabilities admits a restriction in the sub-space of physical states which does satisfy the unitarity relation. It is a necessary condition for the theory to be physically acceptable. We noted that the free-field equation (14.30) makes this result plausible for a linear gauge condition in quantum electrodynamics, a result which we shall prove order by order in perturbation theory in Chapter 18. This success made us believe that the 'minimal' choice of a Fock space, namely the one containing transverse, longitudinal, and scalar excitations, will be always adequate. Therefore, it came as a surprise when people realised that, for a non-linear theory, either in the gauge condition or the interaction, this is no longer true. It turned out that in order to satisfy both Lorentz and unitarity conditions, we should start from an even larger unphysical space. This counter intuitive conclusion was first reached by Feynman by trial and error. We shall briefly present Feynman's argument in the next section and we shall develop a systematic way for constructing this large Fock space using symmetry arguments. In Chapter 18 we shall prove that physical transition probabilities are indeed gauge independent and satisfy the unitarity relation.

14.4   Gauge Fixing and BRST Invariance

14.4.1   Introduction

This section will be devoted to the explicit construction of the large Fock space which will turn out to be necessary for the consistent quantisation of gauge theories with a non-Abelian invariance. In the process of doing so we shall uncover a new symmetry of the theory which is called *Becchi–Rouet–Stora–Tyutin (BRST) symmetry*. It is realised non-linearly and it will turn out to have unsuspected and very deep consequences. Inserted in the path integral approach, it will provide for a powerful method, not only to obtain the correct Feynman rules for all gauge choices, but also to answer all important technical questions such as renormalisation, unitarity of the S-matrix, and the problem of quantum anomalies. The power of the method is such that once we understand its origin, it will become clear that we could have postulated it and used it in order to *define* the theory. In fact it provides for a possible axiomatic definition of gauge theories because it introduces at once all fields and symmetries which are necessary in the quantisation process. In this section we shall not follow this axiomatic approach, but we shall instead *derive*, by trial and error, the quantum theory.

14.4.2   The Traditional Faddeev–Popov Method

14.4.2.1   *Early Attempts*

Already during the nineteenth century, people faced the problem of formulating the classical mechanics of a dynamical system whose variables were not all independent but were satisfying some constraint equations. We have seen already that gauge theories belong to this class. A real history of the subject should start from these problems, but here we will not go that far and we will restrict ourselves to the more recent attempts to quantise non-Abelian gauge theories.

In the 1950s, by using a heuristic method[3] that generalises the quantisation of classical constrained systems, Dirac wrote a formula for the gauge-fixed path integral of a

---

[3] Here the word heuristic means that the mathematics is only justified for a system with a finite number of degrees of freedom, for instance a mechanical system of particles where the canonical variables are submitted to constraints.

gauge invariant Lagrangian $\mathcal{L}_{\text{classical}}$ with a gauge function $\mathcal{M}$. He was using a Hamiltonian approach which is not very convenient for a relativistic system, but, rewritten in a Lagrangian language and inserted in a path integral, it becomes, formally,

$$Z(\mathcal{J}) \sim \int [\text{d}\varphi] \delta\Big(\mathcal{M}(\varphi)\Big) \det\Big(\sum_{\varphi} \frac{\delta\mathcal{M}}{\delta\varphi} \frac{\delta_{\epsilon}^{\text{gauge}}(\varphi)}{\delta\epsilon} \Big) \exp \int \text{d}x (\mathcal{L}_{\text{classical}} + \mathcal{J}\varphi), \qquad (14.32)$$

where $Z(\mathcal{J})$ is the generating functional of Green functions and $\mathcal{J}(x)$ is a set of external sources. We note already the appearance of a determinant which originates from the gauge-fixing condition $\mathcal{M} = 0$ and which will play an important role in the discussion.

Dirac's work did not attract much attention, since, at the time, no interesting gauge symmetry was really used, besides QED, for which the gauge-fixing problem had been solved by brute force in a situation where the determinant in Eq. (14.32) is an irrelevant constant.[4] We found it previously in Eq. (12.158). The non-Abelian gauge invariance was poorly known, and the only non-trivial gauge symmetry studied was gravity, and even that only as a classical theory. In fact, Dirac's work looked at the time as purely academic. He had in mind the quantisation of gravity, but it did not look very promising because he was working in a non-explicitly covariant Hamiltonian framework, not the best starting point for a quantum theory of general relativity.

Let us anticipate what we shall present in the next sections and note that Dirac could have actually accelerated history, had he invented the ghost fields to express the determinant he introduced. He could have made perhaps contact with a BRST invariant local action by using the formal identity that was introduced later,

$$\delta\Big(\mathcal{M}\Big) \det \Big(\sum_{\varphi} \frac{\delta\mathcal{M}}{\delta\varphi} \frac{\delta_{\epsilon}^{\text{gauge}}(\phi)}{\delta\epsilon} \Big)$$
$$= \int [\text{d}\Omega][\text{d}\bar{\Omega}][\text{d}b] \exp i \int \text{d}x \Big(-\bar{\Omega}\Big(\frac{\delta\mathcal{M}}{\delta\varphi} \frac{\delta_{\epsilon}^{\text{gauge}}(\phi)}{\delta\epsilon} \Big) \Omega + b\mathcal{M} \Big)$$
$$= \int [\text{d}\Omega][\text{d}\bar{\Omega}][\text{d}b] \exp i \int \text{d}x s\Big(\bar{\Omega}\mathcal{M}(\varphi) \Big), \qquad (14.33)$$

where s acts as a differential operator acting with a graded Leibniz rule $s(XY) = (sX)Y \pm XsY)$ and the $-$ sign occurs if X has odd grading, with

$$s\varphi = \frac{\delta_{\epsilon}^{\text{gauge}}(\phi)}{\delta\epsilon}\Omega \quad s\bar{\Omega} = b \quad sb = 0. \qquad (14.34)$$

The first part of Eq. (14.33) is a straightforward application of the integration rules for anticommuting variables we introduced in Chapter 11. We shall come back to the s operation later. Of course history did not follow this route, and it took a long time before

---

[4] Dirac's method inspired B. DeWitt, who was among the first to be interested in a quantum theory of gravity.

it became possible (i) to set up the Dirac formalism in a practical setting to allow for explicit computations, and (ii) to discover at last the BRST symmetry with the nilpotent operation *s*, which will be the main object of this chapter.

The next player in this game whose work has been very influential is Feynman, who, in the 1960s, looked at the Yang–Mills theory. He was not interested in the theory per se, but mainly as a laboratory for a quantum theory of gravity. As we have seen, Yang–Mills theories share with general relativity the problem of non-linearity, but they are simpler because the Lagrangian is still polynomial in the fields. Feynman tried to compute the amplitude for the scattering of two gauge bosons as an example for the calculation of the graviton–graviton amplitude. The calculation is simple. The Lagrangian is given by Eqs. (14.18) and (14.20). The interaction contains three- and four-boson vertices of the form

$$\mathcal{L}_I = 2g f^{abc} [\partial_\mu A_\nu^a - \partial_\nu A_\mu^a] A^{\mu b} A^{\nu c} + g^2 f^{abc} f^{ab'c'} A_\mu^b A_\nu^c A^{\mu b'} A^{\nu c'}. \tag{14.35}$$

Let us choose the Feynman gauge in which the propagator is

$$G_{\mu\nu}^{ab}(k) \sim \frac{\eta_{\mu\nu}}{k^2} \delta^{ab}, \tag{14.36}$$

with $\eta_{\mu\nu}$ the Minkowski metric. It propagates all four components of the vector field, but G_{00}^{ab} has the wrong sign. Let us look, for example, at the 1-loop diagram of Fig. 14.1(a). Feynman noted that the unphysical components with the zero index can circulate around the loop and they will contribute to the scattering amplitude. It is easy to check that this contribution will not be cancelled by that of any other diagram to that order. Therefore, a *physical* quantity, such as a scattering amplitude, would depend on the propagation of the *unphysical* components.[5] Note that this problem does not appear in the Feynman gauge of quantum electrodynamics because there are no diagrams with only photon lines which form a loop. We see that the problem is due to the non-linearity which gives the self-couplings of the gauge bosons.

In order to solve this problem Feynman made a bold suggestion. He was probably the only man at that time who understood that a quantum field theory is *defined* through the Feynman rules, so he decided to arbitrarily modify the naïve rules and supplement them with a new one: when computing closed loops of gauge bosons, we should add a new diagram, in which the gauge boson circulating in the loop is replaced by a fictitious scalar field whose couplings are arranged so that the unwanted unphysical contributions of the propagator (14.36) exactly cancel, Fig. 14.1(b). We call these scalar fields *ghosts* and we shall see shortly that the term is fully justified.

---

[5] Feynman made the computation in the 'Feynman' gauge with the propagator given by Eq. (14.36). In a more general way, we could choose a family of gauges depending on a parameter α, as we did for QED, compute the sum of all 1-loop diagrams for the two-boson scattering amplitude (they include the three- and four-boson vertices), and show that the resulting expression for the amplitude is still α-dependent. Hence, the need to introduce the ghosts.

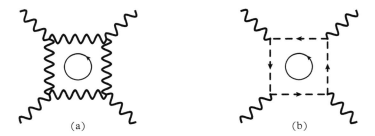

Figure 14.1 *A 1-loop diagram for the scattering amplitude of two Yang–Mills bosons. (a) The gauge boson contribution, (b) the ghost contribution.*

14.4.2.2 The Faddeev–Popov Formula

The Feynman rule of adding the ghost contribution sounds like a recipe. First, it is not clear how to generalise it to other gauges. Second, if we can check its effect for the 1-loop diagrams, it is not easy to guess why it should work at higher orders. Feynman conjectured that this was indeed the case but the argument, based on symmetry properties, was not fully convincing.

The correct version of Dirac's formula was found in 1967 by Ludvig Dmitrievich Faddeev and Victor Nikolaevich Popov from St Petersburg (then Leningrad) who succeeded to translate in the path integral language the intuitive idea of ghosts of Feynman. Moreover, they were able to replace the δ function for imposing the gauge condition $\mathcal{M}(\phi) = 0$ in the measure by a quadratic gauge-fixing term in the action, the way we used in QED in Eq. (12.160). This gave a formal justification for the Feynman recipe and offered the basis for a rigorous proof of the consistency of the method. But probably even more important, it uncovered a new symmetry which we will present in the following sections.

The way Faddeev and Popov proceeded differed from the way Dirac did by choosing to work directly in the manifestly covariant path integral formalism, rather than the Hamiltonian approach. Both methods are formally equivalent, but the approach of Faddeev and Popov provided eventually a local and covariant method for the perturbative gauge fixing of Yang–Mills theories.

More specifically, the general question, which we already asked in the particular case of QED, is the following: in quantum field theory the mean value of an observable represented by an operator $\mathcal{O}$ is given by the ratio of two functional integrals

$$< \mathcal{O} > = \frac{\int [\mathrm{d}\varphi] \mathcal{O}(\varphi) \exp[\mathrm{i} \int \mathrm{d}x (\mathcal{L}_{\mathrm{classical}})]}{\int [\mathrm{d}\varphi] \exp[\mathrm{i} \int \mathrm{d}x (\mathcal{L}_{\mathrm{classical}})]} \equiv \frac{\mathcal{N}}{\mathcal{D}}, \tag{14.37}$$

where the integration is taken over all classical field configurations. As we have often explained, the problem arises when the classical theory has a gauge invariance, i.e. the classical Lagrangian remains invariant under the transformation

$$\varphi \to \, {}^{g}\varphi. \tag{14.38}$$

Here g is an arbitrary gauge group element. We see that to every field configuration $\varphi(x)$ there corresponds a whole family of configurations labelled by g, which leave the Lagrangian invariant. We call this family *the orbit* of the gauge group. The g's include the infinitesimal transformations $\delta^{\text{gauge}}(\varphi)$, when $g \sim 1 + \delta^{\text{gauge}}$, but also all large gauge transformations and, for a general group, elements which may be disconnected from the identity. This invariance creates two problems. First, the differential operator which appears in the quadratic approximation of the classical Lagrangian has zero modes and, therefore, it is not invertible. As a result we cannot compute the Green functions and the perturbation expansion cannot be defined. Second, let us choose $\mathcal{O}$ to be a gauge invariant operator representing some physical observable. Since the measure of integration $[d\varphi]$, and $\mathcal{L}_{\text{classical}}$ are both invariant,[6] we see that the path integral for $< \mathcal{O} >$ counts many times (in fact, an infinite number of times) the same value, for all configurations of φ that belong to the same gauge group orbit. Of course, the same reasoning can be applied to any symmetry group $\mathcal{G}$, not only a gauge symmetry. For a finite system, we can argue that we get the relevant result times the volume of the symmetry group.[7] It will appear as a factor in both the numerator and the denominator of Eq. (14.37) and will cancel in the ratio. For a system with an infinite number of degrees of freedom we expect something analogous, but it is difficult to define precisely the meaning of the volume of the gauge group, which, as we have seen, should be viewed as $\mathcal{G}^{\infty}$.

For all these questions, the gauge-fixing problem is quite a puzzling one. It leads us to all sorts of infinities which may be hard to control. The Faddeev–Popov construction offered the first systematic way to address it. Let us reproduce the historical path, and blankly use formulae that are only correct for finite integrals. The justification, whenever possible, will come later.

We have the following formally obvious identity where the functional integration extends over the gauge group $\mathcal{G} \equiv \{g(x)\}$, and ${}^{g}\varphi$ is the gauge transformation of an arbitrary given field configuration φ

$$\int_{\mathcal{G}} [dg] \delta[\mathcal{M}({}^{g}\varphi)] \det \frac{\delta \mathcal{M}({}^{g}\varphi)}{\delta g} = 1. \tag{14.39}$$

We call the choice of the function $\mathcal{M}$ *the gauge choice*. This equality assumes that $\mathcal{M}$ is such that the condition $\mathcal{M}(\varphi) = 0$ has a unique solution in φ. In other words, we assume the existence of a gauge function, such that in the space of gauge field configurations, the curve $\mathcal{M}(\varphi) = 0$ cuts once and only once every given orbit $\{{}^{g}\varphi\}$,

---

[6] The gauge invariance of the measure $[d\varphi]$ is trivial at the tree level, but must be checked order by order in perturbation theory. This non-trivial property will be studied in a later chapter using the BRST symmetry

[7] All these arguments can be made rigorous by defining the system in a finite, Euclidean space–time lattice with N points and a lattice spacing a. The functional integrals become ordinary integrals and all steps can be fully controlled. The problems will appear when we take the double limit of large volume $N \to \infty$ and continuum space–time $a \to 0$.

where φ is a given field configuration and g runs over the gauge group. We will not discuss this condition for the function $\mathcal{M}$ in any detail here. We will briefly come back to it at the end of this chapter. We only mention that it is a very complex one from a global point of view (when we consider large gauge transformations) and it is generally wrong (as first shown by Vladimir Naumovich Gribov for the non-Abelian Yang–Mills theory in 1977). However, we can easily verify it for the class of gauge choices we will consider and restricting ourselves to gauge transformations in the vicinity of the identity.

As a first step, let us slightly generalise the gauge choice from the condition $\mathcal{M} = 0$ to the one $\mathcal{M} = \lambda(x)$ with $\lambda(x)$ being a given function. The identity (14.39) is still valid:

$$\int_{\mathcal{G}} [\mathrm{d}g] \delta[\mathcal{M}(^{g}\varphi - \lambda)] \det \frac{\delta \mathcal{M}(^{g}\varphi)}{\delta g} = 1. \tag{14.40}$$

Faddeev and Popov noted that we can choose any λ and integrate over all such choices with a Gaussian measure. This gives us the formulae

$$\int [\mathrm{d}\lambda] \exp[-\mathrm{i} \int \mathrm{d}x \frac{\lambda^2}{2\alpha}] \int_{\mathcal{G}} [\mathrm{d}g] \delta[\mathcal{M}(^{g}\varphi - \lambda)] \det \frac{\delta \mathcal{M}(^{g}\varphi)}{\delta g} = C(\alpha), \tag{14.41}$$

where α is an arbitrary parameter which takes real values and $C(\alpha)$ a constant whose precise value will not be needed. This formula can be rewritten as

$$\int_{\mathcal{G}} \int [\mathrm{d}\lambda][\mathrm{d}g] \delta[\mathcal{M}(^{g}\varphi - \lambda)] \det \frac{\delta \mathcal{M}(^{g}\varphi)}{\delta g} \exp[-\mathrm{i} \int \mathrm{d}x \frac{\lambda^2}{2\alpha}] = C(\alpha). \tag{14.42}$$

These formulae can be inserted into both the numerator $\mathcal{N}$ and the denominator $\mathcal{D}$ in the definition of the mean value of the gauge invariant observable $\mathcal{O}$ in Eq. (14.37). For example, for $\mathcal{N}$ we find that

$$\mathcal{N} = \frac{1}{C(\alpha)} \int_{\mathcal{G}} [\mathrm{d}g] \int [\mathrm{d}\varphi] \mathcal{O}(\varphi)$$
$$\times \det \frac{\delta \mathcal{M}(^{g}\varphi)}{\delta g} \exp[\mathrm{i} \int \mathrm{d}x(\mathcal{L}_{\mathrm{classical}}(\phi) - \frac{(\mathcal{M}(^{g}\varphi))^2}{2\alpha})]. \tag{14.43}$$

Due to the gauge invariance of the measure, the Lagrangian, and the observable, we can perform the integral over φ by setting everywhere $g = 1$, (we must change variables $\varphi \to {}^{g}\varphi$ and use the cyclicity property of the determinant). The last formula gives

$$\mathcal{N} \equiv \frac{1}{C(\alpha)} (\int_{\mathcal{G}} [\mathrm{d}g]) \int [\mathrm{d}\varphi] \mathcal{O}(\varphi) \det \frac{\delta \mathcal{M}(^{g}\varphi)}{\delta g} \Big|_{g=1}$$
$$\times \exp[\mathrm{i} \int \mathrm{d}x(\mathcal{L}_{\mathrm{classical}}(\varphi) - \frac{(\mathcal{M}(\varphi))^2}{2\alpha})] \tag{14.44}$$

or, equivalently,

$$\mathcal{N} \equiv \frac{1}{C(\alpha)} \left(\int_{\mathcal{G}} [\mathrm{d}g] \right) \int [\mathrm{d}\varphi] \mathcal{O}(\varphi) \det\left[\frac{\delta \mathcal{M}}{\delta \varphi} \frac{\delta_{\epsilon}^{\mathrm{gauge}}(\varphi)}{\delta \epsilon} \right]$$
$$\times \exp[\mathrm{i} \int \mathrm{d}x (\mathcal{L}_{\mathrm{classical}}(\varphi) - \frac{(\mathcal{M}(\varphi))^2}{2\alpha})]. \tag{14.45}$$

This is the desired result: we succeeded in factorising the volume of the gauge group $\int_{\mathcal{G}}[\mathrm{d}g]$, which is field-independent, and the constant C times a well-defined gauge-fixed path integral. In computing the ratio $\mathcal{N}/\mathcal{D}$ the field independent factors will cancel and we are left with a gauge-fixed action. The price is the appearance of the determinant, but because of it, the mean value of $\mathcal{O}$ cannot (formally) depend on the choice of the gauge function $\mathcal{M}$.

In order to obtain a path integral for a local quantum field theory, we use the integration rules for anticommuting Grassmann algebra valued fields and replace the determinant by a fermionic integral. We get

$$< \mathcal{O} > \sim \int [\mathrm{d}\varphi][\mathrm{d}\Omega][d\bar{\Omega}] \; \mathcal{O}(\varphi)$$
$$\times \exp[\mathrm{i} \int \mathrm{d}x (\mathcal{L}_{\mathrm{classical}}(\varphi) - \frac{(\mathcal{M}(\varphi))^2}{2\alpha} - \bar{\Omega} \left(\frac{\delta \mathcal{M}}{\delta \varphi} \frac{\delta_{\epsilon}^{\mathrm{gauge}}(\varphi)}{\delta \epsilon} \right) \Omega)]. \tag{14.46}$$

The fields Ω and $\bar{\Omega}$ are Lorentz scalars but they are quantised using anti-commutation rules. As a result they cannot represent physical degrees of freedom; otherwise, they would violate the spin-statistics theorem. Hence the term 'ghosts'. They should not appear in the asymptotic states; therefore, we do not need to compute amplitudes with external ghost lines.

We can reintroduce an auxiliary field, which will denote by b, and get instead

$$< \mathcal{O} > \sim \int [\mathrm{d}\varphi][\mathrm{d}\Omega][d\bar{\Omega}][\mathrm{d}b] \; \mathcal{O}(\varphi)$$
$$\times \exp[i \int \mathrm{d}x (\mathcal{L}_{\mathrm{classical}}(\varphi) + \frac{b^2}{\alpha} + b\mathcal{M}(\varphi) - \bar{\Omega} \left(\frac{\delta \mathcal{M}}{\delta \varphi} \frac{\delta_{\epsilon}^{\mathrm{gauge}}(\varphi)}{\delta \epsilon} \right) \Omega)]. \tag{14.47}$$

This formula is called the Faddeev–Popov formula. It tells us that all this business of gauge fixing amounts to replacing the original gauge invariant Lagrangian by an effective one which contains the gauge-fixing function, but also new fields, the Faddeev–Popov ghosts. It answers the question we asked at the beginning of this section, namely it gives the Fock space, including physical as well as unphysical degrees of freedom, which is necessary for the consistent quantisation of a gauge theory. In an axiomatic formulation we should start from this space. For a covariant formalism ghosts are needed for a correct balance of propagating fields and we must therefore include them in the theory from the beginning.

Before closing this section, let us look at some simple cases.

Everything we have said applies to any gauge invariant theory, including QED. For example, let us choose the usual gauge-fixing function $\mathcal{M} = \partial^\mu A_\mu$. The term depending on the ghost fields is just $\bar{\Omega}\square\Omega$. It follows that the ghost fields are free fields and we can integrate them away. We thus obtain the usual gauge-fixed effective Lagrangian. If, however, we decide to use a more complicated gauge condition, such as $\mathcal{M} = \partial_\mu A^\mu + A^\mu A_\mu$, this is no more true and we must include the ghost fields in our computations.

Of course, we should be perverse to choose such a non-linear gauge condition in QED, but if we move to a Yang–Mills theory, we cannot get rid of the non-linearity. Choosing $\mathcal{M} = \partial^\mu A_\mu^a$, we see that $\square$ is replaced by $\partial^\mu D_\mu$ with D_μ the covariant derivative. It follows that we still obtain a ghost term given by $\bar{\Omega}\square\Omega$, but, in addition, we obtain an extra term proportional to $f^{abc}\bar{\Omega}^a A_\mu^b \partial^\mu \Omega^c$. The ghost fields are coupled to the gauge fields and should be included in the calculation. They give the loop terms conjectured by Feynman. We see that in the Yang–Mills theory, the simplest set of ghost fields that Feynman thought of turns out to be the same pair of anticommuting scalar fields, a ghost $\Omega^a(x)$ and an anti-ghost $\bar{\Omega}^a(x)$, that occur in the Faddeev–Popov formula. Ω and $\bar{\Omega}$ are valued in $\mathcal{G}$, as the gauge field A.[8] Their unphysical statistics (unphysical means here that it is opposite to that given by the usual spin-statistics relation) and their Feynman rules must allow for the above-mentioned compensations in all physical amplitudes, at least at any finite order of perturbation theory.

Since we have more fields than gauge fields A, to maintain the general idea of gauge covariance for physics, it is natural to look for a symmetry that encodes the definition of the gauge symmetry acting on A, but also generalises it in a consistent way on the new fields $\Omega, \bar{\Omega}$, and b. This symmetry must interchange the gauge fields and the ghosts, in order to ensure the compensations between these fields in closed loops for any choice of gauge. Such a symmetry between fields of different statistics is the BRST symmetry. It was actually found at the end of a rather long and intricated process, by Carlo Maria Becchi, Alain Rouet, and Raymond Stora, on the one hand, and Igor Viktorovich Tyutin on the other, as a symmetry of the Faddeev–Popov effective action. The initial motivation was to give a compact form of the complicated relations among the couplings of gauge fields and ghosts.

Since that time, the mathematics of the space of gauge field configurations has been developed in such a way that it incorporates the ghosts as well-defined mathematical entities. The BRST symmetry can be directly constructed, before even thinking that it can be used for a gauge-fixing problem. As we will see, the ghosts and the gauge fields can be unified as an extended geometrical object.

14.4.3 Graded Notation for the Classical and Ghost Yang–Mills Fields

In order to construct the symmetry between classical fields and ghosts in the easiest way, it is convenient to introduce some compact notations.

---

[8] The notation is slightly misleading. The 'bar' does not mean that Ω and $\bar{\Omega}$ are related by an operation of complex conjugation. They are two independent fields, but, as we shall explain shortly, they carry opposite quantum numbers.

We shall adopt a matrix notation for the Yang–Mills field and we shall saturate its Lorentz indices with differential forms dx^μ. We end up with formulae well adapted for using the formalism of differential algebra.

The field

$$A = \mathcal{A}_\mu dx^\mu = A_\mu^a T^a dx^\mu \tag{14.48}$$

is called the Yang–Mills connection. It is an anticommuting (odd) quantity.

We will shortly define a generalised grading that determines the statistics of all objects, including the ghosts. If it is even, they are commuting objects; if it is odd, they anticommute.

In our case, this grading will be the sum of the ordinary form degree and a new quantum number, which we will discuss in more detail later on, called the ghost number.

For instance, the scalar ghost and antighost have ghost numbers 1 and −1, respectively. Therefore they anticommute. The auxiliary field b has ghost number 0 and commutes.

We introduce a universal graded bracket [,]: given two elements X and Y valued in the Lie algebra, we have

$$[X, Y] \equiv XY \pm YX, \tag{14.49}$$

where the + sign only occurs if both X and Y are odd, that is, when their matrix elements are anticommuting objects. We have, for instance,

$$F = dA + gAA = dA + \frac{g}{2}[A, A] \tag{14.50}$$

and the 2-form curvature F satisfies the following Bianchi identity, due to $d^2 = 0$ and the Jacobi identity in the Lie algebra $\mathcal{G}$:

$$dF = -g[A, F], \quad \text{which implies that} \quad DF = 0. \tag{14.51}$$

We have $g[F, Y] = DDY$, where Y is any given function of the fields. We can write the covariant derivative of Ω as $D\Omega = d\Omega + g(A\Omega + \Omega A) = d\Omega + g[A, \Omega]$, we have $DD\Omega = g[F, \Omega]$, and so on.

When Y belongs to a given representation of the gauge algebra which is not the adjoint, we can conveniently define the notation $[X, Y]^i \equiv gX^a T_{aj}^i Y^j$. It allows us to write in a unified way equations for matter fields in arbitrary representations of the gauge group.

An infinitesimal gauge transformation is given by

$$\delta A = D\epsilon = d\epsilon + g[A, \epsilon], \tag{14.52}$$

where $\epsilon = \epsilon^a(x) T^a$ is now a $\mathcal{G}$-valued matrix of infinitesimal parameters.

Consistency requires that the commutator of two infinitesimal gauge transformations is a gauge transformation. This is automatically fulfilled because $\mathcal{G}$ is a Lie algebra and the structure constants satisfy the Jacobi identity. This implies that

$$[A, [A, A]] = 0 \qquad 2[A, [A, \epsilon]] + [\epsilon, [A, A]] = 0, \qquad (14.53)$$

which makes it possible to prove the closure of gauge transformations.

Using this notation, we have for the Yang–Mills Lagrangian:

$$\mathcal{L}_{\mathrm{YM}} = -\frac{1}{4} F^{\mu\nu a} F^a_{\mu\nu} = -\frac{1}{4} \mathrm{Tr} F^{\mu\nu} F_{\mu\nu} = -\frac{1}{4} \mathrm{Tr} F \wedge^* F. \qquad (14.54)$$

$\mathcal{L}_{\mathrm{YM}}$ is gauge invariant because F transforms covariantly under gauge transformations:

$$\delta F = -g[\epsilon, F]. \qquad (14.55)$$

Note that[9]

$$\mathcal{L}_{\mathrm{YM}} = -\frac{1}{8} \mathrm{Tr}(F +^* F) \wedge (F +^* F) + \frac{1}{4} \mathrm{Tr} F \wedge F. \qquad (14.56)$$

The last term is locally a pure derivative term because the Bianchi identity implies that $d(\mathrm{Tr} F \wedge F) = 2\mathrm{Tr}(DF \wedge F) = 0$, and $\mathrm{Tr}(F \wedge F) = d(A \wedge F - \frac{1}{6} A \wedge A \wedge A)$. Therefore, it does not contribute to the equations of motion. The classical equations of motion

$$D^* F = 0 \qquad (14.57)$$

are complicated non-linear equations, but they can be solved when the self-duality condition

$$F +^* F = 0 \qquad (14.58)$$

is satisfied in Euclidean space, since the first term of the action is a positive square. This equation has very interesting solutions, called instantons, which will be discussed briefly later.

14.4.4 Determination of the BRST Symmetry as the Extension of the Gauge Symmetry for the Classical and Ghost Fields

Coming back to our problem, we are looking for a new symmetry, which encodes the gauge symmetry and will leave invariant the full local action, after it has been gauge-fixed, with any possible choice of the gauge function, as follows:[10]

$$\mathcal{L} = \mathrm{Tr}\left(-\frac{1}{4} F^{\mu\nu} F_{\mu\nu} + b\mathcal{M}(A) + \cdots\right). \qquad (14.59)$$

Here, the terms ... are yet unknown, but may depend on $A, \Omega, \bar{\Omega}$, and b.

---

[9] The notation is slightly misleading. A factor $d^4 x$ should be understood in the first three terms.

[10] We follow the approach developed in: L. Baulieu and J. Thierry-Mieg, *Nucl. Phys.* **B197**, 477 (1982); L. Alvarez-Gaumé and L. Baulieu, *Nucl. Phys.* **B212**, 255 (1983). For a pedagogical review see: L. Baulieu, *Perturbative Gauge Theories*, lectures given at the Cargèse Summer School (Plenum Press, 1983), and *Phys. Rept.* **129**, 1 (1985).

$\mathcal{M}(A)$ is a function that characterises the gauge. More precisely we shall choose $\mathcal{M}$ with the following properties: it must be a polynomial in the fields and their derivatives of canonical dimension equal to 2 and it must vanish sufficiently rapidly when the Euclidean point x goes to infinity. This is the usual assumption concerning the terms in the Lagrangian in perturbation theory.

Furthermore, let $\{g\}$ denote an element of the gauge group and $A \to {}^{\{g\}}A$ the corresponding gauge transformation. We consider the equation $\mathcal{M}({}^{\{g\}}A) = C(x)$, with $C(x)$ a given function belonging to the adjoint representation of the gauge group. $\mathcal{M}(A)$ must be such that this equation, considered as an equation for $\{g\}$, admits a unique solution for any A and any C, at least when $\{g\}$ is restricted to be in the neighbourhood of the identity element. We shall examine later the case of large gauge transformations.

Although non-covariant gauges can be, and often are, used, we shall restrict here to functions $\mathcal{M}(A)$ which are Lorentz scalars. In this case the term $\mathcal{M}(A) = \partial_\mu A^\mu$, which is the only one which is linear in the gauge field, is necessary in order to have a gauge fixing in the quadratic part of the Lagrangian and obtain well-defined propagators. It gives the Landau–Feynman-type gauges.

Since the symmetry that acts on A, Ω, $\bar{\Omega}$, and b must in particular relate the longitudinal part of A to the ghosts (we have in mind the compensations in closed loops predicted by Feynman), the parameters of the symmetry must be anticommuting parameters and the simplest possibility is that it involves a single constant anticommuting parameter η. If we define

$$\delta_{\mathcal{BRST}} = \eta s, \qquad (14.60)$$

s will be called the generator of the BRST symmetry. We shall derive the action of s on all fields, from general arguments.

Let us first define a grading which will be used to define the different statistics of all fields. It is the sum of the usual form degree and the ghost number g, defined as follows

fields	A	Ω	$\bar{\Omega}$	b
form degree	1	0	0	0
ghost number	0	1	-1	0
grading	1	1	-1	0

The grading of all fields is defined modulo two. When it is even, it denotes a commuting object and when it is odd an anticommuting one.

Because of the definition $\delta_{\mathcal{BRST}} = \eta s$, the BRST operator s acts as a left-differential operator, with the graded Leibniz rule

$$s(XY) = (sX)Y \pm XsY. \qquad (14.61)$$

Here the minus sign occurs only if X has odd grading. The same graded rule holds for the exterior derivative d. Thus, both s and d are differential operators with grading one. The exterior derivative d increases the form degree by one unit and leaves the ghost number unchanged; in contrast, the BRST operator increases the ghost number by one unit, but leaves fixed the form degree.

The grading of the differential operators that act on the fields is

operators	d	s
form degree	1	0
ghost number	0	1
grading	1	1

To determine the way s will transform all the fields (modulo rescaling of the fields and the coupling constant), the following property is in fact sufficient:

$$s^2 = 0 \tag{14.62}$$

$$sd + ds = 0. \tag{14.63}$$

The second property is the only possible one, because gauge transformations must commute with d. The first property, $s^2 = 0$, is the simplest one that makes sense for a graded differential operator. It can be enforced because the fields are valued in a Lie algebra, which implies that the graded bracket $[.,.]$ satisfies the Jacobi identity.

The symmetry operators we have studied so far were all chosen to act linearly on the fields. It will turn out that this is impossible for s, so we shall make the next simplest assumption: the s-transform of every field will be a local polynomial in the fields and their derivatives. In this case, ordinary dimensional analysis and the conservation of the ghost number uniquely determine the action of s, modulo a redefinition of the field b, and rescaling of the other fields and g.

Let us recall that the canonical dimension of A equals 1 and that of b, derived from (14.59), equals 2. Since the ghost fields Ω and $\bar{\Omega}$ have spin 0, we expect them to have each dimension equal to 1 and this is indeed the natural choice. However, because of the conservation of ghost number, they will only appear in the Lagrangian in the combination $\bar{\Omega}\Omega$, which means that only the sum of their dimensions is relevant.

It will turn out that the choice $Dim[\Omega] = 0$ and $Dim[\bar{\Omega}] = 2$ will be slightly more convenient. Under these conditions, the most general solution for the action of s is (see Problem (14.2))

$$\begin{aligned}
sA &= -d\Omega - g[A, \Omega] \\
s\varphi &= -g[\Omega, \varphi] \\
s\Omega &= -\frac{g}{2}[\Omega, \Omega] \\
s\bar{\Omega} &= b \\
sb &= 0,
\end{aligned} \tag{14.64}$$

where φ is a matter field that transforms covariantly under the gauge symmetry. Note that s acts on A as well as the matter fields as a gauge transformation with the ghost field

$\Omega(x)$ playing the role of the gauge parameter. The dependence on the coupling constant g can be absorbed in a redefinition of the structure coefficients f_{bc}^a.

The properties $s\bar{\Omega} = b$, $sb = 0$ imply that any function that is s-invariant and depends on b and $\bar{\Omega}$ must be s-exact. This identifies the pair $(\bar{\Omega}, b)$ as a so-called 'trivial BRST doublet'.

We can in fact verify that the property of $(\bar{\Omega}, b)$ to be a trivial BRST doublet remains true if we redefine b by a rescaling and the addition of any combination of the fields $A, \varphi, \Omega, \bar{\Omega}$. This of course changes the definition of sb and, accordingly, that of $s\bar{\Omega}$. The important point is that sb must be the s-transform of some combination of the fields F, in which case $s\bar{\Omega}$ will be given by $b - F$, so that we have both $s^2\bar{\Omega} = 0$, and $s^2b = 0$.

Suppose now that we adopt a general definition of the bracket $[\bullet, \bullet]$ namely that $[X, Y]^a \equiv F_{bc}^a X^b Y^c$, where the coefficients F_{bc}^a are matrix elements, which are yet undetermined. Then, we can verify that the property $s^2\Omega = 0$ for $s\Omega = -\frac{g}{2}[\Omega, \Omega]$ implies that the F_{bc}^a's satisfy the Jacobi identity, and correspond therefore to Lie algebra coefficients, and the property $s^2A = 0$ implies in turn that $sA = -d\Omega - g[A, \Omega]$.

We can also verify that the latter property $s^2A = 0$ is equivalent to the closure of the gauge transformations on A.

The proof that $s^2 = 0$ completely determines the BRST equations, modulo rescaling factors and possible redefinitions of b, relies only on Lie algebra identities , which imply that, e.g., $[\Omega, [\Omega, \Omega]] = 0$, $[\Omega, [A\Omega]] = 2[A, [\Omega, \Omega]]$.

In fact, the intuitive property $s^2 = 0$ summarises all properties of infinitesimal gauge transformations.

Before closing this section we want to point out an elegant way to unify, in a compact notation, the ghost and the gauge fields. We can define $\hat{d} = d + s$, with $\hat{d}^2 = 0$ and $\hat{A} = A + \Omega$. Then the BRST equations for A and Ω can be elegantly written as

$$\hat{F} \equiv (d + s)(A + \Omega) + \frac{g}{2}[A + \Omega, A + \Omega] = F. \tag{14.65}$$

The property that $s^2 = 0$, that is $\hat{d}^2\hat{A} = 0$, is guaranteed by the Bianchi identity

$$\hat{D}\hat{F} = (d + s)\hat{F} + g[A + \Omega, \hat{F}] = 0. \tag{14.66}$$

The physical interpretation of this unification between the ghost and the gauge field is as follows.

There is a generalised one form $\hat{A} = A + \Omega$, where the classical part is $\hat{A}_1^0 = A$ and the ghost part is $\hat{A}_0^1 = \Omega$. Here the upper index is the ghost number and the lower one the ordinary form degree, and the degree of a generalised form is the sum of both.

It is important to realise that the Faddeev–Popov ghost can be considered as a generalised one form, thanks to the introduction of this bi-grading.

We understand that $\hat{A}$ is not observable, so it can include a ghost part as well as a longitudinal part.

On the other hand its curvature $\hat{F}$ must be physical, which justifies the property that the components with ghost number 1 and 2, namely $\hat{F}_1^1$ and $\hat{F}_0^2$, are zero. These vanishing conditions are precisely the BRST equations.

Eq. (14.65) is in fact the tip of an iceberg. It is the sign of a geometrical origin for the BRST equations and of the depth of the notion of the BRST symmetry.

However, its interpretation involves the fibre bundle formalism, which allows us to associate in a single entity the space–time and a Lie group, but goes beyond the scope of this book.

In practice, it will be most useful in our further study of the question of internal consistency of a gauge theory.

14.4.5   General BRST Invariant Action for the Yang–Mills Theory

Now comes the determination of the Lagrangian that we will insert in the path integral formalism of the Yang–Mills theory. Although we have the answer for a particular class of gauge-fixing functions in the Faddeev–Popov formula (14.47), we want to derive it here in a more general way using only the BRST symmetry. We postulate that the action depends on all fields in the quartet $(A, \Omega, \bar{\Omega}, b)$ and is invariant under the BRST transformations. Using this symmetry we shall be able to fix the dots we left undetermined in (14.59). We are looking for an action $\mathcal{L}$ that must satisfy the following requirements:

- $\mathcal{L}$ is a local polynomial in the fields and their derivatives.
- $\mathcal{L}$ is BRST invariant.
- $\mathcal{L}$ has dimension 4.
- $\mathcal{L}$ has ghost number 0.
- $\mathcal{L}$ is Lorentz invariant.

Since the BRST invariance is such that $s^2 = 0$, we can express $\mathcal{L}$ as

$$\mathcal{L} = \mathcal{L}_{cl} + sK, \tag{14.67}$$

where $\mathcal{L}_{cl}$ is s-invariant, but does not contain terms which are s-exact. In other words, $\mathcal{L}_{cl}$ represents the cohomology with ghost number 0 of s with the appropriate requirements listed above. Since s increases the ghost number by one unit, K must have ghost number -1, and, thus, it must be of the form

$$K = \mathrm{Tr}\big(\bar{\Omega} K_{gf}\big). \tag{14.68}$$

Here K_{gf} is a polynomial of all possible fields, with ghost number 0 and dimension 2. For convenience, we shall restrict ourselves here to gauge-fixing terms that preserve Lorentz invariance as well as the global symmetry of the gauge group G, although non-covariant and/or asymmetric gauges can be used. The only terms which are absolutely essential are *b* and ∂A because they will give well-defined propagators in perturbation theory. They both belong to the adjoint representation of the group, which means that all terms in K_{gf} should be chosen to be Lorentz scalars and members of the adjoint representation of G. Furthermore, they must satisfy all the other requirements we stated in the discussion following Eq. (14.59).

Going over all possible monomials that depend on $A, \Omega, \bar{\Omega}$, and b, we can verify that the general solution for $\mathcal{L}$, which satisfies the above requirements, is, modulo an overall renormalization factor and boundary terms

$$\mathcal{L} = \text{Tr}\left(-\frac{1}{4}F^{\mu\nu}F_{\mu\nu} + s\bar{\Omega}\left(\frac{\alpha}{2}b + \partial_\mu A^\mu + \frac{\beta}{2}[\bar{\Omega}, \Omega]\right)\right), \tag{14.69}$$

where α and β are arbitrary real numbers. They are called gauge parameters.[11]

If we expand (14.69), using the definition of s, we find that

$$\mathcal{L} = \text{Tr}\left(-\frac{1}{4}F^{\mu\nu}F_{\mu\nu} + \frac{\alpha}{2}b^2 + b(\partial_\mu A^\mu + \beta[\bar{\Omega}, \Omega]) - \bar{\Omega}\partial_\mu D^\mu \Omega + \frac{\beta}{2}[\bar{\Omega}, \Omega]^2\right). \tag{14.70}$$

We can eliminate the field b by performing a Gaussian integration, which amounts to replacing b by its algebraic equation of motion,

$$b = -\frac{\partial_\mu A^\mu}{\alpha} - \frac{\beta}{\alpha}[\bar{\Omega}, \Omega]. \tag{14.71}$$

Thus, we get

$$\mathcal{L} = \text{Tr}\left(-\frac{1}{4}F^{\mu\nu}F_{\mu\nu} - \frac{1}{2\alpha}(\partial_\mu A^\mu)^2 - \bar{\Omega}\partial_\mu D^\mu \Omega - \frac{\beta}{\alpha}\partial_\mu A^\mu[\bar{\Omega}, \Omega] \right.$$
$$\left. - \frac{\beta}{2}\left(1 - \frac{\beta}{\alpha}\right)[\bar{\Omega}, \Omega]^2\right). \tag{14.72}$$

This Lagrangian has a complicated form, but it is conceptually very simple.

The Dirac formula, completed by the work of Faddeev–Popov, when the gauge function is a linear function of the physical fields A, φ, is just a particular case of our general formula when $\alpha = \beta = 0$. We can prove that these values of the gauge parameters form a fixed point under the renormalisation process which we shall develop in a later chapter.

Let us rest for a while, and contemplate the results. The gauge invariance has been fixed in a BRST invariant way, since $\mathcal{L}$ is by construction s-invariant and contains the term $(\partial_\mu A^\mu)^2$ that determines an invertible quadratic form for A. It also depends on

---

[11] In many cases we could add more terms. For example, if the gauge group is either $U(1)$ or $SU(N)$ with $N > 2$, we could add a term proportional to $A_\mu A^\mu$. Indeed, for these groups we can build an element of the adjoint representation by taking the symmetric product of two adjoints. Apart from curiosity, there are no compelling reasons to add such terms.

the ghosts and, in the family of gauges we have used, it contains quadratic and quartic terms. This is the price for maintaining the BRST invariance. We can simplify it by choosing special gauges, such as $\beta = 0$, which is perfectly adequate for most practical computations.

In the case of electrodynamics $D^\mu \Omega = \partial^\mu \Omega$ and, in the $\beta = 0$ gauge, the ghost fields decouple. We thus recover the usual form of the Lagrangian. The physical observables of the theory are defined as functions of **E** and **B**, i.e. of the transverse part of the gauge field A. In the Abelian case, we have $sA_\mu = \partial_\mu \Omega$. Thus, the s-invariance of observables coincides with the intuitive requirement that A must be transverse. This definition becomes elusive in the non-Abelian case.

The more precise—and in fact correct—definition of observables that we will shortly adopt is that they are functionals with ghost number 0 of the fields and are s-invariant and defined modulo s-exact terms.

In short, observables are defined as the elements of the cohomology of the BRST symmetry with ghost number 0.

We see that they cannot depend on b and $\bar\Omega$, since these fields build a BRST-trivial doublet. But since they cannot depend on $\bar\Omega$, they cannot depend also on Ω, because of the conservation number. We have thus that the observables are gauge invariant functionals of the gauge field.

The subtlety that occurs in quantum field theory is that the renormalisation can mix their mean values with those of BRST-exact functionals of the fields with the same dimensions, which justifies the definition of observables as the elements of the cohomology of the BRST operator s. A good example of an observable is in fact the classical Yang–Mills action.

Let us now verify that the quadratic approximation of the Lagrangian truly gets invertible propagators. Keeping only the quadratic terms in our action, we get

$$\mathcal{L}_{\text{quad}} = \text{Tr}\left(-\frac{1}{4}A^\mu\left((g_{\mu\nu} - \partial^\rho\partial_\rho\partial_\mu\partial_\nu) - \frac{1}{2\alpha}\partial_\mu\partial_\nu\right)A^\nu - \bar\Omega\partial_\mu\partial^\mu\Omega\right). \tag{14.73}$$

We easily get that the free propagators are in momentum space

$$\langle A_\mu^a(q), A_\nu^b(-q)\rangle = \delta^{ab}\left(\frac{1}{q^2 + i\varepsilon}\left(g_{\mu\nu} - \frac{q_\mu q_\nu}{q^2}\right) + \frac{1}{\alpha}\frac{1}{q^2 + i\varepsilon}\frac{q_\mu q_\nu}{q^2}\right) \tag{14.74}$$

$$\langle \bar\Omega^a(q), \Omega^b(-q)\rangle = \frac{\delta^{ab}}{q^2 + i\varepsilon}. \tag{14.75}$$

Note that the ghost propagator connects a field $\bar\Omega$ to a field Ω, although, as we pointed out already, the first is not the complex conjugate of the second. In a Feynman diagram a ghost line will have an arrow showing the flow of the ghost number.

So far we have not considered the possible coupling of matter, for spinor or scalar fields. The BRST transformation of a scalar Φ or a spinor Ψ reproduces the action of an infinitesimal gauge transformation, when we replace the parameter of the transformation by the ghost Ω:

$$s\Psi = -[\Omega, \Psi] \qquad s\Phi = -[\Omega, \Phi]. \tag{14.76}$$

(We already introduced the convenient notation that when φ^i belongs to any given representation, $[\Omega, \varphi]^i \equiv T_j^{ai} \Omega^a \varphi^j$.)

Thus, asking for the most general BRST invariant matter action, we get that

$$\mathcal{L}_{\text{matter}} = \Phi D_\mu D^\mu \Phi + V(\Phi) + \bar{\Psi}(\gamma^\mu (1 + a\gamma^5) D_\mu - m)\Psi, \tag{14.77}$$

where D is the covariant derivative, m is a mass for the spinor, and V an invariant self-interacting potential for the scalar field. In practice it is chosen as a polynomial in Φ of maximum degree four. When the coupling constant a is not zero, the Lagrangian violates parity.

The presence of a scalar field may allow for a new phenomenon which will turn out to be very important for physical applications. Depending on the choice of the potential V, the value of the field Φ which corresponds to the minimum energy solution of the classical equations of motion may be non-zero.

This phenomenon will be studied in detail in Chapter 15. In quantum language we say that Φ acquires a non-zero vacuum expectation value (v.e.v.) $v = \langle \Phi \rangle$. In this case, the BRST exact part of the Lagrangian can contain extra-terms, such as

$$\mathcal{L} = \text{Tr}\Big(-\frac{1}{4}F^{\mu\nu}F_{\mu\nu} + \mathcal{L}_{\text{matter}} + s\bar{\Omega}\big(\frac{\alpha}{2}b + \partial_\mu A_\mu + \gamma[v, \Phi] + \frac{\beta}{2}[\bar{\Omega}, \Omega]\big)\Big), \tag{14.78}$$

where γ is a new gauge parameter. We shall see that such gauges are very convenient in practical calculations.

The Lagrangian $\mathcal{L}$, with a suitable choice of the gauge group and of the v.e.v. of Φ, turns out to be the right one for describing, in a unified framework, the strong, electromagnetic, and weak interactions that we observe in nature.

14.5 Feynman Rules for the BRST Invariant Yang–Mills Action

We just proved that after all this gauge-fixing procedure is finished, we are left with an effective action with which computations should be performed. The original gauge invariance is explicitly broken by the gauge-fixing term, but a residue remains, the invariance under the BRST transformations which act non-linearly on the fields, including the ghosts. In the usual family of gauges, given by the function $\partial^\mu A_\mu$, the effective action is shown in Eq. (14.72) with $\beta = 0$. In this section we want to write explicitly the resulting set of Feynman rules:

- The propagator for the gauge field is given in Eq. (14.74). As expected, it depends on the gauge parameter α.

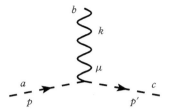

Figure 14.2 *The gauge boson-ghost-antighost vertex.*

- The ghost propagator is given in Eq. (14.75). It corresponds to a contraction between $\bar{\Omega}$ and Ω. Although these two must be treated as two independent fields, there are no $\langle \Omega, \Omega \rangle$, or $\langle \bar{\Omega}, \bar{\Omega} \rangle$ contractions.

- The interaction terms between the gauge fields give the three- and four-boson vertices obtained in the expressions (13.163) and (13.164).

- The new term, resulting from the gauge-fixing process, comes from the term $\bar{\Omega}\partial_\mu D^\mu \Omega$ in the effective Lagrangian which gives a gauge boson – ghost – anti ghost vertex. After partial integration it becomes

$$\mathcal{L}_{\text{YMgh}} = -g f_{abc} \partial^\mu (\bar{\Omega}^a) A^b_\mu \Omega^c. \tag{14.79}$$

Note that the derivative operator ∂^μ applies to $\bar{\Omega}$ but not to Ω: another indication that these fields should not be considered as complex conjugates to each other. This vertex is represented by Fig. 14.2 and it is given by $g p^\mu f_{abc} (2\pi)^4 \delta^4 (k + p - p')$.

- A further rule should be added to those listed above: it stems from the fact that $\bar{\Omega}$ and Ω are anti-commuting ghost fields. It follows that ghosts do not appear in the external lines of an S-matrix diagram and there is a minus sign for every closed ghost loop.

14.6   BRST Quantization of Gravity Seen as a Gauge Theory

We have seen in the Chapter 4 that gravity is invariant under the gauge symmetry acting on fields by means of the Lie derivative:

$$g_{\mu\nu} \to g_{\mu\nu} q + \mathcal{L}_\xi g_{\mu\nu}. \tag{14.80}$$

Like every gauge invariance, (14.80) implies that the quadratic part of the Einstein–Hilbert action $\int \mathrm{d}^4 x \sqrt{-\det(g)} R(g_{\mu\nu})$ has zero modes and it is not invertible. It follows that we cannot compute the propagator for the metric field.

Indeed, if we do a perturbative expansion of the metric $g_{\mu\nu} = \eta_{\mu\nu} + \kappa G_{\mu\nu}$ and of $\int \mathrm{d}^4 x \sqrt{-\det(g)} R(g_{\mu\nu})$ in powers of κ, we find that the quadratic approximation of the action in the field $G_{\mu\nu}$ is invariant under $G_{\mu\nu} \to G_{\mu\nu} + \partial_{\{\mu} \xi_{\nu\}}$, where $\xi_\mu(x)$ is an infinitesimal local vector parameter. This is a gauge invariance analogous to the one that occurs in the

Yang–Mills theory, but in the case of gravity the infinitesimal parameter of the symmetry is a vector.

Thus, for the quantisation, when we try to build a path integral of the type

$$\int \mathrm{d}[g_{\mu\nu}]\mathrm{d}[\varphi]\exp i \int\int \mathrm{d}^4x\sqrt{-\det(g)}(R(g_{\mu\nu}) + L_{\mathrm{matter}}(\varphi)), \qquad (14.81)$$

we must do the path integration over the metric, while remaining orthogonal to the space of the zero modes of Eq. (14.80).

This is analogous to the problem we encountered—and solved—in the case of the gauge fields, where the zero modes come from the invariance of the action under $A_\mu \to A_\mu + D_\mu\epsilon$.

In the case of gravity, there are good gauge choices at the classical level. One of them is given by the harmonic, or de Donder, gauge condition

$$\partial_\mu g^{\mu\nu} = 0. \qquad (14.82)$$

This condition obviously breaks reparametrisation invariance. If one succeeds in adding consistently a term like

$$\frac{1}{\alpha}\int \mathrm{d}^4x\sqrt{-\det(g)}\partial_\mu g^{\mu\nu}\partial_\rho g^{\rho\sigma}g_{\nu\sigma} \qquad (14.83)$$

to the action, it will ensure the propagation of zero modes that are not present in the quadratic field approximation of the Einstein–Hilbert action $\int \mathrm{d}^4x\sqrt{-\det(g)}R(g_{\mu\nu})$. Clearly, the word 'consistently' means that this term must be accompanied by other terms that ensure a BRST invariance associated with the gauge symmetry for reparametrisation. Indeed the physical quantities will be defined from the cohomology of the BRST symmetry of reparametrisation invariance. Formally, their mean values will depend neither on the choice of the gauge condition nor on the value of whichever parameter α we introduce for the gauge fixing, due to the presence of ghost terms that the BRST invariance determines.

We will thus generalise what we did for the case of the Yang–Mills symmetry, by replacing the algebra of Yang–Mills transformations by that of infinitesimal diffeomorphisms.

How to construct the BRST symmetry for reparametrisation symmetry? We start from the gauge transformations

$$\delta g_{\mu\nu} = \mathcal{L}_\xi g_{\mu\nu}. \qquad (14.84)$$

We associate to the vector parameter ξ^μ an anti-commuting vector ghost field $c^\mu(x)$.

We need an anti-ghost field $\bar{c}_\mu(x)$ and its Lagrange multiplier field b_μ. We keep in mind that the latter will be used to enforce a gauge condition on $g_{\mu\nu}$.

The nilpotent BRST symmetry operation is easy to construct as a generalisation of the Yang–Mills BRST invariance. It is

$$sg_{\mu\nu} = \mathcal{L}_\xi g_{\mu\nu} \quad sc^\mu = \mathcal{L}_\xi c^\mu = c^\nu \partial_\nu c^\mu \tag{14.85}$$

$$s\bar{c}_\mu = b_\mu \quad sb_\mu = 0. \tag{14.86}$$

The nilpotency property of s, $s^2 = 0$, is obvious when it acts on $\bar{c}_\mu$ and b_μ. The property $s^2 = 0$ on $g_{\mu\nu}$ and c^μ must be verified by computation. It is actually a direct consequence of the property that the Lie derivative operation $\mathcal{L}$ generates a closed Lie algebra.

Obviously we have

$$s \int d^4x \sqrt{-\det(g)} R(g_{\mu\nu}) = 0 \tag{14.87}$$

since the Einstein–Hilbert action is reparametrisation invariant and the action of s on $g_{\mu\nu}$ is identical to a gauge transformation when we replace ξ by c.

Then the following gauge-fixing action

$$I_{gf} = s \int d^4x \, s(\bar{c}_\mu (\partial_\nu g^{\mu\nu} + \alpha g^{\mu\nu} b_\nu)), \tag{14.88}$$

where α is a gauge parameter, is s-exact and thus s invariant.

It is what we need for a BRST invariant gauge fixing of the Einstein–Hilbert action. Indeed, by expansion of the s-exact term, it gives

$$I_{gf} = s \int d^4x (\alpha g^{\mu\nu} b_\nu b_\mu + b_\mu (\partial_\nu g^{\mu\nu} + \alpha \bar{c}_\nu \mathcal{L}_c g^{\mu\nu}) + \bar{c}_\mu \partial_\nu \mathcal{L}_c g^{\mu\nu}. \tag{14.89}$$

By Gaussian integration

$$\alpha g^{\mu\nu} b_\nu b_\mu + b_\mu (\partial_\nu g^{\mu\nu} + \alpha \bar{c}_\nu \mathcal{L}_c g^{\mu\nu}) \sim -\frac{1}{\alpha} |\partial_\nu g^{\mu\nu} + \alpha \bar{c}_\nu \mathcal{L}_c g^{\mu\nu}|^2. \tag{14.90}$$

we get the desired gauge-fixing term

$$-\frac{1}{\alpha} \partial_\nu g^{\mu\nu} \partial_\rho g^{\rho\sigma} g_{\mu\sigma}. \tag{14.91}$$

The remaining terms are rather complicated, but easy to compute, ghost-dependent terms by expanding the BRST-exact term. We must note that there are ghost-dependent terms with interactions of a higher degree, as a consequence of the definition of the BRST symmetry.

Once we have a gauge-fixed action with a BRST invariance, we can proceed to try to define a perturbation expansion. The method consists of expanding the metric $g_{\mu\nu}$ around a fixed background $g_{0\mu\nu}$, for instance

$$g_{\mu\nu} = g_{0\mu\nu} + \kappa h_{\mu\nu}. \tag{14.92}$$

We can then define a BRST invariant and gauge-fixed perturbation theory in κ for the field $h_{\mu\nu}$, which fluctuates around its classical value $g_{0\mu\nu}$, as predicted by the path integral. This is an important physical difference between gravity and another field theory. For field theories we compute the quantum fluctuations of the fields around a classical value which is often chosen to be zero. For gravity the field is the metric of space-time and the quantum fluctuations are always considered around a given classical background.

This analysis determines, in principle, the rules for computing physical quantities in a perturbation expansion of quantum gravity. However, we understand that in practice the application of these rules will not be easy. The reason is that the effective Lagrangian is not a polynomial in the quantum field $h_{\mu\nu}$ and when expanded in powers of the coupling constant, will yield an infinite number of terms. We shall see in a later chapter that this problem is so severe that it does not allow us to build a consistent quantum field theory of general relativity.

14.7 The Gribov Ambiguity: The Failure of the Gauge-Fixing Process beyond Perturbation Theory

14.7.1 A Simple Example

Already at Chapter 3 on the classical electromagnetic field we encountered the problem of gauge fixing. It was fully developed in section 14.4 for a general gauge theory. The underlying assumption was the existence of a function $\mathcal{M}(\varphi)$ such that in the space of gauge field configurations φ, the curve $\mathcal{M}(^g\varphi) = 0$ cuts once and only once every given orbit, when g runs over the gauge group. We postponed the discussion of this condition and here we want to show two things: first that, in general, this condition is not satisfied and second that this failure will not affect the actual computations we shall present in this book.

The first result follows from elementary functional analysis. It is easy to see that if the gauge variation of the determinant of the functional derivative of the gauge function has at least one zero mode, every orbit will be cut more than once by the gauge function. This phenomenon was discovered as early as 1977 by Gribov[12] and it is known as *the Gribov ambiguity*. He studied the simple example of the three-dimensional Yang–Mills theory in the Landau gauge in which the Faddeev–Popov determinant is $\partial(\partial + A)$. He found many normalisable zero modes of the form

$$\partial(\partial + A)\omega = 0 \quad ; \quad \int \mathrm{Tr}(\omega\omega) < \infty \tag{14.93}$$

and related gauge field configurations with finite norm $\int A^2 < \infty$.

---

[12] V. N. Gribov, *Nucl. Phys.* **B139**, 1 (1978).

This gives the so-called Gribov copies of the gauge field configuration A, i.e.

$$A' = A + (\partial + A)\omega, \tag{14.94}$$

It is clear that A' will contribute as well as A to the path integral. This shows the ambiguity of the gauge-fixing process. Note that even in this very simple case, it is still unknown whether the number of Gribov copies is the same for all orbits and we have no way to classify the orbits that behave differently. So Gribov's observation reveals a general loophole in the gauge-fixing process, but neither its details are fully understood, nor its consequences fully appreciated.

14.7.2   The Gribov Question in a Broader Framework

Gauge invariance is a key feature not only for electrodynamics and its non-Abelian Yang–Mills generalisation, but also for the theory of gravitation where the role of gauge invariance is played by the invariance under general changes of coordinates. In all these cases, we understand very precisely the way unphysical modes disappear perturbatively from the observables of the theory. Gauge fixing was an essential step in this process. The Feynman rules we derived depend on the gauge choice. What Gribov discovered was that, at least in the cases he studied, this process was ambiguous, gauge fixing was never complete. In the previous section we announced two results and here we want to complete the presentation. We shall show that the gauge ambiguity *is* a general phenomenon for gauge theories and we shall discuss its importance for physical computations.

In what follows, A can denote any kind of gauge field (photon, Yang–Mills field, graviton, etc.). Perturbatively, we explained that getting one representative per orbit of a gauge field around a given background is solved by adding to the classical Lagrangian a gauge-fixing term with gauge function $\mathcal{F}(A)$, and a Faddeev–Popov term given by

$$L_{\text{gauge-fixing}} = \int d^4x \, \frac{\mathcal{F}(A(x))^2}{2\alpha} - \bar{\Omega}(x)D_{FP}\Omega(x), \tag{14.95}$$

where the Fadeev–Popov functional operator is[13]

$$D_{\text{FP}} \equiv \frac{\delta\mathcal{F}}{\delta A(x)} \frac{\delta}{\delta_{\text{gauge}}}. \tag{14.96}$$

We showed that the mean values of gauge invariant observables are independent of α at any given order of the perturbation expansion. When $\alpha \to 0$, the condition $\mathcal{F}(A) = 0$ is enforced in the quantum field theory.

---

[13] In fact, we can occasionally encounter situations in which higher order ghost interactions must be added in order to get a fully consistent perturbation theory. This can be seen by the correct application of the BRST formalism and does not affect the question of whether the gauge function $\mathcal{F}$ systematically cuts once and only once each orbit.

All these questions of gauge fixing in perturbation theory were fully understood already in 1974, but three years later, Gribov produced the example we showed in Eqs. (14.93) and (14.94), indicating that the programme failed at the non-perturbative level. The general mathematical answer was found by I. M. Singer in 1978, who showed that the Gribov ambiguity is not an exception, but a genuine property of all gauge theories. We shall not reproduce the proof here, which is quite involved, but we shall try to explain the physical content.

Let us rephrase the question of gauge fixing using a sufficiently abstract language, which is suitable to any type of gauge symmetry, in particular those studied extensively in this book.

Given a gauge group $\mathcal{G}$, we have the (huge) space $\{A\}$ of all possible gauge field configurations which contribute to the path integral. Many of these fields are equivalent modulo gauge transformations and, since the integration measure is assumed to be gauge invariant, this leads to multiple counting. Therefore, it is intuitively justified to reorganise the elements of the space $\{A\}$ in a more structured space, called *the orbit space $\mathcal{A}$*, defined as follows: Given a gauge field configuration A, its orbit is made out of all other elements A^g that can be connected to A by all possible elements $g(x)$ of the gauge group $\mathcal{G}$. It is obvious from the multiplication property of gauge transformations that two different orbits cannot intersect. We call *moduli space* the space $\mathcal{A}/G$ of equivalence classes of gauge field configurations defined modulo all possible gauges transformations. Ideally we would have liked to define the path integral over the elements of $\mathcal{A}/G$ rather than those of $\{A\}$ with a well-defined path integral measure. This would eliminate the necessity of gauge fixing. Unfortunately, this operation is impossible—rather, no one has as yet been able to define it concretely—within the context of a Lorentz invariant local quantum field theory, the framework we have been using all along this book. Hence the need for gauge fixing.

In order to better understand the structure of the space of orbits we shall introduce the topological notion of *small* and *large* gauge transformations.[14] A gauge transformation is called 'small' if it is continuously connected to the identity transformation by the exponential of a succession of infinitesimal gauge transformations in the Lie algebra; otherwise, it is called 'large', which means that it is disconnected from the identity. We can understand this way that the representations of the space of gauge orbits may be non-trivial and the space of all orbits presents a lot of intertwinings.[15]

---

[14] The terminology may be misleading since it refers to topological properties of the orbits and not to any intuitive notion of *size*.

[15] A concrete example of small and large transformations is obtained by studying the reparametrisation symmetry of the two-dimensional metric tensor. In the simplest case of the torus the explicit calculation is not very difficult and yields the following results: the orbits can be materialised in various representations, e.g. as curves piercing the upper complex plane or the Riemann disk, etc. The set of 2D-metrics defined modulo 'small diffeomorphisms' can be visualised as the coordinates of an upper complex plane. Then the remaining infinite discrete degeneracy, due to large gauge transformations that connect them, can be computed to form an $SL(2,Z)$ group, such that fundamental modular domains—where no more points exist that can be related by a 'large', often called a 'modular' transformation,—build an infinite set of non-degenerated tiling patches, each one with a curved triangular shape, in the upper complex plane. A complete gauge fixing over

Mathematically, the idea of gauge fixing sounds very simple. We wish to pick up one representative for each orbit, using a gauge fixing local functional $\mathcal{F}(\mathcal{A})$, which is globally well defined over the space of gauge field configurations and cuts each orbit only once. After the work of Gribov in 1977, it became clear that the gauge fixing with a local gauge function is perfectly well defined perturbatively, since, by construction, perturbation theory only deals with infinitesimal variations of the fields around a fixed configuration and the range of variations is too small to give a second intersection with each orbit. However, it may be not well defined non-perturbatively, since many gauge equivalent representatives will contribute to the path integral. This way the path integral will produce many copies of the true partition function, and will give a wrong answer. Moreover, the complexity of the space of orbits is such that the number of intersection points of the gauge function might vary depending on the orbit, and the factorisation of the 'volume' of the gauge group due to the infinite repletion of copies will not be achieved.

Shortly after the work of Gribov, I. M. Singer gave a complete mathematical proof that all these problems do occur. He showed the general impossibility of a non-perturbative gauge fixing with a local gauge function, no matter how ingenious the expression of the functional $\mathcal{F}$ is. Singer showed that the topological properties of the space of gauge orbits imply that it has a non-trivial curvature, and there exists no local functional $\mathcal{F}(\mathcal{A})$ which cuts each orbit only once; the orbits have such a complicated intertwining that any gauge fixing sub-manifold may not intersect some gauge orbits at all, or it may intersect some others more than once.

This result not only settles the mathematical question, but also it indicates why it is not relevant to the numerical computations we may be willing to perform. Indeed, these computations will be essentially of two sorts. First, there will be perturbation expansions around a given field configuration. By definition, they involve only infinitesimal gauge transformations, continuously connected to the identity and, as we just pointed out, the range of variations is too small to give a second intersection with any given orbit. Second, we shall perform direct non-perturbative numerical computations of the path integral by approximating the four-space by a discrete finite lattice. In this case we shall prove that for a compact group, no gauge fixing is required. Nevertheless, in our present understanding of the theory, we are still facing the embarrassing situation of being unable to give, even in principle, let alone in practical computations, an analytic expression of the partition function of a gauge theory in a continuous formulation. There have been several attempts to find criteria to choose a unique representative of the Yang–Mills gauge field per orbit, but, despite of their ingenuity, they only provide 'improved' non-perturbative formulations, but no rigorous proofs.[16]

---

all 2-D tori metrics amounts to restrict the path integral to one arbitrarily chosen triangle. See L. Baulieu and D. Zwanziger, *Phys. Rev.* **D87**, 086006 (2013).

[16] For a review see: N. Vandersickel and D. Zwanziger, *Phys.Rept.* **520**, 175 (2012).

14.8   Historical Notes

Although many versions of the history of gauge theories exist already in the recent literature,[17] the message has not yet reached the textbooks students usually read. We quote a comment from the review by J. D. Jackson and L. B. Okun: 'it is amusing how little the authors of textbooks know about the history of physics.' Here we present a very brief historical note on the evolution of the concept of gauge symmetry.

The vector potential was introduced in classical electrodynamics during the first half of the nineteenth century, either implicitly or explicitly, by several authors independently. It appears in some manuscript notes by Carl Friedrich Gauss as early as 1835 and it was fully written by Gustav Kirchhoff in 1857, following some earlier work by Franz Neumann. It was soon noted that it carried redundant variables and several 'gauge conditions' were used. The condition, which in modern notation is written as $\partial_\mu A^\mu = 0$, was proposed by the Danish mathematical physicist Ludvig Valentin Lorenz in 1867. Incidentally, most physics books misspell Lorenz's name as *Lorentz*, thus erroneously attributing the condition to the famous Dutch H. A. Lorentz, of the Lorentz transformations.[18] However, for internal symmetries, the concept of gauge invariance, as we know it today, belongs to quantum mechanics. It is the phase of the wave function, or that of the quantum fields, which is not an observable quantity and produces the internal symmetry transformations. The local version of these symmetries are the gauge theories we study here. The first person who realised that the invariance under local transformations of the phase of the wave function in the Schrödinger theory implies the introduction of an electromagnetic field was Vladimir Aleksandrovich Fock in 1926,[19] just after Schrödinger wrote his equation. In other words, Fock formulated and solved the problem we propose in Problem 14.1. Naturally, we would expect non-Abelian gauge theories to be constructed following the same principle immediately after isospin symmetry was established in the 1930s, following Heisenberg's original article of 1932. This is the method we followed in section 14.2. But here history took a totally unexpected route.

The development of the general theory of relativity offered a new paradigm for a gauge theory. The fact that it can be written as the theory invariant under local translations was certainly known to Hilbert,[20] hence the name of *Einstein–Hilbert action*. For the next decades it became the starting point for all studies on theories invariant under local transformations, such as the electromagnetic and the gravitational ones, which were the only fundamental interactions known at that time. It was, therefore, tempting to look for a unified theory, namely one in which both interactions follow from the same gauge principle. Today we know the attempt by Theodor Kaluza, completed by

---

[17] See, among others, O. Darrigol, *Electrodynamics from Ampère to Einstein'* (Oxford University Press, Oxford, 2000); L. O'Raifeartaigh, *The Dawning of gauge theory*, (Princeton University Press, Princeton, NJ 1997); L. O'Raifeartaigh and N. Straumann, *Rev. Mod. Phys.* **72**, 1 (2000); J. D. Jackson and L. B. Okun, *Rev. Mod. Phys.* **73**, 663 (2001)

[18] In French: On ne prête qu'aux riches.

[19] V. Fock, *Z. Phys* **39**, 226 (1926); Translation: *Physics-Uspekhi* **53**, 839 (2010).

[20] D. Hilbert, *Gött. Nachr.* 395 (1915).

Oscar Benjamin Klein,[21] which is often used in supergravity and superstring theories. These authors consider a theory of general relativity formulated in a five-dimensional space–time (1+4). They remark that if the fifth dimension is compact, the components of the metric tensor along this dimension may look to a four-dimensional observer as those of an electromagnetic vector potential. What is less known is that the idea was introduced earlier by the Finnish Gunnar Nordström,[22] who had constructed a scalar theory of gravitation. In 1914 he wrote a five-dimensional theory of electromagnetism and showed that if we assume that the fields are independent of the fifth coordinate, the assumption made later by Kaluza, the electromagnetic vector potential splits into a four-dimensional vector and a four-dimensional scalar, the latter being identified with his scalar field of gravitation, in some sense the mirror theory of Kaluza and Klein. An important contribution from this period is due to Hermann Klaus Hugo Weyl.[23] He is more known for his 1918 unsuccessful attempt to enlarge diffeomorphisms to local scale transformations, but, in fact, a byproduct of this work was a different form of unification between electromagnetism and gravitation. In his 1929 paper, which contains the gauge theory for the Dirac electron we saw in section 14.1, he introduced many concepts which have become classic, such as the Weyl two-component spinors and the vierbein and spin-connection formalism. Although the theory is no more scale invariant, he still used the term *gauge invariance*, a term which has survived ever since.

In particle physics we put the birth of non-Abelian gauge theories in 1954, with the fundamental paper of Chen Ning Yang and Robert Laurence Mills.[24] It is the paper which introduced the $SU(2)$ gauge theory and, although it took some years before interesting physical theories could be built, it is from that date that non-Abelian gauge theories became part of high energy physics. It is not surprising that they were immediately named *Yang–Mills theories*. The influence of this work in high energy physics has often been emphasised, but here we want to mention some earlier and little known attempts which, according to present views, have followed a quite strange route.

The first is due to Oscar Klein. In an obscure conference in 1938 he presented a paper with the title: '*On the Theory of Charged Fields*',[25] in which he attempts to construct an $SU(2)$ gauge theory for the nuclear forces. This paper is amazing in many ways. First, of course, because it was done in 1938. He started from the discovery of the muon, misinterpreted as the Yukawa meson, in the old Yukawa theory in which the mesons were assumed to be vector particles. This provides the physical motivation. The aim is to write an $SU(2)$ gauge theory unifying electromagnetism and nuclear forces. Second, and even more amazing, because he followed an incredibly circuitous road: he considered general relativity in a five-dimensional space, he compactified *à la* Kaluza–Klein, but he took the $g_{4\mu}$ components of the metric tensor to be 2×2 matrices. He wanted to describe the $SU(2)$ gauge fields but the matrices he used, although they depend on three fields, are

---

[21] Th. Kaluza, *K. Preuss. Akad. Wiss.* p 966 (1921); O. Klein, *Z. Phys.* **37**, 895 (1926).
[22] G. Nordström, *Phys. Z.* **15**, 504 (1914).
[23] H. Weyl, *Deutsch Akad. Wiss.* Berlin p 465 (1918); *Z. Phys.* **56**, 330 (1929).
[24] C. N. Yang and R. L. Mills, *Phys. Rev.* **96**, 191 (1954). It seems that similar results were also obtained by R. Shaw in his thesis.
[25] O. Klein, in *Les Nouvelles Théories de la Physique* (Paris, 1939), 81. Report in a Conference organised by the Institut International de Coopération Intellectuelle, Warsaw, 1938.

not traceless. Despite this problem he found the correct expression for the field strength tensor of $SU(2)$. In fact, answering an objection by Møller, he added a fourth vector field, thus promoting his theory to $U(1) \times SU(2)$. He added mass terms by hand and it is not clear whether he worried about the resulting breaking of gauge invariance. It is not known whether this paper has inspired anybody else's work and Klein himself mentioned it only once in a 1955 conference in Berne.[26]

The second work in the same spirit is due to Wolfgang Pauli,[27] who in 1953, in a letter to Abraham Pais, as well as in a series of seminars, developed precisely this approach: the construction of the $SU(2)$ gauge theory as the flat space limit of a compactified higher dimensional theory of general relativity. He was closer to the approach followed today because he considered a six-dimensional theory with the compact space forming an S_2. He never published this work and we do not know whether he was aware of Klein's 1938 paper. He had realised that a mass term for the gauge bosons breaks the invariance and he had an animated argument during a seminar by Yang in the Institute for Advanced Studies in Princeton in 1954.[28] What is certainly surprising is that Klein and Pauli, fifteen years apart from one another, decided to construct the $SU(2)$ gauge theory for strong interactions and both choose to follow this totally counterintuitive method. It seems that the fascination which general relativity had exerted on this generation of physicists was such that, for many years, local transformations could not be conceived independently of general coordinate transformations. Yang and Mills were the first to understand that the gauge theory of an internal symmetry takes place in a fixed background space which can be chosen to be flat, in which case general relativity plays no role.

With the work of Yang and Mills gauge theories entered particle physics. Although the initial motivation was a theory of strong interactions, the first semi-realistic models aimed at describing weak and electromagnetic interactions. We shall present later on in this book the present state of what is known as the *standard model of elementary particle physics*. To complete the story, we mention only a paper by Sheldon Lee Glashow and Murray Gell-Mann[29] which is often left out from the history articles. In this paper the authors extend the Yang–Mills construction, which was originally done for $SU(2)$, to arbitrary Lie algebras. The well-known result of associating a coupling constant with every simple factor in the algebra result, which we presented in section 14.2, appeared for the first time there. We can find the seed for what we will call later *a grand unified theory*. In a footnote they say:

> The remarkable universality of the electric charge would be better understood were the photon not merely a singlet, but a member of a family of vector mesons comprising a simple partially gauge invariant theory.

[26] O. Klein, *Helv. Phys. Acta Suppl.* **IV**, 58 (1956).

[27] W. Pauli, unpublished. It is summarised in a letter to A. Pais, dated 22–25 July 1953. A. Pais, *Inward Bound*, (Oxford University Press, Oxford, 1986), 584.

[28] C. N. Yang, *Selected Papers 1945–1980 with Commentary* (Freeman, San Francisco, 1983), 525.

[29] S. L. Glashow and M. Gell-Mann, *Ann. of Phys.* **15**, 437 (1961); see also: R. Utiyama, *Phys. Rev.* **101**, 1597 (1956).

14.9 Problems

Problem 14.1 The Schrödinger equation, together with the normalisation condition, is invariant under the $U(1)$ group of constant phase transformations of the wave function $\Psi(x, t) \to e^{i\theta}\Psi(x, t)$. Show that extending the invariance to local transformations, i.e. θ replaced by $\theta(x, t)$, implies the introduction of a new interaction. The resulting theory describes the motion of a non-relativistic charged particle in an external electromagnetic field.

Problem 14.2 Determination of the action of BRST transformations, using power counting.

Let us first assume that the gauge group is semi-simple. Using the fact that the ghost numbers and mass dimensions of $A_\mu, \Omega, \bar{\Omega}, b, \phi$ are, respectively $0, 1, -1, 0, 0$ and $1, 0, 2, 2, 1$, show that the nilpotency of the BRST symmetry s and its grading properties imply that the most general s-transformations of the fields are

$$sA = Z_\Omega(-\mathrm{d}\Omega - gZ_A[A, \Omega])$$
$$s\phi = Z_\Omega(-gZ_A[\Omega, \phi])$$
$$s\Omega = Z_\Omega(-\frac{g}{2}Z_A[\Omega, \Omega])$$
$$s\bar{\Omega} = Z_{\bar{\Omega}}(b - \alpha\frac{g}{2}Z_A Z_\Omega[\Omega, \bar{\Omega}])$$
$$sb = -\alpha\frac{g}{2}Z_A Z_\Omega[\Omega, b]),$$

where all Z-factors and α are arbitrary dimensionless constants.

Show that we can set $Z_\Omega = Z_A = Z_{\bar{\Omega}} = 1$ and $\alpha = 0$ by multiplicative and linear field redefinitions. Show also that g can be absorbed into field redefinitions.

Prove the curvature condition Eq. 1.48 for $A + c$, and show that for a scalar field, we have an analogous condition,

$$(s + d) + [A + c, \phi] = D\phi.$$

Write the Bianchi identity for this equation. Conversely, verify directly from this Bianchi identity that $s\phi = -[c, \phi]$.

Generalise the exercise when the gauge group includes Abelian $U(1)$ factors.

Problem 14.3 A more general renormalisable gauge-fixing term.

Generalise the Faddeev–Popov formula (14.47), using the gauge function

$$\mathcal{F}(A, \phi)^a = \partial^\mu A_\mu^a + \mathrm{d}^{abc}A_b^\mu A_{\mu c} + F_{ij}^a \phi^i v^j$$

when the Lie algebra of the gauge group possesses a symmetric invariant tensor d_{abc} and there is a scalar field ϕ^i valued in a given representation of the Lie algebra, with structure coefficients F_{ij}^a. v^j defines a constant direction in the Lie algebra.

Application: Compute the ghost interactions in quantum electrodynamics, using the non-linear gauge function

$$\partial^\mu A_\mu + \alpha A_\mu A^\mu,$$

where α is a real massive parameter.

Compute the ghost part of the action with the non-covariant gauge function $\mathcal{F}_\beta(A) = \partial^i A_i^a + \beta \partial^0 A_0^a$. Is it a well-defined gauge for the parameter $\beta \neq 0$? The Coulomb gauge is the limit $\beta \to 0$. Is it a well-defined limit?

15

Broken Symmetries

15.1 Introduction

Already at the classical level, an infinite system may exhibit the phenomenon of phase transitions. This is often accompanied by a change of the symmetry of the ground state. For example, in the liquid–solid phase transition, the translational invariance of the liquid phase is reduced to the discrete subgroup which leaves invariant the lattice of the solid. A field theory describes a system with an infinite number of degrees of freedom, so we expect to find here also the phenomenon of phase transitions. In this chapter we will study this phenomenon for field theories having both global and local symmetries. We will see that, in many cases, we encounter at least two phases.

In one of them, whichever symmetry is present is manifest in the spectrum of the theory whose excitations appear to form irreducible representations of the symmetry group. This is called *the symmetric phase*, or *the Wigner phase*. For a gauge theory this implies that the vector gauge bosons are massless and belong to the adjoint representation. But, as we will argue later, we have good reasons to believe that for non-Abelian gauge theories, a strange phenomenon occurs in this phase: all physical states are singlets of the group. All non-singlet states, such as those corresponding to the gauge fields, are supposed to be *confined*, in the sense that they do not appear as physically realisable asymptotic states. Although there are several indications in the direction of confinement, a rigorous proof is still missing and this constitutes one of the central unsolved problems in quantum field theory today. Nevertheless, such a gauge theory, based on the group $SU(3)$ in this confined phase, provides the theory for strong interactions. It is called *quantum chromodynamics* and it will be presented in a later chapter.

In the other phase, part of the symmetry is hidden from the spectrum. It is called *spontaneously broken phase* and, for gauge theories, we will show presently that the corresponding gauge bosons become massive. Such a gauge theory, based on the group $U(2)$ spontaneously broken to $U(1)$, describes in a unified way the electromagnetic and weak interactions. In this chapter we will study these phenomena of phase transitions in some simple examples.

The realisation that a physical problem possesses a certain symmetry often simplifies its solution considerably. For example, let us calculate the electric field produced by a uniformly charged sphere at a point A outside the sphere. We could solve the problem

From Classical to Quantum Fields. Laurent Baulieu, John Iliopoulos and Roland Sénéor.
© Laurent Baulieu, John Iliopoulos and Roland Sénéor, 2017. Published 2017 by Oxford University Press.

the hard way by considering the field created by a little volume element of the sphere and then integrating over. But everyone knows that it is sufficient to realise that the problem has a spherical symmetry and then Gauss's theorem for the surface through A gives the answer immediately. In this reasoning we have implicitly assumed that symmetric problems always possess symmetric solutions. Stated in this form the assumption sounds almost obvious. However, in practice, we need a much stronger one. Indeed, a real sphere is never absolutely symmetric and the charge is never distributed in a perfectly uniform way. Nevertheless, we still apply the above reasoning, hoping that small deviations from perfect symmetry will induce only small departures from the symmetric solution. This, however, is a much stronger statement, which is far from obvious, since it needs not only the existence of a symmetric solution but also an assumption about its stability. Let us investigate the conditions for such broken symmetries and some of their consequences.

15.2   Global Symmetries

15.2.1   An Example from Classical Mechanics

A very simple example is provided by the problem of the bent rod. Let a cylindrical rod be charged as in Fig. 15.1. The problem is obviously symmetric under rotations around the z axis. Let z measure the distance from O, and $X(z)$ and $Y(z)$ give the deviations, along the x and y directions, respectively, of the axis of the rod at the point z from the symmetric position. The general equations of elasticity are non-linear but, for small deflections, they can be linearised as

$$IE\frac{\mathrm{d}^4 X}{\mathrm{d}z^4} + F\frac{\mathrm{d}^2 X}{\mathrm{d}z^2} = 0; \quad IE\frac{\mathrm{d}^4 Y}{\mathrm{d}z^4} + F\frac{\mathrm{d}^2 Y}{\mathrm{d}z^2} = 0, \tag{15.1}$$

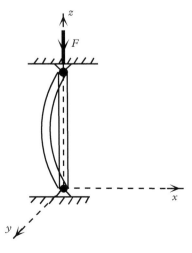

Figure 15.1 *A cylindrical rod bent under a force F along its symmetry axis. (© F. Boudjema)*

where $I = \pi R^4/4$ is the moment of inertia of the rod and E is the Young modulus. It is obvious that the system (15.1) always possesses a symmetric solution $X = Y = 0$. However, we can also look for asymmetric solutions of the general form $X = A + Bz + C \sin kz + D \cos kz$ with $k^2 = F/EI$, which satisfy the boundary conditions $X = X'' = 0$ at $z = 0$ and $z = l$. We find that such solutions exist, $X = C \sin kz$, provided $kl = n\pi$; $n = 1, \ldots$. The first such solution appears when F reaches a critical value F_{cr} given by

$$F_{cr} = \frac{\pi^2 EI}{l^2}. \tag{15.2}$$

The appearance of these solutions is already an indication of instability and, indeed, a careful study of the stability problem proves that the non-symmetric solutions correspond to lower energy. From that point Eqs. (15.1) are no longer valid, because they only apply to small deflections, and we must use the general equations of elasticity. The result is that this instability of the symmetric solution occurs for all values of F larger than F_{cr}.

What has happened to the original symmetry of the equations? It is still hidden in the sense that we cannot predict in which direction in the $x - y$ plane is the rod going to bend. They all correspond to solutions with precisely the same energy. In other words, if we apply a symmetry transformation (in this case, a rotation around the z axis) to an asymmetric solution, we obtain another asymmetric solution which is degenerate with the first one.[1]

We call such a symmetry 'spontaneously broken', and in this simple example we see all its characteristics. There exists a critical point, i.e. a critical value of some external quantity which we can vary freely (in this case the external force F; in several physical systems it is the temperature), which determines whether spontaneous symmetry breaking will take place. Beyond this critical point:

(i)   the symmetric solution becomes unstable; and

(ii)   the ground state becomes degenerate.

15.2.2   Spontaneous Symmetry Breaking in Non-relativistic Quantum Mechanics

These considerations apply also to quantum physics with some new features. A symmetry transformation is represented by an operator which commutes with the Hamiltonian

---

[1] It is instructive to look at this problem also from a slightly different point of view. We can argue that the rod has imperfections which break the rotational symmetry. Let us model these imperfections by a small lateral force f, which, for simplicity, we assume to be applied at the centre of the rod and along the x direction. For non-zero f, the rod will bend in the direction of f and we can compute the maximum deviation $\Delta - X(l/2)$ as a function of f. The problem can be solved exactly and we obtain the function $\Delta(f)$. The important point is that for $F < F_{cr}$, we find that $\Delta(f)$ goes to 0 with f but for $F \geq F_{cr}$, $\lim_{f \to 0} \Delta(f) \neq 0$. In other words, in this case, the symmetry breaking effects remain even in the absence of symmetry breaking forces.

$$[H, Q] = 0. \tag{15.3}$$

As a result, if $|\Psi>$ is an eigenstate of the Hamiltonian, so is $Q|\Psi>$ with the same eigenvalue, unless it happens that the state $|\Psi>$ is annihilated by Q, $Q|\Psi>= 0$.

Let us start with the trivial example of a particle moving in a symmetric double well square potential. The symmetry is the discrete parity transformation P because the potential satisfies $V(-x) = V(x)$. Let us first neglect the tunnelling probability. This means that we consider the potential barrier to be infinite. The spectrum of states consists of two separated Hilbert spaces $|n, L>$ and $|n, R>$, where n labels the energy level and L or R indicate whether the particle is in the left, or right, well. $H|n, L>= E_n|n, L>$ and the same for R. All energy levels are doubly degenerate, but this is trivial, since the two spaces are totally disconnected. All wave functions are strictly localised in one of the two regions and $<n, L|m, R>= 0$ for all n and m. The parity operator connects the two spaces $P|n, L>= |n, R>$, but the matrix elements $<n, L|P|m, L>$ all vanish.

As we have learned in quantum mechanics, the situation becomes more interesting if we lower the potential barrier and allow for the tunnelling probability of the particle to move from one well to the other. The states now become common eigenstates of H and P, the latter having eigenvalues ± 1. In the intermediate region between the two wells the wave function is a superposition of real exponentials of the form $e^{\pm Kx}$ because the motion of the particle in this region is classically forbidden. In the path integral this tunnelling corresponds to trajectories in Euclidean (i.e. imaginary) time.

This picture can be generalised in many ways, leading to interesting phenomena, some of which are related to spontaneous symmetry breaking. First, we can consider a potential with N identical wells and take the large N limit. We have studied this problem of a periodic one-dimensional potential in quantum mechanics and we know that it leads to the propagation of Bloch waves. The corresponding Euclidean time trajectories in the path integral will appear again in quantum field theory and we shall study them in a later chapter. A different generalisation consists of considering a system with many degrees of freedom and/or enlarging the symmetry to a continuous Lie group of transformations. The Heisenberg ferromagnet is a good example in which this kind of phenomena can be studied numerically (and in simple cases even analytically).

It is known that several materials exhibit a phase transition and, at low temperature, they develop a spontaneous magnetisation. Heisenberg proposed a simple model which captures the essential physics of the phenomenon without inessential complications. For simplicity, let us consider a regular cubic lattice. The main assumptions are the following. (i) We neglect all atomic degrees of freedom and keep only a spin variable S_i, $i = 1, 2, ..., N$, at each lattice point i. (ii) The spin variables are subject to short-range forces and we can keep only nearest neighbour interactions. (iii) The interaction is assumed to be rotationally invariant. We are thus led to an effective interaction Hamiltonian of the form

$$H = -\frac{1}{2}\mathscr{J}\sum_{i,j} S_i \cdot S_j, \tag{15.4}$$

where the sum extends over all pairs of nearest neighbouring points (in d dimensions each point has $2d$ nearest neighbours) and $\mathcal{J}$ is a positive coupling constant. For a three-dimensional lattice the system can be studied numerically and shows the phenomenon of spontaneous symmetry breaking of the $O(3)$ symmetry.

Following the same terminology we used for the classical bent rod, we can say the following:

1. The system has two phases: (i) the symmetric phase with the full $O(3)$ symmetry, where the spins are randomly oriented and there is no privileged direction in space; and (ii) the spontaneous broken phase in which the spins are oriented parallel to each other and a macroscopic magnetisation M appears. The symmetry is broken from the initial $O(3)$ down to $O(2)$, the subgroup of rotations around the axis defined by M.

2. The external control parameter, the analogue of the external force, is the temperature. There is a critical value T_{cr}, called *the Curie temperature*, which determines in which phase the system will be found. At any temperature there are two competing mechanisms. The interaction privileges parallel spin configurations in order to minimise the energy in (15.4). Its strength grows with the number of nearest neighbours, i.e. with d. Thermal fluctuations, on the other hand, tend to destroy any long-range order. For most values of d (except $d = 1$ and $d = \infty$), there is a phase transition at a critical temperature T_{cr}. At $T > T_{cr}$ thermal fluctuations win and the system is in a disordered, or symmetric, phase. At $T < T_{cr}$ the interaction wins and an ordered phase with spontaneous symmetry breaking appears.

3. The order parameter, the analogue of the parameter $\Delta = X(z = l/2)$ of the bent rod, is the magnetisation M. It is a d-dimensional vector, so the manifold of ground states will span the surface of a $(d$-$1)$-dimensional sphere. The new feature here is that the initial symmetry is not completely broken. A subgroup remains exact even in the low-temperature phase.

The interesting point with this model is that we can study the physics starting from a system with a finite number of spins N and then take the limit $N \to \infty$. In the three-dimensional case this can only be done numerically, but some features can be understood qualitatively. Let us call $\mathcal{J}_i$, $i = 1, 2, 3$, the three generators of $O(3)$. They all commute with the Hamiltonian (15.4). For simplicity we shall choose the spins S_i to be equal to $\frac{1}{2}$. If we fix the z direction, at every lattice site we have a two-dimensional space of states with basic vectors given by

$$|+> = \begin{pmatrix} 1 \\ 0 \end{pmatrix} \qquad |-> = \begin{pmatrix} 0 \\ 1 \end{pmatrix}. \tag{15.5}$$

At low temperature the ground state consists of all spins parallel, giving a total magnetisation M. It is infinitely degenerate because we can rotate M in any direction. Starting from (15.5), the rotated spin state will be given by

$$|i, \pm >_\theta = e^{iJ \cdot \theta} |i, \pm >. \tag{15.6}$$

For spin $\frac{1}{2}$ we have $J = \sigma/2$. We can compute the overlap between two such states as the matrix element of the rotation operator between them,

$$_{\theta_2} < i, +|e^{iJ \cdot \theta_{12}} |i, + >_{\theta_1} = \cos \theta_{12}, \tag{15.7}$$

where θ_{12} is the angle between the two directions defined by θ_1 and θ_2. The important observation is that this overlap is smaller than 1, so when $N \to \infty$, it goes to 0. For a system with an infinite number of degrees of freedom the various ground states are orthogonal and so are the Hilbert spaces which we can build above each one of them. Physically this means that although it costs no energy to turn a finite number of spins, it is impossible to turn simultaneously an infinite number of them. This is the origin of the phase transition.

15.2.3 A Simple Field Theory Model

Let $\phi(x)$ be a complex scalar field whose dynamics is described by the Lagrangian density

$$\mathcal{L}_1 = (\partial_\mu \phi)(\partial^\mu \phi^*) - M^2 \phi \phi^* - \lambda (\phi \phi^*)^2, \tag{15.8}$$

where $\mathcal{L}_1$ is a classical Lagrangian density and $\phi(x)$ is a classical field. No quantisation is considered for the moment. Equation (15.8) is invariant under the group $U(1)$ of global transformations:

$$\phi(x) \rightarrow e^{i\theta} \phi(x). \tag{15.9}$$

To this invariance corresponds the current $j_\mu \sim \phi \partial_\mu \phi^* - \phi^* \partial_\mu \phi$ whose conservation can be verified using the equations of motion.

We are interested in the classical field configuration which minimises the energy of the system. We thus compute the Hamiltonian density given by

$$\mathcal{H}_1 = (\partial_0 \phi)(\partial_0 \phi^*) + (\partial_i \phi)(\partial_i \phi^*) + V(\phi) \tag{15.10}$$

$$V(\phi) = M^2 \phi \phi^* + \lambda (\phi \phi^*)^2. \tag{15.11}$$

The first two terms of $\mathcal{H}_1$ are positive definite. They can only vanish for $\phi = $ constant. Therefore, the ground state of the system corresponds to $\phi = $ constant $= $ minimum of $V(\phi)$. V has a minimum only if $\lambda > 0$. In this case the position of the minimum depends on the sign of M^2. (Note that we are still studying a classical field theory and M^2 is just a parameter. We should not be misled by the notation into thinking that M is a 'mass' and M^2 is necessarily positive.) For $M^2 > 0$ the minimum is at $\phi = 0$ (symmetric solution, shown in Fig. 15.2, left side), but for $M^2 < 0$ there is a whole circle of minima at the

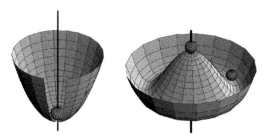

Figure 15.2 *The potential $V(\phi)$ with $M^2 \geq 0$ (left) and $M^2 < 0$ (right).*

complex ϕ-plane with radius $v = (-M^2/2\lambda)^{1/2}$ (Fig. 15.2, right side). Any point on the circle corresponds to a spontaneous breaking of (15.9).

We see that:

the critical point is $M^2 = 0$;

for $M^2 > 0$ the symmetric solution is stable; and

for $M^2 < 0$ spontaneous symmetry breaking occurs.

Let us choose $M^2 < 0$. In order to reach the stable solution we translate the field ϕ. It is clear that there is no loss of generality by choosing a particular point on the circle, since they are all obtained from any given one by applying the transformations (15.9). Let us, for convenience, choose the point on the real axis in the ϕ-plane. We thus write

$$\phi(x) = \frac{1}{\sqrt{2}} \left[v + \psi(x) + i\chi(x) \right]. \tag{15.12}$$

Bringing (15.12) into (15.8) we find that

$$\begin{aligned}
\mathcal{L}_1(\phi) \;\rightarrow\; \mathcal{L}_2(\psi, \chi) &= \tfrac{1}{2}(\partial_\mu \psi)^2 + \tfrac{1}{2}(\partial_\mu \chi)^2 - \tfrac{1}{2}(2\lambda v^2)\psi^2 \\
&\quad -\lambda v \psi(\psi^2 + \chi^2) - \tfrac{\lambda}{4}(\psi^2 + \chi^2)^2.
\end{aligned} \tag{15.13}$$

Note that $\mathcal{L}_2$ does not contain any term proportional to χ^2, which is expected since V is locally flat in the χ direction. A second remark concerns the arbitrary parameters of the theory. $\mathcal{L}_1$ contains two such parameters, a mass M and a dimensionless coupling constant λ. In $\mathcal{L}_2$ we have again the coupling constant λ and a new mass parameter v which is a function of M and λ. It is important to note that although $\mathcal{L}_2$ contains also trilinear terms, their coupling strength is not a new parameter but is proportional to $v\lambda$.

The term *spontaneously broken symmetry* is slightly misleading because the invariance is not broken. $\mathcal{L}_2$ is still invariant under the transformations with infinitesimal parameter θ,

$$\delta\psi = -\theta\chi; \qquad \delta\chi = \theta\psi + \theta v, \tag{15.14}$$

to which corresponds a conserved current

$$j_\mu \sim \psi \partial_\mu \chi - \chi \partial_\mu \psi + v \partial_\mu \chi. \tag{15.15}$$

The last term, which is linear in the derivative of χ, is characteristic of the phenomenon of spontaneous symmetry breaking.

It should be emphasised here that $\mathcal{L}_1$ and $\mathcal{L}_2$ are completely equivalent Lagrangians. They both describe the dynamics of the same physical system and a change of variables, such as (15.12), cannot change the physics. However, this equivalence is only true if we can solve the problem exactly. In this case we shall find the same solution using either of them. However, we do not have exact solutions and we intend to apply perturbation theory, which is an approximation scheme. Then the equivalence is no longer guaranteed and, in fact, perturbation theory has much better chances to give sensible results using one language rather than the other. In particular, if we use $\mathcal{L}_1$ as a quantum field theory and we decide to apply perturbation theory taking, as the unperturbed part, the quadratic terms of $\mathcal{L}_1$, we immediately see that we shall get nonsense. The spectrum of the unperturbed Hamiltonian would consist of particles with negative square mass, and no perturbation corrections, at any finite order, could change that. This is essentially due to the fact that, in doing so, we are trying to calculate the quantum fluctuations around an unstable solution and perturbation theory is just not designed to do so. In contrast, we see that the quadratic part of $\mathcal{L}_2$ gives a reasonable spectrum; thus, we hope that perturbation theory will also give reasonable results. Therefore, we conclude that our physical system, considered now as a quantum system, consists of two interacting scalar particles, one with mass $m_\psi^2 = 2\lambda v^2$ and the other with $m_\chi = 0$. We believe that this is the spectrum we would have found also starting from $\mathcal{L}_1$, if we could solve the dynamics exactly. The remark we made on the Heisenberg ferromagnet applies here as well. We can choose any point on the circle of minima as our vacuum state, but the Hilbert spaces we build above each one of them are orthogonal to each other and there is no transition between states belonging to different spaces.

The appearance of a zero-mass particle in the quantum version of the model is an example of a general theorem due to J. Goldstone. To every generator of a spontaneously broken symmetry there corresponds a massless particle, called the Goldstone particle. This theorem is just the translation, into quantum field theory language, of the statement about the degeneracy of the ground state. The ground state of a system described by a quantum field theory is the vacuum state, and you need massless excitations in the spectrum of states in order to allow for the degeneracy of the vacuum. We will give a more general proof of this theorem in a next section.

15.2.4 The Linear σ-Model

This model was introduced in 1960 by M. Gell-Mann and M. Lévy.[2] Let us first look at the bosonic sector. It consists of four real spin-0 fields $\phi^i(x)$, $i = 1, ..., 4$. The Lagrangian

---

[2] An earlier version was first proposed by J. Schwinger.

density is assumed to be invariant under $O(4)$ transformations:

$$\mathcal{L} = \frac{1}{2}(\partial^\mu \phi^i)(\partial_\mu \phi^i) - \frac{1}{2}M^2 \phi^i \phi^i - \frac{\lambda}{4!}(\phi^i \phi^i)^2. \tag{15.16}$$

The model is just a generalisation of the previous one, Eq. (15.8), from $O(2)$ to $O(4)$; therefore, the same analysis applies. We can compute the Hamiltonian density $\mathcal{H}$,

$$\mathcal{H} = \frac{1}{2}(\partial^0 \phi^i)(\partial^0 \phi^i) + \frac{1}{2}(\nabla \phi^i)(\nabla \phi^i) + V(\phi) \tag{15.17}$$

$$V(\phi) = \frac{1}{2}M^2 \phi^i \phi^i + \frac{\lambda}{4!}(\phi^i \phi^i)^2, \tag{15.18}$$

and we see that for $M^2 < 0$, we have the phenomenon of spontaneous symmetry breaking. The set of minima of the potential forms the surface of a four-dimensional hypersphere with $\phi^i \phi^i = v^2$. If we choose a point in the ϕ^4 direction we break the symmetry from $O(4)$ to $O(3)$. We thus obtain three Goldstone bosons[3] corresponding to the three fields ϕ^i, $i = 1, 2, 3$.

The model becomes richer if we add fermions. There exists an alternative way to write (15.16) taking advantage of the fact that $O(4)$ is locally isomorphic to $SU(2) \times SU(2)$. The four-vector ϕ^i can be written as a 2×2 matrix Φ:

$$\Phi(x) = \frac{i\boldsymbol{\pi}(x) \cdot \boldsymbol{\tau} + \sigma(x)\mathbb{1}}{\sqrt{2}}, \tag{15.19}$$

where $\boldsymbol{\tau}$ and $\mathbb{1}$ are the three Pauli matrices and the 2×2 unit matrix, respectively, and we have renamed the fields $\phi^{1,2,3}(x) = i\pi^{1,2,3}(x)$ and $\phi^4(x) = \sigma(x)$. The factor i has been introduced in order to make the π-fields pseudoscalar, as we shall see presently. The $O(4)$ transformations act on Φ as left- and right-$SU(2)$ transformations:

$$\Phi(x) \to e^{i\boldsymbol{\tau}\cdot\boldsymbol{\theta}_L} \Phi(x) e^{-i\boldsymbol{\tau}\cdot\boldsymbol{\theta}_R}. \tag{15.20}$$

The Lagrangian (15.16) can be rewritten as

$$\mathcal{L} = \frac{1}{2}\text{Tr}[(\partial^\mu \Phi)(\partial_\mu \Phi^\dagger)] - \frac{1}{2}M^2 \text{Tr}[\Phi\Phi^\dagger] - \frac{\lambda}{4!}\text{Tr}[\Phi\Phi^\dagger]^2. \tag{15.21}$$

We add a doublet of Dirac fermion fields $\Psi(x) = \begin{pmatrix} p(x) \\ n(x) \end{pmatrix}$ which we split into the left- and right-hand components $\Psi_{\text{L,R}}(x) = (1 \pm \gamma_5)\Psi(x)$. Under $SU(2) \times SU(2)$ Ψ_L and Ψ_R are assumed to transform as members of the $(2,1)$ and $(1,2)$ representation,

---

[3] $O(4)$ has six generators and $O(3)$ has three.

respectively. When the model was originally proposed it was meant to describe the pion–nucleon strong interactions and $p(x)$ and $n(x)$ were the proton and neutron Dirac fields in Heisenberg's isospin notation. We keep this notation, although the motivation is no more valid. The $SU(2) \times SU(2)$ invariant Lagrangian (15.21) can then be completed as

$$\mathcal{L} = \bar{\Psi}_L \mathrm{i}\, \partial\!\!\!/\, \Psi_L + \bar{\Psi}_R \mathrm{i}\, \partial\!\!\!/\, \Psi_R + g\bar{\Psi}_L \Phi \Psi_R + h.c.$$
$$+ \frac{1}{2}\mathrm{Tr}[(\partial^\mu \Phi)(\partial_\mu \Phi^\dagger)] - \frac{1}{2}M^2 \mathrm{Tr}[\Phi\Phi^\dagger] - \frac{\lambda}{4!}\,\mathrm{Tr}[\Phi\Phi^\dagger]^2 \qquad (15.22)$$

where *h.c.* stands for 'Hermitian conjugate'.

Some remarks: first, the fermions are massless. Indeed, a mass term proportional to $\bar{\Psi}_L \Psi_R + h.c.$ is forbidden by the invariance under $SU(2) \times SU(2)$. Second, if we expand the Yukawa interaction term we find that the σ-field is coupled to $\bar{\Psi}\Psi$ and the π-fields to $\bar{\Psi}\gamma_5\boldsymbol{\tau}\Psi$. So σ is a scalar and the three pions are pseudoscalars.

As we said previously, spontaneous symmetry breaking occurs if $M^2 < 0$. It is an easy exercise to translate the σ-field $\sigma \to \sigma + v$ with the following results:

1. The symmetry is broken $SU(2) \times SU(2) \to SU_V(2)$. This $SU_V(2)$, which is left intact, is the diagonal subgroup of the initial $SU(2) \times SU(2)$ with pure vector currents. It is the well-known isospin group. This result is due to our choice of translating only the σ-field which is scalar under parity. Under $SU_V(2)$, i.e. isospin, σ is a singlet and the three pions form a triplet.

2. The triplet of pions is massless. They are the Goldstone bosons of the axial part of the symmetry which is spontaneously broken.

3. The fermions acquire a mass. The Yukawa term gives, after the σ translation, a term proportional to $gv\bar{\Psi}\Psi$ which implies a common mass for both proton and neutron. This, of course, is a consequence of the invariance of the translated Lagrangian under isospin.

4. The original Lagrangian (15.22) has a further $U(1)$ invariance under which $\Psi \to \mathrm{e}^{\mathrm{i}\theta}\Psi$ and $\Phi \to \Phi$. This invariance remains unbroken and the corresponding conserved charge is the fermion number.

It is easy now to introduce a small explicit breaking the way we did for the problem of the bent rod. (See footnote 1 in section 15.2.1.) We add a term linear in the σ-field which breaks $SU(2) \times SU(2)$ to $SU_V(2)$,

$$\mathcal{L} = \bar{\Psi}_L \mathrm{i}\, \partial\!\!\!/\, \Psi_L + \bar{\Psi}_R \mathrm{i}\, \partial\!\!\!/\, \Psi_R + g\bar{\Psi}_L \Phi \Psi_R + h.c. - c\sigma$$
$$+ \frac{1}{2}\mathrm{Tr}[(\partial^\mu \Phi)(\partial_\mu \Phi^\dagger)] - \frac{1}{2}M^2 \mathrm{Tr}[\Phi\Phi^\dagger] - \frac{\lambda}{4!}\mathrm{Tr}[\Phi\Phi^\dagger]^2, \qquad (15.23)$$

where c is a constant with dimensions [Mass]^3. In order to reach the minimum of the potential, we still translate σ. Keeping only terms of first order in c, we obtain, for $M^2 > 0$, $\sigma' = \sigma + c/M^2$; in other words, when there is no underlying spontaneous symmetry breaking, the shift of the field goes to 0 together with c. The same happens to all symmetry breaking effects, such as the fermion mass which is given by $m_\Psi = gc/M^2$, again proportional to c and the $\pi - \sigma$ mass difference: $m_\sigma^2 - m_\pi^2 = \lambda c^2/6M^4$. On the other hand, if $M^2 < 0$, the shift is $\sigma' = \sigma + v - c/2M^2$ with $v^2 = -6M^2/\lambda$. In this case, the symmetry breaking effects remain, even when c goes to 0. In particular, the pions, which were massless Goldstone bosons in the absence of c, acquire a mass $m_\pi^2 = \lambda c^2/6M^4$. We call them *pseudo-Goldstone bosons* and we believe that such a mechanism is a good model for real pions. As a final remark, we compute the triplet of axial currents in this model. Noether's theorem gives

$$j_\mu^5 = \bar{\Psi}\gamma_\mu\gamma^5\boldsymbol{\tau}\Psi + \mathrm{i}(\boldsymbol{\pi}\,\partial_\mu\sigma - \sigma\,\partial_\mu\boldsymbol{\pi}). \tag{15.24}$$

It is conserved in the absence of c. Using the equations of motion of the Lagrangian (15.23) we obtain

$$\partial^\mu j_\mu^5 \sim c\boldsymbol{\pi}. \tag{15.25}$$

In other words, the divergence of the current is proportional to the field of the pseudo-Goldstone boson. We call this relation *P.C.A.C.*, for 'Partially Conserved Axial Current' and it has played an important role in the development of the theory of both strong and weak interactions.

15.2.5   The Non-linear σ-Model

As we mentioned earlier, the σ model was introduced as a quantum field theory for the interactions among pions and nucleons. Although today neither the pions nor the nucleons are considered as elementary particles, we still need an effective field theory to describe them. Such a model could be useful at relatively low energies, $E \leq 1\,\mathrm{GeV}$ (i.e. relatively large distances), where the composite nature of the particles is not manifest. Indeed, this model has many attractive features: it has a built-in isospin symmetry. The pions are almost massless and they are the pseudo-Goldstone bosons of the spontaneously broken axial symmetry. This 'explains' an experimental puzzle, namely the fact that the pion masses (~ 140 MeV) are very small compared to a typical hadronic mass, such as the nucleon mass, which is of order 1 GeV. Furthermore, the pions satisfy the P.C.A.C. relation (15.25), which is also phenomenologically very successful. The only drawback is the appearance of the σ-field because there is no physical scalar boson in this mass range. A possible solution, already proposed in the original paper of Gell-Mann and Lévy, is to make σ very heavy. But, as we saw, the various parameters of the model are not all independent. Indeed, we have

$$m_\sigma^2 \sim \lambda v^2; \quad m_\Psi \sim gv, \tag{15.26}$$

where g is the pion–nucleon coupling constant and λ that of the pion self-interaction. We know experimentally that g is of order 1, so v is of order 1 GeV. We see that a heavy σ implies very strongly interacting pions. On the other hand, we want to keep the approximate invariance of the Lagrangian under $O(4)$ even in the absence of the σ-field.

Since there is no linear representation of $O(4)$ of dimension 3, it follows that the triplet of pions will transform non-linearly. Gell-Mann and Lévy proposed to use the simplest $O(4)$ invariant condition among the fields σ and $\boldsymbol{\pi}$ which is

$$\boldsymbol{\pi}^2 + \sigma^2 = v^2, \tag{15.27}$$

i.e. to promote the relation among the minimum energy classical field configurations to a relation among the quantum fields. We can use (15.27) to eliminate σ and obtain a non-linear model. Keeping only the kinetic energy terms we obtain

$$\mathcal{L}_{\mathrm{nl}} = \frac{1}{2} \sum_{i,j=1}^{3} \left[\delta_{ij} + \frac{\pi_i \pi_j}{v^2 - \sum_k \pi_k^2} \right] \partial_\mu \pi_i \partial^\mu \pi_j. \tag{15.28}$$

The kinetic energy term of σ is replaced by a term proportional to the kinetic energy of π with a coefficient depending non-linearly on the pion field.

So far the discussion was classical. To obtain the quantum theory we apply the standard recipe and put everything under a functional integral. We can integrate over the σ-field using the condition (15.27) and we obtain a functional integral involving only the pion fields. The generating functional is given by

$$Z[\mathcal{J}_i] = \int \left[\frac{\mathcal{D}[\pi_i]}{(v^2 - \boldsymbol{\pi}^2)^{1/2}} \right] \mathrm{e}^{-S[\pi] + \int \mathrm{d}^n x J(x) \cdot \boldsymbol{\pi}(x)}, \tag{15.29}$$

where $S[\pi]$ is the resulting action involving only the pion fields obtained from the Lagrangian (15.28) with the addition of the interaction terms. The integration measure is the $O(4)$ invariant functional measure on the three-sphere. The result reminds us of the one we obtained in the previous chapter on the quantisation of gauge theories with the Faddeev–Popov determinant. For future convenience, we wrote $\mathrm{d}^n x$ for the Euclidean space integration measure, leaving its dimensionality free.

Because of its non-linearity, this model is not easy to analyse using the techniques we have developed so far. Nevertheless, the idea of enforcing a symmetry in a non-linear way turned out to be quite powerful and we will come back to it in a later chapter.

We can generalise the model to any group $\mathcal{G}$ spontaneously broken to a subgroup $\mathcal{H}$. If the dimension of the Lie algebra of $\mathcal{G}$ is d_G and that of $\mathcal{H}$ is d_H, we will obtain $n = d_G - d_H$ massless Goldstone bosons, which we call π_i, $i=1,...,n$, all other fields being massive. In the same way we considered the limit $m_\sigma \to \infty$ for the $O(4)$ sigma model, we can consider the limit when all massive degrees of freedom become very heavy. This way we expect to obtain an effective theory describing the interaction among the massless Goldstone bosons alone. The experience from the simple model shows that this theory will be strongly interacting. We hope that such an effective theory may be relevant for low-energy phenomena.

There is a formal way to describe this infinite mass limit. Following the previous steps, let us distinguish in the original theory two sets of fields: $\pi_i(x)$, the fields that become massless after spontaneous symmetry breaking, and $\sigma(x)$, which denotes, collectively, all the others. In particular, the set $\sigma(x)$ may contain fields of any spin. In this theory the vacuum-to-vacuum amplitude is given by a functional integral of the form

$$Z = \int \mathcal{D}[\sigma] \Pi_{i=1}^n \mathcal{D}[\pi_i] e^{i \int d^4 x \mathcal{L}(\sigma, \pi_i)}. \tag{15.30}$$

Following a suggestion by K. Wilson, let us imagine that we integrate over all heavy degrees of freedom and keep only the set of massless fields which will transform non-linearly under $\mathcal{G}$. We shall obtain an effective theory describing the strong interaction among the Goldstone modes:

$$Z = \int \Pi_{i=1}^n \tilde{\mathcal{D}}[\pi_i] e^{i \int d^n x \mathcal{L}_{\text{eff}}(\pi_i)}. \tag{15.31}$$

$\tilde{\mathcal{D}}[\pi_i]$ is the new integration measure which takes into account the fact that the π fields satisfy a non-linear relation. In the general case we do not know how to perform such an integration, so we cannot compute $\mathcal{L}_{\text{eff}}(\pi_i)$ explicitly. We can only guess some general properties. It does not have to be polynomial in the π fields and their derivatives and it will realise the original symmetry under $\mathcal{G}$ non-linearly. Since it can only be used as an effective theory valid at low energies, we need only to keep the terms with the smallest number of derivatives. We need a term with two derivatives in order to define a Lorentz invariant dynamical theory, so the general form of the effective Lagrangian density will be

$$\mathcal{L}_{\text{eff}}(\pi_i) = \sum_{i,j=1}^n F_{ij}(\pi) \partial_\mu \pi_i(x) \partial^\mu \pi_j(x) + ..., \tag{15.32}$$

which resembles (15.28). F_{ij} is some matrix-valued function of the fields π and the dots stand for potential energy terms which do not contain derivatives of the fields. So the effective theory appears as a field theory with non-canonical, field-dependent, kinetic energy terms. We shall call all such theories *non-linear σ-models*. We can interpret them as follows: the fields $\pi_i(x)$ span a manifold isomorphic to the coset space $\mathcal{G}/\mathcal{H}$. So $\mathcal{L}_{\text{eff}}(\pi_i)$ can be viewed as a dynamical theory in this manifold with a metric given by F_{ij}. We propose a problem in the next chapter to study a particular example of such a theory with $\mathcal{G} = O(N)$ and $\mathcal{H} = O(N-1)$.

15.2.6   Goldstone Theorem

In section 15.2.3 we saw an example of Goldstone's theorem which states that to every generator of a spontaneously broken symmetry corresponds a massless particle. We want to give here a general proof which will not depend on a particular field theory model and which will illustrate the underlying assumptions for its validity.[4]

---

[4] This argument is due to W. Gilbert, *Phys. Rev. Lett.* **12**, 713 (1964), who presented it as a proof against the possibility of spontaneous symmetry breaking without massless particles. As we shall see, the argument fails in the presence of gauge interactions.

Let us consider a theory with the following properties:

1. Lorentz covariance;
2. It is defined in a Hilbert space with only positive definite norm states;
3. There exists a minimum energy 'vacuum' state $|0>$, although we make no assumption about its unicity; and
4. The theory is invariant under a Lie group of transformations G and this invariance implies the existence of a set of conserved currents $j_\mu^i(x)$ $i = 1, 2, ..., N$, the dimension of the Lie algebra of G. The current conservation implies the time-independence of the corresponding charges:[5]

$$\partial^\mu j_\mu^i(x) = 0 \quad \Rightarrow \quad Q^i \equiv \int d^3x j_0^i(x) \quad \dot{Q}^i = 0. \tag{15.33}$$

We assume that the symmetry is not trivial; in other words, there exists at least one operator $A(x)$ which is not a singlet of G:

$$\delta A(x) \equiv [Q^i, A(x)] \neq 0. \tag{15.34}$$

Assumptions 1–4 just given are the ones we usually make in local quantum field theory. Now comes the one about spontaneous symmetry breaking:

5. We assume that for some Q^i, the vacuum expectation value of δA is non-zero:

$$< 0|\delta A|0 > = < 0|[Q^i, A]|0 > \neq 0. \tag{15.35}$$

Note that this implies that the vacuum state is not invariant under G and, therefore, it is not annihilated by all the generators Q^i. This is the formal version of the degeneracy of the vacuum we found in the previous examples. We saw there that applying a symmetry transformation to a minimum energy field configuration (a 'vacuum' state), we obtained another field configuration with the same energy. Using these assumptions we will prove that the spectrum of states contains a massless particle.

Let us consider the Fourier transform of the vacuum expectation value of the commutator between the current $j_\mu^i(x)$ and the operator A:

$$G_\mu \equiv \int d^4x e^{ikx} < 0|[j_\mu^i(x), A(0)]|0 >. \tag{15.36}$$

---

[5] At this point our proof is only heuristic. The existence of the integrals in (15.33) assumes that all fields vanish sufficiently rapidly at infinity so that surface terms can be dropped. However, in the presence of massless particles this is not always true. See, for example, the Coulomb scattering amplitude in quantum mechanics. Note also that it is precisely the existence of such massless particles that we want to prove. A more rigorous way would be to define the charges as integrals inside a given volume V, $Q_V^i \equiv \int_V d^3x j_0^i(x)$ and study carefully the limit $V \to \infty$. The result is that although the relations (15.33) are not valid in a strong sense, i.e. as operator equations, they can still be used in a weak sense as matrix elements between suitably localised states.

G_μ is a function of the four-vector k and, by assumptions 1. and 5., it is of the general form

$$G_\mu(k) = k_\mu f(k^2); \quad f(k^2) \neq 0. \tag{15.37}$$

We now compute

$$k^\mu G_\mu(k) = k^2 f(k^2) = -i \int d^4 x (\partial^\mu e^{ikx}) < 0|[j_\mu^i(x), A(0)]|0 >= 0, \tag{15.38}$$

where we used partial integration (see footnote 5) and the conservation of the current. The only solution of the two equations (15.37) and (15.38) is

$$f(k^2) = C\delta(k^2) \tag{15.39}$$

with C some constant; in other words, the correlation function $G_\mu(k)$ has a delta-function singularity at $k^2 = 0$. Notice that $\delta(k^2) \sim \mathrm{Im}(k^2 - i\epsilon)^{-1}$ which is the propagator of a massless particle. To finish the proof we must show that this propagator corresponds to a massless physical state and it is not cancelled by some other massless state. This is guaranteed by assumption 2). Since all states have positive norm they all contribute with the same sign and no cancellations can occur. *QED*

A final remark: Goldstone's theorem tells us that to every generator of a spontaneously broken symmetry corresponds a massless particle. This Goldstone (or pseudo-Goldstone) particle has the same quantum numbers as the divergence of the corresponding symmetry current. This is obvious from (15.38) and it is exemplified in the P.C.A.C. relation (15.25). Therefore, it can be used to answer the opposite question: if we observe a massless, or nearly massless, particle in nature, how could we know whether it is the Goldstone, or pseudo-Goldstone, particle of some spontaneously broken symmetry? The answer is to look for the existence of a conserved, or nearly conserved, symmetry current with the right quantum numbers. We will give a more operational version of this property in a later chapter.

15.3   Gauge Symmetries

In the previous chapter we introduced the concept of a gauge symmetry. We saw that it follows from a fundamental geometric principle, but it seems to imply the existence of massless vector particles. Since, apart from the photon, such particles are not present in physics, we concluded that gauge theories and, in particular, non-Abelian gauge theories cannot describe the interactions among elementary particles. In this chapter we studied the phenomenon of spontaneous symmetry breaking. We saw that it is also associated with the existence of massless particles, the Goldstone particles. We found one possible application in the pion system but, again, this requirement of massless particles seems to severely limit their applicability in particle physics. In this section we want to study the consequences of spontaneous symmetry breaking in the presence of a gauge symmetry. We will find a very surprising result. When combined together the two problems will solve each other. It is this miracle that we want to present here. We start with the Abelian case.

15.3.1 The Abelian Model

We look at the model of section 15.2.3 in which the $U(1)$ symmetry (15.9) has been promoted to a local symmetry with $\theta \to \theta(x)$. As we explained already, this implies the introduction of a massless vector field, which we can call the 'photon' and the interactions are obtained by replacing the derivative operator ∂_μ by the covariant derivative D_μ and adding the photon kinetic energy term:

$$\mathcal{L}_1 = -\frac{1}{4}F_{\mu\nu}^2 + |(\partial_\mu + ieA_\mu)\phi|^2 - M^2\phi\phi^* - \lambda(\phi\phi^*)^2. \tag{15.40}$$

$\mathcal{L}_1$ is invariant under the gauge transformation

$$\phi(x) \to e^{i\theta(x)}\phi(x); \quad A_\mu \to A_\mu - \frac{1}{e}\partial_\mu\theta(x). \tag{15.41}$$

The same analysis as before shows that for $\lambda > 0$ and $M^2 < 0$ there is a spontaneous breaking of the $U(1)$ symmetry. Substituting (15.12) into (15.40) we obtain

$$\begin{aligned}
\mathcal{L}_1 \to \mathcal{L}_2 = &-\frac{1}{4}F_{\mu\nu}^2 + \frac{e^2v^2}{2}A_\mu^2 + evA_\mu\partial^\mu\chi \\
&+ \frac{1}{2}(\partial_\mu\psi)^2 + \frac{1}{2}(\partial_\mu\chi)^2 - \frac{1}{2}(2\lambda v^2)\psi^2 + \cdots,
\end{aligned} \tag{15.42}$$

where the dots stand for coupling terms which are at least trilinear in the fields.

The surprising term is the second one which is proportional to A_μ^2. It looks as though the photon has become massive. Note that (15.42) is still gauge invariant since it is equivalent to (15.40). The gauge transformation is now obtained by substituting (15.12) into (15.41):

$$\begin{aligned}
\psi(x) &\to \cos\theta(x)[\psi(x) + v] - \sin\theta(x)\chi(x) - v \\
\chi(x) &\to \cos\theta(x)\chi(x) + \sin\theta(x)[\psi(x) + v] \\
A_\mu &\to A_\mu - \frac{1}{e}\partial_\mu\theta(x).
\end{aligned} \tag{15.43}$$

This means that our previous conclusion, that gauge invariance forbids the presence of an A_μ^2 term, was simply wrong. Such a term can be present; only the gauge transformation is slightly more complicated; it must be accompanied by a translation of the field.

The Lagrangian (15.42), if taken as a quantum field theory, seems to describe the interaction of a massive vector particle (A_μ) and two scalars, one massive (ψ) and one massless (χ). However, we can see immediately that something is wrong with this counting. A warning is already contained in the non-diagonal term between A_μ and $\partial^\mu \chi$. Indeed, the perturbative particle spectrum can be read from the Lagrangian only after we have diagonalised the quadratic part. A more direct way to see the trouble is to count the apparent degrees of freedom before and after the translation:

Lagrangian (15.40):

(i) One massless vector field: 2 degrees

(ii) One complex scalar field: 2 degrees

Total: 4 degrees

Lagrangian (15.42):

(i) One massive vector field: 3 degrees

(ii) Two real scalar fields: 2 degrees

Total: 5 degrees

Since physical degrees of freedom cannot be created by a simple change of variables, we conclude that the Lagrangian (15.42) must contain fields which do not create physical particles. This is indeed the case, and we can exhibit a transformation which makes the unphysical fields disappear. Instead of parametrising the complex field ϕ by its real and imaginary parts, let us choose its modulus and its phase. The choice is dictated by the fact that it is a change of phase that describes the motion along the circle of the minima of the potential $V(\phi)$. We thus write

$$\phi(x) = \frac{1}{\sqrt{2}}[v + \rho(x)]e^{i\zeta(x)/v}; \quad A_\mu(x) = B_\mu(x) - \frac{1}{ev}\partial_\mu \zeta(x). \tag{15.44}$$

In this notation, the gauge transformation (15.41) or (15.44) is simply a translation of the field ζ: $\zeta(x) \rightarrow \zeta(x) + v\theta(x)$. Substituting (15.44) into (15.40) we obtain

$$\mathcal{L}_1 \rightarrow \mathcal{L}_3 = -\frac{1}{4}B_{\mu\nu}^2 + \frac{e^2 v^2}{2}B_\mu^2 + \frac{1}{2}(\partial_\mu \rho)^2 - \frac{1}{2}(2\lambda v^2)\rho^2$$
$$-\frac{\lambda}{4}\rho^4 + \frac{1}{2}e^2 B_\mu^2(2v\rho + \rho^2) \tag{15.45}$$
$$B_{\mu\nu} = \partial_\mu B_\nu - \partial_\nu B_\mu.$$

The field $\zeta(x)$ has disappeared. Formula (15.46) describes two massive particles, a vector (B_μ) and a scalar (ρ). It exhibits no gauge invariance, since the original symmetry $\zeta(x) \rightarrow \zeta(x) + v\theta(x)$ is now trivial.

We see that we obtained three different Lagrangians describing the same physical system. $\mathcal{L}_1$ is invariant under the usual gauge transformation, but it contains a negative square mass and, therefore, it is unsuitable for quantisation. $\mathcal{L}_2$ is still gauge invariant, but the transformation law (15.44) is more complicated. It can be quantised in a space containing unphysical degrees of freedom. This, by itself, is not a great obstacle and it occurs frequently. For example, ordinary quantum electrodynamics is usually quantised in a space involving unphysical (longitudinal and scalar) photons. In fact, it is $\mathcal{L}_2$, in a suitable gauge, which is used for general proofs of renormalisability as well as for practical calculations. Finally, $\mathcal{L}_3$ is no longer invariant under any kind of gauge transformation, but it exhibits clearly the particle spectrum of the theory. It contains only physical particles and they are all massive. This is the miracle that was announced earlier. Although we start from a gauge theory, the final spectrum contains massive particles only.

Actually, $\mathcal{L}_3$ can be obtained from $\mathcal{L}_2$ by an appropriate choice of gauge. Indeed, let us choose to quantise $\mathcal{L}_2$ in a Landau–Feynman gauge. In the notation of the previous chapter, we add a term proportional to $\frac{1}{2}\mathcal{F}^2$ with $\mathcal{F} = \partial_\mu A^\mu$. Since it is a linear gauge and the model is Abelian, the ghost fields are decoupled. The quadratic part of the $A-\chi$ system now becomes

$$\mathcal{L}_{A-\chi} = -\frac{1}{4}F_{\mu\nu}^2 - \frac{1}{2\alpha}(\partial_\mu A^\mu)^2 + \frac{e^2 v^2}{2}A_\mu^2 + ev A_\mu \partial^\mu \chi + \frac{1}{2}(\partial_\mu \chi)^2. \tag{15.46}$$

The propagators are well defined but there is still a non-diagonal $A-\chi$ term. A more convenient gauge choice is given by $\mathcal{F} = \partial_\mu A^\mu - ev\alpha\chi$. In this case $\mathcal{L}_{A-\chi}$ becomes

$$\mathcal{L}_{A-\chi} = -\frac{1}{4}F_{\mu\nu}^2 - \frac{1}{2\alpha}(\partial_\mu A^\mu)^2 + \frac{e^2 v^2}{2}A_\mu^2 + \frac{1}{2}(\partial_\mu \chi)^2 - \frac{\alpha e^2 v^2}{2}\chi^2. \tag{15.47}$$

The two propagators decouple and we find, in momentum space, that

$$G_{\mu\nu}(k) = \frac{-1}{k^2 - e^2 v^2}\left[g_{\mu\nu} - (1-\alpha)\frac{k_\mu k_\nu}{k^2 - \alpha e^2 v^2}\right]$$

$$G(k) = \frac{1}{k^2 - \alpha e^2 v^2} \tag{15.48}$$

for the A and χ propagators, respectively. We note that we find the same singularity for both the χ-propagator and the $k_\mu k_\nu$ part of the A-propagator. We can check (see Problem 15.2) that these singularities cancel for every physical amplitude, as they should, since physical amplitudes must be independent of the gauge parameter α. Some special values for α: (i) $\alpha = 0$: both χ and the $k_\mu k_\nu$ part of A correspond to massless propagators. (ii) $\alpha = 1$: they all have the same mass ev. (iii) $\alpha \to \infty$: $G_{\mu\nu}$ becomes the propagator of a massive vector field. χ becomes infinitely heavy and decouples. The Lagrangian becomes $\mathcal{L}_3$. This choice is often referred to as *the unitary gauge*. Only physical degrees of freedom propagate.

The conclusion of this section can now be stated as follows.

In a spontaneously broken gauge theory the gauge vector bosons acquire a mass and the would-be massless Goldstone bosons decouple and disappear. Their degrees of freedom are used in order to make possible the transition from massless to massive vector bosons. This phenomenon has a complicated history. It was implicit in the first phenomenological description of superconductivity by F. and H. London as well as in the L. D. Landau and V. L. Ginzburg theory of 1950. In the framework of the BCS theory it was studied by Ph. Anderson in 1962. J. Schwinger was the first to understand the physical principles in particle physics. In four-dimensional field theory it was first introduced by R. Brout and F. Englert as well as by P. Higgs in 1964. It was further studied by G. S. Guralnik, C. R., Hagen, and T. W. B. Kibble. It is commonly known as the 'Higgs mechanism', but we will call it in this book the 'BEH mechanism'.

A final remark: The BEH phenomenon seems to violate Goldstone's theorem: there is a spontaneous symmetry breaking and no massless Goldstone particle. The reason is that in quantising a gauge theory, we do not respect the assumptions we stated in section 15.2.6. In particular, assumptions 1 (explicit Lorentz invariance) and 2 (positivity in Hilbert space) cannot be enforced simultaneously. As we saw in the example of quantum electrodynamics, if we choose a covariant gauge we must introduce unphysical degrees of freedom, such as scalar photons, which come with negative metric; otherwise, if we want to have only physical, positive metric degrees of freedom, we must choose a non-covariant gauge, such as the Coulomb gauge.

15.3.2   The Non-Abelian Case

The extension to the non-Abelian case is straightforward. Let us consider a gauge group G with m generators and, thus, m massless gauge bosons. The claim is that we can break part of the symmetry spontaneously, leaving a subgroup H with h generators unbroken. The h gauge bosons associated with H remain massless while the $m - h$ others acquire a mass. In order to achieve this result we need $m - h$ scalar degrees of freedom with the same quantum numbers as the broken generators. They will disappear from the physical spectrum and will re-appear as zero helicity states of the massive vector bosons. As previously, we will see that we need at least one more scalar state which remains physical.

We introduce a multiplet of scalar fields ϕ_i which transform according to some representation, not necessarily irreducible, of G of dimension n. According to the rules we explained in the last section, the Lagrangian of the system is given by

$$\mathcal{L} = -\frac{1}{4}\mathrm{Tr}(F_{\mu\nu}F^{\mu\nu}) + (D_\mu\Phi)^\dagger D^\mu\Phi - V(\Phi). \tag{15.49}$$

In component notation, the covariant derivative is, as usual, $D_\mu\phi_i = \partial_\mu\phi_i - ig^{(a)}T^a_{ij}A^a_\mu\phi_j$, where we have allowed for the possibility of having arbitrary coupling constants $g^{(a)}$ for the various generators of G because we do not assume that G is simple or semi-simple. $V(\Phi)$ is a polynomial in Φ invariant under G of degree equal to 4. As before, we assume that we can choose the parameters in V such that the minimum is not at $\Phi = 0$ but rather at $\Phi = v$ where v is a constant vector in the representation space of Φ. v is not unique. The m generators of G can be separated into two classes: h generators, which annihilate v and form the Lie algebra of the unbroken subgroup H, and $m - h$

generators, represented in the representation of Φ by matrices T^a, such that $T^a v \neq 0$ and all vectors $T^a v$ are independent and can be chosen orthogonal. Any vector in the orbit of v, i.e. of the form $e^{iw^a T^a} v$ is an equivalent minimum of the potential. As before, we should translate the scalar fields Φ by $\Phi \to \Phi + v$. It is convenient to decompose Φ into components along the orbit of v and orthogonal to it, the analogue of the χ and ψ fields of the previous section. We can write

$$\Phi = i \sum_{a=1}^{m-h} \frac{\chi^a T^a v}{|T^a v|} + \sum_{b=1}^{n-m+h} \psi^b u^b + v, \tag{15.50}$$

where the vectors u^b form an orthonormal basis in the space orthogonal to all $T^a v$'s. The corresponding generators span the coset space G/H. As before, we will show that the fields χ^a will be absorbed by the BEH mechanism and the fields ψ^b will remain physical. Note that the set of vectors u^b contains at least one element since, for all a, we have

$$v \cdot T^a v = 0, \tag{15.51}$$

because the generators in a real unitary representation are antisymmetric. This shows that the dimension n of the representation of Φ must be larger than $m - h$ and, therefore, there will remain at least one physical scalar field.

Let us now bring in the Lagrangian (15.49) the expression of Φ from (15.50). We obtain

$$\begin{aligned} \mathcal{L} = {} & \frac{1}{2} \sum_{a=1}^{m-h} (\partial_\mu \chi^a)^2 + \frac{1}{2} \sum_{b=1}^{n-m+h} (\partial_\mu \psi^b)^2 - \frac{1}{4} \mathrm{Tr}(F_{\mu\nu} F^{\mu\nu}) \\ & + \frac{1}{2} \sum_{a=1}^{m-h} g^{(a)2} |T^a v|^2 A_\mu^a A^{\mu a} - \sum_{a=1}^{m-h} g^{(a)} T^a v \partial^\mu \chi^a A_\mu^a \\ & - V(\Phi) + \cdots, \end{aligned} \tag{15.52}$$

where the dots stand for coupling terms between the scalars and the gauge fields. In writing (15.52) we took into account that $T^b v = 0$ for $b > m - h$ and that the vectors $T^a v$ are orthogonal.

The analysis that gave us Goldstone's theorem shows that

$$\frac{\partial^2 V}{\partial \phi_k \partial \phi_l}|_{\Phi=v} (T^a v)_l = 0, \tag{15.53}$$

which shows that the χ-fields would correspond to the Goldstone modes. As a result, the only mass terms which appear in V in Eq. (15.52) are of the form $\psi^k M^{kl} \psi^l$ and do not involve the χ-fields.

As far as the bilinear terms in the fields are concerned, the Lagrangian (15.52) is the sum of terms of the form found in the Abelian case. All gauge bosons which do not correspond to H generators acquire a mass equal to $m_a = g^{(a)} |T^a v|$ and, through their mixing with the would-be Goldstone fields χ, develop a zero helicity state. All other gauge bosons remain massless. The ψ's represent the remaining physical BEH fields.

15.4 Problems

Problem 15.1 The patterns of spontaneous symmetry breaking of $SU(5)$. It will be useful in constructing grand unified theories in a later chapter. We consider a model with a 24-plet of scalar fields belonging to the adjoint representation of $SU(5)$, which we write as a 5×5 traceless matrix $\Phi(x)$. Write the most general polynomial Lagrangian with terms of dimension smaller than or equal to 4, invariant under global $SU(5)$ transformations. (As usual, via discrete transformations, we can ignore terms trilinear in the field Φ and its derivatives. The answer is given in Eq. (26.15).) Study the various patterns of spontaneous symmetry breaking for this model which result from different choices of the coupling constants appearing in the potential of the scalar field. (The answer is given in section 26.2.)

Problem 15.2 We consider a field theory model with the following fields. (i) A Dirac spinor $\Psi(x)$ which we split into its left- and right-handed parts: $L = \frac{1}{2}(1 + \gamma_5)\Psi$, $R = \frac{1}{2}(1 - \gamma_5)\Psi$. (ii) A complex scalar field $\phi(x)$. (iii) A real massless vector field $A_\mu(x)$. We assume that the theory is invariant under the Abelian group of gauge transformations given by

$$L \to e^{ie\theta} L; \quad R \to e^{-ie\theta} R; \quad \phi \to e^{2ie\theta} \phi; \quad A_\mu \to A_\mu + \partial_\mu \theta. \tag{15.54}$$

1. Write the most general gauge invariant Lagrangian density with terms of dimension smaller than or equal to 4. Show that the fermion field is massless and the gauge field couples to the axial fermionic current.

2. Choose the mass-square of the scalar field negative and study the model in the phase with spontaneous symmetry breaking. Find the mass spectrum in this phase.

3. In the gauge given by the function $\mathcal{F} = \partial^\mu A_\mu - ev\alpha\chi$ which we studied in section 15.3.1 compute the boson propagators and verify Eqs. (15.47) and (15.48).

4. Compute the amplitude for the elastic scattering of two Dirac particles in the one-particle exchange approximation and show that the gauge-dependent poles cancel.

Problem 15.3 Consider a theory invariant under $O(3)$ gauge transformations involving, in addition to the gauge bosons, three real scalar fields forming a triplet of $O(3)$.

1. Write the gauge invariant Lagrangian density containing all terms with dimension smaller than or equal to 4.

2. Choose the parameters in order to induce a spontaneous breaking of the symmetry.

3. Describe the spectrum of one-particle states in the broken phase.

16

Quantum Field Theory at Higher Orders

16.1 Existence of Divergences in Loop Diagrams. Discussion

The language of Feynman diagrams influenced considerably our approach to quantum field theory by providing a space–time image for the various terms in the perturbation expansion of Green functions. At this level, the Lagrangian is viewed as just a shorthand encoding of the Feynman rules. As we saw already, there exists a one-to-one correspondence between the diagrams we can draw and the terms which appear using the Wick theorem. The rules of the previous chapters offer a precise algorithm for all calculations at any given order of perturbation.

However, a cursory look at these rules shows that they often involve expressions which are mathematically ill-defined. As an example, let us consider the simplest quantum field theory, that of a single scalar field $\phi(x)$ interacting through a ϕ^4 interaction. If we compute the 4-point Green function at second order of perturbation, we find the diagram of Fig. 16.1.

A straightforward application of the Feynman rules leads in Euclidean momentum space to an integral of the form

$$I = \int \frac{\mathrm{d}^4 k}{(k^2 + m^2)[(k-p)^2 + m^2]}, \tag{16.1}$$

which diverges logarithmically at large k.

Similar divergent terms can be found in all quantum field theories. A simple example in quantum electrodynamics is given by the photon self-energy diagram of Fig. 16.2, which is traditionally called 'vacuum polarisation'.

The corresponding integral is

$$I_{\mu\nu} = \int \mathrm{d}^4 k \mathrm{Tr} \left(\gamma_\mu \frac{1}{k+m} \gamma_\nu \frac{1}{k-q+m} \right), \tag{16.2}$$

From Classical to Quantum Fields. Laurent Baulieu, John Iliopoulos and Roland Sénéor.
© Laurent Baulieu, John Iliopoulos and Roland Sénéor, 2017. Published 2017 by Oxford University Press.

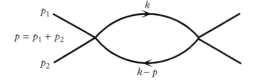

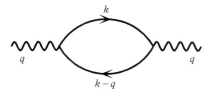

Figure 16.1 *A 1-loop divergent diagram in the ϕ^4 theory.*

Figure 16.2 *The 1-loop vacuum polarisation diagram.*

which is quadratically divergent. Such divergences occur when the momenta of the virtual particles appearing in the loops become very large and are called *ultraviolet* or *short-distance* singularities.

Let us first recall that it is not the first time we encounter such divergences. When we computed the energy of the vacuum state using free fields we found an infinite contribution corresponding to the sum of the zero-point energies for the infinite number of harmonic oscillators which describe each quantum field. In this chapter we shall present a specific method for avoiding these difficulties.

16.2 Connected and 1-PI Diagrams

The description of the perturbation series in terms of Feynman diagrams simplifies the introduction of some notions from elementary topology. We have already encountered in Chapter 12 the notion of *connected* and *disconnected* Green functions. In terms of Feynman diagrams these definitions are obvious. We could even introduce some further simplifications: an *amputated* diagram is a connected diagram multiplied by an inverse free propagator for every external line. Since these factors do not introduce any integrations, they do not affect the convergence properties of the diagram. A connected, amputated diagram will be called *one-particle irreducible* (1-PI), if it cannot be separated into two disconnected sub-diagrams by cutting a single internal line. One-particle reducible diagrams are obtained as products of 1-PI ones multiplied by the propagators of the connecting lines whose momenta are fixed by energy and momentum conservation. We conclude that if we want to study the convergence properties of the integrals involved in the calculation of Feynman diagrams, it is sufficient to restrict ourselves to the subset of the 1-PI ones. The sum of all 1-PI diagrams gives the 1-PI Green functions.

This diagrammatic analysis can also be obtained in an abstract way using the general formalism we developed in the previous chapters. Let $\mathcal{L}(\phi_i(x))$, $i = 1, ..., N$, be the

Lagrangian density describing the system of N interacting fields and let $\mathcal{J}_i(x)$ be N c-number functions of space–time which transform with respect to all the symmetries of $\mathcal{L}$ in such a way that $\sum_{i=1}^{N} \mathcal{J}_i(x)\phi_i(x)$ is an invariant. They are often called *classical sources*. If we consider $\mathcal{L} + \sum \mathcal{J}_i\phi_i$ and calculate the vacuum-to-vacuum transition amplitude in the presence of the sources, we obtain a functional $\mathcal{F}[\mathcal{J}_i]$, which, if expanded in powers of $\mathcal{J}_i$, gives all the Green functions of the theory, including the disconnected pieces. Using the path integral, it can be written as

$$\mathcal{F}[\mathcal{J}_i] = \frac{\int \mathcal{D}\phi_i e^{\int d^4x(\mathcal{L}+\sum \mathcal{J}_i\phi_i)}}{\int \mathcal{D}\phi_i e^{\int d^4x\mathcal{L}}} \tag{16.3}$$

and the disconnected, n-point Green function $G_{i_1,...,i_n}(x_1,...,x_n)$ is given by

$$G_{i_1,...,i_n}(x_1,...,x_n) = \frac{\delta^n \mathcal{F}[\mathcal{J}_i]}{\delta \mathcal{J}_{i_1}(x_1),...,\delta \mathcal{J}_{i_n}(x_n)}\bigg|_{\mathcal{J}_i=0}. \tag{16.4}$$

Similarly, we can construct generating functionals for the connected and the 1-PI Green functions. We define

$$\mathcal{F}[\mathcal{J}_i] = e^{W[\mathcal{J}]}. \tag{16.5}$$

In Chapter 12 we showed that $W[\mathcal{J}]$, expanded in powers of $\mathcal{J}$ around $\mathcal{J} = 0$, generates the connected Green functions. The generating functional for the 1-PI functions is obtained from $W[\mathcal{J}]$ by functional Legendre transformation the way the Hamiltonian is obtained from the Lagrangian in classical mechanics. We define the *classical field* $\phi_i^{(cl)}(x)$ as the conjugate variable to $\mathcal{J}_i(x)$,

$$\phi_i^{(cl)}(x) = \frac{\delta W[\mathcal{J}]}{\delta \mathcal{J}(x)}, \tag{16.6}$$

and the *effective action* by

$$\Gamma[\phi_i^{(cl)}] = W[\mathcal{J}_i] - \sum_{i=1}^{N} \int d^4x \mathcal{J}_i(x)\phi_i^{(cl)}(x). \tag{16.7}$$

In (16.7) we are supposed to express $\mathcal{J}$ as a functional of $\phi^{(cl)}$ through the inverse of (16.6). We obtain

$$\mathcal{J}_i(x) = -\frac{\delta \Gamma}{\delta \phi_i^{(cl)}(x)}. \tag{16.8}$$

If we functionally expand the effective action in powers of $\phi^{(cl)}$, the coefficient functions are precisely the 1-PI Green functions

$$\Gamma_{i_1,\dots,i_n}(x_1,\dots,x_n) = \frac{\delta^n \Gamma[\phi_i^{(\text{cl})}]}{\delta\phi_{i_1}^{(\text{cl})}(x_1),\dots,\delta\phi_{i_n}^{(\text{cl})}(x_n)}\Big|_{\phi_i^{(cl)}=0}, \tag{16.9}$$

i.e. the sum of all 1-PI Feynman diagrams with n external lines corresponding to the fields $i_1,\dots,i_n$. This property is well known in statistical mechanics. Let us give a simple proof here. Consider a theory with a single, neutral, scalar field $\phi(x)$. The extension to other theories is straightforward. We start with the 2-point function. Taking the functional derivative of both sides of Eq. (16.6) with respect to $\phi^{cl}(y)$, we obtain

$$\delta^4(x-y) = \frac{\delta}{\delta\phi^{cl}(y)}\frac{\delta W[\mathcal{J}]}{\delta\mathcal{J}(x)} = \int d^4z \frac{\delta\mathcal{J}(z)}{\delta\phi^{cl}(y)}\frac{\delta^2 W[\mathcal{J}]}{\delta\mathcal{J}(z)\delta\mathcal{J}(x)}. \tag{16.10}$$

We remind that in this equation we are supposed to express the source $\mathcal{J}$ as a functional of the classical field ϕ^{cl} through Eq. (16.8). The first factor under the integral is the second derivative of Γ with respect to ϕ^{cl}. If now we put all external sources equal to 0, we obtain the result that the connected 2-point function is the inverse of the amputated one-particle irreducible 2-point function. In momentum space this tells us that

$$G_c^{(2)}(p^2)\Gamma^{(2)}(p^2) = 1. \tag{16.11}$$

Let us look at this result in lowest order perturbation theory. The complete propagator is given by $G^{(2)}(p^2) = i/(p^2-m^2)$, where the $+i\epsilon$ prescription is understood. So, in the tree approximation Eq. (16.11) tells us that $\Gamma^{(2)}(p^2) = -i(p^2-m^2)$. Let us verify it explicitly: to this order the 1-PI 2-point function equals the amputated 2-point function. In order to obtain the latter we multiply the two external lines by the inverse propagator, which, to this order, is the free propagator. So we obtain $\Gamma^{(2)}(p^2) = G_{c,\text{amp}}^{(2)}(p^2) = -i(p^2-m^2)$, in agreement with Eq. (16.11). It follows that the perturbation expansion of $\Gamma^{(2)}(p^2)$ can be written as

$$\Gamma^{(2)}(p^2) = -i(p^2-m^2-\Sigma^{(2)}(p^2)), \tag{16.12}$$

where $\Sigma^{(2)}(p^2)$ is a formal power series expansion in the coupling constant with the loop contributions.

This result has a simple graphical representation shown in Fig. 16.3. The round bulb represents the connected 2-point function $G_c^{(2)}(p^2)$. The elliptical bulb is the expression for $-i\Sigma^{(2)}(p^2)$. It gives

$$G_c^{(2)}(p^2) = \frac{i}{p^2-m^2} + \frac{i}{p^2-m^2}\frac{\Sigma^{(2)}(p^2)}{i}\frac{i}{p^2-m^2} + \cdots . \tag{16.13}$$

By summing the series we obtain

$$G_c^{(2)}(p^2) = \frac{i}{p^2-m^2-\Sigma^{(2)}(p^2)} = \left[\Gamma^{(2)}(p^2)\right]^{-1}, \tag{16.14}$$

the result of Eq. (16.11).

Figure 16.3 *The 1-PI decomposition of the 2-point function.*

We can continue with higher Green functions. If the 3-point function $\Gamma^{(3)}$ does not vanish by some symmetry property, it is given by differentiating once more Eq. (16.10) with respect to $\phi^{cl}(w)$ and then setting the external sources equal to 0. We obtain two terms: the first comes from differentiating the factor $\delta \mathcal{J}(z)/\delta \phi^{cl}(y) \sim \delta^2 \Gamma/\delta \phi^{cl2}$ and the second from the differentiation of the $\delta^2 W/\delta \mathcal{J}^2$, which gives a term of the form $G_c^{(3)} \Gamma^{(2)}$. More explicitly we get

$$0 = \int \mathrm{d}^4 z \left[\Gamma^{(3)}(z, y, w) G^{(2)}(z, x) + \int \mathrm{d}^4 u \Gamma^{(2)}(z, y) G_c^{(3)}(z, x, u) \Gamma^{(2)}(u, w) \right], \quad (16.15)$$

which expresses the 1-PI 3-point function in terms of the connected 3-point function. Using the relation (16.11) we can rewrite it as

$$\begin{aligned}
\Gamma^{(3)}(x, y, w) &= \int \mathrm{d}^4 x_1 \mathrm{d}^4 y_1 \mathrm{d}^4 w_1 \, G_c^{(3)}(x_1, y_1, w_1) \Gamma^{(2)}(x_1, x) \Gamma^{(2)}(y_1, y) \Gamma^{(2)}(w_1, w) \\
&= G_{c,\mathrm{amp}}^{(3)}(x, y, w), \quad (16.16)
\end{aligned}$$

which means that the 1-PI 3-point function is just the amputated 3-point function, a result we can easily check by drawing the corresponding diagram, Fig. 16.4.

We can continue with all the higher order functions. By successive differentiations we can generalise the relation (16.16) and express any $\Gamma^{(n)}$ function as a sum of terms: the first is the corresponding n point amputated connected function, the others contain combinations of $G^{(m)}$ with $m < n$ and $\Gamma^{(2)}$'s. The proof that the resulting $\Gamma^{(n)}$ is the corresponding 1-PI proceeds by induction, although the combinatorics is, as usual, quite lengthy. Note also that if the 3-point function vanishes, as it is the case for a ϕ^4 theory, the analogue of Eq. (16.16) is valid for the 4-point function: $\Gamma^{(4)}(x, y, w, z) = G_{c,\mathrm{amp}}^{(4)}(x, y, w, z)$ (Fig. 16.5).

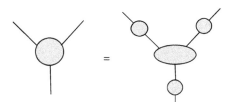

Figure 16.4 *The 1-PI decomposition of the 3-point function.*

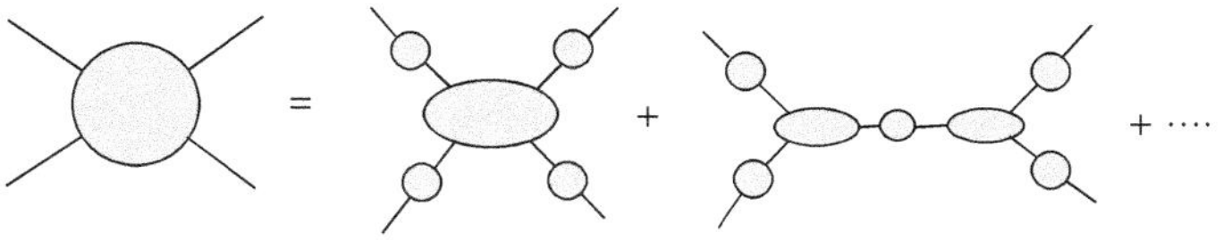

Figure 16.5 *The 1-PI decomposition of the 4-point function. The dots stand for crossed terms. We see that if the 3-point function vanishes by symmetry, $\Gamma^{(4)} = G^{(4)}_{c,amp}$.*

16.3 Power Counting. Definition of Super-Renormalisable, Renormalizable, and Non-renormalisable Quantum Field Theories

We saw already that a straightforward application of Feynman rules may lead to divergent integrals. In this section we want to introduce a systematic method for finding all such divergences. As we noted earlier, it is sufficient to restrict ourselves to the study of the 1-PI diagrams. We want to emphasise that the entire approach will be perturbative, i.e. all calculations are supposed to be performed order by order in a power series expansion of the coupling constant. The calculation of the nth order cannot be envisaged unless the contributions of all lower orders have been successfully completed.

A single-loop integral will be ultravioletly divergent if and only if the numerator is of degree equal or higher in the loop momentum than the denominator. For multiloop diagrams this may not be the case, since the divergence may be entirely due to a particular sub-diagram; however, as we emphasised earlier, in the spirit of perturbation theory, the divergent sub-diagram must be treated first. We thus arrive at the notion of *superficial degree of divergence d* of a given 1-PI diagram, defined as the difference between the degree of integration momenta of the numerator minus that of the denominator. The diagram will be called *primitively divergent* if $d \geq 0$. Let us compute d for the diagrams of some simple field theories.

We start with scalar field theories with interaction of the form ϕ^m with m integer larger than 2. (For $m = 2$ we have the trivial case of a free field theory.) Let us consider a 1-PI diagram of nth order in perturbation with I internal and E external lines. Every internal line brings four powers of k in the numerator through the d^4k factor and two powers in the denominator through the propagator. Every vertex brings a δ^4-function of the energy-momentum conservation. All but one of them can be used to eliminate one integration each, the last reflecting the overall conservation which involves only external momenta. Therefore, we obtain

$$d = 2I - 4n + 4. \tag{16.17}$$

This expression can be made more transparent by expressing I in terms of E and m. A simple counting gives $2I + E = mn$ and (16.17) becomes

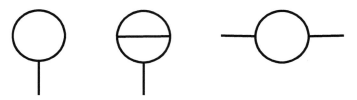

Figure 16.6 *The primitively divergent 1-PI diagrams in the four-dimensional ϕ^3 theory.*

$$d = (m - 4)n - E + 4. \tag{16.18}$$

This is the main result. We see that $m = 4$ is a critical value and we can distinguish three cases:

1. $m = 3$, $d = 4 - n - E$. d is a decreasing function of n, the order of perturbation theory. Only a limited number of diagrams are primitively divergent. Above a certain order they are all convergent. The divergent ones are shown in Fig. 16.6. For reasons that will be clear soon, we shall call such theories *super-renormalisable*.

2. $m = 4$, $d = 4 - E$. d is independent of the order of perturbation theory. If a Green function is divergent at some order, it will be divergent at all orders. For the ϕ^4 theory we see that the primitively divergent diagrams are those with $E = 2$, which have $d = 2$ and are quadratically divergent and those with $E = 4$, which have $d = 0$ and are logarithmically divergent. (Note that for this theory, all Green functions with odd E vanish identically because of the symmetry $\phi \rightarrow -\phi$.) We shall call such theories *renormalisable*.

3. $m > 4$, d is an increasing function of n. Every Green function, irrespectively of the number of external lines, will be divergent above some order of perturbation. We call such theories *non-renormalisable*.

This power-counting analysis can be repeated for any quantum field theory. As a second example, we can look at quantum electrodynamics. We should now distinguish between photon and electron lines which we shall denote by I_γ, I_e, E_γ, and E_e for internal and external lines, respectively. Taking into account the fact that the fermion propagator behaves like k^{-1} at large momenta, we obtain for the superficial degree of divergence of a 1-PI diagram:

$$d = 2I_\gamma + 3I_e - 4n + 4 = 4 - E_\gamma - \frac{3}{2}E_e. \tag{16.19}$$

We see that d is independent of the order of perturbation theory and, therefore, the theory is renormalisable. The 1-PI Green functions which may be divergent are shown in Fig. 16.7. They are those with $E_\gamma = 2$, $E_e = 0$, the photon self-energy diagrams; $E_\gamma = 0$, $E_e = 2$, the electron self-energy; $E_\gamma = 1$, $E_e = 2$, the vertex diagrams, and, finally

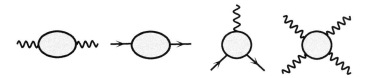

Figure 16.7 *The primitively divergent 1-PI Green's functions of quantum electrodynamics. The last one, the light-by-light scattering, is, in fact, convergent as a consequence of gauge invariance, as we shall see later.*

the $E_\gamma = 4$, $E_e = 0$, the scattering of light by light. We will show shortly that the last one is, in fact, convergent, because of the gauge invariance of quantum electrodynamics. A final remark: the theory is invariant under the operation of charge conjugation which changes the sign of the photon field. It follows that a Green function with no external electron lines and an odd number of photon lines vanishes identically. This property is known as *Furry's theorem*.

We leave as an exercise to the reader to establish the renormalisation properties of other field theories (see, Problem 16.1). In four dimensions of space–time, the result is the following:

1. There exists only one super-renormalisable field theory with interaction of the form ϕ^3.
2. There exist five renormalisable ones:
    2a. ϕ^4.
    2b. Yukawa $\bar\psi\psi\phi$
    2c. Quantum electrodynamics $\bar\psi\gamma_\mu A^\mu\psi$
    2d. Scalar electrodynamics, which contains two terms: $[\phi^\dagger\partial_\mu\phi - (\partial_\mu\phi^\dagger)\phi]A^\mu$ and $A^\mu A_\mu\phi^\dagger\phi$
    2e. Yang–Mills $\mathrm{Tr}\,G_{\mu\nu}G^{\mu\nu}$
3. All other theories are non-renormalisable.

For ϕ^3 the energy will turn out to be unbounded from below, so this theory alone cannot be a fundamental theory for a physical system. A most remarkable fact is that, as we will see later, nature uses *all* renormalisable theories to describe the interactions among elementary particles.

Before closing this section we want to make a remark which is based on ordinary dimensional analysis. In four dimensions a boson field has dimensions of a mass (remember, we are using units such that the speed of light c and Planck's constant h are dimensionless) and a fermion field that of a mass to the power $\frac{3}{2}$. Since all terms in a Lagrangian density must have dimensions equal to 4, we conclude that the coupling constant of a super-renormalisable theory must have the dimensions of a mass, that of a renormalisable theory must be dimensionless, and that of a non-renormalisable theory must have the dimensions of an inverse power of mass. In fact we can rephrase the power-counting argument for the superficial degree of divergence of a 1-PI diagram as an argument based on dimensional analysis. The result will be this connection between the dimensions of the coupling constant and the renormalisation properties of the theory. There is, however, a fine point: for this argument to work we must assume that all boson propagators behave like k^{-2} at large momenta and all fermion ones like k^{-1}. So the argument will fail if this behaviour is not true. The most important example of such failure is a theory containing massive vector fields whose propagator goes like a constant at large k. As a result such theories, although they may have dimensionless coupling constants, are in fact non-renormalisable.

16.4   Regularisation

Since, as we just saw, the perturbation expansion often contains divergent integrals, we must introduce some prescription to deal with them. This means that we must modify the theory somehow in order to improve its short distance behaviour. A first idea could be to modify the rules of quantum field theory and obtain propagators which decrease, at large momenta, faster than the canonical powers we encountered so far. We can show that this is impossible without violating some of the fundamental principles we introduced before. In Problems 16.2 and 16.3 we present two such methods and their shortcomings. Alternatively, we can act on the short distance structure of space–time. The simplest such scheme, both conceptually and physically, is to replace the Euclidean space–time continuum by a regular lattice with spacing a. a gives the shortest possible distance and similarly $\Lambda \sim a^{-1}$ the largest possible momentum. Poincaré invariance is broken to a discrete subgroup, the symmetry group of the lattice. For fixed a there are no ultraviolet divergences since we integrate only up to momenta of order of Λ. At the limit $a \to 0$ we recover the usual continuum theory, including its divergences. We call this procedure *regularisation*. It consists of introducing a new parameter into the theory; here it is the lattice spacing, called *cut-off*, at the price of violating some principle, in this example the symmetries of space–time. For finite values of the cut-off all calculations yield convergent, but cut-off dependent, expressions. The limit when the cut-off disappears will be studied in the next section.

The formulation of quantum field theory on a space–time lattice is interesting in its own right because, on top of providing a cut-off for the calculation of Feynman diagrams, it offers a universal, non-perturbative definition of the theory. The space–time point x is replaced by an index i which takes integer values and labels the lattice site. The fields $\phi(x)$ become ϕ_i.[1] Derivatives of fields are replaced by finite differences between neighbouring points of the form $\phi_i - \phi_{i\pm 1}$ divided by the lattice spacing. The action is

---

[1] Gauge fields $A_\mu(x)$ depend on both the point i and the direction in space given by the index μ. In other words, they depend on the link joining the point i and its neighbour in the direction μ. This is why they are called 'connections'.

now given by a discrete sum over all lattice points and the functional integral reduces to a multiple integral in which, for every lattice site, we integrate all fields from -∞ to +∞. If we introduce also an infrared cut-off by considering only a finite number of lattice sites, we obtain an ordinary finite-dimensional integral which can be evaluated numerically. Every quantity will depend on both, the lattice spacing and the lattice size. If we repeat the calculation by varying these parameters we can, in principle, study the continuum, infinite space limit. The important point is that the result will be independent of the properties of the perturbation expansion, thus providing a way to study field theory in the strong coupling region, where perturbation methods are unreliable. We shall come back to this point in a later chapter.

On the other hand, if our purpose is not to give a non-perturbative definition of field theory but rather to perform computations of Feynman diagrams, we may choose any method that renders these diagrams finite. There exists a plethora of such methods and there is no need to give a complete list. The two examples we presented in Problems 16.2 and 16.3 do that by modifying the Lagrangian density and, hence, the field propagators. Although conceptually simple, they are not very convenient to use.

A different choice is to act directly on the divergent expressions at the level of each individual diagram. A simple prescription consists of expanding the integrand of every diagram in powers of the external momenta. Let us consider the divergent integral of (16.1) and expand it in powers of p.

$$
\begin{aligned}
I &= \int \frac{\mathrm{d}^4 k}{(2\pi)^4} \frac{1}{(k^2+m^2)[(k-p)^2+m^2]} \\
&= \int \frac{\mathrm{d}^4 k}{(2\pi)^4} \left(\frac{1}{(k^2+m^2)^2} + p_\mu \frac{\partial}{\partial p_\mu} \frac{1}{(k^2+m^2)[(k-p)^2+m^2]} \Big|_{p=0} + \dots \right),
\end{aligned}
\tag{16.20}
$$

where the dots stand for higher order terms. It is clear that only the first term is divergent; therefore, we can *define* the finite part of the diagram of Fig. 16.1 as the one we obtain by dropping the first divergent term. This prescription can be generalised to any diagram and yields finite expressions. For more divergent diagrams we may have to drop also higher terms in the expansion, but we shall always be left with convergent integrals. The method looks rather arbitrary; for example, instead of expanding in powers around zero external momentum, we could choose to do so around any other fixed value, provided it is a point of analyticity of the integrand. We shall come back to this arbitrariness in the next section; here, we only note that the terms we drop are polynomials in the external momenta.

For practical calculations it is clear that we must choose a cut-off procedure that renders these computations as simple as possible. In the next section we will argue that physical quantities are independent of the particular choice of the cut-off.

The simplest regularisation scheme turned out to be, by trial and error, a quite counterintuitive one. We start by illustrating it in the simple example of the divergent integral of (16.1). Since we are interested only in the divergent part, we can simplify the discussion by considering the value of I at $p = 0$. We thus obtain

$$I = \int \frac{\mathrm{d}^4 k}{(2\pi)^4} \frac{1}{(k^2 + m^2)^2}. \tag{16.21}$$

Ignoring for the moment the divergence, we note that the integrand depends only on k^2, so we choose spherical coordinates and write $\mathrm{d}^4 k = k^3 dk d\Omega^{(3)}$, where $d\Omega^{(3)}$ is the surface element on the three-dimensional unit sphere. We further note that I would have been convergent if we were working in a Euclidean space–time of three, two, or one dimension. The crucial observation is that in all three cases we can write the result in a compact form,

$$I^{(d)} = \int \frac{\mathrm{d}^d k}{(2\pi)^d} \frac{1}{(k^2 + m^2)^2} = \frac{1}{(4\pi)^{d/2}} \frac{\Gamma(2 - d/2)}{(m^2)^{(2-d/2)}}; \quad d = 1, 2, 3 \tag{16.22}$$

where $\Gamma(z)$ is the well-known special function which generalises for a complex z the concept of the factorial. The important values for (16.22) are given by

$$\Gamma(n) = (n-1)!; \quad \Gamma(n + 1/2) = \frac{(\pi)^{1/2}}{2^n}(2n - 1)!!; \quad n = 1, 2, \dots. \tag{16.23}$$

And now comes the big step. Nothing on the right-hand side of (16.22) forces us to consider this expression only for $d = 1, 2$, or 3. In fact Γ is a meromorphic function in the entire complex plane with poles whenever its argument becomes equal to an integer $n \leq 0$. For the integral $I^{(d)}$, using the identity $n\Gamma(n) = \Gamma(n + 1)$, we see that when $d \to 4$, the Γ function behaves as $\Gamma(2 - d/2) \sim 2/(4 - d)$. So, we can argue that, at least for this integral, we have introduced a regularisation, i.e. a new parameter, namely $\epsilon = 4 - d$, such that the expression is well defined for all values in a region of ϵ and diverges when $\epsilon \to 0$.

Before showing how to generalise this approach to all other integrals we may encounter in the calculation of Feynman diagrams, let us try to make the logic clear by emphasising what this regularisation does not claim to be. First, it does not claim to be the result we would have obtained by quantising the theory in a complex number of dimensions. In fact we do not know how to consistently perform such an operation. In this sense, dimensional regularisation does not offer a non-perturbative definition of the field theory. The prescription applies directly to the integrals obtained order by order in the perturbation expansion. Second, it cannot even be viewed as the analytic continuation to the complex d plane of the results we obtain in performing the integral for $d = 1, 2, 3$. Indeed, the knowledge of the values of a function on a finite number of points on the real axis does not allow for a unique analytic continuation. Instead, the claim is that (16.22), appropriately generalised, offers an unambiguous prescription to obtain a well-defined answer for any Feynman diagram as long as ϵ stays away from 0.

The observation which allows for such a generalisation is that Feynman rules always yield a special class of integrals. In purely bosonic theories, whether renormalisable or not, they are of the form

$$I(p_1, p_2, ...p_n) = \int \prod_i \left(\frac{d^d k_i}{(2\pi)^d} \right) \frac{N(k_1, k_2, ...)}{D(k_1, k_2, ...)} \prod_r \left((2\pi)^d \delta^d(k, p) \right), \tag{16.24}$$

where the k's and the p's are the momenta of the internal and external lines, respectively, the product over i runs over all internal lines, that of r over all vertices, the δ functions denote the energy and momentum conservation on every vertex, and N and D are polynomials of the form

$$N(k_1, k_2, ...) = k_1^{\mu_1} k_1^{\mu_2} ... k_2^{\nu_1} k_2^{\nu_2} ... \tag{16.25}$$

$$D(k_1, k_2, ...) = \prod_i (k_i^2 + m_i^2). \tag{16.26}$$

D is just the product of all propagators and m_i is the mass of the ith line. N appears through derivative couplings and/or the $k^\mu k^\nu$ parts of the propagators of higher spin bosonic fields. It equals 1 in theories with only scalar fields and non-derivative couplings, such as ϕ^4. All scalar products are written in terms of the d-dimensional Euclidean metric $\delta_{\mu\nu}$ which satisfies

$$\delta_\mu^\mu = \text{Tr}\mathbf{1} = d. \tag{16.27}$$

The dimensional regularisation consists in giving a precise expression for $I(p_1, p_2, ...p_n)$ as a function of d which coincides with the usual value whenever the latter exists and is well defined for every value of d in the complex d plane except for those positive integer values for which the original integral is divergent.

At 1 loop the integral (16.24) reduces to

$$I(p_1, p_2, ...p_n) = \int \frac{d^d k}{(2\pi)^d} \frac{N(k)}{D(k, p_1, p_2, ...)} \tag{16.28}$$

with k being the loop momentum. The denominator D is of the form

$$D(k, p_1, p_2, ...) = \prod_i [(k - \Sigma_{(i)} p)^2 + m_i^2], \tag{16.29}$$

where $\Sigma_{(i)} p$ denotes the combination of external momenta which goes through the ith internal line. This product of propagators can be cast in a more convenient form by using a formula first introduced by Feynman

$$\frac{1}{P_1 P_2 ... P_\eta} = (\eta - 1)! \int_0^1 \frac{dz_1 dz_2 ... dz_\eta \delta(1 - \Sigma_i z_i)}{[z_1 P_1 + z_2 P_2 + \cdots + z_\eta P_\eta]^\eta}. \tag{16.30}$$

With the help of (16.30) and an appropriate change of variables, all 1-loop integrals become of the general form

$$\hat{I}(p_1, p_2, ...p_n) = \int \frac{\mathrm{d}^d k}{(2\pi)^d} \frac{k_{\mu_1} k_{\mu_2} ... k_{\mu_l}}{[k^2 + F^2(p, m, z)]^\eta} \tag{16.31}$$

with F being some scalar function of the external momenta, the masses, and the Feynman parameters. F has the dimensions of a mass. $I(p_1, p_2, ...p_n)$ is obtained from $\hat{I}(p_1, p_2, ...p_n)$ after integration with respect to the Feynman parameters z_i of (16.30). For odd values of l $\hat{I}$ vanishes by symmetric integration. For l even, it can be easily computed using spherical coordinates. Some simple cases are as follows:

$$\int \frac{\mathrm{d}^d k}{(2\pi)^d} \frac{1}{[k^2 + F^2(p, m, z)]^\eta} = \frac{1}{(4\pi)^{d/2}} \frac{\Gamma(\eta - d/2)}{\Gamma(\eta)} [F^2]^{(d/2-\eta)} \tag{16.32}$$

$$\int \frac{\mathrm{d}^d k}{(2\pi)^d} \frac{k_\mu k_\nu}{[k^2 + F^2(p, m, z)]^\eta} = \frac{1}{(4\pi)^{d/2}} \frac{\delta_{\mu\nu}}{2} \frac{\Gamma(\eta - 1 - d/2)}{\Gamma(\eta)} [F^2]^{(d/2+1-\eta)}. \tag{16.33}$$

At the end we are interested in the limit $d \to 4$. The first integral (16.32) diverges for $\eta \leq 2$ and the second (16.33) for $\eta \leq 3$. For $\eta = 2$ and $d = 4$ (16.32) is logarithmically divergent and our regularised expression is regular for $\mathrm{Re}\, d < 4$ and presents a simple pole $\sim 1/(d-4)$. For $\eta = 1$ it is quadratically divergent but our expression has still a simple pole at $d = 4$. The difference is that now the first pole from the left is at $d = 2$. We arrive at the same conclusions looking at the integral of (16.33): By dimensionally regularising a 1-loop integral corresponding to a Feynman diagram which, by power counting, diverges as Λ^{2n}, we obtain a meromorphic function of d with simple poles starting at $d = 4 - 2n$. By convention, $n = 0$ denotes a logarithmic divergence.

The generalisation to multiloop diagrams is straightforward: starting from (16.24) and using the Feynman formula (16.30) we write, in a compact notation, a general multi-loop diagram as

$$I(p_1, p_2, ...p_n) = \int \prod_l \left(\frac{\mathrm{d}^d k_l}{(2\pi)^d} \right) \int \mathrm{d}z \frac{N(k_1, k_2, ...)}{[D_2(k, z) + D_1(k, p, z) + D_0(p, z, m)]^\eta}, \tag{16.34}$$

where, as before, η is the total number of internal lines, k_l ($l = 1, 2, ..L$) are the independent loop momenta, the integral over z denotes the multiple integral over the Feynman parameters, and the denominator is a quadratic form in the loop momenta which we have split according to the degree in k. In a vector notation in which $\mathbf{k} = (k_1, k_2, ..., k_L)$ we write

$$D_2 = (\mathbf{k}, \mathbf{Ak}) \tag{16.35}$$

$$D_1 = 2(\mathbf{k} \cdot \mathbf{q}) \tag{16.36}$$

with $\mathbf{A}$ an $L \times L$ z-dependent matrix and $\mathbf{q}$ L linear combinations of the external momenta p with z-dependent coefficients. D_0 is the k-independent part of the denominator.

Diagonalising $\mathbf{A}$, rescaling the momenta and shifting the integration variables we can write (16.34) as

$$I(p_1, p_2, \ldots p_n) = \int \prod_l \left(\frac{\mathrm{d}^d K_l}{(2\pi)^d} \right) \int \mathrm{d}z (\det \mathbf{A})^{-\frac{d}{2}} \frac{N(k_1, k_2, \ldots)}{[\mathbf{K}^2 + (\mathbf{q}, \mathbf{A}^{-1} \mathbf{q}) + D_0]^{\eta}}. \qquad (16.37)$$

Here the L momentum integrations have been separated and can be performed one after the other using the formulae (16.32), (16.33), or, appropriate generalisations. However, this is not the end of the story. We are still left with the integration over the Feynman parameters. There exists no closed formula for the general multiloop integral. The z integration may be singular because $\det \mathbf{A}$ often vanishes at the edge of the integration region. A useful method is to write $(\det \mathbf{A})^{\alpha} \sim (\alpha + 1)^{-1} \partial (\det \mathbf{A})^{\alpha+1}$, where ∂ means the partial derivative with respect to one of the z's. By repeated partial integrations we can isolate and calculate explicitly all pieces which diverge when d goes to four. The finite pieces often require numerical integrations.

Although formula (16.37) is general, it is not always the most convenient. The reason is that it treats all internal lines equally and does not distinguish possible divergent subdiagrams. We will come back to this point in the next section.

So far we have considered only bosonic theories. The introduction of fermions presents a new feature, namely the presence of γ-matrices. In d Euclidean dimensions we write

$$\{\gamma_\mu, \gamma_\nu\} = -2\delta_{\mu\nu} \qquad (16.38)$$

with the Euclidean metric $\delta_{\mu\nu}$ satisfying (16.27). All d-dimensional properties of the Clifford algebra follow from (16.38). For example, we have

$$\mathrm{Tr}\gamma_\mu\gamma_\nu = \mathrm{Tr}\gamma_\nu\gamma_\mu = -d\delta_{\mu\nu}$$
$$\mathrm{Tr}(\gamma_{\mu_1}\gamma_{\mu_2}\ldots\gamma_{\mu_{2k+1}}) = 0 \qquad (16.39)$$
$$\gamma_\mu\gamma_\mu = -d\mathbb{1}.$$

Similar expressions can be written for all products of γ-matrices or traces of such products. The only exception is γ_5 which, in four dimensions, is defined as $\gamma_5 = (1/4!)\epsilon_{\mu\nu\rho\sigma}\gamma_\mu\gamma_\nu\gamma_\rho\gamma_\sigma$. The ϵ symbol with four indices can only be defined in four dimensions and there is no appropriate generalisation. γ_5 can only appear in a Feynman diagram if it is present in the interaction, so we expect that for such theories, diagrams with an odd number of γ_5 vertices will not be dimensionally regularised. This sounds like a technical remark but we shall see in a later section that it is in fact the sign of a far-reaching property of quantum field theory.

16.5 Renormalisation

16.5.1 1-Loop Diagrams

In the last section we saw that by introducing an appropriate regularisation scheme, we obtain well-defined expressions for any Feynman diagram which, however, often depend on a new parameter, the *cut-off*. In the dimensional regularisation it is given by $\epsilon = 4 - d$.

The limit $\epsilon \to 0$ cannot be taken directly without encountering the divergences of the original theory. On the other hand, it is precisely this limit which is physically interesting. In this section we want to address this question, namely, under which circumstances a meaningful four-dimensional theory can be recovered from the regularised ϵ-dependent expressions? As we could have anticipated, the answer will turn out to be that this is only possible for the renormalisable (and super-renormalisable) theories we introduced earlier. The procedure to do so is called *renormalisation*. In this section we will present it for some simple examples.

Let us start with the simplest four-dimensional renormalisable theory given by our already familiar Lagrangian density:

$$\mathcal{L} = \frac{1}{2}\partial_\mu \phi(x)\partial^\mu \phi(x) - \frac{1}{2}m^2\phi^2(x) - \frac{1}{4!}\lambda\phi^4(x). \tag{16.40}$$

In $d = 4$ the field ϕ has the dimensions of a mass and the coupling constant λ is dimensionless. Since we intend to use dimensional regularisation, we introduce a mass parameter μ and write the coefficient of the interaction term $\lambda \to \mu^\epsilon \lambda$, so that the coupling constant λ remains dimensionless at all values of ϵ. We shall present the renormalisation programme for this theory at the lowest non-trivial order, that which includes all diagrams up to and including those with one closed loop.

The power-counting argument presented in section 16.3 shows that, at 1 loop, the only divergent 1-PI diagrams are those of Fig. 16.8.

The 2-point diagram is quadratically divergent and the 4-point diagram logarithmically. We could prevent the appearance of the first diagram by normal ordering the ϕ^4 term in the interaction Lagrangian, but we prefer not to do so. Normal ordering is just a particular prescription to avoid certain divergences but it is not the most convenient one. First, it is not general; for example, it will not prevent the appearance of the 4-point diagram of Fig. 16.8, or even the 2-point one at higher orders and, second, as we will see in a later section, its use may complicate the discussion of possible symmetries

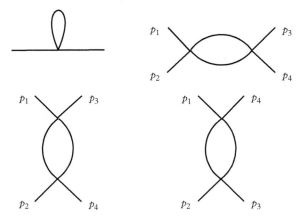

Figure 16.8 *The 1-loop primitively divergent diagrams of the ϕ^4 theory.*

of $\mathcal{L}$. We choose to work entirely with dimensional regularisation and we obtain for these diagrams in Minkowski space–time, using (16.32) at the limit $d \to 4$,

$$\Gamma_1^{(2)} = \frac{\lambda \mu^\epsilon}{2} \int \frac{d^d k}{(2\pi)^d} \frac{1}{k^2 - m^2} = \frac{i\lambda m^2}{16\pi^2} \frac{1}{\epsilon} \tag{16.41}$$

$$\Gamma_1^{(4)}(p_1, .., p_4) = \frac{1}{2} \lambda^2 \mu^{2\epsilon} \int \frac{d^d k}{(2\pi)^d} \frac{1}{(k^2 - m^2)[(k-P)^2 - m^2]} + \text{crossed}$$

$$= \frac{1}{2} \lambda^2 \mu^{2\epsilon} \int_0^1 dz \int \frac{d^d k}{(2\pi)^d} \frac{1}{[k^2 - m^2 + P^2 z(1-z)]^2} + \text{crossed} \tag{16.42}$$

$$= \frac{3i\lambda^2}{16\pi^2} \frac{1}{\epsilon} + \text{finite terms},$$

where $P = p_1 + p_2$, 'crossed' stands for the contribution of the two crossed diagrams in Fig. 16.8 and 'finite terms' represent the contributions which are regular when $d = 4$. Some remarks follow.

1. The divergent contributions are constants, independent of the external momenta. We will see shortly, in the example of quantum electrodynamics, that this is a particular feature of the ϕ^4 theory. In fact, even for ϕ^4, it is no more true when higher loops are considered. For example, the 2-loop diagram of Fig. 16.9 contributes a divergent term proportional to p^2. However, we can prove the following general property. All divergent terms are proportional to monomials in the external momenta. We have already introduced this result in the last section. For 1-loop diagrams the proof is straightforward. We start from the general expression (16.31) and note that we can expand the integrand in powers of the external momenta p taken around some fixed point. Every term in this expansion increases the value of η, so, after a finite number of terms, the integral becomes convergent. We will use this argument later on. It takes some more work to generalise the proof to multi-loop diagrams, but it can be done.

2. The dependence of the divergent terms on m^2 could be guessed from dimensional analysis. This is one of the attractive features of dimensional regularisation.

3. The finite terms in (16.42) depend on the parameter μ. The Laurent expansion in ϵ brings terms of the form $\ln\{[m^2 - P^2 z(1-z)]/\mu^2\}$.

The particular form of the divergent terms suggests the prescription to remove them. Let us start with the 2-point function. In the loop expansion we write

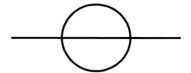

Figure 16.9 *A 2-loop diagram for the 2-point function with a divergent term proportional to p^2.*

$$\Gamma^{(2)}(p^2) = \sum_{l=0}^{\infty} \Gamma_l^{(2)}(p^2) = \Gamma_0^{(2)}(p^2) + \Gamma_1^{(2)}(p^2) + ...,\tag{16.43}$$

where the index l denotes the contribution of the diagrams with l loops. In the tree approximation we have

$$\Gamma_0^{(2)}(p^2) = -\mathrm{i}(p^2 - m^2).\tag{16.44}$$

The 1-loop diagram adds the term given by (16.41). Since it is a constant, it can be interpreted as a correction to the value of the mass in (16.44). Therefore, we can introduce a *renormalised* mass m_R^2, which is a function of m, λ, and ϵ. Of course, this function can only be computed as a formal power series in λ. Up to and including 1-loop diagrams we write

$$m_R^2(m, \lambda, \epsilon) = m^2 \left(1 + \frac{\lambda}{16\pi^2} \frac{1}{\epsilon} \right) + O(\lambda^2).\tag{16.45}$$

A formal power series whose zero-order term is non-vanishing is invertible in terms of another formal power series. So, we can write m as a function of m_R, λ, and ϵ,

$$m^2(m_R, \lambda, \epsilon) = m_R^2 \left(1 - \frac{\lambda}{16\pi^2} \frac{1}{\epsilon} \right) + O(\lambda^2) \equiv m_R^2 Z_m + O(\lambda^2),\tag{16.46}$$

where we have defined the function $Z_m(\lambda, \epsilon)$ as a formal power series in λ with ϵ-dependent coefficients.

The parameter m is often called *bare* mass. Replacing in the Lagrangian (16.40) the bare mass m with the help of (16.46) results in (i) changing in the Feynman rules m by m_R and (ii) introducing a new term in $\mathcal{L}$ of the form

$$\delta\mathcal{L}_m = m_R^2 \frac{\lambda}{32\pi^2} \frac{1}{\epsilon} \phi^2(x).\tag{16.47}$$

Since $\delta\mathcal{L}_m$ is proportional to the coupling constant λ, we can view it as a new vertex in the perturbation expansion which, to first order, gives the diagram of Fig. 16.10. In this case the complete 2-point function to first order in λ is given by

$$\begin{aligned}
\Gamma^{(2)}(p^2) &= -\mathrm{i}(p^2 - m_R^2) + \frac{\mathrm{i}\lambda m_R^2}{16\pi^2} \frac{1}{\epsilon} - \frac{\mathrm{i}\lambda m_R^2}{16\pi^2} \frac{1}{\epsilon} + O(\lambda^2) \\
&= -\mathrm{i}(p^2 - m_R^2) + O(\lambda^2),
\end{aligned}\tag{16.48}$$

Figure 16.10 *The new diagram resulting from $\delta\mathcal{L}_m$ of Eq. (16.47).*

which means that if we keep fixed m_R and λ instead of m and λ, we can take the limit $d \to 4$ and find no divergences up to and including 1-loop diagrams for the 2-point function.

Let us now consider the 4-point function. In the same spirit we write

$$\Gamma^{(4)}(p_1, .., p_4) = \sum_{l=0}^{\infty} \Gamma_l^{(4)}(p_1, .., p_4) = \Gamma_0^{(4)}(p_1, .., p_4) + \Gamma_1^{(4)}(p_1, .., p_4) + ... \,. \quad (16.49)$$

In the tree approximation $\Gamma_0^{(4)}(p_1, .., p_4) = -i\lambda$. Including the 1-loop diagrams we obtain

$$\Gamma^{(4)}(p_1, .., p_4) = -i\lambda \left(1 - \frac{3\lambda}{16\pi^2} \frac{1}{\epsilon} + \text{finite terms} \right) + O(\lambda^3). \quad (16.50)$$

The procedure is now clear. We change from the *bare* coupling constant λ to the *renormalised* one λ_R by writing

$$\lambda_R(\lambda, \epsilon) = \lambda \left(1 - \frac{3\lambda}{16\pi^2} \frac{1}{\epsilon} + O(\lambda^2) \right), \quad (16.51)$$

or, equivalently,

$$\lambda(\lambda_R, \epsilon) = \lambda_R \left(1 + \frac{3\lambda_R}{16\pi^2} \frac{1}{\epsilon} + O(\lambda_R^2) \right) \equiv \lambda_R Z_\lambda. \quad (16.52)$$

Again, replacing λ with λ_R in $\mathcal{L}$ produces a new 4-point vertex which cancels the divergent part of the 1-loop diagrams of Fig. 16.8. Let us also note that we can replace λ with λ_R in (16.46) since the difference will appear only at higher order.

Until now we have succeeded in building a new, *renormalised* Lagrangian and the resulting theory is free from divergences up to and including 1-loop diagrams. It involves two new terms which change the coefficients of the ϕ^2 and ϕ^4 terms of the original Lagrangian. These terms are usually called *counter-terms*. Before looking at higher orders, let us see the price we had to pay for this achievement. It can be better seen at the 4-point function. Looking back at the expression (16.42) we make the following two observations. First, as we noted already, the finite part seems to depend on a new arbitrary parameter with the dimensions of a mass μ. Second, the definition of Z_λ in (16.52) seems also arbitrary. We could add to it any term of the form $C\lambda_R$ with C any arbitrary constant independent of ϵ. Such an addition would change the value of the coupling constant at the 1-loop order. The two observations are not unrelated. Indeed, changing the parameter μ from μ_1 to μ_2 in (16.42) adds a constant term proportional to $\lambda \ln(\mu_1/\mu_2)$ which, as we just saw, can be absorbed in a redefinition of Z_λ and, thus, of the value of the coupling constant. In a later section we will study the μ-dependence of the theory more deeply, but for the moment we note that at 1 loop, all this renormalisation procedure for the 4-point function amounts to choosing the value of the coupling constant. Since the latter was always considered as an arbitrary parameter of the theory, this does not seem to be a high price. We shall postpone a more elaborate discussion of this point to a later section.

Let us summarise the discussion so far. At the 1-loop level all arbitrariness of the renormalisation programme consists in assigning prescribed values to two parameters of the theory, which can be chosen to be the mass and the coupling constant. A convenient choice is given by two conditions of the form.

$$\Gamma^{(2)}(p^2 = m_R^2) = 0 \tag{16.53}$$

$$\Gamma^{(4)}(p_1, .., p_4)|_{\text{point } M} = i\lambda_R^{(M)}. \tag{16.54}$$

The first one, Eq. (16.53), defines the physical mass as the pole of the complete propagator. Although this choice is the most natural for physics, from a purely technical point of view, we could use any condition assigning a prescribed value to $\Gamma^{(2)}(p^2)$ at a fixed point $p^2 = M^2$, provided it is a point in which $\Gamma^{(2)}(p^2)$ is regular. Similarly, in the second condition Eq. (16.54), by 'point M' we mean some point in the space of the four-momentum p_i, $i = 1, .., 4$, provided it is a point in which $\Gamma^{(4)}$ is regular. For a massive theory the point $p_i = 0$ is an example. Once these conditions are imposed all Green functions at 1 loop are well defined and calculable.

A final remark: To be precise, we must include in our list of counter-terms an additive constant in the Lagrangian which ensures that the vacuum state has zero energy. It was discussed in Chapter 8 as a normal ordering prescription, but it is not different to any of the counter-terms we encountered here. Like any of the Z's introduced so far, it has an expansion in perturbation theory and must be determined order by order. It corresponds to the divergences of diagrams with no external lines, but since we rarely need to compute such diagrams, this counter-term is traditionally left out of the renormalisation programme.

As a second example, we will present the renormalisation for the 1-loop diagrams of quantum electrodynamics. The primitively divergent diagrams are shown in Fig. 16.7. We start with the first one, the photon self-energy diagram given by

$$\Gamma_{\mu\nu}^{(2,0)}(q) = -e^2 \mu^\epsilon \int \frac{d^d k}{(2\pi)^d} \frac{\text{Tr}[\gamma_\mu (\not{k} + m)\gamma_\nu (\not{k} - \not{q} + m)]}{(k^2 - m^2)[(k - q)^2 - m^2)]}, \tag{16.55}$$

which, by power counting, is quadratically divergent. Using Feynman's formula (16.30) and the integrals (16.32) and (16.33), we obtain

$$\begin{aligned}
\Gamma_{\mu\nu}^{(2,0)}(q) &= -ie^2 \mu^\epsilon \frac{2d}{(4\pi)^{d/2}} (q_\mu q_\nu - q^2 g_{\mu\nu}) \Gamma(2 - d/2) \int_0^1 dz z(z - 1)(F^2)^{\frac{d}{2}-2} \\
&\Rightarrow \frac{2i\alpha}{3\pi} \frac{1}{\epsilon} (q_\mu q_\nu - q^2 g_{\mu\nu}) + ...,
\end{aligned} \tag{16.56}$$

where $F^2 = m^2 + q^2 z(1 - z)$, $\alpha = e^2/4\pi$ is the fine structure constant and the dots stand for finite terms.

The surprise in (16.56) is that the pole at $d = 2$ which signals the quadratic divergence has disappeared. The $1/\epsilon$ part of $\Gamma_{\mu\nu}^{(2,0)}(q)$ is proportional to $q_\mu q_\nu - q^2 g_{\mu\nu}$ and satisfies $q^\mu \Gamma_{\mu\nu}^{(2,0)}(q) = 0$. This is a welcome result. Indeed, were $\Gamma^{(2,0)}$ quadratically divergent, the corresponding coefficient, by dimensional analysis, would have been a constant independent of the momentum q. In order to absorb the divergence we would have to introduce a counter-term in the Lagrangian of quantum electrodynamics proportional to $A_\mu A^\mu$, i.e. a photon mass term. Such a term is absent in the original Lagrangian because it is forbidden by gauge invariance. We see in this example the advantage of using a cut-off procedure, such as dimensional regularisation, which respects the invariance properties of the theory. If we had used a more direct cut-off, for example cutting all momentum integrals at a scale Λ, we would have been obliged to introduce such a counter-term with a new constant. Such a term would have been a nuisance rather than a catastrophe. We could always recover the usual theory by imposing a renormalisation condition implying the vanishing of the renormalised value of the photon mass. However, for all intermediate calculations, we would have to keep the corresponding counter-term.

The tensor structure of the divergence in (16.56) implies that we need a counter-term for the photon kinetic energy part of the QED Lagrangian. In the same spirit as in (16.66), we write

$$A^\mu(x) = Z_3^{1/2} A_R^\mu(x) = \left(1 - \frac{\alpha}{3\pi}\frac{1}{\epsilon} + O(\alpha^2)\right) A_R^\mu(x). \tag{16.57}$$

Similarly, we compute the electron self-energy diagram and obtain

$$\Gamma^{(0,2)}(p) = -e^2 \mu^\epsilon \int \frac{d^d k}{(2\pi)^d} \frac{\gamma_\mu(\not{p} - \not{k} + m)\gamma^\mu}{k^2[(p-k)^2 - m^2]}$$
$$\Rightarrow \frac{i\alpha}{2\pi}\frac{1}{\epsilon}\not{p} - \frac{2i\alpha}{\pi}\frac{1}{\epsilon}m + ..., \tag{16.58}$$

where we have suppressed spinor indices and, again, the dots stand for finite terms. We now introduce two new counter-terms, one for the electron kinetic energy and another for its mass:

$$\psi(x) = Z_2^{1/2}\psi_R(x) = \left(1 - \frac{\alpha}{4\pi}\frac{1}{\epsilon} + O(\alpha^2)\right)\psi_R(x) \tag{16.59}$$

$$m = Z_m m_R = \left(1 - \frac{2\alpha}{\pi}\frac{1}{\epsilon} + O(\alpha^2)\right)m_R. \tag{16.60}$$

It is easy to check that replacing the $\bar{\psi}(i\not\partial - m)\psi$ part of the Lagrangian density with $Z_2\bar{\psi}_R(i\not\partial - Z_m m_R)\psi_R$, produces precisely the counter-term required to cancel the two divergent contributions of the electron self-energy.

The vertex diagram of Fig. 16.7 gives

$$\Gamma_\mu^{(1,2)}(p,p') = -e^3 \mu^{3\epsilon/2} \int \frac{d^d k}{(2\pi)^d} \frac{\gamma_\nu (\not{p} - \not{k} + m)\gamma_\mu (\not{p} - \not{k} + m)\gamma^\nu}{k^2[(p-k)^2 - m^2][(p'-k)^2 - m^2]}$$
$$\Rightarrow \frac{i\alpha}{2\pi}\frac{1}{\epsilon}e\gamma_\mu + \dots . \tag{16.61}$$

The complete computation of the finite terms is rather lengthy and will not be presented here. However, some parts of this diagram have important physical interpretation and are shown in Problem 16.4. For the renormalisation programme only the divergent part is needed. Since the tree approximation vertex is just $-ie\gamma_\mu$, it follows that the complete vertex, including the 1-loop correction, is given by

$$\Gamma_\mu^{(1,2)}(p,p') = -ieZ_1\gamma_\mu + \dots = -ie\gamma_\mu \left(1 - \frac{\alpha}{2\pi}\frac{1}{\epsilon}\right) + \dots, \tag{16.62}$$

where the dots stand again for finite contributions. In order to cancel this divergence we introduce a coupling constant renormalisation $e = Z_e e_R$ with $Z_e = 1 + O(\alpha)$. Putting all counter-terms together, the interaction Lagrangian becomes

$$-e\bar\psi\gamma_\mu\psi A^\mu = -Z_e Z_2 Z_3^{1/2} e_R \bar\psi_R \gamma_\mu \psi_R A_R^\mu. \tag{16.63}$$

It follows that the condition which determines the charge renormalisation constant Z_e is

$$Z_e Z_2 Z_3 = Z_1. \tag{16.64}$$

By comparing (16.62) and (16.59), we see that at least at this order, $Z_1 = Z_2$; therefore, the entire charge renormalisation is determined by the photon self-energy diagram. In the next section we will show that this property is valid for all orders of perturbation theory and is a consequence of gauge invariance.

The last diagram of Fig. 16.7 represents the scattering of light by light. By power counting, it is logarithmically divergent, so, in order to compute the coefficient of the $1/\epsilon$ part, it is sufficient to set all external momenta $q_i = 0$ and keep the highest power of the loop momentum in the numerator. We thus obtain

$$\Gamma_{\mu\nu\rho\sigma}^{(4,0)}(q_1,q_2,q_3,q_4) = -e^4 \int \frac{d^d k}{(2\pi)^d} \frac{\mathrm{Tr}(\gamma_\mu \not{k}\gamma_\nu \not{k}\gamma_\rho \not{k}\gamma_\sigma \not{k})}{(k^2+m^2)^4} + \dots, \tag{16.65}$$

where the dots stand for permutations in the indices and finite terms. We leave the explicit calculation as an exercise (see Problem 16.4), but it will not come as a surprise to the reader the result that $\Gamma^{(4,0)}$ is, in fact, convergent. The residue of the $1/\epsilon$ pole vanishes. The reason is the same which guaranteed the absence of the quadratically divergent part of the photon self-energy diagram. Indeed, were $\Gamma^{(4,0)}$ divergent, we would have to introduce a counter-term in the Lagrangian of quantum electrodynamics proportional to $(A^\mu A_\mu)^2$ which is forbidden by gauge invariance. Therefore, in a regularisation scheme such as dimensional regularisation, this diagram must be convergent. The explicit calculation shows that $\Gamma^{(4,0)}_{\mu\nu\rho\sigma}(q_1, q_2, q_3, q_4)$ satisfies the same transversality condition we found for $\Gamma^{(2,0)}_{\mu\nu}(q)$, namely $q_1^\mu \Gamma^{(4,0)}_{\mu\nu\rho\sigma}(q_1, q_2, q_3, q_4) = q_2^\nu \Gamma^{(4,0)}_{\mu\nu\rho\sigma}(q_1, q_2, q_3, q_4) = \ldots = 0$.

We conclude that the 1-loop renormalisation properties of quantum electrodynamics resemble those of ϕ^4. All divergences can be absorbed in four counter-terms: the electron wave function and mass counter-terms, the photon wave function counter-term, and the vertex counter-term. Furthermore, we found a relation among the electron wave function and the vertex counter-terms, so that the entire electric charge renormalisation is due to the photon wave function counter-term. This property results from the gauge invariance of the theory and has an important physical consequence: if we consider two charged fields and we assume that they both have the same unrenormalised electric charge, this property guarantees that the renormalised values of the charge will remain the same, irrespectively of whichever other interactions these fields may have. On the other hand, a very well established empirical fact is that all particles in nature have electric charges which are integer multiplets of an elementary charge. Strictly speaking, the gauge invariance of quantum electrodynamics does not explain this fact because it does not prevent us from using different unrenormalised charges. However, it does solve part of the problem because it guarantees their equality after renormalisation if, for some reason, the unrenormalised ones are chosen equal. We will develop this point further in a later chapter.

As we did for the $\lambda\phi^4$ theory, we note that all arbitrariness concerning the finite parts of the counter-terms can be determined by imposing four renormalisation conditions. One of them can be used to determine the physical mass of the electron and a second the value of the coupling constant, i.e. the electric charge. In contrast to ϕ^4, there exists a 'physical' value of α, the one which corresponds to the classically measured electric charge. It is the condition in which the electron is on its mass shell and the photon carries zero momentum. We shall have a closer look at the renormalisation of quantum electrodynamics soon, so we will not elaborate on these points here.

16.5.2 Some 2-Loop Examples

Let us have a quick look at 2 loops. Some examples are presented in Fig. 16.11. They include genuinely 2-loop diagrams, but also ones we generate by inserting the new vertices we introduced at the previous order to the old 1-loop diagrams. For example, the diagram of Fig. 16.11 (c) is, by power counting, logarithmically divergent. However, a closer look reveals that the divergence is quadratic but it is entirely due to the insertion of the 1-loop subdiagram. If we add to it the diagram of Fig 16.11 (d) with the counter-term (16.47), which is on the same order in λ_R, the divergence disappears. In fact, if the counter-term is just the $1/\epsilon$ part we computed in (16.47), the entire contribution vanishes. This gives us an example of a superficially divergent diagram which is not

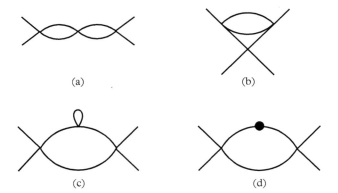

Figure 16.11 *Examples of 2-loop divergent diagrams in the* ϕ^4 *theory.*

primitively divergent and requires no new counter-term. We see, in this example, the importance of treating sub-diagrams first.

The other diagrams bring new terms proportional to $1/\epsilon$ and $1/\epsilon^2$. For most of them the extension of the renormalisation programme presented previously is obvious. The new terms can again be absorbed in a redefinition of the constants Z_m and Z_λ which acquire new terms proportional to λ_R^2. The only qualitatively new feature comes from the diagram of Fig. 16.9. The explicit calculation of the divergent contributions is left as an exercise (see Problem 16.6). It is instructive because it offers a non-trivial example of a multi-loop diagram. It yields divergent contributions which are constant, are independent of the external momentum p, and can be absorbed in Z_m, and a new one, which, as we noted previously, contributes an $1/\epsilon$ divergent term proportional to $ip^2 C\lambda_R^2$ with C a numerical constant. Such a term cannot be absorbed in a redefinition of either Z_m or Z_λ and requires a new renormalisation of the 2-point function. In the same spirit we change now the coefficient of the kinetic energy term in the Lagrangian. It is convenient to define a renormalised field by

$$\phi(x) = Z_\phi^{1/2}\phi_R(x) = \left(1 + \frac{C\lambda_R^2}{2}\frac{1}{\epsilon} + O(\lambda_R^3)\right)\phi_R(x), \tag{16.66}$$

which introduces a kinetic energy counter-term of the form

$$\delta\mathcal{L}_{k.e.} = \frac{1}{2}C\lambda_R^2\partial_\mu\phi_R(x)\partial^\mu\phi_R(x). \tag{16.67}$$

This counter-term is often called *wave function renormalisation*. The presence of Z_ϕ requires a new renormalisation condition in addition to those of (16.53) and (16.54). It can be expressed as a condition on the first derivative of the 2-point function with respect to p^2,

$$\frac{\partial \Gamma^2(p^2)}{\partial p^2}\Big|_{p^2=M^2} = -\mathrm{i},\qquad\qquad (16.68)$$

where, again, M^2 is some conveniently chosen value of p^2. Note that the values of M appearing in (16.54) and (16.68) need not be the same.

16.5.3   All Orders

The philosophy remains the same at higher loops but, as the number of diagrams increases very fast, the correct bookkeeping becomes essential. At each order we can distinguish diagrams which diverge because of insertions of divergent subdiagrams, in which case they require no special treatment, from those which are primitively divergent and introduce a new counter-term. One of the complications comes from diagrams, like that of Fig. 16.12, which can be viewed as containing divergent subdiagrams in more than one way. Making sure that all these *overlapping divergences* are included with the correct combinatoric weight requires some care.

An important ingredient is the concept of a *forest*, first introduced by W. Zimmermann. If g_1 and g_2 are two subdiagrams of a given diagram G, we will say that g_1 and g_2 are *disjoint* if their intersection is empty and *non-overlapping* if they are either disjoint, or if one is a sub-diagram of the other. A forest U of G is a set of non-overlapping 1-PI subdiagrams of G, including the extreme cases of the empty one and G itself.

For a diagram g_i which has no divergent sub-diagrams, we introduce the operation $FP(g_i)$, for *finite part* as follows: if g_i is convergent, $FP(g_i) = g_i$ If it is divergent, write g_i using the dimensional regularisation formula (16.37). Compute the Laurent series in ϵ and drop the terms which are singular for $\epsilon \to 0$. This operation is straightforward if g_i has no divergent sub-diagrams. Note that the so computed $FP(g_i)$ is precisely the result of including the necessary counter-terms to remove the divergent pieces of g_i. For a diagram G whose all divergent sub-diagrams are non-overlapping the generalisation is obvious: $FP(G)$ is obtained by starting from the innermost divergent sub-diagram and working outwards. Zimmermann's forest formula generalises this procedure for a diagram G which has overlapping divergent sub-diagrams. It reads

$$FP(G) = \sum_U \prod_{g \in U} FP(g).\qquad\qquad (16.69)$$

The sum runs over all forests and the product over all sub-diagrams in each forest. Zimmermann proved this formula using his subtraction scheme which yields the finite

Figure 16.12 *A 2-loop diagram of quantum electrodynamics with overlapping divergences.*

part of a diagram by expanding the integrand in powers of the external momenta and dropping as many terms as necessary to make the integral convergent. It is easy to adjust the proof for the finite part obtained using the dimensional regularisation scheme.

We will not present here the complete proof to all orders of perturbation. It can be found in a rigorous way in several specialised references.[2] We give only the result: all divergences, at any finite order of perturbation, for the ϕ^4 quantum field theory can be absorbed by introducing counter-terms in the original Lagrangian. There are three such terms, Z_m, Z_λ, and Z_ϕ, which are determined as formal power series in the coupling constant. All ambiguities in their determination are fixed by three renormalisation conditions of the form (16.53), (16.54), and (16.68). Once these conditions are imposed, all Green functions, at any order of perturbation, are finite and calculable. Among the various choices for these renormalisation conditions there is one which we will call *physical*. It consists of choosing the value of the physical mass as the pole of the propagator in (16.53) and the same value $M = m_R$ for the wave function renormalisation condition (16.68). It is clear that it is this choice that yields the S-matrix elements through the reduction formula we presented in Chapter 12. There is no corresponding 'physical' choice for Z_λ because the coupling constant is not a directly measurable quantity in ϕ^4 quantum field theory. We often use the value of the on-shell elastic scattering amplitude at threshold, i.e. $p_i^2 = m_R^2$, $i = 1, .., 4$, and $s = (p_1 + p_2)^2 = 4m_R^2$, but there is nothing wrong with using any other point to define (16.54).

It must be clear by now how to apply this programme to any quantum field theory. At every order of perturbation we will need a counter-term for every new divergence, one which is not simply due to a divergent sub-diagram. Using the classification we introduced in section 16.3, we see that we have three distinct cases.

For a super-renormalisable theory, this programme terminates at some finite order of perturbation. All divergent contributions to the counter-terms are explicitly known. In a four-dimensional space–time the only such theory is ϕ^3. For a renormalisable theory, such as ϕ^4 or QED, we have a finite number of counter-terms, but each one is given by an infinite series in the coupling constant. Finally, for a non-renormalisable theory, we must introduce new counter-terms at every order. Eventually, every Green function will be primitively divergent. Since for every counter-term we need a renormalisation condition to determine the corresponding arbitrary constant, these theories have no predictive power.

This last point may sound like a technicality, something related to our inability to handle ill-defined expressions. We want to argue here that this is not the case; non-renormalisable theories cannot provide for a fundamental theory and the appearance of divergences growing without limit is a sign of a deep lying difficulty. In Problem 16.6 we consider a famous example, the Fermi current $\times$ current theory. Following K. Wilson, we argue that all field theories, whether renormalisable or not, should be viewed as effective theories describing physics up to a given scale. No theory can claim accuracy

---

[2] See, for example, John H. Lowenstein, 'BPHZ Renormalization', 1975 Erice Conference in Mathematical Physics, NYU-TR11-75; O. Piguet and S. P. Sorella, '*Algebraic Renormalization*', Lecture Notes in Physics, Monographs (Springer, 1995).

to all scales, since we cannot possibly know the physics at arbitrarily high energies. In this view, a cut-off Λ denotes the scale up to which the theory can be trusted. For a typical non-renormalisable theory, such as the Fermi theory, see Problem 16.6, the effective coupling constant is not $G/\sqrt{2}$, the one that appears in the Lagrangian, but the dimensionless quantity $G\Lambda^2$. Since experimentally $G/\sqrt{2} = 10^{-5}$ GeV^{-2}, it follows that for perturbation theory to make sense, Λ must be smaller than $O(10^2)$GeV. In fact, as we show in Problem 16.7, this estimate can be considerably lowered. Precision measurements can bring it down to $O(1)\,GeV$. It was this argument that forced theorists to look beyond Fermi theory and led to the construction of the standard model, which we will study in a later chapter. This argument can be generalised. In Problem 16.7 we study a theorem, known as 'decoupling theorem'. For a field theory to make physical sense particles with arbitrarily high masses should decouple and have no effect in low-energy dynamics. However, as we show in Problem 16.7, this is only true for renormalisable theories. It is this lack of the decoupling property that makes non-renormalisable theories unacceptable, since any prediction depends crucially on necessarily unknown physics at unaccessible energies.

A final point concerning renormalisable theories. Let us consider the example of the Yukawa interaction between a spin-$\frac{1}{2}$ field $\psi(x)$ and a scalar $\phi(x)$ given by $\mathcal{L}_Y \sim \bar{\psi}\psi\phi$. It is a renormalisable theory and we can determine, order by order, the necessary counterterms. At 1 loop we find that the four-ϕ diagram of Fig. 16.13 is divergent and requires a counter-term of the form $\delta\mathcal{L} \sim \phi^4$, although such a term was not present in the Lagrangian we started from. This term will not spoil the renormalisability of the theory, since ϕ^4 is also a renormalisable theory; it only shows that the Yukawa and ϕ^4 theories should be renormalised together, as we could have guessed by power counting.

This example allows us to introduce the notion of *stability*, sometimes also called *completeness*. A renormalisable Lagrangian field theory will be called *stable* if it contains all monomials in the fields and their derivatives that are needed for the consistent implementation of the renormalisation programme to all orders of perturbation. It is obvious that a non-renormalisable theory is never stable. The stability requirement is of fundamental importance, as it tells us how many parameters are left free by the renormalisation procedure. This, in turn, tells us how many renormalisation conditions we must impose in order to uniquely determine these parameters, and thus obtain a well-defined theory.

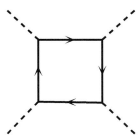

Figure 16.13 *A 1-loop primitively divergent diagram with four external boson lines in the Yukawa theory.*

16.5.4   An 'Almost' Renormalisable Theory

We just saw that power counting classifies all polynomial quantum field theories as super-renormalisable, renormalisable, and non-renormalisable. We also explained that the latter have no predictive power. Here we want to present an example which, in a certain sense, contradicts this statement.

In section 14.3 we discussed briefly a model of quantum electrodynamics with a massive photon. We wrote the Lagrangian density as

$$\mathcal{L}_1 = -\frac{1}{4}(\partial_\mu A_\nu - \partial_\nu A_\mu)^2 + \frac{1}{2}M^2 A_\mu A^\mu + \bar{\psi}\,(i\partial\!\!\!/ - m)\,\psi + e\bar{\psi}\gamma_\mu\psi A^\mu. \tag{16.70}$$

Following the discussion of section 14.3 we note that the vector field is coupled to a conserved current and the condition $\partial_\mu A^\mu = 0$ is a consequence of the equations of motion. The propagator is given by Eq. (14.25) (with M^2 replacing m^2) and the theory is non-renormalisable by power counting. Explicit computations confirm this prediction. For example, computing either of the two diagrams of Fig. 16.14 which represent the 1-loop contribution to the four-fermion correlation function, we find quadratically divergent results. However, if we put the two calculations together (see Problem 16.9), we find a surprise. The divergent contributions of the two diagrams cancel if the external fermion lines are on the mass-shell. It turns out that this miraculous cancellation persists for all Green functions and to all orders. If we compute S-matrix elements, the theory behaves as a renormalisable theory, although all off-shell Green functions are hopelessly divergent. In other words, the conclusion that such a theory is useless for physics is erroneous, although, technically, all steps of the argument are correct. The reason is that we do not necessarily need the full Green functions for arbitrary values of the external momenta in order to compute physical quantities. We want to explain this result, which was first obtained by E. C. G. Stückelberg, because it is an excellent introduction to the gauge model which describes the electromagnetic and weak interactions.

When computing S-matrix elements, external fermion lines of momentum p satisfy $(p\!\!\!/ - m)u = 0$ and external boson lines with momentum k and polarisation ϵ satisfy $k^2 = M^2$ and $k_\mu\epsilon^\mu = 0$. Let us split the boson propagator into $g_{\mu\nu}/(k^2 - M^2)$ and $k_\mu k_\nu/M^2(k^2 - M^2)$. We want to argue that the last term does not contribute. The splitting is done, formally, by considering the Lagrangian

Figure 16.14 *The divergent contributions of the two diagrams cancel when the fermions are on the mass shell.*

$$\mathcal{L}_2 = -\tfrac{1}{2}(\partial_\mu A^\nu)^2 + \tfrac{1}{2}M^2 A_\mu A^\mu + \tfrac{1}{2}(\partial_\mu \phi)^2 - \tfrac{1}{2}M^2 \phi^2 + \bar{\psi}\,(\mathrm{i}\partial\!\!\!/ - m)\psi$$
$$+\mathrm{e}\bar{\psi}\gamma_\mu\psi\,(A^\mu + \tfrac{1}{M}\partial^\mu\phi). \tag{16.71}$$

Comparing $\mathcal{L}_1$ and $\mathcal{L}_2$ we note the following. First, $\mathcal{L}_2$ contains two new degrees of freedom, namely the scalar field ϕ, which has a derivative coupling to the fermion current, and the scalar component of A_μ. Indeed the equation $\partial_\mu A^\mu = 0$ does not follow any more from the equations of motion. Furthermore, as we explained in section 14.3, this new degree of freedom corresponds to a negative metric state. Second, the A_μ propagator is now just proportional to $g_{\mu\nu}/(k^2 - M^2)$. Third, although $\mathcal{L}_1$ and $\mathcal{L}_2$ describe totally different field theories, if we restrict ourselves to diagrams with only external ψ lines and/or external A-lines satisfying $k_\mu \epsilon^\mu = 0$, they give identical results. This is true for every diagram, at any order of regularised but unrenormalised perturbation theory. The $k_\mu k_\nu$ part of the vector propagator is reproduced by the internal ϕ-lines. Of course, $\mathcal{L}_2$ is still a non-renormalisable theory although all propagators are well behaved. The trouble now arises from the derivative coupling of the ϕ-field with the fermions.

Let us now consider a field transformation of the form

$$\psi' = \exp\left[-\mathrm{i}\frac{e}{M}\phi\right]\psi. \tag{16.72}$$

Replacing ψ by ψ' in $\mathcal{L}_2$ we obtain the Lagrangian

$$\mathcal{L}_3 = -\tfrac{1}{2}(\partial_\mu A^\nu)^2 + \tfrac{1}{2}M^2 A_\mu A^\mu + \tfrac{1}{2}(\partial_\mu \phi)^2 - \tfrac{1}{2}M^2 \phi^2$$
$$+\bar{\psi}'(\mathrm{i}\partial\!\!\!/ - m)\psi' + e\bar{\psi}'\gamma_\mu\psi' A^\mu. \tag{16.73}$$

The ϕ-field has decoupled and is now a harmless free field. $\mathcal{L}_3$ is still quantised in a space containing unphysical negative norm states, but it is renormalisable by power counting. Since the transformation (16.72) is of the form, perturbatively, $\psi' = \psi + \mathcal{O}(e)$, the formal asymptotic theory we developed in Chapter 12 tells us that the S-matrices of $\mathcal{L}_2$ and $\mathcal{L}_3$ are identical. In particular, even if computed with $\mathcal{L}_2$, ϕ would turn out to be a free field. But, as we said previously, the S-matrix of $\mathcal{L}_2$, suitably restricted, is that of $\mathcal{L}_1$. We conclude that the latter can be computed using a renormalisable theory, namely $\mathcal{L}_3$.

Let us summarise and state the result in a more precise way. We started with the Lagrangian $\mathcal{L}_1$ which correctly describes the physical degrees of freedom. It is quantised in a space with only physical states but it is non-renormalisable by power counting and cannot be used to define a field theory order by order in perturbation. For this reason we turned to $\mathcal{L}_3$, which is quantised in a larger Hilbert space containing that of $\mathcal{L}_1$, but also unphysical, negative norm states. It is renormalisable by power counting and we can compute, order by order, an S-matrix. It is pseudo-unitary in the indefinite metric large space. The formal argument shows that this S-matrix admits a restriction in the subspace of physical states which is unitary. It is this restriction which defines the S-matrix of the original theory.

16.5.5   Composite Operators

Before closing this section we want to go one step further. Up to now we have been concerned with Green functions consisting of vacuum expectation values of time-ordered products of any number of fields. It is for such quantities that the Feynman rules and the renormalisation programme were set up. However, in perturbation theory we can construct local operators as functions of the elementary fields.[3] It will be useful to extend the programme to cover Green functions of such objects. We will do it here for operators which are monomials in the fields and their derivatives. Examples of such operators for the scalar theory (16.40) are $\phi^2(x)=\phi(x)\cdot\phi(x)$, $\phi^4(x)$, $\phi(x)\Box\phi(x)$, etc. For quantum electrodynamics a particularly interesting operator is the current $j_\mu(x) = \bar{\psi}(x)\gamma_\mu\psi(x)$. We will be interested in Green functions of the form

$$
\begin{aligned}
&\mathcal{G}(y_1,y_2,...,y_n;x_1,x_2,...,x_m)\\
&=<0|T(\mathcal{O}_1(y_1)\mathcal{O}_2(y_2)...\mathcal{O}_n(y_n)\phi_1(x_1)\phi_2(x_2)...\phi_m(x_m))|0>,
\end{aligned}
\tag{16.74}
$$

where $\mathcal{O}_a$ denotes any composite operator of the type we are discussing and ϕ_b any of the elementary fields of the theory. We assume that the fields ϕ_b interact through a Lagrangian density $\mathcal{L}$ and we know how to compute and renormalise any Green function involving only products of fields. A formal way to obtain the Feynman rules for $\mathcal{G}$ is to change $\mathcal{L}$ by adding new terms

$$
\mathcal{L} \to \mathcal{L} + \sum_a \mathcal{J}_a(x)\mathcal{O}_a(x),
\tag{16.75}
$$

with $\mathcal{J}_a(x)$ denoting classical, c-number external sources chosen such that the new term in (16.75) respects all symmetries of $\mathcal{L}$. The derivatives of the corresponding generating functional with respect to $\mathcal{J}_a(y_a)$ taken at $\mathcal{J}_a = 0$ are precisely the Green functions we are interested in,

$$
\begin{aligned}
&Z[j_1,j_2,...,j_l;\mathcal{J}_1,\mathcal{J}_2,...,\mathcal{J}_k]\\
&= \frac{\int \prod_{b=1}^{l}\mathcal{D}[\phi_b]e^{i\int d^4x\left(\mathcal{L}+\sum_{b=1}^{l}j_b(x)\phi_b(x)+\sum_{a=1}^{k}\mathcal{J}_a(x)\mathcal{O}_a(x)\right)}}{\int \prod_{b=1}^{l}\mathcal{D}[\phi_b]e^{i\int d^4x\mathcal{L}}}
\end{aligned}
\tag{16.76}
$$

$$
\begin{aligned}
&\mathcal{G}(y_1,y_2,...,y_n;x_1,x_2,...,x_m)\\
&= \frac{\delta^{n+m}Z}{\delta\mathcal{J}_1(y_1)...\delta\mathcal{J}_n(y_n)\delta j_1(x_1)...\delta j_m(x_m)}\bigg|_{j=\mathcal{J}=0},
\end{aligned}
\tag{16.77}
$$

where l is the number of elementary fields of the theory and k that of the independent composite operators we want to consider. Since all the $\mathcal{O}_a$'s are monomials in the fields and to their derivatives, they can be viewed as new vertices, thus providing the Feynman

---

[3] In a later chapter we will introduce a particular form of a non-local operator but, for the general case, there is no renormalisation theory describing them.

rules for the calculation of $\mathcal{G}$ at any order in an expansion in powers of the coupling constants of $\mathcal{L}$. We emphasise this point: the perturbation series is always determined by the expansion in powers of the original coupling constants. The new vertices generated by the composite operators will be taken only at fixed finite order. This is the reason we do not restrict them to those corresponding to renormalisable theories. It follows that the calculation of $\mathcal{G}$ in a regularised version of the theory is straightforward. We expect, however, new problems to appear when we attempt to take the limit $\epsilon \to 0$. Indeed, these operators are formal products of fields taken at the same space–time point and such products are, in general, singular. Therefore, we expect to require new renormalisation counter-terms in order to render the Green functions (16.74) finite. We will study here the renormalisation properties of Green functions involving a single composite operator and a string of fields. The general case is not essentially different.

Let us start with the simplest example, that of the operator $\phi^2(x)$ in a theory with ϕ^4 interaction. The new vertex is shown on the left of Fig. 16.15. At 1 loop the only 1-PI divergent Green function is the one with two fields $< 0|T(\phi^2(y)\phi(x_1)\phi(x_2))|0 >$ shown on the right of Fig. 16.15. (Strictly speaking, there is also the vacuum Green function $< 0|\phi^2|0 >$ which is also primitively divergent, but we will rarely need to consider it. A simple rule is to fix all such counter-terms by requiring the vacuum expectation value of any such composite operator to vanish.) The analytic expression for the diagram of Fig. 16.15 is in fact the same with the one we computed for the 4-point function of Fig. 16.8. The divergence will be removed by a counter-term Z_{ϕ^2} given by

$$Z_{\phi^2} = 1 + \frac{\lambda_R}{16\pi^2} \frac{1}{\epsilon}, \tag{16.78}$$

which multiplies the $\mathcal{J}_{\phi^2}\phi^2$ term in (16.75).

More complicated operators can be treated following the same lines. The required counter-terms are obtained by power counting and the general property of renormalisation applies, namely they are given by monomials in the fields and their derivatives. For example, let us look at $\phi(x)\Box\phi(x)$. The divergent 1-loop diagrams are shown in Fig. 16.16 and contribute to the Green functions

$$< 0|T(\phi(y)\Box\phi(y)\phi(x_1)\phi(x_2))|0 > \quad \text{and} \quad < 0|T(\phi(y)\Box\phi(y)\phi(x_1)...\phi(x_4))|0 > .$$

The first is quadratically divergent and the second one logarithmically. We leave the complete calculation as a problem but we see immediately a new feature. We cannot

Figure 16.15 *The vertex for the insertion of the operator ϕ^2 (left) and the 1-loop divergent diagram (right).*

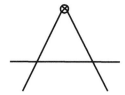

Figure 16.16 *The primitively divergent 1-loop diagrams for the insertions of the operator $\phi\Box\phi$.*

renormalise the Green functions of the operator $\phi(x)\Box\phi(x)$ without simultaneously looking at those of $\phi^2(x)$ and $\phi^4(x)$. We will call a set of composite operators *complete* under renormalisation if all Green functions with one operator from the set and a string of fields can be renormalised without the need to introduce operators outside the set. A complete set is called *irreducible* if it cannot be split into two disjoint complete sets. We saw already that $\phi^2(x)$ by itself forms a complete and irreducible set and so do the three operators $\phi^2(x)$, $\phi^4(x)$, and $\phi(x)\Box\phi(x)$. In Problem 16.10 we establish, by power counting, the rule to find such sets for any renormalisable field theory. Given an operator $\mathcal{O}_a$ with dimensions d_a, the corresponding complete and irreducible set contains all operators with the same quantum numbers and dimensions smaller than or equal to d_a. If $\mathcal{O}_a$, $a = 1, 2, ..., k$, form such a set, a simple way to organise the counter-terms is as a matrix Z_{ab} with $a, b = 1, 2, ..., k$. The finite parts of these matrix-valued renormalisation counter-terms can be determined by appropriately chosen renormalisation conditions (see Problem 16.10). There is no difficulty in extending the programme to the general case of Green functions with several composite operators.

16.6   The Renormalisation Group

16.6.1   General Discussion

In the previous section we presented the renormalisation programme. We saw that for every renormalisable field theory, there exists a well-defined prescription which makes it possible, order by order in perturbation theory, to obtain finite expressions for all the Green functions. They are formal power series in the expansion parameter, for example the number of closed loops, and each term depends on the external momenta, the parameters such as the masses and coupling constants, but also the particular renormalisation scheme we have adopted. In this section we want to study this dependence more closely.[4]

In order to be specific, let us choose the ϕ^4 theory we studied earlier, (16.40). A general renormalisation scheme is defined through three conditions, two for the 2-point function, such as (16.53) and (16.68) we used in the previous section, and one for the 4-point function. For the sake of generality, we may not introduce the physical mass and use instead three conditions of the form,

---

[4] The concept of the renormalisation group was introduced formally in quantum field theory by E. C. G. Stueckelberg and A. Petermann, *Helv. Phys. Acta* **26**, 499 (1953). The first application to the asymptotic behaviour of Green functions is due to M. Gell-Mann and F. E. Low, *Phys. Rev.* **95**, 1300 (1954).

$$\Gamma^{(2)}|_{p^2=M^2} = i(m_R^{(M)})^2 \tag{16.79}$$

$$\frac{\partial \Gamma^2(p^2)}{\partial p^2}\Big|_{p^2=M^2} = -i \tag{16.80}$$

$$\Gamma^{(4)}(p_1,..,p_4)|_{\text{sym. point } M} = i\lambda_R^{(M)}, \tag{16.81}$$

where by 'sym. point M' we mean the point in momentum space in which we consider all momenta p_i ($i = 1, 2, 3, 4$) entering the graph with $\Sigma_i p_i = 0$, $p_i^2 = M^2$, and $(p_i + p_j)^2 = 4M^2/3$ for all pairs $i \neq j$. M is an arbitrary mass parameter which is assumed to lie in the analyticity domains of both $\Gamma^{(2)}$ and $\Gamma^{(4)}$. Note that the parameter $m_R^{(M)}$ in (16.79) is not the position of the pole of the propagator and is not equal to the physical mass.[5] At lowest order the latter is given by $(m_R^{ph})^2 = M^2 + (m_R^{(M)})^2$.

Note also that for the physical choice $M = m_R^{ph}$, the two conditions (16.79) and (16.80) imply that $\Gamma^{(2)}(p^2)$ has the form

$$\Gamma^{(2)}(p^2) = i(p^2 - (m_R^{ph})^2)[1 + (p^2 - (m_R^{ph})^2)\hat{\Gamma}^{(2)}(p^2)], \tag{16.82}$$

with $\hat{\Gamma}^{(2)}(p^2)$ a function of p^2 such that the product $(p^2 - (m_R^{ph})^2)\hat{\Gamma}^{(2)}(p^2)$ vanishes at $p^2 = (m_R^{ph})^2$.

As we saw in the previous section, these three renormalisation conditions uniquely determine, order by order in perturbation theory, the three counterterms Z_ϕ, Z_m, and Z_λ. The renormalisability of the theory then guarantees that all Green functions $\Gamma^{(2n)}$, for all n, are finite and calculable as formal power series. They are of the form $\Gamma^{(2n)}(p_1, ..., p_{2n}; m_R^{(M)}, \lambda_R^{(M)}; M)$, where the external momenta p_i are subject to the condition $\Sigma_{i=1}^{2n} p_i = 0$. The two parameters, the mass and the coupling constant, have a direct physical meaning and are related to measurable quantities. By varying them we map one physical theory to another. In contrast, the subtraction point M was introduced only in order to remove the ambiguities in the finite parts of the divergent diagrams. If we change from one point M_1 to another M_2 we will obtain a different set of Green functions, but we expect them to describe the same physical theory. In this section we want to prove this property. In other words, we want to prove that the renormalisation programme provides us with a family of Green functions $\Gamma^{(2n)}$ labelled by the point M, but physical quantities are independent of M.

Let us consider the transformation

$$M_1 \rightarrow M_2 = \rho M_1, \tag{16.83}$$

with ρ being a numerical coefficient which can take any value in a domain including the value $\rho = 1$ and such that all resulting points M_2 belong to the domain of analyticity of $\Gamma^{(2)}$ and $\Gamma^{(4)}$ and yield well-defined sets of Green functions. The question we want to answer is under which conditions all these different sets describe one and the same physical theory.

---

[5] We could have chosen $m_R^{(M)} = 0$. This would only add an irrelevant constant to the 2-point function.

To illustrate the point, let us look first at the 4-point function and ignore, for the moment, the other two conditions. For the two choices M_1 and M_2 the renormalisation condition (16.81) gives

$$\Gamma_1^{(4)}|_{\text{sym. point } M_1} = i\lambda_R^{(M_1)}; \quad \Gamma_2^{(4)}|_{\text{sym. point } M_2} = i\lambda_R^{(M_2)}, \tag{16.84}$$

where the subscripts 1 and 2 denote the Green functions obtained using the points M_1 and M_2, respectively. It is clear that if the coupling constants $\lambda_R^{(M_1)}$ and $\lambda_R^{(M_2)}$ are totally unrelated the two sets of Green functions will be unrelated as well. Indeed, in the scheme based on M_1 the value of the 4-point function at the symmetric point M_2 is well defined and calculable as a formal power series in powers of $\lambda_R^{(M_1)}$. For the two sets to describe the same theory, $\Gamma_1^{(4)}$ and $\Gamma_2^{(4)}$ must take the same value when computed at the same point; therefore, the λ's must satisfy

$$i\lambda_R^{(M_2)} = \Gamma^{(4)}|_{\text{sym. point } M_2} = i\lambda_R^{(M_1)} + iC_2(\lambda_R^{(M_1)})^2 + iC_3(\lambda_R^{(M_1)})^3 + \dots. \tag{16.85}$$

The C's are finite and calculable and depend on $m_R^{(M)}$, M_1, and M_2. It is important to note that the first coefficient C_1 equals 1, because, at lowest order, the two schemes must be identical.

It is now clear how to describe the general case for any Green function and taking into account all renormalisation conditions. The transformation $M_1 \rightarrow M_2$ implies different choices for the three counterterms Z_ϕ, Z_m, and Z_λ. The two sets of Green functions will be physically equivalent if we can find three functions $Z_\phi(m_R^{(M_1)}, \lambda_R^{(M_1)}, M_1, M_2)$, $m_R^{(M_2)}(m_R^{(M_1)}, \lambda_R^{(M_1)}, M_1, M_2)$, and $\lambda_R^{(M_2)}(m_R^{(M_1)}, \lambda_R^{(M_1)}, M_1, M_2)$ satisfying

$$Z_\phi(m_R^{(M_1)}, \lambda_R^{(M_1)}, M_1, M_2) = 1 + O(\lambda_R^{(M_1)})$$
$$M_2^2 + (m_R^{(M_2)})^2(m_R^{(M_1)}, \lambda_R^{(M_1)}, M_1, M_2) = M_1^2 + (m_R^{(M_1)})^2 + O(\lambda_R^{(M_1)}) \tag{16.86}$$
$$\lambda_R^{(M_2)}(m_R^{(M_1)}, \lambda_R^{(M_1)}, M_1, M_2) = \lambda_R^{(M_1)} + O((\lambda_R^{(M_1)})^2)$$

and such that

$$M_1 \rightarrow M_2$$
$$\Gamma^{(2n)}(p_1, ..., p_{2n}; m_R^{(M_1)}, \lambda_R^{(M_1)}; M_1) \tag{16.87}$$
$$= Z_\phi^{-n}(m_R^{(M_1)}, \lambda_R^{(M_1)}, M_1, M_2)\Gamma^{(2n)}(p_1, ..., p_{2n}; m_R^{(M_2)}, \lambda_R^{(M_2)}; M_2).$$

In other words, the change $M_1 \rightarrow M_2$ should be absorbed in a change of the values of the mass, the coupling constant and the normalisation of the field.

The transformations (16.87) form a group, called *the renormalisation group*. It describes the change in the Green functions induced by a change of the subtraction point. It is instructive to write also the corresponding infinitesimal transformations with ρ of Eq. (16.83) close to 1. For that we take the derivative of both sides of (16.87) with respect to M_2 and set $M_1 = M_2 = M$:

$$[M\frac{\partial}{\partial M} + \beta\frac{\partial}{\partial\lambda} + \gamma_m m\frac{\partial}{\partial m} - n\gamma]\Gamma^{(2n)}(p_1,...,p_{2n};m,\lambda;M) = 0, \tag{16.88}$$

where we have dropped the subscripts R and the reference to the subtraction point. The functions β, γ_m, and γ are defined as

$$\beta(\lambda,\frac{m}{M}) = M\frac{\partial\lambda_R^{(M_2)}}{\partial M_2}|_{M_1=M_2=M}$$

$$\gamma_m(\lambda,\frac{m}{M}) = \frac{M}{m}\frac{\partial m_R^{(M_2)}}{\partial M_2}|_{M_1=M_2=M} \tag{16.89}$$

$$\gamma(\lambda,\frac{m}{M}) = M\frac{\partial lnZ_\phi}{\partial M_2}|_{M_1=M_2=M}.$$

These functions are dimensionless and they can only depend on the ratio m/M. By applying the general equation (16.88) for $n = 1$ and 2 and using the renormalisation conditions (16.79), (16.80), and (16.81), we can express them as combinations of Green functions, or derivatives of Green functions, thus proving that they have a well-defined expansion in powers of the coupling constant. We obtain

$$\gamma(\lambda,\frac{m}{M}) = i[M\frac{\partial}{\partial M}\frac{\partial\Gamma^{(2)}}{\partial p^2}]|_{p^2=M^2}$$

$$\gamma_m(\lambda,\frac{m}{M}) = \frac{1}{2}\gamma + \frac{i}{2m^2}M\frac{\partial\Gamma^{(2)}}{\partial M}|_{p^2=M^2} \tag{16.90}$$

$$\beta(\lambda,\frac{m}{M}) = 2\lambda\gamma + iM\frac{\partial\Gamma^{(4)}}{\partial M}|_{\text{sym. point } M}.$$

At lowest order β and γ vanish because there is no dependence on M at that order. This is not true for γ_m, but this is only due to our peculiar renormalisation condition (16.79) which shifts the value of the mass by M^2 even in the tree approximation.

It is clear how to generalise this analysis to renormalisable theories with several fields ϕ_i with masses m_i, $i = 1,...,N$, and several coupling constants λ_j, $j = 1,...,k$. We will need N γ and γ_m functions and k β functions. In general, they will all depend on the k coupling constants. We can also extend the analysis to describe the subtraction point dependence of Green functions involving one composite operator. Since, as we saw, the renormalisation programme may introduce mixing among different operators, the corresponding γ functions will be matrix-valued functions of the coupling constants.

16.6.2   The Renormalisation Group in Dimensional Regularisation

Equations (16.90) allow for the explicit calculation of the functions β, γ, and γ_m to any given order in perturbation theory. However, as we have said already, they do not provide for the most convenient way to perform such calculations. We want to re-express them using the full power of the method of dimensional regularisation.

Let us assume that we compute all Green functions in d dimensions and extract the four-dimensional finite part by dropping the $1/\epsilon$ terms. As we explained in section 16.5, the Green functions depend now on a parameter μ, with the dimensions of a mass, which plays the same role as the subtraction point in more conventional schemes. As we showed in the previous section, a change in μ induces a change in the three counter-terms. Therefore, we can immediately write the renormalisation group equation (16.88), with μ replacing M. It is, however, instructive to re-derive the equation following the steps of section (16.5). We start again from the bare Lagrangian

$$\mathcal{L} = \frac{1}{2}\partial_\mu\phi_B(x)\partial^\mu\phi_B(x) - \frac{1}{2}m_B^2\phi_B^2(x) - \frac{1}{4!}\lambda_B\phi_B^4(x), \tag{16.91}$$

where the subscript B stands for 'bare'. Green functions computed with (16.91) depend on the cut-off parameter ϵ and they diverge when $\epsilon \to 0$ if the bare parameters are kept fixed. The renormalisability of the theory guarantees that we can introduce renormalisation counterterms Z_ϕ, Z_m, and Z_λ, as shown in section (16.5), functions of the cut-off, such that

$$\Gamma_R^{(2n)}(p_1, .., p_{2n}; \lambda_R, m_R, \mu) = Z_\phi^n \Gamma_B^{(2n)}(p_1, .., p_{2n}; \lambda_B, m_B, \epsilon) + ... \tag{16.92}$$

where the dots stand for terms which vanish when $\epsilon \to 0$. As shown in section 16.5, Eqs. (16.45), (16.52), and (16.66), the connection between the bare and the renormalised quantities is given by

$$\begin{aligned}
\phi(x)_B &= Z_\phi^{1/2}(\lambda_R, \epsilon)\phi_R(x) \\
m_B^2 &= Z_m(\lambda_R, \epsilon)m_R^2 \\
\lambda_B &= \mu^\epsilon Z_\lambda(\lambda_R, \epsilon)\lambda_R,
\end{aligned} \tag{16.93}$$

where we have introduced explicitly the factor μ^ϵ to make the coupling constant dimensionless, as we explained in the previous section. The important point is that in the minimal subtraction scheme, in which the Z's are determined only by the residues of the $1/\epsilon$ poles, they cannot depend on either m or μ.

The renormalisation group equation is obtained by taking the derivative of both sides of (16.92) with respect to μ, keeping the bare quantities and ϵ fixed,

$$[\mu\frac{\partial}{\partial\mu} + \beta\frac{\partial}{\partial\lambda} - n\gamma + \gamma_m m\frac{\partial}{\partial m}]\Gamma^{(2n)}(p_1, .., p_{2n}; \lambda, m, \mu) = 0 \tag{16.94}$$

with

$$\beta(\lambda, \epsilon) = \mu\frac{\partial}{\partial\mu}\Big|_{\lambda_B, \epsilon} \lambda(\lambda_B\mu^{-\epsilon}, \epsilon) = -\epsilon\lambda - \lambda\mu\frac{\partial}{\partial\mu}\Big|_{\lambda_B, \epsilon} \ln Z_\lambda$$

$$\gamma(\lambda, \epsilon) = \mu \frac{\partial}{\partial \mu}\bigg|_{\lambda_B, \epsilon} \ln Z_\phi \tag{16.95}$$

$$\gamma_m(\lambda, \epsilon) = -\frac{1}{2}\mu \frac{\partial}{\partial \mu}\bigg|_{\lambda_B, \epsilon} \ln Z_m.$$

Three points should be noted here. First, these functions appear to depend on both λ and ϵ. However, the renormalisation group equation (16.94) shows that they can be expressed in terms of the renormalised Green functions. Therefore, they have well-defined limits when $\epsilon \to 0$. Second, the functions defined in (16.95) are not identical with those introduced previously in (16.90) because the latter depend on m/M. We will argue in a subsequent section that, nevertheless, they both contain all the physically interesting information. Third and most important, these functions can be computed very easily. Indeed, in order to compute the functions of (16.90) we need to compute the entire 2- and 4-point functions, including the finite parts. In contrast, for those of (16.95), only the pole terms in the Z's are needed. Let us consider, as an example, the β function. The first of Eqs. (16.95) can be rewritten as

$$[\beta + \epsilon\lambda + \beta\lambda \frac{\partial}{\partial\lambda}]Z_\lambda = 0. \tag{16.96}$$

Z_λ has an expansion in powers of ϵ^{-1},

$$Z_\lambda = \sum_{i=0}^{\infty} \frac{Z_\lambda^{(i)}(\lambda)}{\epsilon^i}, \tag{16.97}$$

with $Z_\lambda^{(0)}(\lambda) = 1$. The residues $Z_\lambda^{(i)}(\lambda)$ for $i \geq 1$ are formal power series in λ starting at λ^i. On the other hand, $\beta(\lambda, \epsilon)$, which is regular at $\epsilon = 0$, admits an expansion in positive powers of ϵ:

$$\beta(\lambda, \epsilon) = \sum_{i=0}^{\infty} \beta^{(i)}(\lambda)\epsilon^i. \tag{16.98}$$

Since Eq. (16.96) is an identity in ϵ we can compute the coefficients $\beta^{(i)}(\lambda)$ by matching the powers of ϵ. We see immediately that $\beta^{(1)}(\lambda) = -\lambda$ and all $\beta^{(i)}$'s with $i \geq 2$ vanish. From now on, we will write $\beta(\lambda, \epsilon) = \beta(\lambda) - \epsilon\lambda$, dropping the index 0. Substituting in (16.96) we obtain the recursive relation

$$\lambda^2 \frac{\partial Z_\lambda^{(i+1)}(\lambda)}{\partial\lambda} = \beta(\lambda)\frac{\partial(\lambda Z_\lambda^{(i)}(\lambda))}{\partial\lambda}. \tag{16.99}$$

This is a very important relation. First, it makes it possible to compute $\beta(\lambda)$ order by order in perturbation theory. Writing (16.99) for $i = 0$ we obtain

$$\beta(\lambda) = \lambda^2 \frac{\partial Z_\lambda^{(1)}(\lambda)}{\partial \lambda}. \tag{16.100}$$

In other words, $\beta(\lambda)$ is entirely determined by the residues of the *simple* poles of Z_λ. Since $Z_\lambda^{(1)}$ starts with a term proportional to λ (see Eq. (16.52)), it follows that $\beta(\lambda)$ starts as λ^2. Comparing with (16.52), we find that

$$\beta(\lambda) = b_0\lambda^2 + b_1\lambda^3 + ...; \quad b_0 = \frac{3}{16\pi^2}. \tag{16.101}$$

A second important consequence of the recursive relation (16.99) is that although in the nth order of perturbation theory we encounter singularities up to ϵ^{-n}, only the simple poles are really new. The residues of the poles of higher order are uniquely determined by the lower order ones. At this point this is a technical property of the structure of the perturbation series and can be traced to the forest formula (16.69). In a later section we will use it to obtain physically interesting results concerning the asymptotic behaviour of Green functions for large values of the external momenta.

16.6.3   Dependence of the β and γ Functions on the Renormalization Scheme

The way we obtained the renormalisation group functions shows that, in general, they depend on the particular renormalisation scheme we used to compute them. Since we intend to use them in order to extract physical results, we want to study the extend of such scheme dependence. We will prove that the first non-vanishing term in the perturbation expansion is universal.

Let us start with a theory with only one coupling constant. In one renormalisation scheme we have the function $\beta(\lambda)$ with an expansion

$$\beta(\lambda) = b_0\lambda^2 + b_1\lambda^3 + ... \tag{16.102}$$

By changing the renormalisation scheme we obtain a new coupling constant λ' and a new β-function $\beta'(\lambda')$. The two are related by

$$\lambda' = F(\lambda) = \lambda + f_1\lambda^2 + O(\lambda^3) \tag{16.103}$$

$$\beta'(\lambda') = \mu\frac{\partial}{\partial\mu}\lambda' = \frac{\partial F}{\partial\lambda}\beta(\lambda). \tag{16.104}$$

Note the crucial fact that the first term in the expansion of $F(\lambda)$ in (16.103) is λ. It is this property that makes F a new acceptable coupling constant.

We expand now both sides of (16.104) and find that

$$\beta'(\lambda') = (1 + 2f_1\lambda + O(\lambda^2))(b_0\lambda^2 + b_1\lambda^3 + O(\lambda^4))$$
$$= b_0\lambda^2 + (b_1 + 2f_1b_0)\lambda^3 + O(\lambda^4) = b_0\lambda'^2 + b_1\lambda'^3 + O(\lambda'^4), \tag{16.105}$$

where we have used the inverse relation implied by (16.103), namely $\lambda = \lambda' - f_1\lambda'^2 + O(\lambda'^3)$. We have thus established the universality of b_0 and b_1. The generalisation to several coupling constants is straightforward. The result is that the first non-vanishing term is always scheme independent. Note that the same proof shows that the first non-vanishing term in the expansion of the anomalous dimension of the field $\gamma(\lambda)$ is also scheme independent. At the end of the chapter we propose a problem in which we ask the reader to prove that there exists a renormalisation scheme in which the β function of the $\lambda\phi^4$ theory is exactly given, to all orders of perturbation theory, by its first two non-vanishing terms. All higher order corrections can be chosen to vanish.

16.7 Problems

Problem 16.1 The power counting in various field theories.

1. Repeat the power-counting computation and prove that the four-dimensional field theories mentioned in section 16.3 are indeed renormalisable.

2. Prove that in a two-dimensional space–time all field theories involving only scalar fields with non-derivative polynomial interactions are super-renormalisable.

3. Are there any renormalisable theories in a six-dimensional space–time?

Problem 16.2 A simple way to regularise the ultraviolet behaviour of a quantum field theory is to introduce higher derivative terms in the quadratic part of the Lagrangian density. Consider the example of a single scalar field ϕ with $\mathcal{L}$ given by

$$\mathcal{L} = \frac{1}{2}(\partial_\mu\phi)(\partial^\mu\phi) - \frac{1}{2}m^2\phi^2 + \frac{\xi}{2}(\partial_\mu\phi)\Box(\partial^\mu\phi) \qquad (16.106)$$

with ξ a constant with dimensions $[m^{-2}]$.

Show that the propagator behaves like k^{-4} at large momenta, but this has been achieved at the price of introducing degrees of freedom with negative energy.

Hint: Show that the propagator in this theory is equivalent to that of a theory with two scalar fields ϕ_1 and ϕ_2 given by a Lagrangian density:

$$\mathcal{L} = \frac{1}{2}(\partial_\mu\phi_1)(\partial^\mu\phi_1) - \frac{1}{2}m_1^2\phi_1^2 - \frac{1}{2}(\partial_\mu\phi_2)(\partial^\mu\phi_2) - \frac{1}{2}m_2^2\phi_2^2. \qquad (16.107)$$

Express the parameters of (16.106) in terms of those of (16.107) and study the limit $m_2^2 \to \infty$.

This shows that introducing auxiliary fields, in this case the ϕ_2 field, with negative kinetic energy, or, equivalently, quantised in a space with negative norm, results into a regularisation of the theory.

Problem 16.3 The Pauli–Villars regularisation method. In 1949 Wolfgang Pauli and Felix Villars generalised the previous method and adapted it to quantum electrodynamics.

Consider the QED Lagrangian density, for example, in the Feynman gauge:

$$\mathcal{L} = -\frac{1}{2}(\partial_\mu A_\nu)(\partial^\mu A^\nu) + \bar{\psi}(i\partial\!\!\!/ - m)\psi - e\bar{\psi}A\!\!\!/\psi. \tag{16.108}$$

Introduce a set of n_γ auxiliary photon fields and n_f auxiliary fermion fields and generalise (16.108) to

$$\begin{aligned}
\mathcal{L}_{PV} &= -\tfrac{1}{2}\sum_{i=0}^{n_\gamma}\left[(\partial_\mu A_{\nu(i)})(\partial^\mu A_{(i)}^\nu) + \tfrac{1}{2}M_i^2 A_{\nu(i)}A_{(i)}^\nu\right] \\
&\quad + \sum_{j=0}^{n_f}\left[\bar{\psi}_{(j)}(i\partial\!\!\!/ - m_j)\psi_{(j)}\right] - e\sum_{i=0}^{n_\gamma}\sum_{j=0}^{n_f}c_{ij}\bar{\psi}_{(j)}A\!\!\!/_i\psi_{(j)},
\end{aligned} \tag{16.109}$$

where $A_{(0)}$ and $\psi_{(0)}$ are the usual photon and electron fields with masses $M_0 = 0$ and $m_0 = m$, respectively. The c_{ij}'s are numerical coefficients with $c_{00} = 1$. Every QED Feynman diagram will generate, under (16.109), a whole series of diagrams with identical structure in which every photon line is replaced by each one of the photon auxiliary fields and every fermion loop by the corresponding loop of the fermion auxiliary fields. It is also obvious that gauge invariance is preserved because a local change of phase of any fermion field can be compensated by the corresponding change of the photon field, with all massive photon auxiliary fields remaining unchanged. It is also clear that at the limit $M_i \to \infty$ and $m_j \to \infty$, $i = 1, ..., n_\gamma$ and $j = 1, ..., n_f$, we recover the original theory (16.108). Show that if the parameters in (16.109) satisfy the relations

$$\forall j \;\; \sum_{i=0}^{n_\gamma} c_{(ij)}^2 = 0; \quad \forall i \;\; \sum_{j=0}^{n_f} c_{(ij)}^2 = \sum_{j=0}^{n_f} c_{(ij)}^2 m_j^2 = 0, \tag{16.110}$$

all Green's functions are finite. Note, however, that this gauge invariant regularisation has been achieved at the price of sacrificing the hermiticity of the Lagrangian because all c coefficients cannot be chosen real. Pauli and Villars chose to introduce imaginary couplings for the auxiliary fields, rather than negative signs in the kinetic energy. The practical result at the level of the Feynman diagrams is the same. Find the minimum number of auxiliary fields necessary to satisfy the conditions (16.110)

Problem 16.4 The anomalous magnetic moment of the electron. We saw in Chapter 7 that one of the triumphs of the Dirac theory has been the accurate prediction of the electron gyromagnetic ratio $g = 2$. One of the first triumphs of quantum electrodynamics was the computation of the radiative corrections to this value. Compute these corrections at the 1-loop order.

Hint: Compute the finite part of the vertex diagram of Fig. 16.7 and isolate the term proportional to $q_\mu = (p - p')_\mu$.

Problem 16.5 Compute the 1-loop contribution to the light-by-light scattering amplitude (Fig. 16.7) and show that it is finite and satisfies the requirements of gauge invariance.

Problem 16.6 Compute the $1/\epsilon$ part of the 2-loop diagram of Fig. 16.9 and isolate the term proportional to p^2.

Problem 16.7 The effective coupling constant in the Fermi theory of the weak interactions. Already in 1934, Enrico Fermi proposed a simple model to describe the amplitude of neutron β-decay. This model proved to be very general encompassing the entire field of low-energy weak interactions. It describes the point interaction among four fermion fields with an interaction Lagrangian density of the form

$$\mathcal{L} = \frac{G_F}{\sqrt{2}} \left(\bar{\psi}_1 \gamma_\mu (1 + \gamma_5) \psi_2 \right) \left(\bar{\psi}_3 \gamma^\mu (1 + \gamma_5) \psi_4 \right) + h.c., \tag{16.111}$$

where ψ_i, $i = 1, ..., 4$, denote four fermion fields (in β-decay they are the fields of the neutron, the proton, the electron, and the neutrino) and G_F is a coupling constant with dimension $[m^{-2}]$. $h.c.$ stands for 'Hermitian conjugate'. We will have a closer look at this Lagrangian in a later chapter.

1. Show that the Fermi theory is non-renormalisable.

2. Show that a Green function in the nth order of perturbation expansion has a leading divergence proportional to $(G\Lambda^2)^n$ with Λ a cut-off with the dimensions of a mass. We want to interpret the cut-off as the scale up to which the model can be trusted.

3. The unrenormalised perturbation theory will break down when the $n + 1$ order becomes equal to the n order. Using the experimental value of the Fermi coupling constant $G_F/\sqrt{2} \approx 10^{-5} m_p^{-2}$ with m_p the proton mass $m_p \approx 1$ GeV, derive a first estimation of Λ.

4. As Joffe and Shabalin have pointed out, precision measurements can improve this bound considerably. The weak interactions are known to violate many quantum numbers, such as parity and strangeness, which are conserved by

strong interactions. Experimentally, we know that such violations in hadronic or nuclear physics are suppressed by at least eight orders of magnitude. Using this information, show that the limit on Λ can be reduced to a few GeV.

Problem 16.8 Consider a model with two neutral scalar fields, $\phi(x)$ and $\Phi(x)$, whose interaction is described by the Lagrangian density

$$\mathcal{L} = \frac{1}{2}(\partial_\mu\phi)(\partial^\mu\phi) + \frac{1}{2}(\partial_\mu\Phi)(\partial^\mu\Phi) - \frac{1}{2}m^2\phi^2 - \frac{1}{2}M^2\Phi^2 - \frac{\lambda_1}{4!}\phi^4 - \frac{\lambda_2}{4!}\Phi^4 - \frac{2g}{4!}\phi^2\Phi^2.$$
(16.112)

We consider the limit $M \to \infty$ keeping m and the interaction energy finite. Show that in this limit all effects of the Φ field in the dynamics of ϕ can be absorbed in the renormalisation conditions of the effective Lagrangian obtained from (16.112) by putting $\Phi = 0$

$$\mathcal{L}_{\text{eff}} = \frac{1}{2}(\partial_\mu\phi)(\partial^\mu\phi) - \frac{1}{2}m^2\phi^2 - \frac{\lambda_1}{4!}\phi^4.$$
(16.113)

This is an example of 'the decoupling theorem'. It states that if we have a renormalisable field theory and we take the limit in which some degrees of freedom become infinitely massive, the resulting theory is described by an effective field theory obtained by setting all heavy fields equal to 0, provided the resulting theory is also renormalisable.

Problem 16.9 Compute the divergent contributions for the two diagrams of Fig. 16.14 and show that they cancel when the external fermion lines are on the mass-shell.

Problem 16.10 Renormalisation of composite operators. In this problem we want to establish the renormalisation properties of the Green functions we considered in section 16.5.5.

1. In a quantum field theory of a single scalar field with interaction term proportional to ϕ^4, consider the set of operators

$$\mathcal{O}_1 = \phi^2; \quad \mathcal{O}_2 = \phi^4; \quad \mathcal{O}_3 = \phi\Box\phi.$$
(16.114)

Compute the matrix Z_{ab}, $a, b = 1, 2, 3$ of the counterterms necessary to renormalise all Green functions consisting of one operator O_a and a string of fields ϕ at 1-loop. Establish a set of renormalisation conditions in order to determine this matrix completely.

2. Using a power-counting argument show the completeness property announced in section 16.5.5. Given an operator $\mathcal{O}_a$ with dimensions d_a, the corresponding complete and irreducible set contains all operators with the same quantum numbers and dimensions smaller than or equal to d_a.

Problem 16.11 Compute the first non-vanishing term in the expansion of the β-function for the standard renormalisable field theories, i.e. $\lambda\phi^4$, Yukawa, spinor electrodynamics, scalar electrodynamics, and Yang–Mills theories.

Problem 16.12 Consider our usual $\lambda\phi^4$ theory. At a given renormalisation scheme the perturbation expansion of the β-function can be written as

$$\beta(\lambda) = b_0\lambda^2 + b_1\lambda^3 + b_2\lambda^4 + b_3\lambda^5 + \dots. \tag{16.115}$$

Prove that we can construct a renormalisation scheme which makes the coefficient $b_2 = 0$. By induction, prove that the scheme can be adapted order by order in perturbation theory in order to make *all* coefficients b_i with $i \geq 2$ equal to 0.

17

A First Glance at Renormalisation and Symmetry

17.1 Introduction

The results of the renormalisation programme can be summarised as follows. Given a classical Lagrangian density which is polynomial in a set of fields and their first derivatives and corresponds, by power counting, to a renormalisable theory, there exists a well-defined prescription to obtain a unique finite expression for every Green function at any order of perturbation theory. By Green function we mean the vacuum expectation value of the time-ordered product of any number of fields, as well as local operators which are functions of them. In this section we want to address a further question. We have often seen the crucial role played by symmetries in classical field theories. Does this role survive the procedure of quantisation and renormalisation? In other words, under which conditions can we find the consequences of the symmetries of the classical Lagrangian in the renormalised Green functions? We have already anticipated the answer to this question in a number of cases. For example, when we were studying the renormalisation properties of the ϕ^4 theory we argued that all renormalised Green functions with an odd number of fields vanish because of the discrete symmetry $\phi \rightarrow -\phi$. However, this symmetry is imposed at the level of the classical Lagrangian and, strictly speaking, we must prove that its consequences remain valid after renormalisation. For this example, the proof is obvious since there are no Feynman diagrams with an odd number of external lines. A less trivial example is that of Poincaré invariance. We have always worked with Lorentz invariant Lagrangians and we assumed that after renormalisation, the resulting Green functions still have the correct transformation properties. This is less obvious, because, although the Feynman rules manifestly respect Lorentz invariance, we still must prove that it is not spoiled by regularisation and renormalisation. A simple proof consists of noting that there exist many regularisation schemes which explicitly respect Lorentz invariance. Therefore, the invariance will be valid at any intermediate step of the calculation and for all values of the regulator parameter. In particular, it will be valid separately for the divergent and the finite parts of any diagram and, consequently, for the renormalised expressions. Note however, that the proof would have been more

From Classical to Quantum Fields. Laurent Baulieu, John Iliopoulos and Roland Sénéor.
© Laurent Baulieu, John Iliopoulos and Roland Sénéor, 2017. Published 2017 by Oxford University Press.

complicated had we used a Lorentz non-invariant regulator, such as a space-time lattice. Even in this case we can prove that Lorentz invariance can still be recovered after renormalisation when the lattice spacing goes to 0 although it is broken at every intermediate step, but we will not present the proof here.

In this chapter we want to study the general properties of symmetries in renormalised quantum field theory and, in particular, the non-trivial case of continuous symmetries.

17.2 Global Symmetries

Let us start by the simple example of internal global symmetries.[1] They were first introduced by W. Heisenberg in 1932 when he postulated an approximate invariance of nuclear forces under the internal symmetry $SU(2)$, which we call today *isospin symmetry*. With the discovery of the π-mesons it gave rise to the pion–nucleon interaction, the first field-theory model for strong interactions. The two nucleons, p and n, are assumed to form a doublet of $SU(2)$ and the pions in the three charge states π^+, π^-, and π^0 a triplet. Since the pions are pseudoscalar, the most general renormalisable Lagrangian invariant under isospin transformations is given by

$$\mathcal{L} = \bar{\Psi}(x)(i\partial\!\!\!/ - m)\Psi(x) + \tfrac{1}{2}\partial_\mu \boldsymbol{\pi}(x) \cdot \partial^\mu \boldsymbol{\pi}(x) - \tfrac{1}{2}\mu^2 \boldsymbol{\pi}(x) \cdot \boldsymbol{\pi}(x)$$
$$+ \quad g\bar{\Psi}(x)\gamma_5 \boldsymbol{\tau}\Psi(x) \cdot \boldsymbol{\pi}(x) - \tfrac{\lambda}{4!}\left(\boldsymbol{\pi}(x) \cdot \boldsymbol{\pi}(x)\right)^2 . \tag{17.1}$$

It describes all pion–nucleon scattering amplitudes for all charge states in terms of a single coupling constant g. It follows that the $SU(2)$ symmetry implies relations among Green functions such as $\pi^+ + p \to \pi^+ + p$, $\pi^+ + n \to \pi^+ + n$, $\pi^+ + n \to \pi^0 + p$, etc. It is relations of that kind that we want to investigate and see whether they may remain valid after renormalisation.

Let us study the general case of a Lagrangian density $\mathcal{L}$, which is a polynomial in a set of fields $\phi^i(x)$ and their first derivatives $\partial_\mu \phi^i(x)$, $i = 1, ..., m$. $\mathcal{L}$ is supposed to be invariant under a group of transformations $\mathcal{G}$ under which the fields $\phi^i(x)$ transform linearly

$$\delta\phi^i(x) = \epsilon^\alpha (T_\alpha)^i_j \phi^j(x), \tag{17.2}$$

where α runs from 1 to N, the dimension of the Lie algebra of $\mathcal{G}$, ϵ^α are N, constant, infinitesimal parameters, one for every generator, and the T's are N numerical matrices which characterise the representation, not necessarily irreducible, in which the fields belong. Since the ϵ's are x-independent, the derivatives of the fields transform the same way:

$$\delta\partial_{\mu_1}...\partial_{\mu_n}\phi^i(x) = \partial_{\mu_1}...\partial_{\mu_n}\delta\phi^i(x) = \epsilon^\alpha (T_\alpha)^i_j \partial_{\mu_1}...\partial_{\mu_n}\phi^j(x). \tag{17.3}$$

---

[1] In this section we follow the presentation by S. Coleman, 'Renormalization and Symmetry: A Review for Non-specialists', Erice Lectures 1971, reprinted in S. Coleman, *Aspects of Symmetry* (Cambridge University Press, Cambridge, 1985).

We have already shown that the invariance of the Lagrangian under (17.2) implies the existence of N conserved currents $j_\mu^\alpha(x)$. They are operators of dimension equal to 3 and are constructed as second-degree polynomials in the unrenormalised, classical fields and their first derivatives. The corresponding charges Q^α are the generators of the infinitesimal transformations (17.2),

$$Q^\alpha = \int d^3x j_0^\alpha(x); \quad [\epsilon^\alpha Q_\alpha, \phi^i(x)] = i\delta\phi^i(x); \quad [Q_\alpha, Q_\beta] = if_{\alpha\beta\gamma} Q_\gamma, \tag{17.4}$$

where $f_{\alpha\beta\gamma}$ are the structure constants of the Lie algebra of $\mathcal{G}$.

Since all composite operators $\mathcal{O}_i$ we consider in this chapter are monomials in the fields and their derivatives, we can obtain, at the classical level, similar commutation relations between them and the Q's.

It is straightforward to transform, always at the classical level, these commutation relations into relations among Green functions. We can follow the general method we used in Chapter 14, but, for global symmetries, a simpler, formal way is to start from the expression.

$$G_\mu^\alpha(k_\mu; x_1, ..., x_n) = \int d^4y e^{-iky} < 0|T(j_\mu^\alpha(y)\phi(x_1)...\phi(x_n))|0>, \tag{17.5}$$

where we have suppressed the Lorentz and internal symmetry indices of the fields. We now compute the divergence $k^\mu G_\mu^\alpha$ and obtain

$$\begin{aligned}
k^\mu G_\mu^\alpha =& i \int d^4y(\partial^\mu e^{-iky}) < 0|T(j_\mu^\alpha(y)\phi(x_1)...\phi(x_n))|0> \\
=& -i \int d^4y e^{-iky} < 0|T(\partial^\mu j_\mu^\alpha(y)\phi(x_1)...\phi(x_n))|0> \\
&+ i \sum_{i=1}^n \int d^4y e^{-iky}\delta(y^0 - x_i^0) < 0|T(\phi(x_1)...[j_0^\alpha(y), \phi(x_i)]...\phi(x_n))|0>.
\end{aligned} \tag{17.6}$$

We have used partial integration and the sum of terms containing the equal time commutators in the last line are obtained when the $\partial/\partial y$ derivative hits the θ-functions of the time-ordered product. We can simplify this expression, first by using current conservation $\partial^\mu j_\mu^\alpha(y) = 0$ and, second, by going to the $\mathbf{k} = 0$ frame:

$$\begin{aligned}
k^0 G_0^\alpha =& i \sum_{i=1}^n e^{-ik^0 x_i^0} < 0|T(\phi(x_1)...[Q^\alpha, \phi(x_i)]...\phi(x_n))|0> \\
=& - \sum_{i=1}^n e^{-ik^0 x_i^0} < 0|T(\phi(x_1)...\delta^\alpha\phi(x_i)...\phi(x_n))|0>.
\end{aligned} \tag{17.7}$$

Since, by the assumption (17.2), $\delta^\alpha\phi(x_i)$ is a linear combination of the elementary fields $\phi(x)$, (17.7) expresses the divergence of the Green function G_μ^α as a linear combination of Green functions of the elementary fields. Relations of this form are called *Ward–Takahashi identities* and they were first derived for quantum electrodynamics. They encode all the information coming from the symmetry of the Lagrangian at the level of the Green functions and, from there, at all measurable quantities. If the current is not exactly conserved the corresponding term in (17.6) does not vanish but it is given by the variation of the Lagrangian density under the symmetry transformation:

$$k^0 G_0^\alpha = i \sum_{i=1}^{n} e^{-ik^0 x_i^0} < 0| T(\phi(x_1)...[Q^\alpha, \phi(x_i)]...\phi(x_n))|0 >$$

$$= - \sum_{i=1}^{n} e^{-ik^0 x_i^0} < 0| T(\phi(x_1)...\delta^\alpha \phi(x_i)...\phi(x_n))|0 > \qquad (17.8)$$

$$- i \int d^4 y e^{-ik^0 y^0} < 0| T(\delta\mathcal{L}(y)\phi(x_1)...\phi(x_n))|0 >.$$

As a first application we can consider the identities (17.7) and (17.8) at the limit $k^0 \to 0$. If we assume that our theory does not contain any massless field with the same quantum numbers as j_μ^α or $\partial^\mu j_\mu^\alpha$, then it is easy to see that the left-hand side vanishes. Indeed, in perturbation theory, pole singularities in the external momenta arise only from the propagators of massless particles. Therefore, (17.7) shows that a sum of Green functions of the elementary fields of the Lagrangian vanishes, as a consequence of the invariance of $\mathcal{L}$ under $\mathcal{G}$. In Problem 17.1 we show that for the case of (17.1), these identities imply, in particular, the same mass for protons and neutrons, as well as for charged and neutral pions and a single value for all pion–nucleon coupling constants. At finite k, (17.7) gives a relation between the Green functions of the symmetry currents and those of the elementary fields of the theory.

All these conclusions are valid at the classical level and the question we want to address is what happens to the quantum theory. Can we obtain a set of renormalised Green functions still satisfying the relation (17.7)? For the case of global symmetries the answer is yes and the simplest way to prove it is to exhibit a regulator which is compatible with any global symmetry. We want to emphasise, however, that the existence of such an invariant regulator is a sufficient condition for the symmetries to be preserved at the quantum level but it is not at all necessary. The symmetry could be recovered when the regulator is removed, but, in this case, we must give a proof that this is possible.

The simplest, conceptually, regulator which respects all global symmetries is the one we presented in Problem 16.2. It is based on the relation (17.3), which implies that terms with higher derivatives in the kinetic energy of the fields are still invariant under $\mathcal{G}$. Since adding such terms can render all Feynman diagrams convergent for all values of the regulator parameter, the Ward identities can be maintained.[2]

This analysis has an important, although very simple, consequence for the renormalisation programme of a quantum field theory. In Chapter 16 we introduced

---

[2] To be precise, we must note that such a regulator is acceptable only when applied to massive fields; otherwise, some diagrams may become infrared divergent.

the concept of stability, or completeness. The results of this Chapter can be used to sharpen this concept. A Lagrangian field theory contains a certain number of parameters, masses, and coupling constants. At the classical level these parameters are arbitrary and we can impose any set of relations among them. A symmetry can be viewed as a particular set of such relations. An arbitrary relation which reduces the number of parameters in a field theory will not, in general, give a stable Lagrangian; the renormalisation programme will not respect the relation. What we have shown here is that relations which increase the global symmetries of a theory give stable Lagrangians. In fact, they are the only relations which are compatible with the requirement of stability, with the exception of the trivial case of a relation which yields a free-field theory. The example of the pion–nucleon interaction (17.1) will illustrate this point better. Let us consider some special relations.

Setting $\lambda = 0$ does not increase the symmetry of the Lagrangian. The four-pion interaction term will appear as a counter-term at higher orders. The same is true with relations such as $\lambda = g$, or, more general, $f(\lambda, g) = 0$. The trivial exception is the relation $g = 0$ for which the fermions become free. Looking at the mass terms we note that $\mu = 0$ is not a stable relation. A theory with massless pions is not more symmetric than that with massive ones. In contrast, $m = 0$ is stable. The kinetic energy term of massless fermions has a larger symmetry because it is invariant under chiral transformations. This can be easily seen by separating the right and left components of the fermion field $\psi = \psi_R + \psi_L$ and noting that the kinetic energy splits into $\bar{\psi} i \partial\!\!\!/ \psi = \bar{\psi}_R i \partial\!\!\!/ \psi_R + \bar{\psi}_L i \partial\!\!\!/ \psi_L$. Therefore, we can perform independent phase transformations of the right and left components. The mass term breaks half of this symmetry because it mixes ψ_R and ψ_L: $\bar{\psi} \psi = \bar{\psi}_R \psi_L + \bar{\psi}_L \psi_R$. So does the Yukawa term, but a discrete subgroup remains because we can absorb a change of sign of the fermionic part with a corresponding change of the pion field. We can verify these properties by explicit 1-loop calculations.[3]

The formalism we set up to study the symmetry properties of a field theory allows us to go one step further and examine theories with broken symmetries. Let us consider again a Lagrangian density $\mathcal{L}$ invariant under a group of global transformations $\mathcal{G}$ and add to it a symmetry breaking term which we assume, as usual, to be a polynomial in the fields and their first derivatives of degree no higher than 4:

$$\mathcal{L} = \mathcal{L}_{\text{inv}} + \mathcal{L}_{\text{br}}. \tag{17.9}$$

$\mathcal{L}_{\text{br}}$ is assumed to be invariant only under a subgroup $\mathcal{G}_0$ of $\mathcal{G}$. Obviously, this statement is meaningless unless we specify the symmetry breaking term. We can do so by performing first the renormalisation of $\mathcal{L}_{\text{inv}}$ and consider $\mathcal{L}_{\text{br}}$ as insertion of a composite operator. We saw in Chapter 16 how Green functions are renormalised in this case: renormalisation forces us to include *all* terms with dimensions less than or equal to $\mathcal{L}_{\text{br}}$ and compatible with the symmetries which are still left unbroken, in this case $\mathcal{G}_0$. This allows us to distinguish between *soft* and *hard* breaking terms. We will call soft breaking

---

[3] The relation $\mu = m = 0$ looks like a stable relation because the Lagrangian is left with no dimensionful parameter and becomes, classically, invariant under scale transformations. We will show, however, shortly that scale invariance is always broken by quantum corrections.

Figure 17.1 *The 1-loop diagram for the insertion of the operator* $j_\mu^\alpha(x)$.

a term whose dimension is less than or equal to 3 and hard breaking one with dimension equal to 4. The origin of the terminology is clear: adding a soft breaking term with dimension $d \leq 3$ will force us to break $\mathcal{G}$ to $\mathcal{G}_0$ for all terms with dimensions less than or equal to d, but we can still use all interaction terms with dimension equal to 4 invariant under the full group $\mathcal{G}$. On the other hand, a hard breaking term with $d = 4$ will break $\mathcal{G}$ to $\mathcal{G}_0$ everywhere. In our example of the pion–nucleon interaction (17.1), introducing a mass splitting between the pions, which has $d = 2$, will not affect any other term, one between the nucleons, $d = 3$, will force us to split also the pions and any breaking in the interaction terms which have $d = 4$ will break the symmetry completely. The first two are soft breakings and the last is hard.

Before closing this section, we want to point out a further, very important consequence of the Ward–Takahashi identities (17.7). As we said earlier, the current operator $j_\mu^\alpha(x)$ is a monomial in the unrenormalised fields and, for scalar fields, their derivatives. For quantum electrodynamics it is our familiar $\bar{\psi}(x)\gamma_\mu\psi(x)$. We emphasise here the fact that the fields ψ are the unrenormalised operators. Indeed, the commutation relations (17.4) which show that the charges are the generators of the symmetry transformations of $\mathcal{G}$ are obtained using the canonical commutation relations. *A priori*, we would expect the Green functions of $j_\mu^\alpha(x)$ (17.5) to be renormalised in two ways: first, by the Z factors that relate the unrenormalised to the renormalised fields and, second, by the counter-term which removes the divergence of the Green functions of the composite operator

$$G_\mu^\alpha(k_\mu; x_1, ..., x_n) = Z_j^{-1}Z_2...G_{\mu R}^\alpha(k_\mu; x_1, ..., x_n),$$ (17.10)

where Z_j is the renormalisation factor for the composite operator $j_\mu^\alpha(x)$, Z_2 is the square of the wave function renormalisation for the field $\psi(x)$, and the dots stand for whichever other renormalisation counter-terms are needed for the rest of the Green function. For quantum electrodynamics at 1 loop, Z_2 is determined by the electron 2-point function and is given in Eq. (16.59). Z_j is determined by the diagram of Fig. 17.1. The important point is that these counter-terms spoil the Ward–Takahashi identity (17.7) because they multiply the left-hand side by $Z_j^{-1}Z_2$. However, we have just shown that renormalisation will never force us to violate any global internal symmetry. It follows that we can always choose

$$Z_j = Z_2.$$ (17.11)

This relation will be automatically satisfied if we use a regularisation scheme which respects the symmetry under $\mathcal{G}$, but even if we do not, we will always be able to enforce it. It follows that the Green functions of a symmetry current require no renormalisation. We could arrive at the same conclusion by noting that the commutation relations of the symmetry generators (17.4) are non-homogeneous and, therefore, they fix uniquely the absolute scale of the Q's.

This is not a new result, it is the one we obtained in (16.64). Indeed it is easy to check diagrammatically that Z_j is equal to our old Z_1 of the vertex function renormalisation of (16.62). Although quantum electrodynamics has a local $U(1)$ symmetry, for the relation $Z_1 = Z_2$ only the global part is needed.

17.3　Gauge Symmetries: Examples

Gauge symmetries change the picture in two ways. First, for the non-Abelian case, the Ward identities are more complicated. As we showed in Chapter 14, the Lagrangian with which we actually derive the Feynman rules is not invariant under local transformations, because of the necessity to choose a particular gauge. The resulting Ward identities, which are called *Slavnov–Taylor identities*, are obtained using the BRST global symmetry under which the fields transform non-linearly. Second, the regularisation scheme of higher derivative kinetic terms is no more available because the relation (17.3) is no more true. Only the covariant derivatives of the fields transform like the fields themselves but they introduce new couplings and do not regularise all the Feynman diagrams. Dimensional regularisation offers a convenient alternative whenever applicable and, indeed, using this scheme we can prove in a straightforward way that we can enforce the Slavnov–Taylor identities for the renormalised Green functions at every order of perturbation theory. The next chapter will be entirely devoted to this question. Here we will only look at some specific cases. The results will serve as introduction to the general discussion in the following chapter.

It follows from the previous discussion that the only cases where trouble may arise are the cases where dimensional regularisation cannot be used. As we pointed out in section 16.4, this is the case of chiral transformations involving the matrix γ_5. It is this case that we want to investigate in this section.

17.3.1　The Adler–Bell–Jackiw Anomaly

Let us start with the simplest gauge theory, namely quantum electrodynamics. It has a conserved vector current $j_\mu(x)$:

$$j_\mu(x) = \bar{\psi}(x)\gamma_\mu\psi(x); \quad \partial^\mu j_\mu(x) = 0. \tag{17.12}$$

In the limit of vanishing electron mass it has a second current which is also conserved as a consequence of the classical equations of motion. It is the axial current:

$$j_\mu^5(x) = \bar{\psi}(x)\gamma_\mu\gamma^5\psi(x); \quad \partial^\mu j_\mu^5(x) = 0 \tag{17.13}$$

The existence of the two conserved currents is a consequence of the invariance of the classical Lagrangian of massless electrodynamics under the group of chiral $U(1) \times U(1)$ transformations. When the electron mass is non-zero the axial symmetry is broken by the mass term, and the conservation equation (17.13) changes into

$$\partial^\mu j_\mu^5(x) = 2im_e \bar{\psi}(x)\gamma^5\psi(x) \equiv 2im_e j^5(x). \tag{17.14}$$

It is the validity of Eq. (17.12) and (17.13), or (17.12) and (17.14), that we want to investigate order by order in perturbation theory. A naïve application of the method used in deriving Eq. (17.6) gives

$$\begin{aligned}
k^\mu G_\mu^5 &= i \int d^4y (\partial^\mu e^{-iky}) < 0|T(j_\mu^5(y)\phi(x_1)...\phi(x_n))|0> \\
&= -i \int d^4y e^{-iky} < 0|T(\partial^\mu j_\mu^5(y)\phi(x_1)...\phi(x_n))|0> \\
&\quad + i \sum_{i=1}^{n} \int d^4y e^{-iky}\delta(y^0 - x_i^0) < 0|T(\phi(x_1)...[j_0^5(y),\phi(x_i)]...\phi(x_n))|0>,
\end{aligned} \tag{17.15}$$

where the fields $\phi(x)$ represent collectively the fields of electrons and/or photons. Let us choose, for simplicity, only photon fields, in which case the canonical commutation relations imply that all the equal time commutators on the right-hand side vanish. Taking the simplest non-trivial case with $n=2$ we rewrite (17.15) as

$$\begin{aligned}
k^\mu G_{\mu\nu\rho}^5 &= i \int d^4y (\partial^\mu e^{-iky}) < 0|T(j_\mu^5(y)A_\nu(x_1)A_\rho(x_2))|0> \\
&= -i \int d^4y e^{-iky} < 0|T(\partial^\mu j_\mu^5(y))A_\nu(x_1)A_\rho(x_2))|0> \\
&= 2m_e \int d^4y e^{-iky} < 0|T(j^5(y)A_\nu(x_1)A_\rho(x_2))|0> \equiv 2m_e G_{\nu\rho}^5.
\end{aligned} \tag{17.16}$$

Let us try to verify this relation order by order in perturbation theory. To lowest order the only diagrams are those of Fig. 17.2, together with those obtained by crossing the two external photon lines. By power counting, the diagrams are linearly divergent, so we assume that some suitable regulator has been introduced. The calculation of $G_{\mu\nu\rho}^5$ involves the following trace of γ matrices and fermion propagators:

$$k^\mu G_{\mu\nu\rho}^5 \Rightarrow \text{Tr}[\frac{1}{\slashed{p} - m}\gamma_\rho \frac{1}{\slashed{p} + \slashed{q}_2 - m}(\slashed{q}_1 + \slashed{q}_2)\gamma^5 \frac{1}{\slashed{p} - \slashed{q}_1 - m}\gamma_\nu]. \tag{17.17}$$

We now write $\slashed{q}_1 + \slashed{q}_2 = (\slashed{p} + \slashed{q}_2 - m) - (\slashed{p} - \slashed{q}_1 + m) + 2m$ and use the fact that γ^5 anticommutes with all four γ matrices to get

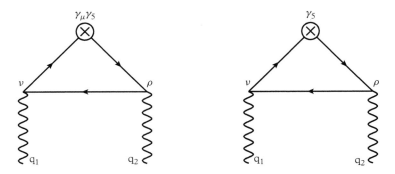

Figure 17.2 *The two lowest order diagrams contributing to the axial current Ward identity.*

$$k^\mu G^5_{\mu\nu\rho} \Rightarrow \mathrm{Tr}[\frac{1}{\not{p}-m}\gamma_\rho\gamma^5\frac{1}{\not{p}-\not{q}_1-m}\gamma_\nu]$$

$$+ \mathrm{Tr}[\frac{1}{\not{p}-m}\gamma_\rho\frac{1}{\not{p}+\not{q}_2-m}\gamma^5\gamma_\nu] \qquad (17.18)$$

$$+ 2m\mathrm{Tr}[\frac{1}{\not{p}-m}\gamma_\rho\frac{1}{\not{p}+\not{q}_2-m}\gamma^5\frac{1}{\not{p}-\not{q}_1-m}\gamma_\nu].$$

The last term is the right-hand side of (17.16) we want to compute. We can rewrite the first two terms using the cyclic property of the trace and adding the two terms obtained from the crossed diagram ($\nu \leftrightarrow \rho$ and $q_1 \leftrightarrow q_2$) as

$$\mathrm{Tr}[\frac{1}{\not{p}-\not{q}_1-m}\gamma_\nu\frac{1}{\not{p}-m}\gamma_\rho\gamma^5] - \mathrm{Tr}[\frac{1}{\not{p}-m}\gamma_\nu\frac{1}{\not{p}+\not{q}_1-m}\gamma_\rho\gamma^5]$$

$$+\mathrm{Tr}[\frac{1}{\not{p}-\not{q}_2-m}\gamma_\rho\frac{1}{\not{p}-m}\gamma_\nu\gamma^5] - \mathrm{Tr}[\frac{1}{\not{p}-m}\gamma_\rho\frac{1}{\not{p}+\not{q}_2-m}\gamma_\nu\gamma^5]. \qquad (17.19)$$

Now, we remember that p is an integration variable. If the integrals were convergent, we would have been allowed to shift integration variables. Changing, for example, in the first term $p' = p - q_1$ will make the first two terms cancel. Similarly, the change $p' = p - q_2$ makes the sum of the third and fourth terms vanish. In this case the Ward identity (17.16) is satisfied.

The trouble is that shifting integration variables is *not* a legitimate operation for linearly divergent integrals. For example, if we use a cut-off Λ, the integral will behave like Λ. The shifted integral will give Λ + a constant. So, the difference will not give 0.

The simplest way to proceed is indeed to introduce a cut-off. Dimensional regularisation does not respect the anti-commutation properties of γ^5 with all γ matrices in d dimensions, but, following 't Hooft and Veltman, we can fix γ^5 in four dimensions as $i\gamma^0\gamma^1\gamma^2\gamma^3$. As a result, γ^5 will anti-commute with the four-dimensional γ matrices but it will commute with all the others in $d-4$ dimensions. In our diagram only the integration variable p lives in d dimensions, so we split it as

$$p = p_{\parallel} + p_{\perp}, \tag{17.20}$$

where $p_{\parallel}$ has components only in four dimensions and $p_{\perp}$ only in the other $d-4$. The previous argument will apply only to $p_{\parallel}$; therefore, we will obtain the identity (17.16) with an extra term proportional to $2\gamma^5 p_{\perp}$. The trace with the factor γ^5 requires the presence of four other four-dimensional γ matrices. We have already the factors γ_ν and γ_ρ, and, in order to obtain a non-zero trace, we need one factor $\not{q}_1$ and one $\not{q}_2$. The symmetric integration forces us to keep another factor $p_{\perp}$. The calculation is now straightforward, introducing Feynman parameters and replacing $p_{\perp}p_{\perp}$ by $p^2\epsilon/d$. The divergent integral gives a pole term of ϵ^{-1} and, therefore, we obtain an extra *finite* term in (17.16):

$$k^\mu G^5_{\mu\nu\rho} = 2m_e G^5_{\nu\rho} + \frac{e^2}{2\pi^2}\epsilon_{\alpha\beta\nu\rho}q_1^\alpha q_2^\beta. \tag{17.21}$$

This is an astonishing result. It is the first time we encounter a case in which the consequences of the classical equations of motion and the canonical commutation relations are not valid in perturbation theory. It seems that the quantum theory does not respect the full $U(1) \times U(1)$ chiral symmetry of the classical Lagrangian. The second term in (17.21), which is responsible for such a violation, is called an *anomaly*.

Several questions should be answered before assessing the importance of this result. First, although the above derivation is unambiguous, we would like to verify the result using a different method. Second, we want to know whether this breakdown of the classical symmetry depends on the particular matrix element we considered and how it changes in higher orders of perturbation. Third, we must study the possibility of absorbing the extra term in a redefinition of the axial current. Last, but not least, comes the question of the physical significance of the result. In this chapter we will answer the first three questions. The last one will be addressed later, when quantum field theory will be applied to describe the physics of fundamental interactions.

17.3.2 A Path Integral Derivation

It is easy to answer the first question. In Problem 17.2 we propose an independent calculation of the triangle diagram which assumes only the conservation of the vector current. The result is, of course, the same. Here we want to present a method which does not require the calculation of any Feynman diagram. It was introduced in 1979 by Kazuo Fujikawa and uses directly the path integral quantisation of gauge theories we studied in Chapter 14. In fact the method applies equally well to the Abelian gauge theory of quantum electrodynamics as well as to any non-Abelian Yang–Mills theory.

Let us start with the path integral representation of the generating functional of the Green functions. We consider, as an example, a gauge theory based on a compact Lie group $\mathcal{G}$ with fermions belonging to an n-dimensional representation,

$$Z[\eta, \bar{\eta}, \mathcal{J}] = \frac{\int [\mathcal{D}\mu] e^{i\int d^4x\left[\bar{\psi}(i\not{D}\ -m)\psi+\frac{1}{2g^2}\,TrF_{\mu\nu}F^{\mu\nu}+\bar{\psi}\eta+\bar{\eta}\psi-A_\mu\mathcal{J}^\mu\right]}}{\int [\mathcal{D}\mu] e^{i\int d^4x\left[\bar{\psi}(i\not{D}\ -m)\psi+\frac{1}{2g^2}\,TrF_{\mu\nu}F^{\mu\nu}\right]}}, \tag{17.22}$$

where, η, $\bar{\eta}$, and $\mathcal{J}^\mu$ are the classical sources for the quantum fields $\bar{\psi}$, ψ, and A_μ and $\mathcal{D}\mu$ is the measure of integration given by

$$\mathcal{D}\mu = \prod_x [\mathcal{D}A_\mu(x)][\mathcal{D}\bar{\psi}(x)][\mathcal{D}\psi(x)]. \tag{17.23}$$

In the gauge field part $[\mathcal{D}A_\mu(x)]$ we include the Faddeev–Popov determinant, which, as usual, we assume to be independent of the fermion fields.

Under a γ_5 transformation the fields transform as

$$\psi(x) \to e^{i\theta(x)\gamma_5}\psi(x); \quad \bar{\psi}(x) \to \bar{\psi}(x)e^{i\theta(x)\gamma_5}. \tag{17.24}$$

The Lagrangian density in (17.22) transforms as

$$\mathcal{L} \to \mathcal{L} - \partial_\mu\theta(x)\bar{\psi}\gamma^\mu\gamma_5\psi - 2im\theta(x)\bar{\psi}\gamma_5\psi. \tag{17.25}$$

Using these transformation properties and, assuming the invariance of the measure (17.23), we derive, in the usual way, the classical Ward identities. The only step which is not obvious is, precisely, the invariance of the fermion part of the measure under (17.24). Let us examine it more closely.

We expand the classical fermion fields in a complete set of eigenfunctions which we choose to be those of the operator $\slashed{D}$. In Euclidean space it is a Hermitian operator with real eigenvalues. We write

$$\slashed{D}\,\phi_n(x) = \lambda_n\phi_n(x); \quad \int \phi_n(x)^\dagger\phi_m(x)d^4x = \delta_{n,m} \tag{17.26}$$

This expression is formal because in R^4 the spectrum of eigenvalues is continuous. We can consider a finite box with some kind of boundary conditions, or a compactified version of R^4, for example a stereographic projection of Euclidean four space onto a four-sphere S^4. The fermion fields can be expanded in the basis of $\phi_n(x)$ as

$$\psi(x) = \sum_n a_n\phi_n(x); \quad \bar{\psi}(x) = \sum_n \phi_n(x)^\dagger \bar{b}_n, \tag{17.27}$$

where, as we explained in Chapter 11, the coefficients a_n and $\bar{b}_n$ are independent elements of a Grassmann algebra. The integration measure (17.23) can be written as

$$\mathcal{D}\mu = \prod_x [\mathcal{D}A_\mu(x)] \prod_n d\bar{b}_n \prod_m da_m. \tag{17.28}$$

Under the chiral transformation (17.24) the coefficient a_n transforms as

$$a'_m = \sum_n \int d^4x\phi_m(x)^\dagger e^{i\theta(x)\gamma_5}\phi_n(x)a_n \equiv \sum_n C_{m,n}a_n, \tag{17.29}$$

and similarly for $\bar{b}_n$. In the transformation $a_n \rightarrow a'_n$ the measure picks up a factor $\det[C]^{-1}$, where -1 appears because of the anti-commutation properties of the a's. Since we have the matrix elements of C, we can compute the determinant as the exponential of the trace of $\ln C$. For infinitesimal θ we obtain

$$[\det C]^{-1} = e^{-i\sum_n \int d^4 x \theta(x) \phi_n(x)^\dagger \gamma_5 \phi_n(x)} \equiv e^{-i\int d^4 x \theta(x) A(x)}, \tag{17.30}$$

where

$$A(x) = \sum_n \phi_n(x)^\dagger \gamma_5 \phi_n(x). \tag{17.31}$$

The classical chiral Ward identity (17.14) or (17.15) is obtained by noting that $A(x)$ is formally proportional to the trace of γ_5 and, therefore, it vanishes. This implies the invariance of the measure of the functional integral. Of course, this argument is only formal because the expression (17.31) is, in fact, divergent. By considering the system in a finite volume we have regulated the infrared behaviour. We will need also an ultraviolet cut-off for the large eigenvalues. K. Fujikawa used a factor $e^{-(\lambda_n/M)^2}$, which regulates eigenvalues such that $\lambda_n^2 \gg M^2$:

$$A(x) = \lim_{M\to\infty} \left(\sum_n \phi_n(x)^\dagger \gamma_5 e^{-(\lambda_n/M)^2} \phi_n(x) \right)$$
$$= \lim_{M\to\infty} \left(\sum_n \phi_n(x)^\dagger \gamma_5 e^{-(\slashed{D}/M)^2} \phi_n(x) \right). \tag{17.32}$$

If, for simplicity, we assume that the gauge group is $SU(N)$ and the fermions belong to the fundamental representation, we can write $\slashed{D}^2 = D^2 + \frac{1}{4}[\gamma^\mu, \gamma^\nu] F_{\mu\nu}$ with $F_{\mu\nu} = \partial_\mu A_\nu - \partial_\nu A_\mu + [A_\mu, A_\nu]$. We can also choose plane wave solutions of the free Dirac operator as the basis for our expansion. When $M \rightarrow \infty$ the dominant contributions will come from the large momentum solutions and we can expand in powers of the gauge field. Since we have the factor γ_5, we will need a minimum of four γ matrices in order to obtain a non-zero trace. Therefore, we obtain

$$A(x) = \lim_{M\to\infty} \text{Tr} \int \frac{d^4 k}{(2\pi)^4} \gamma_5 e^{-ikx} e^{-(\slashed{D}/M)^2} e^{ikx}$$
$$= \lim_{M\to\infty} \text{Tr} \gamma_5 \left([\gamma^\mu, \gamma^\nu] F_{\mu\nu} \right)^2 \frac{1}{4M^4} \frac{1}{2} \int \frac{d^4 k}{(2\pi)^4} e^{-k^2/M^2} \tag{17.33}$$
$$= \frac{-1}{16\pi^2} \text{Tr} \tilde{F}_{\mu\nu}(x) F^{\mu\nu}(x),$$

where $\tilde{F}_{\mu\nu} = \frac{1}{2}\epsilon_{\mu\nu\rho\sigma} F^{\rho\sigma}$. The trace is taken in both Dirac and $SU(N)$ indices, whenever applicable.

The non-vanishing of $A(x)$ is the manifestation of the anomaly. Indeed, going back to the functional integral (17.22) and performing the γ_5 transformation (17.24) and (17.25), we obtain an extra term on the right-hand side of (17.25):

$$\mathcal{L} \to \mathcal{L} + \theta(x) \left(\partial_\mu j_5^\mu - 2imj_5 + \frac{1}{8\pi^2} \text{Tr}\tilde{F}_{\mu\nu}(x)F^{\mu\nu} \right) \tag{17.34}$$

in complete agreement with Eq. (17.21). The factor of 2 in front of the anomaly between Eq. (17.33) and (17.34) comes from the two factors $\mathcal{D}\bar{\psi}$ and $\mathcal{D}\psi$ of the functional measure (17.23), and the absence of a factor proportional to the square of the gauge coupling constant is due to our normalisation of the gauge fields. This formula for the anomaly applies to the Abelian theory we examined before, in which case the extra term reproduces the triangle diagram in perturbation theory, as well as to non-Abelian Yang–Mills theories. In this case the $\text{Tr}\tilde{F}_{\mu\nu}(x)F^{\mu\nu}(x)$ term will produce a triangle diagram, but also a square and a pentagon. However, these are not new anomalies because gauge invariance fixes their coefficients uniquely in terms of that of the triangle. In a later chapter we shall study the precise form of the anomaly in some physically interesting Yang–Mills theories. Let us also note here that we can define a current $G_\mu(x)$ as

$$G_\mu = 2\epsilon_{\mu\nu\rho\sigma} \text{Tr} \left(A^\nu \partial^\rho A^\sigma + \frac{2}{3} A^\nu A^\rho A^\sigma \right). \tag{17.35}$$

It is straightforward to verify that G_μ is not gauge invariant but its divergence is

$$\partial^\mu G_\mu = \text{Tr}\tilde{F}_{\mu\nu} F^{\mu\nu}. \tag{17.36}$$

Before going further, let us look again at (17.31). Since $\not{D}$ and γ_5 anticommute, if ϕ_n is an eigenfunction of $\not{D}$ with eigenvalue $\lambda_n \neq 0$, $\gamma_5\phi_n$ is also an eigenfunction with eigenvalue $-\lambda_n$. Since eigenfunctions corresponding to different eigenvalues are orthogonal, it follows that in the integral, only terms corresponding to zero eigenvalues survive. In the subspace of these zero modes we can choose a basis with eigenvectors common to both $\not{D}$ and γ_5. Because $\gamma_5^2 = 1$, its eigenvalues are ± 1. Let us denote the number of such eigenvalues by $n_\pm$. If chiral symmetry is exact, $n_+ = n_-$ and the anomaly vanishes. Therefore, with the normalisation (17.26) and setting $\nu = n_+ - n_-$, we obtain an integrated form of the anomaly:

$$n_+ - n_- = \nu = -\frac{1}{32\pi^2} \int d^4x \text{Tr}\tilde{F}_{\mu\nu}(x)F^{\mu\nu}(x). \tag{17.37}$$

At first sight, given the relation (17.36), we would conclude that $\nu = 0$, since it is given by the integral over all space of the divergence of the current G_μ. Indeed, in our calculations we often assumed that the classical fields vanish at infinity. However, for gauge fields this is not necessarily a reasonable assumption. Even if we want to assume that asymptotically we reach the vacuum state, this will only force the gauge fields to vanish up to gauge transformations. We can find field configurations which give a non-vanishing integral in (17.37), although the fields at infinity are gauge equivalent to 0. But then we obtain another unexpected result. The left-hand side is necessarily an integer; therefore, the integral on the right must also be an integer. It turns out that this is guaranteed by a deep mathematical theorem. We have seen that $F^{\mu\nu}$ has a very precise geometrical meaning; it is the curvature in the fibre bundle defined by the gauge theory. There exists a topological theorem, due to A. Pontryagin, which, in our notation, states precisely this property. The integral is a topological invariant of the theory and the integer ν is called the *Pontryagin index*, or *the winding number*. We shall explain the origin of this name in a later chapter. Equation (17.37) relates a topological quantity related to the gauge theory with the chirality deficit, i.e. the difference between the number of positive and negative chirality zero modes of the Dirac operator. This relation, which we have derived heuristically here, has been proven rigorously in mathematics and it is called *the Atiyah–Singer index theorem*. It is an amazing theorem because it relates seemingly independent properties of the manifold. The winding number is a topological quantity and, as such, it describes the global properties. The chirality deficit, on the other hand, is given by the zero eigenvalues of the Dirac operator, which is a first-order linear differential operator. Normally, we expect this quantity to reflect the local properties of the manifold. This relation between global and local properties shows the special role played in mathematics by the Dirac operator, which was introduced in physics, roughly as a square root of the four-dimensional Laplace operator, by the need to write a relativistic wave equation for the electron. Another beautiful example of the inter-relation between physics and mathematics.

The second nice feature of (17.33) is that it generalises immediately to any space–time with an *even* number of dimensions $d = 2n$ because we can always write an epsilon symbol with $2n$ indices. This will determine the number of gauge tensors we will use. For example, in two dimensions the anomaly will be proportional to $\epsilon_{\mu\nu}F^{\mu\nu}$.

This calculation answers our first question, namely the independence of the anomaly on the particular way we computed the triangle diagram. In fact, K. Fujikawa has gone one step further and shown that, in his method, we obtain the same anomaly factor with any kind of smooth regularisation of the large eigenvalues.

The second question concerns the degree of generality of the anomaly, regarding both its operator form and the value of its coefficient. There is a precise answer to this question in perturbation theory for Abelian and non-Abelian gauge theories, known as *the Adler–Bardeen theorem*. It states that the axial anomaly is universal. The violation of the axial current conservation equation has the form we obtained here for all Green functions. Even more surprisingly, the numerical coefficient we found, by performing either the triangle diagram calculation or the semi-classical one of the fermionic measure, remains unchanged at every order in the perturbation expansion. It is one of the rare examples in quantum field theory in which a 1-loop effect remains unchanged at all orders.

This theorem was proven by a careful analysis of the divergence structure of the diagrams that could contribute to the anomaly to all orders. We will not reproduce this proof here. We only note that by relating the anomaly to a topological invariant of the theory, the path integral derivation makes this result plausible. Indeed, the presence of non-trivial renormalisations would make the anomaly a formal power series in the coupling constant and it does not seem possible to reconcile such a structure with the fact that the right-hand side of (17.37) must always remain an integer equal to the chirality deficit of the Dirac operator. We repeat, however, that in the form presented here, this is not a proof, only a plausibility argument.

17.3.3 The Axial Anomaly and Renormalisation

In this section we will answer our third question. We will prove that the anomaly cannot be absorbed by any gauge invariant redefinition of the axial current. Let us look again at Eq. (17.16) and (17.21). The first is the consequence of the classical Ward identity and the second the result we found by an explicit calculation of the triangle diagram. We want to prove that there is no acceptable subtraction of the triangle diagram which absorbs the anomaly.

In order to be acceptable a subtraction term must satisfy the following requirements: (i) it must have the correct Lorentz transformation properties, i.e. it must be a three-index axial tensor $S_{\mu\nu\rho}$, function of the two independent momenta q_1 and q_2. (ii) It must be symmetric under the combined exchange ($\nu \leftrightarrow \rho$ and $q_1 \leftrightarrow q_2$), a consequence of Bose statistics for photons. (iii) As we have explained in the previous chapter, subtraction terms must be polynomials in the external momenta. By power counting we see that, for the triangle, this polynomial must be of maximum degree equal to 1. (iv) It must satisfy the requirements of vector current conservation:

$$q_1^\nu S_{\mu\nu\rho}(q_1, q_2) = q_2^\rho S_{\mu\nu\rho}(q_1, q_2) = 0. \tag{17.38}$$

It is easy to see that there is no tensor satisfying all these requirements. Condition (i) imposes the use of the totally antisymmetric tensor $\epsilon_{\mu\nu\rho\sigma}$. This leaves one index to be saturated with a linear combination of q_1 and q_2. Bose statistics implies that this combination is $(q_1 - q_2)^\sigma$. But then the vector current conservation conditions (17.38) are not satisfied.

This argument shows that the anomaly is not intrinsically linked to the axial current. It is rather due to the incompatibility between the conservation of the two currents, the vector and the axial. We can subtract the triangle diagram and enforce the axial current Ward identity, at the price of violating the corresponding one for the vector current. And this conclusion has nothing to do with the choice of any conceivable regularisation scheme. A widespread misunderstanding on this point has resulted in many erroneous claims in the literature which proposed 'miraculous' regulators aiming at renormalising the anomaly away.

17.3.4 A Consistency Condition for Anomalies

As we just noted, we have a certain freedom in defining what we will call 'anomaly'. Even in the Abelian case we can choose the current whose conservation will be violated.

J. Wess and B. Zumino have derived the consistency conditions that any anomaly choice should satisfy for the general case of non-Abelian Yang–Mills theories.

Let us consider, for simplicity, a field theory with fermions belonging to the fundamental representation of some group $SU(N)$. We can construct N^2 vector currents and N^2 axial vector ones. Since the anomaly is a 1-loop phenomenon, we do not need to worry about the gauge interactions. If we want to study the Ward identities for $SU(N) \times SU(N)$, it is enough to consider free fermions in external, classical vector and axial vector fields V_μ^i and A_μ^i, respectively, with $i = 1, 2, ..., N^2 - 1$. The Ward identities can be derived by considering the variation of the generating functional of connected diagrams with respect to infinitesimal vector or axial gauge transformations. If $v(x)$ and $a(x)$ are the infinitesimal parameters, the variations are

$$
\begin{aligned}
\delta V_\mu &= v \times V_\mu - \partial_\mu v; \quad \delta A_\mu = v \times A_\mu \\
\delta A_\mu &= a \times V_\mu - \partial_\mu a; \quad \delta V_\mu = a \times A_\mu,
\end{aligned}
\tag{17.39}
$$

where we have used a vector product notation for $SU(N)$, i.e. $(v \times V_\mu)^i = f^{ijk} v^j V_\mu^k$, with f^{ijk} the structure constants. For a complete list of all Ward identities, for example if we have massive fermions, we may have to introduce also external pseudoscalar, or even scalar, fields. The classical Ward identities will express the invariance of the generating functional under the transformations (17.39) while the anomaly will appear as a particular form of the variation. The important point is that the variations will satisfy the $SU(N) \times SU(N)$ algebra and they will be restricted by the Jacobi identity. Following the notation of Wess and Zumino, let us define the variations under the gauge transformations as differential operators of the form

$$
\begin{aligned}
-X &= \partial_\mu \frac{\delta}{\delta V_\mu} + V_\mu \times \frac{\delta}{\delta V_\mu} + A_\mu \times \frac{\delta}{\delta A_\mu} + ... \\
-Y &= \partial_\mu \frac{\delta}{\delta A_\mu} + A_\mu \times \frac{\delta}{\delta V_\mu} + V_\mu \times \frac{\delta}{\delta A_\mu} + ...,
\end{aligned}
\tag{17.40}
$$

where the dots refer to the other fields. They satisfy the commutation relations of the vector and axial charges:

$$
\begin{aligned}
[X_i(x), X_j(y)] &= f_{ijk} \delta(x - y) X_k(x) \\
[X_i(x), Y_j(y)] &= f_{ijk} \delta(x - y) Y_k(x) \\
[Y_i(x), Y_j(y)] &= f_{ijk} \delta(x - y) X_k(x).
\end{aligned}
\tag{17.41}
$$

We assume that we have renormalised the theory keeping the vector currents conserved. Therefore, the X-variations will be zero

$$
X_i W \equiv F_i = 0.
\tag{17.42}
$$

The anomaly will appear in the axial transformations:

$$Y_i W \equiv G_i[V_\mu, A_\mu].$$ (17.43)

We have assumed here that the anomaly G_i will be a functional only of the external vector and axial vector fields and not of the other external fields we could have introduced. Combining (17.41), (17.42), and (17.43), we find that

$$X_i(x) G_j(y) = f_{ijk}\delta(x-y) G_k(x)$$
$$Y_i(x) G_j(y) - Y_j(y) G_i(x) = 0.$$ (17.44)

The first one shows that the anomaly belongs to the adjoint representation. The second is the real consistency condition. Already for practical purposes, the knowledge of this condition is very important because it tells us precisely how much freedom we really have. For example, it has been used to check detailed computations. But, most importantly, it offers the framework to go beyond the standard perturbation theory. Wess and Zumino asked the question of constructing an *effective* Lagrangian density, or, an effective generating functional, which would reproduce the 1-loop Ward identities, including the anomaly, in the classical approximation. We know that this is impossible with an effective Lagrangian which is polynomial in the fields and their derivatives. This is the meaning of the impossibility of finding a subtraction term which is a polynomial in the external momenta. If we are willing to enlarge the scene and consider non-polynomial Lagrangians, the consistency condition (17.44) will constitute an important guide. We will come back to this point in the next chapter.

17.4   The Breaking of Conformal Invariance

At the beginning of this section we noted that we can always renormalise a quantum field theory in a way that respects the classical transformation properties of Green functions under the Poincaré group. The argument was based on the fact that we can always choose a Poincaré invariant regulator. Here we want to study other space–time symmetries for which we cannot find invariant regulators.[4]

It is known from classical mechanics that there exist geometrical transformations, beyond those of rotations and translations, which leave the dynamics of some physical systems invariant. The example often given is the three-dimensional Laplace equation which is left invariant under space inversions $x \to x/(x)^2$.[5] They are discrete transformations which leave the points on the surface of the unit sphere invariant and map the points in the interior outside and vice versa. The origin is mapped at infinity, which

---

[4] We follow the presentation given in S. Coleman, 'Dilatations', Erice Lectures 1971; Reprinted in S. Coleman, *Aspects of Symmetry* (Cambridge University Press, Cambridge, 1985).

[5] The terminology could be misleading because we have used the term "inversion" earlier to denote a parity transformation.

means that a completion of R^3 with a cone at infinity is necessary. Another (continuous) transformation is space dilatation, also called *scale transformation*, which multiplies all distances by a constant. It leaves invariant every system which has no intrinsic scale. Considering the four-dimensional Minkowski space, we can search for the smallest connected group which contains the Poincaré transformations and the inversions. It turns out that it is a fifteen-parameter group called *the conformal group*. In infinitesimal form it contains the ten Poincaré generators, four generators of special conformal transformations obtained by combining translations and inversions, and one generator of dilatations. In this section we want to study the consequences of such transformations in quantum field theory.

Let us consider our simple field theory model, that of a self-interacting real scalar field:

$$\mathcal{L} = \frac{1}{2}\partial_\mu\phi(x)\partial^\mu\phi(x) - \frac{1}{2}m^2\phi(x)^2 - \frac{\lambda}{4!}\phi(x)^4. \tag{17.45}$$

A space dilatation, or scale transformation, acts on the coordinate x as a multiplication by a constant,

$$x \rightarrow e^\alpha x, \tag{17.46}$$

where α is a constant parameter. If we assume that the field $\phi(x)$ transforms linearly under (17.46), we can write

$$\phi(x) \rightarrow e^{\alpha d}\phi(e^\alpha x) \tag{17.47}$$

with d being some constant which we will call *the scale dimension* of the field ϕ. For infinitesimal transformations, (17.47) gives

$$\delta\phi(x) = \left(d + x_\mu \frac{\partial}{\partial x_\mu}\right)\phi(x). \tag{17.48}$$

Is there a choice of d for which this transformation leaves invariant the Lagrangian (17.45)? Applying (17.48) to every term in (17.45) and performing an integration by parts, we find that the variations of the kinetic energy and the interaction terms are proportional to $(d - 1)$, while that of the mass term is proportional to $(d - 2)$. This is not surprising. We expect scale transformations to be symmetries in the absence of any dimensionful parameter because, in this case, there is no *a priori* scale. So if we assign $d = 1$ to our scalar field, scale invariance is broken only by the mass term. It will not appear as a surprise if we say that had we done the analysis with a Lagrangian involving fermion fields, the result would have been $d = \frac{3}{2}$. Following Sidney Coleman, we want to stress, however, that despite appearances, this is *not* simple dimensional analysis! These are real *scale* transformations. The proof is that mass terms break the symmetry while nothing could conceivably violate dimensional analysis. The only result so far is that, classically, the scale dimensions of the various fields should be chosen equal to their canonical dimensions in the sense of dimensional analysis for the theory to be invariant under dilatations, up to terms with dimensionful parameters. It is easy to check that for a renormalisable Lagrangian, the invariance covers the whole conformal group.

The canonical Ward identities are obtained through Eq. (17.8) setting $k^0 = 0$. We write them for the 1-PI Green functions in momentum space using the transformation properties of the field (17.48) and the fact that the dilatation symmetry breaking term is the mass term in (17.45),

$$\left(\sum_{i=1}^{2n-1} p_i \frac{\partial}{\partial p_i} + 2n - 4 \right) \Gamma^{(2n)}(p_1, p_2, ..., p_{2n}) + im^2 \Gamma_{\phi^2}^{(2n)}(0, p_1, p_2, ..., p_{2n}) = 0, \quad (17.49)$$

where the last term, due to the symmetry breaking, is the $2n$-point 1-PI Green function with an insertion of a ϕ^2 operator at zero momentum. We have put the scale dimension of the scalar field d equal to 1. We have also used the conservation of energy and momentum which gives a delta function $\delta^{(4)}(\sum_{i=1}^{2n} p_i)$.

This is the classical Ward identity of broken scale invariance for the scalar field theory (17.45). Let us try to check it in perturbation theory. For example, let us take the 4-point function, $n = 2$. For the massless theory the Ward identity becomes very simple: the 4-point function is independent of the external momenta! This is indeed correct in the tree approximation, where $\Gamma^{(4)}$ is a constant, but it is obviously false at higher orders, as we have seen in the calculations of the previous sections. We can reach the same conclusion by looking at the full massive theory. The consequences of (17.49) are not satisfied in perturbation theory beyond the tree approximation. In our terminology, the Ward identities have anomalies.

With the experience we have gained so far, we should not be very surprised by this result. It is obvious that any conceivable regularisation introduces, one way or another, a dimensionful parameter which breaks scale and conformal invariance. Although this does not prove that the classical Ward identities *must* be violated, it raises the suspicion that they may be. In this section we want to derive the correct Ward identities, those that are satisfied order by order in perturbation theory. They are known as *the Callan–Symanzik equations*.

Let us go back to our Lagrangian (17.45). The parameters which appear are the un-renormalised ones. We know that Green functions calculated from (17.45) are divergent, but they can be defined if we introduce a suitable cut-off which we shall call Λ. In this case the unrenormalised Green functions will depend on

$$\Gamma^{(2n)}(p_1, p_2, ...p_{2n}; m, \lambda, \Lambda). \quad (17.50)$$

Since we are interested in the mass dependence, it is natural to compute the derivative of $\Gamma^{(2n)}$ with respect to m. This is easy because $\Gamma^{(2n)}$ is given by a sum of terms, each one being a multiple integral of a product of propagators $(k^2 - m^2)^{-1}$. The derivative will be a sum of terms in each of which one propagator is replaced by its square. But this is, up to a multiplicative constant, just the unrenormalised Green function with a zero momentum ϕ^2 insertion:

$$\frac{\partial}{\partial m^2} \Gamma^{(2n)}(p_1, p_2, \dots p_{2n}; m, \lambda, \Lambda) = -i\Gamma^{(2n)}_{\phi^2}(p_1, p_2, \dots p_{2n}; m, \lambda, \Lambda). \tag{17.51}$$

The renormalisability of the theory implies the existence of functions

$$m_R(m, \lambda, , \Lambda); \quad \lambda_R(m, \lambda, \Lambda); \quad Z(m, \lambda, \Lambda); \quad Z_{\phi^2}(m, \lambda, \Lambda) \tag{17.52}$$

such that

$$
\begin{aligned}
\Gamma^{(2n)}(p_i; m, \lambda, \Lambda) &= Z^n \Gamma^{(2n)}_R(p_i; m_R, \lambda_R) + O(1/\ln\Lambda) \\
\Gamma^{(2n)}_{\phi^2}(p_i; m, \lambda, \Lambda) &= Z_{\phi^2} Z^n \Gamma^{(2n)}_{\phi^2 R}(p_i; m_R, \lambda_R) + O(1/\ln\Lambda).
\end{aligned}
\tag{17.53}
$$

The renormalised Green functions $\Gamma^{(2n)}_R$ and $\Gamma^{(2n)}_{\phi^2 R}$ are independent of Λ; m_R and λ_R are the physical values of the mass and the coupling constant, respectively.

The Callan–Symanzik equation is obtained by combining (17.51), (17.52) and (17.53),

$$\left[m_R \frac{\partial}{\partial m_R} + \beta(\lambda_R) \frac{\partial}{\partial \lambda_R} + n\gamma(\lambda_R) \right] \Gamma^{(2n)}_R = m_R^2 \delta(\lambda_R) \Gamma^{(2n)}_{\phi^2 R}, \tag{17.54}$$

where we have defined the functions

$$
\begin{aligned}
\beta(\lambda_R) &= \frac{m_R^2 (\partial\lambda_R/\partial m^2)}{2(\partial m_R^2/\partial m^2)}; \quad \gamma(\lambda_R) = \frac{m_R^2 (\partial \ln Z/\partial m^2)}{2(\partial m_R^2/\partial m^2)} \\
\delta(\lambda_R) &= -i\frac{Z_{\phi^2}}{2(\partial m_R^2/\partial m^2)}.
\end{aligned}
\tag{17.55}
$$

It is easy to express the functions β, γ, and δ in terms of particular renormalised Green functions of the theory. Therefore, they approach well-defined limits when $\Lambda \to \infty$ and since they are dimensionless, they can only depend on λ_R. γ is often called *the anomalous dimension* of the field. This terminology may be misleading, since, obviously, the canonical dimension of any quantity cannot change. What it means is that γ describes the difference between the scaling dimension of the field, which we called d, and the canonical dimension, which, for a scalar field, equals 1.

Equation (17.54) is the Callan–Symanzik equation which involves only renormalised quantities. It gives the response of the Green functions under a change of the physical mass. Since, for a ϕ^4 theory, the mass is the only parameter that breaks scale invariance, this equation is also the broken scale invariance Ward identity. A final point: we have renormalised the Green functions on the mass shell and all quantities entering (17.54) are the physical ones. It follows that the limit $m_R \to 0$ does not exist because we have not introduced any independent subtraction point. It is with this price that the Callan–Symanzik equation has this simple form.

We want to make two remarks before closing this section: first, what went wrong with the derivation of (17.49) and which are the differences with (17.54). The answer is simple: in deriving (17.49) we forgot everything about renormalisation. We assumed that the mass term which appears in the Lagrangian, in other words the unrenormalised mass, is the only source of violation of scale invariance. As we see in (17.50) and (17.51) this is not true because the unrenormalised quantities depend also on the cut-off Λ through the loops of perturbation theory diagrams in a complicated way. The unrenormalised Green functions have a simple dependence on the bare mass m but a complicated dependence on the cut-off. The renormalised Green functions have no dependence on the cut-off, but the dependence on the physical mass is complicated because it enters under the renormalisation conditions. It is this complicated dependence which is expressed through the Callan–Symanzik equation.

The second remark is that, at first sight, this equation looks similar to the renormalisation group equation we derived in Chapter 16. But, in fact, their only similarity lies in the fact that they are both based on the renormalisability of the theory. Otherwise, they express different properties. We will come back to this point in a later chapter.

17.5 A Non-Perturbative Anomaly

This section does not belong to the mainline of this chapter. It attempts to study a phenomenon which is not present at any order of perturbation theory. It was first discovered by Edward Witten and we present it here because it is simple, interesting, and instructive.

The problem concerns a Yang–Mills theory based on the group $SU(2)$ defined in four space–time dimensions. As usual, we define first the Euclidean version. It is given formally by a functional integral,

$$ Z = \int [\mathcal{D}A_\mu] \mathrm{e}^{-\frac{1}{2g^2} \int \mathrm{d}^4 x \mathrm{Tr} F_{\mu\nu} F^{\mu\nu}}, \tag{17.56} $$

where, again, we assume that all gauge-fixing terms and Faddeev–Popov determinants are included in the functional measure $[\mathcal{D}A_\mu]$. A gauge transformation will be denoted by $U(x)$, where U is an element of $SU(2)$ and x a point in the four-dimensional Euclidean space. We have often considered a simplification in the form of a finite volume approximation. Let us choose, for example, a projection of E_4 to S_4. The limit of infinite radius can be taken later. Therefore, a gauge transformation can be viewed as a mapping from S_4 to the gauge group. $SU(2)$ is the group of unitary and unimodular 2×2 matrices. Any such matrix can be written as a superposition of the unit matrix and the three Pauli matrices $U = u_0 \mathbb{1} + i\boldsymbol{u} \cdot \boldsymbol{\sigma}$ with the four real numbers u_0 and $\boldsymbol{u}$ satisfying the condition $\sum_{i=0}^{3} u_i^2 = 1$. This shows that $SU(2)$ has the topology of a three-sphere S_3. In other words, these gauge transformations involve mappings among hyperspheres. In mathematics there is a standard way to divide such mappings into equivalence classes, according to whether they can be continuously deformed to each other. It is called *homotopy theory*. Since we are interested in the mappings from S_4 we must study what mathematicians call *the fourth homotopy group* of $SU(2)$, $\pi_4[SU(2)]$. It is hard to visualise mappings among hyperspheres, but all these groups have been computed and the important result for our problem is that $\pi_4[SU(2)]$ is non-trivial:

$$\pi_4[SU(2)] = Z_2. \tag{17.57}$$

This means that, in four-dimensional space, there is a gauge transformation $U(x)$ such that $U(x) \to \mathbb{1}$ as $|x| \to \infty$ but it cannot be continuously deformed to the trivial transformation which is the identity everywhere. If we want to have a geometric picture we can think of U as being wrapped once around the three-dimensional hypersphere representing $SU(2)$. Equation (17.57) tells us that we do not have to worry about transformations wrapped several times around $SU(2)$ because all those with an odd number of wrappings are deformable to our U and all those with an even number to the identity.

As Witten pointed out, the existence of this topologically non-trivial mapping means that our expression (17.56) involves a double counting, because, to every gauge field A_μ there is a second one, obtained by gauge transforming A by U:

$$A_\mu^U = U^{-1} A_\mu U - \mathrm{i} U^{-1} \partial_\mu U. \tag{17.58}$$

Two remarks may be useful here. First, note that U is not what we called previously *a small* gauge transformation. Therefore, our method of gauge fixing does not resolve this problem. A and A^U give precisely the same contribution to the functional integral since they are gauge transformed to each other and give the same F. But because in perturbation theory we study only small deviations from a given field configuration, usually the zero field configuration, we are not sensitive to problems of this kind which give no contribution to any order of perturbation theory. The second remark concerns, in fact, the relation with a subject we will study in a later chapter. We will encounter situations in which the gauge field configurations can be separated into classes and we will argue there that it is enough to integrate inside each class and then sum over the classes. But nothing like this can work here. U is a gauge transformation everywhere, not only at infinity. In going from A to A^U we do not encounter any singularities or divergences in the functional integral. They both belong to the same sector in field space. We are facing a real ambiguity resulting from an incomplete gauge-fixing procedure.

As long as we are studying a pure Yang–Mills theory this double counting is harmless because the effect will cancel between the numerator and the denominator in any Green function calculation. The situation changes when we introduce massless fermions. Let us consider, as usual, that they belong to the fundamental, doublet, representation of $SU(2)$. The functional integral becomes

$$Z = \int [\mathcal{D}A_\mu][\mathcal{D}\psi][\mathcal{D}\bar\psi]\mathrm{e}^{-\int \mathrm{d}^4 x \left[\frac{1}{2g^2}\mathrm{Tr}F_{\mu\nu}F^{\mu\nu} + \bar\psi \mathrm{i}\slashed{D}\psi\right]}. \tag{17.59}$$

Since the Lagrangian density is bilinear in the fermion fields, we can integrate them out and obtain an effective theory including only the gauge fields. For a doublet of Dirac fermions this integration gives

$$\int [\mathcal{D}\psi][\mathcal{D}\bar\psi]\mathrm{e}^{-\int \mathrm{d}^4 x \bar\psi \mathrm{i}\slashed{D}\psi} = \det(\mathrm{i}\slashed{D}). \tag{17.60}$$

The determinant is given, formally, by the product of the infinite number of the eigenvalues of $\not{D}$. This is the expression we considered in (17.22). In deriving the Adler–Bell–Jackiw anomaly we were interested in axial transformations. We found that, in certain cases, we could not define this operator respecting gauge invariance. Here this problem does not arise because a doublet of Dirac fermions admits a gauge invariant Dirac mass. Therefore, this theory can be regularised by Pauli–Villars regulators (a doublet of negative metric massive Dirac fields), which implies that $\not{D}$ can be defined in a gauge invariant way, both under small gauge transformations and under the U transformation.

Let us now consider, instead, a doublet of Weyl fermions. Since they have only one half the degrees of freedom of Dirac fermions, we expect to obtain the square root of the determinant $[\det(i\not{D})]^{1/2}$. Let us assume that we have defined the sign of the square root, for example by choosing it to be positive for a given field configuration A_μ. Then $[\det(i\not{D})]^{1/2}$ will be invariant under infinitesimal gauge transformations, since they cannot change the sign. But what about the topologically non-trivial gauge transformation U? The claim is that the square root is odd under U:

$$[\det(i\not{D}\,(A_\mu))]^{1/2} = -[\det(i\not{D}\,(A_\mu^U))]^{1/2}. \tag{17.61}$$

This equation should be understood as follows: if we vary the gauge field continuously from A to A^U we will find the opposite sign for the square root. Since in the functional integral we must integrate over all field configurations, it follows that Z vanishes and the theory cannot be defined.

The proof of (17.61) follows the arguments we used earlier, except that now we do not care about the zero modes of $\not{D}$ because they satisfy (17.61). As we noted, the non-vanishing eigenvalues come in pairs of opposite sign. In the square root we want to keep only half of them, so let us assume that we have arranged, for a given field A, to keep for every pair $\pm\lambda$ only $+\lambda$. Now we vary A continuously, for example following the path $A_\mu^t = (1 - t)A_\mu + tA_\mu^U$ for t varying continuously from 0 to 1. The spectrum of $\not{D}$ is the same for $t = 0$ and $t = 1$ but there may be a rearrangement and at some intermediate value $0 < t_0 < 1$ there may be a crossing. One of the eigenvalues which was positive at $t = 0$ becomes negative at $t = 1$ and one which was negative becomes positive. The two will cross 0 at t_0 where the determinant vanishes. If, at $t = 0$, we had defined the square root by keeping only the positive eigenvalues and then follow their evolution continuously, we will end up at $t = 1$ with one negative eigenvalue.

The Atiyah–Singer index theorem which we mentioned previously allows us to follow the flow of the eigenvalues as we vary the gauge field. We will not present the proof here, which consists of studying the spectrum of the Dirac operator in five dimensions with the fifth dimension compactified in a circle. The result is what we indicated earlier. An odd number of pairs of eigenvalues cross 0 between A_μ and A_μ^U and $[\det(i\not{D})]^{1/2}$ changes sign.

As announced, this is a non-perturbative anomaly. The theory is very well defined at the classical level, it has a well-behaved, renormalisable perturbation expansion, but it is ill-defined when non-perturbative effects are taken into account. It shows that quantum field theories and, in particular, gauge theories are much richer than perturbation theory may lead us to believe. We will examine some more interesting cases from the physical point of view in a later chapter.

17.6 Problems

Problem 17.1 Show that the validity of the Ward–Takahashi identities derived from the invariance of the Lagrangian (17.1) under isospin transformations imply that the renormalised masses of the proton and the neutron are equal and that the same holds true for those of the charged and neutral pions. Similarly, show that a single, renormalised coupling constant describes all pion–nucleon vertices.

Problem 17.2 Check the validity of Eq. (17.21) by a direct computation of the triangle diagrams of Fig. 17.2 using the Pauli–Villars regularisation method introduced in Problem 16.2.

18

Renormalisation of Yang–Mills Theory and BRST Symmetry

18.1 Introduction

In the previous chapter we presented the renormalisation programme for various Lagrangian field theories. We showed that it applies to Lagrangians which are polynomials in the fields and their derivatives and satisfy the power-counting requirement. The four-dimensional Yang–Mills theory, in the gauges we considered in Chapter 14, falls into this category; therefore, it is a renormalisable theory. We also studied the consequences of symmetries of the classical action in the renormalised Green functions. In this chapter we want to repeat this exercise for the particular case of the BRST symmetry and show how it can be preserved by renormalisation. It will not be a trivial exercise because the BRST transformations of (14.64) act non-linearly on the fields, but it will be essential because it is only through this symmetry that we will be able to guarantee the gauge independence of physical quantities.

The BRST invariant Lagrangian $\mathcal{L} + \mathcal{L}_{\mathrm{matter}}$ encodes the gauge symmetry under the form of the BRST symmetry. It involves ghost fields and contains a gauge-fixing term and a ghost term. The part of the action that enforces the gauge fixing is a BRST-exact term and the classical action and the observables are BRST-invariant expressions, defined modulo BRST-exact terms. In other words, the gauge-fixing action belongs to the trivial cohomology of the differential operator s and the physical quantities belong to the non-trivial part of the cohomology of s.

As we said previously, the theory is renormalisable by power counting. Therefore, by imposing a finite number of renormalisation conditions, any Green function is finite and calculable at any finite order of perturbation theory. The general theorem we stated in the previous chapter applies and there is not much we could add. Nevertheless, many new questions arise. Some sound technical and some are physical. In this chapter we will address them in the following order.

The first question is whether there exist renormalisation conditions which preserve the consequences of BRST symmetry for the renormalised Green functions, or, equivalently, whether the theory contains anomalies, in which case the loop corrections will

From Classical to Quantum Fields. Laurent Baulieu, John Iliopoulos and Roland Sénéor.
© Laurent Baulieu, John Iliopoulos and Roland Sénéor, 2017. Published 2017 by Oxford University Press.

break the relations among the Green functions implied by BRST symmetry in the tree approximation. We will show that, within the BRST formalism, the possibility of anomalies for the gauge symmetry is related to the existence of a non-trivial cohomology of s in the sector with ghost number one. Once this result is established, we can classify all possible anomalies that can occur in gauge theories, and check whether they cancel. For example, the axial anomaly found in the previous chapter will appear as a special case of this general result.

The second question concerns the structure of counter-terms. Assuming the theory has no anomalies, we will determine the precise form of the counter-terms needed in order to ensure finite Green functions at any order of perturbation theory which satisfy the BRST Ward identities.

The third question is the most important for physics. Gauge theories are formulated in a large space with many unphysical degrees of freedom (longitudinal and scalar gauge bosons, ghost-fields...) Furthermore, they contain unphysical parameters which determine the gauge. Green functions depend on all that. The physical question is to construct well-defined expressions which are calculable in perturbation theory and are suitable for representing physically measurable quantities. Intuitively, we expect such expressions to depend on the physical degrees of freedom only and to be independent of the choice of gauge. If we can define S-matrix elements, we expect them to satisfy a unitarity relation. However, the physical theories we will consider will always contain sectors in which the gauge symmetry will remain unbroken. Consequently, the corresponding gauge bosons will be massless and, as we will see in Chapter 21, the definition of a physically measurable quantity is more subtle.

18.2 Generating Functional of BRST Covariant Green Functions

We introduced the BRST symmetry as the graded operator s for the generator of a scalar global symmetry with $sd + ds = s^2 = 0$. We also introduced a constant anticommuting parameter η, for the transformation

$$\delta_{\mathrm{BRST}} = \eta s. \tag{18.1}$$

As usual, this symmetry implies the existence of Ward identities among the Green functions. The latter are obtained by computing the path integral of products of fields taken at different space–time points times the exponential of the BRST invariant action:

$$G_N = \int [\mathcal{D}\phi]\phi(x_1)\dots\phi(x_i)\dots\phi(x_N)\ \exp(-S). \tag{18.2}$$

Here ϕ stand for all fields $(A, \Omega, \bar{\Omega}, b, \Phi, \Psi)$, and we understand that a Wick rotation has been performed (we define the path integral in Euclidean space). This expression is still formal because no regularisation and renormalisation prescription has been specified.

As a consequence of the BRST symmetry of the action, we have the following relations among the tree-level Green functions, where one field $\phi(x_i)$ has been replaced by its BRST transform $s\phi(x_i)$:

$$0 = \sum_i \pm \int [\mathcal{D}\phi]\phi(x_1)\dots s\phi(x_i)\dots\phi(x_N)\ \exp(-S). \tag{18.3}$$

This tree-level identity can be proven by doing the change of variable $\phi \to \phi + \delta_\eta\phi$ in the path integral, and using the s invariance of the action and of the path integral measure.

Equation (18.3) is the Ward identity for the BRST symmetry.

We must prove that all Ward identities of this type hold true order by order in perturbation theory. In other words, we must prove that the renormalisation programme can be done while preserving the Ward identities of the BRST symmetry.

The path integral allows for a compact way of formulating all Ward identities. We introduce sources $\mathcal{J}_\phi$ for every field $\phi = (A, \Omega, \bar{\Omega}, b, \Phi, \Psi)$, as well as sources v_ϕ for every operator $s\phi$. We must be careful and avoid double counting. Indeed, we have $s\bar{\Omega} = b$, and thus there is no need to introduce a source $v_{\bar{\Omega}}$. In our notation, we set $v_{\bar{\Omega}} = 0$ and use $\mathcal{J}_b$ whenever a source for $s\bar{\Omega}$ is needed. Moreover, since $sb = 0$, we set $v_b = 0$.

As we explained in the previous chapter, the fact that some of the operators $s\phi$ are non-linear functions of the fields means that they must be independently renormalised. All superficially divergent correlators of the type $\langle\phi(x_1)\dots s\phi(x_i)\dots\phi(x_N)\rangle$ need a renormalisation. In this way the fields and their s-transforms acquire anomalous dimensions, which must be perturbatively computed by inserting the operators in all possible divergent 1-PI Green functions of the theory. The Ward identities will give relations among all these anomalous dimensions. In fact, the operators $s\phi$ can mix by renormalisation with all other operators with the same quantum numbers and equal or smaller dimensions, but the BRST symmetry will restrict the possibilities for such mixings.

Let us first define the effective action that includes the sources

$$S_{\text{eff}}(\phi, \mathcal{J}_\phi, v_\phi) = \int \mathrm{d}^4x\left(\mathcal{L} + \mathcal{L}_{\text{matter}} + \mathcal{J}_\phi(x)\phi(x) + v_\phi(x)s\phi(x)\right) \tag{18.4}$$

with the obvious notation that the expression $\mathcal{J}_\phi(x)\phi(x)$ contains a summation over all fields ϕ, that is $\mathcal{J}_\phi(x)\phi(x) \equiv \sum_{\phi=(A,\Omega,\bar{\Omega},b,\Phi,\Psi)} \mathcal{J}_\phi(x)\phi(x)$. We have also consistently $v_\phi(x)s\phi(x) \equiv \sum_{\phi=(A,\Omega,\Phi,\Psi)} v_\phi(x)s\phi(x)$.

We also define

$$\Sigma(\phi, v_\phi) = \int \mathrm{d}^4x\left(\mathcal{L} + \mathcal{L}_{\text{matter}} + v_\phi(x)s\phi(x)\right). \tag{18.5}$$

The generating functional $Z_c(\mathcal{J}_\phi, v_\phi)$ of all possible connected Green functions of fields and their BRST transforms is thus $\log Z(\mathcal{J}_\phi, v_\phi)$, where

$$Z(\mathcal{J}_\phi, v_\phi) = \frac{\int [\mathcal{D}\phi]\ \exp[-S_{\text{eff}}(\phi, \mathcal{J}_\phi, v_\phi)]}{\int [\mathcal{D}\phi]\ \exp[-S_{\text{eff}}(\phi, 0, 0)]}. \tag{18.6}$$

We define as usual the generating functional of 1-PI Green functions $\Gamma(\phi, v_\phi)$ by Legendre transformation of $Z_c(\mathcal{J}_\phi, v_\phi)$ in the couple of variables $(\phi, \mathcal{J}_\phi)$.[1]

$$\Gamma(\phi, v_\phi) = Z_c(\mathcal{J}_\phi, v_\phi) - \int \mathrm{d}^4x \mathcal{J}_\phi(x)\phi(x) \tag{18.7}$$

It follows that

$$\phi(x) = \frac{\delta Z}{\delta \mathcal{J}_\phi(x)}, \tag{18.8}$$

which allows us to express the source $\mathcal{J}$ in terms of the field ϕ, as explained in the previous chapters.

18.2.1 BRST Ward Identities in a Functional Form

By doing again the change of variables $\phi \to \phi + \delta_\eta \phi$ in the path integral definition of $Z_c(\mathcal{J}_\phi, v_\phi)$, it is easy to show that all tree-level Ward identities as in Eq. (18.3) can be put under the compact equivalent forms

$$\int \mathrm{d}^4x \ (\mathcal{J}_\phi(x)\frac{\delta Z_c}{\delta v_\phi(x)} + \mathcal{J}_{\bar{\Omega}}(x)b(x)) = 0 \tag{18.9}$$

for the connected Green functions, and

$$\int \mathrm{d}^4x(\frac{\delta^R\Gamma}{\delta\phi(x)}\frac{\delta^L\Gamma}{\delta v_\phi(x)} + \frac{\delta^R\Gamma}{\delta\bar{\Omega}(x)}b(x)) = 0 \tag{18.10}$$

for the 1-PI Green functions, where δ^R and δ^L stand for right and left differentiations, respectively.

In order to go on in a simplest notational way, we define a graded symplectic bracket for the fields ϕ and sources v_ϕ, as follows: given any pair of functionals $A(\phi, v_\phi)$ and $B(\phi, v_\phi)$, their graded bracket is

$$\{A, B\} \equiv \int \mathrm{d}^4x\left(\frac{\delta^R A}{\delta\phi(x)}\frac{\delta^L B}{\delta v_\phi(x)} - \frac{\delta^R A}{\delta v_\phi(x)}\frac{\delta^L B}{\delta\phi(x)}\right). \tag{18.11}$$

This definition for the graded bracket $\{A, B\}$ takes care of all possibilities for the ghost numbers of the fields. Note the absence of the fields $\bar{\Omega}$ and b in the definition of the bracket. In this notation the Ward identity is

---

[1] ϕ is what we called $\phi^{(\mathrm{cl})}$ in Chapter 16. In order to lighten the notation we will drop the indication 'classical' whenever there is no risk of confusion. This is the case when ϕ appears as an argument of the effective action Γ.

$$\{\Gamma, \Gamma\} = 0 \tag{18.12}$$

Suppose now that we have constructed a functional Γ which satisfies Eq. (18.12). Then, we can define the nilpotent transformation $\mathbf{s}_\Gamma$

$$\mathbf{s}_\Gamma = \{\Gamma,\ \} = \int \mathrm{d}^4 x\, \frac{\delta^R \Gamma}{\delta\phi(x)} \frac{\delta^L}{\delta v_\phi(x)} - \frac{\delta^R \Gamma}{\delta v_\phi(x)} \frac{\delta^L}{\delta\phi(x)}. \tag{18.13}$$

We have $\mathbf{s}_\Gamma^2 = 0$ on all fields $\phi = (A, \Omega, \Phi, \Psi)$ because of the Jacobi identity of the bracket $\{\ ,\ \}$ and the property (18.12).

These definitions will help us understand the renormalisation of the Yang–Mills theory by simplifying the algebraic manipulations which will determine the renormalised Lagrangian and what will be called 'the renormalised symmetry'.

18.3 Anomaly Condition

Since the action $S = \int d^d x\, \mathcal{L}$ of Eq. (14.72) is renormalisable by power counting it can be used to perturbatively define the Green functions of the fields and their BRST transformations with the following recursion technique. We assume that the theory has been renormalised at order n, yielding $\Gamma_n(\phi, v_\phi)$, which satisfies the Ward identity at order n, that is

$$\{\Gamma_n(\phi, v_\phi), \Gamma_n(\phi, v_\phi)\} = O(\hbar^{n+1}). \tag{18.14}$$

The renormalised nilpotent BRST symmetry is at this order

$$\mathbf{s}_n \equiv \{\Gamma_n(\phi, v_\phi),\ \}. \tag{18.15}$$

We have the expansion

$$\Gamma_n(\phi, v_\phi) = \Sigma + \hbar\Gamma_{(1)} + \cdots + \hbar^n \Gamma_{(n)}. \tag{18.16}$$

We can now proceed and compute the radiative corrections to order $n + 1$. The result involves further counter-terms of order $O(\hbar^{n+1})$ and yields a renormalised $\Gamma_{n+1}(\phi, v_\phi)$.

The first question is to find out whether $\Gamma_{n+1}(\phi, v_\phi)$ can be computed such that it satisfies the Ward identity up to terms of order $O(\hbar^{n+2})$.

Since we do not want to assume that we use a regularisation that preserves the BRST symmetry (and, actually, when there are chiral fermions, such a regularisation does not exist), we expect that the Ward identity for $\Gamma_{n+1}(\phi, v_\phi)$ may be violated at order $\hbar^{n+1}$, as follows:

$$\{\Gamma_{n+1}(\phi, v_\phi), \Gamma_{n+1}(\phi, v_\phi)\} = \hbar^{n+1} \int \Delta_4^1(\phi, v_\phi) + O(\hbar^{n+2}). \tag{18.17}$$

The upper and lower indices remind us that $\Delta_4^1(\phi, v_\phi)$ has ghost number 1 and has dimension 4.

For any given method of regularisation and subtraction, and after a given choice of renormalisation conditions has been adopted for the new divergent diagrams at order $\hbar^{n+1}$, the properties of Feynman diagrams imply that $\Delta_4^1(\phi, v_\phi)$ is a *local* functional of the fields and sources ϕ, and v_ϕ.

This last property is called quantum action principle. It is a general property valid order by order in the perturbation expansion of a quantum field theory that satisfies power-counting requirements as was explained in the previous chapter. Note that the terms $O(\hbar^{n+2})$ generally contain terms that are non-local functionals of the fields.

It may happen that the origin of the apparently anomalous term $\Delta_4^1(\phi, v_\phi)$ is just a choice of renormalisation conditions that were not well adjusted in going from order $\hbar^n$ to $\hbar^{n+1}$. In this case, $\Delta_4^1(\phi, v_\phi)$ can be eliminated by adjusting the local counter-terms, and we can proceed to order $\hbar^{n+2}$, and so on, which means that the theory can be renormalised while preserving all the Ward identities. In this case, $\Delta_4^1(\phi, v_\phi)$ is a spurious anomaly.

Otherwise, the theory is called anomalous, which means that the Ward identity breaking term $\Delta_4^1(\phi, v_\phi)$ cannot be eliminated perturbatively. In this case gauge invariant observables cannot be computed order by order in perturbation theory. In fact the consequences of an anomaly depend on the choice of gauge.

18.3.1   General Solution for the Anomalies of the Ward Identities

Determining the general expression of $\Delta_4^1(\phi, v_\phi)$ may look as an impossible task. However, the algebraic properties of the BRST symmetry simplify tremendously the determination of the most general expression of the possible anomalous term $\Delta_4^1(\phi, v_\phi)$. Then, the explicit computation of its coefficient indicates whether the theory is anomalous.

The functional $\Delta_4^1(\phi, v_\phi)$ is in fact severely restricted. We have indeed

$$0 \equiv \{\Gamma_{n+1}, \{\Gamma_{n+1}, \Gamma_{n+1}\}\} = \hbar^{n+1} \mathbf{s}_\Sigma \int \Delta_4^1(\phi, v_\phi) + O(\hbar^{n+2}). \tag{18.18}$$

Thus, $\Delta_4^1(\phi, v_\phi)$ must satisfy the so-called consistency condition:

$$\mathbf{s}_\Sigma \int \Delta_4^1(\phi, v_\phi) = 0. \tag{18.19}$$

This equation generalises the Wess and Zumino consistency condition found for the case of the current algebra anomaly for theories with a global symmetry; see Eq. (17.44). As we will shortly see, it is quite simple to solve it, using power counting, since $\mathbf{s}_\Sigma$ is basically the BRST symmetry operator. Equation (18.19) is quite similar to that which defines the BRST invariant Lagrangian, except for the fact that $\Delta_4^1(\phi, v_\phi)$ has ghost number 1.

The consistency equation (18.19) is trivially satisfied if $\Delta_4^1(\phi, v_\phi) = \mathbf{s}_\Sigma \Delta_4^0(\phi, v_\phi)$, in which case the anomaly is spurious, since $\hbar^{n+1} \Delta_4^0(\phi, v_\phi)$ can be added as a local counter-term which is just a mere modification of the tree-level part of Γ.

Thus, non-spurious anomalies are the solutions $\Delta_4^1(\phi, v_\phi)$ that are defined modulo BRST-exact terms, that is, the elements $\Delta_4^1(\phi, v_\phi)$ of the non-trivial cohomology with ghost number equal to 1 of the BRST symmetry.

Equation (18.19) can be written as the local equation

$$\mathbf{s}_\Sigma \Delta_4^1(\phi, v_\phi) + d\Delta_3^2(\phi, v_\phi) = 0 \tag{18.20}$$

$$\Delta_4^1(\phi, v_\phi) \sim \Delta_4^1(\phi, v_\phi) + \mathbf{s}_\Sigma \Delta_4^0(\phi, v_\phi). \tag{18.21}$$

This equation can be solved by various techniques, using power counting, and writing all possible local polynomials of dimension 4 and ghost number 1 for $\Delta_4^1(\phi, v_\phi)$.

We first show that $\Delta_4^1(\phi, v_\phi)$ can only depend on A and Ω. In fact, the dimensions of the sources v imply quite trivially that the possible v dependence can only be trivial (that is BRST exact). It is slightly more involved to show that any possible dependence on $\bar{\Omega}, b, \Phi, \Psi$ is also trivial, but it can be done by inspection.

Then, we are led to solve the consistency equation for a function Δ_4^1 that depends only on A and Ω.

Consider the case when the gauge group is simple, say $SU(N)$ or $SO(N)$. We will show shortly that the solution for the anomaly $\Delta_4^1(\Omega, A, F)$, defined modulo d- and s-exact terms, is the following 4-form with ghost number 1,

$$\Delta_4^1 = \mathrm{Tr}(d\Omega \wedge (A \wedge dA + \frac{1}{3} A \wedge A \wedge A)), \tag{18.22}$$

where the trace is for the matrices in the Lie algebra. This expression does not look very suggestive. However, it can be written under the following amazingly simple as well as mathematically appealing expression

$$\Delta_4^1 = \mathrm{Tr}\left(\Omega \frac{\delta}{\delta A}\Big|_F \Delta_5(A, F) \right), \tag{18.23}$$

where $\Delta_5(A, F)$ is the Chern–Simons term of order 5, which is the following 5-form, defined in dimension $d \geq 5$:

$$\Delta_5(A, F) = \mathrm{Tr}(A \wedge F \wedge F - \frac{1}{2} A \wedge A \wedge A \wedge F + \frac{1}{10} A \wedge A \wedge A \wedge A \wedge A). \tag{18.24}$$

Moreover, in dimension $d \geq 6$, we have the so-called Chern–Simons identity[2]

$$\mathrm{Tr}(F \wedge F \wedge F) = d\Delta_5(A, F). \tag{18.25}$$

This last equation is, in fact, implied by the property $d\mathrm{Tr}(F \wedge F \wedge F) = 3\,\mathrm{Tr}(DF \wedge F \wedge F) = 0$, due to the Bianchi identity $DF = 0$ and the cyclicity of the trace. Then, the local Poincaré lemma, which states that any quantity f that satisfies $df = 0$ must be locally

---

[2] For $d \leq 6$ this equation is trivial and encodes no information, since both sides vanish identically, due to the fact that we cannot write a non-vanishing 6-form with less than 6 independent forms dx^μ. Therefore, when A is a genuine 1-form, to get non trivial information from this equation the space must have at least six dimensions in order that $dx^{\mu_1} \wedge \ldots \wedge dx^{\mu_6} \neq 0$.

d-exact, implies the existence of the form $\Delta_5(A, F)$ within the identity (18.25). Note that the Chern–Simons term $\Delta_5(A, F)$ is defined only modulo d-exact terms, since the change $\Delta_5(A, F) \to \Delta_5(A, F) + \mathrm{d}(..)$ leaves Eq. (18.25) unchanged. Once the existence of $\Delta_5(A, F)$ has been proven, it is a simple exercise to compute one of its representatives as a function of A and F, as in Eq. (18.24).

Let us now prove the result of Eq. (18.23). From now on we will omit the $\wedge$ symbol for writing products of forms. The beautiful idea is to use the Chern–Simons identity (18.25) for extended forms, when the form degree is equal to the sum of the ordinary form degree and the ghost number. Indeed, since $(s + \mathrm{d})^2 = 0$, we can do the following substitutions into Eq. (18.25),

$$A \to A + \Omega, \quad F \to \hat{F} = (\mathrm{d} + s)(A + \Omega) + (A + \Omega)^2 \quad \mathrm{d} \to \mathrm{d} + s,$$

(18.26)

so that

$$\Delta_5(A + \Omega, \hat{F}) = \mathrm{Tr}\left((A + \Omega)\hat{F}\hat{F} - \frac{1}{2}(A + \Omega)^3 \hat{F} + \frac{1}{10}(A + \Omega)^5\right)$$

(18.27)

with

$$\mathrm{Tr}(\hat{F}\hat{F}\hat{F}) = (\mathrm{d} + s)\Delta_5(A + \Omega, \hat{F}).$$

(18.28)

This identity now carries non-trivial information, even when A is a 1-form in four dimensions, since s and Ω have both ordinary form degree equal to 0, so that $(\mathrm{d} + s)\Delta_5(A + \Omega, \hat{F})$ has non-vanishing components of form degree 4, 3, 2, 1, and 0. On the other hand, since the BRST equations are $\hat{F} = F$, we have identically

$$\mathrm{Tr}(\hat{F}\hat{F}\hat{F}) = \mathrm{Tr}(FFF),$$

(18.29)

and since any given ordinary 6-form vanishes in four dimensions, we have $\mathrm{Tr}(FFF) = 0$ and thus

$$\mathrm{Tr}(\hat{F}\hat{F}\hat{F}) = 0.$$

(18.30)

To summarise this chain of arguments, we have

$$(\mathrm{d} + s)(\Delta_5(A + \Omega, F) = \mathrm{Tr}(\hat{F}\hat{F}\hat{F}) = \mathrm{Tr}(FFF) = 0$$

(18.31)

so that

$$(\mathrm{d} + s)(\Delta_5(A + \Omega, F) = 0.$$

(18.32)

By isolating in this equation the terms with the same form degree and ghost number, we get a 'tower' of solutions of the consistency equations

$$s\Delta^g_{5-g} + \mathrm{d}\Delta^{g+1}_{4-g} = 0 \tag{18.33}$$

for $1 \le g \le 5$, with

$$\Delta^g_{5-g} = (\Delta_5(A + \Omega, F))^g_{5-g}. \tag{18.34}$$

This proves Eq. (18.23), as a particular case for $g = 1$. Equations. (18.33) are often called descent equations.

By using the expression of $\Delta_5(A, F)$, computed in $d \ge 5$, it is then quite easy to deduce from Eq. (18.23) that the constant anomaly Δ^1_4, defined modulo d-exact and s-exact terms, is the expression $\Delta^1_4 = \mathrm{Tr}(\mathrm{d}\Omega\,(A\mathrm{d}A + \frac{1}{3}AAA))$ given in Eq. (18.22). It will be shortly used to select all possible anomalous vertices of the theory.

When the Lie algebra has several factors, the methodology is the same, since the Chern–Simons formula holds true for any given invariant polynomial of curvatures. In this case there are several possible terms for $\Delta_5(A + \Omega, F)$, which we may label by an index i, and the only possible arbitrary parameters are real numbers a_i, which multiply each independent factor of the Chern–Simons term.

For instance, if the Lie algebra is $U(1) \times SU(N)$, the anomaly contains three possible terms, corresponding to the following Chern classes in $d \ge 6$ dimensions, $A_{U(1)}\mathrm{d}A_{U(1)}\mathrm{d}A_{U(1)}$, $\mathrm{d}A_{U(1)}\mathrm{Tr}_{SU(N)}FF$, and $\mathrm{Tr}_{SU(N)}FFF$. The consistent anomalies are thus

$$a_1\mathrm{d}\Omega_{U(1)}\mathrm{d}A_{U(1)}\mathrm{d}A_{U(1)} + a_2\mathrm{d}\Omega_{U(1)}\mathrm{Tr}_{SU(N)}FF$$

$$+ a_3\mathrm{Tr}_{SU(N)}(\mathrm{d}\Omega\,(A\mathrm{d}A + \frac{1}{3}AAA)). \tag{18.35}$$

The second term $\mathrm{d}\Omega_{U(1)}\mathrm{Tr}_{SU(N)}FF$ is called a mixed anomaly.

Note that in order for a purely non-Abelian anomaly to exist, the corresponding six-dimensional invariant $\mathrm{Tr}_{SU(N)}FFF$ must be non-vanishing. For this, we need an invariant symmetrical 3-tensor d_{abc} in the Lie algebra, in order that $\mathrm{Tr}_{SU(N)}FFF = d_{abc}F^aF^bF^c \ne 0$ in six dimensions. The existence of such an invariant tensor depends on the gauge group (for instance, it exists for $SU(3)$ but not for $SU(2)$). We see therefore that the algebraic determination of the consistent anomaly straightforwardly predicts that certain gauge groups cannot have anomalies, simply because their Lie algebra does not admit certain constant tensors. It is also a striking feature that anomalies in four dimensions are somehow related to the existence of Chern classes in dimension 6, which establishes a link between the chiral anomaly in four dimensions and topological invariants in higher dimensions.

For the theory to be anomaly-free, it is necessary and sufficient that, order by order in perturbation, all coefficients a_i vanish. We will shortly explain how we select the 1-PI Green functions that determine the values of these coefficients, and how we perturbatively compute them. Moreover, the Adler–Bardeen theorem we mentioned in the previous chapter shows that for the anomalies associated with chiral fermions, if a coefficient vanishes at 1 loop, it vanishes to all orders.

18.3.2   The Possible Anomalous Vertices and the Anomaly Vanishing Condition

Assuming that the BRST Ward identity has been enforced at $O(\hbar^n)$ for the renormalised order n-generating functional Γ_n, the only possible way the BRST identity can be broken by radiative corrections is thus

$$\{\Gamma_{n+1}, \Gamma_n\} = \hbar^{n+1} a \int \mathrm{Tr}(\,\mathrm{d}\Omega\,(A\mathrm{d}A + \frac{1}{3}AAA)\,) + O(\hbar^{n+2}). \qquad (18.36)$$

This equation allows us to perturbatively compute the *finite* real coefficient a, by selecting the renormalised 1-PI Green function that uniquely determines it. The right-hand side is an operator composite out of the fields and their derivatives and its Green functions can be computed as we explained in Chapter 16. For the sake of simplicity, we consider the case of a simple Lie algebra, with only one possible coefficient a, the generalisation to a general Lie algebra being obvious.

To identify the anomaly coefficient, we must differentiate both sides of (18.36) with respect to the appropriate fields and identify a as the numerical coefficient of given tensorial component of a particular renormalised 1–PI Green function. Let us see how it works.

Since $\Delta_4^1 \propto \int \mathrm{Tr}\mathrm{d}\Omega\,(A\mathrm{d}A + \frac{1}{3}AAA)$, there are two possibilities. We should differentiate the generating functional with respect to Ω, and then two or three times with respect to A. Then we set all fields and sources equal to 0. The left-hand side of the resulting equation will identify the possible non-spurious anomalous 1-PI Green function, and the right-hand side will give a times some space–time distributions and gauge group and Lorentz invariant tensors. By computing the 1-PI Green function, the value of a will follow, with the same value, no matter which one of the two possibilities we follow.

Let us choose the first one, with only three differentiations. We obtain[3]

$$\hbar^n \mathrm{Tr}(T_a T_b T_c) a \epsilon_{\mu\nu\rho\sigma} \frac{\partial}{\partial y_\rho} \delta^4(x-y) \frac{\partial}{\partial z_\sigma} \delta^4(x-z)$$
$$= \frac{\delta^L}{\delta\Omega^a(x)} \frac{\delta^L}{\delta A_\mu^b(y)} \frac{\delta^L}{\delta A_\nu^c(z)} \int \mathrm{d}^4 t \frac{\delta^R \Gamma}{\delta A_\sigma^d(t)} \frac{\delta^L \Gamma}{\delta v_{A\,d}^\sigma(t)}. \qquad (18.37)$$

Note that only a symmetric tensor d^{abc} can occur, simply by the fact that we have a symmetric derivation in A. Moreover, many terms in $\{\Gamma_n, \Gamma_n\}$ have disappeared, after application of the operation $\frac{\delta^3}{\delta\Omega^a(x)\delta A_\mu^b(y)\delta A_\nu^c(z)}$ and taking the sources equal to 0, simply by conservation of the ghost number.

---

[3] In the formulae of this section all fields and sources are set equal to 0, after the functional derivatives are taken, although we do not write it explicitly.

We must have therefore

$$\hbar^n \text{Tr}(T_a T_b T_c) \, \boldsymbol{a} \, \epsilon_{\mu\nu\rho\sigma} \frac{\partial}{\partial y_\rho} \delta^4(x-y) \frac{\partial}{\partial z_\sigma} \delta^4(x-z)$$

$$= \int d^4 t (\langle A_\mu^b(y), A_\nu^c(z), A_\sigma^d(t) \rangle \langle v_{Ad}^\sigma(t), \Omega^a(x) \rangle$$

$$+ \langle A_\mu^b(y), A_\sigma^d(t) \rangle \langle A_\nu^c(z), v_{Ad}^\sigma(t), \Omega^a(x) \rangle$$

$$+ \langle A_\nu^c(z), A_\sigma^d(t) \rangle \langle A_\mu^b(y), v_{Ad}^\sigma(t), \Omega^a(x) \rangle). \tag{18.38}$$

The last two terms must vanish because of the symmetries of the covariant tensors on the left-hand side of the equation. By multiplication by the inverse tensors of the group and of $\epsilon_{\mu\nu\rho\sigma}$, we get (assuming a suitable normalisation for the Lie algebra generators T_a)

$$\hbar^n \, \boldsymbol{a} \, \frac{\partial}{\partial y^{[\mu}} \delta^4(x-y) \frac{\partial}{\partial z^{\nu]}} \delta^4(x-z)$$

$$= \frac{1}{4} \text{Tr}(T_a T_b T_c) \epsilon_{\mu\nu}^{\rho\sigma} \int d^4 t \langle A_\sigma^b(y), A_\rho^c(z), A_\lambda^d(t) \rangle \langle v_{Ad}^\lambda(t), \Omega^a(x) \rangle. \tag{18.39}$$

Since we assumed that the Ward identity is satisfied at order $n-1$, we have

$$\langle v_{Ad}^\lambda(t), \Omega^a(x) \rangle = \frac{\partial}{\partial x_\lambda} \delta^4(x-t) \delta^{ad}. \tag{18.40}$$

Thus, we finally have the identification of the anomaly coefficient as

$$\hbar^n \, \boldsymbol{a} \, \frac{\partial}{\partial y^{[\mu}} \delta^4(x-y) \frac{\partial}{\partial z^{\nu]}} \delta^4(x-z) = \frac{1}{4} \text{Tr}(T_a T_b T_c) \epsilon_{\mu\nu}^{\rho\sigma} \frac{\partial}{\partial x^\lambda} \langle A_\sigma^b(y), A_\rho^c(z), A_\lambda^a(x) \rangle. \tag{18.41}$$

A necessary and sufficient vanishing condition of the anomaly is thus that the 1-PI Green function $\langle A_\sigma^b(y), A_\rho^c(z), A_\lambda^a(x) \rangle$ has no structure coefficient proportional to the fully antisymmetric tensor $\epsilon_{\mu\nu\rho\sigma}$.

We leave as an exercise the determination of the value of $\boldsymbol{a}$ from the computation of the 4-point function.

The coefficient of the 1-loop anomaly is, therefore, a rather easy computation, which is actually the same as that of the Adler–Bell–Jackiw anomaly, which we discussed in the previous chapter. Indeed, at the 1-loop level, only a closed loop of chiral fermions can contribute to the value of the gauge anomaly coefficient $\boldsymbol{a}$. Therefore, we can ignore at this level the possible propagation of gauge fields, and the same diagrams contribute to the calculation of the chiral, as well as the gauge anomaly.

In our discussion we considered the logical possibility that the anomaly coefficient could vanish up to a certain number of loops, and only get a non vanishing value at the next order. Remarkably enough, the Adler–Bardeen theorem we mentioned earlier states that if $\boldsymbol{a}$ vanishes at 1 loop, it vanishes to all orders.

As said earlier, in order to get a non-vanishing anomaly coefficient, the theory must involve the tensor $\epsilon_{\mu\nu\rho\sigma}$ and/or chiral fermions. Thus, a pure Yang–Mills theory, or one involving only Dirac fermions, is non-anomalous. Indeed, in this case, very effective symmetry preserving regularisations exist, such as the dimensional regularisation, and the previous discussion may appear as superfluous, since the counter-terms will be by construction BRST invariant, and no anomaly can occur.

In contrast, in the presence of chiral fermions, the formula involves the tensor $\epsilon_{\mu\nu\rho\sigma}$. 1-loop computations indicate the possibility of anomalies, as seen in the previous chapter (which actually proves that there exists no invariant regularisation). In particular, this is the case of the theory which describes the weak interactions among elementary particles which are known to violate parity and are formulated in terms of chiral fermions. The consistency of the theory at the quantum level requires the vanishing of the coefficient of the anomaly, which must be achieved by compensations among the contributions of various fermion multiplets. Since, for the actual calculations, we must use a non-invariant regularisation scheme, we need non-invariant counter-terms in order to restore the Ward identities. The method developed in this section is essential for their determination. In Chapter 25 the physical example of the theory of weak, electromagnetic, and strong interactions will be treated in detail. Let us only note here that it is based on the gauge group $SU(2) \times U(1) \times SU(3)$, but only the $SU(2) \times U(1)$ sector contains chiral couplings. The condition of having no anomaly is thus that the 3-point function

$$\epsilon^{\mu\nu\rho\sigma}\langle \partial_x^\tau A_\tau(x), W^+_\mu(y), W^-_\nu(z)\rangle \tag{18.42}$$

vanishes to all orders in perturbation theory. This amplitude corresponds to the invariant generalised 6-form $F^{photon}\mathrm{Tr}(F^{su(2)}F^{su(2)})$. The 1-loop condition $a = 0$ leads to a consistency condition which determines the fermion content of the theory and will be discussed in Chapter 25.

18.4   Dimensional Regularisation and Multiplicative Renormalisation

18.4.1   Introduction

We have shown that when quantising a classical Yang–Mills theory we may encounter one of the following three cases.

The theory may have anomalies whose coefficients a do not vanish. In this case, the situation is hopeless and, in general, we will not be able to prove that the resulting perturbation theory defines a consistent quantum field theory. In particular, we won't be able to introduce gauge invariant physical observables. Fortunately, no physically interesting theories seem to fall into this class.[4]

---

[4] In this book we have studied only field theories formulated on Minkowski flat space. The possible existence of defects, external fluxes, or curved backgrounds are not considered, unless otherwise mentioned.

The second case concerns the theories for which anomalies are, in principle, possible, but their coefficients vanish as a result of compensations among different diagrams. It is the case of the theory which describes the electromagnetic and weak interactions. It contains chiral fermions whose γ_5 couplings prevent us from using an invariant regularisation scheme, such as dimensional regularisation. In such a case, the way to enforce the Ward identities going from one order of perturbation theory to the next may be quite tricky. Indeed, the regularisation breaks the BRST symmetry, and we must fulfil the Ward identities by a recursive adjustment of counter-terms, whose structure violates the BRST symmetry in such a way that it compensates for the violations introduced by the non-symmetric regularisation. The general situation is thus that non-symmetric counter-terms are needed in order to enforce the symmetry at the renormalised level. Such symmetry violating counter-terms cannot be obtained in an automatic way, although the methodology is straightforward. At every given order, depending on the theory under consideration and the chosen regularisation scheme, we isolate by power counting the primitively divergent 1-PI diagrams. As we explained in the previous section, for every divergent counter-term we have an arbitrary finite constant which we are free to adjust. Then we proceed by counting. We prove that, at every order, we have enough free parameters to allow us to enforce all BRST Ward identities.

The third case is the simplest, both conceptually and technically. It is the one in which there is no possibility of an anomaly. The most common example is a gauge theory with only Dirac fermions and no γ_5 couplings. The simplicity here is due to the possibility of using dimensional regularisation which guarantees the validity of the BRST Ward identities, order by order in perturbation theory. In what follows, we will explain how the so-called 'multiplicative renormalisation' can be performed for the particular choice of linear gauges, which greatly simplify the proofs, as well as the practical computations. Note, however, that we solved the question of anomalies for general renormalisable gauges, and our proof can be extended to the case of more general non-linear gauges.

18.4.2   Linear Gauges and Ward Identities for the BRST Symmetry and Ghost Equations of Motion

The class of linear gauges are gauges for which the gauge function is linear in the gauge fields and possibly the scalar fields. Such gauges are often called Feynman–t'Hooft gauges. They define the theory by the BRST invariant local effective action

$$
\begin{aligned}
&\Sigma_0(A, \Phi, \Psi, \Omega, \bar{\Omega}, b, v_A, v_\Phi, v_\Psi, v_\Omega, ; g) \\
&\equiv \int \mathrm{d}^4x \Big(\mathcal{L}_{\mathrm{cl}}(A, \Phi, \Psi) + \frac{\alpha}{2}b^2 + b(\partial A + \beta[v, \Phi]) \\
&\quad + (v_{A_\mu} + \partial_\mu \bar{\Omega})D^\mu \Omega - (v_\Phi - \beta[v, \bar{\Omega}])[\Omega, \Phi] - v_\Psi[\Omega, \Psi] - v_\Omega \frac{1}{2}[\Omega, \Omega] \Big).
\end{aligned}
\tag{18.43}
$$

It is convenient to isolate the gauge-fixing term from the rest of the action, including its ghost dependence, denoted by the superscript ˆ

$$\Sigma_0(A, \Phi, \Psi, \Omega, \bar{\Omega}, b, v_A, v_\Phi, v_\Psi, v_\Omega; g)$$
$$\equiv \hat{\Sigma}_0(A, \Phi, \Psi, \Omega, v_A - \mathrm{d}\bar{\Omega}, v_\Phi - \beta[v, \bar{\Omega}], v_\Psi, v_\Omega; g)$$
$$+ \int \mathrm{d}^4 x \left(\frac{\alpha}{2} b^2 + b(\partial A + \beta[v, \Phi]) \right). \tag{18.44}$$

The BRST symmetry of the action and its dependence on b and $\bar{\Omega}$ imply the non-linear equations

$$\{\hat{\Sigma}_0, \hat{\Sigma}_0\} \equiv \mathcal{B}_{\hat{\Sigma}_0} \hat{\Sigma}_0$$
$$\equiv \frac{\delta \hat{\Sigma}_0}{\delta A} \frac{\delta \hat{\Sigma}_0}{\delta v_A} + \frac{\delta \hat{\Sigma}_0}{\delta \Omega} \frac{\delta \hat{\Sigma}_0}{\delta v_\Omega} + \frac{\delta \hat{\Sigma}_0}{\delta \Phi} \frac{\delta \hat{\Sigma}_0}{\delta v_\Phi} + \frac{\delta \hat{\Sigma}_0}{\delta \Psi} \frac{\delta \hat{\Sigma}_0}{\delta v_\Psi} = 0 \tag{18.45}$$

and the linear ones

$$\frac{\delta \Sigma_0}{\delta \bar{\Omega}} = \partial_\mu \frac{\delta \hat{\Sigma}_0}{\delta v_{A_\mu}} + \beta[v, \frac{\delta \Sigma_0}{\delta \Phi}]$$
$$\frac{\delta \Sigma_0}{\delta b} = \alpha b + \partial A + \beta[v, \Phi]. \tag{18.46}$$

Both Ward identities (18.46) are specific to the case of linear gauges. We can show that they cannot be anomalous.

18.4.3 Inverting the Ward Identities in Linear Gauges for a Local Field and Source Functional

The renormalisation programme in linear gauges consists of renormalising the theory in such a way that the quantum action Γ that we can compute from the Feynman rules stemming from the action (18.44) satisfies the same functional identities as Σ_0 in (18.45) and (18.46). The idea will be to incorporate the needed local counter-terms in a local action Σ_R that can be identified with Σ_0 in the 0-th order approximation.

It will be instructive to solve the inverse problem of determining the most general local solution that satisfies Eqs. (18.45) and (18.46), when Σ_0 is replaced by a yet unknown local functional Σ_R of the same fields and parameters, with the same canonical dimension and global symmetries, such as Lorentz invariance.

We can, in fact, expand Σ_R as a Taylor series in the sources v. Then, using power counting, we find that the dependence in the v's is at most linear. By imposing that all terms in this expansion satisfy the Ward identities, we obtain that Σ_R must be of the form

$$\Sigma_R(A, \Phi, \Psi, \Omega, \bar{\Omega}, b, v_A, v_\Phi, v_\Psi, v_\Omega; g)$$
$$= \hat{\Sigma}_R(A, \Phi, \Psi, \Omega, v_A - \mathrm{d}\bar{\Omega}, v_\Phi - \beta[v, \bar{\Omega}], v_\Psi, v_\Omega; g) + \frac{\alpha}{2} b^2 + b(\partial A + \beta[v, \Phi]), \tag{18.47}$$

where

$$\hat{\Sigma}_{\mathrm{R}} \equiv S_{\mathrm{R}}(A, \Phi, \Psi) + \sum_{\varphi = A, \Phi, \Psi} v_{\varphi} s_{\mathrm{R}} \varphi \tag{18.48}$$

and the differential operator s_{R} is such that

$$s_{\mathrm{R}}^2 = 0 \tag{18.49}$$

and

$$s_{\mathrm{R}} S_{\mathrm{R}}(A, \Phi, \Psi) = 0. \tag{18.50}$$

We found in an exercise of the previous chapter on the definition of the BRST symmetry that the nilpotency property of s_{R} implies that its action on the fields must be

$$\begin{aligned} s_{\mathrm{R}} A &= Z_c(-\mathrm{d}\Omega - g Z_A[A, \Omega]) \\ s_{\mathrm{R}} \phi &= Z_c(-g Z_A[\Omega, \phi]) \\ s_{\mathrm{R}} c &= Z_c(-\frac{g}{2} Z_A[\Omega, \Omega]), \end{aligned} \tag{18.51}$$

where Z_A and Z_c are arbitrary dimensionless constants.

Then the most general possible form for the local action $S_R(A, \Phi, \Psi)$ is

$$\begin{aligned} S_{\mathrm{R}} = \int \mathrm{d}^4 x \mathrm{Tr} \, \frac{1}{Z^2} \Big(&-\frac{1}{4} \big(Z_A \partial_{[\mu} A_{\nu]} + g Z_A^2 [A_{[\mu}, A_{\nu]}]\big)^2 \\ &-\frac{1}{2} Z_\Phi^2 \big(\partial_\mu \Phi + g Z_A [A_\mu, \Phi]\big)^2 - \frac{1}{2} Z_\Psi^2 \Psi \big(\partial_\mu \Psi + g Z_A [A_\mu, \Psi]\big)\Big) + \mathcal{V}_{\mathcal{R}}(\Phi). \end{aligned} \tag{18.52}$$

All Z factors are arbitrary constants and $\mathcal{V}_{\mathcal{R}}(\Phi)$ is a globally invariant potential which is a polynomial of degree four of the scalar field with arbitrary coefficients representing the mass and coupling constants.

It follows that the most general local solution of the Ward identity is in fact linked to Σ_0 as follows:

$$\hat{\Sigma}_{\mathrm{R}} = \frac{1}{Z^2} \hat{\Sigma}_0 (Z_A A, Z_\Phi \Phi, Z_\Psi \Psi, Z_\Omega \Omega, Z^2 v_A, Z_A Z^2 Z_\Phi^{-1} v_\Phi, Z_A Z^2 Z_\Psi^{-1} v_\Psi, Z_A Z^2 v_\Omega; g). \tag{18.53}$$

In fact, the difference between $\hat{\Sigma}_{\mathrm{R}}$ and $\hat{\Sigma}_0$ is just a matter of multiplicative renormalisations of fields and constants. The overall constant Z can be interpreted as a further rescaling of the coupling constant. In fact, both operations s_{R} and s are identical, modulo these rescalings. The constants Z_A, etc., are usually called wave function renormalisation constants, and we have used the notation Z_A instead of the usual $Z_A^{\frac{1}{2}}$ we used in Chapter

16 in order to simplify the form of the equations. s_R can be called the renormalised BRST symmetry.

It is worth noting that the linear Ward identities corresponding to the equations of motion of $\bar{\Omega}$ and b have completely fixed the dependence on these fields, a property that is called 'a non-renormalisation theorem'.

18.4.4 The Structure of the Counter-terms within the Dimensional Regularisation Method

Let us now show how the Ward identities determine the structure of the local counter-terms of the renormalisation programme and parametrise their form, in the case where we use the BRST symmetry preserving dimensional regularisation. As said earlier, this is possible only if no chiral couplings exist in the theory.

The symmetry preserving property means that when we use the Feynman rules from the local action Σ_R (Eq. 18.52), with arbitrary given constants Z, we get after an n-loop computation, where all loops have been dimensionally regularised but yet not renormalised, a generating functional $\Gamma_n^{R,\epsilon}(\phi, v_\phi\,;\, g)$ of the 1-PI Green functions which exactly satisfies the Ward identities (18.45) and (18.46):

$$\{\hat{\Gamma}_n^{R,\epsilon}, \hat{\Gamma}_n^{R,\epsilon}\} \equiv \frac{\delta \hat{\Gamma}_n^{R,\epsilon}}{\delta A}\frac{\delta \hat{\Gamma}_n^{R,\epsilon}}{\delta v_A} + \frac{\delta \hat{\Gamma}_n^{R,\epsilon}}{\delta \Omega}\frac{\delta \hat{\Gamma}_n^{R,\epsilon}}{\delta v_\Omega} + \frac{\delta \hat{\Gamma}_n^{R,\epsilon}}{\delta \Phi}\frac{\delta \hat{\Gamma}_n^{R,\epsilon}}{\delta v_\phi} + \frac{\delta \hat{\Gamma}_n^{R,\epsilon}}{delta\Psi}\frac{\delta \hat{\Gamma}_n^{R,\epsilon}}{\delta v_\psi} = 0 \quad (18.54)$$

$$\frac{\delta \Gamma_n^{R,\epsilon}}{\delta \bar{\Omega}} = \partial_\mu \frac{\delta \hat{\Gamma}_n^{R,\epsilon}}{\delta v_{A_\mu}} + \beta[v, \frac{\delta \Gamma_n^{R,\epsilon}}{\delta \Phi}]$$

$$\frac{\delta \Gamma_n^{R,\epsilon}}{\delta b} = \alpha b + \partial A + \beta[v, \Phi]. \quad (18.55)$$

Each term in the $\hbar$ expansion of $\Gamma_n^{R,\epsilon}$ can be written as a Laurent series in ϵ. Since $\hat{\Gamma}_n^{R,\epsilon}$ contains terms up to order $\hbar^n$, the identity (18.54) involves terms up to order $\hbar^{2n}$. Moreover, it depends on the constants Z, which can be thought of as Laurent series in ϵ whose residues are Taylor expansions in g. All three Ward identities can be expanded in a Laurent series in ϵ, with complicated identities satisfied among the coefficients.

We proceed by induction. We suppose that the factors Z, Z_φ in the effective action $\hat{\Sigma}_{R,n}$ (18.53) have been computed at order n, yielding

$$\hat{\Sigma}_{R,n} == \equiv \frac{1}{Z_n}\hat{\Sigma}_0(\varphi, v_\varphi, Z_\varphi^n;\, g). \quad (18.56)$$

The constants Z and Z_φ are themselves Laurent series in ϵ and polynomials of order n in $\hbar$. From our recursion hypothesis, they are such that the theory has been renormalised at order $\hbar^n$. Thus, using the Feynman rules stemming from $\hat{\Sigma}_{R,n}$, we get a regularised generating functional of 1-PI Green functions that can be computed at any given finite order of perturbation theory, giving

$$\Gamma_n^{R,\epsilon} = \text{ terms of order up to } \hbar^n \text{ that are finite when } \epsilon \to 0$$
$$+ \text{ terms of order higher than } \hbar^n \text{ that are possibly divergent when } \epsilon \to 0. \tag{18.57}$$

We define $\Gamma_{(n)}^{R,\epsilon}$ as the truncation of $\Gamma_n^{R,\epsilon}$ when we only retain the terms up to order $O(\hbar^n)$. The definition of the renormalised generating functional Γ_n^R at order n is

$$\Gamma_n^R = \lim_{\epsilon \to 0} \Gamma_{(n)}^{R,\epsilon}. \tag{18.58}$$

By assuming that the BRST symmetry is preserved by dimensional regularisation, we have

$$\{\hat{\Gamma}_n^{R,\epsilon}, \hat{\Gamma}_n^{R,\epsilon}\} = 0 \tag{18.59}$$

and

$$\{\hat{\Sigma}_{R,n}, \hat{\Sigma}_{R,n}\} = 0, \tag{18.60}$$

because of the definition of $\hat{\Sigma}_{R,n}$ in (18.53). These properties are of course verified for $n = 0$, taking all Z factors equal to 1.

Let us now specify in more detail the terms of order $O(\hbar^{n+1})$ in $\Gamma_n^{R,\epsilon}$. They are a mixture of divergent and non-divergent terms when $\epsilon \to 0$. Since the use of $\hat{\Sigma}_{R,n}$ renormalises the theory up to $O(\hbar^n)$, the quantum action principle implies that the divergent terms form a *local* functional $\Gamma_{n+1,\,\text{div}}^{\text{local}}$ of the fields φ and the sources v_φ. Thus, Eq. (18.57) can be made more precise as follows:

$$\Gamma_n^{R,\epsilon} = \Gamma_{(n)}^{R,\epsilon} + \hbar^{n+1}(\Gamma_{n+1,\,\text{div}}^{\text{local}} + \Gamma_{n+1,\,\text{finite}}) + O(\hbar^{n+2}). \tag{18.61}$$

$\Gamma_{n+1,\,\text{finite}}$ is finite and generally contains non-local terms. Since the Z's have been computed only up to order $O(\hbar^n)$, the terms $O(\hbar^{n+2})$ generally contain local and non-local terms which may diverge when $\epsilon \to 0$.

When we insert Eq. (18.61) in Eq. (18.59), the only sources of divergent terms of order $\hbar^{n+1}$ are such that we must have identically

$$\{\hat{\Sigma}_0(\varphi, v_\varphi, Z = 1 \,;\, g), \Gamma_{n+1,\,\text{div}}^{\text{local}}\} = 0. \tag{18.62}$$

As we will see shortly, this fixes the expression of $\Gamma_{n+1,\,\text{div}}^{\text{local}}$.

Moreover, if we compute $\Gamma_{n+1}^{R,\epsilon}$ out of

$$\hat{\Sigma}_{R,n+1} = \hat{\Sigma}_{R,n+1} - \hbar^{n+1}\Gamma_{n+1,\,\text{div}}^{\text{local}} \tag{18.63}$$

we will find that $\Gamma_{n+1}^{R,\epsilon}$ is finite up to terms of order $O(\hbar^{n+2})$,

$$\Gamma_{n+1}^{R,\epsilon} = \text{finite} + O(\hbar^{n+2}) \tag{18.64}$$

with

$$\{\Gamma_{n+1}^{R,\epsilon}, \Gamma_{n+1}^{R,\epsilon}\} = O(\hbar^{n+2}) \tag{18.65}$$

because of Eq. (18.62).

To go further, we must determine the general structure of $\Gamma_{n+1,\,\text{div}}^{\text{local}}$.

Dimensional analysis indicates that it can be at most linear in the sources v_φ, since it has dimension 4, and the sources $v_A, v_\Omega, v_\Phi, v_\Psi$ have respectively dimensions $3, 4, 3$, and $5/2$, which excludes any dependence that is quadratic or higher.

In fact, Eq. (18.62) implies that

$$\Gamma_{n+1,\,\text{div}}^{\text{local}} \delta Z^{(n+1)} \hat{\Sigma}_0(\varphi, v_\varphi, Z = 1 \,;\, g) + \{\hat{\Sigma}_0(\varphi, v_\varphi, Z = 1 \,;\, g), X\}. \tag{18.66}$$

We have indeed $\{\hat{\Sigma}_0, \hat{\Sigma}_0\} = 0$ and $\{\hat{\Sigma}_0, \{\hat{\Sigma}_0, X\}\} \frac{1}{2} \{X, \{\hat{\Sigma}_0, \hat{\Sigma}_0\}\} = 0$, for any given functional X. Then, dimensional analysis implies that

$$X = \sum_{\varphi = A, \Omega, \Phi, \Psi} \delta Z_\varphi^{(n+1)} \varphi v_\varphi. \tag{18.67}$$

$\delta Z^{(n+1)}$ and $\delta Z_\varphi^{(n+1)}$ are constants that will be identified from the computation of $\Gamma_{n+1,\,\text{div}}^{\text{local}}$.

We have

$$\hbar^{n+1}\{\hat{\Sigma}_0, X\} = \hbar^{n+1} \delta Z_\varphi^{n+1} \Big(\varphi \frac{\delta}{\delta\varphi} + v_\varphi \frac{\delta}{\delta v_\varphi}\Big) \hat{\Sigma}_0(\varphi, v_\varphi, Z_\varphi = 1 \,;\, g)$$

$$\hat{\Sigma}_0(\varphi, v_\varphi, Z_\varphi = 1 + \hbar^{n+1} c_\varphi^{n+1} \,;\, g) - \hat{\Sigma}_0(\varphi, v_\varphi, Z = 1 \,;\, g) + O(\hbar^{n+2}), \tag{18.68}$$

where the constant c_φ^{n+1} and $\delta Z^{(n+1)}$ are simply related, since $\varphi \frac{\delta}{\delta\varphi} + v_\varphi \frac{\delta}{\delta v_\varphi}$ is nothing but the dilatation operator.

It follows that we can express $\Sigma_{R,n+1}$ as

$$\hat{\Sigma}_{R,n+1} = \frac{1}{Z_{n+1}} \hat{\Sigma}_0(\varphi, v_\varphi, Z_\varphi^{n+1} \,;\, g) + O(\hbar^{n+2}). \tag{18.69}$$

The last term is irrelevant since we are only interested by the renormalisation at order $\hbar^{n+1}$, so that our induction hypothesis will be satisfied by choosing the following 'multiplicatively renormalised effective action' at order $\hbar^{n+1}$:

$$\hat{\Sigma}_{R,n+1} = \frac{1}{Z_{n+1}} \hat{\Sigma}_0(\varphi, v_\varphi, Z_\varphi^{n+1} \,;\, g). \tag{18.70}$$

We have therefore shown that a given set of constants Z that parametrise $\hat{\Sigma}$ can always be determined in such a way that $\lim_{\epsilon \to 0} \Gamma_n^{R,\epsilon}$ is finite. This defines a renormalised functional $\Gamma^{R,n}$ that corresponds to certain values of the renormalisation conditions of the Green functions. These conditions can always be further adjusted by finite changes in the factors Z, as we just explained.

Since the number of independent Z factors is equal to the number of fields and parameters of the theory, this proves the property of stability of the theory under renormalisation.

Let us also note that when the Lie algebra is such that we have group factors that can mix, the renormalisation remains multiplicative, but in a matricial way. We will see an example in the next chapter.

18.5   Observables

Classically, gauge invariant quantities are functions of the matter fields, which transform tensorially under gauge transformations, of their covariant derivatives, and of the Yang–Mills field-strength $F_{\mu\nu}$.

For the quantum theory, we will define the observables as mean values of field functionals that build the cohomology with ghost number 0 of the BRST symmetry.

Within the dimensional regularisation method an observable is thus defined as

$$\langle \mathcal{O} \rangle \lim_{\epsilon \to 0} \int [\mathcal{D}\phi] \, \mathcal{O} \exp - \int \mathrm{d}^{4-\epsilon} x (\mathcal{L}_r + \mathcal{L}_{r\text{matter}}), \tag{18.71}$$

where all relevant Z factors have been defined in such a way that the theory is renormalised at the order n we are interested in.

Any given observable is independent of variations of all gauge parameters α, β, γ, which we have introduced to obtain a gauge-fixed BRST invariant action. Indeed, the dependence on any parameters which belong to an s_r-exact term, for instance α, is such that

$$\frac{\partial}{\partial \alpha} \langle \mathcal{O} \rangle \lim_{\epsilon \to 0} \int [\mathcal{D}\phi] \, s_r(\bar{K}) \mathcal{O} \exp - \int d^{4-\epsilon} x (\mathcal{L}_r + \mathcal{L}_{r\text{matter}}). \tag{18.72}$$

Using the s_r invariance of the measure $[\mathcal{D}\phi]\mathcal{O}$ and $\mathcal{L}_r + \mathcal{L}_{r\text{matter}}$ immediately implies that

$$\frac{\partial}{\alpha} \langle \mathcal{O} \rangle = 0. \tag{18.73}$$

This property is obviously valid order by order in perturbation theory. Note that it indicates that physical observables only depend on the parameters that are contained in $\mathcal{L}_r + \mathcal{L}_{r\text{matter}}$.

Local observables are thus the same as those we expect from the classical argument, independently of the subtleties of the gauge-fixing process.

18.6   Problems

Problem 18.1 Anomaly cocycles.

Compute the explicit expression of Δ_{5-g}^{g} in Eq. (18.34) as a function of A and Ω. Check the validity of Eq. (18.33), using directly the BRST transformation laws of A and Ω.

Problem 18.2 Antighost independence of Lagrangians and consistent anomalies.

Show that any given functional of the fields that is s-invariant and depends either on the antighost $\bar{\Omega}$ or on b must be s-exact. Deduce from this that any given part of a Lagrangian that depends on the ghost Ω must be part of an s-exact functional. Use power counting to get Eq. (14.69).

Problem 18.3 Mixed Abelian–non-Abelian anomalies.

Since the anomaly is defined modulo s- and d-exact terms, show that the mixed anomaly can be also written as $\Delta_4^1 = \int dA_{U(1)} \mathrm{tr}_{SU(N)} (d\Omega A)$. Show that the mixed anomaly can be therefore interpreted as either a $U(1)$ anomaly or an $SU(N)$ anomaly, with the same coefficient.

Problem 18.4 Anomaly compensating Green–Schwartz counterterms.

We can extend the BRST transformations of the Yang–Mills theory by introducing an 'Abelian 2-form gauge field' $B_{\mu\nu}$, transforming under the BRST symmetry s as

$$sB_2 = dB_1^1 - \mathrm{Tr}(\Omega dA)$$

where

$$sB_1^1 = dB_0^2 - \mathrm{Tr}(\Omega\Omega A)$$

$$sB_2^0 = -\frac{1}{6}\mathrm{Tr}(\Omega\Omega\Omega)$$

and $B_2 = \frac{1}{2}B_{\mu\nu}dx^\mu dx^\nu$ and $B_1^1 = \frac{1}{2}B_\mu^1 dx^\mu$. B_μ^1 is called a vector anti-commuting ghost and B_2^0 a scalar commuting ghost of ghost.

Check that $s^2 = 0$. Compute the s-variation of the 3-form curvature $G_3 = dB_2 + \mathrm{Tr}(AF - \frac{1}{6}AAA)$.

Find a BRST invariant action depending on B_2 and A.

Consider now an action that has an $SU(2) \times U(1)$ anomaly. Show that its consistent anomaly $\int \Omega_{U(1)} \text{Tr}_{SU(2)} FF$ can be expressed as a BRST exact-term, which is the s variation of $\int B_2 \text{Tr} FF$.

Interpret the result, and explain the physical origin of the disappearance of the anomaly when the 2-form gauge field B_2 is introduced.

Hint: Show that the BRST transformations can be written as descent equations, using the curvature equation $\hat{G}_3 \equiv (\mathrm{d} + s)\hat{B}_2 + \text{Tr}(\hat{A}\hat{F} - \frac{1}{6}\hat{A}\hat{A}\hat{A}) = F$, *where* $\hat{A} = A + \hat{A}$ *and* $\hat{B}_2 = B_2 + B_1^1 + B_0^2$.

19

Some Consequences
of the Renormalisation Group

19.1 Introduction

We found already two sets of equations, (16.88) or (16.94) and (17.54). The first two are different forms of the renormalisation group equation and the third is the equation of Callan and Symanzik. They look similar, so we start by pointing out what each one represents. First of all, as we explained earlier, these equations have similar form because they are obtained using the same framework of renormalised perturbation theory.

The Callan–Symanzik equation is the correct Ward identity of broken scale invariance. It expresses the fact that for a massive theory, scale invariance is broken. When we vary the physical mass we obtain a family of physically different field theories. The equation tells us that this dependence of the Green functions on the value of the physical mass is of two kinds: the classical one through the propagators which describes the range over which the interaction is spread. A scale invariant interaction is spread over all distances and the range is a measure of the breaking. But for a renormalisable theory, there is a second dependence. The physical mass enters, through the renormalisation conditions, on the definition of the physical coupling constant and the scale of the field.

On the other hand, the renormalisation group equation describes the dependence of the Green functions on an *unphysical* parameter. When we vary the value of the subtraction point we obtain a family of Green functions and the renormalisation group tells us how they all describe one and the same physical theory.

The resemblance comes from the fact that, here also, the value of the subtraction point enters, through the renormalisation conditions, on the definition of the parameters of the theory. Stated this way, they both sound of limited physical interest. Nobody cares much about the dependence on an unphysical parameter and nobody knows how to design an experiment in which the mass of a particle varies continuously.

Now that we explained in what these equations are different, we can turn into what they have in common. They both describe the response of the Green functions under a change of a dimensionful parameter. Therefore, by ordinary dimensional analysis, we hope to be able to relate this response with that of the Green functions under the change

From Classical to Quantum Fields. Laurent Baulieu, John Iliopoulos and Roland Sénéor.

of other dimensionful quantities. Of particular interest are, of course, quantities related to the incident energy because it can be varied in physical experiments. In this chapter we want to obtain this information.

19.2   The Asymptotic Behaviour of Green Functions

Let us start with the Callan–Symanzik equation. In this section we shall only deal with renormalised quantities, so, in order to keep the notation light, we will drop the subscript R everywhere. The equation simplifies in a particular kinematical region, called *the deep Euclidean region*. We first take all momenta Euclidean, i.e. $p_i^2 < 0$ with real space parts and imaginary time parts. The reason for this strange choice is technical. In this book we will present only a brief discussion of the analyticity properties of Green functions as functions of the external momenta. The result will be that in the Euclidean region we stay away from all singularities. Here we can give only a plausibility argument by noticing that the Euclidean propagators are given by the inverse of $(k^2 + m^2)$ which does not vanish.

Let us now multiply all p_i's with a real parameter ρ. The deep Euclidean region is reached by choosing ρ very large keeping all partial sums among different p_i's different from 0. It is a very unphysical region in which all masses and all momentum transfers become large and negative.

The first step is to simplify the equation by using a theorem originally due to S. Weinberg. It is a rigorous version of dimensional analysis. Let us first remark that $\Gamma^{(2n)}/\Gamma^{(2n)}_{\phi^2}$ has dimension equal to $[M^{-2}]$. Therefore, we expect the ratio $\Gamma^{(2n)}(\rho p_i)/\Gamma^{(2n)}_{\phi^2}(\rho p_i)$ to behave at large ρ as ρ^{-2}. The theorem states that this naïve expectation is in fact correct at any finite order of perturbation theory, with corrections which behave at most as a power of $\ln \rho$. Therefore, we can write the asymptotic form of (17.54) by omitting the right-hand side:

$$\left[m \frac{\partial}{\partial m} + \beta(\lambda) \frac{\partial}{\partial \lambda} + n\gamma(\lambda) \right] \Gamma^{(2n)}_{as}(\rho p_i; m, \lambda) = 0. \tag{19.1}$$

We can use again dimensional analysis to trade the derivative with respect to m for that with respect to ρ. Γ^{2n} has dimensions $[M^{4-2n}]$ and it can be written as

$$\Gamma^{(2n)}(\rho p_i; m, \lambda) = m^{4-2n} F^{(2n)} \left(\frac{\rho p_i}{m}; \lambda \right). \tag{19.2}$$

Therefore, the function $\Phi^{(2n)} = \rho^{2n-4} \Gamma^{(2n)}$ satisfies

$$\left[-\rho \frac{\partial}{\partial \rho} + \beta(\lambda) \frac{\partial}{\partial \lambda} + n\gamma(\lambda) \right] \Phi^{(2n)}_{as} = 0. \tag{19.3}$$

This equation can be solved by using Monge's standard method. We define $\bar{\lambda}$ as a function of λ and ρ through the equation

$$\left[-\rho\frac{\partial}{\partial\rho} + \beta(\lambda)\frac{\partial}{\partial\lambda}\right]\bar{\lambda}(\lambda,\rho) = 0; \quad \bar{\lambda}(\lambda,\rho = 1) = \lambda, \tag{19.4}$$

which is equivalent to

$$\rho\frac{\partial\bar{\lambda}}{\partial\rho} = \beta(\bar{\lambda}). \tag{19.5}$$

The general solution of (19.3) is now given by

$$\Phi^{(2n)}(\rho p_1, ..., \rho p_{2n}; m, \lambda) = \Phi^{(2n)}(p_1, ..., p_{2n}; m, \bar{\lambda})\exp\left\{n\int_1^\rho \mathrm{d}\rho'\gamma[\bar{\lambda}(\lambda,\rho')]\right\}. \tag{19.6}$$

The physical meaning of (19.6) is clear. Scaling all momenta of a Green function by a common factor ρ and taking ρ large has the following effects: (i) it multiplies every external line by a factor, the exponential of (19.6), and (ii) it replaces the physical coupling constant λ by an effective one, $\bar{\lambda}$, which is the solution of (19.5). In the deep Euclidean region the effective strength of the interaction is not determined by the physical coupling constant λ but by $\bar{\lambda}$, which is a function of ρ and λ; i.e. it depends on how far in the deep Euclidean region we are. For this reason $\bar{\lambda}$ is sometimes called *running* coupling constant.

The fact that the effective coupling constant of a renormalisable field theory depends on the scale of the external momenta can be understood by looking at simple Feynman diagrams. A Feynman diagram of nth order is proportional to λ^n. On the other hand, if it is a multiloop diagram, it contains logarithms of the external momenta of degree up to $\ln^n(k^2/m^2)$, where k^2 is the square of a typical momentum transfer. For k^2 large compared to the mass these factors become very important. Summing up these logarithms introduces an effective coupling constant which depends on the scale. The differential equation (19.5) provides a formal way to take into account this dependence. It shows that if $\beta > 0$, $\bar{\lambda}$ increases with increasing ρ and it will continue to increase as long as β remains positive. The limit of $\bar{\lambda}$ when $\rho \to \infty$ will be the first zero of β on the right of the initial value λ. Similarly, for negative β $\bar{\lambda}$ decreases with increasing ρ and, for $\rho \to \infty$, it goes to the first zero of $\beta(x)$ for $x < \lambda$. Finally, if $\beta(\lambda) = 0$, $\bar{\lambda}$ is independent of ρ.

This analysis shows that we can classify the zeros of β in two classes: Those of Fig. 19.1(a) are attractors; i.e. if we start somewhere in their neighbourhood, $\bar{\lambda}$ approaches them for $\rho \to \infty$. Those of Fig. 19.1(b) are repulsors; i.e. $\bar{\lambda}$ goes further away from them as $\rho \to \infty$. An attractor is always followed by a repulsor (multiple zeros must be counted accordingly). The conclusion is that the asymptotic behaviour of a field theory depends on the position and nature of the zeros of its β-function.

As long as perturbation theory is our only guide, we cannot say anything about the properties of $\beta(\lambda)$ for arbitrary λ. We do not know whether it has any zeros, let alone their

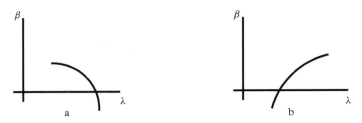

Figure 19.1 *The zeros of the β function.*

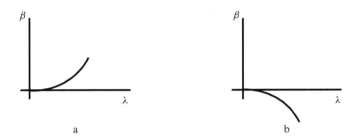

Figure 19.2 *The nature (repulsor or attractor) of the free field theory 0 of the β function.*

nature. The only information perturbation theory can hopefully provide is the behaviour of $\beta(\lambda)$ at the vicinity of $\lambda = 0$. We know that $\beta(0) = 0$, because $\lambda = 0$ is a free field theory. The nature of this zero (attractor or repulsor), will depend on the sign of the first non-vanishing term in the expansion of $\beta(\lambda)$ in powers of λ. But this expansion is precisely perturbation theory. If β starts as in Fig. 19.2(a) i.e. from positive values, the origin is a repulsor. The effective coupling constant will be driven away to larger values as we go deeper and deeper into the Euclidean region. In contrast, if the first term of β is negative, Fig. 19.2(b), the origin is an attractor. If we start somewhere between the origin and the next zero of β, the effective coupling constant will become smaller and smaller and it will vanish in the limit. Such a theory is called *asymptotically free*.

And now we can state the following, very important theorem.

Out of all renormalisable field theories, only the non-Abelian gauge theories are asymptotically free.

This theorem was first proven by David Gross, David Politzer, and Franck Wilczek. The proof consists of a simple, straightforward calculation. We compute the first term in the expansion of the β-function for all renormalisable field theories, i.e. ϕ^4, Yukawa, spinor electrodynamics, scalar electrodynamics, non-Abelian Yang–Mills theories, and any combination of these. Only for the pure non-Abelian Yang–Mills theory, which has a single coupling constant, the β-function is negative. If we add any other, independent pieces, we lose asymptotic freedom.

We reach the same conclusion by looking at the renormalisation group equation (16.88) or (16.94). For the Callan–Symanzik equation we had to rely on Weinberg's theorem to prove that in the deep Euclidean region, the right-hand side can be neglected. For the renormalisation group equation we must use another theorem, sometimes referred to as Kinoshita–Lee–Nauenberg theorem, which we can again prove order by order in perturbation theory. It states that the asymptotic behaviour of the Green functions in the deep Euclidean region is given by that of the corresponding massless theory. Although we will not prove any of these theorems here, we indicate that they both have to do with the behaviour of the theory when the mass goes to 0. The Green functions for the Callan–Symanzik equation are renormalised on the mass shell, which means that the subtraction point is chosen equal to the physical mass. This means that the limit $m \to 0$ is singular; otherwise, we would have obtained a theory with no dimensionful parameter whatsoever. What the theorem shows is that this singularity is no worse than a power of $\ln m^2$; no pole singularities, such as m^{-2}, appear. If this is correct, then ordinary dimensional analysis shows that we can drop the right-hand side of the equation. Similarly for the renormalisation group equation. If we keep the subtraction point fixed, we can put the mass equal to 0, and since there are no terms going like inverse powers of the mass, the asymptotic behaviour at large momenta remains the same. From this point the analysis of (16.88) or (16.94) is the same as that of the Callan–Symanzik equation. We obtain Eq. (19.3) and, finally, Eq. (19.6).

We started this section in order to draw physically relevant conclusions out of seemingly technical equations. We are only halfway through. We obtained an equation which gives the dependence of Green functions on the external momenta, but only in the unphysical deep Euclidean region. We still have some work to do in order to extract results which can be directly compared to experiment. We will do this in a later chapter.

19.3 Stability and the Renormalization Group

In Chapter 17 we noted that relations among the parameters of a field theory which increase the global symmetries of the Lagrangian are stable under renormalisation. Here we want to ask the opposite question: given a renormalisable field theory, can we find all possible stable relations among its parameters? Remarkably enough, the answer turns out to be very simple and provides for another interesting application of the concepts of the renormalisation group.

In order to illustrate the argument let us show a simple field theory example. It consists of two scalar fields $\phi_1(x)$ and $\phi_2(x)$ with quartic interactions. To keep the example as simple as possible, let us impose the discrete symmetries $\phi_i(x) \to -\phi_i(x)$, $i = 1, 2$ and $\phi_1(x) \leftrightarrow \phi_2(x)$. The most general renormalisable Lagrangian for this model is

$$\mathcal{L} = \frac{1}{2}(\partial_\mu \phi_1)^2 + \frac{1}{2}(\partial_\mu \phi_2)^2 - \frac{1}{2}m^2(\phi_1^2 + \phi_2^2) - \frac{\lambda}{4!}(\phi_1^4 + \phi_2^4) - \frac{2g}{4!}\phi_1^2\phi_2^2. \tag{19.7}$$

The model contains two arbitrary coupling constants, λ and g, and we can ask the question whether there exists a relation among them

$$\eta = \frac{g}{\lambda} = C \qquad (19.8)$$

with a constant C, which is stable under renormalisation. By inspection, we see immediately that here the answer is yes. In fact $C = 0$ is an obvious such relation, since, at this point, the model describes a system of two uncorrelated scalar fields. Are there any other, less trivial, relations? The renormalisation group offers a systematic way to address this question.

The coupling constants λ and g and their ratio η, depend on the renormalisation scheme. This dependence is perturbative, which means that the values at any scheme can be computed as a formal power series in the values at any other. Let us consider two such schemes with coupling constants $\lambda_{(1)}$, $g_{(1)}$ and $\lambda_{(2)}$, $g_{(2)}$. We have

$$\begin{aligned}
\lambda_{(1)} &= \lambda_{(2)} + a_1\lambda_{(2)}^2 + a_2 g_{(2)}^2 + a_3\lambda_{(2)}g_{(2)} + \ldots \\
g_{(1)} &= g_{(2)} + b_1\lambda_{(2)}^2 + b_2 g_{(2)}^2 + b_3\lambda_{(2)}g_{(2)} + \ldots,
\end{aligned} \qquad (19.9)$$

where the a's and the b's are calculable numbers which relate the two schemes (for example, they may depend on the ratio μ_1/μ_2 of the subtraction points in the two schemes) and the dots stand for higher loop terms. The important, although trivial, observation, is that at lowest order we must always have $\lambda_{(1)} = \lambda_{(2)}$ and $g_{(1)} = g_{(2)}$, since there is no renormalisation in the classical approximation. Furthermore, (19.9) exhausts all possible choices in the sense that, given a set of coupling constants λ and g obtained in a scheme (1), any other possible set will be given in terms of a set of numbers a and b. The scheme dependence (19.9) is governed by the renormalisation group. A straightforward calculation gives for the β-functions associated with the coupling constants λ and g at the 1-loop level:

$$\begin{aligned}
16\pi^2\beta_\lambda &= 3\lambda^2 + \frac{1}{3}g^2 \\
16\pi^2\beta_g &= \frac{4}{3}g^2 + 2\lambda g.
\end{aligned} \qquad (19.10)$$

A relation invariant under renormalisation is a fixed point of the renormalisation group, i.e. a zero of the corresponding β-function. For example, we find immediately the trivial fixed point $g = 0$, a zero of β_g, which means that if we start with no g term in the Lagrangian, no such term will be generated by renormalisation. If we are interested in the behaviour of the ratio $\eta = g/\lambda$ we compute the β-function

$$\beta_\eta = \frac{\lambda\beta_g - g\beta_\lambda}{\lambda^2} = \frac{-1}{48\pi^2}\eta\lambda(\eta^2 - 4\eta + 3). \qquad (19.11)$$

Equation (19.11) shows that we obtain three fixed points, namely $\eta=0$, 1, and 3. The first is the trivial one we saw earlier. The second, $\eta=1$, gives a more symmetric theory in which the discrete symmetries of (19.7) are promoted into a continuous $O(2)$ group with ϕ_1 and ϕ_2 becoming the two components of an $O(2)$ vector. Finally, the third fixed point, $\eta=3$, through the transformation $\psi_1 = \phi_1 + \phi_2$ and $\psi_2 = \phi_1 - \phi_2$ reduces again to a system of two uncoupled fields with interaction $\psi_1^4 + \psi_2^4$. Furthermore, Eq. (19.11) shows that these are the only possible values of the ratio of the coupling constants which are respected by renormalisation.

For a theory with more than two coupling constants the solution requires the study of all possible relations which are admissible in perturbation theory, a task whose weight increases very fast with the number of coupling constants. For every relation, we compute the corresponding β-function and look for possible fixed points. Although an exhaustive study may be quite tedious, the answer is always unambiguous. The moral of the story is that the reducibility of a model, in other words, the possibility of imposing renormalisation stable relations among its parameters, can be studied by studying the zeros of the β-functions. We may ask the question: how reliable is such a computation? Answer: as reliable as perturbation theory can possibly be. The reason is the result we proved in section 16.6.3 showing that the first non-vanishing terms in the expansion of the β-functions are independent of the renormalisation scheme we use to compute them.

The reader may wonder whether all this is worth the trouble, since the results of this example could be easily guessed by inspection. To show that such is not the case, we turn now to a less obvious example.

We consider a model involving a Majorana spinor $\psi(x)$ and two spin zero fields, a scalar $A(x)$ and a pseudoscalar $B(x)$. They interact through the Lagrangian density

$$\mathcal{L} = \tfrac{1}{2}(\partial_\mu A)^2 + \tfrac{1}{2}(\partial_\mu B)^2 + i\bar{\psi}\,\partial\!\!\!/\psi \tag{19.12}$$
$$+ f\bar{\psi}\psi A + if\bar{\psi}\gamma_5\psi B - \tfrac{1}{2}\lambda(A^2 + B^2)^2 - gA^2B^2,$$

where we have left out possible mass terms which will not affect our discussion at the 1-loop order. The model is renormalisable by power counting and contains three independent coupling constants, a common Yukawa f and the two boson self-interactions λ and g. It is easy to show that due to discrete symmetries, it is stable. It is also easy to guess, following the results of the previous model, that the point $\lambda = g$ gives also a stable theory. The question is whether there are other stable points, for example a relation between all three couplings f, λ, and g. Unless we know already the answer, we do not believe it can be guessed by inspection. So let us go on with the calculation.

The β functions are

$$\pi^2 \beta_{f^2} = \tfrac{2}{3}f^4 \tag{19.13}$$
$$\pi^2 \beta_\lambda = f^2\lambda + \tfrac{9}{4}\lambda^2 + \tfrac{1}{4}g^2 - 2f^4$$
$$\pi^2 \beta_g = f^2\lambda + \tfrac{3}{2}\lambda g + g^2 - 2f^4.$$

We are interested in the variables $\eta_1 = \lambda/f^2 - 1$ and $\eta_2 = g/f^2 - 1$. The linear combinations $\delta_1 = \eta_1 - \eta_2$ and $\delta_2 = \eta_1 + \tfrac{1}{5}\eta_2$ satisfy, to first order in η_1 and η_2, the relations

$$\beta_{\delta_1} \sim \frac{3}{2}\delta_1; \quad \beta_{\delta_2} \sim \frac{9}{2}\delta_2, \tag{19.14}$$

which show that the point $\lambda = g = f^2$ is a fixed point and gives a stable theory. It is not easy to guess the underlying symmetry because it transforms fermions into bosons and vice versa. It is called *supersymmetry* and we will introduce it in the last chapter.

19.4   Dimensional Transmutation

In the examples studied in Chapter 15, the critical point for the phase transition between the unbroken and the broken phase was found to be $M^2 = 0$, the vanishing of the mass of a scalar field. Sidney Coleman and Erick Weinberg asked the question of how this value is affected by quantum corrections. By doing so, they discovered an interesting physical phenomenon, that of *dimensional transmutation*.

Let us look at the Abelian example of section 15.3.1. At the classical level we can choose the value $M^2 = 0$. In this case the model contains two arbitrary parameters, the coupling constants e and λ. They are both dimensionless and the theory exhibits classical scale invariance. We have already noted that the $M^2 = 0$ choice is not stable under renormalisation but let us ignore this point for the moment. For positive values of λ the potential $V(\phi)$ has only one minimum, the point $\phi = 0$, which is invariant under the $U(1)$ transformations (15.41). We conclude that the $U(1)$ symmetry is unbroken. However, the absence of a ϕ^2 term makes the potential very flat around the $\phi = 0$ point and it is legitimate to ask the question whether quantum corrections may change the situation. It turns out that this question can be answered using perturbation theory and the renormalisation group. We will follow the original discussion of Coleman and Weinberg. Before going to a specific model, we introduce some new field theory concepts which we will find useful.

In section 16.2 we introduced the *effective action* $\Gamma[\phi^{(\mathrm{cl})}]$ as the Legendre transform of the generating functional of the connected Green functions, Eq. (16.7). When expanded in powers of the classical field $\phi^{(cl)}$, Eq. (16.9), it gives the one-particle irreducible Green functions. On the other hand, we noted in Chapter 15 that the ground state of the system corresponds to a constant field configuration. This suggests an alternative expansion of $\Gamma[\phi^{(cl)}]$, namely in the number of derivatives of the classical field:

$$\Gamma[\phi^{(\mathrm{cl})}] = \int \mathrm{d}^4x [-V(\phi^{(\mathrm{cl})}) + \frac{1}{2}(\partial_\mu \phi^{(\mathrm{cl})})^2 Z(\phi^{(\mathrm{cl})}) + ...]. \tag{19.15}$$

$V(\phi^{(\mathrm{cl})})$ is called *the effective potential*. In the classical theory it coincides with what we have used so far, namely the sum of all non-derivative terms in the Lagrangian. Its higher order corrections are given by the sum of all 1-PI diagrams with zero external momenta. The renormalisation conditions, which we have expressed in terms of the 1-PI Green functions appearing in the expansion (16.9) of $\Gamma[\phi^{(\mathrm{cl})}]$, can be expressed equally well in terms of the functions V, Z, etc. of the expansion (19.17). For example, the conditions (16.53), (16.54), and (16.68) for the ϕ^4 theory can be rewritten in terms of V and Z as

$$\frac{\mathrm{d}^2 V}{\mathrm{d}\phi^{(\mathrm{cl})2}}\big|_{\phi^{(cl)}=0} = m^2; \quad \frac{\mathrm{d}^4 V}{\mathrm{d}\phi^{(\mathrm{cl})4}}\big|_{\phi^{(\mathrm{cl})}=0} = \lambda; \quad Z(0) = 1. \tag{19.16}$$

After these preliminaries, we consider the Coleman–Weinberg model, that of a massless complex scalar field $\phi(x)$ interacting with the electromagnetic field $A_\mu(x)$, i.e. Eq. (15.40) with $M = 0$:

$$\mathcal{L}_1 = -\frac{1}{4}F_{\mu\nu}^2 + |(\partial_\mu + \mathrm{i}eA_\mu)\phi|^2 - \lambda(\phi\phi^*)^2. \tag{19.17}$$

The analysis presented in section 15.3.1 shows that, classically, the ground state is given by the symmetric configuration $\phi = 0$. Let us compute the 1-loop quantum corrections. We write $\phi = (\phi_1 + \mathrm{i}\phi_2)/\sqrt{2}$ and compute V as the sum of all 1-loop 1-PI diagrams with zero external momenta. Because of the $U(1)$ symmetry, it is sufficient to consider only external ϕ_1 lines. We need a gauge-fixing term and we choose to work in the Landau gauge in which the photon propagator is transverse, given by $-\mathrm{i}[g_{\mu\nu} - k_\mu k_\nu/k^2]/(k^2 + \mathrm{i}\epsilon)$. We have three types of diagrams to consider: (i) those of Fig. 19.3(a) with only ϕ-lines in the loop, (ii) those of Fig. 19.3(b) with only photon lines, and (iii) the mixed ones of Fig. 19.3(c). It is easy to see that in the Landau gauge and taking into account the fact that all external lines carry zero momentum, the diagrams of Fig. 19.3(c) vanish. We are left with the first two classes. The computation is straightforward and we find that

$$V_1 = \mathrm{i}\int \frac{\mathrm{d}^4 k}{(2\pi)^4} \sum_{n=1}^{\infty} \frac{1}{2n}\left[\left(\frac{\lambda\phi^2}{2(k^2 + \mathrm{i}\epsilon)}\right)^n + \left(\frac{\lambda\phi^2}{6(k^2 + \mathrm{i}\epsilon)}\right)^n + 3\left(\frac{e^2\phi^2}{4(k^2 + \mathrm{i}\epsilon)}\right)^n\right], \tag{19.18}$$

where we have dropped the superscript (cl) for the classical field. The first term comes from the diagrams with only ϕ_1 internal lines, the second with only ϕ_2, and the third with only photons. Apart from combinatoric factors, they give identical contributions. Each one of the terms in the series is divergent. The first two have the usual ultraviolet divergences which we know how to remove by renormalisation. We must introduce a mass and a coupling constant counterterm. Note that the absence of a mass term $M^2\phi\phi^*$ in the Lagrangian (19.17) does not prevent the appearance of a mass counter-term. This is in agreement with our conclusions in chapter 17 in which we showed that setting the

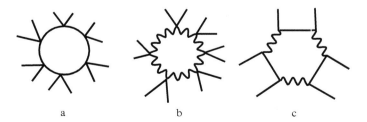

Figure 19.3 *The three types of 1-loop diagrams contributing to* $V(\phi)$.

mass of a scalar field equal to 0 does not increase the symmetries of the theory. However, the absence of a mass term in the propagators makes all terms with $n > 1$ in the series (19.18) infrared divergent. However, if we introduce some infrared cut-off, we can sum the series and obtain, after rotation to Euclidean space,

$$V_1 = \frac{1}{2} \int \frac{\mathrm{d}^4 k}{(2\pi)^4} \left[\ln\left(1 + \frac{\lambda\phi^2}{2k^2}\right) + \ln\left(1 + \frac{\lambda\phi^2}{6k^2}\right) + 3\ln\left(1 + \frac{e^2\phi^2}{4k^2}\right) \right]. \qquad (19.19)$$

In order to obtain the total effective potential to this order we add the 0-loop contribution V_0, given, including the counter-terms, by

$$V_0 = \frac{\lambda Z_\lambda}{4!}\phi^4 + \frac{\delta M^2}{2}\phi^2, \qquad (19.20)$$

where, as usual, the counter-terms are given by

$$Z_\lambda = 1 + z_\lambda^1 \lambda + z_e^1 e^2; \quad \delta M^2 = \delta_\lambda^1 \lambda + \delta_e^1 e^2 \qquad (19.21)$$

with the z's and the δ's divergent constants to be determined by the renormalisation conditions. In our case we want to use conditions of the form

$$\frac{\mathrm{d}^2 V}{\mathrm{d}\phi^2}\Big|_{\phi=0} = 0, \qquad (19.22)$$

which ensures that the renormalised mass of the scalar field is 0. This condition will fix the counter-terms δ_λ^1 and δ_e^1. For the coupling constant renormalisation we need a condition involving the fourth derivative of V with respect to ϕ. We cannot impose the one used in (19.16) because of the infrared divergence at $\phi = 0$. So we introduce a subtraction point μ and we impose

$$\frac{\mathrm{d}^4 V}{\mathrm{d}\phi^4}\Big|_{\phi=\mu} = \lambda. \qquad (19.23)$$

We can now perform the momentum integrations in (19.19) and find that:

$$V = \frac{\lambda}{4!}\phi^4 + \left(\frac{5\lambda^2}{1152\pi^2} + \frac{3 e^4}{64\pi^2}\right)\phi^4 \left(\ln\frac{\phi^2}{\mu^2} - \frac{25}{6}\right). \qquad (19.24)$$

This is our final answer for the effective potential at this order. Note that because of the vanishing of the diagrams of Fig. 19.3 (c), there is no λe^2 term. Higher orders will bring corrections proportional to λ^3, $\lambda^2 e^2$, and e^6. The consistency of the perturbation expansion requires that all coupling constants are $\ll 1$, so we must drop the λ^2 term and keep only the terms λ and e^4. The effective potential (19.24) has a minimum away from the origin in classical field space. If we call v the value of the field at the minimum, we obtain

$$\frac{\mathrm{d}V}{\mathrm{d}\phi} = 0 \quad \to \quad \left(\frac{\lambda}{6} - \frac{11e^4}{16\pi^2}\right)v^3 = 0, \tag{19.25}$$

so we may have $v \neq 0$, provided the coupling constants satisfy the relation

$$\lambda = \frac{33}{8\pi^2}e^4. \tag{19.26}$$

From this point the analysis parallels that of Chapter 15. Translating the field $\phi \to \phi + v$ breaks the gauge symmetry spontaneously and gives a mass to the photon $m_\gamma^2 = e^2 v^2$. The would-be Goldstone boson decouples and we are left with a single neutral scalar field with mass $m_S^2 = 3e^4 v^2/8\pi^2$.

We have a few remarks before closing this section.

1. What we have established is an extension of the domain where spontaneous symmetry breaking occurs by considering the 1-loop radiative corrections to the effective potential. This result can be easily extended to the non-Abelian case.

2. In principle, we can repeat the calculation for the simple ϕ^4 model of section 15.2.3. The radiative corrections to the effective potential are given by (19.24), setting $e = 0$. However, here the analogy stops. Finding a non-zero v would require a balance between a term of order λ and one of order λ^2 and such a relation makes no sense in perturbation theory. In our case we are saved because of the presence of two coupling constants. The relation (19.26) can be correct even if both are very small.

3. In the symmetric phase the model is described by two dimensionless parameters, the coupling constants λ and e. In the broken phase the two are related by (19.26), but we now have a new *dimensionful* parameter v. In other words, we have traded a dimensionless parameter for one bringing a mass scale. This phenomenon, which Coleman and Weinberg called *dimensional transmutation*, is expected to be quite general, for all theories which are classically scale invariant. Whether its study is possible by perturbation depends on the presence of more than one coupling constant.

4. In our calculations we have worked in the Landau gauge. The effective potential is not a physically measurable quantity and, in general, it is not gauge invariant. This raises the question of the physical relevance of the result. The answer, which we do not prove here, is that although all intermediate steps, in particular the expression for the effective potential, are indeed gauge dependent, the physically measurable conclusions, such as the scalar to vector mass ratio m_S/m_γ, are not.

5. The relation (19.26) looks rather peculiar; in particular, it does not seem to result from any symmetry of the Lagrangian. We have learned in this book to be suspicious against such arbitrary relations. The parameters of a Lagrangian are supposed to take generic values, unless otherwise dictated by an underlying symmetry. How general is an effect which seems to depend on a specific relation?

The answer turns out to be surprisingly simple. Let us start with a couple of generic values for λ and e with the only restriction of both being much smaller than 1 for perturbation theory to be reliable. Since we have a massless theory, these coupling constants should be defined by renormalisation conditions imposed at a given scale μ. The dependence on μ can be studied using the equations of the renormalisation group. The result (see Problem 19.2) is that no matter what the initial values are, by a change of scale we can always bring the effective values of the coupling constants in agreement with the relation (19.26). The phenomenon of dimensional transmutation is indeed a genuine physical phenomenon.

19.5   Problems

> **Problem 19.1** In section 19.3 we studied the stability properties of a field theory involving two scalar fields. The purpose of this problem is to generalise the analysis to the case of a theory invariant under $O(N) \times O(N)$, as a function of N.
>
> 1. The case $N = 2$. Consider a couple of two-component fields $\Phi_1^i(x)$ and $\Phi_2^i(x)$, $i=1,2$. We use a vector notation and write the fields as $\boldsymbol{\Phi}_1$ and $\boldsymbol{\Phi}_2$. The scalar product $\boldsymbol{\Phi}_1 \cdot \boldsymbol{\Phi}_1$ denotes the sum $\Phi_1^1 \Phi_1^1 + \Phi_1^2 \Phi_1^2$. The Lagrangian is
>
> $$\mathcal{L} = \tfrac{1}{2}(\partial_\mu \boldsymbol{\Phi}_1) \cdot (\partial^\mu \boldsymbol{\Phi}_1) + \tfrac{1}{2}(\partial_\mu \boldsymbol{\Phi}_2)(\partial^\mu \boldsymbol{\Phi}_2) - \tfrac{1}{2}m^2(\boldsymbol{\Phi}_1 \cdot \boldsymbol{\Phi}_1 + \boldsymbol{\Phi}_2 \cdot \boldsymbol{\Phi}_2)$$
> $$- \tfrac{\lambda}{4!}[(\boldsymbol{\Phi}_1 \cdot \boldsymbol{\Phi}_1)^2 + (\boldsymbol{\Phi}_2 \cdot \boldsymbol{\Phi}_2)^2] - \tfrac{2g}{4!}(\boldsymbol{\Phi}_1 \cdot \boldsymbol{\Phi}_1)(\boldsymbol{\Phi}_2 \cdot \boldsymbol{\Phi}_2). \quad (19.27)$$
>
> The discrete invariance we studied in section 19.3 has now been promoted to $O(2) \times O(2) \times P$, where P denotes the exchange transformation $\boldsymbol{\Phi}_1 \leftrightarrow \boldsymbol{\Phi}_2$.
> (i) Compute again the 1-loop β-functions β_λ and β_g.
> (ii) Find the fixed points of the ratio $\eta = g/\lambda$ and explain their physical meaning.
> (iii) What can we say about their stability properties?
> 2. We study now the general case of N-component fields where the scalar product denotes the sum $\boldsymbol{\Phi}_1 \cdot \boldsymbol{\Phi}_1 = \sum_{i=1}^{N} \Phi_1^i \Phi_1^i$. The Lagrangian density has the same form as (19.27). The invariance is $O(N) \times O(N) \times P$.
> (i) Compute again the 1-loop β-functions β_λ and β_g as functions of N.
> (ii) Find the fixed points of the ratio $\eta = g/\lambda$.
> (iii) Study their stability properties.
> (iv) Consider the limit $N \to \infty$ keeping $N\lambda = \lambda^*$ fixed. Explain the meaning of the fixed points.

The answer to the last question is as follows. At the limit $N \to \infty$ with λ^* fixed, the fixed points are $\eta=1,0$, and -1. They correspond to the uncoupled case, the symmetric case and one in which the interaction becomes proportional to $(\lambda^*/N)(\Phi_1^2-\Phi_2^2)^2$. By a change of variables of the form $X = (\Phi_1 + \Phi_2)/\sqrt{2}$ and $Y = (\Phi_1 - \Phi_2)/\sqrt{2}$ it can be written as $(\lambda^*/2N)(X \cdot Y)^2$. It is invariant under a subgroup of the original $O(N) \times O(N)$ symmetry, namely its diagonal subgroup $O(N)$ which rotates simultaneously X and Y. This coupling alone does not correspond to any fixed point for finite N, because the $O(N)$ symmetry allows for other couplings, such as X^4, Y^4, and $X^2 Y^2$. However, we can easily check that the diagrams which would give counter-terms away from this point are suppressed by inverse powers of N, so at the limit $N \to \infty$ this becomes a fixed point. Indeed, the counter-term for the $(X \cdot Y)^2$ interaction can be isolated by looking at a 1-loop diagram with $X_1 Y_1 X_2 Y_2$ external lines. There are N such diagrams because we can circulate all N components in the loop. So the counter-term is proportional to $((\lambda^*)^2/N)$. On the other hand, the counter-term for the X^4 term is obtained by looking at a diagram with four X_1 external lines. If we start with only a $(\lambda^*/2N)(X \cdot Y)^2$ interaction we have only one such diagram, so we obtain a counter-term proportional to $((\lambda^*)^2/N^2)$.

Problem 19.2 Compute the β-functions for the coupling constants e and λ of the Coleman–Weinberg model and show that the relation (19.26) which exhibits the phenomenon of dimensional transmutation is always reached for generic, but small, values of the coupling constants.

Problem 19.3 The purpose of this problem is to study the renormalisation properties of the non-linear σ-model we introduced in section 15.2.5. Consider the N component scalar field $\phi_i(x)$, $i = 1, 2, ..., N$, interacting through an $O(N)$ invariant interaction. Following the analysis of section 15.2.4, we assume that $O(N)$ is spontaneously broken to $O(N-1)$ and introduce the notation $\phi = (\sigma, \pi_j)$, $j = 1, 2, ..., N-1$. We eliminate the field σ through the condition

$$v^2 = \sigma^2 + \sum_{j=1}^{N-1} \pi_j^2, \tag{19.28}$$

and we rescale the π field through $\pi_j \to \sqrt{\lambda}\pi_j$. We want to study the perturbation expansion in powers of λ of the resulting non-linear model.

(i) Show that at any order of perturbation the effective Lagrangian density is given by

$$\mathcal{L}_{\text{eff}} = \frac{1}{2}\left[\partial_\mu \boldsymbol{\pi} \cdot \partial^\mu \boldsymbol{\pi} + \lambda \frac{(\boldsymbol{\pi} \cdot \partial^\mu \boldsymbol{\pi})^2}{v^2 - \lambda \boldsymbol{\pi} \cdot \boldsymbol{\pi}}\right] + \ln \det(v^2 - \lambda \boldsymbol{\pi} \cdot \boldsymbol{\pi}) \tag{19.29}$$

and we can ignore the bound induced by the condition (19.28). The non-linear terms in (19.29) are understood as producing an infinite power series in λ.

(ii) Find the Feynman rules.

(iii) Ignoring possible infrared divergences, use a power-counting argument to prove that the model is renormalisable in a two-dimensional space–time.

(iv) Show that the linear term $c\sigma$ of Eq. (15.23) regularises the infrared behaviour. By power counting show that it does not impose a breaking of the $O(N)$.

(v) Find the counter-terms.

The model is analysed in detail in 'Quantum Field Theory and Critical Phenomena', by Jean Zinn-Justin (Oxford Science).

20

Analyticity Properties of Feynman Diagrams

20.1 Introduction

In Chapter 12 we saw that the axioms of relativistic quantum field theory, in particular that of locality and causality, imply certain analyticity properties for the Wightman functions. They are analytic in an extended tube and their physical values are obtained as boundary values of these analytic functions when we approach the physical region. In this chapter we want to see how this programme is realised in the case of simple Feynman diagrams. In doing so we will discover a surprisingly simple physical picture of the unitarity property of the S-matrix.

For simplicity, we will consider only scalar field theories and, in this chapter, only massive particles. The extension to higher spins brings only inessential complications. In contrast, those due to the presence of massless particles are highly non-trivial, both technically and physically. We will look at them very briefly in a later chapter.

We start with a simple kinematical analysis. The n-point Green function depends on the n external momenta. By Lorentz invariance it really depends on the scalar products $p_i \cdot p_j$, whose number is restricted by the overall energy–momentum conservation: $\sum_i p_i = 0$. For simplicity we consider all momenta incoming. If some are outgoing we must change their signs. For example, the 2-point function is a function of only one variable, p^2. For the 4-point function (Fig. 20.1), we introduce the Mandelstam variables

$$s = (p_1 + p_2)^2; \quad t = (p_1 + p_3)^2; \quad u = (p_1 + p_4)^2. \tag{20.1}$$

Energy–momentum conservation implies that they satisfy

$$s + t + u = \sum_{i=1}^{4} p_i^2 \tag{20.2}$$

which means that the 4-point function depends on six scalar variables, which can be chosen to be the four squares of the external momenta and two scalar products. For a

From Classical to Quantum Fields. Laurent Baulieu, John Iliopoulos and Roland Sénéor.

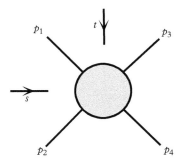

Figure 20.1 *The Mandelstam variables.*

scattering amplitude, where the external momenta are on the mass shells of the corresponding particles, they reduce to two variables, which, in a particular reference frame, correspond to the incident energy and the scattering angle. Note also that for a process $p_1 + p_2 \rightarrow (-p_3) + (-p_4)$ $s = (p_1 + p_2)^2$ equals the square of the total energy in the centre-of-mass frame and t and u are the squares of the two momentum transfers. This counting can be extended to higher n-point functions, each additional momentum bringing four more variables, its square and three scalar products. Obviously, for a physical amplitude, all these variables take real values, but it will be instructive to consider the extension to the complex domain and study the analyticity properties. Note also that a fundamental property of quantum field theory which we call *crossing symmetry*[1] implies that when we compute a 4-point Green function, we can obtain the scattering amplitudes for three physical processes, depending on which momenta we consider as incoming and which as outgoing. We call these processes *channels*. Thus we speak about the s-channel $(p_1 + p_2 \rightarrow (-p_3) + (-p_4))$, the t-channel $(p_1 + p_3 \rightarrow (-p_2) + (-p_4))$, or the u-channel $(p_1 + p_4 \rightarrow (-p_2) + (-p_3))$. It is easy to check that the physical regions for these three processes do not overlap. For example, when all masses are equal, the physical region for the s-channel is $s \geq 4m^2$, t and u are negative, and, similarly, that for the t-channel is $t \geq 4m^2$, while s and u are negative. The fact that the same Green function describes all three scattering amplitudes according to the domain of the variables implies that we can relate these amplitudes by analytic continuation. This is another motivation to study the analyticity properties.

20.2 Singularities of Tree Diagrams

We start with the trivial case of tree diagrams. The only singularities are the poles of the propagators at $k^2 = m^2$ with the Feynman prescription $m^2 \rightarrow m^2 - i\epsilon$. If we consider a two-particle scattering amplitude (Fig. 20.2), the singularity appears at $s = (p_1 + p_2)^2 = m^2$. The physical region for this amplitude is $s \geq (m_1 + m_2)^2$, the square

---

[1] Crossing symmetry is obvious in perturbation theory when we compute a scattering amplitude using Feynman diagrams.

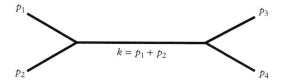

Figure 20.2 *The s-channel pole of a scattering amplitude.*

of the sum of the two incoming particles. On the other hand, we recall that the asymptotic theory which allowed us to relate scattering amplitudes to Feynman diagrams applies, strictly speaking, only to stable particles. Unstable particles do not correspond to *in*, or *out*, states. The stability of the exchange particle implies both $m < m_1 + m_2$ and $m < m_3 + m_4$, the masses of the outgoing particles. In other words, the singularity is always below the threshold of the physical region. The same is true if we look at a pole in a momentum transfer variable t or u. The physical region is $t \leq (m_1 - m_3)^2$ but the stability of the outgoing particle implies $m_3 > m + m_1$, which means again that the pole singularity is outside the physical region.

This analysis applies also to 'tree-like' lines in a diagram, i.e. lines that, if cut, separate the diagram into two disconnected pieces (Fig. 20.3). The pole singularity appears at $\left(\sum_{i=1}^{n} p_i\right)^2 = m^2$ but again stability implies that m is smaller than the sum of the masses of all the particles on the left or the right. We can find connected but one particle reducible Feynman diagrams whose tree-like lines appear to generate poles of higher order; see, for example the diagram of Fig. 20.4. The internal line with momentum k will give a factor

$$\frac{i}{k^2 - m^2 + i\epsilon} \Sigma(k^2) \frac{i}{k^2 - m^2 + i\epsilon}, \tag{20.3}$$

where $\Sigma(k^2)$ is the 1-PI part of the 2-point function of the field whose propagator we are considering. For an arbitrary choice of the renormalisation condition used to define $\Sigma(k^2)$, this diagram does indeed produce a double pole, $(k^2 - m^2)^{-2}$. However, for a scattering amplitude we must use physical, on-shell, renormalisation conditions which

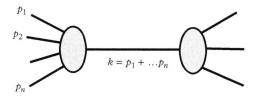

Figure 20.3 *The pole generated by a tree-like internal line.*

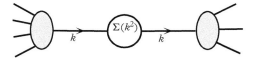

Figure 20.4 *An example of a double pole.*

force $\Sigma(k^2)$ to vanish when k^2 equals m^2, the value of the physical mass. As a result, scattering amplitudes have only simple poles. We will come back to this point when discussing unstable particles.

20.3   Loop Diagrams

In Contrast, to the tree diagrams, the singularity structure of loop diagrams is much more involved. Let us look at the 1-PI ones. The expression for a general diagram was given at (16.24) with N and D given in (16.25) and (16.26). Singularities will appear only when D vanishes. It is clear that when all momenta are Euclidean the propagators for massive particles are strictly positive definite and never vanish. It follows that the Green functions are analytic in any finite region of Euclidean space. It is not difficult to prove that this conclusion survives renormalisation for some set of Euclidean space renormalisation conditions, because, as we explained already, order by order in perturbation theory, the counter-terms introduce only polynomials in the external momenta. Therefore, singularities will appear only when we move to the physical Minkowski space and, in particular, when we consider scattering amplitudes. A zero of D will be a singular point of the Green function if it appears in the domain of integration of the loop momenta and cannot be avoided by suitably deforming the integration contour. Using the general formula (16.37), we can trade the momentum integrations for those over the Feynman parameters z which vary from 0 to 1. Singularities will appear in two cases: (i) when, for some values of the external momenta, the denominator vanishes at the edge of the domain of the z integration (end-point singularities); it is clear that such a point cannot be avoided by any kind of deformation of the z integration contour and we will obtain a singularity; and (ii) when the contour of the z integration is pinched between two, or more, singularities (*pinching singularities*). L. D. Landau has derived a set of equations which exhibit the necessary conditions for such singularities to occur for a multiloop diagram, but we will not discuss them here. We will limit ourselves to showing explicitly the singularity structure of some simple 1-loop diagrams.

We start with the simplest one, the 1-PI diagram of Fig. 20.5. It can be viewed either as the 1-loop contribution to the 2-point function in a ϕ^3 theory or as part of the 1-loop contribution to the 4-point function in a ϕ^4 theory. We started its computation in (16.42). Here we will need also the finite part. We write that

$$I(s) = \int_0^1 dz \int \frac{d^d k}{(2\pi)^d} \frac{1}{[k^2 - m^2 + sz(1-z)]^2}, \tag{20.4}$$

where s is either p^2, for a 2-point function, or $(p_1 + p_2)^2$ for a 4-point one. We are interested in the analytic properties of $I(s)$ in the complex s plane. Performing the

Figure 20.5 *The 1-loop singularities of the 2-point function.*

d-dimensional momentum integration and dropping terms which vanish when ϵ goes to 0, we obtain

$$-\mathrm{i}16\pi^2 I(s) = \frac{2}{\epsilon} - \gamma + \ln(4\pi) - \int_0^1 \mathrm{d}z \ln[m^2 + sz(z-1)]. \tag{20.5}$$

We have used the expansions $\Gamma(\epsilon/2) = 2/\epsilon - \gamma + O(\epsilon)$ and $\alpha^\epsilon = 1 + \epsilon \ln \alpha + O(\epsilon^2)$. γ is the Euler constant. The analytic properties in s are contained in the last term. We write

$$sz^2 - sz + m^2 = s(z - z_+)(z - z_-); \quad z_\pm = \frac{1 \pm \sqrt{1 - 4\eta}}{2}; \quad \eta = \frac{m^2}{s}. \tag{20.6}$$

With this notation the integral in (20.5) gives

$$\int_0^1 \mathrm{d}z \ln[m^2 + sz(z-1)] = \ln s + \ln(z_+ z_-) + \sqrt{1 - 4\eta} \ln \frac{z_+}{z_-} - 2. \tag{20.7}$$

Keeping only the principal definition of the logarithm, often called *the physical sheet*, we obtain

$$\int_0^1 \mathrm{d}z \ln[m^2 + sz(z-1)] = \ln m^2 - 2 + \sqrt{1 - 4\eta} \ln \frac{1 + \sqrt{1 - 4\eta}}{1 - \sqrt{1 - 4\eta}}. \tag{20.8}$$

The singularity structure can be read from this expression. We see, in particular, that: (i) for real $s < 4m^2$, z_+ and z_- are complex conjugate to each other and the amplitude is real and analytic on the physical sheet.[2] (ii) The threshold of the physical region $s = 4m^2$ is a branch point with a cut running from $s = 4m^2$ to infinity. The discontinuity across the cut can be computed as the difference $2\mathrm{i}\Delta I(s) \equiv I(s + \mathrm{i}\epsilon) - I(s - \mathrm{i}\epsilon)$:

$$2\mathrm{i}\Delta I(s) = \frac{\mathrm{i}}{8\pi} \sqrt{\frac{s - 4m^2}{s}} \theta(s - 4m^2). \tag{20.9}$$

In principle, if a function $I(s)$ is analytic in the cut plane, the knowledge of the discontinuity across the cut allows us to reconstruct the function by writing the Cauchy theorem along a contour C going around the cut, as shown in Fig. 20.6. Such a relation was first used by R. Krönig and H. A. Kramers in order to express the forward scattering amplitude of light of frequency ω in a medium as an integral over the imaginary part of the scattering amplitude at all frequencies. By the optical theorem this equals the total absorption cross section of light in the medium. These relations are called *dispersion relations*. However, in our case, $\Delta I(s)$ goes to a constant when s goes to infinity and the integral does not converge. We can go around this difficulty by introducing a *subtraction*, namely by writing the Cauchy integral not for the function $I(s)$ but for the difference

---

[2] Note, however, a singularity at $s = 0$ on the second sheet.

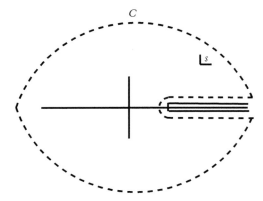

Figure 20.6 *The contour for the s-channel dispersion relation.*

$[I(s) - I(s_0)]/(s - s_0)$. s_0 is an arbitrary point provided it is a point of analyticity of $I(s)$. We thus obtain a *once subtracted dispersion relation* which takes the form

$$I(s) = I(s_0) + \frac{s - s_0}{\pi} \int_{4m^2}^{\infty} ds' \frac{\Delta I(s')}{(s' - s_0)(s' - s)}. \tag{20.10}$$

It determines the value of $I(s)$, in particular the value of its real part at any given point s, in terms of the imaginary part along the cut. The price we pay is the introduction of an unknown constant, the value of I at the point s_0. But this is precisely the arbitrariness we always had in going through the renormalisation programme. For example, if I is the scalar field 2-point function in a ϕ^3 theory, this will correspond to the renormalisation condition which determines the mass. If it is the 4-point function in a ϕ^4 field theory, we can choose $s_0 = 4m^2/3$, in which case $I(s_0)$ is just the coupling constant.

For the ϕ^4 theory we have two more identical diagrams, one for each crossed channel, t and u. It follows that for this theory and at 1-loop order, the two-particle elastic scattering amplitude satisfies independent dispersion relations in each of the variables. In each of them, the imaginary part is independent of the other variables. This simple property ceases to be true for other field theories, or for ϕ^4 at higher orders. The simplest example is the box diagram in a ϕ^3 theory, Fig. 20.7. Apart from trivial numerical factors, it is given by the integral

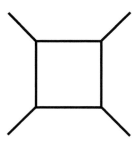

Figure 20.7 *The box diagram.*

$$I^{(4)}(p_1, ..., p_4) =$$

$$\int \frac{\mathrm{d}^4 k}{(2\pi)^4} \frac{1}{(k^2 - m^2)[(k+p_1)^2 - m^2][(k+p_1+p_3)^2 - m^2][(k-p_2)^2 - m^2]}, \quad (20.11)$$

where, as usual, by m^2 we mean $m^2 + i\epsilon$. There is no reason to introduce a regulator to an already nicely convergent integral. Using the Feynman parameters z_i, $i = 1,2,3,4$, Eq. (16.30), we rewrite (20.11) as

$$I^{(4)}(p_1, ..., p_4) = 3! \int \frac{\mathrm{d}^4 k}{(2\pi)^4} \int_0^1 \frac{\prod_{i=1}^4 dz_i \, \delta(1 - \sum_{i=1}^4 z_i)}{[k^2 + F^2]^4} \quad (20.12)$$

with F^2 given by

$$F^2 = z_2 z_4 s + z_1 z_3 t - m^2 [z_1 + z_3 - (z_2 + z_4)^2]. \quad (20.13)$$

We have put the external momenta on the mass shell, $p_i^2 = m^2$. s and t are the Mandelstam variables, Eq. (20.1). The four-dimensional momentum integral can be performed and gives

$$I^{(4)}(s, t) \sim \int_0^1 \frac{\prod_{i=1}^4 dz_i \, \delta(1 - \sum_{i=1}^4 z_i)}{(F^2)^2}. \quad (20.14)$$

To this expression we must add the crossed diagram, which is the same as (20.14) with s and u interchanged. The analyticity properties of the scattering amplitude $I^{(4)}(s, t)$ for complex values of the squares of the centre-of-mass energy s and the momentum transfer t can be obtained by looking at the singularities of (20.14). The physical region of the amplitude is given by $s \geq 4m^2$, $t, u \leq 0$ restricted by the condition (20.2), $s + t + u = 4m^2$. The new feature is that the discontinuity across the cut in the s-plane is a function of t which has its own singularities in the complex t-plane. A straightforward calculation gives the discontinuity in s for fixed t, defined as $2i\Delta_s I^{(4)}(s, t) = I^{(4)}(s + i\epsilon, t) - I^{(4)}(s - i\epsilon, t)$, as

$$I^{(4)}(s + i\epsilon, t) - I^{(4)}(s - i\epsilon, t) = \frac{i}{2\pi} \frac{1}{D} \ln \frac{A + 2sD\sqrt{s(s - 4m^2)}}{A - 2sD\sqrt{s(s - 4m^2)}}, \quad (20.15)$$

where a factor $\theta(s - 4m^2)$ is understood. A and D are functions of the Mandelstam variables

$$A = s^2(s - 2m^2)^2 - s^2(s - 4m^2)(t - u) \quad (20.16)$$

$$D = \sqrt{st[(s - 4m^2)(t - 4m^2) - 4m^4]}. \quad (20.17)$$

Note also that the centre-of-mass scattering angle θ is given in terms of the Mandelstam variables as

$$\cos\theta = \frac{t-u}{s-4m^2}.$$ (20.18)

Using (20.15), we can write an (unsubtracted) fixed t dispersion relation for $I^{(4)}(s,t)$:

$$I^{(4)}(s,t) = \frac{1}{\pi} \int_{4m^2}^{\infty} \frac{\Delta_s I^{(4)}(s',t)}{s'-s} \mathrm{d}s'.$$ (20.19)

On the other hand, $\Delta_s I^{(4)}$ has a discontinuity in t for fixed s, given again by $2i\rho_{st}(s,t) = \Delta_s I^{(4)}(s,t+i\epsilon) - \Delta_s I^{(4)}(s,t-i\epsilon)$. It follows that we can express $I^{(4)}(s,t)$ in terms of a double dispersion relation of the form

$$I^{(4)}(s,t) = \frac{1}{\pi^2} \int \int \frac{\rho_{st}(s',t')}{(s'-s)(t'-t)} \mathrm{d}s' \mathrm{d}t'.$$ (20.20)

Adding the crossed box diagram gives a similar expression with s and u interchanged, i.e. a double dispersion relation in t and u. The sum is

$$\pi^2 I^{(4)}(s,t) = \int \int \frac{\rho_{st}(s',t')}{(s'-s)(t'-t)} \mathrm{d}s' \mathrm{d}t' + \int \int \frac{\rho_{ut}(u',t')}{(u'-u)(t'-t)} \mathrm{d}u' \mathrm{d}t'.$$ (20.21)

The double discontinuities ρ_{st} and ρ_{ut} are called *double spectral functions*. They vanish outside a certain region in their variables which we can determine by looking at (20.15).

The relation (20.21) is a particular example of a general expression which is assumed to describe any scattering amplitude, called *the Mandelstam representation*. We will see it again in a later section.

20.4   Unstable Particles

The restriction to stable particles, although theoretically justified, is too strong for physical applications. The large majority of physical particles whose interactions we describe by quantum field theory are in fact unstable.[3] Nevertheless, we expect this formalism to be still applicable, provided the lifetime of the unstable particle is long compared to the typical time during which the interaction takes place. However, this raises the spectrum of having singularities in the physical region of scattering amplitudes, which is obviously unacceptable. We will address this question here.

Let us consider a very simple model consisting of two real scalar fields, $\phi(x)$ and $\Phi(x)$. Their interaction is assumed to be of the form

$$\mathcal{L} = \frac{1}{2}(\partial\phi)^2 + \frac{1}{2}(\partial\Phi)^2 - \frac{1}{2}m^2\phi^2 - \frac{1}{2}M^2\Phi^2 - \frac{\lambda}{4!}\phi^4 - \frac{gm}{2}\Phi\phi^2.$$ (20.22)

---

[3] The only elementary particles which appear to be absolutely stable are the photon, since it is massless, the neutrinos, as the lightest fermions, the electron, as the lightest electrically charged particle, and, possibly, the proton, if baryon number turns out to be absolutely conserved.

We have explicitly included the factor m to make the coupling constant g dimensionless. In the quantum theory these fields describe two spin-0 particles which we call A, for Φ, and a, for ϕ. We will assume $M > 2m$, which means that A can decay into two a's. More realistic models will be examined in a later chapter.

The physical region for the scattering amplitude of the process $a + a \to a + a$ is $s > 4m^2$. In lowest order (tree approximation) this amplitude receives a contribution from the diagram of Fig. 20.8(a) given by

$$I^{(4)}(s, t) = ig^2 m^2 \frac{1}{s - M^2} + ...,\tag{20.23}$$

where the dots stand for the crossed terms with the A exchange in the t and u channels. The contribution (20.23) clearly exhibits the $s = M^2$ pole in the physical region. The same interaction which is responsible for this singularity causes the A particle to decay with an amplitude given, to lowest order, by the diagram of Fig. 20.9:

$$I^{(3)} = igm.\tag{20.24}$$

We obtain the decay probability by integrating over the two-particle phase-space[4],

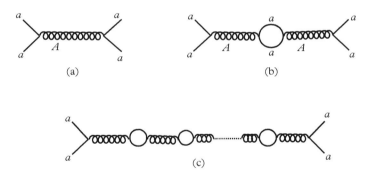

(a)   (b)

(c)

Figure 20.8 *The contribution of an unstable intermediate particle: in the tree approximation (a), including higher orders (b) and (c).*

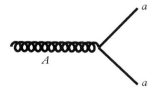

Figure 20.9 *The decay diagram of the unstable particle A to two a's.*

---

[4] In our model the two particles in the final state are identical, so we integrate over half the phase space.

$$\mathrm{d}\Gamma = \frac{g^2 m^2}{4M} \, \mathrm{d}\tilde{q}_1 \, \mathrm{d}\tilde{q}_2 \, (2\pi)^4 \delta^4(P - q_1 - q_2), \qquad (20.25)$$

where $\mathrm{d}\tilde{q}$ denotes the invariant measure on the hyperboloid $q^2 = m^2$ given by

$$\mathrm{d}\tilde{q} = \frac{\mathrm{d}^4 q}{(2\pi)^3} \delta(q^2 - m^2)\theta(q^0) = \frac{\mathrm{d}^3 q}{(2\pi)^3 2q^0}; \quad q^0 = \sqrt{|\boldsymbol{q}|^2 + m^2}. \qquad (20.26)$$

The result is

$$\Gamma = \frac{g^2 m^2}{2M} \frac{1}{16\pi} \sqrt{\frac{M^2 - 4m^2}{M^2}}. \qquad (20.27)$$

We want to argue that if the lifetime of the particle, which is the inverse of Γ, is *long*, the field-theoretic description should be adequate. We want to make this statement more precise.

The tree diagrams of Fig. 20.8(a) give the order g^2 contribution to the $a + a \to a + a$ scattering amplitude. At order g^4 we have several new diagrams, but, among them, we want to single out the one of Fig. 20.8(b). It is the g^2 correction to the A propagator. We have already computed this diagram and let us call the result $ig^2 m^2 \Sigma(p^2)$. At this order the A propagator is given by

$$G_A(p^2) = \frac{i}{p^2 - M^2} \left[1 + g^2 m^2 \Sigma(p^2) \frac{1}{p^2 - M^2} \right]. \qquad (20.28)$$

Let us assume, for the moment, that this procedure can be iterated; in other words, let us assume that we are allowed to isolate the set of diagrams of Fig. 20.8(c) and sum the resulting series. We will try to justify this assumption presently. Under this assumption, the A propagator becomes

$$G_A(p^2) = \frac{i}{p^2 - M^2 - g^2 m^2 \Sigma(p^2)}. \qquad (20.29)$$

The important point is that as we have computed already for $p^2 > 4m^2$, $\Sigma(p^2)$ has both a real and an imaginary part. The ultraviolet divergence $1/\epsilon$ belongs to the real part, while the imaginary part (20.9) is well defined and calculable. Therefore, the renormalisation condition which determines the mass of A can only be enforced on $\mathrm{Re}\Sigma(p^2)$:

$$\mathrm{Re}\Sigma(p^2)|_{p^2 = M^2} = 0. \qquad (20.30)$$

It follows that the position of the pole of the propagator (20.29) is shifted to the complex p^2-plane with an imaginary part proportional to $\sqrt{(M^2 - 4m^2)/M^2}$, leaving the real axis free of singularities. This mechanism seems to solve the problem of the $s = M^2$ pole in the $a + a$ scattering amplitude, but it appears as an artefact. Let us have a closer look.

The first observation is that the imaginary part of $\Sigma(p^2)$ is given, apart from a trivial kinematical factor, by the decay probability (20.27). We will argue in the next section that this is a general result in quantum field theory, but here this observation allows us to write the propagator (20.29), at the vicinity of the pole, as

$$G_A(p^2) = \frac{i}{p^2 - M^2 + 2iM\Gamma}. \tag{20.31}$$

This resembles the Breit–Wigner formula we obtain in non-relativistic potential scattering for a resonance. Because of that we often call unstable particles *resonances*. In the $a + a$ scattering amplitude the A exchange represented by (20.31) will give the dominant contribution if the position of the pole is not too far from the real axis. This enables us to sharpen the long lifetime criterion for an unstable particle. The necessary condition for the applicability of this formalism is $\Gamma \ll M$. We conclude that switching on the interaction which makes a particle to decay moves the pole originally situated at $p^2 = M^2$ to the complex plane and, in fact, it produces a pair of complex conjugate poles on the unphysical sheet because the imaginary part of Σ changes sign when we turn around the branch point $p^2 = 4m^2$. It is in this sense that we often say that unstable particles have *complex masses*.

The last point we want to make in this section concerns the summation of the series of diagrams in Fig. 20.8(c). In principle, the perturbation series is an expansion in powers of the coupling constant and each term should contain all diagrams of a given order. It is only in this case that formal properties, such as the unitarity of the S-matrix or its gauge invariance, are expected to hold. Here we have not followed this rule; we have selected a certain class of diagrams at each order. Formally, it is just a rearrangement of the terms in the perturbation expansion which produces a new series in which the zero order A propagator is a sum containing arbitrary powers of g. This raises the question of the consistency of the scheme. Rather than starting a lengthy discussion on the properties of the perturbation expansion, discussion in which we are in no position to reach meaningful conclusions, we note that it is just a formal series and the only thing we need is a parameter allowing us to identify successive orders. The coupling constant is one such parameter, but it is by no means the only possible one. In order to justify our expansion it is sufficient to exhibit a new parameter, let us call it η, such that each term in the expansion of the $a + a$ scattering amplitude in powers of η automatically contains the infinite series of diagrams of Fig. 20.8(c). There is a simple formal way to obtain such an η-expansion which is useful in many other applications of quantum field theory.

We modify our Lagrangian (20.22) by introducing N copies of the field ϕ. The interaction term becomes $\frac{gm}{2}\Phi\sum_i\phi_i^2$. We will consider a double expansion in powers of g and $1/N$ for the scattering of $a_1 + a_1 \rightarrow a_2 + a_2$, where a_i is the particle associated with the field ϕ_i. The tree diagram of Fig. 20.8(a) is of order g^2 and the box diagram of Fig. 20.10 of order g^4. The 1-loop correction of Fig. 20.8(b) gives g^4N because we can circulate all the N components in the loop. Every subsequent term in the series of Fig. 20.8(c) brings a factor g^2N. This suggests we should consider the expansion in powers of $1/N$ while keeping $\hat{g} = g\sqrt{N}$ fixed. This will make all terms in the series of

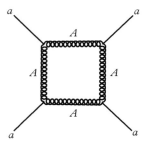

Figure 20.10 *The box diagram contribution of the unstable particle A to the scattering amplitude.*

Fig. 20.8(c) of order $1/N$, while the box diagram and all the others are of order $1/N^2$ and higher. Intuitively, we expect such an expansion to make sense for large N, but we want to emphasise here that for the purposes of defining a formal power series, any value will do. It defines a theory in which it is legitimate to replace the real value of the mass in the propagator of an unstable particle by its complex value. The $1/N$ expansion provides a powerful method for studying quantum field theory in many applications which go far beyond the unstable particles we are studying here. We will see some examples in a later chapter.

20.5   Cutkosky Unitarity Relations

In section 20.3 we derived dispersion relations for the scattering amplitude in quantum field theory. We made the analogy with the Krönig–Krammers dispersion relation for the scattering of light. However, the analogy is still incomplete. In the Krönig–Krammers relation the discontinuity across the cut, *i.e.* the spectral function, was related by the optical theorem to the total cross section. It is this kind of correspondence that we want to establish in this section for the amplitudes of quantum field theory. We have already done so for the special case of the imaginary part of the 2-point function, which was found to be proportional to the total decay probability of the unstable particle. We want to generalise this relation for the imaginary part of every amplitude.

Let us go back to the derivation of the relation between $\mathrm{Im}\,\Sigma(p^2)$ and Γ. Using our previous notation we can start from the expression of the 2-point function:

$$I^{(2)}(P^2) = \int \frac{\mathrm{d}^4 k}{(2\pi)^4} \frac{\mathrm{d}^4 q}{(2\pi)^4} \frac{1}{(k^2 - m^2 + \mathrm{i}\epsilon)(q^2 - m^2 + \mathrm{i}\epsilon)} (2\pi)^4 \delta^4(P - k - q)$$

$$= \int \frac{\mathrm{d}^4 k}{(2\pi)^4} \frac{1}{(k^2 - m^2 + \mathrm{i}\epsilon)[(k - P)^2 - m^2 + \mathrm{i}\epsilon]}. \tag{20.32}$$

As we have pointed out already, this expression is real and analytic when $P^2 = s < 4m^2$. An imaginary part will appear when s reaches the physical threshold $s = 4m^2$. Let us see how: with P^2 real and positive we choose the $\mathbf{P} = 0$ frame, i.e. the centre-of-mass frame in the case of a two-particle scattering amplitude. The product of the two propagators can be split into four factors as

$$\frac{1}{(k_0 - K_0 + i\epsilon)(k_0 + K_0 - i\epsilon)(k_0 - P_0 - K_0 + i\epsilon)(k_0 - P_0 + K_0 - i\epsilon)}, \qquad (20.33)$$

with $K_0 = \sqrt{k^2 + m^2} \geq m$.

In the complex k^0 plane this expression has four poles which, with Feynman's $i\epsilon$ prescription, have moved into the complex plane, two into the upper half and two into the lower half. We perform the k_0 integration by closing the contour into the lower half plane and the result is the sum of the two residues: $[2K_0 P_0(2K_0 - P_0)]^{-1}$ for the $k_0 = K_0 - i\epsilon$ pole and $[2K_0 P_0(2K_0 + P_0)]^{-1}$ for the $k_0 = P_0 + K_0 - i\epsilon$ pole. We still have to integrate over k, which is equivalent, up to the angular part, to integrating over K_0 from m to infinity. The denominator of the second residue never vanishes during this integration, so this pole will give a real contribution to $I^{(2)}(P^2)$. Only the first residue will contribute to the imaginary part because it has a singularity at $K_0 = P_0/2$ which reaches the region of integration when P_0 reaches the physical threshold $2m$. The result is a branch cut starting at $s = 4m^2$ to infinity. The discontinuity across the cut will be given by the difference between the values at $P_0 + i\epsilon$ and $P_0 - i\epsilon$. For the distribution $[x - x_0 \pm i\epsilon]^{-1}$ we have

$$\frac{1}{x - x_0 \pm i\epsilon} = P\frac{1}{x - x_0} \mp i\pi\delta(x - x_0), \qquad (20.34)$$

where P denotes the principal value. It follows that the discontinuity across the cut is obtained by first choosing the $k_0 = K_0 - i\epsilon$ pole in the k_0 integration and, second, by replacing the $P_0 = 2K_0$ singularity by $\delta(P_0 - 2K_0)$. The important point is that these operations have both a very simple interpretation. The first is obtained by replacing the propagator $(k^2 - m^2 + i\epsilon)^{-1}$ in the diagram by the mass-shell condition $\delta(k^2 - m^2)\theta(k_0)$ under the k_0 integration and the second by doing the same with the second propagator, namely replacing $(q^2 - m^2 + i\epsilon)^{-1}$ with $\delta(q^2 - m^2)\theta(q_0)$. But this is precisely the expression for the total decay probability if $I^{(2)}(P^2)$ is the 1-loop contribution to the propagator of an unstable particle. Obviously, this result remains true if (20.32) is a 4-point function in a ϕ^4 theory. The discontinuity across the cut is again obtained by putting the two internal lines on their mass-shell and integrating over the phase space. We can represent this result diagrammatically in Fig. 20.11. We see that it resembles the unitarity relation. The imaginary part of an amplitude from an initial to a final state $\mathcal{M}(a_{in} \to a_f)$ at 1 loop is given in terms of the product of amplitudes in the tree approximation $\mathcal{M}(a_{in} \to a_n)$

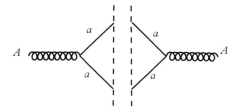

Figure 20.11 *The cutting rule for the 2-point function.*

and $\mathcal{M}^*(a_n \rightarrow a_f)$, where a_n is an intermediate state. It is an example of the so-called *cutting rules*, which yield the *Cutkosky unitarity relation*.

The rules can be stated as the following: the imaginary part of a Feynman diagram corresponding to a singularity in the physical region may be obtained by the following

(i)   Cutting the diagram in all possible ways;

(ii) For each cut replacing the cut propagators by the mass-shell delta functions;

(iii) Integrating each time over the corresponding phase-space; and

(iv) Summing over all possible cuts.

It is straightforward to repeat the proof for any 1-loop diagram. For example, an interested and courageous reader can compute this way the discontinuities of the box diagram and verify the results of section 20.3 (see Problem 20.1). As usual, the extension to multiloop diagrams is possible, albeit rather complicated. It requires a careful analysis of all the Landau singularities of a general diagram and the use of the theory of functions of several complex variables.

Two remarks: First, the Cutkosky rules can be viewed as a generalisation of the unitarity relation and the optical theorem for any individual Feynman diagram. Second, they can be used to redefine the Feynman rules for the computation of loop diagrams without the use of regularisation. Indeed, since the tree diagrams can be computed without integrations, they can be used to define the imaginary part of 1-loop diagrams because the phase-space integrals are always convergent. Then the real part of the 1-loop diagram can be obtained by writing a dispersion relation. This relation may require subtractions if the imaginary part does not vanish at infinity. Thus, the power counting and the renormalisation programme for 1-loop diagrams can be re-expressed in terms of the asymptotic behaviour of tree diagrams. Then the programme can be extended to higher loops by iteration.

20.6   The Analytic *S*-Matrix Theory

The rules of quantum field theory allow the computation of Green functions defined as vacuum expectation values of time-ordered products of operators at arbitrary external momenta. Out of these expressions we can construct *S*-matrix elements by appropriate choice of the external lines and taking the boundary value in the physical momentum region. This programme often appears as both unnecessarily general and unnecessarily heavy. It is unnecessarily general because the basic axioms are formulated in terms of objects, the quantum fields, which are not directly related to measurable quantities. It is also unnecessarily heavy because the calculation often goes through unphysical quantities, such as the Green functions of longitudinal, scalar, or ghost degrees of freedom in gauge theories. For practical purposes, this makes sometimes the calculation prohibitively lengthy. For these reasons, the formulation of alternative approaches has been attempted. They are all based, in one way or another, on the unitarity of the *S*-matrix.

We will present here briefly the conceptual part of a programme which, although not actively pursued at present, has given some important results in quantum scattering theory.[5]

The original idea of an S-matrix theory goes back to Wheeler in 1937 and Heisenberg in 1943, but it was only in the late 1950s and early 1960s that it was developed as an alternative to quantum field theory. It is motivated by the fact that in elementary particle physics, practically all physical information is obtained through scattering experiments, so one should formulate a theory in which the scattering amplitudes are the fundamental objects. This approach was developed by Geoffrey F. Chew, Stanley Mandelstam, Tullio Regge, et al. The basic object is the S-matrix, which is assumed to satisfy certain axioms. They are extracted from the properties one finds in local quantum field theory order by order in perturbation expansion, but here no underlying field theory is assumed. These axioms include the following

(i) *Unitarity.* S is supposed to be a unitary matrix. As usual, it is convenient to subtract the no-scattering part and define the matrix T as $S = \mathbb{1} + \mathrm{i}T$. S and T are assumed to satisfy the unitarity relations

$$SS^\dagger = S^\dagger S = \mathbb{1}; \quad 2\mathrm{Im}\,T = TT^\dagger. \tag{20.35}$$

(ii) *Poincaré invariance.* The S-matrix elements are functions of the external momenta. Invariance under space and time translations implies that they are proportional to $(2\pi)^4 \delta^4(\sum p_i)$. Invariance under Lorentz transformations implies that up to kinematical factors related to the normalisation of the states, they are functions of the independent scalar products $p_i \cdot p_j$. For spinless external particles the amplitude is a scalar function. For particles with spin it can be developed in a standard basis of helicity amplitudes with scalar coefficient functions. In particular, the two-particle to two-particle amplitude is a function of the Mandelstam variables s, t, and u, subject to the condition $s + t + u = \sum m^2$.

(iii) *Crossing symmetry.* The same set of functions describe the amplitudes for all processes with the external momenta considered as incoming or outgoing. In the presence of spin and/or internal quantum numbers, this rule is understood for linear combinations of partial amplitudes and with the possible exchange of particles and anti-particles.

(iv) *Internal symmetries.* Invariance under possible internal symmetries, such as isospin for strong interactions, is implemented in the usual way by applying the Wigner–Eckart theorem. For example, all pion–nucleon processes are described by two basic amplitudes, which can be viewed, in the $\pi - N$ channel, as corresponding to $I = \frac{1}{2}$ and $I = \frac{3}{2}$.

---

[5] Many books have been published on this subject, mainly during the nineteen sixties. See, for example, R. J. Eden, P. V. Landshoff, D. I. Olive, and J. C. Polkinghorne, '*The Analytic S-Matrix*' (Cambridge University Press, 1966); G. F. Chew, '*S-Matrix Theory of Strong Interactions*', A Lecture Note and Reprint Volume (W. A. Benjamin, 1962).

(v) *Maximum analyticity.* This is the main *new* axiom in the theory. It can be formulated intuitively as the requirement that the Lorentz invariant functions representing a scattering amplitude are analytic functions of the external momenta with only those singularities imposed by unitarity. The trouble is that although it is easy to give specific examples of this axiom, the precise mathematical formulation for the general case, as well as its self-consistency, has remained somehow vague. Let us look at the unitarity relation (20.35) for a 'generic' case of a two-particle to two-particle amplitude $A + A \rightarrow A + A$, ignoring spin and isospin complications. We assume that all particles have the same mass m and carry no extra conserved quantum numbers. Introducing a complete set of intermediate states $|n >$ we rewrite (20.35) as

$$< a_f |\text{Im}\, T |a_{in} > \sim \sum_n < a_f | T |a_n > < a_n | T^\dagger |a_{in} > (2\pi)^4 \delta^4 (P_i - P_n). \qquad (20.36)$$

This is taken to mean that the singularities of the amplitude correspond to the possible intermediate states in (20.36).

The first such state is the one-particle state, which, although it lies below the physical region of the scattering amplitude, can be reached by analytic continuation. It gives an imaginary part proportional to $\delta(s - m^2)$, which implies a single pole of the form $1/(s - m^2)$ for the amplitude. This gives the first rule: stable particle intermediate states give singularities in terms of simple poles.

The next intermediate state is the two-particle state which starts at $s = 4m^2$. It gives a branch cut from $s = 4m^2$ to infinity. The next state is the three-particle state with a cut from $s = 9m^2$ to infinity.

We continue with multiparticle states with branch points at $s = (nm)^2$, etc. We conclude with our second rule, namely that multiparticle intermediate states give an infinite number of branch cuts, as shown in Fig. 20.12. As usual in the theory of functions of a complex variable, introducing these cuts we make the amplitude a single-valued function on a Riemann surface.

It is straightforward to extend these considerations to more realistic scattering amplitudes, such as the pion + nucleon $\rightarrow$ pion + nucleon. If we take the pion–nucleon channel as the s-channel, the pole lies at $s = M^2$, with M the nucleon mass, and the cuts start at $s = (M + m)^2$, $s = (M + 2m)^2$,..., $s = (M + nm)^2$, until we reach the nucleon–antinucleon threshold which gives a branch point at $s = 9M^2$, etc. In

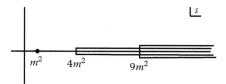

Figure 20.12 *The single-particle pole and the multiparticle thresholds of the scattering amplitude in the complex s plane keeping t fixed.*

the absence of massless particles (or accidental degeneracies), all branch points are separated.[6]

Many amplitudes in strong interactions are known to present several resonances. For the pion–nucleon system the first is called N^* and has spin and isospin equal to $\frac{3}{2}$. Although, strictly speaking, it is not an intermediate state, we postulate, according to our previous analysis, that it gives a pair of complex conjugate poles in the second sheet. This is our third rule. In principle, it should be considered as simply empirical because, in a complete theory, resonances should be calculable from the dynamics, and in the S-matrix theory, the knowledge of the physical singularities should be sufficient to determine those in the unphysical sheets.

So far we have looked at the s-channel keeping t fixed. We can repeat the analysis at the t-channel. In our toy example it is again an $A + A \to A + A$ amplitude and the same analysis applies. We obtain a pole at $t = m^2$ and branch points at $t = (nm)^2$. For fixed $u = u_0$, these t-singularities are reflected into s-singularities through the relation $s + t + u = 4m^2$. The resulting picture is shown in Fig. 20.13. It represents the singularity structure of the physical sheet for the $A + A \to A + A$ amplitude. By convention, we decided to draw the cuts along the real s axis. Obviously, we can deform them anywhere in the complex plane. Only the end points are fixed for fixed u_0.

We still must decide which is the physical value of the $A + A \to A + A$ amplitude. Taking $u_0 < 0$ and s real $s > 4m^2 - u_0$, we see that all three variables s, t, and u take values which correspond to the physical region of the s-channel amplitude. By analogy with the Feynman $i\epsilon$ prescription which we saw in perturbation theory, we decide to define the physical region as the boundary value of the analytic function $F(s, u_0)$ shown in Fig. 20.13 when s becomes real, larger than $4m^2$, reaching the real axis from above: $\lim F(s + i\epsilon, u_0)$. This is our fourth rule. From this *physical sheet*, going through one- or more of the cuts, we reach the *unphysical sheets*.

We are now in a position to give a precise meaning to our axiom (iii) of crossing symmetry. Starting from the general analytic function $F(s, u_0)$ we saw that the physical amplitude in the s-channel is obtained by taking $s > 4m^2$, $\lim F(s + i\epsilon, u_0)$. If we want the physical amplitude in the t-channel, we must take $t > 4m^2$, $t + i\epsilon \to t$. With u_0 fixed

Figure 20.13 *The singularities (poles and cuts) in the complex s plane for the two-particle scattering amplitude, coming from intermediate states in the s-channel (on the right) and the t-channel (on the left). The arrows show the approach to the s-channel physical region.*

[6] This discussion is restricted to hadronic physics where there are no massless particles. In particular, we neglect photons and electromagnetic interactions. The argument is that the amplitude for the emission of n photons will be suppressed by a factor e^n. In other words, we assume that in contrast to the hadronic interactions, the electromagnetic ones can be treated in perturbation. Of course, this was the 'historical' line. We will see in a later chapter that what we consider today as the fundamental strong interactions involve a non-Abelian Yang–Mills theory with massless gauge bosons called 'gluons'. The multi-gluon intermediate states will produce branch points all piling up at the same point, thus complicating the picture considerably. Although we will argue that perturbation theory is still applicable, the techniques of the S-matrix theory will turn out to be very useful.

and negative this is reached by analytically continuing F to $s < 4m^2$ without crossing any cuts, through the gap between the right- and left-hand cuts to the lower half s-plane, and approaching the negative real axis from below.

This simple picture gets more complicated for the case of unequal masses and conserved quantum numbers, like the pion–nucleon system. If the t-channel is the $\pi + \pi \to N + \bar{N}$, the physical region starts at $t = 4M^2$, but the first intermediate state and, hence, the first branch point start at the two-pion intermediate state at $t = 4m^2$. By G-parity[7] there is no three-pion intermediate state, so the next branch point is at $t = 16m^2$; etc. Obviously, all these points are reached by analytic continuation.

(vi) *Polynomial boundedness.* Our sixth axiom is the one necessary to write multidimensional dispersion relations. It states that the amplitude, considered as a function of several complex variables, is bounded at infinity by a polynomial at any complex direction. Although easy to formulate, it is not easy to study mathematically. In particular, it is not known under which conditions it may follow from the general axioms of quantum field theory.

The combined use of these axioms allows us to write general dispersion relations. For the function $F(s, u_0)$ of Fig. 20.13 the Cauchy contour is shown in Fig. 20.14.

Considering the analytic function $\mathcal{M}(s, t, u)$ as a function of two complex variables we can write a generalised Cauchy theorem expressing the value of the amplitude at any given point as a multiple integral over the discontinuities. Since, according to axiom (v), they correspond to the possible intermediate states, they are, in principle, measurable. This is the spirit of the approach: unitarity and analyticity completely determine the scattering amplitude, up to possible subtractions. Axiom (vi) guarantees that we will encounter only a finite number of them.

The typical example is the Mandelstam representation which generalises the relation we found in (20.21). For the pion–nucleon scattering amplitude it can be written as

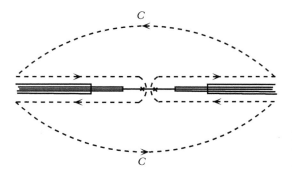

Figure 20.14 *The contour for the Cauchy integral in the general s-channel dispersion relation.*

[7] G-parity generalises the concept of charge conjugation C for particles with non-zero electric charge. It is defined as the product of C with a 180° rotation around the second axis in isospin space: $G = Ce^{i\pi T_2}$. Since strong interactions are invariant under both charge conjugation and isospin transformations, they preserve G. It is easy to see that an n-pion state is an eigenstate of G with eigenvalue equal to $(-)^n$.

$$\mathcal{M}(s,t,u) = \frac{1}{\pi} \int ds' \frac{\rho_s(s')}{s'-s} + \frac{1}{\pi} \int dt' \frac{\rho_t(t')}{t'-t} + \frac{1}{\pi} \int du' \frac{\rho_u(u')}{u'-u}$$
$$+ \frac{1}{\pi^2} \int \int ds' dt' \frac{\rho_{st}(s',t')}{(s'-s)(t'-t)} + \frac{1}{\pi^2} \int \int ds' du' \frac{\rho_{su}(s',u')}{(s'-s)(u'-u)}$$
$$+ \frac{1}{\pi^2} \int \int dt' du' \frac{\rho_{tu}(t',u')}{(t'-t)(u'-u)}. \tag{20.37}$$

Up to possible subtractions, this is the most general double dispersion relation we can write. The spectral functions can be computed by considering all possible intermediate states. As we saw in the analysis of the box diagram, a double spectral function, such as ρ_{st}, will be non-vanishing if, for a given physical multiparticle intermediate state in the s-channel, there are physical, multiparticle intermediate states in the t-channel. Figure 20.15 shows the result for the case of the nucleon + nucleon → nucleon + nucleon amplitude which is simple enough because all external masses are equal and there are no single particle poles, but it presents the complication of having the two-pion intermediate states in the $N + \bar{N} \to N + \bar{N}$ channel.

The basic assumption for applying such relations in practical calculations is that if we are interested in the value of the amplitude at low energies, the dispersion integrals are dominated by the closest singularities, so only low-lying intermediate states need to be considered. In Problem 20.1 we show this property in a simple example. The exchange of the lightest particle in the cross channel gives the dominant contribution

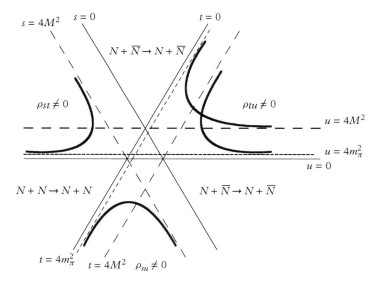

Figure 20.15 *The Mandelstam diagram for the reaction $N + N \to N + N$. In every channel the physical region starts at $4M^2$ but the regions of non-vanishing double spectral functions extend down to $4m_\pi^2$ because of the two-pion intermediate state in the $\bar{N} + \bar{N} \to \bar{N} + \bar{N}$ channel.*

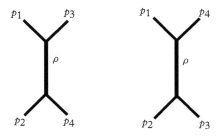

Figure 20.16 *The one ρ exchange diagrams in the t and u channel, respectively.*

to the long-range forces in the direct channel. This observation gave rise to a practical attempt to illustrate the *S*-matrix programme known as *the bootstrap hypothesis*. As we just said, the exchange of a virtual particle in the crossed channel generates forces in the direct channel. These in turn could produce bound states, or resonances, which should be identified with those whose exchange we considered. Hence the name *bootstrap* from the well-known Münchausen tale: a particle produces the forces which create it. Historically, this was first applied to the ρ resonance in pion–pion scattering with only qualitative success. Let us give a brief description.

We consider the amplitude $A(s, t)$ which describes the pion–pion scattering. We will try to compute it using the axioms of *S*-matrix under certain simplifying approximations. (i) In the unitarity condition (20.35) and (20.36) we will keep only the two-pion intermediate state. This approximation is often called *the elastic unitarity approximation*. This means that in the physical region of the *s*-channel we keep only the cut starting at $s = 4m_\pi^2$ and we ignore all the higher ones. We expect such an approximation to be justified at low energies. (ii) We approximate all singularities in the cross channels by the ones created by the 1-ρ exchange diagrams of Fig. 20.16. As we explained earlier, this exchange creates a left-hand singularity in the *s*-channel. We can now describe the various steps of the calculation.

Step 1: Compute the two diagrams of Fig. 20.16. The spin and parity properties of the pions and the ρ are 0^- and 1^-, respectively, and both have isospin equal to 1. So, the $\pi - \pi - \rho$ vertex is described by an effective Lagrangian density of the form $\mathcal{L} = g_{\pi\pi\rho} \boldsymbol{\pi} \times \partial_\mu \boldsymbol{\pi} \cdot \boldsymbol{\rho}^\mu$, where the three-vector notation refers to isospin. The same Lagrangian describes the decay $\rho \to 2\pi$, so the coupling constant $g_{\pi\pi\rho}$ is given in terms of the ρ width Γ_ρ. A straightforward calculation gives

$$A = \frac{g_{\pi\pi\rho}^2}{16(2\pi)^2} f_{\alpha\gamma i} f_{\beta\delta i} \frac{2q^2(1 + \cos\theta) + s}{2q^2(1 - \cos\theta) + m_\rho^2}, \tag{20.38}$$

where we have used the pion mass as the unit of masses, q is the momentum in the centre-of-mass frame, $s = 4(q^2 + 1)$, and $f_{\alpha\gamma i}$ are isospin coefficients.

Step 2: According to the bootstrap hypothesis, we expect this exchange to create the ρ resonance in the s-channel, so we project the previous result in the $L = 1$ and $I = 1$ partial wave. This gives us a function $A_{11}^L(s)$, where the subscript 11 means that the amplitude is in the $L = 1$ and $I = 1$ s-channel partial wave and the superscript L that, in our approximation, it contains all the left-hand cuts. $A_{11}^L(s)$ depends on two parameters, m_ρ and Γ_ρ, the mass and the width of the ρ resonance.

The isospin projection gives $< I = 1|f_{\alpha\gamma i}f_{\beta\delta i}|I = 1 >= 1$ and the projection on the $L = 1$ partial wave is obtained by integrating A with the $L = 1$ Legendre polynomial

$$A_{11}^L(s) = \int_{-1}^1 P_1(\cos\theta)A(s, \cos\theta)\mathrm{d}\cos\theta, \tag{20.39}$$

which, using the expression for the amplitude given by (20.38), yields

$$A_{11}^L(s) = \frac{g_{\pi\pi\rho}^2}{16\pi}\frac{4q^2 + m_\rho^2 + s}{2q^2}\left(\frac{2q^2 + m_\rho^2}{2q^2}\ln\frac{|2q^2 + m_\rho^2|}{m_\rho^2} - 2\right). \tag{20.40}$$

Step 3: We write the partial wave amplitude $A(s)$ (we drop the subscript 11) as

$$A(s) = \frac{N(s)}{D(s)}, \tag{20.41}$$

where $D(s)$ is supposed to have only the right-hand ('physical') singularities and $N(s)$ only the left-hand ones. In our approximation $N(s)$ has the singularities of $A_{11}^L(s)$.[8] On the right-hand cut N is real and on the left-hand cut D is real, so we can write

$$\mathrm{Im}D(s) = N(s)\mathrm{Im}A^{-1}(s); \quad \mathrm{Im}N(s) = D(s)\mathrm{Im}A(s). \tag{20.42}$$

These equations are exact but not very tractable. We will use our approximations to simplify them and bring them to a solvable form.

(i) The discontinuity of $A(s)$ on the left-hand cut equals that of $A^L(s)$, so the second of Eq. (20.42) can be written as

$$\mathrm{Im}N(s) = D(s)\mathrm{Im}A^L(s). \tag{20.43}$$

(ii) According to the elastic unitarity approximation the discontinuity of $A^{-1}(s)$ on the right-hand cut is given by the two-pion intermediate state in the $L = 1$, $I = 1$

---

[8] This separation is not unique, since we can multiply both the numerator and the denominator of (20.41) by a common factor, provided we do not introduce extra singularities. This kind of ambiguity has been discussed extensively in the literature (it is called *the Castillejo, Dalitz, Dyson*, or CDD, ambiguity), but it will not affect directly our calculation.

partial wave. This implies that the right-hand cut discontinuity $\operatorname{Im} A^{-1}(s)$ is given by a kinematical factor $\operatorname{Im} A^{-1}(s) = -q/\sqrt{s}$ and the first of Eq. (20.42) simplifies as

$$\operatorname{Im} D(s) = -\frac{q}{\sqrt{s}} N(s). \tag{20.44}$$

(iii) We will try to solve the system of the two equations (20.43) and (20.44) by iterations. We start with $D = 1$, in which case (20.43) implies that $N(s) = A^L(s)$, up to possible subtractions. Bringing this result to (20.44) we determine the first iteration of $D(s)$ through a dispersion relation. And we start all over again.

This method allows us to construct an approximate form of the pion–pion scattering amplitude in which the one ρ exchange is the only singularity in the cross channel. The amplitude depends on two parameters, m_ρ and Γ_ρ. According to the bootstrap hypothesis, we must identify the same ρ resonance in the s-channel. It will appear as a complex pole in the amplitude of the Breit–Wigner form we found in Eq. (20.31). By construction, $N(s)$ has no singularities in the physical region, so this pole will correspond to a zero of $D(s)$ in the complex plane. Thus, the equation we must solve is

$$D(s) = 0. \tag{20.45}$$

A solution of this equation will be some complex number $s = s_0(m_\rho, \Gamma_\rho)$. The real and imaginary parts of s_0 should be identified with the mass and the width of the resonance, according to (20.31), i.e. m_ρ and Γ_ρ. This gives us a set of consistency equations which can be solved numerically in order to determine the only free parameters of the problem, m_ρ and Γ_ρ. The results are not spectacular: the mass turns out to be around 320 MeV and the width around 50 MeV, compared to the experimental values of 750 and 100 MeV, respectively. We could argue that this is not unexpected since the basic assumption was that the one ρ singularity in the cross channels dominates the two-pion elastic cuts. The latter have a threshold of $2m_\pi = 280$ MeV. Therefore, it is not surprising that the calculation gives a very light ρ.

The programme has been applied also to other problems, such as the π-K scattering with the K^* resonance replacing ρ. The results are always, at best, qualitative. It is clear that the one-particle exchange approximation is a very crude one but anything beyond it yields much more complicated equations.

This calculation illustrates the use of the analytic properties of the S-matrix elements to determine the two particle-to-two particle scattering amplitude. In principle, it is possible to write higher dimensional dispersion relations for multiparticle amplitudes but they have not been analysed in any detail.

20.7   Problems

Problem 20.1 Compute the box diagram of Fig. 20.7 using the cutting rules of section 20.5.

Problem 20.2 The purpose of this problem is to study the inter-relation between the singularities of a scattering amplitude in the various channels.

Consider the scattering of two spinless particles $a + a \to a + a$. Show that the t-channel exchange of a particle A with mass m_A induces in the s-channel a Yukawa potential of the form $e^{-m_A r}/r$. In other words, the large distance interaction is dominated by the t-channel singularity which is closest to the physical region.

21

Infrared Singularities

21.1 Introduction. Physical Origin

We encountered already the notion of *infrared singularities*, first in Chapter 3 in the study of the classical electromagnetic field and later in many other places where we had to deal with particles of zero mass. We had postponed the discussion of these singularities and here we want to fill partially this gap. We have seen clearly that the origin of these divergences is the zero mass of a certain particle, often the photon, which makes it possible to consider excitations with arbitrarily low frequency and prevents the appearance of a mass gap in the energy spectrum. In this sense, the infrared singularities are the opposite to the ultraviolet ones we studied so far. The important difference between the two is that the second was due to a mathematical mistake in our formalism, namely the multiplication of the field operator-valued distributions at the same space–time point. They were correctly handled in perturbation theory by the process of renormalisation. On the other hand, the infrared divergences have a clear physical origin, the appearance in the spectrum of states of excitations with arbitrarily large wavelength which produce long-range forces. Therefore, we do not expect a technical solution but a physical one; we must correctly define the physically measurable quantities in the presence of long-range forces. This will require some modification of the scattering formalism we presented in Chapter 12. Indeed, in formulating the asymptotic condition which led us to the definition of the scattering amplitude, we explicitly assumed that the interaction dies off sufficiently fast at spatial infinity, or, equivalently, that we had no zero mass particles. The problem appears already in non-relativistic quantum mechanics where the scattering by a Coulomb potential is more complicated than that from a short-range one. It is also evident by looking at the analyticity properties of the scattering amplitude we studied in Chapter 20. In the presence of a zero mass particle, all branch points for an arbitrary number of such intermediate particles collapse at $s = 0$ which becomes an essential singularity of the amplitude. In this chapter we will show, in some simple examples, how to handle these complications.

From Classical to Quantum Fields. Laurent Baulieu, John Iliopoulos and Roland Sénéor.
© Laurent Baulieu, John Iliopoulos and Roland Sénéor, 2017. Published 2017 by Oxford University Press.

Figure 21.1 *Diagrams contributing to the forward (backward) electron–electron scattering.*

21.2 The Example of Quantum Electrodynamics

The first example of a field theory showing infrared singularities is that of quantum electrodynamics due to the vanishing mass of the photon. Because of that, if we consider a graph with a photon line, internal or external, with momentum k, the point $k = 0$ may belong to the physical region and may create singularities in certain kinematical configurations. Some examples follow.

A trivial example is the elastic electron–electron scattering amplitude. At lowest order we find the two diagrams of Fig. 21.1 Both contain the $1/k^2$ photon propagator. Using the Mandelstam variables, the first is proportional to $1/t$ and the second to $1/u$. Both singularities belong to the edge of the physical region. Going to the centre-of-mass frame we find that

$$t = -\frac{1}{2}(s - 4m^2)(1 - \cos\theta); \quad u = -\frac{1}{2}(s - 4m^2)(1 + \cos\theta), \tag{21.1}$$

where θ is the scattering angle. We see that $t = 0$ and $u = 0$ correspond to forward and backward scattering, respectively. Due to the identity of the particles in the initial and final states, we expect the cross section to be invariant under the exchange $\theta \leftrightarrow \pi - \theta$ or, equivalently, $t \leftrightarrow u$. In problem 13.3 we compute the differential cross section for this process and find the well-known Møller formula

$$\frac{d\sigma}{d\Omega}(e^- + e^- \to e^- + e^-) = \frac{4\alpha^2}{s}\left[\frac{(s - 2m^2)^2}{(s - 4m^2)^2}\left(\frac{4}{\sin^4\theta} - \frac{3}{\sin^2\theta}\right) + \frac{4}{\sin^2\theta} + 1\right]. \tag{21.2}$$

We recover the singularities at θ equal 0 and π.

A second example is the Bhabha scattering amplitude for $e^- + e^+ \to e^- + e^+$ which we computed again in problem 13.2. Here the one photon exchange diagram contributes only in the t channel. The result is given by

$$\frac{d\sigma}{d\Omega}(e^- + e^+ \to e^- + e^+) = \frac{\alpha^2}{2s}\left[u^2\left(\frac{1}{s} + \frac{1}{t}\right)^2 + \left(\frac{t}{s}\right)^2 + \left(\frac{s}{t}\right)^2\right], \tag{21.3}$$

which is still singular at θ equal 0 and π.

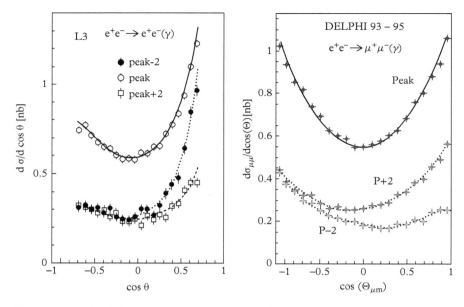

Figure 21.2 *The differential cross sections for $e^- + e^+$ annihilation as measured at LEP. The three curves correspond to the cross section at the peak of the Z^0 pole and at ± 2 GeV above and below it. On the left-hand side we see the Bhabha scattering with the steep rise in the forward direction. For comparison, we show on the right-hand side the same quantity for the $e^- + e^+ \rightarrow \mu^- + \mu^+$ process in which there is no one-photon exchange contribution in the crossed channel and, as a result, there is no such phenomenon. CERN, LEP collaborations, L3 (left), DELPHI (right).*

For comparison, in section 13.3 we computed also the amplitude for the electron–positron annihilation into $\mu^- + \mu^+$ which has no one photon exchange diagrams in the cross channels and no singularity in the physical region; see Eq. (13.157).

Before discussing the physical significance of such singularities, we want to point out that the above expressions are, experimentally, correct. Fig. 21.2 shows the measured e^- e^+ cross section both for $e^- + e^+$ and $\mu^- + \mu^+$ final states[1] in the vicinity of the Z^0 pole. We see the rise of the cross section at small angles. The theoretical curves are those given in Eqs. (13.159) and (13.157), but they include the radiative corrections. It is because of such corrections in the initial and final states that the curves do not look symmetric. We see that the agreement between theory and experiment is perfect.

The usual answer to the problem of forward scattering is the remark that it is not a measurable quantity, since we cannot put a detector there, because it will be saturated by the incoming particles which did not scatter. Although this is, essentially, correct, it does not answer all the questions because the singularity in (13.160) is not integrable. It

---

[1] There are no good measurements for the Møller amplitude at very small angles.

follows that if we want to compute the total cross section by integrating (13.160) over all angles (with a factor $\frac{1}{2}$ because of the identity of the particles), we will find a divergent result.

Similar infrared singularities appear in all tree-level QED scattering amplitudes. They all correspond to a corner of phase space where the momentum of a photon line goes to 0.

At 1 loop the situation may become worse. The classical example is the vertex function $\Gamma_\mu^{(1,2)}(p, p')$ of Fig. 16.7. For p and p' on the mass-shell, it gives the 1-loop correction to the scattering of an electron from an external classical electromagnetic potential. For $t = (p - p')^2 = 0$ it gives the static properties of the electron, namely its charge and its magnetic moment. In Problem 16.4 we computed the latter and found a spectacular agreement with experiment. In contrast, the value of the charge contained the ultraviolet divergence and had to be imposed by a renormalisation condition. We computed the $1/\epsilon$ part in (16.58). In order to impose a renormalisation condition we need also the finite part. Among the various choices, the one which leads to the physical S-matrix is the one in which the external momenta are on-shell. For $\Gamma_\mu^{(1,2)}(p, p')$ this means that $p^2 = p'^2 = m^2$, $\not{p}u(p) = mu(p)$, $\not{p}'u(p') = mu(p')$, and $t = 0^2$. Let us try to compute the corresponding counter-term here. In the notation we used in Problem 16.4 we must compute the quantity $F_1(t)$, which we had called 'the charge form factor', at $t = 0$. We find that

$$F_1(0) = 1 + \frac{\alpha}{2\pi}\frac{1}{\epsilon} + \cdots + 4 \int_0^1 \frac{\mathrm{d}z_1}{1 - z_1}. \qquad (21.4)$$

The first term is the classical value. The next is the ultraviolet divergence found in (16.58). The dots stand for terms which are finite and contain, in particular, a term proportional to $\ln(m/\mu)$; i.e. they contain the dependence on the scale parameter introduced by the dimensional regularisation. The trouble comes from the last term. z_1 is one of the Feynman parameters and the integral is divergent when $z_1 \to 1$. We can convince ourselves that this is an infrared divergence because we can repeat the calculation replacing the photon propagator $1/k^2$ with $1/(k^2 + m_\gamma^2)$, assuming a fictitious mass for the photon, and the term is replaced by $\ln(m/m_\gamma)$. Note that in our on-shell renormalisation scheme, we must introduce a counter-term such that $F_1(0) = 1$. Equation (21.4) shows that this is impossible.

What is the meaning of this divergence? We can look at it in two ways. The first tells us that what we called 'physical' renormalisation conditions are impossible to enforce for this theory. Since they were the ones leading to the definition of the S-matrix, this means that *scattering amplitudes do not exist in quantum electrodynamics*. Although this result may sound catastrophic, we cannot pretend it was totally unexpected. In setting up the framework for the theory of scattering we had repeatedly assumed the absence of zero mass particles. A second way to look at this divergence is to view it as part of the 1-loop correction to the amplitude of scattering of an electron from an external

---

[2] This last choice, $t = 0$, is not important for what follows.

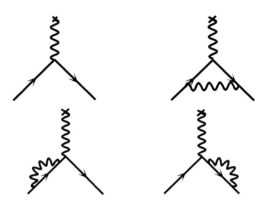

Figure 21.3 *The diagrams contributing to the scattering of an electron in an external electromagnetic field, in the Born approximation and the 1-loop corrections.*

potential, the philosophy we had in Problem 16.4. It is in this framework that F. Bloch and A. Nordsieck found the solution in 1937. Let us illustrate their argument.

For simplicity let us keep the infrared cut-off m_γ. We want to compute the amplitude for the scattering with a momentum transfer $t = (p - p')^2$. The physical region corresponds to $t \leq 0$. The total amplitude for our process will be given, in perturbation theory, by the diagrams of Fig. 21.3. The first is the Born approximation followed by the 1-loop corrections. The ultraviolet renormalisation is straightforward to perform at some value of t, for example $t = 0$, and all the dependence on ϵ and μ will disappear. The total amplitude will be of the form

$$\mathcal{M}(t) = \mathrm{i}e A_\mu^{\mathrm{ext}} \bar{u}(p') \left[\gamma^\mu \left(1 + \alpha C(t) \ln \frac{-t}{m_\gamma^2} + \cdots \right) + \cdots \right] u(p), \qquad (21.5)$$

where $C(t)$ is a calculable function of t which behaves at large momentum transfers as $\ln(-t/m^2)$. The dots stand again for terms which have finite limits when m_γ goes to zero. The differential cross section, summed and averaged over the electron final and initial spin, can be written as

$$\frac{\mathrm{d}\sigma}{\mathrm{d}\Omega} = \left(\frac{\mathrm{d}\sigma}{\mathrm{d}\Omega} \right)_{cl} \left[1 + 2\alpha C \ln \frac{-t}{m_\gamma^2} + \cdots + \mathcal{O}(\alpha^2) \right]. \qquad (21.6)$$

$(\mathrm{d}\sigma/\mathrm{d}\Omega)_{cl}$ is the cross section in the classical approximation which is proportional to e^2. We have written explicitly only the interference term between the 0- and 1-loop amplitudes which diverges when $m_\gamma \to 0$. The $\mathcal{O}(\alpha^2)$ terms contain the square of the 1-loop contributions, but also the interference between the 0- and the 2-loop diagrams and, at this stage of our calculation, cannot be included.

The expression (21.6) is still infrared divergent but Bloch and Nordsieck remarked that in the limit $m_\gamma \to 0$, it is not a physically measurable quantity. To understand the reason let us think of an experimental setup. We need a source of electrons with momentum p and a detector capable of detecting electrons of momentum p' which we

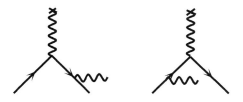

Figure 21.4 *Bremsstrahlung processes.*

can place at an angle Ω with respect to the incident beam. In order to compare with (21.6) we must be sure that we detect an electron and nothing else; for example, we must not count events in which the final electron is accompanied by one or more, photons. Examples of such processes are shown in the diagrams of Fig. 21.4. In the presence of the external field, a photon is radiated by the initial, or the final, electron. It is our familiar bremsstrahlung process. The key point is that every detector has a finite resolution, both in energy, i.e. it cannot detect photons with an energy smaller than some threshold δ, and in angle, i.e. it automatically integrates over an angle $\Delta\Omega$. In other words, our experimental setup does not measure the cross section for the process of Fig. 21.3 but rather the incoherent sum of all processes representing the electron scattering with 0, 1, 2, etc. soft photons integrated over a certain angular region. In perturbation theory every emitted photon carries a factor of e, so the $\mathcal{O}(\alpha)$ cross section is the one of Fig. 21.4. The amplitude is proportional to

$$\mathcal{M}(t,\omega) \sim (eA_\mu^{\text{ext}})\mathrm{e}\,\bar{u}(p')\mathbb{E}\left[\frac{\not{p}'+\not{k}+m}{(p'+k)^2-m^2}\gamma^\mu + \gamma^\mu\frac{\not{p}-\not{k}+m}{(p-k)^2-m^2}\mathbb{E}\right]u(p), \qquad (21.7)$$

where ω is the photon energy and E its polarisation. The important thing is that the electron propagators behave, at small ω, like $1/\omega$. Taking the square and integrating over the phase space for $m_\gamma < \omega < \delta$ we find the quantity we should add to (21.6) in order to obtain the $\mathcal{O}(\alpha)$ measurable quantity. The result, which we could have anticipated, is that the sum is finite when we take the limit $m_\gamma \to 0$, keeping δ fixed. The exact computation is rather lengthy, but it simplifies in the physically interesting case when the momentum transfer $-t$ is much larger than m^2 (see Problem 21.1). The result takes the form

$$\left(\frac{\mathrm{d}\sigma}{\mathrm{d}\Omega}\right)_{\text{eff}} \Rightarrow \left(\frac{\mathrm{d}\sigma}{\mathrm{d}\Omega}\right)_{cl}\left[1 - \frac{\alpha}{\pi}\ln\left(\frac{-t}{m^2}\right)\ln\left(\frac{-t}{\delta^2}\right) + \mathcal{O}(\alpha^2)\right], \qquad (21.8)$$

where we have called 'effective differential cross section' the quantity which is actually measured by our detector. We have used the asymptotic form of the function $C(t)$. We note that the result contains a double logarithm, one which we can call 'ultraviolet' and the second which we call 'infrared'. This double logarithm structure was first discovered by V. V. Sudakov.

21.3 General Discussion

The result of Eq. (21.8) answers the technical part of the infrared divergence problem in QED. We defined a physically measurable quantity in which the infrared cut-off disappeared and is replaced by a physical scale δ provided by our detector. However, this answer is not yet totally satisfactory. First, it is only a first-order result. This can, in principle, be remedied by investing more work in the computation of higher orders. However, there is a second point: as Sudakov has shown, these double logarithms will appear at every order, so the leading behaviour of the nth order will be proportional to $\left[\alpha \ln \left(\frac{-t}{m^2} \right) \ln \left(\frac{-t}{\delta^2} \right) \right]^n$. So at large momentum transfer, the effective expansion parameter is not α but rather $\left[\alpha \ln \left(\frac{-t}{m^2} \right) \ln \left(\frac{-t}{\delta^2} \right) \right]$. α is a very small number ($\sim 1/137$), but each one of the logs can also become quite important. The situation is much more critical for quantum chromodynamics, the theory that describes the strong interactions, which we will study shortly. In this theory the corresponding α_{strong} is of order $\frac{1}{5}$ to $\frac{1}{10}$. For all these reasons, we want to go beyond the lowest order result. This was already done in the original work of Bloch and Nordsieck, but it has been considerably improved since.

The principle of the computation is rather simple, but the practical details are not. Let us stick to our previous problem of the scattering of an electron from an external potential. We will sketch only the calculation for the leading infrared singularity, namely, in each order of perturbation the term with the highest power of the infrared logarithm. We can follow the electron line. There will be several photon lines attached to it. For our argument it does not matter whether they are external or internal, the only important point is whether they are 'hard' or 'soft'. We call 'hard' a photon which is detected in our detector, i.e. whose energy is larger than δ. By contrast, a 'soft' photon line is one whose momentum k can go to 0. Following the incoming electron line we can mark the first hard photon. We are interested only in the photons which are emitted *before* this one because the electron propagator which follows such an emission behaves like $1/k$. Those which are emitted after the hard photon, even if they are soft, they could not produce such an infrared divergence. Similarly, we are interested in the photon lines which are emitted *after* the last hard photon from the outgoing electron line. The typical diagram is shown in Fig. 21.5. We intend to sum all these diagrams.

The typical diagram with n soft photons will contribute to the sum a term proportional to

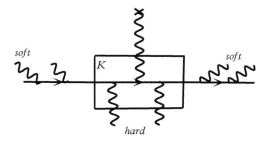

soft   *soft*

K

hard

Figure 21.5 *A Bremsstrahlung diagram with multiple soft photon emission.*

$$M^{(n)}(p, p') \sim e^n \bar{u}(p') \left[\not{E}_n \frac{\not{p}'+\not{k}_n+m}{(p'+k_n)^2-m^2} \not{E}_{n-1} \frac{\not{p}'+\not{k}_n+\not{k}_{n-1}+m}{(p'+k_n+k_{n-1})^2-m^2} \cdots \right.$$

$$eA_\mu^{\text{ext}} K^\mu \left(p' + \sum'\{k\}, p - \sum\{k\}\right)$$

$$\left. \cdots \frac{\not{p}-\not{k}_1-\not{k}_2+m}{(p-k_1-k_2)^2-m^2} \not{E}_2 \frac{\not{p}-\not{k}_1+m}{(p-k_1)^2-m^2} \not{E}_1 \right] u(p), \qquad (21.9)$$

where K is some kernel which represents the box part of the diagram. $\sum'\{k\}$ and $\sum\{k\}$ represent the sums of the momenta of the soft photons emitted from the outgoing and incoming electron line, respectively. Some of these photon lines may recombine to form photon loops. We must sum all these diagrams, something which sounds like a formidable task. The calculation is simplified if we remark that since we are interested only in the limit when all k's go to 0, we can drop the k factors in the numerators, the k^2's in the denominators and all the k-dependence in the kernel. Furthermore, we can bring the p's and p''s next to the spinors and use the Dirac equation. We are left with scalar products $p \cdot E$ in the numerators and $p \cdot k$ in the denominators. Then we must sum over all possible permutations of the k's, square the matrix element, sum over the photon polarisations, and integrate over the phase space with all photon energies smaller than δ. All these operations are superficially easy, although, if we want to be careful, they are rather involved. As it was the case with the lowest order expression (21.8), the final result can be written in a simple form at the limit $-t \gg m^2$. In that limit we find that all the Sudakov double logarithms exponentiate and we obtain the simple answer

$$\left(\frac{d\sigma}{d\Omega}\right)_{\text{eff}} \Rightarrow \left(\frac{d\sigma}{d\Omega}\right)_{\text{cl}} \times \left| \exp\left[-\frac{\alpha}{\pi} \ln\left(\frac{-t}{m^2}\right) \ln\left(\frac{-t}{\delta^2}\right) \right] \right|^2. \qquad (21.10)$$

The exponential of the double logarithm is called 'the Sudakov form factor'. This expression gives the sum of all the leading infrared terms and answers all our previous questions.[3] We see that the effective probability goes to 0 when δ goes to 0, a result easy to understand: the probability of observing a bremsstrahlung process with no emitted photons vanishes. There exist similar, but more complicated expressions for the next-to-leading terms, the next-to-next-to-leading terms, etc.

21.4 Infrared Singularities in Other Theories

In the previous section we showed the appearance of the infrared singularities in QED and the way to treat them. Our result can be phrased by saying that scattering amplitudes cannot be defined but measurable transition probabilities can. It is obvious that similar results can be obtained for every theory which has massless particles which can be radiated incoherently. The most interesting case for high energy physics is quantum chromodynamics (QCD). It is an unbroken Yang–Mills theory based on the group

---

[3] The simple expression of Eq. (21.10) is valid only in the region $-t \gg m^2$; therefore, it does not cover the $t \to 0$ region. However, we can repeat the analysis without the large t simplification and show that the physically measurable integrated effective cross section is indeed finite.

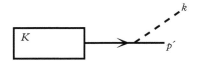

Figure 21.6 *Emission of a soft pion from an external nucleon line.*

$SU(3)$ with fermions, the quarks belonging to the fundamental, triplet, representation. The gauge bosons form an octet of massless spin-1 fields, called gluons. They are like photons except that they are self-coupled according to the Yang–Mills Lagrangian. In perturbation theory we can show that the infrared properties are similar to those of QED but the phenomenon is overshadowed by a non-perturbative property of the theory, the one we called 'confinement' in Chapter 15.

There are no other massless bosons in particle physics, but we can address the question of the low-energy behaviour of theories with almost massless particles. In section 15.2.4 we saw that at the limit in which chiral symmetry is broken only spontaneously, the pions are massless Goldstone bosons. Which are the infrared properties of the σ-model with zero mass pions? The answer turns out to be very simple, at least in the leading order approximation. Let us consider the radiation of a soft pion from an external nucleon line, the part of the diagram we show in Fig. 21.6. As we saw in section 15.2.4, the pions are pseudoscalars, so the amplitude for this emission is proportional to

$$M^{(\pi)}(p,p') \sim g\bar{u}(p')\gamma_5 \frac{p' + k + m_N}{(p' + k)^2 - m_N^2} K(p' + k, p), \tag{21.11}$$

where we have suppressed isospin indices. $K(p' + k, p)$ is some kernel, which transforms like a spinor and represents the rest of the diagram. The important point is that because of the presence of the γ_5 matrix, the numerator gives $\bar{u}(p')[-p' - k + m_N] = -\bar{u}(p')k$. In other words, the amplitude is suppressed by a factor proportional to k. We can still study the phenomenon of the radiation of soft pions but it will not play a role as important as that of soft photons.

In particle physics we do not have any other candidates of field theories which are liable to exhibit the phenomenon of soft quanta radiation. If gravitons exist they may have such a behaviour, but we do not know whether quantum gravity will be described by a quantum field theory. However, scalar field theories are often used to describe phase transitions in models of statistical mechanics and it is important to know their infrared properties.

For a general field theory we can perform a power-counting argument for the infrared, similar to the one we made for the ultraviolet. The results are, obviously, inverted. Non-renormalisable theories are infrared-safe but super-renormalisable ones are singular. Let us illustrate it for a massless ϕ^3 theory in four dimensions. The 1-loop, 2-point function, shown in Fig. 21.7(a), is given by the integral

$$\Gamma^{(2)}(p^2) = g^2\mu^2 \int \frac{\mathrm{d}^4 k}{(2\pi)^4} \frac{1}{k^2(k-p)^2}, \tag{21.12}$$

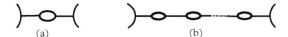

Figure 21.7 *The 1-loop correction to the scalar propagator in ϕ^3 theory (a). An internal line with several self-interaction bubbles (b).*

where $g\mu$ is the coupling constant. The integral is ultravioletly divergent and needs a subtraction. To ensure the masslessness of the theory, we should impose the condition $\Gamma^{(2)}(p^2 = 0) = 0$. But this is impossible because at $p = 0$ the integral is infrared divergent. So we must subtract it at some other value. We can interpret this result in two ways. We can say that the massless theory does not exist mathematically because, at higher orders, we may encounter a diagram of the form shown in Fig. 21.7(b), where an internal line has several self-interaction bubbles. It gives a contribution to the diagram of the form

$$I \sim \int \frac{d^4 k}{(2\pi)^4} \frac{1}{k^2} \left[\frac{\Sigma(k^2)}{k^2} \right]^n. \tag{21.13}$$

$\Sigma(k^2)$ is the subtracted 2-point function, but as we just explained, $\Sigma(k^2 = 0) \neq 0$. As a result the integral (21.13) is violently infrared divergent and the theory makes no sense. A second interpretation is to perform first the summation of the bubble diagrams following the method we explained in section 20.4. The result is that the pole of the propagator will be displaced from $p^2 = 0$ to some other point $p^2 = m^2$. Since we have no instability, the value of the mass will be real but the theory will describe a massive particle.

Super-renormalisable massless theories in less than four dimensions present similar problems. K. Symanzik, using an expansion in powers of $\epsilon = d - 4$, has shown that a massless ϕ^4 theory can be consistently defined in a space with dimensions between three and four. The worse case is in $d = 2$. Even a free massless scalar field in two dimensions has a singular 2-point function of the form

$$G(x) = \int \frac{d^2 k}{(2\pi)^2} \frac{e^{ikx}}{k^2}. \tag{21.14}$$

Before closing this section we want to point out that massless fields are often ill-defined if we quantise the theory in a compact space. In all cases evaluating the Feynman propagator involves computing the inverse of the d-dimensional Laplace operator. This operator has always a zero mode, the one corresponding to the constant function. In a non-compact space this eigenfunction can be ignored because it is non-normalisable and the zero-eigenvalue does not belong to the spectrum of the operator.[4] In a compact space this is no more true and the Laplacian is not invertible. The most important application of this remark is the d-dimensional de Sitter space whose Euclidean version is S_d, the

---

[4] The case of the two-dimensional flat space is marginal. The zero eigenvalue does not belong to the spectrum but it is the accumulation point of a series of eigenvalues.

d-dimensional compact sphere. It follows that a massless scalar field cannot be consistently defined in de Sitter space without introducing some important modifications in the quantum field theory axioms we have considered so far.

21.5 Problems

Problem 21.1 Compute the order α^2 effective differential cross section for the scattering of an electron on an external electromagnetic field and verify Sudakov's formula (21.8).

22

Coherent States and Classical Limit of Quantum Electrodynamics

22.1 Introduction

In this chapter we want to address the question of the classical limit in quantum field theory. The main concepts and applications in this book have been developed with the perspective of elementary particle physics. In particular, the space of physical states, to which we are mainly referring, is the Fock space. Every state in this space contains a well-defined number of particles, and states containing different numbers of particles are orthogonal to each other. It is perfectly suitable to describe scattering experiments, even more so, if the particles in question are massive, but it is not the best starting point to consider the classical limit. When computing scattering amplitudes we often take the limit of plane waves, while classical particles should be localised in space. We could, and should, work with wave packets, which are characterised by square integrable functions. The well-known problem in this case is that a wave packet describing a free particle, since it corresponds to a superposition of waves with different frequencies, spreads in time. It was Schrödinger, already in 1926, at the very early days of quantum mechanics, who first asked the question to find the 'best' wave function for a free particle, meaning the one with the minimum spread. We shall give the answer in this chapter.

The question of the classical limit becomes more complex if we consider massless particles. We saw already that the concept of a one-particle asymptotic state is meaningless when massless particles are involved and we called these problems, collectively, 'infrared problems'. If we ignore the graviton, which has never been detected, the only known massless particles are the gluons and the photons. Both are kept massless because they are the gauge bosons of unbroken local symmetries. The first mediate the strong interactions between the quarks and the resulting gauge theory is called 'quantum chromodynamics'. It will be studied in a later chapter. The infrared problems in this theory are very important but, as we have already noted, their macroscopic effects are overshadowed by a different, poorly understood, phenomenon, that of 'confinement'. Gluons are not present in asymptotic states. We are left with the photons whose infrared problems we addressed in the previous chapter. Here we want to ask the question of the classical

From Classical to Quantum Fields. Laurent Baulieu, John Iliopoulos and Roland Sénéor.

limit of an *n*-photon state. How is it related to the classical state of a given configuration of electric and magnetic fields?

It turns out that the answer to these questions involves the Bargmann space of coherent states, which we introduced in Chapter 8. Here we shall describe their properties and argue that they offer the natural basis to discuss the classical limit.

22.2   The Definition of Coherent States

When we studied the harmonic oscillator and determined the spectrum of the theory, we found that the construction of a Fock space is a very convenient and powerful tool that gives an insight which is complementary to that given by the set of wave functions.

In Chapter 8 we introduced another basis of states, called coherent states or Bargmann representation, which, at the time, looked mostly as an interesting exercise. One motivation was that the Bargmann representation is particularly well suited for the definition of the path integral formula, but we want to show here that these states have a deep physical meaning going much beyond the technical aspects we have seen.

The history of coherent states is intimately related to the connection between classical and quantum physics. In the historical introduction we presented in Chapter 8 for the approach which gave rise to the path integral formalism, we mentioned the work of Schrödinger, who invented the coherent states as early as 1926 in quantum mechanics–he called them 'states of minimal dispersion'. He wanted to find wave functions that maximally concentrate around a classical trajectory. We will present this property shortly. However, the use of this concept was largely ignored till the second half of the twentieth century. It is only that late that John Klauder, followed by Valentine Bargmann, Roy Glauber, George Sudarshan, and others, generalised in depth the idea to quantum field theory. The reader will not be surprised to learn that the importance of this work grew with the fast development in the last decades of the technologies in lasers and photonics, to which developments the concepts of coherent states played an important role. Today, they form one of the pillars of quantum optics.

Coherent states can be constructed for any quantum theory with a vacuum $|0>$ and a Fock space with an orthonormal basis. The latter involves a creation and a destruction operator $a^\dagger$ and a whose actions allow us to connect all possible elements of the basis, using the relation $[a, a^\dagger] = \hbar$. In what follows we will often adopt the usual convention $\hbar = 1$, since the only thing that matters is that the commutator never vanishes. Once the coherent states have been understood for the harmonic oscillator, not much must be changed for considering the situation when the Hamiltonian is perturbed by non-quadratic interactions.

In Chapter 8 the Bargman states were introduced by the relation $|z>= \exp[za^\dagger]|0>$, where z is a complex number and we showed that $|z>$ is an eigenvector of the destruction operator a with eigenvalue z. We will use this property here as a definition and *define* the coherent states as the set of all eigenvectors of a.

A side remark: there is a strong asymmetry in choosing to diagonalise a rather than $a^\dagger$. In fact, $a^\dagger$ has no eigenvector, but the null vector, because the Fock basis only involves integers $n \geq 0$, since, by construction, it involves states with positive energy. Note that the states that diagonalise the Hamiltonian are not coherent states, with the exception of the vacuum.

The coherent states $|z>$ are in fact labelled by a complex number z and defined as the general solution of the eigenvector equation

$$a|z> = z|z> . \tag{22.1}$$

z is complex because a is not a Hermitian operator, and the vacuum $|0>$ is the coherent state for $z = 0$. By doing an expansion of the eigenvalue equation for $|z>$ on the standard orthogonal basis $\{|n>\}$ of the Fock space of the harmonic oscillator, we find recursion relations. They can be easily solved, and we find the following equivalent definitions for all possible coherent states,

$$|z> = \exp-\frac{|z|^2}{2} \sum_{n=0}^{n=\infty} \frac{z^n}{\sqrt{n!}} |n> = \exp-\frac{|z|^2}{2} \exp za^\dagger |n>$$
$$= \exp(za^\dagger - \bar{z}a)|0> \tag{22.2}$$

with $\bar{z}$ the complex conjugate of z.

The unitary operator

$$D(z) = \exp(za^\dagger - \bar{z}a) \tag{22.3}$$

is called the displacement operator. It creates the coherent state $|z>$ out of the vacuum. We changed here the definition given in Chapter 8 by choosing a normalisation such that $|z>$ has norm 1 for all values of z.

The operator $D(z)$ makes it possible to connect to each other any pair of coherent states, using the composition relation

$$D(z)D(z') = \exp\frac{z\bar{z}' - z'\bar{z}}{2} D(z + z'). \tag{22.4}$$

Indeed the factor $\exp(z\bar{z}' - z'\bar{z})$ is a pure phase, and the last formula implies that

$$D(z)|z'> = \exp\frac{z\bar{z}' - z'\bar{z}}{2} |z + z'> . \tag{22.5}$$

The coherent states build an overcomplete basis of the Hilbert space, as can be seen from these formulas: two elements are never orthogonal, since

$$<z|z'> = \exp-\frac{|z - z'|^2}{2}. \tag{22.6}$$

We have, however, an orthogonality relation for the displacement operator, meaning that

$$\text{Tr}[D(z)D(z')] = \pi\delta^2(z - z').$$ (22.7)

The fact that the basis of coherent states is overcomplete is obvious. It is labelled by the ensemble of complex numbers, the cardinal of which is much larger than that of the ensemble of integers that label the states of a Fock basis, which is complete.

Getting an overcomplete non-orthogonal basis is a source of enrichment for describing the system in its various states. In fact, orthogonality is not a necessary demand for a basis. It is no more than a mathematically convenient property for a set of basic vectors in a Hilbert space, but using such a basis sometimes hides the geometrical nature of objects we wish to describe.

If we restore the dependence on $\hbar \neq 1$ in Eq. (22.6), we find that it appears in the denominator of the exponential, $< z|z' > = \exp{-\frac{|z-z'|^2}{2\hbar}}$. Thus, as soon as $z \neq z'$, the overlap goes very fast to 0 in the limit $\hbar \to 0$. The classical case can be somehow seen as the limiting case where two different coherent states become orthogonal.

What truly matters for describing quantum physics is the existence of the completeness relation

$$\sum_z |z > < z| \equiv \int dz |z > < z| = 1.$$ (22.8)

When a basis is overcomplete, part of the projectors may contribute negatively, to get some compensations.

The completion relation makes it possible to define density matrices as

$$\rho \equiv \int d^2z P(z) |z > < z|.$$ (22.9)

Systems that behave almost classically correspond to cases where the function $P(z)$ is almost never negative. If it is the case, the quantum matrix density ρ can be interpreted as a probability density for states that look almost classical.

When systems have many degrees of freedom, as in QED, we index the states by further labels such as the momentum k and the spin s, and do appropriate summations, as we will see shortly, $|z > \to |z_{k,s} >$.

For lasers, the generalised occupation number mean value

$$< n_{k,s} > = \text{Tr}\rho a^\dagger a$$ (22.10)

vanishes almost completely outside a small volume around the value of k (corresponding to the value of the frequency of the monochromatic laser), which leads to an almost positive function $P(z)$, which explains the almost classical aspect of the laser beam.

More refined examples exist in QED with an almost positive function $P(z)$, which looks classical, but the evolution of the corresponding states cannot be described by the classical equations of motion. The squeezed states we will shortly describe are examples of this, and have been materialised by producing the so-called laser beams with 2-photon states. Note that in this case

$$< E > \sim \hbar < n_{k,s} > \omega. \tag{22.11}$$

The richness and the predictive power of the quantum world appears when we observe that there are many limits leading to the same energy density, since the right-hand side involves the product $\hbar < n_{k,s} >$ and we can take different limits of both $\hbar$ and $< n_{k,s} >$ that lead to the same value of $< E >$.

A given coherent state $|z>$ can be projected on any given element of the Fock space basis and satisfies

$$< n|z > = \frac{z^n}{\sqrt{n!}} \exp{-\frac{|z|^2}{2}}, \tag{22.12}$$

and, thus (with the exception $|z>$ being the vacuum), it contains all possible states of the Fock basis, distributed with a Poisson probability distribution. In other words, the occupation number for the eigenvalues of the Hamiltonian is as random as it can be in any given coherent state, through a Poisson distribution. In fact, for the mean value for the occupation number for $N = a^{\dagger}a$, we have

$$\bar{N} = < z|N|z > = |z|^2. \tag{22.13}$$

It is easy to see that, depending on the value of $|z|$, $\bar{N}$ can be as small or as large as possible, and never zero but for the vacuum. Whatever the value of $|z|$ is, the coherent states exhibit a maximally random distribution over all states of an orthogonal basis.

This has physically interesting consequences. For instance, a 'soft' laser beam with a greatly attenuated beam behaves qualitatively exactly as a very powerful one, which explains the interference pattern experiments where the fringes appear at a very slow rate, with the wrong (although intuitive) interpretation that photons arrive one by one.

Interestingly enough, the probability amplitude of getting the vacuum in any given coherent state is

$$< 0|z > = \exp{-|z|^2}. \tag{22.14}$$

22.3   Fluctuations

What is really the advantage of using such an overcomplete basis? It was in fact inferred to by Schrödinger, since he was looking for wave functions that disperse minimally in time, both for the mean value of the position operator $x = a + a^{\dagger}$ and for that of the momentum $p = -i(a - a^{\dagger})$.

He was concerned with the dissipation properties of waves, in particular those of the 'natural' basis that diagonalise the Hamiltonian, which can give huge values of the quadratic fluctuations between some states $< |\Delta x|^2 >$ and $< |\Delta p|^2 >$, allowed by the Heisenberg inequalities $< |\Delta x|^2 > < |\Delta p|^2 > \geq \frac{\hbar^2}{4}$.

The coherent states are such that the Heisenberg relation gives an equality, instead of a minoration. For that reason they are the 'best' approximation to a classical state, and as we will see shortly, they remain so when time evolves.

Coming back to the q and p representation in the case of the harmonic oscillator , we easily compute the following normalised to one coherent state wave functions

$$\Psi_z(x) \equiv\; < x|z > = (\frac{\omega}{\pi\hbar})^{\frac{1}{4}} \exp \operatorname{Im}(z)^2 \exp-\frac{\omega}{2\hbar}[x - (\frac{2\hbar}{\omega})^{\frac{1}{2}}z]^2. \tag{22.15}$$

Under this form, discovered by Schrödinger, there is not much intuition of why this wave packet dissipates in a minimal way. We will show why it is so in a direct way.

22.3.1   Time Evolution of Coherent States

The time-evolution is given by the evolution operator $\exp-\frac{i}{\hbar}Ht$. Eigenstates of a are not eigenstates of H. So they have a non-trivial time evolution. For the harmonic oscillator, using the previous equations and $H = \omega(aa^\dagger + \frac{1}{2})$, we have

$$|z > \;\rightarrow\; |z(t) > = \exp-i\frac{\omega t}{2\hbar}\; |\exp-i\frac{\omega t}{\hbar}z > . \tag{22.16}$$

So $|z(t) >$ remains a coherent state, and, moreover, it comes back equal to itself for every time interval equal to an integer multiple of its period. In fact, for a high enough frequency, a coherent state has basically no dispersion, and it is as near as possible to a classical state. This explains the fact that its wave function in position space appears as describing an object concentrated around the trajectory of a classical motion.

22.3.2   Dispersion of Coherent States

Because coherent states z are eigenvectors of a, and they evolve with time towards other coherent states, the quadratic fluctuations of both operators a and $a^\dagger$ vanish identically, as well as those of their time evolution. Thus,

$$< z(t)|(\Delta a)^2|z(t) > \equiv\; < z(t)|a^2|z(t) > -(< z(t)|a|z(t) >)^2 = 0 \tag{22.17}$$

and

$$< z(t)|(\Delta a^\dagger)^2|z(t) > \equiv\; < z(t)|a^{\dagger 2}|z(t) > -(< z(t)|a^\dagger|z(t) >)^2 = 0. \tag{22.18}$$

In contrast, the quadratic fluctuation of $x = a + a^\dagger$ and its momentum $p = -i(a - a^\dagger)$ don't vanish. An easy computation, using $H = \hbar\omega(aa^\dagger + \frac{1}{2})$, gives

$$< z(t)|(\Delta x)^2|z(t) > = \frac{\hbar}{2\omega} \qquad < z(t)|(\Delta p)^2|z(t) > = \frac{\hbar\omega}{2} \tag{22.19}$$

and thus

$$< z(t)|(\Delta x)|^2|z(t) > < z(t)|(\Delta p)^2|z(t) > = \frac{\hbar^2}{4}. \tag{22.20}$$

This holds true for all values of z. This is the typical property of coherent states. Being eigenvalues of the annihilation operator, coherent states (i) minimise the quadratic fluctuation of the position operator x and its canonical momentum p, and (ii) keep them equal for the whole time evolution. Other sets only satisfy the Heisenberg uncertainty principle, $< \Delta x^2 > < \Delta p^2 > > \frac{\hbar^2}{4}$.

Coherent states give position and momentum fluctuations at the lowest possible equal values allowed by the Heisenberg uncertainty relation. We, thus, find the general result that coherent states, defined by diagonalising destruction operators, approach classical states in the nearest possible way.

We see that the quadratic fluctuations are independent of z, in contrast to the mean values $< z(t)|x^2|z(t) >$ or $< z(t)|p^2|z(t) >$, which depend quadratically on z. Furthermore, by comparing $< x^2 >$ and $< \Delta x^2 >$, we find that the most obvious departure from classical physics is for the eigenstates with the smallest eigenvalues, when $|z| << 1$.

22.4 Coherent States and the Classical Limit of QED towards Maxwell Theory

In QED, the free field A is a linear superposition of oscillators a_k and $a^\dagger_k$ indexed by the continuous three-dimensional momentum k and the value of the spin index s. The vacuum is for zero photons, and the Fock space is made of states with an arbitrary number of real photons with any given momentum and spin.

The basis of coherent states is defined as for an infinite collection of harmonic oscillators,

$$a_{k,s}|z_{k,s} > = z_{k,s}|z_{k,s} >, \tag{22.21}$$

for all values of k and s. We have a continuum of complex eigenvalues, indexed by the three-vector k and the polarisation index s.

Once coherent states are defined for the operator of the vector potential A_μ, we get other coherent states for the electric and magnetic fields, the components of the field strength $F_{\mu\nu}$, which are linear in A_μ.

We decompose the field operators A and F on components with positive and negative energy $\pm\omega$, with $\omega = |k|$, with the obvious notation $A = A^+ + A^-$. We have for the monochromatic electric field E

$$\begin{aligned} E &= I(\omega)\epsilon \cdot [a \exp i(k \cdot r - \omega t) + a^\dagger \exp -i(k \cdot r - \omega t)] \\ &= I(\omega)\epsilon \cdot [x \cos(k \cdot r - \omega t) + p \sin(k \cdot r - \omega t)]. \end{aligned} \tag{22.22}$$

In a problem at the end of this chapter we ask the result for the coherent state describing the electric field in a rectangular cavity.

We are now ready to understand the limiting case of quantum electrodynamics and the way a classical configuration is created by the interaction of the vacuum of quantum electrodynamics with a set of classical macroscopic charges producing a classical current:

$$\mathscr{J}^{\mu}(X) = (\rho(X), j(X)). \tag{22.23}$$

The vacuum of the theory is defined by a state with zero real photons. Such states can carry energy.

From the basic rules of perturbation theory, the time-evolution operator of the theory is

$$\exp -\frac{i}{\hbar} \int dt H_I(t), \tag{22.24}$$

where $H_I(t)$ is the linear interaction Hamiltonian between the field and the current

$$H_I(t) = -\int_V d^3 r \mathscr{J}(r, t) \cdot A(r, t). \tag{22.25}$$

After a Fourier transformation from the space of positions $\{r\}$ to that of momenta $\{k\}$, $H_I(t)$ can be expanded in creation and annihilation operators as

$$H_I(t) = -C \sum_{k,s} \sqrt{\frac{\hbar}{2\omega}} (\epsilon_{k,s} \cdot \mathscr{J}(k, t) a_{k,s} e^{-i\omega t} + \epsilon_{k,s}^* \cdot \mathscr{J}(k, t)^* a_{k,s}^\dagger e^{i\omega t}), \tag{22.26}$$

where $\mathscr{J}(k, t)$ is the Fourier transform of the macroscopic classical current density $\mathscr{J}(r, t)$, $\mathscr{J}(k, t)^*$ its complex conjugate, and C a normalisation factor, which depends on the volume of the space.

Perturbation theory implies that after the current $\mathscr{J}$ is switched on adiabatically, the probability amplitude that the vacuum state at $t = -\infty$ evolves to a vacuum state with zero real photons at time $t = \infty$ is

$$< 0| \exp -\frac{i}{\hbar} \int_{-\infty}^{\infty} H_I(t) |0 >, \tag{22.27}$$

where, as usual, a time ordering is understood.

Now, we observe that the time-evolution operator $\exp -\frac{i}{\hbar} \int_{-\infty}^{\infty} H_I(t)$ is nothing but a displacement operator in the space of coherent states with eigenvalues of order $\epsilon_{k,s} \cdot \mathscr{J}(k, t)$. Therefore, the final state, that is, $\exp -\frac{i}{\hbar} \int_{-\infty}^{\infty} H_I(t) |0 >$, must be a coherent state.

We conclude with two remarks. Starting from a coherent state with a large eigenvalue (because the sources are macroscopic), the final coherent state has very small fluctuations, with a very small dispersion in the occupation number of each Fock space element over which the coherent state decomposes; it is therefore as close to a classical state as it can be. It is a priori different from the vacuum at $t = -\infty$, which persisted before the sources were adiabatically turned on.

Furthermore, a formula can be given for the value of the final coherent state, showing that the field is given by the Maxwell theory with the classical current $\mathcal{J}(r, t)$. Indeed, using the explicit form of the evolution operator, and applying the decomposition of the Feynman propagator in positive and negative frequency parts, the standard rules of Green functions imply that the final value of $A(r, t)$ at $t = \infty$ can be computed by the Lienard–Wiechert formula using the retarded Green functions (assuming that the classical field and the current are zero at time $t = -\infty$ and that the current has been switched on adiabatically at $t - \infty$).

It must be noted that in standard quantum mechanics and for certain specific potentials, we can precisely and often non-perturbatively compute the evolution of the coherent states. This is also the case in quantum field theory.

22.5 Squeezed States

The squeezed states are a generalisation of coherent states. Their existence was pointed out by theorists, in their quest of finding interesting basis for the Hilbert space. They first attracted attention in quantum optics in the late 1970s. In the early development, their principal potential applications were for the field of optical communications and quantum non-demolition experiments designed for the detection of gravity waves. Later on, because of their capacity for optimising quantum fluctuations, squeezed states have been used in various subjects, such as quantum measurement theory, quantum non-linear dynamics, molecular dynamics, dissipative quantum mechanics, condensed matter physics, and gravity.

In quantum optics, the uncertainty principle is a fundamental obstruction that engineers encounter for coding and transmitting information by optical means. In fact, the quantum noise of light beams places a limit on the information capacity of an optical beam since the uncertainty principle is a statement about areas in phase space; noise levels in different quadratures are statements about intersections of uncertainty ellipses with these axes. Any procedure which can deform or squeeze the uncertainty circle for Δx and Δp to an ellipse can in principle be used for noise reduction in one of the quadratures.

So the idea of squeezed states is to isolate states such that the dispersion of canonical variables has been distributed in such a way that the dispersion of one variable is reduced, at the cost of increasing that of the rest of the variables in the phase space.

We will illustrate it in the case of two canonical variables x and p we defined as (putting $\hbar = 1$)

$$x = a^\dagger + a \qquad p = \mathrm{i}(a^\dagger - a). \tag{22.28}$$

We have a circular symmetry for the uncertainty dispersion relations since we have seen that

$$< (\Delta x)^2 >^{\frac{1}{2}} = < (\Delta p)^2 >^{\frac{1}{2}} = 1, \tag{22.29}$$

so that the uncertainty product is at its minimum value.

Squeezed states are just defined by doing a unitary transformation, with

$$x_s = a^\dagger \exp{-i\beta} + a \exp{i\beta} \qquad p_s = i(a^\dagger \exp{i\beta} - a \exp{-i\beta}) \tag{22.30}$$

where β is a real number.

It is easy to show that

$$\Delta x_s = \Delta x \cos\beta + \Delta p \sin\beta$$
$$\Delta p_s = -\Delta x \sin\beta + \Delta p \cos\beta \tag{22.31}$$

and that we can find angles β, such that

$$\Delta x_s < 1 \qquad \Delta p_s > 1. \tag{22.32}$$

The uncertainty dispersion has now an ellipsoidal symmetry, and we see that the quantum noise for x_s can be much less than that for x.

We can choose a rotated base

$$a_s = r(\cos\beta a + \sin\beta a^\dagger) \tag{22.33}$$

and define directly squeezed states as the eigenstates of a_s.

Given a generic coherent state $|z>$, the so-called '2-photons' squeezed states $|v, z>_s$ are obtained from the vacuum by first making a coherent state $|z>$ using the displacement operator $D(z)$ and then by squeezing it with the unitary operator

$$S(v) = \exp \frac{1}{2}(v^* a^2 - v a^{\dagger 2}), \quad \text{where} \quad v \equiv \exp{i\beta}, \tag{22.34}$$

so that

$$|v, z>_s \equiv S(v) \exp(z a^\dagger - z^* a)|0>. \tag{22.35}$$

The time evolution of such states has been often mimicked by introducing some non-linearity in the Hamiltonian for absorbing the z dependence. In fact, squeezed states can be experimentally produced by forcing coherent states to propagate through a non-linear medium, which reproduces the effect of the non-linearity of the Hamiltonian.

We can apply this idea to a monochromatic electric field E by applying unitary operations to the creation and destruction operators, yielding $E \rightarrow E_s$. 'Two-photon laser experiments' allow us to realise experimentally the change $E \rightarrow E_s$, and we can observe that the quantum fluctuation occur in an ellipsoidal way for the two transverse components of E_s instead of in a circular way for the two components in E. The production of squeezed states using media with non-linear optical responses has become an art, which can be even extended to the case of condensed matter physics, by replacing photons by phonons. There is a wealth of experiments, where the construction of squeezed states has been achieved, optimising the production of quantum states that defeat as much as possible the uncertainty principle. They realise a macroscopic quasi-classical limit of electrodynamics, which cannot be interpreted from the purely classical theory.

22.6 Problems

Problem 22.1 Coherent states of the electromagnetic field
  Consider the electromagnetic field quantised in a cubic box of dimension L.

1. Define the coherent states for such an electromagnetic field.

2. If $|\Psi(0)>$ is the coherent state describing the state of the field at $t = 0$, find its state $|\Psi(t)>$ at a later time t.

3. Study the quantum fluctuations of the energy $\Delta H/ < H >$ of the electromagnetic field which is described by a coherent state. Give an estimation of this quantity for the case of the radiation in a resonant cavity of $L{=}10$ cm, which is tuned to a frequency of 1 MHz. The average external field in the cavity equals 1 Gauss. What can we deduce for the validity of the classical approximation for this system?

23

Quantum Field Theories with a Large Number of Fields

23.1 Introduction

In section 20.4, in the study of the properties of unstable particles, we introduced the notion of the $1/N$ expansion, where N is the number of fields in the theory. This was used as an artefact to justify a particular reordering of terms in the infinite series of the perturbation expansion. In this section we want to study this concept a little more systematically.

Almost all the results we have obtained so far, and most of those we will obtain later, in quantum field theory are based on renormalised perturbation theory, in which, for practical reasons, we keep only the first few terms. This approach has two important limitations. The first concerns the numerical results. At best, they can be useful at the weak coupling limit when the value of the effective coupling constant is small. We say 'at best' because, in the absence of any mathematical understanding of the convergence properties of the series, we cannot guarantee that even weak coupling results are reliable. In a later chapter we will expose briefly the few rigorous results which have been obtained in this subject. On the other hand, as we will see in the next chapter, the results obtained using the perturbation expansion are often in spectacular agreement with experiment. The second limitation concerns the general properties of the theory. We have seen in previous sections that the first terms in the perturbation expansion have rather simple properties as analytic functions and this does not change radically as we go to higher orders. In particular, effects such as the presence of bound states will appear only if we can sum an infinite number of terms. For all these reasons, and others we can easily imagine, it is interesting to invent a different approach. Since summing the entire series does not appear to be feasible, people have tried to invent methods of partial summations. The $1/N$ expansion is one of them. It has several advantages. First, it is systematic. Under some well-defined conditions, it gives an unambiguous rule to perform these partial summations. It offers a formal parameter, namely $1/N$, which can be used to label the terms in the new expansion. In this respect, it is as 'consistent' as the ordinary perturbation series. Second, in some simple cases, the summations can be

From Classical to Quantum Fields. Laurent Baulieu, John Iliopoulos and Roland Sénéor.
© Laurent Baulieu, John Iliopoulos and Roland Sénéor, 2017. Published 2017 by Oxford University Press.

performed explicitly. Third, the new series has a richer structure and we can hope that it may capture some of the non-perturbative properties of the underlying theory. In this section we will study some of these aspects.

23.2 Vector Models

By vector models we mean theories with N-component fields which form a vector representation of a group $O(N)$. An example was the model we used in section 20.4. It is the simplest large N model and we shall use it as an example. It is described by a Lagrangian density of the form

$$\mathcal{L} = \frac{1}{2}(\partial_\mu \phi_i)(\partial^\mu \phi_i) - \frac{1}{2}m^2(\phi_i\phi_i) - \frac{\lambda}{4!}(\phi_i\phi_i)^2, \tag{23.1}$$

where the index i runs from 1 to N and a summation over repeated indices is understood. Following the arguments of section 20.4, let us consider the scattering amplitude for the process $a_1 + a_1 \rightarrow a_2 + a_2$. The first diagrams in the perturbation expansion are shown in Fig. 23.1. The tree diagram (a) is of order λ and the 1-loop diagrams are of order λ^2. However, diagram 23.1(b) has an extra factor N relative to those in 23.1(c) and (d) because we can circulate all N components in the first loop, but not in the others. This argument can be extended to diagrams with any number of loops and we see that at any order, diagrams, even with the same topology, will come with various powers of N depending on their particular group theory structure. If we try to take the large N limit keeping λ fixed we obtain divergent expressions reflecting the simple fact that at $N \rightarrow \infty$, we obtain an infinite number of graphs. In section 20.4 we saw that the correct limit was to first rescale the coupling constant and define

$$\tilde{\lambda} = \lambda N \tag{23.2}$$

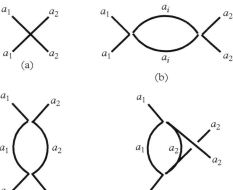

Figure 23.1 *The first diagrams in the perturbation expansion for the $a_1 + a_1 \rightarrow a_2 + a_2$ scattering amplitude. The lines are drawn in a way to show the factor of N in the loop of the s-channel diagram (b).*

and take the limit $N \to \infty$ keeping $\tilde{\lambda}$ fixed. There is a systematic way to prove this result and keep track of all the powers of N in the perturbation expansion by rewriting the Lagrangian (23.1) in a way which resembles the one we used in section 20.4:

$$\mathcal{L} = \frac{1}{2}(\partial_\mu \phi_i)(\partial^\mu \phi_i) - \frac{1}{2}m^2 \phi_i \phi_i + \frac{3}{2}\frac{N}{\tilde{\lambda}}\Phi^2 - \frac{1}{2}\Phi\,\phi_i\phi_i. \tag{23.3}$$

The difference with the case we studied in section 20.4 where Φ was the field of an unstable particle is that in the Lagrangian (23.3) it is an auxiliary field. This Lagrangian is in fact strictly equivalent to (23.1). Indeed, since the field Φ does not have a kinetic energy term, its equation of motion is $6\Phi = \lambda \phi_i \phi_i$ and if we replace it in (23.3) we recover (23.1). This simply reflects the fact that the Lagrangian (23.3) is quadratic in Φ with no kinetic energy and the functional integration is trivial. Note that the field Φ has canonical dimension equal to 2, so no terms of the form Φ^n with $n > 2$ will ever be generated as counter-terms in perturbation theory. On the other hand, the Lagrangian has no symmetry $\Phi \to -\Phi$; therefore, we expect to find a term linear in Φ. Indeed, the first non-trivial diagram in the perturbation expansion of (23.3) with Φ external lines is the one shown in Fig. 23.2. It contributes to the 1-point function and it gives a contribution proportional to N from the ϕ loop. Since it is divergent we must introduce a counter-term of the form $Nc\Phi$. In the terminology we used in section 16.5.3 the Lagrangian (23.3) is not stable. We restore stability by adding such a term already in the classical theory and we complete (23.3) as

$$\mathcal{L} \to \mathcal{L} - Nc\Phi. \tag{23.4}$$

This has no real effect on the dynamics because we can eliminate it by a translation of the field $\Phi \to \Phi + V^2$ with V a constant. As usual, we can determine V by a renormalisation condition and we choose the simplest one, namely

$$< 0|\Phi|0 > = 0. \tag{23.5}$$

In the tree approximation $V^2 = c\tilde{\lambda}/3$. The condition (23.5) eliminates all one-Φ Green's functions and only changes the value of the mass of ϕ: $\tilde{m}^2 = m^2 + V^2$. At 1 loop, V receives the contribution of the diagram of Fig. 23.2, which is divergent. The divergence is eliminated through a counterterm δc determined by the

Figure 23.2 *The 1-loop diagram contributing to the 1-point Green's function.*

renormalisation condition (23.5). In the original Lagrangian (23.1) it corresponds to mass renormalisation.

What we gained with all this rearrangement is that in the new Lagrangian, the powers of N are easier to follow because every term in it is, effectively, proportional to N. This is true for the Φ^2 term, but also for all the others because of the summation over i. In the Feynman rules N appears explicitly in the Φ propagator which is a constant proportional to $1/N$.

We can go one step further by noting that (23.3) is also quadratic in the fields ϕ_i, so we can formally integrate over them. The operation will be only formal because we will obtain a Φ-dependent determinant of the form $[\det(\Box + m^2 + \Phi)]^{-1/2}$, which we can only compute in perturbation. The result will be an effective theory depending only on Φ,

$$
\mathrm{e}^{\mathrm{i}S_{\mathrm{eff}}[\Phi,N]} = \int \prod_i \mathcal{D}[\phi_i] \mathrm{e}^{\mathrm{i}S[\phi_i,\Phi,N]}, \tag{23.6}
$$

where we have indicated explicitly the N-dependence. As we noted earlier, every term in (23.3) is proportional to N, so we obtain

$$
S_{\mathrm{eff}}[\Phi,N] = NS_{\mathrm{eff}}[\Phi,1]. \tag{23.7}
$$

In other words, N plays the same role as $1/\hbar$ in the usual perturbation expansion. A simple counting shows that a one-particle irreducible diagram with E external Φ lines and L loops has a power of N given by N^{1-E-L}. This shows that the leading term in $1/N$ will be given by the diagrams in our effective field theory with no loops (remember, the effective field theory has only Φ lines) and the minimum number of external Φ lines. Since we have eliminated the one-Φ Green's function, the leading term will be $1/N$ and corresponds to tree diagrams with two external Φ lines. This is very simple. It is a 1-PI diagram, computed with S_{eff} with no internal Φ lines, so it is given by the sum of the free contribution plus the 1-loop correction of Fig. 23.3:

$$
\Gamma_\Phi^{(2)}(k^2) = -\mathrm{i}\frac{3N}{\tilde{\lambda}} + \frac{1}{4}\int \frac{\mathrm{d}^4 q}{(q^2 - \tilde{m}^2)[(k-q)^2 - \tilde{m}^2]}. \tag{23.8}
$$

The momentum integral in this expression is logarithmically divergent but the divergence can be absorbed in the renormalisation of the coupling constant. So the result can be written as

Figure 23.3 *The exact 1-PI 2-point function to leading order in $1/N$.*

$$\Gamma_{\Phi}^{(2)}(k^2) = -i\frac{3N}{\tilde{\lambda}} + i\Sigma(k^2),\tag{23.9}$$

where the constants $\tilde{m}$ and $\tilde{\lambda}$ are assumed to represent the renormalised values and $i\Sigma(k^2)$ is the finite part of the 1-loop contribution. The complete Φ propagator is the inverse of $\Gamma_{\Phi}^{(2)}(k^2)$ given by

$$G_{\Phi}^{(2)}(k^2) = \frac{i}{\frac{3N}{\tilde{\lambda}} - \Sigma(k^2)}.\tag{23.10}$$

Diagrammatically this is equivalent to the series of diagrams given in Fig. 23.4. We emphasise that this result is exact to leading order in $1/N$ but to all orders in the perturbation expansion.

For example, the $a_1 + a_1 \to a_2 + a_2$ scattering amplitude we considered at the beginning of this section is obtained, in leading $1/N$ order, but to all orders in perturbation, by the one-Φ exchange diagram shown in Fig. 23.5(a) with the complete Φ propagator given in (23.10). We can arrive at the same conclusion if we eliminate the auxiliary field and go back to the original variables ϕ_i. The amplitude is now given in terms of the sum of the bubble diagrams shown in Fig. 23.5(b). We can again sum the infinite series and verify that the two expressions, Fig. 23.5(a) and Fig. 23.5(b), are indeed identical.

Let us summarise. In our $O(N)$ vector model we can compute exactly any scattering amplitude in the leading $1/N$ order but to all orders in perturbation by summing an infinite series of Feynman diagrams. The model remains renormalisable but we need only a mass and a coupling constant renormalisation. The wave function renormalisation, which is normally associated with the diagram of Fig. 16.9, appears only in the next order in the $1/N$ expansion.

In a four-dimensional space–time the scalar $O(N)$-invariant Lagrangian (23.1) we studied here is essentially the only vector model which yields a renormalisable quantum

Figure 23.4 *The diagrammatic expansion of the Φ propagator.*

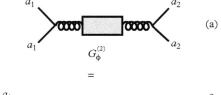

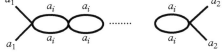

(a)

(b)   **Figure 23.5** *The $a_1 + a_1 \to a_2 + a_2$ amplitude in terms of the Φ propagator (a), and the sum of the bubble diagrams (b).*

field theory. As we explained earlier, we can sum the infinite series of diagrams explicitly and obtain expressions in closed form for the leading terms in $1/N$. This is obtained by rescaling the coupling constant as shown in Eq. (23.2). The limit $N \to \infty$ gives well-defined expressions only if we keep $\tilde{\lambda}$ of Eq. (23.2) fixed. In Problem 23.1 we compute various quantities, such as the effective potential, in this theory, but no essentially new phenomena appear. The situation changes if we consider theories in three or two space–time dimensions, and in Problem 23.2 we consider a very interesting case of a spinor theory of the general form $(\bar{\psi}_i \psi_i)^2$, the so-called *Gross–Neveu model*. It is easy to show that it is renormalisable in two space–time dimensions and in the large N limit it presents many interesting features. Since we have not studied field theories in other than four dimensions in this book, we will not go to further details here. Some rigorous results concerning this model will be presented later.

23.3 Fields in the Adjoint Representation

In the previous section we studied the large N limit of models in which the fields belong to the N-dimensional representation of the symmetry group (the vector representation for the $O(N)$ group or the fundamental representation for the $U(N)$ group in the Gross–Neveu model). When N goes to infinity the number of fields grows linearly with N. We showed that we can sum the series of dominant diagrams and obtain explicit expressions for quantities such as scattering amplitudes or the effective potential. In this section we want to address the same question but for theories in which the fields belong to the adjoint representation of the symmetry group. The application we have in mind is, naturally, the Yang–Mills theories. The obvious difference is that now the number of fields grows quadratically with N.

Let us first consider a pure Yang–Mills theory on the group $SU(N)$. The gauge potentials A_μ are matrix valued, so in order to keep track of factors of N we put the matrix indices explicitly: $A_{\mu i}^{j}$. In fact, the Lorentz index μ will not play any role, so we will drop it in what follows. The analysis could be applied equally well to any spin fields, provided they belong to the adjoint representation of $SU(N)$. From the group theory point of view this matrix representation reflects the fact that the adjoint of $SU(N)$ can be written as the traceless product of a fundamental and an anti-fundamental representation: $A_i^j \sim q^j \bar{q}_i$, where the q's belong to the fundamental representation. The A-propagator will have a group theory structure given by

$$< A_i^j \, A_k^l > \, \sim \, < (q^j \bar{q}_i) \; (q^l \bar{q}_k) > \, \sim \, \delta_k^j \delta_i^l - \frac{1}{N} \delta_i^j \delta_k^l, \tag{23.11}$$

while that of a q will be

$$< q^j \, \bar{q}_i > \, \sim \, \delta_i^j. \tag{23.12}$$

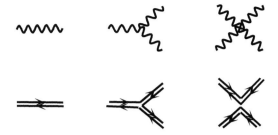

Figure 23.6 *'t Hooft's double-line representation for the diagrams of a Yang–Mills theory.*

The second term on the right-hand side of Eq. (23.11) is due to the fact that we considered an $SU(N)$ theory, rather than a $U(N)$ one. It is clear that in the $N \to \infty$ limit the difference will appear only in the sub-dominant terms. These relations suggest a new way of writing the Feynman diagrams, first proposed by G. 't Hooft. We write an oriented line to follow every fundamental representation index, as shown in Fig. 23.6. The vector bosons are represented by a double line with opposite orientations.[1] It is now straightforward to represent the vertices with three or four gauge bosons, always ignoring the space–time indices and momentum factors.

The Yang–Mills action is proportional to the trace of the product FF, so, instructed by our experience with the vector model, we rescale the fields and write the action as

$$S = \frac{N}{\tilde{g}^2} \int d^4x \left(-\frac{1}{4}\mathrm{Tr}FF\right). \tag{23.13}$$

The large N limit will be taken keeping fixed the *'t Hooft coupling constant*:

$$\tilde{g}^2 = Ng^2. \tag{23.14}$$

We are now in a position to count the factors of N in a given diagram. Let us start with the vacuum to vacuum diagrams. In the double-line representation of Fig. 23.6 all indices must be contracted to make loops. The lines which form every loop will close the perimeter of an oriented polygon. Every loop, i.e. every face of a polygon, will give a factor of N from the summation over the indices. Every propagator, i.e. every edge of the polygon, will give a factor $1/N$, while every vertex will give a factor of N. These factors come from our rescaled action of Eq. (23.13). Putting together all polygons we obtain a two-dimensional surface. Since all polygons are oriented and the lines in the edges go in opposite directions, we can define an orientation for the entire surface. A diagram with F polygon faces, E edges, and V vertices has a factor of N given by N^{F+V-E}. The exponent is known in topology and it is called the *Euler characteristic* χ. An oriented surface is topologically equivalent to a sphere which may have a number B

---

[1] Here the fact that we considered an $SU(N)$ group is important. Had we considered an orthogonal group, in which the fundamental and the anti-fundamental representations are equivalent, we would not have obtained oriented lines.

of boundaries; in other words, holes cut on the surface and a number H of handles. The Euler characteristic is given by

$$\chi = 2 - 2H - B. \tag{23.15}$$

It follows that the leading behaviour of the connected vacuum to vacuum amplitude is N^2 and it is given in terms of diagrams with no boundaries which have the topology of a sphere. In our double-line notation a boundary is formed by a line with an index going only in one direction. This line does not correspond to a vector boson propagator since the latter comes always with a double line with indices going in opposite directions. It corresponds to a field belonging to the fundamental representation. In chapter 25 we will study an $SU(3)$ gauge theory for the strong interactions with an octet of gauge bosons which we call gluons and a set of spinor fields in the fundamental triplet representation which describe the quarks. We conclude that at large N, the leading vacuum to vacuum diagrams are made out of gluons with no quark lines. This result is intuitively obvious, because at every order in perturbation theory we can create either a pair of gluons or a quark–antiquark pair and we have N^2 gluons and only N quarks. Going back to the traditional single line representation of the Feynman diagrams we lose the orientation of the surface. Since we have no handles, the resulting diagram can be drawn on a plane with no line crossings in which one line must go on top of the other, i.e. a planar diagram. So our previous result can be phrased in the following form: *In an $SU(N)$ gauge theory, taking the limit $N \rightarrow \infty$ with $\tilde{g}$ fixed, the leading behaviour of the connected vacuum to vacuum amplitude is given by the sum of all planar diagrams with no quark lines.* Quarks will appear only in the subleading behaviour. The topological relation (23.15) shows that it is given by diagrams with no handles ($H = 0$), but one boundary. So they are planar diagrams with a single quark loop which forms the boundary of the graph. They give a contribution to the amplitude which grows like N.

The planar diagrams form a much larger set than the bubble diagrams we found in the vector models and nobody has been able to compute the sum. Therefore, the large N behaviour of these theories is not known explicitly. There have been some interesting attempts to extract physical consequences for the theory of strong interactions, especially meson phenomenology, but they have not given any quantitative results.

Before closing this section, let us draw an analogy between the $N \rightarrow \infty$ limit and the $\hbar \rightarrow 0$ limit. Formally, they are identical. Let us write Feynman's path integral for some gauge invariant quantity

$$< \mathcal{O} > \sim \int \mathcal{D}[A]\mathcal{O}[A] e^{\frac{iN}{\hbar \tilde{g}^2} S[A]}, \tag{23.16}$$

where we have restored the factor $\hbar$ and we denote all fields collectively by A. The N factor is taken from our expression (23.13). Taking the classical limit means that as $\hbar$ goes to 0 the measure is concentrated on fields which are solutions of the classical equations of motion A_{cl}. The value of the gauge-invariant quantity is the one it obtains for this classical field configuration $\mathcal{O}[A_{\mathrm{cl}}]$, or any gauge equivalent of it. Since the group is $SU(N)$, the gauge potential is given by some $N \times N$ matrix. Similarly, as N goes to infinity we expect the measure to be concentrated on some classical field configuration, except that this time it will be an infinite-dimensional matrix, $A_{\text{cl}}^{(\infty)}$. Summing the planar diagrams amounts into finding this unique field configuration. If we have it we can compute the large N value of any observable, simply as $\mathcal{O}[A_{\text{cl}}^{(\infty)}]$. For this reason $A_{\text{cl}}^{(\infty)}$ is called *the master field*. The problem is that although in simple cases we can solve the classical equations of motion and find A_{cl}, we have no practical way of computing, even approximately, the master field. If we make the reasonable assumption that, at large N, the solution will respect symmetries such as translational invariance, we expect the master field to be gauge equivalent to a constant field configuration, so it will be an infinite-dimensional, constant matrix.

23.4   The Large *N* Limit as a Classical Field Theory

In the previous sections we studied various field theory models in the limit when the number of fields goes to infinity. We used the language of Feynman diagrams and we found that we must rescale the coupling constant to take into account the fact that in this limit we have an infinite number of diagrams at every order of perturbation. When N goes to infinity the bare coupling constant must go to 0. In this section we will try to study an alternative limit which will give us a classical field theory formulated in a space with extra dimensions. The result will be particularly interesting for gauge theories.

Let us start with the vector model. The field is $\phi^i(x)$, $i = 1, 2, ..., N$. When N goes to infinity we can replace $\phi^i(x)$ by a single field $\phi(x, \sigma)$, $0 \leq \sigma < 2\pi$, using any kind of discrete transformation. For example, we can expand $\phi(x, \sigma)$ in Fourier series, or in the basis of Legendre polynomials, etc. In an orthonormal basis we find that

$$\sum_i \phi^i(x)\phi^i(x) \;\rightarrow\; \int_0^{2\pi} \mathrm{d}\sigma\, \phi(x, \sigma)\phi(x, \sigma), \qquad (23.17)$$

i.e. we obtain a single field in a five-dimensional space where the fifth dimension is compact. The problem with the vector model of Eq. (23.1) is that the interaction term $(\phi^i\phi^i)^2$ is no more local in σ. The surprising feature is that for a Yang–Mills theory, the resulting expression turns out to be local.

Let us consider an $SU(N)$ Yang–Mills theory in a d-dimensional space with potentials

$$A_\mu(x) \;=\; A_\mu^a(x)\, t_a, \qquad (23.18)$$

where t_a are the standard $SU(N)$ matrices. The statement is that in an appropriate large N limit there exists a reformulation of the theory in which the gauge fields and the gauge potentials are replaced by

$$A_\mu(x) \to \mathcal{A}_\mu(x, z_1, z_2) \qquad F_{\mu\nu}(x) \to \mathcal{F}_{\mu\nu}(x, z_1, z_2), \qquad (23.19)$$

where $\mathcal{A}$ and $\mathcal{F}$ are c-number functions in a $(d+2)$-dimensional space, Greek indices still run from 0 to $d-1$ and z_1 and z_2 are local symplectic coordinates on a two-dimensional closed surface. So far there is no surprise. We obtain two extra compact dimensions because the gauge fields are represented by matrices which have two indices. The surprise comes when we express the Yang–Mills action in this $(d+2)$-dimensional space. The result is

$$\int \mathrm{d}^4 x \mathrm{Tr} F_{\mu\nu} F^{\mu\nu} \;\rightarrow\; \int \mathrm{d}^4 x \mathrm{d}\Omega\, \mathcal{F}_{\mu\nu}(x, z_1, z_2) \mathcal{F}^{\mu\nu}(x, z_1, z_2), \qquad (23.20)$$

where $\mathrm{d}\Omega$ is the measure on the surface (for a sphere it is the familiar $\sin\theta \mathrm{d}\phi \mathrm{d}\theta$) and $\mathcal{F}_{\mu\nu}(x, z_1, z_2)$ is given by

$$\mathcal{F}_{\mu\nu}(x, z_1, z_2) = \partial_\mu \mathcal{A}_\nu(x, z_1, z_2) - \partial_\nu \mathcal{A}_\mu(x, z_1, z_2) + \{\mathcal{A}_\mu(x, z_1, z_2), \mathcal{A}_\nu(x, z_1, z_2)\}. \quad (23.21)$$

$\{f, g\}$ is the classical Poisson bracket defined by

$$\{f, g\} = \frac{\partial f}{\partial z_1} \frac{\partial g}{\partial z_2} - \frac{\partial f}{\partial z_2} \frac{\partial g}{\partial z_1}. \qquad (23.22)$$

We see that, as announced, in the action (23.20) all terms are local in the $(d+2)$-dimensional space.

The proof of this statement is basically algebraic. We can show that the $SU(N)$ algebra, at the limit $N \to \infty$, with the generators appropriately rescaled, becomes isomorphic to the algebra of symplectic, or area preserving diffeomorphisms of a two-dimensional surface. The generators of symplectic transformations of a surface, locally have the form

$$L_f = \frac{\partial f}{\partial z_2} \frac{\partial}{\partial z_1} - \frac{\partial f}{\partial z_1} \frac{\partial}{\partial z_2} \qquad (23.23)$$

and satisfy the algebra

$$[L_f, L_g] = L_{\{f,g\}}. \qquad (23.24)$$

The direct way to prove this result is to compute the structure constants of the $SU(N)$ algebra and show that at the limit when $N \to \infty$ they go to the corresponding structure constants of [SDiff(S)].[2] We will present here a simpler proof which applies to the case when the two-dimensional surface is a sphere. In this case there exists a globally defined pair of symplectic coordinates, such as $z_1 = \cos\theta$ and $z_2 = \phi$. In the relation (23.23) we can form a basis of generators choosing f to be the spherical harmonics: $f = Y_{l,m}(\theta, \phi)$. The generators of the area-preserving diffeomorphisms can then be expressed as

[2] See J. Hoppe, MIT PhD thesis, 1982.

$$L_{l,m} = \frac{\partial Y_{l,m}}{\partial \cos\theta} \frac{\partial}{\partial\phi} - \frac{\partial Y_{l,m}}{\partial\phi} \frac{\partial}{\partial\cos\theta}. \tag{23.25}$$

They satisfy the Lie algebra

$$[L_{l,m}, L_{l',m'}] = f^{l'',m''}_{l,m;\, l',m'} L_{l'',m''} \tag{23.26}$$

with the structure constants given by

$$\{Y_{l,m}, Y_{l',m'}\} = f^{l'',m''}_{l,m;\, l',m'} Y_{l'',m''}, \tag{23.27}$$

where the curly bracket represents the Poisson bracket with respect to ϕ and $\cos\theta$.

After these preliminaries, we turn to the proof of our statement. We first remark that the spherical harmonics $Y_{l,m}(\theta,\phi)$ are harmonic homogeneous polynomials of degree l in three Euclidean coordinates x_1, x_2, x_3,

$$x_1 = \cos\phi\sin\theta, \quad x_2 = \sin\phi\sin\theta, \quad x_3 = \cos\theta \tag{23.28}$$

$$Y_{l,m}(\theta,\phi) = \sum_{\substack{i_k=1,2,3 \\ k=1,\dots,l}} \alpha^{(m)}_{i_1\dots i_l} x_{i_1}\dots x_{i_l}, \tag{23.29}$$

where $\alpha^{(m)}_{i_1\dots i_l}$ is a symmetric and traceless tensor. For fixed l there are $2l+1$ linearly independent tensors $\alpha^{(m)}_{i_1\dots i_l}$, $m = -l, \dots, l$.

Let us now choose, inside $SU(N)$, an $SU(2)$ subgroup by choosing three $N \times N$ Hermitian matrices which form an N-dimensional irreducible representation of the Lie algebra of $SU(2)$:

$$[S_i, S_j] = i\epsilon_{ijk} S_k. \tag{23.30}$$

The S matrices, together with the α tensors introduced earlier, can be used to construct a basis of $N^2 - 1$ matrices acting on the fundamental representation of $SU(N)$,

$$S^{(N)}_{l,m} = \sum_{\substack{i_k=1,2,3 \\ k=1,\dots,l}} \alpha^{(m)}_{i_1\dots i_l} S_{i_1}\dots S_{i_l}$$

$$[S^{(N)}_{l,m}, S^{(N)}_{l',m'}] = i f^{(N)l'',m''}_{l,m;\, l',m'} S^{(N)}_{l'',m''}, \tag{23.31}$$

where the f's appearing on the right-hand side of (23.31) are just the $SU(N)$ structure constants in a somehow unusual notation. Their normalisation is given by

$$\mathrm{Tr}\left(S^{(N)}_{l,m} S^{(N)}_{l',m'}\right) = \frac{1}{4\pi} N \left(\frac{N^2-1}{4}\right)^l \delta_{ll'}\delta_{mm'}. \tag{23.32}$$

The important, although trivial, observation is that the three $SU(2)$ generators S_i, rescaled by a factor proportional to $1/N$, will have well-defined limits as N goes to infinity:

$$S_i \rightarrow T_i = \frac{2}{N} S_i. \tag{23.33}$$

Indeed, all matrix elements of T_i are bounded by $|(T_i)_{ab}| \leq 1$. They satisfy the rescaled algebra

$$[T_i, T_j] = \frac{2\mathrm{i}}{N} \epsilon_{ijk} T_k \tag{23.34}$$

and the Casimir element

$$T^2 = T_1^2 + T_2^2 + T_3^2 = 1 - \frac{1}{N^2}. \tag{23.35}$$

In other words, under the norm $\|x\|^2 = \mathrm{Tr} x^2$, the limits as N goes to infinity of the generators T_i are three objects x_i which commute by (23.34) and are constrained by (23.35):

$$x_1^2 + x_2^2 + x_3^2 = 1. \tag{23.36}$$

If we consider two polynomial functions $f(x_1, x_2, x_3)$ and $g(x_1, x_2, x_3)$ the corresponding matrix polynomials $f(T_1, T_2, T_3)$ and $g(T_1, T_2, T_3)$ have commutation relations for large N which follow from (23.34):

$$\frac{N}{2\mathrm{i}} [f, g] \rightarrow \epsilon_{ijk} x_i \frac{\partial f}{\partial x_j} \frac{\partial g}{\partial x_k}. \tag{23.37}$$

If we replace now in the $SU(N)$ basis (23.31) the $SU(2)$ generators S_i by the rescaled ones T_i, we obtain a set of $N^2 - 1$ matrices $T_{l,m}^{(N)}$ which, according to (23.29), (23.31), and (23.37), satisfy

$$\frac{N}{2\mathrm{i}} [T_{l,m}^{(N)}, T_{l',m'}^{(N)}] \rightarrow \{Y_{l,m}, Y_{l',m'}\}. \tag{23.38}$$

The relation (23.38) completes the algebraic part of the proof. It shows that the $SU(N)$ algebra, under the rescaling (23.33), does go to that of $[\mathrm{SDiff}(S^2)]$. It is now straightforward to obtain the limit for the Yang–Mills action. We expand the classical fields $SU(N)$ on the basis of the matrices $T_{l,m}^{(N)}$,

$$A_\mu(x) = \sum_{\substack{l=1,\ldots,N-1 \\ m=-l,\ldots,l}} A_\mu^{l,m}(x) T_{l,m}^{(N)}, \tag{23.39}$$

which, according to (23.38), implies that the rescaled commutator has as a limit the Poisson bracket:

$$N[A_\mu, A_\nu] \rightarrow \{A_\mu(x, \theta, \phi), A_\nu(x, \theta, \phi)\}. \tag{23.40}$$

Combining (23.40) with the fact that the trace of the T matrices is obtained from (23.32) by rescaling

$$\text{Tr}\left(T^{(N)}_{l,m} \, T^{(N)}_{l',m'}\right) \rightarrow N \, \delta_{ll'} \delta_{mm'}, \tag{23.41}$$

we obtain the result for the case of the sphere. The torus can be treated similarly because it also admits a pair of globally defined symplectic coordinates.

Putting now all N factors together we see that the correct rescaling to obtain this result is given by the limit of the commutator (23.40). On the left-hand side the A's are $N \times N$ matrices and on the right-hand side, they are c-number functions. It implies that in this large N limit we keep the combination $g^{-2}N^3$ fixed. In contrast to what happens in the 't Hooft limit, here the bare coupling constant goes to infinity for large N. This is reflected to the fact that we cannot use ordinary perturbation expansion starting from the action (23.20) because the quadratic part has no derivatives with respect to z_1 and z_2. Every term in this expansion is divergent, as it should since the rescaling did not absorb the infinite number of diagrams in the original Yang–Mills theory. The only perturbation which could be meaningful would be around a non-trivial classical solution.

23.5 Problems

Problem 23.1 In section 19.4 we introduced the concept of the effective potential $V(\phi)$ as a function of the classical field. We saw that its knowledge allows us to study the various phases of the theory. In the vector model defined by the Lagrangian density (23.1), compute the effective potential in the leading $1/N$ approximation. Use this expression in order to investigate whether this theory exhibits, in this approximation, the phenomenon of spontaneous breaking of the $O(N)$ symmetry.

Problem 23.2 The Gross–Neveu model. We consider a field theory in two space–time dimensions defined by the Lagrangian density

$$\mathcal{L} = \bar{\psi}_i i\partial\!\!\!/ \psi_i + g(\bar{\psi}_i \psi_i)^2, \tag{23.42}$$

where ψ is a fermion field, $i = 1, ..., N$, and a summation over repeated indices is understood. The Lagrangian is invariant under a $U(N)$ symmetry.

(i) Show by power counting that in two space–time dimensions the theory is renormalisable.

(ii) Show that despite the absence of a mass term, the model is stable under renormalisation because of a discrete chiral symmetry.

(iii) We want to study the large N limit of this theory. Find the effective coupling constant $\tilde{g}$ which should be kept fixed in order to obtain finite expressions when $N \to \infty$.

(iv) Use the methods we introduced in the study of the $O(N)$ vector model, section 23.2, in order to compute the effective potential as a function of the classical auxiliary field.

(v) Show that the theory is asymptotically free in the sense we explained in section 19.2.

(vi) Study the phenomena of spontaneous symmetry breaking and dimensional transmutation.

The model is analysed in the lectures by S. Coleman at the 1979 Erice Summer School.

24

The Existence of Field Theories beyond the Perturbation Expansion

24.1 Introduction

All throughout this book we have emphasised the deep connection between the sophisticated mathematical description of the fundamental laws of nature and the detailed experimental results. It is this connection which characterises modern physics. Quantum field theory is the best example of it. As we have seen in the previous chapters, quantum field theories are the natural extension of classical field theories (the main one being electromagnetism), with the integration of the ideas of special relativity and quantum mechanics. A given field theory, whether classical or quantum, is supposed to describe a particular class of physical phenomena. This is reflected by the choice of the variables, the fields, and the particular form of the interaction we assume taking place among them. Historically, the early formulations were following the Hamiltonian, approach and we were splitting the Hamiltonian into the 'free' part, which contained only the kinetic energies and the mass terms, and the interaction terms. In the formulation we have presented so far the strength of the latter was taken to be 'weak'. This was understood either in absolute terms, a coupling constant whose numerical value was much smaller than 1, or relative to the contribution of the other terms, such as the kinetic energies and masses. This way we can imagine that a physical quantity can be computed as a series of 'elementary' processes involving more and more times the interaction. This was the perturbation expansion whose properties we studied in the previous chapters. The basic assumption is that a term of order n, i.e. one involving the interaction n times, should be smaller than the corresponding one of order m with $m < n$. It yields the expansion in terms of Feynman diagrams and, after several formal manipulations, the renormalised perturbation series.

A conceptually new step was the path integral approach introduced to its full generality by Feynman. As we explained in Chapters 8 to 12, it contains two fundamental ingredients: Dirac's visionary focus on the importance of this new formulation of quantum mechanics, on the one hand, and the introduction by Wiener of an integral

From Classical to Quantum Fields. Laurent Baulieu, John Iliopoulos and Roland Sénéor.
© Laurent Baulieu, John Iliopoulos and Roland Sénéor, 2017. Published 2017 by Oxford University Press.

description of the Brownian motion (the so-called Wiener integral), on the other. Feynman's path integral formulation, the functional integral defined in Chapter 8 as

$$\int e^{i \int [\phi(x)j(x) - \mathcal{L}(x)] d^4 x} \prod_{y \in \mathbb{R}^4} d\phi(y), \tag{24.1}$$

was the first global form of the generating functional of a physical theory, the theory being defined by a Lagrangian $\mathcal{L}(\phi(x)) = \mathcal{L}(x)$. Naturally, as for the Hamiltonian, it is possible to split the Lagrangian $\mathcal{L}$ into a quadratic part $\mathcal{L}_0$ corresponding to a free system, i.e. without interaction, and an interactive part, $\mathcal{L}_I$, usually describing the self couplings of the fields (non-quadratic terms) or the couplings between different fields. Then the integral (24.1) is formally proportional to

$$\int e^{i \int [\phi(x)j(x) - \mathcal{L}_I(x)] d^4 x} e^{-i \int \mathcal{L}_0(x) d^4 x} \prod_{y \in \mathbb{R}^4} d\phi(y) \propto \int e^{i \int [\phi(x)j(x) - \mathcal{L}_I(x)] d^4 x} \mathcal{D}\mu(\phi) \tag{24.2}$$

with the introduction of the formal Gaussian measure $\mathcal{D}\mu(\phi)$ built from the free Lagrangian $\mathcal{L}_0$.

After normalisation, we obtain this way the generating functional $Z[j]$ given by formula (12.136).

If the object of interest is the estimation of some given physical quantity, such as, for example, the T-product $< T(\phi(x_1) \cdots \phi(x_n)) >$ of n fields, then its formal perturbative expansion is obtained by expanding in the functional integrals of (12.131) the exponentials of the interaction as a sum of powers of $\mathcal{L}_I$ as given by formula (12.135). We then recover the formal perturbative expansion to arbitrary order as given by the Feynman recipe. This approach has been incredibly efficient with great successes in numerical predictions. The intermediate step to get these results was the renormalisation procedure that we have explained in the previous chapters.

What is missing is the intrinsic consistency of this complex mathematical machinery: functional integration, perturbation expansion, and renormalisation. We can express a physical theory through compact formulae. But these formulae are of the form of functional integrals which imply integration on a space of functions, a space of infinite dimensions. The existence of these functional integrals is very hard to prove because the argument is oscillatory and very little is known about the functional property of the measure of integration. Moreover, many of the main results were obtained from perturbation theory and, apart from the formal relation, nothing is really known between the global theory and the perturbation expansion. The question we want to address in this chapter is how to reassemble in a rational way the global aspect, the perturbation theory, and the renormalisation procedure.

During the past forty years considerable progress has been made on the existence and the properties of the physical theories. We will show, admittedly on some simplified models, that it is possible to understand precisely how the different practical approaches can be made coherent.

24.2 The Equivalence between Relativistic and Euclidean Field Theories

The basic ingredient that makes it possible to give a mathematical meaning to field theories is an equivalence relation between field theories in Minkowski space and their corresponding theories in the Euclidean $\mathbb{R}^4$ space. The metric in Minkowski space is $(+,-,-,-)$. As we have done several times previously, we perform a $-90°$ rotation of the time coordinate (this is often called the Wick rotation), i.e. we change $t \to -it$ in order to arrive at the metric $(-,-,-,-)$ or, equivalently, $(+,+,+,+)$. We have presented this idea in the path integral framework in Chapter 10. Formally, this has the effect of changing the exponent in the functional integral (24.2) from $-i\int \mathcal{L}_I(x)\mathrm{d}^4x$ where $x = (t,x_1,x_2,x_3)$ to $-\int \mathcal{L}_I(x)\mathrm{d}^4x$ where now $x = (x_0,x_1,x_2,x_3)$. It results from this complex rotation that in the integrals the exponential is now $\exp -\int \mathcal{L}_I(x)\mathrm{d}^4x$; thus, if the interaction Lagrangian has some positivity property, the function to integrate will not blow up when the fields (as integration variables) go to infinity and we can expect to control the integrals. It also reduces the singularity due to the metric: in Minkowski space $x.x \equiv t^2 - x_1^2 - x_2^2 - x_3^2 = 0$ means that x is on the light cone while in Euclidean space $x.x \equiv x_0^2 + x_1^2 + x_2^2 + x_3^2 = 0$ means that $x \equiv 0$, i.e. a point, and also simplifies the invariance properties since we go from Lorentz invariance $SL(2,\mathbb{C})$ to the Euclidean $SO(4)$ invariance.

In 1972, Konrad Osterwalder and Robert Schrader have proposed a set of axioms for Euclidean field theories.[1] These axioms are expressed as necessary and sufficient conditions for Euclidean Green functions to define a unique Wightman field theory. More precisely, they proved that under these conditions, the Euclidean Green functions have analytic continuation whose boundary values define a unique set of Wightman functions. These conditions are, except for one, very similar to the Wightman axioms. They set five axioms for Euclidean Green functions $\mathcal{E}_n(x_1, x_2, \cdots, x_n)$, $n = 0, 1, 2\ldots$, as follows.

Let $f_1, \cdots, f_n \in \mathcal{S}(\mathbb{R}^4)$. We define $f = (f_1, f_2, \cdots f_n) \in \mathcal{S}(\mathbb{R}^{4n})$

$$\mathcal{E}_n(f) = \int \mathcal{E}_n(x_1, x_2, \cdots, x_n) f_1(x_1) \cdots f_n(x_n) \mathrm{d}^4x_1 \cdots \mathrm{d}^4x_n. \tag{24.3}$$

We introduce the following transformations for all functions $f(x_1, x_2, \cdots, x_n) \in \mathcal{S}(\mathbb{R}^{4n})$:

$$
\begin{array}{llll}
f \to & f^* & \text{by} & f_n(x_1, x_2, \cdots, x_n) = \bar{f}(x_n, \cdots, x_1) & (24.4) \\
f \to & \Theta f & \text{by} & (\Theta f)_n(x_1, \cdots, x_n) = f_n(\theta x_1, \cdots, \theta x_n) & (24.5) \\
f \to & f^\pi & \text{by} & f^\pi(x_1, \cdots, x_n) = f(x_{\pi(1)}, \cdots, x_{\pi(n)}) & (24.6)
\end{array}
$$

with $\theta x = (-x_0, x_1, x_2, x_3)$ and $\pi(1, 2, \cdots n) = (\pi(1), \cdots, \pi(n))$.

The list of axioms is the following.

---

[1] K. Osterwalder and R. Schrader, *Comm. Math. Phys.* **31**, 83 (1973).

- E0 : *Distributions*

$$\mathcal{E}_0 \equiv 1 \qquad \mathcal{E}_n \in \mathcal{S}'(\mathbb{R}^{4n}), \qquad (24.7)$$

  i.e. the Euclidean Green functions are tempered distributions.

- E1 : *Euclidean invariance*
  For all $a \in \mathbb{R}^4$ and $R \in SO(4)$, with obvious notations, we have

$$\mathcal{E}_n(f) = \mathcal{E}_n(f_{(a,R)}), \qquad (24.8)$$

  where for a function $g : \mathbb{R}^4 \to \mathbb{C}$, $g_{(a,R)}(x) = g(a + Rx)$.

- E2 : *Positivity*

$$\sum_{n,m} \mathcal{E}_{n+m}(\Theta f_n^* \times f_m) \geq 0 \qquad \forall f_n \in \mathcal{S}(\mathbb{R}^{4n}), \qquad n = 1, 2, \cdots . \quad (24.9)$$

- E3 : *Symmetry*

$$\mathcal{E}_n(f) = \mathcal{E}_n(f^\pi) \qquad \text{for all permutations} \qquad \pi \qquad \text{of} \qquad (1, 2, \cdots, n). \tag{24.10}$$

- E4 : *Cluster property*

$$\lim_{\lambda \to \infty} \sum_{n,m} [\mathcal{E}_{n+m}(\Theta f_n^* \times g_{m(\lambda a,1)}) - \mathcal{E}_n(\Theta f_n^*)\mathcal{E}_m(g_m)] = 0, \qquad (24.11)$$

where a is a space-like vector.

Remark: The Euclidean space is symmetric under the Euclidean group of rotations. It is the axiom E2 on positivity which points a particular direction, through the function θ, which will be the time direction in the Minkowski space.

From these axioms we can prove the following theorem.

Theorem 18 (Equivalence between Euclidean and Relativistic Field Theories). *To a given sequence of Euclidean Green functions satisfying conditions E0, E1, E2, E3, and E4 there corresponds a unique sequence of Wightman functions $\mathcal{W}_n$, $n = 1, 2 \cdots$, satisfying the properties (a)–(e) of Theorem 12 and Theorem 13.*

The method for proving this theorem is to perform the analytic continuation corresponding to the Wick rotation, i.e. the complexification of time. Euclidean world and Minkowski world are related by analytic continuation. That this is possible is due to the fact that the Wightman functions are boundary values of analytic functions which are analytic in tubes. These tubes can be extended as we have seen, since from Lorentz covariance follows the invariance under the complex Lorentz group, to a much larger complex region: the extended tubes. We have seen that unlike the forward tubes, the extended tubes contain real points, the Jost points, which are space-like to each other; hence by locality, the Wightman functions are symmetric functions of their complex arguments. This in turn makes it possible to extend the domain of analyticity (more precisely the domain of *holomorphy*),[2] to a larger one, the so-called Bargmann–Hall–Wightman domain of holomorphy, which contains Euclidean points, i.e. points of the form $z = (it, x_1, x_2, x_3)$. This analytic continuation is something complicated since for an n-point Wightman function, it is n complex time variables which must be analytically continued. Usually the Euclidean functions resulting from the Wick rotation of Wightman functions are called Schwinger functions.

The main difference between the Euclidean world and the Minkowski one is that in the Euclidean world no direction of space plays a special role, in contrast to what happens in Minkowski space. As we pointed out earlier, the choice of this direction results from the axiom of positivity E2.

Most of the main results concerning properties and existence of quantum field theories were obtained in the Euclidean framework. All of the following sections of this chapter are written in this framework. By Osterwalder–Schrader axioms they can be understood as results obtained in Minkowski space.

24.3   Construction of Field Theories

In the following sections of this chapter we will show that, thanks to the Osterwalder–Schrader axioms, we can prove the global existence and some properties of a relativistic field theory by proving the existence of the corresponding global Euclidean field theory. This is immediate for a theory quadratic in the fields since, by Minlos' theorem, we know how to prove the existence of infinite-dimensional Gaussian integrals. Therefore, the goal is to find systematically under which conditions the existence of a global measure or, equivalently, the existence of its moments can be proven for the case where the interaction is a polynomial of order greater than 2. Usually, the difficulty of the proof increases with the dimensions of the underlying space over which the theory is defined, but other parameters, such as the nature of fields, fermionic or bosonic, massive or massless, scalars spinors or vectors, simple polynomial or gauge interactions, can affect the difficulties considerably.

We will start with simplified models and then go gradually to models with more physical significance. In this section we will only present the problems and the results, with no proofs. A sample of these proofs will be given in the following sections.

Some properties characterising quantum field theories (QFT) can be easily understood by using simplified models. For example, the divergence of the perturbation series

---

[2] In contrast to what happens in one dimension, in higher dimensions a domain of holomorphy is a maximal domain in the sense that there exists an analytic (or holomorphic function) which cannot be extended to a larger domain. In our case, the extension is possible because of the symmetry property of the Wightman functions for space-like points and the Lorentz invariance.

and the relation between its properties and those of the underlying global expression, the so-called Borel summability, appear in low dimensions of space–time and are independent of the Lorentz transformation properties of the fields. Already in zero dimensions we see easily that a polynomial interaction of order $2k$, $k \geq 2$, when expanded in a power series, generates numerical coefficients growing up as fast as a factorial to the power k. The resulting series is divergent, no matter how small the coefficient in front of this polynomial is (i.e. of the smallness of the coupling constant). The same problem appears in higher space–time dimensions: the number of graphs at order n in the perturbation expansion increases as a power of $n!$. This is the fundamental difficulty in studying the convergence properties of any perturbation series. Each term of the series can be small, provided the coupling constant is small, but the number of such small terms grows too fast. At any given order n of perturbation the estimates on both the values of the graphs produced and the coupling constants grow geometrically with n, while the number of graphs grows like a power of $n!$. So even if the coupling constants are very small, they cannot control this type of divergence. We will be able to overcome this difficulty in some specific cases by introducing new types of expansions in which, at every order, we produce enough 'small' factors able to *dominate* the factorial increase in the number of graphs. For the rest of this chapter *domination* will be the name of the game. This process will grow more and more sophisticated as we study models of increasing complexity. The technical level of the detailed proofs becomes soon very involved so, in most cases, only the main ideas and the results will be presented. For simplicity, we will start with a scalar field interacting through a ϕ^4 term, because this choice is the simplest one which does not correspond to a free-field theory and offers the guarantee that the energy is bounded from below. We will consider this model embedded in a Euclidean space of d dimensions[3] and we will denote it by ϕ_d^4. At the end we will study a model with fermions and one with gauge bosons. We give here the list of the models and a short discussion for each of them. This list does not exhaust the models which have been studied, but it helps introducing all the essential techniques and results.

- *The zero-dimensional model ϕ_0^4 (see section 24.4).* In some sense it is a trivial case because the corresponding QFT is described by usual functions, the field ϕ being now a single scalar variable x. The functional integral reduces to a single one-dimensional integral, from $-\infty$ to $+\infty$, of the exponential of $-(\frac{1}{2}x^2 + \lambda x^4)$. It is a nicely convergent integral, as long as Re $\lambda > 0$, but it diverges for all Re $\lambda < 0$. This shows that an expansion in powers of λ cannot be absolutely convergent and the series will turn out to be only Borel summable. We will establish this result in the next section.

- *The two-dimensional model ϕ_2^4 (see section 24.6).* We then prove the same results in dimension 2. The theory will be scalar and massive and the interaction quartic. We will start by defining it in a finite volume. As for the zero-dimensional case, the

---

[3] We recall that a d-dimensional Euclidean space corresponds to a Minkowski space with 1 time and $d-1$ space dimensions.

quadratic part will make it possible to define the Gaussian measure, which will be the reference measure, and the quartic part will be the interaction. The perturbation expansion is obtained by expanding the interaction part. But then, due to the increase in dimension, we will encounter the first new difficulty: there are infinities which appear in the graphs of the perturbative expansion. Nevertheless, this theory is superrenormalisable, i.e. there is a finite number of divergent graphs, the ones we called 'tadpoles', resulting from the contraction of two fields at the same point. This is the first appearance of a renormalisation problem. As we saw in Chapter 10, this particular problem can be algebraically solved by the Wick ordering, but to do it properly we need to introduce an ultraviolet cut-off. In turn this cut-off breaks the boundedness from below of the energy; as a result, the exponential of the interaction part of the Lagrangian is blowing up like a power of the logarithm of the cut-off. We will call this estimate a 'Wick bound'. Thus, in order to remove the ultraviolet cut-off, we must do something. It will be an 'improved' expansion. Roughly speaking, which will be made more precise in section 24.6, we will have to expand enough to have small constants, but not too much to avoid the number of terms produced to explode. All these arguments are related to a given scale which in this case is made of unit squares. This is well defined since the action is local and the exponential of the action in a given volume can be written as the product of exponentials, one per square, each of them being the exponential of the action restricted to this square. The idea is, starting from the theory with a given ultraviolet cut-off, to try to lower this cut-off to another value (in a suitable decreasing sequence of cut-offs). This can be done by creating enough perturbative terms which are small (at least because the coupling constants are small), but not too many because the number of these small terms is growing like factorials. As a side result, we check at the order at which we stop the expansion that the overall small coefficient produced is small enough not only to still control the number of terms but also to dominate the Wick bound. We can bound the sequence of terms produced, and the last term is a bounded path integral corresponding to a finite ultraviolet cut-off, both estimates being independent of the initial ultraviolet cut-off. This expansion makes it possible to bound any expectation value by an exponential of the volume, independently of the initial ultraviolet cut-off. As a last step, the existence of the thermodynamic, or infinite volume, limit must be proven. This is done by mimicking statistical mechanics via two expansions: a cluster expansion, based on the sufficiently rapid decrease of the correlation between two distant clusters, and a Mayer expansion, making it possible to perform the division of two cluster expansions, one for the expectation value of some quantity (like a product of fields) and the other for the normalisation factor or, in the statistical mechanics language, the partition function. The basic ideas justifying this result are the following.

1. The theory is massive, from which follows that for not too large coupling constant, the interaction is of short range.

2. The final measure is well approached by a Gaussian measure built from the quadratic part of the Lagrangian.

3. The lowering of the ultraviolet cut-off is made by an expansion producing small momentum factors (depending on the cut-off). The coupling constant must be small to ensure a small parameter when there are no small momentum dependent factors.

- *The three-dimensional model ϕ_3^4 (see section 24.7).* At first sight this sounds like a straightforward generalisation of ϕ_2^4. Both models are superrenormalisable, so we expect the same techniques to apply. However, there are important differences which create two new, essential complications.

  1. The theory is still superrenormalisable, i.e. there is a finite number of perturbative graphs which are infinite (tadpole, vacuum, and mass counter-terms). Tadpoles can still be eliminated by Wick ordering, but the other ones need to be renormalised, and this cannot be done algebraically.

  2. We cannot dominate the Wick bound which is proportional to the square of the cut-off for a quartic interaction.

  Because of this last condition, we must refine the space localisation and introduce momentum slices. Suppose we have a sequence $M, M^2, \ldots, M^j \cdots$, with $M > 1$. We can consider that the ultraviolet cut-off of the theory is M^ρ, for some large integer ρ. Then we will introduce coverings indexed by i with cubes of size M^{-3i} and the expansion for the removal of the ultraviolet cut-off will be related to the size of the cubes in a given momentum slice. Having to do an estimate on cubes of size smaller than 1 gives a supplementary small factor and if we have optimised suitably the number of terms produced, this will be enough to control the Wick bound and to give a small factor allowing the resummation of all the terms generated by the expansion. This is what is called *a phase space cell expansion*. To summarise, the principle of the phase space cell expansion is to apply a truncated perturbation expansion on the exponential in smaller and smaller cubes in such a way that at the end of the expansion a field localised in Δ and of high momentum M remains in the exponent if $M^2|\Delta| \leq O(1)$ (this will give in Δ a finite Wick bound).

Since there are some divergent graphs, the starting expression for the theory with a cut-off must be done in such a way that when expanding in powers of the coupling constant, i.e. producing perturbative graphs, each term must have a limit which is the expected renormalised graph. Hence, we must introduce the corresponding counter-terms, with their right coefficients. These counter-terms are finite since we have a cut-off and they have the right coefficients to exactly compensate the part going to infinity when the cut-off is removed. It will turn out that the introduction of these counter-terms in the exponential changes the Wick bound, but not significantly.

As was said previously, there are small factors due to the momentum expansion, but a given vertex localised in the cube whose size is related to the momentum expansion has more than one field and each field has a range of momentum values running from the highest momentum M_h to the lowest one M_l. Given a cube Δ of a given size, a field in this cube such that $M_l^3|\Delta| < 1$ does not produce a small factor when contracted (or integrated with the Gaussian measure). This generates a new difficulty: the domination of badly localised fields which must be estimated in another way. Nevertheless, we can prove, even in this case, the same results as those we proved for the two-dimensional case.

- *The Gross–Neveu model (see section 24.8)*. We saw in the comparison between the ϕ_2^4 and the ϕ_3^4 models that the increase in the number of primitively divergent diagrams created non-trivial complications. So it is not surprising that these methods will break down if we consider a renormalisable theory, like ϕ_4^4. Indeed, the infinite number of primitively divergent diagrams prevents us from finding enough small factors to dominate the Wick bound. This seems to be the case for all renormalisable theories, with some notable exceptions, to wit *asymptotically free theories*. As we can show, in such theories the decrease of the strength of the coupling at high momenta will provide the necessary convergence factors.

  The first model of this kind we will study is the massive version of the two-dimensional Gross–Neveu model we analysed in Problem 23.2. This model is very interesting in the sense that it has a convergent perturbation expansion. It is a little more different technically than what we studied previously since the fields are fermions (which make the perturbative expansion to converge). Nevertheless, in the discussion of this theory, we encounter the same type of difficulties as those we found in the ϕ_3^4 field theory: we need to do a phase space cell expansion, but there is a drastic difference since now there are infinitely many counter-terms. But here asymptotic freedom comes to the rescue. It tells us that the parameters of the theory, the coupling constant, the mass, and the wave function, obey the renormalisation group flow equations. This means that the coupling constant is decreasing with the increase of the ultraviolet momentum cut-off and that the leading terms can be computed from 1-loop perturbative graphs with, if necessary, 2-loop corrections. Hence at a given momentum we must put as counterterms in the exponential only the most divergent ones (or their most divergent parts). We then must perform a phase space cell expansion. Since there are divergences at any order of perturbation theory, we must modify the momentum expansion part of the phase space cell expansion: some of the divergent graphs appearing in these expansions will not be compensated if there are enough convergent factors already produced which can be associated with them. Furthermore, there are also low momentum fields which must be treated differently than by Gaussian integration. Despite these new difficulties, we are again able to prove the existence of the theory with all the desired properties.

- *The four-dimensional Yang–Mills theory (see section 24.9)*. This is the last model we will attempt to present. As we have seen already, it is renormalisable and asymptotically free. We will see in the next chapter that this kind of theory, based on the group $SU(3)$, describes the fundamental strong interactions. So this theory has a capital physical importance; unfortunately, it is much more complicated than all those we have studied so far. At the end, the same tools will be applied, but after a tedious preparation of the theory and for results much less general. This type of theory is more singular for at least three reasons.

1. The Lagrangian is positive, but if we use the quadratic part to define the Gaussian measure, the remaining terms are no more positive. Therefore, there is a lack of positivity, an ingredient which is essential for the proof of the existence of the global measure.

2. The theory is gauge invariant. We must fix the gauge to have a well-defined integral. As we showed in Chapter 14, this gauge fixing introduces other complications such as the Faddeev–Popov ghosts (or equivalently a field-dependent determinant).

3. The theory is massless. We must introduce an infrared cut-off. To remove the infrared cut-off means to solve the confinement problem and nobody has obtained any rigorous results in this direction. At present, all available results for a Yang–Mills theory are restricted to a finite volume.

The proof we could present is the existence of the theory without ultraviolet cut-off in a finite volume. Even that turns out to be quite complicated. First, we must define the precise model we start with. We choose, because it preserves positivity, to fix the initial expression to be in the axial gauge. The price to be paid is a breaking of explicit Euclidean invariance. Computations of Feynman diagrams in this gauge are awkward, so, in order to perform such computations, we must change the gauge and pass to a gauge like the Feynman one.

The most important choice will be that of the momentum cut-off. The momentum cut-off we will chose is not gauge invariant but Euclidean invariant and, therefore, it generates Euclidean invariant counter-terms such as A^2 and A^4, where A is the Euclidean gauge field. In fact, we will take a class of cut-offs which has the property that the coefficient of A^4 is positive (the A^2 term has a negative coefficient but this can be controlled). As a consequence, the action is positive definite even if the coefficient of the highest degree term is vanishing when the cut-off is removed. This positivity, which is of the form $g^4 A^4$, is enough to ensure the domination of all the terms which need to be dominated.

We have a final remark before leaving this section. All the results we present deal with the rigorous definition of the functional integral measure starting from the Gaussian measure. They can be viewed as results on the summability properties of the perturbation expansion. As such they say nothing about the possible existence of phases away from the weak coupling regime. Our experience, theoretical as well as numerical, from models in statistical mechanics tells us that a system often has a phase diagram with regions not accessible to a weak coupling expansion. If such phases exist for the models we have studied, our approach will miss them. This could be the case for sufficiently strong coupling in the low-dimensional models, but it is certainly true for the gauge theories. We will see in the next chapter that both phenomenological and numerical evidence show that, at zero temperature, the system is in a confining phase (i.e. a phase in which the only physical states are bound states not visible in the perturbation expansion), for all values of the coupling. There is experimental evidence based on heavy ion collisions that a phase transition to a 'deconfining phase' (namely a phase in which the physical states are those corresponding to free gauge bosons) may appear, but only at sufficiently high temperature.

In the rest of this chapter we present these results, including, in some cases, the main steps of the proofs, only for the ϕ_0^4 and ϕ_2^4 models. Although we will not show all the details, the next sections, especially the last one, will be quite technical. The reader who is not interested in these mathematical aspects can skip them. They will not be needed in the following chapters. They have been included for completion, but also, because, as it has been said, the models we study introduce progressively some very powerful specific techniques. These techniques are always based on physical intuition or on subtle analysis of the relationships between different technical objects or ideas. Some have already been introduced, but some others are new. They all form a powerful arsenal which has been proven essential in our struggle to understand quantum field theory beyond the perturbation expansion. However, the most important physical problems, the rigorous definition of four-dimensional gauge theories with broken, or unbroken, symmetry, are still in front of us and the road to their solution is not clear.

24.4   The Zero-Dimensional $\lambda\phi^4$ Model

Let us consider the zero-dimensional quantum field theory given by the Lagrangian $\mathcal{L}(x) = \mathcal{L}_0(x) + \mathcal{L}_I(x) = \frac{1}{2}ax^2 + \lambda x^4$, $a > 0$, where $\mathcal{L}_0 = \frac{1}{2}ax^2$. It will be the toy model corresponding to the one described by (10.65). The associated free measure, built up to a normalisation from the free Lagrangian part, is the Gaussian measure $\mathrm{d}\mu_{a^{-1}}(x)$ given by

$$\mathrm{d}\mu_{a^{-1}}(x) = \sqrt{\frac{a}{2\pi}}\,\mathrm{e}^{-\frac{1}{2}ax^2}\,\mathrm{d}x,$$

satisfying

$$\int_{-\infty}^{+\infty}\mathrm{d}\mu_{a^{-1}}(x) = 1 \tag{24.12}$$

and the integration by parts formula is given by

$$\int F(x)x\,\mathrm{d}\mu_{a^{-1}}(x) = a\int \frac{\mathrm{d}F(x)}{\mathrm{d}x}\,\mathrm{d}\mu_{a^{-1}}(x). \tag{24.13}$$

The generating functional of expectation values $Z[j]$ is

$$Z[j] = \frac{1}{Z}\int_{-\infty}^{+\infty}\mathrm{e}^{jx}\mathrm{e}^{-\mathcal{L}_I(x)}\,\mathrm{d}\mu_{a^{-1}}(x) = \frac{1}{Z}\int_{-\infty}^{+\infty}\mathrm{e}^{jx}\mathrm{e}^{-\lambda x^4}\,\mathrm{d}\mu_{a^{-1}}(x), \tag{24.14}$$

where the partition function is given by

$$Z[\lambda] = \int_{-\infty}^{+\infty} e^{-\lambda x^4} d\mu_{a^{-1}}(x). \tag{24.15}$$

This insures that the global or interaction measure $d\mu$ defining the theory is given by

$$d\mu = \frac{e^{-\lambda x^4} d\mu_{a^{-1}(x)}}{Z[\lambda]}. \tag{24.16}$$

.

We will show that already from such a simple model we can explore the relationship between a global theory and its perturbation expansion.

24.4.1   The Divergence of the Perturbation Series

As we have noted already, a Taylor expansion in powers of λ cannot be convergent because the integral clearly diverges for $\text{Re}\lambda < 0$, but it will be instructive to see it explicitly.

In the sequel we will encounter a lot of numerical constants. From now on, unless explicitly stated, we will use equivalently for them the notation K or $O(1)$.

Lemma 1. *For the moments of order m of the interaction measure given by $< x^m > [\lambda] = \frac{1}{Z[\lambda]} \int_{-\infty}^{+\infty} x^m e^{-\lambda x^4} d\mu_{a^{-1}}(x)$, $m = 0, 1, \cdots$, the power series in λ resulting from the expansion of $e^{-\lambda x^4}$ are not summable.*

Proof.   Let us consider the numerator

$$\int_{-\infty}^{+\infty} x^m e^{-\lambda x^4} d\mu_{a^{-1}}(x) = S_m[\lambda]. \tag{24.17}$$

After expanding the exponential in powers of λ, we get

$$S_m[\lambda] = \sum_{n=0}^{+\infty} (-1)^n \frac{\lambda^n}{n!} \int_{-\infty}^{+\infty} x^{4n+m} d\mu_{a^{-1}}(x). \tag{24.18}$$

The integral term can be easily calculated and gives

$$\int_{-\infty}^{+\infty} x^{4n+m} d\mu_{a^{-1}}(x) = a^{2n+\frac{m}{2}} \frac{(4n+m)! 2^{-(2n+\frac{m}{2})}}{(2n+\frac{m}{2})!} = a_{4n+m} \tag{24.19}$$

if m is even and $a_{4n+m} = 0$ if m is odd.

Since we are interested in the convergence of (24.18), we look at the behaviour of a given term for large values of n, large with respect to m, m being fixed. Using the Stirling formula, $n! \simeq (\frac{n}{e})^n \sqrt{2\pi n}$, we get that

$$a_{4n+m} \simeq \sqrt{2}a^{\frac{m}{2}}(\frac{e}{2})^{\frac{m}{2}}a^{2n}(\frac{n}{e})^{2n}2^{4n} \simeq C^n a^{2n}(2n)! \tag{24.20}$$

for some constant C. Thus, the generic term in the alternating series $S_m[\lambda]$ behaves as

$$\frac{\lambda^n}{n!}a_{4n+m} \simeq \frac{1}{\sqrt{\pi n}}(\frac{ae}{2})^{\frac{m}{2}}(16\lambda a^2)^n(\frac{n}{e})^n$$
$$= C(m)(a^2 C_2 \lambda)^n(n-1)! \tag{24.21}$$

for some other constant C, which means, by application of the Leibniz test, that the series is not convergent.

A similar result can be proved for $Z[\lambda]$, which corresponds to $S_m[\lambda]$ for $m = 0$. It is easy to see that the ratio of the two power series in λ cannot, as a formal power series in λ, be convergent.

24.4.2   The Borel Summability

If we have a function of $z \in \mathbb{C}$, $A(z)$ which has an asymptotic expansion $A(z) = \sum a_k z^k$ as $z \to 0$, then we wish to determine conditions under which this function $A(z)$ can be uniquely reconstructed from its asymptotic expansion. One method of reconstruction is given by the Borel summation.

We say that the formal power series $\sum a_k z^k$ is Borel summable if

- Its Borel transform

$$\mathcal{B}A(z) = \sum \frac{a_k}{k!}z^k \tag{24.22}$$

converges in some circle $|z| < R$.

- $\mathcal{B}A(z)$ has an analytic continuation to a neighbourhood of the positive real axis.

$$B(z) = \frac{1}{z}\int_0^{+\infty} e^{-t/z}\mathcal{B}A(t)\mathrm{d}t \tag{24.23}$$

converges for some $z \neq 0$.

Remark: If the integral (24.23) converges for some $z_0 \neq 0$, then (because it is a Laplace transform) it converges for all z with $\Re z^{-1} > \Re z_0^{-1}$. This domain is a circle tangent to the imaginary axis.

$\mathcal{B}A(z)$ is the Borel transform of the series $\sum a_k z^k$ and $B(z)$ is its Borel sum.

A theorem of G. N. Watson gives a sufficient condition for the function $A(z)$ to be equal to the Borel sum of its asymptotic series. It was improved by A. Sokal in a way which was particularly useful for quantum field theories, i.e. by restricting the domain of analyticity of $A(z)^4$.

Let us define, for some positive number R, the disc $C_R = \{z \in \mathbb{C} \mid \Re z^{-1} > R^{-1}\}$.

Theorem 19 (Sokal). *Let $A(z)$ be analytic in the disc C_R and has in this domain an asymptotic expansion*

$$A(z) = \sum_{k=0}^{N-1} \frac{a_k}{k!} z^k + R_N(z) \quad with \quad | R_N(z) | \leq AC^N |z|^N N! \tag{24.24}$$

uniformly in N for some positive constants A, C and for all $z \in C_R$.

Then $\mathcal{B}A(z) = \sum \frac{a_k}{k!} z^k$ converges for $|z| < C^{-1}$ and has an analytic continuation to the striplike region $S_C = \{z \in \mathbb{C} \mid dist(z, \mathbb{R}) < C^{-1}\}$ where it satisfies the bound

$$| \mathcal{B}A(z) | < K e^{|z|/R} \tag{24.25}$$

uniformly in every $S_{C'}$ with $C' > C$. Furthermore, $A(z)$ can be represented by the absolutely convergent integral

$$A(z) = \frac{1}{z} \int_0^{+\infty} e^{-t/z} \mathcal{B}A(t) dt \quad for\ any \quad z \in C_R. \tag{24.26}$$

There is a converse part in this theorem that we have not stated.

It is now easy to establish for $\lambda \in \mathbb{C}$ using $\left| e^{-\lambda x^4} \right| < e^{-\Re \lambda x^4}$ that

$$\left| \int_{-\infty}^{+\infty} x^m e^{-\lambda x^4} d\mu_{a^{-1}}(x) \right| < \int_{-\infty}^{+\infty} x^m e^{-\Re \lambda x^4} d\mu_{a^{-1}}(x) \tag{24.27}$$

and that the functions $Z[\lambda]$ and $Z[\lambda] S_m[\lambda]$ are analytic for $\lambda \in \mathbb{C}$ in some disc C_R of the complex plane.

Furthermore, with the estimates of 24.4.1, starting from

$$e^{-\lambda x^4} = 1 - \lambda x^4 + \frac{1}{2!}(-\lambda x^4)^2 + \cdots + \frac{1}{N!}(-\lambda x^4)^N + R_N[\lambda](x) \tag{24.28}$$

with

$$R_N[\lambda](x) = \frac{1}{N!} \int_0^\lambda \frac{d^{N+1} e^{-x^4 t}}{dt^{N+1}} (\lambda - t)^N dt = (-x)^{4(N+1)} \frac{1}{N!} \int_0^\lambda e^{-x^4 t} (\lambda - t)^N dt$$

$$= (-\lambda x^4)^{N+1} \frac{1}{N!} \int_0^1 e^{-\lambda u x^4} (1 - u)^N du \tag{24.29}$$

---

[4] A. Sokal, *J. Math. Phys.* **21**, 261 (1980).

we get that

$$\left|R_N[\lambda](x)\right| \leq |\lambda|^{N+1} x^{4(N+1)} \frac{1}{N!}. \tag{24.30}$$

We can now introduce these results in $S_m[\lambda]$ as given by (24.17)

$$S_m[\lambda] = \int_{-\infty}^{+\infty} x^m e^{-\lambda x^4} d\mu_{a^{-1}}(x) = \sum_{n=0}^{N}(-1)^n \frac{\lambda^n}{n!} \int_{-\infty}^{+\infty} x^{4n+m} d\mu_{a^{-1}}(x)$$
$$+ \int_{-\infty}^{+\infty} x^m R_N[\lambda](x) d\mu_{a^{-1}}(x) = \sum_{n=0}^{N}(-1)^n \frac{\lambda^n}{n!} \int_{-\infty}^{+\infty} x^{4n+m} d\mu_{a^{-1}}(x) + R_N(\lambda)$$

and the remainder $R_N(\lambda)$ is bounded by

$$\left|R_N(\lambda)\right| = |\lambda|^{N+1} \frac{1}{N!} \int_{-\infty}^{+\infty} x^{4(N+1)+m} d\mu_{a^{-1}}(x) \leq A(C|\lambda|)^N N!, \tag{24.31}$$

thus showing that this zero-dimensional QFT is Borel summable. Moreover, we have defined the 'interacting' measure for the x^4 interaction.

24.5   General Facts about Scalar Field Theories in *d* = 2 or *d* = 3 Dimensions

We will mainly use the formalism introduced in chapter 10. The field theory we want to construct is the one given by the Lagrangian (10.65) except that it is the theory of a scalar field with a quartic self-interaction in a space of arbitrary dimensions d, $d \in \mathbb{Z}_+$, i.e. $x \in \mathbb{R}^d$.

The Lagrangian is made of two parts $\mathcal{L}(x) = \mathcal{L}_0(x) + \mathcal{L}_I(x)$, the free Lagrangian corresponding to the quadratic part and the interacting Lagrangian $\mathcal{L}_I$ to the interacting part. The corresponding action is $S = \int_{\mathbb{R}^d} \mathcal{L}(x) d^d x = S_0 + S_I$ and the formal generating functional is given by

$$Z[j] = \int e^{\int \phi(x)j(x)d^d x} e^{-S} \prod_{x \in \mathbb{R}^d} \mathcal{D}\phi(x), \tag{24.32}$$

which can be written using the free part of the action to define, up to a normalisation factor, the Gaussian measure $d\mu_C$ of covariance

$$C(x-y) = (-\Delta + m^2)^{-1}(x-y) = \frac{1}{(2\pi)^d} \int e^{ip.(x-y)} \frac{1}{p^2 + m^2} d^d p, \tag{24.33}$$

$$Z[j] = \int e^{\int \phi(x)j(x)d^d x} e^{-S_{\text{int}}} d\mu_C = \int e^{\int \phi(x)j(x)d^d x} e^{-\lambda \int \phi(x)^4 d^d x} d\mu_C. \tag{24.34}$$

We get then the vacuum expectation values (v.e.v.)

$$< \phi(x_1)\ldots\phi(x_n) >= \frac{1}{Z}\int \phi(x_1)\ldots\phi(x_n)e^{-\lambda\int\phi(x)^4 d^d x}d\mu_C \qquad (24.35)$$

with the partition function Z given by

$$Z = Z[j]\mid_{j=0} = \int e^{-\lambda\int\phi(x)^4 d^d x}d\mu_C$$

$$Z < \phi(x_1)\ldots\phi(x_n) > = \int \phi(x_1)\ldots\phi(x_n)e^{-\lambda\int\phi(x)^4 d^d x}d\mu_C. \qquad (24.36)$$

In fact

$$d\mu = \frac{1}{Z}e^{-\lambda\int\phi(x)^4 d^d x}d\mu_C$$

defines a normalised measure; the measure of the interacting theory and the vacuum expectation values are nothing else than the moments of this measure.

We note that for coinciding arguments the covariance

$$C(x-x) = C(0) = \frac{1}{(2\pi)^d}\int_{\mathbb{R}^d}\frac{1}{p^2+m^2}d^d p \qquad (24.37)$$

is infinite for $d \geq 2$.

For this reason and also for other ones which will become clear later on, we will introduce momentum cut-offs.

For the rest of this section, given an increasing sequence of positive numbers $\{M_i \mid i \in \mathbb{Z}\}$ with $M_i = 0$ for $i = 0$, $M_1 > 1$ and $M_\infty = +\infty$, we will use the fact that

$$\frac{1}{p^2+m^2} = \int_0^\infty d\alpha e^{-\alpha(p^2+m^2)} = \sum_{i=0}^\infty \int_{M_{i-1}^{-2}}^{M_i^{-2}} d\alpha e^{-\alpha(p^2+m^2)} \qquad (24.38)$$

and define the cut-off covariances (24.33) for some positive integer ρ

$$C_\rho = \sum_{i=0}^{\rho-1} C^i \quad \text{with} \quad C^i = \frac{1}{(2\pi)^2}\int e^{ip(x-y)}\int_{M_{i-1}^{-2}}^{M_i^{-2}}\cdot d\alpha e^{-\alpha(p^2+m^2)}d^2 p. \qquad (24.39)$$

We can prove the following in d dimensions.

Lemma 2. $\exists K > 0$ *such that*

$$C^i(x-y) \leq K(M_{i+1})^{d-2}e^{-M_i|x-y|} \qquad i = 1, 2, \cdots \qquad (24.40)$$

- *for d = 2, the coefficient $(M_{i+1})^{d-2}$ is replaced by $\ln(M_{i+1}/M_i)$*
- *for i = 0, the exponential decrease is given by $e^{-m|x-y|}$.*

According to Minlos' theorem, it follows from (24.39), that the Gaussian measure is now a product of measures

$$d\mu_{C_\rho} = \prod_i d\mu_{C^i}, \tag{24.41}$$

which means that the field ϕ is a discrete sum

$$\phi = \sum_{i=0}^{\rho-1} \phi^i \tag{24.42}$$

of independent Gaussian variables ϕ^i of covariance C^i.

The usual tool to compute expectation values is the integration by parts formula (24.13)

$$\int \phi(x)F(\phi)d\mu_C = \int_{R^d} d^d y C(x-y) \int \frac{\delta}{\delta\phi(y)} F(\phi)d\mu_C. \tag{24.43}$$

Since we want to compute the v.e.v. of product of fields and since the interaction is a polynomial in the field, the Gaussian integration generates a sum of products of covariances with all the internal arguments (the vertices) being integrated. We have seen in section 10.3 that the result can be expressed as diagrams, the so-called Euclidean Feynman diagrams, with precise rules to compute their associated values. These rules were given for fields over $\mathbb{R}^4$ and we have seen that for the ϕ^4 interaction in four dimensions some diagrams are divergent. To study the perturbative expansion, we need to know which diagrams are divergent. In 16.3, a formula is given expressing the superficial degree of divergence of a 1-PI diagram G for a ϕ^m theory in four dimensions. From section 10.3 and the previous formula, we can easily deduce that, in d dimensions, a diagram G has a superficial degree of divergence $D(G)$ given by

$$D(G) = VD - E(d/2 - 1) + d \tag{24.44}$$

with E the number of external lines, $2n$ is the power of the monomial of ϕ defining the interaction, V the number of vertices, and $D = 2n(d/2 - 1) - d$ the power counting of a vertex. When $D(G) < 0$ for all diagrams except a finite number, the theory is super-renormalisable. It is the case for $n = 2$ and $n = 1$ in three dimensions and the trivial case $n = 1$ in four dimensions. If $D(G)$ is independent of V, the theory is just renormalisable. This is the case for $n = 3$ in three dimensions and $n = 2$ in four dimensions.

Finally, we will also cut off the space of integration of the Lagrangian density. This will be necessary in order to control the expressions occurring in the computation of the v.e.v.. The space cut-off will be denoted here by Λ, a volume in $\mathbb{R}^d$, and the interaction part of the action is

$$\int_\Lambda \mathcal{L}_{\text{int}}(x)\mathrm{d}^d x. \tag{24.45}$$

All v.e.v. will now be indexed by the two cut-offs ρ and Λ, ρ meaning that the free measure $\mathrm{d}\mu_C$ is replaced by $\mathrm{d}\mu_{C_\rho}$, and Λ meaning that the space integration will be restricted to Λ.

We are now ready to apply this formalism to the $\lambda\phi^4$ theory in two and three dimensions and transpose to these cases the results of the previous section.

24.6   The $\lambda\phi^4$ Theory in $d = 2$ Dimensions

In this section we want to illustrate some of the methods which were described only in qualitative terms in the previous sections, in the particular example of the two-dimensional, massive, $\lambda\phi^4$ field theory. We will assume that the coupling constant λ is small enough[5] and we start with the study of this theory when the underlying space has two dimensions. In fact, the results can be extended to the case where the interaction is a polynomial of even degree $P(\phi)$ with small coefficients.

24.6.1   The Divergence of the $\lambda\phi_2^4$ Perturbation Series

We have done the power counting of the theory in Problem 16.7. We start by doing the expansion with respect to λ in the normalisation term Z. We suppose λ to be small

$$
\begin{aligned}
Z(\rho, \Lambda) &= \int \mathrm{d}\mu_{C_\rho} - \lambda \int \left(\int \phi(x)^4 \mathrm{d}^2 x \right) \mathrm{d}\mu_{C_\rho} + O(\lambda^2) \\
&= 1 - \lambda \int C_\rho(0)^2 \mathrm{d}^2 x + O(\lambda^2)
\end{aligned}
\tag{24.46}
$$

but

$$C_\rho(0) = \frac{1}{(2\pi)^2} \int_0^{M_\rho^2} \frac{1}{p^2 + m^2} \mathrm{d}^2 p \simeq C \ln M_\rho. \tag{24.47}$$

Taking the limit of $\rho \to \infty$, we find the divergent tadpole. Obviously, the presence of a divergent diagram is not acceptable. If we want to prove the existence of any theory, we need to have at least all terms of its perturbative expansion to be finite. This is in dimension 2 the first trace of the need to renormalise the theory. We already have seen this necessity in Chapter 10 and in this chapter we got the answer namely, we had to replace $\phi(x)^4$ with its Wick ordered form $: \phi(x)^4 :$, where formally (see 10.61)

$$: \phi(x)^4 := \phi(x)^4 - 6C(0)\phi(x)^2 + 3C(0)^2 \tag{24.48}$$

---

[5] See, however, a remark at the end of this section.

and with the cut-off

$$:\phi(x)^4 := \phi(x)^4 - 6C_\rho(0)\phi(x)^2 + 3C_\rho(0)^2 \tag{24.49}$$

with the property that

$$\int :\phi(x)^4 : \mathrm{d}\mu_{C_\rho} \equiv 0.$$

According to the formulae related to the power counting, $D = -2$ and $D(G) = 2(1 - V)$. Thus, the only primitively divergent diagrams in $\lambda\phi_2^4$ have one vertex, i.e. the tadpoles. They correspond to $D(G) = 0$. The diagrams are divergent but as a logarithm rather than a power. So removing all the tadpoles with the Wick ordering ensures the existence of a well-defined perturbation expansion.

For our studies in the previous chapters, this was the end of the game. We proved that every term in the perturbation expansion is well defined. Here, however, we want to go further: we want to prove that there exists a way to sum the series and define a measure for the interacting theory. The starting expressions to define this interacting measure corresponding to a $\lambda\phi_2^4$ theory are

$$\mathrm{d}\mu_{\rho,\Lambda} = \frac{1}{Z(\rho,\Lambda)} e^{-\lambda \int_\Lambda :\phi(x)^4 : \mathrm{d}^2 x} \mathrm{d}\mu_{C_\rho}. \tag{24.50}$$

To prove the existence of this measure for $\rho \to \infty$, we will have to estimate its moments, i.e. the various v.e.v.'s, and prove that they have a well-defined limit when we remove the cut-off. Then we must study their properties in order to prove, by the reconstruction and Osterwalder–Schrader theorems, that this measure corresponds to a two-dimensional quantum field theory with quartic interaction.

Let us take the example of the partition function

$$Z(\rho,\Lambda) = \int e^{-\lambda \int_\Lambda :\phi(x)^4 : \mathrm{d}^2 x} \mathrm{d}\mu_{C_\rho}. \tag{24.51}$$

The expansion in power of λ is given by

$$Z(\rho,\Lambda) = \sum_{n=0}^{\infty} \frac{(-\lambda)^n}{n!} \int \left(\int_\Lambda :\phi(x)^4 : \mathrm{d}^2 x \right)^n \mathrm{d}\mu_{C_\rho}. \tag{24.52}$$

We will now give an estimate of the generic term

$$\int \left(\int_\Lambda :\phi(x)^4 : \mathrm{d}^2 x \right)^n \mathrm{d}\mu_{C_\rho}. \tag{24.53}$$

We will give some arguments showing that this term behaves as the same one in zero dimensions. Roughly speaking we must estimate the Gaussian integration of $4n$ fields, more precisely the number of vacuum diagrams we can draw with n 4-vertices. For that we can use an $L_1 - L_\infty$ estimate. Let us explain how this method[6] can be used to get estimates on sums.

Suppose we must estimate

$$\sum_{i \in A} a_i, \tag{24.54}$$

where A is a set of denumerable indices and a_i a function on this set. If we can find positive numbers c_j such that

$$\sum_{j \in A} c_j^{-1} \leq 1 \tag{24.55}$$

then we can replace the sum over a set of indices by taking the supremum of a suitable expression:

$$\sum_{i \in A} a_i = \sum_{i \in A} c_i c_i^{-1} a_i \leq \sum_{i \in A} c_i^{-1} \sup c_i |a_i| \leq \sup c_i |a_i|. \tag{24.56}$$

The coefficients c_i are called combinatoric factors.

We give some examples.

- $A = \mathbb{N}^n$. We can take $c_i = K|i|^{n+\epsilon}$ where $|i|^2 = i_1^2 + \cdots + i_n^2$ for $(i_1, \cdots, i_n) \in (\mathbb{N}^n)$ and K is such that

$$\frac{1}{K} \sum_{i \in \mathbb{N}^n} \frac{1}{|i|^{n+\epsilon}}. \tag{24.57}$$

- $A = \mathbb{D}$ where $\mathbb{D}$ is a partition of $\mathbb{R}^n$ by unit cubes Δ. The cube Δ can be the location of some field and we are interested in the choice of the cube Δ' in which this field will contract another field. The choice of the cube Δ' is done by the factor $c_{\Delta,\Delta'} = K d(\Delta, \Delta')^{n+\epsilon}$, $\epsilon > 0$ where $d(\Delta, \Delta') = \sup(1, \text{dist}(\Delta, \Delta'))$, $\text{dist}(\Delta, \Delta')$ being the Euclidean distance between the two cubes and K being such that

$$\frac{1}{K} \sum_{\Delta' \in \mathbb{D}} \frac{1}{d(\Delta, \Delta')^{n+\epsilon}} \leq 1. \tag{24.58}$$

In order to estimate the number of terms produced by the Gaussian integration in (24.53) we take one of the fields in one of the interaction packages $\int_\Lambda : \phi(x)^4 : \mathrm{d}^2 x$. Because of the Wick ordering, this field cannot contract with another field in the same interaction package. It will therefore contract to another of the remaining $n-1$ packages.

---

[6] This was introduced by J. Glimm and A. Jaffe, *Fortschr. Phys.*, **21**, 327 (1973).

There are $n-1$ choices. Now, there are four fields in each interaction which means a factor 4 to choose which of the four fields will be hit by the contraction. Thus the complete contraction of one of the n components gives $(4(n-1))^4$. Now it remains at most $n-1$ packages. Repeating the process we can estimate the number of terms as

$$(4(n-1))^4(4(n-3))^4 \cdots 1 = \left(\frac{4n4(n-1)\cdots 4(2)4}{4n4(n-2)\cdots 4} \right)^4$$

$$= \left(\frac{4^n n!}{8^{n/2} \frac{n}{2}!} \right)^4 \simeq 4(2^{2n}) \left(\frac{2n}{e} \right)^{2n} \simeq \frac{2^{(2n+2)}}{\sqrt{4\pi n}} (2n)!. \tag{24.59}$$

This is of the same order as the result obtained in the case of the zero-dimensional model (24.20). The complete bound on (24.53) is now obtained by an a priori estimate of a diagram with n-vertices. We now give a simple argument showing that a diagram G with n vertices of a $\lambda\phi_2^4$ theory is bounded by C^n for some positive constant C.

The argument is the following: given an n-vertex diagram, we can number its vertices by $1, 2, \cdots, n$. Each vertex is an elementary diagram g_j with initial and final lines. The initial lines of g_j are those joining vertices g_k with $k < j$. The final lines of g_j are those joining vertices g_l with $l > j$. If we think of each vertex g_j as an operator acting on the initial lines to give the final ones, then the diagram G can be bounded by the product of the Hilbert–Schmidt norms of each g_j. With a $\lambda\phi^4$ theory, the four lines of each vertex can be oriented in three different types:

- Four initial lines and no final lines or no initial lines and four final;
- One initial line and three final ones or three initial lines and one final; and
- Two initial lines and two final lines.

Thus, the contribution of the nth-order term is bounded for some constants K and C by

$$\int \left(\int_\Lambda : \phi(x)^4 : \mathrm{d}^2 x \right)^n \mathrm{d}\mu_{C_\rho} \leq K C_3^n \lambda^n n!. \tag{24.60}$$

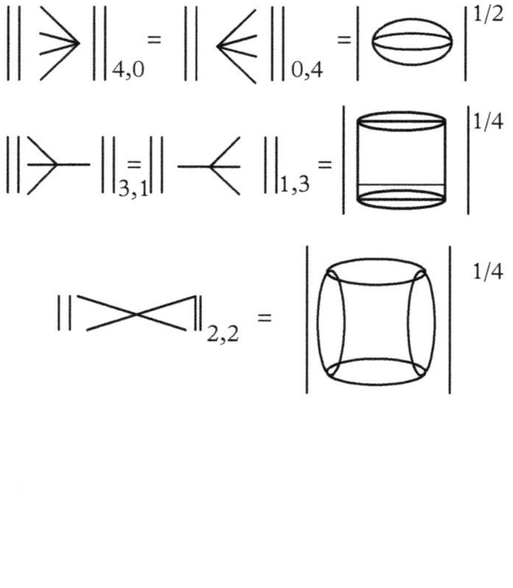

Figure 24.1 *The Hilbert–Schmidt norms.*

This is a bound on the nth-order term and not an estimate, but the dominant contribution is the number of terms. This contribution behaves as an $n!$ as for the zero-dimensional model, a contribution which cannot be beaten by a constant to the nth power. We thus conclude that the power expansion in λ of the partition function is divergent. It is easy to extend this result for any moment of the theory, i.e. terms of the form $< \phi(x_1) \cdots \phi(x_m) >$.

24.6.2 The Existence of the $\lambda\phi_2^4$ Theory

In the case of the zero-dimensional field theory, the existence of the measure, or, equivalently, of its various moments, was obvious because the exponential of the interaction reinforces the convergence of the measure and all the integrals were absolutely convergent. We will see that for the $\lambda\phi_2^4$ theory the situation is more complicated. Although the renormalisation we had to perform in order to obtain, order by order, a well-defined perturbation expansion was a very simple one, namely the Wick ordering of the interaction monomial, we will see that, by doing so, we lose the positivity property of the interaction. In fact, with the Wick order defined by the cut-off covariance C_ρ, we have that

$$
\begin{aligned}
: \phi(x)^4 : &= \phi(x)^4 - 6C_\rho(0)\phi(x)^2 + 3C_\rho(0)^2 \\
&= (\phi(x)^2 - 3C_\rho(0))^2 - 6C_\rho(0)^2 \geq -6C_\rho(0)^2;
\end{aligned} \tag{24.61}
$$

thus,

$$
\begin{aligned}
\left| Z(\rho, \Lambda) \right| = \left| \int e^{-\lambda \int_\Lambda :\phi(x)^4:d^2x} d\mu_{C_\rho} \right| &\leq \int e^{6\lambda C_\rho(0)^2|\Lambda|} d\mu_{C_\rho} \\
= e^{6\lambda C_\rho(0)^2|\Lambda|} &\leq e^{K(\ln M_\rho)^2 \lambda |\Lambda|}
\end{aligned} \tag{24.62}
$$

for some constant K. We thus see that when $\rho \to \infty$ we do not have an a priori finite bound on Z.

To simplify the notations we will note that

$$
V_\Lambda = \int_\Lambda : \phi(x)^4 : d^2x. \tag{24.63}
$$

We will also choose the sequence of momentum cut-offs $\{M_i\}$ to be of the form $M_i = L^i$, for some $L > 1$.

We will now give some elements showing how we prove the existence of the $\lambda\phi_2^4$ theory. It will be done in two steps. For the first step we will prove that $Z(\rho, \Lambda)$ has a limit when $\rho \to \infty$ and moreover

$$
\lim_{\rho \to \infty} Z(\rho, \Lambda) < Ke^{|\Lambda|} \tag{24.64}
$$

for some positive constant K result, which shows that following a statistical mechanics interpretation (Z being the partition function) the energy per unit volume is bounded

$$\frac{1}{|\Lambda|} \lim_{\rho \to \infty} \ln Z(\rho, \Lambda) < C \tag{24.65}$$

for another positive constant C.

Then, for the second step we will apply a classical method from statistical mechanics (the so-called Mayer expansion) where the v.e.v.'s have a limit when $\Lambda \to \mathbb{R}^2$. This property results from the fact that the covariance linking two distant vertices decreases exponentially when the distance between these two vertices increases with a rate bounded by the mass m.

We will mainly explain the way to get the bound (24.64) and only give some arguments for the Mayer expansion, then the analyticity property in λ and the estimate of the reminder in the power expansion of the interaction part will be easy to prove.

To prove the estimate (24.64) and more generally a similar bound for any v.e.v., i.e. Schwinger functions, we will introduce a covering of $\mathbb{R}^2$ with unit squares $\Delta_1, \Delta_2, \Delta_3, \cdots$. We then try to lower the cut-off index ρ in each unit square. This will generate small terms due to the coupling constant λ, which is supposed to be small, and convergent factors coming from the lowering of momentum cut-off. With enough small factors we can hope to beat the blowing up of the bound on the exponential of the interaction.

To lower the cut-off index in Z, we associate with each square $\Delta \in \Lambda$ a parameter $t_{\rho, \Delta}$ which will ensure the interpolation between ϕ_ρ and $\phi_{\rho-1}$ in Δ. Therefore, the interaction V, which is

$$V = \lambda \int_\Lambda : \phi(x)^4 : \mathrm{d}^2 x = \sum_{\Delta \subset \Lambda} \lambda \int_\Delta : \phi(x)^4 : \mathrm{d}^2 x = \sum_{\Delta \subset \Lambda} V_\Delta, \tag{24.66}$$

becomes in each square

$$\lambda \int_\Delta : \left(t_{\rho,\Delta} \phi^\rho(x) + \phi_{\rho-1}(x) \right)^4 : \mathrm{d}^2 x = V_\Delta(t_{\rho,\Delta}) \tag{24.67}$$

and $Z(\rho, \Lambda)$ becomes

$$Z(\rho, \Lambda; t_\rho) = \int \prod_{\Delta \subset \Lambda} \mathrm{e}^{-\lambda \int_\Delta :\phi(x, t_{\rho,\Delta})^4 :\mathrm{d}^2 x} \mathrm{d}\mu_{C_\rho} \tag{24.68}$$

with an obvious definition of $\phi(x, t_{\rho,\Delta})$ and $t_\rho = \{t_{\rho,\Delta} \mid \Delta \subset \Lambda\}$.

The value of $Z(\rho, \Lambda; t_\rho)$ at $t_\rho \equiv 0$ is $Z(\rho - 1, \Lambda)$.

We then write for each Δ

$$\mathrm{e}^{-V_\Delta} = \mathrm{e}^{-V_\Delta(t_{\rho,\Delta})}\Big|_{t_{\rho,\Delta}=1}$$

$$= \int_0^1 \frac{\mathrm{d}}{\mathrm{d}t_{\rho,\Delta}} \mathrm{e}^{-V_\Delta(t_{\rho,\Delta})} \mathrm{d}t_{\rho,\Delta} + \mathrm{e}^{-V_\Delta(t_{\rho,\Delta})}\Big|_{t_{\rho,\Delta}=0} \tag{24.69}$$

and we define the operators $I_{\rho,\Delta}$ and $P_{\rho,\Delta}$

$$I_{\rho,\Delta}e^{-V_\Delta(t_{\rho,\Delta})} = e^{-V_\Delta(t_{\rho,\Delta})}\Big|_{t_{\rho,\Delta}=0}$$

$$P_{\rho,\Delta}e^{-V_\Delta(t_{\rho,\Delta})} = \int_0^1 \frac{d}{dt_{\rho,\Delta}}e^{-V_\Delta(t_{\rho,\Delta})}dt_{\rho,\Delta}. \tag{24.70}$$

The operator P acts as a derivative $\dfrac{d}{dt_{\rho,\Delta}}$ on $V_\Delta(t_{\rho,\Delta}) = \int_\Delta : \phi(x,t_{\rho,\Delta})^4 : d^2x$, thus on the field

$$\frac{d}{dt_{\rho,\Delta}}\phi(x,t_{\rho,\Delta}) = \phi^\rho, \tag{24.71}$$

then when integrated, we will see that the field ϕ^ρ will produce a small factor.

We now define an expansion by applying n_i times formula (24.69) in each square $\Delta \subset \Lambda$ and for each index i, $1 \leq i \leq \rho$. The operators I and P have the following properties:

- $I_{\rho,\Delta}$ and $P_{\rho',\Delta'}$ commute if the pair $(\rho,\Delta) \neq (\rho',\Delta')$; and
- $P_{\rho,\Delta}I_{\rho,\Delta} \equiv 0$ and $[I_{\rho,\Delta}]^n = I_{\rho,\Delta}$.

Therefore, we can write

$$e^{-V} = \prod_{\Delta \subset \Lambda}\left\{\prod_{i\geq 0}\left(I_{i,\Delta} + P_{i,\Delta}\right)^{n_i}e^{-V_\Delta}\right\}. \tag{24.72}$$

A generic term of the expansion has therefore the form

$$T(\{\nu_{i,\Delta}\}) = \prod_\Delta\left\{\prod_i[P_{i,\Delta}]^{\nu_{i,\Delta}}e^{-V_\Delta(\{\nu_{i,\Delta}\})}\right\}, \tag{24.73}$$

where $\nu_{i,\Delta} \in \mathbb{N}, 0 \leq \nu_{i,\Delta} \leq n_i$ and

$$V_\Delta(\{\nu_{i,\Delta}\}) = V_\Delta|_{t_{i,\Delta}=0} \quad if \quad \nu_{i,\Delta} \neq n_i. \tag{24.74}$$

For each Δ, we define that $i(\Delta) = \sup\{i|t_{i,\Delta} \neq 0\}$, then we can prove the following lemma.

Lemma 3. *There exist constants K and ϵ such that*

$$\left|\int T(\{\nu_{i,\Delta}\})d\mu_{C_\rho}\right| \leq \prod_\Delta\left\{e^{K(\log L^{i(\Delta)})^2}\prod_{i>0}\frac{(4\nu_{i,\Delta})^{2\nu_{i,\Delta}}}{(\nu_{i,\Delta})!}[Ki^8L^{-i\epsilon}]^{\nu_{i,\Delta}}\right\}. \tag{24.75}$$

We can also prove the following corollary.

Lemma 4. *There exists an ϵ such that*

$$\left| \int T(\{v_{i,\Delta}\}) \mathrm{d}\mu_{C_\rho} \right| \leq \prod_{\substack{\Delta \\ \text{such that} \\ v_{i,\Delta}=0 \; \forall i}} K \prod_{i,\Delta} [L^{-i\epsilon}]^{v_{i,\Delta}} \frac{1}{(v_{i,\Delta})!}. \tag{24.76}$$

This corollary expresses the fact that when no vertex has been produced in a square Δ, the price to pay is a factor K and we get a factor $[L^{-i}]^\epsilon$ each time a ϕ^4 is hit by the action of a P operator.

The explanation of why we get a small factor for each $P_{i,\Delta}$ comes from the fact that $P_{i,\Delta}$ will produce a ϕ^4-vertex at a point x integrated in a unit square Δ with one of the four fields being of covariance C^i. From (24.40), we get a factor $\ln(M_{i+1}/M_i)$, i.e. $\ln L$, with our choice of cut-off and an exponential decrease $\mathrm{e}^{M_i|x-.|}$, thus integrating over x

$$\int_\Lambda \mathrm{e}^{-M_i|x-y|} \mathrm{e}^{-M_{\alpha_1}|x-y_1|} \mathrm{e}^{-M_{\alpha_2}|x-y_2|} \mathrm{e}^{-M_{\alpha_3}|x-y_3|} \mathrm{d}^2 x$$

$$\leq \int_\Lambda \mathrm{e}^{-M_i|x-y|} \mathrm{d}^2 x \leq K|\Lambda|M_i^{-2} = K|\Lambda|L^{-2i}. \tag{24.77}$$

This is the small factor per vertex which will ensure, besides λ, the convergence of the expansion. We extract from it the convergent bound $L^{-i\epsilon}$ we use for the proof of Lemma 4.

From the corollary, we get immediately a bound for (24.62)

$$|Z(\rho, \Lambda)| \leq \prod_{\Delta \subset \Lambda} \left\{ K + \prod_{i>0} (L^{-i\epsilon} + \cdots + \frac{[L^{-i\epsilon}]^{n_i}}{n_i!}) \right\} \tag{24.78}$$

and since

$$L^{-i\epsilon} + \cdots + \frac{[L^{-i\epsilon}]^{n_i}}{n_i!} \leq \mathrm{e}^{L^{-i\epsilon}} \tag{24.79}$$

we have

$$K + \prod_{i>0} (L^{-i\epsilon} + \cdots + \frac{[L^{-i\epsilon}]^{n_i}}{n_i!}) \leq K + \prod_{i>0} \mathrm{e}^{L^{-i\epsilon}} \leq K + K(\epsilon) \leq K \tag{24.80}$$

and we finally get the bound

$$\left| \int \mathrm{e}^{-V_\Lambda} \mathrm{d}\mu_{C_\rho} \right| \leq K^{|\Lambda|} \tag{24.81}$$

uniformly in ρ.

It is easy to check from these estimates that $Z(\rho, \Lambda)$ has a limit when $\rho \to \infty$ and that this limit satisfies (24.81).

As a result of the Wick ordering in the interaction term, we also have, using Jensen's inequality, a very straightforward lower bound

$$Z(\rho, \Lambda) = \int e^{-\lambda \int_\Lambda :\phi(x)^4: d^2x} d\mu_{C_\rho} \geq e^{-\lambda \int \left(\int_\Lambda :\phi(x)^4: d^2x \right) d\mu_{C_\rho}} = 1. \tag{24.82}$$

We will now give some arguments for the proof of Lemma 3 and Lemma 4. First we split in two parts the integral to estimate

$$\int T(\{v_{i,\Delta}\}) d\mu_{C_\rho} = \int \prod_\Delta \left\{ \prod_i [P_{i,\Delta}]^{v_{i,\Delta}} e^{-V_\Delta} \right\} d\mu_{C_\rho}$$

$$\leq \left(\prod_\Delta \sup e^{-V_\Delta} \right) \left[\int \prod_\Delta \prod_i \left| [P_{i,\Delta}]^{v_{i,\Delta}} \right|^2 d\mu_{C_\rho} \right]^{\frac{1}{2}}, \tag{24.83}$$

where in the second part of the formula $[P_{i,\Delta}]^{v_{i,\Delta}}$ is the polynomial in the fields resulting from the action of the derivatives on the exponential.

Now using the bound (24.61),

$$V_\Delta(\{v_{i,\Delta}\}) \geq -K(\ln L^{i(\Delta)})^2. \tag{24.84}$$

The first term in the right-hand side product of (24.83) gives the required exponential bound of the lemma.

It remains to compute the Gaussian integral

$$I = \int \prod_\Delta \prod_i \left| [P_{i,\Delta}]^{v_{i,\Delta}} \right|^2 d\mu_{C_\rho}. \tag{24.85}$$

In order to estimate the Gaussian integral I we will use the combinatoric factors. We know that to compute the Gaussian integral we will have to contract all the fields and we know also that the result is independent of the order we choose to perform the contraction. We will use this possibility and order the fields before performing the contraction. Let us set $m_{i,\Delta} = 2v_{i,\Delta}$. We order the pairs (i, Δ) by

$$(i, \Delta) \geq (i', \Delta') \quad \text{if} \quad m_{i,\Delta} \geq m_{i',\Delta'}$$

and we start by contracting the fields corresponding to the highest values of the numbers $m_{i,\Delta}$.

So let us suppose we want to contract a field indexed by (i, Δ) with a field indexed by (i', Δ'). By definition of the process $m_{i,\Delta} \geq m_{i',\Delta'}$, to control the number of term produced, we have the following combinatoric factors (the c_i of formula (24.57) and (24.58)):

- $Kd(\Delta, \Delta')^3$ to choose in which square the contracted field is located;
- $K[i']^2$ to choose the momentum index; and
- $4m_{i',\Delta'}$ to choose which field characterised by the pair (i', Δ') is contracted.

Because of the ordering we have chosen, we can symmetrise the estimate by replacing

$$m_{i',\Delta'} \leq (m_{i',\Delta'})^{1/2}(m_{i,\Delta})^{1/2}.$$

The factor to bound the number of terms produced is

$$\prod_{\text{Propagator}} d(\Delta, \Delta')^3 \prod_{\text{Vertex} P_{i,\Delta}} K i^8 (4m_{i,\Delta})^{2m_{i\Delta}}. \tag{24.86}$$

We now must estimate a given diagram. A diagram is made of an ensemble of vertices localised in squares of $\mathcal{R}^2$ and lines joining these vertices to these lines are associated propagators with an exponential decrease. We will take a fraction δ of this exponential decrease, $e^{-\delta L^i d(\Delta,\Delta')}$ which is bounded by $e^{-\delta d(\Delta,\Delta')}$, to control the factor per line in (24.86). We must therefore finally estimate a diagram with propagator of the form

$$e^{\delta L^i d(\Delta,\Delta')} \chi(x) C^i(x-y) \chi'(y),$$

χ and χ' being the characteristic functions of the unit square Δ and Δ'. The remaining part of the diagram is estimated as in section 24.4.1. We have one more piece of information: if a vertex g_j has a line of lower momentum index i, we can extract from the integrals a factor $(L^i)^{-(1-\eta)}$. This gives a bound

$$\prod_{\text{line}} e^{-\delta d} \prod_{P_{i,\Delta}} K(L^i)^{-(1-\eta)}. \tag{24.87}$$

Collecting all these results (24.86) and (24.87) we obtain the proof of Lemma 3 with $1 - \eta = \epsilon$ which can be taken close to 1 and the $1/(v_{i,\Delta})!$ comes from the integration on the $t_{i,\Delta}$.

To prove Lemma 4, we will use the convergent factor of Lemma 3, $(L^{-i})^{-\epsilon}$ divided into four pieces $(L^{-i})^{-\epsilon/4}$, and use the fact that

1. $(4v_{i,\Delta})^{2v_{i,\Delta}}[L^i]^{-\epsilon v_{i,\Delta}/4} \leq 1$ for $i \geq 1$, $v_{i,\Delta} \leq n_i$ with $n_i \simeq [L^i]^{\epsilon}$;
2. $K e^{K(\ln L)^2} i^8 [L^i]^{-\epsilon/4} \leq 1$ for $i \geq 1$; and
3. $K e^{K(\ln L^{i(\Delta)})^2} [L^{i(\Delta)}]^{-n_{i(\Delta)}\epsilon/4} \leq 1$ for $i(\Delta) \geq 1$.

The last condition to be satisfied shows the minimal growth on n_i, which has to grow faster than $(\ln L^i)^2$. The remaining $L^{-i\epsilon/4}$ factor is the convergent factor appearing in Lemma 4.

Doing the same analysis, we can also prove the following bound, uniformly in ρ, for the Schwinger functions , with K_1, K_2, and C being some positive constants

$$Z(\rho, \Lambda) < \phi(x_1) \cdots \phi(x_m) > = \int \phi(x_1) \ldots \phi(x_n) e^{-\lambda : \int_\Lambda \phi(x)^4 : d^d x} d\mu_C$$

$$\leq K_1 C^m m! K_2^{|\Lambda|}. \tag{24.88}$$

In all these cases, we can prove the existence of the limit of all these quantities when $\rho \to \infty$. To do it, for example for the partition function, we prove that $Z(\rho, \Lambda)$ satisfies a Cauchy sequence in the variable ρ. For this, it is enough to study the difference $\Delta Z = Z(\rho, \Lambda) - Z(\rho-1, \Lambda)$ using the expansion. Introducing the parameters $t_{\rho,\Delta}$, for all squares Δ, we see that the difference ΔZ contains at least one term with a derivative with respect to $t_{\rho,\Delta}$ for at least one Δ. This means that a field ϕ^ρ will be contracted, thus producing a convergent factor $L^{-\rho}$, a part of which can be used as the factor showing that the difference ΔZ tends to 0 as $L^{-\rho\epsilon}$.

In order to prove the infinite volume limit, the thermodynamic limit in the statistical mechanics language, we perform a cluster (or Mayer or Kirkwood–Salzburg) expansion.

24.6.3   The Cluster Expansion

We will now describe a very important tool in the study of rigorous results for large physical systems: the cluster expansion. It is presented here for a massive scalar field theory in two dimensions, but these results can be extended easily to higher dimensions and to other types of field theories.

The cluster expansion is based on the sufficiently rapid decrease with the distance of the connected Schwinger functions as a function of their arguments.[7] Technically the method has its origin in statistical mechanics to prove the infinite volume limit of thermodynamic quantities.

We will study Schwinger Ω-functions where the Ω fields are localised in disjoint unit squares[8] $\Delta_\alpha, \Delta_\beta, \cdots, \Delta_\Omega$:

$$S_\Lambda(\Delta_\alpha, \cdots, \Delta_\Omega) \tag{24.89}$$

$$= \frac{1}{Z_\Lambda} \lim_{\rho \to \infty} \int_{\Delta_\alpha} \cdots \int_{\Delta_\Omega} \int \phi(z_1) \ldots \phi(z_n) e^{-\lambda \int_\Lambda : \phi(x)^4 : d^2 x} d\mu_\rho d^2 x_1 \ldots d^2 x_n,$$

where

$$Z_\Lambda = \lim_{\rho \to \infty} \int e^{-\lambda \int_\Lambda : \phi(x)^4 : d^2 x} d\mu_\rho. \tag{24.90}$$

---

[7] The exponential decrease is due to the fact that this is a massive field theory, the range of the forces being on the order of m^{-1}. It translates, in the rigorous quantum field theory language, the Yukawa potential we find in the old meson theory. The strength of the exponential decrease has a value which is close to the mass m, the bare mass, initially introduced in the theory. The fact that the coupling constant λ is small leads to the fact that the physical mass is close to the bare one. For massless theories, there are long-range effects and the clustering depends strongly on the way the covariance is decreasing, usually like a power, like the Coulomb potential in electrodynamics.

[8] The disjointness condition is not essential

We also introduce the unnormalised function

$$I_\Lambda(\Delta_\alpha, \cdots, \Delta_\Omega) = \int \left(\prod_{i=\alpha}^{\Omega} \int_{\Delta_i} \phi(z_i) \mathrm{d}^2 z_i\right) e^{-\lambda \int_\Lambda :\phi(x)^4: \mathrm{d}^2 x} \mathrm{d}\mu_\rho. \tag{24.91}$$

The squares $\Delta_\alpha, \ldots, \Delta_\Omega$ will be taken as squares of reference; they are elements of a network of unit squares $\mathbb{D}_0$. Because the theory is a massive theory, there is an exponential decrease in the distance of the covariances with a strength close to the value of the mass (the bare mass). The idea of the expansion is to test the link between any square of $\mathbb{D}_0$ and the set of reference squares with the effect that only the aggregation of squares close to the squares of reference will count significantly. This is a direct consequence of the fact that because the coupling constant is small enough, the global measure of the theory is correctly approximated by the free measure.

We thus introduce a set of parameters $\{s_i | i \in (1, 2, \ldots,)\}$ with values $0 \le s_i \le 1$. These parameters will make it possible to decouple a given square from the other ones. Thus let us choose $\Delta_1 \in \mathbb{D}_0$ and consider the covariance

$$\begin{aligned} C(x - y; s_1) &= C(x - y)[s_1 + (1 - s_1)(\chi_{\Delta_1}(x)\chi_{\Delta_1}(y) \\ &\quad + (1 - \chi_{\Delta_1}(x))(1 - \chi_{\Delta_1}(y)))] \\ &= C(s_1)(x - y), \end{aligned} \tag{24.92}$$

where $\chi_{\Delta_1}(x)$ is the characteristic function of the square Δ_1. When $s_1 = 1$ we get the usual covariance and when $s = 0$, the covariance is different from 0 in two cases: either when both x and y are in Δ_1 or when both x and y are in the complementary of Δ_1. Moreover, the derivative of the covariance with respect to s_1 has the effect of coupling Δ_1 to its complementary set.

Looking at the connected part of Schwinger functions S_Λ^c we will prove the following.

Lemma 5. *Let $\phi(x_\alpha)$ be a field localised in Δ_α, $\phi(x_\beta)$ a field localised in Δ_β up to $\phi(x_\Omega)$, a field localised in Δ_Ω, $\Delta_\alpha, \Delta_\beta \cdots \subset \Lambda$. Then the connected Schwinger function $S_\Lambda^c(\Delta_\alpha, \cdots, \Delta_\Omega)$ has a limit $S^c(\Delta_\alpha, \cdots, \Delta_\Omega)$ when $\Lambda \to \mathbb{R}^2$, and given ϵ, there exist two constants, λ_ϵ and K_ϵ, independent of Λ such that*

$$S^c(\Delta_\alpha, \cdots, \Delta_\Omega) \le K_\epsilon e^{-(m-\epsilon)T(\Delta_\alpha, \cdots, \Delta_\Omega)} \tag{24.93}$$

for any $\lambda \le \lambda_\epsilon$, $T(\Delta_\alpha, \cdots, \Delta_\Omega)$ being the length of the smallest tree linking all the Δ_i's, $i = \alpha, \ldots, \Omega$.

The proof goes as follows. We replace in the definition of (24.90) and (24.91), the covariance C by $C(s_1)$ and thus define $I(s_1)$ and $Z(s_1)$. Then the first step of the expansion is defined, starting from the expression when $s_1 = 1$ by

$$I(1) = I(0) + \int_0^1 \frac{\mathrm{d}}{\mathrm{d}s_1} I(s_1). \tag{24.94}$$

Since the measure dμ factorises for $s_1 = 0$, we have that

$$I(0) = I_{\Lambda \backslash \Delta_1} I_{\Delta_1} \tag{24.95}$$

and

$$\frac{\mathrm{d}}{\mathrm{d}s_1} I(s_1) = \sum_{\Delta_2 \neq \Delta_1} I_{\Delta_1, \Delta_2}(s_1) \tag{24.96}$$

with

$$I_{\Delta_1, \Delta_2}(\Delta_\alpha, \dots \Delta_\Omega; s_1) = \int \mathrm{d}u \mathrm{d}v [\chi_{\Delta_1}(u)\chi_{\Delta_2}(v) + \chi_{\Delta_1}(v)\chi_{\Delta_2}(u)] \tag{24.97}$$

$$C(u,v)\frac{\delta^2}{\delta\phi(u)\delta\phi(v)}\left\{\left(\prod_{i=\alpha}^{\Omega}\phi(\Delta_i)\right)e^{-\lambda \int_\Lambda :\phi(x)^4:\mathrm{d}^2 x}\right\}\mathrm{d}\mu(s_1)$$

with $\phi(\Delta)$ the integral of $\phi(z)$ over Δ. The right-hand side of (24.97) shows that a propagator is connecting the square Δ_2 to the square Δ_1. We then repeat the same procedure for $I_{\Lambda \backslash \Delta_1}$ starting from an arbitrary square of $\Lambda \backslash \Delta_1$ and for $I_{\Delta_1, \Delta_2}(s_1)$ by introducing a new parameter s_2 which interpolates between $\Delta_1 \cup \Delta_2$ and $\Lambda \backslash (\Delta_1 \cup \Delta_2)$.

The expansion will terminate when all the squares of the space Λ have been tested with at least another square. We then get

$$I_\Lambda(\Delta_\alpha, \dots, \Delta_\Omega) = \sum_{q \geq 1} \sum_{X_1, \dots, X_q} \prod_{i=1}^{q} I(X_i; \{\Delta\}_i), \tag{24.98}$$

where the sums run over all partitions of Λ in disjoint sets. $I(X_i; \{\Delta\}_i)$ represents I restricted to X_i where the $\phi(\Delta_\gamma)$ are on square $\Delta_\gamma \in X_i$. We can rewrite $I(X_i; \{\Delta\}_i)$ as a sum of connected graphs joining the squares of X

$$I(X) = \sum_G I(X, G). \tag{24.99}$$

From the way the development has been defined these graphs have a tree structure based on the squares of X with, if necessary, some intermediate squares. Among all these graphs, some are particular: they are the ones built over a unique square not containing any $\phi(z)$. We will note a the value of $I(X)$ in this case. Using Jensen's inequality and the bound on Z_Δ, (24.64), we get that a is finite and different from 0.[9] This number a, will play the role of the activity in condensed matter physics.

---

[9] It can be taken close to 1 by letting the coupling constant go to 0.

We therefore replace (24.98) with

$$a^{-|\Lambda|}I_\Lambda(\Delta_\alpha,\ldots,\Delta_\Omega) = \sum_{q \geq 1} \sum_{X_1,\ldots,X_q} \prod_{i=1}^{q}\left(I(X_i;\{\Delta\}_i)a^{-|X_i|}\right), \tag{24.100}$$

where the second sum on the right-hand side runs over all the partitions $X_1,\ldots,X_q$ of Λ such that either $|X_i| \geq 2$ or $|X_i| = 1$, but then there is at least one $\phi(z)$ in X_i.

Each $I(X,G)$ is a sum of elementary terms generated by the expansion. We will now explain how these terms are made. For this purpose we need to look in more detail at the form taken by the interpolating covariances.

At the first step starting from Δ_1, we introduce a covariance linking Δ_1 to a square Δ_2 different from Δ_1. We note this covariance $C(s_1)$. We then test the coupling of $\Delta_1 \cup \Delta_2$ with a square Δ_3 in Λ different from Δ_1 and Δ_2:

$$C(s_1,s_2) = s_2 C(s_1) + (1-s_2)C(s_1)_1 \tag{24.101}$$

with

$$\begin{aligned}
C(s_1)_1(x,y) &= \chi_{\Delta_1\cup\Delta_2}(x)C(s_1)(x,y)\chi_{\Delta_1\cup\Delta_2}(y) \\
&+ \chi_{(\Lambda-\Delta_1\cup\Delta_2)}(x)C(s_1)(x,y)\chi_{(\Lambda-\Delta_1\cup\Delta_2)}(y)
\end{aligned} \tag{24.102}$$

or more formally (with Δ standing for χ_Δ, the characteristic function of Δ)

$$C(s_1)_1 = \Delta_1 \cup \Delta_2 C(s_1)\Delta_1 \cup \Delta_2 + (\Lambda - \Delta_1 \cup \Delta_2)C(s_1)(\Lambda - \Delta_1 \cup \Delta_2)$$

without the explicit x and y dependance.

By the definition of $C(s_1,s_2)$, the derivative of $C(s_1,s_2)$ with respect to s_2 will link a square Δ_3 in the complementary of $\Delta_1 \cup \Delta_2$ to either Δ_1 or Δ_2.

We then introduce more generally

$$C(s_1,s_2,\ldots,s_i) = s_i C(s_1,\ldots,s_{i-1}) + (1-s_i)C(s_1,\ldots,s_{i-1})_i \tag{24.103}$$

with

$$\begin{aligned}
C(s_1,\ldots,s_{i-1})_i &= \Delta_1 \cup \ldots \Delta_i C(s_1,\ldots,s_{i-1})\Delta_1 \cup \ldots \Delta_i \\
&+ (\Lambda - \Delta_1 \cup \ldots \Delta_i)C(s_1,\ldots,s_{i-1})(\Lambda - \Delta_1 \cup \ldots \Delta_i).
\end{aligned} \tag{24.104}$$

We can now express the complete cluster expansion introducing the operator

$$P(\Delta,\Delta') = \int \frac{dC(s)}{ds}(x,y)\frac{\delta^2}{\delta\phi(x)\delta\phi(y)}\chi_\Delta(x)\chi_{\Delta'}(y)d^2x d^2y \tag{24.105}$$

and this gives on $Q = \phi(\Delta_\alpha) \ldots \phi(\Delta_\Omega) e^{-\int :\phi(x)^4: d^2 x}$:

$$I = \int Q d\mu = \int Q d\mu \Big|_{\Delta_1} I_{\sim \Delta_1} \qquad (24.106)$$

$$+ \sum_{\Delta_2 \neq \Delta 1} \int ds_1 \int P(\Delta_1, \Delta_2) Q d\mu(s_1, s_2) \Big|_{\substack{s_2 = 0 \\ \Delta_1 \cup \Delta 2}} I_{\sim \Delta_1 \cup \Delta 2}$$

$$+ \sum_{i=3}^{\infty} \sum_{\eta} \sum_{\substack{\{\Delta_j\} \\ i \geq j \geq 1}} \int P(\Delta_{\eta(i)}, \Delta_i) \ldots$$

$$\ldots P(\Delta_2, \Delta_1) Q d\mu(s_1, \ldots, s_i) \Big|_{\substack{s_i = 0 \\ \cup \Delta_i}} ds_1 \ldots ds_{i-1} I_{\sim \cup \Delta_i}. \qquad (24.107)$$

We now must estimate the derivatives with respect to s_k of the covariances $C(s_1, \ldots, s_k)$ appearing in the definition of the P's. By this same definition of the P's, a square Δ_{k+1} not in $\Delta_1 \cup \cdots \cup \Delta_k$ will be linked to a square of $\Delta_1 \cup \cdots \cup \Delta_k$ whose index is given by the value of the function $\eta(k)$ taken in the set $(1, \ldots, k)$.

We have the following lemma.

Lemma 6.

$$\chi_{\Delta_{\eta(i)}} \frac{d}{ds_{i-1}} C(s_1, \ldots, s_{i-1}) \chi_{\Delta_i} = \left(\prod_{k=\eta(i)}^{i-2} s_k \right) \chi_{\Delta_{\eta(i)}} C \chi_{\Delta_i}. \qquad (24.108)$$

Proof. Let us give the proof.

$$\chi_{\Delta_{\eta(i)}} \frac{d}{ds_{i-1}} C(s_1, \ldots, s_{i-1}) \chi_{\Delta_i} \qquad (24.109)$$

$$= \chi_{\Delta_{\eta(i)}} [C(s_1, \ldots, s_{i-2}) - C(s_1, \ldots, s_{i-2})_{i-1}] \chi_{\Delta_i}$$

$$= \chi_{\Delta_{\eta(i)}} C(s_1, \ldots, s_{i-2}) \chi_{\Delta_i}$$

$$= \chi_{\Delta_{\eta(i)}} \begin{Bmatrix} C \\ s_{i-2} C(s_1, \ldots, s_{i-3}) \end{Bmatrix} \chi_{\Delta_i} \quad \text{if} \quad \begin{cases} \eta(i) = i-1 \\ \eta(i) = i-2 \end{cases}$$

and the proof results from repeating this analysis.

We then define a function

$$f(\eta, s) = \prod_i s_{\eta(i)} \ldots s_{i-2}. \qquad (24.110)$$

We will show some examples of the structures which are produced by the expansion by drawing some trees and the associated η values.

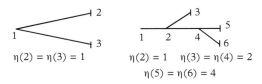

$\eta(2) = \eta(3) = 1$ $\eta(2) = 1$ $\eta(3) = \eta(4) = 2$

$\eta(5) = \eta(6) = 4$

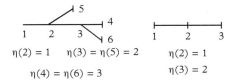

$\eta(2) = 1$ $\eta(3) = \eta(5) = 2$ $\eta(2) = 1$

$\eta(4) = \eta(6) = 3$ $\eta(3) = 2$

Figure 24.2 *Some examples of trees and η.*

Figure 24.3 *The graph G.*

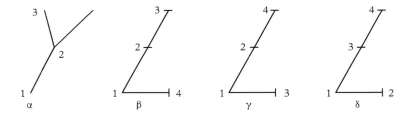

Figure 24.4 *The link $\Delta_1 \Delta_2$.*

Figure 24.5 *The four possible graphs resulting from G.*

In Fig. 24.2, we can see that different η functions can lead to the same graph.

Let us now consider the graph G shown in Fig. 24.3 with a given square Δ_1.

We will see how it can be built following the procedure outlined in this section.

From Δ_1, we first choose a square Δ_2 (see Fig. 24.4). We have then two possible choices for the positions of Δ_3 and of Δ_4. This leads to graphs shown in Fig. 24.5. We finally add Δ_5 and then in order to get G, there is only one choice for each of the four graphs drawn in Fig. 24.5.

By introducing now the short-hand notation $C(i,j)$ for $C(x,y)[\chi_i(x)\chi_j(y) + \chi_i(y)\chi_j(x)]$, then for each of the four graphs

- α corresponds to $s_1 s_2^2 s_3 \, C(1,2)C(2,3)C(2,4)C(1,5)$;
- β corresponds to $s_1 s_2^2 s_3 \, C(1,2)C(2,3)C(1,4)C(2,5)$;

- γ corresponds to $s_1 s_2^2 s_3 \; C(1,2)C(1,3)C(2,4)C(2,5)$; and
- δ corresponds to $s_1 s_3 \; C(1,2)C(1,3)C(3,4)C(3,5)$.

We then have for $I(X, G)$, $X = \cup_i^n \Delta_i$ and G given,

$$I(X, G) = \sum_{\eta(G)} \int f(\eta, s) \left[\int \dots \int \prod C(x_{\eta,i}, x_i) \right. \tag{24.111}$$

$$\left. \times \prod \chi_{\Delta_{\eta(i)}}(x_{\eta,i}) \chi_{\Delta_i}(x_i) \frac{\delta^2}{\delta\phi(x_{\eta,i})\delta\phi(x_i)} Q \Big|_X d\mu(s) \prod d^2 x_i \right] ds_1 \dots ds_n.$$

A graph G is a tree whose initial vertex, Δ_1, and the topological structure are given. Moreover it is characterised, for given indices of the other vertices, i.e. η, by a sequence n_i where

$$n_i = \{\text{number of} \quad k \quad \text{such that} \quad \eta(k) = i\} \qquad i = 1, \dots, n-1. \tag{24.112}$$

We remark that for different sequences $\{n_i\}$ can correspond to graphs having the same topological structure.

Introducing

$$q_j = \{\text{number of } k \text{ such that } \eta(k) \le j \le k-2\}, \quad j = 1, \dots, n-1 \text{ to} \tag{24.113}$$

we can show easily that

$$n_1 + \dots + n_{n-1} = n - 1$$

and that for the corresponding η-function

$$f(\eta, s) = s_1^{q_1} \dots s_{n-1}^{q_{n-1}}. \tag{24.114}$$

Lemma 7. *Given a function η, we get the following identities relating the ns, the qs, and $f(\eta, s)$,*

$$q_j = \sum_{k \le j} n_k - j \qquad\qquad j = 1, \dots, n-1$$
$$n_j = q_j - q_{j-1} + 1 \quad j = 1, \dots, n-1 \quad \text{with by convention } q_0 = 0$$

and

$$\sum_{\eta} \int_0^1 \dots \int_0^1 f(\eta, s) ds_1 \dots ds_{n-1} \prod n_i! = 1, \tag{24.115}$$

the sum being over all the η-functions compatible with the given choice of ns.

If we now go back to the four graphs α, β, γ, and δ, we remark that all these graphs have $n_1 = 2$, and that three, α, β, and γ, have $n_2 = 2$ although the fourth, δ, has $n_3 = 2$, the other n's being zero. The lemma can be checked for these examples, the proof resulting from the fact that the integration over the s's gives the combinatoric factors related to the weight of the graph.

Lemma 8. *Given X and G, there exists constants $K(\epsilon)$, C, and K independent of X, G, and Λ such that*

$$|I(X, G)| \leq CK^{|X|} \prod_{p \in G} [K(\epsilon)e^{-(m-\epsilon)d(p)}]\lambda^{[n-l]_+}, \tag{24.116}$$

where $|X|$ is the number of unit squares in X, $d(p)$ is the Euclidean distance between the centres of the squares in which are the end points of the propagator p, l is the number of fields $\phi(z)$, and

$$[a]_+ = a \text{ for } a \in \mathbb{N}$$
$$= 0 \text{ elsewhere.}$$

Replacing in the formula

$$\int \frac{dC(s)}{ds}(x, y) \frac{\delta^2}{\delta\phi(x)\delta\phi(y)} \chi_\Delta(x)\chi_{\Delta'}(y) d^2x d^2y \tag{24.117}$$

with

$$\sup_{x \in \Delta\, y \in \Delta'} \left(\frac{dC(s)}{ds}(x, y)\right) \int \frac{\delta^2}{\delta\phi(x)\delta\phi(y)} \chi_\Delta(x)\chi_{\Delta'}(y) d^2x d^2y \tag{24.118}$$

we must, η given, estimate

$$\left[\prod_{j, \eta(j)=i} \frac{\delta}{\delta\phi(x_{\eta,j})}\right] \frac{\delta}{\delta\phi(x_i)} [\phi(\Delta_i)]^{l_i} e^{-\lambda \int_{\Delta_i} :\phi(x)^4: d^2x}, \tag{24.119}$$

the product being on derivatives with respect to fields localised at points $x_{\eta,j}$ contained in Δ_i. The number of derivatives with respect to ϕ is by definition equal to $r_i = n_i + l_i$ (when $i = 1$, $r_1 = n_1$). Some of these derivatives will hit the exponential. They give

$$\frac{\delta}{\delta\phi(x)} e^{-\lambda \int_{\Delta_i} :\phi(x)^4: d^2x} = -4\lambda : \phi(x)^3 : e^{-\lambda \int_{\Delta_i} :\phi(x)^4: d^2x}. \tag{24.120}$$

Let p be the number of derivatives acting on the exponential; we have

$$[(r_i - l_i)/4]_+ \leq p \leq r_i; \tag{24.121}$$

thus, $r_i - p$ act on the field $\phi(z)$ or on the fields produced by derivatives which have acted on the exponential. The most singular term (i.e. producing monomials harder to control) comes when among the r_i derivatives, the p derivatives act first. The other derivatives act therefore on ϕ^{3p+l_i} (p derivatives having acted on ϕ^4 plus the l_i fields) and give

$$\frac{(3p + l_i)!}{(4p + l_i - r_i)!}\phi^{3p+l_i-r_i+p}, \tag{24.122}$$

where the product of the fields is in reality a product of integrals in Δ_i of monomials $: \phi^\alpha :$ with $1 \leq \alpha \leq 3$. Such a term is bounded by

$$K^{3p+l_i-r_i+p}[(3p + l_i - r_i + p)!]^{l/2}. \tag{24.123}$$

The case where all the p derivatives don't act first on the exponential gives factors all bounded by $(3p + l_i)!/(4p + l_I - r_i)!$. The sum over all possible choices of derivatives acting on the exponential is bounded by a factor 2 per derivative to decide whether a derivative acts on the exponential or on already produced or existing fields. Adding the numerical factors coming from the derivative of a monomial (typically a ϕ^4), we get

$$K^{r_i} \sum_p K^{l_i} \sup \frac{(3p+l_i)!}{(4p+l_i-r_i)!}[(4p + l_i - r_i)!]^{1/2}$$

$$\leq K^{r_i}K^{l_i}[l_i!]^{1/2}[r_i!]^{3/2}. \tag{24.124}$$

We can thus estimate the contribution of these factors on the product of derivatives by

$$K^n K^l [l!]^{1/2} \prod_{\Delta \in G} [n(\Delta)!]^{3/2} \tag{24.125}$$

with $n(\Delta_i) = r_i$.

The λ-dependance is given by $\lambda^{[n-l]_+}$ since if there are more squares than fields $\phi(z)$, at least $n - l$ derivations will hit the exponential. This is the factor which will make the thermodynamic limit (infinite volume limit) to converge.

We now give some arguments justifying why derived fields can be bounded by square roots of factorials. We start with (24.111). The integration over the position x_i is done by taking the supremum for each propagator. It remains to bound the absolute value of an expression like (24.119). From the bound on the propagators we extract a factor $e^{-md(l)}$ where $d(l)$ is the Euclidean distance between the centres of the end squares of the line l. A part of this decrease $e^{-(m-\epsilon)d(l)}$ will be used to prove the decrease involved in Lemma 5, formula (24.93). After having performed the derivations we must estimate

$$\int \prod_\Delta [\prod \int_\Delta : \phi(x)^\alpha : e^{-\lambda \int_\Delta :\phi(x)^4:}] d\mu \tag{24.126}$$

that we bound by

$$\left[\int \prod_\Delta [\prod \int_\Delta : \phi(x)^\alpha :]^2 d\mu\right]^{1/2} \left[\int [\prod_\Delta e^{-\lambda \int_\Delta :\phi(x)^4:}] d\mu\right]^{1/2}, \tag{24.127}$$

the second term being the partition function Z_X which is bounded by K^n.

We now prove the following.

Lemma 9.

$$\int \prod_\Delta \left[\prod \int_\Delta : \phi(x)^{\alpha(\Delta)} : \right]^2 d\mu \le K^p \prod_\Delta [p(\Delta)!]^{1/2}, \tag{24.128}$$

where $p = \sum_\Delta p(\Delta)$ *and* $p(\Delta) = \sum \alpha(\Delta)$.

The proof of this lemma is very simple. The left-hand side of formula (24.127) is a sum of products of Gaussian contractions. These contractions are obtained by the integration by parts formula from Chapter 10 and are independent of the order in which the integration by parts is applied, and in particular, from which field we start first, which one is second, and so on. The idea is therefore to introduce the partial order relation on the squares Δ given by the number of fields in a given square $n(\Delta) : n(\Delta_1) \ge n(\Delta_2) \ge \ldots$. We start the contractions with fields in Δ_1. We choose by a factor $[d(\Delta_1, \Delta')]^{2+\delta}$ the square Δ' in which the contraction will hold and by $n(\Delta')$ which fields in Δ' will be contracted. Since $n(\Delta_1) \ge n(\Delta')$, a combinatoric factor $n(\Delta_1)^{1/2}$ will be attributed to the contacting square and a factor $n(\Delta')^{1/2}$ to the contracted one, and the factor $[d(\Delta_1, \Delta')]^{2+\delta}$ will be attributed to the propagator between the two fields. This last factor will be controlled by the decrease of the propagator (which must be, as we have seen, integrable) and the result will be a constant K.

Finally, to conclude the proof of Lemma 8, we must bound the right-hand side of (24.111):

$$\sum_{\eta(G)} \int f(\eta, G)|\ldots| \le \left|\sum_{\eta(G)} \int f(\eta, G) \prod n_i!\right| \cdot \sup_{\{n\}} \frac{1}{n_i!} |\ldots| \le \sup_{\{n\}} \frac{1}{n_i!} |\ldots|. \tag{24.129}$$

This means to estimate

$$\lambda^{n-l} \prod_p e^{-(m-\epsilon)d(p)} K^n K^l [l!]^{1/2} \left[\prod_p e^{-\epsilon d(p)} \prod_\Delta \frac{[n(\Delta)!]^{3/2}}{n(\Delta)!}\right]. \tag{24.130}$$

The rules to construct the tree will help us to beat the factorial factors per square using the exponential decrease of the propagators. Indeed, let us consider a square Δ_i; it is related to r_i other squares $\Delta_i(1), \ldots, \Delta_i(r_i)$, which by construction are distinct squares. If all these squares were regularly laid around Δ_i, they would form a disk centred around

Δ_i and of radius r such that $\pi r^2 \sim r_i$. Let now δ, $0 < \delta < 1$. It follows that $r_i \delta$ of the squares are at least at a distance d from Δ_i such that

$$d \geq K[r_i(1-\delta)]^{1/2}, \tag{24.131}$$

from which it follows that

$$\prod_p e^{-\epsilon d} \leq \prod_i \left[e^{-\epsilon K[r_i(1-\delta)]^{1/2}/2} \right]^{r_i \delta} \leq K(\epsilon, k)^n \prod_i [r_i!]^{-k} \tag{24.132}$$

for any $k \in \mathbb{N}$. This completes the proof of lemma 8.

The bound on $I(X)$ is similar to that given in Lemma 8 since the sum over all possible graphs G having X for support is equivalent to sum up all possible values of the n_i's. In fact, we have

$$\sum_{\substack{n_1, \dots, n_{n-1} \\ \sum n_i = n-1}} = 2^{n-1} \sum_{\substack{n_1, \dots, n_{n-1} \\ \sum n_i = n-1}} \prod \frac{1}{2^{n_i}} \leq 2^{n-1} 2^{n-1} = K^n. \tag{24.133}$$

24.6.4   The Mayer Expansion

In the previous section we introduced the cluster expansion. It is an expansion which, starting from the expectation value with respect to an interacting measure localised in some finite volume Λ of some function Q, localised in space, expresses it as a convergent expansion. These unnormalised quantities blow up as an exponential of the volume $|\Lambda|$. The Mayer expansion deals with the normalised quantities. It was introduced in statistical mechanics to prove the existence of macroscopic quantities in the thermodynamic limit. More precisely, it shows that all the thermodynamical ensembles have the same (and unique) infinite volume limit. In fact, what the cluster expansion proves is that in the absence of long-range forces, the pressure of the system $\lim_{\Lambda \to \mathbb{R}^2} \frac{1}{|\Lambda|} \log Z_\Lambda$ exists. It is the Mayer expansion which shows that the normalised quantities, such as the normalised correlation functions, exist. They correspond in field theory to normalised expectation values. Again, as for the cluster expansion, the Mayer expansion will be described in the framework of the $\lambda\phi_2^4$ quantum field theory but extended to more general situations, i.e. higher dimensions or other types of theories.

The first idea is to rewrite an expression like (24.98) without the disjointness conditions of the X_i's. Let $\widetilde{\Delta}_\alpha \dots \widetilde{\Delta}_\Omega$ be the squares of Λ which contain the fields $\phi(\Delta)$ and let us call Ω their union. In addition, let us introduce a potential $V_{\alpha,\beta}$

$$V_{\alpha,\beta} = \begin{cases} 0 & \text{if the squares } \Delta_\alpha \text{ and } \Delta_\beta \text{ are different} \\ +\infty & \text{if the squares } \Delta_\alpha \text{ and } \Delta_\beta \text{ are identical.} \end{cases} \tag{24.134}$$

Then (24.98) can be written as

$$I(\widetilde{\Delta}_\alpha \ldots \widetilde{\Delta}_\Omega) = a^{|\Lambda|} \sum_{n \leq |\Lambda| - \Omega} \frac{1}{n!} \sum_{\substack{\Delta_1 \ldots \Delta_n \\ \Delta_i \in \Lambda}} \prod_{\alpha < \beta} e^{-V_{\alpha,\beta}}$$

$$\times \sum_{q \geq 1} \sum_{\substack{q\text{-partitions} \\ X = \cup X_i \\ X \supset \Omega}} \prod_1^q \widetilde{I}(X_j; \{\Delta\}_j), \tag{24.135}$$

where

$$X = \bigcup_\alpha^\Omega \widetilde{\Delta}_j \bigcup_1^n \Delta_i.$$

An order has been put on the union of $\widetilde{\Delta}$ and Δ and the q-partitions are partitions $\omega_1, \ldots, \omega_q$ of the indices $\alpha, \beta, \ldots, \Omega, 1, \ldots, n)$ in q sets not reduced to a single square. The notation $\widetilde{I}$ stands for the product of I by the associate a^{-I}. The main difference between formula (24.98) and formula (24.135) is that in X the square can now overlap with, as a counterpart, a symmetry factor $1/n!$.

Let us remark that since by convention $I(X)$ is zero as soon as two squares coincide, it is possible to suppress the condition $n \leq |\Lambda| - \Omega$.

Now, the next step is to replace everywhere $e^{V_{\alpha,\beta}}$ (except for the squares containing the fields $\phi(z)$) by

$$e^{V_{\alpha,\beta}} = 1 + (e^{V_{\alpha,\beta}} - 1) = 1 + M_{\alpha,\beta}. \tag{24.136}$$

If two squares Δ_α and Δ_β are disjoint, then $M_{\alpha,\beta} = 0$ and nothing is changed. If these two squares coincide then $M_{\alpha,\beta} = -1$ and if $\Delta_\alpha \in X_i$ and $\Delta_\beta \in X_j$ with $i \neq j$, then $I(X_i)$ and $I(X_j)$ are said to be linked by a Mayer link and the result is called a polymer (i.e. a connected set made with one layer or more). We will note $\widetilde{I}(X_i; \{\Delta\}_j)$, the connected set built from the $\widetilde{I}(X_i)$'s linked by Mayer links. The support of $\widetilde{I}$ is the union of the support of the $\widetilde{I}(X_i)$'s, which means that squares belonging to the support can have multiplicities. Thus,

$$I(\widetilde{\Delta}_\alpha \ldots \widetilde{\Delta}_\Omega) = a^{|\Lambda|} \sum_{n \geq 0} \frac{1}{n!} \sum_{\substack{\Delta_1 \ldots \Delta_n \\ \Delta_i \in \Lambda}} \sum_{q \geq 1} \sum_{\substack{q\text{-partitions} \\ X = \cup X_i \\ X \supset \Omega}} \prod_1^q \widetilde{I}(X_j; \{\Delta\}_j). \tag{24.137}$$

Among the connected sets X_i some contain some of the $\widetilde{\Delta}_j$'s. We will note still X for these and use the notation Y for those which don't contain any of the $\widetilde{\Delta}_j$'s. The sum on the partitions factorises into a product of partitions of X-type times partitions of Y-type. Let n_1 be the set of indices corresponding to X-type connected sets. Thus, the sum over

n is replaced by sums over n_1 and $n_2 = n - n_1$. Since the number of partition of n as a sum $n_1 + n_2$ is given by the binomial coefficient, we have

$$I(\widetilde{\Delta}_\alpha \ldots \widetilde{\Delta}_\Omega) = a^{|A|} \sum_{\substack{n_1 \geq 0}} \frac{1}{n_1!} \sum_{\substack{\Delta_1 \ldots \Delta_n \\ \Delta_i \in A \\ X_i \cap \Omega \neq \emptyset}} \sum_{q \geq 1} \sum_{\substack{q\text{-partitions} \\ X = \cup X_i \\ X \supset \Omega}} \prod_1^q \widetilde{I}(X_i; \{\Delta\}_i)$$

$$\times \Bigg[\sum_{n_2 \geq 0} \frac{1}{n_2!} \sum_{\substack{\Delta_1 \ldots \Delta_n \\ \Delta_i \in A \\ X_i \cap \Omega \neq \emptyset}} \sum_{q \geq 1} \sum_{\substack{q\text{-partitions} \\ Y = \cup Y_i}} \prod_1^q \widetilde{I}(Y_j; \{\Delta\}_j) \Bigg]. \quad (24.138)$$

It is not difficult to see that the term in between the brackets is nothing else than the partition function Z_A; therefore,

$$S_A(\widetilde{\Delta}_\alpha, \ldots, \widetilde{\Delta}_\Omega) = \sum_{n \geq 0} \frac{1}{n!} \sum_{\substack{\Delta_1 \ldots \Delta_n \\ \Delta_i \in A \\ X_i \cap \Omega \neq \emptyset}} \sum_{q \geq 1} \sum_{\substack{q\text{-partitions} \\ X = \cup X_i \\ X \supset \Omega}} \prod_1^q \widetilde{I}(X_i; \{\Delta\}_i). \quad (24.139)$$

Now from Lemma 8, we see that $\forall A$, $A > 1$, there exists $\lambda(A, K)$ such that for $\lambda \leq \lambda(A, K)$

$$|I(X, G)| \leq CK^{|X \cap \Omega|} \prod_{p \in G} [K(\epsilon)e^{-(m-\epsilon)d(p)}] A^{-|X \setminus \Omega|}. \quad (24.140)$$

This bound extends obviously to the polymers $\widetilde{I}$'s provided $|X|$ is now the number of squares with their multiplicities.

We remark also that the sum over q is bounded since $q \leq |\Omega|$. The sum over the squares Δ_i is as before, starting from the trees attached to the graphs G_j appearing in the decomposition $I(X) = \sum I(X, G)$, a decomposition which also extends to Mayer graphs $\widetilde{I}(X) = \sum \widetilde{I}(X, G)$. As before, the summation over the positions of the squares can be done using $e^{-\epsilon d(p)}$ and the decrease between the squares of the $\phi(z)$ is extracted in the same way up to a redefinition of ϵ.

The connected functions with respect to the $\phi(z)$'s are defined by

$$S(\Omega) = \sum_{\substack{\text{non-trivial partitions} \\ \Omega = \Omega_1 \cup \cdots \cup \Omega_p}} \prod_i S^c(\Omega_i) \quad (24.141)$$

and reciprocally

$$S^c(\Omega) = - \sum_{\substack{\text{non-trivial partitions} \\ \Omega = \Omega_1 \cup \cdots \cup \Omega_p}} (-1)^p (p-1)! \prod_i S(\Omega_i). \quad (24.142)$$

We have therefore

$$S_\Lambda^c(\tilde{\Delta}_\alpha, \ldots, \tilde{\Delta}_\Omega) = \sum_{n \geq 0} \frac{1}{n!} \sum_{\substack{\Delta_1 \ldots \Delta_n \\ \Delta_i \subset \Lambda}} \sum \tilde{I}(X_i; G_i) \tag{24.143}$$

and the proof of Lemma 5.

There are many terms which are produced during the Mayer expansion. Let us explain how they are controlled. A connected Mayer graph is a union of graphs $G_1, \ldots, G_k$ linked by Mayer links. We begin by constructing a skeleton of this polymer: we start with G_1 giving an order to the squares of G_1. If more than one link connect G_1 to G_j, we keep only the Mayer link starting from the square of G_1 having the lowest index. Then we do the same analysis for each of the G_j linked to G_1, testing their Mayer links with the graphs G_i not already linked to G_1. Since in each G_j the squares are distinct, two Mayer links cannot start from the same square. The summation is then essentially reduced to the estimation of the number of graphs having the same skeleton. We must define combinatoric factors for the number of Mayer links leaving a graph G_i, the number of squares n_j in each of the graphs G_j linked by Mayer links to G_i, the number of external indices in G_i (i.e. the number of $\phi(z)$ in G_i), and the index of the square which is an end point to the link. The link of G_1 to different G_i's is obtained by the binomial law, i.e. replacing $1/n!$ with $1/n_i!$ for each G_i, and the choices of the external indices are bounded by C^{n_i}. It remains for each G_i to control

$$\sum_{n \geq 1} \frac{C^n}{n!} \sum_{\Delta_1 \ldots \Delta_n} \sum_G I(X, G). \tag{24.144}$$

But the $I(X, G)$'s are bounded by $\prod_{\text{prop}} e^{-\epsilon d}$, the sum being on all G having different indices n_i. Using

$$\sum_{\Delta_1 \ldots \Delta_n} \prod e^{-\epsilon d} \leq K^n \tag{24.145}$$

We can sum over all possible n_i and each skeleton is multiplied by a factor $K^n n!$, which concludes the sketch of the proof.

Note that it is possible using these methods to replace the initial $\phi(\Delta)$ by $\int :\phi^\alpha(x): \chi_\Delta(x)\mathrm{d}^2 x$ for any $\alpha \in \mathbb{Z}_+$ and to get a similar result.

A remark before closing this section: we have chosen to give a proof of the convergence of the Mayer expansion following the spirit in which the cluster expansion was presented, despite the fact that more generally, the existence (and the convergence) of the Mayer expansion can be shown to result from a condition expressed in a theorem of D. Brydges.[10]

---

[10] D. Brydges, 'A Short Course on Cluster Expansions', in *Critical Phenomena, Random Systems, Gauge Theories*, Les Houches session XLIII, 1984 (Elsevier Science, 1986).

Theorem 20 (Brydges' bound). *Suppose that the polymers in the cluster expansion formula (24.98) satisfy the bound*

$$\sum_{0 \in X} |I(X)| e^{|X|} < 1. \tag{24.146}$$

Then the Mayer expansion converges.

Where $0 \in X$ means that the union of squares forming X contains the origin 0. This point, 0, can be replaced by any point of $\mathbb{R}^2$, this condition being only necessary to break the translational invariance.

We remark that the Brydges' condition can be replaced by the following.

Lemma 10. *Given K, if for λ small enough we have*

$$\sum_{0 \in X} |I(X)| K^{|X|} < 1; \tag{24.147}$$

then the Mayer expansion converges.

This condition can be checked during our proof.

24.6.5   The Infinite Volume Limit of $\lambda\phi_2^4$

The infinite volume limit results immediately for ϕ^4-field theory in two dimensions from the cluster and the Mayer expansions.

Let us consider an imbedded sequence of domains Λ_i, $\Lambda_i \subset \Lambda_{i+1}$ with $\lim_{i \to \infty} \Lambda_i = \mathbb{R}^2$. They can be obtained by adding at each step a square Δ_{i+1} to Λ_i in order to construct Λ_{i+1}. We assume that $\Omega \subset \Lambda_i$. Thus, $S_{\Lambda_{i+1}}$ appears as a perturbation of S_{Λ_i} by $e^{-\int_{\Delta_{i+1}} :\phi^4:}$. We then introduce a dependance on ζ, replacing this term with $e^{-\zeta \int_{\Delta_{i+1}} :\phi^4:}$. We thus have

$$S_{\Delta_{i+1}}(\zeta)\big|_{\zeta=1} = S_{\Lambda_{i+1}} \quad S_{\Delta_{i+1}}(\zeta)\big|_{\zeta=0} = S_{\Lambda_i}$$

$$\frac{d}{d\zeta} S_{\Delta_{i+1}}(\zeta) = \frac{d}{d\zeta} I_{\Lambda_{i+1}}/Z_{\Lambda_{i+1}} = Z_{\Lambda_{i+1}}^{-1}(\zeta) \frac{d}{d\zeta} I_{\Lambda_{i+1}} - S_{\Lambda_{i+1}} Z_{\Lambda_{i+1}}^{-1} \frac{d}{d\zeta} Z_{\Lambda_{i+1}}(\zeta)$$

$$= -S_{\Lambda_{i+1}}(\Omega \cup \Delta; \zeta) + S_{\Lambda_{i+1}}(\Omega; \zeta) S_{\Lambda_{i+1}}(\Delta; \zeta). \tag{24.148}$$

As for the connected graphs, only graphs joining one square of Ω to Λ are different from 0. Thus extracting from the tree associated with each graph a factor linking Ω to Δ we get

$$\left| \frac{d}{d\zeta} S_{\Lambda_{i+1}}(\zeta) \right| \leq \text{(usual bound)}.e^{-\epsilon \text{dist}(\Omega, \Delta_i)} \tag{24.149}$$

and thus

$$\left| S_{\Lambda_{i+1}} - S_{\Lambda_i} \right| \le K e^{-\epsilon \operatorname{dist}(\Omega, \Delta_i)}, \tag{24.150}$$

which proves the limit by extracting a Cauchy sequence.

24.6.6   The Borel Summability of the $\lambda \phi_2^4$ Theory

The proof of the Borel summability of $\lambda \phi_2^4$ theory is very similar to the results we got in the two preceding sections. It proceeds by following two steps: (1) bound on the remainder of the perturbation expansion; (2) analyticity of the global expression in the Sokal sector.

We have shown in section 24.6.1 on the divergence of the perturbation series that a generic term of order n behaves as

$$KC^n n!$$

From this result we can infer that its Borel transform is convergent in a disc of radius C, centred at the origin, and define in this way an analytic function in this domain.

As for the zero-dimensional model, we can expand in powers of λ the exponential of the interaction up to the Nth order. The rest of order N of the Taylor formula is given by

$$R_N[\lambda \mid \rho, \Lambda] = \frac{1}{N!} \int_0^\lambda \left(\int \frac{\mathrm{d}^{N+1}}{\mathrm{e} t^{N+1}} \mathrm{e}^{-t \int_\Lambda : \phi(x)^4 : \mathrm{d}^2 x} \mathrm{d}\mu_{C_\rho} \right) (\lambda - t)^N \mathrm{d}t. \tag{24.151}$$

We must estimate $R_N[\lambda \mid \rho, \Lambda]$.
Each derivative will produce a $\int_\Lambda : \phi(x)^4 : \mathrm{d}^2 x$; thus,

$$R_N[\lambda \mid \rho, \Lambda] \tag{24.152}$$
$$= \frac{1}{N!} (-1)^{N+1} \int_0^\lambda \left(\int \left(\int_\Lambda : \phi(x)^4 : \mathrm{d}^2 x \right)^{N+1} \mathrm{e}^{-t \int_\Lambda : \phi(x)^4 : \mathrm{d}^2 x} \mathrm{d}\mu_{C_\rho} \right) (\lambda - t)^N \mathrm{d}t.$$

First, we must estimate the Gaussian integral

$$I(\rho, \Lambda) = \int \left(\int_\Lambda : \phi(x)^4 : \mathrm{d}^2 x \right)^{N+1} \mathrm{e}^{-t \int_\Lambda : \phi(x)^4 : \mathrm{d}^2 x} \mathrm{d}\mu_{C_\rho}. \tag{24.153}$$

As we did in the previous section, we will introduce in the field the interpolation parameters $t_{\rho, \Delta}$ and the operators $I_{\rho, \Delta}$ and $P_{\rho, \Delta}$.
We get

$$I(\rho, \Lambda) = \int \prod_{\Delta \subset \Lambda} \left\{ \prod_{i \ge 0} (I_{\rho, \Delta} + P_{\rho, \Delta})^{n_i} \right\} \left(\int_\Lambda : \phi(x)^4 : \mathrm{d}^2 x \right)^{N+1} \mathrm{e}^{-t \int_\Lambda : \phi(x)^4 : \mathrm{d}^2 x} \mathrm{d}\mu_{C_\rho}. \tag{24.154}$$

It is easy to see that the existence of $4(N + 1)$ fields will produce in the bound (24.81) an overall factor $C^{(2(N+1)}(2(N + 1))!$

The first Gaussian integral in (24.154) can be estimated by

$$\int \left(\int_\Lambda : \phi(x)^4 : \mathrm{d}^2 x \right)^{2(N+1)} \mathrm{d}\mu_{C_\rho} \leq |\Lambda| K C^{2(N+1)} (2(N + 1))!, \qquad (24.155)$$

from which follows that after the change of variable $t \rightarrow \lambda t$ in the last integral over t, we get

$$|R_N[\lambda \mid \rho, \Lambda]| \leq C |\lambda|^N C^N N! \qquad (24.156)$$

for some constants C.

These estimates on the perturbation expansion show that the Borel transform is analytic in a disc centred at the origin.

It remains now to look at the analyticity. A first remark is that the exponential of the interaction is an entire function of λ. Moreover in the estimate over the v.e.v. nothing is changed if we replace λ with its real part; thus the various integrals giving the partition function and the other v.e.v. are analytic in a neighbourhood of 0 in the half-space $\mathbb{R}\lambda \geq 0$, thus showing by the Borel transformation the relation between a global theory and its perturbation expansion.

24.6.7 The Mass Gap for ϕ_2^4 in a Strong External Field

In two dimensions the simple quartic interaction term can be replaced with a polynomial of even degree, since the only thing that entered the proofs we presented was that this polynomial had to be bounded from below, with the bound independent of the size of the integration variable ϕ. The idea of the proof reflects the fact that the dominating part in the Lagrangian function is the quadratic part, the free part, the interaction appearing in some way only as a small perturbation of a Gaussian system. This condition was essential in the proofs we presented earlier. It is also easy to understand it intuitively. Consider a system in statistical mechanics. It is well known that if we start with a free system and increase progressively the strength of the interaction there will appear some critical values of the parameters. In the case of $\lambda \phi^4$, this means explicitly that there is a critical value λ_c such that for $\lambda \ll \lambda_c$ we are in a single phase region, clearly dominated, if λ is small enough, by the free underlying system. In this case we showed the existence of an exponential clustering with a range related to the bare mass (m^{-1}). When λ increases and approaches the critical value $\lambda = \lambda_c$, we are at the critical point where there is a long-range order. As a result the previous expansion fails to converge and this signals the appearance of a phase transition.

In this section, we want to point out that our previous results could be extended to cover values of the coupling constant much larger than the critical value, provided some care is taken to ensure that the system is in a single phase. This is because, by going through the critical value, we are experiencing a phase transition and beyond this critical values there may exist more than one phase. We need therefore to find a way to be in one of these phases. Take the example of the Ising model; we can single out a phase by introducing a coupling with a strong magnetic field. When this field is large enough, we can again be in a system where the main behaviour is that of a free (Gaussian) theory with a well-defined mass, which, however, can be different from the one we started with.[11] For the $\lambda\phi_2^4$ theory this can be technically achieved by choosing to study an interaction of the form $\lambda P(\phi) + \mu\phi^k$, with P even and $1 \leq k <\deg P$. In this case we can prove that, for μ large enough, there is a mass gap (thus, a non-zero minimal mass) in the energy spectrum. We will not go through all the details, but the essential steps are the following.

Let us take $P(\phi) = \phi^4$ in two dimensions, i.e. an interaction $\lambda\phi^4 + \mu\phi$. If $a = a(\lambda, \mu)$ is the minimum of $\lambda y^4 + \mu y$ and m_0, the mass of the Gaussian measure, we perform the following transformations:

- Translation
  We translate the fields : $\phi(x) \to \phi(x) + a$; thus an expectation value is changed as

$$< F(\phi) >_{\Lambda,P,m_0^2} =< F^a(\phi) >_{\Lambda,P^a+am_0^2 y,m_0^2}$$

  with $F^a(\phi) = F(\phi + a)$ and $P^a(y) = P(y + a) - P(a)$.

- Scaling

$$< F >_{\Lambda,P,m_0^2} =< F_s >_{s\Lambda,s^{-2}P,s^{-2}m_0^2},$$

  where $F_s(\phi(x)) = F(\phi(sx))$ and $s^2 = a^2/\sigma$ with σ a constant independent of a to be determined later.

- Mass shift

$$< F >_{\Lambda,P+\frac{1}{2}b^2,m_0^2} =< F >_{\Lambda,P^\star,m_0^2+b^2},$$

  where $P^\star$ is characterised by a change of Wick ordering,

$$: P :_{m_0^2} =: P :_{m_0^2+b^2},$$

  with $\frac{1}{2}b^2 = \sigma b_2$, b_2 being the coefficient of y^2 in the development of $P(y) - P(a)$ in power of y.

After these transformations the theory is characterised by a mass $m_1^2 = m_0^2 a^{-2}\sigma + 2b_2\sigma$, all the coefficients of the interaction polynomial being small with respect to the bare transformed mass m_1^2, and we are under the conditions of a weakly coupled ϕ^4 theory.

---

[11] T. Spencer, *Comm. Math. Phys.* **39**, 63 (1974).

24.7   The λ(φ)⁴ Theory in *d* = 3 Dimensions

In the previous section we established the existence of the $\lambda\phi_2^4$ theory. In this section we want to extend this result to three dimensions. In order to emphasise the role of the dimension, we will first comment on the results of the preceding section for the ϕ^4 interaction in an arbitrary dimension d and then sketch the main ideas of the proof on the existence and on the Borel summability of the $\lambda\phi_3^4$ theory. In some special cases the proofs given for $d = 3$ can be extended to four dimensions (for example, the planar or the infrared ϕ_4^4).[12]

In $\lambda\phi_2^4$, the existence of the v.e.v.'s in a finite volume was obtained by introducing an increasing sequence of cut-offs $M_0 = 0 < M_1 < M_2,\ldots, < M_i,\ldots \quad < M_{i-1} < M_i$, $i \in Z^+$, a global cut-off M_ρ, defined by the index ρ, the largest index, a cover of the space of interest Λ by squares of unit size and by creating an expansion whose aim was, in each square of the cover, to progressively lower this cut-off value, going first from ρ to $\rho - 1$ up to the momentum cut-off being equal to 1 for which the bound was obvious. This way we had to compare the interaction measure with a cut-off M_ρ to an interaction with cut-off $M_{\rho-1}$ and so on. This comparison was obtained by performing a truncated perturbation expansion which produced interaction vertices with at least one field having a momentum M_ρ. For a ϕ^4 interaction, this leads in d-dimensions to a convergence factor

$$[M_{\rho-1}]^{-(4-d)} \tag{24.157}$$

per ϕ^4 vertex.

In fact, redoing the analysis leading to formula (24.77), we know that a vertex produced in the expansion has one field of high lower index i. In d dimensions, the covariances obey the estimate (24.40). This means that we get a factor $M_{i+1}^{(d-2)}$. For a covariance linking two vertices, only half of it (the square root) will be attributed to each of the two vertices. By the nature of the expansion, all the remaining fields will have smaller lower cut-offs; therefore, we will bound the contribution of the four propagators by $K[M_i^{\frac{(d-2)}{2}}]^4$ (if the interaction is ϕ^{2n}, we will have to replace it by $K[M_i^{\frac{(d-2)}{2}}]^{2n}$). Integrating over a unit volume in dimension d (or a larger one) gives a factor M_i^{-d}; therefore, we got very generally for each ϕ^{2n}-produced vertex a bound

$$K[M_i^{\frac{(d-2)}{2}}]^{2n}M_i^{-d} = KM_i^{(n(d-2)-d)}, \tag{24.158}$$

---

[12] We will follow the references:

1.   J. Glimm and A. Jaffe, *Fortschr. Phys.* **21**, 327 (1973).

2.   J. Magnen and R. Sénéor, *Comm. Math. Phys.* **56**, 237 (1977).

3.   R. Sénéor, 'Théorie constructive des champs.', Troisième cycle de la physique en Suisse Romande, Semestre d'hiver (1986/1987).

which gives $KM_i^{(d-4)}$ for a ϕ^4 interaction, i.e. $M_i^{(-1+d/4)}$ per field. It is a small factor provided $d < 4$.

From now on, except when specifically stated, we will discuss only the case of a ϕ^4 theory.

We also know that by doing a perturbative expansion of the interaction term we will get a small factor λ (if it has a small value). This factor is independent of the scale of momenta. In addition, by the way the perturbation is built, we will get, for all dimensions less than four, momentum-dependent small factors. If enough perturbation vertices are produced, we can hope in this way to beat the divergence of the Wick bound. However, we also know, from the analysis of the perturbation expansion, that the resulting series is divergent. More specifically, when we have produced k vertices, i.e. we have done an expansion up to the kth order and $4k$ fields have been produced, we end up, by Gaussian integration, with $2k!$ contractions, i.e. a blowing up of the number of diagrams produced. At the kth order, there is a convergent $1/k!$ factor which takes account of the combinatoric of the expansion, but it is not enough to compensate the number of diagrams produced. Thus, we need to optimise the value of k between the divergence of the perturbation series and the convergent factors generated to stop optimally the perturbation expansion, and this choice must be such that at the end of the estimate there is enough convergence to control the Wick bound and to resum each term of the expansion.

Let us take, as an example, what happens going from M_ρ to $M_{\rho-1}$. In a unit volume, we obtain, using the estimate on the number of terms produced given by (24.19),

$$[M_{\rho-1}]^{-k(4-d)} \frac{2k!}{k!} 2^{-k} \simeq e^{k[-(4-d)\ln M_{\rho-1} + \ln 2k - 1]}. \tag{24.159}$$

The minimum is obtained when $k = 1/2[M_{\rho-1}]^{(4-d)}$ and the value is $e^{-1/2[M_{\rho-1}]^{4-d}}$. This factor is enough to dominate the Wick bound if

$$[M_{\rho-1}]^{4-d} \gg \begin{cases} [\ln M_\rho]^2 & \text{if } d = 2 \\ [M_\rho]^{2(d-2)} & \text{if } d > 2. \end{cases} \tag{24.160}$$

The first inequality is possible as soon as the sequence of cut-off increases at least as fast as $\rho^{O(1)}$. However, the last inequality is impossible as soon $4 - d < 2(d-2)$, i.e. $d > 8/3$. In other words, for dimension $d > 2$ we cannot control the Wick bound by doing an analysis in unit volumes. We will have to adapt the size of the volumes in which an expansion will be done according to the index of the momenta we are considering. This gives rise to an expansion known as *a phase space*, or a *multiscale* expansion (depending on the point of view).

Let us now apply what we previously learned in dimensions 0 and 2 to the $\lambda\phi^4$-theory in three dimensions. The degree of divergence D of a vertex of this theory is -1. The theory is still superrenormalisable, but the Wick ordering of the interaction which suppresses the tadpoles does not remove all the divergences. It remains three divergent diagrams: two vacuum energy diagrams, and one mass diagram.

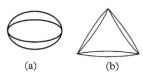

(a)   (b)   **Figure 24.6** *The divergent vacuum diagrams.*

Among the 2-vertex vacuum diagrams of Fig. 24.6, (a) is linearly divergent, $D(G) = -1$, and (b) is logarithmically divergent, $D(G) = 0$. There is also a mass correction given by the diagram of Fig. 16.9 for which $D(G) = 0$, thus corresponding to a logarithmic divergence.

In order to prove results identical to those proven for $\lambda\phi^4$ in two dimensions, we first introduce an increasing sequence of momentum cut-offs with an upper cut-off of index ρ. Then we add to the interaction Lagrangian the counter-terms corresponding to the divergent diagrams with their proper coefficients, in such a way that, when performing a truncated perturbation expansion, to the divergent diagrams will be associated their cut-off counter-terms.[13]

The renormalised action in the volume Λ is given by

$$V(\Lambda, g, \rho; \phi) = \lambda \int_\Lambda : \phi(x)^4 : \mathrm{d}^3x - \frac{\lambda^2}{2} \int_\Lambda \delta m_\rho^2(x) : \phi(x)^2 : \mathrm{d}^3x$$
$$+ \frac{\lambda^2}{2} \int \left(\int_\Lambda : \phi(x)^4 : \mathrm{d}^3x \right)^2 \mathrm{d}\mu_\rho$$
$$- \frac{\lambda^3}{3!} \int \left(\int_\Lambda : \phi(x)^4 : \mathrm{d}^3x \right)^3 \mathrm{d}\mu_\rho, \tag{24.161}$$

where $\mathrm{d}\mu_\rho$ stands for $\mathrm{d}\mu_{C_\rho}$, where the mass correction $\delta m_\rho^2(x)$ corresponding to the diagram of Fig. 16.9 is given by

$$\delta m_\rho^2(x) = -4^2 \int \left(: \phi(x)^3 : \int : \phi(y)^3 : \mathrm{d}^3y \right) \mathrm{d}\mu_\rho, \tag{24.162}$$

and where the third and fourth terms of the interaction (24.161) correspond to the vacuum diagrams of Fig. 24.6.

The partition function is given by

$$Z(\Lambda, \lambda, \rho) = \int \mathrm{e}^{-V(\Lambda, \lambda, \rho; \phi)} \mathrm{d}\mu_\rho. \tag{24.163}$$

---

[13] This is in principle the same mechanism as that corresponding to replace monomials in the field by the corresponding Wick ordered ones. For example, when one Wick orders ϕ^4, this means that everytime the ϕ^4 vertex will be produced, there will be associated with the right combinatoric coefficient a ϕ^2 counter-term corresponding to the contraction of two of the fields of the vertex, forming a tadpole, and a constant one corresponding to the complete self-contractions of all four fields.

To summarise, the general principle of all the expansions is that by adding the counter-terms in the exponent, the divergent diagrams minus their counter-terms will converge to finite quantities when $\rho \to \infty$. We remark that in contradistinction with the Wick order mechanism, these counter-terms are not necessarily linear in the coupling constant λ. For example, the mass counter-term we obtained in Eq. (24.162) is proportional to λ^2. This non-linearity of the counter-terms has for consequence that at any given order in the perturbation expansion there remain some divergent terms which are not compensated. The compensation is complete only if we look at the expansion to all orders. Regarding the Wick bound, this will not be a problem if, by the nature of the expansion, we have produced enough small factors to compensate this divergence.

As an example, let us show this compensation mechanism for the mass counter-term.

Suppose that two functional derivatives $\delta/\delta\phi(x)\delta/\delta\phi(y)$ act on the exponential of the interaction. The following two terms will be produced,

$$(4\lambda)^2 \int \,:\phi^3(x)::\phi^3(y): \,\mathrm{d}\mu_{C_\rho} - 2\delta m_\rho^2(x)\delta^{(3)}(x-y)$$
$$= (4\lambda)^2 3![C_\rho(x-y)]^3 - 2\delta m_\rho^2(x)\delta^{(3)}(x-y)$$
$$= 2F_\rho(x-y) - 2\delta m_\rho^2(x)\delta^{(3)}(x-y) \equiv F_\rho^{\mathrm{ren}}(x-y) \tag{24.164}$$

with $F_\rho(x-y) \equiv \frac{(4\lambda)^2 3!}{2}[C_\rho(x-y)]^3$ and $\delta m_\rho^2(x) \equiv \frac{(4\lambda)^2 3!}{2}\int [C_\rho(x-y)]^3 \mathrm{d}^3y$. It is easy to check that $F_\rho^{\mathrm{ren}}(x-y)$ is not singular (we can integrate with some regular function $f(y)$), although $C(x)^3 \simeq 1/|x|^3$ is not integrable.

A consequence of this renormalisation procedure is that the Fourier transform $\widetilde{F_\rho^{\mathrm{ren}}}(p)$ of F_ρ^{ren} is finite for all p and all ρ such that $\widetilde{F_\rho^{\mathrm{ren}}}(0) = 0$.

More concretely, in the general series resulting from expansions and contractions, we will obtain expressions like

$$\int \left[C(.,x)C(.,y)\frac{\delta}{\delta\phi(x)}\frac{\delta}{\delta\phi(y)}e^{-V(\Lambda,\lambda,\rho;\phi)} \right] \mathrm{d}^3x\mathrm{d}^3y\mathrm{d}\mu_\rho. \tag{24.165}$$

Among the terms produced by the application of the functional derivatives, we will get

$$\int \left[2\, C(.,x)C(.,y)\left[F_\rho(x-y) - \left(\int F_\rho(x-z)\mathrm{d}^3z\right)\delta^3(x-y)\right]\mathrm{d}^3x\mathrm{d}^3y\right]e^{-V(\Lambda,\lambda,\rho;\phi)}\right]\mathrm{d}\mu_\rho$$
$$= -\int \left[\int [C(.,x) - C(.,y)]^2 F_\rho(x-y)\mathrm{d}^3x\mathrm{d}^3y\right]e^{-V(\Lambda,\lambda,\rho;\phi)}\right]\mathrm{d}\mu_\rho$$
$$= -\int \left[\int [\vec{\nabla}C(.,x)]^2 (x-y)^2 F_\rho(x-y)\mathrm{d}^3x\mathrm{d}^3y\right]e^{-V(\Lambda,\lambda,\rho;\phi)}\right]\mathrm{d}\mu_\rho,$$

showing that renormalisation amounts to the replacement of F by $(x-y)^2 F$ and the application of a gradient on the two external legs. Vacuum energy terms can also be produced during the expansion, but since they are constant terms they will be identically cancelled by the counter-terms.

Now we give some elements on how, in order to prove the existence of the limit when $\rho \to \infty$, the expansion must be modified because the underlying space is of dimension three.

We know from the commentaries made at the beginning of this section that we cannot hope to dominate the Wick bound by producing enough perturbation terms in volumes of unit size. If a perturbation term means small factors on one side, it also means increasing divergent factorials in the number of terms produced on the other side. Therefore, in order to control the Wick bound, if some amount of perturbation terms are produced in volumes of size 1, then other perturbation terms must be produced in covers with cubes of smaller and smaller sizes.

Let us start by giving a more precise definition of the sequence of cut-offs. Choosing M, $M > 1$, large enough, we define the sequence of momenta

$$M_0 = 0, M_1 = M, \ldots, M_i = M^i, \ldots, M_\rho = M^\rho \tag{24.166}$$

with ρ being the index of the largest cut-off. Next, we introduce the notion of momentum slices. To each slice i, defined by momenta in the interval $[M_{i-1}, M_i]$, we associate a cover $\{\mathcal{D}_i\}$, partition of $\mathcal{R}^3$, made of cubes $\Delta \in \mathcal{D}_i$ of size $\Delta^{1/3} = M_i^{-1}$. Each $\{\mathcal{D}_i\}$ is a refinement of the cover $\mathcal{D}_{i-1}$, the sequence of volumes satisfying

$$|\Delta_0| = 1 > |\Delta_1| > \cdots > |\Delta_j| > \ldots, \quad \text{with} \quad |\Delta_{j-1}| > |\Delta_j|, \quad j \in \mathbb{Z}^+. \tag{24.167}$$

We define $\widetilde{\mathcal{D}} = \bigcup_{0 \leq l \leq \rho} \mathcal{D}_l$ and the elements of phase space by the following.

Definition 1. *Cell of phase space*
A cell of the phase space is a pair $[M_{i-1}, M_i] \times \Delta$, $\Delta \in \mathcal{D}_i$.

We also define the following.

Definition 2. *Properly localised field*

$$M_i^{-1} \leq |\Delta|^{1/3}. \tag{24.168}$$

A field ϕ localised in a cube Δ and with lower momentum M_i satisfying (24.168) is said to be properly localised.

The introduction of this notion of properly localised fields relies on the fact that it is impossible to simultaneously localise a function and its Fourier transform with an arbitrary precision.

Because of the local nature of the interaction, for any index i, introducing $\phi = \sum_{\Delta \in \mathcal{D}_i} \phi \chi_\Delta = \sum_{\Delta \in \mathcal{D}_i} \phi_\Delta$, with $\phi_\Delta = \phi \chi_\Delta$, we have

$$V(\lambda, \Lambda, \rho; \sum_{\Delta \in \mathcal{D}_i} \phi_\Delta) = \sum_{\Delta \in \mathcal{D}_i \bigcap \Lambda} V(\lambda, \Lambda, \rho; \phi_\Delta) = \sum_{\Delta \in \mathcal{D}_i \bigcap \Lambda} V_\Delta. \qquad (24.169)$$

24.7.1 The Expansion: Definition

24.7.1.1 The Factors Which Will Guarantee the Convergence

In dimension $d > 2$ we learned that we cannot hope to beat the Wick bound in cubes of unit size and we need to do expansions in cubes of smaller and smaller sizes. This has led to the introduction of momentum slices and associated covers and thus to phase space.

The existence of the theory in an infinite volume (the thermodynamic limit) will rely on the following:

- The smallness of the coupling constant λ
- A sufficiently rapid decrease of the propagators[14]

$$c_n d(\Delta, \Delta')^{-n} \qquad (24.170)$$

with $n \in \mathbb{Z}^+$, where $d(\Delta, \Delta')$ is the scaled distance:

$$d(\Delta, \Delta') = \sup(1, \text{dist}(\Delta, \Delta'), M_l \text{dist}(\Delta, \Delta'), M_{l'} \text{dist}(\Delta, \Delta')) \qquad (24.171)$$

and $\text{dist}(\Delta, \Delta')$ is the Euclidean distance between Δ and Δ'. We remark that in case both lower momentum cut-offs are zero, the scaled distance reduces to $\sup(1, \text{dist}(\Delta, \Delta'))$.

This behaviour of a propagator, $c_n d(\Delta, \Delta')^{-n}$, $n \in \mathbb{Z}^+$, corresponds to the contractions of two fields, one ϕ with low momentum M localised in Δ, with another one ϕ' localised in Δ' with low momentum M'. In fact, if ϕ is properly localised and if $\Delta \in \mathcal{D}_l$ and $\Delta' \in \mathcal{D}_{l'}$ with $l' \leq l$, the factor $d(\Delta, \Delta')^{-m}$ satisfies

$$\sum_{\Delta' \in \mathcal{D}_{l'}} d(\Delta, \Delta')^{-m} \leq O(1) \quad \forall m > 3 \qquad (24.172)$$

with some $O(1)$ depending on m and showing that it is a good combinatoric factor for the sum, Δ resulting from all possible $\Delta' \in \mathcal{D}_{l'}$. This factor was already noted in section 24.6 when we introduced the notion of combinatoric factors.

- The boundedness from below of the interaction

---

[14] For the theories we are considering, the decrease of the propagators is exponential, although a power like one with $n > 3$ would have been sufficient.

and the facts that at scale i and depending on the dimension.

- A propagator behaves like

$$K_n M_i^2 d(\Delta, \Delta')^{-n} \quad in \quad d = 4 \quad \text{dimensions}$$
$$K_n M_i d(\Delta, \Delta')^{-n} \quad in \quad d = 3 \quad \text{dimensions.}$$

- The integration is done on cells of size $|\Delta| = M_i^{-d}$

from which follows that the following convergent factors can be attributed

$$M_i^{(d-2)/2} \quad \text{per leg or per field of index} \quad i$$
$$M_i^{-d} \quad \text{per integration.}$$

24.7.1.2   *The Expansions*

The phase space expansion is then defined as the result, for each momentum slice, of two successive expansions:

- A horizontal expansion
  In each slice i, a cluster expansion is done to test the coupling between the different cubes of $\mathcal{D}_i$ as it was done in section 24.6.3 on $\lambda\phi^4$ in two dimensions by introducing interpolated covariances, depending on parameters $s_1, s_2, \ldots$ as in (24.103) and (24.105).
- A vertical expansion
  From a cluster expansion at scale i we cannot hope to perform a thermodynamic limit since the interaction still couples the various horizontal slices; therefore, in each cell (i, Δ), an expansion is done to test the coupling between fields of indices larger than or equal to i and those of indices smaller than i by introducing in the interaction new interpolating parameters $\{t_{i,\Delta}\}$ different from the $\{s_i\}$'s introduced for the cluster expansion.
  This is done in modifying each V_Δ by

$$\int_{\Delta \in \mathcal{D}_i} \left[\left(\phi_h(x) + t_{i,\Delta}\phi_l(x) \right)^4 + (1 - t_{i,\Delta}^4)\phi_l(x)^4 \right] \mathrm{d}^3 x \equiv V_\Delta(\{t_\Delta\}) \qquad (24.173)$$

with the splitting of the field into two parts $\phi(x) = \phi_h(x) + \phi_l(x)$ with

$$\phi_h = \sum_{j \geq i} \phi^j \quad \text{the well-localised part or high momentum part}$$

$$\phi_l = \sum_{j < i} \phi^j \quad \text{the low momentum part.} \qquad (24.174)$$

We remark that the parameters $t_{i,\Delta}$ interpolate the interaction, going from $t_{i,\Delta} = 1$ to $t_{i,\Delta} = 0$. This interpolation can be seen as connecting two situations: an initial expression with an interaction where in $\Delta \in \mathcal{D}_i$ there are the full components of the fields, their high momentum as well as their low momentum parts, and a final expression where the interaction is now the sum of two ϕ^4 interactions in Δ, one with high momentum fields (from which a single cluster and Mayer expansions make it possible to perform the thermodynamic limit), and another one with low momentum fields, i.e. with no couplings between high and low momentum fields, for which we will have to repeat the procedure.

A Taylor expansion is done up to the fifth order on each $e^{-V_\Delta(t)}$. The reason to expand to this order is that each time a ϕ^4 is derived, at least one high momentum leg and one low momentum leg will be simultaneously produced. The existence of the high momentum leg will guarantee the convergence. If in a cube Δ five derivations are done, this implies that in this cube there are five low momentum legs (that we considered at scale i as external legs) and thus the corresponding sub-diagrams, with at least five external legs, will be convergent according to the general power counting of the theory.

This expansion is expressed algebraically, for each $t_{i,\Delta}$, by

$$e^{-V_\Delta(t)}|_{t_{i,\Delta}=1} = (I_{i,\Delta}^{(4)} + R_{i,\Delta}^{(4)})e^{-V_\Delta(t)}, \tag{24.175}$$

where $I_{i,\Delta}^{(4)}e^{-V_\Delta(t)}$ are the first four Taylor expansion terms around $t_{i,\Delta} = 0$ and $R_{i\Delta}^{(4)}e^{-V_\Delta(t)}$ is the remainder term (which involves five derivatives with respect to $t_{i,\Delta}$).

In each slice, the two expansions will generate graphs the same way as in ϕ_2^4, for example through formula (24.98) or through the Mayer links. Some of these graphs are connected. The horizontal connectivity is the usual one as it appears in cluster expansions. The vertical connectivity results from the notion of connectivity for cubes belonging to different covers.

Definition 3 (Connectivity). *$\Delta_i \in \mathcal{D}_i$ and $\Delta_j \in \mathcal{D}_j$, with $j \le i$, are connected if:*

- *There is a propagator connecting them (this is an element of the cluster expansion at level i). This implies that $i = j$;*
- *$\Delta_i \subset \Delta_j$ and there is a vertex localised in Δ with two fields or more, one of index i and one of index j;*
- *$i = j + 1$, $\Delta_i \subset \Delta_{i-1}$ and the ith vertical expansion has produced a remainder term $R_{\Delta_i}^{(4)}$ in Δ_i, i.e. there are five low momentum legs produced by a vertex in Δ_i.*

24.7.2 The Expansion Completed

The expansion, as defined by a succession of cluster and Mayer expansions, must be completed. For example, we have in the exponential of the interaction a mass counter-term of second order in λ which behaves as $\ln M^\rho$. This mass counter-term must cancel the divergence part of a mass term formed by the contraction of some vertices generated during the expansion. On the other hand, we know that unless expanding an infinite number of terms, there will remain some non-cancelled divergences. We will have to explain what to do in terms of contractions when a vertex is produced by a cluster expansion or when it is produced by a vertical expansion. This will lead to a combination of perturbation and contraction formulae which will form what is called a 'partly renormalised expansion'. It is partly renormalised because in the later stages of the expansion some logarithmic divergences will remain uncancelled. They will occur only in terms where convergent factors produced earlier in the expansion are strong enough to dominate these divergences. Finally, we must explain what we should do with the low momentum fields. Low momentum fields cannot be integrated with Gaussian measure because they will produce too many terms. Thus, for $\Delta \in \mathcal{D}_i$, we split each low momentum field part of index l as defined in (24.174) into two parts,

$$\phi_{l,\Delta}(x) = \bar{\phi}_{l,\Delta}(x) + \delta\phi_{l,\Delta}(x), \tag{24.176}$$

where

$$\bar{\phi}_{l,\Delta}(x) = \frac{\chi_\Delta(x)}{|\Delta|} \int \phi_{l,\Delta}(y)\mathrm{d}^3y. \tag{24.177}$$

We will show that $\delta\phi_{l,\Delta}$ behaves as a high momentum field. To see this behaviour, we decompose the field over the momentum slices and prove this property for each component: $\phi_l = \sum_j \phi_l^j$; thus $\delta\phi_{l,\Delta}(x) = \sum_j \delta\phi_{l,\Delta}^j(x)$ with $i > j$ and

$$\delta\phi_{l,\Delta}^j(x) = \phi^j(x)\chi_\Delta(x) - \frac{\chi_\Delta(x)}{|\Delta|} \int \phi_{l,\Delta}^j(y)\mathrm{d}^3y; \tag{24.178}$$

now introducing in the integral part a formal Taylor expansion of $\phi^j(y)$ around x

$$\phi^j(y) = \phi^j(x) + \int_0^1 \sum_\nu (y-x)^\nu \partial_\nu \phi^j(x + s(y-x))\mathrm{d}s \tag{24.179}$$

we get that

$$\delta\phi_{l,\Delta}^j(x) = -\frac{\chi_\Delta(x)}{|\Delta|} \int \chi_\Delta(y)\left(\int_0^1 \sum_\nu (y-x)^\nu \partial_\nu \phi^j(x + s(y-x))\mathrm{d}s\right)\mathrm{d}^3y. \tag{24.180}$$

As a result, with the arguments given at the beginning of this section, we see that with respect to the low momentum component the $y-x$ coefficient gives a M^{-i} and the derivative acting on the field gives a M^j; thus the whole right-hand side gives an extra factor M^{j-i} which transforms this low momentum component into a high momentum one.

What remains of the low momentum part will not, as it has been said, be integrated with the Gaussian measure and will be controlled by the positivity of the interaction using

$$|\bar{\phi}_{l,\Delta}|^4 = \left|\frac{1}{|\Delta|} \int_\Delta \phi_l(x)\mathrm{d}^3x\right|^4 \le \frac{\lambda^{-1}}{|\Delta|} \int_\Delta \lambda\phi(x)^4\mathrm{d}^3x. \tag{24.181}$$

Now, each term of the expansion will be of the form of a sum of products of high momentum fields and $\delta\phi$-fields times the exponential of the interaction and will be estimated using

$$\int \prod_h (\phi_h, \delta\phi) \prod_\Delta |\phi_{l,\Delta}| e^{-\lambda \int (:\phi^4:+CT)} d\mu_\rho$$
$$\leq \left[\int \left[\prod_h (\phi_h, \delta\phi) \right]^2 d\mu_\rho \right]^{1/2} \left(\sup \prod_\Delta |\phi_{l,\Delta}| e^{-\int (:\phi_l^4:+CT)} \right), \tag{24.182}$$

where CT stands for the counter-terms and the supremum is bounded by

$$|\bar\phi_\Delta|^p \leq [2\lambda^{-1/4}|\Delta|^{-1/4}]^p p! e^{\frac{\lambda}{2} \int_\Delta \phi^4}. \tag{24.183}$$

In order to control this last exponential term, we use the fact that in $-\int_\Delta :\phi^4:$, the low momentum part can be written as

$$-\int_\Delta :\phi_l^4: := -\frac{1}{2} \int_\Delta \phi_l^4 - \int_\Delta \left(\frac{1}{2}\phi_l^4 - 6c_l\phi_l^2 + 3c_l^2 \right). \tag{24.184}$$

The first term of the right-hand side will be used to dominate the exponential factor introduced in the bound (24.183) for a product of low momentum fields localised in a cube Δ. The second term is bounded from below by $Kc_l^2|\Delta|$ for some constant K. This term can be estimated as follows: since we are dealing with the low momentum part at level i, in d dimensions, $\Delta \in \mathcal{D}_i$, $|\Delta| = M_{-i}^d$ and $c_l^2 = [M_i^{(d-2)}]^2$; thus

$$Kc_l^2|\Delta| \leq [M_i]^{2(d-2)-d} = M_i^{d-4}, \tag{24.185}$$

thus showing that for slices of high momentum i in each cube Δ (as they are generated by the vertical expansion) of a cover $\mathcal{D}_i$, the bound of the exponential of the interaction (or Wick bound, it will be the same bound for the high momentum field part) corresponds to an estimate (for dimensions less than 4) of a constant per cube.

The expansion thus completed is called 'a partly renormalised phase space expansion' (PRPSE).

We will not develop more these points here.[15] We go directly to the results.

24.7.3   The Results

As for the ϕ_2^4 case, we have the following theorems.

Theorem 21 (I). *Let* $\lambda \in \mathbb{C}$, $|Arg\lambda| < \frac{\pi}{2}$, *then there exists* $C(\lambda) \geq 0$ *such that*

$$\left| \int e^{-V(\lambda,\Lambda,\rho)} d\mu \right| \leq e^{C(\lambda)|\Lambda|}. \tag{24.186}$$

---

[15] See J. Magnen and R. Sénéor, *Comm. Math. Phys.*, **56**, 237 (1977).

Theorem 22 (II). *In the weak coupling region, i.e. λ/m small enough, the Euclidean vacuum expectation values (Schwinger functions) satisfy the Osterwalder–Schrader axioms with an exponential clustering. They can be analytically continued in λ in the domain defined in Theorem I and are C^∞ at $\lambda = 0$.*

Theorem 23 (III). *Under the conditions of Theorems I and II, the perturbation series of these functions at $\lambda = 0$ are Borel summable.*

We will not prove these theorems. Their proofs result from the same arguments used to prove the existence and the cluster property of the theory in a finite volume and is similar to that given for ϕ_2^4. The infinite volume limit is then proved exactly as for the two-dimensional case.

24.8   The Massive Gross–Neveu Model in *d* = 2 Dimensions

The massive Gross–Neveu field theory is probably the simplest renormalisable theory: it is two dimensional, it is massive, i.e. the infrared difficulties are avoided, it has a quartic interaction, its perturbation series is, with finite ultraviolet cut-off, convergent in a disk, and the diagrams of the theory are topologically the same as those of a ϕ^4 bosonic theory. It has an infinite number of divergent diagrams but it is asymptotically free. It was introduced in Problem 23.2.

The interest of the model stems from the fact that it mimics many properties which, as we will see in the next chapter, are found in the real world. Both massless or massive Gross–Neveu models are asymptotically free. By massive we mean having a non-zero bare mass. We will deal with massive N-component fermion fields. We will construct the Schwinger functions of the model using the property of asymptotic freedom together with the existence of a non-zero bare mass to get a small parameter of expansion. The model being just renormalisable there are infinitely many primitively divergent diagrams requiring coupling constant, mass, and wave function counter-terms.

We will first introduce the model with ultraviolet cut-offs and finite volume and show that the unnormalised and unrenormalised Schwinger functions are entire functions of the coupling constant. The next step will be to define a partly renormalised phase space expansion (PRPSE) with given bare coupling and a correct bare mass ansatz. This PRPSE is shown to be absolutely convergent if the bare constants are suitably chosen following the renormalisation group expressions.

24.8.1   Definition of the Model

The model is defined by the Lagrangian density

$$\mathcal{L}(\overline{\psi}, \psi) = \overline{\psi}(\mathrm{i}\zeta\,\slashed{\partial} + m)\psi - g(\overline{\psi}.\psi)^2 \tag{24.187}$$

with $\overline{\psi}.\psi = \sum_{a,\alpha} \overline{\psi}_a^\alpha \psi_a^\alpha$. $a, b, \ldots$ are spinor indices taking the values $0, 1$ and $\alpha, \alpha', \ldots$ are internal indices; we will call them 'colour indices', taking the values $1, \ldots, N$[16]; we may also note $A, B, \ldots$ pairs of colour and spinor indices: $A = (a, \alpha), B = (b, \beta), \ldots$; m is the mass and ζ the wave function renormalisation; and finally $g > 0$. The finite volume Λ will be a square box in $\mathbb{R}^2$ of area $|\Lambda|$. Given $M > 1$, we define a sequence of momentum cut-offs $M, M^2, \cdots, M^\rho$; the starting ultraviolet cut-off will be referred to as corresponding to the index ρ.

The model has a perturbation expansion similar to a $(\bar{\phi}.\bar{\phi})^2$ field theory with an N-component vector field $\bar{\phi}$. Formally the $2p$-point functions are defined by

$$S_{2p}^{\gamma_1, \cdots, \gamma_p; \delta_1, \cdots, \delta_p} (y_{1,c_1}, \cdots, y_{p,c_p}; z_{1,d_1} \cdots, z_{p,d_p}) \tag{24.188}$$

$$= \frac{1}{Z} \int \psi_{c_1}^{\gamma_1}(y_1) \cdots \psi_{c_p}^{\gamma_p}(y_p) \overline{\psi}_{d_1}^{\delta_1}(z_1) \cdots \overline{\psi}_{d_p}^{\delta_p}(z_p) \mathrm{e}^{-\int \mathcal{L}(\overline{\psi}, \psi)} \prod_x \mathrm{d}\overline{\psi}(x) \mathrm{d}\psi(x)$$

$$Z = \int \mathrm{e}^{-\int \mathcal{L}(\overline{\psi}, \psi)} \prod_x \mathrm{d}\overline{\psi}(x) \mathrm{d}\psi(x), \tag{24.189}$$

where the fermionic integration follows the Berezin rules.

With cut-offs, we can rewrite this formula as;

$$S_{2p, \Lambda, \rho}^{\gamma_1, \cdots, \gamma_p; \delta_1, \cdots, \delta_p} (y_{1,c_1}, \cdots, y_{p,c_p}; z_{1,d_1} \cdots, z_{p,d_p}) \tag{24.190}$$

$$= \frac{1}{Z_{\Lambda, \rho}} \int \psi_{c_1}^{\gamma_1}(y_1) \cdots \psi_{c_p}^{\gamma_p}(y_p) \overline{\psi}_{d_1}^{\delta_1}(z_1) \cdots \overline{\psi}_{d_p}^{\delta_p}(z_p) \mathrm{e}^{-\int_\Lambda \mathcal{L}_\rho(\overline{\psi}, \psi)} \mathrm{d}\mu_\rho(\overline{\psi}, \psi)$$

with

$$\mathcal{L}_{I,\rho}(\overline{\psi}, \psi) = -g_\rho (\overline{\psi}.\psi)^2,$$

m_ρ, ζ_ρ, and g_ρ being respectively, with cut-off ρ, the bare mass, the bare wave function renormalisation, and the bare coupling constant. Also from now on we will omit the colour indices. We have also introduced, using the quadratic part of the cut-off Lagrangian $\mathcal{L}_{0,\rho}(\overline{\psi}, \psi) = \overline{\psi}(i\zeta_\rho \slashed{\partial} + m_\rho)\psi$, the fermion Gaussian measure defined by

$$\int \psi_a(x) \bar{\psi}_b(y) \mathrm{d}\mu(\overline{\psi}\psi) = C_{a,b}(x - y), \tag{24.191}$$

where $C_{a,b}(x-y)$ is the fermion propagator.

With a cut-off $\eta_\rho(p)$, it is defined by

$$C_\rho(x - y) = \int \mathrm{e}^{\mathrm{i}p.(x-y)} \frac{\zeta_\rho \slashed{p} + m_\rho}{\zeta_\rho^2 p^2 + m_\rho^2} \eta_\rho(p) \mathrm{d}^2 p \tag{24.192}$$

---

[16] We consider $N \geq 2$ since the Gross–Neveu model corresponding to $N = 1$ can be solved completely.

with

$$\gamma_0 = \begin{pmatrix} i & 0 \\ 0 & -i \end{pmatrix}, \qquad \gamma_1 = \begin{pmatrix} 0 & 1 \\ -1 & 0 \end{pmatrix}$$

$$\slashed{p} = p_0\gamma_0 + p_1\gamma_1 \qquad \gamma_a = -\gamma_a^* \qquad \{\gamma_a, \gamma_b\} = -2\delta_{a,b}. \tag{24.193}$$

The theory is invariant under a change of scale

$$\begin{aligned} \zeta &\to s\zeta \\ m &\to m/s \\ g &\to g/s^2, \end{aligned} \tag{24.194}$$

which means that we can choose ζ to be equal to 1.

Let us start for simplicity with a Gross–Neveu model where η_ρ is an exponential cut-off of index ρ, m depends on ρ, and $\zeta = 1$.

Now we rewrite C_ρ as the convolution of two terms

$$C_\rho(x-y) = \int e^{ip.(x-y)} \frac{\slashed{p}+m}{p^2+m^2} e^{-2p^2 M^{-2(\rho+1)}} d^2p = \int A_\rho(x-t)B_\rho(t-y)dt, \tag{24.195}$$

where

$$\tilde{A}(p) = \frac{\slashed{p}+m}{(p^2+m^2)^{3/4}} e^{-p^2 M^{-2(\rho+1)}} \quad \text{and} \quad \tilde{B}(p) = \frac{e^{-p^2 M^{-2(\rho+1)}}}{(p^2+m^2)^{3/4}}. \tag{24.196}$$

In this case, the unnormalised 2-point Schwinger function of formula (24.190), omitting all colour and spinor indices, is given, after a complete expansion in powers of g_ρ of the exponential of the interaction, by

$$\mathbf{S}_{2p,\Lambda,\rho}(\{y\}, \{z\}) \tag{24.197}$$

$$= \sum_{n=0}^{+\infty} \frac{(g_\rho)^n}{n!} \int \psi(y_1) \cdots \psi(y_p)\overline{\psi}(z_1) \cdots \overline{\psi}(z_p) \left[\int_\Lambda (\overline{\psi}\psi)^2 d^2x \right]^n d\mu_\rho(\overline{\psi}, \psi)$$

$$= \sum_{n=0}^{+\infty} \frac{(g_\rho)^n}{n!} \int_\Lambda d^2x_1 \cdots d^2x_n \sum_{a_i,b_i} \cdots \begin{bmatrix} y_{1,c_1} & \cdots & y_{p,c_p} & x_{1,a_1} & x_{1,b_1} & \cdots & x_{n,a_n} & x_{n,b_n} \\ z_{1,d_1} & \cdots & z_{p,d_p} & x_{1,a_1} & x_{1,b_1} & \cdots & x_{n,a_n} & x_{n,b_n} \end{bmatrix},$$

where g_ρ is the bare coupling constant and

$$\int \psi_a(x)\bar{\psi}_b(y)d\mu_\rho = C_{\rho;a,b}(x-y). \tag{24.198}$$

We use the Cayley's notation for determinants

$$\binom{u_{i,a}}{v_{j,b}} = \det C_{\rho;a,b}(u_i - v_j). \tag{24.199}$$

Now we can estimate the determinant using Gram's inequality which says that

$$\left| \det \left(\int f_i(x) g_j(x) dx \right) \right| \leq \prod_i \|f_i\|_2 \|g_i\|_2. \tag{24.200}$$

Applying it to the right-hand side. of (24.197) we get

$$|\mathbf{S}_{2p,\Lambda,\rho}(\{y\}, \{z\})| \leq \sum_n \int_\Lambda d^2x_1 \cdots d^2x_n \frac{(g_\rho)^n}{n!} \tag{24.201}$$

$$\times \left\{ \left(\sum_a \|A_{\rho;a}\|_2 \right) \left(\sum_b \|B_{\rho;b}\|_2 \right) \right\}^{2n} \cdot \prod_{i=1\ldots p} \|A_{\rho;c_i}\|_2 \|B_{\rho;d_i}\|_2$$

with

$$\|A_{\rho;a}\|_2 = \left\{ \sum_b \int_\Lambda A_{\rho;a,b}(x) \bar{A}_{\rho;b,a}(x) d^2x \right\}^{1/2} \leq O(1) M^{\rho/2} \tag{24.202}$$

and similar formula for B. Therefore,

$$|\mathbf{S}_{2p,\Lambda,\rho}(\{y\}, \{z\})| \leq \sum_n \int_\Lambda d^2x_1 \cdots d^2x_n \frac{(g_\rho)^n}{n!} [O(1) M^\rho]^{p+2n} |\Lambda|^n, \tag{24.203}$$

which shows that the radius of convergence of $S_{2p,\Lambda,\rho}$ as a power series in g_ρ is infinite. This proves easily (reintroducing ζ).

Lemma 11. *The unnormalised Schwinger functions $S_{2p,\Lambda,\rho}(\{y\}, \{z\})$, in a finite volume Λ and with a finite ultraviolet cut-off M^ρ, are entire functions of the bare coupling constant g_ρ for any finite value of the bare mass m and of the wave function coefficient ζ (both of them can depend on the ultraviolet cut-off).*

This result is in strong contrast with the case of the similar bosonic ϕ^4- field theory for which the radius of convergence is zero. This is a consequence of the Pauli principle which is expressed by the fact that the interaction is the logarithm of a determinant in which, by its structure, many cancellations are hidden. It is reflected in the Feynman rule according to which diagrams with an even, or odd, number of fermionic loops have opposite signs.

In order to go further, we must face two main problems: the infinite volume limit and the removal of the ultraviolet cut-off.

24.8.2   The Infinite Volume Limit

As we did in the previous section, ρ being given we choose $\zeta_\rho = 1$ and m_ρ fixed. We will prove that with fixed ultraviolet cut-off, the bare expansion of the normalised Schwinger functions $S_{2p,\rho}$, which are the limits of $S_{2p,\Lambda,\rho}$ when Λ tends to $\mathbb{R}^2$, still has a non-zero radius of convergence.

The proof of the existence of the infinite volume limit results from a cluster expansion followed by a Mayer expansion. It follows mainly what was described in sections 24.6.3 and 24.6.4 and will not be reproduced here. It relies on the fact that the theory is massive and propagators fall off rapidly with the distance. The result is expressed in the following lemma.

Lemma 12. *The expansions for the normalised, as well as the truncated Schwinger functions S and S^T, for fixed bare mass m_ρ, considered as power series in g_ρ, have a non-zero radius of convergence $r_\rho(\Lambda, m_\rho)$, which satisfies*

$$[r_\rho(\Lambda, m_\rho)]^{-1} \le K(m_\rho)M^{10\rho} = C_\rho \tag{24.204}$$

with a bound uniform in Λ. The thermodynamic limit of S and S^T exists, uniformly in g_ρ for $|g_\rho| \le (C_\rho)^{-1}$.

Therefore as power series in g_ρ, these normalised functions have a non-zero radius of convergence $r_\rho(m_\rho)$, with $r_\rho(m_\rho) \ge (C_\rho)^{-1}$.

24.8.3   The Removal of the Ultraviolet Cut-off

To remove the ultraviolet cut-off we need to refine the analysis, as it was done for ϕ_3^4, by introducing momentum slices given by the scale of momenta $1, M, M^2, \ldots, M^\rho$, accompanied by a corresponding sequence of covariances $C^0(x - y), \ldots, C^\rho(x - y)$ with $C_\rho(x - y) = \sum_{k=0,\ldots,\rho} C^k(x - y)$. The functions A and B are decomposed in a similar way such that $C^k(x - y) = \int A^k(x - t)B^k(t - y)\mathrm{d}^2 t$. The techniques are the same as before, with one important difference: the theory is renormalisable and, therefore, we have an infinite number of primitively divergent diagrams requiring an infinite number of counter-terms. We saw in the ϕ_3^4 theory that in order to define the measure of the interacting theory, we had to raise the counter-terms in the exponent of the exponential, each counter-term with the right coefficient, in order to cancel the divergent graphs of the perturbation expansion. We explained there that this had to be done up to a given order, thus producing what we called 'a partly renormalised phase space expansion' (PRPSE). This was achieved for ϕ_3^4 because the theory was superrenormalisable and there were only a finite number of divergent diagrams. In the case of the Gross–Neveu model the need to introduce counter-terms at any order in the perturbation expansion makes the construction of a PRPSE more difficult, but nevertheless possible, because of the property of the theory to be asymptotically free. In fact, in a given momentum slice, it will be necessary to introduce only a finite number of counter-terms. We also see why these methods do not apply to theories like ϕ_4^4, QED, etc.

24.8.4 The Behaviour of the Effective Constants and the Approximate Renormalization Group Flow

Our aim is to study the evolution of the effective constants as functions of the cut-off index ρ. The main behaviour of g_ρ is governed by the first non-vanishing term of the corresponding β-function, although the subleading term also plays a role. For the coupling constant we choose as an ansatz the formula obtained by retaining the first orders of perturbation theory

$$g_\rho(C) = f(\rho, C)^{-1} \quad \text{where} \quad f(\rho, C) \equiv -\beta_2(\ln M)\rho + (\beta_3/\beta_2)\ln \rho + C \qquad (24.205)$$

with C being a large enough positive constant.

By parity considerations, the mass term is only logarithmically divergent and because, except for the initial tadpole which is linear in the coupling constant, all these 2-point graphs have at least two vertices, we obtain, by asymptotic freedom 'a convergent logarithmic power counting' for large energy.

Let us define

$$g_{\rho-1} = g_\rho + \delta g_\rho \quad m_{\rho-1} = m_\rho + \delta m_\rho \quad \zeta_{\rho-1} = \zeta_\rho + \delta\zeta_\rho, \qquad (24.206)$$

where δg_ρ, δm_ρ, and $\delta\zeta_\rho$ are given by the sum of the counter-terms corresponding to the 4- and 2-point functions at level ρ computed with effective coupling g_ρ and effective propagators C^ρ.

The transformation $\rho \to \rho - 1$ is a renormalisation group transformation

At lowest order in g_ρ, we obtain

$$g_{\rho-1} \simeq g_\rho\{1 - (\ln M)[\beta_2(g_\rho/\zeta^2) + (-\beta_2^2(\ln M + \gamma_3)(g_\rho/\zeta^2)^2] + O(g_\rho^3)\}$$
$$m_{\rho-1} \simeq m_\rho\{1 - \gamma(\ln M)(g_\rho/\zeta^2) + O(g_\rho^2)\}, \qquad (24.207)$$
$$\zeta_{\rho-1} \simeq \zeta_\rho\{1 + \gamma_2(\ln M)(g_\rho/\zeta^2)^2 + O(g_\rho^3)\},$$

where

$$\beta_2 = -2N(N-1)/\pi \quad \gamma = -(2N-1)/\pi \quad \gamma_2 = (2N-1)/(2\pi)^2. \qquad (24.208)$$

The fact that β_2 is negative is the characteristic of an asymptotically free theory.

We define $\beta_3 = \gamma_3 + 2\gamma_2$ and, since $\beta_3 = 2(N-1)/\pi^2$ has already been computed, we get $\gamma_3 = (N-3/2)/\pi^2$.

The Gaussian measure at level $\rho - 1$ corresponds to the covariance

$$C^*_{\rho-1}(p) \equiv \eta_{\rho-1}(p)\frac{\zeta_{\rho-1}(p)\ \not{p} + m_{\rho-1}(p)}{\zeta^2_{\rho-1}(p)p^2 + m^2_{\rho-1}(p)}, \qquad (24.209)$$

where

$$
\begin{aligned}
\zeta_{\rho-1}(p) &\equiv \zeta_\rho(1 - \eta_{\rho-1}(p)) + \zeta_{\rho-1}(\eta_{\rho-1}(p)) \\
m_{\rho-1}(p) &\equiv m_\rho(1 - \eta_{\rho-1}(p)) + m_{\rho-1}(\eta_{\rho-1}(p)).
\end{aligned}
\tag{24.210}
$$

More generally, we can prove the following. We define inductively g_k, m_k, and ζ_k by

$$
g_k = g_{k+1} + \delta g_{k+1} \quad m_k = m_{k+1} + \delta m_{k+1} \quad \zeta_k = \zeta_{k+1} + \delta \zeta_{k+1}
\tag{24.211}
$$

with δg_{k+1}, δm_{k+1}, and $\delta \zeta_{k+1}$ given by the sum of the counter-terms corresponding to the 4- and 2-point functions at level $k+1$ computed with effective coupling g_k and effective propagators C^{*q}, $q \geq k+1$ given by

$$
C^{*q}(p) \equiv \eta_q(p) \frac{\zeta_q(p)\, \slashed{p} + m_q(p)}{\zeta_q^2(p)p^2 + m_k^2(p)},
\tag{24.212}
$$

where

$$
\begin{aligned}
\zeta_q(p) &\equiv \zeta_\rho(1 - \eta_{\rho-1}(p)) + \sum_{q<i<\rho} [\zeta_i(\eta_i(p) - \eta_{i-1}(p))] + \zeta_q \eta_q(p) \\
m_q(p) &\equiv m_\rho(1 - \eta_{\rho-1}(p)) + \sum_{q<i<\rho} [m_i(\eta_i(p) - \eta_{i-1}(p))] + m_q \eta_q(p).
\end{aligned}
$$

Then, starting from the ansatz (24.205), we obtain the following.

Lemma 13.

$$
\begin{aligned}
\delta g_k = &-(\ln M)\,[\beta_2 g_k^2 + (\beta_3 - \beta_2^2(\ln M)g_k^3 - \delta_3 g_k^2/\rho(-\beta_2 \ln M) \\
&+ O(\ln k/k^4) + (g_k^2)(e^{-O(1)(\rho-k)})],
\end{aligned}
\tag{24.213}
$$

where by definition $\beta_3 = \gamma_3 + \delta_3$ is the usual third-order term in the β-function used in the previous formulae.

This makes it possible, starting from the ansatz on g_ρ, to prove some estimates on the values taken by the constants at step k.

Lemma 14. *There exists some constant $O(1)$ such that if C is a positive number satisfying $C \geq O(1)$, and g_ρ is defined by (24.205), then g_k is positive and behaves according to the bounds*

$$-\beta_2(\ln M)k + (\beta_3/\beta_2)\ln k + C/2 \le g_k^{-1}$$
$$\le -\beta_2(\ln M)k + (\beta_3/\beta_2)\ln k + 2C, \tag{24.214}$$

and choosing for the mass and the wave function, the g_ρ-dependent ansatz

$$m_\rho = m.\{1 - [\beta_2(\ln M) + f(\rho, C)]\lambda_\rho\}^{(N-1/2)/(N-1)} \tag{24.215}$$
$$\zeta_\rho = 1, \tag{24.216}$$

the running mass and the wave function satisfy

$$(m/2).k^{-(N-1/2)/(N-1)} \le m_k \le 2m.k^{-(N-1/2)/(N-1)} \tag{24.217}$$
$$\zeta_k - \zeta_\rho = (\gamma_2/\beta_2^2 \ln M)[(1/k) - (1\rho)] + O[(\ln k)/k^2]. \tag{24.218}$$

We will first prove formula (24.214) of this lemma.
Assuming (24.214) at order k, we rewrite (24.206) and (24.213) as

$$g_{k-1}^{-1} - g_k^{-1} = (\ln M)[\beta_2 + \frac{\beta_3}{(-\beta_2(\ln M)k}$$
$$- \frac{\delta_3}{\rho(-\beta_2 \ln M)} + O\left(\frac{\ln k}{k^2}\right) + e^{-O(1)(\rho-k)}] \tag{24.219}$$

and (24.214) will follow from the logarithmic divergence of $1/k$, the summability of $(\ln k)/k^2$, of $e^{-O(1)(\rho-k)}$, and the obvious bound

$$\sum_{j=k,\ldots,\rho} (|\delta_3|/\rho) = |\delta_3|(\rho-k)/\rho \le |\delta_3|, \tag{24.220}$$

which ends the proof of this formula.

The rest of this lemma relies essentially on the definition of δg_k taking account only of the most divergent terms and on the following lemmas with obvious graphical notations.

Lemma 15.

$$\delta g_k = -(\ln M)[\beta_2^{k,\rho} g_k^2 + (\gamma_3^{k,\rho} - (\beta_2^{k,\rho})^2(\ln M))g_k^3 + O(g_k^4)], \tag{24.221}$$

where $\beta_2^{k,\rho}$ and $\gamma_3^{k,\rho}$ are defined by the diagrammatic expressions shown in Fig. 24.7. Furthermore the following limits exist,

$$\beta_2 = \lim_{k\to\infty} \beta_2^{k,\infty}, \quad \gamma_3 = \lim_{k\to\infty} \gamma_3^{k,\infty} \tag{24.222}$$

$$(\log M)\beta^{k,\rho} = \sum_{i_1,i_2 = k,...,\rho \text{ with inf } (i_1,i_2) = k} \int \quad \raisebox{0pt}{[diagram]} \quad d^2y$$

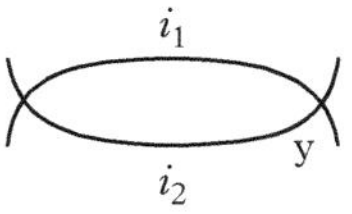

$$(\log M)\gamma_3^{k,\rho} = \sum_{i_1,i_2,i_3,i_4 = k,...,\rho \text{ with inf } (i_1,i_2,i_3,i_4) = k} \int \quad \raisebox{0pt}{[diagram]} \quad i_4 d^2y\,d^2z$$

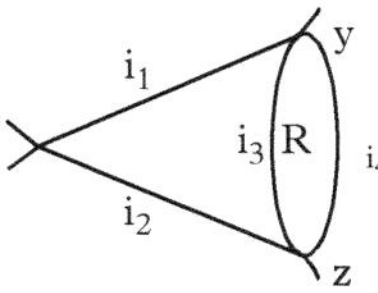

Figure 24.7 *The diagrammatic expressions for $\beta_2^{k,\rho}$ and $\gamma_3^{k,\rho}$. The mark 'R' in the second diagram means that the inner bubble is renormalised.*

and β_2 is the usual first coefficient of the β function, computed with the wave function $\zeta = 1$ introduced in sections 24.8.1 and 24.8.2. We remark that with standard convention β_2 is negative, which corresponds to asymptotic freedom, hence to $\delta\lambda_k > 0$ for small $g_k > 0$.

This lemma can be proven by inspection on the diagrams.

Now it is easy to show that (24.207) leads for large ρ, starting from finite ζ_0 and positive g_0, to

$$g_\rho \simeq \zeta^2[-\beta_2(\ln M)\rho + (\beta_3/\beta_2)\ln\rho + C]^{-1} \equiv f(\rho,C)^{-1} \qquad (24.223)$$
$$\zeta_\rho \simeq \zeta$$

with C and ζ two constants. Finally, using the scale invariance (24.194) we will choose a g_ρ-dependent mass ansatz for large values of ρ

$$m_\rho = m.\{1 - [\beta_2(\ln M) + f(\rho,C)]g_\rho\}^{(N-1/2)/(N-1)} \qquad (24.224)$$
$$\zeta_\rho = 1.$$

These are the essential ingredients. We see that asymptotic freedom provides the necessary logarithmic convergent factors, so we do not have to carry on the expansions any further. Roughly speaking, for the purposes of constructing a partly renormalised phase space expansion, an asymptotically free renormalisable theory behaves, effectively, as a superrenormalisable theory.

We are now in a position to formulate the basic results, under the form of a theorem.

Theorem 24. *Under the ansatz (24.205), the PRPSE is absolutely and uniformly convergent for C positive large enough. The normalised Schwinger functions $S_{2p,\rho}(C)$, sums of the PRPSE, have limits $S_{2p}(C) = \lim_{\rho\to\infty} S_{2p,\rho}(C)$ and are the Schwinger functions of a non-trivial Euclidean theory satisfying the Osterwalder–Schrader axioms.*

The limits $S_{2p}(C)$ are analytic in C for C complex with $\Re C \geq O(1)$.

The renormalised coupling constant g_{ren} defined at zero momenta lies in a disc $D_C = \{g|\Re(1/g) > C'\}$. For C' large enough and for $g_{\mathrm{ren}} \in D_{C'}$, the map from C to g_{ren} may be inverted and C becomes an analytic function of $g_{\mathrm{ren}} \in D_{C'}$, with $\Re C \geq O(1)$. The theory can therefore be parametrised by g_{ren} instead of C, is analytic in $D_{C'}$, and is the Borel sum of the usual renormalised perturbation series in g_{ren} with renormalised mass and g_{ren} determined by the renormalisation conditions

$$g_{\mathrm{ren}} = S_4^A(C, m)(0, 0, 0, 0)$$
$$(m_{\mathrm{ren}})^{-1} = S_2(C, m)(0, 0)$$
$$\zeta_{\mathrm{ren}} = (m_{\mathrm{ren}})^2 \left[-i\frac{d}{dp_0} S_2(C, m)(p) \right]\Big|_{p=0}, \qquad (24.225)$$

where S_4^A is the amputated 4-point function.

We will not prove this theorem. A large part of the proof is based on the same type of estimates as those given to prove the convergence of the $g\phi^4$-theory in three dimensions.

24.9   The Yang–Mills Field Theory in *d* = 4

As we said earlier, this is by far the most complicated among all models for which rigorous results have been obtained. It is also the most interesting one, since it is close to the theory which describes the fundamental strong interactions among elementary particles. We have studied its properties in perturbation theory in the previous chapters, where we have established that it is: (i) renormalisable and (ii) asymptotically free. Therefore, we would expect that the results obtained in the previous section concerning the Gross–Neveu model should be applicable also to the four-dimensional Yang–Mills gauge theory. It will turn out that this is partly true, but only after several crucial questions, both physical and technical, have been answered. We will give here a list of the most important among these difficulties and how they have been bypassed.

24.9.1   A Physical Problem

We start with a physical problem, which, of course, is not unrelated to many technical ones: the gauge bosons associated with an unbroken gauge symmetry are massless.[17] We recall that in all our previous estimates a non-zero mass gap was essential for the convergence of the cluster expansion and the existence of the infinite volume limit. It follows that in the case of the Yang–Mills theory, we will be unable to take this limit and all our results will be valid only inside a fixed finite volume. Does this mean that they are uninteresting for physics? Fortunately no, and we want to explain here why.

---

[17] We have explained already that breaking the symmetry spontaneously gives masses to the gauge bosons, but at the price of losing the property of asymptotic freedom.

In section 24.3 we explained that in the theory for strong interactions which we call 'quantum chromodynamics', all available experimental results as well as numerical estimations point to the existence of a confining phase for all values of the coupling, which means that at infinite volume the system is not described by the states we find in perturbation theory, namely free, massless gauge bosons, but instead a collection of bound states not present in perturbation. The gauge bosons are 'confined' inside a small volume. Experimentally the size of this confining volume is found to be of order of 1 f, or 10^{-13} cm. Scattering experiments in present accelerators, like the LHC, probe distances down to $10^{-16} - 10^{-17}$ cm, much smaller than the confining volume. Because of asymptotic freedom, the effective coupling constant inside this volume is small and we perform perturbation theory calculations. It is therefore important to know the properties of the perturbation expansion, even inside the confining volume.

24.9.2   Many Technical Problems

As we have said already, the physical problem of the zero mass gauge bosons creates many technical problems, which we will try to indicate qualitatively in this section. They are mostly related to the resulting gauge invariance of the theory we start with.

The proof of the existence of the Yang–Mills theory without ultraviolet cut-off will rely essentially on the same constructive machinery which has been developed in the previous sections. Several elements were essential in this process:

1. In order to start with well-defined initial expressions, we must introduce a sequence of ultraviolet cut-offs.

2. We know, and that was the reason to shift to Euclidean space, that we need some control of the exponential of the Lagrangian, the so-called positivity of energy.

3. Similarly we need to define a reference measure as this was done previously in the scalar ϕ^4 theories with weak coupling. In these cases this measure was the Gaussian measure and it was extracted from the quadratic part of the Lagrangian. This Gaussian measure played two roles, one as a reference measure, the remaining of the Lagrangian assumed to be a small perturbation of this measure, and second this measure had to be the right tool to generate the perturbative diagrams of the expansion.

The trouble is that the gauge structure of the theory, and in particular the need for a gauge fixing, makes this machinery much more complex to set up, requiring many artificial-looking choices.

Let us start with the ultraviolet cut-off. Ideally, we would have liked to use one which respects gauge invariance; otherwise, we would produce gauge non-invariant counterterms, not present in our initial Lagrangian. When we studied the renormalisation properties of the Yang–Mills theories in Chapter 18, we used dimensional regularisation. But this is a prescription applicable only to the calculation of Feynman diagrams and it is not obvious how to adapt it to the definition of a global measure in a functional integral. As we explained in section 24.3, we will use a Euclidean invariant cut-off with the consequence of producing counterterms proportional to A^2 and A^4. This will not create too many problems and, in fact, it will be essential to the definition of a global measure. By a suitable choice of a class of Euclidean invariant cut-offs, it will be possible to get the coefficient of the A^4 term positive. This will be the key property which will help controlling the positivity properties, although the practical computations will not be that simple.

Then we come to the gauge choice. It is easy to see that there exists no global choice which satisfies all our requirements. In the previous chapters we often used gauges belonging to the family $\partial^\mu A_\mu = C(x)$, with C some given function, for example $C = 0$ for the Feynman gauge. It has the advantage of producing well-defined and easy-to-use Feynman rules, but it breaks every one among our other requirements: (i) it suffers from the Gribov ambiguity we explained in section 14.7 and this makes a global definition of the path integral measure problematic. (ii) It breaks positivity. We saw this already in the Abelian case where we had to use a space of states with negative norm, as well as in the Yang–Mills case where we had to introduce the Faddeev–Popov ghosts. In this section we will use an axial gauge, in which one of the components of the Euclidean gauge field is set equal to 0. It has many attractive properties, such as to be free of Gribov copies and Faddeev–Popov ghosts. For these reasons it is the one used exclusively in the numerical computations of Yang–Mills theories on a space–time lattice we will present in the next chapter. It respects positivity of the Gaussian measure, but it is singular when used in perturbation theory calculations. For this reason, in order to recover our usual Feynman rules of perturbation theory, we will have to make a local change of gauge and pass to one of the families mentioned earlier.

The real story is even more complicated. As we mentioned earlier, in order to define the theory we introduce a momentum cut-off of a certain type which breaks gauge invariance and requires gauge restoring counter terms. These gauge-dependent counter-terms proportional to A^4 will stabilise the field variables in the Lie algebra at sizes of order g^{-1}. Combined with the positivity of the axial gauge, then the field variables in the Lie algebra become, in probability, of order $g^{-1/2}$. To exploit this fact, we must make a division of the phase space for the field into small field regions and large field regions and expect that the large field regions can be resumed being more 'rare' than the small field regions. An explicit change of gauge can then be done in the small field regions, making it possible to use the usual perturbation theory. However, large field regions and small field regions are connected by vertices connecting high momentum and low momentum fields. Usually, it is possible to dominate the low momentum fields, as it was done for $g\phi_3^4$, using the positivity of the interaction and thus to free the small field regions. This can be done in some specific cases using the positivity due to the ad hoc positivity of the gauge-restoring A^4-counter-term, but it is not always possible. In these cases, we must use a background-dependent gauge and a background-dependent propagator to analyse the small field perturbative regions. The background field at a given scale and position is roughly made of all the large fields of low frequencies located at this position. The use of the background-dependent gauge and the resulting propagator makes the cluster expansion and the evaluation of the large field region more complicated than how it is done for simpler models. In these regions, the functional integrals are renormalised by their coupling to higher momenta in small field regions. This coupling results in a determinant which reflects the difference in normalisation between the Gaussian measure with a given background field and the ordinary Gaussian measure with no background field. This determinant must be controlled. The proof that the construction is correct is given by the fact that the Schwinger functions which are obtained satisfy the appropriate Ward identities, which are the remnant of gauge invariance under small gauge transformations.

Other effects, such as the invariance under large gauge transformations or non-trivial topological effects such as instantons, will not be discussed. Also the complete Osterwalder–Schrader axioms have not been proven, although the Osterwalder–Schrader positivity is, probably, correct.[18]

The final result, whose proof we will not be presented here,[19] is that in the ultraviolet limit, keeping the infrared cut-off fixed, the Schwinger functions exist and satisfy the corresponding Ward identities.

---

[18] An indication is that this positivity has been proven for the theory with a lattice ultraviolet cut-off and we expect to be able to link the two by introducing an overall momentum cut-off (of our type) large enough with respect to the lattice spacing.

[19] See J. Magnen, V. Rivasseau and R. Sénéor, *Comm. Math. Phys.* **155**, 325 (1993). For an alternative approach using a lattice ultraviolet cut-off, see T. Balaban, *Comm. Math. Phys.* **95**, 17 (1984); **96**, 223 (1984); **98**, 17 (1985); **99**, 75 (1985); **99**, 329 (1985); **102**, 277 (1985); **109**, 249 (1987); **116**, 1 (1988); **119**, 243 (1988); **122**, 175 (1989); **122**, 355 (1989).

25

Fundamental Interactions

25.1 Introduction. What Is an 'Elementary Particle'?

The modern era of theoretical high energy physics has a precise starting date: 2 June 1947, the date of the Shelter Island Conference. The most important contribution that was presented at this conference was not a theoretical breakthrough but an experimental result. Willis Lamb, of Columbia University, reported the measurement of an energy difference of about 1000 MHz between the $2S_{1/2}$ and $2P_{1/2}$ levels of the hydrogen atom. We recall that, as we saw in Chapter 7, the Dirac theory predicts an exact degeneracy for these two levels. The importance of the 'Lamb shift' is not that it was unexpected. As Steven Weinberg puts it, 'It was not so much that it forced us to change our physical theories, as that it forced us to take them seriously.' In the months which followed, Richard Feynman and Julian Schwinger set the foundations of the theory of renormalisation, the well-defined prescription to compute the higher order terms in the perturbation series of quantum electrodynamics, which we saw in Chapter 16. As it became known a bit later, similar ideas had been developed independently in Japan by Sin-Itiro Tomonaga, who was the first to give the precise computation of the Lamb shift to first order. The entire programme was formally described by Freeman Dyson in 1949. Quantum electrodynamics, the theory of interacting photons and electrons, supplemented with the programme of renormalisation, is one of the most successful theories in physics. Its agreement with experiment is spectacular. But it was also the first successful quantum field theory, the quantum mechanics of a relativistic system with an infinite number of degrees of freedom.

In the previous chapters we set the formalism for such a theory. In this chapter we will show that it provides the unifying language to describe physical phenomena at the fundamental level. Through the successful use of these ideas there has been spectacular progress during the past 40 years which has resulted in a much deeper comprehension of the laws of nature.

The purpose of elementary particle physics is twofold: to discover the fundamental constituents of matter, on the one hand, and to understand the nature of their interactions, on the other. A chapter on this subject should normally begin with a definition of what is an elementary particle. The trouble is that we have no such definition. We can

From Classical to Quantum Fields. Laurent Baulieu, John Iliopoulos and Roland Sénéor.

Table 25.1 *This table shows our present ideas on the structure of matter. Quarks and gluons do not exist as free particles and the graviton has not been observed.*

TABLE OF ELEMENTARY PARTICLES

Quanta of Radiation

Strong interactions	Eight gluons
Electromagnetic interactions	Photon (γ)
Weak interactions	Bosons W^+, W^-, Z^0
Gravitational interactions	Graviton (?)

Matter Particles

	Leptons	Quarks
1st family	ν_e, e^-	u_a, d_a, $a = 1, 2, 3$
2nd family	ν_μ, μ^-	c_a, s_a, $a = 1, 2, 3$
3rd family	ν_τ, τ^-	t_a, b_a, $a = 1, 2, 3$

BEH Boson

only give a table whose entries evolve with time and represent, at any given moment, the current state of our knowledge (or ignorance) of the structure of matter. The idea that the structure of matter is discontinuous is very old and goes back to Democritus, a Greek philosopher who, around 400 BC, postulated the existence of fundamental building blocks of all matter which he called 'atoms', i.e. 'unbreakables'. Today we know that atoms are not elementary but have instead a complex internal structure and can be broken. However, they are 'unbreakable' in the sense that the pieces of an iron atom once broken are no more iron. The physical existence of atoms was established only by the beginning of the previous century and during the past hundred years we have uncovered deeper and deeper layers of the cosmic structure:

atoms → electrons, nuclei → electrons, protons, neutrons → electrons, quarks →??

There is no reason to believe that we have reached the innermost layer, if such a thing exists at all. Therefore, to the question 'what is an elementary particle?' we can only answer by showing the Table of Elementary Particles known today (Table 25.1). A good introduction to the laws of fundamental physics is to go through the various entries of this table. This will lead us to introduce the kinds of interactions that these particles undergo.

25.2   The Four Interactions

Although our perception on the identity of elementary particles has changed radically many times during the past century, our knowledge of their interactions has remained remarkably stable. We know that the structure of matter at all scales known today, from the clusters of galaxies down to the shortest distances explored ($\sim 10^{-16}$ cm), can be qualitatively understood by the action of four types of interaction. In order of decreasing strength, they are the following:

- *Strong interactions.* They are mainly responsible for nuclear structure. Indeed, we know experimentally that the various nuclei are bound states of protons and neutrons. Protons carry each a unit of positive electric charge and, therefore, they are subject to electrostatic repulsion. The remarkable cohesion of nuclei shows the existence of attractive forces among protons and neutrons which must be much stronger and outweigh the electromagnetic repulsive forces. Strong interactions have a short range, typically on the order of 10^{-13} cm, and their effects are not manifest at everyday life. Not all known particles are subject to strong interactions. Those that are, are called 'hadrons'. Among the particles of Table 25.1, the quarks and the gluons are hadrons but all other particles, such as the electrons and the neutrinos, are not. Understanding the nature of the strong interactions has been a long-lasting problem in high-energy physics. Its solution has offered deep insight into the laws of nature, insight that far exceeds the domain of nuclear forces. We will give a brief description of these ideas later on in this chapter.

- *Electromagnetic interactions.* They are responsible for atomic and molecular structure. They have a long range and their effects are observable at macroscopic scales. We learned in a previous chapter that the electromagnetic interactions are the results of the exchange of virtual photons, the quanta of the electromagnetic field, between electrically charged particles. The long range is due to the zero mass of the photon.

- *Weak interactions.* They are responsible for nuclear β-decay as well as the decays of other unstable particles. To give a measure of how weak interactions are, we note that, experimentally, neutrinos, which have no strong or electromagnetic interactions, are produced at the centre of the sun and escape nevertheless without being absorbed. For a very massive star, neutrino radiation is the only known cooling mechanism. We know today that weak interactions share many common features with electromagnetic ones. They are also due to the exchange of virtual quanta, the weak vector bosons W^+, W^-, and Z^0. However, these quanta are massive and, consequently, the weak interactions are short ranged. In this chapter we will explore this weak–electromagnetic analogy much further.

- *Gravitational interactions.* Their importance in both terrestrial and cosmic phenomena is well known but, at the microscopic level, their effects are too small to be observable. We presented already their description at the classical level, but we will mention only very briefly their quantum effects at the end of this book.

A final look at Table 25.1 reveals some remarkable regularities. (i) All interactions are produced by the exchange of virtual quanta. They are the eight gluons for the strong interactions, the photon for the electromagnetic ones, and the bosons W and Z for the weak, and we believe the graviton plays that role for the gravitational interactions. For the first three, they are vector (spin-1) fields, while the graviton is assumed to be a tensor, spin-2 field. (ii) The constituents of matter appear to be all spin-$\frac{1}{2}$ particles. They are divided into quarks, which are hadrons, and 'leptons' which have no strong interactions. (iii) In the table we see six quark species, u, d, c, s, t, and b. In the physicists' jargon they are called 'flavours'. In turn, each one of them appears under three forms, often called 'colours' (the terms flavour and colour have no relation with the ordinary sense of these words). This triplet structure will turn out to be very important later. (iv) Quarks and gluons do not appear as free particles. They form a large number of bound states, the hadrons. (v) Quarks and leptons seem to fall into three distinct groups, or 'families'. This family structure is one of the great puzzles in elementary particle physics. We believe we understand the importance of the first family. It is composed of an electron and its associated neutrino as well as up- and down-quarks. These quarks are the constituents of protons and neutrons. The role of each member of this family in the structure of matter is obvious. In contrast, the role of the other two remains obscure. Muon and tau leptons seem to be heavier versions of the electron but they cannot be viewed as excited states of it because they seem to carry their own quantum numbers. The associated quarks with exotic names, such as charm, strange, top, and bottom, form new, unstable hadrons which are not present in ordinary matter. Why does nature produce three similar copies of apparently the same structure? (vi) The sum of all electric charges inside any family is equal to 0.

In this chapter we will present the successful application of the concepts and techniques of quantum field theory to these interactions. These efforts gave rise to one of the most exciting and rewarding periods in modern physics. It is usually said that progress in science occurs when an unexpected experimental result contradicts current theoretical beliefs. This forces scientists to change their ideas and leads to a new theory. This has often been the case in the past, but the revolution we are going to describe here had a theoretical, rather an aesthetic motivation. It was a triumph of abstract theoretical thought which brought geometry into physics. The road has been long and circuitous and many a time it gave the impression of leading to a dead end. The detailed history of this revolution has not yet been written and we will not attempt here to follow the historical order. This would had led to a too lengthy presentation, mainly because the theoretical developments often anticipated the experimental results. In many cases theorists had only their intuition for a guide and their deductions were necessarily speculative. We will use instead all the experimental results known today and show that they point unmistakably to what is called 'the standard model' of elementary particle interactions.

25.3 The Standard Model of Weak and Electromagnetic Interactions

25.3.1 A Brief Summary of the Phenomenology

Before going into model building, let us first recall the essential phenomenological properties of weak interactions. It is instructive to go back to the late sixties when the essential features of the model were found.

Weak interaction phenomena were, until 1971, well described by a simple phenomenological model involving an operator $\mathcal{J}_\lambda(x)$, the 'weak current', which is the analogue of the familiar electromagnetic current. In terms of it we can build an effective Lagrangian density of the form

$$\mathcal{L}_F = \frac{G}{\sqrt{2}} \mathcal{J}^\lambda(x) \mathcal{J}_\lambda^\dagger(x), \tag{25.1}$$

where † means Hermitian conjugation.

This is the famous current×current theory and $G/\sqrt{2}$ is the Fermi coupling constant, which is equal to $10^{-5} m_{\text{proton}}^{-2}$. The weak current $\mathcal{J}_\lambda(x)$ is a sum of two parts, a leptonic part $l_\lambda(x)$ and a hadronic one $h_\lambda(x)$. They are both of the $V - A$ form, i.e. they can be written as a particular superposition of a vector and an axial part and satisfy simple algebraic relations. We will come back to this point presently. It is straightforward to write the leptonic current in terms of the fields of known leptons as

$$l_\lambda(x) = \bar{e}(x)\gamma_\lambda(1 + \gamma_5)\nu_e(x) + \bar{\mu}(x)\gamma_\lambda(1 + \gamma_5)\nu_\mu(x), \tag{25.2}$$

where we have used a notation in which the particle symbols e, ν_e, μ, and ν_μ denote also the corresponding Dirac fields. In 1971 the leptonic current had only these two terms; today we know that we must add a third one to describe the τ family: $\bar{\tau}(x)\gamma_\lambda(1 + \gamma_5)\nu_\tau(x)$. The $V - A$ structure we mentioned in the first part of this paragraph is shown by the presence of the $1 + \gamma_5$ projector. The vector and the axial currents enter with exactly the same coefficient. This property, equal strength for all currents, is often called 'universality'. As we saw in Chapter 7, $1 + \gamma_5$ projects the four-component Dirac spinors into two-component left-handed Weyl spinors. In other words, $V - A$ implies that only the left-handed leptons participate in weak interactions. Another remarkable property of the $V - A$ current is that it generates an $SU(2)$ algebra.

The origin of this algebra is simple to trace. Let us introduce a compact notation which will be useful later. We put every charged lepton, together with the associated neutrino, into a doublet,

$$\Psi_{\text{L,R}}^i(x) = \begin{pmatrix} \nu_i(x) \\ \ell_i^-(x) \end{pmatrix}_{\text{L,R}}, \tag{25.3}$$

where the index i runs over the three families, $\nu_i = \nu_e$, ν_μ, and ν_τ and $\ell_i^- = e^-$, μ^-, and τ^- for $i=1,2$, and 3, respectively. L and R denote the left- and right-handed spinors $\Psi_{\text{L,R}} = \frac{1}{2}(1 \pm \gamma_5)\Psi$. For each family we can construct eight currents, four involving only Ψ_{L} and four only Ψ_{R},

$$j_{\lambda\text{L,R}}^i(x) = \bar{\Psi}_{\text{L,R}}^i(x)\gamma_\lambda\Psi_{\text{L,R}}^i(x) \quad ; \quad \vec{j}_{\lambda\text{L,R}}^i(x) = \bar{\Psi}_{\text{L,R}}^i(x)\gamma_\lambda\boldsymbol{\tau}\Psi_{\text{L,R}}^i(x), \tag{25.4}$$

where $\boldsymbol{\tau}$ are the usual Pauli matrices. All currents we have used so far can be written as linear combinations of these j's. For example, the electromagnetic current is given by

$$j_\lambda^{\text{em}} = \sum_i \bar\Psi_{\text{L}}^i \gamma_\lambda \frac{1-\tau^3}{2} \Psi_{\text{L}}^i \; + \; \sum_i \bar\Psi_{\text{R}}^i \gamma_\lambda \frac{1-\tau^3}{2} \Psi_{\text{R}}^i. \tag{25.5}$$

The summations extend over the three lepton families. Similarly, the charged weak current (25.2) is given by

$$l_\lambda = \sum_i \bar\Psi_{\text{L}}^i \gamma_\lambda \tau^- \Psi_{\text{L}}^i. \tag{25.6}$$

Two remarks are in order here. First, Eqs. (25.5) and (25.6) show that electromagnetic and weak interactions, being vectorial do not mix right- and left-handed components. Second, in 1971 the charged current (25.6) described all weak interactions known at the time. There was no evidence for a need to introduce neutral currents involving the unit matrix, or the τ^3 matrix in (25.6).

Since leptons have no strong interactions, the complete leptonic Lagrangian, in the limit of vanishing lepton masses, could be written as

$$\mathcal{L}_{\text{lept}} = \sum_i \bar\Psi_{\text{L}}^i i\partial\!\!\!/ \, \Psi_{\text{L}}^i + \sum_i \bar\Psi_{\text{R}}^i i\partial\!\!\!/ \, \Psi_{\text{R}}^i - \frac{1}{4} F_{\mu\nu}^2 - e j_\lambda^{\text{em}} A^\lambda + \frac{G}{\sqrt{2}} l_\lambda l^{\lambda\dagger}. \tag{25.7}$$

It is invariant under $U(2) \times U(2)$ transformations[1] which rotate separately right- and left-handed doublets, as well as several $U(1)$ phase transformations which correspond to the conservation of the various lepton numbers. A mass term, which is proportional to $\bar\Psi_{\text{L}}\Psi_{\text{R}} + \bar\Psi_{\text{R}}\Psi_{\text{L}}$, breaks the separate L and R invariance into the diagonal subgroup leaving only the vector currents conserved. Note also that if neutrinos are massless, the right-handed neutrino fields are free fields and can be omitted. As we will see later, recent experimental data show that neutrinos do have non-zero masses but it is still unclear whether their right-handed components are present. Furthermore, experiments show that all three lepton numbers are not separately conserved and neutrinos oscillate among the different species, but let us ignore this complication for the moment.

The generators of the $U(2) \times U(2)$ symmetry are precisely the space integrals of the time components of the $V - A$ and $V + A$ currents

$$Q_{\text{L,R}}^i = \sum_l \int \mathrm{d}^3 x j_{0\,\text{L,R}}^{\,i}(x) \tag{25.8}$$

and satisfy the $U(2) \times U(2)$ algebra, i.e. the two $U(1)$ charges commute among themselves as well as with all other charges and the six $SU(2)$ ones satisfy

$$[Q_{\text{L,R}}^{i,a}, \; Q_{\text{L,R}}^{j,b}] = \delta^{ij} i\epsilon_{abc} Q_{\text{L,R}}^{i,c}; \quad [Q_{\text{L}}^i, \; Q_{\text{R}}^j] = 0, \tag{25.9}$$

---

[1] As we explained in section 17.3, the axial part of the $U(1)$ current will be subject to the Adler–Bell–Jackiw anomaly condition, but this fact will play no role in our present discussion which, for the moment, concerns only the classical symmetries.

where both the family index i, j and the $SU(2)$ index a, b, c run through 1,2,3. Current conservation guarantees the time independence of the charges Q^i.

It is not so easy to write such an explicit form for the weak hadronic current $h_\lambda(x)$. First of all, it will depend on which particles we consider as elementary. But even if today there is a general consensus on this question in favour of the quarks, we still need to guess the form of the current using experimental data which involve hadrons. To do so we need to know some of the properties of strong interactions. It was a very important discovery, which took several years and is due to the work of several people, when it was finally established that, at the limit of vanishing quark masses, strong interactions are invariant under a chiral symmetry group $U(f) \times U(f)$, where f is the number of quark flavours, always modulo the Adler–Bell–Jackiw anomaly condition. The weak hadronic current $h_\lambda(x)$ can be identified with the currents of this symmetry group. Today we know that $f = 6$ and the symmetry between hadrons and leptons is striking. In the late sixties, however, we only knew the up, down, and strange quarks and this analogy was far from obvious. A further complication was related to the conservation of the various flavour numbers. In the leptonic world there was good evidence that the electronic and muonic numbers were separately conserved.[2] This was mainly due to the absence of the decay mode $\mu \to e + \gamma$, which is allowed by angular momentum and electric charge conservation. In contrast, strangeness, which is the quantum number associated with the s quark, was not even approximately conserved. Strange particles, such as K-mesons or Λ and Σ baryons, were known to decay via weak interactions to non-strange particles, such as π-mesons and protons and neutrons. The hadronic weak current which described all data at that time was of the form

$$h_\lambda = \bar{u}_L \gamma_\lambda (\cos \theta \; d_L + \sin \theta \; s_L), \tag{25.10}$$

with θ a phenomenological parameter known as 'the Cabibbo angle'. We see that only a linear combination of the down- and strange quarks participated in the weak interactions, while the orthogonal one ($\cos \theta \; s_L - \sin \theta \; d_L$) did not. Let us note also that in writing the current (25.10), a summation over the three colour indices of the quarks is understood. Unless absolutely necessary, we will avoid writing this colour summation explicitly, in order to keep the notation simple.

The Lagrangian (25.1), with the total weak current $\mathcal{J}_\lambda(x)$ given by the sum of (25.6) and (25.10), despite its phenomenological character, is an extremely elegant structure. This simple and compact form could not only fit a large variety of data, but also it incorporated fundamental physical principles, such as universality and current algebra, that we mentioned earlier. Thus, at the phenomenological level, we had a perfectly working scheme; there was no compelling experimental reason to try to change it. It described correctly all experimental results which were inside its natural domain, namely all data which, at the time, could definitely be attributed to weak interactions. And yet we were not happy. What we wanted was not a phenomenological scheme, but a physical theory. Yang–Mills theories were studied not because they fitted the data better, but

---

[2] Although today we know that this is not absolutely true (see the discussion in the section on neutrino masses), it is still a very good approximation.

rather because of their aesthetic beauty and, more important, their mathematical consistency. The problem with (25.1) is that because of its dimensionful coupling constant, it is hopelessly non-renormalisable. Only tree-level diagrams can be computed.

25.3.2   Model Building

In this section we want to apply all the powerful machinery of gauge theories to the real world and construct a realistic gauge theory. Following the historical order we start with the weak and electromagnetic interactions. The idea of putting these two fundamental forces together and treating them on equal footing was not obvious. Indeed, as we just saw, weak interactions violate parity and involve $V - A$ currents while electromagnetic ones conserve parity and involve vector currents. The first suggestion along this line goes back to the work of Sheldon Lee Glashow in 1961. Here let us try to be general and develop a step-by-step approach to model building. The essential steps are the following.

1. Choose a gauge group G.
2. Choose the fields of the 'elementary' particles whose interactions you want to describe and assign them to representations of G. Include scalar fields to allow for the Brout–Enblert–Higgs (BEH) mechanism.
3. Write the most general renormalisable Lagrangian invariant under G. At this stage gauge invariance is still exact and all gauge vector bosons are massless.
4. Choose the parameters of the BEH potential so that spontaneous symmetry breaking occurs. In practice, this often means to choose a negative value for a parameter μ^2.
5. Translate the scalars and rewrite the Lagrangian in terms of the translated fields. Choose a suitable gauge and quantise the theory.

Remark: Gauge theories provide only the general framework, not a detailed model. The latter will depend on the particular choices made in steps 1 and 2.

The construction of the standard model has been one of the most exciting and most rewarding periods of modern physics. Although the motivation was entirely theoretical, both theory and experiment made spectacular discoveries which were parallel and complementary. It would have been instructive to follow in detail the historical order, precisely in order to show this close cooperation, but it would lengthen the exposition considerably. Therefore, we choose to go directly to the model in its final form, as it has been established today and we will only comment on the highlights of the interrelation between theory and experiment. We will follow step by step the programme we set up.

25.3.3   The Lepton World

The leptonic part of this model was first proposed by Steven Weinberg and Abdus Salam in 1967 and 1968.

- *Step 1:* Looking at Table 25.1 we see that for the combined electromagnetic and weak interactions, we have four gauge bosons, namely $W^\pm$, Z^0, and the photon. As we explained in Chapter 14, each one of them corresponds to a generator of the group G. The only non-trivial group with four generators is $U(2) \approx SU(2) \times U(1)$. Here comes the first remark about the theory–experiment connection: when the model was proposed the three intermediate vector bosons were still out of reach of any accelerator. Furthermore, the weak neutral currents, such as $(\bar{e}\gamma_\lambda(1 \pm \gamma_5)e)$ and $(\bar{\nu}\gamma_\lambda(1 \pm \gamma_5)\nu)$, had not yet been observed and their very existence was in doubt. Consequently, several attempts were made to build models which avoided them. None were particularly attractive and the subsequent discovery of the neutral currents at CERN by the Gargamelle group in 1973 established the $SU(2) \otimes U(1)$ model unambiguously and offered the first triumph of these ideas. A naïve identification would have been to assign the photon to the generator of $U(1)$ and the other three to those of $SU(2)$. Glashow's remark was that a much richer structure is obtained if you allow for a mixing between the two neutral generators, that of $U(1)$ and the neutral component of $SU(2)$. Following the notation which was inspired by the hadronic physics, we call T_i, $i = 1, 2, 3$, the three generators of $SU(2)$ and Y that of $U(1)$. Then the electric charge operator Q will be a linear combination of T_3 and Y. By convention, we write

$$Q = T_3 + \frac{1}{2}Y. \tag{25.11}$$

The coefficient in front of Y is arbitrary and only fixes the normalisation of the $U(1)$ generator relatively to those of $SU(2)$. We will come back to this point in a later chapter. This ends our discussion of the first step.

- *Step 2:* Leptons have always been considered as elementary particles, both in 1967, when the model was initially proposed, and today. Their number has increased from 4 to 6 with the discovery of the τ and its associated neutrino ν_τ, so we must look for $SU(2)$ representations of dimension 6. However, as we noted already, a striking feature of the data is the phenomenon of family repetition. We do not understand why nature chooses to repeat itself three times, but the simplest way to incorporate this observation with the model is to use three times the same representations, one for each family. This leaves $SU(2)$ doublets and/or singlets as the only possible choices. A further experimental input we will use is the fact that the charged W's couple only to the left-handed components of the lepton fields, in contrast to the photon, which couples with equal strength to both right and left. These considerations lead us to assign the left-handed components of the lepton fields to doublets of $SU(2)$. In the notation of section 25.3.1, we write

$$\Psi_L^i(x) = \frac{1}{2}(1 + \gamma_5) \begin{pmatrix} v_i(x) \\ \ell_i^-(x) \end{pmatrix}; \quad i = 1, 2, 3. \tag{25.12}$$

The right-handed components are assigned to singlets of $SU(2)$:

$$\nu_{iR}(x) = \frac{1}{2}(1 - \gamma_5)\nu_i(x) \ (?); \quad \ell_{iR}^-(x) = \frac{1}{2}(1 - \gamma_5)\ell_i^-(x). \tag{25.13}$$

The question mark next to the right-handed neutrinos means that the presence of these fields is not confirmed by the data. We will drop them in this chapter, but we will come back to this point later. We will also simplify the notation and put $\ell_{iR}^-(x) = R_i(x)$. The resulting transformation properties under local $SU(2)$ transformations are

$$\Psi_L^i(x) \rightarrow e^{i\boldsymbol{\tau}\boldsymbol{\theta}(x)}\Psi_L^i(x); \quad R_i(x) \rightarrow R_i(x) \tag{25.14}$$

with $\boldsymbol{\tau}$ the three Pauli matrices. This assignment and the Y normalisation given by Eq. (25.11) fix also the $U(1)$ charge and, therefore, the transformation properties, of the lepton fields. For all i we find that

$$Y(\Psi_L^i) = -1; \quad Y(R_i) = -2. \tag{25.15}$$

If a right-handed neutrino exists, it has $Y(\nu_{iR}) = 0$, which shows that it is not coupled to any gauge boson.

We are left with the choice of the BEH scalar fields. Although one of them has been discovered, we still have no precise information concerning their number, so we will choose the minimal solution. We must give masses to three vector gauge bosons and keep the fourth one massless. The latter will be identified with the photon. We recall that for every vector boson acquiring mass, a scalar with the same quantum numbers decouples. At the end we will remain with at least one physical, neutral, scalar field. It follows that the minimal number to start with is four, two charged and two neutral. We choose to put them, under $SU(2)$, into a complex doublet:

$$\Phi = \begin{pmatrix} \phi^+ \\ \phi^0 \end{pmatrix}; \quad \Phi(x) \rightarrow e^{i\boldsymbol{\tau}\boldsymbol{\theta}(x)}\Phi(x), \tag{25.16}$$

with the conjugate fields ϕ^- and ϕ^{0*} forming $\Phi^\dagger$. The $U(1)$ charge of Φ is $Y(\Phi) = 1$.

This ends our choices for the second step. At this point the model is complete. All further steps are purely technical and uniquely defined.

- *Step 3:* What follows is straightforward algebra. We write the most general, renormalisable, Lagrangian, involving the fields (25.12), (25.13), and (25.16) invariant under gauge transformations of $SU(2) \times U(1)$. We will also assume the separate conservation of the three lepton numbers, but we will further discuss this point in a later section. The requirement of renormalisability implies that all terms in the Lagrangian are monomials in the fields and their derivatives and their canonical dimension is smaller than or equal to 4. The result is

$$\mathcal{L} = -\frac{1}{4}W_{\mu\nu} \cdot W^{\mu\nu} - \frac{1}{4}B_{\mu\nu}B^{\mu\nu} + |D_\mu\Phi|^2 - V(\Phi)$$

$$+ \sum_{i=1}^{3}\left[\bar{\Psi}_L^i\,\mathrm{i}\slashed{D}\Psi_L^i + \bar{R}_i\mathrm{i}\slashed{D}R_i - G_i(\bar{\Psi}_L^i R_i\Phi + h.c.)\right]. \tag{25.17}$$

If we call W and B the gauge fields associated with $SU(2)$ and $U(1)$, respectively, the corresponding field strengths $W_{\mu\nu}$ and $B_{\mu\nu}$ appearing in (25.17) are given by (14.16) and (3.6)

$$W_{\mu\nu} = \partial_\mu W_\nu - \partial_\nu W_\mu + gW_\mu \times W_\nu \qquad B_{\mu\nu} = \partial_\mu B_\nu - \partial_\nu B_\mu. \tag{25.18}$$

Similarly, the covariant derivatives in (25.17) are determined by the assumed transformation properties of the fields, as shown in (14.13):

$$D_\mu\Psi_L^i = \left(\partial_\mu - \mathrm{i}g\tfrac{\tau}{2}\cdot W_\mu + \mathrm{i}\tfrac{g'}{2}B_\mu\right)\Psi_L^i; \quad D_\mu R_i = \left(\partial_\mu + \mathrm{i}g'B_\mu\right)R_i$$
$$D_\mu\Phi = \left(\partial_\mu - \mathrm{i}g\tfrac{\tau}{2}\cdot W_\mu - \mathrm{i}\tfrac{g'}{2}B_\mu\right)\Phi. \tag{25.19}$$

The two coupling constants g and g' correspond to the groups $SU(2)$ and $U(1)$, respectively. The most general potential $V(\Phi)$ for the scalar fields compatible with the transformation properties of the field Φ is

$$V(\Phi) = \mu^2\Phi^\dagger\Phi + \lambda(\Phi^\dagger\Phi)^2. \tag{25.20}$$

The last term in (25.17) is a Yukawa coupling term between the scalar Φ and the fermions. Since we have assumed the absence of right-handed neutrinos, this is the most general term which is invariant under $SU(2) \times U(1)$. As usual, *h.c.* stands for 'Hermitian conjugate'. G_i are three arbitrary coupling constants.

A final remark: As expected, the gauge bosons W_μ and B_μ appear to be massless. The same is true for all fermions. This is not surprising because the assumed different transformation properties of the right- and left-handed components forbid the appearance of a Dirac mass term in the Lagrangian. On the other hand, the assumption about the conservation of the three leptonic numbers forbids the appearance of a Majorana mass term. In fact, the only dimensionful parameter in (25.17) is μ^2, the parameter in the BEH potential (25.20). Therefore, the mass of every particle in the model is expected to be proportional to $|\mu|$.

- *Step 4:* The next step of our programme consists in choosing the parameter μ^2 of the scalar potential negative in order to trigger the phenomenon of spontaneous symmetry breaking and the BEH mechanism. The minimum of the potential occurs at a point $v^2 = -\mu^2/\lambda$. As we explained in Chapter 15, we can choose the direction of the breaking to be along the real part of ϕ^0.

- *Step 5:* Translating the BEH field by a real constant

$$\Phi \to \Phi + \frac{1}{\sqrt{2}} \begin{pmatrix} 0 \\ v \end{pmatrix} \qquad v^2 = -\frac{\mu^2}{\lambda} \tag{25.21}$$

transforms the Lagrangian and generates new terms, as was explained in Chapter 15. Let us look at some of them.

(i) *Fermion mass terms.* Replacing ϕ^0 by v in the Yukawa term in (25.17) creates a mass term for the charged leptons, leaving the neutrinos massless.

$$m_e = \frac{1}{\sqrt{2}} G_e v \qquad m_\mu = \frac{1}{\sqrt{2}} G_\mu v \qquad m_\tau = \frac{1}{\sqrt{2}} G_\tau v. \tag{25.22}$$

Since we had three arbitrary constants G_i, we can fit the three observed lepton masses.

(ii) *Gauge boson mass terms.* They come from the $|D_\mu \Phi|^2$ term in the Lagrangian. A straight substitution produces the following quadratic terms among the gauge boson fields:

$$\frac{1}{8} v^2 [g^2 (W_\mu^1 W^{1\mu} + W_\mu^2 W^{2\mu}) + (g' B_\mu - g W_\mu^3)^2]. \tag{25.23}$$

Defining the charged vector bosons as

$$W_\mu^\pm = \frac{W_\mu^1 \mp i W_\mu^2}{\sqrt{2}} \tag{25.24}$$

we obtain their masses:

$$m_W = \frac{vg}{2}. \tag{25.25}$$

The neutral gauge bosons B_μ and W_μ^3 have a 2×2 non-diagonal mass matrix. After diagonalisation, we define the mass eigenstates,

$$\begin{aligned} Z_\mu &= \sin \theta_W B_\mu - \cos \theta_W W_\mu^3 \\ A_\mu &= \cos \theta_W B_\mu + \sin \theta_W W_\mu^3 \end{aligned} \tag{25.26}$$

with $\tan \theta_W = g'/g$. They correspond to the mass eigenvalues

$$m_Z = \frac{v(g^2 + g'^2)^{1/2}}{2} = \frac{m_W}{\cos \theta_W}. \tag{25.27}$$

$$m_A = 0.$$

As expected, one of the neutral gauge bosons is massless and will be identified with the photon. The BEH mechanism breaks the original symmetry according to $SU(2) \times U(1) \rightarrow U(1)_{\text{em}}$ and θ_W is the angle between the original $U(1)$ and the one left unbroken. It is the parameter first introduced by S. L. Glashow, although it is often referred to as 'Weinberg angle'.

(iii) *Physical scalar mass.* Three out of the four real fields of the Φ doublet will be absorbed by the BEH mechanism in order to allow for the three gauge bosons $W^{\pm}$ and Z^0 to acquire a mass. The fourth one, which corresponds to $(|\phi^0 \phi^{0\dagger}|)^{1/2}$, remains physical. Its mass is given by the coefficient of the quadratic part of $V(\Phi)$ after the translation (25.21) and is equal to

$$m_S = \sqrt{-2\mu^2} = \sqrt{2\lambda v^2}. \tag{25.28}$$

Note that the mechanism of spontaneous symmetry breaking is at the origin of the creation of all masses of elementary particles, fermions as well as gauge bosons, with the exception of the remaining physical scalar.

In addition, we produce various coupling terms which we will present, together with the hadronic ones, in the next section.

25.3.4 Extension to Hadrons

Introducing the hadrons into the model presents some novel features. They are mainly due to the fact that the individual quark quantum numbers are not separately conserved and we have the phenomenon of flavour mixing. As regards to the second step, today there is a consensus regarding the choice of the 'elementary' constituents of matter. Besides the six leptons, there are six quarks. They are fractionally charged and come each in three 'colours'.

Let us pause here for a moment and make a second history comment. In the sixties only three quark flavours were known, the ones we present in Table 25.1 as u, d, and s. Their electric charges are $+\frac{2}{3}, -\frac{1}{3}$, and $-\frac{1}{3}$, respectively, and the weak current which was known at the time was the one given in Eq. (25.10). Trying to extend the gauge theory ideas to this three quark hadronic world we were faced with the following difficulty. The commutator of h and $h^{\dagger}$, which gives the neutral component of $SU(2)$, contains pieces of the form $\bar{d}s$ and $\bar{s}d$, i.e. flavour-changing neutral currents. Their presence would induce decays of the type $K^0 \rightarrow \mu^+\mu^-$, or $K^0 \rightarrow \nu\bar{\nu}$, both of which were absolutely excluded experimentally. The solution to this puzzle was found in 1970 by S. L. Glashow, J. Iliopoulos, and L. Maiani. It consisted of proposing the existence of a fourth quark flavour, named c for 'charm', which made possible the addition of a second piece to the charged weak current:

$$h_\lambda = \bar{u}_L \gamma_\lambda (\cos\theta \, d_L + \sin\theta \, s_L) + \bar{c}_L \gamma_\lambda (\cos\theta \, s_L - \sin\theta \, d_L). \tag{25.29}$$

It is easy to check now that the resulting neutral current is diagonal in flavour space. The introduction of a fourth quark implied the prediction on the existence of an entire

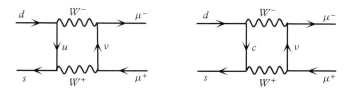

Figure 25.1 *1-loop contributions to a strangeness changing neutral current process.*

new hadronic sector, namely the hadrons containing c among their constituents. Since, experimentally, the value of $\cos\theta$ is close to 1, the 'charmed' particles are predicted to decay predominantly to strange particles. We could even go a step further. At the limit of exact flavour symmetry, when all quarks have the same mass, the processes involving flavour changing neutral currents were forbidden by the symmetry. But we know that flavour symmetry is broken and the quarks do not have the same mass. If we look at these processes at higher orders we find, for example, the square diagrams of Fig. 25.1. They induce amplitudes for processes such as $K^0 \rightarrow \mu^+\mu^-$, or the $\Delta S = 2$ transition $K^0 - \bar{K}^0$, which gives the $K_s - K_L$ mass difference. The u and c quark contributions have opposite signs, so the resulting amplitudes are proportional to $(m_c^2 - m_u^2)/m_W^2$. This property implied the translation of the experimental upper limits on these processes to an upper limit on m_c, i.e. upper limits of a few GeV on the masses of the charmed particles. The subsequent experimental discovery of these new particles in the predicted mass range and with precisely the correct decay signature was a second great success of these ideas.

Coming back to present day, we see that in order to explore the lepton–hadron universality property, we must use also doublets and singlets for the quarks. The first novel feature we mentioned in the previous paragraph is that all quarks appear to have non-vanishing Dirac masses, so we must introduce both right-handed singlets for each family. A naïve assignment would be to write the analogue of Eqs. (25.12) and (25.13) as

$$Q_L^i(x) = \frac{1}{2}(1 + \gamma_5) \begin{pmatrix} U^i(x) \\ d^i(x) \end{pmatrix}; \quad U_R^i(x); \quad D_R^i(x) \tag{25.30}$$

with the index i running over the three families as $U^i = u, c, t$ and $D^i = d, s, b$ for $i = 1, 2, 3$, respectively.[3] This assignment determines the $SU(2)$ transformation properties of the quark fields. It also fixes their Y charges and, hence, their $U(1)$ properties. Using Eq. (25.11), we find that

$$Y(Q_L^i) = \frac{1}{3}; \quad Y(U_R^i) = \frac{4}{3}; \quad Y(D_R^i) = -\frac{2}{3}. \tag{25.31}$$

---

[3] An additional index a, running also through 1,2, and 3 and denoting the colour, is understood.

The presence of the two right-handed singlets has an important consequence. Even if we had only one family, we would have two distinct Yukawa terms between the quarks and the scalar fields of the form

$$\mathcal{L}_{\text{Yuk}} = G_d(\bar{Q}_{\text{L}} D_{\text{R}} \Phi + h.c.) + G_u(\bar{Q}_{\text{L}} U_{\text{R}} \tilde{\Phi} + h.c.). \tag{25.32}$$

$\tilde{\Phi}$ is the doublet made out of ϕ^{0*} and ϕ^-. It has the same transformation properties under $SU(2)$ as Φ, but the opposite Y charge (see Problem 25.1).

If there were only one family, this would have been the end of the story. The hadron Lagrangian $\mathcal{L}_{\text{h}}^{(1)}$ is the same as (25.17) with quark fields replacing leptons and the extra term of (25.32). The complication we alluded to before comes with the addition of more families. In this case the total Lagrangian is not just the sum over the family index. The physical reason is the non-conservation of the individual quark quantum numbers we mentioned previously. In writing (25.30), we implicitly assumed a particular pairing of the quarks in each family, u with d, c with s, and t with b. In general, we could choose any basis in family space and, since we have two Yukawa terms, we will not be able to diagonalise both of them simultaneously. It follows that the most general Lagrangian will contain a matrix with non-diagonal terms which mix the families. By convention, we attribute it to a different choice of basis in the $d - s - b$ space. It follows that the correct generalisation of the Yukawa Lagrangian (25.32) to many families is given by

$$\mathcal{L}_{\text{Yuk}} = \sum_{i,j} \left[(\bar{Q}_{\text{L}}^i G_d^{ij} D_{\text{R}}^j \Phi + h.c.) \right] + \sum_i \left[G_u^i (\bar{Q}_{\text{L}}^i U_{\text{R}}^i \tilde{\Phi} + h.c.) \right], \tag{25.33}$$

where the Yukawa coupling constant G_d has become a matrix in family space. After translation of the scalar field, we will produce masses for the up-quarks given by $m_u = G_u^1 v$, $m_c = G_u^2 v$- and $m_t = G_u^3 v$, as well as a 3×3 mass matrix for the down-quarks given by $G_d^{ij} v$. As usually, we want to work in a field space where the masses are diagonal, so we change our initial $d - s - b$ basis to bring G_d^{ij} into a diagonal form. This can be done through a 3×3 unitary matrix $\tilde{D}^i = U^{ij} D^j$ such that $U^\dagger G_d U = \text{diag}(m_d, m_s, m_b)$. In the simplest example of only two families, it is easy to show that the most general such matrix, after using all freedom for field redefinitions and phase choices, is a real rotation

$$C = \begin{pmatrix} \cos\theta & \sin\theta \\ -\sin\theta & \cos\theta \end{pmatrix}, \tag{25.34}$$

with θ being our familiar Cabibbo angle. For three families we can show (see Problem 25.2) that the matrix has three angles, the three Euler angles, and an arbitrary phase. It is traditionally written in the form

$$CKM = \begin{pmatrix} c_1 & s_1 c_3 & s_1 s_3 \\ -s_1 c_3 & c_1 c_2 c_3 - s_2 s_3 e^{i\delta} & c_1 c_2 s_3 + s_2 c_3 e^{i\delta} \\ -s_1 s_2 & c_1 s_2 c_3 + c_2 s_3 e^{i\delta} & c_1 s_2 s_3 - c_2 c_3 e^{i\delta} \end{pmatrix}, \tag{25.35}$$

with the notation $c_k = \cos\theta_k$ and $s_k = \sin\theta_k$, $k = 1, 2, 3$. The novel feature is the possibility of introducing the phase δ. This means that a six-quark model has a natural source of CP, or T, violation, while a four-quark model does not. This brings us to a third historical remark. It has been known since 1965 that weak interactions are not invariant under time reversal. The absence of a natural source for such violation in a two-family model prompted M. Kobayashi and T. Maskawa to postulate the existence of a third family, prediction, which was again verified with the discovery of τ, b, and t. So we see that the consistency and the simplicity of the electroweak theory repeatedly determined the spectrum of elementary particles.

The total Lagrangian density, before the translation of the scalar field, is now

$$
\begin{aligned}
\mathcal{L} = &-\frac{1}{4} W_{\mu\nu} \cdot W^{\mu\nu} - \frac{1}{4} B_{\mu\nu} B^{\mu\nu} + |D_\mu \Phi|^2 - V(\Phi) \\
&+ \sum_{i=1}^{3} \left[\bar{\Psi}_{\rm L}^i {\rm i}\slashed{D} \Psi_{\rm L}^i + \bar{R}_i {\rm i}\slashed{D} R_i - G_i (\bar{\Psi}_{\rm L}^i R_i \Phi + h.c.) \right. \\
&+ \bar{Q}_{\rm L}^i {\rm i}\slashed{D} Q_{\rm L}^i + \bar{U}_{\rm R}^i {\rm i}\slashed{D} U_{\rm R}^i + \bar{D}_{\rm R}^i {\rm i}\slashed{D} D_{\rm R}^i + G_u^i (\bar{Q}_{\rm L}^i U_{\rm R}^i \tilde{\Phi} + h.c.) \Big] \\
&+ \sum_{i,j=1}^{3} \left[(\bar{Q}_{\rm L}^i G_d^{ij} D_{\rm R}^j \Phi + h.c.) \right].
\end{aligned}
\tag{25.36}
$$

The covariant derivatives on the quark fields are given by

$$
D_\mu Q_{\rm L}^i = \left(\partial_\mu - {\rm i}g\frac{\boldsymbol{\tau}}{2} \cdot W_\mu - {\rm i}\frac{g'}{6} B_\mu \right) Q_{\rm L}^i
\tag{25.37}
$$

$$
D_\mu U_{\rm R}^i = \left(\partial_\mu - {\rm i}\frac{2g'}{3} B_\mu \right) U_{\rm R}^i
$$

$$
D_\mu D_{\rm R}^i = \left(\partial_\mu + {\rm i}\frac{g'}{3} B_\mu \right) D_{\rm R}^i.
$$

The classical Lagrangian (25.36) contains 17 arbitrary real parameters. They are the following:

- The two gauge coupling constants g and g'.
- The two parameters of the BEH potential λ and μ^2.
- Three Yukawa coupling constants for the three lepton families, $G_{e,\mu,\tau}$.
- Six Yukawa coupling constants for the three quark families, $G_u^{u,c,t}$ and $G_d^{d,s,b}$.
- Four parameters of the Cabibbo–Kobayashi–Maskawa (CKM) matrix, the three angles and the phase δ.

A final remark: Fifteen out of these 17 parameters are directly connected with the BEH sector.

Translating the field Φ by Eq. (25.21) and diagonalising the resulting down-quark mass matrix produces the mass terms for fermions and bosons which we introduced before as well as several coupling terms. We will write them here in the unitary gauge we introduced in section 15.3 in which only the fields representing physical particles appear. In other gauges we must also include the ghosts and the three unphysical BEH fields.

(i) *The gauge boson–fermion couplings.* They are the ones which generate the known weak and electromagnetic interactions. A_μ is coupled to the charged fermions through the usual electromagnetic current

$$\frac{gg'}{(g^2 + g'^2)^{1/2}} \left[\bar{e}\gamma^\mu e + \sum_{a=1}^{3} \left(\frac{2}{3}\bar{u}^a\gamma^\mu u^a - \frac{1}{3}\bar{d}^a\gamma^\mu d^a \right) + ... \right] A_\mu, \tag{25.38}$$

where the dots stand for the contribution of the other two families $e \to \mu, \tau$, $u \to c, t$ and $d \to s, b$ and the summation over a extends over the three colours. Equation (25.38) shows that the electric charge e is given, in terms of g and g' by

$$e = \frac{gg'}{(g^2 + g'^2)^{1/2}} = g\sin\theta_W = g'\cos\theta_W. \tag{25.39}$$

Similarly, the couplings of the charged W's to the weak current are

$$\frac{g}{2\sqrt{2}} \left(\bar{\nu}_e\gamma^\mu(1 + \gamma_5)e + \sum_{a=1}^{3} \bar{u}^a\gamma^\mu(1 + \gamma_5)d^a_{KM} + ... \right) W^+_\mu + h.c. \tag{25.40}$$

As expected, only left-handed fermions participate. d_{CKM} is the linear combination of $d - s - b$ given by the *CKM* matrix (25.35). By diagonalising the down-quark mass matrix we introduced the off-diagonal terms into the hadron current. When considering processes, like nuclear β-decay, or μ-decay, where the momentum transfer is very small compared to the W mass, the W propagator can be approximated by m_W^{-2} and the effective Fermi coupling constant is given by

$$\frac{G}{\sqrt{2}} = \frac{g^2}{8m_W^2} = \frac{1}{2v^2}. \tag{25.41}$$

In contrast to the charged weak current (25.40), the Z^0-fermion couplings involve both left- and right-handed fermions:

$$-\frac{e}{2}\frac{1}{\sin\theta_W\cos\theta_W} \left[\bar{\nu}_L\gamma^\mu\nu_L + (\sin^2\theta_W - \cos^2\theta_W)\bar{e}_L\gamma^\mu e_L \right.$$
$$\left. +2\sin^2\theta_W\bar{e}_R\gamma^\mu e_R + ... \right] Z_\mu \tag{25.42}$$

$$\frac{e}{2} \sum_{a=1}^{3} \left[\left(\frac{1}{3} \tan \theta_W - \cot \theta_W \right) \bar{u}_L^a \gamma^\mu u_L^a + \left(\frac{1}{3} \tan \theta_W + \cot \theta_W \right) \bar{d}_L^a \gamma^\mu d_L^a \right.$$
$$\left. + \frac{2}{3} \tan \theta_W (2 \bar{u}_R^a \gamma^\mu u_R^a - \bar{d}_R^a \gamma^\mu d_R^a) + \dots \right] Z_\mu. \tag{25.43}$$

Again, the summation is over the colour indices and the dots stand for the contribution of the other two families. We verify in this formula the property of the weak neutral current to be diagonal in the quark flavour space. Another interesting property is that the axial part of the neutral current is proportional to $[\bar{u} \gamma_\mu \gamma_5 u - \bar{d} \gamma_\mu \gamma_5 d]$. This particular form of the coupling is important for the phenomenological applications, such as the induced parity violating effects in atoms and nuclei.

(ii) *The gauge boson self-couplings.* One of the characteristic features of Yang–Mills theories is the particular form of the self-couplings among the gauge bosons. They come from the square of the non-Abelian curvature in the Lagrangian, which, in our case, is the term $-\frac{1}{4} W_{\mu\nu} \cdot W^{\mu\nu}$. Expressed in terms of the physical fields, this term gives

$$- ig(\sin \theta_W A^\mu - \cos \theta_W Z^\mu)(W^{\nu-} W_{\mu\nu}^+ - W^{\nu+} W_{\mu\nu}^-)$$
$$- ig(\sin \theta_W F^{\mu\nu} - \cos \theta_W Z^{\mu\nu}) W_\mu^- W_\nu^+$$
$$- g^2 (\sin \theta_W A^\mu - \cos \theta_W Z^\mu)^2 W_\nu^+ W^{\nu-}$$
$$+ g^2 (\sin \theta_W A^\mu - \cos \theta_W Z^\mu)(\sin \theta_W A^\nu - \cos \theta_W Z^\nu) W_\mu^+ W_\nu^- \tag{25.44}$$
$$- \frac{g^2}{2} (W_\mu^+ W^{\mu-})^2 + \frac{g^2}{2} (W_\mu^+ W_\nu^-)^2,$$

where we have used the notation $F_{\mu\nu} = \partial_\mu A_\nu - \partial_\nu A_\mu$, $W_{\mu\nu}^\pm = \partial_\mu W_\nu^\pm - \partial_\nu W_\mu^\pm$, and $Z_{\mu\nu} = \partial_\mu Z_\nu - \partial_\nu Z_\mu$ with $g \sin \theta_W = e$. Let us concentrate on the photon-$W^+ W^-$ couplings. If we forget, for the moment, about the $SU(2)$ gauge invariance, we can use different coupling constants for the two trilinear couplings in (25.44), say e for the first and $e\kappa$ for the second. For a charged, massive W, the magnetic moment μ and the quadrupole moment Q are given by

$$\mu = \frac{(1+\kappa)e}{2m_W} \qquad Q = -\frac{e\kappa}{m_W^2}. \tag{25.45}$$

Looking at (25.44), we see that $\kappa = 1$. Therefore, $SU(2)$ gauge invariance gives very specific predictions concerning the electromagnetic parameters of the charged vector bosons. The gyromagnetic ratio equals 2 and the quadrupole moment equals $-em_W^{-2}$.

(iii) *The scalar–fermion couplings.* They are given by the Yukawa terms in (25.17). The same couplings generate the fermion masses through spontaneous symmetry breaking. It follows that the physical BEH scalar couples to quarks and leptons with strength proportional to the fermion mass. Therefore, the prediction is that it will decay predominantly to the heaviest possible fermion compatible with phase space. This property provides a typical signature for its identification.

(iv) *The scalar–gauge boson couplings.* They come from the covariant derivative term $|D_\mu \Phi|^2$ in the Lagrangian. In the unitary gauge only one neutral scalar field survives:

$$\frac{1}{4}(v + \phi)^2 \left[g^2 W_\mu^+ W^{-\mu} + (g^2 + g'^2) Z_\mu Z^\mu \right]. \tag{25.46}$$

(v) *The scalar self-couplings.* They are proportional to $\lambda(v + \phi)^4$. Equations (25.28) and (25.41) show that $\lambda = G m_S^2 / \sqrt{2}$, so this coupling has been measured together with m_S, the BEH mass. The latter is around 126 GeV, so λ is on the order of $\frac{1}{6}$, appreciable, but according to our experience, still in the perturbative regime. On the other hand, this relation shows that were the physical scalar particle very heavy, it would have been also strongly interacting and this sector of the model would have been non-perturbative.

The five-step programme is now complete for both leptons and quarks. Although the number of arbitrary parameters seems very large, we should not forget that they are all mass and coupling parameters, like the electron mass and the fine structure constant of quantum electrodynamics. The reason we have more of them is that the standard model describes in a unified framework a much larger number of particles and interactions.

Just a few words concerning the fermion masses. Leaving aside the neutrinos, the other fermion masses are shown in Table 25.2.

Some remarks:

- Even without the neutrinos, these values are spread over a huge range. The ratio between the electron and the *t*-quark masses is larger than 10^5. Il we include the limits on the neutrino masses it exceeds 10^{11} and, probably, much higher. It is hard to imagine that they all come from the same spontaneous symmetry breaking mechanism.

Table 25.2 *Fermion masses are shown in MeV. The lepton masses are directly measurable but those of the quarks are only indirectly estimated. This is particularly significant for the u and d quarks. The uncertainties, experimental or estimated, are also shown. For the electron and the muon the uncertainties are too small to be included.*

FERMION MASSES					
Leptons			Quarks		
e	μ	τ	u/d	c/s	t/b
0.51100	105.6584	1776.82 ± 0.16	2.3 (+0.7, −0.5) 4.8 (+0.5, −0.3)	1275 ± 25 95 ± 5	173070 ± 520 4180 ± 30

- For the two heavy quark families the up-type quarks are heavier than the down-type ones. This pattern is reversed for the light quarks of the first family.

- If we restrict ourselves to the first family we see that the mass ratio of the two quarks equals a factor of 2, so, at the quark level, isospin symmetry is very badly broken. If it appears to be a good approximate symmetry in hadronic physics, it is because the quark masses contribute very little to the masses of hadrons. The scale of the first is a few MeV while that of the second is 1 GeV. The global isospin symmetry seems to be accidental. We will come back to the origin of hadron masses later.

- Despite their small values, the mass pattern of the light quarks seems to have very important consequences. The fact that the down-quark is much heavier than the up one results, probably, in the neutron being slightly heavier than the proton. Hydrogen stability, at the origin of all matter creation, is due to this accident.

Our confidence in this model is amply justified on the basis of its ability to accurately describe the bulk of our present-day data and, especially, of its enormous success in predicting new phenomena. We have mentioned already some of them. Let us give here a brief summary.

1. The discovery of weak neutral currents by Gargamelle in 1973:

$$\nu_\mu + e^- \rightarrow \nu_\mu + e^-; \quad \nu_\mu + N \rightarrow \nu_\mu + X.$$

   Not only their existence, but also their detailed properties were predicted. In general we would expect, for every lepton and every quark flavour a parameter that determines the strength of the neutral current relatively to the charged one and another to fix the ratio of the vector and axial parts. In the standard model, in which the breaking comes through an isodoublet scalar field, they are all expressible in terms of the angle θ_W. This is brilliantly confirmed by the data.

2. The discovery of charmed particles at SLAC in 1974–1976. Their presence was essential to ensure the absence of strangeness changing neutral currents, ex. $K^0 \rightarrow \mu^+ + \mu^-$. Their characteristic property is to decay predominantly into strange particles.

3. As we will see shortly, a necessary condition for the consistency of the model is that $\sum_i Q_i = 0$ inside each family. When the τ lepton was discovered the b and t quarks were predicted with the right electric charges.

4. The discovery of the W and Z bosons at CERN in 1983 involved a brilliant innovation in accelerator technology. The characteristic relation of the standard model with an isodoublet BEH mechanism $m_Z = m_W/\cos\theta_W$ is checked with very high accuracy (including radiative corrections).

5. The t-quark was *seen* at LEP through its effects in radiative corrections before its actual discovery at Fermilab.

6. The final touch: the recent discovery of the Brout–Englert–Higgs scalar at CERN.

A brief comparison with all available experimental data will be presented later. All 17 parameters of the model have been determined experimentally.

At lowest order the parameters g, g', and v enter into the expressions for the electric charge (25.39), the Fermi coupling (25.41), the W and Z masses (25.25) and (25.27), and the couplings of the weak neutral current (25.42) and (25.43). This gives us a large number of predictions, all of which have been verified experimentally.

The Yukawa couplings and the parameters of the *CKM* matrix are determined through the fermion spectrum and decay properties. An important experimental effort is actually devoted to the precise determination of the mixing angles and the phase δ connected with *CP* violation.

The last parameter λ has been determined through the measurement of the mass of the recently discovered scalar boson. At lowest order it enters only into the BEH mass and self-coupling.

25.3.5 The Neutrino Masses

We have postponed the discussion concerning the masses of the neutrinos. This was done for simplicity, but also because, until recently, all experimental results were compatible with vanishing neutrino masses. The situation changed radically with a new generation of experiments including solar, atmospheric, and terrestrial neutrinos. We will present here these results, which are among the most remarkable discoveries of high-energy physics in recent years.

The first question concerns the total number of neutrino species. In the standard model it is related to the total number of families. A new neutrino, even with very low, or vanishing, mass, would still remain undetectable in ordinary experiments, provided its companion charged lepton is sufficiently heavy. This problem was brilliantly solved by the first LEP experiment. The clue was to measure very precisely the decay parameters of the Z^0 boson. Equations (25.42) and (25.43) show the contribution of each elementary fermion in the model. The decays into quarks and charged leptons can be measured separately and their sum constitutes the visible part of the Z^0 width.[4] The neutrino final states cannot be measured directly but their total contribution can be determined by the difference between the experimentally measured total and visible widths. In turn this influences the value of the cross section at the peak of the Z^0 curve. Using (25.42) we can express it in terms of N_ν, the number of neutrino species contributing to the decay, i.e. those with masses less than half the mass of Z^0. In practice, in order to reduce the errors, we define the 'invisible ratio' $R^0_{\text{inv}} = \Gamma_{\text{inv}}/\Gamma_{ll}$, i.e. the ratio of the invisible width over the purely leptonic width. We will assume, first, that all invisible width is due to standard model neutrinos and, second, that universality for all families holds. Then we obtain

$$R^0_{\text{inv}} = N_\nu \left(\frac{\Gamma_{\nu\bar{\nu}}}{\Gamma_{ll}} \right)_{\text{SM}} , \qquad (25.47)$$

---

[4] By 'visible', we mean the decays whose final states are visible in the detector.

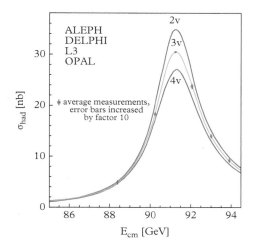

Figure 25.2 *The hadronic production cross section in the vicinity of Z^0. The predicted curves for two-, three-, and four-neutrino species are shown under the assumption of standard model couplings and negligible masses. Note that the error bars have been multiplied by 10 for illustration purposes (© CERN, LEP collaborations).*

where the ratio which multiplies N_ν is the one we compute in the standard model. We compare the measured value of R^0_{inv} with the standard model calculation and we obtain

$$N_\nu = 3. \tag{25.48}$$

We can visualise the result in Fig. 25.2 which shows also the strong dependence of the hadronic peak cross section on the number of neutrinos. The great precision of the measurements makes it possible to put very strict limits on any other conceivable contributions of unknown neutral particles in the Z^0 decay width. If new families exist, their neutrinos are very heavy.

Let us come now to the problem of masses. We have two classes of experiments. The first aims at a direct measurement of the mass of each particular species. They have given only upper limits so far. They are, including the corresponding confidence levels,

$$m_{\nu_e} < 2.2 \ eV \ (95\% \ \text{CL}) \tag{25.49}$$

$$m_{\nu_\mu} < 170 \ \text{keV} \ (90\% \ \text{CL}); \quad m_{\nu_\tau} < 18.2 \ \text{MeV} \ (95\% \ \text{CL}). \tag{25.50}$$

The best limit is the one on m_{ν_e}. It comes from the study of the electron energy spectrum in the β-decay of tritium. In Problem 25.6 we compute this spectrum for the simple case of neutron β-decay and we show that the form of the end-point, i.e. the point where the electron has the maximum possible kinetic energy, is very sensitive to m_{ν_e}. A new generation of experiments along the same lines are expected to improve the limit (25.49) by an order of magnitude. A better, but indirect, limit is obtained by the absence of neutrinoless double β-decay. An ordinary double β-decay process is one in which two neutrons in a nucleus decay simultaneously produce two electrons and two anti-neutrinos: $N_1 \rightarrow N_2 + 2e^- + 2\bar{\nu}_e$. Being double weak, these processes are very rare, but have been observed for some isotopes for which all other decay channels are energetically

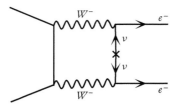

Figure 25.3 *Neutrinoless double beta decay. The cross represents the presence of a Majorana mass.*

forbidden. In 1939 W. Furry suggested the existence of a neutrinoless double β-decay, a process of the form shown in the Feynman diagram of Fig. 25.3. Such decay violates lepton number conservation and, as can be seen from Fig. 25.3, requires the neutrino to be identical with its own anti-particle and, therefore, it tests the presence of a Majorana mass. The resulting limit is on the order of 0.4 eV, which again may be substantially improved in the near future.

All these are negative results and support the view that neutrinos could indeed be massless. However, in recent years, a different series of experiments changed radically this picture. We know today that neutrinos do have non-zero masses, although we do not know their precise values. They are all based on the observation that with massive neutrinos, the lepton sector of the standard model is expected to exhibit the phenomenon of flavour mixing, as is the case with the quark sector. In analogy with the down-quarks, neutrino masses could be described by a general 3×3 matrix, in which case each one of the three lepton numbers would not be separately conserved. It follows that neutrinos could undergo quantum mechanical oscillations among the three species. In Problem 25.7 we compute the amplitude for such oscillations and show that it depends on the mass differences Δm_ν. For example, a ν_e, produced in β decay could change into a ν_μ or a ν_τ in the course of its propagation. Such oscillations, which are impossible if all three neutrinos have degenerate, in particular zero, masses, have been detected in three types of experiments. The first used solar neutrinos, i.e. neutrinos produced by the nuclear reactions in the interior of the Sun. The second used the neutrinos found in the decay products of π or K mesons produced by cosmic rays in the upper atmosphere, and the third used the neutrinos coming from nuclear reactors in power stations or specifically produced using particle accelerators. We are still far from determining the entire neutrino mass matrix accurately, but we know some general features: the mass differences are very small, on the order of 10^{-3} and 10^{-5} eV, and the mixing angles among the flavours are larger than those we encounter in the quark sector. Most experimental results are compatible with a minimal neutrino content in which the neutrinos are described by Majorana fermions, but further studies are necessary in order to confirm this hypothesis. Note that if the actual mass values are on the same order as the mass differences, their direct kinematical determination will be impossible in the near future. On the other hand, such tiny values may indicate that their origin lies in physics beyond the standard model.

25.3.6   Some Sample Calculations

The standard electroweak model is a renormalisable quantum field theory and, consequently, the technology of Feynman diagrams we developed in the previous chapters allows us to compute any correlation function in the perturbation expansion. Here we will present only some calculations which will help us to understand better the intricacies of spontaneously broken gauge theories.

We start with a series of tree diagram computations, first performed by Christopher Llewellyn Smith in 1973. By that time the standard model was known, but the calculation could have been done any time previously. The purpose is to show that the requirement of renormalisability points unambiguously to a spontaneously broken gauge theory and exhibits clearly the role of each piece of the puzzle.

Let us consider the Fermi theory of weak interactions, Eq. (25.1). We showed already that it is the low-energy approximation of the standard model with the coupling strength given by eq. (25.41), but let us suppress this information for the moment. It is a non-renormalisable theory, so we can only compute tree-level diagram-s, but, at that level, it is phenomenologically successful. The purpose of the exercise is to seek possible modifications of the theory in order to improve its convergence properties, still keeping the agreement with experiment.

We look at the electron-neutrino system. For example, the cross section for the elastic electron–anti-neutrino scattering of Fig. 25.4(a) in the four-fermion theory (25.1) is given by the square of the amplitude:

$$M_{\text{el}} = i\frac{G}{\sqrt{2}}[\bar{u}_e(p_1')\gamma_\mu(1+\gamma_5)v_\nu(p_2')][\bar{v}_\nu(p_2)\gamma^\mu(1+\gamma_5)u_e(p_1)]. \tag{25.51}$$

The Fermi coupling constant has dimensions $[M]^{-2}$. At high energies we can neglect the fermion masses and the cross section will behave as

$$\sigma_{\text{el}} \sim G^2 s, \tag{25.52}$$

where s is the total energy in the centre of mass, $s = (p_1+p_2)^2$. Remember, however, that in section 20.5 we showed that this cross section is proportional to the imaginary part of

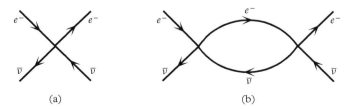

(a)   (b)

Figure 25.4 *The elastic electron–anti-neutrino scattering at first order in the Fermi coupling constant (a) and at second order (b).*

the corresponding 1-loop amplitude of Fig. 25.4(b). It follows therefore that if we write a dispersion relation for the latter of the form

$$M_{el}(s) \sim \int \frac{\sigma_{el}(s')}{s' - s} ds', \tag{25.53}$$

it will require two subtractions. Of course, this is in agreement with the fact that the diagram of Fig. 25.4(b) is quadratically divergent. It is not difficult to guess a first modification of (25.1) to improve this result. We assume the existence of a massive, charged, intermediate vector boson $W^{\pm}$ and write (25.1) in the form of (25.40). g is dimensionless, so we expect any lepton cross section to behave at high energies not worse than a constant. In fact, the suggestion for the existence of such an intermediate boson was made much before the construction of the standard model.

Now we have enlarged our set of particles and we must study all tree amplitudes in the e–v–W system. We choose the inelastic process $v + \bar{v} \rightarrow W^{+} + W^{-}$. In the tree approximation it is given by the diagram of Fig. 25.5(a). We may think that a W, being a vector particle, is like a photon, but it is massive. In the diagram of Fig. 25.5(a) this difference manifests itself only in the fact that a W may have three degrees of polarisation, in contrast to a photon which has only two. So we look at the cross section to produce a pair of longitudinally polarised W's. We obtain

$$M(v\bar{v} \rightarrow WW) \sim g^2 \bar{v}(p_1) \not{E}^{-}(1 + \gamma_5) \frac{\not{q} + m_e}{q^2 - m_e^2} \not{E}^{+}(1 + \gamma_5)u(p_2), \tag{25.54}$$

where $E^{\pm}$ are the polarisation vectors for the pair of W's. A longitudinal polarisation is proportional to the W momentum k, $E_L \sim k/m_W$, so at large k, the cross section will behave as

$$\sigma(v\bar{v} \rightarrow W_L W_L) \sim \frac{g^4}{m_W^4} s. \tag{25.55}$$

In other words, we are back to the problem we had in (25.52). This again is compatible with the fact that the propagator of a massive vector boson behaves at large momenta like a constant and the diagram of Fig. 25.5(b) is quadratically divergent. Adding a massive W did not make the Fermi theory renormalisable.

We must further enlarge our physical system to produce new tree diagrams to dump the bad asymptotic behaviour found in Eq. (25.55). Note that the trouble appears in the p-wave scattering, since both W's are longitudinally polarised. Looking at the diagram of Fig. 25.5(a) we see that we have the following choices: we can add a contribution to the s channel, or one in the u channel, or a combination of the two. The first leads to the diagram of Fig. 25.6(a), which implies the addition of a new neutral vector particle Z. The second is shown in Fig. 25.6(b) and requires a new positively charged lepton. The second choice leads to the Georgi–Glashow model we introduce in Problem 25.8, but here we assume only the first since we know experimentally of the existence of Z.

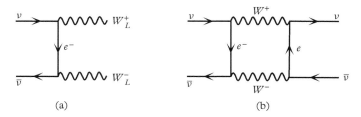

Figure 25.5 *The $\nu\,\bar{\nu} \rightarrow W^+ W^-$ amplitude in the tree approximation (a) and the 1-loop contribution to the $\nu\,\bar{\nu}$ elastic amplitude (b).*

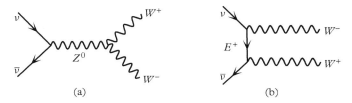

Figure 25.6 *Possible corrections to the $\nu\,\bar{\nu} \rightarrow W^+ W^-$ amplitude in the s-channel (a) and the u-channel (b).*

We want to have a cancellation in the asymptotic behaviour between the two diagrams of Figs. 25.5(a) and 25.6(a), so the Z exchange must be also proportional to g^2. The neutrinos are left-handed, so the ν–$\bar{\nu}$–Z vertex is essentially unique, up to an arbitrary coupling constant: $\sim \bar{\nu}\gamma_\mu(1 + \gamma_5)\nu Z^\mu$. We must still determine the Z–W^+–W^- vertex. We have three vector fields, so we need one derivative. We also note that W^+ and W^- are complex conjugate to each other and the effective interaction Lagrangian must be Hermitian. This leaves us with three possible dimension 4 terms, the ones we have in Eq. (25.44) with arbitrary coefficients. The cancellation in the cross section for longitudinally polarised W's gives a relation between these coupling constants. We can go one step further because the presence of electrically charged particles forces us to enlarge our system and include the photon. We repeat the exercise for the $e^+ + e^- \rightarrow W^+_L + W^-_L$, as well as the amplitudes for producing pairs of W–Z and W–γ. We thus arrive at the solution of (25.44) which describes the Yang–Mills couplings.

We are almost home, but we are still missing the scalar particle. For that we must look at the scattering of two vector particles, for example the one shown in Fig. 25.7(a). We find that a good high-energy behaviour requires the addition of a scalar field (Fig. 25.7(b)). We thus arrive at the standard model without assuming a gauge theory to start with.

We can object that in our search we succeeded because we knew the answer and this is certainly true. However, now that we have the machinery, we can ask the more general question. Consider a set of fields, spin 0, spin 1/2, and spin 1. Among the latter, some

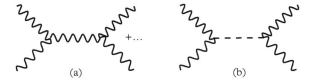

Figure 25.7 *The high-energy behaviour in the tree diagram describing the scattering of two vector bosons (a) is tamed by the scalar boson contribution (b).*

are assumed massless and some are massive. Write the most general Hermitian, Lorentz invariant, interaction Lagrangian among them with terms of dimension smaller than or equal to 4. We can simplify it by imposing invariance under some global symmetry. Our theory so far is characterised by a large set of parameters, masses, and coupling constants. Impose 'tree unitarity' for all amplitudes, i.e. no cross section computed in the tree approximation should grow at infinity.[5] This gives relations among the parameters. The resulting theory is equivalent to a Yang–Mills gauge theory. The relations among the parameters simply express the BRST symmetry in the unitary gauge; in other words, if the BRST identities are not satisfied the theory cannot be renormalisable. All massive vector bosons correspond to spontaneously broken generators via a BEH mechanism with two possible exceptions: (i) the trivial one of free fields, and (ii) massive neutral vector bosons coupled to conserved currents following the Stückelberg formalism we exposed in section 16.5.4.

As a second example of calculations in the framework of the standard model we want to compute some static quantities of boson and fermion fields, the analogue of the *g-2* computation we indicated in Problem 16.4. We want to show explicitly the fundamental importance of the underlying BRST symmetry, so we choose to work in the unitary gauge we introduced in section 15.3 which contains only physical degrees of freedom. The strange property of this gauge is that the theory is non-renormalisable by power counting; nevertheless, physical quantities will turn out to be finite.

Let us start with the electrostatic properties of the charged vector boson $W^{\pm}$; in other words, we consider the scattering of a W from an external electromagnetic field. We write the amplitude for this process as $\mathcal{M}_{\mu\nu\rho}(p_1, p_2)E_1^\nu E_2^\rho$, where p_1, E_1 and p_2, E_2 are the momenta and the polarisation vectors of the incoming and outgoing W, respectively. We must study the vertex function $\mathcal{M}_{\mu\nu\rho}(p_1, p_2)$. In order to simplify the calculations we can ignore the lepton and quark fields and restrict ourselves to the bosonic sector of the standard model. At 1 loop the diagrams which contribute to this process are shown in Figs. 25.8 and 25.9. We analyse $\mathcal{M}_{\mu\nu\rho}(p_1, p_2)$ in form factors. The initial and final W's are on the mass shell, so we must have

$$q^\mu \mathcal{M}_{\mu\nu\rho} = 0; \quad p_1^2 = p_2^2 = m_W^2; \quad p_1 \cdot E_1 = p_2 \cdot E_2 = 0, \qquad (25.56)$$

---

[5] We want to stress here that this argument is by no means a proof of renormalisability. A real proof requires the steps we went through in Chapter 18. It is, however, a proof of the opposite statement, namely that a massive, non-Abelian Yang–Mills theory is not renormalisable if the vector bosons do not acquire their mass through the mechanism of spontaneous symmetry breaking. Tree unitarity is a necessary ingredient for a renormalisable theory but not a sufficient one. See, for example, the theories with anomalous Ward identities, whose presence is manifest only in loop diagrams. We will address this question for the standard model in the next section.

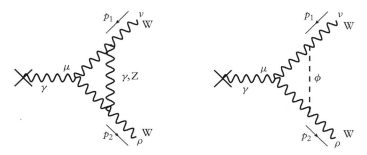

Figure 25.8 *Diagrams contributing to the scattering amplitude of a charged W in an external electromagnetic field.*

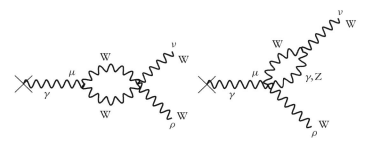

Figure 25.9 *Diagrams contributing to the scattering amplitude of a charged W in an external electromagnetic field.*

where $q = p_2 - p_1$. The most general form of $\mathcal{M}$ compatible with these constraints can be written as

$$\mathcal{M}_{\mu\nu\rho}(p_1, p_2) = F_1(q^2)\left[(p_1 + p_2)_\mu g_{\nu\rho} + 2(q_\nu g_{\mu\rho} - q_\rho g_{\mu\nu})\right]$$
$$+ F_2(q^2)\left(q_\nu g_{\mu\rho} - q_\rho g_{\mu\nu}\right) + F_3(q^2)\frac{1}{m_W^2}(p_1 + p_2)_\mu q_\nu q_\rho, \tag{25.57}$$

where F_1, F_2, and F_3 are three form factors which, in the tree approximation, are given by $F_1(0) = ie$ and $F_2(0) = F_3(0) = 0$. The last two are the anomalous magnetic moment and the anomalous quadrupole moment of the W, respectively, and, in a renormalisable theory, they must be both finite and calculable. It is this property which we want to verify for the spontaneously broken $SU(2) \times U(1)$ theory using the unitary gauge.

Let us start with the remark that only the three diagrams of Fig. 25.8 contribute to the anomalous quadrupole moment $\Delta Q = F_3(0)$. Indeed, the four-boson vertices of the Lagrangian (25.44) have no derivatives; therefore, each one of the diagrams of Fig. 25.9 will give a tensor depending on either p_1 or p_2 but not on both, so it will contribute to F_1 and F_2 but not to F_3.

The explicit computation of all these diagrams by hand is too lengthy and tedious. However, there exist today several efficient computer programs which compute symbolically any set of Feynman diagrams at 1 loop (for some processes even higher loops). For pedagogical purposes we will present here the explicit calculation of the simplest among them, the scalar exchange of Fig. 25.8 and show that its contribution to $F_2(0)$ and $F_3(0)$ is finite.

Using the Lagrangian (25.44) and (25.46) we obtain

$$
\begin{aligned}
\mathcal{M}_{\mu\nu\rho}^{(S)}(p_1, p_2) = \mathrm{i} e m_W^2 g^2 \int \frac{\mathrm{d}^d k}{(2\pi)^d} &\Big\{ \big[(p_1 - k)_\mu g_{\rho'\sigma'} - (p_1 - k)_{\sigma'} g_{\mu\rho'} \\
&+ (p_2 - k)_\mu g_{\rho'\sigma'} - (p_2 - k)_{\rho'} g_{\mu\sigma'} + q_{\sigma'} g_{\mu\rho'} - q_{\rho'} g_{\mu\sigma'} \big] \\
&\left[g^{\rho\rho'} - \frac{(p_1-k)^\rho (p_1-k)^{\rho'}}{m_W^2} \right] \left[g^{\sigma\sigma'} - \frac{(p_2-k)^\sigma (p_2-k)^{\sigma'}}{m_W^2} \right] \\
&\frac{1}{(k^2 - m_S^2)[(p_1-k)^2 - m_W^2][(p_2-k)^2 - m_W^2]} \Big\},
\end{aligned}
\tag{25.58}
$$

where we have used the coupling constant of the scalar-W vertex given by (25.46):

$$
\frac{1}{2} g^2 v = g m_W. \tag{25.59}
$$

We want to compute the $1/\epsilon$ contributions to $F_2(0)$ and $F_3(0)$. The calculation is straightforward. In order to obtain a divergent contribution, we must have at least two powers of k in the numerator. It follows that the term proportional to $g^{\rho\rho'} g^{\sigma\sigma'}$ is finite.[6]

25.3.7   Anomalies in the Standard Model

In the previous section we saw in the explicit calculations the crucial role played by the BRST Ward identities in the consistency of the standard model. The renormalisable gauges contain unphysical degrees of freedom and it is only by virtue of this symmetry that they decouple from physical quantities. Therefore, we expect that a failure to enforce these identities may result in a physically unacceptable theory. In section 18.3 on the other hand, we saw that the algebra $SU(2) \times U(1)$ is not automatically anomaly-free. The purpose of this section is twofold. We first verify explicitly that, in general, the standard electroweak theory does contain anomalies and, what is worse, they do prevent us from defining a physically relevant theory. Second, we show that for the particular family structure we observe in nature, the coefficient of the anomaly vanishes. These results were obtained by Cl. Bouchiat, J. Iliopoulos, and Ph. Meyer as well as D. Gross and R. Jackiw.

---

[6] In a renormalisable gauge the W propagator has only this term and the contribution of this particular diagram is finite. This shows, if need there is, that the contributions of individual Feynman diagrams are gauge dependent.

In the general analysis presented in section 18.3 we saw that for the $SU(2) \times U(1)$ algebra, the dangerous piece is the axial $U(1)$. Therefore, in order to avoid lengthy calculations, we first start with a simplified Abelian model consisting of a fermion field Ψ, two massless neutral vector fields A_μ and B_μ, and a complex scalar field ϕ. We define, as usual, the right- and left-handed spinors $L = \frac{1}{2}(1 + \gamma_5)\Psi$ and $R = \frac{1}{2}(1 - \gamma_5)\Psi$. We require the Lagrangian to be invariant under two Abelian gauge groups under which the fields transform as

$$I. L \to e^{ie\theta}L; \quad R \to e^{ie\theta}R; \quad A_\mu \to A_\mu + \partial_\mu\theta; \quad B_\mu \to B_\mu; \quad \phi \to \phi \tag{25.60}$$

and

$$II. L \to e^{ig\theta}L; \quad R \to e^{-ig\theta}R; \quad A_\mu \to A_\mu; \quad B_\mu \to B_\mu + \partial_\mu\theta; \quad \phi \to e^{2ig\theta}\phi. \tag{25.61}$$

The Lagrangian density which is invariant under the $U(1) \times U(1)$ gauge transformations (25.60) and (25.61) is given by

$$\mathcal{L} = -\tfrac{1}{4}A_{\mu\nu}A^{\mu\nu} - \tfrac{1}{4}B_{\mu\nu}B^{\mu\nu} + \bar{L}i\gamma^\mu(\partial_\mu - ieA_\mu - igB_\mu)L \tag{25.62}$$
$$+ \bar{R}i\gamma^\mu(\partial_\mu - ieA_\mu + igB_\mu)R + |(\partial_\mu - 2igB_\mu)\phi|^2 - \mu^2\phi\phi^* - \lambda(\phi\phi^*)^2$$
$$- \sqrt{2}G(\bar{R}L\phi^*) - \sqrt{2}G(\bar{L}R\phi)$$

with $A_{\mu\nu} = \partial_\mu A_\nu - \partial_\nu A_\mu$ and $B_{\mu\nu} = \partial_\mu B_\nu - \partial_\nu B_\mu$.

We now assume that $\mu^2 < 0$ and the vacuum expectation value $< \phi >_0 = v/\sqrt{2}$ is different from 0. We can always choose v real and write

$$\phi = \frac{\phi_1 + v + i\phi_2}{\sqrt{2}}. \tag{25.63}$$

In terms of the translated fields the quadratic and the interaction parts of the Lagrangian are

$$\mathcal{L} = \mathcal{L}_0 + \mathcal{L}_I \tag{25.64}$$

$$\mathcal{L}_0 = -\tfrac{1}{4}A_{\mu\nu}A^{\mu\nu} - \tfrac{1}{4}B_{\mu\nu}B^{\mu\nu} + \tfrac{1}{2}(\partial_\mu\phi_1)^2 + \tfrac{1}{2}(\partial_\mu\phi_2)^2 + \bar{\Psi}i\partial\!\!\!/\Psi \tag{25.65}$$
$$+ \tfrac{1}{2}(4g^2v^2)B_\mu B^\mu - \tfrac{1}{2}(2\lambda v^2)\phi_1^2 - 2gvB_\mu\partial^\mu\phi_2 - Gv\bar{\Psi}\Psi$$

$$\mathcal{L}_I = -2gB^\mu(\phi_1\partial_\mu\phi_2 - \phi_2\partial_\mu\phi_1) + 4g^2vB_\mu^2\phi_1 + 2g^2B_\mu^2(\phi_1^2 + \phi_2^2) \tag{25.66}$$
$$+ e\bar{\Psi}\gamma_\mu\Psi A^\mu - g\bar{\Psi}\gamma_\mu\gamma_5\Psi B^\mu - G\bar{\Psi}\Psi\phi_1 + iG\bar{\Psi}\gamma_5\Psi\phi_2$$
$$- \lambda v\phi_1(\phi_1^2 + \phi_2^2) - \tfrac{1}{4}\lambda(\phi_1^2 + \phi_2^2)^2.$$

The constant v is chosen such that the coefficient of the term linear in ϕ_1 vanishes. As expected, the B field appears massive.

The Feynman rules are readily obtained. The important point is that in a gauge such that $\partial_\mu B^\mu = 0$, the propagators of the B and ϕ_2 fields in momentum space are

$$< B_\mu, \, B_\nu > = \frac{-i}{k^2 - (2gv)^2} \left[g_{\mu\nu} - \frac{k_\mu k_\nu}{(2gv)^2} \right] - \frac{i k_\mu k_\nu}{k^2 (2gv)^2} \qquad (25.67)$$

$$< \phi_2, \, \phi_2 > = \frac{i}{k^2} \qquad (25.68)$$

$$< B_\mu, \, \phi_2 > = 0. \qquad (25.69)$$

Equations (25.67) to (25.69) show two things: first, that the large k behaviour of all propagators is that of a renormalisable theory and second, that this has been achieved at the price of introducing an unphysical zero-mass pole in the last term of (25.67). This reminds us of the case we studied in section 15.3.1 and, as it happened there, we expect this pole to cancel against the corresponding one of the ϕ_2 propagator for all physical quantities. Indeed, it is easy to verify that such a cancellation does occur, for example, at lowest order in the fermion–fermion scattering amplitude given by the two diagrams of Fig. 25.10. We studied a simplified version of this model in Problem 15.2.

Looking closer at this cancellation we see that it is the result of a very special relation among the four constants which enter in the calculation: the $B - \Psi$ coupling constant g, the $\phi_2 - \Psi$ Yukawa coupling constant G, the B mass gv, and the Ψ mass Gv. This relation is the lowest order version of the BRST Ward identity. We can now easily guess in which case this cancellation will fail. We look at the forward scattering amplitude of two A vector bosons. The lowest order diagrams which exhibit the zero-mass pole are shown in Fig. 25.11. We compute the residue of the pole. The first diagram with the B contribution gives

$$R(B) = \frac{-i}{4v^2} (k_1 + k_2)_\alpha (k_1 + k_2)_\beta \, T^\alpha_{\mu\nu}(k_1, k_2) \, T^\beta_{\rho\sigma}(k_1, k_2) \qquad (25.70)$$

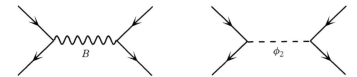

Figure 25.10 *The fermion–fermion scattering amplitude in the tree approximation in the simplified Abelian version of the standard model.*

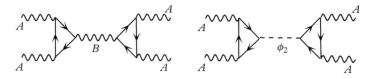

Figure 25.11 *The triangle anomaly in the simplified standard model.*

and the second one with ϕ_2

$$R(\phi_2) = iG^2 T_{\mu\nu}(k_1, k_2) T_{\rho\sigma}(k_1, k_2), \tag{25.71}$$

where $T^\alpha_{\mu\nu}(k_1, k_2)$ and $T_{\mu\nu}(k_1, k_2)$ are defined by the corresponding triangle diagrams. We see now that in order for the cancellation to occur, we need an identity of the form $-(k_1 + k_2)_\alpha T^\alpha_{\mu\nu} = 2Gv T_{\mu\nu}$. But the T's are precisely the triangle diagrams we computed in section 17.3. We learnt that the correct Ward identity is instead

$$- (k_1 + k_2)_\alpha T^\alpha_{\mu\nu} = 2Gv T_{\mu\nu} + 8\pi^2 \epsilon_{\mu\nu\rho\sigma} k_1^\rho k_2^\sigma. \tag{25.72}$$

We see that the residue of the $(k_1 + k_2)^2 = 0$ pole does not vanish, which means that the unphysical degrees of freedom do not decouple. It follows that this model is unacceptable for physical applications.

In order to remedy to this difficulty we first observe that the anomaly is independent of the fermion mass. Therefore, we introduce another fermion field ψ which has a new Yukawa coupling constant G' but carries the opposite charges under both gauge groups. It follows that Ψ and ψ have different masses but the fermion part of both currents is odd under the exchange $\Psi \to \psi$. Therefore, the anomalies disappear and the residues of the $(k_1 + k_2)^2 = 0$ poles associated with the B and the ϕ_2 propagators cancel.

The extension of this anomaly cancellation mechanism to the real world of the standard $SU(2) \times U(1)$ model is immediate. Considering only one family of fermions, we first see that a purely leptonic model would indeed be anomalous. Adding the other leptons only multiplies the coefficient of the anomaly by 3. The cancellation comes between leptons and hadrons. For the first family of the electron and its neutrino, adding n hadron doublets with electric charges Q_i and $Q_i - 1$, $i = 1, ..., n$, we see immediately that the requirement of vanishing anomaly coefficient (18.42) implies the relation

$$\sum_{i=1}^{n} Q_i - 1 = 0. \tag{25.73}$$

Since the electron charge equals -1, this condition can be expressed by saying that the sum of all electric charges in a family must vanish. It is satisfied, in particular, by the three-colour model with $Q_i = \frac{2}{3}$, but also by other, integer charge models.

In fact, the anomaly vanishing condition (25.73) has a wider application. The standard model could have been invented after the Yang–Mills theory was written, much before the discovery of the quarks. At that time the 'elementary' particles were thought to be the electron and its neutrino, the proton and the neutron, so we would have used one lepton and one hadron doublet. The condition (25.73) is satisfied. When quarks were discovered we changed from nucleons to quarks. The condition is again satisfied. If tomorrow we find that our known leptons and/or quarks are composite, the new building blocks will be required to satisfy this condition again. Since the contribution of a chiral fermion to the anomaly is independent of its mass, it must be the same no matter which mass scale we are using to compute it.

This anomaly cancellation condition ensures that the coefficient of the axial anomaly vanishes at 1 loop. Then the Adler–Bardeen theorem we have mentioned previously implies the vanishing at all orders and guarantees the consistency of the theory. Families must be complete. Thus, the discovery of a new lepton, the tau, implied the existence of two new quarks, the b and the t, a prediction which was again verified experimentally.

Before closing this section we want to point out that the discussion on the anomaly cancellation mechanism we presented here is not limited to the $U(1) \times SU(2)$ gauge theory of the standard model. Mathematical consistency requires that the cancellation condition should be imposed in any gauge theory. In particular, this includes several models we will study in the following sections which include a gauge theory for the strong interactions, as well as more general models, which we call 'grand unified theories'. H. Georgi and S. L. Glashow found the generalisation of the anomaly equation (25.73) for a gauge theory based on any Lie algebra with a given fermion content. It takes a surprisingly simple form (see Problem 25.9)

$$\mathcal{A}_{abc} = \mathrm{Tr}\left(\gamma^5 \{\Gamma_a, \Gamma_b\} \Gamma_c\right), \tag{25.74}$$

where Γ_a denotes the Hermitian matrix which determines the coupling of the gauge field W_a^μ to the fermions through the interaction $\bar\Psi \gamma_\mu \Gamma_a \Psi W_a^\mu$. As we see, Γ_a may include a γ^5. Georgi and Glashow showed that the anomaly is always a positive multiplet of $\mathcal{A}_{abc}$, so this quantity should vanish identically for all values of the Lie algebra indices a, b, and c.

Since gauge theories are believed to describe all fundamental interactions, the anomaly cancellation condition plays an important role not only in the framework of the standard model, but also in all modern attempts to go beyond, from grand unified theories to superstrings. It is remarkable that this seemingly obscure higher order effect dictates to a certain extent the structure of the world.

25.4   A Gauge Theory for Strong Interactions

25.4.1   Are Strong Interactions Simple?

The discovery of the fundamental theory of strong interactions was made as a response to an experimental challenge. For many years the efforts to understand the nature of strong interactions were concentrated in the study of the experimental results from hadronic collisions. The resulting picture invariably appeared to be too complicated to allow for a simple interpretation. We understand now that this complexity should not be attributed to the fundamental interactions themselves, but is instead due to the fact that the objects we are dealing with, namely the hadrons, are themselves too complicated. It is as if we were trying to discover quantum electrodynamics by studying the interactions among complex molecules. There were already several experimental hints pointing towards a composite structure for the hadrons, but the decisive progress came with a set of experiments from the Stanford Linear Accelerator Centre (SLAC) by the late sixties and early seventies. They were studying the large momentum transfer scattering, often called

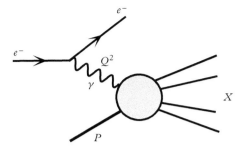

Figure 25.12 *Deep inelastic electron-nucleon scattering with the production of a final state $|X>$.*

deep inelastic scattering, of electrons off nucleons.[7] The process is shown schematically in Fig. 25.12 in the one-photon exchange approximation. It is an 'inclusive process', in the sense that in the final state, only the momentum of the electron was measured. The hadronic state, denoted by X, was not. Therefore, what was measured corresponds to a sum over all possible X. This inclusive cross section may be written as

$$\frac{d\sigma}{d\Omega' dE'} \sim \sum \frac{1}{q^4} |j_\mu^e(0) < X|\mathcal{J}^\mu(0)|P > |^2 (2\pi)^4 \delta^4(k' + P_X - k - P), \qquad (25.75)$$

where $\sum$ denotes a sum over all possible final states $|X>$ and an average over the initial spins. $j_\mu^e(0) = \bar{u}^e(k')\gamma_\mu u^e(k)$ is the electron current and $\mathcal{J}_\mu(0)$ the hadron current. We can rewrite the cross section as

$$\frac{d\sigma}{d\Omega' dE'} \sim \frac{1}{q^4} m_{\mu\nu} W^{\mu\nu} \qquad (25.76)$$

with

$$m_{\mu\nu} = 2(k_\mu k'_\nu + k_\nu k'_\mu - g_{\mu\nu} k \cdot k')$$

$$W_{\mu\nu} = \frac{1}{2} \sum < P|\mathcal{J}_\mu^\dagger(0)|X><X|\mathcal{J}_\nu(0)|P > (2\pi)^3 \delta^4(k' + P_X - k - P) \qquad (25.77)$$

$$= \int \frac{d^4x}{4\pi} e^{-iqx} < P|\mathcal{J}_\mu^\dagger(x)\mathcal{J}_\nu(0)|P >$$

and $q = k - k' = P_X - P$. In computing $m_{\mu\nu}$ we have taken into account that the spinors $u^e(k)$ and $u^e(k')$ satisfy the free Dirac equation but we have put $m_e = 0$. In the last line we have omitted the sign showing the average over the spin of the nucleon. Note that $W_{\mu\nu}$ can be rewritten as

---

[7] If we use a hydrogen target we obtain the scattering on protons. If we combine the results with those on other nuclei, for example deuteron, we can extract the scattering cross section on neutrons.

$$W_{\mu\nu} = \int \frac{\mathrm{d}^4 x}{4\pi} \mathrm{e}^{-iqx} < P|[\mathcal{J}_\mu^\dagger(x), \mathcal{J}_\nu(0)]|P > \ \sim \mathrm{Im}\, T_{\mu\nu}$$

$$T_{\mu\nu} = \int \frac{\mathrm{d}^4 x}{4\pi} \mathrm{e}^{-iqx} < P|T(\mathcal{J}_\mu^\dagger(x)\mathcal{J}_\nu(0))|P > . \tag{25.78}$$

In replacing the matrix element of the product of the two currents with that of their commutator, we have added a term proportional to $< P|\mathcal{J}_\nu(0)\mathcal{J}_\mu^\dagger(x)|P >$. It is easy to see that this term gives a vanishing contribution. Indeed, the intermediate states $|n >$ which are liable to contribute must have the quantum numbers of the target nucleon and an energy E_n equal to $M_N - q^0$. But q^0 is positive and there are no hadronic states with the quantum numbers of a nucleon and with energy smaller than the nucleon mass. The relation (25.78) is an example of the optical theorem; it states that $W_{\mu\nu}$ equals the imaginary part of the forward Compton amplitude of a virtual photon with mass q^2.

Taking into account the conservation of the electromagnetic current as well as invariance under parity (see Problem 25.10 for parity violating processes), we can write the most general form of $W_{\mu\nu}$ as

$$W_{\mu\nu} = -\left(g_{\mu\nu} - \frac{q_\mu q_\nu}{q^2}\right) W_1 + \frac{1}{M_N^2} \left(P_\mu - q_\mu \frac{q \cdot P}{q^2}\right)\left(P_\nu - q_\nu \frac{q \cdot P}{q^2}\right) W_2 \tag{25.79}$$

with $W_{1,2}$ two scalar functions which can depend only on $\nu = 2q \cdot P$ and q^2. We can choose two dimensionless variables, $x = Q^2/\nu$ and M_N^2/Q^2, with $Q^2 = -q^2$. After trivial rescaling it has become customary to define two dimensionless functions $F_1 = M_N W_1$ and $F_2 = \nu W_2/M_N$, called *structure functions*.

The surprising result of the SLAC experiment was that when both ν and Q^2 become large with fixed ratio x, these structure functions were, to a good approximation, functions only of x. Here 'large' means large compared to the nucleon mass. This property became known as *scale invariance*; changing the energy scale of the experiment, Q^2 and ν with x fixed, does not change the cross section.

This result is very interesting because it is very easy to understand it using a naïve and wrong reasoning. When $Q^2 \to \infty$ with fixed x, the second variable M_N^2/Q^2 goes to 0. Naïvely, a function $F(x, M_N^2/Q^2)$ can be approximated by

$$F(x, M_N^2/Q^2) = F(x, 0) + \frac{M_N^2}{Q^2} F^{(1)}(x, 0) + ..., \tag{25.80}$$

where $F^{(1)}$ is the first derivative of F with respect to M_N^2/Q^2 keeping x fixed. So we expect to be left with only the x dependence. This argument expresses the intuitive idea that at very high energies and momentum transfers, the masses are unimportant and the theory exhibits scale invariance. Feynman had even built a simple model which had this scale behaviour. Let us assume that the target nucleon is made out of elementary constituents which interact with the incident photon as point-like particles. We will call them collectively *partons*. If we neglect all interactions among the partons, we can easily reproduce this scaling property. The trouble, of course, is that the assumption of no interaction is meaningless. The partons cannot be free and still bind strongly to form a nucleon. Nevertheless, it was such a schizophrenic behaviour that was indicated by the data. The partons were almost free when probed by a virtual photon in the deep inelastic region and still very strongly bound in ordinary hadronic experiments. The answer was obtained by using the Callan–Symanzik equation obtained in section 17.4, which is the correct form of the scale invariance Ward identity. We saw that there is a class of theories, those we called *asymptotically free*, which have precisely this property, i.e. the effective strength of the interaction may be large at low energies, but it goes to 0 at high energies. We learnt that only pure non-Abelian gauge theories are asymptotically free. If we add any other pieces this property is lost. Such is the case, for example, for the gauge theory of the standard electroweak model. The coupling constants g' for the $U(1)$ piece of the gauge group, G_i for the Yukawa terms, and λ for the scalar field self-coupling, all become large in the deep Euclidean region. Since at higher orders the β-functions mix the evolutions of the coupling constants, asymptotic freedom is entirely lost.

The message of this theorem is clear. If we want to understand the success of the naïve parton model in a quantum field theory language, we must assume that strong interactions are described by an unbroken non-Abelian gauge theory. In the deep inelastic region asymptotic freedom has already set in and the effective strength of strong interactions has become small. At small momentum transfers, on the other hand, where most of the data of hadronic reactions come from, Eq. (19.5) shows that the effective coupling constant increases. We enter the strong coupling regime and perturbation theory breaks down.

25.4.2   Quantum Chromodynamics

In principle, there are several ways to realise a non-Abelian gauge theory for strong interactions, but, in practice, if we assume the standard quark model with fractional electric charges and three colours, there is one natural scheme. The quark fields can be written as a matrix with f rows and c columns, where f and c denote the number of flavours and colours, respectively. In the usual quark model $f = 6$ and $c = 3$. Therefore, there is a natural non-Abelian group $SU(6) \otimes SU(3)$. The $SU(6)$ piece mixes quark fields having the same colour and different flavours, while the $SU(3)$ piece does the opposite, mixes colours and leaves flavours unchanged. We know experimentally that the flavour group is badly broken because the quark masses are spread over a very wide range and there exist no massless vector bosons with flavour quantum numbers. As we explained already, a spontaneous breaking destroys asymptotic freedom and we know that an explicit breaking is incompatible with renormalisability. This leaves the colour group $SU(3)$ as the natural choice. The resulting theory is called *quantum chromodynamics*, or QCD.[8] It has eight ($3^2 - 1$) massless gauge bosons, which we call *gluons*. For massless quarks the Lagrangian density is given by

---

[8] We follow the presentation of D. Gross, 'Applications of the Renormalization Group to High-Energy Physics', in *Les Houches 1975, Proceedings, Methods in Field Theory* (North-Holland/World Scientific, 1976).

$$\mathcal{L}_{\mathrm{QCD}} = -\frac{1}{2}\mathrm{Tr}\mathcal{G}_{\mu\nu}\mathcal{G}^{\mu\nu} + \sum_{f=1}^{6}\sum_{i,j=1}^{3} \bar{q}_f^i \, i\mathcal{D}_{ij} \, q_f^j. \tag{25.81}$$

The field strength $\mathcal{G}_{\mu\nu}$ can be written in terms of the eight gluon gauge fields $G_\mu^\alpha(x)$ $\alpha = 1, ..., 8$ as

$$\mathcal{G}_{\mu\nu} = \partial_\mu \mathcal{G}_\nu - \partial_\nu \mathcal{G}_\mu - \mathrm{i}g_s \left[\mathcal{G}_\mu, \mathcal{G}_\nu\right] \tag{25.82}$$

with

$$\mathcal{G}_\mu(x) = \sum_{\alpha=1}^{8} G_\mu^\alpha(x)\lambda^\alpha \tag{25.83}$$

and λ^α the eight 3×3 traceless Gell–Mann matrices. g_s is the strong interaction coupling constant. The covariant derivative $\mathcal{D}_\mu$ is a 3×3 matrix given by

$$\mathcal{D}_\mu = \partial_\mu + \mathrm{i}g_s\mathcal{G}_\mu. \tag{25.84}$$

The 1-loop β-function of this theory receives two contributions. The first comes from the non-Abelian self-coupling among the vector bosons, including the Faddeev-Popov ghosts, and it is negative, as expected. The second comes from the diagrams with one fermion loop. Since each flavour contributes independently, the result is proportional to N_f, the total number of flavours. They have exactly the same structure as the vacuum polarisation term in quantum electrodynamics, so they give a positive contribution to the β-function. For a general $SU(N)$ group with fermions in the fundamental representation, the result is

$$\beta(g_s) = -\frac{g_s^3}{(4\pi)^2}\left(\frac{11}{3}N - \frac{2}{3}N_f\right) = -b_0 g_s^3. \tag{25.85}$$

We see that for $SU(3)$, we still have asymptotic freedom, provided $N_f < 17$. Following the analysis of section 19.2, we conclude that the effective strength of the interaction in the deep Euclidean region will be given by the running coupling constant $\bar{g}_s$, solution of the equation

$$t\frac{\partial \bar{g}_s}{\partial t} = \beta(\bar{g}_s), \tag{25.86}$$

which, for t large enough, may become sufficiently small for perturbation theory to be applicable. Indeed, solving (25.86) with β given by (25.85), we find that

$$\bar{g}^2(t) = \frac{g^2}{1 + 2b_0 g^2 t}, \tag{25.87}$$

where the momenta are scaled according to $p_i \rightarrow \rho p_i$ and $t = \ln \rho$. As long as b_0 remains positive, i.e. $N_f < 17$, the denominator does not vanish. For large ρ $\bar{g}$ goes to 0 logarithmically. As in quantum electrodynamics, it is customary to define $\alpha_s = g^2/4\pi$. If Q denotes the typical momentum which grows large and μ the initial subtraction point where $\rho = 1$, we have $2t = \ln Q^2/\mu^2$ and the relation (25.87) can be written as

$$\alpha_s(Q^2) = \frac{\alpha_s(\mu^2)}{1 + 4\pi b_0 \alpha_s(\mu^2) \ln Q^2/\mu^2}. \tag{25.88}$$

We introduce a parameter Λ with the dimensions of a mass through the relation $\ln \Lambda^2 = \ln \mu^2 - [4\pi b_0 \alpha_s(\mu^2)]^{-1}$ and rewrite (25.88) as

$$\alpha_s(Q^2) = \frac{1}{4\pi b_0 \ln Q^2/\Lambda^2} \tag{25.89}$$

in which all reference to the initial value of the coupling constant $\alpha_s(\mu^2)$ has disappeared in favour of a scale parameter Λ. It shows clearly the behaviour of the effective coupling constant. At scales Q^2 much larger than Λ, $\alpha_s(Q^2)$ decreases according to the property of asymptotic freedom. On the other hand, when Q^2 decreases, $\alpha_s(Q^2)$ increases and it diverges when $Q = \Lambda$. Of course, the perturbation theory on which this analysis is based cannot be trusted when α_s becomes large but we expect this equation to describe the behaviour of the effective strength of the theory for $Q^2 > \Lambda^2$. Figure 25.13 shows a comparison with the available data. The agreement between theory and experiment is impressive.

What about the region $Q^2 \leq \Lambda^2$? Perturbation theory breaks down, in particular because (25.81) predicts the existence of asymptotic states corresponding to massless quarks and gluons. None of them have ever been seen in hadronic collisions. We get out of this difficulty by assuming that this breakdown of perturbation is not only quantitative, but also qualitative, in the sense that even the space of states is not correctly described by it. We have previously called this property 'confinement' and argued that only colour singlet states appear as free particles. In the absence of a real proof of this property, we will only try to explain this concept a bit further.

Let us go back to quantum electrodynamics. Consider an electron and a positron, created at point x_0 and pulled a distance L apart. For large L the electric field between them falls like L^{-2} because the field lines are spread over all space. Imagine that we had a way to confine the electric field lines in a thin tube starting at the position of the electron and ending at that of the positron. The field would remain constant, independent of L, and we would need an infinite amount of energy to separate the two particles. The QED vacuum does not have this property and, therefore, in physics electric charges are not confined. The closest we can think of is a perfect superconductor, except that it is not electric but magnetic lines that would form a thin tube. If magnetic monopoles exist, they would be confined in a superconducting medium.

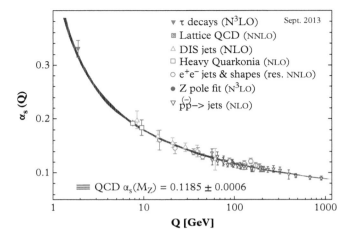

Figure 25.13 *The evolution of the QCD effective coupling constant and comparison with experimental measurements. The theoretical precision with which α_s has been extracted from the data is indicated in parentheses: NLO means 'next-to-leading order', NNLO 'next-to-next-to-leading order', etc. The width of the line gives the QCD prediction including the theoretical uncertainties. Experimentally, the most precise determination of α_s comes from LEP measurements on the Z^0 mass. The lowest energy point, which gives a very significant contribution to the overall precision, comes from an accurate measurement of the τ hadronic decays. The LHC collaborations have recently presented data extending the Q range to 1 TeV. They are included in the graph (© CERN, CMS collaboration).*

What we need is to show that QCD vacuum confines the colour electric lines. K. Wilson gave a simple criterion for such confinement. Consider, for QCD, the analogue of the Bohm–Aharonov phase along a closed path.

$$W[C] = \mathrm{Tr}\left[P \exp\left(ig_s \oint_C dx^\mu \mathcal{G}_\mu(x) \right) \right]. \tag{25.90}$$

Here $\mathcal{G}_\mu(x)$ is the 3×3 matrix valued gluon field and the integral is taken along C, a closed path in space–time. The symbol P denotes a particular ordering, which we will call *path ordering*. This is necessary because the field is matrix valued and the matrices in different points do not commute. A simple ordering choice is to parametrise the path C in terms of a parameter s, $0 < s \leq 1$, and decide to place the matrices with the smaller value of s to the left. For a line along the time axis, this convention coincides with the time ordering we introduced earlier. Equation (25.90) defines the operator W as a functional of the path. It is easy to show (see Problem 25.11) that, after taking the trace, it is gauge invariant. In fact, people have argued that all gauge invariant quantities can be expressed in terms of such Wilson loops, but we will not develop this point here.

We do not know how to compute a Wilson loop for large paths because, at large separations, g_s becomes large and perturbation theory does not apply. Wilson gave a prescription of how to evaluate the vacuum expectation value of W numerically by approximating the continuum space–time by a four-dimensional lattice. What is relatively simple to prove is that this quantity offers a criterion for confinement. It can be expressed as

$$< W[C] >_0 ~ \sim ~ \begin{cases} e^{-KL(C)} & \text{no confinement} \\ e^{-K'\Sigma(C)} & \text{confinement.} \end{cases} \tag{25.91}$$

Here $L(C)$ denotes the length of the perimeter of C and $\Sigma(C)$ the area enclosed by it. K and K' are constants. In other words, the behaviour of the vacuum expectation value of a Wilson loop for large paths, area or length law, determines the phase of the theory, confining or not.

A simple argument in favour of (25.91) is the following. Let us consider again the pair creation process we introduced previously. We assume that at time $-T/2$ a quark–antiquark pair is created at the point $x = 0$ and then pulled a distance L apart. Let us call $V(L)$ the resulting potential energy. We let them stay at the points $\pm L/2$ for time T and then they come together again and annihilate. If the quarks are very heavy, we can consider that during the time T they act as external sources $\mathcal{J}_\mu^\alpha(x)$ given by

$$\mathcal{J}_\mu^\alpha(x) = \delta_{\mu 4}\lambda^\alpha \left[\delta^3\left(x + \frac{L}{2}\right) - \delta^3\left(x - \frac{L}{2}\right) \right], \tag{25.92}$$

where λ^α are the colour $SU(3)$ matrices. When T becomes very large the process is just the vacuum-to-vacuum amplitude in the presence of the sources:

$$< 0|0 > = \frac{\int \mathcal{D}U e^{(S_E + \int \mathcal{J} \cdot A d^4 x)}}{\int \mathcal{D}U e^{S_E}} = < e^{\int \mathcal{J} \cdot A d^4 x} >. \tag{25.93}$$

In Euclidean space this amplitude is given by $e^{[E_0(\mathcal{J}) - E_0]T}$ where $E_0(\mathcal{J})$ and E_0 are the ground state energies in the presence and absence of the source, respectively. The difference is precisely the potential energy V. On the other hand, with the sources given by (25.92), we can choose a gauge in which the right-hand side of (25.93) is proportional to a Wilson loop spanned by T and L. We thus obtain the relation

$$V(L) \sim \lim_{T \to \infty} \frac{1}{T} \ln < W[C] >. \tag{25.94}$$

Since the area of the loop is LT, we see that if we have an area law, the potential is linear in L and we have a constant confining force, while if we have a length law it goes like constant $+ \mathcal{O}(1/L)$. The force falls like $1/L^2$ and the charges are free. We will see in a later section how this conclusion is affected by the presence of dynamical quarks. It is customary to introduce the quantity

$$\sigma = \lim_{L \to \infty} \frac{V(L)}{L}, \tag{25.95}$$

which, for historical reasons, it is called *the string tension*. The lattice simulations of quantum chromodynamics performed so far, admittedly at rather small lattices, support the confinement hypothesis.

In writing (25.81), we left the flavour group $SU(6)$ as a global symmetry. In fact, (25.81) with massless quarks has a much larger flavour symmetry, namely $U(6) \otimes U(6)$ because we can perform $U(6)$ rotations independently to the left and right components of the fermion fields. Since the zero-mass approximation cannot be justified for all quark flavours, we expect at least part of this symmetry to be badly broken. It is instructive to see the fate of the various pieces.

The $U(1)$ vector symmetry, associated with a common phase change of all quark flavours, right and left, remains as an exact symmetry and corresponds to the conservation of baryon number. The axial $U(1)$ part (opposite phase changes for right and left fields) is expected to be broken by quantum effects, the same way that we found the anomalous conservation law of the axial current in quantum electrodynamics. We will come back to this point at the end of this section.

Let us come now to the $SU(6) \otimes SU(6)$ part. The massless approximation is very good for u and d, questionable for s, and meaningless for all higher quark flavours.[9] If it were not for the weak interactions, we would still have the conservation of the vector currents which correspond to the diagonal generators of vector $SU(6)$ describing the separate conservation of each quark species. But weak interactions break these symmetries and cause the decay of strange, charm, b– and t– hadrons. We are thus left with $SU(2) \otimes SU(2)$, which, if our theory is right, must be a good approximate symmetry of the real hadronic world. It is precisely the symmetry we studied in section 15.2.4. The diagonal subgroup of this symmetry is vector $SU(2)$ and corresponds to the well-known isospin symmetry of strong interactions. Indeed, all states, hadrons as well as nuclear levels, can be classified according to $SU(2)$ representations with very small mass splittings among the members of an isospin multiplet. For example, the proton–neutron mass difference is measured to be $(m_n - m_p)/(m_n + m_p) \sim 7 \times 10^{-4}$. Similarly, hadronic processes can be described by isospin invariant amplitudes.

What about the axial part of $SU(2) \otimes SU(2)$? At first sight there is no such symmetry in the spectrum of hadron states. The nucleon does not have an almost degenerate partner with opposite parity. Could nevertheless the axial currents be still conserved? A physical process involving the matrix elements of this current is the decay of the charged pion: $\pi^{\pm} \to l^{\pm} + \nu_l(\bar{\nu}_l)$, where l stands for e or μ. In units of the W mass, it is a very low

---

[9] Naturally, the validity of this statement depends on the energy scale we are considering. At extremely high energies, much higher than 100 GeV, even the heavy quark flavours can be considered as approximately massless. We will encounter some theoretical speculations along these lines in the next chapter.

momentum transfer process and, as we explained before, we can use the effective Fermi current×current theory. The decay amplitude is given by

$$
\begin{aligned}
\mathcal{A}_l &= \; < l(k_1), \nu_l(k_2)|\mathcal{H}_F(0)|\pi(q) > \\
&= \frac{G}{\sqrt{2}} \; < l, \nu_l|l_\lambda(0)|0 > < 0|h^\lambda(0)|\pi(q) > \\
&= \frac{G}{\sqrt{2}} \; [\bar{u}_\nu(k_2)\gamma_\lambda(1+\gamma_5)v_l(k_1)] \; < 0|h^\lambda(0)|\pi(q) >,
\end{aligned}
\tag{25.96}
$$

where $\mathcal{H}_F$ is the Fermi interaction Hamiltonian given in (25.1), $|l(k_1), \nu_l(k_2) >$ is the state with a charged lepton l with momentum k_1 and the associated neutrino with momentum k_2, and $|\pi(q) >$ is the one-pion state with momentum q. u and v are the Dirac spinors for the leptons. Since the pion has intrinsic parity -1, only the axial part of the hadronic current contributes to the decay. The most general form of this matrix element, consistent with Lorentz invariance, is

$$
< 0|h^\lambda(0)|\pi(q) > = < 0|A^\lambda(0)|\pi(q) > = f_\pi q^\lambda.
\tag{25.97}
$$

A^λ denotes the axial part of the hadronic current and f_π is a constant whose value is determined by the observed pion decay rate. Multiplying both sides of (25.97) by q_λ and taking into account that $q^2 = m_\pi^2$, we obtain the desired result

$$
< 0|\partial_\lambda A^\lambda(0)|\pi(q) > = \mathrm{i}f_\pi m_\pi^2,
\tag{25.98}
$$

which reminds us of the P.C.A.C. relation we found in section 15.2.4. Two conclusions can be drawn from this equation. First the axial current cannot be exactly conserved, since this would imply a stable and/or massless pion. Second, the conservation could be approximate, valid only at the limit of vanishing pion mass. We recall that, experimentally, $m_\pi \approx 140$ MeV, a very low value in the scale of hadron masses, so this approximation may be reasonable.

If this partial conservation of the axial hadronic current is indeed an approximate symmetry of the strong interactions, we must explain why we see no trace of it in the particle spectra. Here the analysis we made in Chapter 15 will be very useful. We learnt that a symmetry of the equations of motion may not appear as a symmetry of the solution. We called this phenomenon 'spontaneous symmetry breaking' and, for a global symmetry, we argued that it is accompanied by the appearance of massless particles, the so-called Goldstone particles, which have the quantum numbers of the divergence of the corresponding current. In the study of the σ-model of sections 15.2.4 and 15.2.5 we saw a model which exhibits this phenomenon of partial conservation of the axial current. Since in $SU(2) \otimes SU(2)$ we have a triplet of axial currents, we expect a triplet of pseudo-scalar Goldstone bosons. We conclude that, at least as far as its symmetry properties, the Lagrangian (25.81) can provide an approximate description of the dynamics of strong interactions, provided we assume that the chiral $SU(f) \otimes SU(f)$ symmetry is spontaneously broken to its diagonal vector subgroup $SU(f) \otimes SU(f) \to SU(f)_{\text{vector}}$. Since we do not have the analogue of the σ field, we do not know the detailed mechanism which is responsible for this breaking. Looking at the spectrum of hadrons we infer that it occurs at an energy scale on the order of a few hundred MeV, where the effective coupling strength of QCD is large and perturbation theory does not apply. We believe it to be dynamical, triggered by the formation of a scalar quark–anti-quark bound state which acquires a non-zero vacuum expectation value. If this is correct, it means that the role of the σ field is played by a bound state. The observable signal of this mechanism is the appearance of almost massless pseudoscalar bosons. They are the Goldstone bosons of the spontaneous symmetry breaking. They are not exactly massless because the original symmetry was only approximate. Lattice computations support this view but it has also several observable consequences. Let us derive the simplest of them, originally due to Stephen Adler. We consider the matrix element $M_\mu(k)$ of the axial current between two hadronic states $|a>$ and $|b>$.

$$M_\mu(k) = \int d^4x\, e^{-ikx} < a|A_\mu(x)|b > . \tag{25.99}$$

If we compute $k^\mu M_\mu$ we obtain the matrix element of the divergence of the axial current. We may be tempted to put the result equal to 0, and indeed this would have been the case if the symmetry was not spontaneously broken. But Goldstone's theorem tells that there exists precisely a massless particle whose quantum numbers are those of the divergence of the current. In this case we have an almost Goldstone particle with a mass m_π close to 0. The resulting one-pion intermediate state, represented by the diagram of Fig. 25.14, still dominates the process at low momenta because of the one-pion propagator. Its contribution is equal to

$$k^\mu M_\mu(k) \sim \frac{f_\pi m_\pi^2}{k^2 - m_\pi^2} M_\pi(k), \tag{25.100}$$

where f_π is the constant introduced in Eq. (25.98) and $M_\pi(k)$ is the amplitude for emission (or absorption) of a pion of momentum k in the process $a \rightarrow b + \pi$ (or $a + \pi \rightarrow b$). In the numerator we have put $k^2 = m_\pi^2$ but the assumption of the pion pole dominance is valid only as long as the pion mass is close to 0. For an exactly massless pion, i.e. an exactly conserved axial current, the factor in front is just k^2/k^2. At low momenta we obtain the relation

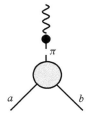

Figure 25.14 *One-pion pole in the matrix element of the divergence of the axial current.*

$$\lim_{k \to 0} M_\pi(k) \sim f_\pi^{-1} \lim_{k \to 0} k^\mu M_\mu(k). \tag{25.101}$$

The relation (25.101) expresses Adler's condition. The amplitude for emission or absorption of a Goldstone particle vanishes at low energy with a term linear in the momentum of the Goldstone particle. Although we derived it for pions, it is clearly valid for any Goldstone particle. The only ingredient we used was the fact that the quantum field of the Goldstone particle and the divergence of the corresponding current had exactly the same quantum numbers and we could use the second, appropriately rescaled, in place of the first in the reduction formula.

We can obtain similar *low-energy theorems* for amplitudes involving any number of pions. In Problem 25.12 we derive the *Adler–Weisberger relation* which relates the low-energy pion–nucleon scattering amplitude to the weak coupling constant of the axial current in nuclear beta-decay. The method is always the same. We write the chiral symmetry Ward identity we obtained in Eq. (17.8) and replace the divergence of the axial current with the pion field.

The spontaneous breaking of the chiral symmetry offers an explanation for the hadron masses. As we said, the breaking occurs at a scale on the order of the QCD parameter Λ, i.e. a few hundred MeV. Since a nucleon is made out of three quarks, its mass is on the order of $3 \times \Lambda \sim 1$ GeV.

We must make a final remark concerning the axial $U(1)$ symmetry. The conservation of the corresponding gauge invariant current is broken by the Adler–Bell–Jackiw anomaly, as we explained in section 17.3. Does this settle the question of this symmetry? Not completely, because, as we showed in Eq. (17.35), we can add to the gauge invariant current the gauge dependent piece G_μ and obtain a conserved current. Since it is not a gauge invariant operator we cannot use it directly in the analogue of Eq. (25.97) to create a physical state out of the vacuum. However, in any covariant gauge, we can consider the vacuum expectation value of the time-ordered product of this gauge dependent but conserved current with a string of gauge invariant operators, for example appropriate composite operators of quark-anti-quark fields. We can now repeat the procedure we used in Chapter 17 and obtain Ward identities because the commutation relations between the zero component of the current and the quark fields remain unchanged. Then, following the same argument which we presented after Eq. (17.8), we conclude that either we have a symmetry relating the Green functions of gauge invariant operators or we have a zero-mass pole with the quantum numbers of the divergence of the current. The first part of the alternative is excluded by the data because there is no trace of an axial symmetry in the spectrum of hadrons. On the other hand, there is no massless, or nearly massless pseudoscalar particle with the $U(1)$ quantum numbers, such as an isoscalar partner of the pion. In other words, this symmetry does not appear either as a manifest or as a spontaneously broken symmetry. This puzzle was known as the *missing Goldstone boson*. The only possible answer would be that although a zero-mass pole appears in matrix elements of the gauge-dependent current, it does not appear in physical, gauge invariant, quantities. This is not impossible because, as we have seen, gauge theories quantised in a covariant gauge always contain unphysical degrees of freedom, some of which may appear as massless. In a later section we will show that this is indeed the case.

We see that testing the quantitative predictions of quantum chromodynamics against experimental results will not be an easy task. We must adopt different strategies according to the energy scale we want to probe. In order to use the techniques of perturbation theory we have studied so far we must make sure that the running coupling constant $\bar{g}_s$ is small and this happens only at large energy scales. On the other hand, at low energies the theory becomes strongly interacting and perturbation theory is unreliable. We must develop non-perturbative methods, such as the study of phenomenological models, or, more directly, computations on a space–time lattice. In the following sections we will give very brief presentations of these two methods.

25.4.3   Quantum Chromodynamics in Perturbation Theory

In section 19.2 we showed that the knowledge of the 1-loop β-function allows us to compute the effective value of the running coupling constant in the deep Euclidean region where all masses can be neglected. We also noted that this is not the region in which real experiments are done. Here we want to complete the analysis and bridge the gap between theory and experiment. We will first consider the structure functions W_i $i = 1, 2$, appearing in the expression (25.79).

25.4.3.1   *Kinematics*

We start with some kinematics. Since $q = k - k'$, we have $q^2 = 2EE'(\cos\theta - 1)$, where E and E' are the energies of the initial and final electron and θ is the scattering angle. In the deep inelastic region the energy of the incident electron is large, θ is large; therefore, q^2 is large and negative. However, this does not bring $T_{\mu\nu}$ of (25.78) in the deep Euclidean region because the nucleon is on its mass shell, $P^2 = M_N^2$, and, furthermore, $T_{\mu\nu}$ is a forward scattering amplitude of a virtual photon on the target nucleon, which means that the momentum transfer is zero. Looking closer at the kinematics we see that the inequality $P'^2 = (P + q)^2 \geq M_N^2$ implies that the physical region for the ratio $x = Q^2/\nu$ is given by $0 \leq x \leq 1$. Since $W_{\mu\nu}$ is the matrix element of the commutator of two current operators, it vanishes when their separation is space-like. Let us choose a frame in which the target nucleon is at rest and the z-axis is along the vector $\mathbf{q}$. When we let $Q^2 \to \infty$ keeping x fixed, we have $q_0 = Q^2/(2M_N x)$, $q_z^2 = Q^2 + q_0^2 \sim q_0^2$, which implies that $q_+ = q_0 + q_z \sim \mathcal{O}(Q^2)$ and $q_- = q_0 - q_z$ finite. It follows that in the space–time integral of Eq. (25.78) the dominant contributions will come from the region in which the exponent is stationary; in other words, the region

$$x_- = t - z \sim \mathcal{O}(1/Q^2); \quad x_+ = t + z \sim \text{finite}; \quad x_\mu x^\mu = x_- x_+ \to 0. \qquad (25.102)$$

We thus arrive at the conclusion that in deep inelastic scattering we probe the commutator of two current operators near the light cone.

25.4.3.2   *The Short Distance Expansion*

We start by analysing a simpler situation, that of the corresponding commutator at short distances $x_\mu \to 0$, i.e. at the tip of the light cone, rather than the entire light cone. This is based on the intuitive picture that equal time commutation relations will be simple even in presence of interactions.

Let us consider two local operators $A(x)$ and $B(y)$. In a quantum field theory they may be monomials in the fundamental fields and their derivatives. In 1969 Kenneth Wilson postulated a simple form for the product $A(x)B(y)$ when the difference $x-y$ goes to 0,

$$A(x)B(0)|_{x_\mu \to 0} \sim \sum_i C_i(x) O^i(0), \tag{25.103}$$

where the sum extends over the infinite set of all local operators O^i which have the same quantum numbers as the product on the left. The C_i's are c-number coefficient functions which may be singular when x goes to 0. Two points are important in this expansion. First, it is assumed to be an operator equation. The coefficient functions C_i do not depend on the particular matrix element we may consider.[10] Second, the behaviour of $C_i(x)$ at the origin can be determined by dimensional analysis. If d_A, d_B, and d_i are the dimensions of the operators A, B, and O^i, respectively, the coefficient functions C_i are assumed to behave at short distances, up to logarithmic corrections, as

$$C_i(x)|_{x_\mu \to 0} \sim |x|^{d_i - d_A - d_B}, \tag{25.104}$$

where $|x|$ denotes the modulus of x_μ. In fact, it is simpler to consider this relation in Euclidean space. In Minkowski space we must let x_μ go to 0 staying in the interior of the light cone in which $x_\mu x^\mu$ is positive. Since we will often apply this relation to the matrix element of a commutator which vanishes outside the light cone, this is not any important restriction. The expansion (25.104) shows that the product of the two operators A and B will be dominated at short distances by the operators with the lowest dimension d_i. In a four-dimensional space–time, all canonical fields, as well as the derivative operators, have strictly positive dimensions. Therefore, for any d_i we can construct only a finite number of local operators O_i. It follows that any product (25.103) will be approximated, with any desired accuracy, by a finite number of terms.[11]

The validity of this expansion can be verified in quantum field theory order by order in perturbation. It is obviously correct for free-field theories. For example, if we consider

---

[10] This should not be understood as an assumption about the convergence of the expansion (25.103) in a strong sense. Although we write it this way, we will only use it inside a Green function and it is under this form that we can prove it in perturbation theory, as we will explain shortly.

[11] This statement should be made more precise. For operators made out of free fields the numbers which appear in (25.104) are the canonical dimensions. When interactions are included they change according to the discussion in section 16.6 on the renormalisation group. These changes induce corrections which are formal series in powers of the coupling constant. The assumption is that for small values of the coupling, these changes do not induce large deviations from the canonical values. In particular, $|d - d^{can}| \ll 1$ for all operators.

a neutral scalar field $\phi(x)$, the trivial case to check (25.103) is to look at the vacuum expectation value of the product of two fields:

$$< 0|\,T(\phi(x)\phi(0))|0 >\,|_{x_\mu \to 0} = iG_F(x)|_{x_\mu \to 0} \sim \frac{-1}{4\pi^2} \frac{1}{x^2 - i\epsilon}. \tag{25.105}$$

As announced, this is a trivial case. Taking the vacuum expectation value has killed all but the unit operator on the right-hand side of the expansion (25.103). Since $d_I = 0$, the singularity is $|x|^{-2}$. Had we looked at the product $T(\phi(x)\phi(0))$ we would have found an extra term equal to $:\phi^2(0):$ whose vacuum expectation value vanishes because of the normal ordering. Staying with free fields, we can consider the product of two composite operators of the form $T(:\phi^2(x):::\phi^2(0):)$. We have normal ordered the composite operators because we are interested in the singularity when x goes to 0 and not in those induced by taking the product of two fields at the same point. Using Wick's theorem and the singularity of the Feynman propagator function (25.105) we obtain

$$T(:\phi^2(x):::\phi^2(0):)_{x_\mu \to 0} = \frac{1}{8\pi^4(x^2 - i\epsilon)^2} - \frac{1}{\pi^2(x^2 - i\epsilon)} :\phi^2(0): + :\phi^4(0): \tag{25.106}$$

in agreement with the expansion (25.103) and the singularity structure given by (25.104).

We can now switch on the interactions and check these relations order by order in perturbation. For the scalar theory we can consider a $\lambda\phi^4$ interaction. The simplest case is the product of two fields $T(\phi(x)\phi(0))$. Generalising (25.105) we expect to find the unit operator with a singular function $1/x^2$ and a $:\phi^2(0):$ operator with a coefficient which has at most a logarithmic singularity of the form $\ln(|x|m)$ to some power. The unit operator will contribute only to the disconnected part of whichever Green function we are considering. Diagrammatically, the expansion is shown in Fig. 25.15 where we have considered the matrix element of $T(\phi(x)\phi(0))$ between two states $|a>$ and $|b>$. The left-hand side is the matrix element $< a|\,T(\phi(x)\phi(0))|b >$ and can be computed in perturbation theory using the Feynman rules. The right-hand side represents the insertion of the unit operator, which contributes only in the trivial case with $|a>=|b>$,

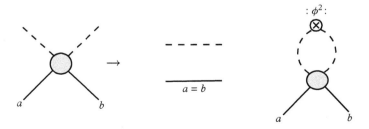

Figure 25.15 *The diagrammatic representation of the short distance expansion of Eq. (25.103).*

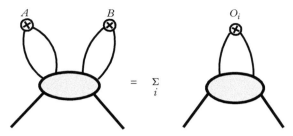

Figure 25.16 *The short distance expansion for two operators A and B.*

and the insertion of the composite : $\phi^2(0)$: operator which we have studied in section 16.5.5. At the limit $x_\mu \to 0$ it is, indeed, logarithmically divergent and matches the divergence on the left-hand side.

Figure 25.16 generalises this picture for two arbitrary local operators A and B.

We see immediately that we will need two sorts of counter-terms: our usual Z_A and Z_B in order to render well defined the Green functions involving either of these operators and a string of fields. These counter-terms are often matrix valued. However, they may be altogether absent if the operators satisfy a conservation equation. In this case, as we have shown in Chapter 16, the corresponding anomalous dimensions vanish. This is the case of the electromagnetic, or the weak, currents which are exactly, or partially, conserved. The second type of counter-term is the one we may need in order to render well-defined Green functions involving the product AB and a string of fields. We will call this counter-term Z_{AB} and it may also be matrix valued. It may be present even for conserved currents. If A and B are both equal to the electromagnetic current $\bar{\psi}\gamma_\mu\psi$, simple power counting shows that the only Green function which may become divergent is the vacuum expectation value $< 0|T(\mathcal{J}_\mu(x)\mathcal{J}_\nu(0))|0 >$. We have computed this diagram in Chapter 16. We saw that although it is potentially quadratically divergent if we use a gauge invariant regularisation, such as the dimensional regularisation, the result is proportional to $(g_{\mu\nu}q^2 - q_\mu q_\nu)$ and it becomes only logarithmically divergent. This counter-term will give a new anomalous dimension $\gamma_{\mathcal{J}\mathcal{J}}$ but its effect is to give a multiplicative factor to the matrix element and, as we will see shortly, it will not affect our calculation. Using the formalism we developed in Chapter 16 we can extend the proof of the validity of the short-distance expansion (25.103) to all orders. The power-counting argument tells us that the singular terms will involve all operators with dimensions smaller than or equal to $d_A + d_B$. The detailed proof involves some rather complicated combinatorics to show that the coefficients of the singular terms precisely match on the right- and left-hand sides of Fig. 25.16.

25.4.3.3 Light Cone Operator Product Expansion

In the previous section we studied the short distance singularities for a product of two local operators. The expansion was simple because the short distance limit allowed us to go to Euclidean space in which $x_\mu x^\mu \to 0$ implies that $x_\mu \to 0$. This is no more true in Minkowski space. For the applications we have in mind in the deep inelastic region we need the commutator of two current operators with separations in which $x_\mu x^\mu \to 0$ while the components of x_μ may remain finite. This changes the expansion (25.103) because its basic ingredient was the fact that the building blocks of a composite local operator, i.e. powers of the fields or derivative operators, have all positive dimensions. At the light cone, however, we can use also powers of x_μ which remains finite but has dimension equal to -1. Therefore, operators with high dimension will contribute to the expansion provided they have a high spin. This yields a light cone operator product expansion of the form

$$A(x)B(0)|_{x_\mu x^\mu \to 0} \sim \sum_{\mathcal{J},i} C_i^{\mathcal{J}}(x^2) x^{\mu_1} x^{\mu_2} ... x^{\mu_{\mathcal{J}}} O_{\mu_1...\mu_{\mathcal{J}}}^{i,\mathcal{J}}(0), \tag{25.107}$$

where the operators $O_{\mu_1...\mu_{\mathcal{J}}}^{i,\mathcal{J}}$ are symmetric, traceless tensors of spin equal to $\mathcal{J}$. Dimensional analysis now modifies Eq. (25.104) and tells that the expected behaviour of the coefficient functions $C_i^{\mathcal{J}}(x^2)$ will be given by

$$C_i^{\mathcal{J}}(x^2)|_{x^2 \to 0 \, , \, x^\mu : \text{finite}} \sim (x^2)^{(d_i^{\mathcal{J}} - \mathcal{J} - d_A - d_B)/2}. \tag{25.108}$$

We see that the relevant quantity now is the difference between the dimension $d_i^{\mathcal{J}}$ of the operator and its spin $\mathcal{J}$. We will call this quantity *twist* and we conclude that operators with the lowest twist dominate. In a scalar field theory the lowest twist operators have twist equal to 2 and they are of the form $\phi \overset{\leftrightarrow}{D}_{\mu_1} ... \overset{\leftrightarrow}{D}_{\mu_{\mathcal{J}}} \phi$. With fermion fields we can again construct twist two operators of the form $\bar{\psi} \gamma_{\mu_1} \overset{\leftrightarrow}{D}_{\mu_2} ... \overset{\leftrightarrow}{D}_{\mu_{\mathcal{J}}} \psi$. Similarly with gauge fields we have the operators $G_{\mu_1 \nu} \overset{\leftrightarrow}{D}_{\mu_2} ... \overset{\leftrightarrow}{D}_{\mu_{\mathcal{J}-1}} G_{\mu_{\mathcal{J}}}^{\nu}$. In all these formulae D denotes an ordinary, or a covariant, derivative. In quantum chromodynamics we will encounter the last two sets of operators where ψ are the quark fields which carry both colour and flavour indices and $G_{\mu\nu}$ is the gluon field strength which carries only colour indices. To give a non-vanishing matrix element between one-nucleon states the colour indices should combine to make a colour singlet. The operator with only gluon fields will be always flavour singlet while the one having also quark fields can be singlet or non-singlet. Since the quarks are assumed to belong to the fundamental representation of $SU(N_f)$, the non-singlet piece belongs to the adjoint. We can separate singlet and non-singlet by looking at particular combinations of structure functions involving protons or neutrons. For example, it is easy to see that the combination $F_2^P - F_2^N$, i.e. the difference of the F_2 structure functions measured in the deep inelastic scattering of electrons off protons and neutrons, receives contributions only from the flavour non-singlet operators.

In perturbation theory we can prove the expansion (25.107) and (25.108) using the same techniques as for (25.103) and (25.104), order by order in the coupling constant expansion. Here also the underlying assumption, which is common to all perturbation theory calculations, is that summing the higher orders will not upset the lower order results. In particular, the contribution of operators with higher twist will be suppressed, relative to those of twist two, by powers of $M^2 x^2$, where M is a typical mass in the problem, for example, the nucleon mass in deep inelastic scattering. In the nth order of perturbation theory this power suppression is corrected by $\ln^n(M^2x^2)$ and the assumption is that these logarithms do not sum up to powers. At any order of perturbation theory renormalisation will simply introduce anomalous dimensions for the various operators. Here the difference between flavour singlet and non-singlet operators becomes important because it implies that the non-singlet operators will be multiplicatively renormalised at 1-loop order while we expect to have a mixing between the two operators which contribute to the flavour singlet quantities.

25.4.3.4 *Renormalisation Group Equations for the Coefficient Functions*

We saw the important role played by the coefficient functions in the operator product expansion. In this section we want to show that they satisfy the same renormalisation group, or Callan–Symanzik, equations as the Green functions.

Let us start with the short distance expansion (25.103). We can consider the right- and left-hand sides as kernels in an n-point function of the underlying field theory and we can transform the expansion to a relation among Green functions involving insertions of composite operators:

$$< 0|T(A(x)B(0)\phi_1(y_1)...\phi_n(y_n))|0 > |_{x_\mu \to 0}$$
$$\sim \sum_i C_i(x) < 0|T(O^i(0)\phi_1(y_1)...\phi_n(y_n))|0 > . \qquad (25.109)$$

To simplify the notation we have written, collectively, ϕ_k to denote all elementary fields of the theory. The index k can be used to distinguish among them. For quantum chromodynamics they are the gluons and the various quark fields. We assume that the underlying field theory is renormalisable and the Green functions appearing in (25.109) are renormalised and depend on the coupling constant (or constants) g, the various masses which we call M and the parameter μ of the renormalisation scheme we are using. In a perturbation expansion of quantum chromodynamics the coefficient functions will depend on the quark masses. At very high energies all but the top quark can be assumed massless. On the other hand, if we start from a nucleon target, the probability of exciting a $t - \bar{t}$ pair is very small. Therefore, the same asymptotic theorems we used in section 19.2 allow us to compute the large Q^2 behaviour using a massless theory. Let us now apply to both sides of (25.109) the renormalisation group differential operator $\mathcal{D}$:

$$\mathcal{D} = \mu^2 \frac{\partial}{\partial \mu^2} + \beta \frac{\partial}{\partial g}. \qquad (25.110)$$

Since the Green functions satisfy the renormalisation group equation, we obtain

$$\sum_i [\mathcal{D}C_i]\, \Gamma_{O_i}^{(n)} = \sum_i C_i(\gamma_i - \gamma_A - \gamma_B - \gamma_{AB})\Gamma_{O_i}^{(n)}, \qquad (25.111)$$

where γ_i, γ_A, and γ_B are the anomalous dimensions of the operators O_i, A, and B, respectively, and we have simplified the form of the equation by assuming that they do not mix under renormalisation, either among themselves or with other operators in the theory. γ_{AB} is the anomalous dimension which comes from the counter-term Z_{AB} which is often required to make the vacuum expectation value of the product AB well defined. The generalisation to more realistic cases is straightforward and amounts to writing a matrix equation with matrix-valued anomalous dimensions.

The important point is that the operator product expansion (25.103) is assumed to be an operator equation (although in a weak sense), which implies that the renormalisation group equation (25.111) is valid for every Green function $\Gamma^{(n)}$. It follows that the C_i's must satisfy the equation

$$[\mathcal{D} - \gamma_i + \gamma_A + \gamma_B + \gamma_{AB}] \, C_i(x, g, \mu) = 0. \tag{25.112}$$

In the realistic cases where the operators A and B are conserved, or partially conserved, currents, we have $\gamma_A = \gamma_B = 0$ and γ_{AB} is a constant independent of i, so it can be absorbed in a multiplicative factor common to all C_i's. To simplify the notation, we will drop it in the following.

25.4.3.5 *Application to Deep Inelastic Scattering*

We are now ready to apply this formalism to the experimental results in deep inelastic scattering. We want, in particular, to obtain a detailed description of the observed approximate scale invariance of the structure functions.

We start with a simple toy case in which we use scalar, rather than vector, currents which means that W and T have no tensor indices:

$$W(x, M_P^2/Q^2) = \int \frac{\mathrm{d}^4 y}{4\pi} \mathrm{e}^{-iqy} < P|[\mathcal{J}(y), \mathcal{J}(0)]|P > \ \sim \mathrm{Im}\, T$$
$$T(x, M_P^2/Q^2) = \int \frac{\mathrm{d}^4 y}{4\pi} \mathrm{e}^{-iqy} < P| T(\mathcal{J}(y)\mathcal{J}(0))|P > . \tag{25.113}$$

W and T are related by a dispersion relation in ν with fixed Q^2, as we showed in section 20,

$$T(Q^2, \nu) = \nu^s \int \frac{\mathrm{d}\nu'}{\nu'^s(\nu' - \nu)} W(Q^2, \nu') + P_{s-1}(Q^2, \nu), \tag{25.114}$$

where we have assumed that the dispersion relation requires s subtractions. $P_{s-1}(Q^2, \nu)$ is a polynomial of $s - 1$ degree in ν whose coefficients are functions of Q^2. The integral extends over the discontinuities in ν shown in Fig. 25.17. The right-hand cut goes from Q^2 to ∞ and corresponds to the intermediate states in the s-channel. The left-hand cut goes from $-Q^2$ to $-\infty$ and corresponds to the u-channel intermediate states. But T is a forward scattering amplitude and (25.113) shows that it is symmetric when q goes into $-q$, i.e. $T(Q^2, \nu) = T(Q^2, -\nu)$. Indeed, changing the sign of q can be absorbed into changing the sign of y in the integral. Because of translational invariance the matrix element of the time-ordered product satisfies

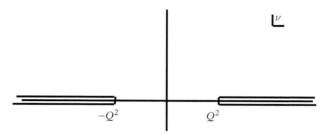

Figure 25.17 *The cut ν-plane.*

$< P | T(\mathcal{J}(-y)\mathcal{J}(0)) | P >=< P | T(\mathcal{J}(0)\mathcal{J}(y)) | P >=< P | T(\mathcal{J}(y)\mathcal{J}(0)) | P >.$[12] The data suggest that for large Q^2 with x fixed, both W and T satisfy approximate scale invariance, which implies that the polynomial P_{s-1} satisfies $P_{s-1}(Q^2, \nu) \rightarrow P_{s-1}(x^{-1})$. Under this assumption, the dispersion relation (25.114) becomes a dispersion relation in x

$$T(Q^2, x) = x^{1-s} \int_{-1}^{+1} \frac{x'^{s-1} dx'}{x' - x} W(Q^2, x') + P_{s-1}(x^{-1}), \tag{25.115}$$

where again, up to a factor of 2, the integral can be restricted to positive x. At large Q^2 T is dominated by the light cone, so we can use the expansion (25.107),

$$T \rightarrow \sum_{i,\mathcal{J}} \int d^4 x e^{-iqx} x^{\mu_1} ... x^{\mu_\mathcal{J}} C_i^{\mathcal{J}}(x^2) < P | O_{\mu_1 ... \mu_\mathcal{J}}^{i,\mathcal{J}} | P >$$

$$= \sum_{i,\mathcal{J}} (2i)^{\mathcal{J}} (Pq)^{\mathcal{J}} \left(\frac{\partial}{\partial q^2} \right)^{\mathcal{J}} \left[\int d^4 x e^{-iqx} C_i^{\mathcal{J}}(x^2) \right] O^{i,\mathcal{J}} \tag{25.116}$$

$$= \sum_{i,\mathcal{J}} x^{-\mathcal{J}} \tilde{C}_i^{\mathcal{J}}(Q^2) O^{i,\mathcal{J}},$$

where we have defined

$$\tilde{C}_i^{\mathcal{J}}(Q^2) = (iQ^2)^{\mathcal{J}} \left(\frac{\partial}{\partial q^2} \right)^{\mathcal{J}} \int d^4 x e^{-iqx} C_i^{\mathcal{J}}(x^2) \tag{25.117}$$

and

$$< P | O_{\mu_1 ... \mu_\mathcal{J}}^{i,\mathcal{J}} | P >= O^{i,\mathcal{J}} [P_{\mu_1} ... P_{\mu_\mathcal{J}} - \text{trace terms}]. \tag{25.118}$$

The trace terms are subtracted to make the operator a pure spin $\mathcal{J}$. They contain terms of the form $P_{\mu_i} P_{\mu_j} g^{\mu_i \mu_j} = M_N^2$, so they are sub-dominant in the deep inelastic region. Combining with the dispersion relation (25.115) and expanding in powers of the scale variable x, we finally obtain that

---

[12] For the physical case of vector currents this property implies that $T_{\mu\nu}(Q^2, \nu) = T_{\nu\mu}(Q^2, -\nu)$.

$$\int_0^1 \mathrm{d}x x^{\mathcal{J}-1} W(Q^2, x) \to \sum_i \tilde{C}_i^{\mathcal{J}}(Q^2) O^{i,\mathcal{J}}; \quad \mathcal{J} \geq s. \tag{25.119}$$

Because of the subtractions, the first s terms are undetermined but, for $\mathcal{J} \geq s$, the moments of the structure function, which can be extracted from the data, determine the Fourier transform of the corresponding coefficient functions in the light cone expansion. $O^{i,\mathcal{J}}$ are just numerical coefficients. They are given by the matrix elements of the local operators $O^{i,\mathcal{J}}(0)$ between one-nucleon states. They depend on the nucleon mass but not on Q^2. All dependence on Q^2 is contained in the coefficient functions $\tilde{C}_i^{\mathcal{J}}(Q^2)$. Thus, the operator product expansion which gave the relation (25.119) achieves our goal: it factors the short distance properties which can be studied by perturbation theory and the large distance ones which give only the unknown numerical coefficients $O^{i,\mathcal{J}}$. The experimental results on a particular value of Q^2 can be used to fit these numbers and then the renormalisation group and perturbation theory will determine the Q^2 dependence of the structure functions.

Extending this analysis to the real world of deep inelastic scattering requires only technical changes. Some are simplifications and some are complications. The simplifications come from the fact that the electromagnetic current of Eq. (25.78), or the weak current we saw in Section 25.3.1, are conserved, or partially conserved. According to our discussion in Chapter 17 their anomalous dimensions vanish. The same is true for any operator $O^{i,\mathcal{J}}$ which satisfies a conservation equation. This is the case for the energy-momentum tensor which will appear on the right-hand side of the light cone expansion for $\mathcal{J} = 2$. The complications are of two kinds. The first comes from the presence of the two tensor indices which means that we must write the dispersion relations separately for every structure function F_i. The second comes from the mixing of the flavour singlet operators as a result of renormalisation which means that the anomalous dimensions which determine the scaling properties of the corresponding coefficient functions are matrix valued.

Putting all this together we obtain the Q^2 dependence of the moments of the structure functions. A typical behaviour, for a non-singlet combination, is given by

$$
\begin{aligned}
M_{\mathcal{J}}(Q^2) &= \int_0^1 \mathrm{d}x x^{\mathcal{J}-2} F(Q^2, x) = \sum_i \tilde{C}_i^{\mathcal{J}}(Q^2/\mu^2, g) O_{i\mathcal{J}} \\
&= \sum_i \tilde{C}_i^{\mathcal{J}}(1, \bar{g}(t)) \exp[-\int_0^t \gamma_i^{\mathcal{J}}(\bar{g}(t')) \mathrm{d}t'] O_{i\mathcal{J}},
\end{aligned} \tag{25.120}
$$

where μ is some reference scale which must be already at the perturbative regime of quantum chromodynamics. In practice a scale of a few GeV is sufficient. For the scale parameter t we take $2t = \ln(Q^2/\mu^2)$ and the anomalous dimension of the operator $O_{i\mathcal{J}}$ is given, at 1-loop order, by $\gamma_i^{\mathcal{J}}(g^2) = r_i^{\mathcal{J}} g^2$. We took into account the fact that the anomalous dimensions of the currents vanish. When Q^2 goes to infinity the effective coupling constant $\bar{g}(t)$ goes to 0, as shown in Eq. (25.87). It follows that at this limit the moment $M_{\mathcal{J}}(Q^2)$ behaves like

$$M_{\mathcal{J}}(Q^2)|_{Q^2 \to \infty} \sim \sum_i K_i^{\mathcal{J}} \left(\ln \frac{Q^2}{\mu^2} \right)^{-(r_i^{\mathcal{J}}/4\pi b_0)}, \tag{25.121}$$

with $K_i^{\mathcal{J}}$ a constant to be determined by the data at the reference scale μ. The logarithmic factor is the correction to the free-field result.

It is also useful to rewrite Eq. (25.120) as an evolution equation. We have

$$Q^2 \frac{\partial}{\partial Q^2} M_{\mathcal{J}}(Q^2) = \sum_i \left[-\mu^2 \frac{\partial}{\partial \mu^2} \tilde{C}_i^{\mathcal{J}}(Q^2/\mu^2, g) \right] O_{i\mathcal{J}}$$

$$= \sum_i \left[\left(\beta(g) \frac{\partial}{\partial g} - \gamma_i^{\mathcal{J}}(g) \right) \tilde{C}_i^{\mathcal{J}}(Q^2/\mu^2, g) \right] O_{i\mathcal{J}}. \tag{25.122}$$

In QCD $\beta(g)$ starts as g^3 (see Eq. (25.85)), and the first-order corrections to $\tilde{C}_i^{\mathcal{J}}$ are of order g^2. It follows that the term proportional to the β function is of order g^4 and can be dropped if we are interested only at the 1-loop corrections. At 1 loop and for a non-singlet combination of structure functions, we have only one operator for a given spin and this allows us to write the solution of (25.122) in a very simple form:[13]

$$M_{\mathcal{J}}(Q^2) = M_{\mathcal{J}}(\mu^2) \left[\frac{\alpha_s(\mu^2)}{\alpha_s(Q^2)} \right]^{r^{\mathcal{J}}/4\pi b_0}. \tag{25.123}$$

By inverse Mellin transformation we can reconstruct the structure functions themselves from their moments and compare with the measurements at any value $Q^2 > \mu^2$. Alternatively, we can extract $M_{\mathcal{J}}(Q^2)$ from F. Note that these operations require the precise measurement of F at small x, especially for large $\mathcal{J}$. On the other hand, in order to reach small x keeping Q^2 large we need very high energies. In recent years a dedicated electron–proton (or positron–proton) collider (HERA) at DESY near Hamburg has provided high-quality measurements fully confirming quantum chromodynamics in deep inelastic scattering.

25.4.3.6 *Connection with the Parton Model Formalism. The A.P.D.G.L. (Altarelli–Parisi, Dokshitzer–Gribov–Lipatov) Equations*

In the previous sections we developed two pictures to describe the composite structure of hadrons. The first, the parton model, is intuitively very simple. Hadrons, such as protons or neutrons, are made out of elementary constituents, the partons. At very high energies, in the frame in which the hadron moves very fast, it is seen by the virtual photon as a thin pancake because of Lorentz contraction. The basic assumption is that interactions among partons, which in this frame involve exchange forces in the

---

[13] For the flavour singlet combinations we must compute the matrix of the anomalous dimensions. At higher orders the computations become more complicated because the operators $O_{i\mathcal{J}}$ mix also with gauge non-invariant operators, such as those made out of the Faddeev–Popov ghost fields.

transverse plane, are very weak. So partons interact with the incident photon as quasi-free particles. The strong point of this picture is its simplicity. Among the weak points we can mention are the lack of a theoretical basis for the free-particle assumption, the absence of a systematic way to estimate the corrections to this 'zero-order' approximation, and the difficulty to understand why partons are not kicked out of the hadron after they are hit by the virtual photon. The second picture is based on the property of non-Abelian gauge theories to be asymptotically free. It identifies the partons with quarks, anti-quarks, and gluons. It explains the quasi-free particle behaviour and offers a consistent expansion scheme to compute the logarithmic corrections to scale invariance. Finally, through the property of the effective coupling constant to increase with distance, it provides for an intuitive, albeit neither rigorous nor quantitative, understanding of confinement. However, all this is achieved at the price of using a rather heavy formalism based on the operator product expansion in which the simple intuitive picture of the photon interacting with individual point-like partons is lost. Here we want to combine the two pictures and obtain a generalised parton model which includes the QCD corrections.

We start with a more precise formulation of the parton picture. Let k_i^μ denote the four-momentum of the parton i in the rest frame of the nucleon. We have

$$P^\mu = \sum_i k_i^\mu = \left(\sum_i k_i^0, \mathbf{0} \right) = (M_N, \mathbf{0}). \tag{25.124}$$

We can perform a Lorentz boost in the z direction with parameter $\omega = \sqrt{\frac{1+v_z}{1-v_z}}$ and define $P^+ = \frac{P^0 + P^z}{2}$ and $P^- = P^0 - P^z$. Under the boost the momenta become

$$\begin{aligned}
P^\mu &\longrightarrow (P^+, P^-, P^\perp) = \left(\tfrac{1}{2} M e^\omega, M e^{-\omega}, 0^\perp \right) \\
k_i^\mu &\longrightarrow (k_i^+, k_i^-, k_i^\perp) = \left(k_i^+ e^\omega, k_i^- e^{-\omega}, k_i^\perp \right),
\end{aligned} \tag{25.125}$$

where $\perp$ denotes the momentum in the $x - y$ plane. The parton picture becomes simple in the infinite momentum frame, $\omega \to \infty$. The key concept of the model is the parton distribution function $f_i(z)$ defined as the probability to find inside the nucleon a parton of type i carrying a fraction z of the nucleon's longitudinal momentum. In the limit of large ω we have

$$z = \frac{k^L}{P^L} \simeq \frac{k^+}{P^+} + O\left(\frac{M_N}{P^+} \right). \tag{25.126}$$

The basic assumptions are that $0 \leq z \leq 1$; in other words, at large ω there are no partons going in the opposite direction and, furthermore, the parton transverse momenta $k_i^\perp$ are bounded. The process is presented diagrammatically in Fig. 25.18. The photon–nucleon interaction is presented as the incoherent sum of photon–parton interactions.

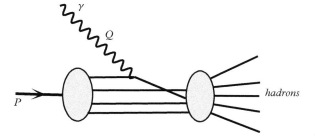

Figure 25.18 *The photon–nucleon interaction as the incoherent sum of photon–parton interactions.*

In quantum chromodynamics the partons are the various quark flavours with distribution functions $u(z)$, $d(z)$, $s(z)$, etc., the anti-quarks $\bar{u}(z)$, etc., and the gluons $G_i(z)$. In the usual quark model with fractionally charged quarks we obtain immediately the sum rules

$$\sum_i \int_0^1 dz\, z\, [q_i(z) + \bar{q}_i(z) + G_i(z)] = 1 \qquad (25.127)$$

$$\int_0^1 dz\, [u(z) - \bar{u}(z)] = 2$$

$$\int_0^1 dz\, [d(z) - \bar{d}(z)] = 1$$

$$\int_0^1 dz\, [s(z) - \bar{s}(z)] = 0$$

$$\int_0^1 dz\, [c(z) - \bar{c}(z)] = 0 \qquad (25.128)$$

$$\text{etc.}$$

$$\sum_q \int_0^1 dz\, [q(z) - \bar{q}(z)] = 3, \qquad (25.129)$$

where we have called collectively $q(z)$ the distribution function of any quark species. The first one (25.127) expresses the conservation of the total momentum, those of Eqs. (25.128) refer to a proton target and express the conservation of the electric charge, and the last one (25.129) express that of the baryon number. The notation is slightly misleading. The range of the index i in the summation in (25.127) is not the same in the three terms.

We can now compute the diagram of Fig. 25.18. The gluons are neutral, so only quarks and anti-quarks interact with the incident virtual photon,

$$W_{\mu\nu}(x, Q^2) = \sum_q \int_0^1 \frac{dz}{z} q(z)\, W_{\mu\nu}^{(q)}\left(\frac{Q^2}{2k_q \cdot q}, Q^2\right), \qquad (25.130)$$

where $W_{\mu\nu}^{(q)}$ is given by the same expression as in (25.78) with a quark, or anti-quark, with momentum k_q replacing the nucleon. The explicit $1/z$ factor takes into account the difference in the incident fluxes between a nucleon of momentum P and a quark of momentum k. $W_{\mu\nu}^{(q)}$ is the imaginary part of the forward Compton scattering amplitude of a virtual photon of mass $-Q^2$ on a free on-shell quark (or anti-quark), satisfying the free Dirac equation. So, apart from kinematic factors, we have

$$W_{\mu\nu}^{\prime(q)} \sim \text{Im} \sum_{\text{spins}} \frac{\bar{u}(k_q)\gamma_\mu (\not{q}+\not{k_q}+m_q)\gamma_\nu u(k_q)}{(q+k_q)^2-m_q^2+i\epsilon}$$
$$\sim \delta(x-z)\text{Tr}\left[\not{k_q}\gamma_\mu (\not{q} + \not{k_q})\gamma_\nu\right]. \tag{25.131}$$

For simplicity, we will neglect the quark masses since the probability of finding inside the nucleon a very heavy quark flavour is negligible. Replacing (25.131) in (25.130) and using (25.78), we finally obtain that

$$F_1(x) = \sum_i e_i^2\, q_i(x); \quad F_2(x) = 2\sum_i e_i^2\, x\, q_i(x), \tag{25.132}$$

where e_i is the charge of the ith quark. The scaling property of the parton model is manifest and, furthermore, we obtain an identity, the *Callan–Gross* relation

$$2\, x\, F_1(x) \;=\; F_2(x), \tag{25.133}$$

which translates the fact that the charged partons have spin $1/2$.

Obviously the parton model can be re-derived in the language of field theory by using a free quark model. The proton is a collection of three quarks and the electromagnetic current is given by the sum of terms $e_q\bar{q}\gamma_\mu q$. Using free-field theory to compute the coefficient functions in the light cone expansion of the product of two currents, we immediately obtain the parton model results as expressed in the relations (25.132) (see Problem 25.13). On the other hand, we saw in the previous section that taking into account the leading QCD corrections introduces a t dependence of the structure functions F_i which can be computed by inverse Mellin transformation starting from the moments given in Eqs. (25.121). Therefore, we can *define* quark and gluon distribution functions $q(z,t)$, $\bar{q}(z,t)$, and $G(z,t)$ such that they reproduce in a parton model language the QCD results. Staying with the non-singlet combinations, we write

$$M_{\mathcal{J}}(t) = \int_0^1 dx\, x^{\mathcal{J}-1}[q(x,t) - \bar{q}(x,t)]. \tag{25.134}$$

Following APDGL, we note that the 1-loop evolution equations (25.123) can be reproduced if we postulate that the quark distribution functions satisfy a master equation of the form

$$\frac{dq(x,t)}{dt} = \frac{\alpha_s(t)}{4\pi} \int_x^1 \frac{dy}{y} q(y,t)P(x/y); \quad \int_0^1 dz\, z^{\mathcal{J}-1}P(z) = r_{\mathcal{J}} \tag{25.135}$$

or, equivalently,

$$q(x,t) + dq(x,t) = \int_0^1 dy \int_0^1 dz \delta(zy - x) q(y,t) \left[\delta(z-1) + \frac{\alpha_s(t)}{4\pi} P(z) dt \right]. \quad (25.136)$$

This equation incorporates the 1-loop QCD corrections but it admits a parton model interpretation: $q(y,t)$ is the probability of finding inside the nucleon a quark with a fraction y of the total longitudinal momentum. This quark can radiate a gluon and reduce its part of the longitudinal momentum from y to x. Thus, the quantity inside the square bracket $\delta(z-1) + \frac{\alpha_s(t)}{4\pi} P(z) dt$ can be interpreted as the probability density of finding inside a quark another quark with fraction z of the parent momentum. The second of the equations (25.135) shows that the moments of this probability density are given by the 1-loop anomalous dimensions of the lowest twist operators in the light cone expansion.

The generalisation to cover also the singlet case is now straightforward. We will need the gluon distribution function $G(z,t)$ and write a system of two coupled integro-differential equations of the form

$$\frac{dq_i(x,t)}{dt} = \frac{\alpha_s(t)}{4\pi} \int_x^1 \frac{dy}{y} \left[\sum_{j=1}^{2f} q_j(y,t) P_{q_i q_j}(x/y) + G(y,t) P_{q_i G}(x/y) \right] \quad (25.137)$$

$$\frac{dG(x,t)}{dt} = \frac{\alpha_s(t)}{4\pi} \int_x^1 \frac{dy}{y} \left[\sum_{j=1}^{2f} q_j(y,t) P_{G q_j}(x/y) + G(y,t) P_{GG}(x/y) \right], \quad (25.138)$$

where the summation goes up to twice the number of flavours because we have used q for both quarks and anti-quarks. Concerning the probability functions we note that gluons are flavour singlets; therefore, we must have $P_{q_i q_j} = \delta_{ij} P_{qq}$. Similarly, in the limit of exact flavour symmetry, i.e. in the limit of vanishing quark masses, $P_{q_i G} = P_{qG}$ and $P_{G q_i} = P_{Gq}$. In this case the APDGL equations (25.137) and (25.138) take the simpler form

$$\frac{dq_i(x,t)}{dt} = \frac{\alpha_s(t)}{4\pi} \int_x^1 \frac{dy}{y} \left[q_i(y,t) P_{qq}(x/y) + G(y,t) P_{qG}(x/y) \right] \quad (25.139)$$

$$\frac{dG(x,t)}{dt} = \frac{\alpha_s(t)}{4\pi} \int_x^1 \frac{dy}{y} \left[\sum_{j=1}^{2f} q_j(y,t) P_{Gq}(x/y) + G(y,t) P_{GG}(x/y) \right]. \quad (25.140)$$

The probability density is now a matrix and, again, its moments are given by the matrix of the anomalous dimensions.

Equations (25.139) and (25.140), although formulated using the concepts of the parton model, are strictly equivalent to the 1-loop QCD equations we derived in the previous section. The connection with the operator product expansion is provided by the relation among the moments of the probability density and the anomalous dimensions of the lowest twist operators. Their advantage is that they give a simple picture of the process. At large Q^2 the scattering of the virtual photon with the charged constituents takes place at a time scale which is much shorter than the characteristic time of the interactions among the constituents. This is manifest in the infinite momentum frame. Once this basic property is understood, it is easy to apply quantum chromodynamics to processes other than the deep inelastic lepton–nucleon scattering.

Obviously, the equivalence between the APDGL equations and QCD is not limited to the 1-loop approximation.[14] We can generalise Eqs. (25.139) and (25.140) by including the evolution of the effective coupling constant and obtain an effective parton model which includes all logarithmic corrections predicted by quantum chromodynamics.

25.4.3.7 *Application to Other Processes*

The experiments of deep inelastic scattering of leptons (electrons, muons, or neutrinos) on nucleons played an important role in discovering the nature of the fundamental strong interactions because, historically, they were the first ones in which the properties of the parton model and the QCD corrections were studied and understood. As we explained in the previous sections, this was due to the fact that they were the only processes in which we could isolate a piece belonging to the deep Euclidean kinematic region. We remind that this was the region we felt safe to apply the asymptotic theorems which allowed us to extract physical results out of the renormalisation group equations. The reason was that in this region we were sure to stay away from infrared singularities and we could study the limit of a massless theory. However, after the discussion we had in Chapter 21, we know that going to the deep Euclidean region is certainly a sufficient condition to be infrared safe, but it is by no means a necessary one. The appearance of infrared singularities is limited to certain well-defined kinematic regions, those we called 'exceptional momenta', which are generically characterised by the property that the square of at least one partial sum of external momenta remains finite. The results of the renormalisation group can be safely applied to all other regions. This result enlarged considerably the domain of applicability of QCD and allowed us to compute many processes in hadronic physics. We will mention briefly some of these applications in this section.

We start with the simplest process, that of the total electron–positron annihilation cross section in hadrons. The diagram is shown in Fig. 25.19 in the one-photon approximation. Following the same analysis that yielded Eq. (25.78), we obtain

$$\sigma(e^+ + e^- \to \text{hadrons}) = \frac{8\pi^2\alpha^2}{3Q^4} \int \mathrm{d}^4x e^{iQ\cdot x} < 0|[\mathcal{J}_\mu(x), \mathcal{J}^\mu(0)]|0 > . \tag{25.141}$$

Q is the virtual photon momentum and Q^2 is the square of the total energy in the centre-of-mass system. By the optical theorem we have

$$\sigma = \frac{32\pi^3\alpha^2 \mathrm{Im}\Pi(Q^2)}{Q^2}, \tag{25.142}$$

[14] See L. Baulieu and C. Kounnas, *Nucl. Phys.* **B141**, 423 (1978) ; **B155**, 429 (1979).

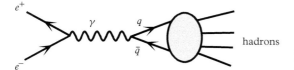

hadrons

Figure 25.19 *The process $e^+ e^- \to$ hadrons in the one-photon approximation.*

where $\Pi(Q^2)$ is given by the vacuum expectation value of the T-ordered product of two currents:

$$\Pi_{\mu\nu}(Q^2) = (g_{\mu\nu}Q^2 - Q_\mu Q_\nu)\Pi(Q^2) = \int \mathrm{d}^4 x \mathrm{e}^{iQ \cdot x} < 0| T(\mathcal{J}_\mu(x)\mathcal{J}_\nu(0))|0 > . \quad (25.143)$$

$\Pi_{\mu\nu}(Q^2)$ represents the QCD corrections to the photon propagator. When $Q^2 \to \infty$ the dominant contributions to the integral come from the tip of the light cone, so it is the short distance expansion of Eq. (25.103) which is relevant. This simplifies the analysis tremendously because the entire process takes place at short distances and, in contrast to what happened at the deep inelastic scattering, we do not have to factor out any large distance contributions. Furthermore, when Q^2 is far from all thresholds, we expect to be able to apply the renormalisation group analysis (we will come back to this point shortly), which tells us that after renormalisation, $\Pi(Q^2)$ can be computed in QCD perturbation theory, using the running coupling constant $\alpha_s(Q^2)$. Figure 25.20 shows the zero-order (parton model) result as well as the order α_s corrections. It is instructive to define the ratio of the total hadronic cross section divided by that of $e^+ + e^- \to \mu^+ + \mu^-$ at the same energy because the latter is a pure QED process involving only point-like particles:

$$R(Q^2) = \frac{\sigma(e^+ + e^- \to \text{hadrons})}{\sigma(e^+ + e^- \to \mu^+ + \mu^-)}. \quad (25.144)$$

In R all kinematical factors as well as the photon propagator cancel. At lowest order in QCD the hadronic system consists of a single quark–anti-quark pair and we must sum over all quarks whose production is energetically possible. We thus obtain the very simple result

$$R(Q^2) = \sum_i e_i^2, \quad (25.145)$$

where e_i is the electric charge, measured in units of the electron charge, of the ith type of quark. The sum extends over the three quark colours as well as the quark flavours. In this approximation R is a constant. The Q^2 dependence comes with the α_s corrections which are given by the diagrams of Figs. 25.20(b), (c), and (d). They are identical with the 2-loop vacuum polarisation diagrams in QED. The special QCD features, such as the three or four gluon couplings, will appear only at higher orders. Figure 25.21 shows the value of R, as measured in electron–positron colliders. The comparison with Eq. (25.145) requires some comments. At very low values of $\sqrt{Q^2}$, below 1 GeV, the

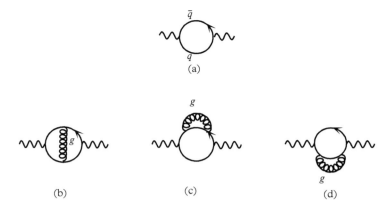

Figure 25.20 *The hadronic contributions to the photon propagator at zero order in α_S (a) and the one-gluon corrections (b), (c), and (d).*

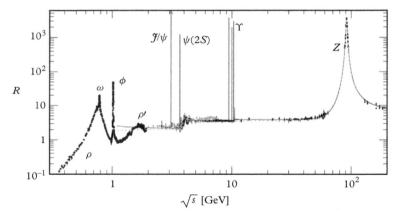

Figure 25.21 *The ratio R from low energies, up to and above the Z^0 mass. The gray dotted lines is the parton model prediction and the red one includes QCD corrections (© Particle Data Group).*

effective QCD coupling constant is large and we do not expect perturbation theory to apply. Indeed the data show large variations in R due to resonance production. Surprisingly, the asymptotic regime of Eq. (25.145) seems to be reached quite soon, at $\sqrt{Q^2} \sim 2$ GeV, although the corresponding value of α_s is not that small (see at Fig. 25.13). At this energy only u, d, and s quarks can be produced and Eq. (25.145) gives $R = 3(4/9 + 1/9 + 1/9) = 2$ in rather good agreement with the data. At $\sqrt{Q^2} \sim 3$–4 GeV we cross the charm threshold. As expected, perturbation theory breaks down because at this range Q^2 cannot be considered larger than all relevant masses. We observe instead the charm–anti-charm resonances, such as the J/Ψ, the Ψ', etc. However, as soon as we pass the thresholds, R settles down to a new constant value, in agreement with the parton model result. The new value is obtained by adding the charm quark contribution which equals $3 \times 4/9 = 4/3$ bringing the new value of R to $10/3$. The same happens again at $\sqrt{Q^2} \sim 10$ GeV when we cross the $b - \bar{b}$ threshold. Above it R increases by $3 \times 1/9 = \frac{1}{3}$ to a total of $\frac{11}{3}$. We predict a new jump by $4/3$ when we cross the $t - \bar{t}$ threshold, but no electron–positron collider has reached this energy yet. At energies close to the Z^0 mass the one-photon approximation of Eq. (25.141) breaks down and the total cross section is dominated by the Z^0 pole.

This discussion illustrates an approximate method for treating threshold effects, called 'the step function approximation'. In computing R we included the diagrams of Fig. 25.20 for all quark flavours satisfying $2m_q < \sqrt{Q^2}$ for which we set $m_q = 0$. Flavours with $2m_q > \sqrt{Q^2}$ were completely left out, as if their masses were infinite. Hence the name 'step function'. Quark masses are either zero, if below $\sqrt{Q^2}$, or infinite if they are above. We can compute the corrections to this approximation at higher orders in α_s, but, as the data show, these corrections do not seem to be numerically very important.

The parton model with the foundations provided by QCD changed radically our views on hadronic collisions. Hadrons are viewed as bags full of partons and a hadronic collision is the result of collisions among individual partons. Most of the time the latter are peripheral processes in which the momentum transfer Q between the colliding partons is low. The characteristic scale of the interaction, given by the inverse of the momentum transfer, is large, comparable to that of the interaction among the constituents. In this case we expect the free parton model to be a bad approximation. Alternatively, we can say that the effective strong interaction coupling constant $\alpha_s(Q^2)$ is large and perturbation theory does not apply. These events constitute the bulk of the measurements and are beyond the domain of perturbative QCD. However, we expect that, from time to time, events occur, similar to those which led Rutherford to discover the existence of the atomic nucleus. They are hard collisions in which the momentum transfer between the colliding partons is large. In this case we expect $\alpha_s(Q^2)$ to be small and perturbation theory to apply.

Let us consider, as an example, a proton–proton collision in which a quark of the first proton makes a hard scattering with an anti-quark of the second. It may be a quark-anti-quark annihilation producing a final state F with a large invariant mass Q^2. Using the parton model notation we write in the centre-of-mass system:[15]

$$\sigma\left(p(P_1) + p(P_2) \to F(Q) + X\right)$$
$$= \int_0^1 dx_1 dx_2 \sum_{ij} q_i(x_1) \bar{q}_j(x_2) \sigma\left(q_i(x_1 P) + \bar{q}_j(x_2 P) \to F(Q)\right). \tag{25.146}$$

The distribution functions q_i and $\bar{q}_j$ can be extracted from the deep inelastic scattering data, so we need only to compute the cross section $q_i + \bar{q}_j \to F$. A simple case, which has a clear experimental signature, is the one in which $i = j$ and the state F consists of a virtual photon of positive invariant mass Q^2 which gives a lepton–anti-lepton pair, such as $e^+ e^-$, or $\mu^+ \mu^-$. So the total process appears as $p + p \to l^+ + l^- + X$. It is an inclusive process,

---

[15] The symbols q and $\bar{q}$ are used with two different meanings: $q_i(x_1)$ and $\bar{q}_j(x_2)$ are the parton model distribution functions to find a quark (or anti-quark) of flavour i (or j) carrying a fraction x_1 (or x_2) of the proton momentum. $\sigma\left(q_i(x_1 P) + \bar{q}_j(x_2 P) \to F(Q)\right)$ is the cross section for the quark–anti-quark annihilation to the final state F.

in the sense that we sum over all hadronic states X. The lepton pair has a large invariant mass. These reactions are called *Drell–Yann processes*. The subprocess $q + \bar{q} \to l^+ + l^-$ is just the one we considered in Fig. 25.19 with the initial and final states interchanged. At zero order in α_s it is given by the corresponding QED process, averaged over the quark colour indices:

$$\sigma(q_i + \bar{q}_i \to l^+ + l^-) = \frac{4\pi \alpha_{em}^2 e_i^2}{9Q^2}. \tag{25.147}$$

In actual experiments one usually measures the momenta k_1 and k_2 of the final leptons, so we measure the momentum of the virtual photon $Q = k_1 + k_2$. A useful quantity is the photon's *rapidity*, denoted by Y. It is defined, in the proton–proton centre-of-mass frame, as

$$Q^0 = \sqrt{Q^2} \cosh Y. \tag{25.148}$$

If $s = 4E^2$ is the square of the total energy in the centre-of-mass system of the collision, the momenta of the initial protons are $P_1 = (E, 0, 0, E)$ and $P_2 = (E, 0, 0, -E)$. Assuming, in the spirit of the parton model, that the transverse momenta of the partons are negligible, we obtain immediately the relations

$$Q^2 = x_1 x_2 s; \quad e^Y = \left[\frac{x_1}{x_2}\right]^{1/2}$$
$$x_1 = \left[\frac{Q^2}{s}\right]^{1/2} e^Y; \quad x_2 = \left[\frac{Q^2}{s}\right]^{1/2} e^{-Y}. \tag{25.149}$$

It follows that the differential cross section for a Drell–Yann process producing a lepton–anti-lepton pair with invariant square mass Q^2 and rapidity Y is given by

$$\frac{d^2\sigma}{dQ^2 dY}(p + p \to l^+ + l^- + X) = \sum_{i=1}^{f} x_1 q_i(x_1) x_2 \bar{q}_i(x_2) \frac{4\pi \alpha_{em}^2 e_i^2}{9Q^2} \tag{25.150}$$

with x_1 and x_2 given by (25.149).

The generalisation of these ideas to describe any hard collision among partons is straightforward. In QCD partons are quarks, anti-quarks, or gluons. If we denote, collectively, their distribution functions $f_i(x)$, we can compute the differential cross section for the reaction shown in Fig. 25.22. Two hadrons, for example two protons, collide at high centre-of-mass energy involving a hard subprocess in which a parton i of the first hadron and a parton j of the second scatter give two partons i' and j'. We assume that all partons are massless and the total energy $\bar{s}$ and momentum transfer $\bar{t}$ of the subprocess are both large. In the parton model the differential cross section is given by

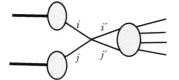

Figure 25.22 *The parton picture of the collision between two hadrons.*

$$\frac{\mathrm{d}\sigma}{\mathrm{d}x_i\mathrm{d}x_j\mathrm{d}\bar{t}}(p+p \rightarrow i'+j'+X) = \sum_{i,j} f_i(x_i)f_j(x_j)\frac{\mathrm{d}\sigma}{\mathrm{d}\bar{t}}(i+j \rightarrow i'+j'). \qquad (25.151)$$

It is a simple kinematical exercise to rewrite this cross section in terms of the parton rapidities $y_{i'}$ and $y_{j'}$ as well as the transverse momentum $p^{\perp}$ of the final partons in the centre-of-mass frame of the collision

$$\pi\frac{\mathrm{d}^4\sigma}{\mathrm{d}y_{i'}\mathrm{d}y_{j'}\mathrm{d}^2p^{\perp}} = \sum_{i,j} x_i f_i(x_i)x_j f_j(x_j)\frac{\mathrm{d}\sigma}{\mathrm{d}\bar{t}}(i+j \rightarrow i'+j'), \qquad (25.152)$$

where the variables x_i and x_j are given by

$$x_i = \frac{2p^{\perp}}{\sqrt{s}}\cosh y\, \mathrm{e}^{Y}; \quad x_j = \frac{2p^{\perp}}{\sqrt{s}}\cosh y\, \mathrm{e}^{-Y} \qquad (25.153)$$

with $y = (y_{i'} - y_{j'})/2$ and $Y = (y_{i'} + y_{j'})/2$. The summation extends over all pairs of partons i and j.

In the Drell–Yan process i' and j' were a pair of leptons. In general, they could be two partons, quarks, anti-quarks, or gluons. They will not appear as such in the final state of the experiment because of the phenomenon of confinement. Each one will turn to a number of hadrons. The details of this hadronisation involve soft processes which are not describable by perturbative QCD. But, if they are soft, they will not change the large transverse momentum of the parent parton i' or j'. Therefore, we expect to observe two jets of collimated hadrons as shown in Fig. 25.23. At higher orders we can consider subprocesses of the form $i+j \rightarrow i'+j'+k'$, such as a hard quark–quark scattering with a radiation of an extra hard gluon. This will result in an event with three large angle jets in the final state.

This phenomenon of large angle jet production in hadronic collisions has been extensively studied in recent years. It turns out to be even more important in LHC because of its higher energy and luminosity. Apart from providing a confirmation of the QCD picture, it threatens to give the main background in the research for eventual new physics. Indeed, a new heavy particle produced at the LHC will decay promptly, often giving hadronic jets at large angles. For this reason, a considerable theoretical effort has been invested in the calculation of this background as accurately as possible.

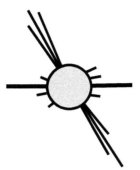

Figure 25.23 *A hard collision with two jets in the final state.*

25.4.4   Quantum Chromodynamics on a Space–Time Lattice

The formulation of quantum field theory on a space–time lattice may have several motivations. For a physical condensed matter system the lattice represents the physical reality, the underlying crystal structure of the solid. For such a system, the description in terms of a continuum field theory is an approximation, valid when a typical correlation length is much larger than the lattice spacing. Such is the case, for example, for temperatures near the critical point of a phase transition. At the opposite side, in the relativistic quantum field theories we studied in this book, we assumed a physical, continuum Minkowski space–time. For these theories the lattice is an approximation. In section 16.4 we introduced this concept as a means to regulate the short distance behaviour of the theory. We noted that this method does provide for an ultraviolet cut-off, but it is not very convenient for performing actual calculations of Feynman diagrams. Since we have been mainly concerned with quantum field theory in perturbation expansion, we adopted other, more efficient, regularisation schemes. However, we also noted that a discrete space–time offers a way to define directly the path integral without reference to perturbation expansion. Therefore, it makes it possible, in principle, to perform approximate calculations in the strong coupling regime where perturbation theory is unreliable. In this section we will try to use this approach for the physically interesting case of quantum chromodynamics where, as we saw earlier, new physical phenomena, such as quark and gluon confinement, are expected to appear for sufficiently strong coupling.

Let us start by rewriting our familiar expressions for the various field theory models we studied so far for the case where the continuum space–time is replaced by a Euclidean lattice which, for simplicity, we assume to have a hypercubic symmetry. The space–time point x_μ is replaced by

$$x_\mu \to n_\mu a, \tag{25.154}$$

where a is a constant length (the lattice spacing), and n_μ is a d-dimensional vector with components $n_\mu = (n_1, n_2, ..., n_d)$ which take integer values $0 \le n_\mu \le N_\mu$. N_μ is the number of points of our lattice in the direction μ. The total number of points, i.e. the volume of the system, is given by $V \sim \prod_{\mu=1}^{d} N_\mu$. The presence of a introduces an ultraviolet, or short distance, cut-off because all momenta are bounded from above by $2\pi/a$. The presence of N_μ introduces an infrared, or large distance cut-off because the momenta are also bounded from below by $2\pi/Na$, where N is the maximum of N_μ. The infinite volume continuum space is recovered at the double limit $a \to 0$ and $N_\mu \to \infty$. From our experience with systems in statistical mechanics, we suspect that these limits should not be taken independently. For example, in classical thermodynamics we obtain finite values for quantities such as the pressure or the temperature only if the number of particles N and the volume V go to infinity with fixed ratio $N/V = \rho$, the particle density. The particular limit we should consider in our field theory examples will be studied in this section.

25.4.4.1 *Scalar Fields on a Lattice*

Let us start with the simplest field theory, that of a self-interacting scalar field $\phi(x)$. In the continuum Euclidean space the action is written as

$$S_E[\phi] = \int \left[\frac{1}{2} \phi(x) \Box \phi(x) - V(\phi) \right] \mathrm{d}^d x, \tag{25.155}$$

where $\Box = \sum_{\mu=1}^{d} \frac{\partial^2}{\partial x_\mu^2}$ and, as usual, we assume that the potential $V(\phi)$ is a polynomial in the field $\phi(x)$.[16] On the lattice the point x is replaced by the lattice site n according to the relation (25.154) and we will drop the factor a to simplify the notation. The derivatives are replaced by the corresponding finite differences. So the action becomes

$$S_E(\phi) = a^d \sum_n \left[\phi_n \sum_{r=1}^{d} \frac{\phi_{n+\hat{r}a} - 2\phi_n + \phi_{n-\hat{r}a}}{a^2} - V(\phi_n) \right], \tag{25.156}$$

where $\hat{r}$ denotes the unit vector in the direction r. Note that for fixed lattice spacing, a is the minimum length of the theory; therefore, we can only consider quantities whose characteristic size is larger than a. In particular, this applies to the wave length of a particle, i.e. if m is the mass, we must always keep $ma \ll 1$.

The action given by (25.156) has the correct continuum limit but it is certainly not unique. We can add any term provided it vanishes when a goes to 0. In the notation we used in section 16.5.5, let $\mathcal{O}(x)$ denote any local operator, which, in the continuum space–time, is a monomial in the field ϕ and its derivatives of canonical dimension D. Its lattice version will contain products of the field ϕ at the same, or neighbouring, points. We can determine its $a \to 0$ behaviour by ordinary dimensional analysis. In a space–time of dimension d it behaves like a^{D-d}. For this reason, in $d = 4$ dimensions, operators with dimension $D \geq 5$ are called *irrelevant operators*. They give vanishing contributions in the continuum limit, although they affect the physics at finite lattice spacing. We will make use of this freedom to add such irrelevant operators in the next section.

---

[16] With this phase convention the exponential factor in the path integral is given by $\exp(S_E/\hbar)$.

Starting from the classical action (25.155) we obtain the continuum quantum field theory by a formal functional integration. In practice however, for a four-dimensional theory, we can compute this path integral only in perturbation. We expect that the discrete version of the path integral will give us the corresponding lattice quantum field theory. We want to know in which sense the continuum theory is obtained as a limit of the discrete one and whether we can use the second to define the first non-perturbatively. On the lattice the functional integral reduces to a multiple ordinary integral

$$\mathcal{D}[\phi] \rightarrow \prod_n \mathrm{d}\phi_n, \tag{25.157}$$

where the product extends over all lattice points. If we take, for simplicity, $N_\mu = N$ for all μ, we must perform an N^d-dimensional integral, each one from $-\infty$ to $+\infty$.

In order to get some intuition let us study the simplest case of the free-field theory with a potential $V(\phi) = m^2 \phi^2 / 2$. We have seen that in the continuum theory the only non-trivial 1-PI correlation function is the 2-point function which, in momentum space, is given by $\Gamma^{(2)}(k) \sim (k^2 + m^2)$. In Minkowski space it gives the usual dispersion relation $k^0 = \pm\sqrt{(\mathbf{k}^2 + m^2)}$. Let us compute the corresponding quantity on the lattice. We define the momentum on the lattice by

$$p = \frac{2\pi P}{Na} \tag{25.158}$$

with $P = (P_1, P_2, ..., P_d)$, a vector whose components are integers modulo N. Using periodic boundary conditions, we diagonalise the kinetic term with a discrete Fourier transformation:

$$\phi_n = \frac{1}{N^{d/2}} \sum_p \mathrm{e}^{iap \cdot n} \tilde{\phi}_p; \quad \tilde{\phi}_p = \frac{1}{N^{d/2}} \sum_n \mathrm{e}^{-iap \cdot n} \phi_n. \tag{25.159}$$

The lattice action (25.156) in momentum space becomes

$$S_E(\tilde{\phi}) = a^d \sum_p \tilde{\phi}_p \left[\frac{1}{a^2} \sum_{r=1}^d \cos p_r a - \frac{1}{2}\left(m^2 + \frac{2d}{a^2} \right) \right] \tilde{\phi}_{-p} \tag{25.160}$$

and the 2-point function is given by

$$\Gamma^{(2)\mathrm{Lat}}(p) \sim \frac{1}{a^2} \sum_{r=1}^d \cos p_r a - \frac{1}{2}\left(m^2 + \frac{2d}{a^2} \right). \tag{25.161}$$

Let us first remark that we obtain an additive mass term equal to $2d/a^2$. It is sometimes called 'lattice mass counter-term', but this terminology is slightly misleading. Strictly speaking there is no mass renormalisation in a free-field theory. The mass, defined by the zero of the 1-PI 2-point function, is always equal to m^2. The constant d/a^2 just cancels the one we obtain from the $\cos pa$ term in Eq. (25.161). However, this constant is divergent when a goes to 0, so, for practical computations, it is convenient to define an effective mass on the lattice:

$$m_{\text{Lat}}^2 = m^2 + \frac{2d}{a^2}. \tag{25.162}$$

In particular, if we want to describe a massless field in the continuum limit, we should start with a $2d/a^2$ effective mass term on the lattice. The use of periodic boundary conditions implies that the momentum p takes values inside the Brillouin zone $[-\pi/a, +\pi/a)$. We obtain the dispersion relation by looking at the zeros of the 2-point function (25.161) inside the Brillouin zone. Going back to Minkowski space we obtain

$$\frac{4}{a^2} \sinh^2 \frac{p_0 a}{2} = m^2 + \frac{4}{a^2} \sum_{i=1}^{d-1} \sinh^2 \frac{p_i a}{2}, \tag{25.163}$$

which, for $a \to 0$, reduces to the usual $p_0^2 = \mathbf{p}^2 + m^2$. We can solve for p_0 as a function of p_i even for finite a with the result

$$\frac{p_0 a}{2} = \ln \left[\pm \sqrt{\frac{m^2 a^2}{4} + \sum_{i=1}^{d-1} \sin^2 \frac{p_i a}{2}} + \sqrt{1 + m^2 a^2 + \sum_{i=1}^{d-1} \sin^2 \frac{p_i a}{2}} \right]. \tag{25.164}$$

Inside the Brillouin zone, the right-hand side is a monotonically increasing function of p_i. Therefore, this relation has the right properties for a dispersion relation. In particular, for $p_i = 0$ and $ma \ll 1$, the two solutions are $p_0 = \pm m$.

Let us summarise. In a space–time lattice scalar fields live on the lattice sites and, as it happens in the continuum, the kinetic energy term couples fields at nearby points. The lattice offers a good regularisation scheme, and the resulting theory has the correct free-field limit.

25.4.4.2 *Fermion Fields on a Lattice*

Let us now turn to spin-1/2 fields. The obvious difference is that they obey Fermi–Dirac statistics, so the integrations over the field variables must be performed in a Grassmannian manifold with the Berezin rules we explained in Chapter 11. This sounds very serious because computers do not usually handle such integrals, but, in fact, this is not a real problem. All theories we may be interested in are quadratic in the fermion fields, so the integrations can be done explicitly and they just produce determinants which depend only on boson fields. We will never have to do anything else. The second difference comes from the kinetic energy term which, for fermions, has first-order derivatives. Therefore, the lattice action will be given by

$$S_E(\psi) = a^d \sum_n \left[\bar{\psi}_n \sum_{\mu=1}^{d} \gamma_\mu \frac{\psi_{n+a\mu} - \psi_{n-a\mu}}{2a} - im\bar{\psi}_n\psi_n + \dots \right], \qquad (25.165)$$

where the dots stand for coupling terms which will be of the general form $\bar{\psi}_n A(\phi)\psi_m$ with A some matrix-valued function of boson fields. We can repeat the exercise we did for the scalar fields and obtain the free-field 2-point function on the lattice, the analogue of Eq. (25.161):

$$\Gamma_{\psi\bar{\psi}}^{(2)\mathrm{Latt}}(p) \sim \sum_{\mu=1}^{d} \gamma_\mu \frac{\sin p_\mu a}{a} - im, \qquad (25.166)$$

which should be compared with the corresponding expression in the continuum: $\Gamma_{\psi\bar{\psi}}^{(2)}(p) \sim \not{p} - im$.

Two remarks are in order here. The first is the absence of an additive mass counter-term, in contradistinction to what we found for the scalar field. This result is not unexpected. From dimensional analysis, an additive mass counter-term should be proportional to a^{-1}. But we know that fermion masses are protected from such counter-terms because of chiral symmetry. This is also in agreement with the fact that in the 2-point function, Eq. (25.166), we find a sine rather than a cosine. The second remark is more subtle. As we did for the scalar case, let us compute the dispersion relation in Minkowski space for finite lattice spacing and compare with the corresponding result in the continuum. To simplify the result, let us choose $m = 0$. We find that

$$p_0 a = \ln \left[\pm \sqrt{\sum_{i=1}^{d-1} \sin^2 p_i a} + \sqrt{1 + \sum_{i=1}^{d-1} \sin^2 p_i a} \right]. \qquad (25.167)$$

The important difference with the corresponding result in Eq. (25.164) is that we have $\sin^2 pa$ rather than $\sin^2 pa/2$. This is due to the fact that the fermion kinetic energy has first-order derivatives, while that of the scalar field has second order. As a result, the right-hand side of Eq. (25.167) is no more a monotonically increasing function of p_i inside the Brillouin zone. In addition to the expected solution $p_0 = 0$ for $p_i = 0$, we also obtain $p_0 = 0$ for $p_i = \pm\pi/a$. It seems that on the lattice we describe 2^d fermionic states, all having the same, in this example zero, mass.

We will argue shortly that this result too is not unexpected and it is also due to chiral symmetry, but, for the moment, we remark that, in contrast to the scalar field case, the lattice regularisation does not correctly describe the physical fermionic states. This is a serious problem because we wanted to use the lattice formulation to describe theories like QCD.

Several solutions to this problem have been proposed and we will show here one, which is most commonly used and is due to Kenneth Wilson. In his 1976 Cargèse lectures Wilson proposed to use a Euclidean lattice regularisation in order to study

QCD numerically. To solve the problem of the extra fermionic states he suggested the introduction of a new term in the action with the following properties: (i) it must be an irrelevant operator so that it vanishes in the continuum limit. (ii) For finite a it must lift the degeneracy between the 'normal' fermionic state and its lattice partners in such a way that the latter decouple in the continuum limit. These requirements do not determine this term uniquely and, in the notation of Eq. (25.156), a simple choice is given by

$$S_{\text{Wilson}} \sim a^d \sum_n \frac{\mathrm{i} f}{2a} \bar{\psi}_n \sum_{r=1}^d (\psi_{n+\hat{r}a} - 2\psi_n + \psi_{n-\hat{r}a}) \tag{25.168}$$

with f an adjustable constant. Note that this term mimics the second-order derivative of the scalar field action of Eq. (25.156). It satisfies the first requirement because, for small a, it goes like

$$S_{\text{Wilson}} \rightarrow \frac{af}{2} \int \mathrm{d}^d x \bar{\psi}(x) \Box \psi(x). \tag{25.169}$$

In other words, it vanishes with a. It satisfies also the second requirement because it has no γ matrices, so, on the lattice, it contributes a p-dependent mass term. Since it comes from a second derivative, it contributes to the dispersion relation terms proportional to $(f/a) \sum_i \sin^2 p_i a/2$. It gives a vanishing contribution to the mass of the state with $p_i = 0$, but it moves that of all states which live on the boundary of the Brillouin zone, where at least one of the components of p_i equals $\pm\pi/a$, by an amount proportional to $1/a$. In this way the new, unwanted, states will become infinitely heavy in the continuum limit. In Problem 16.8 we presented the so-called 'decoupling theorem' which states under which conditions an infinitely massive state decouples from low-energy dynamics. We will come back to this point shortly.

With the addition of the Wilson term the fermionic lattice action becomes[17]

$$S_E(\psi) = a^d \sum_n \left[\bar{\psi}_n \sum_{\mu=1}^d \left(\frac{\gamma_\mu - \mathrm{i}f}{2a} \psi_{n+\mu} - \frac{\gamma_\mu + \mathrm{i}f}{2a} \psi_{n-\mu} \right) + \mathrm{i}\bar{\psi}_n \left(m - \frac{df}{a} \right) \psi_n \right]. \tag{25.170}$$

We first note that the Wilson term does introduce a lattice mass counter-term which, when a goes to 0, is linearly divergent. In the terminology we used for the scalar fields, in order to describe a zero-mass fermion we must choose an effective mass on the lattice equal to df/a. This shows that the Wilson term breaks explicitly chiral invariance. This is not without consequences because one of the properties of quantum chromodynamics in the strong coupling regime we wanted to study using the lattice formulation was precisely the spontaneous breaking of chiral symmetry. We will postpone the discussion of this point until we have a complete formulation of the theory.

---

[17] We have chosen all four γ matrices in Euclidean space to be anti-Hermitian.

The conclusion of this section is that we can use a lattice regularisation for field theories involving fermions which, like scalar fields, are localised on the lattice sites. However, chiral symmetry is explicitly broken by the lattice.

25.4.4.3   *Gauge Fields on a Lattice*

We have learnt how to introduce matter fields, both spin 0 and spin 1/2, on a space–time lattice. In order to complete the picture, we should also study gauge fields. We will follow the same reasoning presented in Chapter 14 which led us to their introduction in the continuum.

Let us consider a lattice field theory containing a set of fields ψ_n^i, which, in the notation of Eq. (14.8), we write as a multiplet Ψ_n. As before, n labels the lattice site and i runs from 1 to r. The fields can be scalars or spinors and the dynamics is given by some lattice action $S(\Psi)$. We assume that the ψ's transform linearly according to an r-dimensional representation, not necessarily irreducible, of a group G which leaves S invariant,

$$
\Psi_n = \begin{pmatrix} \psi_n^1 \\ \vdots \\ \psi_n^r \end{pmatrix} \qquad\qquad \Psi_n \to \Omega(\omega)\Psi_n \qquad\qquad \omega \in G, \qquad (25.171)
$$

where $\Omega(\omega)$ is the matrix of the representation of G corresponding to the group element ω. This is just the lattice transcription of (14.8). If G is a Lie group we can write the analogue of (14.9),

$$
\Psi_n \to e^{i\Theta}\Psi_n \qquad\qquad \Theta = \sum_{\alpha=1}^{m} \theta^\alpha T^\alpha, \qquad (25.172)
$$

where the θ^α's are a set of m constant parameters, and the T^α's are m $r \times r$ matrices representing the m generators of the Lie algebra of G which satisfy the commutation relations (14.10).

These are global transformations. As before, we obtain the corresponding gauge ones by taking the group elements ω of (25.171), or the parameters θ^α of (25.172), to depend on the lattice site n. The terms in the action which are non-invariant under these local transformations are those which contain products of fields in neighbouring points coming from the lattice version of the derivative operators in Eq. (25.156) or (25.165). Under the transformations (25.171) these terms transform as

$$
\Psi_n^\dagger \Psi_{n+\hat{r}a} \to \Psi_n^\dagger \Omega^{-1}(\omega_n)\Omega(\omega_{n+\hat{r}a})\Psi_{n+\hat{r}a}. \qquad (25.173)
$$

As we did in the continuum, we restore invariance by introducing gauge fields U which are matrix valued in the group G and correspond to every ordered pair of neighbouring points n and $n + \hat{r}a$. All local terms in the action are unaffected, but the products on different points become

$$\Psi_n^\dagger \Psi_{n+\hat{r}a} \rightarrow \Psi_n^\dagger U_{n,n+\hat{r}a} \Psi_{n+\hat{r}a}. \tag{25.174}$$

They are invariant under (25.171) provided the U's transform as

$$U_{n,n+\hat{r}a} \rightarrow \Omega(\omega_n) U_{n,n+\hat{r}a} \Omega^{-1}(\omega_{n+\hat{r}a}) \tag{25.175}$$

and we will impose on U the constraint $U_{n,m} = U_{m,n}^{-1}$. We see that, in contrast to matter fields, gauge fields do not live on the lattice sites but on oriented links. It is a clear way to exhibit the fact that they correspond to the connections we introduced in the previous chapters. The terms (25.174) give on the lattice the couplings between matter and gauge fields.

If the group G is a Lie group we can write the matrices U as

$$U_{n,n+\hat{r}a} = e^{iga \sum_{\alpha=1}^m A_{n,n+\hat{r}a}^\alpha T^\alpha}, \tag{25.176}$$

where, in order to make contact with the expressions we used in the continuum, we introduced a constant g, which will become the coupling constant, and the lattice spacing a to give A the dimensions of a mass. Note that U takes values in the group G while A takes values in the corresponding Lie algebra. For the continuum limit it is convenient to introduce the notation $U_{n,n+\mu} \rightarrow U_\mu(n)$ and similarly for A. It is easy to check that, in the continuum limit, the A's have the right transformation properties for gauge fields we found in Eq. (14.15). Indeed, let us expand (25.176) and (25.175) in powers of a and θ, keeping only first-order terms:

$$U_\mu(n) = \mathbb{1} + iag A_\mu^\alpha(n) T^\alpha + \dots.$$
$$e^{i\theta^\alpha T^\alpha} = \mathbb{1} + i\theta^\alpha T^\alpha + \dots. \tag{25.177}$$

Inserting these expressions into the expansion of the transformation properties of U, Eq. (25.175), and restoring $x = na$, we obtain that

$$A_\mu^\alpha(x) \rightarrow A_\mu^\alpha(x) - \frac{1}{g}\partial_\mu \theta^\alpha(x) + A_\mu^\beta(x)\theta^\gamma(x) f^{\alpha\beta\gamma} \tag{25.178}$$

in agreement with Eq. (14.15). The derivative on θ comes from the expansion of $\theta(n+\mu)$.

We can now write the gauge invariant action of scalar or fermion fields and study the continuum limit. Let us take, for example, a charged scalar field and the group $G = U(1)$. The coupling terms will come from products of the scalar fields in neighbouring points. The action becomes

$$S[\phi, U] = a^d \sum_n \phi_n^\dagger \sum_{\mu=1}^d \frac{U_\mu(n)\phi_{n+\mu} + U_\mu^\dagger(n-\mu)\phi_{n-\mu}}{a^2} - \dots, \tag{25.179}$$

where the dots stand for terms not involving the gauge field U. In order to obtain the continuum limit we must develop the fields and keep terms up to order a^2:

$$\phi_{n\pm\mu} = \phi(x) \pm a\partial_\mu\phi(x) + \frac{a^2}{2}\Box\phi(x) + \dots$$
$$U_\mu(n) = 1 + iagA_\mu(x) - \frac{a^2g^2}{2}A_\mu(x)A^\mu(x) + \dots$$
$$U_\mu^\dagger(n-\mu) = 1 - iagA_\mu(x) + a^2\left(ig\partial_\mu A^\mu(x) - \frac{g^2}{2}A_\mu(x)A^\mu(x)\right) + \dots.$$

Substituting (25.4.4.3) into (25.179), we obtain the action of scalar electrodynamics of Eq. (15.40) with the exception of the F^2 term.

The procedure we developed so far gives us the lattice formulation of gauge theories where the gauge fields are external classical fields without dynamical degrees of freedom. To obtain the full theory we need the lattice version of the F^2 terms. As for the continuum, the guiding principle is gauge invariance. Let us consider two points on the lattice n and m. We will call a path $p_{n,m}$ on the lattice a sequence of oriented links which join continuously the two points. Consider next the product of the gauge fields U along all the links of the path $p_{n,m}$:

$$P^{(p)}(n, m) = \prod_p U_{n,n+\mu}\dots U_{m-\nu,m}. \tag{25.180}$$

Using the transformation rule (25.175), we see that $P^{(p)}(n, m)$ transforms as

$$P^{(p)}(n, m) \rightarrow \Omega(\omega_n)P^{(p)}(n, m)\Omega^{-1}(\omega_m). \tag{25.181}$$

It follows that if we consider a closed path $c = p_{n,n}$ the quantity $\text{Tr}P^{(c)}$ is gauge invariant. The simplest closed path for a hypercubic lattice has four links and it is called *plaquette*. We introduce the notation

$$\mathcal{P}_{\mu\nu}(n) = U_\mu(n)U_\nu(n + \mu)U_\mu^\dagger(n + \nu)U_\nu^\dagger(n). \tag{25.182}$$

It is now straightforward to show (see Problem 25.11), that the action

$$S_g = \frac{1}{g^2} \sum_{n,\mu<\nu} \left[1 - \text{Re}(\text{Tr}\mathcal{P}_{\mu\nu}(n))\right] \tag{25.183}$$

is gauge invariant and, in $d=4$, goes to the space–time integral of $\text{Tr}(F^2)$ when a goes to 0. It is the term introduced by K. Wilson to describe the pure gauge part of the action on the lattice and it is called *the Wilson action*. In section 25.4.2 we introduced the concept of a Wilson loop, Eq. (25.90). We see that $\text{Tr}\mathcal{P}_{\mu\nu}(n)$ is the simplest Wilson loop on the lattice. So the Wilson action involves the sum over all elementary Wilson loops. The restriction $\mu < \nu$ avoids double counting.

This concludes our discussion on the lattice description of gauge fields. We have the choice to consider either the group-valued fields U or the Lie algebra valued ones A as independent variables. For the continuum limit the A's are clearly more convenient. On the other hand, for purely lattice calculations, the U's present several advantages.

First, they allow us to study numerically non-Lie groups. The simplest example is the gauged version of Z_2, the group of the Ising model. The gauge fields take the values $U = \pm 1$ and, for the pure gauge theory, the functional integral is reduced to a sum. Independently of its significance in statistical mechanics, the model is interesting as a toy model for more complicated cases.

Second, for a compact Lie group G, the use of the variables U which take values in the group allows us to write the lattice version of the path integral without the need to add a gauge-fixing term. The reason is that the integration over the U's produces a harmless factor given by the volume of the group which is assumed to be compact. This finite factor is common to both the numerator and the denominator in the expression for the generating functional and it safely cancels. In contradistinction, the continuum space formulation uses the A's as dynamical variables which take values in the Lie algebra of the group. As we noted already, the factor which results from the integration is infinite, corresponding, formally, to the volume of the group G^∞. It was the necessity to deal with such situations that led us to the gauge-fixing procedure and the BRST symmetry. We conclude that as long as we stay on the lattice keeping a finite, gauge fixing is not compulsory.

25.4.4.4 QCD on a Lattice. Formal Discussion

We are now in a position to formulate and study any gauge theory on a space–time lattice. The obvious choice is quantum chromodynamics because it presents a rich variety of interesting non-perturbative phenomena. We would like to address issues such as the spontaneous breaking of chiral symmetry, the confinement of colour non-singlet degrees of freedom, and, of course, the spectrum of hadronic states. Before looking at the results of the numerical simulations, let us present some formal arguments related to these topics.

We saw already that the Wilson term (25.168) breaks explicitly chiral invariance for any finite value of the lattice spacing. The introduction of such a term was necessary to ensure that the continuum limit contains only the desired fermionic degrees of freedom. We want to argue here that the resulting breaking of chiral symmetry was not accidental, due to some peculiarity of the Wilson term, but it was dictated by the known properties of this symmetry. The space–time lattice can be viewed as an efficient cut-off which regulates all short distance singularities. Therefore, all formal manipulations are legitimate as long as a is kept different from 0. Furthermore, we learned that this regulator respects whichever gauge invariance we want to impose. If we could find a prescription which respects chiral invariance and has the correct continuum limit, we would have obtained a way to write a gauge theory without chiral anomaly. But we proved in section 17.3 that this is impossible, which means that a regulator respecting both gauge and chiral symmetry does not exist. This result can be proven directly on the lattice without reference to the proof we presented in the continuum and is known as *the Nielsen–Ninomiya theorem*. For quantum chromodynamics it implies that the study of the spontaneous breaking of chiral invariance will not be a straightforward exercise. We will see the results in the next section. Note also that if we wanted to study on the lattice a theory like the standard model of the electroweak interactions in which both the vector and the axial currents are coupled to gauge fields, we could not expect to obtain the correct continuum limit unless we used the full fermion content of a family for which, as we proved in section 25.3.7, the coefficient of the anomaly vanishes. Indeed, we pointed out in the formulation of the decoupling theorem in Problem 16.8 that the heavy modes decouple only if the initial and the final theory are both renormalisable. This is not the case for the standard model if the families are not complete because the axial anomaly induces a non-renormalisable counter-term.

Before going to the second point in our list, namely the phenomenon of quark and gluon confinement, it is instructive to discuss the properties of the phase diagram of a lattice field theory and exhibit the differences between global and local symmetries. Let us first note that there exists a formal analogy between a quantum field theory formulated on a Euclidean lattice and the partition function of a statistical system. The coupling constant g^2 of the first plays the role of the temperature in the second. Let us look at the simplest model based on the group Z_2. For a global symmetry we know that a suitable order parameter is the average magnetisation. Keeping a and the volume finite the average magnetisation is 0 in the absence of an external magnetic field because the configurations with spins plus or minus 1 have equal probabilities. At the infinite volume limit, however, the system has two phases. At sufficiently high temperature it is in a symmetric or disordered phase in which the magnetisation vanishes. Below T_c we are in a phase of broken symmetry with non-zero magnetisation. We can probe the order parameter by introducing an external source coupled to it, in this case a magnetic field h. In the presence of h the basic configurations $\sigma_n = \pm 1$ are separated by an energy difference $\Delta E = 2hN$ which diverges when N goes to infinity with fixed h. The external source should be put to 0 after the infinite volume limit has been taken. All this is well known from both exact solutions, whenever available, and numerical simulations. We want to know how much of this analysis survives when the underlying Z_2 symmetry becomes local. The answer is contained in a theorem proven by Shmuel Elitzur who showed that there exists no local order parameter for a gauge symmetry in which the fields take values in a compact manifold. The proof goes as follows.

Let us consider a local function of the gauge fields $f(U)$. We compute the mean value $<f(U)>$ by

$$<f(U)> = \lim_{\mathcal{J}\to 0} \lim_{N\to\infty} <f(U)>_{\mathcal{J},N}$$
$$= \lim_{\mathcal{J}\to 0} \lim_{N\to\infty} Z_{\mathcal{J},N}^{-1} \int \exp[S(U) + \mathcal{J} \cdot U] f(U) \mathcal{D}U, \qquad (25.184)$$

where the limits are taken in the right order, as explained earlier. The action and the integration measure are supposed to be invariant under the gauge transformations (25.175), which we write, formally, as $U \to {}^\omega U$. If f is going to be a candidate for an order parameter it should not be gauge invariant, i.e. we must have

$$\int f({}^\omega U) \mathrm{d}\omega = 0, \qquad (25.185)$$

where the integral is taken independently in every ω_n. Since the symmetry is local, we can consider a subgroup of the transformations which act only on a finite subset of the U's, which we call U', leaving the rest $\hat{U}$ invariant: $\hat{U} = {}^\omega\hat{U}$. If now we change the integration variables in (25.184) from U to ${}^\omega U$, we obtain

$$< f(U) >_{\mathcal{J},N} = Z_{\mathcal{J},N}^{-1} \int \exp\left[S(U) + \mathcal{J}' \cdot {}^\omega U' + \hat{\mathcal{J}} \cdot \hat{U}\right] f({}^\omega U)\mathcal{D}U\mathrm{d}\omega. \qquad (25.186)$$

For a gauge theory we can always choose the set U' finite and independent of N, such that if the source $\mathcal{J}$ is bounded from below by ϵ we have

$$|\exp(\mathcal{J}' \cdot U') - 1| \le \eta(\epsilon) \qquad (25.187)$$

with η staying positive for positive ϵ and going to 0 uniformly in ϵ, independently of $\mathcal{J}$ or N. Note that such an inequality does not hold for the global Ising symmetry we considered before because all the spins transform simultaneously and we cannot find transformations which act only on a finite subset of them. Going back to (25.186) we write $\exp\mathcal{J}' \cdot U' = 1 + (\exp\mathcal{J}' \cdot U' - 1)$. The first term vanishes because of (25.185) and the second can be bounded using (25.187). So we obtain the bound

$$| < f(U) >_{\mathcal{J},N} | \le 2\eta(\epsilon)\mathrm{Sup}f, \qquad (25.188)$$

which, in the ordered limit $\mathcal{J} \to 0$, $N \to \infty$, gives

$$< f(U) > = 0 \qquad (25.189)$$

for all local functions f satisfying (25.185).

This result is sometimes interpreted as saying that a gauge symmetry cannot be spontaneously broken because only gauge invariant local functions of the fields can have a non-zero vacuum expectation value. Although such an interpretation may be misleading because the theorem does not imply anything directly for the continuum theory,[18] it is certainly true that in order to study the phase diagram of the lattice theory, we must use a non-local order parameter. The Wilson loop was introduced for this purpose. There is a simple argument, first developed by Alexander Polyakov, which shows that a lattice gauge theory based on a compact group at the strong coupling limit exhibits always the property of confinement. It uses the fact that a Wilson loop is made out of the sum of all plaquettes which span it, so if it were not for the integration measure, it would

---

[18] A non-zero mean value for a gauge non-invariant quantity would be a signal for a phase transition resulting in a spontaneous breaking of the gauge symmetry. Elitzur's theorem shows that this does not happen on the lattice. As we just saw, it is based on the fact that on the lattice the group can be taken to be compact. On the other hand, since it is a gauge group, it can be chosen to act on a fixed, finite subset of variables. So a gauge group on a lattice with a fixed lattice spacing a behaves effectively like a system with a finite number of degrees of freedom and cannot have any phase transitions. But such transitions may occur when we take the limit $a \to 0$ because in the continuum, even an arbitrarily small but finite region of space contains always an infinite number of degrees of freedom.

always have an area law, which, as we showed in section 25.4.2, implies confinement. But at the strong coupling limit $1/g^2 \rightarrow 0$, $\exp(g^{-2}S_E)$ goes to 1 and the integration over the compact group gives only a constant. This argument applies to all compact groups, Abelian as well as non-Abelian, at the infinite coupling limit. The next question is to examine whether there is a phase transition when we go towards the weak coupling region and/or the continuum theory. We can prove that such a deconfining phase transition is indeed present at weak coupling for the Abelian lattice theory, but we have only numerical evidence concerning its absence for the non-Abelian case. Note that the presence of dynamical fermions affects these conclusions in an important way. We argued previously that a perimeter law for the Wilson loop indicates that the force between two opposite charges vanishes at large distances. We interpreted this result as showing the presence of free quarks. This, however, is true only if quarks are infinitely heavy and we do not consider the possibility of creating quark–anti-quark pairs from the gluon field out of the vacuum. In the real world such pair creations are possible and the phenomenon can occur in the following way. We consider a quark–anti-quark pair which we attempt to separate. Let us assume that the force between them remains constant, which, for classical quarks, would indicate confinement. However, in reality, after a certain distance it becomes energetically advantageous to create a new pair. The newly created quark (anti-quark) can bind with the initial anti-quark (quark), resulting in the appearance of two mesons. The force between them goes to 0 with distance, but this does not imply deconfinement, but screening. We still do not observe free quarks or gluons although the effective forces between them vanish at large distance. The presence of screening makes the study of the phase diagram, both analytical and numerical, more complicated.

25.4.4.5 *QCD on a Lattice. Numerical Results*

There are certainly many more analytical studies we can perform on lattice gauge theories, but the most important results are those obtained by direct numerical integrations. Indeed, it is for this purpose that lattice gauge theories were introduced in the first place. We will briefly present the ideas and discuss some of the results. Let us only warn the reader that it is an extremely active field of research covering topics not only from physics, but also from numerical analysis and computer science and cannot be adequately summarised in one section.

It is clear that had we unlimited computing power, we could compute any expectation value, or any correlation function, on any given lattice characterised by a lattice spacing a and a total volume N. We could repeat the calculation by varying a and N and obtain the results as functions of the lattice parameters. With enough accuracy we could study the limit of infinite volume and vanishing lattice spacing. In order to understand the non-perturbative aspects of QCD we would like to compute quantities such as the expectation value of the Wilson loop and the correlation functions of composite operators. The first would show us directly whether the theory exhibits the property of confinement. The second would give us the spectrum of physical states. Let us, for example, consider the 2-point function of an operator $\mathcal{O}$: $G^{(2)}(x) \sim < T(\mathcal{O}(-x/2), \mathcal{O}(x/2)) >$ and let $\tilde{G}^{(2)}(k^2)$ be its Fourier transform. We found in Problem 20.2 that the large distance behaviour of $G^{(2)}$ is given by the nearest singularity of $\tilde{G}^{(2)}(k^2)$ in the complex k^2 plane. If this singularity is a simple pole on the real axis at $k^2 = m^2$, showing the existence of a stable particle of mass m with the quantum numbers of the operator $\mathcal{O}$, $G^{(2)}$ behaves as e^{-mr}. Therefore, from the study of the 2-point functions we can infer the particle spectrum of the theory. If we choose $\mathcal{O} \sim \bar{q}\gamma_5\tau^i q$ with q the doublet of u and d quarks, we obtain the quantum numbers of the pion. A nearly zero value of its mass would indicate the spontaneous breaking of approximate chiral symmetry.

Of course, in the real world we do not possess unlimited computing power and the programme presented earlier can be realised only to a certain degree of approximation. In fact, it is not surprising that reliable results to a fraction of these questions have only recently been obtained and the programme is still going on.

It must be clear by our previous discussion that a direct computation of the multidimensional integral to which the Feynman path integral is reduced on the lattice is beyond any available, or even imaginable, computing system.[19] New algorithms which are inspired by the well-known method of steepest descent have been developed. The most commonly used, *the Metropolis algorithm,* was invented in the very early days of the first computers (1953) by Nicholas Metropolis, but it is more often known as *the Monte Carlo algorithm.* Strictly speaking it is not an integration programme but rather a sampling algorithm, suitable to study the landscape of hyper-surfaces. It is a Markov chain Monte Carlo method which provides efficient samplings from a probability distribution for which direct methods are impossible, or extremely time-consuming. In the course of the years it has grown into a full branch of computational science incorporating many rigorous results from numerical analysis and the theory of stochastic processes. Let us illustrate the principle in a simple example from lattice gauge theories. We choose to compute numerically the mean value of some quantity $f(U)$,

$$< f(U) > = \frac{\int \mathcal{D}U \, f(U) \, e^{S[U]}}{\int \mathcal{D}U \, e^{S[U]}}. \tag{25.190}$$

The Metropolis algorithm consists of starting from an initial field configuration chosen randomly. For Z_2 a simple choice is $U_\mu(n) = \pm 1$ with equal probability. For $U(1)$ we can write $U_\mu(n) = e^{i\theta_\mu(n)}$ and choose θ uniformly in the interval $[0, 2\pi)$. This is similar for any other group. We will call this configuration C_0. We will generate a new configuration by applying the following recipe. We go through all the links, one after the other. Each time we attempt to change the field on that link. For Z_2 we just flip the spin. For a continuous group we replace $U_\mu(n)$ by $U_\mu(n)V_\mu(n)$, no summation in n or μ, with V uniformly chosen in the group. For example, for $U(1)$ we take $V = e^{i\phi}$ and we choose ϕ uniformly inside $[0, 2\pi)$. Each time we compute the resulting variation of the action δS and we keep the new value of the field in the link n to $n + \mu$ if δS satisfies

---

[19] A simple order of magnitude estimation. Let us assume that the computation of a single integral from $-\infty$ to $+\infty$ requires K simple operations. For simple well-behaved functions K may be of order 10^2 to 10^3. For a four-dimensional theory a rather low value for N is of order 10^6, which gives around 30 points in each direction. The resulting number of elementary operations is K^N, clearly beyond any system.

$$e^{\delta S} > r, \tag{25.191}$$

where r is a number uniformly distributed in the interval 0 to 1 and independently chosen for each link. After sweeping this way through all the links we obtain a new configuration, C_1. We start all over again and we generate a sequence of configurations C_2, C_3,..., C_M,... . For each of them we compute the corresponding value of f: $f_M(U)$. We expect the average we are looking for to be given by

$$< f(U) >= \lim_{M \to \infty} \frac{1}{M} \sum_{m=0}^{M-1} f_m(U). \tag{25.192}$$

Under some specific assumptions we can prove the existence of the limit in (25.192) and its independence on the initial configuration C_0. In practice, of course, we stop the sweep at a certain step M and, again under some assumptions, we can estimate the error δf. There is no universal, easily implementable criterion to decide when to stop, but we expect to have a good approximation if the resulting field configurations C_M are statistically independent, for example, when the correlation between C_M and $C_{M'}$ goes to 0 for large $|M - M'|$.

This is the principle of the algorithm in its simplest form. Several improvements have been incorporated in order to speed up the convergence (for example, instead of updating one link at a time we proceed to change a sizable fraction of them simultaneously) or, to obtain a more reliable estimate of the uncertainties. For the case of QCD, people have also used the possibility of adding irrelevant operators to the lattice action, in addition to the one introduced by Wilson to solve the fermion problem, see Eq. (25.168). They aim at improving the convergence and/or minimising the finite size effects. Another obvious addition is related to the question of gauge invariance. As we noted before, on the lattice we do not have to fix the gauge because the volume of the gauge group is finite. However, when applying the Metropolis algorithm, we do not want to spend computational time on configurations which are gauge equivalent, only to find out that δS is zero. We want to be sure that every updating of the value of the field in a link yields an inequivalent configuration and this requires a gauge fixing. A commonly used gauge is given by imposing on all U's the condition $U_4(n) = \mathbb{1}$. It corresponds, in the continuum, to the familiar gauge $A_0(x) = 0$, for which the Faddeev–Popov determinant is trivial.

An important bottleneck in the calculation has been the computation of the determinant of the huge matrix resulting from the integration over the fermion fields for each field configuration. For this reason, all early results from lattice simulations were treating quarks as external sources with no dynamical degrees of freedom. This was called *the quenched approximation*. It is only recently, with the increasing computing power, that the influence of fermion loops is included.

Before going into the actual calculations, let us estimate the lattice parameters we should ideally need in order to simulate the physically interesting, non-perturbative region of QCD. In units of mass, we want to cover the area from a few MeV, the order of magnitude of the light quark masses, to a few GeV, above which perturbation theory becomes reliable. The upper end tells us that we need to consider lattice spacings a

smaller than one-tenth of a fermi.[20] The lower end requires a total lattice size larger than 100 fermi, which means at least 1000 points in each direction. For the moment we do not have the means to consider so large lattices and, as we will see, the low mass region where the spontaneous breaking of chiral symmetry is expected to occur can be approached only by extrapolation. State of the art computations use lattices of $2^6 = 64$ points in each direction.

Let us now present some of the results that have been obtained and briefly comment on them.

Figure 25.24 gives a general view of the spectrum of hadrons obtained in various lattice simulations. The different colours refer to different groups who performed the calculations. The *b*-states are displaced by 4 GeV in order to show them in the same figure. Isospin symmetry has been assumed by taking the masses of the two light quarks m_u and m_d equal. The input parameters are the common mass $m_u = m_d$ and the masses of the *s*, *c*, and *b* quarks. To them we must add a third one which corresponds to the value of the strong interaction coupling constant, or, equivalently, a scale parameter similar to the one we introduced in the continuum in Eq. (25.89). The question of the spontaneous breaking of the chiral symmetry will be shown independently.

Putting all these results in the same figure may be misleading, because they come from different simulations and the assumptions and the input parameters are not always the

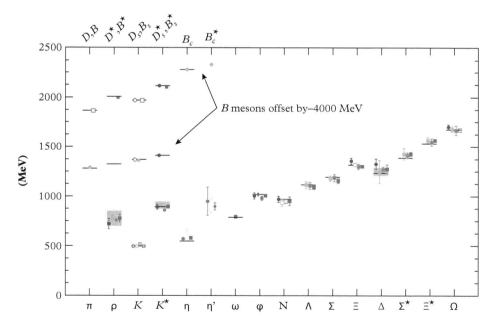

Figure 25.24 *The hadron spectrum (©A.S. Kronfeld).*

---

[20] 1 fermi equals 10^{-13} cm, which, in our units, corresponds to $(200 \text{ MeV})^{-1}$.

same. However, it offers a comprehensive picture and shows that the lattice calculations have reached maturity. The fit is quite impressive.

In the previous results the low value of the pion mass is an input. The phenomenon of the spontaneous breaking of chiral symmetry is the object of an independent study and we show the results in Figs. 25.25 and 25.26. The idea is to compute the pion mass M_π^2 as a function of the quark mass m. Although we must use rather large values of m, if the idea of spontaneous breaking is correct, we must find, by extrapolation, that M_π^2 goes to 0 with m. We remind that in the lattice chiral symmetry must be explicitly broken. In Fig. 25.25 the breaking is given by the Wilson fermionic term of Eq. (25.168). In Fig. 25.26 a different method is used. We start by writing QCD in a five-dimensional lattice. Then we consider a layered phase in which the five-dimensional space is split into layers of four-dimensional sublattices. We can show that the fermionic zero modes become localised in these subspaces. In the continuum limit we can obtain a four-dimensional space with the correct fermion degrees of freedom. Both calculations are consistent with spontaneous breaking of the chiral symmetry.

There are many more results which have been obtained in lattice QCD including the spectrum of gluon bound states, called *glueballs*, the hadronic matrix elements in weak interaction transitions, or the evidence for confinement for all couplings in non-Abelian theories. The work is still in progress but the results presented here show already the power of the method.

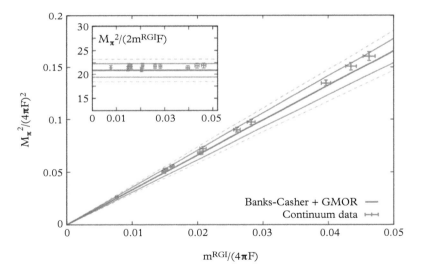

Figure 25.25 *The relation between the pion square mass and the quark mass, both normalised to $4\pi F \sim 1\,GeV$. F is the pion decay constant f_π of Eq. (25.97). The line is the theoretical extrapolation to the GMOR (Gell-Mann – Oakes – Renner) relation based on the chiral current algebra Eq. (25.9). The calculation uses Wilson fermions. The lattice parameters are indicated (© G.P. Engel et al, arXiv:1406.4987).*

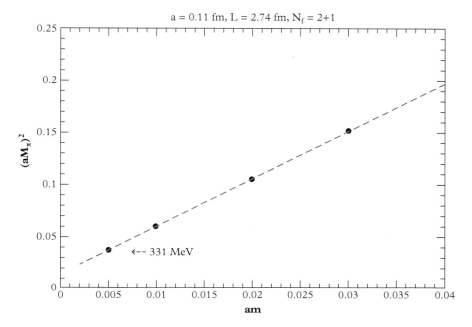

a = 0.11 fm, L = 2.74 fm, N_f = 2+1

<--- 331 MeV

Figure 25.26 *The relation between the pion square mass and the quark mass. The calculation uses domain wall fermions. The lattice parameters are indicated (© C. Allton et al, RBC-UKQCD Collaborations, arXiv: 0804.0473).*

25.4.5   Instantons

25.4.5.1   *Perturbation Expansion and Semi-classical Approximation*

Formulated in terms of Feynman's path integral, perturbation is an expansion around a background field configuration, often chosen to be solution of the classical equations of motion. If we write the classical action $S[\phi]$, we set $\phi(x) = \phi_0(x) + \tilde{\phi}(x)$. If the Lagrangian density is polynomial in ϕ the translated action $S[\phi_0(x) + \tilde{\phi}(x)]$ will be again polynomial of the same degree in $\tilde{\phi}(x)$ with $\phi_0(x)$-dependent coefficients. For example, for the ϕ^4 theory we obtain

$$S[\tilde{\phi}] = S[\phi_0] + \frac{1}{2}(\partial_\mu \tilde{\phi})^2 + \frac{1}{2}\left(m^2 + \frac{1}{2}\lambda\phi_0^2\right)\tilde{\phi}^2 + \frac{\lambda}{3!}\phi_0\tilde{\phi}^3 + \frac{\lambda}{4!}\tilde{\phi}^4. \qquad (25.193)$$

If $\phi_0(x)$ does not satisfy the classical equations of motion, we should add the terms linear in $\tilde{\phi}$. In general, an expansion around a field configuration which is a solution of the classical equations of motion amounts to computing small fluctuations around the classical solution, hence the name *semi-classical approximation*. It is easy to see that this is in fact equivalent to perturbation expansion. The simplest way is to note that we can rescale the field $\phi \to \sqrt{\lambda}\phi$. The path integral becomes, in the Euclidean space,

$$\int \mathcal{D}[\phi] e^{-S[\phi]} \rightarrow \int \mathcal{D}[\phi] e^{\frac{-1}{\lambda} S[\phi]}. \tag{25.194}$$

The coupling constant λ does not change the classical equations of motion, so, for the classical theory, it can be set to any value. In the quantum theory on the other hand, the value of λ determines how much configurations which are not classical solutions contribute to the integral. We can say, naïvely, that if $\lambda = 0$, only the classical solution contributes. So for small λ we can keep only configurations close to it. So expansion in powers of λ appears to be equivalent to semi-classical approximation.[21] $\phi_0 = 0$ is a solution of the classical equations of motion in the absence of sources. An expansion around it corresponds to our usual perturbation expansion. We have also encountered the case $\phi_0 = v$ with v a constant, given by the position of the minimum of the classical potential. It led us to the study of the phenomenon of spontaneous symmetry breaking. The general case of including in the functional integral several field configurations ϕ_0^i was avoided so far, first for reasons of simplicity and, second, because we expect their contribution to be exponentially small compared to that of the configuration which gives the absolute minimum of the action. However, we know that in quantum mechanics there are phenomena, such as those related to penetration through a potential barrier, which are not accessible to any finite order of perturbation theory precisely because they are suppressed by exponential factors. In this section we want to study these questions with particular emphasis on gauge theories and QCD.

25.4.5.2   *The Winding Number Rediscovered*

For simplicity, let us consider first a pure Yang–Mills theory based on some compact, simple group G. The invariant action is proportional to the integral over all Euclidean space of $\mathrm{Tr}F^2$. The perturbation theory we have considered so far is the semi-classical expansion around the zero field configuration. The corresponding value of the classical action is $S = 0$. We want to study here the possibility of taking into account the expansion around other finite action field configurations which may be solutions of the classical equations of motion.[22]

The Euclidean action is $O(4)$ invariant, so, in the integral over the four-space E_4, we can use four-dimensional spherical coordinates. We can imagine performing first the integration over the angular variables, so we are left with an integration over the radial variable r. We will obtain a finite action if F vanishes at infinity faster than $1/r^2$. Since

---

[21] The argument is, at best, intuitive and it makes sense only to the extent that perturbation expansion makes sense. It is based on the observation that the contribution of each field configuration to the integral is weighted by $e^{-S[\phi]}$, but this does not necessarily mean that we need to keep only small action configurations because the argument ignores the question of how many such configurations exist, in other words their measure in the functional space. In particular, it may lead us to conclude that in the calculation of the path integral, only field configurations of finite action need to be considered. In fact, it is the opposite that may be true. In simple toy models we can find examples in which finite action field configurations form a set of measure zero in the functional space, so we would conclude that they can be omitted. The configurations of finite action and, in particular, *small* finite action become important only in perturbation theory because it is only around them that we can meaningfully expand.

[22] We follow the presentation given by S. Coleman, in 'The Uses of Instantons', Erice Summer School 1977, reprinted in *Aspects of Symmetry* (Cambridge University Press, 1985).

$F \sim dA$, a naïve conclusion would be that A should go to 0 at infinity faster than $1/r$. But this reasoning forgets gauge invariance. F is gauge invariant but A is not. So the correct statement is that A should go to 0 at infinity faster than $1/r$, *up to gauge transformations*,

$$A_\mu(x) = \Omega(x)\partial_\mu\Omega^{-1}(x) + ..., \tag{25.195}$$

where the dots stand for terms which vanish at infinity faster than $1/r$. $\Omega(x)$ is a function from the Euclidean four-space to the gauge group G and, for the purposes of this discussion, we can choose it to be of order 1, i.e. a function of the angular variables only. Thus, $\Omega(x)$ establishes a mapping from S^3, the angular part of E_4, into G. Any gauge potential in the family (25.195) will give a finite action field F. Obviously, $\Omega(x)$ is not unique. We can change it by a gauge transformation $\Omega \to g\Omega$ with $g(x)$ an element of G. Since we are interested in the asymptotic behaviour of the fields for large r, we can ask the question whether we can always choose $g(x) = \Omega^{-1}$ for $r \to \infty$. We want g to be a regular gauge transformation everywhere in E_4; otherwise, we may introduce singularities in the action integral. This implies that $g(0)$ must be a non-singular constant gauge transformation. All such transformations are continuously connected to the identity transformation because our gauge groups do not have disconnected pieces. So the question is whether the gauge transformation Ω, which is defined on the hypersphere at infinity, is continuously connected to the identity transformation. This question is answered precisely by the study of the homotopy classes of the mapping $S^3 \to G$. In Problem 25.15 we study explicitly a toy model in which G is replaced by the Abelian group $U(1)$ and the space is the Euclidean plane, so S^3 is replaced by the circle S^1. Topologically $U(1)$ is equivalent to a circle, so in the problem we study the mappings $S^1 \to S^1$. Armed with this experience we turn here to the physically more interesting case of the mappings $S^3 \to G$ and we choose first the simple case $G = SU(2)$. We can write every $SU(2)$ matrix in terms of the unit 2×2 matrix and the three Pauli matrices as

$$g(a_\mu) = a_4\mathbb{1} + i\boldsymbol{a} \cdot \boldsymbol{\sigma}; \quad |a|^2 = \sum_{i=1}^{4} a_i^2 = 1, \tag{25.196}$$

so a group element g is parametrised by the four real numbers satisfying the unit modulus condition (25.196); in other words, $SU(2)$ is topologically equivalent to S^3, so we must study the mappings $S^3 \to S^3$. We can follow the same intuitive chain of arguments that led us to the solution of the toy model, but, in fact, the problem has been solved mathematically and the homotopy classes of all such mappings are known. For the case of $S^3 \to S^3$ there exist an infinite number of homotopy classes labelled by an integer number ν, called *the winding number*. Every mapping is homotopic to one in the family

$$g^{(\nu)}(x) = [g(x_\mu)]^\nu \tag{25.197}$$

with $g(x_\mu)$ given by (25.196) for $a_\mu = x_\mu$. We can view ν as the number of times the hypersphere at infinity is wrapped around $SU(2)$, which justifies the term 'winding'. In fact, we can show that the winding number is our old Pontryagin index we found in section 17.3 which is given by the integral (17.37). This result can be generalised to arbitrary groups as follows: (i) the maps $S^3 \to U(1)$ are all homotopic to the trivial one, i.e. they are continuously deformable to the identity. (ii) The maps $S^3 \to G$, with G any simple Lie group, can be continuously deformed into a mapping of $S^3 \to SU(2)$, a subgroup of G. So, for any simple gauge group, they are characterised by a winding number. An arbitrary gauge group can be written locally as $G \approx [U(1)]^n \times G_1 \times \ldots \times G_k$ with the G_i's being simple groups, so it will be characterised by k independent winding numbers, one for each G_i.

This answers our first question, namely we found that finite action field configurations form homotopy classes, those of the mappings $S^3 \to G$. For non-Abelian groups they are labelled by a set of winding numbers. The $A_\mu = 0$ configuration we have been using in perturbation theory belongs to the class with $\nu = 0$.

25.4.5.3   *An example of a non-trivial configuration*

The most important result of the previous section is the existence of a non-trivial winding number. Before studying the consequences for our favourite gauge theories let us show an explicit example of a finite action classical field configuration belonging to a winding number $\nu = 1$. As we explained earlier, there is no essential loss of generality by restricting ourselves to a gauge group $SU(2)$.

Any finite action gauge potential A_μ belonging to the $\nu = 1$ class will be asymptotically gauge equivalent to

$$A_\mu = g^{(1)}(x)\partial_\mu[g^{(1)}(x)]^{-1}; \quad g^{(1)}(x) = \frac{x_4 \mathbb{1} + \mathrm{i}x \cdot \boldsymbol{\sigma}}{r} \tag{25.198}$$

with the obvious notation $x = (x_4, x)$, $r = (x_4^2 + x^2)^{1/2}$. We can define $x^* = (x_4, -x)$ and obtain $[g^{(1)}(x)]^{-1} = g^{(1)}(x^*)$. This leads us to look for solutions of the equations of motion of the form

$$A_\mu = f(r)g^{(1)}(x)\partial_\mu[g^{(1)}(x)]^{-1}. \tag{25.199}$$

$f(r)$ must be regular everywhere, which implies the boundary condition $f(0) = 0$. We must now obtain (and solve) a differential equation for f. The task is rendered easier by the following remark. We start from the definition $\tilde{F}_{\mu\nu} = \frac{1}{2}\epsilon_{\mu\nu\rho\sigma}F^{\rho\sigma}$. It follows that $FF = \tilde{F}\tilde{F}$ and this gives us the Schwarz inequality,

$$\int \mathrm{d}^4x FF = \left[\int \mathrm{d}^4x FF \int \mathrm{d}^4x \tilde{F}\tilde{F}\right]^{\frac{1}{2}} \geq \left|\int \mathrm{d}^4x F\tilde{F}\right| = 32\pi^2|\nu|, \tag{25.200}$$

where we have used the index theorem of (17.37). We thus obtain a bound for the classical action

$$S = \frac{1}{4g^2}\int \mathrm{d}^4x FF \geq \frac{8\pi^2|\nu|}{g^2}, \tag{25.201}$$

where g is the Yang–Mills coupling constant. The bound is saturated only for field configurations which are self-, or anti-self-, dual:

$$F = \pm \tilde{F}. \tag{25.202}$$

We will look for solutions of the self-duality equation (25.202), which is a first-order differential equation, because, if we can find them, we know that they will be also solutions of the full equations of motion since they saturate the bound (25.201) and consequently they give local minima of the action. Furthermore, they will give the minimum action solutions inside a given homotopy class, thus providing the dominant contributions in a semi-classical expansion.

Using the ansatz (25.199) in the self-duality equation (25.202), A. A. Belavin, A. M. Polyakov, A. S. Schwartz, and Yu. S. Tyupkin found the simple solution $f(r) = r^2/(r^2 + \rho^2)$. ρ is an arbitrary scale and reflects the fact that the initial classical theory is scale invariant. This solution is centred at $x = 0$. Using the invariance under translations we can find a family of such solutions by just replacing everywhere $x \to x - x_0$ with x_0, another free parameter. Finally, we can perform gauge transformations which do not change the homotopy class.

A simpler and more suggestive form of this solution is given by

$$A_\mu^a = \frac{2}{g} \frac{\eta_{a\mu\nu}(x - x_0)^\nu}{(x - x_0)^2 + \rho^2}, \tag{25.203}$$

where η is antisymmetric in μ and ν and is given by

$$\eta_{aij} = \epsilon_{aij}; \quad \eta_{ai4} = \delta_{ai}. \tag{25.204}$$

Although the solution (25.203) looks strange, it has a simple interpretation. The space E_4 has an $O(4) \approx SU(2) \times SU(2)$ invariance. The sphere at infinity has an $SU(2)$ invariance. The internal symmetry is also $SU(2)$. Since we want a solution that maps $SU(2)$ into $SU(2)$, it is natural to establish a correspondence between one of the subgroups of $O(4)$ and the internal symmetry group. The tensor η plays that role. So the solution (25.203) is $O(4)$ invariant because any $O(4)$ rotation can be compensated by an internal symmetry transformation. We can verify by explicit computation of the integral of $F\tilde{F}$ that this solution has winding number equal to 1. In fact, Eq. (25.203) represents an eight-parameter family of solutions with $\nu=1$ which corresponds to the symmetries of the original equations of motion: four parameters for space–time translations given by x_0, one parameter for dilatations given by ρ, and three parameters for constant gauge transformations which are still global symmetries of the theory, even after gauge fixing. A mathematical theorem, due to M. Atiya and R. Ward, shows that every finite action solution with $\nu = 1$ belongs to this family of solutions.

The solution is localised around $x = x_0$ not only in space but also in time. For this reason it is called *an instanton*. It goes to 0 everywhere at infinity, but it is not equivalent to the zero field solution. If we go back to Minkowski space, it interpolates between the zero field configuration and one which is obtained by a gauge transformation which is not continuously connected to the identity. Let us also note that by space inversion, we obtain a new solution which has a winding number equal to −1. This is obvious since $F\tilde{F}$ is a pseudoscalar. We call this new solution *an anti-instanton*.

A final remark: Suppose that we could find a similar instanton solution in E_3 rather than E_4. E_3 is the space part of four-dimensional Minkowski space. An E_3 instanton would be a time-independent, space-localised solution of the Yang–Mills equations; therefore, we would expect such a solution to describe, in the quantum theory, a 'particle', but one which is not present in perturbation theory. Are there such 'particles'? The answer is no, for a pure Yang–Mills theory and the reason is very simple. In order to find the corresponding winding number we should study the mappings of S^2 (the sphere at infinity of E_3) into S^3. All these mappings are trivial, in the sense that they can be continuously deformed to the mapping on the identity.[23] Therefore, such a 'particle' would not be topologically stable. However, the situation will change if we consider a Yang–Mills theory with BEH fields, as we will show in the next chapter.

25.4.5.4   *The Structure of the Ground State in Gauge Theories*

The existence of a non-trivial winding number shows the existence of an infinity of finite action field configurations for the classical Yang–Mills theory. All these configurations fall into homotopy classes. Here we want to study the consequences for the quantum theory.[24]

In the quantum theory we must first fix the gauge. Until now we have worked in covariant gauges because they give the simplest Feynman rules to compute loop diagrams. However, here we are not going to compute any diagrams; we will only evaluate the functional integral around a given field configuration. It turns out that in this case it is simpler to use a ghost-free gauge, i.e. a gauge in which the Faddeev–Popov ghosts decouple. We choose, following S. Coleman, the so-called 'axial gauge' and we set $A_3 = 0$. The calculation becomes more transparent if we start by quantising the theory in a finite volume and take the limit of infinite volume at the end. We consider a box of volume V ($\pm X, \pm Y, \pm Z$) in E_3 and a Euclidean time interval $\pm T$. We must impose boundary conditions at the edges of the box as well as the initial and final times $\pm T$ and a convenient choice is to ensure that all surface terms when we perform partial integrations vanish. Since such terms are given by

$$\delta S = \frac{1}{g^2} \int \mathrm{d}^3 S \, n^\mu F_{\mu\nu} \delta A^\nu, \qquad (25.205)$$

where n^μ is a unit vector normal to the surface and $F_{\mu\nu}$ is antisymmetric in μ and ν, we conclude that δS depends only on the tangential components of the gauge field. Obviously, we must choose conditions such that $A_3 = 0$, but also such that when the size

---

[23] In fact all mappings of S^n into S^m are trivial for $n < m$.
[24] In this section we summarise the discussion presented by Sidney Coleman in his 1977 Erice lectures.

of the box goes to infinity, the resulting classical field configuration gives a finite action. This is guaranteed if we choose boundary conditions at $\pm T$ corresponding to definite values of the winding number, say n_i at $-T$ and n_f at $+T$. This in turn implies that the field inside the box corresponds to a winding number $n = n_f - n_i$. This is possible because Eq. (17.36) shows that the winding number, being proportional to the integral over all space of $\partial_\mu G^\mu$, depends only on the normal components of G_μ, which are given in terms of the tangential components of A_μ (see Eq. (17.35)). It follows that the winding number will remain constant as the box goes to infinity. We can show (see Problem 25.16) that with our choice of fixing the tangential components of the gauge potential at the boundaries of the box corresponding to a definite value of the winding number, we can take the limit of large V and T provided we integrate over field configurations inside this homotopy class,

$$M(V, T, n) \sim \int \mathcal{D}[A_\mu] e^{S_{cl}[A]} \delta_{\nu n}, \tag{25.206}$$

where the Kronecker symbol $\delta_{\nu n}$ means that we integrate only over field configurations with winding number $\nu = n$. As usual, $M(V, T, n)$ represents a transition matrix element between an initial and a final state determined by the boundary conditions and is given by

$$M(V, T, n) \sim\, <n_f| e^{-2HT} |n_i>. \tag{25.207}$$

$|n_i>$ and $|n_f>$ are states defined by field configurations with the corresponding winding number. Equation (25.207) shows that when $T \to \infty$, we pick up the eigenstate of H with the lowest eigenvalue. The expression becomes more transparent if we Fourier transform M with respect to n:

$$\tilde{M}(V, T, \theta) = \sum_n e^{in\theta} M(V, T, n) \sim \int \mathcal{D}[A_\mu] e^{S_{cl}[A]} e^{i\nu\theta}. \tag{25.208}$$

The important point is that the winding number is given by Eq. (17.37), which is an integral over all space of a local density. The size of an instanton is determined by the parameter ρ which is independent of T or V. Therefore, it remains fixed when the volume goes to infinity. It follows that we can consider the configuration with total winding number n as the result of an arbitrary number of instantons and anti-instantons, provided their difference equals n. For example, if we split the box T into $T = T_1 + T_2$, we obtain, approximately,

$$M(V, T_1 + T_2, n) = \sum_{n_1+n_2=n} M(V, T_1, n_1) M(V, T_2, n_2); \tag{25.209}$$

in other words, we put a field configuration with winding number n_1 in the first half of the box and one with winding number n_2 in the second. By doing so, we ignore field configurations which have a winding number n in the entire box but do not correspond to a definite winding number in any of the two parts. Since the size of the instanton configurations is very small compared to the sizes of the boxes, we expect the contribution of such configurations to go to 0 as the box goes to infinity, roughly as a surface-to-volume effect.

The composition law (25.209) implies that for large T_1 and T_2, $\tilde{M}(V, T, \theta)$ satisfies

$$\tilde{M}(V, T_1 + T_2, \theta) = \tilde{M}(V, T_1, \theta)\tilde{M}(V, T_2, \theta), \qquad (25.210)$$

which means that $\tilde{M}$ satisfies a composition law of a simple exponential, as expected from the expectation value of the operator e^{-HT} in an energy eigenstate. As $T \to \infty$, we select the lowest energy eigenstate corresponding to winding number ν. Since it is a state with the lowest energy we will call it θ-*vacuum* and denote it by $|\theta >$. In this section we will compute the energy corresponding to each θ-state.

In the spirit of the previous discussion, we approximate the configuration with winding number ν by an arbitrary number of instanton and anti-instanton configurations satisfying $\nu = n - \bar{n}$. They are centred at points x_0^i, which are separated by distances large compared to their sizes given by the parameters ρ^i. This is often called *the dilute gas approximation*, by analogy to the corresponding one used in statistical mechanics. We will calculate the functional integral for such a configuration.

Let us call S_0 the value of the classical action for one instanton, or anti-instanton. Computing the quadratic fluctuations around this solution yields an expression of the form $[\det A]^{-1/2}e^{-S_0}$. Using the explicit form of the solution (25.203) we can compute the values of S_0 and $\det A$, but we will not need these values here, so let us call the one-instanton contribution to the functional integral Ke^{-S_0}. In the dilute gas approximation the classical value for n such configurations equals nS_0 and the quadratic fluctuations give $K^n e^{-nS_0}$. This is the contribution of n configurations centred around the points x_0^i which are held fixed. We must still integrate over these positions. In each of the four dimensions of the box V, T we can place the centres at $-X < x_0^1 < x_0^2 < ... < x_0^n < +X$. Integrating over all positions gives

$$\int_{-X}^{+X} \mathrm{d}x_0^n \int_{-X}^{x_0^n} \mathrm{d}x_0^{n-1} ... \int_{-X}^{x_0^2} \mathrm{d}x_0^1 = \frac{(2X)^n}{n!}. \qquad (25.211)$$

Putting all these together we write Eqs. (25.207) and (25.208) as

$$< \theta|e^{-2HT}|\theta > \sim \frac{1}{n!\bar{n}!} \sum_{n,\bar{n}} (Ke^{-S_0})^{n+\bar{n}} e^{i(n-\bar{n})\theta}(2VT)^{n+\bar{n}} = e^{4KVT e^{-S_0}\cos\theta}, \qquad (25.212)$$

which shows that the θ-state has an energy density per unit volume given by

$$E_\theta = -2K\cos\theta e^{-S_0}. \qquad (25.213)$$

The precise value in Eq. (25.213) is not important; what *is* important, however, is that E_θ does depend on θ, which means that the θ-vacua are indeed different states.

The expression (25.208) has a very simple interpretation. Taking into account field configurations with non-vanishing winding number in the Yang–Mills functional integral results in a continuum of vacuum states, labelled by a parameter θ, in complete analogy with the Bloch waves in a periodic potential. The perturbation theory around each one of these vacuum states is determined by an effective action given by Eq. (25.208):

$$S_{\text{eff}} = S_{\text{YM}} + \frac{i\theta}{32\pi^2} \int d^4x F\tilde{F}, \qquad (25.214)$$

where we have used the expression (17.37) for the winding number ν. Our old perturbation theory corresponds to the particular value $\theta = 0$. In other words, a Yang–Mills theory has a more complex structure than naïvely anticipated; namely, it contains a second 'hidden' coupling constant, the parameter θ. We could be tempted to ignore the new term in (25.214) because, as shown in Eq. (17.36), it is equal to the integral over all space of a four-divergence. However, our instanton analysis proved that the assumption that all fields always vanish at infinity is not correct for a gauge theory.

A final remark: $F\tilde{F}$ is a pseudoscalar; therefore, for $\theta \neq 0$ this term violates parity. Since it conserves charge conjugation, it violates CP and, consequently, time-reversal invariance. This is also obvious from the presence of the i-factor. Since strong interactions are known to conserve CP and T, it means that experiments should put severe bounds on the allowed values of θ for QCD. We will discuss this point shortly.

25.4.5.5 *Instantons in QCD*

The only new element that QCD introduces in the discussion presented so far is the presence of quark fields. They do not affect the instanton solutions because configurations such as the one shown in Eq. (25.203) are still finite action solutions of the classical equations of motion corresponding to zero values of the quark fields. However, they do affect the conclusions we reached concerning the structure of the vacuum state as we will explain here.

Let us first consider the simple case of exactly massless quarks. The important points were already presented in the discussion of the chiral anomaly in section 17.3. They are summarised in Eqs. (17.24), (17.30), (17.33), and (17.37). The first three show that a chiral transformation on a massless fermion field results in the addition in the effective Lagrangian density of a term proportional to the trace of $F\tilde{F}$. Since this is precisely the effect of taking into account the functional integral of field configurations with non-vanishing winding number, as shown in Eq. (25.214), we conclude that in QCD with massless quarks, the effective θ coupling constant can be chosen equal to 0 and all instanton effects disappear. How is this compatible with the result shown in Eq. (25.213) which led us to conclude that the $|\theta >$ states were all different? The answer is given by Eq. (17.37), which relates the winding number with the zero modes of the covariant Dirac operator. It shows that for $\nu \neq 0$, the covariant Dirac operator in a gauge field with non-vanishing winding number has at least one zero mode. We have already seen that the constant K which appears in the energy density of the $|\theta >$ vacua is related to the determinants we obtain when computing the quadratic fluctuations in the functional integral. For fermion fields this determinant appears in the numerator. The presence of a zero mode means that the determinant vanishes and hence, $K = 0$. All $|\theta >$ vacua have vanishing energy density. Note that for this result it suffices to have just one quark flavour with vanishing mass.

We are tempted to interpret this result as meaning that the theory possesses a family of ground states with degenerate energies. To prove that this is indeed the case we must show that the $|\theta >$ states are genuinely different states and they are not just proportional to each other. The simplest way to show it, which will also exhibit the underlying physics, is to compute the expectation value $< \theta|\sigma_\pm(x)|\theta >$ where we choose $\sigma_\pm$ to be gauge invariant chiral operators. For a single-quark field with vanishing mass the simplest choice is

$$\sigma_\pm(x) = \frac{1}{2}\overline{\psi}(x)(1 \pm \gamma_5)\psi(x). \qquad (25.215)$$

Under a chiral transformation they satisfy

$$\delta\sigma_\pm(x) = \mp 2i\sigma_\pm(x). \qquad (25.216)$$

In Problem 25.17 we compute the expectation value $< \theta|\sigma_\pm(x)|\theta >$ and show that it is different from 0. We conclude that this theory satisfies all the conditions we set in Chapter 15 for the phenomenon of spontaneous symmetry breaking. The $U(1)$ chiral symmetry is spontaneously broken and the $|\theta >$ vacua form a family of degenerate ground states. What is remarkable is that we discovered this phenomenon by performing a first-order semi-classical calculation. The classical theory shows no sign of it. It is the first time that we encounter a field theory which exhibits the phenomenon of spontaneous symmetry breaking and this phenomenon is manifest in finite order perturbation theory. By contrast, we can think of the $SU(2)$ chiral symmetry of QCD. We argued in section 25.4.2 that it is spontaneously broken and the pions are the approximate Goldstone bosons. But there is no way to study this phenomenon at any finite order of perturbation theory. We tried to explain this failure by saying that it happens at a low mass scale in which the effective coupling constant is large and perturbation theory breaks down. The interaction becomes strong and produces quark–anti-quark bound states, a phenomenon clearly outside perturbation theory. Therefore, we must admit that, in contrast to the Abelian part, the spontaneous breaking of the non-Abelian chiral symmetry remains a conjecture.

This brings us to the question we asked in 25.4.2: granted that the $U(1)$ chiral symmetry is spontaneously broken, why there is no physical Goldstone boson? We have now the tools to answer this question quantitatively. Let us start by looking at the gauge invariant Green function $< \theta|\sigma_+(x)\sigma_-(0)|\theta >$. In the functional integral we must include the $A_\mu = 0$ configuration, as well as the one with one instanton and one anti-instanton. The first gives the ordinary perturbation theory, which, at lowest order, is the diagram with one quark loop. It has a cut but no poles. The second factors into the product

$< \theta | \sigma_+ | \theta > < \theta | \sigma_- | \theta >$. We have computed these expressions in Problem 25.17 and, again, they have no zero-mass pole. On the other hand, if we look at the gauge-dependent matrix element $< \theta | G_\mu(x) \sigma_-(0) | \theta >$, the only configurations that contribute are those with $\nu = 1$. Therefore, taking into account Eqs. (17.36) and (17.37), we obtain

$$\int d^4x \partial^\mu < \theta | G_\mu(x) \sigma_-(0) | \theta > = 32\pi^2 < \theta | \sigma_- | \theta > \neq 0. \qquad (25.217)$$

Since this relation is valid for all k, we conclude that the Fourier transform of the matrix element $< \theta | G_\mu(x) \sigma_-(0) | \theta >$ does have a $k^2 = 0$ pole whose residue is proportional to the expectation value $< \theta | \sigma_- | \theta >$. This analysis answers our main question and solves the mystery of the missing Goldstone boson. A $k^2 = 0$ pole does appear but only in gauge-dependent Green functions.

If QCD had at least one massless quark flavour, this would have been the end of the story. The θ-term in the effective Lagrangian (25.214) could be put equal to 0 and no observable *CP* violation would result. The lightest quark seems to be the u-quark and this is the natural candidate for being massless. In the framework of the standard model this would mean that the parameter G_u^1 in the Yukawa Lagrangian (25.33) is equal to 0. In reality, the mass of a light quark is not a directly measurable physical quantity because the quarks do not appear as physical states. The masses that appear in the Lagrangian are parameters which can be determined only indirectly. The trouble is that all indirect determinations, such as lattice simulations, chiral symmetry perturbation approximations, current algebra calculations, all seem to indicate a non-vanishing mass for all quark flavours. If this is the case, the θ-vacua are all different and the θ-term is observable. This is known as *the strong CP problem*. Can we set by hand $\theta = 0$? If QCD was the only interaction the quarks had, this would have been an acceptable choice because radiative corrections would not force us to violate this symmetry. However, weak interactions are known to violate *CP* and as a result, they produce an effective θ-term. It turns out that the strongest constraint comes from the absence of a neutron electric dipole moment. It can be translated as an upper bound for θ on the order of $\theta < 10^{-9}$, several orders of magnitude smaller than what we could naïvely expect from the electroweak radiative corrections. A deeper explanation is clearly needed.

Roberto Peccei and Helen Quinn proposed to enlarge the model in such a way as to restore a $U(1)$ axial symmetry, even in the presence of massive quarks. For the standard model a simple way to achieve this goal is to consider two BEH doublets, one which is coupled only to up-type quarks and a second one for the down-type. Now we have an extra freedom, namely to perform phase changes of the scalar fields. It is easy to check that the θ-term can be absorbed and no *CP* violation is induced. However, as Steven Weinberg and Franck Wilczek have observed, this $U(1)_{PQ}$ is spontaneously broken at the classical level, since both BEH doublets acquire a non-zero vacuum expectation value. The resulting Goldstone boson does not remain massless, because of the instanton effects. It follows that such a theory contains a pseudo-scalar pseudo-Goldstone boson which is called *an axion*. Its detailed properties, mass and couplings, depend on the particular model. The simplest two-doublet model is already ruled out by experiment, but other models are not. An active experimental programme is being currently pursued aiming to discover and, possibly, study axions.

25.5   Problems

Problem 25.1 The properties of $SU(2)$ spinors. Show that if Ψ denotes an $SU(2)$ doublet, the complex conjugate spinor Ψ^* belongs to an equivalent representation of $SU(2)$ and find the equivalence matrix. This is a particular example of the general property of $SU(2)$ for which all representations are real. Such a property is lost for higher unitary groups.

Problem 25.2 The general Kobayashi–Maskawa matrix. Given a theory with n_f quark families find the number of physically measurable arbitrary parameters which appear in the resulting Kobayashi–Maskawa matrix. Verify that for n_f=3, we obtain the three Euler angles and a phase.

Problem 25.3 Stability of the $SU(2) \times U(1)$ electroweak theory. The standard electroweak model has 17 parameters. An interesting question is to examine whether they are all truly independent, or whether we can impose some relation among them. In particular, people have looked at the possibility of determining the ratio m_Z/m_S, i.e. expressing the mass of the BEH scalar in terms of that of the gauge bosons. Using the renormalisation group analysis we presented in section 19.3, show that a relation of the form $m_Z/m_S = C$, with C a fixed constant, will not be stable under renormalisation.

Problem 25.4 Higgs hunting. One of the main items in the agenda of experimental high energy physics has been the study of the mechanism underlying the spontaneous symmetry breaking of the theory.

1. In the standard model find the dominant decay modes of the BEH particle as a function of its mass.
2. Assuming $m_S \approx 126$ GeV compute the branching ratio for the two-photon decay mode $B_{\gamma\gamma} = \frac{\Gamma(S \to \gamma\gamma)}{\Gamma(S \to all)}$, where the total width can be estimated by the dominant decay mode.

Problem 25.5 In the symmetry breaking pattern of the standard model we found that, in the classical approximation, we have the relation (25.27) between the gauge boson masses and the mixing angle. Find how this relation is modified if we add a second scalar field, belonging to a triplet of weak $SU(2)$, which takes a vacuum expectation value v'.

Problem 25.6 The end-point spectrum in β-decay. Compute the energy spectrum of the emitted electron in neutron β-decay in the vicinity of the 'end-point', i.e. in the region in which the electron takes almost all the available energy. Show that the form of the spectrum is very sensitive to a possible non-vanishing value of the neutrino mass m_ν. Given the very low value of the momentum carried by the leptons compared to the W mass, it is sufficient to use the Fermi theory.

Problem 25.7 Neutrino oscillations.

1. We start with a two-flavour model, ν_e and ν_μ. We assume that neutrinos are massive Dirac particles and the flavour numbers are not separately conserved. Let us call the mass eigenstates $|\nu_1 >$ and $|\nu_2 >$ with mass eigenvalues m_1 and m_2, respectively. At $t = 0$ a ν_e is produced with momentum p and travels through the vacuum. Find the probability that it interacts like a ν_μ at time t. Show that it exhibits an oscillatory behaviour which is absent if the two masses m_1 and m_2 are equal. Show that the oscillation amplitude depends on a single mixing parameter.

2. Repeat the exercise with the physical case of three flavours. Find the number of parameters which determine the mixing, when the neutrinos are described by (i) Dirac fields and (ii) Majorana fields.

   Hint: The difference between the two is that Majorana fields are real and we cannot redefine their phases.

3. The solar neutrinos which were detected on earth and first pointed to the phenomenon of neutrino oscillations are produced by nuclear reactions in the interior of the sun. The first such reaction, which triggers all the others, is hydrogen fusion: proton+proton $\rightarrow$ deuteron+positron+neutrino. All these are low-momentum transfer reactions, so only electron neutrinos are produced. On the other hand, the neutrinos travel a long way inside the sun, which is a relatively high density medium, so the approximation we made in the first part of this problem of vacuum propagation is not justified. The purpose of this question is to model the presence of matter.

   (i) Show that in the standard model, electron neutrinos interact with matter differently from the other two species and draw the corresponding diagrams. What is the difference between neutrinos and anti-neutrinos?

   (ii) Let us first make the simplifying assumption of a medium with a constant matter density. If H is the total Hamiltonian which governs neutrino propagation, we can write $H = H_V + H_m$. H_V is the part which describes the vacuum propagation and contains the kinetic energy and mass terms. As we noted earlier, it is diagonal in the basis of the mass eigenstates $|\nu_1 >$, $|\nu_2 >$ and $|\nu_3 >$. Show that H_m is diagonal in the basis of flavour eigenstates $|\nu_e >$, $|\nu_\mu >$, and $|\nu_\tau >$. Show that on this basis and in the Fermi theory approximation, in which the vector boson propagator is replaced by a point interaction, it can be written, up to an overall factor, as $H_m \propto \text{diag}\,[(1 + C\rho_e),\ 1,\ 1]$ with ρ_e the density of electrons and C a numerical constant which changes sign between neutrinos and anti-neutrinos.

   (iii) Study the phenomenon of oscillations in this medium.

Remark: *The constant density approximation is not justified for the sun where the density changes from a maximum in the centre to 0 at the surface. The general case can only be solved numerically since it depends on the solar model. There is, however, an approximation, which is quite realistic and allows for exact solutions. It is the adiabatic approximation valid when density variations are small over one oscillation length. In this case the problem can be solved going to a variable basis of eigenstates, exactly analogous to the phenomenon of spin precession in a slowly varying magnetic field.*

Problem 25.8 The Georgi–Glashow model. In the very early seventies, the dawn of the gauge theory era, the existence of neutral currents was not yet experimentally confirmed. So theorists looked for models, alternative to the standard one, with no observable neutral currents. The first, and the simplest, was proposed by Howard Georgi and Sheldon Lee Glashow. Although it was very soon ruled out by experiment, it still serves as a toy model for spontaneously broken gauge theories because it has several attractive features. It unifies weak and electromagnetic interactions in a simple gauge group. The electric charge is one of the generators and, since the group is simple, the values of electric charges are automatically quantised. It exhibits phenomena such as the existence of magnetic monopoles and the electric–magnetic duality which we find typically in models going beyond the standard model, some of which we will study in the next chapter.

The model is based on the gauge group $O(3)$. We studied the pattern of spontaneous symmetry breaking in Problem 2.3. Here we want to introduce fermions.

1. We restrict to the lepton world. The first idea is to consider the group $SU(2)$, which is locally isomorphic to $O(3)$, and put the leptons of each family in a doublet. Show that this model is incompatible with the observed weak and electromagnetic interactions of the leptons.

2. Following Georgi and Glashow we enlarge the lepton spectrum by assuming the existence of heavy, as yet unobserved, leptons. For each family we introduce one positively charged and one neutral heavy lepton. Show that we can obtain an acceptable, by 1970 standards, model in which left-handed leptons are put in $O(3)$ triplets and right-handed ones in $O(3)$ singlets. Exhibit the lepton–vector boson couplings after spontaneous symmetry breaking.

It turns out that there is no simple acceptable extension of this model to include quarks. Furthermore, the experimental discovery of weak neutral currents, followed by that of the neutral vector boson, put an end to the search of this kind of models.

Problem 25.9 Using the general results of section 18.3 prove the Georgi–Glashow formula (25.74).

Problem 25.10 Deep inelastic neutrino–nucleon scattering. Repeat the analysis which gave us the expression (25.79) for $W_{\mu\nu}$ in the case of electron–nucleon deep inelastic scattering to the case when the incident lepton is a neutrino.

Hint: Neutrino–nucleon interactions do not conserve parity.

Problem 25.11 Show that the Wilson loop defined in Eq. (25.90) is gauge invariant.

Problem 25.12 Chiral symmetry and β-decay.

1. We write the contribution of the weak axial current in the neutron β-decay as

$$< p(q_2)|A^\lambda|n(q_1) > \; \sim \; \bar{u}_p(q_2)\left[g_A(q^2)\gamma^\lambda + g_P(q^2)q^\lambda\right]\gamma_5 u_n(q_1), \quad (25.218)$$

where u_n and u_p are the Dirac spinors of the initial neutron and the final proton with momenta q_1 and q_2, respectively, and $q = q_1 - q_2$ is the momentum transfer. g_A and g_P are two form factors which, from Poincaré invariance, depend only on q^2. We remind that the proton–neutron mass difference is very small, so the approximation $q^2 \approx 0$ is very good. Apply the principle of P.C.A.C. of Eq. (15.25) to the axial current and find a relation of the form

$$Mg_A(0) \; \sim \; f_\pi g, \quad (25.219)$$

where M is the mass of the nucleon, f_π the pion decay constant given in Eq. (25.97), and g, the pion–nucleon coupling constant defined in Eq. (17.1).

   This relation, known as the Goldberger–Treiman relation, played an important role in the development of the ideas related to chiral symmetry.

2. In studying the non-relativistic scattering theory we found the useful concept of the *scattering length*, which is defined, up to a kinematical factor, as the value of the elastic scattering amplitude at threshold. In this problem we want to compute such quantities for the physically interesting case of the elastic scattering of pions on nucleons.

   We start by introducing the Green functions

$$G^{ab;ij} = \frac{1}{f_\pi^2 m_\pi^4} \int d^4x d^4y < p_2,j|T\left(\partial^\mu A_\mu^a(x)\partial^\nu A_\nu^b(y)\right)|p_1,i >$$
$$(k_1^2 - m_\pi^2)(k_2^2 - m_\pi^2)e^{-ik_1 x}e^{ik_2 y}, \quad (25.220)$$

where a and b are the axial current isotopic spin indices, $a,b = 1,2,3$, i and j those of the initial and final nucleon states, $i,j = 1,2$, and p_1 and p_2 the momenta of the initial and final nucleons. G is a function of the external momenta k_1, k_2, p_1, and p_2, subject to the overall energy–momentum conservation and we have chosen the normalisation factor such that at the limit $k_l^2 \to m_\pi^2$, $l = 1, 2$, G equals the corresponding pion–nucleon scattering amplitude.

Similarly, we define the Green functions

$$G_{\mu\nu}^{ab;ij} = \frac{1}{f_\pi^2 m_\pi^4} \int d^4x d^4y < p_2, j | T \left(A_\mu^a(x) A_\nu^b(y) \right) | p_1, i >$$
$$(k_1^2 - m_\pi^2)(k_2^2 - m_\pi^2) e^{-ik_1 x} e^{ik_2 y} \tag{25.221}$$

$$G_1^{ab;ij} = \frac{-ik_1^\mu}{f_\pi^2 m_\pi^4} \int d^4x d^4y < p_2, j | [A_\mu^a(x), A_0^b(y)] | p_1, i > \delta(x^0 - y^0)$$
$$(k_1^2 - m_\pi^2)(k_2^2 - m_\pi^2) e^{-ik_1 x} e^{ik_2 y} \tag{25.222}$$

$$G_0^{ab;ij} = \frac{1}{f_\pi^2 m_\pi^4} \int d^4x d^4y < p_2, j | [A_0^a(x), \partial^\nu A_\nu^b(y)] | p_1, i > \delta(x^0 - y^0)$$
$$(k_1^2 - m_\pi^2)(k_2^2 - m_\pi^2) e^{-ik_1 x} e^{ik_2 y}. \tag{25.223}$$

(i) Show that they satisfy a Ward identity of the form

$$G^{ab;ij} = k_1^\mu k_2^\nu G_{\mu\nu}^{ab;ij} + G_0^{ab;ij} + G_1^{ab;ij}. \tag{25.224}$$

We want to use this identity to obtain the value of the s-wave pion–nucleon scattering amplitude at threshold, neglecting terms of order m_π/m_N.

(ii) Prove that the first two terms give a vanishing contribution when $\mathbf{k}_1 = \mathbf{k}_2 = 0$.

Hint: For the first term show that the contribution of the pole diagrams to the s-wave amplitude goes to 0 and for the second use P.C.A.C.

(iii) Use the $SU(2) \times SU(2)$ current algebra to show that the identity (25.224) determines the combination of the pion–nucleon scattering lengths which is antisymmetric in the pion isospin indices.

(iv) In the pion–nucleon channel the total isospin can take two values, $|\mathbf{I}| = \frac{1}{2}$ and $|\mathbf{I}| = \frac{3}{2}$. We call the corresponding scattering lengths $a_{1/2}$ and $a_{3/2}$. Show that the previous results can be combined to give

$$a_{1/2} - a_{3/2} \approx 0.3 m_\pi^{-1}; \quad a_{1/2} + 2a_{3/2} \approx 0 \tag{25.225}$$

in good agreement with experiment.

(v) Since the scattering lengths are the values of the corresponding scattering amplitudes at threshold, use the forward dispersion relations and the optical theorem to express the first of these relations in terms of an integral over the difference of the pion-nucleon total cross sections.

In this form, as an integral over cross sections, this relation was derived by Stephen Adler as well as William Weisberger. It is called the Adler–Weisberger relation and gave the first great success of the current algebra ideas. The derivation we present here, which is expressed as a low-energy theorem, is due to Steven Weinberg.

Problem 25.13 Apply the light-cone operator product expansion formalism we developed in section 25.4.3 to the free quark parton model and derive the relations (25.132).

Problem 25.14 Show, by explicit computation, that the continuum limit of the Wilson action of Eq. (25.183) is the Yang–Mills action.

Problem 25.15 Study the homotopy classes of the mapping between two circles, $S^1 \to S^1$, and show that they are labelled by an integer winding number.

Problem 25.16 Study the large volume and large time limits of the Euclidean functional integral we considered in section 25.4.5 and derive the energy of the θ-states given in Eq. (25.213).

Problem 25.17 Compute the expectation value $< \theta | \sigma_\pm(x) | \theta >$, where the chiral operator σ is given in terms of a massless quark field ψ by Eq. (25.215), $\sigma_\pm(x) = \frac{1}{2}\overline{\psi}(x)(1 \pm \gamma_5)\psi(x)$, and $|\theta >$ is the vacuum state defined in section 25.4.5.

26

Beyond the Standard Model

This chapter follows a different spirit from the one that has prevailed in the rest of the book. The physical concepts and models we have presented so far, although often quite abstract, were solidly anchored in experimental results. The mathematical methods we used were motivated to serve a well-defined phenomenological purpose. In this chapter we will break away from this philosophy and we will attempt to present a panorama of theoretical speculations which have dominated research in theoretical high-energy physics over the past decades. The style of presentation will change accordingly, from deductive and pedagogical, appropriate to a text book, to more descriptive, like that of a general review. The purpose is to stimulate interest for further reading rather than to build a convincing sequence of arguments. We will not present a well-established scientific doctrine but a line of research in progress. The reader should be more critical than before.

In high-energy physics we have been extremely fortunate to construct a fundamental theory in remarkable agreement with experiment. In contrast, over the past decades theorists have often worked with very little experimental input. The enormous complexity of modern high-energy or astrophysics experiments has stretched the time between the conception, the design, and the completion of an experiment to decades. As a result, our ideas are necessarily tentative and seem to go in many directions. We try to cover as much of these ideas as possible, but many results will be presented without a full justification and the reader will be referred to the original literature. The choice of the material is, obviously, subjective, but we believe it accurately reflects the general opinions prevailing in the scientific community. With one notable exception. We do not cover at all the subject of quantum gravity and, consequently, all the extremely rich and exciting line of research in local supersymmetry and superstring theory. To present even a partial view of this subject would have required a second volume.

We are aware of the fact that coming experiments may change completely our theoretical prejudices. However, we feel that many of these speculations are sufficiently attractive to deserve our attention. Either in their present form or embedded in a larger scheme, they may form the basis of the theory of tomorrow, the one that experiments will establish. We have been extremely frustrated during all these years with experiments presenting no new directions to explore and we cannot hide our excitement now that the

From Classical to Quantum Fields. Laurent Baulieu, John Iliopoulos and Roland Sénéor.
© Laurent Baulieu, John Iliopoulos and Roland Sénéor, 2017. Published 2017 by Oxford University Press.

long-awaited results are at last in sight. Never in the past were new experimental facilities loaded with so many expectations. We are confident that great and exciting discoveries lay ahead.

26.1 Why

26.1.1 The Standard Model Has Been Enormously Successful

As we have already emphasised, the agreement between the standard model and experiment is truly remarkable. Many new phenomena were discovered following its predictions: the weak neutral currents, the charmed particles as well as those of the heavier flavours, the intermediate vector bosons, the effective strength of the interactions as predicted by the renormalisation group, the presence of gluons among the constituents of hadronic matter, and, finally, the recent discovery of an 'elementary' scalar particle. All these spectacular successes of the standard model are in fact successes of renormalised perturbation theory. Indeed what we have learnt was how to apply the methods which had been proven so powerful in quantum electrodynamics to other elementary particle interactions. But in some cases, we have been able to go beyond perturbation. The beautiful results on the spectrum of light hadrons which were obtained by the QCD calculations on a space–time lattice show clearly that what we call 'the standard model' is, in fact, *a fundamental theory*. The remarkable quality of modern high-energy physics experiments has provided us with a large amount of data of unprecedented accuracy. All can be fit using the standard model with no free parameters beyond those which are directly determined by experiment. Figure 26.1 indicates the overall quality of such a fit. It shows a variety of physical quantities, masses, decay widths, asymmetries, etc., which can be measured experimentally and determined theoretically in the framework of the standard model. The theoretical determination can be either part of the global fit or indirect through radiative corrections. There is one measurement which lays beyond two standard deviations away from the theoretical predictions, but it is not sufficient to conclude that there is evidence for new physics rather than an accident, or the result of incorrectly combining incompatible experiments. To this list we should add some low-energy high precision measurements, such as the anomalous magnetic moments of the electron and the muon, which could also indicate the presence of possible new physics beyond the standard model.

Another impressive fit concerns the strong interaction effective coupling constant as a function of the momentum scale (Fig. 25.13). This fit already shows the importance of taking into account the radiative corrections, since, in the tree approximation, α_s is, obviously, a constant. Similarly, Fig. 26.2 shows the importance of the weak radiative corrections in the framework of the standard model. Because of the special Yukawa couplings, the dependence of these corrections on the fermion masses is quadratic, while it is only logarithmic in the BEH mass. The ϵ parameters are designed to disentangle the two. The ones we use in Fig. 26.2 are defined by

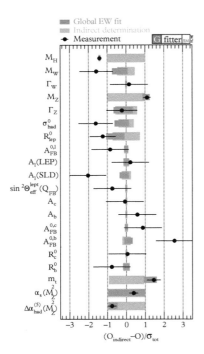

Figure 26.1 *Various physical quantities and their agreement with the Standard Model computations (©: M. Baak et al, G-Fitter Group Collaboration, arXiv: 1407.3792).*

$$\epsilon_1 = \frac{3 G_F m_t^2}{8\sqrt{2}\pi^2} - \frac{3 G_F m_W^2}{4\sqrt{2}\pi^2} \tan^2\theta_W \ln\frac{m_S}{m_Z} + \dots \tag{26.1}$$

$$\epsilon_3 = \frac{G_F m_W^2}{12\sqrt{2}\pi^2} \ln\frac{m_S}{m_Z} - \frac{G_F m_W^2}{6\sqrt{2}\pi^2} \ln\frac{m_t}{m_Z} + \dots, \tag{26.2}$$

where the dots stand for subleading corrections. As you can see, the ϵ's vanish in the absence of weak interaction radiative corrections; in other words, $\epsilon_1 = \epsilon_3 = 0$ are the values we get in the tree approximation of the standard model but including the purely QED and QCD radiative corrections. We see clearly in Fig. 26.2 that this point is excluded by the data. The latest values for these parameters are $\epsilon_1 = (5.21 \pm 0.08) \times 10^{-3}$ and $\epsilon_3 = (5.279 \pm 0.04) \times 10^{-3}$.

26.1.2 Predictions for New Physics

This great success of the standard model can be interpreted by saying that perturbation theory is very reliable outside the region where strong interactions become important. We have seen in section 25.4.3 that in all known physics this happens only in hadronic processes in which the relevant energy and momentum transfer is comparable to the corresponding typical mass scale. We can use this property to make a qualitative prediction for the expected results of future experiments in the multi-hundred GeV range, such as

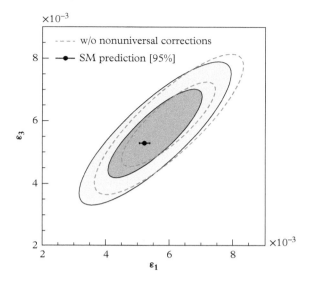

Figure 26.2 *The importance of the weak interaction radiative corrections. The standard model prediction, at 95% confidence level, is included with the corresponding error bar (©: M. Ciuchini et al, arXiv:1306.4644).*

the LHC. The new information is that LHC has indeed found a 'light' scalar particle, in agreement with the prediction based on previous experiments, implying a value of the coupling constant λ on the order of $\frac{1}{6}$. This value lies inside the classically 'acceptable' region:

$$1 > \lambda > 0. \tag{26.3}$$

We have already discussed these limits in the previous chapter. The lower limit for λ comes from the classical stability of the theory. If λ is negative the potential is unbounded from below and there is no ground state. The upper limit comes from the requirement of keeping the theory in the weak coupling regime. If $\lambda \geq 1$ the scalar sector of the theory becomes strongly interacting and we expect to see plenty of resonances and bound states rather than a single elementary particle.

Going to higher orders is straightforward, using the renormalisation group equations. The running of the effective mass is determined by that of λ. Keeping only the dominant terms and assuming $\rho = \log(v^2/\mu^2)$ is small ($\mu \sim v$), we find that

$$\frac{d\lambda}{d\rho} = \frac{3}{4\pi^2}[\lambda^2 + 3\lambda h_t^2 - 9h_t^4 + ...], \tag{26.4}$$

where h_t is the coupling of the scalar boson to the top quark. The dots stand for less important terms, such as the other Yukawa couplings to the fermions and the couplings with the gauge bosons. This equation is correct as long as all couplings remain smaller than 1, so that perturbation theory is valid, and no new physics beyond the standard model becomes important. Now we can repeat the argument on the upper and lower

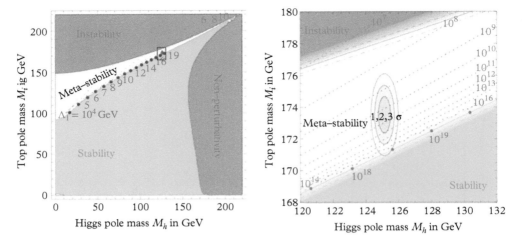

Figure 26.3 *The Standard Model phase diagram in terms of the BEH scalar and top quark masses. It shows the regions of stability, meta-stability and instability of the broken phase vacuum. The dotted lines show the scale in which instability occurs. The diagram assumes no new physics up to the Planck scale. At the right we zoom at the region around the experimental values of M_h and M_t. We see that we are in the meta-stability region. The plot includes radiative corrections. The grading of the colours indicates the estimated theoretical uncertainties (© D. Buttazzo et al arXiv: 1307.3536).*

bounds for λ but this time taking into account the full scale dependence $\lambda(\mu)$. We thus obtain for the BEH mass an upper bound given by the requirement of weak coupling regime ($\lambda(\mu) < 1$) all the way up to the scale μ, and a lower bound by the requirement of vacuum stability ($\lambda(\mu) > 0$), again up to μ. Obviously, the bounds will be more stringent the larger the assumed value of μ. Figure 26.3 shows the regions of stability, meta-stability and instability assuming no new physics up to the Planck scale. It is interesting to notice that the experimental point lies in the meta-stability region, although the estimated life-time of the Standard Model vacuum state exceeds the age of the Universe.

How reliable is this conclusion? These are the standard model predictions, assuming no new physics shows up to the scale μ. We want to argue here that, in fact, the opening of new physical thresholds can be safely predicted already at a scale of a few TeV, although their precise nature cannot.

With the experimental discovery of the BEH scalar boson the electroweak theory is complete. All coupling constants have been measured and they belong to the perturbative regime. Following K. Wilson, we can fix a scale μ and imagine that we integrate over all degrees of freedom with energies above μ. It is the scale up to which the theory can be trusted; in other words, we assume that physics at scales below μ is accurately described by the standard model. We thus obtain an effective field theory describing the light, meaning lighter than μ, degrees of freedom. We can compute this theory in perturbation, but, even non-perturbatively, we can guess its general form. Integrating over the heavy degrees of freedom does not break any symmetry, so the effective theory will be a sum over all operators built out of the light fields consistent with the symmetries of the standard model. The dependence of the coefficients on μ can be deduced from dimensional analysis. If μ is larger than any other mass scale of the theory, for the standard model this means μ much larger than the top quark mass, the coefficient of an operator $\mathcal{O}_i$ with dimensions d_i, is proportional to $\mu^{(4-d_i)}$, up to logarithmic corrections. Therefore, we can distinguish three classes of operators:

(i) Operators whose dimension d_i is larger than 4. Their contribution decreases as a power of μ, so they become irrelevant for large μ.

(ii) Operators with $d_i = 4$. They are precisely the operators appearing in the standard model Lagrangian and they receive logarithmic corrections in μ. These corrections are described by the equations of the renormalisation group. They are sometimes called 'marginal operators'.

(iii) Operators with $d_i < 4$. Their coefficients grow like positive powers of μ, so they become dominant at large scales. In the standard model there are only two such operators: the unit operator **1** with dimension equal to 0 and the operator Φ^2, where Φ is the field of the BEH scalar, with dimension equal to 2. The first contributes only to the induced cosmological constant which, in the absence of gravitational interactions, is not observable.[1] We conclude that *the only dominant operator of the standard model is the mass term of the scalar boson.* It receives corrections which grow quadratically with the energy scale.

This problem is often referred to as *the hierarchy problem* because it was first analysed in the framework of the so-called 'grand unified theories', which we will present in the next section. However, we want to emphasise here that it is a genuine instability of all generic quantum field theories involving scalar fields. This quadratic dependence on the scale should not be confused with the quadratic divergence, which is present in scalar theories. The latter is removed by renormalisation, while the former affects the finite mass counter-term, which must be adjusted, order-by-order, in order to cancel the scale dependence. This adjustment will be more and more severe the higher the assumed scale is.

This argument allows us to introduce the concept of *naturalness*. The underlying idea is that all physical theories are effective theories valid up to a certain scale, because we can never assume that we know physics at all scales. A quantum field theory will be called natural if the values of its parameters depend only logarithmically on this large energy scale. According to this definition, the standard model is not natural. It must be replaced by a different theory above a certain scale μ. Given the fact that the BEH scalar has a rather strong coupling to the top quark, the μ^2 dependence induced by a t-quark loop will require very fine adjustments if the scale is much larger than 1 TeV,

$$(m_S^2)_{\text{eff}} = m_S^2 + C\alpha_{\text{eff}}\mu^2, \tag{26.5}$$

---

[1] We have also the fermion mass terms which are dimension three operators, but they are protected by the symmetries of the model.

where C is a calculable numerical coefficient of order 1 and α_{eff} some effective coupling constant. In practice it is dominated by the large coupling to the top quark. The moral of the story is that the BEH particle cannot remain light unless there is a precise mechanism to cancel this quadratic dependence on the high scale. We will see later how such cancellation mechanisms can be implemented and what kind of new physics they imply.

26.1.3   Unsolved Problems of the Standard Model

Despite its enormous success, there are several reasons to suspect that the gauge theory of the standard model cannot be considered as a truly fundamental theory. We have already mentioned some of its shortcomings. One was the family problem: why do we observe three, apparently similar, families of elementary fermions? Another is the problem of masses. It is hard to believe that all these widely spread mass values are arbitrary parameters in a fundamental theory. This problem existed already with mass values such as m_e and m_t. It is accentuated with the values of the neutrino masses. A third very important problem is that $U(1) \times SU(2) \times SU(3)$ is not a unified theory at all. Each group factor comes with its own coupling strength. Even worse is the presence of $U(1)$ because it allows for any number of coupling constants. We have already explained that in a non-Abelian group the coupling constant is fixed by the gauge boson self-coupling and it must be the same, except for Clebsch–Gordan coefficients, for every matter multiplet. For $U(1)$, however, this is not so. In other words, the present theory does not explain why electric charge appears to be quantised and we do not see particles with charge πe. For the standard model the observed very precise equality (up to one part in 10^{20}) of the electric charges of the proton and the positron seems to be accidental. Last, but not least, the standard model leaves out the gravitational forces. Although, at present energies, the latter are very weak, we expect a fundamental theory to describe *all* fundamental interactions.

For all these reasons, and others that we can easily add, theorists tried to imagine schemes that go beyond the standard model. There exists an old theoretical prejudice which says that a better theory is a more symmetric one, so it is not surprising that most of these extensions involve some enlargement of the symmetries of the standard model. In this chapter we will briefly present some of these speculations.

26.2   Grand Unified Theories

26.2.1   Generalities

The hypothesis of grand unification states that $U(1) \times SU(2) \times SU(3)$ is the remnant of a larger, simple, or semi-simple group G, which is spontaneously broken at very high energies. The scheme looks like

$$G \xrightarrow{M} U(1) \times SU(2) \times SU(3) \xrightarrow{m_W} U(1)_{e.m.} \times SU(3), \tag{26.6}$$

where the breaking of G may be a multistage one and M is one (or several) characteristic mass scale(s).

Two questions immediately arise concerning this idea:

(i) Is it possible? In other words, are there groups which contain $U(1) \times SU(2) \times SU(3)$ as a subgroup and which can accommodate the observed particles?

(ii) Does it work? i.e. is the observed dynamics compatible with this grand unification idea?

We will try to answer each one of these questions separately following the approach introduced first by Howard Georgi and Sheldon Glashow.

We first observe that G must contain electromagnetism; i.e. the photon must be one of the gauge bosons of G. This is part of the requirement that G contains the group of the standard model. Another way to say the same thing is to say that the electric charge operator Q must be one of the generators of the algebra of G. Since G is semi-simple, all its generators are represented by traceless matrices. It follows that in any representation of G, we must have

$$\text{Tr}(Q) = 0. \tag{26.7}$$

In other words, the sum of the electric charges of all particles in a given representation vanishes.

For simplicity, let us make a first assumption. The 15 (or 16) spinors of a family fill a representation, not necessarily irreducible, of G; i.e. we assume that there are no other, as yet unobserved, particles which sit in the same representation. Property (26.7), together with the above assumption, have a very important consequence. As we have remarked, the members of a family satisfy (26.7) because the sum of their charges vanishes. This, however, is not true if we consider leptons or quarks separately. Therefore, each irreducible representation of G will contain both leptons and quarks. This means that there exist gauge bosons of G which can change a lepton into a quark, or vice versa. We conclude that a grand unified theory, which satisfies our assumption, cannot conserve baryon and lepton numbers separately. This sounds disastrous, since it raises the spectrum of proton decay. The amplitude for such a decay is given by the exchange of the corresponding gauge boson and, therefore, it is of order M^{-2}, where M is the gauge boson's mass. The resulting proton lifetime will be of order

$$\tau_p \sim \frac{M^4}{m_p{}^5}. \tag{26.8}$$

Using the experimental limit (for particular decay modes) of a few times 10^{33} years, we can put a lower limit on M:

$$M \geq 10^{16} \text{GeV}. \tag{26.9}$$

Grand unification is not a low-energy phenomenon!

26.2.2   The Simplest GUT: $SU(5)$

In this section we will answer the first question by giving a specific example of a group G which satisfies our requirements.

$U(1) \times SU(2) \times SU(3)$ corresponds to a Lie algebra of rank 4 (i.e. there are four generators which commute: one of $U(1)$, one of $SU(2)$, and two of $SU(3)$). Therefore, let us first look for a grand unification group of rank 4. We list all possible candidates:

$$[SU(2)]^4, [SO(5)]^2, [G_2]^2, SO(8), SO(9), Sp(8), F_4, [SU(3)]^2, SU(5).$$

The first two are excluded because they have no $SU(3)$ subgroup. The next five admit no complex representations; therefore, they cannot accommodate the observed families where, as we already saw, the right- and left-handed particles do not transform the same way. (We again assume that no unobserved fermions will complete a given representation). Finally, in $SU(3) \times SU(3)$ quarks and leptons must live in separate representations because the leptons have no colour. But $\Sigma Q_{\text{quarks}} \neq 0$ and the same is true for leptons. This leaves us with $SU(5)$ as the only candidate of a grand unified theory (GUT) group of rank 4. It is the simplest and, although, as we will see, it has many shortcomings, it can be considered as the 'standard model' of grand unification.

The gauge bosons belong to the 24-dimensional adjoint representation. It is useful to decompose it into its $SU(2) \times SU(3)$ content. We find that

$$24 = [(2,3) \oplus (2,\bar{3})] \oplus (1,8) \oplus [(3,1) \oplus (1,1)]$$
$$\begin{array}{ccc} \downarrow & \downarrow & \downarrow \\ \begin{pmatrix} X \\ Y \end{pmatrix} \begin{array}{l} Q = 4/3 \\ Q = 1/3 \end{array} & \text{gluons} & W^\pm, Z^0, \gamma, \end{array} \tag{26.10}$$

where the first number denotes the $SU(2)$ and the second the $SU(3)$ representation. The known vector bosons can be identified with the eight gluons of QCD in the $(1,8)$ piece (a singlet of $SU(2)$ and an octet of $SU(3)$), as well as the electroweak gauge bosons W, Z, and γ in the $(3,1) \oplus (1,1)$ piece. We are left with 12 new ones, called X and Y, with electric charges $\frac{4}{3}$ and $\frac{1}{3}$, respectively, which transform as a doublet of $SU(2)$ and a triplet and anti-triplet of $SU(3)$. They must be heavy, according to the limit (26.9).

Let us now come to the matter-field assignment. We will try to put all the two-component spinors of a family in a representation (not necessarily irreducible) of $SU(5)$. But before doing so, we observe that all gauge couplings, being vectorial, conserve helicity. Therefore, we cannot put right- and left-handed spinors in the same representation. We go around this problem by replacing all right-handed spinors with the corresponding left-handed charge conjugate ones. A quick glance at the representation table of $SU(5)$ suggests to use each family in order to fill two (or three, if a right-handed neutrino exists) distinct, irreducible, representations: the $\bar{5}$ and the 10. Their $SU(2) \otimes SU(3)$ content is

$$\bar{5} = (1,\bar{3}) \oplus (2,1) \tag{26.11}$$

$$10 = (2,3) \oplus (1,\bar{3}) \oplus (1,1). \tag{26.12}$$

The identification is now obvious. We often write these representations as a five-vector and a 5×5 antisymmetric matrix. Comparing the quantum numbers in a family with (26.11) and (26.12) and using (26.7), we find that

$$\bar{5} = \begin{pmatrix} d^c_1 \\ d^c_2 \\ d^c_3 \\ e^- \\ -\nu_e \end{pmatrix}_L = \psi_{L_a} \qquad 10 = \begin{pmatrix} 0 & u^c_3 & -u^c_2 & -u_1 & -d_1 \\ & 0 & u^c_1 & -u_2 & -d_2 \\ & & 0 & -u_3 & -d_3 \\ & & & 0 & -e^c \\ & & & & 0 \end{pmatrix}_L = \psi^{ab}_{\ L}. \tag{26.13}$$

If we have a right-handed neutrino, it must be assigned to the singlet representation of $SU(5)$. This is an unpleasant feature. In the absence of a ν_R we could say that the choice of $SU(5)$ was 'natural', in the sense that it is the only group with an acceptable 15-dimensional representation (although not an irreducible one). As we will see, with 16 dimensions, other choices are aesthetically more appealing. A technical remark is that it is important to note that the sum of these representations is anomaly-free.

Let us finally study the symmetry breaking system. The first step goes through a 24-plet of scalars $\Phi(x)$. It is convenient to represent the 24 as a 5×5 traceless matrix. The vacuum expectation value which breaks $SU(5)$ down to $U(1) \times SU(2) \times SU(3)$ is proportional to the diagonal matrix

$$\lambda_{24} = \frac{1}{\sqrt{15}} \begin{pmatrix} 1 & & & & \\ & 1 & & & \\ & & 1 & & \\ & & & -3/2 & \\ & & & & -3/2 \end{pmatrix}. \tag{26.14}$$

$SU(3)$ is defined to act on the upper three components of the five-dimensional space and $SU(2)$ on the lower two. The potential for the $\Phi(x)$ field can be written as

$$V(\Phi) = -\frac{1}{2}m^2 \text{Tr}(\Phi^2) + \frac{h_1}{4}[\text{Tr}(\Phi^2)]^2 + \frac{h_2}{2}\text{Tr}(\Phi^4). \tag{26.15}$$

The vacuum expectation value of Φ is determined by the minimum of $V(\Phi)$. It is easy to show (see Problem 15.1) that for h_1 and h_2 positive, this minimum is precisely $V\lambda_{24}$ with

$$V^2 = m^2 \left[h_1 + \frac{7}{15}h_2 \right]^{-1}. \tag{26.16}$$

Note that if $h_2 < 0$ with $h_1 > 0$ the direction of breaking is instead $SU(5) \rightarrow U(1) \times SU(4)$.

Can we use the same 24-plet in order to obtain the second breaking of the standard model? The answer is no for two reasons. First, the 24 does not contain any $(2, 1)$ piece (see Eq. (26.10)) which is the one needed for the $U(1) \times SU(2) \rightarrow U(1)_{e.m.}$ breaking. Second, the 24 does not have the required Yukawa couplings to the fermions. Indeed with the $\bar{5}$ and 10 assignments the fermions can acquire masses through Yukawa couplings with scalars belonging to one of the representations in the products

$$\bar{5} \otimes 10 = 5 \oplus 45 \tag{26.17}$$

$$10 \otimes 10 = \bar{5} \oplus \bar{45} \oplus \bar{50}. \tag{26.18}$$

We see that the 24 is absent while the 5 looks promising. If $H(x)$ is a five-plet of scalars, the complete potential of the scalar fields is

$$V_S = V(\Phi) + V(H) + V(\Phi, H) \tag{26.19}$$

with $V(\Phi)$ given by (26.15) and

$$V(H) = -\frac{1}{2}\mu^2 H^\dagger H + \frac{1}{4}\lambda(H^\dagger H)^2 \tag{26.20}$$

$$V(\Phi, H) = \alpha H^\dagger H \mathrm{Tr}(\Phi^2) + \beta H^\dagger \Phi^2 H. \tag{26.21}$$

We can show that for an appropriate range of the parameters m^2, μ^2, h_1, h_2, λ, α, and β, we obtain the desired breaking

$$< \Phi >_0 \sim V \begin{pmatrix} 1 & & & & \\ & 1 & & & \\ & & 1 & & \\ & & & -3/2 - \epsilon/2 & \\ & & & & -3/2 + \epsilon/2 \end{pmatrix} \tag{26.22}$$

$$< H >_0 \sim v \begin{pmatrix} 0 \\ 0 \\ 0 \\ 0 \\ 1 \end{pmatrix}. \tag{26.23}$$

The small number ϵ in (26.19) is due to the mixed terms $V(\Phi, H)$ in the potential and it causes a breaking of $SU(2) \times U(1)$ already from the vacuum expectation value of Φ. We must have $\epsilon \ll 1$; otherwise the breaking of the standard model would have been of the same order as that of $SU(5)$. Using the potential (26.19) we find that

$$\epsilon = \frac{3\beta v^2}{20 h_2 V^2} + O\left(\frac{v^4}{V^4}\right), \tag{26.24}$$

which means that ϵ must be of the order 10^{-28}. It is hard to see how such a number may come out for generic values of the coupling constants. This is part of the naturalness problem which we presented in the previous section. As we will explain in the next section, it plagues all grand unified theories. In the case of $SU(5)$ this problem has two aspects. The first is the general problem of the two widely separated symmetry breaking scales. We expect to have $V^2 \sim m_{\Phi}^2$ and $v^2 \sim m_H^2$. But the presence of the mixed terms in $V(\Phi, H)$ induces a $(m_H^2)_{\text{eff}}$ on the order of V^2, unless the parameters of the potential are very finely tuned. The second is related to the five-plet of scalars H which, under $U(1) \times SU(2) \times SU(3)$, is split as shown in Eq. (26.11). The $SU(2)$ doublet is used for the electroweak breaking and must have a mass on the order of v^2. The $SU(3)$ triplet components, however, can mediate baryon number violating transitions and should be superheavy on the order of V^2. Again there is no natural way to obtain such a doublet–triplet splitting without a fine tuning of the parameters in the potential.

The fermion masses are due to the vacuum expectation value of H. Looking back at the assignment (26.13) we see that the up-quarks take their masses through (26.18) while the down-quarks and the charged leptons through (26.17).

This discussion answers the first question; namely it shows that there exist groups which have the required representations to be used for grand unification. Let us now turn to the second question, namely the study of the dynamical consequences of GUTs.

26.2.3 Dynamics of GUTs

26.2.3.1 Tree-Level SU(5) Predictions

Let us first examine the dynamical predictions of $SU(5)$ at the Lagrangian level without taking into account higher order effects. There are several such predictions:

(i) The first concerns the coupling constants. $SU(5)$ is a simple group and hence it has only one coupling constant g. On the other hand in nature we observe three distinct ones, g_1, g_2, and g_3, corresponding to each one of the factors of the standard model $U(1) \times SU(2) \times SU(3)$. The naive prediction would be $g_1 = g_2 = g_3$. However, we must be more careful with the relative normalisations. For non-Abelian groups, like $SU(2)$, $SU(3)$, or $SU(5)$, the normalisation of the generators is fixed by the algebra, which is a non-linear relation. So the question arises only for $U(1)$. In the standard model the $U(1)$ generator Y is related to the electric charge and the third component of weak isospin T_3 by $Q = T_3 + 1/2 Y$; see the relation (25.11). The factor $\frac{1}{2}$ was chosen for historical reasons. For the embedding of $U(1) \times SU(2) \times SU(3)$ into $SU(5)$, all generators must be normalised the same way. Let us choose the normalisation by requiring $\text{Tr}(\mathcal{T}_i \mathcal{T}_j) = R\delta_{ij}$, where R is a constant which may depend on the representation we use to compute the trace, but it is independent of i and j. Let us now compute $\text{Tr}(T_3^2)$ using, for example, the electron family. We find $\text{Tr}(T_3^2) = 2$. Similarly we find $\text{Tr}[(1/2 Y)^2] = \frac{10}{3}$. Therefore, we see that for the embedding, the $U(1)$

generator must be rescaled by $1/2 Y \to c Y$ with, $c^2 = \frac{5}{3}$. Therefore, the tree-level prediction of any grand unified theory based on a simple group is

$$g_2 = g_3 = g \qquad \sin^2\theta_W = \frac{g_1{}^2}{g_1{}^2 + g_2{}^2} = \frac{g^2/c^2}{g^2/c^2 + g^2} = \frac{3}{8}. \qquad (26.25)$$

(ii) The second concerns Fermion masses. Fermion masses are generated through the same mechanism as in the standard model, i.e. through Yukawa couplings with scalar fields. Therefore, they depend on the particular BEH system one assumes. In the minimal $SU(5)$ model with only a 5-plet of scalars we see in Eqs. (26.17) and (26.18) that we have two independent coupling constants for each family. The up-quarks take their masses through (26.18), while the down-quarks and the charged leptons through (26.17). This last property implies the relations

$$m_d = m_e \qquad\qquad m_s = m_\mu \qquad\qquad m_b = m_\tau. \qquad (26.26)$$

It is obvious that these predictions are lost if we assume a more complicated scalar system, for example by including higher dimensional representations.

(iii) The third concerns baryon and lepton number violation. X, Y, or heavy scalar boson exchanges lead to baryon and lepton number violation. In Fig. 26.4 we depict some diagrams contributing to proton decay. In $SU(5)$ the main decay mode is expected to be $p \to \pi^0 e^+$ with a branching ratio on the order of 30–40% followed by $p \to \omega e^+$ or $p \to \rho^0 e^+$. The neutrino modes, such as $p \to \pi^+ \bar{\nu}$, are expected to be rare ($\sim 10\%$ or less). Bound neutrons are also expected to decay with $n \to \pi^- e^+$ being the dominant mode. All these decay modes are easily detectable. The overall lifetime depends on the masses of the superheavy gauge bosons X and Y (see Eq. (26.8)).

As we noted already, baryon and lepton number violation is a general feature of all grand unified theories. A large experimental effort has been concentrated to detect any trace of proton instability. The result is a higher limit on its lifetime. At present, for the easily detectable decay modes, such as the $\pi^0 + e^+$ one, this limit

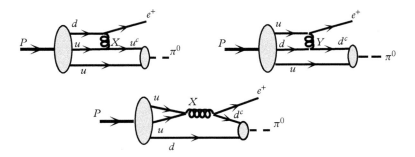

Figure 26.4 *Some diagrams contributing to proton decay.*

is close to 10^{33} years. In the absence of any direct observation of baryon number non-conservation, physicists have tried to see its possible effects in cosmology.

In traditional cosmological models baryons and antibaryons were assumed to have been created in pairs since the Hamiltonian conserved baryon number. The only way to obtain a non-zero baryon number was to put it in by hand as an initial condition. In the so-called 'symmetric' cosmologies it was argued that within some range of temperatures (~ 1 GeV), a phase transition occurs which results in a spontaneous symmetry breaking and thermal radiation becomes unstable against separation of nucleons from antinucleons. The situation was compared to what happens in a ferromagnet where a domain structure appears. According to this view the observed predominance of matter over antimatter is a local effect. The trouble with this theory is that there is no evidence for the presence of large amounts of antimatter anywhere in the universe. The rare traces of antinucleons detected in cosmic rays are compatible with the estimated production of antimatter in particle collisions and no large-scale annihilations have been observed. Nevertheless, this was the accepted doctrine for many years. The reason is that in a symmetric cosmological model, where no net baryon number is put in by hand under the initial conditions, the eventual appearance of baryon excess requires (i) the violation of C and CP invariance, (ii) the violation of baryon number conservation, and (iii) the departure from thermal equilibrium. The necessity for the first two conditions is obvious since otherwise there is no distinction between baryons and antibaryons. The significance of the third one is also very simple. In a stationary universe, where all interactions are in thermal equilibrium, the particle abundances are given by Boltzmann's law which involves only the particle masses. But PCT invariance guarantees that baryons and antibaryons have equal masses and, therefore, no net baryon number can possibly be produced.

Charge conjugation is known to be maximally violated in weak interactions. The violation of CP has been observed already in 1964 in the decays of neutral kaons. The expansion of the universe provides the necessary departure from thermal equilibrium. So out of the three necessary conditions only the second one, the violation of baryon number, has not yet been experimentally verified. Grand unified theories provide a theoretical framework for such a violation and offer the possibility for an estimation of the resulting number of baryons in the universe. It is worth noting that the suggestion for a baryon–antibaryon asymmetry was first made by A. D. Sakharov in 1966, much before the advent of gauge theories. Several detailed calculations have been performed in order to reproduce the observed ratio of baryon number density to entropy density $kn_B/s \sim 10^{-10}$, or, equivalently the baryon to photon ratio $n_B/n_\gamma \sim 10^{-9}$. The results depend on the particular model and the symmetry breaking pattern we assume. In recent models one takes into account also the violation of lepton number. Despite the fact that no absolutely convincing model has been produced, it is remarkable that even a qualitative agreement between theory and observation can be reached.

(iv) Finally we remark that the $SU(5)$ Lagrangian, including the Yukawa couplings, is invariant under a $U(1)$ group of global phase transformations

$$\psi^{ab} \to e^{i\theta}\psi^{ab} \qquad \psi_a \to e^{-3i\theta}\psi_a \qquad H_a \to e^{-2i\theta}H_a, \qquad (26.27)$$

with all other fields left invariant. We can verify that this global symmetry is also anomaly free. The non-zero vacuum expectation value of H seems to break this symmetry spontaneously. This sounds disastrous since it normally leads to the appearance of a truly massless Goldstone boson. However, we are saved because the symmetry is not really broken; it is simply changed. We can check immediately that even after the translation of the scalar fields, the linear combination $\mathcal{J} + 4Y$ remains as a global symmetry, where $\mathcal{J}$ is the generator of (26.27) and Y the $U(1)$ part of $SU(5)$ given by (26.14). The conserved charge of this symmetry is the difference $B - L$ of baryon and lepton numbers. This conservation has some very important consequences. In particular, it gives some precise predictions for the nucleon decay properties. For example, $p \to e^+\pi^0$ or $n \to e^+\pi^-$ are allowed but $n \to e^-\pi^+$ is not. The same is true for $n - \bar{n}$ oscillations which violate $B - L$. As we will see, this property remains true (or nearly true) in many grand unified models.

26.2.3.2   *Higher Order Effects*

The tree-level predictions (26.25) and (26.26) are in violent disagreement with experiment. However, before throwing away the entire scheme, we must understand that these predictions are consequences of the full symmetry and can only be true at energies well above M where the $SU(5)$ breaking can be neglected. In order to compare these predictions with the real world we must extrapolate to present day energies. These extrapolations can be performed using the equations of the renormalisation group, as was first suggested by Howard Georgi, Helen Quinn, and Steven Weinberg. Several assumptions enter in this procedure. The most important one is connected with the very idea of grand unification. We must assume that we know all fundamental physics, in particular the entire spectrum of elementary particles, from our accelerators, to energies of 10^{16} GeV or higher. As we will see, the results are sensitive to the possible existence of new thresholds at high energies.

For simplicity let us ignore the standard model symmetry breaking and study the case $SU(5) \to U(1) \times SU(2) \times SU(3)$. It will be clear that the generalisation to more realistic cases is considerably longer, but straightforward. Since we have only one breaking we will keep only the 24-plet of scalar fields with the vacuum expectation value given by (26.14). Furthermore, all fermions will remain massless. We write the Lagrangian density as

$$\mathcal{L} = -\frac{1}{4}\text{Tr}G_{\mu\nu}G^{\mu\nu} + \bar{\psi}i\slashed{D}\psi + \frac{1}{2}\text{Tr}(D_\mu\Phi D^\mu\Phi) - V(\Phi) \qquad (26.28)$$

with the potential $V(\Phi)$ given by Eq. (26.15). The Lagrangian (26.28) contains four independent parameters, namely the gauge coupling constant g and the three parameters of the scalar potential m^2, h_1, and h_2. After spontaneous symmetry breaking via the translation $\Phi \rightarrow \Phi + (V/\sqrt{2})\lambda_{24}$, we obtain the following mass spectrum:

(i) $M^2 = (5/3)V^2 g^2$ for the vector gauge bosons $G_9,...,G_{20}$.

(ii) $M_3^2 = (1/3)V^2 h_2$ for the scalars $\Phi_1,...,\Phi_8$.

(iii) $M_2^2 = (4/3)V^2 h_2$ for the scalars $\Phi_{21},...,\Phi_{23}$.

(iv) $M_1^2 = 2V^2[h_1 + (7/15)h_2]$ for Φ_{24}, with the vacuum expectation value V given by Eq. (26.16). The eight gauge bosons $G_1,...,G_8$ (the gluons), as well as $G_{21},...,G_{24}$ (the photon, the Z^0 and the W's), remain massless. The scalar components Φ_9 to Φ_{20} will be absorbed by the BEH mechanism.

We choose to renormalise the broken theory by imposing the following seven renormalisation conditions:

(i) Three conditions for the three independent masses M^2, M_1^2, and M_2^2, which will be chosen physical, as the poles of the corresponding propagators.

(ii) Three wave function renormalisation conditions for fermions, vector bosons and scalars, respectively.[2] For example, the condition on the scalars reads

$$\frac{\mathrm{d}}{\mathrm{d}p^2}\Gamma_A^{(2)}(p^2)|_{p^2=-\mu^2} = 1 \tag{26.29}$$

and similarly for fermions and gauge bosons. μ is a subtraction point. However, since the SU(5) symmetry is broken, we must specify on which component of Φ we impose the condition. This is done with the index A which runs from 1 to 24. The conservation of $U(1) \times SU(2) \times SU(3)$ implies that we need only to distinguish four distinct cases ($A=1,...,8; 9,...,20; 21,...,23; 24$). Once the condition is imposed on any of the components, all the other Green's functions become finite and calculable. The same remarks apply to the vector bosons.

(iii) Finally, we must impose one renormalisation condition for the gauge coupling constant. Let us choose it to be the value of the fermion–vector boson 3-point function at $p^2 = -\mu^2$

$$\Gamma_{Aab}^{(3)}(p_1, p_2)|_{p^2=-\mu^2} = ig_A\gamma^\mu T_{ab}^A \quad \text{no summation over } A, \tag{26.30}$$

where p_1 and p_2 are the momenta of the external fermion lines which carry SU(5) indices a and b with $p = p_1 - p_2$, T is the corresponding SU(5) matrix, and, again, the index A denotes the particular vector boson we have used.

---

[2] In fact, for fermions we need as many conditions as the number of different irreducible representations that appear in (26.28).

We want to emphasise that once these three kinds of renormalisation conditions have been imposed, the perturbation theory is completely defined and all Green functions are finite and calculable as formal power series in g_A. It is not possible to impose any further conditions. Note, in particular, that there exists only one gauge coupling constant, as we should expect from a grand unified theory based on a simple group, like $SU(5)$. Does this mean that we can compute a QCD process as a power series in the weak interaction coupling constant? Formally, the answer is yes, but in practice this is not so. Formal perturbation theory guarantees that if in (26.30) we choose A to denote one of the $SU(2)$ gauge bosons, all other 3-point functions are finite and calculable. In particular, if $\Gamma_3^{(3)}$ is the 3-point function with a gluon external gauge boson we can write[3]

$$\Gamma_3^{(3)}(p^2 = -\mu^2) = g_2 + C_1 g_2^3 + C_2 g_2^5 + ..., \tag{26.31}$$

where the coefficients C_n are finite and calculable functions of μ and the masses of the theory. However, it is easy to check that C_n is of the form $C_n \sim [\ln(M^2/\mu^2)]^n$, which, for $M/\mu \sim 10^{15}$, gives $[70]^n$. In other words, although the series (26.31) is well defined, it is useless for practical computations.

The remedy to this difficulty is easy to guess. We renormalise the same broken $SU(5)$ theory in three different ways by choosing the index A in the conditions (26.29) and (26.30) corresponding to the bosons of $U(1)$, $SU(2)$, or $SU(3)$. This gives us three perturbation expansions in powers of g_1, g_2, or g_3 always of the same theory, but each one is suited to particular processes. The values of the g_i's will be fixed by experiment, but we must always remember that all three describe the same theory, so we write the generalisation of (26.31),

$$g_i = F_{ij}(g_j, M^2/\mu^2, \alpha), \tag{26.32}$$

where α denotes, collectively, the ratios M_1/M and M_2/M. In the limit of exact $SU(5)$ symmetry, i.e. when $M^2/\mu^2 \to 0$, all coupling constants must be equal. This happens only at infinite energy. So the functions F_{ij} must satisfy

$$F_{ij}(g_j, 0, \alpha) = g_j. \tag{26.33}$$

We can rewrite (26.32) as renormalisation group equations by taking the total derivative with respect to μ^2 of both sides of (26.32). In the Landau gauge we obtain

$$\beta^{(i)}(g_i, \rho, \alpha) = \left[-\rho \frac{\partial}{\partial \rho} + \beta^{(j)}(g_j, \rho, \alpha) \frac{\partial}{\partial g_j} \right] F_{ij}, \tag{26.34}$$

where

$$\beta^{(k)}(g_k, \rho, \alpha) = \mu^2 \frac{\mathrm{d} g_k}{\mathrm{d}\mu^2}; \quad \rho = \frac{M^2}{\mu^2}. \tag{26.35}$$

---

[3] We will use the following notation: for $A=1,...,8$ we call $g_A = g_3$; $A=21,...,23$ $g_A = g_2$; $g_{(A=24)} = g_1$.

The differential equations (26.34) with the boundary conditions (26.35) are our basic equations. The β-functions are calculable at any given order of perturbation theory as power series of the form

$$\beta^{(k)}(g_k, \rho, \alpha) = b_0^{(k)}(\rho, \alpha)g_k^3 + \ldots \ . \tag{26.36}$$

Note that the b coefficients, unlike those of F_{ij} itself, do not contain large logarithms. This can be easily understood since the β-functions, as defined by Eqs. (26.35), possess well-defined limits for both $\rho \to \infty$, when they become the β-functions of $SU(3)$, $SU(2)$, or $U(1)$, and $\rho \to 0$, when they become all equal to that of $SU(5)$. In contrast, F_{ij} has no limit when $\rho \to \infty$ with g_j kept fixed.

The system of Eqs. (26.34) can be solved with the standard method of characteristics which we saw in section 19.2. The solution expresses any coupling constant in terms of any other,

$$g_i = F_{ij} = \eta(g_j, \rho\alpha) \tag{26.37}$$

with η a known function which can be determined order by order in perturbation theory by solving the system of the renormalisation group equations, provided we have computed the β-functions up to that order. In the classical approximation, when all β-functions vanish, we have the tree-level results of Eq. (26.25). At 1 loop we obtain

$$\frac{1}{g_i^2} = \frac{1}{g_j^2} + 2 \int_0^\rho \frac{\mathrm{d}x}{x} [b_0^{(i)}(x, \alpha) - b_0^{(j)}(x, \alpha)]. \tag{26.38}$$

The integral is well defined at $x = 0$ because the difference of the β-functions vanishes when the scale goes to infinity.

These equations contain an important prediction. We can use as input the experimentally measured effective strengths of strong, electromagnetic, and weak interactions at moderate (let us say ~ 10 GeV) energies. The two independent equations (26.37) contain three unknown parameters, namely ρ and the two masses M_1/M and M_2/M of the scalar fields denoted, collectively by α. However, it turns out that the dependence on α is very weak, even when higher orders are taken into account. If we ignore it, we can use the two equations to fix ρ, which means the grand unification scale M, and obtain a relation between measured values of the three coupling constants g_i. This relation is usually presented as a prediction for the value of $\sin\theta_W$. For a comparison with experiment we must make a more complete analysis and include the breaking of the standard model with a 5-plet of scalars. A pictorial way often used to present the result is to write the renormalisation group equations in the step function approximation we introduced in section 25.4.3 in which, at any given scale, we put all heavier masses at infinity and all lower masses at zero. In this approximation for all energies up to the scale M, the three coupling constants g_1, g_2, and g_3 evolve independently, each one following its own renormalisation group equation. The prediction is correct if the three curves meet at the scale M. Figure 26.5 shows the result for the group $SU(5)$. As we can see the curves do not really meet. We will see later that this can be considerably improved if we extend

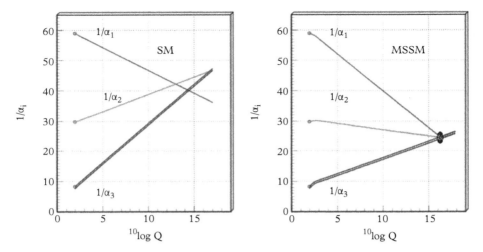

Figure 26.5 *The renormalisation group evolution of the inverse coupling constants* $\alpha_i^{-1}, i = 1, 2, 3$ *of the three Standard Model groups* $U(1)$, $SU(2)$ *and* $SU(3)$ *without supersymmetry, (left) and with supersymmetry, (right). The uncertainties are also shown. The calculations include 2-loop effects (© Particle Data Group).*

the model. Here we want only to stress that Fig. 26.5 is only valid in the step function approximation. In the real analysis we presented earlier, the curves meet only at infinite energy, as shown in the boundary conditions (26.33). Indeed, if we plot the real solutions and expand around the scale M we obtain Fig. 26.6.

This renormalisation group analysis can be repeated for the fermion masses of Eq. (26.26), since they are proportional to the scalar-fermion coupling constants. At 1 loop we find the ratio

$$\frac{m_b}{m_\tau} = \left[\frac{\alpha_s(\mu = 2m_b)}{\alpha_s(\mu = M)} \right]^{\frac{12}{33-6f}}, \qquad (26.39)$$

where α_s is the strong interaction coupling constant evaluated at the corresponding scale and f is the number of families. The agreement is very good precisely for $f = 3$. Note that this result was obtained by A. Buras, J. Ellis, M. K. Gaillard, and D. V. Nanopoulos before LEP had shown that there exist only three light neutrinos.

We might be tempted to apply the same analysis to the other two mass ratios, for example m_s/m_μ, where the agreement with experiment is not good, but we will not discuss this point here.

26.2.4 Other Grand Unified Theories

In the previous section we examined in some detail the grand unified model based on the group $SU(5)$. The main reason for this choice was its simplicity. In fact, as we mentioned

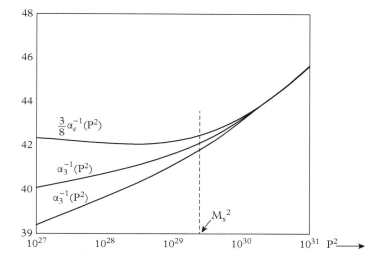

Figure 26.6 *Detail of the region around the grand unification scale using exact 2-loop β-functions (© I. Antoniadis et al, LPTENS 81–4).*

already, this simplest model does not quite fit the experimental data. However, the general properties remain the same in practically all models and the methods we developed can be applied in a straightforward way to every other model, although the detailed numerical results may differ. In this section we will briefly present some other 'classical' grand unified theories and we will try to explain their respective merits.

26.2.4.1 A Rank 5 GUT: SO(10)

The $SU(5)$ model, in its simple and most attractive version, has no natural place for a right-handed neutrino. We must add it as an extra singlet. The only simple group which can be used for grand unification without need for a singlet representation and without introducing exotic fermions is $SO(10)$. It is a group of rank 5, which means that the corresponding algebra has five commuting generators. For $SU(5)$ we had proven that it was the only acceptable group of rank 4. Similarly, we can prove that $SO(10)$ is the only one of rank 5. The proof goes along the same lines: we list all possible candidates and we eliminate the unacceptable ones. The only candidate which almost makes it is $SU(6)$. It has a 15-dimensional representation but its decomposition in $SU(2) \times SU(3)$ shows that it cannot accommodate the members of a family. We find that

$$15 = (2,3) \oplus (1,\bar{3}) \oplus (1,3) \oplus (2,1) \oplus (1,1)$$

The troublesome piece is the $(1,3)$ which is a singlet of $SU(2)$ and triplet of colour rather than an anti-triplet.

The only successful candidate is $SO(10)$ which has a 16-dimensional irreducible representation. $SO(10)$ contains $SU(5)$ as a subgroup and the 16-plet decomposes under $SU(5)$ into

$$16 = 10 \oplus \bar{5} \oplus 1, \tag{26.40}$$

i.e. we find our old $\bar{5}$ and 10 as well as a singlet. The obvious interpretation of this last one is a right-handed neutrino (or $\nu^c{}_L$).

The salient features of this GUT are the following. The gauge bosons belong to a 45-dimensional (adjoint) representation which under $SU(5)$ decomposes as

$$45 = 24 \oplus 10 \oplus \bar{10} \oplus 1. \tag{26.41}$$

An interesting property of the model is that all members of a family, enlarged with a right-handed neutrino, belong to a single irreducible representation, the 16-dimensional spinorial representation we mentioned earlier (Eq. 26.40). In this respect the family structure seems more natural in $SO(10)$ than in $SU(5)$. On the other hand, again no explanation is offered for the observed number of families. It is also interesting to point out that $SO(10)$ has no anomalies. Another interesting feature is that $B-L$ is now a gauge generator. It must be spontaneously broken; otherwise, there would remain a massless photon coupled to it. However, this violation does not lead to any observable effects in nucleon decay because the branching ratio of forbidden to allowed decays is predicted to be very small.

In the long journey from $SO(10)$ down to $U(1) \times SU(2) \times SU(3)$, nature may choose various paths. She can take the direct road (just one big break) or she may decide to go through one, or more, of the intermediate subgroups:

$$SO(10) \rightarrow \left\{ \begin{array}{c} \rightarrow \\ \rightarrow SU(5) \rightarrow \\ \rightarrow SU(5) \times U(1) \rightarrow \\ \rightarrow SU(4) \times SU(2) \times SU(2) \rightarrow \\ \rightarrow SU(4) \times SU(2) \times U(1) \rightarrow \\ \cdots \end{array} \right\} \rightarrow U(1) \times SU(2) \times SU(3). \tag{26.42}$$

The BEH system depends on the breaking pattern we choose, but, in any case, it is more complex than that of $SU(5)$. Several representations are necessary.

The main experimental prediction of $SO(10)$, which differs substantially from that of $SU(5)$, concerns the neutrino mass. The presence of ν_R allows for a Dirac mass $\bar{\nu}_R \nu_L$ and the violation of $B - L$ allows for a Majorana term. The Dirac mass term comes presumably from a Yukawa coupling to a BEH scalar and, therefore, it is an adjustable parameter, like any other fermion mass in the theory. A priori we expect a term on the same order of magnitude as the up-quark masses. The problem then is how to make sure that the physical neutrino masses are sufficiently small. The main remark here is that the Majorana mass, which comes from the superheavy breaking, is expected to be large, on the order of the $SO(10)$ symmetry breaking scale. The resulting neutrino mass spectrum will depend on the details of the BEH system. For example, if $SO(10)$ is broken through the vacuum expectation value of a 126-plet of scalars, the neutrino mass matrix for one family takes the form

$$(\bar{v}_L, \bar{v}_L^c) \begin{pmatrix} 0 & m_D/2 \\ m_D/2 & M \end{pmatrix} \begin{pmatrix} v_R^c \\ v_R \end{pmatrix}, \tag{26.43}$$

where m_D and M are the Dirac and Majorana masses. As we just explained, we expect $m_D \ll M$. Then the $SU(2)$ doublets and singlets will be approximate mass eigenstates with masses $m_D^2/4M$ and M, respectively. For $m_D \sim 1$ GeV and $M \sim 10^{16}$ GeV we find a negligibly small mass for the doublet neutrino of order 10^{-7} eV. Of course one also expects mixings among the three families but they are very model dependent. Let us also mention that even if the Majorana mass term is forbidden in the tree approximation (for example, if $B - L$ is not broken through the 126 but through a 16 representation), it may be generated in higher orders through a particular 2-loop diagram. The value of M is suppressed in this case by coupling constants and the resulting neutrino mass may be on the order of 1 eV or higher.

The moral of the story is twofold. First, the theory offers, through the spontaneous breaking of $B - L$, a natural mechanism to obtain very light neutrinos, and second almost any desired pattern of masses and mixings can be reproduced by adjusting the parameters of an already rather complicated BEH system.

26.2.4.2 *Other Models*

It is by now obvious that a large variety of grand unified theories can be obtained by playing around with elementary group theory and scalar field representations. This partly explains the popularity that GUTs have enjoyed for more than 30 years. There is no point in giving a complete list of all models which claim some agreement with data, which is usually the case for published models. We will only mention two examples of models based on exceptional groups.

An aesthetically unattractive feature of all models based on unitary, orthogonal, or symplectic algebra is that these form infinite series, so it may be hard to understand why any one in particular would provide the basis for a fundamental theory. Exceptional groups, on the other hand, are unique; they are just $G_2, F_4, E_6, E_7,$ and E_8. The first one is excluded because it is too small to contain $SU(3)$ as a subgroup. The others could, in principle, be used as candidates for GUTs. We present here only E_6 and E_8.

E_6: It is the most attractive exceptional group for grand unification. It is the only one which admits complex representations and, from the group theory point of view, it can be considered as the natural extension of $SU(5)$ and $SO(10)$. Indeed, based on the Dynkin diagram pattern, we could define exceptional algebras E_4 and E_5 as isomorphic to $SU(5)$ and $SO(10)$, respectively. Of course, E_6 contains $SO(10)$ and *a fortiori* $SU(5)$, as subgroups. Furthermore it is 'safe', i.e. it is automatically anomaly-free. Its fundamental representation has 27 dimensions and, under $SO(10)$ and $SU(5)$, decomposes as

$$27 \xrightarrow{SO(10)} 16 \oplus 10 \oplus 1 \xrightarrow{SU(5)} 10 \oplus \bar{5} \oplus 1 \oplus 5 \oplus \bar{5} \oplus 1. \tag{26.44}$$

The 78-dimensional adjoint representation decomposes as

$$78 \stackrel{SO(10)}{\longrightarrow} 45 \oplus 16 \oplus \overline{16} \oplus 1$$
$$\stackrel{SU(5)}{\longrightarrow} 24 \oplus 10 \oplus \overline{10} \oplus 1 \oplus 10 \oplus \bar{5} \oplus 1 \oplus \overline{10} \oplus 5 \oplus 1 \oplus 1. \tag{26.45}$$

There are several inequivalent possibilities of constructing grand unified theories based on E_6 and, in this section, we will mention only one example which satisfies the following requirements:

   (i)   All fermions of one family belong to the same irreducible representation.

  (ii)   All unobserved, fermions get naturally superheavy masses.

 (iii)   All required scalars belong to representations appearing in the product of two fermion representations. This last requirement means that the resulting scalar particles can be viewed as fermion–anti-fermion bound states.

We will assign the fermions of each family to the 27 fundamental representation, which, therefore, contains new, unobserved fermions. The scalar fields must belong, according to (iii), to one, or more, of the representations:

$$27 \otimes 27 = (27 \oplus 351)_S \oplus 351_A. \tag{26.46}$$

The important observation comes from the decomposition (26.44). Out of the 27 fermions of a family, 12 (i.e. $5 \oplus \bar{5} \oplus 1$) can take an $SU(5)$ invariant (and *a fortiori* $U(1) \otimes SU(2) \otimes SU(3)$ invariant) mass. Therefore, these fermions are expected to have masses on the order of 10^{16} GeV (this is sometimes called 'the survival hypothesis'). It is easy to check that with the BEH system (26.46) this indeed happens. This explains why only 15 light fermions are observed in each family.

The simplest symmetry breaking pattern of this model goes through $SO(10)$, although others have also been considered. The detailed predictions, including some interesting speculations concerning the fermion mass spectrum, are model dependent.

E_8: If uniqueness is an important criterion for choosing the group of grand unification, then E_8 is the most prominent candidate. Its unique features include: (i) it is the largest exceptional group with a finite-dimensional Lie algebra. (ii) It contains E_6, and, thus, $SO(10)$, $SU(5)$, etc., as subgroups. (iii) It is the only simple Lie group whose lowest dimensional representation is the adjoint. This offers the possibility of putting fermions and gauge bosons in the same, lowest dimensional, representation. In fact, E_8 has a natural, built-in, fermion–boson symmetry, as we will see later. It is the symmetry group which appears automatically in some modern superstring models. These are the good news. Now the bad news: the adjoint representation has dimension 248, so a large number of new gauge bosons and fermions is required. Similarly, the necessary scalar field representations are enormous, the simplest version using the 3875-dimensional representation. On the other hand, we can put all known fermions, together with many unknown ones, in the same 248 representation. After symmetry breaking, we can arrange to have three light $SU(5)$ families $(\bar{5} \oplus 10)$ and three heavier $(\sim 1 \text{ TeV})$, conjugate ones $(5 \oplus \bar{10})$. All other fermions become superheavy. E_8 has only real representations, so the theory is 'vector-like', i.e. there are equal numbers of right- and left-handed fermions and, before symmetry breaking, we can write the theory using only vector currents.

26.2.5 Magnetic Monopoles

In this section we want to present a general feature of grand unified theories which is interesting in its own right and shows a novel property of gauge theories.[4] This discussion will also allow us to introduce the idea of duality, which plays a crucial role in many modern theories, such as that of strings and branes.

26.2.5.1 *Abelian Magnetic Monopoles*

The empty-space Maxwell's equations possess an obvious invariance under the following interchange of electric and magnetic fields:

$$\mathbf{E} \rightarrow \mathbf{B} \qquad\qquad \mathbf{B} \rightarrow -\mathbf{E}. \qquad (26.47)$$

As a matter of fact, this invariance is much larger and covers the entire $U(1)$ group of transformations $\mathbf{E} + i\mathbf{B} \rightarrow e^{i\phi}(\mathbf{E} + i\mathbf{B})$. However, as we will see presently, only the discreet subgroup (26.47) could possibly survive in the presence of sources.

In a compact relativistic notation the transformation (26.47) can be written as a duality transformation:

$$F^{\mu\nu} \rightarrow \tilde{F}^{\mu\nu} \qquad\qquad \tilde{F}^{\mu\nu} \rightarrow -F^{\mu\nu}. \qquad (26.48)$$

It is an empirical fact that the presence of matter destroys this symmetry. Indeed, we have electric charges and electric currents $j_\mu = (\rho, \mathbf{j})$ but no corresponding magnetic ones. Maxwell's equations are

$$\partial_\nu F^{\mu\nu} = -j^\mu \qquad\qquad \partial_\nu \tilde{F}^{\mu\nu} = 0. \qquad (26.49)$$

We can try to restore the symmetry, but then we must introduce a magnetic current $k_\mu = (\sigma, \mathbf{k})$ and write Eqs. (26.49) as

$$\partial_\nu F^{\mu\nu} = -j^\mu \qquad\qquad \partial_\nu \tilde{F}^{\mu\nu} = -k^\mu, \qquad (26.50)$$

which are invariant under (26.48), provided j^μ and k^μ transform as

$$j^\mu \rightarrow k^\mu \qquad\qquad k^\mu \rightarrow -j^\mu. \qquad (26.51)$$

---

[4] We follow the presentation given in S. Coleman, 'The Magnetic Monopole Fifty Years Later', Erice Summer School 1981 in *The Unity of the Fundamental Interactions*, ed. A. Zichichi (Plenum Press, 1983).

For example, if the electric current results from the motion of point electric charges, the magnetic current k^μ will result from the motion of point magnetic charges, i.e. magnetic monopoles.

The introduction of magnetic charges looks at first sight like a trivial generalisation of quantum electrodynamics. However, it is easy to see that this is not so. The usual quantisation procedure is set up in terms of the vector potential A^μ rather than the electric and magnetic fields $F^{\mu\nu}$. The latter is given by $F^{\mu\nu} = \partial^\mu A^\nu - \partial^\nu A^\mu$. This last relation implies the vanishing of $\partial_\nu \tilde{F}^{\mu\nu}$ and thus the absence of any magnetic current. We conclude that in a theory with magnetic monopoles, the vector potential cannot be a well-defined function of the space–time point x.

In our familiar theory, we can view an isolated magnetic monopole of magnetic charge g sitting at the origin, as one end of a solenoid in the limit when the latter is infinitely long and infinitely thin. The line occupied by the solenoid, say the negative z axis, is called *the Dirac string*. An observer will see a magnetic flux coming out of the origin as if a monopole were present. He won't realise that the flux is coming back through the solenoid, because, by assumption, this is infinitely thin. The total magnetic field is singular on the string and is given by

$$\mathbf{B} = \frac{g}{r^2}\hat{r} + 4\pi g\theta(-z)\delta(x)\delta(y)\hat{z} \tag{26.52}$$

with $\hat{r}$ and $\hat{z}$ being the corresponding unit vectors. The first term represents the field of the monopole which has the usual point-particle singularity, while the second is the singular contribution of the string. We can construct a vector potential $\mathbf{A}$ whose curl is $\mathbf{B}$. Of course, we expect also $\mathbf{A}$ to be singular on the negative z axis. A simple computation, using spherical coordinates, gives

$$\mathbf{A} = \frac{g}{r}\frac{1 - \cos\theta}{\sin\theta}\hat{\boldsymbol{\phi}}, \tag{26.53}$$

where $\hat{\boldsymbol{\phi}}$ is the unit vector in the ϕ-direction. $\mathbf{A}$ can be taken to represent the field of the monopole and indeed this is true everywhere except on the negative z axis. Since, by our previous argument, we know that a string-like singularity must exist, the form (26.53) is the best we can do. Obviously, the choice of the negative z axis as the position of the string is arbitrary and we could have placed the solenoid along any line from the origin to infinity.

So far the discussion was purely classical. Quantum mechanics brings a subtle difference. In classical electrodynamics the vector potential A_μ is not measurable, only the components $\mathbf{E}$ and $\mathbf{B}$ of $F_{\mu\nu}$ are. In quantum mechanics, however, we can detect directly the presence of A_μ by the Bohm–Aharonov effect. By moving around electrically charged test particles we can discover the magnetic flux coming back through the string. The corresponding change in the phase of the wave function will be

$$\psi \to e^{4\pi i e g}\psi, \tag{26.54}$$

where $4\pi g$ is the flux and e the charge of the test particle. The usual interference experiment will detect the phase change and hence the presence of the string, unless $\exp(4\pi ieg)$ equals 1 or

$$eg = 0, \pm\frac{1}{2}, \pm1, \dots . \tag{26.55}$$

Condition (26.55) is the famous Dirac quantisation condition. If it is satisfied, the string is undetectable by any conceivable experiment and we have obtained a real magnetic monopole. On the other hand, it shows that if there exists a magnetic monopole in the world, all electric charges must be quantised; i.e. they must be multiplets of an elementary charge e_0. Similarly all magnetic charges must be multiplets of an elementary magnetic charge g_0 such that $2e_0g_0$ is an integer. A particle that has both electric and magnetic charge is called 'a dyon'.

26.2.5.2 *The 't Hooft–Polyakov Monopole*

In the Abelian case we saw that magnetic monopoles give rise to singular vector potentials. We will now turn to non-Abelian theories. We have good reasons to believe that the electromagnetic gauge group is part of a bigger group which is spontaneously broken through the BEH mechanism. The simplest such theory, although not the one chosen by nature, is the Georgi–Glashow $SO(3)$ model, which we studied in Problems 15.3 and 25.8. It is a theory without weak neutral currents in which the only gauge bosons are W^+, W^-, and γ. We introduce a triplet of scalars $\boldsymbol{\Phi}$ and we write the Lagrangian as

$$\mathcal{L} = -\frac{1}{4}\boldsymbol{G}_{\mu\nu} \cdot \boldsymbol{G}^{\mu\nu} + \frac{1}{2}(\mathcal{D}_\mu\boldsymbol{\Phi}) \cdot (\mathcal{D}^\mu\boldsymbol{\Phi}) - V(\boldsymbol{\Phi}) \tag{26.56}$$

$$\boldsymbol{G}_{\mu\nu} = \partial_\mu\boldsymbol{W}_\nu - \partial_\nu\boldsymbol{W}_\mu - e\boldsymbol{W}_\mu \times \boldsymbol{W}_\nu$$

$$\mathcal{D}_\mu\boldsymbol{\Phi} = \partial_\mu\boldsymbol{\Phi} - e\boldsymbol{W}_\mu \times \boldsymbol{\Phi} \tag{26.57}$$

$$V(\boldsymbol{\Phi}) = \frac{\lambda}{4}(\boldsymbol{\Phi}^2 - v^2)^2.$$

We have written the scalar potential V in a form which exhibits explicitly the minimum away from the origin in field space and we have not included any fermions for simplicity.

From (26.56) we can compute the corresponding Hamiltonian density

$$\mathcal{H} = \frac{1}{2}\sum_{i=1}^{3}[\boldsymbol{E}^i \cdot \boldsymbol{E}^i + \boldsymbol{B}^i \cdot \boldsymbol{B}^i + (\mathcal{D}_0\boldsymbol{\Phi}) \cdot (\mathcal{D}_0\boldsymbol{\Phi}) + (\mathcal{D}^i\boldsymbol{\Phi}) \cdot (\mathcal{D}^i\boldsymbol{\Phi})] + V(\boldsymbol{\Phi}), \tag{26.58}$$

where we have defined the non-Abelian 'electric' and 'magnetic' fields as $\boldsymbol{E}^i = -G^{0i}$ and $\boldsymbol{B}^i = -\epsilon^{ijk}G_{jk}$ and the bold-face vectors refer again to the internal symmetry space. The important point about $\mathcal{H}$ is that it is the sum of positive semidefinite terms. Therefore, the minimum energy solution will be that for which $\mathcal{H} = 0$. On the other hand, $\mathcal{H}$ is invariant under local $SO(3)$ rotations in the internal symmetry space. However, the only symmetric solution, i.e. the field configuration

$$W_\mu = 0 \qquad\qquad \boldsymbol{\Phi} = 0, \qquad\qquad (26.59)$$

gives $\mathcal{H} = v^2$ and thus corresponds to infinite total energy. The zero energy solution must make each term of (26.58) vanish. An example of such a solution is

$$W_\mu = 0 \qquad\qquad \boldsymbol{\Phi} = v\hat{\boldsymbol{k}} \qquad\qquad (26.60)$$

with $\hat{\boldsymbol{k}}$ being the unit vector in the third direction in the internal symmetry space. Obviously, since $\mathcal{H}$ is gauge invariant, any gauge transform of (26.60) will give another zero energy solution. In particular, there is nothing sacred about choosing $\boldsymbol{\Phi}$ to point along the third direction.

The solution (26.60), or any transform of it, exhibits the well-known BEH phenomenon. The symmetry $SO(3)$ is spontaneously broken since the invariance of the solution is reduced to the group of rotations around the third axis, i.e. $U(1)$. Two of the vector bosons acquire a mass and it is natural to identify the third one, which remains massless, with the photon.

Up to now we have found two sets of solutions of the equations of motion given by the Lagrangian density (26.56): one $SO(3)$ symmetric solution given by (26.59) and a whole family of asymmetric ones given by (26.60) and all its gauge transforms. The first corresponds to infinite total energy while all the second ones correspond to zero energy. They describe the family of stable vacuum states. A natural question is the following. Are there any finite, non-zero-energy, non-trivial, particle-like solutions? The condition of finite total energy implies that $\mathcal{H}$ must vanish at large distances; therefore, asymptotically, any such solution will approach one belonging to the family (26.60).

There is no general method for finding the solutions of coupled, non-linear, partial differential equations. What is usually done is to guess a particular form of the solution and to simplify the equations. In doing the guesswork we often try first to guess the symmetries of the solution. Since we are looking for a stable particle-like solution (the magnetic monopole) the solution must be time-independent. Furthermore it will be left invariant by a group of transformations which is a subgroup of the symmetries of the equations of motion. In the rest frame of the particle the latter is $G = SO(3) \otimes SO(3) \otimes P \otimes R$ where the first $SO(3)$ corresponds to spatial rotations, the second corresponds to the internal symmetry, P denotes parity, and R denotes the transformation $\boldsymbol{\Phi} \to -\boldsymbol{\Phi}$. In guessing the form of the monopole solution we will try to enforce as much of the symmetry G as possible. Invariance under spatial rotations will force $\boldsymbol{\Phi}$ to be asymptotically constant and $G^{ij} \sim r^{ij}$. It is easy to verify that this solution has zero total magnetic charge like the vacuum solution of Eq. (26.60). On the other hand, since our solution must approach at large distances one of the internal symmetry breaking ones, we cannot enforce the second $SO(3)$ either. Finally, both P and R change the sign of the magnetic charge and cannot be included. Let us choose, therefore, to impose invariance under $SO(3) \otimes PR$, where $SO(3)$ is the diagonal subgroup of $SO(3) \otimes SO(3)$ and PR is the product of the two. We seek solutions of the general form

$$\Phi_\alpha = H(evr)\frac{x_\alpha}{er^2}$$

$$W^0_\alpha = 0 \tag{26.61}$$

$$W^i_\alpha = -\epsilon_{\alpha ij}\frac{x_j}{er^2}[1 - K(evr)],$$

with H and K functions of a single variable. Space and internal symmetry indices are mixed and the ansatz (26.61) is spherically symmetric in the sense that a spatial rotation can be compensated for by an internal symmetry rotation. The procedure parallels the one we followed in section 25.4.5 when we were discussing instanton solutions. Plugging (26.61) into the equations of motion we obtain a system of coupled ordinary differential equations for H and K which can be solved, at least numerically. It is easy to verify that this solution does describe a magnetic monopole. We can compute the associated magnetic field and we find asymptotically

$$B^i \to -\frac{1}{e}\frac{x^i}{r^3}, \tag{26.62}$$

which corresponds to a magnetic charge,

$$g = -\frac{1}{e}, \tag{26.63}$$

thus obtaining $|eg| = 1$. Note that here e is the charge of the boson which has isospin $= 1$. If we had included isospin $\frac{1}{2}$ fields, for example isospinor fermions, the symmetry group would have been $SU(2)$ and the smallest electric charge in the theory would have been $|e/2|$. We thus recover the Dirac quantisation condition which states that the minimum magnetic charge is given by $|e_{\min}g_{\min}| = \frac{1}{2}$. We can also compute the total energy of the solution, which can be interpreted as the classical approximation to the mass of the monopole. We find that

$$M_{\text{monopole}} = \frac{4\pi v}{e}f(\lambda/e^2) \tag{26.64}$$

with f a given function of the ratio of the coupling constants which is larger than 1 for λ positive. It turns out that for $x = \lambda/e^2$ ranging from 0 to 10 f stays of order 1 ($f(0) = 1$, $f(10) \simeq 1.44$). Since the mass of the massive vector bosons after spontaneous symmetry breaking is of order ev, it follows that

$$M_{\text{monopole}} \sim M_{\text{vector boson}}/\alpha \sim 10^2 M_{\text{vector boson}}. \tag{26.65}$$

B. Julia and A. Zee have generalised the solution (26.61) by choosing a non-vanishing W^α_0. They obtained dyon solutions with masses satisfying

$$M_{\text{dyon}} \geq v(e^2 + g^2)^{1/2}. \tag{26.66}$$

It is interesting to note that the inequality is saturated when the scalar field self-coupling goes to 0. We will see shortly that this limit has other interesting consequences.

In general, we can obtain multi-charged solutions with electric charge ne and magnetic charge mg, with n and m integers. The Dirac quantisation condition is again verified.

The asymptotic form of the scalar field is obtained from (26.61) by solving for H. We find that

$$\Phi_{as}^{\alpha} = v\frac{x^{\alpha}}{r}, \tag{26.67}$$

i.e. Φ_{as} points as in Fig. 26.7. This corresponds to a (singular) gauge transformation of the vacuum configuration (26.60) of Fig. 26.8.

Following the same arguments as in section 25.4.5 we see that topologically Φ_{as} maps the surface at spatial infinity S^2 onto the corresponding surface of internal space which is also S^2. Since it is not possible to continuously deform the map of Fig. 26.7 to the constant map of Fig. 26.8, we conclude that the monopole configuration is topologically stable and cannot decay to the vacuum. We can understand this physically as the consequence of conservation of magnetic charge.

We also note that the solution (26.61) is regular everywhere, including the origin. The presence of the scalar field makes possible the existence of a magnetic monopole solution which not only does not have the Dirac string singularity, but also it is smooth

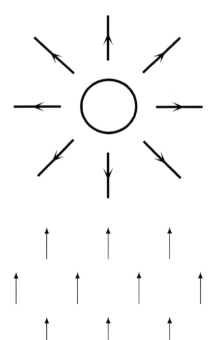

Figure 26.7 *The asymptotic form of the BEH field for the monopole solution.*

Figure 26.8 *The BEH field in the vacuum configuration.*

all the way to $r = 0$. Therefore, the monopoles are non-singular, finite energy, static solutions of the classical equations of motion. It is reasonable to assume, although there is no rigorous proof for this, that they survive quantisation and correspond to real, physical particles. The fact that they have a regular internal structure has a very important physical consequence: the mass of the 't Hooft–Polyakov monopole is calculable, while that of the Dirac one is not. The spectrum of the stable one-particle states will consist of: (i) a massless photon, (ii) a pair of charged vector bosons $W^{\pm}$ with mass of order ev and electric charge $\pm e$, (iii) a neutral scalar boson with mass of order $\sqrt{\lambda}v$, *and* (iv) a monopole and an anti-monopole, each with mass of order v/e and magnetic charge $\pm g = \mp(1/e)$.

An interesting simplification occurs in the so-called 'BPS' (Bogomolny–Prasad–Sommerfield) limit, which consists of $\lambda \rightarrow 0^{+}$ keeping v fixed and maintaining the form of the solution (26.61). In this limit the physical scalar boson becomes massless, like the photon and gives rise to a new long-range force. On the other hand, as we noted already, in this limit the masse of the dyon saturates the bound (26.66). We call this bound 'BPS bound' and we will come back to it shortly.

26.2.5.3 Other Gauge Groups

In quantum electrodynamics magnetic monopoles are a curiosity. They may exist, in which case they have important physical consequences like the quantisation of electric charge, but nobody forces us to introduce them. The theory makes perfect sense without them. The monopole configuration is a singular one for the gauge potential $A^{\mu}(x)$. On the other hand, we just saw that if electromagnetism is part of an $SO(3)$ gauge theory with a triplet of BEH fields, magnetic monopoles are, probably, part of the physical spectrum. Of course, $SO(3)$ and $SU(2)$ are not the right invariance groups for physics because they have no room for weak neutral currents. Could we apply the same reasoning to the gauge group of the standard model which is $SU(2) \otimes U(1)$ broken to $U(1)$? Does the standard model imply the existence of stable magnetic monopoles? The answer is no even if one enlarges the scalar field content. The reason is that the stability argument of the previous section does not apply to the standard model. The argument was based on the fact that there is no $SO(3)$ continuous, non-singular gauge transformation which can deform the monopole configuration of Eq. (26.61) to the vacuum one of Eq. (26.60). However, it is easy to see that this is no more true when the gauge group is $SU(2) \otimes U(1)$. The presence of the $U(1)$ factor makes the 't Hooft–Polyakov magnetic monopole unstable. Therefore, the natural application of these ideas is the framework of grand unified theories. Let us first present briefly the results of the general case which was investigated by P. Goddard, J. Nuyts, and D. Olive.

Let us assume that a gauge theory of a simple group G is spontaneously broken through the BEH mechanism to a subgroup H. (In the 't Hooft–Polyakov case G was $SO(3)$ and H was $U(1)$.) Goddard, Nuyts, and Olive showed that we can again find magnetic monopole and dyon solutions with the following properties: if H is not just $U(1)$, the 'electric' (and 'magnetic') charges are not given by a single number because the fields form multiplets of H. It turns out that the electric charges are described in terms of the group H, while the magnetic ones in terms of its dual, denoted by H^*. The distinction between the two sounds like a mathematical detail because, for all practical purposes, they are locally isomorphic, like $SO(3)$ and $SU(2)$. However, we have seen already that such details may be relevant. For example, $SU(2)$ admits half-integer charges while $SO(3)$ does not and this affects the Dirac quantisation condition. We will encounter this problem again when studying monopoles in the grand unified model $SU(5)$. More precisely, the values of the electric charges can be viewed as vectors q in the weight lattice of H and the magnetic charges as vectors g in that of H^*. The quantisation condition now reads

$$eq \cdot g = 2\pi N; \quad N \in \mathbb{Z}. \tag{26.68}$$

26.2.5.4   *Monopoles and Instantons in Gauge Theories*

Magnetic monopoles are particle-like solutions which are absent from the spectrum of states in perturbation theory. In section 25.4.5 we proved that they do not appear in a pure gauge theory and, indeed, in this section we saw that the presence of the BEH field was essential for their stability. They are classical solutions in four-dimensional Minkowski space, but, since they are time-independent, they can be viewed as solutions in the three-dimensional Euclidean space. In this sense a magnetic monopole in d space-time dimensions is equivalent to an instanton in $d-1$ dimensions. A general gauge theory, in particular a grand unified theory, will contain both such solutions. The monopoles will impose a Dirac-type charge quantisation condition and the instantons will introduce the angle θ as a second effective coupling constant, as shown in Eq. (25.215).

We can repeat the previous analysis and look for monopole and dyon solutions in a gauge theory including the θ term. For a monopole, we obtain a shift in the value of the charge given by

$$q = \frac{\theta e^2}{8\pi^2} g \tag{26.69}$$

and, for a dyon, taking into account the Dirac condition

$$q = ne + \frac{\theta e}{2\pi} m; \quad g = \frac{4\pi}{e} m; \quad n, m \in \mathbb{Z}. \tag{26.70}$$

It is useful to introduce a complex representation and write

$$\tau = \frac{\theta}{2\pi} + i\frac{4\pi}{e^2} \quad ; \quad q + ig = e(n + m\tau). \tag{26.71}$$

The BPS bound can be written using this notation as

$$M \geq ve|n + m\tau|. \tag{26.72}$$

26.2.5.5 *Monopoles in Grand Unified Theories*

In the previous section we argued in favour of grand unified theories in which we start with a simple group G and, after several spontaneous symmetry breakings, we end up with $SU(3) \otimes U(1)$. Since $SU(2)$ can be embedded in any simple G, we expect to find stable monopole solutions in every grand unified theory. This question was first studied by C. P. Dokos and T. N. Tomaras in 1979.

Indeed, we can give a general proof of this statement. The presence of the $U(1)_{e.m.}$ factor in the unbroken gauge subgroup of the simple grand unified group guarantees the existence of smooth, finite-energy, topologically stable, particle-like solutions of the equations of motion with quantised magnetic charge. The proof of stability is the direct generalisation of the simple topological argument given in the previous section. Let us exhibit the grand unified monopoles in the case of the $SU(5)$ prototype model.

Let us try to find the quantum numbers of the $SU(5)$ monopole using only simple symmetry arguments. We start by determining the minimum value of the magnetic charge g_{min}. Since the minimum value of electric charge is $e_{min} = e/3$ (the down-quark) we would expect, naïvely, to have $g_{min} = 3/(2e)$. However, this is incorrect. The reason is, again, our sloppy way in dealing with group theory, not paying much attention to the difference between a group and its Lie algebra. Being more careful, Dokos and Tomaras have established that in the case of $SU(5)$, the unbroken group H is only locally isomorphic to $SU(3) \times U(1)$. The group H is really the set of $SU(5)$ matrices of the form

$$\begin{pmatrix} ue^{-i\alpha} & & \\ & e^{3i\alpha} & \\ & & 1 \end{pmatrix} \tag{26.73}$$

with u an element of $SU(3)$. The mapping from $SU(3) \otimes U(1)$ to H, defined by $(u, e^{i\alpha}) \rightarrow \mathrm{diag}(ue^{-i\alpha}, e^{3i\alpha}, 1)$ is 3 to 1, since the three elements $(u, e^{i\alpha})$, $(ue^{2\pi i/3}, e^{i(\alpha+2\pi/3)})$, and $(ue^{4\pi i/3}, e^{i(\alpha+4\pi/3)})$ of $SU(3) \otimes U(1)$ are mapped to the same element of H. In other words, H is the group $SU(3) \otimes U(1)/Z_3$, with Z_3 being the centre of $SU(3)$. As a consequence of this, using the same phase argument of the previous section, we can show that $g_{min} = [2 \cdot 3 \cdot e_{min}]^{-1} = [2e]^{-1}$ with $|e|$ being again the electron charge. We thus recover the Dirac quantisation condition.

Let us next identify the $SU(2)$ subgroup of $SU(5)$ which will be used to construct the symmetry group of the monopole solution. It is clear that it cannot be a subgroup of $SU(3)_{colour}$ because it must contain electromagnetism. On the other hand it cannot be the $SU(2)$ of the standard model as we have already shown. Therefore, we are led to a choice of the form

$$T = \frac{1}{2} \begin{pmatrix} 0 & & \\ & 0 & \\ & & \boldsymbol{\tau} \\ & & & 0 \end{pmatrix} \tag{26.74}$$

for the $SU(2)$ generators $\boldsymbol{\tau}$. It follows that the monopole magnetic field will have both ordinary magnetic and colour magnetic components.

From now on, the analysis proceeds like in the previous section. We can identify the asymptotic form of the solution and obtain a system of ordinary differential equations. The monopole mass in the classical approximation will be on the order of $10^2 M$, with M being the mass of the vector bosons of $SU(5)$; i.e. the magnetic monopoles of grand unified theories are superheavy with masses of order 10^{18} GeV, far beyond the reach of any accelerator. On the other hand, the lightest monopole is stable. It is amusing to note that the only stable elementary particles in a grand unified theory are the photon, which is massless, the lightest neutrino, the electron, and the monopole. Neutrino decay is forbidden by angular momentum conservation because the neutrino is the lightest spin-$\frac{1}{2}$ particle. The stability of the electron and that of the monopole are guaranteed by the conservation of electric and magnetic charge. Let us also mention that, as shown by C. Callan and V. Rubakov, although baryons in grand unified theories may have long life-times, magnetic monopoles act as catalysers: in their presence baryons decay promptly.

Since monopoles are stable, whichever might have been produced in the course of the evolution of the Universe have a reasonable chance of being around today. The problem of estimating the monopole abundance is therefore reduced into that of estimating the rate of monopole production and that of subsequent monopole–antimonopole annihilation. At the very early Universe, when temperature was sufficiently high, we expect all symmetries, which are spontaneously broken today, to be restored. At $T > T_c \sim M$ we have the full $SU(5)$ symmetry. But for magnetic monopoles to exist $SU(5)$ must be broken. Therefore, there were no monopoles when $T > T_c$. They are produced during the phase transition but the production mechanism is model dependent. If the phase transition is second order the vacuum expectation value of the scalar field undergoes large fluctuations and a domain structure is established. The resulting domain walls contain topological defects in the scalar field orientation which give rise to magnetic monopoles. A rough estimation of their density gives $d_{\text{monopole}} \sim \xi^{-3}$ where ξ is the correlation length. Although its precise estimation is difficult, it is certainly less than the horizon length at temperature T. The latter is of order $C M_P T^{-2}$, where M_P is the Planck mass and C depends on the number of massless degrees of freedom in thermal equilibrium at temperature T. We thus obtain a bound for the initial monopole density on the order $d_{mon} T^{-3} \geq 10^{-10}$. A similar result is obtained even in the case of a first-order phase transition. Order of magnitude estimations show that the annihilation rate in an expanding Universe does not substantially reduce this number. We are thus left with a monopole density $d \sim 10^{-10} T^3$, i.e. comparable to the baryon density. This is obviously absurd if we take into account their enormous mass.

Several mechanisms have been proposed to reduce the number of monopoles surviving but the most attractive proposal today is based on the inflation scenario. During the exponential expansion any initial monopole density is 'inflated away' and reduced to essentially 0. During an eventual subsequent reheating new monopoles could be produced but the production rate depends crucially on the reheating temperature. The final result is model dependent, but acceptably low densities can be obtained.

26.2.5.6 The Montonen–Olive Duality Conjecture

All our quantitative understanding of four-dimensional quantum field theories comes from perturbation theory. The enormous success of the standard model testifies to that. On the other hand we know that such an understanding is necessarily limited. Many exciting physical phenomena are not accessible to perturbation. The obvious example is confinement in QCD. The perturbative spectrum of the theory is described by the Fock space of states of quarks and gluons, but we know that the asymptotic states are those of hadrons. In a qualitative way, we attribute this property to the behaviour of the QCD effective coupling strength as a function of scale. At short distances the coupling is weak and perturbative calculations are reliable. At large distances we enter the strong coupling regime and perturbation theory breaks down. If we had an exact, analytic solution of the quantum field theory, we would have found the hadrons as complicated functionals of the quark and gluon fields, describing static, finite energy, particle-like configurations. Physicists have often tried to guess effective descriptions of QCD in terms of collective variables valid at large distances, where the description in terms of quark and gluon fields is inadequate.

The model with non-Abelian magnetic monopoles may provide a simple example of such a situation. We recall that in the BPS limit, the perturbative, one-particle states of the system, i.e. those corresponding to the elementary fields in the original Lagrangian, are the massless photon, the massless scalar, and the two spin-1 bosons of mass $M = ev$ and electric charge $\pm e$. The strength of the interaction is given by e. In addition the spectrum contains non-perturbative one-particle states, the magnetic monopoles. In contrast to the gauge bosons, they are not point-particles but extended objects. Their masses are $M_{\text{mon}} = gv$ and they have a magnetic charge $g = \pm 1/e$. The electric charge is conserved as a consequence of Noether's theorem, because it corresponds to an unbroken symmetry of the Lagrangian. The magnetic charge is conserved as a consequence of the topological structure of the monopole solution. All this reminds us of the duality symmetry of Maxwell's equations (26.48) and (26.51).

The complete quantum mechanical properties of this model are not known. For small e we can use perturbation theory. Note that in this case g is large and the monopoles are heavy. When e grows large we enter the strong coupling regime and perturbation theory breaks down. In this case g is small and the monopoles are light. C. Montonen and D. Olive made the following conjecture. This model admits two *dual equivalent* field theory formulations in which electric (Noether) and magnetic (topological) quantum numbers exchange roles. The Lagrangian (26.56) in the BPS limit gives the first one. The electrically charged gauge bosons are the elementary fields, e is the coupling constant and the monopoles are the extended objects. Obviously, this description should be useful when e is small. In the dual equivalent description the monopoles are the elementary fields, together with the photon and the massless scalar, g is the new coupling constant, and the electrically charged gauge bosons correspond to extended, soliton-like, solutions. This will be the useful description for large e. This is the conservative version of the conjecture. In fact, Montonen and Olive went a step further. They conjectured that for this particular model, the two field theories are identical; i.e. the new Lagrangian is again given by (26.56) with g replacing e. The monopole fields and the photon form a triplet under a new $SO(3)$ gauge group. This may sound strange because the monopole solution appears to be gauge equivalent to a spherically symmetric one and, naïvely, we expect the monopole to have zero spin. In fact, the situation is more involved and a complete calculation of the total angular momentum of the solution is not easy. We will not develop this point here.

A proof of this conjecture would require a complete solution of the field theory, a rather unlikely state of affairs in any foreseeable future. We can only verify its calculable consequences. So far, such checks have been successful. Let us mention only one. We can generalise the classical solution (26.61) to one describing any number of well-separated monopoles and/or antimonopoles at rest. Studying the instantaneous acceleration of such states, N. S. Manton has found the long-range part of the classical force between two monopoles. The surprising result is

$$ F = -\frac{2g^2}{r^2} \quad \text{opposite charge;} \quad F = 0 \quad \text{same charge,} \quad (26.75) $$

i.e. twice the naively expected attraction for opposite charges and no force at all for same charges. How does this agree with the known forces between the electrically charged particles of the model? It does! The classical force between the two charged vector bosons receives two contributions. The one-photon exchange graph gives a Coulomb force of $\pm e^2/r^2$, attraction for opposite charges, and repulsion for like charges. The one-scalar exchange graph is always attractive, like all even spin exchanges. It is given by the trilinear term in (26.56), after translation of the scalar field. The result is $-e^2/r^2$; i.e. it doubles the attractive part of the photon and cancels the repulsive one. Therefore, in a completely independent calculation, we derive the result (26.75) with e replacing g.

26.3   The Trial of Scalars

The purpose of this section is not to destroy, but to fulfil. It is our firm belief, shared by most physicists, that gauge theories have come to stay. 'Beyond' here does not mean that we propose to replace gauge theories with something else, but rather to embed them into a larger scheme with a tighter structure and higher predictive power. There are several reasons for such a search.

As we said in the previous chapter, gauge theories contain two and possibly three independent worlds. The world of radiation with the gauge bosons, the world of matter with the fermions, and, finally, in our present understanding, the world of BEH scalars. In the framework of gauge theories these worlds are essentially unrelated to each other. Given a group G the world of radiation is completely determined, but we have no way of knowing a priori which and how many fermion representations should be introduced; the world of matter is, to a great extent, arbitrary.

This arbitrariness is even more disturbing if one considers the world of BEH scalars. Not only their number and their representations are undetermined, but their mere presence introduces a large number of arbitrary parameters into the theory. Note that this is independent of our computational ability, since these are parameters which appear in our fundamental Lagrangian. What makes things worse is that these arbitrary parameters appear with a wild range of values. The situation becomes even more dramatic in grand unified theories where we may have to adjust parameters with as many as 28 significant figures. This is the problem of gauge hierarchy which is connected to the enormously different mass scales at which spontaneous symmetry breaking occurs. The breaking of G into $U(1) \otimes SU(2) \otimes SU(3)$ happens at $M \sim 10^{16}$ GeV. This means that a certain BEH field Φ acquires a non-zero vacuum expectation value $V \sim 10^{16}$ GeV. The second breaking, that of $U(1) \otimes SU(2)$, occurs at $\mu \sim 10^2$ GeV; i.e. we must have a second scalar field H with $v \sim 10^2$ GeV. But the combined potential of the scalar fields will contain terms of the form $\Phi^2 H^2$ (see Eq. (26.21)). Therefore, after the first breaking, the H mass will be given by

$$m_H^2 = \mu^2 + O(\alpha V^2), \tag{26.76}$$

where μ is the mass appearing in the symmetric Lagrangian. On the other hand we know that $v^2 \sim m_H^2$, so unless there is a very precise cancellation between μ^2 and $O(\alpha V^2)$, a cancellation which should extend to 28 decimal figures, v will turn out to be of order V and the two breakings will come together; in other words, the theory is not able to sustain naturally a gauge hierarchy. It is the same naturalness problem, except that here it appears already at the classical approximation. This grand-fine tuning of parameters must be repeated order by order in perturbation theory because, unlike fermions, scalar field masses require counter-terms which depend quadratically on the mass scale. The whole structure looks extremely unlikely. The problem is similar to that of the induced cosmological constant in any theory with spontaneous symmetry breaking. We believe that despite its rather technical aspect, the problem is sufficiently important so that some new insight will be gained when it is eventually solved.

One possible remedy is to throw away the scalars as fundamental elementary particles. After all, their sole purpose was to provoke the spontaneous symmetry breaking through their non-vanishing vacuum expectation values. In non-relativistic physics this phenomenon is known to occur but the role of the BEH fields is played by fermion pairs. This idea of dynamical symmetry breaking has been studied extensively, especially under the name of 'technicolour'. The physical idea is very simple and attractive. Let us introduce it by considering first QCD with two massless quark flavours. Although we cannot solve the theory, we believe that it presents the phenomenon of spontaneous chiral symmetry breaking and the ground state contains an arbitrary number of quark–anti-quark pairs. As a result the quarks acquire a mass and a triplet of zero mass Goldstone particles appear, which we identified to the pions.

Let us now introduce the $U(1) \times SU(2)$ gauge theory without any elementary scalar fields. The W boson 2-point function is of the general form

$$G^{\mu\nu}(p) = \left(g^{\mu\nu} - \frac{p^\mu p^\nu}{p^2} \right) \Pi(p^2) \tag{26.77}$$

Figure 26.9 *The one-pion intermediate state in the W propagator.*

and, for a zero mass vector boson, we must have

$$\Pi(0) = 0. \tag{26.78}$$

We do not know how to compute $\Pi(p^2)$ exactly since this would require the complete solution of the QCD dynamics; however, we can estimate the contribution of a particular intermediate state, that of the one-pion state. It is given by the diagram of Fig. 26.9. It gives

$$G^{\mu\nu}(p)|_{\text{1-pion}} = i(igf_\pi p^\mu)\frac{i}{p^2}(-igf_\pi p^\nu), \tag{26.79}$$

where g is the $SU(2)$ coupling constant and f_π the effective constant which couples the pion to the axial current defined in Eq. (25.97). In writing the right-hand side of Eq. (26.79) we made use of the fact that the pion, being a Goldstone particle, is massless. This explains the $1/p^2$ term in the propagator. It follows that the contribution of this particular term in the W propagator is

$$\Pi(p^2) = g^2 f_\pi^2. \tag{26.80}$$

The surprising result is that the two equations (26.80) and (26.78) disagree; in other words, the W has acquired a mass. No other state can cancel the value of (26.80) because no other state has a $1/p^2$ pole. Therefore, we have obtained a W mass given by

$$m_W = gf_\pi. \tag{26.81}$$

What about the neutral vector bosons. Here the situation is more complicated because π^0 couples to both A_μ^3 and B_μ. It follows that there is one linear combination which couples to π^0 and becomes massive and the orthogonal one which remains massless. It is not difficult to recognise Eqs. (25.26) and (25.27) of the Weinberg–Salam model. Even its great success, namely the relation between m_W, m_Z, and θ_W, is there!

Could the pions be the BEH scalars of weak interactions? Unfortunately, the answer is no. First, because there would be no physical pions left. They would have been eaten up by the BEH mechanism. Second, the order of magnitude is not right. Using the experimental value $f_\pi \sim 10^2$ MeV, we find m_W of the same order. But the message of the exercise is clear: we need new interactions which become strong at a scale one thousand times larger than QCD.

Technicolor is the commercial name of the new strong interactions in the multi-hundred GeV region. Let us assume that we have a non-Abelian gauge theory with a certain number of 'techni-flavours', for example one doublet, of massless fermions (the

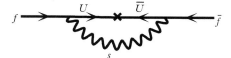

Figure 26.10 *Fermion mass generation in the extended technicolour scheme.*

'techni-quarks'), U_t and D_t, each coming in several 'techni-colours', for example three. The corresponding gauge bosons form a multiplet of 'techni-gluons'. These new strong interactions look like QCD except that we assume that their Λ parameter defined in Eq. (25.89), i.e. the scale in which the interaction becomes strong, is $\Lambda_{\text{Techn}} \sim 10^3 \Lambda_{\text{QCD}}$. All our previous analysis applies verbatim, but now you should read GeV everywhere in the place of MeV. The massless technipions are absorbed through the BEH mechanism, but a whole spectrum of 'techni-hadrons' appear with masses around 1 TeV.

Is this the end of the story? Remember that in the standard model the scalar fields are those which give masses to the fermions. Here who gives masses to ordinary quarks and leptons? We could try to be economical and use the same mechanism at higher orders, but ordinary fermions communicate with the techniquarks only through the electroweak gauge bosons. Since all these couplings are vectorial (vector or axial vector), they cannot change the chirality of the fermions so they will never produce a mass term. We should complicate the model by introducing a third type of gauge interaction which couples directly ordinary fermions to U_t and D_t. In such a scheme of 'extended technicolor' the mechanism for the fermion mass generation is shown in Fig. 26.10. The ordinary fermion f is coupled to U_t or D_t through the extended technicolor gauge boson S. Because of the vacuum condensate of the techniquarks the chirality of the fermion line changes and the net result is an effective $\bar{f}f$ term, i.e. a fermion mass. But now, who gives masses to the new gauge bosons S? We need a new interaction and repeat the same story.

Despite their many attractive features, the technicolor models suffer, up to now, from several important difficulties. First, the available field theory technology does not allow for any precise quantitative computation of bound state effects and everything must be based on analogy with the chiral symmetry breaking in QCD. Second, as we have just seen, although the initial idea is very simple, its detailed application gets quite complicated. Third, and most important, nobody has succeeded in producing an entirely satisfactory phenomenological model. In particular, the simplest models have large flavour-changing neutral currents. Nevertheless there is still hope and the scheme has some predictions which can be tested experimentally in the near future.

27

Supersymmetry, or the Defence of Scalars

27.1 Introduction

The best defence of scalars is the remark that they are not the only ones which reduce the predictive power of a gauge theory. As we have already seen, going through the chain radiation–fermion matter fields–BEH scalars we encounter an increasing degree of arbitrariness. Eliminating the scalars does not eliminate all arbitrariness. A more attractive possibility would be to connect the three worlds with some sort of symmetry principle. Then the knowledge of the vector bosons will determine the fermions and the scalars and the absence of quadratically divergent counter-terms in the fermion masses will forbid their appearance in the scalar masses.

Is it possible to construct such a symmetry? A general form of an infinitesimal transformation acting on a set of fields $\phi^i(x), i = 1, ..., m$, can be written as

$$\delta\phi^i(x) = \epsilon^a (T_a)^i_j \phi^j(x), \tag{27.1}$$

where $a = 1, ..., n$ with n denoting the number of independent transformations; in other words, n is the number of generators of the Lie algebra of the group we are considering. For $U(1)$ $n = 1$, for $SU(2)$ $n = 3$, etc. The ϵ's are infinitesimal parameters and T_a is the matrix of the representation of the fields. Usually the ϵ's are taken to be c-numbers; in which case the transformation (27.1) mixes only fields with the same spin and obeying the same statistics. It is clear that if we want to change the spin of the fields with a transformation (27.1), the corresponding ϵ's must transform non-trivially under rotations. If they have non-zero integer spin they can mix scalars with vectors, or spin-$\frac{1}{2}$ with spin-$\frac{3}{2}$ fields. This was the case with the old $SU(6)$ transformations and their relativistic extensions. If, on the other hand, the ϵ's are anticommuting parameters, they will mix fermions with bosons. If they have zero spin, the transformations (27.1) will change the statistics of the fields without changing their spin; i.e. they will turn a physical field into a ghost. The BRST transformations which we saw in the quantisation of non-Abelian gauge theories belong to this class. Here we want to connect physical bosons

From Classical to Quantum Fields. Laurent Baulieu, John Iliopoulos and Roland Sénéor.
© Laurent Baulieu, John Iliopoulos and Roland Sénéor, 2017. Published 2017 by Oxford University Press.

with physical fermions; therefore, the infinitesimal parameters must be anticommuting spinors. We will call such transformations 'supersymmetry transformations' and we see that a given irreducible representation will contain both fermions and bosons. It is not a priori obvious that such supersymmetries can be implemented consistently, but in fact they can. In the following we will give a very brief description of their properties as well as their possible applications. There exist several specialised textbooks on this subject to which we refer for further reading.

27.2 The Supersymmetry Algebra

We want to find symmetry transformations which generalise (27.1) with anti-commuting ϵ's. First some definitions. Let A_m $m = 1, ..., D$ denote the generators of a Lie algebra and Q_α $\alpha = 1, ..., d$ be the elements of a d-dimensional representation:

$$[A_m, A_n] = f_{mn}^l A_l; \qquad [A_m, Q_\alpha] = s_{m\alpha}^\beta Q_\beta. \tag{27.2}$$

A graded superalgebra is the algebraic scheme which consists of the generators A_m and Q_α if we can find a set of constants $r_{\alpha\beta}^m$ such that

$$[Q_\alpha, Q_\beta]_+ = Q_\alpha Q_\beta + Q_\beta Q_\alpha = r_{\alpha\beta}^m A_m. \tag{27.3}$$

The constraint on the r's is that they must satisfy the corresponding Jacobi identities for the set of Eqs. (27.2) and (27.3) to be self-consistent.

There exist theorems which give a classification of graded superalgebras analogous to the Cartan classification of Lie algebras, but we will not need them here. The only superalgebra we will use is the one in which the Lie algebra is that of the Poincaré group with generators P_μ and $M_{\mu\nu}$ and the grading representation (27.2) is given by a Majorana spinor Q_α:

$$[P_\mu, Q_\alpha] = 0; \qquad [Q_\alpha, M_{\mu\nu}] = i\gamma_{\alpha\beta}^{\mu\nu} Q_\beta \tag{27.4}$$

$$[Q_\alpha, Q_\beta]_+ = 0; \qquad [Q_\alpha, \bar{Q}_\beta]_+ = -2\gamma_{\alpha\beta}^\mu P_\mu, \tag{27.5}$$

in which $\gamma^{\mu\nu} = \frac{1}{4}[\gamma^\mu, \gamma^\nu]$. The defining relations (27.4) and (27.5) admit the Lorentz group $SL(2, C)$ as an automorphism. The components of Q are the generators of the special supersymmetry transformations. The second relation of (27.4) shows only that they have spin $\frac{1}{2}$. The first is more important because it shows that they are translationally invariant. We will come back to this point later. The anticommutation relations (27.5) are the fundamental relations of the new symmetry.

An obvious generalisation of (27.4) and (27.5) consists of starting from the Poincaré algebra $\otimes$ a compact internal symmetry G with generators A_i. If the Q's belong to a certain representation of the internal symmetry, we write them as Q_α^m where the index $\alpha = 1, ..., 4$ labels the components of the Majorana spinor and m is the index of the internal symmetry, $m = 1, 2, ..., d$. The algebra now takes the form

$$[A_i, A_j] = f_{ij}^k A_k; \quad [P_\mu, Q_\alpha^m] = 0; \quad [Q_\alpha^m, M_{\mu\nu}] = i\gamma_{\alpha\beta}^{\mu\nu} Q_\beta^m \qquad (27.6)$$

$$[A_i, Q_\alpha^m] = s_{in}^m Q_\alpha^n; \quad [Q_\alpha^m, Q_\beta^n]_+ = 0; \quad [Q_\alpha^m, \bar{Q}_\beta^n]_+ = -2\delta^{mn}\gamma_{\alpha\beta}^\mu P_\mu. \qquad (27.7)$$

The algebra (27.6) and (27.7) admits $SL(2, C) \otimes G$ as a group of automorphisms.

In the literature both the Majorana and the Weyl representations are frequently used and it is helpful to be able to change from one to the other.

For example, in the Weyl representation (27.5) becomes

$$[Q_\alpha, Q_\beta]_+ = [\bar{Q}_{\dot{\alpha}}, \bar{Q}_{\dot{\beta}}]_+ = 0 \quad ; \quad [Q_\alpha, \bar{Q}_{\dot{\beta}}]_+ = 2(\sigma^\mu)_{\alpha\dot{\beta}}P_\mu. \qquad (27.8)$$

We see that the supersymmetry algebra with N spinorial generators has a natural global $U(N)$ symmetry, the group of unitary transformations that transform the Q's among themselves. Such symmetries were introduced by Pierre Fayet and are called *R-symmetries*.[1] For $N=1$ we have a group of phase transformations

$$Q_\alpha \to e^{i\phi} Q_\alpha. \qquad (27.9)$$

27.3 Why This Particular Algebra; or All Possible Supersymmetries of the *S* Matrix

The superalgebra (27.6) and (27.7) combines in a non-trivial way Poincaré invariance with an internal symmetry. There exists a theorem, known as *the Coleman–Mandula theorem*, which states that for ordinary algebras, such a combination cannot be a symmetry of a unitary S matrix. The only possibility is the trivial case of a direct product 'Poincaré $\otimes$ internal symmetry'. This is what we have been using everywhere in this book. The superalgebras (27.6) and (27.7) seem to violate this theorem. The reason is that, implicit in the Coleman–Mandula proof, there is the assumption that the algebra closes with commutators only. To be more precise, they use the fact that the trace of the operators that appear in the algebraic relations vanishes. This is always true if these relations contain only commutators because of the cyclic property of the trace. But if we allow for anticommutators it is no more true. We will state, without proof, the generalisation of this theorem to include superalgebras, i.e. algebras which close using both commutators and anti-commutators. The remarkable result, obtained by Haag, Lopuszanski, and Sohnius, is that (27.4) and (27.5), or (27.6) and (27.7), is essentially the only admissible one. More precisely, the theorem shows that the grading operators Q should belong to the $(\frac{1}{2}, 0)$, or $(0, \frac{1}{2})$ representation of the Lorentz group. The only possible extension is to generalise the anticommutators in (27.8) by introducing operators which commute with all other operators in the algebra. Thus, we may have

$$[Q_\alpha^m, Q_\beta^n]_+ = \epsilon_{\alpha\beta} Z^{mn}, \qquad (27.10)$$

---

[1] Fayet had introduced the $U(1)$ R-symmetry before the advent of supersymmetry and extended the concept later.

where Z^{mn} are a set of central charges, i.e. operators which commute with every operator in the algebra. Since $\epsilon_{\alpha\beta}$ is antisymmetric in its indices, it follows that the Z's must also be antisymmetric, $Z^{mn} = -Z^{nm}$, which in turn implies that this central charge vanishes for $N=1$. The conclusion is that out of the infinitely many ways we can grade the Poincaré algebra, only the one we introduced, in which we used a spin-$\frac{1}{2}$ operator, may be relevant for physics.

27.4   Representations in Terms of One-Particle States

In order to extract the possible physical consequences of supersymmetry, we must construct the representations of the algebra in terms of one-particle states, i.e. the one-particle 'supermultiplets'. We start by observing that the spinorial charges commute with P_μ and, therefore, they do not change the momentum of the one-particle state. Furthermore, the operator P^2 commutes with all the operators of the algebra, which implies that all the members of a supermultiplet will have the same mass. As it is the case with the Poincaré algebra, we can distinguish two cases: $P^2 \neq 0$ and $P^2 = 0$.

27.4.1   Massive Case

We can go to the rest frame in which the right-hand side of (27.5) or (27.8) becomes a number. Let us first forget about a possible internal symmetry and consider the case $N=1$. Equation (27.8) gives

$$[Q_\alpha, \bar{Q}_{\dot\beta}]_+ = 2M\delta_{\alpha\dot\beta}, \qquad (27.11)$$

where $P^2 = M^2$. Equation (27.11) implies that the operators $Q/\sqrt{2M}$ and $\bar{Q}/\sqrt{2M}$ satisfy the anticommutation relations for creation and annihilation operators of free fermions. Since the index α can take two values, 1 and 2, and $Q_1^2 = Q_2^2 = 0$, starting from any one-particle state with spin S and projection S_z, we can build a four-dimensional Fock space with states

$$|S, S_z; n_1, n_2 >= Q_2^{n_2} Q_1^{n_1} |S, S_z > \qquad n_1, n_2 = 0, 1. \qquad (27.12)$$

We can define a parity operation under which the Majorana spinor Q_α, $\alpha = 1, ...4$, transforms as

$$(Q_\alpha)_P = (\gamma^0 Q)_\alpha. \qquad (27.13)$$

Then the spin-parity content of the representation (27.12) is

$$(S - 1/2)^\eta; \quad S^{i\eta}; \quad S^{-i\eta}; \quad (S + 1/2)^{-\eta}, \qquad (27.14)$$

where $\eta = \pm i, \pm 1$ for S integer or half-integer. Some examples follow:

$S = 0$: a scalar, a pseudoscalar, a spinor.

$S = \frac{1}{2}$: a scalar, a vector, two spinors (or a pseudo-scalar, a pseudo-vector, two spinors).

$S = 1$: a vector, a pseudo-vector, a spinor, a $\frac{3}{2}$ spinor.

$S = \frac{3}{2}$: a vector, a tensor, two $\frac{3}{2}$ spinors.

In this counting the spinors are Weyl or Majorana. We see that the $S=0$ example can be viewed as the simplest supersymmetric extension of the chiral spinors of the standard model, so this multiplet is often called *chiral multiplet*. Similarly, the $S = \frac{1}{2}$ example corresponds to the supersymmetric extension of a massive vector.

The generalisation to include internal symmetries is straightforward. Ignoring, for the moment, the possibility of having central charges in the supersymmetry algebra, we see that the only difference is that now we have more creation operators and the corresponding Fock space has 2^{2N} independent states, where N is the number of spinorial charges. A well-established theoretical prejudice, supported by all available experimental data, is that if we exclude gravitation, there exist no elementary particles with spin higher than 1. This prejudice is based on the great difficulties we encounter if we want to write consistent field theories with high spin particles. Based on this prejudice, $N = 2$ should be the largest supersymmetry for massive particles because it allows precisely the transition from a -1 to a $+1$ spin state in four steps of $\frac{1}{2}$.[2]

The presence of central charges brings an interesting complication. We consider the algebra (27.7) modified by (27.10). We can construct the operator $Q_\alpha^{m(-)}$ as

$$Q_\alpha^{m(-)} = Q_\alpha^m - \sum_{\beta n} \epsilon_{\alpha\beta} U^{mn} \bar{Q}_\beta^n \tag{27.15}$$

with U being a unitary $N \times N$ matrix. Let us compute the anticommutator

$$\sum_{\alpha m} [Q_\alpha^{m(-)}, Q_\alpha^{m(-)\dagger}]_+ = \sum_{\alpha m} [Q_\alpha^m - \sum_{\beta n} \epsilon_{\alpha\beta} U^{mn} \bar{Q}_\beta^n , \bar{Q}_\alpha^m - \sum_{\gamma k} \epsilon_{\alpha\gamma} U^{\dagger mk} Q_\gamma^k]_+$$
$$= 8NH - 2\text{Tr}(UZ^\dagger + ZU^\dagger) \tag{27.16}$$

with H being the Hamiltonian. The left-hand side of (27.16) is a positive-definite operator, so if we take the matrix element between a one-particle state at rest, we obtain the inequality

$$M \geq \frac{1}{4N} \text{Tr}(UZ^\dagger + ZU^\dagger). \tag{27.17}$$

Taking into account the fact that the matrix Z can be written as the product of a positive Hermitian matrix times the unitary matrix U, we can rewrite this inequality as

---

[2] We will see later how this counting may be modified by the presence of possible central charges in the supersymmetry algebra.

$$M \geq \frac{1}{2N} \mathrm{Tr} \sqrt{Z^\dagger Z}, \tag{27.18}$$

which gives an interesting lower bound for the mass of a multiplet in terms of its eigenvalues of the central charges. If the bound is saturated, the operators $Q_\alpha^{m(-)}$ annihilate any state of the multiplet, which justifies the superscript $(-)$. But this means that we have only N creation and annihilation operators, rather than $2N$. We will call such multiplets *short multiplets*, since they have fewer states. We will see that they appear naturally in gauge theories with extended supersymmetry when part of the symmetry is spontaneously broken.[3]

27.4.2   Massless Case

Here we choose the frame $P_\mu = (E, 0, 0, E)$. The relation (27.8) yields

$$[Q_\alpha, \bar{Q}_{\dot\beta}]_+ = 2E(1 - \sigma_z) = 4E\delta_{\alpha 2}\delta_{\dot\beta 2}. \tag{27.19}$$

Only Q_2 and $\bar{Q}_{\dot 2}$ can be considered as creation and annihilation operators. The other two operators Q_1 and $\bar{Q}_{\dot 1}$ annihilate any one-particle state with the momentum $P_\mu = (E, 0, 0, E)$. Starting from a one-particle state with helicity $\pm\lambda$, we obtain the state with helicity $\pm(\lambda + \frac{1}{2})$. It follows that all massless multiplets of $N=1$ supersymmetry contain just two states, one fermionic and one bosonic. As we have shown in Chapter 12 all quantum field theories we are considering satisfy the *CPT* theorem. Therefore, to a state with helicity λ different from 0 corresponds the *CPT*-transformed state with helicity $-\lambda$. For every multiplet we will also have the *CPT*-transformed multiplet. Some interesting examples follow:

$\lambda = -\frac{1}{2}$: one spin $\frac{1}{2}$ and one spin 0 (both massless).

$\lambda = \frac{1}{2}$: one spin $\frac{1}{2}$ and one spin 1 (both massless).

$\lambda = \frac{3}{2}$: one spin $\frac{3}{2}$ and one spin 2 (both massless).

The first is the massless version of the chiral multiplet we found in the massive case. Obviously, we will be tempted to identify the second one with the photon, or any other massless gauge boson, and its supersymmetric partner, often called *photino*, or *gaugino*, and the third with the conjectured mediator of the gravitational forces, *the graviton* and its partner, called *gravitino*. The counting of degrees of freedom that we saw in the BEH effect can be made also in supersymmetry. Including the CPT-transformed multiplets,

---

[3] The mass bound obtained in (27.18) reminds us of a similar one we obtained in section 26.2.5 when we studied the mass spectrum of magnetic monopoles, namely the inequalities (26.66) and (26.72). In the Bogomolny–Prasad–Sommerfield limit the corresponding bound was saturated. As we will see later, the short multiplets of extended supersymmetry appear often in grand unified theories when the spontaneous breaking of the non-Abelian gauge symmetry induces the appearance of magnetic monopoles and in these cases the two bounds coincide. For this reason the short multiplets are also called *BPS multiplets*, although, in general, the two phenomena are unrelated.

we can combine the degrees of freedom of a massless multiplet containing one spin 1 and one spin-$\frac{1}{2}$ particle with those of a multiplet containing a spin-$\frac{1}{2}$ and two spin 0 to obtain a massive multiplet containing a massive spin 1, two spin $\frac{1}{2}$ Majorana, or Weyl (which can be combined into one spin $\frac{1}{2}$ Dirac), and one spin 0.

If we have more than one spinorial charge, i.e. $N > 1$, we obtain N creation and annihilation operators. Again, the operators Q_1^m and $\bar{Q}_1^m$ annihilate the one-particle massless states. Note, that this implies that the same should be true for any central charges; in other words, massless particles must be neutral under central charges in extended supersymmetry. The consequence of the prejudice we mentioned previously is that $N = 4$ is the largest supersymmetry which may be interesting for particle physics without gravitation if we consider massless particles. The reason is that $N = 4$ contains four creation operators and allows us to go from a helicity state $\lambda = -1$ to that of $\lambda = +1$. Any increase in the number of spinorial charges will automatically yield representations containing higher helicities. Finally, if we include gravitation, the same prejudice tells us that we must allow for elementary particles with helicities $|\lambda| \leq 2$. The previous counting argument now gives $N = 8$ as the maximum allowed supersymmetry.

We can easily find examples of massless representations for such extended supersymmetries. Let us assume that we have a state which has the maximum helicity in the representation; we call it λ_{max}. We can apply N annihilation operators, thus obtaining N states with helicity $\lambda_{\text{max}} - \frac{1}{2}$. After n such steps we obtain $N!/n!(N-n)!$ states with helicity $\lambda_{\text{max}} - n/2$ which form a rank n antisymmetric tensor representation of the R-symmetry. We can apply at most N annihilation operators, so the minimum value of the helicity is $\lambda_{\text{min}} = \lambda_{\text{max}} - N/2$. Here again, we must add the CPT-transformed states. Some examples follow: for $N=2$, we have two multiplets not exceeding spin 1.

$\lambda_{\text{max}} = 1$: one spin 1, two spin $\frac{1}{2}$, forming a doublet of R, two spin-0 singlets of R.

$\lambda_{\text{max}} = \frac{1}{2}$: one spin $\frac{1}{2}$, two spin 0, forming a doublet of R.

For obvious reasons, we will call the first *a gauge multiplet* and the second *a matter multiplet*.

Similarly, for $N=4$ we have only one possible supermultiplet containing 1 spin 1, 4 spin $\frac{1}{2}$, and 6 spin 0. It is identical with its CPT-transformed multiplet. In our previous terminology, it is a gauge multiplet and we do not have the analogue of a matter multiplet. In $N=3$, we find the same answer because, for $\lambda_{\text{max}}=1$, we find 1 spin 1 with helicity $+1$, 3 spin $\frac{1}{2}$ with helicity $+\frac{1}{2}$, 3 spin 0 and 1 spin $\frac{1}{2}$ with helicity $-\frac{1}{2}$, to which we should add the CPT-transformed multiplet, thus reconstructing the one we found for $N=4$. As a last example we compute the helicity states of the $N = 8$ multiplet with $\lambda_{\text{max}}=2$. There is only one possibility with 1 spin 2, 8 spin $\frac{3}{2}$, 28 spin 1, 56 spin $\frac{1}{2}$, and 70 spin 0. Again this multiplet is identical with that of $N=7$ taking into account CPT.

A concluding remark: All representations contain equal number of bosonic and fermionic states. All states in an irreducible representation have the same mass.

27.5 Representations in Terms of Field Operators: Superspace

Our aim is to obtain field theoretical realisations of supersymmetry; therefore, we look for representations in terms of local fields. Such representations were first obtained by trial and error, but the most elegant method is to use the concept of superspace.

We want to find a representation of the supersymmetry algebra (27.4) and (27.5) in terms of differential operators. Following the same method we use for the ordinary Poincaré group, we consider the coset space formed by the super-Poincaré group[4] divided by the Lorentz group. This natural parameter space for the operators P_μ, Q, and $\bar{Q}$ has eight dimensions, four (the usual Minkowski space) associated with the operators P_μ and four with the spinors Q and $\bar{Q}$. The last four coordinates, however, are not numbers but elements of a Grassmann algebra. An element of this space will be denoted by $z^M = (x, \theta, \bar{\theta})$. This eight-dimensional space is called 'superspace'. In Chapter 11 we introduced the basic rules of calculus in a Grassmannian manifold. They apply to superspace.

A 'finite' group element can be defined by

$$G(x, \theta, \bar{\theta}) = e^{i[\theta Q + \bar{\theta}\bar{Q} - x \cdot P]}. \tag{27.20}$$

The quotation marks around the word 'finite' mean that although G is formally an exponential in the variables θ and $\bar{\theta}$, its expansion in powers of them terminates after the first terms. In this sense a supersymmetry transformation is always 'infinitesimal'. We can multiply two such group elements and, by Hausdorff's formula, obtain

$$G(y, \xi, \bar{\xi})G(x, \theta, \bar{\theta}) = G(y + x - i\xi\sigma\bar{\theta} + i\theta\sigma\bar{\xi}, \xi + \theta, \bar{\xi} + \bar{\theta}). \tag{27.21}$$

This means that the group induces a motion of the parameter space into itself

$$G(y, \xi, \bar{\xi}) : (x, \theta, \bar{\theta}) \rightarrow (y + x - i\xi\sigma\bar{\theta} + i\theta\sigma\bar{\xi}, \xi + \theta, \bar{\xi} + \bar{\theta}). \tag{27.22}$$

Equation (27.22) shows that supersymmetry transformations act on superspace as generalised translations. The required representation of the algebra (27.4) and (27.5) in terms of differential operators can be read off (27.22):

$$Q_\alpha = \frac{\partial}{\partial \theta^\alpha} - i\sigma^\mu_{\alpha\dot{\alpha}}\bar{\theta}^{\dot{\alpha}}\frac{\partial}{\partial x^\mu}$$

$$\bar{Q}_{\dot{\alpha}} = -\frac{\partial}{\partial \bar{\theta}^{\dot{\alpha}}} + i\sigma^\mu_{\alpha\dot{\alpha}}\theta^\alpha\frac{\partial}{\partial x^\mu} \tag{27.23}$$

$$P_\mu = i\frac{\partial}{\partial x^\mu}.$$

---

[4] That is, the group we obtain by formally integrating the algebra (27.4) and (27.5).

A superfield is a function of the superspace element $z^M : \phi...(x, \theta, \bar{\theta})$, where the dots stand for possible Lorentz tensor or spinor indices. A transformation of the group acts on it as a generalised translation:

$$G(y, \xi, \bar{\xi})\phi(x, \theta, \bar{\theta}) = \phi(y + x - i\xi\sigma\bar{\theta} + i\theta\sigma\bar{\xi}, \xi + \theta, \bar{\xi} + \bar{\theta}). \qquad (27.24)$$

The interest of the superfields derives from the fact that like any function of Grassmann variables, they are polynomials in θ and $\bar{\theta}$,

$$\phi(x, \theta, \bar{\theta}) = A(x) + \theta\psi(x) + \bar{\theta}\bar{\chi}(x) + ... + \theta\theta\bar{\theta}\bar{\theta}R(x), \qquad (27.25)$$

where the coefficient functions $A(x)$ (scalar), $\psi(x)$ (spinor), etc. are ordinary fields, i.e. a superfield is a finite multiplet of fields. Using the transformation property (27.24) and expanding both sides in powers of θ and $\bar{\theta}$ we obtain the transformation properties of the coefficient functions which, under supersymmetry transformations, transform among themselves. In this way we have obtained representations of supersymmetry in terms of a finite number of fields.

It is easy to see that the representation (27.25) is a reducible one. We must be able to impose covariant restrictions on the superfield (27.25) in order to decompose it into irreducible representations. For example, we can verify that the condition on ϕ to be a real function is a covariant one. The systematic way to obtain such covariant constraints is to realise that the algebra (27.4) and (27.5) contains the algebra of the Q's or the $\bar{Q}$'s as subalgebras

$$[Q, Q]_+ = 0; \qquad [\bar{Q}, \bar{Q}]_+ = 0. \qquad (27.26)$$

We can therefore study the motion of the group on the corresponding cosets. We can parametrise the group elements as

$$G_1(x, \theta, \bar{\theta}) = e^{i(\theta Q - x \cdot P)}e^{i\bar{\theta}\bar{Q}}; \qquad G_2(x, \theta, \bar{\theta}) = e^{i(\bar{\theta}\bar{Q} - x \cdot P)}e^{i\theta Q}. \qquad (27.27)$$

The formulae (27.25) and (27.27) give three different but equivalent ways of representing the group elements and, therefore, lead to three different types of super-fields. Of course, by Hausdorff's formula we can shift from one to another, the three representations being equivalent,

$$G(x, \theta, \bar{\theta}) = G_1(x + i\theta\sigma\bar{\theta}, \theta, \bar{\theta}) = G_2(x - i\theta\sigma\bar{\theta}, \theta, \bar{\theta}), \qquad (27.28)$$

and similarly for the corresponding superfields

$$\phi(x, \theta, \bar{\theta}) = \phi_1(x + i\theta\sigma\bar{\theta}, \theta, \bar{\theta}) = \phi_2(x - i\theta\sigma\bar{\theta}, \theta, \bar{\theta}). \qquad (27.29)$$

The generators Q and $\bar{Q}$, which on a superfield of type ϕ were represented by the operators (27.23), when acting on a superfield of type ϕ_1 are given by

$$Q = \frac{\partial}{\partial\theta} - 2i\sigma\bar{\theta}\frac{\partial}{\partial x} \quad ; \quad \bar{Q} = \frac{\partial}{\partial\bar{\theta}} \tag{27.30}$$

and on a superfield of type ϕ_2 by

$$Q = \frac{\partial}{\partial\theta} \quad ; \quad \bar{Q} = -\frac{\partial}{\partial\bar{\theta}} + 2i\theta\sigma\frac{\partial}{\partial x}. \tag{27.31}$$

We see that the same way that we were able to impose a reality constraint which was invariant for a superfield of type ϕ, we can, for example, impose on a superfield of type ϕ_1 to be independent of $\bar{\theta}$, or on ϕ_2 to be independent of θ. In other words, $\partial/\partial\bar{\theta}$ is a covariant derivative when it acts on a superfield of type ϕ_1 and $\partial/\partial\theta$ on ϕ_2. By a shift (27.29) we define covariant derivatives for any type of superfield

$$\mathcal{D}_\alpha\phi = \left(\frac{\partial}{\partial\theta^\alpha} + i\sigma_{\mu\alpha\dot{\alpha}}\bar{\theta}^{\dot{\alpha}}\frac{\partial}{\partial x^\mu}\right)\phi; \quad \bar{\mathcal{D}}\phi = \left(-\frac{\partial}{\partial\bar{\theta}} - i\theta\sigma\frac{\partial}{\partial x}\right)\phi \tag{27.32}$$

$$\mathcal{D}\phi_1 = \left(\frac{\partial}{\partial\theta} + 2i\sigma\bar{\theta}\frac{\partial}{\partial x}\right)\phi_1; \quad \bar{\mathcal{D}}\phi_1 = -\frac{\partial}{\partial\bar{\theta}}\phi_1 \tag{27.33}$$

$$\mathcal{D}\phi_2 = \frac{\partial}{\partial\theta}\phi_2; \quad \bar{\mathcal{D}}\phi_2 = \left(-\frac{\partial}{\partial\bar{\theta}} - 2i\theta\sigma\frac{\partial}{\partial x}\right)\phi_2. \tag{27.34}$$

These differential operators anticommute with the infinitesimal supersymmetry transformations. They will be very useful when we decide to construct Lagrangian field theory models.

A superfield of any of the three types shown in Eqs. (27.32)–(27.34) may have Lorentz indices corresponding to a given spin s. In this case the lowest spin in the representation will be s. We can sharpen this analysis and obtain all linear irreducible representations, but we will not need them here. In this book we have studied only the properties of scalar, spinor, and vector fields and, therefore, we will introduce the superfields that contain only these fields. Obviously, any representation will contain both fermion and boson fields. A superfield of type ϕ_1 will give

$$\phi_1(x,\theta) = A(x) + \theta\psi(x) + \theta\theta F(x). \tag{27.35}$$

It contains spin 0 and spin-1/2 fields, and similarly for $\phi_2(x,\bar{\theta})$. The real superfield ϕ can be expanded as

$$\begin{aligned}
\phi(x,\theta,\bar{\theta}) = {}& C + i\theta\chi - i\bar{\theta}\bar{\chi} + \frac{i}{2}\theta\theta(M+iN) - \frac{i}{2}\bar{\theta}\bar{\theta}(M-iN) \\
& - \theta\sigma_\mu\bar{\theta}v^\mu \\
& + i\theta\theta\bar{\theta}_{\dot{\alpha}}\left(\bar{\lambda}^{\dot{\alpha}} - \frac{i}{2}\sigma^{\dot{\alpha}}_{\mu\alpha}\partial^\mu\chi^\alpha\right) - i\bar{\theta}\bar{\theta}\theta\left(\lambda + \frac{i}{2}\sigma_\mu\partial^\mu\bar{\chi}\right) \\
& + \frac{1}{2}\theta\theta\bar{\theta}\bar{\theta}\left(D + \frac{1}{2}\Box C\right),
\end{aligned} \tag{27.36}$$

where a reality condition on ϕ has been imposed. The supermultiplet (27.35) contains a chiral spinor $\psi(x)$ and it is called 'chiral multiplet' while (27.36) contains a vector field $v_\mu(x)$ and is called 'vector'. The peculiar notation in the coefficients of the expansion (27.36) is used because of historical reasons and also because it leads to simpler transformation properties under infinitesimal transformations.

The global $U(1)$ R transformations we found when studying the algebra have a simple interpretation in superspace. A point $(x, \theta, \bar\theta)$ transforms as

$$(x, \theta, \bar\theta) \rightarrow (x, \theta e^{-i\phi}, \bar\theta e^{i\phi}). \tag{27.37}$$

A vector superfield is 'neutral' under R, while a chiral one may be multiplied by a phase

$$V(x, \theta, \bar\theta) \rightarrow V(x, \theta e^{-i\phi}, \bar\theta e^{i\phi}); \quad S(x, \theta) \rightarrow e^{in\phi} S(x, \theta e^{-i\phi}). \tag{27.38}$$

It is straightforward to write the explicit form of the transformation properties in terms of the component fields. For example, for the chiral multiplet (27.35), we obtain under supersymmetry

$$\delta A = \xi \psi; \quad \delta \psi = 2i\sigma_\mu \bar\xi \partial^\mu A + 2\xi F; \quad \delta F = i\partial^\mu \psi \sigma_\mu \bar\xi, \tag{27.39}$$

where ξ is the parameter of the infinitesimal supersymmetry transformation.

Given a chiral superfield $S(x, \theta)$ of type ϕ_1 and its conjugate $\bar S(x, \bar\theta)$, which is of type ϕ_2, we can construct a vector superfield in the following way. First, through (27.29) we transform both of them into superfields of type ϕ:

$$S(x, \theta) = \phi(x - i\theta\sigma\bar\theta, \theta); \quad \bar S(x, \bar\theta) = \bar\phi(x + i\theta\sigma\bar\theta, \bar\theta). \tag{27.40}$$

This way we can combine them together. The simplest is just to take the difference $i(S - \bar S)$, which we call ∂S. It is a vector multiplet whose components, in the notation of Eq. (27.36), are[5]

$$C = \text{Im}\, A; \quad \chi = \psi; \quad M + iN = F; \quad v_\mu = \partial_\mu \text{Re}\, A; \quad \lambda = D = 0. \tag{27.41}$$

The interest of this special vector multiplet is that it generalises the notion of a gauge transformation in which the vector field is transformed by the derivative of a scalar function. Similarly, given a vector multiplet $V(x, \theta, \bar\theta)$ we can construct chiral multiplets by eliminating the $\bar\theta$ dependence. Of particular interest is the one given by

---

[5] The relation $\chi = \psi$ should be read as χ = the Majorana spinor equivalent to the Weyl spinor ψ; in other words, $\chi = \begin{pmatrix} \psi_\alpha \\ \bar\psi^{\dot\alpha} \end{pmatrix}$. Note also that we could have taken the vector superfield corresponding to the sum $S + \bar S$. It is easy to verify that it is related to the one we consider by a parity transformation.

$$W_\alpha = -\frac{1}{4}\bar{\mathcal{D}}\bar{\mathcal{D}}\mathcal{D}_\alpha V. \qquad (27.42)$$

The chirality condition follows immediately from the fact that the product of three $\bar{\mathcal{D}}$ operators vanishes. It is a superfield with an undotted spinor index and we can verify that among its components it contains the term proportional to $\partial_\mu v_\nu - \partial_\nu v_\mu$. Therefore, it is suitable to describe the 2-form field strength of a vector field.

Two remarks concerning these representations. First, if we compare with the results obtained when we studied the representations in terms of one-particle states, we see that we have more fields than the physical states which are contained in an irreducible representation. Therefore, some of the fields must turn out to be auxiliary fields. Their presence is, however, necessary in order to ensure linear transformation properties. Second, we note that the field F in (27.39) transforms, under supersymmetry, with a total derivative. This is not accidental. It follows from the fact that a supersymmetry transformation acts in superspace as a generalised translation. But the F term is already proportional to the maximum allowed power of θ, so any higher power in the expansion will be necessarily a space derivative. This property will be always true with the last component in the expansion of a superfield, i.e. F for a chiral superfield (27.35), D for a vector (27.36), etc. We will use this property soon.

Before closing this section we must establish a tensor calculus in order to be able to combine irreducible representations together. This is essential for the construction of Lagrangian models. Here again the superfield formalism simplifies our task enormously. All the necessary tensor calculus is contained in the trivial observation that the product of two superfields is again a superfield. For example, let $S_1(x,\theta)$ and $S_2(x,\theta)$ be two superfields of type $\phi_1(x,\theta)$. We form the product: $S_{12}(x,\theta) = S_1(x,\theta)S_2(x,\theta)$. Expanding both members in powers of θ and identifying the coefficients we obtain

$$\begin{aligned} A_{12}(x) &= A_1(x)A_2(x) \\ \psi_{12}(x) &= \psi_1(x)A_2(x) + \psi_2(x)A_1(x) \\ F_{12}(x) &= F_1(x)A_2(x) + F_2(x)A_1(x) - \frac{1}{2}\psi_1(x)\psi_2(x). \end{aligned} \qquad (27.43)$$

Similarly, the product of two vector superfields is again a vector superfield. For example, taking the square V^2 and expanding in powers of θ and $\bar{\theta}$ we find that its D term contains the expression $v_\mu v^\mu$, i.e. the mass of the vector field. We can also multiply superfields upon which we have acted with the corresponding covariant derivatives, Eqs. (27.32) to (27.34) . As we noted already, two, or more, superfields of different types cannot get multiplied together. Rather we should transform them first into superfields of the same type by using the relations (27.29) and then multiply them. For example, we can multiply ϕ and $\bar{\phi}$ of Eq. (27.40) and expand in powers of θ and $\bar{\theta}$. This way we obtain a vector multiplet whose components are bilinears in the components of S. We can verify that the last term in the expansion, i.e. the D component of the vector multiplet, contains

$$\ldots - \theta\theta\bar{\theta}\bar{\theta} \left[A \Box A^* - \frac{i}{2} \psi \sigma^\mu \partial_\mu \bar{\psi} + FF^* \right]. \tag{27.44}$$

The first two terms are recognised as the kinetic energy terms of a complex spin-0 field and a two-component Weyl spinor. The last term has no derivative on the F-field, which shows that F will be, in fact, an auxiliary field.

The last bilinear we will need corresponds to the square of the chiral field (27.42) which we expect to use for the kinetic energy term of a vector field. Indeed, since the $v_{\mu\nu} = \partial_\mu v_\nu - \partial_\nu v_\mu$ term appears in the expansion of W_α in the term proportional to θ_α, the square $v_{\mu\nu} v^{\mu\nu}$ will appear in the F term in the expansion of its square:

$$W_\alpha W^\alpha |_{\theta\theta} + \bar{W}^{\dot\alpha} \bar{W}_{\dot\alpha} |_{\bar\theta\bar\theta} \sim -i\lambda\sigma^\mu \partial_\mu \bar\lambda - \frac{1}{4} v_{\mu\nu} v^{\mu\nu} + \frac{1}{2} D^2 + \frac{i}{4} v_{\mu\nu} \tilde{v}^{\mu\nu}. \tag{27.45}$$

We recognise in this expansion the kinetic energy term of the spinor and the vector fields. D appears with no derivative, so it will be an auxiliary field while the last term is a four derivative. Note that the same term can also be obtained as a D term in the expansion of the vector multiplet $W_\alpha \mathcal{D}^\alpha V$,

$$W_\alpha W^\alpha |_{\theta\theta} + \bar{W}^{\dot\alpha} \bar{W}_{\dot\alpha} |_{\bar\theta\bar\theta} \sim \left[W_\alpha \mathcal{D}^\alpha V + \bar{W}^{\dot\alpha} \bar{\mathcal{D}}_{\dot\alpha} V \right] |_{\theta\theta\bar\theta\bar\theta} + \ldots, \tag{27.46}$$

where the dots stand for terms which are four derivatives.

The final step is to use this tensor calculus and build Lagrangian field theories invariant under supersymmetry transformations. As we said earlier, supersymmetry transformations can be viewed as a kind of generalised translations in superspace. Therefore, the problem is similar to that of constructing translationally invariant field theories. We all know that the only Lagrangian density invariant under translations is a trivial constant. However, what is important is to have an invariant action which is obtained by integrating the Lagrangian density over all four-dimensional space–time. The same must be true for supersymmetry. Now, the Lagrangian density will be a function of some superfields and their covariant derivatives; i.e. it will be a function of the superspace point $(x, \theta, \bar\theta)$. The action will be given by an eight-dimensional integral over superspace:

$$I = \int \mathcal{L}(x, \theta, \bar\theta) \mathrm{d}^4 x \mathrm{d}^2\theta \mathrm{d}^2\bar\theta. \tag{27.47}$$

By construction, this integral is invariant under supersymmetry. This invariance can be verified by noting that only the last term in the expansion of $\mathcal{L}$ in powers of θ and $\bar\theta$, the one proportional to $\theta\theta\bar\theta\bar\theta$, will survive the integration. This follows from the integration rules over Grassmann variables we established in Chapter 11. But, as we noted earlier, the variation of the last term in the expansion of any superfield, such as an F or a D term, is given by a total derivative. Therefore, their integrals over all space–time vanish. We will call *F-terms* those coming from the expansion of a chiral superfield and *D-terms* those coming from vector multiplets. Since chiral and vector superfields are the only ones we will use in this book, all Lagrangian densities we will write will be of one of these two types. In fact, we can always work in superspace in terms of superfields and never write down the component fields explicitly. Feynman rules can be derived and all the results of the next sections can be obtained in a more direct way. We will not use this powerful formalism here for the sake of physical transparency. In this way the next sections can be understood by the reader who has not studied this one very carefully. Another reason is that the superspace formalism cannot be directly generalised to the cases of extended supersymmetries which contain several spinorial generators.

27.6   A Simple Field Theory Model

We will discuss here the simplest supersymmetric invariant field-theory model in four dimensions, that of a self-interacting chiral multiplet S. We introduced already the kinetic term, Eq. (27.44), and the mass term S^2+Hermitian conjugate, Eq. (27.43). For the interaction we choose the term $S^3 + h.c.$ In terms of component fields, the complete Lagrangian, after integration over the Grassmann variables θ and $\bar{\theta}$, reads

$$
\begin{aligned}
\mathcal{L} = {} & \frac{1}{2}[(\partial A)^2 + (\partial B)^2 + i\bar{\psi}\gamma_\mu\partial^\mu\psi + F^2 + G^2] + m\left[FA + GB - \frac{1}{2}\bar{\psi}\psi\right] \\
& + g[F(A^2 - B^2) + 2GAB - \bar{\psi}(A - \gamma_5 B)\psi],
\end{aligned} \tag{27.48}
$$

where we changed the notations in two ways: (i) we separated the real and imaginary parts of the scalar fields $A \to \frac{1}{2}(A + iB)$ and $F \to \frac{1}{2}(F + iG)$ and (ii) we switched to the Majorana representation for the spinor ψ. m is a common mass for all fields and g a dimensionless coupling constant. In terms of superfields the kinetic energy term is the D term in the expansion of the vector multiplet $S\bar{S}$ we obtained in Eqs. (27.40) and (27.44), the mass term is the F term in the expansion of the chiral superfield S^2 of Eq. (27.43), and the interaction term is the F term in the expansion of S^3. As we mentioned earlier, $F(x)$ and $G(x)$ are auxiliary fields and can be eliminated using the equations of motion,

$$
F + mA + g(A^2 - B^2) = 0; \quad G + mB + 2gAB = 0, \tag{27.49}
$$

in which case the Lagrangian takes the form

$$
\begin{aligned}
\mathcal{L} = {} & \frac{1}{2}[(\partial A)^2 + (\partial B)^2 + i\bar{\psi}\gamma_\mu\partial^\mu\psi - m^2(A^2 + B^2) - m\bar{\psi}\psi] \\
& - mgA(A^2 + B^2) - g\bar{\psi}(A - \gamma_5 B)\psi - \frac{1}{2}g^2(A^2 + B^2)^2.
\end{aligned} \tag{27.50}
$$

It describes Yukawa, trilinear, and quartic couplings among a Majorana spinor, a scalar, and a pseudoscalar. The consequence of supersymmetry is that all fields have a common mass and all interactions are described in terms of a single coupling constant. Apart from the mass term, this is the Lagrangian we discovered in section 19.3 as a fixed point of the renormalisation group. Supersymmetry implies the conservation of a spin-$\frac{3}{2}$

current, which is

$$\mathcal{J}^\mu = \gamma^\lambda \partial_\lambda (A - \gamma_5 B) \gamma_\mu \psi - (F + \gamma_5 G) \gamma_\mu \psi. \tag{27.51}$$

The Lagrangian (27.48) is the most general renormalisable supersymmetric invariant theory of one chiral multiplet. Strictly speaking we could add a term linear in the field F:

$$\mathcal{L} \to \mathcal{L} + \lambda F. \tag{27.52}$$

Such a term does not break supersymmetry because, as we said earlier, the variation of F is a total derivative. However, it has no effect on the model because it can be eliminated by a shift in the field A.

The renormalisation of this theory is straightforward. We will go through in some detail, both as an exercise in the technology of Ward identities and because the results will turn out to be unexpected and interesting.

We start by noting that supersymmetry being a global symmetry, we can use several supersymmetric invariant regularisation schemes. A conceptually simple one is to introduce higher derivatives in the kinetic energy the way we saw in section 16.4.[6] It amounts into modifying the kinetic energy part of the Lagrangian (27.48) by adding terms of the form $\xi \frac{1}{2} (\partial_\mu \Box A)^2$ and similarly for all other fields. ξ is a regulator parameter with dimensions $[M]^{-2}$ which will go to 0 at the end of the calculation, after the renormalisation has been performed. The important point is that in the presence of ξ, all diagrams are convergent and, consequently, all formal manipulations are allowed and well defined. Furthermore, the ξ-terms respect supersymmetry.

The first important result is valid for all theories with global supersymmetry and concerns the vacuum energy. Although it is often left out from the renormalisation programme, a generic quantum field theory requires a constant counter-term in order to remove the divergence of the vacuum diagrams. The corresponding renormalisation condition is usually chosen to impose the vanishing of the vacuum energy. In a supersymmetric theory this counter-term is absent. All terms generated by the boson loops in the vacuum diagrams are cancelled by the corresponding ones having fermion loops. Since every supersymmetric theory has equal number of bosonic and fermionic degrees of freedom with degenerate masses, this cancellation is exact. A formal way to obtain this result is to write the local analogue of the supersymmetry algebra (27.5), or (27.8), in the form

$$[Q_\alpha, \mathcal{J}^\mu_\beta(x)]_+ = 2 \left(\gamma^\nu \gamma^0 \right)_{\alpha\beta} T^\mu_\nu(x) + ..., \tag{27.53}$$

where $\mathcal{J}^\mu_\beta(x)$ is the spin-$\frac{3}{2}$ supersymmetry current and $T^\mu_\nu(x)$ is the energy–momentum tensor. The dots stand for terms containing space derivatives whose vacuum expectation value vanishes. The algebra of charges (27.5) or (27.8) is obtained from (27.53) by integration over d^3x. If supersymmetry is exact, the charge Q annihilates the vacuum

---

[6] The use of dimensional regularisation, although possible in this simple case, should be avoided because supersymmetry transformations involve γ_5 in an essential way.

and this implies the vanishing of the vacuum expectation value of the energy–momentum tensor:

$$< 0 | T_\nu^\mu(x) | 0 > = 0, \qquad (27.54)$$

which means that no constant counter-term, or, equivalently, normal ordering of the terms in the Lagrangian, is needed.

The physical significance of this result is understood if we consider a supersymmetric theory in the presence of an external gravitational field. In a generic field theory the vacuum fluctuations induce an infinite vacuum energy, which implies an infinite cosmological constant. In a supersymmetric theory the induced cosmological constant is zero. Note, however, that this is valid only when supersymmetry is exact. In a broken theory, whether explicitly or spontaneously, we obtain a non-zero cosmological constant.

After this general result, let us come back to the particular model of the Lagrangian (27.48). The conservation of the current (27.51) yields Ward identities among different Green functions. A simple way to obtain them is to introduce external sources for each field and write the generating functional as

$$Z[\mathcal{J}] = \frac{\int e^{i \int (\mathcal{L} + \mathcal{J}\Phi) d^4 x} \mathcal{D}[\Phi]}{\int e^{i \int \mathcal{L} d^4 x} \mathcal{D}[\Phi]}, \qquad (27.55)$$

where Φ denotes collectively all fields A, B, F, G, and ψ and the source term is given by $\mathcal{J}\Phi = \mathcal{J}_A A + \mathcal{J}_B B + \mathcal{J}_F F + \mathcal{J}_G G - \bar{\eta}\psi$. The external sources are assumed to form a chiral multiplet so that the part of the action corresponding to $\mathcal{J}\Phi$ is supersymmetric invariant. This means that the sources transform as the fields themselves:

$$\begin{aligned}
\delta \mathcal{J}_A &= -i(\partial_\mu \bar{\eta}) \gamma^\mu \alpha \\
\delta \mathcal{J}_B &= i(\partial_\mu \bar{\eta}) \gamma_5 \gamma^\mu \alpha \\
\delta \mathcal{J}_F &= i \bar{\eta} \alpha \\
\delta \mathcal{J}_G &= i \bar{\eta} \gamma_5 \alpha \\
\delta \eta &= \partial_\mu (\mathcal{J}_F - \gamma_5 \mathcal{J}_G) \gamma^\mu \alpha + (\mathcal{J}_A + \gamma_5 \mathcal{J}_B) \alpha.
\end{aligned} \qquad (27.56)$$

Under the transformation (27.56) $Z[\mathcal{J}]$ is invariant, which gives the Ward identities

$$\begin{aligned}
& -i \frac{\delta Z}{\delta \mathcal{J}_A} (\partial_\mu \bar{\eta}) \gamma^\mu + i \frac{\delta Z}{\delta \mathcal{J}_B} (\partial_\mu \bar{\eta}) \gamma_5 \gamma^\mu + i \frac{\delta Z}{\delta \mathcal{J}_F} \bar{\eta} + i \frac{\delta Z}{\delta \mathcal{J}_G} \bar{\eta} \gamma_5 \\
& - \frac{\delta Z}{\delta \eta} [\partial_\mu (\mathcal{J}_F - \gamma_5 \mathcal{J}_G) \gamma^\mu + \mathcal{J}_A + \gamma_5 \mathcal{J}_B] = 0.
\end{aligned} \qquad (27.57)$$

As usual, we can rewrite (27.57) as Ward identities for the one-particle irreducible Green functions by Legendre transforming the connected Green functions generating functional $W[\mathcal{J}] = -i \ln Z[\mathcal{J}]$,

$$\Gamma[R] = W - \int (\mathcal{J}_A R_A + \mathcal{J}_B R_B + \mathcal{J}_F R_F + \mathcal{J}_G R_G - \bar{\eta} R_\psi)\mathrm{d}^4 x, \tag{27.58}$$

where $R_\phi = \delta W/\delta \mathcal{J}_\phi$ $(\phi = A, B, F, G)$, $R_\psi = \delta W/\delta \bar{\eta}$, and $\mathcal{J}_\phi = -\delta\Gamma/\delta R_\phi$, $\eta = -\delta\Gamma/\delta\bar{R}_\psi$.

$$\mathrm{i}\, R_\psi \frac{\delta\Gamma}{\delta R_A} + \mathrm{i}\gamma_5 R_\psi \frac{\delta\Gamma}{\delta R_B} + \mathrm{i}\gamma^\mu \partial_\mu R_\psi \frac{\delta\Gamma}{\delta R_F} + \mathrm{i}\gamma_5\gamma^\mu \partial_\mu R_\psi \frac{\delta\Gamma}{\delta R_G}$$
$$+ (R_F + \gamma_5 R_G - \partial_\mu R_A \gamma^\mu - \partial_\mu R_B \gamma_5\gamma^\mu)\frac{\delta\Gamma}{\delta\bar{R}_\psi} = 0. \tag{27.59}$$

This equation is the functional form of the Ward identities for the 1-PI Green functions. It states that $\Gamma[R]$ is invariant under supersymmetry transformations, i.e. if we transform the classical fields R as the fields A, B, etc., we obtain $\delta\Gamma = 0$. The existence of a supersymmetric invariant regularisation scheme ensures that these Ward identities can be enforced in the renormalised theory.

Two important consequences follow from these Ward identities. (i) The vacuum expectation values of all fields vanish. (ii) The Lagrangian (27.48) is stable under renormalisation, i.e. no new counter-terms will be needed at every order in perturbation theory.

We can prove these properties very easily. The only fields for which the first does not follow trivially from Lorentz invariance or parity conservation are A and F. We take the functional derivative of the identity (27.57) with respect to $\bar{\eta}$ and then set all sources equal to 0. The result is

$$<F>_0 = 0. \tag{27.60}$$

On the other hand, the equation of motion for F, Eq. (27.49), in the presence of the ξ regulator term, reads $(1 + \xi\Box^2)F + mA + g(A^2 - B^2) = 0$. Taking the vacuum expectation value and using (27.60) and the Ward identity $<AA>_0 = <BB>_0$, we obtain

$$<A>_0 = 0; \tag{27.61}$$

in other words, no counter-terms linear in A or F are needed.

We can now proceed with the renormalisation programme the usual way. We introduce three counter-terms for wave function, mass, and coupling constant renormalisation:

$$\Phi_R = Z^{-\frac{1}{2}}\Phi; \quad m_R = Zm + \delta m; \quad g_R = Z^{\frac{3}{2}}Z'g. \tag{27.62}$$

They are determined by the following three renormalisation conditions:[7]

$$\frac{\mathrm{d}\Gamma_{AA}(p^2)}{\mathrm{d}p^2}\Big|_{p^2=0} = -1; \quad \Gamma_{AF}(p^2 = 0) = m_R$$
$$\Gamma_{FAA}(p_i = 0) = \Gamma_{A\psi\psi}(p_i = 0) = 2g_R. \tag{27.63}$$

---

[7] For simplicity, we choose to impose all renormalisation conditions at zero external momenta. This way the parameter m_R is not the physical mass but the value of the 2-point function at zero momentum. In section 16.6 we showed how the two are related.

The Ward identities tell us that this choice is possible order by order in perturbation. For example, taking different derivatives of (27.59) we obtain for the 2-point functions

$$\Gamma_{AA}(p^2) = \Gamma_{BB}(p^2) = -p^2 \Gamma^{(1)}_{\psi\psi}(p^2) \tag{27.64}$$

$$\Gamma_{AF}(p^2) = \Gamma_{BG}(p^2) = \Gamma^{(2)}_{\psi\psi}(p^2), \tag{27.65}$$

where we have defined $\Gamma_{\psi\psi}(p) = i\not{p}\,\Gamma^{(1)}_{\psi\psi}(p^2) - \Gamma^{(2)}_{\psi\psi}(p^2)$. Equations (27.64) tell us that $\Gamma_{AA}(0) = 0$, which means that no A^2 counter-term needs to be introduced. Similarly for the 3- and 4-point functions we obtain Ward identities of the form

$$\Gamma_{AAA}(p_i = 0) = \Gamma_{ABB}(p_i = 0) = 0; \quad \Gamma_{AAAA}(p_i = 0) = 0, \tag{27.66}$$

meaning that no A^3, AB^2, or A^4 counter-terms are required. This analysis shows that supersymmetry can be maintained in the renormalised theory, a result we knew already from the existence of the supersymmetric invariant ξ regulator.

Let us now have a closer look at the renormalisation and compute the 1-loop mass counter-term. Let us choose, for example, the two diagrams of Fig. 27.1. Since $\gamma_5^2 = -1$, we see immediately that the two contributions cancel and, at this order, $\delta m = 0$. We may think that this is an accident of the 1-loop diagrams but we can also check that we obtain the same result at two loops, although now the cancellation involves diagrams with different topologies, such as those shown in Fig. 27.2. It turns out that this persists to all orders and the way to prove it is to remark that the unrenormalised Lagrangian (27.48) satisfies, in the presence of the ξ regulator, the identity

$$\frac{\partial}{\partial m}\mathcal{L} = \frac{1}{2g}\frac{\partial}{\partial A}\mathcal{L}_g, \tag{27.67}$$

where $\mathcal{L}_g$ denotes the interaction part of $\mathcal{L}$. We use this property and evaluate the derivative with respect to m of a Green function

$$\frac{\partial}{\partial m}Z[\mathcal{J}] = \frac{\partial}{\partial m}\frac{\int e^{i\int(\mathcal{L}+\mathcal{J}\Phi)\mathrm{d}^4x}\mathcal{D}[\Phi]}{\int e^{i\int \mathcal{L}\mathrm{d}^4x}\mathcal{D}[\Phi]}. \tag{27.68}$$

The derivative of the denominator vanishes because of (27.60). For the numerator we find

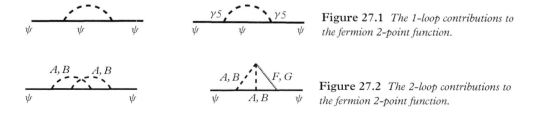

$\psi \quad \psi \quad \psi \qquad \psi \quad \psi \quad \psi$

Figure 27.1 *The 1-loop contributions to the fermion 2-point function.*

Figure 27.2 *The 2-loop contributions to the fermion 2-point function.*

$$\frac{\partial}{\partial m} \int e^{i \int (\mathcal{L} + \mathcal{J}\Phi)\mathrm{d}^4 x} \mathcal{D}[\Phi] = \frac{i}{2g} \int e^{i \int (\mathcal{L} + \mathcal{J}\Phi)\mathrm{d}^4 x} \int \frac{\partial}{\partial A} \mathcal{L}_g \mathrm{d}^4 y \mathcal{D}[\Phi]$$

$$= \frac{1}{2g} \int e^{i \int (\mathcal{L} - \mathcal{L}_g + \mathcal{J}\Phi)\mathrm{d}^4 x} \int \frac{\delta}{\delta A} \int e^{i \int \mathcal{L}_g \mathrm{d}^4 x} \mathrm{d}^4 y \mathcal{D}[\Phi] \tag{27.69}$$

$$= -\frac{im}{2g} \int e^{i \int (\mathcal{L} + \mathcal{J}\Phi)\mathrm{d}^4 x} \int F \mathrm{d}^4 y \mathcal{D}[\Phi] - \frac{i}{2g} \int e^{i \int (\mathcal{L} + \mathcal{J}\Phi)\mathrm{d}^4 x} \int \mathcal{J}_A \mathrm{d}^4 y \mathcal{D}[\Phi].$$

In terms of the connected Green functions this gives

$$\frac{\partial}{\partial m} W[\mathcal{J}] = -\frac{m}{2g} \int \frac{\delta W[\mathcal{J}]}{\delta \mathcal{J}_F(y)} \mathrm{d}^4 y - \frac{1}{2g} \int \mathcal{J}_A(y) \mathrm{d}^4 y, \tag{27.70}$$

which means that for every unrenormalised but regularised connected Green function other than the one $\mathcal{J}_A$ function, the derivative with respect to the unrenormalised mass is equivalent to the insertion of a zero momentum F field. In terms of the 1-PI functions this implies the identity

$$\frac{\partial}{\partial m} \Gamma[R] = -\frac{m}{2g} \int R_F(y) \mathrm{d}^4 y + \frac{1}{2g} \int \frac{\delta \Gamma[R]}{\delta R_A(y)} \mathrm{d}^4 y, \tag{27.71}$$

which means again that for every 1-PI function other than the one R_F, the derivative with respect to the unrenormalised mass is equivalent to the insertion of a zero momentum A field.

Let us now apply Eq. (27.71). We take the functional derivative with respect to R_F and then put all the R's equal to 0. Using (27.60) we find that

$$m = Z^{-1} \Gamma_{FA}(p^2 = 0). \tag{27.72}$$

Comparing with (27.63) we get $m_R = Zm$, or $\delta m = 0$. Similarly, taking the second derivative with respect to R_ψ we obtain

$$\frac{\partial}{\partial m} Z^{-1} \Gamma_{\psi\psi}(p) = \frac{1}{2g} Z^{-\frac{3}{2}} \Gamma_{A\psi\psi}(p_A = 0, p), \tag{27.73}$$

which, for $p = 0$ and using the renormalisation condition (27.63), gives $g_R = Z^{-\frac{3}{2}}g$, i.e. $Z' = 1$.

This completes the proof that for this model, the only counter-term required is merely a common wave function renormalisation. No mass or coupling constant counter-terms are needed.

This result sounds miraculous but it can be understood using the superspace formalism. We noted already that the kinetic energy part of (27.48) is a D-type term while the mass and the interaction terms are of F-type. Going back to the analysis of section 27.5 we remark that the former results from the integrations over both $\mathrm{d}\theta$ and $\mathrm{d}\bar{\theta}$ because a D term comes from a vector multiplet which depends on both. On the other hand, an F term comes from a chiral multiplet which depends on either θ or $\bar{\theta}$. The

Feynman propagators are determined by the kinetic energy terms. It follows that all the counter-terms will involve $d\theta d\bar{\theta}$ integrations, which give a vanishing result if applied to functions of only one of them. This explains the previous result and generalises it to the following *theorem. In a supersymmetric theory only D terms need to be renormalised.* This very peculiar consequence of supersymmetry, known as *non-renormalisation theorem*, was first discovered in this simple model at 1-loop order by Julius Wess and Bruno Zumino and proven to all orders by J. Iliopoulos and B. Zumino. As we will see later, this particular property makes supersymmetry attractive for physical applications.

Before closing this section let us mention an interesting consequence of the simple renormalisation properties of this model. Let us write the Callan–Symanzik equation for the 1-PI Green functions. We noted already in the derivation of Eq. (27.71) that the derivative with respect to the bare mass is equivalent to the insertion of a zero momentum A field, instead of the insertion of a composite operator ϕ^2 which we found in the general case. Since A is an elementary field of the theory, it is renormalised by a $\sqrt{Z}$ factor. It follows that, in the notation we used in section 17.4, the functions $\delta(g)$ and $\gamma(g)$ are proportional to each other. In fact, $\beta(g)$ is also proportional to them because all three come from the single counter-term Z. In Problem 27.1 we derive these proportionality relations. We have often emphasised the importance of the zeros of the β-function. We have also noted that perturbation theory can give us information only about the trivial fixed point $g = 0$. This model is an exception. Let us assume that $\beta(g)$ vanishes for some $g = g^* \neq 0$. Since β and γ are, essentially, the same function, it follows that $\beta(g^*) = \gamma(g^*) = 0$; in other words, at $g = g^*$ we would have a scale invariant theory with all fields having vanishing anomalous dimensions. It is not difficult to convince ourselves that for a field theory quantised in a space with only positive norm states, as is the case for this model, this implies that the theory is free. But the coupling constant was defined as the value of a 3-point function at zero external momenta, see Eqs. (27.63). If this value is non-zero we will have non-trivial scattering amplitudes, which is not possible for a free field theory. So we arrive at the conclusion that the β-function of this model has no fixed points other than the trivial one $g = 0$.

27.7 Supersymmetry and Gauge Invariance

27.7.1 The Abelian Case

A combination of supersymmetry with gauge invariance is clearly necessary for the application of these ideas to the real world. Let us start with an Abelian gauge theory and construct the supersymmetric extension of quantum electrodynamics.

If v_μ is the photon field and ϕ_1 and ϕ_2 are the real and imaginary parts of a charged field, an infinitesimal gauge transformation is given by

$$\delta v_\mu = \partial_\mu \Lambda; \quad \delta \phi_1 = e\Lambda \phi_2; \quad \delta \phi_2 = -e\Lambda \phi_1, \tag{27.74}$$

where Λ is a scalar function. In order to extend (27.74) to supersymmetry we must replace v_μ with a whole vector multiplet $V(x, \theta, \bar{\theta})$. It means that the photon field v_μ will be accompanied by its supersymmetric partners χ, M, N, etc. shown in Eq. (27.36). Let us also assume that the matter fields are given in terms of a charged chiral multiplet $\Psi(x, \theta)$ and $\bar\Psi(x, \bar\theta)$. We expect, therefore, to describe simultaneously the interaction of photons with charged scalars, pseudoscalars, and spinors, together with other interactions involving the supersymmetric partners of the photon. It is obvious that if we keep Λ as a scalar function, the transformation (27.74) is not preserved by supersymmetry because it transforms only the photon but not its partners. We choose to generalise the gauge transformation (27.74) by replacing Λ with a whole chiral multiplet $\Lambda(x, \theta)$ and $\bar\Lambda(x, \bar\theta)$. In Eq. (27.41) we constructed, out of the chiral multiplet $\Lambda(x, \theta)$, a vector multiplet which we called $\partial\Lambda$. Therefore, we write the supersymmetric extension of the gauge transformations (27.74) as

$$\delta V = \partial\Lambda; \quad \delta\Psi_1 = e\Lambda\Psi_2; \quad \delta\Psi_2 = -e\Lambda\Psi_1. \tag{27.75}$$

It is easy to check, using the relations (27.41), that the vector component of the superfield V transforms by the derivative of a scalar function, as it should. On the other hand, we also see that the fields λ and D are gauge invariant.

We are now in a position to write the supersymmetric extension of quantum electrodynamics. We have already the kinetic part of the vector multiplet, Eqs. (27.45) and (27.46), and the kinetic and the mass term of the charged chiral multiplet. We must now construct the gauge invariant interaction term. As usually, the only term which breaks gauge invariance is the free kinetic part of the chiral multiplet. According to Eqs. (27.40) and (27.44), it is given by a bilinear of the form $\Psi(x - i\theta\sigma\bar\theta, \theta)\bar\Psi(x + i\theta\sigma\bar\theta, \bar\theta)$. When Ψ transforms, formally, by a phase with the chiral multiplet Λ the bilinear transforms by a phase

$$\Psi\bar\Psi \rightarrow \Psi\bar\Psi\, e^{i(\Lambda - \bar\Lambda)}. \tag{27.76}$$

We restore gauge invariance by introducing the vector multiplet V which transforms precisely by Eq. (27.75) with $\partial\Lambda$ proportional to $\Lambda - \bar\Lambda$. It follows that the gauge invariant kinetic energy and interaction term is given by

$$\mathcal{L} \sim \left[\Psi e^V \bar\Psi\right]\big|_{\theta\theta\bar\theta\bar\theta}, \tag{27.77}$$

which means that we must take the D term of the vector multiplet obtained by the product of the chiral multiplets Ψ and $\bar\Psi$ and the exponential of the vector multiplet V. This way we have a gauge invariant supersymmetric quantum electrodynamics. The problem is that the interaction terms are non-polynomial in the fields and, as a consequence, it seems that this theory cannot be studied with the methods we developed in this book. Let us remind, however, that this new theory has many more degrees of freedom and we attempt to describe the interactions of a much larger set of fields. On the other hand, this has been achieved by a large extension of gauge invariance. By promoting gauge invariance from (27.74) to the supersymmetric extension (27.75) we introduced a much larger gauge freedom, from that of a single scalar function to that of an entire chiral multiplet. It follows that this new theory contains a large number of unphysical degrees of freedom. We can use the generalised gauge freedom to eliminate as many of them as possible. Looking at the transformation properties (27.41) we see that we can choose a supersymmetric gauge such that all fields of the vector multiplet with the exception of v_μ, λ, and D can be put equal to 0. This is called *the Wess–Zumino gauge* and, by making this choice, the supersymmetric extension of quantum electrodynamics takes the form

$$
\begin{aligned}
\mathcal{L} = & -\frac{1}{4}(\partial_\mu v_\nu - \partial_\nu v_\mu)^2 - \frac{i}{2}\bar{\lambda}\gamma^\mu\partial_\mu\lambda + \frac{1}{2}D^2 \\
& + \frac{1}{2}[(\partial A_1)^2 + (\partial A_2)^2 + (\partial B_1)^2 + (\partial B_2)^2 - F_1^2 - F_2^2 - G_1^2 - G_2^2 \\
& - i\bar{\psi}_1\gamma_\mu\partial^\mu\psi_1 - i\bar{\psi}_2\gamma_\mu\partial^\mu\psi_2] \\
& + m[F_1 A_1 + F_2 A_2 + G_1 B_1 + G_2 B_2 - \frac{i}{2}\bar{\psi}_1\psi_1 - \frac{i}{2}\bar{\psi}_2\psi_2] \\
& + e v_\mu[i\bar{\psi}_1\gamma^\mu\psi_2 - A_1\partial^\mu A_2 + A_2\partial^\mu A_1 - B_1\partial^\mu B_2 + B_2\partial^\mu B_1] \\
& - \frac{1}{2}e^2 v_\mu v^\mu[A_1^2 + A_2^2 + B_1^2 + B_2^2] + eD(A_1 B_2 - A_2 B_1) \\
& - ie\bar{\lambda}[(A_1 + \gamma_5 B_1)\psi_2 - (A_2 + \gamma_5 B_2)\psi_1].
\end{aligned}
\tag{27.78}
$$

The interpretation is straightforward. v_μ is the photon field and λ the field of its supersymmetric partner, *the photino*. The real Majorana spinors ψ_1 and ψ_2 can be combined together to form a complex Dirac spinor, the field of the electron. A_1, A_2, B_1, and B_2 are the real and imaginary parts of two complex, charged, spin-0 fields, a scalar and a pseudoscalar. They are the supersymmetric partners of the electron, sometimes called *selectrons*. As before, the fields F_1, F_2, G_1, G_2, and D are auxiliary. The Lagrangian (27.78) is invariant under ordinary gauge transformations. In fact, if we eliminate the auxiliary fields, we obtain the usual interaction of a photon with a charged scalar, pseudoscalar, and spinor field including the seagull term and the quartic term among the scalar fields. Supersymmetry has introduced only two new elements: (i) the coupling constant in front of the quartic self-interaction of the spin-0 fields is not arbitrary, but it is equal to $e^2/2$, and (ii) new terms, those of the last line in (27.78), appeared which describe a Yukawa-type interaction between the Majorana spinor (the photino) and the spin-$\frac{1}{2}$ and 0 fields of the matter multiplet. The strength of this new interaction is again equal to the electric charge e. Strictly speaking, (27.78) is not invariant under supersymmetry transformations. However, a supersymmetry transformation can be compensated for by a gauge transformation, so all physical results will be supersymmetric.

The theory is renormalisable by power counting and we can easily show that all relations implied by supersymmetry can be maintained. Following our previous discussion, we conclude that there will be no mass counter-term for the chiral multiplet because the latter is an F-type term. In contrast, all other expected counter-terms, i.e. the wave function renormalisation counter-term for the vector multiplet, that for the chiral one, and the coupling constant counter-term, will be present. Here the remark we made in regards to Eq. (27.46) is important. Although the photon kinetic energy term was first written as an F-term, the fact that it can also be written as a D-term implies that it is not protected by a non-renormalisation theorem and it will require its own special counter-term.

27.7.2 The Non-Abelian Case

The extension to the Yang–Mills theories follows the same lines. The only difference is that now the superfields are matrix valued. The vector multiplet, as well as the chiral multiplet of the gauge transformation, belongs to the adjoint representation of a compact Lie group,

$$V = V_i T^i; \quad \Lambda = \Lambda_i T^i, \tag{27.79}$$

where the T's are the matrices that represent the generators of the corresponding Lie algebra in the adjoint representation. The transformation law of Eq. (27.75) becomes

$$e^{V'} = e^{-i\bar{\Lambda}} e^V e^{i\Lambda}, \tag{27.80}$$

where Λ is a left-handed chiral superfield and $\bar{\Lambda}$ is its right-handed Hermitian conjugate. The supersymmetric field strength W_α of Eq. (27.42) generalises into the matrix

$$W_\alpha = -\frac{1}{4} \bar{\mathcal{D}} \bar{\mathcal{D}} e^{-V} \mathcal{D}_\alpha e^V. \tag{27.81}$$

It has the expected transformation properties

$$W'_\alpha = e^{-i\Lambda} W_\alpha e^{i\Lambda}. \tag{27.82}$$

Using W_α we can write the supersymmetric Yang–Mills Lagrangian as

$$\mathcal{L}_{\text{SYM}} = -\frac{1}{4g^2} \text{Tr} \left(W^\alpha W_\alpha|_{\theta\theta} + \bar{W}_{\dot\alpha} \bar{W}^{\dot\alpha}|_{\bar\theta\bar\theta} \right), \tag{27.83}$$

where, as before, the same expression can be written as a D-term. It is again a non-polynomial Lagrangian, but in the Wess–Zumino gauge, after the elimination of the auxiliary fields, it takes the remarkably simple form

$$\mathcal{L} = -\frac{1}{4g^2} \text{Tr}[W_{\mu\nu}^2 - 2i\bar\lambda \gamma^\mu \mathcal{D}_\mu \lambda], \tag{27.84}$$

where

$$W_{\mu\nu} = \partial_\mu v_\nu - \partial_\nu v_\mu + i[v_\mu, v_\nu]; \quad \mathcal{D}_\mu \lambda = \partial_\mu \lambda + i[v_\mu, \lambda]. \tag{27.85}$$

This Lagrangian describes the gauge invariant interaction of m^2-1 massless Majorana fermions belonging to the adjoint representation of $SU(m)$ with the gauge fields. The surprising result is that it is automatically supersymmetric, in the same sense as for (27.76); i.e. a supersymmetry transformation can be compensated for by a gauge transformation. The corresponding spin-$\frac{3}{2}$ conserved current is

$$\mathcal{J}^{\mu} = -\frac{1}{2}\mathrm{Tr}(W_{\nu\rho}\gamma^{\nu}\gamma^{\rho}\gamma^{\mu}\lambda). \tag{27.86}$$

We observe here the appearance of something like a connection between 'radiation', i.e. the gauge fields, and 'matter' multiplets in the sense that the vector gauge field and its supersymmetric partner, the spinor field, give by themselves an interaction invariant under supersymmetry. We will come back to this point later.

In section 25.4.5 we found that the Lagrangian of non-Abelian gauge theories contains an additional effective term proportional to the winding number ν which represents the effects of instantons, Eq. (25.214). It is easy to find its supersymmetric extension. In writing the Yang–Mills Lagrangian, Eq. (27.84), we considered the real part of the superfield $W^{\alpha}W_{\alpha}$. Its imaginary part is given, in the Wess–Zumino gauge, by

$$\frac{1}{2i}\mathrm{Tr}\left(W^{\alpha}W_{\alpha}|_{\theta\theta} - \bar{W}_{\dot{\alpha}}\bar{W}^{\dot{\alpha}}|_{\bar{\theta}\bar{\theta}}\right) = \mathrm{Tr}[W_{\mu\nu}\tilde{W}^{\mu\nu} - 2i\bar{\lambda}\gamma^{\mu}\gamma_{5}\mathcal{D}_{\mu}\lambda]. \tag{27.87}$$

The first term is the instanton Lagrangian. Its supersymmetric partner is the *CP*-violating, gauge invariant interaction of the λ spinors. In Eq. (17.36) we have shown that the first term on the right-hand side of Eq. (27.87) is proportional to a four-derivative. The same is true for the second. Since λ is a Majorana spinor, we find that

$$\mathrm{Tr}\bar{\lambda}\gamma^{\mu}\gamma_{5}\mathcal{D}_{\mu}\lambda \sim \partial_{\mu}\mathrm{Tr}\bar{\lambda}\gamma^{\mu}\gamma_{5}\lambda. \tag{27.88}$$

In all field theories we are considering in this book fermion fields are supposed to fall sufficiently rapidly at infinity so that total derivatives of the fermion terms can be dropped. This is not always the case with boson fields, as we have already seen. Therefore, the instanton term should be maintained.

The introduction of other multiplets in the form of chiral superfields presents no difficulties. Let Ψ ($\bar{\Psi}$) be a left (right)-handed chiral multiplet belonging to some d-dimensional representation of the gauge group. We can rewrite the superfields V and Λ as matrices in the same representation by using the T matrices of Eq. (27.79) that represent the generators in this representation. The transformation properties can be written as

$$\Psi \to e^{-i\Lambda}\Psi; \quad \bar{\Psi} \to \bar{\Psi}e^{i\bar{\Lambda}}. \tag{27.89}$$

Then the superfield $\bar{\Psi}e^{V}\Psi$ is gauge invariant and its D component describes the gauge and supersymmetry invariant interaction of the chiral multiplet to which we can add an invariant mass term as the F component of Ψ^{2}. In the Wess–Zumino gauge we obtain a renormalisable theory. We leave as an exercise the calculation of the β-function for an $SU(m)$ gauge theory interacting with n chiral multiplets belonging to the fundamental representation of $SU(m)$. The result is

$$\beta = -\frac{mg^3}{16\pi^2}(3-n), \tag{27.90}$$

which means that for $n < 3$, the theory is asymptotically free, although it contains scalar and pseudoscalar particles. This is because, in supersymmetric theories, the quartic couplings of the spin-0 fields are not independent but they are determined by the gauge coupling constant, see the remark after Eq. (27.78).

If we have n chiral multiplets Ψ_i, $i = 1, ..., n$, each one belonging to a representation d_i of the gauge group, we can couple them to the gauge multiplet through the D components of the superfields $\bar{\Psi}_i e^V \Psi_i$. In the Wess–Zumino gauge, after the elimination of the auxiliary fields, we obtain the simple Lagrangian

$$\mathcal{L} = -\frac{1}{4g^2}\text{Tr}[W_{\mu\nu}^2 - 2i\bar{\lambda}\gamma^\mu\mathcal{D}_\mu\lambda] + \sum_i |\mathcal{D}_\mu^{(i)}\phi_i|^2 - i\sum_i \bar{\psi}_i\gamma^\mu\mathcal{D}_\mu^{(i)}\psi_i$$
$$- ig\sum_i (t_i^{kra}\bar{\psi}_i^k\phi_i^r\lambda^a + h.c.) - V(\Psi_i) + \mathcal{L}_{inst}, \tag{27.91}$$

where some explanations on the notation are needed: ψ_i and ϕ_i represent the spinor and scalar fields of the chiral multiplets Ψ_i. All terms are supposed to be summed over $i = 1, ..., n$. $D^{(i)}$ is the covariant derivative corresponding to the representation d_i. The Yukawa terms between the gauginos λ and the spinor and scalar fields of the chiral multiplets generalise the corresponding terms we found in Eq. (27.78). k, r, and a are group indices and the constants t_i^{kra} contain the corresponding Clebsch–Gordan coefficients. The last term is the instanton contribution. $V(\Psi_i)$ is the F component of the chiral superfield given by the sum of all the bilinear and trilinear terms, invariant under the gauge group, of the form $m^{ij}\Psi_i\Psi_j + g^{ijk}\Psi_i\Psi_j\Psi_k$. V is the supersymmetric generalisation of the scalar field potential we found in gauge theories and, for this reason, it is often called *the superpotential*. In fact, if we set the spinor fields ψ_i equal to 0, it becomes the classical potential for the scalar fields ϕ_i. The new coupling constants g^{ijk} are not related to the gauge coupling constant, but because all these terms are F terms, they will not be independently renormalised.

27.7.3   Extended Supersymmetries

Supersymmetric gauge theories with more spinorial generators, $2 \leq N \leq 4$, have been only marginally applied to particle physics at currently available energies, because they do not offer an obvious way to distinguish right-handed from left-handed fermions. Nevertheless, they have remarkable properties as quantum field theories, properties that are still under investigation, and some of the results that have been recently obtained are among the most important and most promising in our efforts towards an understanding of quantum field theory beyond the perturbation expansion. We will not give a detailed presentation of these results here, first, because some of the technical methods used for their derivation go beyond what we have studied in this book and, second, because the work is still in progress and it may be too early to assess their real significance. We will only show how to construct the supersymmetric Lagrangians and merely indicate their properties. As we have already noted, we do not have a general superspace formulation applicable to all theories with extended supersymmetry but, in most cases, we can use the $N=1$ superfields and, for the explicit calculations, the usual component formalism.

We start with the $N = 2$ theory. Our construction will be based on two observations. First, we know already the massless states from our analysis in section 27.4, so we know what kind of fields we should include. Since we want to construct a gauge theory, we will use the multiplet which contains helicity equal to 1 states. Second, we know that every theory with $N = 2$ supersymmetry has *also* an $N = 1$ one, with the addition of an $SU(2)$ R-symmetry. So, if a local Lagrangian exhibiting $N = 2$ supersymmetry exists at all, it must be a special case of our general $N = 1$ Lagrangian (27.91), the one which has the right kind of extra fields and upon which we have imposed the R-symmetry. Even if we could not figure out the result immediately, we could always search for the corresponding fixed points of the renormalisation group following the method we developed in section 19.3. In fact, we will not need this computation here because, from symmetry arguments alone, the answer will be unique. In section 27.4 we found that the $N = 2$ multiplet which has a vector field can be decomposed under $N = 1$ supersymmetry, as the sum of one massless vector multiplet (v_μ, λ) and one massless chiral multiplet (ψ, ϕ), both belonging to the adjoint representation of the gauge group. So our Lagrangian will be given by Eq. (27.91) choosing $n = 1$ and d_1 denoting the adjoint representation. This Lagrangian depends on the set of arbitrary parameters of the superpotential $V(\Psi)$, but we must still impose the R-symmetry relating the spinors λ and ψ. In fact, a discrete subgroup of the form $\lambda \to \psi$ and $\psi \to -\lambda$ with all other fields left invariant is sufficient because it implies the vanishing of the superpotential since the latter depends on ψ and not on λ. As a result, the Lagrangian is unique; either it has an $N = 2$ supersymmetry or there is no local Lagrangian which does. To exhibit this extended supersymmetry we note that the Lagrangian we started from has, by construction, an $N = 1$ invariance with multiplets (v_μ, λ) and (ψ, ϕ), to whom we should add the $N = 1$ auxiliary fields D and F. Because of the R symmetry we imposed it has also a second $N = 1$ supersymmetry with multiplets (v_μ, ψ, D) and $(-\lambda, \phi, F)$. Therefore, it is the $N = 2$ theory we are looking for.

This is an $N=2$ supersymmetric 'pure gauge' theory, in the sense that it has only the fields required by supersymmetry and gauge invariance. In particular, it is a trivial free field theory when the gauge group is $U(1)$. We can still add massless matter multiplets, but, as we explained in section 27.4, we do not have $N = 2$ massive matter multiplets, because all such multiplets have spin larger than or equal to 1. So the $N=2$ supersymmetric extension of quantum electrodynamics will describe massless electrons. We will come back to this point shortly.

The $N = 4$ theory can be constructed following the same method. As we showed in section 27.4, the multiplet will contain one spin-1, four spin-$\frac{1}{2}$, and six spin-0 fields, all belonging to the adjoint representation of the gauge group $\mathcal{G}$. The R symmetry is $SU(4) \sim O(6)$. It can be viewed as a set of 4×4 unitary matrices that rotate the four supersymmetry charges. The spinors belong to the fundamental representation (quartet) of $SU(4)$ and the scalars to a sextet, which can be seen as the antisymmetric product of two quartets. We can start again from the general formula (27.91) of an $N=1$ gauge theory. We will need a real vector multiplet V, a left-handed chiral multiplet T_3, and two more left-handed chiral multiplets T_1 and T_2. They are all matrix valued belonging to the adjoint representation of the gauge group. We can construct the Lagrangian by imposing the R symmetry, but it is simpler to pass from the $N=2$ theory using the results we just obtained. Under $N=2$, V and T_3 combine to form an $N=2$ gauge multiplet while T_1 and T_2 combine to form an $N=2$ massless matter multiplet. It is now straightforward to obtain the Lagrangian

$$\mathcal{L}_{N=4} = \mathcal{L}_{N=1}(V) + \sum_{i=1}^{3} \mathrm{Tr}[\bar{T}_i e^V T_i]_{\theta\theta\bar\theta\bar\theta} + 2gf_{abc}[T_1^a T_2^b T_3^c]_{\theta\theta} + \mathcal{L}_{\mathrm{inst}}, \qquad (27.92)$$

where we have included the instanton contribution. In the Wess–Zumino gauge we can express the Lagrangian in terms of the component fields as

$$\mathcal{L}_{N=4} = -\frac{1}{4g^2}\mathrm{Tr}\left(W_{\mu\nu}W^{\mu\nu} + i\sum_{i=1}^{4} \bar\lambda^i \gamma_\mu D^\mu \lambda^i + \sum_{a=1}^{6} D^\mu\phi_a D_\mu\phi_a \right.$$

$$+ \sum_{i,j=1}^{4}\sum_{a=1}^{6} \bar\lambda^i T_{ij}^a \lambda^j \phi_a + \sum_{a,b=1}^{6}[\phi_a,\phi_b][\phi_a,\phi_b]$$

$$\left. + \frac{\theta}{32\pi^2} W_{\mu\nu}\tilde{W}^{\mu\nu} \right), \qquad (27.93)$$

where the T^a's are six antisymmetric 4×4 matrices. The potential for the scalars is given by the trace of the commutator square and vanishes only if ϕ is represented by a diagonal matrix. We thus obtain families of degenerate ground states, all corresponding to zero potential and, therefore, all preserving $N=4$ supersymmetry. They are parametrised by the vacuum expectation values of the scalar fields that correspond to diagonal matrices; in other words, they form a manifold with dimension equal to the rank of the group $\mathcal{G}$. These fields are called *moduli*. Out of all these theories, only that corresponding to $\phi=0$ for all fields leaves the gauge group unbroken. The others will break spontaneously $\mathcal{G}$ to a subgroup $\mathcal{H}$.

As we noted already in section 27.4, this Lagrangian describes also the $N = 3$ theory. All fields belong to the gauge multiplet and there are no separate matter multiplets. In particular, the theory is free if the gauge group is $U(1)$. The connection between radiation and matter we noted in the $N=1$ Yang–Mills theory is here manifest. The theory contains 'matter' (spinor and scalar) fields, but they all belong to the same multiplet as the 'radiation', i.e. the gauge fields.

The remarkable convergence properties of supersymmetric theories, which led to the non-renormalisation theorems we presented earlier, have now even more surprising consequences. The most astonishing result is that the β-function of an $N = 4$ supersymmetric Yang–Mills theory based on any group $SU(m)$ vanishes to all orders of the perturbation expansion. The effective coupling constant is scale independent and does not run. We often say that $N = 4$ is a finite theory with no need of counter-terms, but this may be misleading. The theory is *not* a free-field theory. Gauge invariant composite operators *do* have non-vanishing anomalous dimensions and their correlation functions are non-trivial. For $N = 2$ we have an intermediate result: the β-function receives only

1-loop contributions. These results can be verified by explicit calculations and we can construct all order proofs. They are not straightforward, the way we proved the non-renormalisation theorems in section 27.6, because of two difficulties. First, there is no set of auxiliary fields for $N = 4$ and in order to verify supersymmetry we must use the equations of motion. Second, we have no regularisation scheme that preserves both supersymmetry and gauge invariance. In a later section we will see how we can overcome these technical difficulties and show the vanishing of the β-function order by order in perturbation theory using the general BRST technology. Furthermore, we will indicate how these properties open the way for a non-perturbative understanding of these theories.

27.8   Spontaneous Symmetry Breaking and Supersymmetry

In this section we want to study the phenomenon of spontaneous symmetry breaking in supersymmetric theories. In fact this question has two aspects. The first is the consequences of supersymmetry in the breaking of global or gauge symmetries we have studied so far. The second is the possibility of spontaneous breaking of supersymmetry itself. Indeed, fermions and bosons are not degenerate in nature, so supersymmetry, if it is at all relevant, must be broken. Although explicit but soft, in the sense we used in Chapter 17, breaking mechanisms are used for phenomenological purposes, it is more interesting to study the possibility of spontaneous breaking.

The usual mechanism for spontaneous symmetry breaking of a non-supersymmetric theory is the introduction of some spin-0 field with negative square mass. It is not straightforward to apply this option in supersymmetry because it could imply an imaginary mass for the corresponding fermion. It follows that the entire discussion on spontaneous symmetry breaking we presented in Chapter 15 and we followed until now must be reformulated in the presence of supersymmetry. This is a consequence of the particular relation between the supersymmetry generators and the space–time translations implied by the algebra (27.5). Indeed, this relation implies, in the absence of any central charges, that

$$H = \frac{1}{4} \sum_i Q_i^2, \tag{27.94}$$

where H is the total Hamiltonian and Q_i, $i = 1, ..., 4N$, are the real components of the Majorana spinors representing the N generators of supersymmetry transformations.[8]

---

[8] In this section we will often use the relation between supersymmetry generators and the Hamiltonian, so let us state some of its limitations. The first, namely the absence of central charges, we noted already. A more fundamental one concerns the existence of the charges Q as operators. We write them usually as the integrals over three space of the zero components of the corresponding conserved currents. The convergence of the integral depends on the behaviour of the fields at spatial infinity, which is connected with the range of the interactions. If we assume that all forces are short ranged we can impose on the fields to vanish sufficiently fast as $x \to \infty$. However, the spontaneous breaking of a global symmetry is accompanied by the appearance of massless excitations, the Goldstone particles, which create long-range forces. In this case the charges are

This relation has important consequences. First, it shows that the energy of a supersymmetric system is always positive semi-definite, which excludes the option of a negative square mass for some scalar fields. Second, it shows that if we have a zero energy ground state, we must have

$$H|0> = 0 \quad \Rightarrow \quad <0|Q_i^2|0> = 0 \quad \Rightarrow \quad Q_i|0> = 0. \tag{27.96}$$

In other words, a zero energy eigenstate is always annihilated by all supersymmetry generators and, therefore, it is supersymmetric. This surprising result could be interpreted, naïvely, as implying the impossibility of spontaneous supersymmetry breaking, since it shows that a supersymmetric state is always the ground state. This conclusion is incorrect, as we will show by explicit examples shortly and the loophole lies in the assumption contained in 'if we have a zero energy ground state'. Indeed, the examples will be such that they admit no supersymmetric state at all. In this respect the physics of spontaneous breaking of global supersymmetry is very different from that of other symmetries where one shows that the symmetric state corresponds to a local maximum of the energy and is, therefore, unstable, while the non-symmetric ground state is degenerate. In supersymmetry spontaneous breaking means that there is no supersymmetric state and the ground state does not have to be degenerate. Another conclusion from the global relations (27.94) and (27.96) is that an extended global supersymmetry with N charges cannot be spontaneously broken to one with $N' < N$. Indeed, either the ground state of the system has zero energy, in which case it is annihilated by all supersymmetry generators, or it has positive energy, in which case all are broken. Again, this conclusion depends crucially on the existence of the supersymmetry charges as globally defined operators.

27.8.1   Goldstone and BEH Phenomena in the Presence of Supersymmetry

The phenomenon of spontaneous symmetry breaking, global or local, presents some novel features in the presence of extended supersymmetry as will be shown in some examples. Throughout this section we will use the superfield formalism of $N=1$ supersymmetry, appropriately adapted to higher N.

We start with the simplest example, the $N = 2$ extension of QED. In this language the $N=2$ gauge multiplet can be built from one vector multiplet $V(x, \theta, \bar\theta)$, which, in

---

defined as integrals in a finite volume. In the infinite volume limit the transformations should be interpreted as automorphisms. The possible presence of central charges may be important. Let us consider, as an example, a weaker form of the supersymmetry algebra which consists of writing the local current algebra relation which replaces the global relation (27.8),

$$\int d^3y \left[\mathcal{J}_{0\alpha}^m(y), \bar{\mathcal{J}}_{\mu\dot\beta}^n(x) \right]_+ = 2(\sigma^\nu)_{\alpha\dot\beta} T_{\nu\mu}(x)\delta^{mn} + (\sigma_\mu)_{\alpha\dot\beta} C^{mn}, \tag{27.95}$$

where $\mathcal{J}$ is the supersymmetry current, T is the energy–momentum tensor, and we allowed for the presence of a central charge C. It is clear that if C is different from 0 we cannot integrate this relation over x and the supersymmetry charges will not be defined.

the Wess–Zumino gauge, has components (v_μ, λ, D) and one chiral multiplet $T(x, \theta)$ of $N{=}1$ with components (ζ, t, s, F_T, G_T). The $N{=}2$ matter multiplet is built from one $N{=}1$ chiral multiplet $\Psi_1(x, \theta)$ and one anti-chiral $\Psi_2(x, \bar\theta)$, with components $(\psi_1, A_1, B_1, F_1, G_1)$ and $(\psi_2, A_2, B_2, F_2, G_2)$, respectively, which, as we did in the $N{=}1$ case, we can combine into a charged chiral multiplet $\Psi(x, \theta)$ and $\bar\Psi(x, \bar\theta)$. The resulting Lagrangian is

$$\mathcal{L}_{N=2} = \mathcal{L}_{N=1}(V, \Psi) + \left[(T\bar T)\right]_{\theta\theta\bar\theta\bar\theta} + 2e\left[\Psi_1\bar\Psi_2 T\right]_{\theta\theta}, \qquad (27.97)$$

where $\mathcal{L}_{N=1}(V, \Psi)$ is the $N{=}1$ supersymmetric QED Lagrangian we wrote in the previous section in which the mass of the matter multiplet is set equal to 0. In the Wess–Zumino gauge (27.97) is polynomial and renormalisable by power counting.

The Lagrangian (27.97) is invariant under the following groups of transformations: (i) the $U(1)$ gauge transformations of super-QED. (ii) The $N{=}2$ global supersymmetry. (iii) An R-symmetry of global $U(2) = SU(2) \times U(1)$ transformations. The two supersymmetry charges Q_1 and Q_2 form a doublet of the $SU(2)$ part of R. The Abelian $U(1)$ part has the usual interpretation in $N{=}1$ superspace given in Eqs. (27.37) and (27.38) under which the superfields V, Ψ_1, and Ψ_2 are neutral and T takes the phase 2β. (iv) Since it is a massless theory, it is invariant at the classical level under dilatations and conformal transformations. Note that a mass term for the matter multiplet breaks both dilatation invariance and the $U(1)$ part of the R-symmetry. Under a conformal transformation, the two supersymmetry generators Q_1 and Q_2 transform into a pair of new ones, $\tilde Q_1$ and $\tilde Q_2$, sometimes called 'special' super-charges, so the theory (27.97) is classically invariant under a superconformal algebra.

It is obvious from the Lagrangian (27.97) that if the vacuum expectation values of the scalar fields of the matter multiplet A and B are set equal to 0, the total potential for all the scalar fields vanishes for any value of $< t >_0$ and $< s >_0$. Therefore, a translation of the form

$$t \to t + v, \qquad (27.98)$$

with v an arbitrary mass scale, breaks spontaneously both the $U(1)$ part of the R symmetry and dilatations. The breaking of dilatations implies that of conformal invariance and, therefore, the breaking of the invariance under the special supersymmetry generators $\tilde Q_1$ and $\tilde Q_2$. The resulting potential is still equal to 0, so the theory is invariant under the original $N{=}2$ supersymmetry, $U(1)$ gauge symmetry and $SU(2)$ R symmetry. Under the translation (27.98), the last term in the Lagrangian (27.97) generates a new term proportional to $ev\Psi\bar\Psi$, i.e. a common mass for the matter multiplet, but all the fields of the $N{=}2$ gauge multiplet remain massless.

According to the general theory of spontaneous symmetry breaking of global symmetries we studied in Chapter 15, we expect a massless excitation, the Goldstone mode, for every generator which is broken by the translation (27.98). They are: (i) the charge of the $U(1)$ part of the R-symmetry, which we call R_1, (ii) the generator of the dilatations D, and (iii) the two special supersymmetry charges $\tilde Q_1$ and $\tilde Q_2$. It is not difficult to identify the corresponding massless excitations. We write the transformation properties of every field under these broken symmetries and we identify those that contain, on the right-hand side, the constant v. We thus find

$$[R_1, s] \sim <t>_0; \quad [D, t] \sim <t>_0$$
$$[\tilde{Q}_1, \bar{\xi}]_+ \sim <t>_0; \quad [\tilde{Q}_2, \bar{\lambda}]_+ \sim <t>_0, \tag{27.99}$$

which means that all the partners of the photon in the $N=2$ multiplet are Goldstone particles. At higher orders quantum anomalies are expected to break all these symmetries, so we should call them pseudo-Goldstone particles; however, they are protected from acquiring a mass because the $N=2$ supersymmetry and the $U(1)$ gauge symmetry remain exact.

The net result of all these operations is to give a mass to the matter multiplet, still keeping the extended supersymmetry exact. This sounds strange because, according to our analysis in section 27.4, the smallest massive multiplet of $N = 2$ has a massive spin-1 particle. In other words, the spectrum we found is in disagreement with the $N=2$ algebra (27.7).

It is not difficult to guess the answer to this puzzle. We have seen already that in supersymmetric gauge theories a supersymmetry transformation should be often compensated by a gauge transformation. When we translate the scalar fields the conserved currents get modified and so does the algebra of the corresponding charges. Since $N=2$ supersymmetry remains exact, we must still find two conserved spinorial charges, but now the algebra may be modified. We saw in section 27.3 that the only consistent modification is the appearance of central charges on the right-hand side of (27.7). They must be conserved operators, already existing in the theory because by translating the scalar fields we did not create any new symmetries. Since gauge transformations were necessary to enforce the $N=2$ supersymmetry in the first place, we conclude that the charge operator of the $U(1)$ gauge symmetry will appear as a central charge in the supersymmetry algebra.

This result can be verified by explicit calculation in the present example. In Problem 27.3 we ask the reader to compute the supersymmetry currents and use canonical commutation relations at equal times to evaluate the anti-commutator of the two spinorial charges. The result is given by

$$[Q^1, Q^2]_+ = 2Q^{(em)}(t + \gamma_5 s), \tag{27.100}$$

where Q^1 and Q^2 are the two supersymmetry generators, $Q^{(em)}$ is related to the $U(1)$ gauge generator and the expression $Q^{(em)}(t + \gamma_5 s)$ means that the anticommutator of the two spinorial charges does not vanish as an operator equation but it is given by a field-dependent gauge transformation. In the present case the fields that appear are the scalar and pseudoscalar components of the T multiplet. This implies that the supersymmetry algebra, as it was written in Eq. (27.7) without central charges, can be used only in evaluating gauge invariant quantities which are annihilated by the gauge generator. When we translate the field t by the new mass scale v, the charge $Q^{(em)}$ appears explicitly as a central charge. The massive matter multiplet is an example of a short multiplet like those we introduced in section 27.4. We can easily verify that the mass bound (27.18) is indeed saturated.

A similar phenomenon appears in the spontaneous breaking of a gauge symmetry. Let us stay with the $N=2$ supersymmetric theory and consider the case in which the gauge group is a non-Abelian group $\mathcal{G}$ whose Lie algebra has dimension n. In terms of $N=1$ superfields, the theory contains the vector multiplet V and the chiral multiplet T, both belonging to the adjoint representation of $\mathcal{G}$. We do not need to introduce any matter multiplets Ψ. We can write the Lagrangian as

$$\mathcal{L}_{N=2} = \mathcal{L}_{N=1}(V) + [\bar{T}e^V T]_{\theta\theta\bar{\theta}\bar{\theta}} + \mathcal{L}_{\text{inst}} \tag{27.101}$$

with $\mathcal{L}_{N=1}(V)$ the $N=1$ Yang–Mills Lagrangian given in Eqs. (27.83) and (27.84). The equations that eliminate the auxiliary fields are

$$D = g[t, s]; \quad F = 0; \quad G = 0 \tag{27.102}$$

with g the gauge coupling constant. The potential for the scalar fields t and s is given by

$$\mathcal{V}(t, s) = -\frac{1}{4}g^2 \text{Tr}[t, s]^2. \tag{27.103}$$

As expected, we obtain the symmetric vacuum state in which $\mathcal{V}$ vanishes for $t = s = 0$, but also a whole family of degenerate vacua where the scalar fields take non-zero vacuum expectation values satisfying

$$[< t >, < s >] = 0, \tag{27.104}$$

i.e. the vacuum expectation values belong to the Cartan sub-algebra of $\mathcal{G}$. By translating the superfield T we give mass to some of the components of the vector superfield V, thus breaking the gauge group $\mathcal{G}$ to the subgroup $\mathcal{H}$ which leaves invariant (27.104). The mass spectrum for the components of the gauge superfield V can be read from the second-order expansion of the term $[< \bar{T} > e^V < T >]_{\theta\theta\bar{\theta}\bar{\theta}}$ where the superfield T is replaced by its vacuum expectation value. Let us denote by m the number of generators of $\mathcal{H}$. We see here an example of the supersymmetric extension of the BEH mechanism we explained in section 27.4. A massless vector multiplet combines with a massless chiral multiplet to become a massive vector multiplet. The original n-component V and T superfields, which were in the adjoint representation of $\mathcal{G}$, are now split into two sets: the m corresponding to the unbroken generators of $\mathcal{H}$ and the others $n - m$, which correspond to the broken generators. For the latter the corresponding components of V and T combine to form $N=1$ massive vector multiplets. As a result we obtain: (i) $n - m$ massive vector multiplets corresponding to the broken generators of $\mathcal{G}$, (ii) m massless vector multiplets corresponding to the unbroken generators of $\mathcal{H}$, and (iii) the remaining m massless chiral multiplets which, together with the vector ones, form m massless $N=2$ vector multiplets. The pseudo-Goldstone particles corresponding to the breaking of the global symmetries, such as dilatations, $U(1)$ R symmetry, and special supersymmetry transformations, are linear combinations of the fields contained in these multiplets.

Since the vacuum state, even after the translation of the scalar fields, has zero energy, the $N=2$ supersymmetry remains unbroken. The massive vector multiplets are short multiplets of $N = 2$, so, again, we expect central charges to appear on the right-hand side of the supersymmetry algebra. They appear with the same mechanism which we found in the Abelian case; namely the anticommutator of the two supersymmetry charges produces, on the right-hand side, a field-dependent gauge transformation. In the unbroken phase it can be set equal to 0 but when the scalar fields acquire a non-zero vacuum expectation value, the generators of the corresponding gauge transformations appear as central charges. Let us derive some of their properties which depend only on the symmetry breaking pattern $\mathcal{G} \to \mathcal{H}$. Let us consider the two linear combinations of the $\mathcal{G}$ generators

$$Z_t = \frac{g}{v} < t^a > T^a; \quad Z_s = \frac{g}{v} < s^a > T^a \qquad (27.105)$$

with $< t^a >$ and $< s^a >$ being the vacuum expectation values of the t and s scalar components of T. v is given by $v^2 = \sum_a [(< t^a >)^2 + (< s^a >)^2]$. We saw that both Z's belong to the Cartan sub-algebra of $\mathcal{G}$, so we have

$$[Z_t, Z_s] = 0. \qquad (27.106)$$

If $Y = y^a T^a$ is any generator of $\mathcal{G}$, our first result is as follows.

Y is a generator of $\mathcal{H}$ if, and only if, it commutes with both Z_t and Z_s. Indeed, if f^{abc} are the structure constants of $\mathcal{G}$, for Y to remain unbroken we must have $f^{abc} y^b < t^c >= f^{abc} y^b < s^c >= 0$.

Our second result follows immediately:

Z_t and Z_s belong to the centre of the algebra.

Indeed, from the first result we have that they commute with all unbroken generators of $\mathcal{G}$. Since supersymmetry remains unbroken, it means that they commute with the two spinorial generators. Finally, they commute with the generators of the $SU(2)$ R-symmetry because t and s are $SU(2)$ singlets.

If Z_t and Z_s are proportional, we can use a rotation of the $U(1)$ R-symmetry to set $< s >= 0$. In this case $\mathcal{H}$ contains only one $U(1)$ factor.

An important property of the short multiplets is that their masses are not renormalised at any finite order of perturbation theory. The reason is that they exist only if the mass bound (27.18) is saturated. Any correction would move the masses away from their minimum value. As a result the operators $Q^{(-)}$ we introduced in section 27.4 will no more annihilate the states of the multiplet and we should obtain a regular long multiplet. But we do not expect corrections at any finite order of perturbation theory to create new states. It follows that these mass values are protected against perturbative quantum corrections. It is one of the rare cases in quantum field theory in which a result at the classical level is, in fact, exact.

It is instructive to check all these results by explicit computation in the simple case in which the breaking $\mathcal{G} \to \mathcal{H}$ is $O(3) \to U(1)$.

Starting from the $N = 2$ Lagrangian we can first compute the Noether current of $N = 1$ supersymmetry, the generalisation of Eq. (27.86). Using component fields we find that

$$\mathcal{J}_\mu^{(1)} = -\frac{1}{4} v_{\rho\sigma}^a [\gamma^\rho, \gamma^\sigma] \gamma_\mu \lambda^a - e\epsilon^{abc} \lambda^a \phi^{*b} \phi^c + \frac{1}{\sqrt{2}} \left[(\not{D}\phi)^a \gamma_\mu \zeta_R^a + (\not{D}\phi^*)^a \gamma_\mu \zeta_L^a \right], \quad (27.107)$$

where we have put $\phi = (t + is)/\sqrt{2}$. The subscripts L and R denote the left- and right-handed components of the spinors, written in the Weyl representation. e is the $O(3)$ coupling constant, in anticipation of the fact that under the symmetry breaking $O(3) \rightarrow U(1)$, the unbroken $U(1)$ group is identified with electromagnetism. The second supersymmetry current is now immediately obtained by the substitution $\lambda \rightarrow -\zeta$ and $\zeta \rightarrow \lambda$:

$$\mathcal{J}_\mu^{(2)} = \frac{1}{4} v_{\rho\sigma}^a [\gamma^\rho, \gamma^\sigma] \gamma_\mu \zeta^a + e\epsilon^{abc} \zeta^a \phi^{*b} \phi^c + \frac{1}{\sqrt{2}} \left[(\not{D}\phi)^a \gamma_\mu \lambda_R^a + (\not{D}\phi^*)^a \gamma_\mu \lambda_L^a \right]. \quad (27.108)$$

They are both conserved as a consequence of the equations of motion. We can now compute the anticommutator of the two spinorial charges using canonical commutation relations. After some algebra we can cast the result in the form

$$[Q^{(1)}, Q^{(2)}]_+ \sim \int dS_i R^{0i}, \quad (27.109)$$

where $R^{\mu\nu}$ is a gauge invariant antisymmetric two-index tensor given by

$$R_{\mu\nu} = (v_{\mu\nu}^a - i\tilde{v}_{\mu\nu}^a)\phi^a + \dots \quad (27.110)$$

$\tilde{v}_{\mu\nu}^a$ is the dual of the gauge field strength and the dots stand for terms which involve the fermion fields. Equation (27.109) is exact. In the unbroken phase of the theory, all fields are assumed to go to 0 sufficiently rapidly at infinity and the surface integral vanishes. In the broken phase this is no more true for the boson fields and a central charge appears. We can always choose to put the non-zero vacuum expectation value v in the third component of $t =\text{Re}\phi$ with all other components vanishing. We thus find that

$$v_{0i}^a\phi^a = -vE_i; \quad \tilde{v}_{0i}^a\phi^a = vB_i, \quad (27.111)$$

where E and B are the electric and magnetic fields of the unbroken $U(1)$ subgroup of $O(3)$. The eigenvalues of the central charge operator are given by the surface integrals (27.109) of the radial components of the electric and magnetic fields:

$$\int dS_i E^i = q; \quad \int dS_i B^i = g, \quad (27.112)$$

with q and g being the electric and magnetic charge, respectively. Therefore, we see that the massive multiplets in this model are short $N = 2$ multiplets with electric charges $\pm e$

and no magnetic charge. Their mass equals $\sqrt{2}ev$ and saturates the bound found in Eq. (27.18). As a result, it remains unchanged to all orders of perturbation theory. On the other hand, we saw in section 26.2.5 that this model contains magnetic monopoles and dyons whose masses, in the BPS limit, saturate the bound (26.66). It is remarkable that it is precisely the bound we obtain by combining Eqs. (27.18) with (27.112). The BPS states form short $N=2$ multiplets. This is the reason why short multiplets are often called *BPS multiplets*, although, as we saw, BPS states appear also in non-supersymmetric models.

We can follow the same steps to analyse the spontaneous breaking of the gauge symmetry in the $N=4$ supersymmetric theory. Taking again the example of an $O(3)$ gauge group spontaneously broken to $U(1)$ we find the following.

In the unbroken phase the theory contains three massless vector multiplets of $N=4$ which form a triplet of the $O(3)$ gauge group. The global R-symmetry is $SU(4)$ which is locally isomorphic to $O(6)$. There is no central charge in the supersymmetry algebra.

In the broken phase we obtain the following:

(i) A massive complex vector multiplet which contains the two vector bosons $W_\mu^\pm$ which became massive through the BEH mechanism. Their supersymmetric partners are four Dirac spinors and five complex scalars, all with the same mass $\sqrt{2}ev$. All regular massive multiplets of $N = 4$ have spins ≥ 2, as we saw in section 27.4. Therefore, those we find are short multiplets and a central charge appears in the algebra. It can be again identified with the generator of the unbroken $U(1)$ subgroup.

(ii) A real massless vector multiplet containing a neutral vector boson (the 'photon'), four Majorana spinors (the Goldstone spinors corresponding to the spontaneous breaking of the four special generators of the super-conformal algebra), and six scalars (one dilaton and five Goldstone bosons of the spontaneous breaking of the global symmetry $O(6) \to O(5)$). In addition, we have massive monopole and dyon multiplets, all saturating the BPS bound. Again, all these masses are protected against quantum corrections at any order of perturbation theory.

27.8.2   Spontaneous Supersymmetry Breaking in Perturbation Theory

Until now we looked at the spontaneous breaking of internal bosonic symmetries, but supersymmetry was left unbroken. Let us turn now to the possibility of breaking spontaneously supersymmetry itself. We saw in the general discussion in Chapter 15 that to every generator of a spontaneously broken global symmetry corresponds a zero mass particle with quantum numbers given by the divergence of the associated conserved current. The conserved current of supersymmetry has spin equal to $\frac{3}{2}$, so the divergence has spin equal to $\frac{1}{2}$. It follows that spontaneous supersymmetry breaking will result in the appearance of a zero mass Goldstone fermion for every broken supersymmetry generator. This particle is often called *goldstino*. We want to study this phenomenon of spontaneous supersymmetry breaking and the associated appearance of a goldstino in a simple field theory model.

In our previous analysis we showed that a supersymmetric vacuum state is always stable. Thus, it would have been impossible to break supersymmetry spontaneously if it were not for the strange property, which we mentioned already, namely the possibility of adding to the Lagrangian a term linear in the auxiliary fields without breaking supersymmetry explicitly. If we restrict ourselves to chiral and vector multiplets, in the notation we used previously, the auxiliary fields are F-fields, G-fields, or D-fields. The first are scalars, the other two pseudoscalars. Let ϕ denote, collectively, all other spin-0 fields. We will assume that neither Lorentz nor translational invariance is broken; consequently, only spin-0 fields can acquire non-zero, constant, vacuum expectation values. Let us recall also that the signature of the spontaneous breaking of a symmetry is the appearance of a constant term on the right-hand side of the transformation properties of some field under infinitesimal transformations. Using the transformation properties, such as (27.39), we can easily show that spontaneous symmetry breaking occurs only when one, or more, of the auxiliary fields acquires a non-vanishing vacuum expectation value because they are the only ones which appear without derivatives in the transformation laws of the spinors. The potential of the scalar fields in the tree approximation has the form

$$
-V(\phi) = \frac{1}{2} \left[\sum F_i^2 + \sum G_i^2 + \sum D_i^2 \right]
$$
$$
+ \left[\sum F_i F_i(\phi) + \sum G_i G_i(\phi) + \sum D_i D_i(\phi) \right], \qquad (27.113)
$$

where the functions $F_i(\phi)$, $G_i(\phi)$, and $D_i(\phi)$ are polynomials in the physical fields ϕ of degree not higher than second. The equations which eliminate the auxiliary fields are

$$
F_i = F_i(\phi); \quad G_i = G_i(\phi); \quad D_i = D_i(\phi), \qquad (27.114)
$$

so the potential, in terms of the physical fields, reads

$$
V(\phi) = \frac{1}{2} \left[\sum F_i^2(\phi) + \sum G_i^2(\phi) + \sum D_i^2(\phi) \right]. \qquad (27.115)
$$

The important observation is that V is non-negative and vanishes only for

$$
F_i(\phi) = 0; \quad G_i(\phi) = 0; \quad D_i(\phi) = 0. \qquad (27.116)
$$

This positivity property of the potential is the reflexion of the positivity of the total energy we found in Eq. (27.94). When Eqs. (27.116) are satisfied, Eqs. (27.114) show that all auxiliary fields have zero vacuum expectation values and, as we said already, supersymmetry is unbroken. It follows that the only way to break supersymmetry spontaneously, at least in the classical approximation, is to arrange so that the system of algebraic equations (27.116) has no real solution. In such a case, at least one of the auxiliary fields will have a non-vanishing vacuum expectation value and supersymmetry will be broken.

We saw already in section 27.6 that adding a term linear in F in the simple field theory model in Eq. (27.52) does not break supersymmetry because it can be eliminated by a shift in the field A. In other words, Eqs. (27.116) have a real solution. We can complicate the model by adding more chiral multiplets and in Problem 27.4 we study this case and show that a minimum of three multiplets is needed.

Here we will exhibit a simpler case, namely the supersymmetric extension of quantum electrodynamics, Eq. (27.76), with the addition of a term linear in the auxiliary field D, following the work of P. Fayet and J. Iliopoulos:

$$\mathcal{L} \to \mathcal{L} + \xi D. \qquad (27.117)$$

We repeat that this term does not break supersymmetry explicitly but, D being pseudoscalar, it breaks parity explicitly, but softly. The system of Eqs. (27.116) reads

$$mA_1 = 0; \quad mA_2 = 0; \quad mB_1 = 0; \quad mB_2 = 0$$
$$e(A_1 B_2 - A_2 B_1) + \xi = 0. \qquad (27.118)$$

It is clear that this system has no solution. We therefore expect supersymmetry to be spontaneously broken. Indeed, eliminating the auxiliary fields we find non-diagonal mass terms among the scalars and pseudoscalars. By diagonalisation we obtain the fields $\tilde{A}_1$, $\tilde{A}_2$, $\tilde{B}_1$, and $\tilde{B}_2$ with mass terms

$$-\frac{1}{2}(m^2 + \xi e)(\tilde{A}_1^{\,2} + \tilde{B}_1^{\,2}) - \frac{1}{2}(m^2 - \xi e)(\tilde{A}_2^{\,2} + \tilde{B}_2^{\,2}). \qquad (27.119)$$

This mass spectrum shows clearly that we have obtained a spontaneous breaking of supersymmetry since the masses of the scalar and spinor members of the chiral multiplets are no longer the same. We can easily verify that the corresponding Goldstone particle is the massless spinor λ.

From this point, what follows depends on the sign of the square mass terms $m^2 \pm \xi e$. If they are both positive, the story ends here. If one of them is negative, this in turn depends on the magnitude and sign of the parameter ξ, the corresponding scalar fields become BEH fields for the $U(1)$ gauge symmetry, and the photon becomes massive. Therefore, the introduction of the linear term ξD can trigger the spontaneous breaking of both supersymmetry and gauge symmetry. In this case the Goldstone spinor is a linear combination of λ and the ψ's. It is clear that without loss of generality we can choose $\xi e > 0$. If $m^2 - \xi e < 0$, we can translate the $\tilde{A}_2$ field

$$\tilde{A}_2 \to \tilde{A}_2 + v; \quad v^2 = 2(\xi e - m^2)/e^2, \qquad (27.120)$$

which gives a mass to the photon and the $\tilde{B}_2$ field decouples, as it is the case in the BEH mechanism. The new feature here is that the fermions acquire non-diagonal mass terms which, after diagonalisation, give

$$\eta_1 = \frac{1}{2}[(1 + \cos\beta)\psi_1 - (1 - \cos\beta)\gamma_5\psi_2 - \sqrt{2}\sin\beta\lambda]$$

$$\eta_2 = \frac{1}{2}[(1 - \cos\beta)\gamma_5\psi_1 + (1 + \cos\beta)\psi_2 + \sqrt{2}\sin\beta\gamma_5\lambda] \qquad (27.121)$$

$$\zeta = \frac{1}{\sqrt{2}}\sin\beta(\psi_1 + \gamma_5\psi_2) + \cos\beta\lambda,$$

where $\beta =\arctan(ev/m)$. With the redefinitions (27.121) the fermion mass terms in the Lagrangian take the form

$$\mathcal{L}_m = -\frac{1}{2}\sqrt{m^2 + e^2v^2}(\bar{\eta}_1\eta_1 + \bar{\eta}_2\eta_2), \qquad (27.122)$$

which means that the spinor ζ is massless. We can verify that it has the correct transformation properties of the goldstino. Indeed under an infinitesimal supersymmetry transformation with spinorial parameter α, we find that

$$\delta\eta_1 = \ldots; \quad \delta\eta_2 = \ldots; \quad \delta\zeta = -\frac{m^2}{e\cos\beta}\gamma_5\alpha + \ldots, \qquad (27.123)$$

where the dots stand for terms which are proportional to the other fields. We see that only $\delta\zeta$ contains a constant term which shows that ζ is indeed the goldstino.

This method can be applied to any gauge theory provided the algebra is not semi-simple; otherwise, a linear D term cannot be added.

27.8.3 Dynamical Breaking of Supersymmetry

In studying the standard model we encountered the possibility of a spontaneous breaking of part of the symmetry outside perturbation theory. We envisaged this possibility for the electroweak gauge symmetry breaking in the framework of the technicolor models, but the concrete case we know for sure that such a breaking occurs is the breaking of global chiral symmetry in quantum chromodynamics. The difficulties we found in breaking supersymmetry spontaneously in the classical theory prompted people to consider the spontaneous breaking, either at higher orders in perturbation, or, dynamically, by non-perturbative effects.

If supersymmetry is exact at the classical level, it means that Eqs. (27.116) have a real solution. Since these terms are usually protected by non-renormalisation theorems, this situation is not likely to change when higher order corrections are taken into account. We are left with the possibility of a dynamical breaking by effects which cannot be seen in perturbation theory. Again, explicit calculations in four-dimensional field theories cannot be made, but Witten proposed a simple criterion to decide whether such a breaking could be possible. The argument is based on the following simple observation.

Let us assume that we have a physical system admitting a supersymmetry charge operator Q which commutes with the Hamiltonian H. As we have often done in this book, we start by quantising the system in a finite volume of three-space V. The spectrum of the Hamiltonian is discrete: $H|n> = E_n|n>$. The states of the system will be either bosonic, $|b>$, or fermionic, $|f>$. Taking into account the relation $Q^2 = H$ without any central charges, we obtain, for $E_n \neq 0$,

$$Q|b>= \sqrt{E}|f>; \quad Q|f>= \sqrt{E}|b>. \tag{27.124}$$

In other words, the existence of the charge Q implies that all eigenstates of H with non-vanishing energy come in pairs, one bosonic and one fermionic, with the same energy. This conclusion does not necessarily apply to zero energy eigenstates, because they are all annihilated by Q: $Q|n_0>= 0$. We also recall that when supersymmetry is spontaneously broken, the ground state of the Hamiltonian corresponds to strictly positive energy; therefore, the existence of zero energy eigenstates implies that supersymmetry is unbroken. E. Witten introduced the operator $(-)^F$, where F is the fermionic number. It is equal to 1 for a fermion and 0 for a boson. We can construct the index of this operator as

$$W = \mathrm{Tr}(-)^F. \tag{27.125}$$

W is called *the Witten index*. In computing W we can restrict to the subspace of the zero energy eigenstates because all states which come in fermion–boson pairs give a vanishing contribution to the trace. We see that W measures the difference between the bosonic and fermionic zero energy eigenstates of the Hamiltonian, $W = n_b - n_f$.

In general, we do not know how to compute the spectrum of the Hamiltonian of a given physical system, but we expect the eigenvalues E_n to depend on the volume V and the various parameters of the model, such as masses and coupling constants. In some cases we may be able to compute reliably the energy spectrum only at some limited region of the parameter space, for example when the couplings are very weak. The interesting property of W is that it is largely independent of the particular values of the parameters we used to compute it.

Let us call $\{g\}$, collectively, the entire set of parameters of our model. The energy levels will depend, in general, on $\{g\}$: $E(\{g\})$. As the parameters vary, the energy eigenvalues will change, but the only changes that may affect the value of W are those which change the zero energy eigenstates.[9] They are of two kinds. First, when, for some value $\{g_1\}$, an energy level which was strictly positive goes to 0 $\{g\} \to \{g_1\} \Rightarrow E(\{g\}) \to 0$. But for all values of $E \neq 0$ the level is degenerate corresponding to a boson–fermion pair. Therefore, both states will move to zero energy simultaneously, n_b and n_f will both increase by 1, and the value of W will not be affected. Second, for a different value $\{g_2\}$ a zero energy level may move to a positive value. But again, as soon as the energy becomes different from 0, the level must be degenerate describing a boson–fermion pair. By continuity, it follows that zero energy states can move to positive energy only in pairs, n_b and n_f will both decrease by 1, thus leaving the value of W unchanged. The great interest of the Witten index is precisely this independence of the particular values of the

---

[9] We assume that the deformations of the theory we are considering are sufficiently smooth so that the space of states remains the same. The energy levels may move around but no new states are created. We will not give a precise definition of this 'smoothness' assumption, but an obvious counterexample is provided by the transition from a free to an interacting theory. Setting the coupling constants equal to 0 is not a 'smooth' deformation, as can be easily checked in simple examples.

parameters of the theory. If we succeed in computing it in some region, even away from the physical region, we can trust the result. In fact we can show that W is a topological invariant and this explains why it remains unchanged under a large class of deformations of the underlying theory.

If W is different from 0, it means that there is an unbalance between the bosonic and fermionic zero energy eigenstates and at least one of n_b and n_f is different from 0. This unbalance will not change by varying the parameters of the theory; therefore, the ground state will remain always at zero energy. But a zero energy state is annihilated by the supersymmetry generator; therefore, in this theory supersymmetry will not be spontaneously broken. If, on the other hand, we have $W = 0$, it may happen that for certain values of the parameters, all eigenstates move to positive values of the energy. The ground state of the system will have positive energy and, according to our previous discussion, supersymmetry will be spontaneously broken. However, we cannot draw any firm conclusion because $W = 0$ implies that either $n_b = n_f = 0$, in which case supersymmetry is spontaneously broken, or $n_b = n_f \neq 0$, in which case supersymmetry remains exact.

If $W \neq 0$ supersymmetry cannot be spontaneously broken, perturbatively or non-perturbatively.

If $W = 0$ we cannot predict the outcome for sure. We need a detailed study of the spectrum of the theory.

Let us apply this criterion to the simple supersymmetric field theory model we studied in section 27.6. After elimination of the auxiliary fields, the classical potential for the scalar field A is given by

$$V(A) = \frac{1}{2}g^2 A^2 \left(A + \frac{m}{g} \right)^2 . \tag{27.126}$$

We can study the spectrum of this model in a finite volume in the semi-classical approximation. We find two zero energy ground states corresponding to the values of the scalar field $< A > = 0$ and $< A > = -m/g$. They are both bosonic because, for $m \neq 0$, the fermionic states are massive.[10] So, for this model we find $W = 2$. We knew that this model had no spontaneous supersymmetry breaking at the classical level and the non-renormalisation theorems showed that this remained true at any finite order in perturbation theory, but this argument tells us that it is also true when non-perturbative effects are taken into account.

27.9 Dualities in Supersymmetric Gauge Theories

In section 26.2.5 we introduced the idea of duality in gauge theories. In its simplest form it interchanges electric and magnetic quantities as well as weak and strong coupling

---

[10] Taking the limit $m = 0$ does not change the value of W because the massless fermionic states are accompanied by massless bosonic ones. The important point is that the fermion ψ does not transform like a goldstino.

regimes. We presented the Montonen-Olive conjecture which postulates the actual identity of the two descriptions, at least for the simple Georgi–Glashow model. Strange as it may sound, we saw that this conjecture passed some simple tests. We can now address the following questions: how far can we trust this conjecture? In the absence of any rigorous proof can we, at least, use it as a means to define the theory in the strong coupling region? Does it apply to all gauge theories and, if not, are there models for which it comes closer to the truth? Last but not least, how can we use it in order to extract physically interesting results? In this section we will attempt to give a partial answer to some of these questions.

Let us first note that the identification $g \to 1/g$ cannot be exact everywhere for a generic gauge theory. The reason is that the effective value of the coupling constant depends on the scale and if such an identification can be enforced in one scale, it won't be true in another. However, we have seen in Section 27.7.1 that there is a class of gauge theories for which the running of the effective coupling constant is particularly simple. They are the gauge theories with extended $N = 4$ or $N = 2$ supersymmetries. For these theories the duality conjecture has given novel and interesting results.

Let us illustrate the physical ideas in the simple $N = 2$ Georgi–Glashow model we studied before in which the gauge group is spontaneously broken $O(3) \to U(1)$. The $N = 4$ case can be considered as a field theorist's dream: all anomalies cancel, in both the R-symmetry and the symmetry under dilatations. As a result the theory is conformally invariant and has two well-defined dimensionless coupling constants, the gauge coupling g and a θ angle. In contrast, the $N = 2$ theory shares many essential features with the real world. It is asymptotically free and it has a, presumably, complicated dynamics. After spontaneous symmetry breaking we obtain a massless vector field (the photon), together with its $N = 2$ supersymmetric partners. Since the theory is asymptotically free, at high scales we can use perturbation. At low energies, however, we enter the strong coupling regime. Following the Wilson suggestion we used in section 15.2.5 when we obtained the non-linear σ-model, let us integrate over all massive degrees of freedom and obtain an effective Lagrangian describing the low-energy strong interaction of the massless modes. Note that it is precisely the kind of exercise we would have liked to solve for QCD in order to obtain an effective theory of hadrons. Such a Wilsonian effective Lagrangian is not necessarily renormalisable, in the same sense that the Fermi theory was not. For QCD we do not know how to solve this problem. Here, however, supersymmetry comes to the rescue. Integrating the heavy degrees of freedom does not break supersymmetry, so we expect the effective Lagrangian to be $N = 2$ supersymmetric. In our example the gauge group of the massless modes is $U(1)$, so the theory is expressed in terms of a single massless $N = 2$ neutral vector multiplet. Its $N = 1$ decomposition contains a neutral vector multiplet V and a chiral multiplet T. In the previous sections we have often written the most general $N = 2$ theory for any gauge group, but we were interested in renormalisable theories for which all terms are monomials in the fields and their derivatives with dimension smaller than or equal to 4. For $U(1)$ this gives a trivial free theory; it is the sum of the kinetic energies of the photon and its partners. In an effective theory we can allow for more general terms, but, at low energies, we can expand any amplitude in powers of energy and keep only the first non-vanishing contributions.

As a result the dominant terms are those that contribute the lowest power in energy. We can use a simple power-counting argument, similar to the one we used in section 16.3 to find the degree of divergence of a given diagram, to determine all such terms.

Let us consider a 1-PI diagram with E_f (I_f) external (internal) fermion lines, E_b (I_b) external (internal) boson lines, E_a (I_a) external (internal) lines of auxiliary fields, V_i vertices of type i with d_i derivatives, and L loops. Since all particles are massless, the power of energy associated with any such diagram is given by simple dimensional analysis:

$$D = 4L + \sum_i V_i d_i - 2I_b - I_f. \tag{27.127}$$

Note that the internal lines I_a do not contribute in formula (27.127) because the auxiliary fields have constant propagators. Let us denote by f_i, b_i, and a_i the number of fermion, boson, and auxiliary field lines coming in the vertex V_i. As we did in section 16.3, we can trade the number of internal lines with that of external lines by using the topological properties of the diagram:

$$L = I_b + I_f + I_a - \sum_i V_i + 1; \quad 2I_b + E_b = \sum_i V_i b_i$$
$$2I_f + E_f = \sum_i V_i f_i; \quad 2I_a + E_a = \sum_i V_i a_i. \tag{27.128}$$

The result is

$$D = \sum_i V_i \delta_i + 2L - \frac{1}{2} E_f - E_a + 2; \quad \delta_i = d_i + \frac{1}{2} f_i + a_i - 2. \tag{27.129}$$

We see that we obtain the minimum value of D by: (i) choosing diagrams with zero loops (this is not surprising, since an effective theory is supposed to be used in the tree approximation), and (ii) minimising the factor δ_i. Note that the latter does not depend on b_i, the number of boson lines on the vertex. The vertices V_i will be determined by any combination of integers d_i, f_i, and $a_i \geq 0$ which minimise δ_i, keeping b_i arbitrary. All the terms in the free Lagrangian of the tree approximation give $\delta_i = 0$ The formula still looks complicated but, again, supersymmetry helps to find the answer. We will determine the kind of effective vertices V_i by first considering $N = 1$ supersymmetry and then imposing the $SU(2)$ R-symmetry. $N = 1$ supersymmetry tells us that we will encounter only two types of terms, those which generalise the two terms we found in Eqs. (27.46) and (27.44). To each one we can introduce an arbitrary function of the chiral multiplet T:

$$\mathcal{L}_{\text{eff}} = \frac{1}{2} \mathcal{K}(T, \bar{T})_{\theta\theta\bar{\theta}\bar{\theta}} - \frac{1}{2} \left[\mathcal{T}(T) W_\alpha W^\alpha |_{\theta\theta} + h.c. \right], \tag{27.130}$$

where $\mathcal{K}$ and $\mathcal{T}$ are two arbitrary functions of the chiral superfield T. The important observation is that $\mathcal{T}$ depends on T but not on $\bar{T}$. In expanding the Lagrangian (27.130) in the component fields we must keep only terms that give $\delta_i = 0$.[11] In terms of the component fields the low-energy effective Lagrangian reads

$$
\begin{aligned}
\mathcal{L}_{\text{eff}} = {} & \frac{\partial^2 \mathcal{K}(\phi, \phi^*)}{\partial \phi \partial \phi^*} \left[\partial_\mu \phi \partial^\mu \phi^* + \tfrac{1}{2} (\bar{\psi} \not{p} \psi) - F F^* \right] \\
& + \tfrac{1}{2} \left[\frac{\partial^3 \mathcal{K}(\phi, \phi^*)}{\partial^2 \phi \partial \phi^*} (\bar{\psi} \psi) F^* + \tfrac{1}{2} \frac{\partial^3 \mathcal{K}(\phi, \phi^*)}{\partial^2 \phi \partial \phi^*} (\bar{\psi} \gamma_\mu \gamma_5 \psi) \partial^\mu \phi \right] + h.c. \\
& - \tfrac{1}{4} \frac{\partial^4 \mathcal{K}(\phi, \phi^*)}{\partial^2 \phi \partial^2 \phi^*} (\bar{\psi} \psi)^2 \\
& - \tfrac{1}{4} \left[\tfrac{1}{2} \frac{d^2 \mathcal{T}(\phi)}{d\phi^2} (\bar{\lambda} \lambda)(\bar{\psi} \psi) - \frac{d\mathcal{T}(\phi)}{d\phi} (\bar{\lambda} \lambda) F \right] + h.c. \\
& + \tfrac{1}{4} \mathcal{T}(\phi) \left[(\bar{\lambda} \not{p} \lambda) - \tfrac{1}{2} v_{\mu\nu} v^{\mu\nu} + \tfrac{1}{2} v_{\mu\nu} \tilde{v}^{\mu\nu} - D^2 \right] + h.c. \\
& + \frac{\sqrt{2}}{4} \frac{d\mathcal{T}(\phi)}{d\phi} \left[i(\bar{\lambda} \psi) D - \tfrac{1}{4} (\bar{\lambda} [\gamma_\mu, \gamma_\nu] \psi) v^{\mu\nu} \right] + h.c.,
\end{aligned}
\tag{27.131}
$$

where the components of the vector and chiral multiplets are denoted by (v_μ, λ, D) and (ψ, ϕ, F), respectively. The Lagrangian (27.131) contains all terms with $\delta = 0$. Looking only at the scalar field sector we see that we have obtained an effective theory of the general form shown in Eq. (15.32), i.e. a non-linear σ-model describing the interaction of the moduli fields ϕ. The metric in the moduli space is given by

$$
F_{ij}(\phi) = \frac{\partial^2 \mathcal{K}(\phi, \phi^*)}{\partial \phi \partial \phi^*}.
\tag{27.132}
$$

An even-dimensional manifold in which we can introduce an operation of complex conjugation and choose coordinates such that the relation (27.132) can be enforced locally is called *a Kähler manifold*. By extension, we will call the function $\mathcal{K}(\phi, \phi^*)$ *the Kähler potential.*

We obtain the $N = 2$ theory by imposing the R-symmetry to (27.131). Under $SU(2)$ the two spinors (ψ, λ) transform as a doublet, the three auxiliary fields $(\text{Im}F, \text{Re}F, D/\sqrt{2})$ as a triplet, and all other fields are singlets. As we saw in section 27.7.1, it is sufficient to impose invariance under a subgroup of discrete transformations and we choose a rotation by an angle π around the second axis which gives

$$
\psi \to \lambda; \quad \lambda \to -\psi; \quad D \to -D; \quad F \to F^*.
\tag{27.133}
$$

Invariance under (27.133) implies relations among the functions $\mathcal{K}$ and $\mathcal{T}$. Equating the coefficients of the kinetic energies $\bar{\psi} \not{p} \psi$ and $\bar{\lambda} \not{p} \lambda$ of the spinors we find that

$$
\frac{\partial^2 \mathcal{K}(\phi, \phi^*)}{\partial \phi \partial \phi^*} = \text{Re} \mathcal{T}(\phi).
\tag{27.134}
$$

---

[11] It is obvious that there are no terms in the expansion that give negative δ_i because there are no terms with boson lines and no derivatives.

The absence of a $(\bar{\lambda}\lambda)^2$ term implies that

$$\frac{\partial^4 \mathcal{K}(\phi,\phi^*)}{\partial^2 \phi \partial^2 \phi^*} = 0. \tag{27.135}$$

Finally, separating the right- and left-handed components of ψ and λ and equating the coefficients of the terms $\bar{\psi}\gamma_\mu\gamma_5\psi$ and $\bar{\lambda}\gamma_\mu\gamma_5\lambda$, we obtain, after partial integration,

$$\frac{1}{2}\partial_\mu T(\phi) = \frac{\partial^3 \mathcal{K}(\phi,\phi^*)}{\partial^2 \phi \partial \phi^*}\partial^\mu \phi. \tag{27.136}$$

Under these conditions the Lagrangian (27.131) is invariant under the transformations (27.133) and, as we saw in section 27.7.1, it is invariant under $N = 2$ supersymmetry. It looks as if we have obtained three relations among the functions $\mathcal{K}$ and T, but they are not independent. It is obvious that (27.135) and (27.136) follow in fact from (27.134). This relation implies, in turn, that both $\mathcal{K}$ and T can be expressed in terms of a single function which is usually written as

$$T(\phi) = \frac{1}{4\pi i}\frac{dh(\phi)}{d\phi}; \quad \mathcal{K}(\phi,\phi^*) = \mathrm{Im}\frac{\phi^* h(\phi)}{4\pi}. \tag{27.137}$$

The first of these relations implies that h is an holomorphic function, i.e. it depends on ϕ but not on ϕ^*. This property of holomorphicity is characteristic of supersymmetry and comes from the fact that the corresponding term in the super-potential is expressed as an F-term.[12] We often find the notation

$$h(\phi) = \frac{d\mathcal{F}(\phi)}{d\phi}, \tag{27.138}$$

where the function $\mathcal{F}(\phi)$ is called *the prepotential*.

We understand better this result if we go back to the original theory which is invariant under $N = 2$ supersymmetry and $O(3)$ gauge symmetry. All fields are triplets of $O(3)$ and ϕ is the third component of the scalar field. The spontaneous breaking of the gauge symmetry $O(3) \rightarrow U(1)$ is obtained by giving a non-zero vacuum expectation value to ϕ. We remind that the classical potential is flat in the direction of ϕ, which means that the vacuum expectation value is undetermined. We have called these fields moduli and it is not surprising that the effective theory depends on their values. They have two effects: they produce non-canonical kinetic energy terms for all other fields with a field-dependent coefficient and field-dependent coupling 'constants' for all interaction vertices. The consequence of $N = 2$ supersymmetry is that all this dependence on the moduli can be expressed in terms of a single function $h(\phi)$. In the tree approximation all

---

[12] We remind that an F-term comes from the expansion of a chiral superfield which depends on θ and not on $\bar{\theta}$.

couplings are absent and the kinetic energy is canonical. It follows that the corresponding value of h, with a suitable normalisation of the constants, is

$$h(\phi)|_0 = \left(\frac{4\pi i}{e^2} + \frac{\theta}{2\pi} \right) \phi = \tau_{cl} \phi \tag{27.139}$$

with τ_{cl} given by (26.71).

The remarkable result, first obtained by N. Seiberg and E. Witten, is to show that by performing semi-classical calculations, we can determine $h(\phi)$ exactly. Their method is based on the following observations: (i) the original theory is asymptotically free. So, for large values of ϕ the coupling is weak and low-order perturbation expansion is assumed to be reliable. (ii) The β-function receives only 1-loop contributions, so the evolution of the effective coupling constant is known. These two properties imply that we can express the prepotential $\mathcal{F}(\phi)$ in perturbation theory, including the 1-loop corrections, as

$$\mathcal{F}(\phi) = \frac{i}{2\pi} \phi^2 \ln \frac{\phi^2}{\Lambda^2}, \tag{27.140}$$

where we have used the analogue of Eq. (25.89) to replace the running coupling constant e^2 by a scale Λ. The fact that the $N = 2$ β-function is given by its 1-loop term ensures that this formula does not receive any higher order perturbative corrections. On the other hand, if we had looked at the $N=4$ theory whose β-function vanishes, the formula (27.139) would have been exact.

(iii) The third and most important observation is connected with duality. As we noted already, the effective theory depends on the point in the space of moduli. In the particular example $O(3) \rightarrow U(1)$ we are considering, this space is a two-dimensional manifold and we can choose as a complex co-ordinate the value of the gauge invariant quantity

$$a^2 = \sum_{i=1}^{3} \phi_i \phi_i. \tag{27.141}$$

So, the question is to find the coordinate transformations

$$a \rightarrow f(a) \tag{27.142}$$

that yield physically equivalent theories. Note that since in our effective theory (27.131) the coupling constant is a-dependent, the electromagnetic duality transformation $e \rightarrow 1/e$ belongs to the general class (27.142). However, expressing the Montonen–Olive conjecture is not straightforward for a theory with a running coupling constant. The Seiberg–Witten result is based on the observation that for $N = 2$, duality should be expressed as the invariance under the transformations that respect the Kähler structure in the space of moduli. The full justification of this point is rather lengthy and we will not present all the calculations in detail; we will only indicate the essential steps.

Using the variables (27.141) and the relations (27.137), the metric (27.132) can be written as

$$(\mathrm{d}s)^2 = \mathrm{Im}\frac{\mathrm{d}^2\mathcal{F}(a)}{\mathrm{d}a^2}\mathrm{d}a\mathrm{d}a^*. \tag{27.143}$$

Seiberg and Witten observe that this description can be only locally valid because the harmonic function $\mathrm{Im}\frac{\mathrm{d}^2\mathcal{F}(a)}{\mathrm{d}a^2}$ cannot be positive definite everywhere. This is because a harmonic function attains its extreme values at the boundary of its domain of definition; therefore, if it is defined everywhere in the manifold, it cannot have a minimum.

It will be convenient to introduce the auxiliary variable $a_D = h(a)$ and write the metric in the symmetric form

$$(\mathrm{d}s)^2 = \frac{\mathrm{i}}{2}(\mathrm{d}a\mathrm{d}a_D^* - \mathrm{d}a_D\mathrm{d}a^*). \tag{27.144}$$

We can consider the two complex variables (a, a_D) as coordinates on a two-dimensional complex space X. The form of the metric (27.144) on the moduli space M suggests we introduce on X the symplectic form $\omega = \mathrm{Im}\,\mathrm{d}a_D \wedge \mathrm{d}a^*$. If a complex variable u labels the point on M,[13] the pair $(a(u), a_D(u))$ defines a map $M \to X$ with the induced metric

$$(\mathrm{d}s)^2 = \frac{\mathrm{i}}{2}\left(\frac{\mathrm{d}a}{\mathrm{d}u}\frac{\mathrm{d}a_D^*}{\mathrm{d}u^*} - \frac{\mathrm{d}a_D}{\mathrm{d}u}\frac{\mathrm{d}a^*}{\mathrm{d}u^*}\right)\mathrm{d}u\mathrm{d}u^*. \tag{27.145}$$

The question now is to find the group of transformations that preserve this form of the metric. This is easy to answer. Let us rewrite (27.145) as

$$(\mathrm{d}s)^2 = \frac{\mathrm{i}}{2}\epsilon_{\alpha\beta}\frac{\mathrm{d}a^\alpha}{\mathrm{d}u}\frac{\mathrm{d}a^{*\beta}}{\mathrm{d}u^*}\mathrm{d}u\mathrm{d}u^* \tag{27.146}$$

with the obvious notation in which a^α is the two-component vector (a, a_D) and $\epsilon_{\alpha\beta}$ is the two-index antisymmetric tensor. The metric now is invariant under the special linear group $SL(2, \mathbb{R})$, which leaves $\epsilon_{\alpha\beta}$ invariant and commutes with the operation of complex conjugation. We can also add a constant to either a or a_D, so the transformations are

$$\begin{pmatrix} a \\ a_D \end{pmatrix} \to \mathcal{M}\begin{pmatrix} a \\ a_D \end{pmatrix} + \begin{pmatrix} c_1 \\ c_2 \end{pmatrix} \tag{27.147}$$

with $\mathcal{M}$ a $2{\times}2$ matrix of $SL(2, R)$.

Up to now we have looked only at the scalar field sector of the theory and we found that the Kähler metric on the moduli space is left invariant under the transformations (27.147). We want to understand the physical meaning of these transformations and

---

[13] A convenient choice is $2u = a^2$.

justify the conjecture that they can be used to define the property of duality. For this we must remember that the scalar fields belong to a massless $N=2$ supersymmetry multiplet, together with a pair of spinors and a photon. We know what we should expect from a duality transformation when it acts on the photon field, so we want to see how our $SL(2, \mathbb{R})$ transformations act on photons.

Let us write an element of $SL(2, \mathbb{R})$ in terms of two generators, T_b and S:

$$T_b = \begin{pmatrix} 1 & 0 \\ b & 1 \end{pmatrix}; \quad S = \begin{pmatrix} 0 & -1 \\ 1 & 0 \end{pmatrix} \tag{27.148}$$

with real b. We can immediately read the action of T_b on the photon kinetic energy. We see that it gives $a_D \to a_D + ab$, which implies a shift proportional to ϕb in $h(\phi)$. Using (27.131), (27.137), and (27.139), we conclude that the result of a T_b transformation is to shift the value of θ in the Lagrangian by an amount proportional to b. Our discussion in section 25.4.5 has taught us that θ is an angle. It follows that b must be equal to $2\pi n$ with n being an integer; in other words, only the subgroup $SL(2, \mathbb{Z})$ of $SL(2, \mathbb{R})$ is acceptable. The transformation is of the form $\tau \to \tau + 1$ and it is a symmetry of the Lagrangian.

It takes some more work to find the physical meaning of S. It is a discrete transformation $a_D \to a$, $a \to -a_D$, which implies that $\tau \to -1/\tau$. It reminds us of the electric–magnetic duality transformation we found in section 26.2.5, but it inverts τ rather than just the charge e. As such it is not expected to be a symmetry of the Lagrangian, in the same way that the substitution $e \to 1/e$ is not. It maps one physical theory onto another and, in the case of the electric–magnetic duality, we had conjectured that the two theories describe the same physics. Here supersymmetry helps us to go further and prove this conjecture for the transformation S. Let us go through the main steps of the proof.

The photon field is contained in the chiral superfield W_α in the Lagrangian (27.130). It satisfies the chirality condition $\mathcal{D}W = 0$, where $\mathcal{D}$ is the supercovariant derivative. We can implement this constraint in the functional integral by using a Lagrange multiplier given by a real chiral superfield V_D and add a term proportional to

$$\int \mathrm{d}^4x \mathrm{d}^2\theta \mathrm{d}^2\bar\theta \, V_D \mathcal{D} W \sim \int \mathrm{d}^4x \mathrm{d}^2\theta \mathrm{d}^2\bar\theta \, \mathcal{D} V_D W \sim \int \mathrm{d}^4x \mathrm{d}^2\theta \, W_D W. \tag{27.149}$$

The effective Lagrangian is given by the sum of (27.130) and (27.149). The dependence on W is quadratic and we can perform the Gaussian functional integral to eliminate W in favour of the new field W_D. The result is a new effective Lagrangian $\mathcal{L}_D$ proportional to

$$\mathcal{L}_D \sim \left[\frac{1}{\mathcal{T}(T)} W_D W^D |_{\theta\theta} + h.c. \right]. \tag{27.150}$$

We are halfway through. We need to compute the result of the S transformation on the chiral multiplet T. Using the second of the relations (27.137) we find that

$$\int d^4x d^2\theta d^2\bar\theta h(T)\bar T \to \int d^4x d^2\theta d^2\bar\theta h_D(T_D)\bar T_D \qquad (27.151)$$

with $h_D(h(T)) = -T$. Expressed in terms of superfields, the first of Eqs. (27.137) can be written as $T(T) \sim dh(T)/dT$, which implies that for the coefficient of the transformed kinetic energy term of the gauge field given by Eq. (27.150),

$$-[T(T)]^{-1} = -[dh(T)/dT]^{-1} = dh_D(T_D)/dT_D = T_D(T_D), \qquad (27.152)$$

which shows that the transformation S which, in the space of moduli interchanges $a_D \to a$, $a \to -a_D$, relates two physically equivalent theories. This proves the conjecture that the $SL(2,\mathbb{Z})$ transformations generalise the duality property of electromagnetism in our $N=2$ model. The role of the constants c_1 and c_2 in the general transformation (27.147) will be discussed later.

The next point is to use these properties in order to determine the singularities of the prepotential. This is an essential step because the knowledge of the singularities combined with the asymptotic form determines completely an holomorphic function. An obvious singular point is the point at infinity because we find it in the perturbative result and the large field region corresponds to weak coupling where perturbation can be trusted. In the classical approximation we find also a singularity at $\phi = 0$ because, at this point, the full $O(3)$ gauge symmetry is restored. However, at small values of the field the coupling is strong and the classical picture is not expected to be reliable. The simplest ansatz is to assume that the point at infinity is the only singular point. Looking at the expression (27.140) which is valid for large ϕ, we see that the singularity at infinity is logarithmic. Using our variables a and a_D we can write

$$a_D = h(a) = \frac{d\mathcal{F}}{da} = \frac{2ia}{\pi}\ln\frac{a}{\Lambda} + \frac{ia}{\pi}. \qquad (27.153)$$

The logarithmic singularity implies the existence of a non-trivial monodromy because if we follow a circle around the point at infinity, always staying at large values of the field where the expression (27.153) is correct, the logarithm of ϕ^2 changes by $2\pi i$ and $\ln a \to \ln a + \pi i$. It follows that the action of this monodromy on our (a, a_D) space is given by

$$\mathcal{M}_\infty \begin{pmatrix} a \\ a_D \end{pmatrix} = \begin{pmatrix} -1 & 0 \\ 2 & -1 \end{pmatrix} \begin{pmatrix} a \\ a_D \end{pmatrix} = \begin{pmatrix} -a \\ -a_D + 2a \end{pmatrix}. \qquad (27.154)$$

The interesting point is that the very existence and the nature of this monodromy show that our ansatz, namely the assumption that the logarithmic singularity at infinity is the only singular point in the moduli space, is not correct. The reason is simple: if there is no other singularity all closed paths around the point at infinity would be continuously deformable to each other; in other words, monodromies would form an Abelian representation of the fundamental group (or the first homotopy group), of the moduli space in $SL(2,\mathbb{Z})$. But this implies that the action of *all* monodromies, not only those restricted to large values of the field, would be given by (27.154). Since in this monodromy a goes into $-a$, a^2 would have been a good coordinate defined everywhere in the moduli space. But we argued previously that this is impossible because the induced metric could not be positive definite. The conclusion is that we must have further singularities, or other topological obstructions, in addition to the one we found at infinity.

We could be tempted to consider the point at the origin in field space, which we found in the classical theory, and consider a manifold with two singularities, but the same analysis of the monodromies shows that it does not solve our problem. Furthermore, Seiberg and Witten have argued that this singularity at the origin is washed out in the quantum theory. We conclude that the moduli space must have at least one singular point, somewhere at a value $u = u_0$ different from 0. Since the operation $u \to -u$ is a symmetry of the theory, being part of the unbroken centre of $O(3)$, the point $u = -u_0$ will be a singular point too. We are thus led to considering a manifold with three singularities, $\pm u_0$, and infinity. At this point we must do three things. (i) Find the physical origin of the singularities at $\pm u_0$; (ii) check whether there are other singularities; and (iii) if not, construct the corresponding holomorphic function.

It is easy to complete the first task. We know that this theory contains, in addition to the gauge bosons and their $N = 2$ supersymmetric partners, magnetic monopoles and dyons. They form short $N = 2$ multiplets, so their masses saturate the BPS bound (26.66). At the classical level the value of the central charge Z_{cl} is given by

$$Z_{cl} = a(n_e + \tau_{cl} n_m), \qquad (27.155)$$

where n_e and n_m are the electric and magnetic charges, respectively, which, for $O(3)$, take integer values. As we explained in the previous section, the dyon masses saturate the BPS bound even when all quantum corrections are included because it is only in this case that we can have short multiplets. We can compute the value of the central charge in our effective theory (27.131) by computing the $N = 2$ supersymmetry algebra directly. Alternatively, we can couple the pure gauge theory to a charged matter multiplet. The result is that for electrically charged particles with no magnetic charge the value of Z is an_e. By duality, the corresponding value of the central charge for magnetic monopoles will be $a_D n_m$. So, for a general state we obtain

$$Z = a n_e + a_D n_m. \qquad (27.156)$$

This formula has the correct classical limit and exhibits the property of duality. As we said earlier, it can be verified by direct computation. Before pointing out the consequences for the singularity structure in the moduli space, let us note that it is not invariant under the constant translations of either a or a_D contained in the transformations (27.147). It follows that we must put $c_1 = c_2 = 0$. On the other hand, since n_e and n_m are integers, we obtain an independent derivation of the result that $SL(2, \mathbb{R})$ should be restricted to $SL(2, \mathbb{Z})$. Concerning the singularities, we remind that they occur at points in the space of the moduli in which new massless states appear. Looking at (27.156) we see that given a pair n_e, n_m, Z, and therefore the dyon masses, vanish at points away from the origin $a = 0$, thus creating singularities in the manifold. For example, a magnetic monopole ($n_e = 0, n_m = 1$) becomes massless for $a_D = 0$, although

a is different from 0. A single charge dyon $(1,1)$ becomes massless at $a + a_D = 0$. We could think that if we consider all the series of dyons $(n_e, 1)$ we would find an infinity of singular points, namely the points u such that the equations $a_D + n_e a = 0$ are satisfied. In fact, this is not so. Let us call u_0 the solution of the equation $a_D = 0$. A 1-loop calculation around that point shows that the monodromy is

$$a_D \to a_D; \quad a \to a - 2a_D, \tag{27.157}$$

which shows that the entire series corresponds to the same pair of singularities $u = \pm u_0$.

This discussion completes our first task; namely, it shows the physical mechanism through which singularities, other than the one at infinity, appear.

We will only give a plausibility argument regarding the second task. If the physical origin of singularities is the appearance of massless states, this discussion should exhaust all possibilities because dyons are the only states in the semi-classical spectrum. Seiberg and Witten, by considering the $N = 2$ theory coupled to matter, gave a full justification of this point.

The third task is purely technical. By appropriately rescaling the variable u, we can bring the singularities at $u = \pm 1$. The moduli space of vacua for our theory is a two-dimensional space with singularities at $u = 1$, -1, and ∞ and a $\mathbb{Z}_2$ symmetry $u \leftrightarrow -u$. Since we have three singularities the monodromies form a non-Abelian representation of the fundamental group and the discussion is more complicated than the one we have just presented, but no new physical input is required. The solution determines the holomorphic function $h(a)$. Seiberg and Witten wrote it in a parametric form $a(u)$ and $h = a_D(u)$ as follows

$$a(u) = \frac{\sqrt{2}}{\pi} \int_{-1}^{1} \frac{dx\sqrt{x-u}}{\sqrt{x^2-1}} \tag{27.158}$$

$$a_D(u) = \frac{\sqrt{2}}{\pi} \int_{1}^{u} \frac{dx\sqrt{x-u}}{\sqrt{x^2-1}}. \tag{27.159}$$

This completes the construction and fully determines the low-energy theory (27.131). Similar results can be obtained for the $N = 4$ theory and, to a certain extent, even for $N = 1$. We want to emphasise that although the $N = 2$ model does not resemble anything like the real world, this construction is, nevertheless, a remarkable achievement because it shows that a fully interacting four-dimensional dynamical theory can be determined beyond the perturbation expansion.

27.10 Twisted Supersymmetry and Topological Field Theories

27.10.1 Introduction

In this section, we present a relatively new subject, that of topological quantum field theory (TQFT). The main interest in TQFTs started in 1988, following the work of

E. Witten, who showed that one can use the $N=2$, $d=4$ super-Yang–Mills theory to compute topological invariants on four-dimensional manifolds. In this book we will adopt a rather broad definition of a TQFT as a theory which contains observables whose quantum expectation values are independent of any local changes of the metric which is used to describe the manifold, provided these changes do not affect its topological properties. More precisely, a TQFT is a quantum field theory defined from a local action, whose propagators and possibly the interactions may be metric dependent; however, it allows us to compute expectation values which remain the same when we vary the metric by local deformations. This is a non-trivial programme, since when the metric varies, not only the propagators and vertices change, but also the measure of the path integral. Very subtle compensations must occur for the existence of topological observables.

There are basically two kinds of local actions that can be used to define a TQFT. Some are themselves metric independent, but define a classical propagation of fields, with a celebrated example of the three-dimensional Chern–Simons action. They contain topological observables such as a Wilson loop. Indeed, such observables are classically gauge invariant and metric independent, and changing the metric only influences the standard gauge-fixing sector of the action. The metric independence of the Wilson loop at the quantum level can be simply shown by using standard BRST Ward identities. Other theories are of a more intriguing type, which we will study here. Their action is metric dependent, but possesses a supersymmetry which provides Ward identities, which in turn explain the metric independence of some observables. They are often called TQFTs of the cohomological type.

We may wonder why such theories are of interest for physicists. The obvious answer is that, as we have already seen, topological observables, such as a Wilson loop, often carry important physical information. We will show here a second answer; namely, the study of such theories provides a deep insight into the structure of quantum field theories with extended supersymmetry and, more generally, into the whole quantisation programme.

The addition of topological terms to local actions of various QFTs was done much before the introduction of TQFTs. In section 25.4.5 we saw that many aspects of instanton effects in Yang–Mills theories are taken into account by adding a topological term to the action, which is proportional to the second Chern class. In Eq. (25.214) we saw that the effective theory is given by

$$\mathcal{S}_{\text{eff}} = \int \mathrm{d}^4 x \left(\frac{1}{g^2} \mathrm{Tr}(F_{\mu\nu} F^{\mu\nu}) + \frac{i\theta}{32\pi^2} \epsilon^{\mu\nu\rho\sigma} \mathrm{Tr}(F_{\mu\nu} F_{\rho\sigma}) \right). \tag{27.160}$$

Here θ is a dimensionless constant whose values reflect the definition of the vacuum state. Using this property, we showed that we can answer physical questions such as the famous $U(1)$ problem which is connected with the absence of a light isoscalar pseudo-Goldstone boson. In section 25.4.5 we said that θ is just a second coupling constant and, in most applications, this is indeed correct. However, here we can see immediately that this is true only as long as the first coupling constant g is kept finite and the theory is quantised using the usual Yang–Mills action. Let us consider the formal limit

$g \rightarrow \infty$, keeping θ finite. The Yang–Mills term drops out and the effective action is given entirely by the topological term. What kind of quantum field theory does such an effective theory describe? The immediate answer is a completely trivial one since this term, being topological, is just a constant given by the Pontryagin index. We cannot derive any equations of motion by varying the action with respect to the field A and, therefore, it seems that we have no dynamical degrees of freedom. Note also that this action is independent of the metric we use to describe the geometry of our four-dimensional space–time.[14]

It is not difficult to identify the source of all these problems. The theory given only by the topological term

$$\mathcal{S}_t \sim \int \mathrm{d}^4 x \epsilon^{\mu\nu\rho\sigma} \mathrm{Tr}(F_{\mu\nu} F_{\rho\sigma}) \tag{27.161}$$

has a large gauge invariance, much larger than that of the Yang–Mills theory, because $\mathcal{S}_t$ is invariant under the transformations

$$A_\mu(x) \rightarrow A_\mu(x) + \epsilon_\mu(x) \tag{27.162}$$

with $\epsilon_\mu(x)$ being any $\mathcal{G}$-valued local vector parameter, provided it is globally well defined possibly modulo ordinary gauge transformations, and does not change the winding number of the gauge field configuration. This huge invariance explains why the classical propagation cannot be defined, not only for the longitudinal degrees of freedom of A but also for the transverse ones. It means that the gauge field has local zero modes in all possible directions. This is in contradistinction with the ordinary Yang–Mills theory in which the gauge transformation

$$A_\mu(x) \rightarrow A_\mu(x) + D_\mu^A \epsilon(x) \tag{27.163}$$

affects only the longitudinal component of the field. Thus, from the physicists point of view, the possibility of quantising a topological invariant comes as no surprise. The reader who has followed the BRST method presented in Chapter 14 can guess the procedure which will be developed in this section. We will show that the quite unorthodox question of gauge fixing the huge gauge symmetry (27.162) can be successfully solved by a BRST-like invariant gauge-fixing term of the local invariance (27.162) of $\mathrm{Tr}(F \wedge F)$. The main technical difference is the size of the space of the zero modes, but the methodology is completely analogous. We must build a BRST symmetry that contains ghosts and antighosts and encode the definition of the gauge symmetry on the classical fields and ghosts. Then we must find a practical gauge-fixing action that is BRST exact and check that no further zero modes arise. Then, we can safely define the path integral from this BRST invariant action. Observables will then be defined from the cohomology of the

---

[14] In this book we have studied only quantum field theory in flat Minkowski space. Had we introduced a non-trivial metric $g_{\mu\nu}$ it would have changed the form of the Yang–Mills Lagrangian but not that of the θ term since Eq. (17.37) remains true.

BRST symmetry. If there is no anomaly, the theory is renormalisable, and it is expected that the observables will be independent of the parameters which have been introduced for the gauge-fixing action, through BRST exact terms.

Interesting facts often happen in the process that we just explained for building a TQFT. In the case of the quantisation of the action (27.161), the surprise is that we get, in the simplest case, an action identical to the $N=2$, $d=4$ action, modulo an operation called *twist*,[15] that is, a mere change of variables on the fields that respects a sub-algebra of the global symmetries of the $N=2$, $d=4$ action. In fact, the construction just described introduces fermions that look superficially like ghosts, since they have integer spin, in apparent contradiction with the spin-statistics theorem, and gives an action that looks as a new type of supersymmetric action. Naturally, this contradiction is only apparent because, as we will see, in Euclidean space, the twist operation is just a linear mapping of the two Majorana spinors of $N=2$ to a new set of integer spin fields. So, by the inverse operation, which we can call untwisting, the TQFT action will turn out to be identical with the usual $N=2$, $d=4$ action. A surprising feature that turns out to be quite generic is that the TQFT action is determined by a smaller number of supersymmetries than that given by the number of parameters of the Poincaré supersymmetry of the untwisted model. For instance, in the case of $N=2$, we have a total of 5=1 (scalar) +4 (vector) twisted supersymmetries that determine the theory. But the TQFT action determined by these five supersymmetries turns out to be automatically invariant under three more supersymmetries, with three parameters that are assembled in a self-dual 2-form. These unpredicted extra supersymmetries seem accidental, but the total of the 5+3 supersymmetries is identical to the number of the 8 Poincaré supersymmetry generators of the untwisted $N=2$ action. For the $N=4$ extended supersymmetry, the same kind of phenomenon happens. We find a smaller algebra (with 9 generators) than the super-Poincaré algebra (with 16 generators) that determines the twisted TQFT action. The advantage is that the former constitute an off-shell closed algebra, contrast to what happens to the latter for which we have no finite set of auxiliary fields which can close the algebra. This makes it possible to establish reliable proofs based on Ward identities for the $N=4$ theory. Naturally, the TQFT action turns out again to be automatically invariant under 7 more supersymmetry generators which allow us to recover the 16 generators of the original $N=4$ theory.

To grab some understanding of all this, we first study a very simple case, that of a problem with a finite number of degrees of freedom. We show a method for constructing a TQFT corresponding to the quantum mechanical example where the classical coordinates are those of a point particle $q^i(t)$, $i = 1, 2$, moving in a plane with no origin, and thus with a non-trivial topology. In this case the topological observables are the possible winding numbers of closed trajectories Γ that may enclose the origin, and the classical action is $\int_\Gamma dt \, \frac{\dot{\theta}}{2\pi}$. Expectation values of these observables reproduce the winding numbers of trajectories at the quantum level.

---

[15] No relation with the term introduced in section 25.4.3.

27.10.2 A Quantum Mechanical Toy Model

We start from a classical action equal to the winding number of the trajectory Γ of a particle moving in a plane with no origin,

$$I_t[q^i] = \frac{1}{2\pi} \int_\Gamma dt \, \dot\theta = \frac{1}{2\pi} \int_\Gamma dt \, \epsilon_{ij} \frac{q^i \dot q^j}{q^2} = n_\Gamma, \qquad (27.164)$$

where n_Γ is the winding number of the trajectory Γ around the origin.[16]

We could add this topological term to any given non-relativistic action of quantum mechanics. This would lead to interesting physical effects, analogous to the Aharonov–Bohm effect. However, we will ask ourselves the strange question of defining the path integral

$$< \mathcal{O}([q]) > \equiv \int [dq] \mathcal{O}([q]) \exp I_t[q] =?, \qquad (27.165)$$

where $\mathcal{O}(q)$ is some observable. The question looks quite weird, since we start from an action with no equation of motion and the usual visualisation of quantum mechanical effects being defined as fluctuations around classical trajectories weighted by $\exp I_t[q]$ becomes immaterial. In fact, the question mark in Eq. (27.165) is justified because $I_t[q]$ remains the same for arbitrary local variations of q, so that the path integral is completely degenerate. This observation can be rephrased by saying that we have a gauge invariance of the action, $\delta I_t[q] = 0$, with

$$\delta q(t) = \epsilon(t). \qquad (27.166)$$

The gauge invariance of the action is so large (here the gauge invariance is arbitrary coordinate redefinitions that do not change the winding number) that the path integral is ill-defined. We can, however, quickly recover hope, after a bit of thinking.

This degeneracy of the path integral is in fact analogous to the one we had in the case of the Yang–Mills theory in the sector of the longitudinal components of the gauge fields, yielding an undefined longitudinal propagator. We solved the question in the Yang–Mills case by adding to the action a gauge-fixing term and appropriate ghost-dependent terms, such that the total action becomes BRST invariant, while the added terms were BRST exact and the observables defined as the cohomology of the BRST symmetry.

To define the path integral (27.165), we thus decide to add a gauge-fixing term of the huge symmetry (27.166), while imposing a corresponding BRST-like symmetry. We thus introduce an anti-commuting ghost $\Psi^i(t)$, in correspondence with the local

---

[16] L. Baulieu and E. Rabinovici, *Phys. Lett. B* **316**, 93 (1993).

parameter $\epsilon(t)$, $\epsilon^i(t) \rightarrow \Psi^i(t)$ but the opposite statistics. We must also introduce an antighost $\bar{\Psi}^i(t)$ and a Lagrange parameter $H^i(t)$. We call $q^i(t), \Psi^i(t), \bar{\Psi}^i(t), H^i(t)$ a topological quartet. We thus define the BRST-like symmetry associated with the invariance $\delta q(t) = \epsilon(t)$ as

$$Qq^i = \Psi^i$$
$$Q\Psi^i = 0$$
$$Q\bar{\Psi}^i = H^i$$
$$QH^i = 0. \tag{27.167}$$

We have $Q^2 = 0$. Note the equation $(d + Q)q = dq + \Psi$.

We can use the following q-dependent topological gauge-fixing function, which we will impose in a BRST-like invariant way, that is by adding to the action a Q-exact term,

$$\dot{q}^i + \epsilon_{ij}\frac{q^j}{q^2} = \dot{q}^i + \frac{\delta V[q]}{\delta q^i}, \tag{27.168}$$

where

$$V[q] = \theta. \tag{27.169}$$

This choice of a 'topological gauge function' is a non-trivial one, since θ is multi-valued. In fact, there are instanton-like solutions of the equation $\dot{q}^i + \epsilon_{ij}\frac{q^j}{q^2} = 0$. They are concentric circles on which the particle moves at constant velocity. They have winding number n and their radius goes like $1/\sqrt{n}$.

The local topological action is obtained as

$$I_t[q] \rightarrow I_t[q] + \int_\Gamma dt Q\left(\bar{\Psi}^i g_{ij}\left(\frac{aH^j}{2} + \dot{q}^j + \epsilon^{jk}\frac{q^k}{q^2}\right)\right). \tag{27.170}$$

Here g_{ij} is the metric on the plane without the origin that we supposed constant for simplicity for the computation that we will shortly exhibit.[17]

It is easy to show that with the new action we obtain a well-defined propagator, since the former can be written as

$$I_t[q] + \int_\Gamma dt \left(\frac{a}{2}H^i g_{ij}H^j + H^j g_{ij}\left(\dot{q}^i + \epsilon_{ik}\frac{q^k}{q^2}\right) - \bar{\Psi}_i\left(\dot{\Psi}^i + \frac{\delta^2\theta}{\delta q^i q^j}\Psi^j\right)\right)$$
$$\sim I_t[q] - \int_\Gamma dt \left(\frac{1}{2a}\left(\dot{q}^i + \epsilon_{ij}\frac{q^j}{q^2}\right)^2 + \bar{\Psi}_i\left(\dot{\Psi}^i + \frac{\delta^2\theta}{\delta q^i q^j}\Psi^j\right)\right). \tag{27.171}$$

---

[17] We could in fact had chosen a non-constant metric on the plane, $g_{ij} \rightarrow g_{ij}(q)$, in which case the computation which follows would have yielded an action that is reparametrisation invariant in q and depends on the Christoffel symbols $\Gamma^i_{jk}(g)$ and Riemann tensor $R^{il}_{jk}(g)$.

For $a = 1$ the mixed term in $(\dot{q}^i + \epsilon_{ij}\frac{q^j}{q^2})^2$ cancels the corresponding one in $I_t[q]$, so that we get for the action[18]

$$I[q] = \int_\Gamma \left(\mathrm{d}t \frac{(\dot{q})^2}{2} + \frac{1}{2q^2} + \bar{\Psi}_i \left(\dot{\Psi}^i + \frac{\delta^2 \theta}{\delta q^i q^j} \Psi^j \right) \right). \tag{27.173}$$

This action has the property that a change in the metric, $g + \delta g$, amounts to modifying the action by the addition of a Q-exact term. If we consider Q-invariant observables that are metric independent, such as functionals of Ψ^i and $\mathrm{d}X$ whose Q-variation is a pure derivative, i.e.

$$\epsilon_{ij}\Psi^i\Psi^j, \quad \epsilon_{ij}\mathrm{d}q^i \Psi^j, \quad \epsilon_{ij}\mathrm{d}q^i \mathrm{d}q^j, \tag{27.174}$$

the Ward identities of the Q symmetry imply that their correlators do not depend on variations of the metric. This is a deep result, since we have a quantum mechanical action that depends on the metric, but some expectation values are metric independent, which justifies the denomination of a topological model. Note that we can compute other quantities, using the action I. If they do not belong to the cohomology of the Q symmetry, they will in general depend on variations of the metric. In fact, as seen from the point of view of quantum mechanics, the action is just an ordinary supersymmetric action. It can be interpreted as an action for a particle with spin, as can be shown by a standard Hamiltonian treatment. What is important here is that we have succeeded in building an action for which the metric can be viewed as a sort of gauge parameter, with non-trivial topological observables. We achieved this through a process of gauge fixing of a topological action, in our case a winding number, which preserves the BSRT-like symmetry associated with the general invariance of the topological term.

The process can be clearly generalised. As we will see, in the case of the Yang–Mills field, it will open a new chapter for supersymmetric Yang–Mils theories.

27.10.3 Yang–Mills TQFT

In the previous section we showed in the toy model that a purely topological theory, the one given by the action (27.164), can be reinterpreted at the quantum level as a supersymmetric interacting theory. The generalisation to four dimensions will imply a similar phenomenon for the action given by (27.161). We will show that, in a certain sense which will be made precise shortly, it gives rise to our $N = 2$ supersymmetric Yang–Mills theory. Here we will follow the opposite way; namely, we will start from the $N = 2$ theory and derive the topological term of Eq. (27.161).

---

[18] When we perform a more detailed computation with a non-constant metric, $g_{ij} \to g_{ij}(q)$, nothing is really changed, but the fact that the action (27.173) is generalised as

$$I[q] = \int_\Gamma \mathrm{d}t \left(g_{ij}\frac{\dot{q}^i \dot{q}^j}{2} + \frac{1}{2q^2} + \bar{\Psi}^j g_{ij} \left(\dot{\Psi}^i + \Gamma^I_{kl}(q)\dot{q}^k \Psi^l + \frac{\delta^2 \theta}{\delta q^i q^j} \Psi^j \right) \right. \tag{27.172}$$
$$\left. + R_{ijkl}(g)\Psi^i\Psi^j\bar{\Psi}^k\bar{\Psi}^l \right).$$

27.10.3.1 Notion of Twist for the Yang–Mills Theories, with Emphasis on the d = 4, N = 2 Theory

Depending on the space dimensionality, we have often the possibility of mapping spinors onto tensors, an operation called twist, which has been proven most useful in the context of topological quantum field theory and has interesting applications to supersymmetric theories.

Here we will detail the first historical description of the twist, in the case of the $d = 4$, $N = 2$ theory, starting from our knowledge of supersymmetric Yang–Mills theories. We have seen already that such theories are invariant under both global Lorentz invariance and R-symmetry. The latter is associated with the rank of the supersymmetry. In a Euclidean formulation, the Lorentz symmetry becomes the rotational invariance $SO(d)$. For the $d = 4$, $N = 2$ theory, the R-symmetry is $R = SU(2)$, so the global symmetry is $SO(4) \times SU(2) \sim SU_L(2) \times SU_R(2) \times SU(2)$.[19]

The twist operation consists of breaking this symmetry into one of a lower rank. In practice we arrange for a breaking of $SU_R(2) \times SU(2) \to SU_R'(2) \times U(1)$ with $SU_R'(2)$ the diagonal subgroup of the original product. We are thus left with the symmetry group $SU_L(2) \times SU_R'(2) \times U(1) \approx SO'(4) \times U(1)$. We can now define a new Lorentz group from the analytic continuation of $SO'(4)$. This sounds strange because this new group has a piece coming from our initial internal symmetry group. It follows that the fields will have peculiar transformation properties. The leftover internal symmetry is $U(1)$ and we will call the corresponding conserved charge *shadow*.

To see in more detail how this goes, let us recall that the vector multiplet of the $N = 2$ super-Yang–Mills theory is made out of a gauge field A_μ, a scalar field t, and a pseudo-scalar s, all singlets of the $SU(2)$ R-symmetry, as well as a pair of Majorana fermions λ which form a doublet of $SU(2)$ (in section 27.7.1 they were denoted by λ and ζ). To these fields we must add a triplet of auxiliary fields which we will denote, collectively, G^i (in terms of the $N = 1$ fields they are the F, G, and D auxiliary fields used in the previous sections). All these fields are in the adjoint representation of whichever gauge group $\mathcal{G}$ we are considering. They satisfy the following supersymmetry transformations with a Majorana parameter ϵ which is a doublet of the internal $SU(2)$ symmetry. In the following we will choose ϵ to be a commuting parameter:

$$\delta^{Susy} A_\mu = -i\left(\overline{\epsilon}\gamma_\mu\lambda\right)$$

$$\delta^{Susy} s = -\left(\overline{\epsilon}\gamma_5\lambda\right) \qquad \delta^{Susy} t = -\left(\overline{\epsilon}\lambda\right)$$

$$\delta^{Susy}\lambda = F\epsilon + \gamma_5[s,t]\epsilon - i\left(\gamma_5 Ds + Dt\right)\epsilon + G\epsilon$$

$$\delta^{Susy} G^i = \left(\overline{\epsilon}\tau^i\left(-iD\lambda + \gamma_5[s,\lambda] - [t,\lambda]\right)\right). \tag{27.175}$$

---

[19] The R and L indices refer to the right- and left-handed components of the Lorentz group.

The τ^i's are the usual Pauli matrices and we can easily verify that these transformation properties close an off-shell supersymmetry algebra of the Wess–Zumino type

$$(\delta^{Susy})^2 = \delta^{\text{gauge}}\left((\overline{\epsilon}[\gamma_5 s - t]\epsilon) - i(\overline{\epsilon}\gamma^\mu \epsilon)A_\mu \right) - i(\overline{\epsilon}\gamma^\mu \epsilon)\partial_\mu. \qquad (27.176)$$

Let us write the spinors λ in the Weyl representation in which they carry two types of $SU(2)$ indices: λ^i_α (and $\overline{\lambda}^i_{\dot\alpha}$), with i referring to the internal symmetry $SU(2)$. Altogether, we have eight spinorial, anticommuting, components. The symmetry reduction consists in identifying and summing over indices of type i and α.[20] This operation leads us to decompose the eight spinorial components of any given pair of Majorana spinors, valued in the fundamental representation of the $R = SU(2)$ symmetry, as scalar, vector, and anti-selfdual tensor of $SO'(4))$, and $U(1)$ charges 1, -1, and 1, respectively. The decomposition is $8 = 1 \oplus 4 \oplus 3$. This can be done for all quantities that are organised as pairs of Majorana spinors in the supersymmetric theory, not only the fundamental fields, but also the generators and parameters of supersymmetry.

In practice this is done as follows: the fields of the $N = 2$ gauge multiplet which belong to the adjoint representation of the gauge group $\mathcal{G}$ are twisted into

$$(A_\mu, \lambda^i_\alpha, \overline{\lambda}^i_{\dot\alpha}, t, s, G^i) \rightarrow (A_\mu, \psi_\mu, \chi_{\mu\nu}, \eta, \phi, \overline{\phi}, G_{\mu\nu}), \qquad (27.177)$$

where

$$\begin{aligned} \eta &\equiv \varepsilon^{\alpha\beta}\lambda_{[\alpha\beta]} \\ \psi_\mu &\equiv (\text{-}s_\mu)^{\alpha\dot\alpha}\overline{\lambda}_{\alpha\dot\alpha} \\ \chi_{\mu\nu} &\equiv (s_{\mu\nu})^{\alpha\beta}\lambda_{(\alpha\beta)} \end{aligned} \qquad (27.178)$$

are the anticommuting fields resulting from the twist of the spinor fields $(\lambda^i_\alpha, \overline{\lambda}^i_{\dot\alpha})$ into $(\lambda_{\alpha\beta}, \overline{\lambda}_{\alpha\dot\alpha})$. Note that the Lorentz indices for the scalar fields $(\phi, \overline{\phi})$ and the gauge boson A_μ remain the same after the twisting operation. We see that in flat space, the twist operation looks like a linear change of variables on the fields, and that the lowering of the global symmetry has been obtained by the identification and contraction of the indices of type i and $\dot\alpha$.

The vector multiplet in representations of L' is thus made of the gauge field $A^{(0)}_\mu$, both bosonic scalar fields $\phi^{(2)} = t + s$ and $\overline{\phi}^{(-2)} = t - s$, an anticommuting vector $\psi^{(1)}_\mu$, an anticommuting anti-selfdual 2-form $\chi^{(-1)}_{\mu\nu^-}$, an anticommuting scalar $\eta^{(-1)}$, and a commuting auxiliary field $G^{(0)}_{\mu\nu^-}$ (the $SU(2)$-triplet auxiliary field G^i has been identified with an anti-selfdual 2-form $G^{(0)}_{\mu\nu^-}$ of L') . The superscripts in parenthesis denote the internal symmetry $U(1)$ charges of the fields and they will be dropped in the following.

---

[20] It is obvious that for this operation to be possible, the internal symmetry group must have an $SU(2)$ component. For example, in the $N = 4$, $d = 4$ case the R symmetry is $SU(4)$ and there are three independent possibilities for choosing an $SU(2)$ factor within the R symmetry. Each one gives a different generalisation of the operations we describe in the $N = 2$ case.

The scalar, vector, and tensor anticommuting generators are obtained by the same linear mappings as for the spinorial fields, namely $\delta \equiv \epsilon^{\alpha i} Q_{\alpha i}$, $\delta_{\mu} \equiv i\sigma_{\mu}^{\dot{\alpha} i} Q_{\dot{\alpha} i}$, and $\delta_{\mu\nu^-} \equiv \sigma_{\mu\nu}^{\alpha i} Q_{\alpha i}$.

The interesting result is that the (off-shell) supersymmetry algebra (27.176) greatly simplifies when it is given in twisted variables and can be written, in an obvious notation, as

$$\delta^2 = \delta^{\text{gauge}}(\phi), \quad \{\delta, \delta_{\mu}\} = \partial_{\mu} + \delta^{\text{gauge}}(A_{\mu}), \quad \delta_{\{\mu} \delta_{\nu\}} = g_{\mu\nu} \delta^{\text{gauge}}(\bar{\phi}) \tag{27.179}$$

and

$$\{\delta, \delta_{\rho\sigma}\} = 0, \quad \{\delta_{\mu}, \delta_{\rho\sigma}\} = 4 P_{\rho\sigma\mu}^{-\ \ \nu}(\partial_{\nu} + \delta^{\text{gauge}}(A_{\nu})), \quad \{\delta_{\mu\nu}, \delta_{\rho\sigma}\} = -8 P_{\mu\nu\rho\sigma}^{-} \delta^{\text{gauge}}(\phi). \tag{27.180}$$

The commuting supersymmetry parameter ϵ is twisted into $(\bar{\omega}, \varepsilon^{\mu}, \nu^{\mu\nu})$. The complete supersymmetry generator can be written as

$$\delta^{\text{Susy}} \equiv \bar{\omega} s + \nu^{\mu\nu} s_{\mu\nu} + \varepsilon^{\mu} s_{\mu} = (\epsilon^{\alpha i} + \sigma_{\mu\nu}^{\alpha i}) Q_{\alpha i} + i\sigma_{\mu}^{\dot{\alpha} i} Q_{\dot{\alpha} i} \tag{27.181}$$

and the transformation laws for the twisted $N = 2$, $d = 4$ super-Yang–Mills theory are

$$
\begin{array}{ll}
s A_{\mu} = \psi_{\mu} & s_{\mu} A_{\nu} = -g_{\mu\nu} \eta - \chi_{\mu\nu} \\
s \psi_{\mu} = -D_{\mu} \phi & s_{\mu} \psi_{\nu} = F_{\mu\nu} + G_{\mu\nu} + g_{\mu\nu}[\phi, \bar{\phi}] \\
s \phi = 0 & s_{\mu} \phi = -\psi_{\mu} \\
s \bar{\phi} = \eta & s_{\mu} \bar{\phi} = 0 \\
s \eta = [\phi, \bar{\phi}] & s_{\mu} \eta = D_{\mu} \bar{\phi} \\
s \chi_{\mu\nu} = G_{\mu\nu} & s_{\mu} \chi_{\rho\sigma} = 4 P_{\rho\sigma\mu}^{-\ \ \nu} D_{\nu} \bar{\phi} \\
s G_{\mu\nu} = [\phi, \chi_{\mu\nu}] & s_{\mu} G_{\rho\sigma} = D_{\mu} \chi_{\rho\sigma} - 4 P_{\rho\sigma\mu}^{-\ \ \nu}(D_{\nu} \eta + [\psi_{\nu}, \bar{\phi}])
\end{array} \tag{27.182}
$$

for the scalar and vector pieces, while for the tensor supersymmetry they are

$$
\begin{aligned}
s_{\rho\sigma} A_{\mu} &= -4 P_{\rho\sigma\mu}^{-\ \ \kappa} \psi_{\kappa} \\
s_{\rho\sigma} \psi_{\mu} &= -4 P_{\rho\sigma\mu}^{-\ \ \kappa} D_{\kappa} \phi \\
s_{\rho\sigma} \phi &= 0 \\
s_{\rho\sigma} \bar{\phi} &= \chi_{\rho\sigma} \\
s_{\rho\sigma} \eta &= -G_{\rho\sigma} \\
s_{\rho\sigma} A_{\mu} &= -4 P_{\rho\sigma[\mu}^{-\ \ \kappa}(2 F_{\kappa|\nu]} + G_{\kappa|\nu]}) + 4 P_{\rho\sigma\mu\nu}^{-}[\phi, \bar{\phi}] \\
s_{\rho\sigma} G_{\mu\nu} &= -4 P_{\rho\sigma[\mu}^{-\ \ \kappa}(4 D_{\kappa} \psi_{\nu]} + [\phi, \chi_{\kappa|\nu]}]) - 4 P_{\rho\sigma\mu\nu}^{-}(D_{\kappa} \psi^{\kappa} + [\phi, \eta]). \tag{27.183}
\end{aligned}
$$

We may wonder at this point whether we have done something wrong because we seem to have a theory with integer spin, but anticommuting fields, something which sounds contrary to the spin-statistics theorem. We have learned in Chapter 12 that this is impossible in a theory satisfying the usual quantum field theory axioms and, in particular, positivity of the energy. On the other hand, it is hard to imagine how anything could go wrong with a simple change of variables. The answer, of course, is that these are not canonical fields like those we have studied so far. Indeed, we can rewrite the action in terms of the twisted fields. It is given by

$$\mathcal{S}^{N=2,d=4} = \int \mathrm{d}^4 x \, \mathrm{Tr} \left(\frac{1}{2} \, G^{\mu\nu} \left(F_{\mu\nu} + \frac{1}{2} \, G^{\mu\nu} \right) - \chi^{\mu\nu} \left(D_\mu \psi_\nu + \frac{1}{4} \, [\phi, \chi_{\mu\nu}] \right) + \eta D_\mu \psi^\mu \right.$$
$$\left. + \bar{\phi} \{\psi_\mu, \psi^\mu\} - \bar{\phi} D_\mu D^\mu \phi + [\phi, \bar{\phi}]^2 - \eta[\phi, \eta] \right). \quad (27.184)$$

As we see, the quadratic part of the action is non-diagonal in the fields, so we cannot read immediately the expression for the Hamiltonian and check the positivity of the energy. In fact, in order to do so, we must diagonalise the quadratic part, which, in practice, means to go back to the original variables. So it is legitimate to ask the question whether all this is anything more than a complicated way to obscure the physical content of an otherwise simple theory. To answer this question we must study the theory (27.184) a bit further.

Let us first note that it can be written as an *s*-exact term

$$\mathcal{S}^{N=2,d=4} = s \, \Psi^{(-1)} = s \int \mathrm{d}^4 x \, \mathrm{Tr} \left(\frac{1}{2} \, \chi^{\mu\nu} \left(F_{\mu\nu} + \frac{1}{2} \, G^{\mu\nu} \right) + \bar{\phi} (D_\mu \psi^\mu - [\eta, \phi]) \right). \quad (27.185)$$

In fact $\Psi^{(-1)}$ is the most general local functional of the fields which is renormalisable, of $U(1)$ charge -1 and satisfies the requirement $s_\mu \Psi^{(-1)} = 0$.[21]

We can then check that the action (27.185) is also $s_{\mu\nu}$-invariant. This shows that the invariance under $s_{\mu\nu}$ is rather a consequence of the invariance under the five transformations s and s_μ. This is our first non-trivial result. The twisted formalism shows that the set of eight supersymmetry transformations which describe the invariance of the $N = 2$, $d = 4$ theory is redundant and the invariance depends on five parameters only. Although this sounds like a technical point, it is in fact important for the quantisation programme because it gives a simpler set of Ward identities. And we have seen in Chapter 16 that they play a crucial role in the very definition of the quantum theory. In particular, the proof of the renormalisation properties of theories with extended supersymmetry greatly simplifies if we use the twisted language. For example, we noted in section 27.7.1 that the proof of finiteness of the $N = 4$, $d = 4$ theory was complicated because we have no set of auxiliary fields to realise the supersymmetry off-shell. Now we understand that

---

[21] If we relax the requirement of s_μ invariance, we get the same action, but there are arbitrary coefficients in front of each term in the antecedent of the *s*-exact term. Their variations leave invariant the mean values of observables defined from the cohomology of *s*, according to the general properties of the BRST formalism. In particular, we can get rid of the term $s\bar{\phi}[\eta, \phi]$, which is the source of the quartic interactions on the $N = 2$ action. But the values of the coefficients determined by the demand of s_μ invariance are necessary for the identification of the TQFT action with the untwisted $N = 2$ action. We will come back to this point shortly.

this is due to the fact that we were dealing with a redundant set of symmetry generators. In the twisted formalism the symmetry can be enforced off-shell and the proof of the non-renormalisation theorems follows standard lines.

In fact, the action (27.185) can be written as an ss_μ-exact term, variation of a term involving the Chern–Simons term, as follows:

$$\mathcal{S}^{N=2,d=4}$$

$$= ss_\mu \int \mathrm{d}^4 x \, \mathrm{Tr}\left(\epsilon^{\mu\nu\rho\sigma}(A_\nu \partial\rho A_\sigma - \tfrac{2}{3} A_\nu A_\rho A_\sigma + (g^{\mu\nu}\eta + \chi^{\mu\nu})\psi_\nu\right) \qquad (27.186)$$

We have thus displayed the following novel result: the $N = 2$ Yang–Mills action can be expressed in terms of twisted fields and it is remarkable that it is completely fixed by the requirement of s and s_μ invariance, which means that the theory needs much less symmetry than expected. In particular, the tensor supersymmetry brings no information at the classical level and we can truly let it aside. We can show that the gauge fixing of the action can be done in an s and/or s_μ invariant way, a property that greatly simplifies the classification of observables of the theory at the quantum level.[22]

27.10.3.2   *The Interpretation of Yang–Mills Supersymmetry as a Quantum Topological Symmetry*

The previous results are more than a mathematical curiosity. The algebra (27.179) satisfied by s and s_μ modulo gauge transformations gives a very simple decomposition of the space–time derivative, with no Dirac matrices. The invariance under the action of this simple algebra fully determines the $N = 2$, $d = 4$ theory.

The formulation can be improved to get rid of the gauge transformations, and give a more geometrical interpretation for the fields. By introducing a 'shadow' scalar field, analogous to a Faddeev–Popov field, i.e. an anticommuting scalar field c^a valued in the adjoint representation, we can deduce from s the following scalar graded differential operator, which we call s_{top}:

$$s_{\mathrm{top}} A = \psi - Dc$$
$$s_{\mathrm{top}} \psi = -D\phi - [c, \psi]$$
$$s_{\mathrm{top}} \phi = -[c, \phi]$$
$$s_{\mathrm{top}} c = \phi - \frac{1}{2}[c, c]$$
$$s_{\mathrm{top}} \bar\phi = \eta - [c, \bar\phi]$$
$$s_{\mathrm{top}} \eta = -[\phi, \bar\phi]. \qquad (27.187)$$

The difference between s and s_{top} is basically gauge transformations that involve the fields c. By setting $c = 0$, s_{top} becomes identical to s. Owing to the transformation law of

---

[22] We can verify that there is no anomaly in the tensor symmetry, so that it is also possible to ignore it at the quantum level.

the shadow c, s_{top} is a differential operator that closes exactly (not only modulo gauge transformations). We have on all fields

$$s_{\text{top}}^2 = s_{\text{top}}\mathrm{d} + \mathrm{d}s_{\text{top}} = \mathrm{d}^2 = 0. \tag{27.188}$$

s_{top} is in fact quite analogous to the BRST operator that we used in the quantisation of the Yang–Mills theory. Because it is nilpotent, the classification of invariants can be expressed as a cohomology problem. The analogy with the ordinary BRST symmetry is quite striking, and we can verify from (27.187) the identity

$$(d + s_{\text{top}})(A + c) + \frac{1}{2}[A + c, A + c] = F + \psi + \phi \tag{27.189}$$

with its Bianchi identity

$$(d + s_{\text{top}})(F + \psi + \phi) + [A + c, F + \psi + \phi] = 0. \tag{27.190}$$

This curvature equation gives an interesting geometrical meaning to the twisted fields of supersymmetric theories that appear on the right-hand side of Eq. (27.189).[23] This geometrical equation shows that the twisted fields have a more natural geometrical interpretation as components of a generalised curvature in an extended space, although the untwisted super Poincaré formulation exhibits more clearly the physical degrees of freedom of the theory.

The curvature equation is very suggestive. Exactly as we did in order to properly understand the anomaly equation in the case of the Yang–Mills theory, we can use the Chern–Simons formula to obtain a similar result for the super Yang–Mills theory. For any given invariant polynomial $\mathcal{P}(F)$, we have the identity $d\mathcal{P}(F) = 0$, and thus we have

$$(d + s_{\text{top}})\mathcal{P}(F + \psi + \phi) = 0. \tag{27.191}$$

Taking for instance $\mathcal{P}(F + \psi + \phi) = \text{Tr}(F + \psi + \phi)(F + \psi + \phi)$ we get by expansion in form degree the solutions of the cohomological equation $s_{\text{top}}(...) + d(...) = 0$:

$$s_{\text{top}}\text{Tr}(FF) = -2d\text{Tr}(F\psi)$$

$$s_{\text{top}}\text{Tr}(F\psi) = -d\text{Tr}\left(F\phi + \frac{1}{2}\psi\psi\right)$$

$$s_{\text{top}}\text{Tr}\left(F\phi + \frac{1}{2}\psi\psi\right)) = -d\text{Tr}(2\psi\phi)$$

$$s_{\text{top}}\text{Tr}(\psi\phi)) = -\frac{1}{2}d\text{Tr}(\phi\phi)$$

$$s_{\text{top}}\text{Tr}(\phi\phi)) = 0, \tag{27.192}$$

---

[23] The equation can be improved to include the vectorial s_μ transformations.

where the multiplication is understood as a multiplication of forms. These equations are very similar to the descent equations for the anomaly cocycles. But they have a very different interpretation. If they are integrated over cocycles of appropriate dimensions M_{4-g}, they give the following operators Δ_{4-g}^g that are s_{top} invariant:

$$s_{\text{top}}\Delta_{4-g}^g = 0 \qquad (27.193)$$

with

$$\Delta_4^0 \equiv \int_{M_4} \text{Tr}(FF)$$

$$\Delta_3^1 \equiv \int_{M_3} \text{Tr}(F\psi)$$

$$\Delta_2^2 \equiv \int_{M_2} \text{Tr}\left(F\phi + \frac{1}{2}\psi\psi\right)$$

$$\Delta_1^3 \equiv \int_{M_1} \text{Tr}(\psi\phi)$$

$$\Delta_0^4 \equiv \text{Tr}(\phi\phi). \qquad (27.194)$$

Such operators are called topological observables for the following reason. If we compute the mean values

$$< \Delta_{4-g}^g >^{\mathcal{S}^{N=2,d=4}} \qquad (27.195)$$

by using the (gauge-fixed) $N = 2$, $d = 4$ action for the path integral, the result will be invariant under variations of the metric, i.e.

$$\frac{\delta}{\delta g_{\mu\nu}} < \Delta_{4-g}^g >^{\mathcal{S}^{N=2,d=4}} = 0. \qquad (27.196)$$

The proof of this result is, in principle, quite simple. It is based on the observation that since the action can be written as an s_{top}-exact term (in fact, it is so by construction), its dependence on the metric can only appear through a similar s_{top}-exact term. If we interpret s_{top} as a BRST symmetry, all parameters in the action, and in particular the metric, appear as gauge parameters in the BRST quantisation. The Ward identities for the s_{top} symmetry show that the mean values of all s_{top}-invariant operators are independent of these parameters. The proof follows the same lines as that presented in Chapter 18 for the Yang–Mills theory. We should only check that this prediction of invariance of the observables which belong to the cohomology of the s_{top} symmetry is compatible with renormalisation. Checking that the theory is multiplicatively renormalisable is a straightforward exercise on BRST quantisation,[24] which we leave aside, since the goal of

---

[24] We must in particular determine how the standard Faddeev–Popov ghosts transform under the s_{top} symmetry. Then the observables are defined as the cohomology of s_{top}, within the usual BRST cohomolgy.

this chapter is only to sketch the existence of an unexpected geometrical interpretation of supersymmetry, with the right ingredients.

As we can expect, many other supersymmetries can be twisted, and one can systematically look for topological invariants in analogous ways.[25] For instance, the expectation values $< \Delta^g_{4-g} >$ are related to the so-called Donaldson invariants, which play a profound role in the classification of four-dimensional manifolds.

In deriving the previous result we used only the invariance under s_{top}. Therefore, we may question the role of the vector supersymmetry s_μ. From a pragmatic point of view it is important because it allows us to determine the action uniquely. Indeed, the invariance under s_{top}, together with gauge invariance, implies that the action (modulo trivial field rescalings) is

$$I = \int \mathrm{d}^4 x s_{\text{top}} (\chi_{\mu\nu} \left(\frac{H_{\mu\nu}}{2} + F_{\mu\nu} \right) + \bar{\phi}(D^\mu \psi_\mu + a[\eta, \phi])) \qquad (27.197)$$

with a being an arbitrary parameter. This remaining unique degeneracy of the TQFT action is fixed by imposing the s_μ symmetry which implies $a = 1$. This identifies the action with the $N = 2$, $d = 4$ action, modulo the twist. It must be noted that we can compute the topological observables for all values of a, in particular for $a = 0$. Their expectation values remain the same, due to the Ward identities of the s_{top}-symmetry.

A deeper justification for having the s_μ symmetry is the following. The s_{top} invariance forces the action to be s_{top}-exact. But then the energy–momentum tensor must be also s_{top}-exact,

$$T_{\mu\nu} = s_{\text{top}} Z_{\mu\nu}. \qquad (27.198)$$

In other words, the energy–momentum tensor $T_{\mu\nu}$ has an s_{top}-antecedent. If we demand, as it is natural, that the tensor $Z_{\mu\nu}$ be also conserved, it can be considered as the Noether current of a fermionic symmetry with a vector index, which is precisely the s_μ-symmetry, with $\{s_\mu, s_{\text{top}}\} = \partial_\mu$ (modulo gauge transformations).

Finally, let us summarise by justifying, in a reverse way as compared to the introduction, why the scalar component s_{top} of the twisted supersymmetry can be called a topological symmetry, or rather the BRST symmetry associated with a topological symmetry. By looking at the definition of s_{top}, we can interpret the anticommuting vector field ψ_μ as a generalised Faddeev–Popov ghost associated with the general local transformation of the gauge field, $\delta A_\mu = \epsilon_\mu(x)$, where the local parameter $\epsilon_\mu(x)$ is defined modulo infinitesimal gauge transformations. At the classical level, only a Lagrangian which is locally a pure derivative, such as the second Chern class $\text{Tr}FF$, can be invariant under such a (huge) local gauge symmetry. It gives no equations of motion, so the classical content of such a theory is empty. However, we can perform a BRST invariant

---

[25] See, in particular, L. Baulieu, H. Kanno, and I. M. Singer, *Commun. Math. Phys.* **194**, 149 (1998); L. Baulieu and I. M. Singer, *Nucl. Phys. Proc. Sup.* **15B**, 12 (1988); L. Baulieu, G. Bossard, and A. Tanzini, *JHEP* **0508**, 037 (2005); L. Baulieu and G. Bossard, *Phys. Lett. B* **632**, 131 (2006); L. Baulieu, G. Bossard, and S. P. Sorella, *Nucl. Phys.* **B753**, 273 (2006).

gauge fixing of this action. It must be obtained by adding an s_{top}-exact gauge fixing to the action. This description and Eq. (27.184) indicate that this gauge fixing is nothing but the $N = 2$, $d = 4$ action. This way the latter appears as a gauge-fixed action of the second Chern class $\text{Tr}FF$, if we use as a 'gauge condition' for A the self-duality condition for the curvature F in addition to the Feynman–Landau condition. Then the observables are defined as the cohomology of this BRST symmetry, which yields the cocycles found earlier as the topological observables. In this construction, the mean values of observables are obviously metric independent, which is the definition of a topological observable.

27.11   Supersymmetry and Particle Physics

Since fermions and bosons are not degenerate, supersymmetry, if present in nature, must be broken. Furthermore we saw that a spontaneous breaking results in the appearance of a massless Goldstone fermion. We want to study its properties, independently of any particular model. The first guess would be to try to identify it with one of the neutrinos. It is easy to see that this cannot be true. First, the neutrinos don't seem to have zero masses and, second, they don't even seem to be approximate Goldstone spinors. Indeed, we have shown in section 25.4.2 that a Goldstone particle has very specific properties. They all come from the observation that it has the same quantum numbers as the divergence of the corresponding conserved current. Therefore, it satisfies the low-energy theorems we obtained for pions in section 25.4.2 and, in particular, the Adler relation (25.101). Let $\eta(x)$ be the field associated with the Goldstone fermion and $\mathcal{J}_\mu$, the conserved supersymmetry current. Adler's relation reads

$$\lim_{k_\mu \to 0} M(k) = 0, \tag{27.199}$$

where $M(k)$ is the amplitude for the emission (or absorption) of a Goldstone fermion with momentum k and we have suppressed spinor indices. This is a very powerful prediction and can be checked by studying the end-point spectrum in nuclear β-decay. This is the region in phase space in which the emitted electron takes almost all available energy leaving a neutrino carrying almost zero energy. Unfortunately, the prediction is not verified. Experiments show no such suppression, which means that the electron neutrino is not a Goldstone fermion. The same is true for all other neutrinos.

The question now is the following: if the Goldstone fermion is not one of the neutrinos, then where is it? There are two possible answers to this question which correspond to the two possible ways to implement supersymmetry: (i) as a global symmetry or (ii) as a local, or gauge, symmetry. In the first case the Goldstone fermion is a physical particle and, since it does not seem to appear in our experiments, we must make it 'invisible'. We will discuss the second possibility later.

Several mechanisms have been proposed to hide the zero-mass goldstino, but the simplest is to endow it with a new, conserved quantum number and arrange so that all other particles which share this number are heavy. Such a quantum number appears naturally in the framework of our supersymmetric theories and it is present even in models in which the goldstino problem is absent. It is the R-symmetry we have already introduced in the previous sections. We can define a discrete subgroup of the R-symmetry, which we will call *R-Parity* as

$$(-)^R = (-)^{2S}(-)^{3(B-L)}, \qquad (27.200)$$

where S is the spin of the one-particle state we are considering and B and L are the baryon and lepton numbers, respectively. It is easy to check that Eq. (27.200) gives $R = 0$ for all known particles, fermions as well as bosons, while it will give $R = -1$ for the goldstino. Similarly, the supersymmetric partners of the known particles, such as the photino we introduced in section 27.4, or the selectrons of section 27.7.1, will have $R = \pm 1$. Since R is conserved, the R-particles are produced in pairs and the lightest one is stable. In a spontaneously broken global supersymmetry this is the goldstino, which is massless.

Since the goldstino cannot be a known particle, can it be the partner of such a particle? For example, can it be identified with the photino? As we said earlier, the mechanism of spontaneous symmetry breaking, which is at the origin of the existence of the goldstino, allows us to find some properties of the latter, independently of the details of a particular model. In a spontaneously broken theory the spin-$\frac{3}{2}$ conserved current is given by

$$\mathcal{J}_\mu(x) = d\gamma_\mu\gamma_5\eta(x) + \hat{\mathcal{J}}_\mu(x), \qquad (27.201)$$

where d is a parameter with dimensions (mass)^2, $\eta(x)$ is the goldstino field, and $\hat{\mathcal{J}}_\mu(x)$ is the usual part of the current which is at least bilinear in the fields. In other words, the field of a Goldstone particle can be identified with the linear piece in the current. The conservation of $\mathcal{J}_\mu(x)$ gives

$$d\gamma_5\gamma_\mu\partial^\mu\eta = \partial^\mu\hat{\mathcal{J}}_\mu. \qquad (27.202)$$

This is the equation of motion of the goldstino. In the absence of spontaneous breaking $d = 0$ and $\partial^\mu\hat{\mathcal{J}}_\mu = 0$. In fact, to lowest order, the contribution of a given multiplet to $\partial^\mu\hat{\mathcal{J}}_\mu$ is proportional to the square mass-splitting Δm^2. Thus, the coupling constant of the goldstino to a spin-0–spin-$\frac{1}{2}$ pair is given by

$$f_\eta = \pm\frac{\Delta m^2}{d}, \qquad (27.203)$$

where the sign depends on the chirality of the fermion. It follows that if the goldstino were the photino, $f_\eta \propto e$ and the (mass)^2-splittings would have been proportional to the electric charge. For example, if s_e and t_e were the charged spin-0 partners of the electron, we would have

$$m^2(s_e) + m^2(t_e) = 2m^2(e).$$ (27.204)

This relation is clearly unacceptable. The conclusion is that the photon cannot be the bosonic partner of the goldstino. With a similar argument we prove that the same is true for the Z^0 boson, the scalar boson, or any linear combination of them. This is a model-independent result. Not only can we not identify the goldstino with the neutrino, but also we cannot pair it with any of the known neutral particles. Therefore, strictly speaking, there is no acceptable supersymmetric extension of the standard model with spontaneously broken global supersymmetry. The one that comes closest to it assumes an enlargement of the gauge group to $U(1) \otimes U(1) \otimes SU(2) \otimes SU(3)$, thus involving a new neutral gauge boson. We will not study it in detail here but we will rather extract those features which are model independent and are likely to be present in any supersymmetric theory.

Here is the moment to present the second mechanism to hide the goldstino, namely to promote supersymmetry to a local, or gauge, symmetry. This is not as simple as writing a Yang–Mills theory because supersymmetry is not an internal symmetry. The anticommutator of two supersymmetry charges contains the generator of translations, so gauging supersymmetry will necessarily imply gauging the translations. In other words, at the classical level, a gauge supersymmetry will contain, in the bosonic sector, general relativity. This theory is called *supergravity*. Since in this book we have studied field theory only in flat space, we will not derive its properties in any detail. In the next section we will briefly write the equations and the particle content but here we will only discuss its consequences for the goldstino.

We have already noted that the conserved supersymmetry current has spin equal to $\frac{3}{2}$. In writing a gauge theory we must introduce the gauge fields whose quantum numbers are those of the symmetry currents. It follows that the gauge fields necessary to turn supersymmetry into a local symmetry must be spin-$\frac{3}{2}$ spinors. Similarly, the gauge field necessary to turn translations into a gauge symmetry must have the quantum numbers of the energy–momentum tensor, i.e. it must have spin equal to 2. We have seen that the zero-mass representations contain a multiplet with a spin-2 state, the graviton, and its spin-$\frac{3}{2}$ partner, the gravitino. They are both massless, as expected from gauge fields.

In Yang–Mills theories we noted the BEH phenomenon in which a massless spin-1 particle absorbs a massless spin-0 one and becomes a massive spin-1 vector boson:

$$(m = 0, \text{ spin} = 1 + m = 0, \text{ spin} = 0) = (m \neq 0, \text{ spin} = 1).$$

The corresponding phenomenon in supergravity will involve a massless spin-$\frac{3}{2}$ fermion absorbing a massless spin-$\frac{1}{2}$ goldstino to become a massive spin-$\frac{3}{2}$ particle.

$$\left(m = 0, \text{ spin} = \frac{3}{2} + m = 0, \text{ spin} = \frac{1}{2}\right) = \left(m \neq 0, \text{ spin} = \frac{3}{2}\right).$$

We will call this mechanism naturally *the super-BEH mechanism* and, as it happens in the bosonic case, the goldstino has disappeared from the physical spectrum of states. At low energies, when gravitational interactions are decoupled, the theory will look, presumably, like a model with explicitly broken global supersymmetry. It will contain many arbitrary parameters, usually mass splittings and mixing angles. In principle, they are calculable in terms of the initial supergravity theory, but the relation is not always clear. Most often, they just parametrise our ignorance of the underlying symmetry breaking mechanism. It is this general framework which has been used in most phenomenological studies.

As a final remark, let us have another look at Eq. (27.119). As a result of supersymmetry breaking the masses of the chiral multiplet members are split, but we see that some pattern remains. The masses squared of the spin-0 fields are equally spaced above and below those of the fermions. In other words, we obtained a mass formula of the form

$$\sum_{\mathcal{J}} (-)^{2\mathcal{J}} (2\mathcal{J} + 1) m_{\mathcal{J}}^2 = 0, \qquad (27.205)$$

where $m_{\mathcal{J}}$ is the mass of the particle of spin $\mathcal{J}$. It turns out that such a formula is valid in every spontaneously broken supersymmetric model and even in some explicitly broken ones. We used it already in (27.204) in order to prove the impossibility of pairing together the photon and the goldstino. It plays an important role in model building.

27.11.1   Supersymmetry and the Standard Model

Let us now try to apply these ideas to the real world. We want to build a supersymmetric model which describes the low-energy phenomenology. There may be several answers to this question but, in practice, there is only one class of models which come close to being realistic. They were discovered by Pierre Fayet in the seventies. They assume a super-algebra with only one spinorial generator; consequently, all particles of a given supermultiplet must belong to the same representation of the gauge group. As we explained earlier, this is dictated by chirality. It is easy to see that in a supersymmetric model with two spinorial charges, each supermultiplet will contain fermions with both right and left components and it is not clear how to break this right–left symmetry. In the following we will try to keep the discussion as general as possible, so that our conclusions will be generally valid.

All models of global supersymmetry use three types of multiplets:

(i) Chiral multiplets. As we said already, they contain one Weyl (or Majorana) fermion and two scalars. Chiral multiplets are used to represent the matter (leptons and quarks) fields as well as the BEH fields of the standard model.

(ii) Massless vector multiplets. They contain one vector and one Weyl (or Majorana) fermion, both in the adjoint representation of the gauge group. They are the obvious candidates to generalise the gauge bosons.

(iii) Massive vector multiplets. They are the result of the ordinary BEH mechanism in the presence of supersymmetry. A massive vector multiplet is formed by a vector field, a Dirac spinor, and a scalar. These degrees of freedom are the combination of those of a massless vector multiplet and a chiral multiplet.

The physical degrees of freedom of the particles in the minimal standard model with one scalar multiplet are the following:

Bosonic degrees of freedom = 28

Fermionic degrees of freedom = 90 (or 96, if we include ν_R's).

It follows that a supersymmetric extension of the standard model will necessarily introduce new particles. We can even go one step further. In $N=1$ supersymmetry all the particles of a given supermultiplet must belong to the same representation of the gauge group. For the various particles of the standard model this yields the following:

(i) The gauge bosons are one colour octet (gluons), one $SU(2)$ triplet, and one singlet ($W^{\pm}$, Z^0, γ). No known fermions have these quantum numbers.

(ii) The BEH scalars transform as $SU(2)$ doublets but they receive a non-zero vacuum expectation value; consequently, they cannot be the partners of leptons or quarks. Otherwise, we would have induced a spontaneous violation of lepton or baryon number. Furthermore, we must enlarge the scalar sector by introducing a second complex chiral supermultiplet. This is necessary for several technical reasons which are related to the fact that, in supersymmetry, the scalars must have their own spin-1/2 partners. This in turn creates new problems like, for example, new triangle anomalies which must be cancelled. Furthermore, now the operation of complex conjugation on the scalars induces a helicity change of the corresponding spinors. Therefore, we cannot use the same doublet of scalar fields to give masses to both up- and down-quarks. Finally, with just one scalar supermultiplet, we cannot give masses to the charged partners of the W's. The net result is a richer spectrum of physical scalar particles. Since we start with eight scalars (rather than four), we end up having five physical ones (rather than just one). They are the scalar partners of the massive vector bosons and three neutral ones.

The conclusion is that in the standard model, supersymmetry associates known bosons with unknown fermions and known fermions with unknown bosons. We are far from obtaining a connection between the three independent worlds. For this reason this extension cannot be considered as a fundamental theory. Nevertheless, the phenomenological conclusions we will derive are sufficiently general to be valid, unless otherwise stated, in every theory based on supersymmetry.

We close this section with a table of the particle content in the supersymmetric standard model. Although the spectrum of these particles, as we will see shortly, is model dependent, their very existence is a crucial test of the whole supersymmetry idea. We will argue presently that its experimental verification is expected to be within the reach of LHC.

Strictly speaking, the existence of the particles appearing in Table 27.1 is the only model-independent test of $N = 1$ supersymmetry. It is not a very useful one, however,

Table 27.1 *The assignment is conventional. In any particular model the physical particles may be linear combinations of those appearing here.*

The particle content of the supersymmetric standard model

Spin-1	Spin-1/2	Spin-0	
Gluons	Gluinos	No partner	
Photon	Photino	No partner	
$W^{\pm}$	2 Dirac Winos	$w^{\pm}$	H
Z^0	2 Majorana Zinos	z	i b
		Standard ϕ^0	g o g s s o
	1 Majorana Higgsino	Pseudoscalar $\phi^{0'}$	n s
	Leptons	Spin-0 leptons	
	Quarks	Spin-0 quarks	

without some further theoretical input. Indeed, a prediction about the existence of such a rich spectroscopy of new particles is only meaningful if the masses of the new particles are also predicted, at least to within an order of magnitude. Let us draw here a parallel with the prediction concerning the charmed particles we studied in section 25.3.4. They were introduced in order to suppress the induced flavour changing weak neutral currents, or the $\Delta S = 2$ transitions. This suppression was effective only if the masses of the new particles were not too large, so the prediction could have been falsified by experiment. We will argue here that the situation is similar concerning supersymmetry.

Supersymmetry is the only known scheme which allows, even in principle, a connection between the Poincaré symmetry of space–time and internal symmetries. It provides a framework for the unification of the various worlds of gauge theories. We believe that supersymmetry will turn out to be part of our world; the only question is that of scale. How badly is supersymmetry broken? In other words, how heavy are the supersymmetric partners of the known particles? We claim that under some rather reasonable assumptions, we can answer this question for the supersymmetric standard model.

In section 26.1.2 we pointed out the quadratic dependence of the scalar mass on the high-energy scale μ up to which we assume the theory to be valid. Since the data show that the BEH scalar is light, this scale should be very low, less than 1 or 2 TeV, unless new physics shows up. This new physics should be able to cancel the μ^2 dependence of the scalar mass. Supersymmetry is the only local quantum field theory which has this property. In the presence of broken supersymmetry the scale μ is replaced by Δm, the scale of supersymmetry breaking. Therefore, the previous bound gives an upper bound of a few TeV for the supersymmetric particles. A badly broken supersymmetry is not effective in protecting the small mass scale. If supersymmetry is the mechanism which stabilises the scalar mass in the standard model it has reasonable chances to be discovered at the LHC. We will describe some possible experimental signatures in a later section.

27.11.2   Supersymmetry and Grand Unified Theories

The sensitivity of the BEH mass to the high-energy scale is particularly troublesome in grand unified theories. It is the gauge hierarchy problem which plagues all known GUTs. As we explained already, the problem has two aspects, one physical and one technical. The physical aspect is to understand the profound reason why nature creates the two largely separated mass scales. The technical aspect is related to renormalisation. In the notation of section 26.2, in order for the model to be able to sustain a gauge hierarchy, we must impose a very precise relation among the parameters of the potential. It is this relation which is destroyed by renormalisation effects and must be enforced artificially order by order in perturbation theory. As we will see shortly, supersymmetry may provide the mechanism to answer the physical problem, but it can certainly solve the technical one. The key is the non-renormalisation theorem we mentioned earlier. If supersymmetry is exact, the mass parameters of the potential do not get renormalised.

After these remarks on the gauge hierarchy problem, we can proceed in supersymmetrising our favourite GUT model. The construction parallels that of the low-energy standard model with similar conclusions. Again, no known particle can be the superpartner of another known particle. Furthermore, assuming a spontaneous symmetry breaking, we can repeat the analysis which led us to conclude that $U(1) \otimes SU(2) \otimes SU(3)$ was too small. The corresponding conclusion here will be that $SU(5)$ is too small, since $SU(5)$ does not contain anything larger than the group of the standard model.

Finally, we can repeat the renormalisation group estimation of the grand unification scale and the proton lifetime. We had found in section 26.2.3 that at low energies the effective coupling constants evolve following, approximately, the renormalisation group equations of $U(1)$, $SU(2)$, or $SU(3)$. The same remains true in a supersymmetric theory, but now the values of the β-functions are different. The number of Yang–Mills gauge bosons is the same as before. They are the ones which give rise to negative β-functions. On the other hand, supersymmetric theories have a larger number of 'matter' fields, spinors, and scalars, which give positive contributions. The net result is a smaller, in absolute value, β-function and, therefore, a slower variation of the asymptotically free coupling constants. The agreement now is remarkable (see Fig. 26.5). The three curves appear to come together. Expressed in terms of a prediction for the value of the weak mixing angle $\sin \theta_W$, this agreement is

$$\sin^2 \theta_W (\text{no SUSY}) \sim 0.214; \quad \sin^2 \theta_W (\text{SUSY}) \sim 0.232$$
$$\sin^2 \theta_W (\text{exp}) = 0.23149 \pm 0.00017. \tag{27.206}$$

The resulting value for M is $M \sim 10^{16} - 10^{17}$ GeV. If nothing else contributes to proton decay, it is beyond the reach of experiment. Fortunately, there are other contributions, which, although of higher order, turn out to be dominant. They are due to the exchange of the fermionic partners of the heavy bosons and their contribution is model dependent. Not surprisingly, in many models the result turns out to be on the order of 10^{33} years.

27.11.3   The Minimal Supersymmetric Standard Model

In Table 27.1 we had given the new particles that we expect to find in a supersymmetric extension of the standard model. Note, in particular, a richer system of scalar particles. Since the symmetry is broken, an important element is the breaking mechanism which determines the mass spectrum. Unfortunately, it is the least understood sector of the theory. We believe that it is a spontaneous breaking at a scale where the effective theory is supergravity. In this case the Goldstone fermion is absorbed by the spin-$\frac{3}{2}$ gravitino. At lower energies the theory looks like a model with explicitly, but softly, broken global supersymmetry. It is this framework that has been used in most phenomenological studies so far. The important point is that supersymmetry brings many new particles but no new couplings. At energies lower than the scale of grand unification we still have the three gauge couplings of $U(1)$, $SU(2)$, and $SU(3)$. With no further assumptions we must introduce a set of new arbitrary parameters describing the masses and mixing angles of all new particles. Even with massless neutrinos, this is a very large number. Note that already in the standard model the masses and mixing angles of quarks and leptons are arbitrary parameters to be determined by experiment. But we have seen in Section 26.2.3 that grand unification may reduce this number by providing relations among masses, like, for example, Eq. (26.39), which was the result of the $SU(5)$ relations (26.26). Something similar was applied to supersymmetry by S. Dimopoulos and H. Georgi and, independently, by N. Sakai. In the literature we find many variations of this idea and the most economic one is called the *minimal supersymmetric standard model* (MSSM). The basic assumption is that at the grand unification scale the supersymmetry breaking parameters which determine the mass splittings in the supermultiplets are the simplest possible. From this point we use the renormalisation group equations to derive the spectrum at present energies and compare with experiment. Remember again that these relations involve only the known gauge coupling constants.

In the MSSM the spectrum of the supersymmetric partners of ordinary particles (quarks, leptons, and gauge bosons) at the GUT scale is assumed to be determined by a minimum number of parameters: a common mass parameter $m_{1/2}$ for all gauginos, a corresponding one m_0 for all squarks and sleptons, and a common trilinear coupling among the various scalars, denoted by A. This choices are dictated only by simplicity. In fact, we have no reason to believe that they will be exact in the real world, but they have the obvious merit of providing a simple framework of analysing experimental data. The absence of flavour-changing neutral interactions sets limits on the possible mass differences among squarks and sleptons of different families, but does not force them to be 0. The most interesting sector is the BEH system. We need two doublets, as we explained previously. At the phenomenological level this introduces some new parameters. First, there will be two vacuum expectation values to break $U(1) \otimes SU(2)$, which we will call v_1 and v_2. They are taken by the neutral components of the two doublets, but we have weak isospin $I_z = +\frac{1}{2}$ and the other $I_z = -\frac{1}{2}$. It follows that no CP breaking is introduced because we can rotate the two phases independently and bring both v's to real values. An important parameter for phenomenology is the ratio

$$\tan \beta = v_2/v_1. \tag{27.207}$$

A second parameter is a mixing term between the two scalar multiplets. At the grand unification scale of $SU(5)$ the two doublets are promoted to two chiral supermultiplets belonging to 5 and $\bar{5}$ and the mixing term is written as $\mu H_1 H_2$ with μ a new arbitrary constant. Various versions of the MSSM make different assumptions regarding μ. It is often left arbitrary and it is determined by the requirement that the BEH system produces the correct electroweak symmetry breaking when extrapolated to lower energies using the renormalisation group equations. In all cases this last requirement severely restricts the parameters of the scalar potential. On the other hand, we do not want the electroweak breaking to occur at the grand unification scale, so all corresponding square masses must be positive or zero at that scale. In extrapolating from M_{GUT} down to present energies, we require the correct breaking at m_W, no breaking in between, and the maintaining of the perturbative nature of the theory everywhere. This means that all couplings should remain smaller than 1 and the effective potential bounded from below in the entire region. It turns out that all these requirements leave a relatively narrow window for the possible values of the parameters which can be compared with experiment. The attractive point of this scenario is the introduction in a 'natural' way of the large separation between M_{GUT} and m_W. It is simply due to the logarithmic running of the parameters. In practice the running of the effective mass of the scalars is dominated by the t-quark loop because of the corresponding very large Yukawa coupling. The typical renormalisation group equations give a relation of the form

$$m_W \sim M_{\mathrm{GUT}} \, \mathrm{e}^{-\frac{1}{\alpha_t}}; \quad \alpha_t = \frac{\lambda_t^2}{4\pi}. \tag{27.208}$$

Without any other assumptions the experimental signatures for the discovery of the supersymmetric particles are not very precise. A simple, hand-waving argument shows that ordinary particles are expected to be lighter than their supersymmetric partners. The reason is that the former take their masses solely through the BEH mechanism while the latter through both the supersymmetry breaking and the BEH mechanisms. Similar theoretical arguments almost always predict squarks and gluinos heavier than sleptons and other gauginos. The reason is that in most models the masses are set equal at the grand unification scale and the differences are due to the strong interactions of squarks and gluinos. For the same reason the masses of sneutrinos are predicted to be on the same order as those of the corresponding charged sleptons. Note, however, that the assumption of equal masses at the GUT scale is totally arbitrary with no theoretical, or phenomenological, basis whatsoever. In fact, this simple model is already excluded by the data, so it is only for the simplicity of the presentation that it is still studied, although several of the results we present are quite general.

Some simple relations follow from these general assumptions of the minimal model. The gauginos must obey the renormalisation group equations of the gauge couplings,

$$m_i(\mu) = m \frac{\alpha_i(\mu)}{\alpha_i(M)}, \tag{27.209}$$

where m is the common mass at the GUT scale M and μ is the low scale. This gives a 'prediction' for the three gauginos of $U(1)$, $SU(2)$, and $SU(3)$ at present energies

$$m_1 \ : \ m_2 \ : \ m_3 \ = \ 1 \ : \ 2 \ : \ 7. \tag{27.210}$$

This picture may be slightly complicated because of mixings with higgsinos. Note also that in the MSSM R-parity is conserved; therefore, all new particles are produced in pairs and the lightest among them is stable. It is usually denoted by LSP (lightest supersymmetric particle). In almost all models it is identified with a linear combination of the neutral gauginos and higgsinos. In this case its interactions are comparable to those of the neutrinos and it leaves no trace in the detector. Since all new particles will eventually end up giving LSPs, a precise determination of missing transverse momentum is an essential handle in the search of supersymmetric particles. Furthermore, the LSP offers an excellent candidate for cold dark matter, a necessary ingredient in cosmological models. For all these reasons, m_{LSP} is a very important phenomenological parameter, although no precise predictions for its value can be given. The cosmological arguments mentioned already give a rather loose bound $m_{\mathrm{LSP}} < O(200)$ GeV. Of course, the goldstino, which, if it exists, is massless, is absent from theories derived from supergravity with a super-BEH mechanism.

Let us now briefly discuss some results on masses and decay properties. In the MSSM the analysis is made as a function of the parameters we introduced earlier and the result should be given as a multi-dimensional plot. Some among the most important points follow:

(i) The BEH system. It is probably the most sensitive test of the MSSM. Five scalars are predicted, a pair of charged ones and three neutrals. The requirements of the correct symmetry breaking allow for a rather narrow window in the mass of the lightest neutral, the analogue of the standard model BEH scalar. At the tree level these predictions give a limit for m_ϕ on the order of the Z-mass, already excluded by experiment. Radiative corrections, especially the t-quark loops, raise this limit considerably. In most models we find $m_\phi \leq 130$ GeV, although a small amount of fine tuning is still necessary. The recently discovered scalar particle with a mass of 126 GeV falls remarkably close to this value.

(ii) Scalar partners of quarks and leptons (squarks, sleptons). There exists one such partner for every left- or right-handed quark and lepton. The breaking of $U(1) \otimes SU(2)$ causes mixings among the partners of opposite chirality fermions, so the final mass spectrum is the result of several diagonalisations. For this reason squarks do not necessarily follow the mass hierarchy of their quark partners. Squarks are produced in hadron collisions either in pairs or in association with gluinos (R must be conserved). Their decay modes are of the form $\tilde{q} \to q + \mathrm{LSP}$ (quark + LSP) or, if phase space permits, $\tilde{q} \to q + \tilde{g}$ (quark + gluino). The signature is missing P_T plus jets. Sleptons behave similarly and give $\tilde{l} \to l + \mathrm{LSP}$. The signal is again missing energy and momentum.

(iii) Gluinos. They are an important test of any supersymmetry scheme because they are expected to have substantial production cross sections in hadron colliders. Their specific decay properties are model dependent and the searches allow for various final states. At the end LSPs are expected to be produced.

(iv) Gauginos and higgsinos. They mix among themselves and must be analysed together. The charged ones are the supersymmetric partners of $W^{\pm}$ and $H^{\pm}$ and are described by a 2×2 mass matrix. Even before LHC, LEP had given a limit of 90–100 GeV for the mass of the lighter of the two. Among the neutral ones, the partners of W^3, B, and the two CP-even neutral scalars mix in a 4×4 matrix. The lightest of them is assumed to be the LSP.

The picture that emerges is that supersymmetric particles may be spread all over from a few hundred GeV to 1 or 2 TeV. The most recent LHC data suggest that the lower limit may be as high as 1 TeV.

The mass ratios and decay properties we presented depend often on the minimal hypothesis which was made only for convenience and has no theoretical base. However, if sparticles are discovered and their masses and mixing angles measured, we can easily go back and compute the symmetry breaking pattern at GUT energies. This, in turn, will give us a hint on the breaking mechanism which, as we explained, is probably related to the fundamental way gravity is unified with the other interactions. Looking for supersymmetric particles will be an important part of experimental search in the years to come. We all hope that it will be both exciting and rewarding and, in any case, we will know soon whether supersymmetry is a fundamental symmetry of particle forces at present energies.

27.12   Gauge Supersymmetry

Supergravity is the theory of local supersymmetry, i.e. supersymmetry transformations whose infinitesimal parameters — which are anticommuting spinors — are also functions of the space–time point x. There are several reasons to go from global to local supersymmetry:

(i) We have learned in recent years that all fundamental symmetries in nature are local (or gauge) symmetries.

(ii) The supersymmetry algebra contains the translations. So local supersymmetry transformations imply local translations and we know that invariance under local translations leads to general relativity which, at least at the classical level, gives a perfect description of the gravitational interactions.

(iii) As we already noted, local supersymmetry provided the most attractive explanation for the absence of a physical goldstino.

(iv) In the previous section we saw that in a supersymmetric grand unified theory the unification scale approaches the Planck mass (10^{19} GeV) at which gravitational interactions can no more be neglected.

The gauge fields of local supersymmetry can be easily deduced. Let us introduce an anticommuting spinor ϵ for every spinorial charge Q and write the basic relation (27.7) as a commutator,

$$[\epsilon^m Q^m, \bar{Q}^n \bar{\epsilon}^n] = 2\delta^{mn}\epsilon^m \sigma_\mu \bar{\epsilon}^n P^\mu, \qquad n, m = 1, ..., N, \qquad (27.211)$$

where no summation over m and n is implied. In a local supersymmetry transformation ϵ becomes a function $\epsilon(x)$. Equation (27.211) implies that the product of two super-symmetry transformations with parameters $\epsilon_1(x)$ and $\epsilon_2(x)$ is a local translation with parameter

$$\alpha_\nu(x) = \epsilon_1(x)\sigma_\nu \bar{\epsilon}_2(x). \qquad (27.212)$$

On the other hand, we know that going from a global symmetry with parameter θ to the corresponding local one with parameter $\theta(x)$ results in the introduction of a set of gauge fields which have the quantum numbers of $\partial_\mu \theta(x)$. If $\theta(x)$ is a scalar function, which is the case for internal symmetries, $\partial_\mu \theta(x)$ is a vector and so are the corresponding gauge fields (ex. gluons, $W^\pm$, Z^0, γ). If the parameter is itself a vector, like $\alpha_\nu(x)$ of translations, $\partial_\mu \alpha_\nu(x)$ is a two-index tensor and the associated gauge field has spin 2. In supersymmetry the parameters $\epsilon^m(x)$ have spin 1/2 so the gauge fields will have spin 3/2. We conclude that the gauge fields of local supersymmetry, otherwise called supergravity, are one spin-2 field and N spin-3/2 ones. To those we must add all other fields necessary to complete the multiplet.

27.12.1 *N*=1 Supergravity

This is the simplest supergravity theory and it provides for a good basis for a phenomenological analysis. The gauge fields are the metric tensor $g_{\mu\nu}(x)$, which represents the graviton and a spin-3/2 Majorana 'gravitino' $\psi_\mu(x)$. We can start by writing the Lagrangian of 'pure' supergravity, i.e. without any matter fields. The Lagrangian of general relativity can be written as

$$\mathcal{L}_\mathcal{G} = -\frac{1}{2\kappa^2}\sqrt{-g}R = \frac{1}{2\kappa^2}eR, \qquad (27.213)$$

where $g_{\mu\nu}(x)$ is the metric tensor and $g = \det g_{\mu\nu}(x)$. R is the curvature constructed out of $g_{\mu\nu}(x)$ and its derivatives. We have also introduced the vierbein field e_m in terms of which $g_{\mu\nu}(x)$ is given as $g_{\mu\nu}(x) = e_\mu^m(x)e_\nu^n(x)\eta_{\mu\nu}$ with $\eta_{\mu\nu}$ the Minkowski space metric. It is well known that if we want to study spinor fields in general relativity the vierbein, or tetrad, formalism is more convenient. e equals $-\sqrt{-g}$ and κ^2 is the gravitational coupling constant. Equation (27.213) is the Lagrangian of the gravitational field in empty space. We add to it the Rarita–Schwinger Lagrangian of a spin-3/2 massless field in interaction with gravitation

$$\mathcal{L}_{\mathcal{RS}} = -\frac{1}{2}\sqrt{-g}\epsilon^{\mu\nu\rho\sigma}\bar{\psi}_\mu \gamma_5 \gamma_\nu \mathcal{D}_\rho \psi_\sigma, \qquad (27.214)$$

where $\mathcal{D}_\rho$ is the covariant derivative

$$\mathcal{D}_\rho = \partial_\rho + \frac{1}{2}\omega_\rho^{mn}\gamma_{mn}; \quad \gamma_{mn} = \frac{1}{4}[\gamma_m, \gamma_n] \qquad (27.215)$$

and $\omega_\rho^{mn}(x)$ is the spin connection. Although $\omega_\rho^{mn}(x)$ can be treated as an independent field, its equation of motion expresses it in terms of the vierbein and its derivatives.

The remarkable result is that the sum of (27.213) and (27.214)

$$\mathcal{L} = \mathcal{L}_\mathcal{G} + \mathcal{L}_{\mathcal{RS}} \qquad (27.216)$$

gives a theory invariant under local supersymmetry transformations with parameter $\epsilon(x)$:

$$\delta e_\mu^m = \frac{\kappa}{2}\bar{\epsilon}(x)\gamma^m\psi_\mu$$
$$\delta\omega_\mu^{mn} = 0 \qquad (27.217)$$
$$\delta\psi_\mu = \frac{1}{\kappa}\mathcal{D}_\mu\epsilon(x) = \frac{1}{\kappa}\left(\partial_\mu + \frac{1}{2}\omega_\mu^{mn}\gamma_{mn}\right)\epsilon(x).$$

Two remarks are in order here. First the invariance of (27.216) reminds us of the similar result obtained in global supersymmetry, where we found that the sum of a Yang–Mills Lagrangian and that of a set of Majorana spinors belonging to the adjoint representation were automatically supersymmetric. This means that *all* gauge theories, both of space–time and of internal symmetries, admit a natural supersymmetric extension. This is one of the reasons for which many theorists consider that supersymmetry should be part of our world. The second remark is technical. The transformations (27.217) close an algebra only if we use the equations of motion derived from (27.216). We can avoid this inconvenience by introducing a set of auxiliary fields. In fact, we have partly done so, because the spin connection is already an auxiliary field.

The next step is to couple the $N = 1$ supergravity fields with matter in the form of chiral or vector multiplets. The resulting Lagrangian is quite complicated and will not be given explicitly here. Let us only mention that, in the most general case, it involves two arbitrary functions. If we call z the set of complex scalar fields, the two functions are $G(z, z^*)$, a real function, invariant under whichever gauge group we have used, and $f_{ij}(z)$, an analytic function which transforms as a symmetric product of two adjoint representations of the gauge group.

We may wonder why we have obtained arbitrary functions of the fields, but we must remember that in the absence of gravity, we impose to our theories the requirement of renormalisability which restricts the possible terms in a Lagrangian to monomials of low degree. In the presence of gravity, however, renormalisability is anyway lost, so no such restriction exists. Although supergravity may be a fundamental theory, at our present understanding we can only use it as an effective theory. In view of this, it is quite remarkable that only the two aforementioned functions occur.

27.12.2   *N* = 8 Supergravity

One of the arguments to introduce supersymmetry was the desire to obtain a connection among the three independent worlds of gauge theories, the worlds of radiation, matter, and scalar fields. None of the models presented so far achieved this goal. They all enlarged each world separately into a whole supermultiplet, but they did not put them together, with the exception of an association of some of the scalars with the massive gauge vector bosons. $N = 8$ supergravity is the only one which attempts a complete unification. It is the largest supersymmetry we can consider if we do not want to introduce states with spin higher than 2. We have already constructed the irreducible representation of one-particle states which contain the following:

$$
\begin{array}{ll}
1 & \text{spin-2 graviton} \\
8 & \text{spin-3/2 Majorana gravitini} \\
28 & \text{spin-1 vector bosons} \\
56 & \text{spin-1/2 Majorana fermions} \\
70 & \text{spin-0 scalars.}
\end{array}
\tag{27.218}
$$

The Lagrangian which involves all these fields and is invariant under eight local supersymmetry transformations was constructed by E. Cremmer and B. Julia, who also uncovered its remarkable symmetry properties. In contrast to the $N = 1$ case, there is no known system of auxiliary fields. Since we have 28 vector bosons we expect the natural gauge symmetry to be $SO(8)$. This is bad news because $SO(8)$ does not contain $U(1) \otimes SU(2) \otimes SU(3)$ as subgroup. The remarkable property of the theory, which raised $N = 8$ to the status of a candidate for a truly fundamental theory, is the fact that the final Lagrangian has unexpected symmetries: (i) a global non-compact E_7 symmetry and (ii) a gauge $SU(8)$ symmetry whose gauge bosons are not elementary fields. They are composites made out of the 70 scalars. $SU(8)$ is large enough to contain the symmetries of the standard model, but this implies that all known gauge fields (gluons, $W^\pm$, Z^0, γ) are in fact composite states. The elementary fields are only the members of the fundamental multiplet (27.218). None of the particles we know is among them; they should all be obtained as bound states.

$N = 8$ supergravity promised to give us a truly unified theory of all interactions, including gravitation and a description of the world in terms of a single fundamental multiplet. The main question was whether it defined a consistent field theory.

In pure quantum gravity some higher order computations have been made in order to check the way infinities arise in the perturbation expansion. The task may seem absurd, since the result is known in advance: the quantum theory of gravity is a non-polynomial theory and there is little hope that its perturbation expansion could make sense. However, the way invariant counter-terms are generated by Feynman diagrams order by order is an intrinsically interesting question. Moreover, other theoretical schemes, such as string theory, give a finite theory that includes quantum gravity as a singular limit. It is a challenge to precisely understand how it can be used to regularise these apparently lethal infinities of quantum gravity, provided they are under control.

For supergravity the hope was that the large number of supersymmetries would ensure a sufficient cancellation of the divergences of perturbation theory so that to make the theory finite. We have no clear answer to this question. It may still happen that miraculous compensations due to supersymmetry hold true to all orders, so that, although the theory is not renormalisable by power counting, some, hopefully the physically relevant ones, Green functions are finite to all orders. Checking this property of finiteness has been done for several multiloop diagrams, which is a technical 'tour de force', and the result, although not conclusive, has not yet shown that finiteness definitely fails.

In some sense $N = 8$ supergravity can be viewed as the end of a road, the road of local quantum field theory. The usual response of physicists whenever faced with a new problem was to seek the solution in an increase of the symmetry. This quest for larger and larger symmetry led us to the standard model, to grand unified theories, and then to supersymmetry, to supergravity, and, finally, to the largest possible supergravity, that with $N = 8$. In the traditional framework we are working, that of local quantum field theory, there exists no known larger symmetry scheme. The next step had to be a very radical one. The very concept of point particle, which had successfully passed all previous tests, was abandoned. During the past decades the theoretical investigations have moved towards the theory of interacting extended objects.

27.13 Problems

Problem 27.1 Write the Callan–Symanzik equation for the chiral supersymmetric model of section 27.6 and compute the β, γ, and δ functions at 1 loop.

Problem 27.2 Compute the β-function of the supersymmetric $SU(m)$ gauge theory with n chiral multiplets belonging to the fundamental representation of $SU(m)$ and derive the result of Eq. (27.90).

Problem 27.3 Study the spontaneous symmetry breaking in the $N = 2$ supersymmetric $U(1)$ gauge theory of section 27.8 and derive the central charges in the algebra of Eq. (27.100).

Problem 27.4 Consider the generalisation of the chiral supersymmetric field theory we studied in section 27.6 which involves n chiral multiplets. Add liner terms proportional to the F components and show that a spontaneous supersymmetry breaking can occur when $n \geq 3$.

Appendix A
Tensor Calculus

This appendix develops some of the mathematical tools necessary to understand the first part of this book. It is not a complete mathematical treatise.

A.1 Algebraic Theory of Tensors

In non-relativistic classical physics, it is natural to choose the Euclidean space $\mathbb{R}^3$ as the configuration space and to equip this space with a Cartesian system of coordinates such that a point x of space is represented by (x^1, x^2, x^3). If the system under interest has a spherical symmetry, then it is natural to take spherical coordinates characterised by r, θ, ϕ. The equations of physics can be expressed in any of these systems of coordinates. In relativistic physics, the configuration space is the space–time whose points x are represented in Cartesian coordinates by $x^0 = ct, x^1, x^2, x^3$. Again, this representation is not unique and we can introduce other systems of coordinates taking into account, for example, some particular symmetry of the system, or implying some dynamical aspects, as, for example, the fact that this system is in motion with respect to some observer in the initial frame. In any case, even if the equations of physics are invariant under some transformations of coordinates, they are expressed in terms of quantities, like the velocity or the forces, for which we need to know the way they transform under a change of coordinates.

In the study of physical systems, it appears natural to study quantities, the *tensors*, having well-defined transformation rules with respect to the changes of coordinates.

In what follows, we will define and study the algebraic properties of these objects without emphasising their intrinsic nature. Then we will sketch a more geometric and global presentation of tensors.

A second part of the appendix will be devoted to some elements of differential calculus.

The last part of this appendix, even more sketchy, introduces some basic notions from the theory of groups and Lie algebras.

A.1.1 Definitions

Let us consider an n-dimension space (we may think of $\mathbb{R}^n$). Let q_0 be a point of this space. In the coordinate system x, a point q close to q_0 is given by coordinates $x^1, x^2, \ldots, x^n$. By changing to the system of coordinates $\bar{x}$, we mean that this same point is now represented by the coordinates $\bar{x}^1, \bar{x}^2, \ldots, \bar{x}^n$ and that at least locally, that is to say in a neighbourhood of q_0, we have $x^i = x^i(\bar{x}^1, \ldots, \bar{x}^n)$, for $i = 1, \ldots, n$, $\bar{x}^j = \bar{x}^j(x^1, \ldots, x^n)$ for $j = 1, \ldots, n$ and these functions are sufficiently regular (differentiable). We have, of course, $x = x(\bar{x}(x))$ and $\bar{x} = \bar{x}(x(\bar{x}))$.

Definition 4. *A tensor or a tensor field T of type (r, s) is a set of numbers $T^{i_1, \ldots, i_r}_{j_1, \ldots, j_s}$ such that*

$$T^{i_1, \ldots, i_r}_{j_1, \ldots, j_s}(x) = \sum_{k_1=1}^{n} \cdots \sum_{l_s=1}^{n} T^{k_1, \ldots, k_r}_{l_1, \ldots, l_s}(\bar{x}) \frac{\partial x^{i_1}}{\partial \bar{x}^{k_1}} \cdots \frac{\partial x^{i_r}}{\partial \bar{x}^{k_r}} \frac{\partial \bar{x}^{l_1}}{\partial x^{j_1}} \frac{\partial \bar{x}^{l_s}}{\partial x^{j_s}}, \tag{A.1}$$

where the numbers $T^{i_1, \ldots, i_r}_{j_1, \ldots, j_s}(x)$ are relative to the system of coordinates $(x^1, \ldots, x^n)$ and the numbers $T^{k_1, \ldots, k_r}_{l_1, \ldots, l_s}(\bar{x})$ to the system of coordinates $(\bar{x}^1, \ldots, \bar{x}^n)$.

The lower indices are the covariant indices and the upper indices are the contravariant indices. A tensor of type (r, s) is called r-times contravariant (or contravariant of degree r) and s-times covariant (or covariant of degree s).

A.1.2   Examples

We give some examples of tensors.

1. The velocity.
   The velocity $v(t) = \dfrac{dx(t)}{dt}$ is a $(1, 0)$-tensor. Expressing it indeed as a function of the velocity in the system $\bar{x}$, we get

   $$v^i(t) = \frac{dx^i(t)}{dt} = \frac{\partial x^i}{\partial \bar{x}^j} \frac{d\bar{x}^j}{dt} = \frac{\partial x^i}{\partial \bar{x}^j} \bar{v}^j(t). \tag{A.2}$$

   This is an example of a *vector*.

2. The gradient.
   Let us consider a differentiable function $f(x)$. Its gradient has for components

   $$\nabla f = \left(\frac{\partial}{\partial x^1} f, \ldots, \frac{\partial}{\partial x^n} f \right) \tag{A.3}$$

   and we get

   $$(\nabla f)_i(x) = \frac{\partial \bar{x}^j}{\partial x^i} (\bar{\nabla} f)_j(\bar{x}), \tag{A.4}$$

   where $(\bar{\nabla} f)_j(\bar{x}) = \frac{\partial}{\partial \bar{x}^j} f(x(\bar{x}))$. This shows that ∇f is a covariant tensor. This is an example of a *covector*.

3. A metric.

Let us consider a vector space $\mathbb{R}^n$. A *riemannian metric* over $\mathbb{R}^n$ is defined by giving at each point a positive definite quadratic form g_{ij} on the tangent vectors. It makes it possible to write the infinitesimal element dl of the length of a curve $t \rightarrow x(t)$, $t \in \mathbb{R}$, as

$$dl = \sqrt{g_{ij}v^i(t)v^j(t)}\,dt. \tag{A.5}$$

The properties of transformation of the metric by change of coordinates are defined in such a way that dl is independent of the choice of the system of coordinates.

An example of a metric is the *Euclidean metric*. It is given in Cartesian coordinates by

$$g_{ij} = \delta_{ij}. \tag{A.6}$$

By the invariance property, it must be written in another system of coordinates $\bar{x}$ as

$$g_{ij}(\bar{x}) = g_{kl}(x)\frac{\partial x^k}{\partial \bar{x}^i}\frac{\partial x^l}{\partial \bar{x}^j}. \tag{A.7}$$

This shows that g_{ij} is a tensor of type $(0, 2)$.

Remark that by derivation of $x^i = x^i(\bar{x}(x))$ we get

$$\delta_j^i = \frac{\partial x^i}{\partial \bar{x}^k}\frac{\partial \bar{x}^k}{\partial x^j}, \tag{A.8}$$

which implies that the matrices $\{\frac{\partial x^i}{\partial \bar{x}^j}\}$ and $\{\frac{\partial \bar{x}^i}{\partial x^j}\}$ are inverse of each other.

Exercise 1 Show that in $\mathbb{R}^3$, the metric related to the spherical coordinates $x^1 = \rho \cos\theta, x^2 = \rho \sin\theta \cos\phi, x^3 = \rho \sin\theta \sin\phi$ is given by

$$g_{ij} = \begin{pmatrix} 1 & 0 & 0 \\ 0 & \rho^2 & 0 \\ 0 & 0 & \rho^2 \sin^2\theta \end{pmatrix}. \tag{A.9}$$

Another structure often used by physicists is the structure of *pseudo-Riemannian space*. It is defined by a non-degenerated non-necessarily positive definite metric. According to Silvester's theorem, it is characterised by a signature, i.e. after diagonalisation, by the difference between the number of its positive and negative eigenvalues. The space $\mathbb{R}^4$ equipped with the *Minkowski metric* $g_{\mu\nu}$, $g_{\mu\nu} = \delta_{\mu\nu}$ for $\mu = 1, 2, 3$ and $g_{\mu\nu} = -\delta_{\mu\nu}$ for $\mu = 0$, is the Minkowski space, M^4, of the special relativity.

A.1.3   Algebraic Properties of Tensors

We describe now some operations on tensors that lead to define new tensors.

1. Addition
   Tensors of a given type form at each point a vector space: if T and S are of type (r, s), then $\lambda T + \mu S$ is of type (r, s) and the space is of dimension n^{r+s}.

2. Permutation of indices
   By permutation of indices of the same type, contravariant or covariant, a tensor is transformed into another tensor. Let $\sigma \in S_r$ and $\mu \in S_s$, S_k being the group of permutation of k elements. If $T^{i_1,\ldots,i_r}_{j_1,\ldots,j_s}$ is a tensor of type (r, s), then

$$(T_{\sigma\mu})^{i_1,\ldots,i_r}_{j_1,\ldots,j_s} = T^{\sigma(i_1,\ldots,i_r)}_{\mu(j_1,\ldots,j_s)} \tag{A.10}$$

   is also a tensor of type (r, s).

3. Contraction
   The contraction V^k_l of the kth upper index with the lth lower index is an operation which changes a tensor T of type (r, s) into a tensor $V^k_l T$ of type $(r-1, s-1)$ by

$$(V^k_l T) T^{i_1,\ldots,i_{k-1},i_{k+1},\ldots,i_r}_{j_1,\ldots,j_{l-1},j_{l+1},\ldots,j_s} = T^{i_1,\ldots,i_{k-1},m,i_{k+1},\ldots,i_r}_{j_1,\ldots,j_{l-1},m,j_{l+1},\ldots,j_s}. \tag{A.11}$$

4. Tensor product
   By tensor product, from two tensors, T of type (r, s) and S of type (p, q), we build a new tensor $T \otimes S$ of type $(r + p, s + q)$ whose elements are

$$(T \otimes S)^{i_1,\ldots,i_{r+p}}_{j_1,\ldots,j_{s+q}} = T^{i_1,\ldots,i_r}_{j_1,\ldots,j_s} S^{i_{r+1},\ldots,i_{r+p}}_{j_{s+1},\ldots,j_{s+q}}. \tag{A.12}$$

The proofs are direct applications of the formulae (A.1) and (A.8).

A.1.4   Bases

Vectors (or tensors of type $(1, 0)$ or, in a more standard mathematical language *vector fields*) form, at each point, a space of dimension n. We will note a basis of this space $\{e_i\}_{i=1,\ldots,n}$ and a vector v is written $v = v^i e_i$.

Similarly, the covectors (or tensors of type $(0, 1)$) form, at each point, a space of dimension n. With $\{e^i\}_{i=1,\ldots,n}$ a basis of this space, a covector w is written as $w = w_i e^i$.

A basis of tensors of type (r, s) is given by $e_{i_1} \otimes \cdots \otimes e_{i_r} \otimes e^{j_1} \otimes \cdots \otimes e^{j_s}$ and a tensor T of type (r, s) can be written at each point of space

$$T = T^{i_1,\ldots,i_r}_{j_1,\ldots,j_s} e_{i_1} \otimes \cdots \otimes e_{i_r} \otimes e^{j_1} \otimes \cdots \otimes e^{j_s}. \tag{A.13}$$

Let us return to the definition of a covector given in the examples in section A.1.2. More intrinsically, we can define the gradient of a function f as the components of a linear functional on vectors: the *differential* df of the function. If v is a vector and f a function, we set

$$(\mathrm{d}f, v) = v^i \frac{\partial f}{\partial x^i}. \tag{A.14}$$

With $\{e_i^*\}$ the dual basis of $\{e_i\}$, $\mathrm{d}f$ is written as $\mathrm{d}f = \frac{\partial f}{\partial x^i} e_i^*$. More generally, the covariant tensors are defined as multilinear forms on vectors. It follows that it is natural to take $e^i = e_i^*$ with $(e_i^*, e_j) = \delta_{ij}$. This last relation, however, does not make it possible to identify the space of vectors with the space of covectors because this identification is not invariant by change of coordinates. Indeed, if we want the equality between a vector and a covector to remain after a change of coordinates, it is necessary that the basis $\{e_i\}$ and $\{e_i^*\}$ transform in the same way. But if in the change from x to $\bar{x}$, e_i is transformed by the matrix $A = \{\frac{\partial \bar{x}^j}{\partial x^i}\}$ into $\bar{e}_i = A_i^j e_j$, e_i^* is transformed according to $(A^{-1})^{\mathrm{tr}}$. We thus see that the identification is only possible for orthogonal transformations.

However, for a given choice of basis, we can find simple expressions of the elements of these two bases.

The elements of the basis of vectors can be identified with differential operators. Let, indeed, v be a vector and f be a function. The tensor product of v with ∇f is a $(1, 1)$-tensor. It becomes a scalar by contraction[1] and can be written as

$$\partial_v f = v^i \frac{\partial}{\partial x^i} f. \tag{A.15}$$

∂_v is called the derivative of f along v. To a vector of the basis e_i corresponds by formula $\partial_{e_i} f = \frac{\partial f}{\partial x^i}$ the differential $\frac{\partial}{\partial x^i}$. Through this identification, the operators $\{\frac{\partial}{\partial x^i}\}_{i=1,\ldots,n}$ form a natural basis of vectors.

Similarly, we can exhibit a natural basis for the covectors by using the definition (A.14) of the differential of a function. With the choice of function $f = x^k$, we deduce that $(\mathrm{d}x^k, v) = v^k$, thus $(\mathrm{d}x^k, e_i) = \delta_i^k$, which shows that the $\{\mathrm{d}x^i\}_{i=1,\ldots,n}$ form a basis of the linear forms on the space of vectors. We thus recover the formula of the ordinary differential calculus

$$\mathrm{d}f = \frac{\partial f}{\partial x^i} \mathrm{d}x^i. \tag{A.16}$$

This shows that the differential $\mathrm{d}f$ is in fact the covariant tensor whose components are given by the gradient (one of the examples in the preceding section).

We will see later on that the vectors are elements of the tangent bundle of a manifold M (M is what we call the space up to now) and the covectors are linear forms on the tangent bundle, i.e. elements of the cotangent bundle.

---

[1] From the tensor point of view, a scalar is invariant by changes of coordinates.

A.2 Manifolds and Tensors

In this section, we will define in the more general framework of manifold the notions that were introduced previously. Tensors will appear naturally as intrinsic objects, independent of any system of representation.

A.2.1 Manifolds, Tangent, and Cotangent Bundles

A.2.1.1 *Differentiable Manifolds*

A *differentiable manifold* (this means C^∞ unless explicitly stated otherwise) M of dimension m is a set M with a finite or denumerable collection of charts $\{C_j\}$, each point q of the manifold being in at least one chart.

A *chart* C is a couple (U, Φ) where U is an open set of M (the domain of the chart) and Φ a homeomorphism of U onto an open set of $\mathbb{R}^m$. For $q \in U$, the elements $x \in \mathbb{R}^m$ of the open set, image of U, define a local coordinate system.

Two charts C_1 and C_2 are *compatible* either if $U_1 \cap U_2 = \emptyset$ or if the mappings $\Phi_1 \circ \Phi_2^{-1}$ and $\Phi_2 \circ \Phi_1^{-1}$, suitably restricted, are diffeomorphisms of the corresponding open sets of $\mathbb{R}^m$. Explicitly, this means if $q \in U_1 \cap U_2$, V_1 and V_2 being the image open sets in $\mathbb{R}^m$, that $f = \Phi_2 \circ \Phi_1^{-1} : V_1 \rightarrow V_2$ with $\{x^i\} \in V_1 \mapsto \{f_i(x^1, \ldots, x^n)\} \in V_2$ is a family of m differentiable functions of m variables and the same for $\Phi_2 \circ \Phi_1^{-1}$. We also suppose that the manifold is separable; i.e. any 2 distinct points of M have neighbourhoods which do not intersect.

N is a *submanifold*, $N \subset M$, of dimension n, $m > n$, if $\forall q \in N$, there exists a chart (U, Φ) with $q \in U$ and $\Phi : U \rightarrow (x^1, \ldots, x^m)$, such that $\Phi|_{U \cap N} : U \cap N \rightarrow (x^1, \ldots, x^n, 0, \ldots, 0)$.

An *atlas* is a system of compatible charts covering M.

We define $\mathbb{R}^n_+ = \{x \in \mathbb{R}^n \,|\, x^1 \leq 0\}$ and $\partial \mathbb{R}^n_+ = \{x \in \mathbb{R}^n \,|\, x^1 = 0\}$. A mapping f of an open set U of $\mathbb{R}^n_+$ in $\mathbb{R}^m$ is differentiable if it is the restriction of a differentiable mapping of an open set of $\mathbb{R}^n$ in $\mathbb{R}^m$.

Let M be a separable (and metrisable) space, M is a *manifold with boundary* if there exists a covering with open sets U_i and homeomorphisms Φ_i of the U_i in open sets in $\mathbb{R}^m_+$ such that $\forall i, j, \quad \Phi_i \circ \Phi_j^{-1}|_{\Phi_j(U_i \cap U_j)}$ is C^∞. The boundary of M, $\partial M = \bigcup_i \Phi_i^{-1}(\Phi_i(U_i) \cap \partial \mathbb{R}^m_+)$. $M \setminus \partial M$ and ∂M have the structure of a manifold; ∂M has no boundary.

A.2.1.2 *Tangent Bundle*

Over each point $q \in M$, there exists a vector space, the tangent space $T_q(M)$, whose elements v are the tangent vectors, i.e. the equivalence classes of curves $u(t)$ on M such that $u(0) = q$.

In a chart C, two curves $u(t)$ and $\tilde{u}(t)$ are equivalent if $q = u(0) = \tilde{u}(0)$ and if $\lim_{t \to 0} \frac{(\Phi \circ u(t) - \Phi \circ \tilde{u}(t))}{t} = 0$. We have thus defined a homeomorphism

$$\Theta_C(q) : u(t) \rightarrow D(\Phi \circ u(t)) \in \mathbb{R}^m \tag{A.17}$$

between the equivalence classes of curves on M at q and the vectors of $\mathbb{R}^m$ (if f is a differentiable function of $\mathbb{R}^m$ in $\mathbb{R}^n$, we recall that the derivative, Df, of f, is, in terms of components, the matrix of the partial derivatives $Df_{ij} = \dfrac{\partial f_i}{\partial x^j}$, $i = 1, \ldots, n$ $j = 1, \ldots, m$). We will note by v an equivalent class of curves. If $u(t)$ is a curve, element of the equivalence class $v \in T_q(M)$, we will identify, as much as possible,[2] v to $\dfrac{\mathrm{d}\Phi \circ u(t)}{\mathrm{d}t}\big|_{t=0}$ with components $v^1, \ldots, v^m$.

This complicated formalism used to define the tangent vector of a curve on a manifold is due to the fact that we do not know how to compare two neighbour points in M without referring to the underlying $\mathbb{R}^m$. It makes it possible, however, to understand how a vector $v \in T_q(M)$ is changed after a change of local coordinates for $q \in M$. Let us suppose indeed that q belongs to two charts C and $\bar{C}$ and let Φ and $\bar{\Phi}$ be the corresponding mappings. It is how one moves from the local coordinates x to the local coordinates $\bar{x}$. The bijection mapping Θ_C becomes $\Theta_{\bar{C}} = \Theta_C D(\bar{\Phi} \circ \Phi^{-1})$, but $D(\bar{\Phi} \circ \Phi^{-1})$ is the matrix $\dfrac{\partial \bar{x}^i}{\partial x^j}$ thus for the components of v, $v^i \to \bar{v}^i = v^j \dfrac{\partial \bar{x}^i}{\partial x^j}$ (this is the transformation formula of a vector which was introduced in the preceding section).

In what follows when using local coordinates, we will omit as much as possible to explicitly write the morphisms Φ and Θ_C.

We can then show that the union $T(M)$ of the tangent spaces $\bigcup_{q \in M} T_q(M)$ can naturally be structured as a differentiable manifold of dimension $2m$.

We call Π the natural projection $\Pi : (q, v) \in T(M) \mapsto q \in M$. Its inverse image $\Pi^{-1}(q)$ is the tangent space $T_q(M)$.

Definition 5. *We call vector bundle a quadruplet (X, M, Π, F), where X is the manifold, M a submanifold of X, Π a surjection of X onto M, and F a vector space such that for each point $q \in M$ there exists a neighbourhood U of q whose inverse image $\Pi^{-1}(U)$ is diffeomorphic to $M \times F$. F is called the fibre.*

From this definition results that $T(M)$ is a vector bundle. The fibre F above each point q of M is the tangent space $T_q(M)$. We call $T(M)$ the *tangent bundle*. A local coordinate system for $T(M)$ is given by $(x^1, \ldots, x^m, v^1, \ldots, v^m)$.

A.2.2 Differential of a Mapping

If $f : M_1 \to M_2$ is a differentiable mapping, i.e. if in local coordinates, it is given by differentiable functions, it induces a mapping, the *differential of the mapping*, $T_q(f)$ at each point $q \in M_1$. If $v \in M_1$ and if u is a curve $t \in \mathbb{R} \mapsto u(t) \in M_1$ such that $u(0) = q$ and $\dfrac{\mathrm{d}u(t)}{\mathrm{d}t}\Big|_{t=0} = v$, then[3]

$$T_q(f)(v) = \frac{\mathrm{d}}{\mathrm{d}t}\Big|_{t=0} f(u(t)). \tag{A.19}$$

---

[2] The identification is straightforward only as long as we remain in the same chart.

[3] Rigorously, the differential of the mapping at q should be written as

$$T_q(f) = \Theta_{C_2}^{-1}(f(q)) \circ D(\bar{\Phi}_2 \circ f \circ \Phi_1^{-1}) \circ \Theta_{C_1}(q) \tag{A.18}$$

and is independent of the choices of the charts C_1 and C_2.

It extends to a mapping $T(f) : T(M_1) \to T(M_2)$ given by

$$T(f) : (q, v) \mapsto (f(q), T_q(f) \cdot v). \tag{A.20}$$

We can check that for $h = f \circ g$ then $T(h) = T(f) \circ T(g)$.

A.2.3   Vector Fields

An r-times differentiable *vector field* X is a mapping C^r of M in $T(M)$ such that $\Pi.X = 1$. In a local chart, we have $X : q \to (q, u(q))$. In practice, we often omit to write the reference to the point q of the basis.

A vector field X makes it possible to define, on a manifold, curves whose tangent vectors at each point are given by X. More precisely, if $t \to u(t)$, is a curve on M, it will be called the *integral curve* of X if at each point, the vector tangent to the curve is equal to X.

To each vector field X, we associate a differential equation $\dot{u} = X \circ u$ which can be written in a chart

$$\dot{u}^i = \frac{\mathrm{d}u^i(t)}{\mathrm{d}t} = X^i(u(t)), \quad i = 1, \ldots, m \tag{A.21}$$

with the initial condition $u(0) = q_0$.

Under very general conditions, for example X being a C^∞ vector field, there exists a local solution, i.e. an interval $I = \{-a, a\}$, and a neighbourhood V of q_0 such that

$$\Phi_t^X : q_0 \to u(t, q_0) \tag{A.22}$$

is, for $t \in I$, a diffeomorphism of V into a neighbourhood of M and $u(t, q_0)$, a solution of (A.21).

The diffeomorphisms Φ_t^X form (at least locally) a one-parameter group

$$\Phi_{t_1}^X \circ \Phi_{t_2}^X = \Phi_{t_1+t_2}^X : \tag{A.23}$$

the local *flow* of X.

Reciprocally, if we know a one-parameter group of diffeomorphisms, we can associate with it a vector field

$$X = \left. \frac{\mathrm{d}\Phi_t^X}{\mathrm{d}t} \right|_{t=0}. \tag{A.24}$$

Let now $f \in C_0^\infty$ be an infinitely differentiable function with compact support and consider the function $t \to f \circ \Phi_t^X$. We define the *Lie derivative* L_X associated with the vector field X by

$$L_X f = \left. \frac{\mathrm{d}}{\mathrm{d}t} f \circ \Phi_t^X \right|_{t=0}. \tag{A.25}$$

In terms of coordinates, if $q(t) = u(t, q(0))$ is a solution of (A.21), then $(f \circ \Phi_t^X)(q) = f(q(t))$, and formula (A.25) can be written as

$$L_X f = \frac{\mathrm{d}}{\mathrm{d}t} f(q(t)) \bigg|_{t=0} = \frac{\partial f}{\partial q^i} \frac{\partial q^i}{\partial t} \bigg|_{t=0} = \frac{\partial f}{\partial q^i} X^i(q) = \partial_X f. \tag{A.26}$$

We thus see that on functions, the Lie derivative is nothing else than the derivative along the vector X.

A.2.4 Cotangent Bundle

The dual space of $T_q(M)$ (the set of linear mappings of $T_q(M)$ in $\mathbb{R}$) is the *cotangent space* $T_q^*(M)$.

Remark: At the opposite of all the notions precedingly defined and which are independent of the choice of the atlas, we saw in section A.1.4 that we cannot canonically identify $T_q(M)$ to its dual space $T_q^*(M)$. In contrast, the space of the linear forms on $T_q^*(M)$, i.e. the dual $T_q^{**}(M)$ of $T_q^*(M)$, is identical to $T_q(M)$.

A vector bundle structure can be put on the union $T^*(M) = \bigcup_q T_q^*(M)$.

We saw in section A.1.4 that the differential of a function at the point $q \in M$ was an element of $T_q^*(M)$.

If $f \in C^\infty(M)$, formula (A.14) gives, in local coordinates, the action of the differential $\mathrm{d}f$, at q, on the element $v \in T_q(M)$. Formula (A.26) shows that this is the action of the Lie derivative on f:

$$(\mathrm{d}f, v) = L_v f(q). \tag{A.27}$$

The differential is a vector, i.e. a contravariant tensor of degree 1.

A.2.5 Tensors

We reproduce in this section the content of section A.1 but this time in the framework of the theory of manifolds.

A.2.5.1 *Natural Bases*

We now give a global justification to the content of section A.1.4. For clarification, let $C = C(U, \Phi)$ be a chart and $\Phi(q) = \sum e_i x^i$ where $\{e_i\}$ is a basis of $\mathbb{R}^m$. The inverse mapping $\Theta_C^{-1}(q)$ associates with it a basis in $T_q(U)$ symbolically noted $\{\frac{\partial}{\partial q^i}\}$ or ∂_i. This partial derivative is the Lie derivative with respect to the inverse image of e_i by Θ_C: on a function $g \in C^\infty(M)$, $L_{\Theta_C^{-1}(e_i)}(g) = \frac{\partial g}{\partial q^i}$. To an arbitrary vector $u^i e_i$ of $\mathbb{R}^m$ corresponds the differential operator $u^i \frac{\partial}{\partial q^i}$. To simplify the notations, we will merge any time there are no ambiguities, the basis $\frac{\partial}{\partial q^i}$ with its image under Θ_C: $\frac{\partial}{\partial x^i}$.

By duality, the corresponding basis in $T_q^*(U)$ is noted $\{\mathrm{d}q^i\}$: it is the basis for the linear forms on $T_q(M)$ and $(\mathrm{d}q^i, \partial_j) = \delta_j^i$.

Remarks:

1. For the same reasons as in the previous section, the notation $\mathrm{d}q^i$ is not fortuitous and corresponds to the differential of q^i;

2. For $M = \mathbb{R}^m$ or in an open set $U \subset M$, we will use x instead of q in all these formulae (since $\Theta \equiv 1$).

A.2.5.2 Change of Basis

We now define the effect of changes of basis on $T_q(M)$ and $T_q^*(M)$.

Let us first see how a vector field $X : M \to T(M)$ is transformed by a diffeomorphism $\Phi : M \to N$. It induces on tangent fibres, the mapping $T(\Phi)$ that we saw previously and on X a mapping $\Phi^* X \equiv T(\Phi) \circ X \circ \Phi^{-1}$.

To not repeat, we will suppose that locally Φ is given by the linear mapping $L : x \mapsto Lx \equiv \bar{x}$. It follows that

$$\partial_i = \frac{\partial}{\partial x^i} = \frac{\partial \bar{x}^j}{\partial x^i} \frac{\partial}{\partial \bar{x}^j} = (L^{\mathrm{tr}})^j_i \frac{\partial}{\partial \bar{x}^j} = (L^{\mathrm{tr}})^j_i \partial_j \tag{A.28}$$

and since

$$v^i \partial_i = \bar{v}^i \bar{\partial}_i = v^i L^{\mathrm{tr}j}_i \partial_j = L^j_i v^i \partial_j \tag{A.29}$$

then

$$\bar{v}^i = L^i_j v^j. \tag{A.30}$$

Thus in local coordinates $T(\Phi) : (x^i, v^j) \in T_x(M) \mapsto (L^i_k x^k, L^j_l v^l) \in T_{Lx}(N)$ and

$$\Phi^* X : (x^i, u^j) \mapsto (x^i, L^j_l v^l(L^{-1}x)). \tag{A.31}$$

In the same way, we can study the mapping $T^*(f)$ induced by f on the cotangent bundle. Remarking that $\mathrm{d}\bar{x}^i = L^i_j \mathrm{d}x^j$, we find that

$$T^*(f) : (x^i, v_j) \mapsto (L^i_k x^k, (L^{-1})^l_j v_l). \tag{A.32}$$

A.2.5.3 Definition of Tensors

Definition 6. *A r-times contravariant and a s-times covariant tensor at q is a multilinear mapping*

$$\underbrace{T_q^*(M) \times T_q^*(M) \times \cdots \times T_q^*(M)}_{r-times} \times \underbrace{T_q(M) \times T_q(M) \times \cdots \times T_q(M)}_{s-times} \to \mathbb{R}. \tag{A.33}$$

We note $(T_q)^r_s(M)$ the space of tensors of type (r, s) at $q \in M$.

The union of these spaces, for all points of M, can be given a structure of fibre space which will be noted $T^r_s(M)$.

Examples Multilinear mappings are known when we know them on a basis. Therefore, we will define a basis for contravariant tensors. On $T_q^*(M)$, a basis of the linear forms is given by $e_1, \ldots, e_m$, basis of the dual space $T_q(M)$. Therefore, if $v^{*1}, \ldots, v^{*r}$ are r elements of $T_q^*(M)$, the equation

$$((v^{*1}, \ldots, v^{*r}), e_{i_1} \otimes e_{i_2} \otimes \cdots \otimes e_{i_r}) \equiv (v_1^*, e_{i_1})(v_2^*, e_{i_2}) \ldots (v_r^*, e_{i_r}) \qquad \text{(A.34)}$$

defines a multilinear mapping on

$$\underbrace{T_q^*(M) \times T_q^*(M) \times \cdots \times T_q^*(M)}_{r-times}, \qquad \text{(A.35)}$$

which is noted $e_{i_1} \otimes e_{i_2} \otimes \cdots \otimes e_{i_r}$. Each $e_{i_1} \otimes e_{i_2} \otimes \cdots \otimes e_{i_r}$ is a tensor of order r. The set of these tensors forms a basis of the r-times contravariant tensors.

On this basis, a tensor is written as

$$t = \sum_{\{i\}} t^{i_1, \ldots, i_r} e_{i_1} \otimes e_{i_2} \otimes \cdots \otimes e_{i_r}, \quad t^{i_1, \ldots, i_r} = t^{(i)} \in \mathbb{R}. \qquad \text{(A.36)}$$

Definition 7. *The C^∞ mapping, $t : M \to T_s^r(M)$, such that $\Pi \circ t = 1$ is an r-times contravariant and s-times covariant tensor field (or tensor). We note $\Upsilon_s^r(M)$ the set of all these tensor fields.*

Examples
1. The differential of a function df is a covariant vector field.
2. Locally a tensor field can be written as

$$\sum_{i,j} t^{i_1, \ldots, i_r}_{j_1, \ldots, j_s} \partial_{i_1} \otimes \cdots \otimes \partial_{i_r} \otimes \mathrm{d}q^{j_1} \otimes \cdots \otimes \mathrm{d}q^{j_s}. \qquad \text{(A.37)}$$

The $t^{(i)}_{(j)}$ are the components of the tensor field.

A.2.5.4   *Law of Transformation*

A diffeomorphism $\phi : M_1 \to M_2$ induces an isomorphism $\phi^* : \Upsilon_s^r(M_1) \to \Upsilon_s^r(M_2)$ such that the following diagram is commutative

$$
\begin{array}{ccc}
M_1 \ni q & \xrightarrow{\;\;\phi\;\;} & \phi(q) \\[2pt]
{\scriptstyle t(q)}\downarrow & & \downarrow{\scriptstyle \phi^* t(\phi(q))} \\[2pt]
(T_q)_s^r & \xrightarrow[T_q(\phi)\otimes\cdots\otimes T_q(\phi)\otimes T_q^*(\phi)\otimes\cdots\otimes T_q^*(\phi)]{} & (T_{\phi(q)})_s^r
\end{array}
\qquad \text{(A.38)}
$$

with $t \in \Upsilon^r_s(M)$ and

$$\phi^* t = \underbrace{T(\phi) \otimes \cdots \otimes T(\phi)}_{r\text{--times}} \otimes \underbrace{T^*(\phi) \otimes \cdots \otimes T^*(\phi)}_{s\text{--times}} \circ t \circ \phi^{-1}. \tag{A.39}$$

Here $T(\phi)$ and $T^*(\phi)$ are, respectively,[4] the mappings induced by ϕ between the tangent and cotangent bundles.

Examples

1. If $M = \mathbb{R}^m$ and $\varPhi : \mathbf{x} \to \mathbf{x} + \mathbf{a}, \quad \mathbf{a} \in \mathbb{R}^m$, then

$$T(\varPhi) : (\mathbf{x},\mathbf{u}) \mapsto (\mathbf{x+a},\mathbf{u}) \tag{A.41}$$

and

$$T^*(\varPhi) : (\mathbf{x},\mathbf{v}) \mapsto (\mathbf{x+a},\mathbf{v}). \tag{A.42}$$

A tensor

$$t : \mathbf{x} \mapsto (\mathbf{x}, t^{i_1,\ldots,i_r}_{j_1,\ldots,j_s}(\mathbf{x})\partial_{i_1} \otimes \cdots \otimes \partial_{i_r} \otimes \mathrm{d}x^{j_1} \otimes \cdots \otimes \mathrm{d}x^{j_s}) \tag{A.43}$$

becomes

$$\varPhi^* t : \mathbf{x} \mapsto (\mathbf{x}, t^{i_1,\ldots,i_r}_{j_1,\ldots,j_s}(\mathbf{x-a})\partial_{i_1} \otimes \cdots \otimes \partial_{i_r} \otimes \mathrm{d}x^{j_1} \otimes \cdots \otimes \mathrm{d}x^{j_s}). \tag{A.44}$$

2. If $M = \mathbb{R}^m$ and $\varPhi : x^i \mapsto L^i_k x^k$
   then

$$T^*(\varPhi) : (x^i, \mathrm{d}x^j v_j) \mapsto (L^i_k x^k, \mathrm{d}x^j (L^{-1})^k_j v_k). \tag{A.45}$$

For the covariant indices, L must be replaced by $(L^{-1})^{\mathrm{tr}}$; thus

$$\varPhi^* t : \mathbf{x} \mapsto (\mathbf{x}, \partial_{i_1} \otimes \cdots \otimes \partial_{i_r} \otimes \mathrm{d}x^{j_1} \otimes \cdots \otimes \mathrm{d}x^{j_s} L^{i_1}_{m_1} \ldots L^{i_r}_{m_r} (L^{-1})^{n_1}_{j_1} \ldots \tag{A.46}$$
$$\ldots (L^{-1})^{n_s}_{j_s} t^{m_1,\ldots,m_r}_{n_1,\ldots,n_s}(L^{-1}\mathbf{x})).$$

3. If $g \in C^\infty(M)$:

$$\varPhi^* \mathrm{d}g = \mathrm{d}(g \circ \varPhi^{-1}). \tag{A.47}$$

---

[4] In a chart, we have for $T^*(\phi)$ a formula similar to (A.20):

$$T^*(\phi) : (q,v) \mapsto (\phi(q), (T(\phi^{-1}))^{tr}(q) \cdot v). \tag{A.40}$$

As already remarked, it is possible to add and multiply tensors at the same point. However, this is only valid for tensors of the same type.

We can multiply two arbitrary tensors. This defines a mapping $\otimes$, the tensor product,

$$\Upsilon^{r_1}_{s_1} \times \Upsilon^{r_2}_{s_2} \to \Upsilon^{r_1+r_2}_{s_1+s_2}. \tag{A.48}$$

This mapping is associative and distributive

$$(t_1 \otimes t_2) \otimes t_3 = t_1 \otimes (t_2 \otimes t_3) \tag{A.49}$$

$$t_1 \otimes (\alpha t_2 + \beta t_3) = \alpha t_1 \otimes t_2 + \beta t_1 \otimes t_3. \tag{A.50}$$

A.2.6 Lie Derivative

Let us see now how the one-parameter groups of diffeomorphisms Φ^X_t act on tensors $S \in \Upsilon^r_s$.

The way, using Φ^X_t, to go from x_0 to $x(t)$ is a change of coordinates; we define the law of transformation of the components at x_0 by Φ

$$((\Phi^X_{-t})^* S)^{i_1 \ldots i_r}_{j_1 \ldots j_s} = S^{k_1 \ldots k_r}_{l_1 \ldots l_s} \frac{\partial x^{l_1}}{\partial x^{j_1}_0} \cdots \frac{\partial x^{l_s}}{\partial x^{j_s}_0} \frac{\partial x^{i_1}_0}{\partial x^{k_1}} \cdots \frac{\partial x^{i_r}_0}{\partial x^{k_r}}. \tag{A.51}$$

Remark: since $\Phi^X_t(x_0) = x(t)$, to compare the tensor S at point x_0 with the transformed tensor, we must pull it back from $x(t)$ to x_0. This is done using the property that a diffeomorphism Φ of M induces the diffeomorphisms $T(\Phi)$ and $T^*(\Phi)$ of the tangent and cotangent bundles. Thus, a vector $v \in T_x(M)$ will be compared to an element of $T_{x(t)}(\Phi^X_{-t})$. The transformed tensor at the point q can then be written as

$$(\Phi^X_{-t})^* S = T(\Phi^X_{-t}) \otimes \ldots T(\Phi^X_{-t}) \otimes T^*(\Phi^X_{-t}) \otimes T^*(\Phi^X_{-t}) \circ S \circ \Phi^X_t \tag{A.52}$$

giving formula (A.51).

This makes it possible to define the Lie derivative

$$L_X S = \frac{\mathrm{d}}{\mathrm{d}t} (\Phi^X_{-t})^* S \bigg|_{t=0} \tag{A.53}$$

In coordinates, using the fact[5] that $\frac{\partial x^i}{\partial x_0^j} = \delta_j^i + t\frac{\partial X^i}{\partial x_0^j} + o(t)$ and $\frac{\partial x_0^i}{\partial x^j} = \delta_j^i - t\frac{\partial X^i}{\partial x^j} + o(t)$ we get

$$
\begin{aligned}
(L_X S)_{j_1 \ldots j_s}^{i_1 \ldots i_r} &= X^k \frac{\partial S_{j_1 \ldots j_s}^{i_1 \ldots i_r}}{\partial x^k} \\
&+ S_{kj_2 \ldots j_s}^{i_1 \ldots i_r} \frac{\partial X^k}{\partial x^{j_1}} \cdots + S_{j_1 \ldots j_s k}^{i_1 \ldots i_r} \frac{\partial X^k}{\partial x^{j_s}} - S_{j_1 \ldots j_s}^{ki_2 \ldots i_r} \frac{\partial X^{i_1}}{\partial x^k} \cdots \\
&- S_{j_1 \ldots j_s}^{i_1 \ldots i_r k} \frac{\partial X^{i_r}}{\partial x^k}.
\end{aligned}
\tag{A.54}
$$

Obviously $L_X S \in \Upsilon_s^r$. The Lie derivative L_X is compatible with the algebraic properties of tensors:

1.

$$
L_X(f + g) = L_X(f) + L_X(g).
\tag{A.55}
$$

2.

$$
L_X(f.g) = f.L_X(g) + g.L_X(f).
\tag{A.56}
$$

3. The Lie derivative commutes with contractions.

We recover the preceding definition, applying formula (A.54) to a function. Thus, with a $(1, 0)$-tensor Y, we find that

$$
L_X Y^i = X^j \frac{\partial Y^i}{\partial X^j} - Y^j \frac{\partial X^i}{\partial Y^j}
\tag{A.57}
$$

and thus

$$
L_X Y = -L_Y X.
\tag{A.58}
$$

Theorem 25. *Let f be a function*

$$
\partial_{L_X Y} f = \partial_X(\partial_Y f) - \partial_Y(\partial_X f) = [\partial_X, \partial_Y] f.
\tag{A.59}
$$

The bracket of the operators ∂_X and ∂_Y is a first-order differential operator: the vector field $L_X Y$.

The vector field $[X, Y] = L_X Y$ is called the *Lie bracket* of the fields X and Y.

---

[5] Equation (A.21) can be written as an integral, $x(t) = x_0 + \int_0^t X(x(s))\mathrm{d}s$, with x instead of u. Thus, for t small enough, $x(t) = x_0 + tX(x_0) + o(t)$. By derivation, we get the following two formulae.

A.2.7   Riemannian Structure

Let us introduce a new structure on the manifold M which makes it possible to identify the tangent and the cotangent bundles.

Definition 8. *If on the manifold M, there exists a 2-times covariant symmetric and non-degenerate vector field $g \in \Upsilon_2^0(M)$, then M is said to be a pseudo-Riemannian manifold. If g is positive, M is a Riemannian manifold, and g is its metric.*

In local coordinates $g = g_{ik}\mathrm{d}x^i \otimes \mathrm{d}x^k$, which is generally noted $g_{ij}\mathrm{d}x^i\mathrm{d}x^j$, and non-degenerate means $\det(g_{ik}(q)) \neq 0, \forall q$.

We will suppose from now on that there exists a metric.

Examples On a pseudo-Riemannian space, g defines a linear bijection $T_q(M) \to T_q^*(M)$

$$\forall q \in M : v = v^i e_i \to v_i e^i, \quad v_i = g_{ik}v^k, \tag{A.60}$$

the e^i generating the dual basis: $(e_i, e^j) = \delta_i^j$ and we can define a scalar product $(,)$ between two vectors x, y by

$$(x, y) = x^i g_{ij} x^j. \tag{A.61}$$

This allows us to speak of covariant and contravariant components of a vector and by extension of a tensor. The tensor g can be used to lower or to raise the indices. To this tensor is associated a 2-times contravariant tensor by

$$g^{ik}g_{kj} = \delta_j^i. \tag{A.62}$$

It is also possible to define the scalar product of two covectors by a formula similar to (A.61). If $\hat{x}$ and $\hat{y}$ are two covectors obtained by lowering the indices of two vectors x and y, then the two scalar products are equal:

$$(\hat{x}, \hat{y}) = x_i g^{ij} y_j = x^i g_{ij} y^j = (x, y). \tag{A.63}$$

Because of the existence of a metric, we can speak of the symmetry or of the anti-symmetry of a linear operator A, tensor of type $(1,1)$. In fact, the effect of this operator on vectors is given by the matrix elements A_j^i, and $A_{ij} = g_{ik}A_j^k$ defines a bilinear form on vectors.

Definition 9. *A is symmetrical (respectively antisymmetrical) if the quadratic form A_{ij} is symmetrical: $A_{ij} = A_{ji}$ (respectively antisymmetrical: $A_{ij} = -A_{ji}$).*

Since the eigenvalues of an operator A is a well-defined notion, we can also speak of the eigenvalues of the corresponding bilinear form in a space with metric.

Definition 10. *The eigenvalues of the bilinear form A_{ij} with the metric g_{ij} are the eigenvalues of the operator A.*

Examples

1. If A is a linear operator, $\mathrm{Tr} A^i_j$ and $\det A^i_j$, which are built with the eigenvalues of A, are invariants depending on the metric.

2. The invariants of the electromagnetic field $F_{\mu\nu}$ are the coefficients of the characteristic polynomial

$$P_{\mu\nu} = \det\left(F_{\mu\nu} - \lambda g_{\mu\nu}\right) \qquad (A.64)$$

Appendix B
Differential Calculus

B.1 Differential Form

Let us consider a $(0, 2)$-tensor of components T_{ij}. We can split it into a symmetrical tensor T^s and into an antisymmetrical tensor T^a. The components of T^a are

$$T_{ij}^a = \frac{1}{2}(T_{ij} - T_{ji}). \tag{B.1}$$

We have

$$
\begin{aligned}
T^a = T_{ij}^a \mathrm{d}x^i \otimes \mathrm{d}x^j &= \frac{1}{2} T_{ij}^a (\mathrm{d}x^i \otimes \mathrm{d}x^j - \mathrm{d}x^j \otimes \mathrm{d}x^i) \\
&= \sum_{i<j} T_{ij}^a (\mathrm{d}x^i \otimes \mathrm{d}x^j - \mathrm{d}x^j \otimes \mathrm{d}x^i).
\end{aligned}
\tag{B.2}
$$

Let us introduce the notation

$$\mathrm{d}x^i \wedge \mathrm{d}x^j = \mathrm{d}x^i \otimes \mathrm{d}x^j - \mathrm{d}x^j \otimes \mathrm{d}x^i. \tag{B.3}$$

This example and the notation we introduced can be generalised.

Definition 11. *Let $E_p(M)$ be the space of p-times covariant totally antisymmetric tensor fields; its elements are p-forms.*

We set $E_0(M) = C^\infty(M)$ and $E_1(M) = \Upsilon_1^0(M)$. An element $\omega \in E_p(M)$ can be written in the natural basis of a local chart U as

$$
\begin{aligned}
T = \omega_{i_1 \ldots i_p} \mathrm{e}^{i_1} \otimes \cdots \otimes \mathrm{e}^{i_p} &= \sum_{i_1 < \cdots < i_p} \omega_{i_1 \ldots i_p} \mathrm{d}x^{i_1} \wedge \cdots \wedge \mathrm{d}x^{i_p} \\
&= \frac{1}{p!} \omega_{i_1 \ldots i_p} \mathrm{d}x^{i_1} \wedge \cdots \wedge \mathrm{d}x^{i_p},
\end{aligned}
\tag{B.4}
$$

where the $\omega_{i_1 \ldots i_p} \in C^\infty(U)$ are completely antisymmetric, i.e.

$$\omega_{\sigma(i_1 \ldots i_p)} = (-1)^\sigma \omega_{i_1 \ldots i_p} \tag{B.5}$$

$\sigma \in S_p$ being a permutation of the p numbers $i_1, \ldots, i_p$ and $(-1)^\sigma$ being the sign of this permutation.

The elements

$$dx^{i_1} \wedge \cdots \wedge dx^{i_p}, \qquad i_1 < \cdots < i_p \tag{B.6}$$

form a basis for the p-forms. This notation generalises formula (B.3) and

$$dx^{i_1} \wedge \cdots \wedge dx^{i_p} = \sum_{\sigma \in P_p} e^{\sigma(i_1)} \otimes \cdots \otimes e^{\sigma(i_p)} = (-1)^\sigma dx^{\sigma(i_1)} \wedge \cdots \wedge dx^{\sigma(i_p)} \tag{B.7}$$

where by an abuse of notation we have written $\sigma(i_j)$ the image of i_j by the permutation $\sigma : (i_1, \ldots, i_p) \to \sigma(i_1, \ldots, i_p)$.

We define by $\wedge$ the *exterior product* of 2 forms. It is a mapping of $E_{p_1} \times E_{p_2} \to E_{p_1+p_2}$ associative and distributive which to $\omega_1 \in E_{p_1}$ and to $\omega_2 \in E_{p_2}$ associates $\omega = \omega_1 \wedge \omega_2 \in E_{p_1+p_2}$ locally given by

$$\omega = \sum_{k_1 < \cdots < k_{p_1+p_2}} \omega_{k_1 \ldots k_{p_1+p_2}} dx^{k_1} \wedge \cdots \wedge dx^{k_{p_1+p_2}} \tag{B.8}$$

with

$$\omega_{k_1 \ldots k_{p_1+p_2}} = \sum_{w \in S_{p_1+p_2}} \frac{(-1)^\sigma}{p_1! p_2!} (\omega_1)_{\sigma(k_1) \ldots \sigma(k_{p_1})} (\omega_2)_{\sigma(k_{p_1+1}) \ldots \sigma(k_{p_1+p_2})}. \tag{B.9}$$

We can check that the exterior product of m 1-forms acting on m vectors $v_i \in T_q(M)$,

$$((\omega_1 \wedge \omega_2 \wedge \cdots \wedge \omega_m)(q), v^1 \otimes \cdots \otimes v^m) = \sum_\sigma (-1)^\sigma \prod_{i=1}^m (\omega_i(q), v^{\sigma(i)}) \tag{B.10}$$

$$= \det(\omega_i(q), v^j), \tag{B.11}$$

where the sum is on the $m!$ permutations.

Remark: if in local coordinates $\omega = \omega_i dx^i$ and $v = v^j e_j$ then $(\omega, v) = \omega_i v^i$.
    We easily prove the following.

Lemma 16. $\forall \omega_1 \in E_{p_1}, \omega_2 \in E_{p_2}$

$$\omega_1 \wedge \omega_2 = (-1)^{p_1 p_2} \omega_2 \wedge \omega_1. \tag{B.12}$$

With the use of the metric tensor, we can define an isomorphism

$$* : E_p \to E_{n-p}, \qquad p = 1, \ldots, n, \tag{B.13}$$

which to each p-form ω

$$\omega = \frac{1}{p!}\omega_{i_1,\dots,i_p}\,\mathrm{d}x^{i_1}\wedge\cdots\wedge\mathrm{d}x^{i_p} \tag{B.14}$$

associates an $(m-p)$-form $^*\omega$

$$^*\omega = \frac{1}{p!(m-p)!}\sqrt{|g|}\,\omega_{i_1,\dots,i_p}g^{i_1j_1}\dots g^{i_pj_p}\epsilon_{j_1\dots j_m}\mathrm{d}x^{j_{p+1}}\wedge\cdots\wedge\mathrm{d}x^{j_m}, \tag{B.15}$$

where $|g| = |\det g|$.

We have

$$^*\circ^* = (-1)^{p(m-p)+s}\mathbf{1}, \tag{B.16}$$

where $(-1)^s = \dfrac{\det g}{|\det g|}$

Remark:

1. We check that on 1-forms $\omega_1 = \mathrm{d}x^i$ and $\omega_2 = \mathrm{d}x^j$ the exterior product coincides with the definition introduced at the beginning of the section.

2.

$$\mathrm{d}x^i \wedge \mathrm{d}x^i = 0 \tag{B.17}$$

by antisymmetry.

3. A basis $\mathrm{d}x^{i_1}\wedge\cdots\wedge\mathrm{d}x^{i_p}$ of $E_p(\mathbb{R}^m)$ is made, because of the antisymmetry, of C_m^p independent elements.

4.

$$\mathrm{d}x^{i_1}\wedge\cdots\wedge\mathrm{d}x^{i_p} = 0 \tag{B.18}$$

if $p > m$, m being the dimension of space.

5. On $\mathbb{R}^m$, there is a canonical m-form (the volume form)

$$\mathrm{d}x^1\wedge\cdots\wedge\mathrm{d}x^m = \epsilon_{i_1,\dots,i_m}\mathrm{d}x^{i_1}\otimes\cdots\otimes\mathrm{d}x^{i_m}, \tag{B.19}$$

where $\epsilon_{i_1,\dots,i_m}$ are the components of completely antisymmetric tensors such as $\epsilon_{1,\dots,m} = 1$.

6. On M, of dimension m, every m-form ω is given locally by a number. In fact, there is only one such element in the basis $\mathrm{d}x^1\wedge\cdots\wedge\mathrm{d}x^m$ ($\omega_{\sigma(1,2,\dots,m)} = (-1)^\sigma\omega_{12\dots m}$). By a change of coordinates $x \to \bar{x}$, this $\omega_{1\dots m}$ transforms into $\bar{\omega}_{1\dots m}$ with

$$\omega_{1\dots m} = \bar{\omega}_{1\dots m}\mathfrak{J}, \tag{B.20}$$

where $\mathcal{J}$ is the jacobian of the transformation

$$\mathcal{J} = \det\left(\frac{\partial \bar{x}^i}{\partial x^j}\right). \tag{B.21}$$

7. If ω and ν are 2 p-forms, then $\omega \wedge {}^*\nu = \nu \wedge {}^*\omega$.

Exercise 2 Prove that if g is a $(0, 2)$-tensor generating a non-degenerate quadratic form (g_{ij}), then setting $|g| = |\det g|$,

$$\sqrt{|g|}\, dx^1 \wedge \cdots \wedge dx^m \tag{B.22}$$

is a tensor (it is the volume element for the metric g).

We now define the interior product as the contraction of a differential form with a vector field.

Definition 12. *The interior product by a vector field* $X \in \Upsilon_0^1$ *is a linear mapping* $i_X :$ $E_p(M) \to E_{p-1}(M)$

$$i_X(\omega) = \sum_{j=1}^{p} \frac{1}{p}(-1)^{j+1} V_j^1(X \otimes \omega), \tag{B.23}$$

where the contraction V_j^1 *has been defined in Appendix A.1.3.*

The interior product obeys a Leibniz graded rule. For β a p-form and γ a q-form, we have

$$i_X(\beta \wedge \gamma) = i_X(\beta) \wedge \gamma + (-1)^p \beta \wedge i_X(\gamma). \tag{B.24}$$

It satisfies that

$$i_X(i_Y(\omega)) = -i_Y(i_X(\omega))$$

and we have also

$$[L_X, i_Y](\omega) = [i_X, L_Y](\omega) = i_{[X,Y]}(\omega).$$

Examples For $p = 1$, $i_X(\omega) = (\omega, X) \in C^\infty(M)$.

B.2 Exterior Differential

Definition 13. *The exterior differential is a mapping*

$$d : E_p(M) \rightarrow E_{p+1}(M). \tag{B.25}$$

Let ω be a p-form

$$\omega = \frac{1}{p!} \sum_{(i)} c_{(i)} dx^{i_1} \wedge \cdots \wedge dx^{i_p}, \quad c_{(i)} \in C^{\infty}(M), \tag{B.26}$$

its exterior differential $d\omega$ is given by

$$d\omega = \frac{1}{p!} \sum_{(i)} dc_{(i)} \wedge dx^{i_1} \wedge \cdots \wedge dx^{i_p}. \tag{B.27}$$

It satisfies the properties:

$$d(\omega_1 + \omega_2) = d\omega_1 + d\omega_2, \omega_i \in E_p(M)$$
$$d(\omega_1 \wedge \omega_2) = (d\omega_1) \wedge \omega_2 + (-1)^p \omega_1 \wedge (d\omega_2) \quad \text{if } \omega_1 \text{ is a } p\text{-form}$$

$d(d\omega) = 0, \omega \in E_p(M), \quad p = 0, 1, \ldots, m.$
If Φ is a diffeomorphism, then $\Phi^* d\omega = d\Phi^* w.$

Remark: $d(dx^{i_1} \wedge \cdots \wedge dx^{i_p}) = 0$ for $0 \le p \le m.$

Moreover

$$L_X(\omega) = i_X d\omega + d i_X(\omega).$$

Examples If $M = \mathbb{R}^3$ and $g = 1$, identifying E_0 with E_3 and E_1 with E_2, we recover the elementary expressions of the differential calculus
1. $(df)_i = (\nabla f)_i$
2. $^*(dv)_i = (\nabla \times v)_i = (\text{rot} v)_i$
3. $^*(d^* v) = \nabla . v = \text{div} v.$

Exercise 3 Carry out in $\mathbb{R}^3$ the change to spherical coordinates of the volume form $d^3 x = dx^1 \wedge dx^2 \wedge dx^3$. We find with the polar coordinates (r, θ, ϕ):

$$r^2 \sin \theta \, dr \wedge d\theta \wedge d\phi. \tag{B.28}$$

Definition 14. *A p-form ω is closed if and only if $d\omega = 0$; it is exact if and only if $\omega = dv$ for $v \in E_{p-1}(M)$.*

Lemma 17. *If M is a star-shaped open set of $\mathbb{R}^m$ (a subset Ω of $\mathbb{R}^m$ is star-shaped if there exists a point $p \in \Omega$ such that every segment linking p to another point of Ω is entirely inside Ω), then $d\omega = 0$ implies that ω is an exact form.*

Since in $\mathbb{R}^m$ every neighbourhood contains a convex set, it results that locally every closed form is an exact form.

We can also introduce the *codifferential* $\delta : E_p \to E_{p-1}$ which generalises the notion of divergence. We have

$$\delta \equiv {}^*d\,{}^*(-1)^{m(p+1)+s} \tag{B.29}$$

and

$$d = {}^*\delta\,{}^*(-1)^{pm+1+s}. \tag{B.30}$$

B.2.1 Integration

If M is a manifold of dimension m, an m-form defines a measure on M.

Definition 15. *M, of dimension m, is an orientable manifold, if there exists an m-form Ω which is nowhere zero.*

Examples

1. an open set $\mathbb{R}^m$ is orientable, its m-form is $dx^1 \wedge \cdots \wedge dx^m$.
2. the tangent bundle is orientable, even if M is not orientable.
3. since Ω is different from 0, every m-form can be written as $f\Omega, f \in C^\infty$

Let us suppose M is orientable and let Ω be the associated form. In a local chart (U, Φ), the restriction of Ω to U is, according to 3), of the form $w(x)dx^1 \wedge \cdots \wedge dx^m$. Supposing $w > 0$, one defines the integration of an m-form with support in U by

$$\int \Omega f = \int \Phi^*(\Omega f) = \int_{-\infty}^{\infty} dx^1 \int_{-\infty}^{\infty} dx^2 \ldots \int_{-\infty}^{\infty} dx^m w(x)(f \circ \Phi^{-1})(x). \tag{B.31}$$

The value of the integral is invariant by diffeomorphisms. This makes it possible to give an intrinsic definition of the integral over M by a covering with local charts: if $f = \sum f_i$, each f_i having its support in an open set U_i of the atlas. Thus,

$$\int_M \Omega f = \sum_i \int \Phi_i^*(\Omega f_i). \tag{B.32}$$

The following theorem generalises the Green and Gauss–Ostrogradski formulae.

Theorem 26 (Stokes' theorem). *If M is an orientable manifold with boundary, of dimension m and if ω is an $(m-1)$-form with compact support*

$$\int_M d\omega = \int_{\partial M} \omega. \tag{B.33}$$

Remark: If $\omega = d\upsilon$, another application of Stokes formula leads to $\partial(\partial M) = 0$. The boundary of a manifold has no boundary.

Let us give some examples.

1. Let $M = [a, b)$ be an interval of $\mathbb{R}$, and let $f \in E_0^0$ a function with compact support on M

$$\int_a^b df = -f(a). \tag{B.34}$$

If the support of f is not compact, then $f(b) \neq 0$, and Stoke's formula (B.33) is not true.

2. Let us consider in $\mathbb{R}^2$ a 1-form T and a domain U, with boundary ∂U, we have

$$\int_{\partial U} T = \int_U dT \tag{B.35}$$

or in coordinates

$$\int_{\partial U} (T_1 dx^1 + T_2 dx^2) = \int_U (\frac{\partial T_1}{\partial x^2} - \frac{\partial T_2}{\partial x^1}) dx^1 \wedge dx^2. \tag{B.36}$$

This is Green's formula. When ∂U is a closed curve, i.e. given by $x(s), 0 \leq s \leq 1$ and $x(0) = x(1)$, we note the first integral of formula (B.36) by

$$\oint_{\partial U} T. \tag{B.37}$$

3. Let us consider in the Euclidean space $\mathbb{R}^3$ a 2-form T and let V be a domain with boundary ∂V, then in coordinates

$$\int_{\partial V} \frac{1}{2} T_{ij} dx^i \wedge dx^j = \int_V \left(\frac{\partial T_{12}}{\partial x^3} + \frac{\partial T_{23}}{\partial x^1} + \frac{\partial T_{31}}{\partial x^2} \right) dx^1 \wedge dx^2 \wedge dx^3. \tag{B.38}$$

Remark that introducing the 1-form $B =^* T$ (with the Euclidean metric) the preceding expression can be written as $\mathbf{B} = (B_1, B_2, B_3)$ and the surface element $d\mathbf{S}$, $dS^i = \frac{1}{2}\epsilon_{ijk}dx^j \wedge dx^k$,

$$\int_{\partial V} \mathbf{B}.d\,\mathbf{S} = \int_V (\text{div } \mathbf{B})\, dx^1 \wedge dx^2 \wedge dx^3 = \int_V {}^*\delta B, \tag{B.39}$$

the first equality being the Gauss–Ostrogradski theorem.

4. Let us consider in the Euclidean space $\mathbb{R}^3$ a 1-form T and let U be a surface with boundary ∂U, then in term of coordinates

$$\int_{\partial U} T_i \mathrm{d}x^i = \int_U \mathrm{d}T = \int_U \mathbf{rotT}.\mathrm{d}\mathbf{S}, \tag{B.40}$$

where

$$(\mathrm{rot}\ T)_i = \frac{1}{2}\epsilon_{ijk}\partial_j T_k. \tag{B.41}$$

Remark: in the cases described earlier, the boundary of the domain is not necessarily made out of only one connected part. In these cases one difficulty remains: which sign to give to the integrals on the different connected components of the boundary if there are more than one. An answer can be given in the cases where the domain of integration is a submanifold of an orientable manifold M by following the sign of the normal vector to the curve or the surface.

To conclude, we will give a proof of Cauchy formula as a consequence of Green's formula.

Let us consider the complex plane $\mathbb{C}$ with generic points (x, y) and set $z = x + iy$ and $\bar{z} = x - iy$. Every function of (x, y) can be considered as a function of z and $\bar{z}$, and

$$\frac{\partial}{\partial z} = \frac{\partial}{\partial x} - i\frac{\partial}{\partial y} \qquad \frac{\partial}{\partial \bar{z}} = \frac{\partial}{\partial x} + i\frac{\partial}{\partial y}. \tag{B.42}$$

Let D be a bounded open set of $\mathbb{C}$ whose boundary $\partial D = C$ is sufficiently regular (piecewise C^1). If $f \in C^1(\bar{D})$ ($\bar{D}$ the closure of D) is a complex valued function, by applying Green's formula, we get

$$\int_{\partial D} f\,\mathrm{d}z = \int_D \mathrm{d}f \wedge \mathrm{d}z \tag{B.43}$$

or, noting that $\mathrm{d}f = \frac{\partial f}{\partial \bar{z}}\mathrm{d}\bar{z} \wedge \mathrm{d}z$,

$$\int_{\partial D} f(z, \bar{z})\mathrm{d}z = 2i \int_D \frac{\partial f}{\partial \bar{z}}\mathrm{d}x \wedge \mathrm{d}y = \int_D \frac{\partial f}{\partial \bar{z}}\mathrm{d}\bar{z} \wedge \mathrm{d}z. \tag{B.44}$$

The function f is called *analytic* in D if the Cauchy–Riemann conditions

$$\frac{\partial f}{\partial \bar{z}} = 0 \tag{B.45}$$

or

$$\frac{\partial u}{\partial y} = -\frac{\partial v}{\partial x} \qquad \frac{\partial v}{\partial y} = \frac{\partial u}{\partial x} \tag{B.46}$$

are satisfied, u and v being respectively the real and imaginary parts of f. In this case, according to formula (B.44), the form $f(z, \bar{z})\mathrm{d}z$ is closed. Remark that the

Cauchy–Riemann conditions (B.46) can be written as $\frac{\partial}{\partial\bar{z}}f = \bar{\partial}f = 0$ or equivalently $df = f'dz$ where $f' = \frac{\partial f}{\partial z}$. We deduce the following:

1. every polynomial of z is an analytic function,
2. $\oint_C z^n dz = 0$ for all $n \geq 0$,
3. $\oint_C z^{-n} dz = 0$ for all $n \geq 0$ if C is a contour which does not contain the origin of coordinates,
4. $\oint_C z^{-n} dz$, $n \geq 0$, is independent of the contour C, circling around the origin of coordinates. Therefore taking for C a circle, oriented clockwise and centred at the origin, we have with $z = e^{i\theta}$ that the only non-vanishing integral is for $n = 1$ and

$$\oint_C z^{-1} dz = 2\pi i. \tag{B.47}$$

Lemma 18 (Cauchy formula). *Let be $f \in C^1(\bar{D})$, then for $\zeta \in D$*

$$f(\zeta) = \frac{1}{2\pi i} \left(\int_{\partial D} \frac{f(z)}{z-\zeta} dz + \int_D \frac{\partial f/\partial\bar{z}}{z-\zeta} dz \wedge d\bar{z} \right). \tag{B.48}$$

We prove the lemma by applying Green's formula (B.44) to $f(z)/(z-\zeta)$ in D_ϵ obtained from D by removing a small disc of radius ϵ centred at the origin. We therefore get since $(z-\zeta)^{-1}$ is analytic in D_ϵ

$$\int_{D_\epsilon} \frac{\partial f/\partial\bar{z}}{z-\zeta} d\bar{z} \wedge dz = \int_{\partial D} \frac{f(z)}{z-\zeta} dz - \int_0^{2\pi} f(\zeta + \epsilon e^{i\theta}) i d\theta \tag{B.49}$$

and the lemma is proved by letting ϵ go to 0 and using that f is continuous in a neighbourhood of ζ.

For a function f, analytic in D, formula (B.48) reduces to

$$f(\zeta) = \frac{1}{2\pi i} \int_{\partial D} \frac{f(z)}{z-\zeta} dz \tag{B.50}$$

and consequently the following lemma.

Lemma 19. *If f is analytic in the disc $D_r = \{z; |z| \leq r\}$, then*

$$f(z) = \sum_0^\infty f^n(0) \frac{z^n}{n!}. \tag{B.51}$$

The series is uniformly convergent on every compact subset of D_r and the nth derivative $f^n(0)$ is given by

$$f^n(0) = n! \frac{1}{2\pi i} \oint_\gamma \frac{f(\zeta)}{\zeta^{n+1}} d\zeta \tag{B.52}$$

with the contour $\gamma \subset D_r$ surrounding the origin.

Appendix C
Groups and Lie Algebras

C.1 Lie Groups

C.1.1 Definitions

We recall that a *group* G is a set with a product law such that

1. $g_1 g_2 \in G$ for all $g_1, g_2 \in G$.
2. $(g_1 g_2) g_3 = g_1 (g_2 g_3)$ for all $g_1, g_2, g_3 \in G$.
3. There exists in G a (unique) element e, the neutral element, such that $eg = ge = g$ for all $g \in G$.
4. For each $g \in G$, there exists an inverse g^{-1} such that $g g^{-1} = g^{-1} g = e$.

A subgroup H of a group G is a subset $H \subset G$ such that if $g_1, g_2 \in H$ then $g_1 g_2^{-1} \in H$. A subgroup is a group.

If H is a subgroup, the set of elements $\{Hg\}_{g \in G}$ is the quotient space G/H of G by H. Hg is the right equivalence class of g.

If the subgroup H is distinguished, that is to say, for all $g \in G$, $gHg^{-1} = H$, the quotient space G/H is a group: the *quotient group*.

The set of elements of a group which commutes with the other elements of the group forms a commutative subgroup called the *centre*.

Definition 16. *Let G and G' be two groups. A mapping f of G into G' is a group homeomorphism if*

$$f(g_1 g_2) = f(g_1) f(g_2) \qquad \forall g_1, g_2 \in G. \tag{C.1}$$

If e' is the neutral element of G', its inverse image, $Ker f = f^{-1}(e')$, is the kernel of the mapping. If the mapping is surjective, $f(G) = G'$, and injective, $Ker f = \{e\}$, it is an isomorphism.

We check easily that the kernel of a group homomorphism is a subgroup.

Definition 17. *A group automorphism G is an isomorphism of G on itself.*
It is an inner automorphism if, for some $h \in G$, it is given by

$$G \ni g \mapsto h g h^{-1} \in G. \tag{C.2}$$

An element like $h g h^{-1}$ is called the element of g conjugate by h.

C.1.2   Representations

Let G be a group and V be a vector space on $\mathbb{R}$ or on $\mathbb{C}$.

Definition 18. *A representation of the group G on the vector space V is a mapping π of G into the linear operators of V such that*

1. $\pi(e) = \mathbf{1}$, *is the identity operator on V.*
2. $\pi(g_1 g_2) = \pi(g_1)\pi(g_2)$ $\forall g_1, g_2 \in G$.

A representation is indeed a homomorphism of G in the group of linear operators on V.

If $W \subset V$, W is an invariant vectorial subspace for the representation π if

$$\pi(g)W \subset W \qquad \forall g \in G. \tag{C.3}$$

A *finite-dimensional representation* is a triplet (G, V, π) such that V be of finite dimension. The dimension of V is called the *dimension of the representation*. If $\dim V = n$, the operators $\pi(g)$ are the $n \times n$ matrices .

A representation such that $\pi(g) = \mathbf{1}$, the identity of V, is called *trivial*.

If $\pi(g)$ is a finite-dimensional representation, the traces of the matrix elements $\xi_\pi(g) = \mathrm{Tr}\,\pi(g)$ are called the *characters* of the representation.

Definition 19. *Two representations π and $\tilde{\pi}$ of G, respectively, in V and $\tilde{V}$ are equivalent if there exists a bijection given by the linear operator $A : V \to \tilde{V}$ such that*

$$A\pi(g) = \tilde{\pi}(g)A \qquad \forall g \in G.$$

If π is a representation of G in V whose only invariant subspaces are V and $\{\emptyset\}$ then π is an irreducible representation.

Lemma 20 (Schur's Lemma). *Let π and $\tilde{\pi}$ be irreducible representations of G, respectively, in V and $\tilde{V}$. Suppose there exists an operator A of V in $\tilde{V}$ such that*

$$A\pi(g) = \tilde{\pi}(g)A \qquad \forall g \in G. \tag{C.4}$$

Then either $A = 0$ or A is a bijection of V on $\tilde{V}$.

Proof. Let $W = \{x \in V; Ax = 0\}$, formula (C.4) shows that $\pi W \subset W$. Thus, W is an invariant subspace. If it is an empty set, then A is injective and if it is the whole space V then $A = 0$. Let now $\tilde{W} = AV$, formula (C.4) shows that $\tilde{\pi}\tilde{W} \subset \tilde{W}$. Again, as already argued, the irreductibility of $\tilde{\pi}$ means that either A is surjectif or $A = 0$, which closes the proof of the lemma.

The following lemma, which is the usual application of Schur's lemma, follows from the preceding lemma or can be directly proved.

Lemma 21. *Let π be a irreducible representation of G in V. If A is an operator which commutes with all the elements of the representation, then $A = \lambda 1$, λ being a number.*

Proof. Since A is a linear operator in a finite-dimensional space, it has at least one eigenvalue λ. Then $W = \{x \in V; Ax = \lambda x\}$. The fact that A commutes with the elements of the representation means that Eq. (C.4) is satisfied. Therefore, this means also that W is an invariant subspace and, consequently, because of the irreducibility of π, $W = V$ since $A \neq 0$.

Definition 20. *Let π_1 and π_2 be two representations of a group G in, respectively, V_1 and V_2, vector spaces of finite dimension. We call tensor product $\pi_1 \otimes \pi_2$ of the two representations, a representation π in the vector space $V = V_1 \otimes V_2$, the vector space tensor product,[1] such that for all $v_1 \in V_1$ and $v_2 \in V_2$ and for all $g \in G$*

$$\pi(g)(v_1 \otimes v_2) = \pi_1(g)v_1 \otimes \pi_2(g)v_2. \tag{C.5}$$

The tensor product of two irreducible representations is not an irreducible representation. It splits into a direct sum of irreducible representations, each one with a given multiplicity.

Definition 21. *Let X_i, $i = 1, \ldots, m$, be m vector spaces of finite dimension over the same field K. Their direct sum $X = X_1 \bigoplus \cdots \bigoplus X_m$ is the vector space over K of elements $(x_1, \cdots, x_m)$ obtained by defining on the Cartesian product $X_1 \times X_2 \cdots \times X_m$ two operations*

- $(x_1, \cdots, x_m) + (y_1, \cdots, y_m) = (x_1 + y_1, \cdots, x_m + y_m)$
- $\lambda(x_1, \cdots, x_m) = (\lambda x_1, \cdots, \lambda x_m)$

for any $\lambda \in K$ and for any $(x_1, \cdots, x_m)$ and $(y_1, \cdots, y_m)$ in X.

The dimension $\dim X$ of X is such that $\dim X = \dim X_1 + \cdots + \dim X_m$

Definition 22. *Let X_i, $i = 1, \ldots, m$, be m vector spaces of finite dimension and let π_i be m representations of a group G in the corresponding spaces. The direct sum representation $\pi = \pi_1 \bigoplus \cdots \bigoplus \pi_m$ is defined in the direct sum vector space $X = X_1 \bigoplus \cdots \bigoplus X_m$ by, for all $g \in G$ and for all $x_i \in X_i$, $(x_1, \ldots, x_m) \in X$,*

$$\pi(g)(x_1, \ldots x_m) = (\pi_1(g)x_1, \ldots, \pi_m(g)x_m). \tag{C.6}$$

This representation is highly reducible, since by construction, each X_k is an invariant subspace. A representation π_α appears in the sum with the multiplicity k if k of the representations $\{\pi_i\}_{i=1,\ldots,m}$ are equivalent to π_α. The matrix of the operator $\pi(g)$ in X is a matrix with blocks of dimension $\dim(X_i)$, $i = 1, \ldots, m$, on the diagonal and 0 elsewhere.

---

[1] The tensor product V of two vector spaces E and F is a vector space $V = E \otimes F$ such that there exists a bilinear mapping of $E \times F$ in V given by $(x \times y \mapsto x \otimes y$ for all $x \in E$ and $y \in F)$, the tensor product of the vectors of the basis $e_i \otimes f_j$ generating a basis in V.

Let V be the space of a representation π of a group G. Suppose there is on it a bilinear Hermitian form (,) (if $v, w \in V$, $\overline{(v, w)} = (w, v)$), π is a unitary representation if $\forall g \in G$ $(\pi(g)w, \pi(g)v) = (w, v)$, that is to say the $\pi(g)$'s are unitary matrices.

C.1.3   Lie Groups

We first give a general definition of Lie groups and then restrict to a subcategory: the matrix Lie groups.

Definition 23. *A Lie group G is a group which has the structure of a differentiable manifold such that the mappings*

1. *$\phi : G \to G$ with $\phi(g) = g^{-1}$*
2. *$\psi : G \times G \to G$ with $\psi(g, h) = gh$*

are differentiable.

All the Lie groups we will study are subgroups of the groups $GL(n, \mathbb{R})$ or $GL(n, \mathbb{C})$, the groups of the $n \times n$ matrices, with non-vanishing determinant on $\mathbb{R}$ or $\mathbb{C}$.

Lemma 22. *The groups $GL(n, \mathbb{R})$ and $GL(n, \mathbb{C})$ are Lie groups.*

$A \in GL(n, \mathbb{R})$ if $\det A \neq 0$. We easily check the group structure. The manifold structure can be seen as follows. The fact that the determinant is non-zero means that it is an open set of $\mathbb{R}^{n^2}$, or more precisely of the ring $M(n, \mathbb{R})$ of the $n \times n$ matrices. We can take the matrix elements A_j^i for coordinates in $M(n, \mathbb{R})$. We check easily that to take the inverse of a matrix in $GL(n, \mathbb{R})$ or to make the product of two such matrices are differentiable mappings (indeed C^∞).

We now give some examples of matrix Lie groups.

1. The rotation group $O(n, \mathbb{R})$ of dimension n.
   It is the group of real displacements in the Euclidean space $\mathbb{R}^n$ which leave fixed a point. $A \in O(n, \mathbb{R})$ if $A \in GL(n, \mathbb{R})$ and $A^{\mathrm{tr}}A = 1$. The A's are the orthogonal matrices with determinant ± 1 (because $\det A^{\mathrm{tr}}A = \det A^{\mathrm{tr}} \det A = (\det A)^2 = 1$). The hypersurface generated by the group in $\mathbb{R}^{n^2}$ is given by the system of equations

$$A_j^i A_k^i = \delta_{jk} \tag{C.7}$$

   It corresponds to $n(n + 1)/2$ equations. Therefore, the manifold of the group is of dimension $n^2 - n(n + 1)/2 = n(n - 1)/2$.

2. The group $SO(n, \mathbb{R})$.
   It is the subgroup of the group $O(n, \mathbb{R})$ made of matrices of determinant 1. It has the same dimension as $O(n, \mathbb{R})$. It is the connected component of this group which contains the identity.

3. The group $SL(n, \mathbb{R})$.
   It is the group of order n matrices with determinant equal to 1. The manifold of this group is defined by $\det A = 1$. Its dimension is $n^2 - 1$.

4. The unitary group $U(n)$.
   It is defined in the space of complex matrices of order n by

$$UU^* = 1. \tag{C.8}$$

   In term of coefficients, this is equivalent to

$$U_j^i \bar{U}_k^i = \delta_{jk}. \tag{C.9}$$

   We can check that the system (C.9) corresponding to n^2 real equations. Therefore, the dimension of the manifold[2] is $2n^2 - n^2 = n^2$.

5. The unimodular group $SU(n)$. It is the subgroup of $U(n)$ made of the matrices with determinant 1. It has the dimension $n^2 - 1$.

6. The group $SL(n, \mathbb{C})$.
   It is made of the complex matrices of order n and with determinant 1. Its dimension on $\mathbb{C}$ is $n^2 - 1$ (or on $\mathbb{R}$, $2n^2 - 2$).

7. The group $O((p, q), \mathbb{R})$ (with $p + q = n$).
   It is the group of motions, leaving fixed the origin, in the pseudo-riemannian space $\mathbb{R}^n$ with the metric g. The signature of the metric is (p, q). It can also be defined as the group of transformations leaving invariant the quadratic form $(x, y) = (x^i g_{ij} y^j)$ with $(x, x) = (x^1)^2 + \cdots + (x^p)^2 - \cdots - (x^n)^2$.
   The group $O(3, 1)$ is the Lorentz group.
   If G is the matrix group defined on $\mathbb{R}^n$, we call identical representation, the representation π such that $\pi(g) = g$.

Exercise 4 Check that the identical representations of $GL(n, K)$ and of $SL(n, K)$ are irreducible, K being $\mathbb{R}$ or $\mathbb{C}$.

Some of these groups are studied in detail in this book because they play an important role in physics.

C.1.4   One Parameter Subgroup. Tangent Space

The Lie groups being manifolds, we can study the tangent space at a point. It is usual to take for point the neutral element. In fact, it is possible by a group action to move the tangent space at a point to the tangent space at the neutral element. To study the tangent space, we will consider the one-parameter subgroups of the group.

---

[2] The group manifold is not analytic since the derivatives of Eq. (C.9) with respect to $\bar{U}_j^i$ are not zero.

The one-parameter subgroups are the parametrised curves on the group. A one-parameter subgroup $t \mapsto g(t) \in G$ going through the neutral element is defined by the properties

$g(0) = 1$ (1 is the notation for the neutral element of the group),

$g(t_1 + t_2) = g(t_1)g(t_2)$ which implies that $g(-t) = g(t)^{-1}$.

We associate with $g(t)$ a vector, element of the tangent space, by

$$A = g(t)^{-1} \frac{\mathrm{d}g(t)}{\mathrm{d}t}. \tag{C.10}$$

We check first that A is independent of t. Consider in fact $g(t + s)$, we have

$$\frac{\mathrm{d}g(t)}{\mathrm{d}t} = \left. \frac{\mathrm{d}g(t + s)}{\mathrm{d}s} \right|_{s=0}, \tag{C.11}$$

thus using the group property $g(t + s) = g(t)g(s)$,

$$\frac{\mathrm{d}g(t)}{\mathrm{d}t} = g(t) \left(\left. \frac{\mathrm{d}g(s)}{\mathrm{d}s} \right|_{s=0} \right) \tag{C.12}$$

and then

$$A = \left. \frac{\mathrm{d}g(s)}{\mathrm{d}s} \right|_{s=0}. \tag{C.13}$$

A is the generator of the one-parameter group.

Conversely, the differential equation (C.13), with initial condition $g(0) = 1$, has, for t small enough, a unique solution. It is a one-parameter subgroup $g(t)$. Using the multiplication law of the group, this is in fact a solution for arbitrary t. We then write

$$g(t) = \exp(At), \tag{C.14}$$

a notation justified by the following definition.

Definition 24. *Let $g(t)$ be a one-parameter group with $\left. \dfrac{\mathrm{d}g(t)}{\mathrm{d}t} \right|_{t=0} = A$. The mapping $A \to g(1)$ is called the exponential mapping. It is a mapping from the tangent space of the group to the group.*

In the case of groups of matrices, the exponential mapping is the usual exponential of matrices. We set for $A \in GL(n, \mathbb{R})$

$$\exp(A) = \sum_{p=1}^{\infty} \frac{A^p}{p!}. \tag{C.15}$$

We have thus the following lemma.

Lemma 23. *Introducing the norm* $|A| = \left(\sum_{i,j} (A_j^i)^2 \right)^{1/2}$.

1. *The series (C.15) is convergent.*
2. *If A and B commute*

$$\exp(A + B) = \exp(A) \exp(B). \tag{C.16}$$

3. *If $U = \exp(A)$, then $U^{-1} = \exp(-A)$.*
4. $\exp(A^{\mathrm{tr}}) = (\exp(A))^{\mathrm{tr}}$.
5. $\exp(tA)$ *is solution of the differential equation (C.13).*

We usually write $\exp(A)$ as an ordinary exponential e^A.

The set of the vectors A tangent to the group at the identity forms the tangent space T of the group. It is a vector space of dimension n, if n is the dimension of the group.

We present here another approach, less intrinsic, but more intuitive of tangent space of the group.

Let us study the tangent space of a Lie group G of dimension n. Let $(x^1, \cdots, x^n)$ be a system of coordinates in the neighbourhood of the neutral element of the group 1 of coordinates $(0, \cdots, 0)$. In terms of these coordinates, the two differentiable mappings ϕ and ψ giving the inverse and the product can be written with $g = (x^1, \cdots, x^n)$ and $h = (y^1, \cdots, y^n)$ as

$$\phi^i(g) = \phi^i(x^1, \cdots, x^n) \quad \text{and} \quad \psi^i(gh) = \psi^i(x^1, \cdots, x^n, y^1, \cdots, y^n) \qquad i = 1, \ldots, n. \tag{C.17}$$

Developing these expressions to second order in x and y we obtain that

$$\psi^i(g, h) = ax^i + by^i + a_{jk}^i x^j x^k + b_{jk}^i y^j y^k + c_{jk}^i x^j y^k. \tag{C.18}$$

Since $\psi(g, 1) = \psi(1, g) = g$ we deduce that $a = b = 1$. Since $\psi(g, g^{-1}) = \psi(g, \phi(g)) = 1$, we deduce that at first order $\phi^i(g) = -x^i$, and that a_{jk}^i and b_{jk}^i are antisymmetric in j, k. Formula (C.18) can therefore be written as

$$\psi^i(g, h) = x^i + y^i + c_{jk}^i x^j y^k. \tag{C.19}$$

We define an alternated bilinear operation $[,]$ on T by

$$[x, y]^i = \frac{1}{2} (c_{jk}^i - c_{kj}^i) x^j y^k,$$

where x and y represent, respectively, the points of coordinates $\{x^i\}$ and $\{y^i\}$. From the associativity of the product

$$\psi(g, \psi(h, l)) = \psi(\psi(g, h), l) \tag{C.20}$$

we deduce the formula

$$[x, [y, z]] + [y, [z, x]] + [z, [x, y]] = 0. \tag{C.21}$$

We give some properties of this space.

1. Let $g_1(t)$ be a one-parameter subgroup of generator A_1. Let us thus consider an inner automorphism given by an element $h \in G$: $g_1(t) \mapsto hg_1(t)h^{-1}$. Since $hg_1(t)h^{-1}$ is a one-parameter subgroup of G, we get by derivation that $hA_1h^{-1} \in T$: the group acts by conjugation on the tangent space.

2. Let $g_1(t)$ and $g_2(t)$ be two one-parameter subgroups with, respectively, A_1 and A_2 for generators. The commutator of g_1 and g_2 $F(t, s) = g_1(t)g_2(s)g_1^{-1}(t)g_2^{-1}(s)$ is a two-parameter family of elements of G. Consider the derivative at 0 of $F(t, s)$ with respect to s:

$$H(t) = \left.\frac{\mathrm{d}F(t, s)}{\mathrm{d}s}\right|_{s=0} = g_1(t)A_2g_1(t)^{-1} - A_2. \tag{C.22}$$

It is the sum of two elements of the tangent space, therefore an element of T. More precisely, it is a curve in the tangent space going through 0. Its derivative

$$\left.\frac{\mathrm{d}H(t)}{\mathrm{d}t}\right|_{t=0} \tag{C.23}$$

is an element of the tangent space to T at A. Since $T = \mathbb{R}^n$, T can be identified with its dual space and therefore the derivative (C.23) is an element of T. It can be written as $[A_1, A_2]$.

Exercise 5 Check that

$$\left.\frac{\mathrm{d}^2}{\mathrm{d}t^2}g_1(t)g_2(t)g_1^{-1}(t)g_2^{-1}(t)\right|_{t=0} = [A_1, A_2]. \tag{C.24}$$

The result of this exercise is that the bilinear form $[\,,\,]$ is antisymmetric.

C.2 Lie Algebras

C.2.1 Definition

In section **A.2**, we introduced the notion of a Lie bracket $[X, Y]$ of two vector fields X, Y. The bracket $[X, Y]$ is an alternate bilinear form. We found the same structure when studying the tangent space at a point of a Lie group.

Definition 25. *A vector space V with an alternate bilinear operation, the bracket, $[,]$, is called a Lie algebra if every triplet X, Y and Z of V satisfies the following identity, the Jacobi identity:*

$$[X, [Y, Z]] + [Y, [Z, X]] + [Z, [X, Y]] = 0. \tag{C.25}$$

The Jacobi identity can be obtained as the result of a derivation on the bracket. In fact, for all $X \in V$, let us introduce the linear operator $ad\, X$ defined by $ad\, X(Y) = [X, Y]$. Then formula (C.25) can be rewritten as

$$ad\, X[Y, Z] = [ad\, X(Y), Z] + [Y, ad\, X(Z)], \tag{C.26}$$

that is to say that adX acts as a derivation. We now give some examples of Lie algebras.

Examples

1. The vector fields in a domain of $\mathbb{R}^n$ form a Lie algebra for the Lie bracket. In fact, if $X = X^i \frac{\partial}{\partial x^i}$ and $Y = Y^i \frac{\partial}{\partial y^i}$:

$$[X, Y]^i = L_X Y^i = X^j \frac{\partial Y^i}{\partial x^j} - Y^j \frac{\partial X^i}{\partial y^j}. \tag{C.27}$$

2. The Euclidean vector space $\mathbb{R}^3$ with bracket given by the vector product $\wedge$ is a Lie algebra.
3. An algebra of linear operators is a Lie algebra with a bracket given by the commutator of two operators.
4. The space M_n of order n matrices is a Lie algebra. If $A, B \in M_n$ then $[A, B] = AB - BA$.
5. The tangent space of a Lie group is a Lie algebra.

The interest of the notion of Lie algebra is that because of the exponential mapping, it is possible, knowing the Lie algebra of a Lie group, to recover the connected component of the identity of this group.

We introduced the notion of Lie bracket by studying the vector fields. We just saw that the tangent space of a Lie group has also a structure of Lie algebra. This naturally follows from the theorem

Theorem 27. *If X and Y are two vector fields of the tangent space $T_x(M)$ of a differentiable manifold M, then their Lie bracket is also in this space.*

We give a proof in a simple case. Let us suppose that the manifold M is given by an equation $f(x_1, \ldots, x_n) = 0$. Locally, we can always suppose that by a change of coordinates, it can be written $x^1 = 0$. The fact that the vector fields X and Y are in the tangent space of M is equivalent to

$$\partial_X f|_{f=0} = L_X f|_{f=0} = L_Y f|_{f=0} = 0, \tag{C.28}$$

thus in our case

$$X^1\big|_{x^1=0} = 0; \qquad\qquad Y^1\big|_{x^1=0} = 0. \tag{C.29}$$

Therefore,

$$[X, Y]^1\big|_{x^1=0} = \left(X^i \frac{\partial Y^1}{\partial x^i} - Y^i \frac{\partial X^1}{\partial x^i}\right)\bigg|_{x^1=0} = 0, \tag{C.30}$$

since $X^1\big|_{x^1=0}$ leads that, in the neighbourhood of 0, $X^1 = x^1 g(x)$ and thus

$$\frac{\partial Y^1}{\partial x^i}\bigg|_{x^1=0} = 0 \quad \text{and} \quad \frac{\partial X^1}{\partial x^i}\bigg|_{x^1=0} = 0. \tag{C.31}$$

If π is a representation $\pi : G \to GL(n, \mathbb{R})$ of a group G, then $T_e(\pi)$, the induced tangent mapping at e (the neutral element) maps the tangent space at this point, that is to say the Lie algebra $\mathbf{g}$ of G, into the space of matrices $M(n, \mathbb{R})$. This mapping is a Lie algebra homomorphism, that is to say

$$T_e(\pi)([A, B]) = [T_e(\pi)(A), T_e(\pi)(B)] \qquad \forall A, B \in \mathbf{g}. \tag{C.32}$$

$T_e(\pi)$ is a matrix representation of the Lie algebra $\mathbf{g}$.

C.2.2   Matrix Lie Algebras

We list here the Lie algebras associated with the Lie groups introduced in the preceding section.

1. The Lie algebra of $O(n, \mathbb{R})$.
   We start with a one-parameter subgroup $g(t)$, that is to say an $n \times n$ matrix depending on $t \in \mathbb{R}$. By the definition of the group, we have $g(t)g(t)^{\text{tr}} = 1$. Differentiating this condition with respect to t and taking the value at $t = 0$, we find with $A = \dfrac{\mathrm{d}g(t)}{\mathrm{d}t}\bigg|_{t=0}$

$$A^{tr} + A = 0. \tag{C.33}$$

   The Lie algebra elements are therefore all the antisymmetric matrices with real coefficients. We have that $A \in \mathbf{o(n,R)}$, the Lie algebra of $O(n, \mathbb{R})$. We can check that the dimension of the space generated by these matrices is $n(n-1)/2$. It is as expected the dimension of the group (or of the manifold of the group).

2. The Lie algebra $\mathbf{so(n,R)}$ of $SO(n, \mathbb{R})$.
   It is the same algebra as that of $O(n, \mathbb{R})$. In fact, $SO(n, \mathbb{R})$ is a connected component of $O(n, \mathbb{R})$; therefore, it has the same tangent structure. We can also see this

analytically since the supplementary condition det $g(t) = 1$ leads[3] by derivation to no new constraints.

3. The Lie algebra of $SL(N, \mathbb{R})$.
   From footnote 3 we see that taking the derivative $\det g(t) = 1$ leads to $\mathrm{Tr}\, A = 0$. The Lie algebra is made of traceless matrices.

4. The Lie algebra $\mathbf{u(n)}$ of $U(n)$.
   We easily checks that it is made of the anti-Hermitian matrices, that is to say those which satisfy $A^* + A = 0$.

5. The Lie algebra $\mathbf{su(n)}$ of $SU(n)$ is a Lie subalgebra of $\mathbf{u(n)}$ made of traceless anti-Hermitian matrices.

6. The Lie algebra $\mathbf{sl(n,C)}$ of $SL(n, \mathbb{C})$.
   It is made of traceless complex matrices.

7. The Lie algebra of the group $O((p, q), \mathbb{R})$ is made of the matrices A satisfying $A^{\mathrm{tr}} G + G A = 0$, where G is the matrix associated with the metric g.

To conclude, we will obtain some relations specific to the matrix Lie groups viewed as groups of transformations of Euclidean or pseudo-Euclidean spaces. We take as an example $SO(3, \mathbb{R})$.

An example of a one-parameter subgroup of the rotation group $SO(3, \mathbb{R})$ is given by the rotations $R_3(\theta)$, of angle θ, with z axis of coordinates

$$R_3(\theta) = \begin{pmatrix} \cos\theta & -\sin\theta & 0 \\ \sin\theta & \cos\theta & 0 \\ 0 & 0 & 1 \end{pmatrix}. \tag{C.34}$$

The derivative at point 0 defines an element $\mathcal{J}_3$ of the Lie algebra of dimension 3 of the group. Considering the other rotations, we get the existence of 3 elements $\mathcal{J}_1, \mathcal{J}_2,$ and $\mathcal{J}_3$

$$\mathcal{J}_1 = \begin{pmatrix} 0 & 0 & 0 \\ 0 & 0 & -1 \\ 0 & 1 & 0 \end{pmatrix} ; \quad \mathcal{J}_2 = \begin{pmatrix} 0 & 0 & 1 \\ 0 & 0 & 0 \\ 1 & 0 & 0 \end{pmatrix} ; \quad \mathcal{J}_3 = \begin{pmatrix} 0 & -1 & 0 \\ 1 & 0 & 0 \\ 0 & 0 & 0 \end{pmatrix}, \tag{C.35}$$

satisfying

$$[\mathcal{J}_1, \mathcal{J}_2] = \mathcal{J}_3; \quad [\mathcal{J}_2, \mathcal{J}_3] = \mathcal{J}_1; \quad [\mathcal{J}_3, \mathcal{J}_1] = \mathcal{J}_2. \tag{C.36}$$

These last relations show that $\mathbf{so(3,R)}$ is isomorphic to the Lie algebra of $\mathbb{R}^3$ (the algebra of vectors generated by $\wedge$). The 3 matrices built in this way form a basis of this Lie algebra of dimension 3, that is to say the 3×3 antisymmetrical matrices.

---

[3] We can prove this assertion by using the fact that $\det A = e^{\mathrm{Tr}\,\log A}$.

It is possible to associate with each real (or complex) matrix $\mathcal{J}$ of order n a vector field $T_{\mathcal{J}}$ in $\mathbb{R}^n$ (or $\mathbb{C}^n$) by

$$T_{\mathcal{J}}x = \mathcal{J}x \qquad x \in \mathbb{R}^n. \tag{C.37}$$

The integral curves of these vector fields are defined by the differential equation

$$\dot{x} = -\mathcal{J}x \tag{C.38}$$

of solution

$$x(t) = \exp(-\mathcal{J}t)x_0 \tag{C.39}$$

if $x_0 = x(0)$ is the initial condition.

For the rotations, we check easily that the three vector fields $T_{\mathcal{J}_i}$, $i = 1, 2, 3$, which are generally written L_i, have for components

$$T_{\mathcal{J}_1} = (0, z, -y); \qquad T_{\mathcal{J}_2} = (-z, 0, x); \qquad T_{\mathcal{J}_3} = (y, -x, 0). \tag{C.40}$$

Since we have seen that the Lie bracket of two vector fields is a vector field

$$[T_X, T_Y] = T_{[X,Y]}. \tag{C.41}$$

We deduce in our particular case, by linearity, that

$$[T_{\mathcal{J}_i}, T_{\mathcal{J}_j}] = T_{[\mathcal{J}_i \mathcal{J}_j]} = T_{\epsilon_{ijk}\mathcal{J}_k} = \epsilon_{ijk}T_{\mathcal{J}_k}. \tag{C.42}$$

The differential operators associated with the vector fields $\partial_{T_{\mathcal{J}}}$ (or the Lie derivatives $L_{T_{\mathcal{J}}}$) are the so-called *generators* of the group.

Examples Setting $\partial_{T_{\mathcal{J}_i}} = L_{x^i}$, we have, with $(x^1, x^2, x^3) = (x, y, z)$

$$L_x = z\frac{\partial}{\partial y} - y\frac{\partial}{\partial z}; \qquad L_y = x\frac{\partial}{\partial z} - z\frac{\partial}{\partial x}; \qquad L_z = y\frac{\partial}{\partial x} - x\frac{\partial}{\partial y} \tag{C.43}$$

and the three generators satisfy the commutation relations of the angular momentum.

The generators also make it possible to reconstruct the action of the group on functions.

If $f(x)$ is a function on $\mathbb{R}^n$, under the action of the one-parameter group $\exp(-\mathcal{J}t)$, it becomes $\phi(t, x) = f(\exp(-\mathcal{J}t)x)$. $\phi(t, x)$ is the solution of the differential equation

$$\frac{\mathrm{d}}{\mathrm{d}t}\phi(t, x) = -(\partial_{T_{\mathcal{J}}}\phi)(t, x) \tag{C.44}$$

with the initial condition $\phi(0, x) = f(x)$. We can write this solution as

$$\phi(t, x) = e^{-t\partial T_g} f(x), \tag{C.45}$$

where the exponential is defined by its formal expansion and has a meaning whenever the corresponding series is convergent.

Examples The generator of the translation group in $\mathbb{R}$ is $\frac{d}{dx}$. A translation by a changes a function $f(x)$ in $f(x - a)$. Formula (C.13) can then be written as

$$f(x - a) = e^{-a\frac{d}{dx}} f(x) = f(x) - af^{(1)}(x) + \frac{a^2}{2!} f^{(2)}(x) - \frac{a^3}{3!} f^{(3)}(x) + \cdots, \tag{C.46}$$

which is nothing else than the Taylor expansion of f around x.

Appendix D
A Collection of Useful Formulae

(The numbers in parenthesis refer to the sections in the book in which these concepts are presented.)

D.1 Units and Notations

- The standard high energy unit system $c = \hbar = 1$ is used.
- The Minkowski metric is denoted by $\eta_{\mu\nu}$, or $g_{\mu\nu}$, whenever there is no risk of confusion with the curved space metric.
  It is given by $\eta_{00} = 1$, $\eta_{ii} = -1$, $\eta_{\mu\nu} = 0$ for $\mu \neq \nu$.
- The scalar product of two four-vectors p and q is denoted by $pq = p^0 q^0 - p \cdot q$.
  The mass shell condition for a particle of mass $m \geq 0$ is: $p^2 = m^2$.
- The invariant measure on the positive energy branch of the mass hyperboloid $p^2 = m^2$ is given by Eq. (5.16):

$$d\Omega_m = \frac{d^4 p}{(2\pi)^4} (2\pi)\delta(p^2 - m^2)\theta(p^0) = \frac{d^3 p}{2(2\pi)^3 \sqrt{p^2 + m^2}}. \tag{D.1}$$

- The Fourier transform of a function $f(x)$ in d dimensions is defined by

$$\tilde{f}(p) = \int d^d x e^{-ipx} f(x). \tag{D.2}$$

- The $SU(2)$ Pauli matrices are denoted by either $\boldsymbol{\sigma}$, or $\boldsymbol{\tau}$ and are given by

$$\sigma_1 = \begin{pmatrix} 0 & 1 \\ 1 & 0 \end{pmatrix} \qquad \sigma_2 = \begin{pmatrix} 0 & -i \\ i & 0 \end{pmatrix} \qquad \sigma_3 = \begin{pmatrix} 1 & 0 \\ 0 & -1 \end{pmatrix}. \tag{D.3}$$

- The $SU(3)$ Gell–Mann matrices are denoted by λ^a, $a = 1, ..., 8$ and are given by

$$\lambda_1 = \begin{pmatrix} 0 & 1 & 0 \\ 1 & 0 & 0 \\ 0 & 0 & 0 \end{pmatrix} \lambda_2 = \begin{pmatrix} 0 & -i & 0 \\ i & 0 & 0 \\ 0 & 0 & 0 \end{pmatrix} \lambda_3 = \begin{pmatrix} 1 & 0 & 0 \\ 0 & -1 & 0 \\ 0 & 0 & 0 \end{pmatrix} \lambda_4 = \begin{pmatrix} 0 & 0 & 1 \\ 0 & 0 & 0 \\ 1 & 0 & 0 \end{pmatrix}$$

$$\lambda_5 = \begin{pmatrix} 0 & 0 & -i \\ 0 & 0 & 0 \\ i & 0 & 0 \end{pmatrix} \quad \lambda_6 = \begin{pmatrix} 0 & 0 & 0 \\ 0 & 0 & 1 \\ 0 & 1 & 0 \end{pmatrix} \quad \lambda_7 = \begin{pmatrix} 0 & 0 & 0 \\ 0 & 0 & -i \\ 0 & i & 0 \end{pmatrix} \quad \lambda_8 = \frac{1}{\sqrt{3}} \begin{pmatrix} 1 & 0 & 0 \\ 0 & 1 & 0 \\ 0 & 0 & -2 \end{pmatrix}.$$

$$\tag{D.4}$$

D.2 Free Fields

- **Neutral scalar field**
  - Lagrangian density (section 6.2):

$$\mathcal{L} = \frac{1}{2}[(\partial\phi)^2 - m^2\phi^2] \tag{D.5}$$

  - Equation of motion (section 6.2):

$$(\Box + m^2)\phi = 0 \tag{D.6}$$

  - Feynman propagator (sections 3.8, 6.2):

$$\frac{i}{p^2 - m^2 + i\epsilon} \tag{D.7}$$

  - Expansion in plane waves (section 6.2):

$$\phi(x) = \int \mathrm{d}\Omega_m [a(p)e^{-ipx} + a^*(p)e^{ipx}] \tag{D.8}$$

  - Canonical commutation relations (section 9.4):

$$[a(p), a^*(p')] = (2\pi)^3 2\omega_p \delta^3(\boldsymbol{p}-\boldsymbol{p}') \tag{D.9}$$

- **Complex scalar field**
  - Lagrangian density (section 6.2):

$$\mathcal{L} = (\partial_\mu\phi)(\partial^\mu\phi^*) - m^2\phi\phi^* \tag{D.10}$$

  - Expansion in plane waves (section 6.2):

$$\phi(x) = \int \mathrm{d}\Omega_m [a(p)e^{-ipx} + b^*(p)e^{ipx}] \tag{D.11}$$

  - Canonical commutation relations (section 9.4):

$$[a(p), a^*(p')] = [b(p), b^*(p')] = (2\pi)^3 2\omega_p \delta^3(\boldsymbol{p}-\boldsymbol{p}') \tag{D.12}$$

with all other commutators vanishing.

- **Dirac field**
  - Lagrangian density (section 6.3):

$$\mathcal{L} = \bar{\Psi}\,(i\partial\!\!\!/- m)\,\Psi \tag{D.13}$$

  - Equation of motion (section 6.3):

$$(i\partial\!\!\!/- m)\Psi = 0 \tag{D.14}$$

  - Feynman propagator (section 6.3):

$$\frac{i}{p\!\!\!/- m + i\epsilon} \tag{D.15}$$

  - Expansion in plane waves (section 6.3):

$$\Psi = \int d\Omega_m \sum_{\alpha=1}^{2} \left[a_\alpha(k)u^{(\alpha)}(k)e^{-ikx} + b_\alpha^\dagger(k)v^{(\alpha)}(k)e^{ikx} \right] \tag{D.16}$$

  - The elementary solutions satisfy the orthogonality relations (section 6.3):

$$\begin{aligned} \bar{u}^{(\alpha)}(k)u^{(\beta)}(k) &= 2m\delta^{\alpha\beta} & \bar{u}^{(\alpha)}(k)v^{(\beta)}(k) &= 0 \\ \bar{v}^{(\alpha)}(k)v^{(\beta)}(k) &= -2m\delta^{\alpha\beta} & \bar{v}^{(\alpha)}(k)u^{(\beta)}(k) &= 0 \end{aligned} \tag{D.17}$$

  - Canonical anticommutation relations (section 11.1):

$$\begin{aligned} \{a_s(\boldsymbol{p}), a_{s'}^\dagger(\boldsymbol{p}')\} &= (2\pi)^3 2\omega_p \delta^3(\boldsymbol{p}-\boldsymbol{p}')\,\delta_{ss'} \\ \{b_s(\boldsymbol{p}), b_{s'}^\dagger(\boldsymbol{p}')\} &= (2\pi)^3 2\omega_p \delta^3(\boldsymbol{p}-\boldsymbol{p}')\,\delta_{ss'} \end{aligned} \tag{D.18}$$

  with all other anticommutators vanishing.

- **Massless vector field in a covariant gauge**
  - Lagrangian density (sections 3.6, 6.4):

$$\mathcal{L} = -\frac{1}{4}F_{\mu\nu}F^{\mu\nu} + \frac{1}{2\alpha}(\partial_\mu A^\mu)^2 \quad ; \quad F_{\mu\nu} = \partial_\mu A_\nu - \partial_\nu A_\mu \tag{D.19}$$

  - Equation of motion (sections 3.6, 6.4):

$$\partial^\mu F_{\mu\nu} + \frac{1}{\alpha}\partial_\nu\partial_\mu A^\mu = 0 \tag{D.20}$$

  - Feynman propagator (sections 3.8, 6.4):

$$\frac{i}{k^2 + i\epsilon}\left[g^{\mu\nu} - \frac{k^\mu k^\nu}{k^2(1-\alpha)} \right] \tag{D.21}$$

- Expansion in plane waves (section 6.4):

$$A_\mu(x) = \int d\Omega_0 \sum_{\lambda=0}^{3} \left[a^{(\lambda)}(k) \epsilon_\mu^{(\lambda)}(k) e^{-ikx} + a^{(\lambda)*}(k) \epsilon_\mu^{(\lambda)*}(k) e^{ikx} \right]. \qquad (D.22)$$

The four ϵ vectors are called *polarisation vectors* and satisfy the orthonormality relations (section 6.4):

$$\sum_{\lambda=0}^{3} \frac{\epsilon_\mu^{(\lambda)}(k) \epsilon_\nu^{(\lambda)*}(k)}{\epsilon^{(\lambda)\rho}(k) \epsilon_\rho^{(\lambda)*}(k)} = g_{\mu\nu}; \quad \epsilon^{(\lambda)\rho}(k) \epsilon_\rho^{(\lambda')*}(k) = g^{\lambda\lambda'}, \qquad (D.23)$$

where, again, $*$ means 'complex conjugation'.
- Canonical commutation relations (section 12.5):

$$[a^{(\lambda)}(k), a^{(\lambda')*}(k')] = -2k_0 (2\pi)^3 g^{\lambda\lambda'} \delta^3(\boldsymbol{k} - \boldsymbol{k}'). \qquad (D.24)$$

- **Massive vector field**
  - Lagrangian density (section 6.4):

$$\mathcal{L} = -\frac{1}{4} F^2 + \frac{1}{2} m^2 A^2 \qquad (D.25)$$

  - Equation of motion (section 6.4):

$$\Box A_\mu - \partial_\mu \partial_\nu A^\nu + m^2 A_\mu = 0; \quad m^2 \partial^\mu A_\mu = 0. \qquad (D.26)$$

  - Feynman propagator (section 6.4):

$$\frac{i}{k^2 - m^2 + i\epsilon} \left[g^{\mu\nu} - \frac{k^\mu k^\nu}{m^2} \right] \qquad (D.27)$$

  - Plane wave expansion (section 12.5):

$$A_\mu(x) = \int d\Omega_m \sum_{i=1}^{3} [a^{(i)}(p) \epsilon_\mu^{(i)}(p) e^{-ip.x} + a^{(i)*}(p) \epsilon_\mu^{(i)*}(p) e^{ip.x}]. \qquad (D.28)$$

The three polarisation vectors $\epsilon^{(i)}(p)$ satisfy (section 12.5):

$$p.\epsilon^{(i)}(p) = 0$$
$$\epsilon^{(i)}(p).\epsilon^{(j)}(p) = \delta_{ij}$$
$$\sum_i \epsilon_\varrho^{(i)}(p) \epsilon_\sigma^{(i)}(p) = -(g_{\varrho\sigma} - \frac{p_\varrho p_\sigma}{m^2}) \qquad (D.29)$$

– Canonical commutation relations (section 12.5):

$$[a^{(i)}(p), a^{(j)*}(p')] = \delta_{ij}(2\pi)^3 2\omega_p \delta^{(3)}(\boldsymbol{p} - \boldsymbol{p}').$$ (D.30)

D.3 Feynman Rules for Scattering Amplitudes

A scattering amplitude relating a set of initial to a set of final particles is given, order by order in perturbation theory, by the sum of all connected and amputated Feynman diagrams to that order in the expansion, with external lines corresponding to the initial and final particles of the amplitude. The rules for calculating these diagrams are the following.

- **External lines** (section 13.3)
  - All external momenta are put on the mass shell of the corresponding particle: $p^2 = m^2$.
  - All polarisation vectors correspond to physical particles. In particular, photon polarisations are transverse.
  - A factor, corresponding to the wave function of the external particle, multiplies each external line. They are given in Fig. D.1.
- **Internal lines** (section 13.3)
  - To every internal line corresponds a Feynman propagator, according to the spin of the line. In general, the propagators are matrices carrying Lorentz as well as internal symmetry indices. The expressions are given in section D.2.
  - To every internal line of momentum p there is an integration $\int d^4p/(2\pi)^4$.
- **Vertices** (sections 13.3, 14.5)
  To every vertex there is attached a δ-function of energy–momentum conservation: $(2\pi)^4\delta^4(\Sigma p_i)$, where the momenta p_i of all the lines converging to the vertex are drawn in-coming.
  For a connected diagram the δ-function of energy–momentum conservation in every vertex determines the momenta of all internal lines in terms of the external momenta, leaving only one momentum undetermined for every closed loop. As a

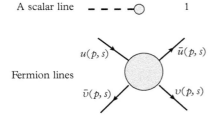

Figure D.1 *The factors applied to every external scalar, spinor or vector line in a Feynman diagram for a scattering amplitude. The momenta and spins are those of the external particle. Possible internal symmetry indices have been suppressed.*

result, we must perform only one integration for every closed loop. An overall δ-function, expressing the total energy–momentum conservation between the initial and final particles, factors out: $(2\pi)^4 \delta^4 (\Sigma(p_i) - \Sigma(p_f))$.

The vertex function is a matrix both in Minkowski space and in whichever internal symmetry space we have. We give here a list of the most commonly used vertices:

− ϕ^4 theory:

$$\mathcal{L}_I = -\frac{\lambda}{4!} \, \phi^4$$

The vertex is represented by

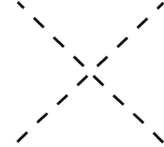

and it is given by $-\frac{\lambda}{4!}$.

− Scalar Yukawa theory:

$$\mathcal{L}_I = g\overline{\psi}\psi\phi$$

The vertex is represented by

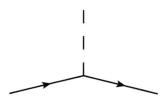

and it is given by g.

− Pseudo-scalar Yukawa theory:

$$\mathcal{L}_I = g\overline{\psi}\gamma_5\psi\phi$$

The vertex is represented by

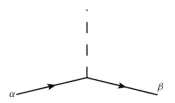

and it is given by $g(\gamma_5)_{\alpha\beta}$.

– $SU(2)$ pseudo-scalar Yukawa theory with a doublet of fermions and a triplet of bosons.

$$\mathcal{L}_I = g\overline{\psi}\gamma_5\boldsymbol{\tau}\psi \cdot \boldsymbol{\phi}$$

The vertex is represented by

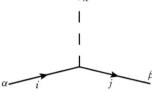

and it is given by $g(\tau^k)_{ij}(\gamma_5)_{\alpha\beta}$.

Here i, j, and k are the $SU(2)$ indices (i,j=1,2 and k=1,2,3) and α and β the Dirac indices (α, β=1,...,4).

– The gluon–quark vertex in QCD. The interaction Lagrangian is given by

$$\mathcal{L}_I = -g_s \sum_{f=1}^{6} \sum_{i,j=1}^{3} \overline{q}_f^i \, \mathcal{G}_{ij} \, q_f^j \qquad \text{(D.31)}$$

It is represented by

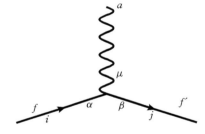

and it is given by $-g_s(\lambda^a)_{ij}(\gamma_\mu)_{\alpha\beta}\delta_{ff'}$, where i and j are the quark colour indices (i,j=1,2,3), a is the gluon colour index (a=1,...,8), α and β are Dirac indices, and λ are the $SU(3)$ Gell-Mann matrices. f and f' are flavour indices and g_s is the strong interaction coupling constant.

– Scalar electrodynamics. It has two couplings:

(i) $\mathcal{L}_I^{(1)} = \mathrm{i}eA_\mu(\phi\partial^\mu\phi^* - \phi^*\partial^\mu\phi)$

The vertex is represented by

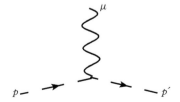

and it is given by $-e(p + p')_\mu$.

(ii) $\mathcal{L}_I^{(2)} = e^2 A_\mu A^\mu \phi \phi^*$

The vertex is represented by

and it is given by $2e^2 g_{\mu\nu}$.

– Yang–Mills theory. In a covariant gauge it has three couplings:

(i) $\mathcal{L}_I^{(1)} = \frac{1}{4} g f_{abc} (\partial_\mu A_\nu^a - \partial_\nu A_\mu^a) A^{\mu b} A^{\nu c}$ where f_{abc} are the structure constants of the Lie algebra of $\mathcal{G}$.

The vertex is represented by

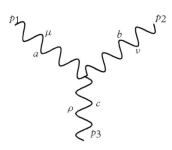

and it is given by

$$gf_{abc}[g_{\mu\nu}(p_1 - p_2)_\rho + g_{\nu\rho}(p_2 - p_3)_\mu \qquad (D.32)$$
$$+ g_{\rho\mu}(p_3 - p_1)_\nu].$$

(ii) $\mathcal{L}_I^{(2)} = -\frac{1}{4} g^2 f_{abc} f_{ab'c'} A_\mu^b A_\nu^c A^{\mu b'} A^{\nu c'}$

The vertex is represented by

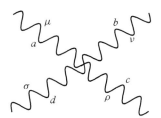

and it is given by

$$-g^2 [f_{eab} f_{ecd} (g_{\mu\rho} g_{\nu\sigma} - g_{\mu\sigma} g_{\nu\rho}) \qquad (D.33)$$
$$+ f_{eac} f_{edb} (g_{\mu\sigma} g_{\rho\nu} - g_{\mu\nu} g_{\rho\sigma}) + f_{ead} f_{ebc} (g_{\mu\nu} g_{\rho\sigma} - g_{\mu\rho} g_{\nu\sigma})]$$

(iii)   The gauge boson–ghost coupling

$$\mathcal{L}_{\text{YMgh}} = -gf_{abc}\partial^{\mu}(\bar{\Omega}^{a})A_{\mu}^{b}\Omega^{c} \tag{D.34}$$

The vertex is represented by

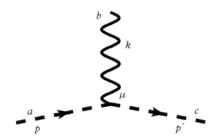

and it is given by $gp^{\mu}f_{abc}$.

- There is an overall numerical factor multiplying every diagram. It is composed by various pieces:
    - A factor i^{n} from the perturbation expansion.
    - A minus sign for every closed fermion loop.
    - A minus sign for every anti-fermion line crossing the diagram from the initial to the final state.
    - Finally, a combinatoric factor coming from all the different contractions in the Wick expansion which give this particular diagram.

D.4   Examples

The only rule which is not spelled out in a totally clear and unambiguous way is the one which gives the final overall combinatoric factor. So, we will end this appendix with a few examples to illustrate the procedure.

- A 1-loop diagram in second-order ϕ^{4} theory. We consider the scattering amplitude of two scalar bosons with momenta p_1 and p_2 going into two bosons with momenta p_3 and p_4.

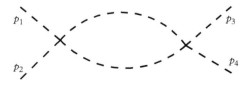

This diagram has a very large combinatoric factor because all lines are identical and can be connected in many different ways. We find the following:

– A factor $i^2 = -1$ from the perturbation expansion.

– A factor $4 \times 3 = 12$ from the ways we can combine the two external lines with momenta p_1 and p_2 to the four lines of the first vertex.

– The same factor $4 \times 3 = 12$ from the second vertex and the lines in the final state.

– A factor of 2 from the two ways to connect the lines in the loop.

The overall factor is -288. This is why we define the coupling constant in the ϕ^4 theory as $\lambda/4!$. If we take this definition into account, the coefficient of this diagram ends up being $-\lambda^2/2$.

If we repeat the analysis for the self-interaction of a charged scalar field with interaction Lagrangian proportional to $(\phi\phi^*)^2$, the diagram looks the same but now the lines have arrows showing the charge flow. In the contractions we must respect charge conservation and we find a factor of 4 (instead of 12) for each vertex and no factor for the loop. The overall factor is -16. In this case we usually define the coupling constant as $\lambda/4$ and the diagram has a resulting factor $-\lambda^2$.

• A 1-loop diagram in fourth-order QED. We chose the one shown here which contributes to the $e^+ + e^- \rightarrow e^+ + e^-$ scattering amplitude.

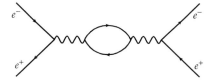

– There is a factor $i^4 = 1$ from the perturbation expansion.

– In QED all lines in a vertex are distinct; therefore, there is only one contraction which produces this diagram. The combinatoric factor equals 1.

– There is a minus sign from the fermion loop.
Therefore, the overall factor is -1.

• The relative sign between two tree diagrams in QED. We consider the same process $e^+ + e^- \rightarrow e^+ + e^-$ and we look at the two tree diagrams

– Both diagrams have a factor $i^2 = -1$ from the perturbation expansion.

– There is no combinatoric factor.

– The second diagram has a minus sign because the positron line goes through the diagram from the initial to the final state.

As a result, the first diagram has an overall factor equal to -1 and the second to +1.

Appendix E
Extract from Maxwell's *A Treatise on Electricity and Magnetism*

These four pages are extracted from Maxwell's *A Treatise on Electricity and Magnetism*. The first two pages show how Maxwell, although he knew both the corpuscular and undulatory theories, is only interested in the adequation of his theory with the latter. The last two pages are concerned with the problem of the action at a distance and the contradiction with Ampere's laws to conclude finally at the necessity of a medium, the aether.

781.] In several parts of this treatise an attempt has been made to explain electromagnetic phenomena by means of mechanical action transmitted from one body to another by means of a medium occupying the space between them. The undulatory theory of light also assumes the existence of a medium. We have now to show that the properties of the electromagnetic medium are identical with those of the luminiferous medium.

To fill all space with a new medium whenever any new phenomenon is to be explained is by no means philosophical, but if the study of two different branches of science has independently suggested the idea of a medium, and if the properties which must be attributed to the medium in order to account for electromagnetic phenomena are of the same kind as those which we attribute to the luminiferous medium in order to account for the phenomena of light, the evidence for the physical existence of the medium will be considerably strengthened.

But the properties of bodies are capable of quantitative measurement. We therefore obtain the numerical value of some property of the medium, such as the velocity with which a disturbance is propagated through it, which can be calculated from electromagnetic experiments, and also observed directly in the case of light. If it should be found that the velocity of propagation of electromagnetic disturbances is the same as the velocity of light, and this not only in air, but in other transparent media, we shall have strong reasons for believing that light is an electromagnetic phenomenon, and the combination of the optical with the electrical evidence will produce a conviction of the reality of the medium similar to that which we obtain, in the case of other kinds of matter, from the combined evidence of the senses.

782.] When light is emitted, a certain amount of energy is expended by the luminous body, and if the light is absorbed by another body, this body becomes heated, showing that is has received energy from without. During the interval of time after the light left the first body and before it reached the second, it must have existed as energy in the intervening space.

According to the theory of emission, the transmission of energy is effected by the actual transference of light-corpuscules from the luminous to the illuminated body, carrying with them their kinetic energy, together with any other kind of energy of which they may be the receptacles.

According to the theory of undulation, there is a material medium which fills the space between the two bodies, and it is by the action of contiguous parts of this medium that the energy is passed on, from one portion to the next, till it reaches the illuminated body.

The luminiferous medium is therefore, during the passage of light through it, a receptacle of energy. In the undulatory theory, as developed by Huygens, Fresnel, Young, Green, &c this energy is supposed to be partly potential and partly kinetic. The potential energy is supposed to be due to the distortion of the elementary portions of the medium. We must therefore regard the medium as elastic. The kinetic energy is supposed to be due to the vibratory motion of the medium. We must therefore regard the medium as having a finite density.

In the theory of electricity and magnetism adopted in this treatise, two forms of energy are recognised, the electrostatic and the electrokinetic (see Arts. 630 and 636), and these are supposed to have their seat, not merely in the electrified or magnetized bodies, but in every part of the surrounding space, where electric or magnetic force is observed to act. Hence our theory agrees with the undulatory theory in assuming the existence of a medium which is capable of becoming a receptacle of two forms of energy*.

865.] There appears to be, in the minds of these eminent men, some prejudice, or *à priori* objection, against the hypothesis of a medium in which the phenomena of radiation of light and heat and the electric actions at a distance take place. It is true that at one time those who speculated as to the causes of physical phenomena were in the habit of accounting for each kind of action at a distance by means of a special aethereal fluid, whose function and property it was to produce these actions. They filled all spaces three and four times over with aethers of different kinds, the properties of which were invented merely to 'save appearances,' so that more rational enquirers were willing rather to accept not only Newton's definite law of attraction at a distance, but even the dogma of Cotes*, that action at a distance is one of the primary properties of matter, and that no explanation can be more intelligible than this fact. Hence the undulatory theory of light has met with much opposition, directed not against its failure to explain the phenomena, but against its assumption of the existence of a medium in which light is propagated.

866.] We have seen that the mathematical expressions for electrodynamic action led, in the mind of Gauss, to the conviction that a theory of the propagation of electric action in time would be found to be the very keystone of electrodynamics. Now we are unable to conceive of propagation in time, except either as the flight of a material substance

---

* 'For my own part, considering the relation of a vacuum to the magnetic force and the general character of magnetic phenomena external to the magnet, I am more inclined to the notion that in the transmission of the force there is such an action, external to the magnet, than that the effects are merely attraction and repulsion at a distance. Such an action may be a function of the other; for it is not at all unlikely that, if there be an other, it should have other uses than simply the conveyance of radiations — Faraday's *Experimental Researches*, 3075.

* Preface to Newton's Principia, 2nd edition.

through space, or as the propagation of a condition of motion or stress in a medium already existing in space. In the theory of Neumann, the mathematical conception called Potential, which we are unable to conceive as a material substance, is supposed to be projected from one particle to another, in a manner which is quite independent of a medium, and which, as Neumann has himself pointed out, is extremely different from that of the propagation of light. In the theories of Riemann and Betti it would appear that the action is supposed to be propagated in a manner somewhat more similar to that of light.

But in all of these theories the question naturally occurs:— If something is transmitted from one particle to another at a distance, what is its condition after it has left the one particle and before it has reached the other? If this something is the potential energy of the two particles, as in Neumann's theory, how are we to conceive this energy as existing in a point of space, coinciding neither with the one particle nor with the other? In fact, whenever energy is transmitted from one body to another in time, there must be a medium or substance in which the energy exists after it leaves one body and before it reaches the other, for energy, as Torriecelli* remarked, 'is a quintessence of so subtile a nature that it cannot be contained in any vessel except the inmost substance of material things.' Hence all these theories lead to the conception of medium in which the propagation takes place, and if we admit this medium as an hypothesis, I think it ought to occupy a prominent place in our investigations, and that we ought to endeavour to construct a mental representation of all the details of its action, and this has been my constant aim in this treatise.

Index